W9-CKI-330

Prefixes used to define multiples of S.I. units. These may be used with any of the basic S.I. units or with units derived from them

Fraction	Prefix	Symbol	Example
10^{-18}	atto	a	
10^{-15}	femto	f	
10^{-12}	pico	p	
10^{-9}	nano	n	1 nanosecond $= 1$ ns $= 10^{-9}$ seconds
10^{-6}	micro	μ	
10^{-3}	milli	m	1 millimetre $= 1$ mm $= 10^{-3}$ metres
10^{-2}	centi	c	1 centimetre $= 1$ cm $= 10^{-2}$ metres
10^{-1}	deci	d	
10	deka	da	
10^2	hecto	h	
10^3	kilo	k	1 kilogram $= 1$ kg $= 10^3$ grams
10^6	mega	M	
10^9	giga	G	
10^{12}	tera	T	

The Greek Alphabet

A	α	alpha
B	β	beta
Γ	γ	gamma
Δ	δ	delta
E	ϵ	epsilon
Z	ζ	zeta
H	η	eta
Θ	θ	theta
I	ι	iota
K	κ	kappa
Λ	λ	lambda
M	μ	mu
N	ν	nu
Ξ	ξ	xi
O	o	omicron
Π	π	pi
P	ρ	rho
Σ	σ	sigma
T	τ	tau
Υ	υ	upsilon
Φ	ϕ	phi
X	χ	chi
Ψ	ψ	psi
Ω	ω	omega

Motion with Constant Acceleration

$$v = v_0 + a\,\Delta t$$
$$\Delta x = v_0\,\Delta t + \tfrac{1}{2}a(\Delta t)^2$$
$$\bar{v} = \tfrac{1}{2}(v_0 + v)$$
$$\Delta x = \tfrac{1}{2}(v_0 + v)\,\Delta t$$
$$v^2 = v_0{}^2 + 2a\,\Delta x$$

SCHAUMBURG TOWNSHIP PUBLIC LIBRARY

3 1257 00486 1398

WITHDRAWN

SCHAUMBURG TOWNSHIP
DISTRICT LIBRARY
130 SOUTH ROSELLE ROAD
SCHAUMBURG, ILLINOIS 60193

GENERAL PHYSICS

SUPPLEMENTS

Student Study Guide and Solutions Manual for General Physics (0-471-92915-3)
Morton M. Sternheim, University of Massachusetts, Amherst
Joseph W. Kane, University of Massachusetts, Amherst

Developed for student use with this text, including objectives, reviews, examples, new concepts and terms, quizzes and exams, as well as solutions for approximately 25% of the problems in the text.

Experiments in Physics—A Laboratory Manual for Scientists and Engineers (0-471-80571-8)
Daryl W. Preston, California State University at Hayward

A laboratory manual designed to be used in the calculus-based general physics course. The experiments follow the order of topics in traditional texts. Optional material is provided as well as flexibility in the choice of experiments.

Essential Mathematics for College Physics with Calculus, A Self-Study Guide (0-471-80876-8)
Michael Ram, SUNY Buffalo

Math supplement for both noncalculus and calculus courses, covering the mathematics used in physics. Uses the tutorial approach. Chapters independent and nonsequential. Emphasis on key points and stumbling blocks, test strategies, memorization tricks, etc.

Concentrated Physics Concepts: Part 1, Mechanics (0-471-82562-X)
David Alexander, Wichita State University
Richard Cornelius, Wichita State University

Interactive computer-aided instruction to help teach problem-solving techniques. Exposes students to a wide range of physics topics such as conversion of units, vectors, significant figures, particle dynamics, rotational dynamics, and many more.

For The Instructor
Instructor's Manual for *General Physics includes solutions to all exercises and problems.*
Transparency Masters for *Physics*, 2nd Edition and *General Physics*

Sixty-six transparency masters from which overhead transparencies may be made, selected from illustration figures in the text.

GENERAL PHYSICS

MORTON M. STERNHEIM
JOSEPH W. KANE

Department of Physics and Astronomy
University of Massachusetts
Amherst, Massachusetts

JOHN WILEY & SONS

New York • Chichester • Brisbane • Toronto • Singapore

SCHAUMBURG TOWNSHIP PUBLIC LIBRARY
32 WEST LIBRARY LANE
SCHAUMBURG, ILLINOIS 60194

Cover photo by David Attie
Cover design by Sheila Granda

Copyright © 1986, by John Wiley & Sons, Inc.

All rights reserved. Published simultaneously in Canada.

Reproduction or translation of any part of
this work beyond that permitted by Sections
107 and 108 of the 1976 United States Copyright
Act without the permission of the copyright
owner is unlawful. Requests for permission
or further information should be addressed to
the Permissions Department, John Wiley & Sons.

Library of Congress Cataloging in Publication Data:

Sternheim, Morton M., 1933–
    General physics.

    Includes index.
    1. Physics.   I. Kane, Joseph W., 1938–
II. Title.

QC23.S84   1986   530   85-12026
ISBN 0-471-80906-3

Printed in the United States of America

10 9 8 7 6 5 4 3 2 1

12|86
ост

530
STERNHEIM, M

3 1257 00486 1398

To my family.   **JWK**

To my wife, Helen, and my children, Laura, Amy, and Jeffrey.   **MMS**

PREFACE

General Physics is an introduction to physics for science majors with some background in calculus. Like our earlier text, *Physics,* it is intended to appeal to students with a wide range of interests and needs. However, it differs in that we have rewritten and added numerous topics that can most effectively be presented and understood with the aid of somewhat more advanced mathematical tools.

Ideally, students using this text will have completed at least one semester of calculus at the outset, but this is not essential. The derivative is introduced in discussing kinematics in Chapter One. Integrals are used only sparingly until Unit Five, Electricity and Magnetism, is reached near the middle of the book. Calculations involving calculus are carried out with a good deal of detail and discussion. A few minor omissions may be appropriate if students begin their calculus along with the physics.

Both books differ in several ways from many other physics texts. First, the specific needs of science majors, including those in the life sciences, have influenced which physics topics are included or emphasized. Thus we cover some topics no longer of great current interest to many physicists, such as geometric optics, the mechanics of fluids, and acoustics, while minimizing historical material and contemporary physics areas that have little direct impact on other sciences. Second, we make extensive use of examples involving biological and chemical systems and alternative energy sources. Finally, we devote entire sections and chapters to applications of physics, covering subjects such as nerve conduction, ionizing radiation, and nuclear magnetic resonance. These features help to motivate students while demonstrating the widespread utility of physics and the unity of science.

General Physics contains 31 chapters, grouped into nine units. In order to accommodate varying needs and tastes, there is more material than can usually be covered in a two-semester or three-quarter course. Chapters that may be treated lightly or omitted entirely include Chapter Eight, Elastic Properties of Materials; Chapter Eighteen, Nerve Conduction; Chapter Twenty-five, Special Relativity; Chapter Twenty-nine, The Structure of Matter; and Chapter Thirty-one, Ionizing Radiation. Most of the chapters end with Supplementary Topics sections containing either applications of physics or traditional topics that can be omitted without loss of continuity. This arrangement assists instructors in selecting what to include or emphasize and also helps students to distinguish the basic principles of physics from more peripheral material.

The changes made in the units on mechanics, thermal physics, and fluids in the present book are relatively minor. We use calculus in our discussions of kinematics, the center of gravity, moments of inertia, work and energy, area and polar moments of inertia, simple harmonic motion, the adiabatic expansion of an ideal gas, and Poiseuille's law. In Electricity and Magnetism, a significant amount of new material has been added, including finding fields and potentials by integration. Gauss' and Ampere's laws are present as Supplementary Topics. In Wave Motion, we use trigonometric functions to represent waves. In Chapter Twenty-eight, Quantum Mechanics and Atomic Structure, we introduce and solve the Schroedinger equation in one dimension, and consider the hydrogen atom wave functions.

Throughout *General Physics,* we present the appropriate calculus-based derivations within the main presentation; the few such derivations included in *Physics* are located in the Supplementary Topics. We have also added a few topics that do not involve calculus. These are the parallel axis theorem for moments of inertia, a derivation of the speed of sound, spherical mirrors, and the twin paradox. SI units are used exclusively.

Each chapter has a checklist of terms to define or explain and exercises keyed to the sections. There are also problems, which are unkeyed; occasional more difficult ones are preceded by an asterisk. Exercises and problems involving calculus are preceded by a [c]. There are also exercises and problems for the Supplementary Topics. About 150 new exercises and problems have been included, most of these involving calculus.

We wish to thank the many students and faculty colleagues who have helped us in so many ways. We are also indebted to the competent and cooperative editorial and production staffs at John Wiley & Sons for their valuable assistance. Most of all, we thank our families for their ongoing patience, help, and encouragement.

MORTON M. STERNHEIM
JOSEPH W. KANE

PROLOGUE

PHYSICS AND THE SCIENCE STUDENT

"Why should I study physics?" Sometimes asked with emotional overtones ranging from anguish to anger, this is one of the questions most frequently heard by physics teachers. It seems appropriate therefore to begin this book by attempting an answer.

One reason this question is asked so often is that many people who have not studied physics—and some who have—lack a clear notion of what physics is. Dictionaries are not much help. A typical short dictionary definition says that physics is the branch of science that deals with matter, energy, and their interactions. This is vague and general enough to include what is usually considered to be chemistry; in any case, it does not give any real feeling for what is involved. Longer dictionary entries usually expand the definition by noting that physics includes subfields such as mechanics, heat, electricity, and so forth. They give no clues as to why some subfields of science are included and others are not.

A better approach to defining physics is to ask what physicists are concerned about. Physicists attempt to understand the basic rules or *laws* that govern the operation of the natural world in which we live. Since their activities and interests evolve with time, the basic science called physics also changes with time. Many of the most active contemporary subfields of physics were undreamed of a generation or two ago. On the other hand, some parts of what are now considered to be chemistry or engineering were once considered to be physics. This is because physicists sometimes gradually abandon a field once the basic principles are known, leaving further developments and practical applications to others.

The fact that physics deals with the basic rules governing how the world works lets us see why people with varied interests may find the study of physics interesting and useful. For example, a historian who wants to understand the origins of our contemporary society will find significance in the story of the development of physics and its relationship to other human activities. Similarly, a philosopher concerned about concepts of space and time will profit greatly from understanding the revolutionary twentieth-century advances in physics. However, since we have written this book primarily for students majoring in the sciences, we have not stressed the historical or philosophical aspects of physics. Instead, we have tried to make clear in every chapter the connection between physics and other areas of science. We have learned that science majors find this approach more appropriate, since it makes clear the relevance and usefulness of studying physics.

An obvious impact of physics on both the life and physical sciences is in the area of instrumentation. Physical principles underlie the operation of light and electron microscopes, of X-ray machines and nuclear magnetic resonance spectrometers, of oscilloscopes and nuclear radiation monitors. Physics is also fundamental to a true understanding of chemistry, biology, and the earth sciences. The physical laws governing the behavior of molecules, atoms, and nuclei are the basis for all of chemistry and biochemistry. At the macroscopic level, the effects of forces of various types strongly influence the shapes of anatomical and human-built structures. Physiology offers many examples of physical processes and principles; diffusion within cells, the regulation of body temperature, the motion of fluids within the circulatory system, and electrical signals in nerves are just a few. In exercise science, activities ranging from running and jumping to karate can be analyzed and sometimes optimized by the application of physical principles. In the course of developing and illustrating the basic principles of physics, we discuss these applications and many others.

A few remarks about how one studies physics may be helpful. More than any other science, physics is a logical and deductive discipline. In any subfield of physics, there are just a few fundamental concepts or laws derived from experimental measurements. Once one has mastered these basic ideas, the applications are usually straightforward conceptually, even though the details may sometimes become complicated. Consequently, it is important to focus one's attention on the basic principles and to avoid memorizing a mass of facts and formulas.

Most of the basic laws of physics can be expressed rather concisely in the form of mathematical equations. This is a great convenience, since a tremendous amount of information is implicitly contained in a single equation. However, this also means that any serious attempt to learn or apply physics necessitates using a certain amount of mathematics. *General Physics* assumes a reasonable level of facility with high school algebra and basic geometry. Also, students should ideally have had some calculus before starting to use this text, although they may take it concurrently. An understanding of what derivatives and integrals mean is important, although not a great deal of skill in applying these concepts is needed. A mathematical review in Appendix B reviews key algebra and geometry topics and also lists the derivatives and integrals required for the examples and problems.

In summary, we believe that science majors will benefit in two major ways from studying physics. They will gain an understanding of the basic laws that govern everything in our world, from the subatomic to the cosmic scale, and will also learn much that will be important in their later work. The study of physics as a basic science is not particularly easy, but we believe it is rewarding, particularly for students planning further training in related sciences. We hope that all who use this book will agree.

M. M. S.
J. W. K.

CONTENTS

GENERAL
PHYSICS

UNIT ONE

THE GENERAL LAWS OF MOTION

(Franklin Wing/Stock Boston).

I N this unit and the next we discuss *mechanics:* the study of the motions of objects and of the forces that affect their motions. Its concepts and principles enter, directly and indirectly, into many areas of the physical and biological sciences. The laws of mechanics allow us to make predictions about such diverse phenomena as the motions of satellites, the movements of animals, and the strength and structure of both living and artificial systems. The laws of mechanics, applied to the motions of large numbers of atoms and molecules, provide an interpretation of the phenomena of heat and temperature. The properties of fluids—at rest and in motion—are understood in terms of the same laws; through them we can understand the flight of airplanes and animals or the flow of rivers and blood. Finally, with some twentieth-century modifications, the same mechanical laws play a central role in our present theories of atomic and nuclear phenomena.

The first two chapters of this unit are devoted to the concepts we require for a quantitative description of motion: *position, velocity,* and *acceleration.* In Chapter One we encounter these in the treatment of motion in a straight line; Chapter Two extends the same ideas to the description of motion in more than one dimension. The general laws of motion introduced by Newton, which relate the motions of objects to their causes, are discussed in Chapters Three to Five.

CHAPTER 1

MOTION IN A STRAIGHT LINE

The most basic and obvious result of physical interactions is motion: a brick falls, an eardrum vibrates, a compass needle swings into line with a magnetic field, a meter needle moves on a scale, a radioactive nucleus emits a beta particle. Most of our understanding of nature is derived from our observations of motions and our efforts to relate them to their causes. Accordingly, we begin our study of physics by developing the ideas needed for a quantitative discussion of motion, starting in this chapter with the case of an object moving in a straight line.

Physics, like many other sciences, is largely based on quantitative measurements. These measurements must be correlated or interpreted in some way; often they are compared with theoretical predictions. To the extent that theory and experiment are in accord, we say that we have some understanding of the phenomena in question. A quantitative discussion of motion requires measurements of times and distances, so we must first consider the *standards, units,* and *errors* involved in physical measurements.

1.1 | MEASUREMENTS, STANDARDS, UNITS, AND ERRORS

Quantitative physical measurements must be expressed by numerical comparison to some agreed-upon set of standards. If you say a lecture lasted 53 minutes, you mean that it went on for the same length of time as it took for the wall clock to make some number of ticks. Here the quantity being measured has the *dimensions* of time, the *unit* for the measurement is the minute, and the clock is the *standard.*

It is a *secondary standard,* since the minute is not defined by the properties of that one clock. All such measuring devices are calibrated directly or indirectly in terms of *primary standards* of length, time, and mass established by the international scientific community.

These primary standards are redefined from time to time as measurements become more precise. For example, the unit of length—the *metre*—was defined in 1889 as the length of a particular platinum-iridium bar kept under controlled conditions. This standard was discarded in 1960 because replication and preservation were inconvenient and subject to inaccuracies. The length standard is now based on the wavelength of the orange-red light emitted by atoms of krypton 86 in an electrical discharge tube. Standards have also been defined for the units of time and mass.

It is not accidental that standards have been set up for length, time, and mass. All mechanical quantities can be expressed in terms of some combination of these three fundamental dimensions, which we will denote at L, T, and M, respectively. For example, a velocity is a distance divided by a time, so its dimensions are L/T.

Systems of Units | *Metric units* have long been used in everyday matters everywhere except in the English-speaking countries, where *British units* were the norm. The British Commonwealth countries recently converted to the metric system, and the United States has slowly begun that complex process. In scientific work, metric units are used worldwide. Accordingly, in this text we will only use the interna-

the British length units (foot, yard, mile) and force unit (pound). Some fairly common non-S.I. units are defined for reference in Appendix C.

Representative lengths and times of various magnitudes are listed in S.I. units in Tables 1.1 and 1.2, respectively. The numbers appear in powers of 10, or "scientific notation," which is reviewed in Appendix B.1. Notice that many of the quantities in these tables look extremely large or small. For this reason, we will often use power-of-10 multiples or submultiples of S.I. units constructed with the aid of standard prefixes, tabulated for easy reference on the inside front cover of this text. For example, the distance between two cities is usually measured in kilometres, where 1 kilometre is 10^3 metres. The dimensions of this book are more conveniently expressed in centimetres than in metres, while the thickness of this page is roughly 0.1 millimetre or 100 micrometres. (1 millimetre = 10^{-3} metre, 1 micrometre = 10^{-6} metre.)

Figure 1.1. The Clinton P. Anderson Los Alamos Meson Physics Facility, a 0.8-kilometre long machine located high in the mountains of northern New Mexico, accelerates large numbers of protons (hydrogen nuclei) to high velocities. When these protons strike a target they produce short-lived particles called mesons used both for basic physics research and for cancer therapy. Measurements made at this "meson factory" employ sophisticated electronic devices, but these are still calibrated indirectly in terms of the basic units of length, time, and mass. (Courtesy of Los Alamos National Laboratory.)

tionally accepted set of metric units called the *Systeme Internationale* (*S.I.*). The *metre, kilogram,* and *second* are its basic units of length, mass, and time, respectively. In older texts, an earlier version of this system is referred to as the *m.k.s. system.* Older texts also sometimes used *c.g.s. units,* built on the *centimetre, gram,* and *second:* 1 centimetre is 0.01 metre, and 1 gram is 0.001 kilogram. The centimetre and gram are considered acceptable submultiples of the basic S.I. units, but most other c.g.s. units are now obsolete. In this text, we mention in passing just a few more commonly encountered remnants of this system. We will also show how to convert into S.I. units

TABLE 1.1

Representative lengths in metres	
Atomic nucleus	10^{-15}
Sodium atom, diameter	10^{-11}
C—C bond	1.5×10^{-10}
DNA, diameter	2×10^{-9}
Microfilament, thickness	4×10^{-9}
Hemoglobin	7×10^{-9}
Cell membrane	10^{-8}
Small virus, diameter	2×10^{-8}
Small bacterium, diameter	2×10^{-7}
Visible light, wavelength	$4 - 7 \times 10^{-7}$
Mitochondria, diameter	$0.5 - 1.0 \times 10^{-6}$
Large bacterium, diameter	10^{-6}
Mammalian liver cell, diameter	2×10^{-5}
Sea urchin egg	7×10^{-5}
Giant amoeba, diameter	2×10^{-4}
Small crustacean	10^{-3}
Ostrich egg, diameter	4×10^{-2}
Mouse	10^{-1}
Human	$1 - 2 \times 10^{0}$
Blue whale	3×10^{1}
Brooklyn Bridge	10^{3}
Earth, diameter	1.3×10^{7}
Sun, diameter	1.2×10^{9}
Earth–sun, distance	1.3×10^{11}
Our galaxy, diameter	10^{22}
Distance to the farthest galaxies ever observed	10^{28}

TABLE 1.2

Representative times in seconds

Nuclear events	$10^{-23} - 10^{-10}$
Atomic events: light absorption, electronic excitation	$10^{-15} - 10^{-9}$
Chemical events	$10^{-9} - 10^{-6}$
Chains of biochemical reactions	$10^{-8} - 10^{2}$
Quick contraction of striated muscle (wink)	10^{-1}
Fastest cell division	5×10^{2}
Typical bacterial generation time	3×10^{3}
Typical protozoan generation time	10^{5}
Small mammal generation time	4×10^{7}
Large mammal lifetime	$4 \times 10^{8} - 4 \times 10^{9}$
Lifetime of a lake	$10^{10} - 10^{12}$
Age of mammals	3×10^{15}
Age of vertebrates	10^{16}
Age of life	$> 10^{17}$
Age of the earth	2×10^{17}

Conversion of Units | Although we mainly use S.I. units, we occasionally need to convert quantities from one set of units to another. It is easy to do this correctly, even in complicated cases, with the aid of a trick involving "multiplying by one." For example, suppose we need to convert 100 feet (100 ft) into the equivalent number of metres (m). From the conversion factors listed on the inside front cover,

$$1 \text{ ft} = 0.3048 \text{ m}$$

Now we divide both sides by 1 ft, just as if the unit (feet) were an algebraic quantity:

$$\frac{1 \text{ ft}}{1 \text{ ft}} = \frac{0.3048 \text{ m}}{1 \text{ ft}}$$

The feet cancel on the left, leaving us a way of writing the quantity 1,

$$1 = \frac{0.3048 \text{ m}}{1 \text{ ft}}$$

If we multiply 100 ft by 1, nothing is changed, so we find

$$100 \text{ ft} = (100 \text{ ft})(1)$$
$$= (100 \text{ ft})\frac{(0.3048 \text{ m})}{1 \text{ ft}}$$
$$= 30.48 \text{ m}$$

Notice that the ft units in the numerator and denominator cancel, leaving the desired unit, m. The

virtue of multiplying by 1 is that this eliminates any doubt as to whether we should multiply or divide by the conversion factor. For instance, we can divide 1 ft = 0.3048 m by 0.3048 m to obtain another way of writing 1,

$$1 = \frac{1 \text{ ft}}{0.3048 \text{ m}}$$

However, if we multiply 100 ft by this factor, the units do not cancel properly.

Sometimes a quantity involves two or more units that must be converted. For example, a volume might be measured in cubic metres or m^3, and a velocity in kilometres per hour or km h^{-1}. (Note that we use negative exponents with units exactly as we would with algebraic quantities, so $1 \text{ h}^{-1} = 1/\text{h}$.) A factor of 1 is used for each conversion in the following examples.

Example 1.1

A small swimming pool is 20 ft long, 10 ft wide, and 5 ft deep. Its volume is the product of these distances, or (20 ft)(10 ft)(5 ft) = 1000 ft³. What is the volume in cubic metres (m^3)?

Here we must convert feet to metres three times, corresponding to changing the units for length, width, and depth. Using 1 ft = 0.3048 m,

$$1000 \text{ ft}^3(1)^3 = 1000 \text{ ft}^3 \frac{(0.3048 \text{ m})^3}{(1 \text{ ft})^3}$$
$$= 1000(0.3048)^3 \text{ m}^3 = 28.3 \text{ m}^3$$

Example 1.2

Convert a velocity of 60 mi h^{-1} (miles per hour) to metres per second (m s^{-1}).

To carry out this conversion, we need a factor of 1 to convert hours to seconds and another to convert miles to metres. Since 1 h = 60 min = 3600 s, dividing by 3600 s gives

$$1 = \frac{1 \text{ h}}{3600 \text{ s}}$$

Also, 1 mi = 1.609 km = 1609 m, so

$$1 = 1609 \text{ m mi}^{-1}$$

Multiplying 60 mi h^{-1} by 1 twice gives

$$(60 \text{ mi h}^{-1})(1)(1) = (60 \text{ mi h}^{-1})\left(\frac{1 \text{ h}}{3600 \text{ s}}\right)(1609 \text{ m mi}^{-1})$$
$$= 60\left(\frac{1609}{3600}\right) \text{ m s}^{-1} = 26.8 \text{ m s}^{-1}$$

Types of Errors | Measurements and predictions are both subject to errors. Measurement errors are of two types, *random* and *systematic*. The meaning of these terms is best understood with the aid of an example: the time T required for a weight on a string to swing back and forth once when released from a given point.

If someone uses a stopwatch to measure T and repeats the experiment several times, each result will be slightly different from the others. Usually, most of the measurements will be close to the average of all the measurements. The variation in results about this average arises from the inability of the observer to start and stop the watch exactly the same way each time. The error introduced by this inability is *random* and can be reduced by taking the average of many measurements.

Even when many measurements are made, the average result for T will be too small if the watch runs slow. This *systematic* error can be reduced by using a better watch or by comparing the watch used with a more accurate one and adjusting the results accordingly.

A systematic error may also result from the observer's reaction time. The observer may systematically start or stop the watch too early or too late. This error can be reduced by doing a more complex experiment. For example, the timing device may be started and stopped using a light beam and a photoelectric cell similar to those used in automatic door openers. Naturally, this apparatus will also have systematic and random errors, but they will be smaller than before.

Both systematic and random errors are present in all experiments. Reducing these errors generally requires increasingly elaborate apparatus and time-consuming procedures. High precision measurements and measurements of small effects require that great attention be given to identifying and reducing these errors.

Theoretical predictions usually have errors arising from various sources. A theoretical formula often contains measured quantities such as the mass of an electron or the speed of light, and there is some error associated with these measurements. For example, we will later obtain a formula for the period of a pendulum, the time T in our discussion above. Theoretical predictions made with this formula depend on the accuracy of our present knowledge of the acceleration due to gravity. Also, like most theoretical formulas, this formula depends on the validity of several approximations. Among these are the absence of friction and air resistance and the assumption that the pendulum does not swing very far from its rest position.

In all careful scientific work the numerical accuracy must be stated with precision. However, it is customary in textbooks to avoid the difficulties of a complete error analysis in doing numerical examples by using the rules for *significant figures*. This means that in the statement "the length of a rod is 2.43 metres" the last digit (3) is somewhat uncertain; the exact length might turn out to be closer to 2.42 or 2.44 m. In the examples, exercises, and problems in this text, all numbers should be treated as known to three significant figures. For example, 2.5 and 3 should be interpreted in calculations as 2.50 and 3.00, respectively. Significant figures are reviewed in Appendix B.3.

1.2 | DISPLACEMENT; AVERAGE VELOCITY

Quantitative discussions of motion are based on measurements and calculations of *positions, displacements, velocities,* and *accelerations*. In this and the following two sections, we use simple examples to introduce these concepts for motion along a straight line; we give their extensions to curved paths in the next chapter. We consider only *translational* motion, in which every part of an object moves in the same direction, and there is no rotation. Rotation will be discussed in Chapter Five.

The average velocity is defined in terms of the *displacement,* or the change in the position of an object that occurs in a specified interval of time. To illustrate what this means, suppose that a car is moving north along a straight highway with marker posts every 100 m, and that it is observed to pass one of these posts every 5 s, as in Fig. 1.2a. During any one of these 5-s intervals, the displacement is 100 m, in a 10-s interval, it is 200 m; and so on. The displacements have a direction as well as a size or magnitude: they are directed along the highway, or north. Specifying the direction is simple for motion along a straight line, but it becomes a bit more complex with curved paths.

The *average velocity* of the car during a specific time interval is the displacement divided by the time elapsed:

$$\text{average velocity} = \frac{\text{displacement}}{\text{time elapsed}}$$

The average velocity is proportional to the displacement, and it has the same direction. This definition is illustrated by the following example.

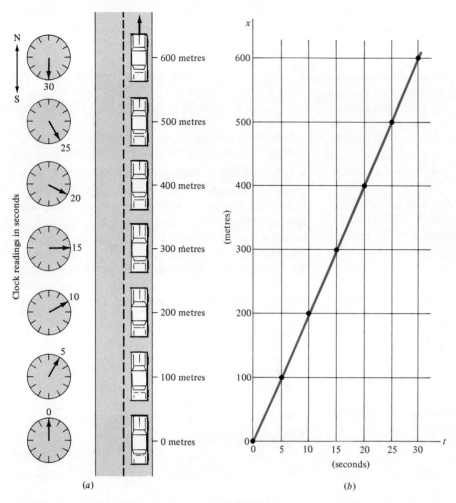

Figure 1.2. (*a*) The position of a car as observed at 5-second intervals. The speedometer reading is constant. (*b*) A graph of the position *x* versus time *t*.

Example 1.3

What is the average velocity of the car in Fig. 1.2*a* during the interval that the clock reading changes from 10 to 25 s?

From the figure, we see that the car travels 500 m − 200 m = 300 m during the 15-s period, so

$$\text{average velocity} = \frac{300 \text{ m}}{15 \text{ s}} = 20 \text{ m s}^{-1}$$

The direction of the average velocity is northward.

Because the car in this example moves equal distances in equal times, the average velocity will be the same no matter what time interval is chosen. In this situation the motion is said to be *uniform,* and the driver will observe the speedometer reading to remain constant. Motion that is not uniform is said to

be *accelerated.* In such cases the average velocity does depend on the time interval chosen. The speedometer reading of an accelerating car will be changing in time.

Often it is convenient to use graphs or algebraic formulas to describe the position and velocity of a moving object. For the car discussed above, we may designate the straight line along which it moves as an "*x* axis," and choose its position *x* to be zero at the post marked 0 metres. Also, we may use the symbol *t* to represent the observation times. If we arbitrarily take *t* = 0 when the first observation is made, then *x* is zero at *t* = 0. Also, *x* is 100 m when *t* = 5 s, *x* is 200 m when *t* = 10 s, and so on. Figure 1.2*b* shows these observations on a graph with time *t* along the horizontal coordinate and with the position *x* along the vertical coordinate. A straight line has been drawn connecting the points corresponding to the observations, since the car presumably moves steadily between observations.

The average velocity can now be defined more symbolically using this notation. Suppose that at some time which we will call t_1 the car is observed to be at position x_1 and that at a later time t_2, it is located at x_2. The displacement is the difference in positions,

$$\Delta x = x_2 - x_1$$

(The symbol Δ is the Greek letter "delta," and Δx is read "delta x." Δ usually represents a difference or change of the quantity that follows it.) The elapsed time between the observations is the difference

$$\Delta t = t_2 - t_1$$

In this notation, the average velocity $\bar{v}$ is the displacement divided by the elapsed time,

$$\bar{v} = \frac{\Delta x}{\Delta t} = \frac{x_2 - x_1}{t_2 - t_1} \tag{1.1}$$

Note that this definition of the average velocity holds whether or not $\bar{v}$ is constant in time.

We will now repeat the preceding example to show how this notation is used.

Example 1.4

Using Eq. 1.1, again find the average velocity of the car in Fig. 1.2 during the period from $t = 10$ s to 25 s.

Here t_1 is 10 s and t_2 is 25 s; from the graph, $x_1 = 200$ m and $x_2 = 500$ m. Thus

$$\bar{v} = \frac{x_2 - x_1}{t_2 - t_1} = \frac{500 \text{ m} - 200 \text{ m}}{25 \text{ s} - 10 \text{ s}} = \frac{300 \text{ m}}{15 \text{ s}} = 20 \text{ m s}^{-1}$$

Describing the straight-line motion of an object in terms of its position along some coordinate axis (here, the highway with its labeled markers is our x axis) automatically keeps track of the directions of displacements and average velocities. In Fig. 1.2, positive displacements are directed toward the north, and negative displacements are in the opposite direction, or south. For example, suppose the car is moving downward (toward the south) in the picture, instead of upward (toward the north). Then x is decreasing with increasing t, and Eq. 1.1 yields a negative average velocity, indicating the object is moving toward the south. This is illustrated numerically in the next example.

Example 1.5

At $t_1 = 5$ s, a car is at $x_1 = 600$ m, and at $t_2 = 15$ s, it is at $x_2 = 500$ m. Find its average velocity.

Using Eq. 1.1,

$$\bar{v} = \frac{x_2 - x_1}{t_2 - t_1} = \frac{500 \text{ m} - 600 \text{ m}}{15 \text{ s} - 5 \text{ s}}$$
$$= \frac{-100 \text{ m}}{10 \text{ s}} = -10 \text{ m s}^{-1}$$

Note that $\bar{v}$ is negative, and the car is moving in the $-x$ direction even though the position x is positive.

When an object undergoes uniform motion, its x-t graph is a straight line, as in the example of Fig. 1.2. If the motion is accelerated, the graph is not a straight line, and the average velocity depends on the particular time interval chosen. For example, in Fig. 1.3, a car starting from rest travels a short distance in the first second and a longer distance in the next second as it speeds up. In this situation the average velocity will be less in the first second of the motion than in later seconds. This is illustrated by the following example.

Example 1.6

A car moves as shown in Fig. 1.3. Find its average velocity from $t = 0$ to $t = 1$ s and from $t = 1$ to $t = 2$ s.

To calculate the average velocities, we need the positions at $t = 0$, 1, and 2 s. From Fig. 1.3, these are 0, 1, and 4 m, respectively. From $t = 0$ to 1 s, the average velocity is

$$\bar{v} = \frac{\Delta x}{\Delta t} = \frac{1 \text{ m} - 0 \text{ m}}{1 \text{ s} - 0 \text{ s}} = 1 \text{ m s}^{-1}$$

From $t = 1$ to 2 s,

$$\bar{v} = \frac{\Delta x}{\Delta t} = \frac{4 \text{ m} - 1 \text{ m}}{2 \text{ s} - 1 \text{ s}} = 3 \text{ m s}^{-1}$$

As anticipated, the average velocity is greater in the later time interval because the car is accelerating.

1.3 | INSTANTANEOUS VELOCITY

Most situations of interest involve accelerated rather than uniform motion. Since the average velocity depends on the time interval if the motion is accelerated, it is often more useful to characterize the motion by the *instantaneous velocity*—the velocity at a particular instant in time. For example, when we say that an accelerating car is moving at 10 m s^{-1}, we are referring to its instantaneous velocity at the present moment. This instantaneous velocity v can be thought of as the average velocity evaluated for an arbitrarily small time interval.

A more rigorous definition of the instantaneous velocity v is that it is the limit of the average velocity $\overline{v} = \Delta x / \Delta t$ as Δt approaches zero. In mathematical notation, this is written as

$$v = \lim_{\Delta t \to 0} \frac{\Delta x}{\Delta t} \qquad (1.2)$$

This equation is the definition of the *derivative* of x with respect to t, dx/dt. *Thus the instantaneous velocity of an object is the derivative of its position with respect to time,*

$$v = \frac{dx}{dt} \qquad (1.2a)$$

The derivative dx/dt is just the limit of the ratio $\Delta x/\Delta t$; Eq. 1.2a is a shorthand form of Eq. 1.2.

In general, derivatives can be interpreted as rates of change. The instantaneous velocity is the rate of change of the position with the time. On a graph of a quantity, a derivative is equal to the *slope*. As we see below, the instantaneous velocity equals the slope of the position-versus-time graph.

If we have an algebraic expression for the variation of the position of an object with the time, we can apply the rules of differentiation to find the instantaneous velocity. A brief list of derivatives is given in Appendix B; longer lists can be found in calculus texts. In the next example, we illustrate how to find v.

Example 1.7

The motion of the car shown in Fig. 1.3 corresponds to the mathematical formula $x = bt^2$, where

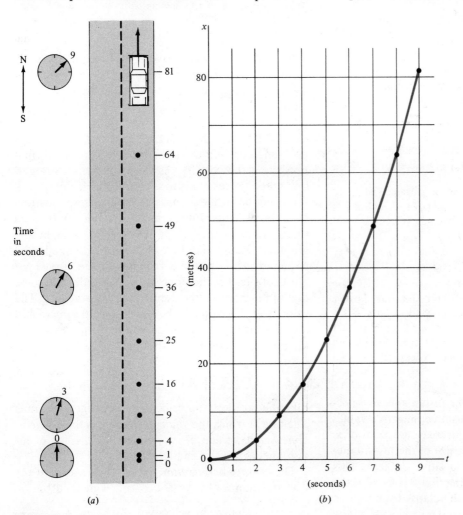

(a)   (b)

Figure 1.3. (a) The positions of an accelerating car at 1-second intervals are represented by solid circles. (b) The position-time graph for the car.

$b = 1$ m s^{-2}. (a) Find the instantaneous velocity of the car at an arbitrary time t. (b) Find the instantaneous velocity at $t = 3$ s. (c) Find the average velocity from 3 to 3.01 s.

(a) The instantaneous velocity v is the derivative of the position with respect to time. At time t, the position of the car is bt^2, and at time $t + \Delta t$ the position is $b(t + \Delta t)^2 = b[t^2 + 2t\Delta t + (\Delta t)^2]$. The displacement Δx in the time interval from t to $t + \Delta t$ is

$$\Delta x = b[t^2 + 2t\Delta t + (\Delta t)^2] - bt^2 = 2bt\Delta t + b(\Delta t)^2$$

Thus the average velocity over this time interval is

$$\overline{v} = \frac{\Delta x}{\Delta t} = \frac{2bt\Delta t + b(\Delta t)^2}{\Delta t} = 2bt + b\Delta t$$

If we take the limit as $\Delta t \to 0$, then $b\Delta t \to 0$, and we find the instantaneous velocity at t is

$$v = 2bt$$

The same result is found immediately if we use Eq. B.22 in Appendix B, $(d/dt)t^n = nt^{n-1}$; with $n = 2$, $(d/dt)t^2 = 2t$. (The factor b just multiplies the derivative of t^2.) From now on we will use differentiation formulas rather than working from the definition of the derivative.

(b) Substituting $t = 3$ s in $v = 2bt$, the instantaneous velocity at that time is

$$v = 2(1 \text{ m s}^{-2})(3 \text{ s}) = 6 \text{ m s}^{-1}$$

(c) When $t = 3$ s, the car is located at $x = bt^2 = (1 \text{ m s}^{-2})(3 \text{ s})^2 = 9$ m. At $t = 3.01$ s, $x = (1 \text{ m s}^{-2})(3.01 \text{ s})^2 = 9.0601$ m. Thus the average velocity $\overline{v}$ from 3 to 3.01 s is

$$\overline{v} = \frac{9.0601 \text{ m} - 9 \text{ m}}{3.01 \text{ s} - 3 \text{ s}} = 6.01 \text{ m s}^{-1}$$

The average velocity for the small time interval of 0.01 s is very close to the instantaneous velocity at $t = 3$ s, $v = 6$ m s^{-1}. In fact, for the shorter time interval from 3 to 3.001 s, we find that $\overline{v}$ is 6.001 m s^{-1}, even closer to the instantaneous value. This is what we expect from the definition of v.

Like the average velocity, the instantaneous velocity can be either positive or negative; negative values correspond to motion toward decreasing values of x. The *speed* is the magnitude or size of the instantaneous velocity, so it is always positive or zero. This definition is equivalent to saying that the speed is the distance traveled divided by the elapsed time.

From now on, when we refer to the velocity or any other rate of change, we mean the instantaneous

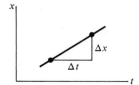

Figure 1.4. The x-t graph for an object undergoing uniform motion. The slope of the line is defined to be $\Delta x / \Delta t$, which is the velocity.

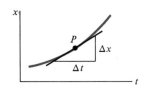

Figure 1.5. The slope of the x-t curve at P is equal to the slope $\Delta x / \Delta t$ of the straight-line tangent to the curve at P. The velocity at P is equal to the slope of the curve at that point.

value unless the word "average" is explicitly used. Note that the distinction between the two kinds of velocities disappears for the special case of uniform motion.

Graphical Interpretation of Velocity | A position-versus-time graph provides direct information about the velocity. As we have seen, a straight-line graph corresponds to a constant velocity or uniform motion, while a curved graph corresponds to a changing velocity. Furthermore, the velocity is equal to the slope of the x-t graph. In the straight-line graph of Fig. 1.4, the slope of the line is defined to be $\Delta x / \Delta t$, which is just the velocity. In the curved x-t graph of Fig. 1.5, the slope at point P is defined to be that of the straight-line tangent to the curve at P. The velocity at P is equal to the slope, since the curve and the tangent line have the same rate of increase with time at that point in the motion.

1.4 | ACCELERATION

Like the position, the velocity can change with time. The rate at which the velocity changes is the *acceleration*. Again we can discuss either the average or instantaneous rate of change.

The *average acceleration* $\overline{a}$ from time t_1 to t_2, if the velocity changes by $\Delta v = v_2 - v_1$, is defined by

$$\overline{a} = \frac{\text{change in velocity}}{\text{time elapsed}} = \frac{\Delta v}{\Delta t} = \frac{v_2 - v_1}{t_2 - t_1} \quad (1.3)$$

If we measure velocity in metres per second and time in seconds, $\bar{a}$ will have the units of metres per second per second (usually abbreviated as m s^{-2} and read as "metres per second squared"). An average acceleration of 1 m s^{-2} corresponds to an average increase in the velocity of 1 m s^{-1} each second. We will illustrate this definition with an example.

Example 1.8

A car accelerates from rest to 30 m s^{-1} in 10 s. What is its average acceleration?

From the definition,

$$\bar{a} = \frac{\Delta v}{\Delta t} = \frac{30 \text{ m s}^{-1} - 0}{10 \text{ s}} = 3 \text{ m s}^{-2}$$

This corresponds to an increase in the velocity of 3 m s^{-1} in each second of the 10-s time interval.

The *instantaneous acceleration a* is defined as a limit much as the instantaneous velocity is defined. The instantaneous acceleration at the time t is the derivative of the velocity with respect to time at that instant,

$$a = \lim_{\Delta t \to 0} \frac{\Delta v}{\Delta t} = \frac{dv}{dt} \tag{1.4}$$

On a graph of the velocity of an object versus time, the instantaneous acceleration is equal to the slope of the curve. These ideas are illustrated in the following examples.

Example 1.9

An object moves according to the formula $x = (b + ct^3)$. What is the instantaneous acceleration at time t?

The instantaneous acceleration is defined as the derivative of the velocity, so we must first find v. From the definition,

$$v = \frac{dx}{dt} = \frac{d}{dt}(b + ct^3) = 3ct^2$$

Then

$$a = \frac{dv}{dt} = \frac{d}{dt}(3ct^2) = 6ct$$

This is the instantaneous acceleration at time t.

Example 1.10

Figure 1.6 shows the velocity-versus-time graph for a car. Describe its acceleration qualitatively.

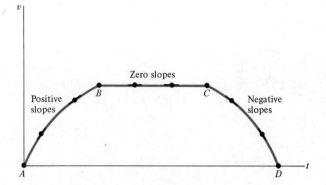

Figure 1.6. The velocity-versus-time graph for a car. The slope and acceleration are positive from A to B, zero from B to C, and negative from C to D.

From A to B, the velocity is increasing, the slope is positive, and the car is accelerating. However, the slope is becoming progressively smaller, so the acceleration is decreasing. From B to C, the velocity is constant and the slope and the acceleration are zero. From C to D, the velocity is decreasing, so the acceleration is negative. This *deceleration* increases in magnitude as the car slows.

1.5 | FINDING THE MOTION OF AN OBJECT

So far we have calculated velocities from position changes and accelerations from velocity changes. However, it is often true that it is the acceleration of an object that is either measured or predicted theoretically, and we wish to know the corresponding velocity and position changes. For example, when an animal jumps vertically from the earth's surface, it is subject to a constant acceleration due to gravity, which determines the motion of the animal while it is off the ground. In this section, we see how to find the subsequent motion of an object given its accleration and its initial position and velocity.

Consider an object initially moving with a velocity v_0. If the object undergoes a constant acceleration a for a time Δt, then from the definition of the acceleration, $a = \Delta v / \Delta t$ or $\Delta v = a \, \Delta t$. Hence the velocity has changed by an amount Δv to a value

$$v = v_0 + a \, \Delta t \tag{1.5}$$

This result has an interesting and useful interpretation in terms of the acceleration-versus-time graph

(Fig. 1.7a). The product of the height a and width Δt of the shaded rectangle is its area $a\,\Delta t$, which is equal to the velocity change. *Thus the change in velocity equals the area under the* a-t *graph over the time interval chosen.* The area is considered to be positive if it lies above the time axis and negative if it lies below. This result is quite general and is not restricted to constant acceleration situations.

If we draw a graph of the velocity versus time as given by Eq. 1.5, we obtain a straight line (Fig. 1.7b). The average velocity for the time interval Δt is $\bar{v} = v_0 + \Delta v/2 = v_0 + (v - v_0)/2$ or

$$\bar{v} = \frac{1}{2}(v_0 + v) \qquad (1.6)$$

The displacement or change in position that occurs during the time interval Δt is related to the average velocity through the definition $\bar{v} = \Delta x/\Delta t$. Thus $\Delta x = \bar{v}\,\Delta t$ or

$$\Delta x = \frac{1}{2}(v_0 + v)\,\Delta t \qquad (1.7)$$

Alternatively, if we substitute $v = v_0 + a\,\Delta t$ in this equation, we find

$$\Delta x = v_0\,\Delta t + \frac{1}{2}a\,(\Delta t)^2 \qquad (1.8)$$

Similarly, we can rewrite Eq. 1.5 as $\Delta t = (v - v_0)/a$ and substitute for Δt in Eq. 1.7 to give

$$\Delta x = \frac{1}{2}(v_0 + v)\frac{(v - v_0)}{a} = \frac{v^2 - v_0^2}{2a}$$

or

$$v^2 = v_0^2 + 2a\,\Delta x \qquad (1.9)$$

Again, our algebraic results have a direct graphical correspondence. The shaded area under the v-t graph

TABLE 1.3
Motion with constant acceleration

$v = v_0 + a\,\Delta t$	(1.5)
$\Delta x = v_0\,\Delta t + \frac{1}{2}a\,(\Delta t)^2$	(1.8)
$\bar{v} = \frac{1}{2}(v_0 + v)$	(1.6)
$\Delta x = \frac{1}{2}(v_0 + v)\,\Delta t$	(1.7)
$v^2 = v_0^2 + 2a\,\Delta x$	(1.9)

in Fig. 1.7b has an average height $\bar{v}$ and a width Δt, so its area is the product $\bar{v}\,\Delta t$, which is equal to Δx as given by Eq. 1.7. (Alternatively, the sum of the triangular and rectangular areas reduces to Eq. 1.8. See Problem 1-70.) *Thus the displacement is equal to the area under the* v-t *graph over the time interval Δt.* This is also true when the acceleration is not constant. As before, areas above the time axis are considered positive, and areas below are negative.

Equation 1.5 for the velocity and Eq. 1.8 for the displacement completely describe the motion of an object with a given initial velocity and position and a constant acceleration. Equations 1.6, 1.7, and 1.9 contain equivalent information, and are sometimes handy for solving problems. For example, Eq. 1.9 is useful when the initial and final velocity and acceleration are given but not the elapsed time. The constant acceleration formulas are listed in Table 1.3 and on the front endpapers for convenient reference. Their use is illustrated by the examples below and in the next section.

Example 1.11

A car initially at rest at a traffic light accelerates at $2\ \mathrm{m\ s^{-2}}$ when the light turns green. After 4 s, what are its velocity and position?

Since we know the acceleration a, the elapsed time Δt,

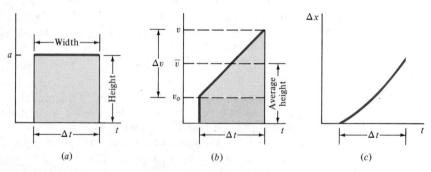

Figure 1.7. Motion with constant acceleration. (*a*) The acceleration a is constant over the time interval Δt. (*b*) The graph of $v = v_0 + a\,\Delta t$. The average velocity is $\bar{v} = (v_0 + v)/2$. (*c*) The displacement-versus-time graph.

and the initial velocity $v_0 = 0$, we can use Eqs. 1.5 and 1.8 to find the velocity and displacement. Thus

$$v = v_0 + a\,\Delta t = 0 + (2\text{ m s}^{-2})(4\text{ s}) = 8\text{ m s}^{-1}$$

$$\Delta x = v_0\,\Delta t + \frac{1}{2}a\,(\Delta t)^2 = 0 + \frac{1}{2}(2\text{ m s}^{-2})(4\text{ s})^2 = 16\text{ m}$$

After 4 s the car has reached a velocity of 8 m s^{-1} and is 16 m from the light.

Note that we could also have found Δx from Eq. 1.7, using our result for v. Constant acceleration problems and physics problems in general can often be solved in more than one way.

Example 1.12

A car reaches a velocity of 20 m s^{-1} with an acceleration of 2 m s^{-2}. How far will it travel while it is accelerating if it is (a) initially at rest; (b) initially moving at 10 m s^{-1}?

(a) Here we know the initial and final velocities as well as the acceleration. Equation 1.9 contains these known quantities plus the displacement Δx. Solving for Δx, with $a = 2$ m s^{-2}, $v = 20$ m s^{-1}, and $v_0 = 0$, we find

$$\Delta x = \frac{(v^2 - v_0{}^2)}{2a} = \frac{(20\text{ m s}^{-1})^2}{2(2\text{ m s}^{-2})} = 100\text{ m}$$

We could also have solved this problem by using $v = v_0 + a\,\Delta t$ to find Δt and then substituting the elapsed time into Eq. 1.8. (The reader should work out this solution as an exercise.)

(b) Proceeding as above but with $v_0 = 10$ m s^{-1},

$$\Delta x = \frac{v^2 - v_0{}^2}{2a}$$

$$= \frac{(20\text{ m s}^{-1})^2 - (10\text{ m s}^{-1})^2}{2(2\text{ m s}^{-2})} = 75\text{ m}$$

The distance required to reach the desired velocity is shorter than in part (a) because the car is initially in motion.

Example 1.13

A car accelerates from rest with a constant acceleration of 2 m s^{-2} onto a highway where traffic is moving at a steady 24 m s^{-1}. (a) How long will it take for the car to reach a velocity of 24 m s^{-1}? (b) How far will it travel in that time? (c) The driver does not want the vehicle behind to come closer than 20 m nor force it to slow down. How large a break in traffic must the driver wait for?

(a) The time needed for the car to reach the velocity $v = 24$ m s^{-1} starting from rest satisfies $v = v_0 + a\,\Delta t$, or

$$\Delta t = (v - v_0)/a = (24\text{ m s}^{-1})/(2\text{ m s}^{-2}) = 12\text{ s}$$

(b) Using Eq. 1.8, the distance traveled by the car in 12 s is

$$\Delta x = v_0\,\Delta t + \frac{1}{2}a\,(\Delta t)^2$$

$$= 0 + \frac{1}{2}(2\text{ m s}^{-2})(12\text{ s})^2 = 144\text{ m}$$

(c) The vehicle behind is moving at a constant velocity $v_0 = 24$ m s^{-1}, so $a = 0$. Using Eq. 1.8, in 12 s it moves a distance

$$\Delta x = v_0\,\Delta t + \frac{1}{2}a\,(\Delta t)^2$$

$$= (24\text{ m s}^{-1})(12\text{ s}) + 0 = 288\text{ m}$$

Since the entering car travels 144 m in this time, the oncoming vehicle gains $(288 - 144)$ m, or 144 m. If it is to come no closer than 20 m, then the break in traffic must be at least $(144 + 20)$ m, or 164 m.

The car in this example reaches 24 m s^{-1} or about 86 km h^{-1} (54 mi h^{-1}) in 12 s, which is a fairly brisk acceleration. A less powerful car would take longer to reach this speed and would require a longer break in traffic.

The above examples illustrate the procedures for solving constant acceleration problems and physics problems in general. We identify which quantities are known and which are to be found and determine the equation or equations relating these quantities. If necessary, we then solve algebraically for one unknown in terms of known quantities. Numerical values are usually best substituted in the final step rather than at an earlier stage. This tends to minimize the arithmetic labor and also facilitates checking for errors.

A useful check on any problem is provided by the dimensions of the final result. If the unknown is a length, then the result should be in length units such as metres or kilometres; if not, an error has been made somewhere. Note, for example, that in parts (b) and (c) of Example 1.13 the time units cancel and the distances are in metres as required.

To summarize, we have found that given the initial position and velocity, we can find the velocity at a later time from the acceleration, and the position from the velocity. The equations listed in Table 1.3 can be used to find the motion of objects undergoing a constant acceleration. When the acceleration is not constant, the average acceleration may often be used in the constant acceleration equations to find an approximate description of the motion. We do this in analyzing vertical jumping in the Supplementary Topics at the end of this chapter. Alternatively, we can use integration techniques if we know an explicit formula for the acceleration. This is also considered in the Supplementary Topics.

1.6 | THE ACCELERATION OF GRAVITY AND FALLING OBJECTS

Until now, our discussion of motion has been based on definitions and their consequences. Specifically, we have considered the mathematical relationships arising from the definitions of velocity and acceleration, but we have made no statements about how the natural world works. However, in order to discuss the motion of falling objects, we must use some information first obtained by Galileo from careful experimental observations.

We know from everyday experience that unsupported objects tend to fall toward the ground. The speed at impact often increases if the distance dropped increases. Thus, it is evident that falling objects undergo an acceleration, which we attribute to *gravity*, the gravitational attraction of the earth. Nevertheless, two essential aspects of this gravitational acceleration are not so readily observed.

Suppose gravity is the only factor affecting the motion of an object falling near the earth's surface, and air resistance is either absent or negligibly small. In this situation, it is found that:

1 *The gravitational acceleration is the same for all falling objects,* no matter what their size or composition.
2 *The gravitational acceleration is constant.* It does not change as the object falls.

Neither of these statements squares entirely with our everyday experience. Coins fall faster than bits of paper, which contradicts statement 1. Objects dropped from great heights reach a maximum or *terminal* velocity, which contradicts statement 2. However, both of these effects result from air resistance. A coin and scrap of paper fall together in a vacuum (Fig. 1.8), and an object falling at high altitude has a constant acceleration until entering the atmosphere.

The acceleration of gravity near the surface of the earth is denoted by g; it is approximately equal to

$$g = 9.8 \text{ m s}^{-2}$$

Small variations in g occur as a result of changes in latitude, elevation, and the density of local geological features.

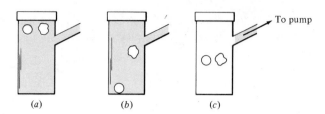

Figure 1.8. (*a*) A coin and a piece of paper are released simultaneously from the top of a container. (*b*) If the container has air in it, the coin hits the bottom first. (*c*) If most of the air is pumped out of the container, both objects reach the bottom at the same time.

In the examples below and in the problems and exercises at the end of the chapter we neglect such variations in g and assume that air resistance is unimportant. This allows us to use the constant acceleration equations to find the motion of falling objects.

Example 1.14

A ball is dropped from a window 84 m above the ground (Fig. 1.9). (a) When does the ball strike the ground? (b) What is the velocity of the ball when it strikes the ground?

In this type of problem we can choose the coordinate system so that the positive x direction is either up or down. Let us choose positive values of x in the upward direction.

Since the ball is dropped with zero initial velocity, $v_0 = 0$. The gravitational acceleration is constant and in the $-x$ direction, so the equations of Table 1.3 can be applied using $a = -g = -9.8 \text{ m s}^{-2}$.

(a) The ball strikes the ground when $\Delta x = -84 \text{ m}$. This happens after a time interval Δt, which satisfies

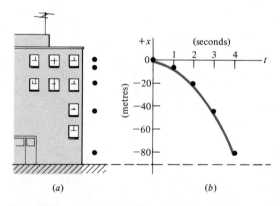

Figure 1.9. (*a*) A freely falling ball released with no initial velocity. Note that it falls through successively larger distances in each second. (*b*) The x-t graph for the ball in Example 1.13, taking $x = 0$ at the window.

$$\Delta x = \frac{1}{2} a (\Delta t)^2$$

or

$$(\Delta t)^2 = \frac{2 \, \Delta x}{a}$$

Thus

$$\Delta t = \sqrt{\frac{2 \, \Delta x}{a}} = \sqrt{\frac{2(-84 \text{ m})}{-9.8 \text{ m s}^{-2}}} = 4.14 \text{ s}$$

The positive root is used because the ball hits the ground after it is released, not before.

(b) Using Eq. 1.5 with $\Delta t = 4.14$ s and $v_0 = 0$,

$$v = a \, \Delta t = (-9.8 \text{ m s}^{-2})(4.14 \text{ s}) = -40.6 \text{ m s}^{-1}$$

This example can also be done when the $+x$ direction is chosen to be downward. Then the acceleration is $a = g = +9.8 \text{ m s}^{-2}$, and the ball hits the ground when $\Delta x = +84$ m. With these values, the same result is found for Δt. The velocity v when the ball hits the ground is positive because $+x$ is now downward. The choice of the positive x direction has no effect on our physical results, but we must correctly interpret the signs in our answers.

Objects Thrown Straight Up | The motion

of a ball or other object thrown straight up raises some interesting and potentially confusing points. The ball experiences the constant downward gravitational acceleration throughout its motion. This is true while the ball is going up, at the instant it reaches its greatest height, and as it is coming down.

However, the velocity is continually changing. Initially it is directed upward. As the ball rises, the magnitude of v uniformly decreases, until the velocity becomes zero at the instant the ball is at its highest point. The ball then begins to move downward, with v steadily increasing in magnitude.

One way to understand the fact that $v = 0$ at the highest point is to note that when a quantity has a maximum or minimum value, its rate of change or derivative is zero. Thus, the maximum value of x occurs where $v = dx/dt = 0$. *Remembering that $v = 0$ at the highest point reached is the key to solving many problems.*

The motion of an object thrown upward is illustrated by the v-t and x-t graphs in Fig. 1.10. At A the object is moving upward, in the $+x$ direction. At B it is instantaneously at rest at the highest point in its motion, and at C it is moving downward. Thus, v is initially positive, then zero for an instant, and nega-

tive afterward. By contrast, *the acceleration is equal to the slope of the v-t graph, or to $-g$, at all times.*

We can see from Fig. 1.10 that the motion is *symmetric* about its highest point (point B). The upward velocity 1 second before the peak and the downward velocity 1 second after the peak have the same magnitude. Consequently the time the object takes to traverse any given portion of the path will be the same going up and coming down.

These features of the motion are illustrated by the next example.

Example 1.15

A ball is thrown upward at 19.6 m s^{-1} from a window 58.8 m above the ground. (a) How high does it go? (b) When does it reach its highest point? (c) When does it strike the ground?

Again choosing $+x$ upward, $v_0 = 19.6$ m s^{-1} and $a = -g = -9.8$ m s^{-2}.

(a) As the ball rises, its velocity decreases uniformly until $v = 0$ at the peak; we want to find the height at which this occurs. From Eq. 1.9, using $v = 0$,

$$\Delta x = \frac{v^2 - v_0^2}{2a} = \frac{0 - (19.6 \text{ m s}^{-1})^2}{2(-9.8 \text{ m s}^{-2})} = 19.6 \text{ m}$$

Thus the ball reaches a maximum height 19.6 m above the window, or $(58.8 + 19.6)$ m $= 78.4$ m above the ground.

(b) Using Eq. 1.5 and the fact that $v = 0$ at the peak height,

$$t = \frac{v - v_0}{a} = \frac{0 - 19.6 \text{ m s}^{-1}}{-9.8 \text{ m s}^{-2}} = 2 \text{ s}$$

The peak is reached in 2 s.

(c) The ball hits the ground when $\Delta x = -58.8$ m, or after a time interval Δt, which satisfies

$$\Delta x = v_0 \, \Delta t + \frac{1}{2} a (\Delta t)^2$$

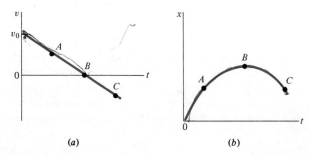

(a)   *(b)*

Figure 1.10. *(a)* v-t and *(b)* x-t graphs for an object thrown straight up from $x = 0$. The velocity is zero at the instant the object reaches its greatest height (point B), but the acceleration is always $-g$.

(New York Public Library Picture Collection.)

GALILEO GALILEI
(1564-1642)

Galileo was born in Pisa. At 17 he began to study medicine, although he had also shown talent in music and art. His interests soon turned toward other branches of science, and he was appointed a professor of mathematics at the University of Pisa. There, between 1589 and 1592, he conducted his investigations of motion which are the basis for this chapter.

The Greek philosopher Aristotle (384–322 B.C.) had taught that heavy objects fall faster than light ones. Galileo performed a series of experiments on objects rolling down smooth inclines and concluded that, in the ideal frictionless case, all objects have the same acceleration. Furthermore, he showed that the distance varies as the time squared, which implies that the acceleration is constant. Galileo is considered to have shown the importance of experimentation in science.

In 1608, Galileo heard that two spectacle lenses used together would magnify a distant object, and he soon built a series of telescopes of increasing magnifying power. He discovered that our moon has mountain ranges, that Jupiter has moons, and that the sun has spots. But his observations also led to problems. Copernicus (1473–1543) had earlier doubted Aristotle's teaching that the earth is the center of the universe. He showed that the apparent motions of the sun, stars, and planets could be most simply explained by supposing that the earth itself is a planet that rotates daily on its axis and revolves annually about the sun. Galileo's observations supported this heretical view that the earth is not unique. Consequently, he ran into serious difficulties with the authorities.

Galileo's clash with Church dogma lasted for nearly two decades. Initially forbidden to discuss his ideas, he was later ordered to describe the Copernican ideas as only a theory. However, Galileo's analysis and presentation of the existing facts were so complete and convincing that, at age 70, he was tried for disregarding the earlier order. After the trial, he was kept under house arrest for the remaining 12 years of his life.

$$-58.8 \text{ m} = (19.6 \text{ m s}^{-1}) \Delta t + \frac{1}{2}(-9.8 \text{ m s}^{-2}) (\Delta t)^2$$

Dividing by 4.9 m s^{-2} and rearranging terms, this becomes

$$(\Delta t)^2 - (4 \text{ s})(\Delta t) - 12 \text{ s}^2 = 0$$

which can be factored to

$$(\Delta t - 6 \text{ s})(\Delta t + 2 \text{ s}) = 0$$

This equation has two solutions, $\Delta t = 6 \text{ s}$ and $\Delta t = -2 \text{ s}$. Since the ball cannot possibly hit the ground before it is thrown, $\Delta t = 6 \text{ s}$ is the correct result. (The extraneous root $\Delta t = -2 \text{ s}$ is not meaningless. A ball thrown upward from the ground level 2 s earlier and passing the window just as the original ball was thrown would follow exactly the same subsequent path.)

1.7 | MODELS IN PHYSICS

Real problems sometimes are so complex that an exact solution is either impossible or else requires very difficult measurements or calculations. A rough estimate of the correct solution can often be found with a *mathematical model* obtained by making simplifying assumptions and approximations.

We used a model in the preceding section when we considered falling objects and neglected air resistance. For a rock or a coin falling at low speeds, this is a good approximation. However, if the speed is high, air resistance is more important, and a model that neglects it can be misleading in its predictions. Also, as we saw, it would be very inaccurate to ignore air resistance for something as light as a scrap of paper.

Mathematical models can show what factors are important in a problem and provide a qualitative understanding that may not be obtainable from a more exact approach. Their predictions may indicate whether a more elaborate effort is worth attempting and may suggest lines of attack for such an effort. However, textbooks—including this one—and other sources are not always fully explicit in stating their assumptions and in distinguishing models from exact treatments. Thus, a moderate amount of skepticism is a healthy thing to acquire.

SUMMARY
Physical quantities are stated in terms of units. To convert units from one system to another, we multiply by 1 and cancel units as though they were algebraic quantities.

The motion of an object along a straight line is described by its position, velocity, and acceleration. The velocity is the time rate of change of the position. The average velocity is the change in position or displacement divided by the elapsed time,

$$\bar{v} = \frac{\Delta x}{\Delta t}$$

The instantaneous velocity is the limit of the average velocity $\bar{v} = \Delta x/\Delta t$ as Δt approaches zero, or the derivative of x with respect to t,

$$v = \frac{dx}{dt}$$

Similarly, the average acceleration is the velocity change divided by the elapsed time,

$$\bar{a} = \frac{\Delta v}{\Delta t}$$

The instantaneous acceleration is the derivative of v with respect to t,

$$a = \frac{dv}{dt}$$

On an x-t graph, the slope is equal to the instantaneous velocity; on a v-t graph, the slope is equal to the instantaneous acceleration.

Often the acceleration can be calculated theoretically or measured experimentally. If the initial position and velocity are known, their later values can then be found from the acceleration. In the special case where the acceleration is constant, we have found the equations of motion. These are listed in Table 1.3 for convenient reference. Whether or not the acceleration is constant, the displacement equals the area under the v-t graph, and the velocity change equals the area under the a-t graph. Areas above the time axis are positive, and areas below the time axis are negative.

Objects falling without appreciable air resistance near the earth's surface all experience a common, constant acceleration g. An object initially thrown upward also has this acceleration. Its velocity steadily decreases in magnitude until it becomes zero at the highest point reached.

Checklist

Define or explain:

dimensions	uniform motion
standards	accelerated motion
units	instantaneous velocity
S.I.	slope
m.k.s. system	average acceleration
c.g.s. system	instantaneous
British units	acceleration
random errors	constant acceleration
systematic errors	formulas
significant figures	gravitational
translational motion	acceleration
displacement	terminal velocity
average velocity	mathematical model

REVIEW QUESTIONS

(Answers are given in this chapter just before the supplementary topics.)

Q1-1   The officially recognized set of units for scientific work is the _____.

Q1-2   Experimental data usually contain _____ and _____ errors.

Q1-3   The best way to convert units is to multiply by a factor of _____ for each unit that must be converted.

Q1-4   The change in position is called the _____.

Q1-5   The average velocity is the ratio _____/ _____.

Q1-6   The instantaneous velocity is the average velocity evaluated for _____.

Q1-7   The average acceleration is the _____ divided by the _____.

Q1-8   On an *x-t* graph, the slope equals the _____.

Q1-9   On a *v-t* graph, the slope equals the _____.

Q1-10   The velocity change equals the area under the _____ graph.

Q1-11   The displacement equals the area under the _____ graph.

Q1-12   In air, a rock falls faster than a feather because of _____.

Q1-13   When an object thrown straight upward reaches its greatest height, its velocity is _____ and its acceleration is _____.

EXERCISES

The student should work out some of the exercises for each section to test his or her general understanding of the concepts before attempting the problems. Answers to most of the odd-numbered exercises and problems are given at the end of the book, usually to 3 significant figures. If your answers differ slightly in the last place, this may be due to differences in rounding off intermediate results, rather than to an error in your work. Exercises and problems preceded by a ᶜ may require the use of calculus.

Section 1.1 | Measurements, Standards, Units, and Errors

1-1   An acre is 43,560 ft^2. How large is this in square metres (m^2)?

1-2   Convert 40 mi h^{-1} to metres per second (m s^{-1}).

1-3   A gallon is 231 cubic inches (in.^{3}), and a litre is 1000 cm^3. How many litres are there in a gallon?

1-4   A furlong is 220 yards, and a fortnight is 14 days. If a snail moves at 2 m h^{-1}, what is this in furlongs per fortnight?

1-5   A cell membrane is 70 angstrom (Å) units thick. If an angstrom unit is 10^{-10} m, what is the membrane thickness in (a) metres; (b) micrometres?

1-6   If two quantities have different dimensions, can they be (a) multiplied; (b) added? Give examples to support your answers.

1-7   In the United States land is measured in acres (1 acre = 43,560 ft^2). In most other countries it is measured in hectares (1 hectare = 10^4 m^2). How large is a 100-acre farm in hectares?

1-8   The volumes of reservoirs are sometimes measured in acre-feet; that is, a lake with an area of 1 acre and an average depth of 1 ft contains 1 acre-foot of water. If a lake has an area of 100 acres and an average depth of 20 feet, find its volume (a) in acre-feet; (b) in cubic feet; (c) in cubic metres. (1 acre = 43,560 ft^2)

1-9   Suppose you wanted to know the area of a rectangular room and had a cloth tape measure available to determine its dimensions. What are some random and systematic errors that might affect your result?

1-10   A driver wishes to check the speedometer by traveling at a constant speed along a highway with markers placed every kilometre and having a passenger note the time intervals with a wristwatch. Discuss the random and systematic errors involved.

Section 1.2 | Displacement; Average Velocity

1-11 A car travels 30 kilometres in 45 minutes on a straight highway. What is its average velocity in kilometres per hour (km h^{-1})?

1-12 A pilot wishes to fly 2000 km in 4 hours. What average velocity in metres per second is required to accomplish this?

1-13 Sketch the position-versus-time graph for a car that starts from rest and is driven 1 kilometre to a store. (Describe the motion in words.)

1-14 A car travels in a straight line at 40 km h^{-1} for 1 hour and at 60 km h^{-1} for 2 hours. (a) How far does it travel? (b) Find the average velocity.

1-15 A woman wants to drive 100 kilometres in 2 hours. If she averages 40 km h^{-1} for the first 1.5 hours, what average velocity must she maintain for the remaining time?

1-16 A man runs a 42-km marathon in 2.5 hours. Find the average velocity in (a) kilometres per hour (km h^{-1}); (b) metres per second (m s^{-1}).

1-17 A sprinter runs the 100-m dash in 9.8 s. (a) What is the sprinter's average velocity? (b) Since the runner starts from rest, the velocity cannot be constant. Sketch the approximate position-versus-time graph for the runner. Explain the assumptions made.

1-18 Light travels at 3×10^8 m s^{-1}. A light-year is the distance light travels in 1 year, 365 days. Find the distance in kilometres to the nearest star, which is 4 light-years away from us.

1-19 A falling object moves so that its height x above the ground at time t is given by the equation $x = 100$ m $- (4.9$ m s$^{-2})t^2$. Find its average velocity from (a) $t = 0$ to $t = 2$ s; (b) $t = 2$ to $t = 4$ s.

1-20 A ball thrown straight upward has a height x above the ground, which is given by the equation $x = (19.6$ m s$^{-1})t - (4.9$ m s$^{-2})t^2$. Find the average velocity from (a) $t = 0$ to $t = 2$ s; (b) $t = 2$ to 4 s.

1-21 From the position-versus-time graph of Fig. 1.11, find the average velocity from $t = 0$ s to (a) $t = 10$ s; (b) $t = 20$ s; (c) $t = 40$ s.

1-22 A baseball reaches a batter at 40 m s^{-1}. If home plate is 0.3 m across, how long will the ball be over the plate?

1-23 In 1970, a record was set when a swimmer

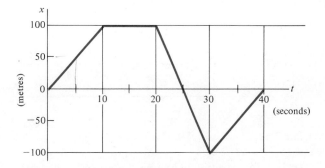

Figure 1.11. Exercises 1-21, 1-24, and 1-25.

swam 100 m in 51.9 s. What was his average velocity in km h^{-1}?

Section 1.3 | Instantaneous Velocity

1-24 In Fig. 1.11, what is the instantaneous velocity at (a) $t = 5$ s; (b) $t = 15$ s; (c) $t = 25$ s; (d) $t = 35$ s?

1-25 Draw the instantaneous velocity-versus-time graph corresponding to Fig. 1.11.

1-26 Figure 1.12 shows the position of a pendulum versus time. In the interval $t = 0$ to T, when is the velocity (a) zero; (b) positive; (c) negative?

1-27 Figure 1.12 shows the position of a pendulum versus time. In the interval from $t = 0$ to T, when does the velocity have its largest positive and negative values?

1-28 In 1875, Matthew Webb became the first man to swim the English Channel without a life jacket. He required 21 hours and 45 minutes to make the 33.8 km crossing. (a) What was his average velocity in km h^{-1}? (b) Since he did not swim in a straight line, he actually covered a distance of 60 km. What was his average speed?

1-29 The maximum speed attained by a runner during a 100-m race is reported as 12.5 m s^{-1}. He completed the race in 9.9 s. Are these numbers consistent? Explain.

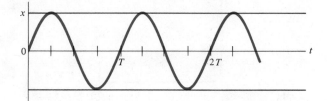

Figure 1.12. The horizontal position of a pendulum versus time. Exercises 1-26 and 1-27; Problem 1-67.

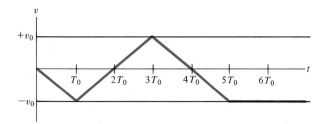

Figure 1.13. Exercises 1-34 and 1-35.

ᶜ1-30 The position of an object is given by $x = bt^3$. (a) If x and t are in metres and seconds respectively, what are the units of b? (b) What is the instantaneous velocity at time T? (c) Find the average velocity from $t = 0$ to $t = T$.

ᶜ1-31 A falling object has a height for $t > 0$ given by $x = (200 \text{ m}) - (4.9 \text{ m s}^{-2})t^2$. (a) Find the instantaneous velocity at $t = 10$ s. (b) Find the average velocity from $t = 0$ to $t = 20$ s.

Section 1.4 | Acceleration

1-32 A car proceeds to pass another. Its speed increases from 50 to 100 km h^{-1} in 4 s. What is the average acceleration?

1-33 A car moves at a constant velocity of 50 m s^{-1} for 20 s. It then slows with a constant acceleration, coming to rest 10 s later. (a) Draw a velocity-time graph for the car. (b) Draw the acceleration-time graph for the car.

1-34 Draw the acceleration-versus-time graph corresponding to Fig. 1.13.

1-35 When does the acceleration corresponding to Fig. 1.13 have (a) a maximum value; (b) a minimum value; (c) a value of zero?

1-36 A car initially moving at 20 m s^{-1} brakes with an acceleration of 2 m s^{-2}. How long will it take for the car to stop?

ᶜ1-37 A car slowing to a stop has a velocity given by $v = v_0(1 - t/T)$ from $t = 0$ to $t = T$. (a) Find the instantaneous acceleration for $0 < t < T$. (b) Find the average acceleration from 0 to T. (c) Explain the relationship between your answers in parts (a) and (b).

ᶜ1-38 A falling object has a height for $t > 0$ given by $x = (200 \text{ m}) - (4.9 \text{ m s}^{-2})t^2$. (a) Find the instantaneous acceleration at $t = 10$ s. (b) Find the average acceleration from $t = 0$ to $t = 20$ s.

Section 1.5 | Finding the Motion of an Object

1-39 A jet plane accelerates on a runway from rest at 4 m s^{-2}. After 5 s, find its (a) distance traveled; (b) velocity.

1-40 In the accelerator shown in Fig. 1.1, protons emerge with a velocity of 2.5×10^8 m s^{-1}. The accelerator is 0.8 km long. (a) If the acceleration is uniform, how large is it? (b) How long does it take for protons to travel the length of the accelerator?

1-41 A train traveling with a velocity of 30 m s^{-1} stops with a uniform acceleration in 50 s. (a) What is the acceleration of the train? (b) What is the distance traveled before coming to rest?

1-42 A car starts from rest at $t = 0$ and accelerates as shown in Fig. 1.14. Find its velocity at (a) $t = 10$ s, and (b) $t = 30$ s. (c) Draw the velocity-versus-time graph.

1-43 Plot the position-versus-time graph corresponding to Fig. 1.15. (Assume the object starts from $x = 0$.)

1-44 Suppose a car can accelerate at 1 m s^{-2}. How large a break in traffic is needed to enter a highway where cars are moving at 200 m s^{-1} if the driver wants to avoid forcing the next car to slow down or to approach closer than 25 m?

1-45 A car moving at 15 m s^{-1} hits a stone wall.

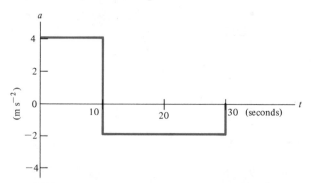

Figure 1.14. Exercise 1-42.

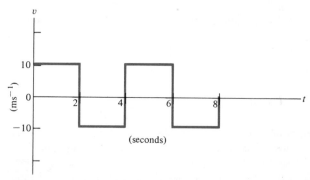

Figure 1.15. Exercise 1-43.

(a) A seat-belted passenger comes to rest in 1 m. What average acceleration does this person experience? (b) Another passenger without a seat belt strikes the windshield and comes to rest in 0.01 m. What average acceleration does this person experience?

1-46 A racing car initially at rest has a constant acceleration for 0.5 kilometres. Its speed at the end of this period is 100 m s^{-1}. (a) What is the acceleration of the car? (b) How long does it require to cover the 0.5-kilometre distance?

1-47 A baseball pitcher throws a ball at 40 m s^{-1}. If the acceleration is approximately constant over a distance of 2 m, how large is it?

1-48 A baseball player catches a ball moving at 30 m s^{-1}. (a) If he does not move his hand, the ball comes to rest in his glove over a distance of 1 cm. What is the average acceleration? (b) If he moves his hand as the ball is caught so that it comes to rest over 10 cm, what is the acceleration?

Section 1.6 | The Acceleration of Gravity and Falling Objects

1-49 From what height must water fall to strike a turbine wheel with a vertical downward velocity of 30 m s^{-1}?

1-50 An antiaircraft shell is fired vertically upward with an initial velocity of 500 m s^{-1}. (a) Compute the maximum height of the shell. (b) How long does it take to reach this height? (c) When will the height be 1000 metres?

1-51 A model rocket is fired straight up from ground level with a constant acceleration of 50 m s^{-2} until the engine runs out of fuel after 4 s. Neglecting air resistance, find (a) the height of the rocket when the engine stops; (b) the maximum height reached; (c) the total time duration of the flight.

1-52 A rock falls from a cliff 60 m high. (a) Find the average velocity during the first 3 seconds of its fall. (b) At what instant of time is the instantaneous velocity equal to the average velocity of part (a)? (c) How long does it take for the rock to fall to the ground?

1-53 A boy standing beside a tall building throws a ball straight up with an initial velocity of 15 m s^{-1}. (a) How high will the ball rise? (b) How long will it take for the ball to reach its maximum height? (c) Another boy reaches out of a window 6 m above the initial position of the ball and

attempts to catch the ball. At what times will the ball pass him?

1-54 A rock dropped from the top of a tower strike the ground in 4 s. (a) Find the velocity of the rock just before it strikes the ground. (b) Find the height of the tower.

1-55 A stone is thrown vertically downward from a bridge with an initial velocity of 100 m s^{-1}. It strikes the water in 3 s. (a) What is the velocity of the stone as it strikes the water? (b) What is the height of the bridge above the water?

1-56 A stone dropped from a bridge strikes the water in 5 s. (a) What is the velocity of the stone when it strikes the water? (b) What is the height of the bridge?

1-57 The hammer of a pile driver strikes the top of a pipe with a velocity of 7 m s^{-1}. From what height did the hammer fall?

1-58 A sandbag dropped from a balloon strikes the ground in 15 s. What was the height of the balloon if it was initially (a) at rest in the air; (b) descending with a velocity of 20 m s^{-1}?

1-59 A box falls from an elevator that is ascending with a velocity of 2 m s^{-1}. It strikes the bottom of the elevator shaft in 3 s. (a) How long will it take the box to reach its maximum height? (b) How far from the bottom of the shaft was the box when it fell off the elevator? (c) What is the height of the elevator when the box is at its highest point?

1-60 Repeat Example 1.14 with the $+x$ direction downward.

1-61 Repeat Example 1.15 with the $+x$ direction downward.

1-62 A car moving at 30 m s^{-1} (108 km h^{-1}) collides head-on with a stone wall. From what height would the car have to fall to achieve the same results?

PROBLEMS

The occasional problems preceded by an asterisk are the most difficult. Answers to most of the odd-numbered problems are given at the end of the book.

1-63 At the instant a bowler releases the ball, her hand is moving horizontally with a speed of 6 m s^{-1} relative to her body. If she is moving forward at a speed of 1 m s^{-1}, what is the velocity of the ball with respect to the floor?

1-64 When a bowling ball is released, a bowler's hand is moving relative to his forearm at

0.82 m s^{-1}. The forearm is moving relative to the upper arm at 0.55 m s^{-1}, and the upper arm is moving at 5.27 m s^{-1} relative to the shoulder. (The velocities refer to the outermost ends of the respective parts of the body and are all horizontal.) If the bowler's shoulder is moving relative to the floor at 1.43 m s^{-1}, how fast is the ball thrown?

1-65   A girl rows 12 km downstream in 2 h. Her return trip takes 3 h. (a) How fast can she row in still water? (b) How fast is the current?

1-66   A boy is on a train moving at 70 km h^{-1}. Relative to the ground, how fast is he moving if he runs at 15 km h^{-1} (a) toward the front of the train; (b) toward the rear?

1-67   The position-versus-time graph for a pendulum is shown in Fig. 1.12. Sketch the velocity-versus-time and acceleration-versus-time graphs for the motion.

1-68   A sled starting from rest slides down a hill with uniform acceleration. It travels 12 m in the first 4 s. When will the sled have a velocity of 4 m s^{-1}?

1-69   A racing car initially at rest accelerates for one-quarter of a kilometre and then brakes to a stop in an additional one-half kilometre? (a) Sketch an approximate velocity-versus-time graph. (b) Sketch an acceleration-versus-time graph.

1-70   Plane A, flying at 500 m s^{-1}, is 10,000 m directly behind plane B, moving in the same direction at 400 m s^{-1}. The pilot of plane A fires a missile that accelerates at 100 m s^{-2}. How long will it take for the missile to reach plane B? (Neglect the effects of the gravitational acceleration.)

1-71   A dog running at 10 m s^{-1} is 30 m behind a rabbit moving at 5 m s^{-1}. When will the dog catch up with the rabbit? (Both velocities remain constant.)

1-72   A rocket-powered experimental sled carrying a test pilot is brought to rest from 200 m s^{-1} over a distance d. If the pilot is not to be subjected to an acceleration greater than six times that of gravity, what is the minimum value of d?

*1-73   The world record for the 100-metre dash is 9.95 s and for the 60-metre dash, it is 6.45 s. Assume a sprinter accelerates at a constant rate up to a maximum velocity that is maintained for the remainder of the race. (a) Find the acceleration. (b) What is the duration of the acceleration period? (c) What is the maximum velocity? (d) The record for the 200-metre dash is 19.83 s, while for the

1000-m run it is about 133.9 s. Are these times consistant with the assumptions made?

1-74   A sandbag is dropped from an ascending balloon that is 300 m above the ground and is ascending at 10 m s^{-1}. (a) What is the maximum height of the sandbag? (b) Find the position and velocity of the sandbag after 5 s. (c) How long does it take for the sandbag to reach the ground from the time it was dropped?

*1-75   In the television show "The Six Million Dollar Man," Colonel Austin had superhuman capabilities. In one episode he tries to catch a man fleeing in a sports car. The distance between them is 100 m when the car begins to accelerate with a constant acceleration of 5 m s^{-2}. Colonel Austin runs at a constant speed of 30 m s^{-1}. Show that he cannot catch the car, and find his distance of closest approach.

1-76   In Fig. 1.7b, the shaded area has a triangular portion (above the v_0 line) and a rectangular portion. Show that the sum of their areas reduces to the right-hand side of Eq. 1.8.

1-77   (a) A magazine article states that cheetahs are the fastest sprinters in the animal world, and that a cheetah was observed to accelerate from rest to 70 km h^{-1} in 2 s. What average acceleration in m s^{-2} does this require? (b) The article also says the cheetah covered 60 metres during that 2-s interval. How large a constant acceleration is implied by this statement? Does it agree with your result in (a)? (c) Accelerations substantially greater than g are difficult for an animal or an automobile to attain, because there is a tendency to slip even on very rough ground with larger accelerations. Given this information, can you guess which number is wrong in the article?

1-78   A rock is dropped into a well and a splash is heard 3 s later. If sound travels in air at 344 m s^{-1}, how deep is the well?

1-79   A lightning flash is seen, and 5 s later thunder is heard. Assuming they are produced simultaneously, how far away is the flash? (In air, sound travels at 344 m s^{-1} and light at $3.00 \times 10^8 \text{ m s}^{-1}$.)

1-80   An estimate of the difficulties involved in space exploration beyond the solar system can be seen from the following calculation. (a) The distance to the moon from the earth is 3.84×10^8 m. Present-day spaceships require about 24 hours to reach the moon. What is the average velocity of these spaceships? (b) The nearest star to our solar system is about 4 light-years away, where a light-year is the distance traveled by light in 1 year (365 days) and the speed of light is 3.00×10^8 m s^{-1}. How many years would it take to reach this star if the spaceship had the velocity calculated in (a)?

^{c}**1-81** A mass on the end of a spring moves so that its position is given by $x = A \sin 2\pi t/T$. (a) Show that the motion repeats after a time T. (T is called the *period*.) (b) What is the greatest distance the mass moves from $x = 0$? (c) Find the instantaneous velocity as a function of time. (d) Find the instantaneous acceleration as a function of time. (e) Show that a is proportional to $-x$ and find the constant of proportionality.

^{c}**1-82** An object falling through a fluid has a velocity at a time t given by $v = v_f(1 - e^{-t/T})$. ($e = 2.718...$ is the base of natural logarithms. See Appendix B.10) (a) Find the velocities at $t = 0$, T, and ∞. (b) Explain why v_f is called the terminal velocity. (c) Show that the instantaneous acceleration is $(v_f - v)/T$. (d) Give a qualitative explanation of your results.

ANSWERS TO REVIEW QUESTIONS

Q1-1, Système Internationale (S.I.); **Q1-2,** random, systematic; **Q1-3,** one; **Q1-4,** displacement; **Q1-5,** displacement, elapsed time; **Q1-6,** an extremely short time; **Q1-7,** velocity change, elapsed time; **Q1-8,** instantaneous velocity; **Q1-9,** instantaneous acceleration; **Q1-10,** a-t; **Q1-11,** v-t; **Q1-12,** air resistance; **Q1-13,** 0, g (downward)

SUPPLEMENTARY TOPICS

1.8 | VERTICAL JUMPING

We can use the equations for constant acceleration to analyze the relative jumping ability of various animals. Table 1.4 lists the heights recorded for some animals jumping vertically. Notice that the jumping height for the human is less than the record for the high jump, which is about 2 m. This is because a man 1.8 m tall is already in a position to clear a bar at about half his height by rotating his body to a horizontal position. The method of human jumping called the Western Roll (Fig. 1.16) is not used by other animals, so the heights listed in Table 1.4 are the appropriate ones to use for comparing jumping abilities.

Figure 1.16. High jumpers using (*a*) the traditional Western Roll and (*b*) the newer Fosbury Flop technique. In both cases the athlete keeps as much of his weight below the bar as possible. ((*a*) United Press International; (*b*) Wide World Photos.)

Animals make standing jumps by folding their legs and extending them rapidly. Usually the acceleration distance d is somewhat shorter than the animal's legs. Once off the ground the animal experiences only the acceleration of gravity, so the constant acceleration formulas apply. We can also analyze the takeoff phase if we make the approximation that the acceleration a_t during takeoff is constant. This approximation is used in the following example.

Example 1.16

Using the data in Table 1.4, find (a) the takeoff velocity v_t for a human and (b) the takeoff acceleration a_t.

(a) Let the coordinate x stand for the position of the person's midpoint, and choose the $+x$ direction upward

TABLE 1.4

Acceleration distances d and vertical heights h for several animals. All distances are in metres.

	Acceleration Distance (d)	Vertical Height (h)
Human	0.5	1.0
Kangaroo	1.0	2.7
Bushbaby	0.16	2.2
Frog	0.09	0.3
Locust	0.03	0.3
Flea	0.0008	0.1

(Fig. 1.17). To find the takeoff velocity, we consider the airborne phase of the jump. During this phase, the acceleration is $-g$ and the velocity changes from $v_0 = v_t$ to $v = 0$. The height changes by $\Delta x = h = 1\,\text{m}$, so $v^2 = v_0^2 + 2a\,\Delta x$ becomes

$$0 = v_t^2 - 2gh$$

or

$$v_t^2 = 2gh$$

Thus

$$v_t = \sqrt{2gh} = \sqrt{(2)(9.8\ \text{m s}^{-2})(1\ \text{m})}$$
$$= 4.4\ \text{m s}^{-1}$$

(b) We will assume that during the takeoff phase the acceleration a_t is constant. The velocity increases from $v_0 = 0$ to $v = v_t$, and the height changes by $\Delta x = d = 0.5\,\text{m}$. Thus, during the takeoff phase, $v^2 = v_0^2 + 2a\,\Delta x$ becomes

$$v_t^2 = 2a_t d$$

Comparing this with $v_t^2 = 2gh$, we find

$$a_t = \frac{h}{d}g$$

$$= \frac{1\ \text{m}}{0.5\ \text{m}}(9.8\ \text{m s}^{-2}) = 19.6\ \text{m s}^{-2}$$

The takeoff accelerations and times are vastly different for different animals. A human jumping with

the same acceleration as a flea would reach a height of more than 50 m! Despite these great differences, in Chapter Six we see that the total energy that a given amount of muscle can supply in a single jump is roughly the same in all animals.

1.9 | FINDING THE MOTION USING INTEGRATION

The acceleration of a swinging pendulum or of a mass oscillating on a spring is continually changing in magnitude and direction. In such situations, we cannot use the uniform acceleration formulas to find the changes in position and velocity even approximately. We show now that if the acceleration is a known function of time, it can be integrated to find the velocity. The velocity can then be integrated to find the position.

From the definition of the average acceleration, $\bar{a} = \Delta v/\Delta t$, the change in the velocity Δv in the time interval between t and $t + \Delta t$ is $\Delta v = \bar{a}\Delta t$. On an a–t graph (Fig. 1.18a), $a\Delta t$ is the area of the colored rectangular strip, where a is the acceleration at t. If Δt is small, the average acceleration $\bar{a}$ approximately equals the acceleration a at the beginning of the interval, so $\Delta v \simeq a\Delta t$. Thus the total change in velocity between the times t_1 and t_2 is the sum of the velocity changes,

$$v_2 - v_1 \simeq \Sigma\, a\Delta t$$

Here Σ (capital Greek sigma) is the conventional symbol for summation.

In the limit where Δt approaches zero, the total area of the strips becomes exactly equal to the area under the a–t curve (Fig. 1.18b). (We saw this earlier for the special case of constant acceleration.) Also, from the definition of an integral as the limit of a sum, we have

$$v_2 - v_1 = \int_{t_1}^{t_2} a\, dt \qquad (1.10)$$

When a is constant, it can be taken out of the integral. Then the integral becomes

$$v_2 - v_1 = a\int_{t_1}^{t_2} dt = a[t]_{t_1}^{t_2} = a(t_2 - t_1)$$

With $\Delta t = t_2 - t_1$ and $\Delta v = v_2 - v_1$, this reduces to $\Delta v = a\Delta t$, the constant-acceleration relationship.

The position can be found from the velocity in a similar fashion. The position change Δx in a time

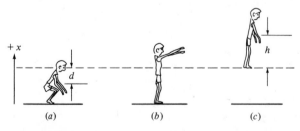

Figure 1.17. Positions during vertical jumping: (a) crouched, with $v = 0$; (b) fully extended at takeoff, with $v = v_t$; and (c) maximum height, with $v = 0$. The coordinate x indicates the height of the person's midpoint.

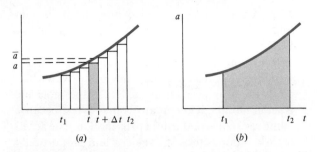

Figure 1.18. (a) The area of the colored strip is $a\Delta t$. In the limit where Δt approaches zero, $\bar{a}$ is equal to the acceleration a at time t. Hence $\Delta v = \bar{a}\Delta t = a\Delta t$, and the velocity change equals the area of the strip. (b) The total velocity change equals the area under the curve from t_1 to t_2.

interval Δt is $\bar{v}\Delta t$. The total position change from t_1 to t_2 is equal to the area under the v–t curve and to the integral

$$x_2 - x_1 = \int_{t_1}^{t_2} v\, dt \qquad (1.11)$$

If v is constant, it can be taken out of the integral. The equation then becomes $x_2 - x_1 = v(t_2 - t_1)$, or $\Delta x = v\Delta t$, as expected.

We illustrate the use of these results with an example of an accelerating car.

Example 1.17

A car starts from rest at time $t = 0$. It moves with an acceleration which diminishes linearly to zero according to the formula $a = a_0(1 - t/T)$, where $a_0 = 2 \text{ m s}^{-2}$ and $T = 10 \text{ s}$ (Fig. 1.19a). (a) What is the velocity of the car when the acceleration reaches zero at $t = T$? (b) How far does the car move during the acceleration period? (c) What is the average velocity from $t = 0$ to $t = T$?

(a) The velocity at time $t = 0$ is zero. At a later time t, the velocity is the integral of the acceleration from 0 to t.

Since we want the velocity at time t, we use t' as the integration variable. Thus

$$v = \int_0^t a_0(1 - t'/T)dt' = a_0 \int_0^t (1 - t'/T)\, dt'$$

Now according to Eq. B.38 in Appendix B.12, $\int t'^n dt' = t'^{n+1}/(n + 1)$. Thus $\int dt' = t'$, $\int t' dt' = t'^2/2$, and v is given by

$$v = a_0[t' - t'^2/2T]_0^t = a_0(t - t^2/2T)$$

This is the velocity at any time t during the acceleration period (Fig. 1.19b). Substituting $t = T$ into this result, the velocity at the end of the acceleration period is

$$v = a_0(T - T^2/2T) = a_0 T/2$$

Using $a_0 = 2 \text{ m s}^{-2}$ and $T = 10 \text{ s}$, we have

$$v = (2 \text{ m s}^{-2})(10 \text{ s})/2 = 10 \text{ m s}^{-1}$$

(b) We find the position change by integrating the velocity. If we measure x from the starting point, then $x = 0$ at $t = 0$, and at later times

$$x = \int_0^t a_0(t' - t'^2/2T)dt'$$

Using Eq. B.38, we have $\int t' dt' = t'^2/2$, $\int t'^2 dt' = t'^3/3$, and

$$x = a_0[t'^2/2 - t'^3/6T]_0^t = a_0(t^2/2 - t^3/6T)$$

This is the distance the car has traveled up to time t (Fig. 1.19c). At $t = T$,

$$x = a_0(T^2/2 - T^3/6T) = a_0 T^2/3$$
$$= (2 \text{ m s}^{-2})(10 \text{ s})^2/3 = 66.7 \text{ m}$$

The car traveled 66.7 m while accelerating.

(c) The average velocity is the displacement divided by the elapsed time,

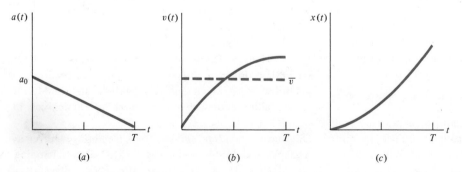

Figure 1.19. Example 1.17. (a) The acceleration of a car satisfies $a = a_0(1 - t/T)$. (b) The velocity of the car versus time if it starts from rest at $t = 0$. Notice the average velocity is more than half the final velocity. (c) Its position versus time.

SCHAUMBURG TWP. PUBLIC LIBRARY

$$\bar{v} = \Delta x / \Delta t = (66.7 \text{ m})/(10 \text{ s}) = 6.67 \text{ m}$$

Note that $\bar{v}$ is *not* equal to the average of the initial and final velocities, $\frac{1}{2}(0 + a_0 T/2) = \frac{1}{2}(10 \text{ m s}^{-1}) = 5 \text{ m s}^{-1}$ (Fig. 1.19b). The average velocity is usually different from the average of the initial and final velocities when the acceleration is not constant.

EXERCISES ON SUPPLEMENTARY TOPICS

Section 1.8 | Vertical Jumping

1-83 A salmon jumps vertically out of the water at an initial velocity of 6 m s^{-1}. (a) How high will it jump? (b) How long will the salmon be out of the water?

1-84 How high will a woman jump if her takeoff velocity is the same as that of the flea?

1-85 From Table 1.4, compute the average takeoff acceleration and takeoff velocity for a locust, assuming the acceleration is constant. Compare these results with that for the human of Example 1.16.

1-86 An astronaut wearing a space suit can jump 0.5 m vertically at the surface of the earth. The gravitational acceleration on Mars is 0.4 times that on the earth. If his takeoff velocity is the same, how high can the astronaut jump on Mars?

1-87 If a human could achieve a takeoff acceleration equal to that of the flea, how high could she jump? (Assume the acceleration distance is still 0.5 metres.)

1-88 Compute the takeoff times for the human, bushbaby, and flea of Table 1.4.

Section 1.9 | Finding the Motion Using Integration

^{c}**1-89** The velocity of an object is $v = bt^2$. If the object is at the origin at $t = 0$, where is it at $t = T$?

^{c}**1-90** An object has an acceleration $a = kt$, where $k = 3$ m s^{-3}. At $t = 0$ s, it is at the origin and has a velocity $v_0 = 10$ m s^{-1}. (a) What is the velocity at $t = 10$ s? (b) What is the position at this time?

PROBLEMS ON SUPPLEMENTARY TOPICS

^{c}**1-91** The velocity of a small object starting from rest at $t = 0$ and falling through a fluid is found to obey the formula $v = v_f(1 - e^{-bt})$, where b is a constant. ($e = 2.718\ldots$ is the base of the natural logarithms. See Appendix B.10.) Find the position at time $t > 0$ if the object is at the origin at $t = 0$.

^{c}**1-92** Using integration techniques, derive the uniform acceleration formulas for v and for Δx, Eqs. 1.5 and 1.8.

^{*c}**1-93** A protein molecule in water is effectively weightless because the buoyancy of the water balances the gravitational attraction. If the molecule is set in motion, it eventually comes to rest because of the resistance of the fluid to its motion. The acceleration is opposite to the velocity and proportional to its magnitude, $a = -cv$. Find the subsequent velocity if the initial velocity at $t = 0$ is v_0. [*Hint:* Find dv/v and integrate.]

Additional Reading

Alfred M. Bork and Arnold B. Arons, Resource Letter Col R-1 on Collateral Reading in Physics, *American Journal of Physics,* vol. 35, 1967, p. 1. A bibliography.

Marjorie Nicholson, Science and Literature, *American Journal of Physics,* vol. 33, 1965, p. 175. A bibliography.

Arnold B. Arons and Alfred M. Bork, *Science and Ideas,* Prentice-Hall, Inc., Englewood Cliffs, N.J., 1964. An anthology.

American Foundation for Continuing Education, *The Mystery of Matter,* Oxford University Press, New York, 1965. An anthology.

National Science Foundation, *Exploring the Universe,* McGraw-Hill Book Co., New York, 1963. An anthology.

W.F. Magie, *Source Book in Physics,* McGraw-Hill Book Co., New York, 1935. An anthology.

L.B. Macurdy, Standards of Mass, *Physics Today,* vol. 4, April 1951, p. 7.

D. Brouwer, The Accurate Measurement of Time, *Physics Today,* vol. 4, August 1951, p. 6.

P. Giacomo, The New Definition of the Meter, *The American Journal of Physics,* vol. 52, 1984, p. 607.

Stillman Drake, *Discoveries and Opinions of Galileo,* Doubleday and Co., Garden City, New York, 1957.

Giorgio de Santillana, *The Crime of Galileo,* University of Chicago Press, Chicago, 1955.

James Hansen, The Crime of Galileo, *Science 81.* March 1981, p. 14.

R.B. Lindsay, Galileo Galilei, 1564–1642, and the Motion of Falling Bodies, *American Journal of Physics,* vol. 10, 1942, p. 285.

Sir James Gray, *How Animals Move,* Cambridge University Press, Cambridge, 1953. Pages 69–80 discuss jumping.

R. McNeill Alexander, *Animal Mechanics,* University of Washington Press, Seattle, 1968. Pages 28–33 discuss jumping.

SCHAUMBURG TWP. PUBLIC LIBRARY

David P. Willoughby, Running and Jumping, *Natural History,* vol. 83, March 1974, p. 68. Comparisons of various animals.

J.B. Rafert and R.N. Nicklin, Velocity Measurement of Humans by Computers, *The Physics Teacher,* vol. 22, 1984, p. 213.

Scientific American articles:

Herbert Butterfield, The Scientific Revolution, September 1960, P. 173.

P.A.M. Dirac, The Physicist's Picture of Nature, May 1963, p. 45.

Freeman J. Dyson, Mathematics in the Physical Sciences, September 1964, p. 128.

Allen V. Astin, Standards of Measurement, June 1968, p. 50.

Lord Ritchie-Calder, Conversion to the Metric System, July 1970, p. 17.

Barry N. Taylor, Donald N. Langenberg, and William H. Parker, The Fundamental Physical Constants, October 1970, p. 62.

J.E. Ravetz, The Origins of the Copernican Revolutions, October 1966, p. 88.

Stillman Drake, Galileo's Discovery of the Law of Free Fall, May 1973, p. 84.

Stillman Drake, The Role of Music in Galileo's Experiments, June 1975, p. 98.

Owen Gingerich, The Galileo Affair, August 1982, p. 132. Galileo's conflict with the church.

Jearl Walker, How to Analyze a City Traffic-Light System From the Outside Looking In, *The Amateur Scientist,* March 1983, p. 138.

CHAPTER 2

MOTION IN TWO DIMENSIONS

Although many of the principles of mechanics can be illustrated by objects moving in a straight line, their applications often involve more complex motions. Animals leaping forward, thrown or struck balls, and figure skaters all move in a vertical or horizontal plane. The definitions of position, velocity, and acceleration and their interrelations can be extended from the one-dimensional case to the more general situation if these quantities are represented by *vectors*. Vectors are mathematical objects having both a magnitude and a direction and are used to represent many other physical quantities in addition to the ones needed to describe motion. By contrast, some physical quantities, such as temperature and time, have no direction and are represented by ordinary numbers or *scalars*.

In the first section we introduce vectors and explain some of the rules for manipulating them. We then restate the definitions of the previous chapter in vector language and show that motion in a plane is equivalent to a pair of one-dimensional motions. This permits us to carry over the results of the preceding chapter and to discuss the motion of projectiles.

2.1 | AN INTRODUCTION TO VECTORS

In this section we introduce vectors and show how to add and subtract them. While vectors are written with an arrow over the symbol ($\vec{A}$), in print they are denoted with boldface type, as in **A**, **s**, and **v**. The magnitude of the vector **A** is written as A or $|\mathbf{A}|$. A vector is pictured in a diagram by an arrow with a length proportional to its magnitude and oriented to indicate its direction. For example, in Fig. 2.1, **A**, **B**, and **C** each have different directions, and **C** is longer or greater in magnitude than **A**.

Addition of Vectors | The concept of vector addition is illustrated by an example involving two vectors representing displacements. Suppose someone walks a certain distance in one direction and then

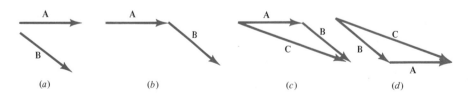

(a) (b) (c) (d)

Figure 2.1. (*a*) The vectors **A** and **B** representing two displacements. (*b*) To add **B** to **A**, we place its tail at the head of **A**. (Moving a vector does not change it so long as the direction and magnitude are not altered.) (*c*) The sum **C** = **A** + **B** is a vector from the tail of the first vector, **A**, to the head of the second vector, **B**. **C** represents the net displacement. (*d*) The order in which the vectors are added does not matter. **A** + **B** = **B** + **A**.

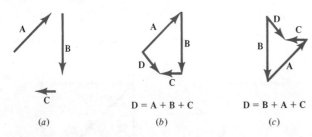

Figure 2.2. (*a*) Three vectors, **A**, **B** and **C**. (*b*) and (*c*) Two possible orderings in computing the sum **D** of the three vectors.

turns and walks another distance in a second direction. The net change in position or displacement will depend on the magnitudes and directions of the two displacements. Let us denote the first displacement by **A** and the second by **B**. The net displacement **C** is the sum of **A** and **B**,

$$C = A + B$$

The procedure for finding **C** is shown in Fig. 2.1. Note that the order of the vectors in the sum is irrelevant: **A** + **B** and **B** + **A** are the same thing. When there are three or more vectors to be summed, two vectors are added first, then the next, and so on. Again, the order of the vectors is irrelevant; for example, in Fig. 2.2, **A** + **B** + **C** = **B** + **A** + **C**.

The following example shows how the magnitude of **A** + **B** depends on the relative directions of **A** and **B**.

Example 2.1

A person walks 1 km due east. If the person then walks a second kilometre, what is the final distance from the starting point if the second kilometre is walked (a) due east; (b) due west; (c) due south?

We will call the first displacement **A** and the second **B**. Using the procedures outlined above, we construct the sum **C** = **A** + **B** for the three cases (Fig. 2.3).

(a) Since **A** and **B** are in the same direction, $C = A + B = 2A = 2$ km. **C** is directed due east.

(b) Here, the vectors are opposite, so $C = A - B = 0$.

(c) From the Pythagorean theorem, $C^2 = A^2 + B^2 = 2A^2$, so

$$C = \sqrt{2}\,A = \sqrt{2}\,\text{km}$$

From Fig. 2.3c, **C** points toward the southeast.

Multiplication of a Vector By a Scalar

Multiplication of a vector by a scalar is defined so that the usual rules of algebra apply. If $2A = A + A$ is to hold, then we must interpret $2A$ as a vector in the same direction as **A** but with twice the length. For example, if **A** is a 2-km displacement due north, $2A$ is a 4-km displacement due north, and $5A$ is a 10-km displacement due north.

Vector Subtraction

Subtraction of vectors is also defined so that the usual algebraic rules apply. In order that $A - A = 0$ holds, $-A$ must be interpreted as a vector equal in magnitude to A but opposite in direction. Thus if **A** is a 2-km displacement due north, $-A$ is a 2-km displacement due south and $-3A$ is a 6-km displacement due south. The vector difference $C = B - A = B + (-A)$ is evaluated by adding the vector $-A$ to the vector **B** as in Fig. 2.4.

Vector Components

A vector in a given plane is specified by two pieces of information, its magnitude and its direction. Equivalently, the vector can be specified by two other quantities, its *components* along a pair of perpendicular axes. These components are often useful in vector calculations.

The procedure for finding the components of a vector is shown in Fig. 2.5. The vector **A** is redrawn from the origin of the *x-y* axes, and the components of **A**, denoted by A_x and A_y, are constructed by drawing the dashed lines at right angles to the axes. Expressions relating the components to the magnitude of **A** and its angle θ (theta) with the *x* axis can be obtained using some properties of right triangles.

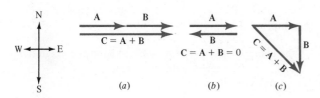

Figure 2.3. The sum of two displacement vectors that are equal in magnitude and (*a*) parallel; (*b*) opposite or antiparallel; and (*c*) perpendicular.

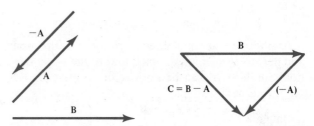

Figure 2.4. The vector **C** = **B** − **A** is found by adding −**A** to **B**.

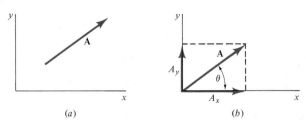
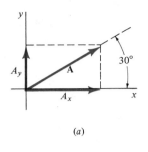

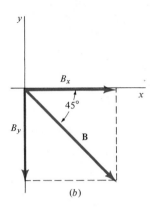

Figure 2.5. (*a*) A vector **A** and perpendicular *x* and *y* axes. (*b*) The components of **A** are A_x and A_y.

Figure 2.7. Example 2.2.

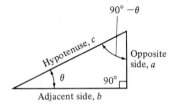

Figure 2.6. A right triangle.

Figure 2.6 shows a right triangle with sides *a*, *b*, and *c*. These satisfy the Pythagorean relation,

$$a^2 + b^2 = c^2 \qquad (2.1)$$

The sine, cosine, and tangent of the angle θ are defined by

$$\sin\theta = \frac{\text{opposite side}}{\text{hypotenuse}} = \frac{a}{c}$$

$$\cos\theta = \frac{\text{adjacent side}}{\text{hypotenuse}} = \frac{b}{c}$$

$$\tan\theta = \frac{\text{opposite side}}{\text{adjacent side}} = \frac{a}{b}$$

Thus, in Fig. 2.5*b*, the components A_x and A_y satisfy

$$A_x^2 + A_y^2 = A^2 \qquad (2.2)$$

Also,

$$\cos\theta = \frac{A_x}{A}, \qquad \sin\theta = \frac{A_y}{A} \qquad (2.3)$$

or

$$A_x = A\cos\theta, \qquad A_y = A\sin\theta \qquad (2.4)$$

A_x is positive when it is directed in the $+x$ direction and negative when it is directed toward the $-x$ direction. Similarly A_y can be positive or negative, as we see in the next example.

Example 2.2
Find the components of the vectors **A** and **B** in Fig. 2.7, if $A = 2$ and $B = 3$.

Using trigonometric tables or an electronic calculator, we find that $\cos 30° = 0.866$ and $\sin 30° = 0.500$. Thus,

$$A_x = A\cos\theta = 2\cos 30° = 2(0.866) = 1.73$$
$$A_y = A\sin\theta = 2\sin 30° = 2(0.500) = 1.00$$

From Fig. 2.7*b*, B_x is positive and B_y is negative. With $\cos 45° = \sin 45° = 0.707$,

$$B_x = 3\cos 45° = 3(0.707) = 2.12$$
$$B_y = -3\sin 45° = -3(0.707) = -2.12$$

The sum of two or more vectors can be calculated conveniently in terms of components. To do this, we define the *unit vector* $\hat{x}$ (read "x-hat"), a vector of length one in the $+x$ direction. Similarly, $\hat{y}$ is a unit vector in the $+y$ direction. With these unit vectors, a vector **A** with components A_x and A_y can be written as

$$\mathbf{A} = A_x\hat{x} + A_y\hat{y}$$

If a vector **B** is also written as

$$\mathbf{B} = B_x\hat{x} + B_y\hat{y}$$

the sum $\mathbf{C} = \mathbf{A} + \mathbf{B}$ is

$$\mathbf{C} = (A_x + B_x)\hat{x} + (A_y + B_y)\hat{y} \qquad (2.5)$$

The components of **C** *are the sums of the components of* **A** *and* **B** (Fig. 2.8). This is illustrated by the following example.

Example 2.3
In Fig. 2.9, $\mathbf{A} = 2\hat{x} + \hat{y}$ and $\mathbf{B} = 4\hat{x} + 7\hat{y}$. (a) Find the components of $\mathbf{C} = \mathbf{A} + \mathbf{B}$. (b) Find the magnitude of **C** and its angle θ with the *x* axis.
(a) Using Eq. 2.5,

$$\mathbf{C} = (2 + 4)\hat{x} + (1 + 7)\hat{y} = 6\hat{x} + 8\hat{y}$$

Thus $C_x = 6$, and $C_y = 8$.

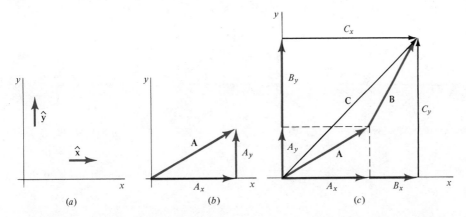

Figure 2.8.   (a) $\hat{x}$ and $\hat{y}$ are unit vectors, vectors of length one directed along the coordinate axes. (b) A vector **A** can be constructed from its components A_x and A_y. (c) **C** = **A** + **B** in components is $C_x = A_x + B_x$, $C_y = A_y + B_y$.

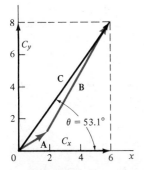

Figure 2.9.   Example 2.3.

(b) From the Pythagorean theorem,

$$C^2 = C_x^{\ 2} + C_y^{\ 2} = 6^2 + 8^2 = 100$$

so $C = 10$. From Fig. 2.9, we see that the angle θ satisfies

$$\tan \theta = \frac{C_y}{C_x} = \frac{8}{6} = 1.333$$

Using a calculator, we find $\theta = 53.1°$.

2.2 | THE VELOCITY IN TWO DIMENSIONS

In two dimensions, position, velocity, and acceleration are represented by vectors. Their definitions are very similar to those for motion along a straight line, and their x components are related to each other in the same way as are x, v, and a for straight-line motion. Since this is also true for the y components, *a problem involving motion in a plane is effectively a pair of one-dimensional motion problems.*

Figure 2.10 shows an object moving in a plane labeled with x and y axes. The object could be a car, an animal, or a red blood cell, and it is represented symbolically by a point. If the displacement in a time interval Δt is denoted by the vector $\Delta \mathbf{s}$, then the *average velocity* of the object is parallel to $\Delta \mathbf{s}$ and is given by

$$\bar{\mathbf{v}} = \frac{\Delta \mathbf{s}}{\Delta t} \tag{2.6}$$

We can write $\Delta \mathbf{s}$ in terms of its components in the x and y direction:

$$\Delta \mathbf{s} = \Delta x \hat{\mathbf{x}} + \Delta y \hat{\mathbf{y}} \tag{2.7}$$

Then the average velocity can also be written in terms of its components.

$$\bar{\mathbf{v}} = \frac{\Delta x}{\Delta t}\hat{\mathbf{x}} + \frac{\Delta y}{\Delta t}\hat{\mathbf{y}} \tag{2.8}$$

Thus the components of $\bar{\mathbf{v}}$ are

$$\bar{v}_x = \frac{\Delta x}{\Delta t} \qquad \bar{v}_y = \frac{\Delta y}{\Delta t} \tag{2.9}$$

Each component of the average velocity looks like a one-dimensional average velocity.

Since the average velocity in a plane is equivalent to 2 one-dimensional average velocities, everything stated in the preceding chapter about the average and instantaneous velocities holds here for each component. For example, the *instantaneous velocity* **v** is the limit of the average velocity as the time interval Δt approaches zero. In component form

$$\mathbf{v} = v_x \hat{\mathbf{x}} + v_y \hat{\mathbf{y}} \tag{2.10}$$

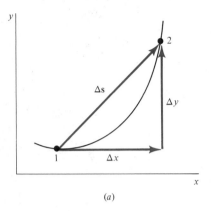
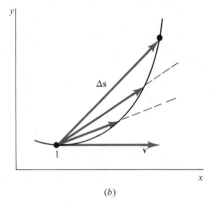

(a)   (b)

Figure 2.10. (*a*) An object moves along a path in a plane. At time t_1, it is at point 1, and at t_2 it is at point 2. The average velocity is parallel to $\Delta\mathbf{s}$. (*b*) As the time interval $t_2 - t_1$ becomes smaller, so does the displacement $\Delta\mathbf{s}$. The average velocity $\overline{\mathbf{v}} = \Delta\mathbf{s}/\Delta t$ approaches the instantaneous velocity $\mathbf{v}$ at t_1, which is tangent to the path at point 1.

Here $v_x = dx/dt$ and $v_y = dy/dt$ are the rates of change of the x and y coordinates of the object with respect to the time. Again, the derivatives dx/dt and dy/dt are shorthand notations for the limits of $\Delta x/\Delta t$ and $\Delta y/\Delta t$ as Δt approaches zero. At any instant, v is directed along the tangent to the path or *trajectory* of the object (Fig. 2.10*b*). These ideas are illustrated by the following examples.

Example 2.4

A car travels halfway around an oval racetrack at a constant speed of 30 m s^{-1} (Fig. 2.11). (a) What are its instantaneous velocities at points 1 and 2? (b) It takes 40 s to go from 1 to 2, and these points are 300 m apart. What is the average velocity of the car during this time interval?

(a) The instantaneous velocity is tangent to the path of

the car, and its magnitude is equal to the speed. Thus, at point 1 the velocity is directed in the $+y$ direction and $\mathbf{v}_1 = 30$ m s$^{-1}\hat{\mathbf{y}}$. Similarly, at point 2 the velocity is along the $-y$ direction, and $\mathbf{v}_2 = (30$ m s$^{-1})(-\hat{\mathbf{y}}) = -30$ m s$^{-1}\hat{\mathbf{y}}$.

(b) The average velocity is the displacement divided by the elapsed time. The displacement is entirely along the x direction, so $\Delta\mathbf{s} = (300$ m$)\hat{\mathbf{x}}$. Since $\Delta t = 40$ s,

$$\Delta\overline{\mathbf{v}} = \frac{\Delta\mathbf{s}}{\Delta t} = \frac{300 \text{ m}}{40 \text{ s}}\hat{\mathbf{x}} = 7.5 \text{ m s}^{-1}\hat{\mathbf{x}}$$

The average velocity during this time interval is directed along the $+x$ axis. Its magnitude is less than the speed of 30 m s^{-1} because the car does not travel in a straight line.

Example 2.5

An object moving in a circle about the origin has coordinates which vary in time according to

$$x = R \cos \omega t \qquad y = R \sin \omega t$$

where R and ω are constants. (a) Find the velocity components. (b) Show that the speed is constant in time.

(a) The velocity component v_x is dx/dt. Using Eq. B.26 in Appendix B, $(d/dt) \cos \omega t = -\omega \sin \omega t$, so

$$v_x = \frac{d}{dt}(R \cos \omega t) = -\omega R \sin \omega t$$

Similarly, $(d/dt) \sin \omega t = \omega \cos \omega t$, and with $v_y = dy/dt$,

$$v_y = \frac{d}{dt}(R \sin \omega t) = \omega R \cos \omega t$$

(b) The speed is the magnitude of the velocity, v. Using the Pythagorean theorem,

$$v^2 = v_x{}^2 + v_y{}^2 = \omega^2 R^2(\sin^2 \omega t + \cos^2 \omega t)$$

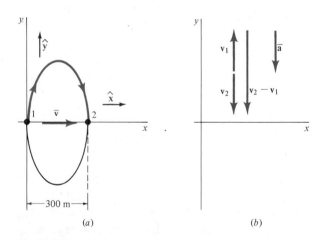

(a)   (b)

Figure 2.11. (*a*) A car goes halfway around an oval track. Its average velocity $\overline{\mathbf{v}}$ during this interval points in the $+x$ direction. (*b*) The average acceleration $\overline{\mathbf{a}}$ points in the $-y$ direction.

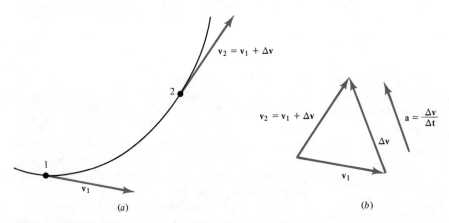

Figure 2.12. (a) The black curve represents the path of an object. The velocities at points 1 and 2 are vectors tangent to the trajectory at times t_1 and t_2. (b) The average acceleration $\bar{\mathbf{a}}$ from t_1 to t_2 is parallel to $\Delta\mathbf{v}$.

Since $\sin^2\theta + \cos^2\theta = 1$ for any argument, $v^2 = \omega^2 R^2$. Thus the velocity has the constant magnitude ωR even though its components are changing in magnitude and sign.

2.3 | THE ACCELERATION IN TWO DIMENSIONS

The acceleration of an object moving in a plane is defined much as in the one-dimensional case. Suppose that in a time interval $\Delta t = t_2 - t_1$ the velocity changes by $\Delta\mathbf{v} = \mathbf{v}_2 - \mathbf{v}_1$. Then the *average acceleration* is

$$\bar{\mathbf{a}} = \frac{\Delta\mathbf{v}}{\Delta t} = \frac{\mathbf{v}_2 - \mathbf{v}_1}{t_2 - t_1} \qquad (2.11)$$

The *instantaneous acceleration* $\mathbf{a}$ is the limit of the average acceleration as the time interval Δt approaches zero (Fig. 2.12). In components,

$$\mathbf{a} = a_x\hat{\mathbf{x}} + a_y\hat{\mathbf{y}} \qquad (2.12)$$

Here $a_x = dv_x/dt$ and $a_y = dv_y/dt$ are the rates of change of v_x and v_y with respect to the time. The following examples illustrate these definitions.

Example 2.6

In Example 2.4 the velocity of the car changed from $\mathbf{v}_1 = 30 \text{ m s}^{-1}\hat{\mathbf{y}}$ to $\mathbf{v}_2 = -30 \text{ m s}^{-1}\hat{\mathbf{y}}$ in 40 s. What was the average acceleration of the car in that time interval?

The average acceleration is defined as the velocity change divided by the elapsed time:

$$\bar{\mathbf{a}} = \frac{\mathbf{v}_2 - \mathbf{v}_1}{\Delta t} = \frac{(-30 \text{ m s}^{-1}\hat{\mathbf{y}}) - (30 \text{ m s}^{-1}\hat{\mathbf{y}})}{40 \text{ s}}$$

$$= -1.5 \text{ m s}^{-2}\hat{\mathbf{y}}$$

Thus the average acceleration during the time the car goes from point 1 to point 2 is directed in the $-y$ direction, or downward in Fig. 2.11*b*.

Example 2.7

In Example 2.5, an object moving in a circle was found to have the velocity components

$$v_x = -\omega R \sin\omega t \qquad \text{and} \qquad v_y = \omega R \cos\omega t$$

Find the acceleration components.

Using the definitions of the acceleration components, we have,

$$a_x = \frac{dv_x}{dt} = \frac{d}{dt}(-\omega R \sin\omega t) = -\omega^2 R \cos\omega t$$

$$a_y = \frac{dv_y}{dt} = \frac{d}{dt}(\omega R \cos\omega t) = -\omega^2 R \sin\omega t$$

Note that if we apply the Pythagorean theorem, we find that a has a constant magnitude, $a = \omega^2 R$.

This example illustrates an important point. A constant *velocity* means that the acceleration is zero, but a constant *speed* may or may not correspond to zero acceleration. An object moving in a curved path with constant speed has its velocity changing direction, so it is accelerating. We feel the effects of this acceleration when a car turns a corner quickly.

To reiterate, the acceleration is zero only when the speed and the direction of motion are both constant.

2.4 | FINDING THE MOTION OF AN OBJECT

As we have stressed, a problem involving motion of an object in a plane is equivalent to a pair of straight-line motion problems. Consequently the x and y components of the position and velocity of an object can be calculated exactly as before if the acceleration and the initial position and velocity are known. In one dimension, when the acceleration is constant, the displacement and velocity are given by $\Delta x = v_0 \, \Delta t + \frac{1}{2} a \, (\Delta t)^2$ and $v = v_0 + a \, \Delta t$. Here, similar equations hold for the x and y motions separately:

$$\Delta x = v_{0x} \, \Delta t + \tfrac{1}{2} a_x \, (\Delta t)^2$$
$$\Delta y = v_{0y} \, \Delta t + \tfrac{1}{2} a_y \, (\Delta t)^2 \qquad (2.13)$$

$$v_x = v_{0x} + a_x \, \Delta t$$
$$v_y = v_{0y} + a_y \, \Delta t \qquad (2.14)$$

Each of the other constant acceleration equations tabulated on the front endpapers also leads to a pair of component equations for two-dimensional motion. As in one dimension, integration methods are required if the acceleration varies in time.

The use of these equations is illustrated in the next section.

2.5 | PROJECTILES

Kicked or thrown balls, jumping animals, and objects dropped from windows are all examples of *projectiles*. *If air resistance is neglected, the motion of a projectile is influenced only by the constant gravitational acceleration.* Given the initial position and velocity, the subsequent position and velocity can be found using the constant acceleration equations.

To use these equations for the motion of a projectile, let us choose the x axis to be horizontal and the $+y$ axis to be directed upward. Then $a_x = 0$ and $a_y = -g$, and Eqs. 2.13 and 2.14 become

$$\Delta x = v_{0x} \, \Delta t, \qquad \Delta y = v_{0y} \, \Delta t - \tfrac{1}{2} g \, (\Delta t)^2 \quad (2.15)$$
$$v_x = v_{0x}, \qquad v_y = v_{0y} - g \, \Delta t \qquad (2.16)$$

The Δx and v_x equations show that the horizontal motion is uniform, and the Δy and v_y equations show the vertical motion is that of an object acting only under the influence of gravity.

The equations for the vertical motion can be used as in straight-line motion to answer various questions. For example, the projectile strikes the ground when y equals the ground height, and the peak height occurs when $v_y = 0$. This approach can be applied to qualitative and quantitative projectile motion problems.

Two interesting examples of projectile motion are often shown as lecture demonstrations. In one of these, two steel balls are simultaneously released from a stand above the floor. Ball 1 is projected horizontally by a spring, while ball 2 is allowed to drop from rest. The problem is to predict which ball will hit the floor first.

Although this problem sounds hard initially, its solution is apparent once we realize that both balls have the same initial vertical velocity and position components and the same acceleration. Consequently, the equations for the vertical motion of the two balls are the same, and they must hit the floor simultaneously (Fig. 2.13). This illustrates clearly the idea that motion in the horizontal and vertical directions are independent.

The second demonstration is a bit more elaborate. A projectile is fired by a "cannon" pointed at a stuffed animal (Fig. 2.14). Just as the projectile leaves the cannon, the stuffed animal is released and falls. Somewhat surprisingly, the projectile strikes the animal in midair. The point is that if there were no gravitational acceleration, the projectile would travel in a straight line. However, since both the projectile and the stuffed animal experience the same gravitational acceleration, both fall at the same rate relative to where they would have been otherwise. Hence, if the cannon is aimed directly at the animal, the only effect

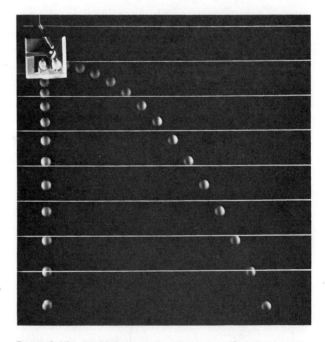

Figure 2.13. Multiple exposure photographs show that a ball dropped from rest and one projected forward fall at the same rate. (Reproduced by permission of the publisher from PSSC Physics, fourth edition. 1976, by D.C. Heath and Company, Lexington, Massachusetts.)

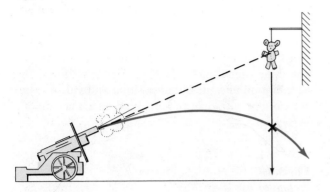

Figure 2.14. The projectile and stuffed animal fall together when simultaneously released. They collide at the point indicated with a cross.

of the gravitational acceleration is that they meet somewhat below where the animal was initially.

The following examples illustrate the solution of quantitative projectile motion problems.

Example 2.8

A ball is kicked from ground level with a velocity of 25 m s^{-1} at an angle of 30° to the horizontal direction. (Fig. 2.15) (a) When does it reach its greatest height? (b) Where is it at that time?

(a) From Fig. 2.15b, the initial velocity has components

$$v_{0x} = v_0 \cos 30° = (25 \text{ m s}^{-1})(0.866) = 21.7 \text{ m s}^{-1}$$
$$v_{0y} = v_0 \sin 30° = (25 \text{ m s}^{-1})(0.500) = 12.5 \text{ m s}^{-1}$$

The greatest height is reached when $v_y = 0$. Using $v_y = v_{0y} - g \Delta t$, this occurs when

$$\Delta t = \frac{(v_{0y} - v_y)}{g} = \frac{(12.5 \text{ m s}^{-1} - 0)}{9.8 \text{ m s}^{-2}} = 1.28 \text{ s}$$

(b) With Eqs. 2.15, the displacement after 1.28 s is given by

$$\Delta x = v_{0x} \Delta t = (21.7 \text{ m s}^{-1})(1.28 \text{ s}) = 27.8 \text{ m}$$
$$\Delta y = v_{0y} \Delta t - \tfrac{1}{2}g (\Delta t)^2$$
$$= (12.5 \text{ m s}^{-1})(1.28 \text{ s}) - \tfrac{1}{2}(9.8 \text{ m s}^{-2})(1.28 \text{ s})^2$$
$$= 7.97 \text{ m}$$

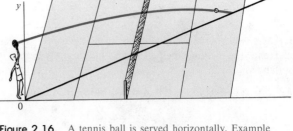

Figure 2.16. A tennis ball is served horizontally. Example 2.9.

Thus the ball is 7.97 m above a point on the ground 27.8 m from where it was kicked.

Example 2.9

A tennis ball is served horizontally from 2.4 m above the ground at 30 m s^{-1} (Fig. 2.16). (a) The net is 12 m away and 0.9 m high. Will the ball clear the net? (b) Where will the ball land?

(a) To find the height of the ball at the net, we must first use the equations for the horizontal motion to find out when it will reach the net, where $\Delta x = 12$ m. From this time, we can determine the height. Solving $\Delta x = v_{0x} \Delta t$ for Δt,

$$\Delta t = \frac{\Delta x}{v_{0x}} = \frac{12 \text{ m}}{30 \text{ m s}^{-1}} = 0.4 \text{ s}$$

Using $\Delta t = 0.4$ s and $v_{0y} = 0$, the vertical displacement is

$$\Delta y = v_{0y} \Delta t - \tfrac{1}{2}g (\Delta t)^2$$
$$= -\tfrac{1}{2}(9.8 \text{ m s}^{-2})(0.4 \text{ s})^2 = -0.78 \text{ m}$$

Since the ball was initially 2.4 m above the ground, it is now $(2.4 - 0.78)$ m $= 1.62$ m above the ground, so it easily clears the net.

(b) The ball lands when $\Delta y = -2.4$ m. Once we have found this time interval, we can find the horizontal displacement from the equation for Δx. Substituting $v_{0y} = 0$ in $\Delta y = v_{0y} \Delta t - \tfrac{1}{2}g (\Delta t)^2$, we find

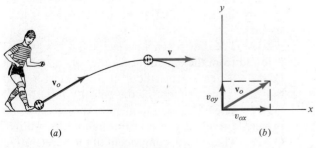

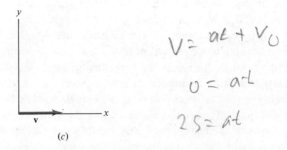

Figure 2.15. (a) A ball kicked from ground level has $v_y = 0$ at its highest point. Example 2.8. (b) Components of the initial velocity. (c) Components of the velocity at the highest point. Notice $v_y = 0$.

$$\Delta y = -\tfrac{1}{2}g\,(\Delta t)^2$$

$$(\Delta t)^2 = \frac{-2\Delta y}{g} = \frac{-2(-2.4 \text{ m})}{9.8 \text{ m s}^{-2}} = 0.490 \text{ s}^2$$

$$\Delta t = 0.700 \text{ s}$$

Hence the distance the ball travels horizontally before it lands is

$$\Delta x = v_{0x}\,\Delta t = (30 \text{ m s}^{-1})(0.700 \text{ s}) = 21.0 \text{ m}$$

Projectiles in Athletics | The preceding example illustrates how the projectile motion formulas can be used to analyze a tennis serve. A serve should clear the net by a small but safe margin and land just inside the service court. If a beginner serves the ball rather slowly, the ball will not clear the net unless it is served slightly above the horizontal direction. More advanced players can serve horizontally or slightly below the horizontal, since the ball moves faster and has a flatter trajectory. While a player can determine his or her own best serving angle by trial and error, the projectile motion formulas can be used to predict this angle given the initial speed. The advice given in textbooks on tennis is sometimes based on this kind of analysis. Many other athletic events involving projectiles that are thrown, kicked, or struck can be discussed using the projectile motion formulas.

SUMMARY

In two dimensions, position, velocity, and acceleration have both magnitude and direction, so they are represented by vectors. In diagrams, vectors are designated by arrows. To find the sum $C = A + B$, we draw the tail of B from the head of A; C is then a vector from the tail of A to the head of B. Multiplying a vector by a positive scalar or ordinary number increases its magnitude by that factor without altering its direction; multiplying by a negative scalar reverses its direction as well as changing the magnitude.

Many calculations involving vectors are simplified by choosing a convenient pair of x-y axes and finding the vector components. If A makes an angle θ with respect to the positive x direction, then its components are $A_x = A \cos\theta$ and $A_y = A \sin\theta$. Adding the x components of two or more vectors gives the x component of their sum. Similarly, adding the y components gives the y component of the sum.

The velocity and acceleration in a plane are defined by restating the one-dimensional definitions in

vector form: $\overline{\mathbf{v}} = \Delta \mathbf{s}/\Delta t$ and $\overline{\mathbf{a}} = \Delta \mathbf{v}/\Delta t$. Also, the x components of these quantities are related to each other exactly as in one dimension, as are the y components, so motion in a plane is effectively a pair of one-dimensional motions. Therefore, if we know the acceleration and the initial position and velocity, the subsequent motion can be found much as in one dimension.

Checklist

Define or explain:

vector
scalar
addition of vectors
subtraction of vectors
vector components
unit vector

velocity in two
　dimensions
acceleration in two
　dimensions
projectile

REVIEW QUESTIONS

Q2-1　What is the difference between a scalar and a vector?

Q2-2　How do we represent vectors in writing and in textbooks?

Q2-3　If A is parallel to B, what is the magnitude of $A + B$? Of $A - B$? Of $2A$?

Q2-4　If A is perpendicular to B, what is the magnitude of $A + B$?

Q2-5　The average velocity in a plane is the _____ divided by the _____.

Q2-6　The average acceleration in a plane is the _____ divided by the _____.

Q2-7　The acceleration is zero when both the _____ and _____ of the velocity are constant.

Q2-8　Motion in a plane is equivalent to a pair of _____.

Q2-9　The motion of a projectile is influenced only by the _____ (assuming air resistance is negligible).

Q2-10　The time when a projectile hits the ground is found from the equations for the _____ motion.

Q2-11　When a projectile is at its greatest height, the _____ component of the velocity is zero.

Q2-12　The _____ component of the velocity of a projectile remains constant throughout its motion.

Q2-13　The vertical component of the acceleration of a projectile is _____, and the horizontal component is _____.

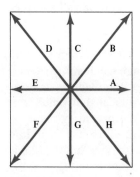

Figure 2.17. Exercise 2-1.

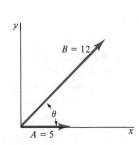

Figure 2.18. Exercises 2-2 and 2-4.

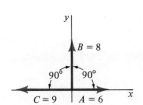

Figure 2.21. Exercise 2-11.

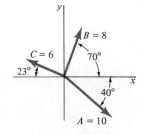

Figure 2.22. Exercise 2-12.

EXERCISES

Section 2.1 | An Introduction to Vectors

2-1 Figure 2.17 shows a collection of vectors that can be combined in various ways. For example, $A + C = B$. Find (a) $E + C$; (b) $A + F$; (c) $A + D$; (d) $E + A$; (e) $E + 2A$; (f) $A - B$; (g) $B - A$; (h) $C - A$.

2-2 In Fig. 2.18, for what value of θ will $C = A + B$ have (a) a minimum magnitude, and (b) a maximum magnitude? (c) Find C when $\theta = 90°$.

2-3 A vector has an x component of -10 and a y component of $+3$. (a) Draw a set of x-y axes and show the vector. (b) Calculate the magnitude and direction of the vector.

2-4 If $\theta = 72°$ in Fig. 2.18, find (a) the direction and magnitude of $C = A + B$ by constructing a drawing using a ruler and a protractor; (b) the direction and magnitude of C using the component method.

2-5 For the vectors A and B in Fig. 2.19, find (a) $A + B$; (b) $B - A$; (c) $A - B$.

2-6 Using components for the vectors in Fig. 2.20, find the direction and magnitude of $E = A + B + C + D$.

2-7 Using components for the vectors in Fig. 2.20, find the direction and magnitude of $F = A - C + B - 2D$.

2-8 A woman walks 10 km north, turns toward the northwest, and walks 5 km further. What is her final position?

2-9 A ship sets out to sail 100 km north but is blown by a severe storm to a point 200 km east of its starting point. How far must it sail, and in what direction, to reach its intended destination?

2-10 One person walks northeast at 3 km h^{-1}, and another heads south at 4 km h^{-1}. How far apart are they after 2 hours?

2-11 For the vectors in Fig. 2.21, find the magnitude and direction of (a) $D = A + B + C$; (b) $E = A - B - C$.

2-12 For the vectors in Fig. 2.22, find the magnitude and direction of (a) $D = A + B + C$; (b) $E = A - B - C$.

Section 2.2 | The Velocity in Two Dimensions

2-13 A car goes around a circular track 500 metres in diameter at a constant speed of 20 m s^{-1}. (a) How long does it take for the car to go halfway around the track? (b) What is its average velocity in that time interval?

2-14 A car goes around a circular track with a radius of 1000 metres at a constant speed of 10 m s^{-1}. (a) How long does it take the car to go once completely around the track? (b) What is the average velocity of the car over this time interval?

2-15 A ball is thrown at 30 m s^{-1} at an angle of 20° to the horizontal direction. Find the horizontal and vertical components of its initial velocity.

2-16 A plane flies for 3 hours and reaches a point 600 km north and 800 km east of its starting point. Find the direction and magnitude of its average velocity.

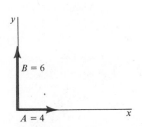

Figure 2.19. Exercise 2-5.

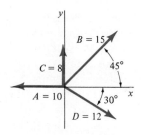

Figure 2.20. Exercises 2-6 and 2-7.

c2-17 An object has its position given by the formula $r = pt\hat{x} + qt^2\hat{y}$, where p and q are constants. Find its instantaneous velocity components as functions of time.

Section 2.3 | The Acceleration in Two Dimensions

2-18 A tennis ball is served by a player, bounces in the opponent's court, and is hit by the opponent toward the first player. Describe the direction and magnitude of the acceleration during each part of the motion.

2-19 A car initially traveling due north goes around a semicircle having a radius of 500 m at a constant speed of 20 m s^{-1}. (a) How long does this take? (b) What is the magnitude and direction of the average acceleration?

2-20 The earth rotates about the sun once each year on an approximately circular path. Find the magnitude of the average acceleration associated with this motion over a 6-month interval. (The average distance from the earth to the sun is 1.50×10^{11} m.)

2-21 A rifle pointed at 30° to the horizontal fires a bullet at 250 m s^{-1}. If the bullet is accelerated uniformly in the barrel for 0.006 s, find (a) the magnitude of the acceleration; (b) its horizontal and vertical components.

c2-22 Find the instantaneous acceleration of the object in Exercise 2-17.

Section 2.4 | Finding the Motion of an Object and Section 2.5 | Projectiles

2-23 A football kicked into the air from ground level hits the ground 30 m from where it started after 4 s. (a) Find its average velocity while in the air. (b) Find its average acceleration while in the air.

2-24 A ball is thrown horizontally at 20 m s^{-1} from a window 15 m above the ground. (a) When will it hit the ground? (b) Where will it land?

2-25 A rifle pointed at 30° to the horizontal fires a bullet at 500 m s^{-1}. The rifle barrel is 0.7 m long. (a) Find the average acceleration in the rifle barrel. (b) Find the horizontal and vertical components of this acceleration.

2-26 A snowball is thrown from 2 m above the ground at a velocity of 10 m s^{-1} directed at 30° above the horizontal. (a) Find its horizontal and vertical position after 1 s. (b) Find its velocity components after 1 s.

2-27 (a) How long will the snowball of the preceding exercise be in the air? (b) Where will it land?

2-28 A baseball is hit at 40 m s^{-1} at angle of 30° to the horizontal. (a) How high will it go? (b) When will it reach that height? (c) What will be its horizontal distance from the batter at that time?

2-29 For the baseball in the preceding exercise, find (a) the distance it will travel; (b) the total time it will be in the air. (Neglect the fact that the ball is struck from slightly above ground.)

2-30 A rifle is aimed directly at a target 200 m away at the same height as the rifle. If the bullet leaves the muzzle at 500 m s^{-1}, by how much will it miss the target?

2-31 The earth revolves about its axis every 24 hours. Find the magnitude of the average acceleration of a point on the equator over a 6-hour time interval. (The radius of the earth is 6.38×10^6 m.)

PROBLEMS

2-32 Suppose that in Fig. 2.14 the stuffed animal is initially 1 m above and 1.5 m to the right of the cannon that is pointed at the animal. The animal starts to fall as the cannonball is fired at 10 m s^{-1}. (a) What are the initial horizontal and vertical velocity components of the cannonball? (b) How long does it take for the horizontal coordinate of the ball to change by 1.5 m? (c) What are the vertical positions of the ball and the animal at that time?

2-33 One ball is thrown horizontally with a velocity v_0 from a height h, and another is thrown straight down with the same initial speed. (a) Which ball will land first? (b) Which ball will have the greater speed as it is about to land?

2-34 A boy throws a ball so that it rises 1 m while travelling 7 m horizontally and then begins to drop. What were the initial speed and direction of the ball?

*2-35 A man wishes to row across a river 0.5 km wide. He points his boat toward the opposite shore and rows at 2 km h^{-1} relative to the water. The current is 4 km h^{-1}. (a) How long does it take him to cross the river? (b) Where does he land?

2-36 A tennis ball is served horizontally at a height of 2.4 m, 12 m from a net that is 0.9 m high. (a) If it is to clear the net by at least 0.2 m, what is its minimum initial velocity? (Neglect air resistance.) (b) If it clears the net by 0.2 m, where will it land?

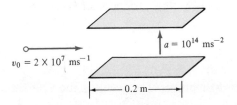

Figure 2.23.   Problem 2-38.

2-37   Some books advise serving a tennis ball at an angle below the horizontal direction. To see if this is sound advice, suppose a ball is struck at an angle of 5° downward at a height of 2.4 m with the relatively high speed of 30 m s^{-1}. How high will it be when it reaches the net 12 m away? (The net is 0.9 m high. Neglect air resistance.)

2-38   The screens on cathode ray tubes in television sets and oscilloscopes emit light when they are struck by rapidly moving electrons. Electrical deflecting plates are used to control where the electrons strike. In Fig. 2.23, electrons with an initial horizontal velocity of 2×10^7 m s^{-1} experience a vertical acceleration of 10^{14} m s^{-2} while they are between the plates, which are 0.2 m long. (a) How long will the electrons be between the plates? (b) In what direction will the electrons be moving after they leave the plates? (c) How far will the electrons be deflected vertically as they leave the plates?

*2-39   A tennis ball is served 2.5 m above the ground at an angle of 5° above the horizontal direction with an initial speed of 30 m s^{-1}. (a) When will it hit the ground? (b) How far will it travel?

2-40   A ski jumper leaves a slope at an angle of 20° above the horizontal direction. She lands 3.5 s later at a point 20 m below her takeoff point. (a) What was her initial speed? (b) How far does she travel horizontally?

2-41   A boy standing 10 m from a building can just barely reach the roof 12 m above him when he throws a ball at the optimum angle with respect to the ground. Find the initial velocity components of the ball.

2-42   Derive a formula for the maximum height reached by a projectile in terms of its initial velocity components.

*2-43   Show that the horizontal and vertical displacement components for a projectile satisfy an equation of the form $\Delta y = a\,\Delta x + b\,(\Delta x)^2$, so that the trajectory is a *parabola*.

^{c}2-44   A point on the rim of a rolling wheel has its position given by $x = R\sin\omega t + \omega R t$, $y =$ $R\cos\omega t + R$. (a) Sketch the path followed by the point. (This curve is called a cycloid.) (b) Find the instantaneous velocity and acceleration components at time t. (c) Find the magnitudes of v and a at time t.

^{c}2-45   A runner goes around a circular track of radius 50 m at a constant speed of 8 m s^{-1} (Fig. 2.24). Assume she starts to run clockwise at $t = 0$ s from $x = 0$ m, $y = 50$ m. For $t > 0$, find the components of her (a) position; (b) velocity; (c) acceleration.

^{c}2-46   (a) Show that the acceleration of an object undergoing circular motion at constant speed is directed toward the center of the circle. (b) Is this true if the speed is not constant? Explain.

^{c}2-47   An object has an acceleration $\mathbf{a} = p\hat{\mathbf{x}} + qt\hat{\mathbf{y}}$, where p and q are constants. It starts from rest at the origin at $t = 0$. Using integration techniques (Sec. 1.9), for $t > 0$ find (a) its velocity; (b) its position.

ANSWERS TO REVIEW QUESTIONS

Q2-1, scalar has magnitude only, vector also has direction; **Q2-2,** arrows over symbols, boldface type; **Q2-3,** $A + B, A - B, 2A$; **Q2-4,** $(A^2 + B^2)^{\frac{1}{2}}$; **Q2-5,** displacement, elapsed time; **Q2-6,** velocity change, elapsed time; **Q2-7,** magnitude, direction; **Q2-8,** one dimensional motions; **Q2-9,** gravitational acceleration; **Q2-10,** vertical; **Q2-11,** vertical; **Q2-12,** horizontal, **Q2-13,** $-g$, 0

SUPPLEMENTARY TOPICS

2.6 | PROJECTILES IN BIOMECHANICS

Many applications of projectile motion occur in athletics and in animal motion. Here we explore some further aspects of this subject.

In applications of projectile motion, it is convenient to have a formula for the horizontal distance traveled

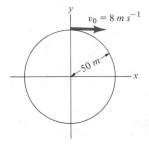

Figure. 2.24.   Problem 2-45.

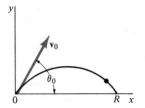

Figure 2.25. A projectile with a velocity $\mathbf{v}_0$ launched at an angle θ_0 to the horizontal has a range R.

or *range, R*. To obtain this formula, consider a projectile launched from a flat surface (Fig. 2.25). The projectile lands after an elapsed time Δt when Δy returns to zero. The range can be found from the equation for Δx once this elapsed time is known.

With $\Delta y = 0$, we can rewrite $\Delta y = v_{0y} \Delta t - \frac{1}{2}g\,(\Delta t)^2$ as

$$(v_{0y} - \tfrac{1}{2}g\,\Delta t)\,\Delta t = 0$$

The solutions of this equation are $\Delta t = 0$, which corresponds to the instant the projectile was launched, and

$$\Delta t = \frac{2v_{0y}}{g} \qquad (2.17)$$

which gives the elapsed time the projectile is in motion.

Using this elapsed time in $\Delta x = v_{0x} \Delta t$, the range is

$$R = \frac{2v_{0x}v_{0y}}{g}$$

If the initial velocity of the projectile is at a *launch angle θ_0* to the ground (Fig. 2.24), $v_{0x} = v_0 \cos\theta_0$ and $v_{0y} = v_0 \sin\theta_0$. Then the range can be expressed as

$$R = \frac{2v_0^2 \sin\theta_0 \cos\theta_0}{g} \qquad (2.18)$$

Alternatively, using the trigonometric identity $\sin 2\theta_0 = 2 \sin\theta_0 \cos\theta_0$

$$R = \frac{v_0^2}{g} \sin 2\theta_0 \qquad (2.19)$$

Two interesting features of projectile motion can be extracted from these equations. First, $\sin\theta_0 = \cos(90° - \theta_0)$, so replacing θ_0 by its complement $90° - \theta_0$ in Eq. 2.18 leaves the range unchanged. For example, footballs punted at 30° or at 60° launch angles will land at the same spot. However, the ball kicked at 60° will have a higher trajectory and will stay in the air longer, giving the kicking team more time to run down the field (Fig. 2.26). Second, the sine has its maximum value of $+1$ when its argument is 90°, so the range is a maximum in Eq. 2.19 when $2\theta = 90°$ or $\theta = 45°$. *The maximum range on flat ground occurs when the launch angle is* 45° (Fig. 2.26).

The calculation of the range is somewhat more complicated when the takeoff and landing heights are not equal. However, we can see from Fig. 2.27 that a shot put thrown from *above* ground level will have its greatest range with a launch angle *less* than 45°. Conversely, an object thrown from *below* its landing point will have its longest range for an angle *greater* than 45° (see Problem 2-61).

The following example illustrates how the range formulas can be used to analyze a situation in athletics.

Example 2.10

A baseball player can throw a ball at 36 m s^{-1}. (a) What is the greatest distance he can throw the ball, assuming it is caught at the same height at which he releases it? (b) If he wishes to throw the ball half this maximum distance in the shortest possible time, at what

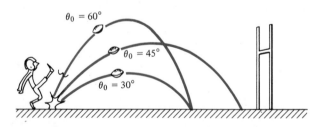

Figure 2.26. Footballs kicked with the same initial speed from ground level with launch angles of 30° and 60° have the same range. The maximum range occurs when the launch angle is 45°.

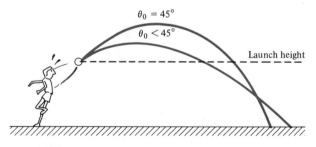

Figure 2.27. A shot put is thrown from above ground level. The trajectories for a launch angle of 45° and a smaller launch angle cross at a point below the launch height. The flatter trajectory has a longer range. In general, the launch angle which will produce the longest range is less than 45° if the landing point is below the launch point.

angle should he throw it? (c) What are the elapsed times in the two situations?

(a) The maximum range occurs for a launch angle of $45°$, or $\sin 2\theta_0 = 1$ in Eq. 2.19. Thus

$$R = v_0^2/g = (36 \text{ m s}^{-1})^2/(9.8 \text{ m s}^{-2}) = 132 \text{ m}$$

(b) Solving Eq. 2.19 for $\sin 2\theta_0$, and using $R = (132 \text{ m})/2 = 66 \text{ m}$,

$$\sin 2\theta_0 = gR/v_0^2 = (9.8 \text{ m s}^{-2})(66 \text{ m})/(36 \text{ m s}^{-1})^2 = 0.5$$

$$2\theta_0 = 30°$$

$$\theta_0 = 15°$$

The same range can also be obtained with $\theta_0 = 90° - 15° = 75°$, but the elapsed time will be longer.

(c) Using Eq. 2.17 and $v_{0y} = v_0 \sin \theta_0$, the times are

$$\Delta t_a = 2v_{0y}/g = 2(36 \text{ m s}^{-1})(\sin 45°)/(9.8 \text{ m s}^{-2}) = 5.20 \text{ s}$$

$$\Delta t_b = 2(36 \text{ m s}^{-1})(\sin 15°)/(9.8 \text{ m s}^{-2}) = 1.90 \text{ s}$$

Notice that the elapsed time in part (b) is less than half that in (a), even though the range is halved, because the trajectory is much flatter. The time for *two* of the shorter throws is less than Δt_a by $(5.20 \text{ s}) - 2(1.90 \text{ s}) = 1.40 \text{ s}$. Rather than throwing directly to the plate, an outfielder frequently throws to a player who relays the ball to home plate. This makes the accuracy less critical and often saves time. (Allowing for the time to make the second throw reduces the savings, while air resistance effects increase them.)

Horizontal Jumping

In Section 1.8, we saw that the constant acceleration formulas can be used to analyze vertical jumping by animals. Similarly, the projectile motion formulas can be used to discuss horizontal jumping, since they accurately describe the motion while the animal is in the air if air resistance is negligible.

Although a $45°$ launch angle produces the maximum range on flat ground for a given initial speed (Fig. 2.28), an animal may customarily jump at some other angle for reasons related to its needs or structure. For example, locusts often jump into the air and then start flying. In this case the range of the jump is clearly irrelevant, but the time duration may be significant. Whether or not they begin to fly, locusts usually jump at about $55°$.

The calculation of the takeoff velocity is illustrated in the following example.

Example 2.11

What is the takeoff speed of a locust if its launch angle is $55°$ and its range is 0.8 m?

Figure 2.28. Frogs frequently jump with a launch angle of approximately $45°$, the angle which produces the maximum range on flat ground.

Since we know R and θ_0, we can find v_0 from

$$R = \frac{2v_0^2}{g} \sin \theta_0 \cos \theta_0$$

With $\sin 55° = 0.819$ and $\cos 55° = 0.574$,

$$v_0^2 = \frac{gR}{2 \sin \theta_0 \cos \theta_0} = \frac{(9.8 \text{ m s}^{-2})(0.8 \text{ m})}{2(0.819)(0.574)}$$

$$= 8.3 \text{ m}^2 \text{ s}^{-2}$$

$$v_0 = 2.9 \text{ m s}^{-1}$$

EXERCISES ON SUPPLEMENTARY TOPICS

Section 2.6 | Projectiles in Biomechanics

2-48 A football is kicked at 20 m s^{-1} from ground level. Find its range if the launch angle is (a) $30°$; (b) $60°$; (c) $45°$.

2-49 In the preceding exercise, how long will the football be in the air in each of the three cases?

2-50 An astronaut wearing his spacesuit can broad jump 2 m on the earth. How far can he jump on a planet where the gravitational acceleration is half that at the surface of the earth?

2-51 A girl wishes to throw a snowball at another child. If she can throw the snowball at 20 m s^{-1}, how far can she stand from the other child and still reach him?

2-52 A rifle is fired horizontally from the top of a tall mountain. Using a sketch, show the effect of the earth's curvature on the range of the bullet.

2-53 A kangaroo can jump 8 m. If it takes off at $45°$ to the horizontal, what is its takeoff speed?

2-54 A baseball thrown at $10°$ to the horizontal returns to its original height after 70 m. What was its original speed?

2-55 A motorcycle stunt rider leaves a ramp at $30°$ to the horizontal, just barely clears a row of trucks 36 m wide, and lands at the same height as his takeoff point. What was his takeoff speed?

2-56 A rifle is aimed slightly above a target 200 m away at the same height as the rifle. The bullet leaves the muzzle at 500 m s^{-1} and strikes the center of the target. At what angle to the horizontal is the rifle barrel?

2-57 A football is kicked 60 metres on a level field. If the launch angle is 60°, how large is its initial velocity?

2-58 A rescue ship is to fire a shell trailing a lifeline to a distressed vessel located at a distance of 300 m. The initial velocity of the shell is 100 m s^{-1}. What are the possible launch angles? (Neglect the effect of the trailing lifeline.)

2-59 A mortar shell is fired at a ground-level target 500 m distant with an initial velocity of 90 m s^{-1}. What is its launch angle? (Mortars are fired at large launch angles.)

2-60 A football is kicked from ground level on a level field with an initial velocity v_0 and a launch angle θ_0. Find the velocity v and the angle θ at which it hits the ground.

PROBLEMS ON SUPPLEMENTARY TOPICS

2-61 With the aid of a sketch, show that the maximum range for an object launched at a given speed from below its landing point occurs for a launch angle greater than 45°.

2-62 (a) Explain why the maximum range for a man doing a standing broad jump is not obtained for a takeoff angle of 45°. (b) Should the angle be less than or greater than 45°? Explain.

2-63 A frog can jump 0.9 m with a takeoff angle of 45°. (a) What initial velocity does this require? (b) With this same initial velocity directed vertically, how high could a frog jump? (c) The maximum height jumped by frogs is 0.3 m. What are some possible explanations for this difference?

2-64 A flea can jump 0.03 m. (a) If the takeoff angle is 70°, what is the initial velocity? (b) If the flea achieves this velocity in a takeoff distance of 8×10^{-4} m, what is its average acceleration during takeoff?

2-65 A boy can throw a ball a maximum horizontal distance of 60 m. Assuming he can throw equally hard in the vertical direction, how high can he throw a ball?

*2-66 Show that for a ball kicked on level ground, the ratio of the maximum height reached to the range is $\frac{1}{4} \tan \theta_0$.

*2-67 The curvature of the earth becomes important in projectile calculations when the distance traveled R is a significant fraction of the radius of the earth, R_E. Ignoring any possible variations in g, show that (a) the extra time the projectile is in motion is approximately given by $\Delta t \simeq \Delta y/v_{0y} \simeq (R^2/2R_E)/v_{0y}$; (b) the fractional error $\Delta R/R$ is given by $\Delta R/R = v_{0x}{}^2/gR_E$.

2-68 Using the equation in the preceding problem, find the fractional error in the range of a projectile due to the curvature of the earth if the projectile is fired with a launch angle of 45° and an intended range of 100 km. (The average radius of the earth is 6.38×10^6 m.)

ᶜ2-69 (a) Explain why the launch angle θ which gives the maximum range must satisfy $dR/d\theta = 0$. (b) Using this requirement, show that a 45° launch angle produces the maximum range on flat ground.

Additional Reading

Sir James Gray, *How Animals Move*, Cambridge University Press, Cambridge, 1953. Pages 69–80 discuss jumping.

R. McNeill Alexander, *Animal Mechanics,* University of Washington Press, Seattle, 1968. Pages 28–33 discuss jumping.

David F. Griffing, *The Dynamics of Sports—Why That's the Way the Ball Bounces,* Mohican, Loudonville, Ohio, 1982. Basic physics applied to track and field, swimming, water skiing, football, etc.

Scientific American articles:

Stillman Drake and James MacLachlan, Galileo's Discovery of the Parabolic Trajectory, March 1975, p. 102.

Cornelius T. Leondes, Inertial Navigation for Aircraft, March 1970, p. 80.

Graham Hoyle, The Leap of the Grasshopper, January 1958, p. 30.

Miriam Rothschild et al., The Flying Leap of the Flea, November 1973, p. 92.

CHAPTER 3

NEWTON'S LAWS OF MOTION

Having learned how to describe motion, we can now turn to the more fundamental question of what causes motion. An object is set into motion when it is pushed or pulled or subjected to a *force*. The discussion of forces and their effects is the central topic in mechanics.

Although there are many kinds of forces in nature, the effects of any force are described accurately by three general laws of motion first stated fully by Sir Isaac Newton (1642–1727). Guided by earlier astronomical observations and making several giant steps of intuition, Newton developed the laws of motion and also the expression for the gravitational attraction between two objects. He then showed that the orbital motions of the planets and the moon were in quantitative agreement with the predictions he made using these laws.

Newton's work represented a tremendous step forward in our understanding of the natural world and exerted a great influence on science and on the way people viewed science. For over two centuries, Newton's laws of motion served as the foundation of mechanics, with later workers finding full agreement between theory and experiment for a wide range of phenomena. Even though twentieth-century advances have shown that Newton's laws are inadequate at the atomic scale and at velocities comparable to the speed of light, 3×10^8 m s^{-1}, they provide an extremely accurate framework for discussing the motions of macroscopic objects at ordinary velocities. Thus, they are fully adequate for most applications in fields such as astronomy, biomechanics, geology, and engineering. The twentieth-century modifications of mechanics are discussed in later units.

3.1 | FORCE, WEIGHT, AND GRAVITATIONAL MASS

If we push or pull an object, we are exerting a force on it. Forces have both magnitudes and directions, so they are vector quantities. It is found that the *net* or *total* force on an object is the vector sum of all the forces acting on the object. For example, if two forces equal in magnitude but opposite in direction act on an object, there is no net force (Fig. 3.1a).

Forces that are exerted only when two objects are in contact are referred to as *contact forces*. Examples are the force exerted by a compressed spring on an attached object, the upward force exerted by a table on a book resting upon it, and the force exerted on a bone by a contracting muscle. Other forces, including gravitational, magnetic, and electric forces, can be exerted between objects that are not in contact. For example, the earth is kept in a nearly circular orbit by the gravitational attraction of the sun. However, the underlying origin of the contact forces that objects

F_1 F_1 F_2

F_2 F
(a)   (b)

Figure 3.1. Forces have both magnitudes and directions and are vector quantities. (*a*) F_1 and F_2 are equal in magnitude but opposite, so their sum is zero. (*b*) Since F_1 and F_2 are equal, the net force is $F = F_1 + F_2 = 2F_1$.

exert on each other lies in electrical and magnetic forces acting among the constituent atomic particles, so this distinction is not clear-cut.

In order to make quantitative statements about forces, we must define a force unit. One way to do this is to use a spring to measure the gravitational force on an object adopted as a standard. If the spring is compressed, a pointer moves along a calibrated dial. When the standard is placed on a plate mounted on the spring, the pointer moves a certain distance. The standard object is then said to exert a force of one unit on the spring. The same reading will always be obtained if this procedure is performed at points on the earth's surface where the gravitational acceleration is the same. Other forces can then be measured by determining the compression of the spring once the dial has been calibrated using multiple copies of the standard object. In Section 3.6, we see that the force unit can also be defined by measuring the acceleration of a standard object.

The S.I. force unit is the *newton* (N), and the British unit is the *pound* (lb). These units are related by

$$1 \text{ N} = 0.225 \text{ lb}$$

We will use the S.I. unit exclusively.

A particularly important force is the gravitational force on an object, which is referred to as its *weight* **w**. The weight of a fairly heavy man might be 1000 N or 225 lb.

Closely related to the weight is the *gravitational mass* m of an object, which is defined as the weight divided by the gravitational acceleration g at the location of the object:

$$m = \frac{w}{g} \qquad (3.1)$$

The gravitational force pulls an object downward along the direction it will fall if it is not supported. Thus, the weight **w** is parallel to the gravitational acceleration **g**, and Eq. 3.1 can be written in vector form as

$$\mathbf{w} = m\mathbf{g} \qquad (3.2)$$

The S.I. unit of gravitational mass is the *kilogram* (kg). For example, a man who weighs 1000 N on the earth has a gravitational mass of $w/g = 1000 \text{ N}/9.8 \text{ m s}^{-2} = 102 \text{ kg}$.

An object with a gravitational mass of 1 kg weighs $w = mg = (1 \text{ kg})(9.8 \text{ m s}^{-2}) = 9.8 \text{ N}$. Since 1 N =

0.225 lb, this weight is 2.2 lb. This is a handy conversion factor to remember: a 1-kg mass has a weight of 2.2 lb.

The mass, denoted dimensionally by M, completes the basic set of physical dimensions. All mechanical quantities can be written in terms of the dimensions of length, time, and mass. Table 3.1 shows the masses of some representative objects.

In this section we have introduced two major concepts, force and gravitational mass. A force is any push or pull on an object. The gravitational force on an object is referred to as its weight, and the gravitational mass of an object is its weight divided by the gravitational acceleration. We explore the relationships among forces and masses in the following sections.

3.2 | DENSITY

When we talk about the properties of materials in a general way, rather than about specific objects, it is often convenient to refer to the mass per unit volume or the *density*. If a sample of a material has a mass m and a volume V, the density is given by the ratio

$$\rho = \frac{\text{mass}}{\text{volume}} = \frac{m}{V} \qquad (3.3)$$

TABLE 3.1

Representative masses in kilograms	
Electron	9×10^{-31}
Proton	2×10^{-27}
Oxygen atom	3×10^{-26}
Insulin molecule (a small protein)	10^{-23}
Penicillin molecule	10^{-18}
Giant amoeba	10^{-8}
Ant	10^{-5}
Hummingbird	10^{-2}
Dog	10^{1}
Human	10^{2}
Elephant	10^{4}
Blue whale	10^{5}
Oil tanker	10^{8}
Moon	7×10^{22}
Earth	6×10^{24}
Sun	2×10^{30}
Our galaxy	2×10^{41}

(The symbol ρ is the Greek letter "rho.") In S.I. units, densities are measured in kilograms per cubic metre (kg m^{-3}). For example, if a block of wood has a mass of 50 kg and a volume of 0.1 m^3, its density is $\rho = m/V = (50 \text{ kg})/(0.1 \text{ m}^3) = 500 \text{ kg m}^{-3}$.

Table 3.2 lists the densities of some materials. Note that the density varies with the temperature and the pressure, especially for gases.

Closely related is the concept of the *relative density* or *specific gravity,* which is defined as the ratio of the density of a substance to that of water at 0° C (Celsius). The relative density can be found readily from Table 3.2, since the density of water at 0° C is 1000 kg m^{-3}. For example, the relative density of mercury is the dimensionless number 13,600/1000 = 13.6. From the conversion factor given in the table, we also see, for example, that a cubic centimetre of mercury has a mass of 13.6 grams.

TABLE 3.2

Some representative densities in kg m^{-3}. Densities of materials are at atmospheric pressure and 0° C, except as noted (at 0° C, 1 cm^3 of water has a mass of 1 gram)

Interstellar space	$10^{-18} - 10^{-21}$
Best laboratory vacuum	10^{-17}
Hydrogen (H$_2$)	0.0899
Air, at 0° C and 1 atmosphere	1.29
at 100° C and 1 atmosphere	0.95
at 0° C and 50 atmospheres	6.5
Water, at 0° C and 1 atmosphere	1000
at 100° C and 1 atmosphere	958
at 0° C and 50 atmospheres	1002
Whole blood, at 25° C	1059.5
Mercury	13,600
Aluminum	2700
Iron, steel	7800
Copper	8900
Lead	11,300
Gold	19,300
The earth, density of core	9500
The sun, density at center	1.6×10^5
White dwarf stars	$10^8 - 10^{15}$
Atomic nuclei	10^{17}
Neutron stars	10^{17}

3.3 | NEWTON'S FIRST LAW

According to the Aristotelian view that dominated medieval ideas about motion, all objects move only if some force causes the motion to occur. Thus, a wagon detached from the horse that was pulling it would quickly come to rest because no force is pulling it along. The modern view is that the cart slows down and stops because of frictional forces acting upon it. This view is summarized in Newton's first law, which states that

Every object continues in a state of rest, or of uniform motion in a straight line, unless it is compelled to change that state by forces acting upon it.

An equivalent statement of the first law is that if there is no force on an object, or if there is no net force when two or more forces act on the object, then

(1) *an object at rest remains at rest, and*
(2) *an object in motion continues to move with constant velocity.*

The first law holds true in the form stated only for measurements made by certain observers. A girl riding on a merry-go-round sees objects not subjected to any net force undergo rather complex motions, while a boy at rest on the ground sees them at rest or moving with constant velocity. Thus, Newton's first law as stated is true for the boy at rest but not for the girl. The point is that the girl is accelerating, since her velocity is changing, and Newton's first law as stated does not hold true for someone who is accelerating.

The first law leads us to define an *inertial* coordinate system or reference frame as one in which Newton's first law holds true as stated. Strictly speaking, the boy in the preceding example is not quite in an inertial frame, since he is standing on the earth, which is revolving daily on its axis, rotating annually about the sun, and moving with the solar system relative to the distant stars. Usually these effects can be ignored, and the earth can be treated as an inertial frame. However, the earth's daily revolution does affect the large-scale motions of the atmosphere and the oceans.

A coordinate system moving at a constant velocity relative to an inertial frame is itself an inertial frame. To see this, consider an observer standing on the ground and another in a car moving at a constant velocity. If both observers measure the velocity of a moving object, their measurements will differ by their *constant* relative velocity. Hence, they will agree on whether the motion of the object is accelerated, and Newton's first law will work equally well for both observers.

From this discussion, we see that cars or airplanes moving at constant velocities relative to the ground are inertial reference frames, while accelerating vehicles, merry-go-rounds, and playground swings are not. As a further illustration, consider the driver of an accelerating car. She feels the back of her seat exerting a force on her. Relative to the ground, she is accelerating, but relative to the car, she is not. The first law is consistent with the measurements of an observer at rest on the ground: the state of motion of the driver is being changed by forces acting on her. The first law does not work in the frame of the accelerating car, since there is a force on the driver, but she remains at rest.

3.4 | EQUILIBRIUM

When the state of motion of an object remains unchanged even though two or more forces are acting upon it, the object is said to be in *equilibrium*. There are three types of equilibrium: *unstable, stable,* and *neutral.* These are best defined with the aid of an example.

Suppose a ball is placed exactly at the crest of a small hill (point *A* in Fig. 3.2). It may remain at rest there briefly, but if a slight breeze moves it a bit, it will rapidly accelerate down the hillside. This is unstable equilibrium: a small displacement leads to an unbalanced force that further increases the displacement from the equilibrium location. Conversely, in the depression at *B*, the ball will roll back if disturbed. This is stable equilibrium: a small displacement leads to an unbalanced force that restores the

body to the equilibrium position. Finally, on the flat level area near *C*, there is still no unbalanced force, even if the ball is displaced slightly. This is neutral equilibrium.

From Newton's first law, the net force on an object must be zero in order that the translational motion of an object remain unchanged. This is one requirement for equilibrium. If we are dealing with an object that can rotate and not a point particle, then there is a second requirement that no twisting effects or *torques* act that would alter the rotational state of motion. This second requirement is discussed in the next chapter.

The simplest kind of equilibrium situation is one where two forces act on an object. A woman standing on a floor is pulled downward by the gravitational attraction of the earth, which is her weight **w** (Fig. 3.3*a*). Since she remains at rest, the first law tells us there is no net force on her, and there must therefore be an equally large upward force exerted on her by the floor. This force is denoted by **N** in Fig. 3.3*a*, since it is *normal* or perpendicular to the floor. (The normal force **N** has a magnitude *N*; this should not be confused with the conventional abbreviation for the newton, N.) Or consider a tug-of-war, with two teams pulling in opposite directions (Fig. 3.3*b*). If the forces they exert have exactly the same magnitude, the knot will remain at rest, but if one team pulls slightly harder, the knot will begin to move in that direction.

Figure 3.4*a* shows a traffic light in equilibrium under the action of three forces. The vector sum of the forces on the traffic light must be zero (Fig. 3.4*b*). Similarly, the sled in Fig. 3.5 is moving with a constant velocity on a flat snow-covered field. Its velocity

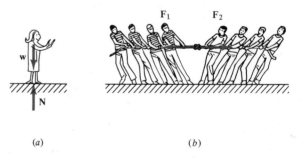

(a)   (b)

Figure 3.3. (*a*) Since the woman is in equilibrium, the force **N** exerted by the floor is equal in magnitude to her weight **w** and opposite in direction. (*b*) If the forces **F**₁ and **F**₂ exerted by the two teams are equal in magnitude, the knot will be in equilibrium or remain at rest.

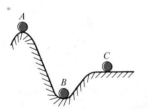

Figure 3.2. A ball at the crest of a hill is in unstable equilibrium. In the depression, it is in stable equilibrium. On the level plain, it is in neutral equilibrium.

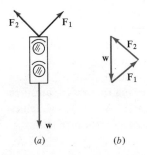

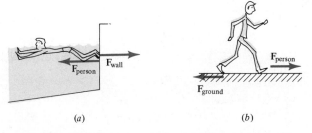

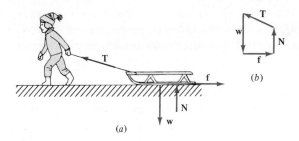

Figure 3.4. (a) A traffic light suspended by two cables is in equilibrium under the action of three forces. (b) The vector sum of the forces is zero.

Figure 3.6. The force on the person is equal in magnitude and opposite in direction to the force on (a) the wall and (b) the ground.

Figure 3.5. (a) A sled moving with a constant velocity. Forces acting on it include the weight **w**, a normal force **N**, a frictional force **f** retarding its motion, and a force **T** exerted by the rope. (b) The vector sum of the forces is zero because the sled is in equilibrium.

is constant, so by the first law the net force on the sled is zero, and it is in equilibrium. The concept of equilibrium is not restricted to objects at rest, since an object that is moving with a constant velocity according to one inertial observer is at rest according to another.

To summarize, an object will be in translational equilibrium if the vector sum of the forces acting upon it is zero. The type of equilibrium is determined by noting how the forces change when the object is disturbed slightly from its original state of rest or of uniform motion.

3.5 | NEWTON'S THIRD LAW

We turn now to Newton's third law, deferring the second law until the next section. The third law relates the forces that two objects exert on each other, and it is familiar in a general way from our everyday experiences. For example, suppose you are at rest in a swimming pool. If you push a wall with your legs, the wall exerts a force that propels you further into the

pool. The *reaction* force the wall exerts on you is opposite in direction to the force you exert on the wall (Fig. 3.6). Similarly, in order for you to start walking forward, your feet must exert a backward force on the ground; the ground, in turn, pushes you forward.

Newton observed that whenever a person exerts a force on an object, the object exerts a force on the person that is *equal in magnitude and opposite in direction*. This relationship, which is referred to as the third law of motion, holds true whether or not the person or the object accelerates. The two forces acting between a person and an object, or between two objects, are called *action* and *reaction* forces.

The general statement of the third law is

If one object exerts a force **F** *on a second, then the second object exerts an equal but opposite force* −**F** *on the first.*

For example, if you push a chair with a horizontal force of 10 N, the chair exerts a force on you of 10 N in the opposite direction. If the earth exerts a downward 5 N gravitational force on a book, then the book exerts an upward 5 N force on the earth.

It is important to realize that in each case the two forces—the action and reaction forces—act on different objects, so their effects do *not* cancel out. When you push the chair, it starts to move unless there is a sufficiently large frictional force exerted by the floor to prevent this. Or if a book is not subjected to any force other than gravity, it will fall toward the earth. In other words, *only the forces acting on a particular object can affect its state of motion;* the forces exerted *by* an object will affect the motion of *other* objects.

There are many situations where forces are equal but opposite, but they are not action-reaction pairs in the sense of the third law. This often happens when an object is in equilibrium and forces therefore cancel

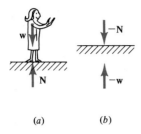

Figure 3.7. (*a*) Forces on a woman in equilibrium. (*b*) The reaction to the weight of the woman is a force −**w** on the earth; the reaction to the force **N** exerted by the floor is a force −**N** she exerts on the floor.

as a consequence of the *first* law. For example, we saw in the preceding section that the forces **w** and **N** acting on the woman are equal in magnitude and opposite in direction (Fig. 3.7). They both act on the same object, and they are *not* an action-reaction pair. Since her weight **w** is a downward force exerted *on her* by the earth, the corresponding reaction force is an upward gravitational force −**w** exerted *by her* on the earth. Similarly, the reaction to the normal force **N** exerted on her by the floor is a downward force −**N** exerted *by the woman* on the floor.

Although it is customary to refer to action and reaction forces, the main point of the third law is that forces between two objects always occur in equal but opposite pairs. Which is the action or reaction is often a matter of choice, as in the case of gravitational forces.

3.6 | NEWTON'S SECOND LAW

When there is a net force acting on an object, the object undergoes an acceleration in the same direction as the force. The acceleration and the force are also proportional in magnitude; if the force on a given object is doubled, so is the acceleration.

If two quantities are proportional, one is equal to some number or *proportionality constant* times the other. Thus, we can relate the net force **F** and the acceleration **a** by *Newton's second law,*

$$\mathbf{F} = m\mathbf{a} \qquad (3.4)$$

The proportionality constant m is called the *mass* of the object. (Strictly speaking, it is the *inertial mass,* but we see in Section 3.10 that the inertial and gravitational masses of an object are equal.)

The mass of an object is a measure of the amount of matter it comprises, or in other words, of its inertia.

The greater the mass of an object, the less effect a given force will have on its motion. Mass is related to but very different from weight. The weight of an object is the gravitational force on the object and is a vector quantity; mass is a scalar quantity.

Newton's second law $\mathbf{F} = m\mathbf{a}$ provides another way of defining the unit of force. A 1-N force acting on a 1-kg mass produces an acceleration of 1 m s^{-2}:

$$1\,\mathrm{N} = (1\,\mathrm{kg})(1\,\mathrm{m\,s^{-2}}) = 1\,\mathrm{kg\,m\,s^{-2}}$$

This definition is equivalent to that given earlier.

To illustrate the use of these units and the second law, we may note that a 1000-kg car accelerating at 2 m s^{-2} must be subjected to a net force $F = ma = (1000\,\mathrm{kg})(2\,\mathrm{m\,s^{-2}}) = 2000$ N. Similarly, if a 50-N force applied to a sled produces an acceleration of 2 m s^{-2}, then its mass must be $m = F/a = (50\,\mathrm{N})/(2\,\mathrm{m\,s^{-2}}) = 25$ kg.

We consider many examples and applications of Newton's second law later in this and other chapters. However, we can see at this point that the second law explains why the equality of action and reaction forces is sometimes not immediately apparent. Since $\mathbf{a} = \mathbf{F}/m$, the acceleration resulting from a given force varies inversely with the mass of an object. A person who steps off a cliff experiences an acceleration g resulting from the gravitational force exerted by the earth on the person. The earth experiences a force of equal magnitude, but its acceleration is very much smaller because of its great mass. Similarly, when a small car and a large truck collide, the forces on each vehicle are equal in magnitude, but the acceleration of the car is much greater than that of the truck.

3.7 | THE SIGNIFICANCE OF NEWTON'S LAWS OF MOTION

Newton's three laws of motion are very fundamental statements about the physical world. All of mechanics may be viewed as Newton's laws applied in direct or indirect ways to a variety of forces and systems. Although we introduce many other quantities and concepts in our discussions of mechanics in the following chapters, none are so basic as Newton's laws of motion. Accordingly, before going on to discuss specific forces and examples of how one uses Newton's laws, we list these laws together here for emphasis and for convenient reference.

(Culver Pictures)

SIR ISAAC NEWTON
(1642–1727)

Born in 1642, the year of Galileo's death, Newton made the crucial advances needed to complete our understanding of motion. He also made major contributions to optics and mathematics.

Newton was a thin frail baby and was raised by his grandmother after his widowed mother remarried when he was 2 years old. His difficult childhood may have contributed to his later psychotic tendencies. Throughout his brilliant career, he was extremely anxious when his work was published and irrationally violent whenever his ideas were challenged. He suffered at least two nervous breakdowns.

As an undergraduate at Cambridge (1661 to 1665), Newton soon mastered the literature of science and mathematics and began to enter unexplored regions. He formulated the binomial theorem and the basic concepts of calculus. During this period and the years immediately following, he also began to do research on optics and on planetary motion. He deduced that the force on a planet due to the sun must vary as $1/r^2$. Some 20 years later, he would extend this idea to the universal law of gravitation.

Although Newton's work was known only to a limited circle because of his reluctance to publish his research, he was appointed a professor at Cambridge in 1669. He developed the first reflecting telescope in order to circumvent the distortions inherent in lenses. When this telescope received an enthusiastic reception from the Royal Society of London, he was encouraged to present his other research in optics to the Society in 1672. Robert Hooke, the leading authority on optics, disagreed with some of Newton's ideas. This led to bitter disputes, with Newton finally withdrawing into isolation for some years.

Newton's greatest achievements were his advances in mechanics. Although many of his results were obtained quite early in his career, he did not present his theory of planetary motion until he was urged to do so in 1684 by Edmond Halley, an astronomer who had heard of Newton's work.

Newton's classic work, *Principia Mathematica,* appeared in 1687. Written in Latin, it contained the three laws of motion and the universal law of gravitation. This treatise constituted one of the foundations of modern science and made Newton internationally famous. It also effectively marked the end of Newton's active research, with his attention gradually turning to politics, theology, and scientific priority disputes.

Newton became master of the mint, a well-paying and normally undemanding job. However, he took the position seriously and was especially zealous in sending counterfeiters to the gallows. He also assumed the role of the leader of English science, becoming president of the Royal Society in 1703; in 1705, he became the first scientist to be knighted. Unfortunately, he repeatedly used his position to carry on bitter arguments with various scientists. The most prolonged of these was a 25-year battle with Leibniz (which ended with Newton's death in 1727) over credit for the development of the calculus. It is now agreed that Leibniz independently developed the calculus after Newton had, but before Newton published his results.

Newton's first law of motion: *When there is no net force on an object, (1) an object at rest remains at rest; and (2) an object in motion continues to move with a velocity that is constant in magnitude and direction.*

Newton's second law of motion: *the force **F** needed to produce an acceleration **a** is*

$$\mathbf{F} = m\mathbf{a} \qquad (3.4)$$

where m is the mass of the object.

Newton's third law of motion: *If one object exerts a force **F** on a second, then the second object exerts an equal but opposite force −**F** on the first.*

Note that in the form given, Newton's laws may be applied only with respect to inertial coordinate systems.

3.8 | SOME EXAMPLES OF NEWTON'S LAWS

We now give several examples of how Newton's laws of motion are used. In each case, we employ a systematic procedure for relating the acceleration of an object or objects to the forces present.

1 For each object, we draw a careful sketch showing the forces acting *on* that object.
2 We then apply Newton's second law $\mathbf{F} = m\mathbf{a}$ to *each object separately.* If there are *n* forces $\mathbf{F}_1$,

$\mathbf{F}_2, \ldots, \mathbf{F}_n$ acting on an object, the net force **F** is the sum of the forces, and we have

$$\mathbf{F} = \mathbf{F}_1 + \mathbf{F}_2 + \cdots + \mathbf{F}_n = m\mathbf{a}$$

In component form, this is

$$F_x = F_{1x} + F_{2x} + \cdots + F_{nx} = ma_x \quad (3.5)$$
$$F_y = F_{1y} + F_{2y} + \cdots + F_{ny} = ma_y \quad (3.6)$$

3 As in the earlier chapters, we usually do not substitute numerical values immediately. Instead we solve the equations for the unknown quantities symbolically and then substitute the numbers if they are given. This procedure facilitates checking for algebra and physics errors and often reduces the amount of arithmetic required.
4 We include the units of the numerical quantities and see that the final answer has the correct dimensions.

We use these steps in the examples below and in later chapters as well.

Example 3.1

An elevator has a mass of 1000 kg. (a) It accelerates upward at 3 m s^{-2}. What is the force *T* exerted by the cable on the elevator? (b) What is the tension *T* if the acceleration is 3 m s^{-2} downward?

(a) The forces on the elevator are its weight **w** and the upward force **T** resulting from the cable (Fig. 3.8). Using Eq. 3.6 with two forces present,

$$F_{1y} + F_{2y} = ma_y$$

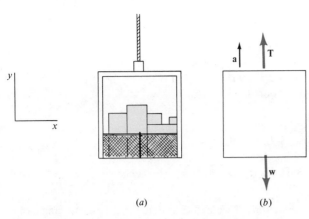

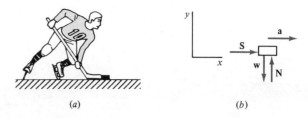

(a)   *(b)*

Figure 3.9. *(a)* A hockey player hits a puck. *(b)* Force diagram for the puck. Since the acceleration is horizontal, the net vertical force component is zero, and the normal force **N** has the same magnitude as the weight **w**.

(a)   *(b)*

Figure 3.8. *(a)* The elevator and cable. *(b)* The force of the cable on the elevator, **T**, and the weight of the elevator **w**. When the elevator is accelerating upward, **T** is greater than **w**.

With $F_{1y} = T$ and $F_{2y} = -w = -mg$, this becomes

$$T - mg = ma_y$$

or

$$T = m(g + a_y) \tag{i}$$

Then, with $a_y = 3$ m s^{-2} and $m = 1000$ kg,

$$T = (1000 \text{ kg})(9.8 \text{ m s}^{-2} + 3 \text{ m s}^{-2})$$
$$= 12{,}800 \text{ N}$$

Note that T is greater than the weight mg. The cable must support the weight of the elevator and also provide the extra force needed for the acceleration.

(b) Equation i holds true no matter what the magnitude or sign of a_y. When **a** is directed downward, $a_y = -3$ m s^{-2}, and Eq. i gives

$$T = m(g + a_y)$$
$$= (1000 \text{ kg})(9.8 \text{ m s}^{-2} - 3 \text{ s}^{-2}) = 6800 \text{ N}$$

Note that T is now less than the weight mg, since the elevator is being allowed to accelerate downward.

Example 3.2

An ice hockey player strikes a puck of mass 0.17 kg with his stick, accelerating it along the ice from rest to a speed of 20 m s^{-1} over a distance of 0.5 m (Fig. 3.9). What force must he exert if the frictional force between the puck and the ice is negligible? (Assume the acceleration is constant.)

To solve this problem, we find the acceleration and then apply Newton's second law. From the constant acceleration formulas in Table 1.3, we have $v^2 = v_0^2 + 2a \Delta x$. With $v_0 = 0$,

$$a = v^2/2 \Delta x = (20 \text{ m s}^{-1})^2/2(0.5 \text{ m}) = 400 \text{ m s}^{-2}$$

Then the net force on the puck is

$$F = ma = (0.17 \text{ kg})(400 \text{ m s}^{-2}) = 68 \text{ N}$$

This net force of 68 N in the direction of the acceleration is the force **S** the player's stick must exert on the puck. This follows since the only other forces acting on the puck are its weight (downward) and a normal force (upward) due to the ice. These two forces must exactly cancel, since the acceleration has no vertical component, and $F_y = ma_y = 0$.

Suppose two people pull the ends of a rope with oppositely directed forces, **F**$_1$ and **F**$_2$. The rope also exerts forces $-$**F**$_1$ and $-$**F**$_2$ on the people, in accordance with the third law of motion. From the second law, we know $F_1 - F_2 = ma$, where m is the mass of the rope and **a** is its acceleration. If that acceleration is zero, or if the mass of the rope is so small that we can idealize the rope as having zero mass, then $F_1 - F_2 = 0$. In this special situation the forces *exerted by the rope* on the two people are equal in magnitude, and the rope can be thought of as simply transmitting a force from one person to the other; it is as though they were holding each other's hands and pulling. The force at any point in the rope is referred to as the *tension*. It is the same everywhere in the rope only if the rope is unaccelerated or if it is idealized as massless. The tension at any point in the rope can be measured in principle by cutting it there and inserting a spring balance.

Example 3.3

A child pulls a train of two cars with a horizontal force **F** of 10 N. Car 1 has a mass $m_1 = 3$ kg, and car 2 has a mass $m_2 = 1$ kg (Fig. 3.10). The mass of the string connecting the cars is small enough so it can be set equal to zero, and friction can be neglected. (a) Find the normal forces exerted on each car by the floor. (b) What is the tension in the string? (c) What is the acceleration of the train?

In this problem, there is a system of several objects. The three objects of interest are the two cars and the string connecting them. When Newton's second law

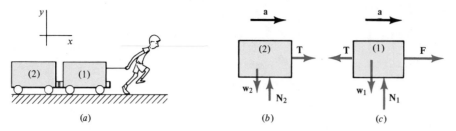

Figure 3.10. (a) The system in Example 3.3. (b) and (c) show the force diagrams of cars (2) and (1), respectively.

$\mathbf{F} = m\mathbf{a}$ is applied to each object, a set of equations results that must be solved simultaneously.

The fact that the string is massless means that the net force $\mathbf{F} = m\mathbf{a}$ on the string must be zero. As we have explained, this implies that the forces *it exerts* on the two cars are equal in magnitude, so we have labeled its tension force by the same symbol $\mathbf{T}$ in the force diagrams for each of the cars.

(a) Since $a_y = 0$ for each car, $F_y = ma_y = 0$, so

$$N_2 - w_2 = 0$$

and

$$N_1 - w_1 = 0$$

Thus the normal forces exerted by the floor are

$$N_2 = w_2 = m_2g = (1\ \text{kg})(9.8\ \text{m s}^{-2}) = 9.80\ \text{N}$$
$$N_1 = w_1 = m_1g = (3\ \text{kg})(9.8\ \text{m s}^{-2}) = 29.4\ \text{N}$$

(b) For both cars, $a_x = a$, so for car 1, $F_x = m_1a_x$ becomes

$$F - T = m_1a \qquad \text{(i)}$$

Both T and a are unknowns, so we cannot find them from this one equation alone; as many equations as there are unknowns are needed to solve a problem. The second equation relating T and a is obtained by applying $F_x = m_2a_x$ to car 2:

$$T = m_2a \qquad \text{(ii)}$$

This gives $a = T/m_2$. Substituting this for a in Eq. i,

$$F - T = m_1\left(\frac{T}{m_2}\right)$$

Solving for T,

$$T = \frac{F}{1 + \dfrac{m_1}{m_2}} = \frac{10\ \text{N}}{1 + \dfrac{3\ \text{kg}}{1\ \text{kg}}} = 2.5\ \text{N}$$

(c) Now we can find a from Eq. ii:

$$a = \frac{T}{m_2} = \frac{2.5\ \text{N}}{1\ \text{kg}} = 2.5\ \text{m s}^{-2}$$

Note that if we had considered the train as a single object of mass $m = m_1 + m_2 = (3 + 1)\ \text{kg} = 4\ \text{kg}$, we could immediately have found the acceleration from $a = F/m = 10\ \text{N}/4\ \text{kg} = 2.5\ \text{m s}^{-2}$. However, the normal and tension forces can only be found by considering the individual cars and not the system as a whole.

Example 3.4

A block of mass $m_1 = 20$ kg is free to move on a horizontal surface. A rope, which passes over a pulley, attaches it to a hanging block of mass $m_2 = 10$ kg (Fig. 3.11). Assuming for simplicity that the pulley and rope masses are negligible and that there is no friction, find (a) the forces on the blocks and (b) their acceleration. (c) If the system is initially at rest, how far has it moved after 2 s?

This problem is similar to the preceding problem of the two cars, although here one mass moves horizontally and one moves vertically. Again, the fact that the rope and pulley are massless implies that the force exerted by the rope on m_1 is equal in magnitude to the force exerted by the rope on m_2. Accordingly we denote both forces by the same symbol $\mathbf{T}$.

(a) We first apply $\mathbf{F} = m\mathbf{a}$ to the block on the surface. Since it has no vertical acceleration component, the net vertical force component must be zero. Thus the normal force N_1 on block 1 due to the surface is

$$N_1 = w_1 = m_1g = (20\ \text{kg})(9.8\ \text{m s}^{-2}) = 196\ \text{N}$$

The system is accelerating with an unknown acceleration a. Block 1 has $a_x = a$, so $F_x = m_1a_x$ becomes

$$T = m_1a \qquad \text{(i)}$$

Note that both T and a are unknown, so we cannot find either until we consider the motion of m_2.

We now apply Newton's second law to block 2. Since it is accelerating downward, $a_y = -a$, and $F_y = m_2a_y$ becomes

$$T - w_2 = -m_2a \qquad \text{(ii)}$$

This equation also involves the two unknowns, T and a. Since the number of equations now equals the number

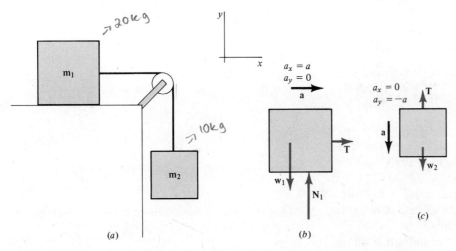

Figure 3.11. (b) and (c) are force diagrams for the two blocks when they are connected as shown in (a).

of unknowns, we can solve for T and a. From Eq. i we have $a = T/m_1$. Substituting this in Eq. ii gives

$$T - w_2 = -m_2 \frac{T}{m_1}$$

Solving for T,

$$T = \frac{w_2}{1 + \dfrac{m_2}{m_1}} = \frac{m_2 g}{1 + \dfrac{m_2}{m_1}} = \frac{(10\ \text{kg})(9.8\ \text{m s}^{-2})}{1 + \dfrac{10\ \text{kg}}{20\ \text{kg}}}$$

$$= 65.3\ \text{N}$$

(b) Using Eq. i, the acceleration is

$$a = \frac{T}{m_1} = \frac{65.3\ \text{N}}{20\ \text{kg}} = 3.27\ \text{m s}^{-2}$$

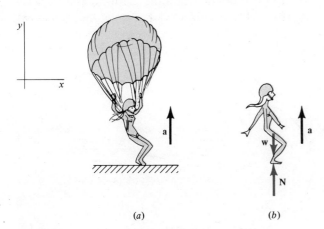

Figure 3.12. (a) A parachutist comes to rest with an acceleration of $3g$. (b) The forces on her are a normal force **N** due to the ground and her weight **w**.

(c) Since the system is initially at rest and uniformly accelerated, the distance it moves in 2 s is

$$\Delta x = \tfrac{1}{2} a\,(\Delta t)^2 = \tfrac{1}{2}(3.27\ \text{m s}^{-2})(2\ \text{s})^2 = 6.54\ \text{m}$$

Thus if we know the initial position and velocity, we can find the subsequent motion from the forces.

Example 3.5

A parachutist of weight **w** strikes the ground with her legs flexed and comes to rest with an upward acceleration of magnitude $3g$. Find the force exerted on her by the ground during landing (Fig. 3.12).

The forces on the woman are her weight **w** and a normal force **N** due to the ground. With $m = w/g$ and $a = 3g$, $F_y = ma_y$ becomes

$$N - w = ma = \left(\frac{w}{g}\right)(3g) = 3w$$

$$N = 4w$$

The force on her feet due to the ground is four times her weight. By contrast, if she is simply standing on the ground, the normal force equals her weight. Note that if she holds her legs more stiffly during landing, she will come to rest with a greater acceleration in a shorter distance, and the force on her feet will be greater.

3.9 | GRAVITATIONAL FORCES

Newton's study of planetary motion led him to infer the formula for the gravitational force between two masses. This formula is referred to as the *law of universal gravitation* and is regarded as a fundamental

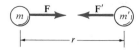

Figure 3.13. Two spheres exert attractive forces on each other. **F** and **F**′ are equal in magnitude but opposite in direction, in accordance with Newton's third law of motion.

law of nature. Using this law along with the three laws of motion, Newton was able to derive the observed laws of planetary motion. In addition, he was able to account accurately for the motion of the moon around the earth and to provide a qualitative explanation of the tides in our oceans.

The law of universal gravitation states that all objects in the universe attract each other. For two uniform spheres, or for two objects of any shape that are so small compared to their separation that they may be considerd as point particles, the law has a simple form. If two spheres or particles have gravitational masses m and m', and their centers are separated by a distance r, the forces between the two spheres have a magnitude

$$F = \frac{Gmm'}{r^2} \tag{3.7}$$

G is called the *gravitational constant* and has a measured value of

$$G = 6.67 \times 10^{-11}\,\text{N}\,\text{m}^2\,\text{kg}^{-2}$$

The gravitational forces are directed along the line connecting the centers of the two spheres (Fig. 3.13). The magnitude of the gravitational force varies as $1/r^2$, so Eq. 3.7 is referred to as an *inverse square law*.

As noted, Eq. 3.7 applies directly to spheres and to particles. For more complex objects, the forces between all the pairs of particles in the objects must be summed to find the net gravitational forces on the objects. However, these net forces are still equal but opposite as required by the third law of motion.

The gravitational force exerted by the earth on an object is relatively large because of the large mass of the earth. In contrast, the gravitational force between two objects of moderate mass is very small and hard to detect, as is shown quantitatively in the next example.

Example 3.6

The centers of two 10-kg spheres are separated by 0.1 m. (a) What is their gravitational attraction?

(b) What is the ratio of this attraction to the weight of one of the spheres?

(a) Using Newton's law of gravitation, the forces between the spheres have a magnitude

$$F = G\frac{mm'}{r^2} = (6.67 \times 10^{-11}\,\text{N}\,\text{m}^2\,\text{kg}^{-2})\,\frac{(10\,\text{kg})(10\,\text{kg})}{(0.1\,\text{m})^2}$$

$$= 6.67 \times 10^{-7}\,\text{N}$$

The forces are directed along the line connecting the centers of the spheres.

(b) The weight of one of the spheres is

$$w = mg = (10\,\text{kg})(9.8\,\text{m}\,\text{s}^{-2}) = 98\,\text{N}$$

Thus the ratio of the gravitational forces between the spheres to the weight of a sphere is

$$\frac{F}{w} = \frac{6.67 \times 10^{-7}\,\text{N}}{98\,\text{N}} = 6.81 \times 10^{-9}$$

The small size of this ratio explains why we do not notice the gravitational attractions among objects of ordinary size. However, the gravitational forces between two such objects can be observed and measured using very sensitive instruments.

In Chapter Five, we will see how Newton's law of universal gravitation leads to an understanding of planetary and lunar motion and of the tides.

3.10 | WEIGHT

The weight of an object is the gravitational force it experiences. For an object near the surface of the earth, this force is mainly due to the earth's attraction.

Let us call the radius of the earth R_E and its mass M_E (Fig. 3.14). An object with a gravitational mass $\tilde{m}$

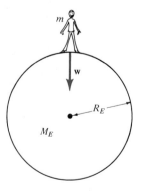

Figure 3.14. A person of mass m has a weight $w = GmM_E/R_E^2$ at the surface of the earth.

at the earth's surface is subjected to a gravitational force. From Eq. 3.7 this force is

$$F = G \frac{\tilde{m} M_E}{R_E{}^2} \tag{3.8}$$

The acceleration g resulting from this force can be found from Newton's second law, $\mathbf{F} = m\mathbf{a}$, which contains the inertial mass m:

$$g = \frac{F}{m} = \frac{1}{m}\left(G \frac{\tilde{m} M_E}{R_E{}^2} \right)$$

If we assume that the inertial and gravitational masses are the same, then $\tilde{m}/m = 1$, and we find

$$g = G \frac{M_E}{R_E{}^2} \tag{3.9}$$

This result says that the gravitational acceleration is the same for all objects. This is in agreement with experiments and justifies the assumption that the gravitational and inertial masses of an object are equal.

Note that the earth's radius R_E is 6400 km. Accordingly, the gravitational acceleration within a few metres or even a few kilometres of the surface will not differ appreciably from the value at the surface, 9.8 m s^{-2}.

As we have seen, mass and weight are related but quite different quantities. The mass of an object is an intrinsic property that is the same whether the object is in Chicago, on the moon, or in interstellar space. It is a measure of the amount of matter in the object and determines its inertia or its response to a force. The weight of an object varies from place to place and is the force resulting from gravity.

The mass of an object is determined *only* by the amount of matter present and is independent of its physical or chemical state. For example, if a cubic metre of oxygen gas at atmospheric pressure is cooled, it will liquefy and fill a volume of about 10^{-3} m^3. Nevertheless, it will still have the same number of molecules and the same mass. Similarly, when a volume of hydrogen gas and another volume of oxygen gas combine to form liquid water, the volume is reduced by a large factor, but again the mass is unchanged. The following example further illustrates the relationship between mass and weight.

Example 3.7

An astronaut weighs 700 N on the earth. What is his weight on planet X, which has a radius $R_X = R_E/2$ and a mass $M_X = M_E/8$?

On the earth, his weight is

$$w_E = \frac{Gm M_E}{R_E{}^2}$$

where m is his mass. On planet X, his mass is the same, but his weight is

$$w_X = \frac{Gm M_X}{R_X{}^2} = \frac{Gm(M_E/8)}{(R_E/2)^2}$$
$$= \frac{4}{8}\frac{Gm M_E}{R_E{}^2} = \frac{1}{2} w_E$$

Thus, his weight on planet X is $(\frac{1}{2})(700 \text{ N}) = 350 \text{ N}$.

3.11 | EFFECTIVE WEIGHT

When an elevator starts to move upward, it accelerates briefly and then moves at a constant velocity until it approaches the desired floor. During the upward acceleration, we feel heavier than usual. Similarly, when the acceleration is downward, we have the feeling that our weight is reduced. Our weight is the gravitational force exerted on us by the earth, and that, of course, is not changed by our being in the elevator. However, our *perception* of our weight is determined by the forces exerted on us by the floor or chair or whatever is supporting us. These forces are not equal to our weight when we are accelerating.

We define the *effective weight* $\mathbf{w}^e$ of a person or an object as the total force that object exerts on a spring scale. According to Newton's third law of motion, this is equal in magnitude and opposite in direction to the force $\mathbf{S}$ the scale exerts on the person or object, so

$$\mathbf{w}^e = -\mathbf{S} \tag{3.10}$$

While such a measurement may not always be feasible, this definition will be useful in understanding and calculating the effective weight of an accelerating object. The following example illustrates how the effective weight is found.

Example 3.8

A woman of mass m stands on a spring scale in an elevator. Find the effective weight of the woman if the elevator is accelerating upward at $0.2g$.

The forces acting on the woman are her weight $\mathbf{w} = m\mathbf{g}$ and the force $\mathbf{S}$ due to the spring scale (Fig. 3.15).

Using Newton's second law, $\mathbf{F} = m\mathbf{a}$,

$$S - mg = ma, \qquad S = mg + ma$$

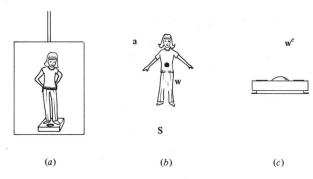

Figure 3.15. (*a*) A woman accelerating upward in an elevator. (*b*) The forces on the woman. (*c*) The woman exerts a force $\mathbf{w}^e$ on the spring scale. By Newton's third law, $\mathbf{S} = -\mathbf{w}^e$.

Since by definition the effective weight is equal in magnitude to the spring scale force S,

$$w^e = m(g + a) = mg(1 + 0.2) = 1.2mg$$

The effective weight of the woman is then 1.2 times her normal weight, *mg*.

If the acceleration were downward, a similar calculation would show that the woman's effective weight is less than *mg*.

From Fig. 3.15, it is clear how to find the effective weight $\mathbf{w}^e$ in general. By definition, $\mathbf{w}^e = -\mathbf{S}$, where $\mathbf{S}$ is the force exerted by the scale. From Newton's second law, $\mathbf{S} + m\mathbf{g} = m\mathbf{a}$, or $-\mathbf{S} = m\mathbf{g} - m\mathbf{a}$. Thus,

$$\mathbf{w}^e = m\mathbf{g} - m\mathbf{a} \qquad (3.11)$$

An object in free fall has an acceleration $\mathbf{a}$ equal to $\mathbf{g}$, so its effective weight is zero. Since an artificial satellite in orbit around the earth is in free fall, an astronaut in such a satellite has zero effective weight, and will float freely about his spaceship if he is not tethered. However, even though his effective weight is zero, his mass has not changed (Fig. 3.16). If he carelessly pushes off hard from one wall and strikes the opposite wall headfirst, he will experience a large and unpleasant force as he abruptly decelerates.

3.12 | FRICTION

Friction is a force that always acts to resist the motion of one object sliding on another. Frictional forces are very important, since they make it possible for us to walk, use wheeled vehicles, and hold books. Microscopically, friction arises from many minute temporary bonds between the contact points of the two surfaces.

Frequently, we try to reduce the frictional forces that oppose some desired motion. This is often done with rollers or with wheels, since rolling peels the surfaces apart with a much smaller force than is required when they are torn apart by sliding.

Frictional forces in fluids are called *viscous forces*. They are often quite small compared to the friction between solid surfaces. Thus the use of lubricating liquids such as oil, which clings to the surfaces of metals, greatly reduces friction. Similarly, an air layer provides an almost frictionless support for a hovercraft vehicle or an air track demonstration apparatus.

When we walk or run, we are not conscious of any friction in our knees or other leg joints. These and many other mammalian joints are well lubricated by *synovial fluid,* which is squeezed through the cartilage lining the joints when they move (Fig. 3.17). This lubricant tends to be absorbed when the joint is stationary, increasing the friction and making it easier to maintain a fixed position. This is an excellent example of the clever biological engineering that nature has employed. Other illustrations of lubrication in our bodies include the saliva we add to food as we chew it, and the mucus coatings in our heart, lungs, and intestines that minimize friction as these organs move in carrying out their functions.

Figure 3.16. An astronaut in free fall has a zero effective weight, and perceives himself to be weightless. (NASA.)

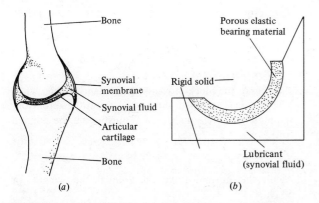

Figure 3.17. Human joints are lubricated by synovial fluid squeezed through the porous cartilage lining the joint. (*a*) A typical human joint. (*b*) A model approximately equivalent to the joint. (From Duncan Dowson, "Lubrication in Human Joints," Verna Wright, (ed), in *Lubrication and Wear in Joints*, Lippincott; Philadelphia, 1969.)

To make quantitative statements about friction, we consider a block at rest on a horizontal surface (Fig. 3.18*a, b*). Since the block is at rest, the first law requires that the net force on the block be zero. The vertical forces are the weight **w** and the normal force **N**, so we must have $N = w$. In the horizontal direction, there is no applied force and no motion, so the frictional force must also be zero, according to the first law.

Now suppose we apply a small horizontal force **T** to the right (Fig. 3.18*c*). If the block *remains at rest*,

the friction force $\mathbf{f}_s$ can no longer be zero, since the first law requires that the net force be zero, or $f_s = T$. If **T** is gradually increased, $\mathbf{f}_s$ increases also. Eventually when **T** becomes large enough, the block begins to slide. *Thus there is a maximum possible static friction force* $\mathbf{f}_s$(max).

Experimentally, it is found that $\mathbf{f}_s$(max) has the following properties:

1 $\mathbf{f}_s$(max) *is independent of the contact area*. For example, if we saw the block in half and stack one piece on the other (Fig. 3.18*d*), then $\mathbf{f}_s$(max) is unchanged.
2 *For a given pair of surfaces*, f_s(max) *is proportional to the normal force N.*
3 The number relating f_s(max) and N, called the *coefficient of static friction*, μ_s, is defined by

$$f_s(\text{max}) = \mu_s N \qquad (3.12)$$

The coefficient μ_s depends on the nature of the two surfaces, and on their cleanliness and smoothness, the amount of moisture present, and so forth. Typically, for metals on metals, μ_s is between 0.3 and 1.0. Values much above 1 are not encountered under ordinary circumstances. However, metals exposed to air form thin layers of oxide on their surfaces. If these are removed and the surfaces are brought into contact in a vacuum, large bonding forces come into play, making the value of μ_s quite large. When lubri-

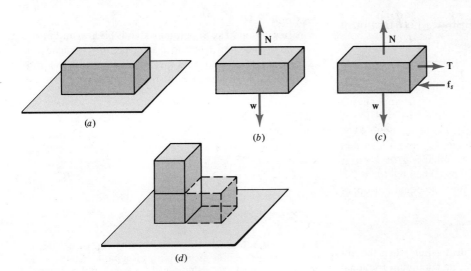

Figure 3.18. (*a*) A block at rest on a horizontal surface. (*b*) The force diagram. (*c*) The force diagram when a force **T** is applied. (*d*) The maximum static friction force is independent of the contact area.

cating oils are used, μ_s is about 0.1 for metals on metals. For Teflon on metals, $\mu_s \simeq 0.04$. In a healthy hip joint, the synovial fluid reduces μ_s to the remarkably low value of 0.003.

4 *The force necessary to keep an object sliding at constant velocity is smaller than that required to start it moving.* For example, it is easier to keep a heavy table or box moving than it is to get it started. Thus the *sliding or kinetic friction force* f_k is less than $f_s(\text{max})$. *It is independent of the contact area* and it satisfies

$$f_k = \mu_k N \qquad (3.13)$$

Here μ_k is the *coefficient of kinetic friction* and is determined by the nature of the two surfaces.

5 μ_k *is nearly independent of the velocity* and, since $f_k < f_s(\text{max})$,

$$\mu_k < \mu_s \qquad (3.14)$$

It is important to note that these properties of frictional forces are *not* fundamental statements about the physical world in the same sense as Newton's laws of motion or the universal law of gravitation. Even though they give a good description of the frictional forces in common situations, they represent only an *approximation* to a complex problem. A more complete discussion would have to include the specific characteristics of the forces among the molecules making up the two surfaces. Friction is still an imperfectly understood phenomenon because of this complexity.

The following examples illustrates how friction is taken into account.

Example 3.9

A 50-newton block is on a flat, horizontal surface (Fig. 3.18). (a) If a horizontal force $T = 20$ N is applied and the block remains at rest, what is the frictional force? (b) The block starts to slide when T is increased to 40 N. What is μ_s? (c) The block continues to move at constant velocity if T is reduced to 32 N. What is μ_k?

(a) Since the block remains at rest when the force **T** is applied, the frictional force $\mathbf{f}_s$ must be equal but opposite to **T**. Consequently

$$f_s = T = 20 \text{ N}$$

(b) Since the block just begins to slide when the applied force is increased to 40 N, the maximum frictional force must be

$$f_s(\text{max}) = 40 \text{ N}$$

The vertical forces must add to zero, so the normal force N is equal to the weight, w, or 50 N. Hence

$$\mu_s = \frac{f_s(\text{max})}{N} = \frac{40 \text{ N}}{50 \text{ N}} = 0.8$$

(c) Since the block moves with constant velocity when a 32-N force is applied, the net force must be zero. Consequently the frictional force f_k must equal the applied force, or

$$f_k = 32 \text{ N}$$

Again the normal force must equal the weight, or 50 N, so

$$\mu_k = \frac{f_k}{N} = \frac{32 \text{ N}}{50 \text{ N}} = 0.64$$

As expected, the coefficient of kinetic friction μ_k is less than the coefficient of static friction μ_s.

As we noted before, the value of μ_s is usually 1 or less. This often limits the force a person or animal can exert. For example, when a man attempts to push or pull horizontally while standing on flat ground, the force he exerts is matched by an equally large reaction force acting on him. He will slip if that reaction force exceeds the maximum friction force, which will equal his weight if $\mu_s = 1$. On soft ground, animals with hooves or claws can dig in and somewhat increase the force they apply without slipping. Locomotives are made very heavy to increase the maximum friction force.

These ideas are used in the next example.

Example 3.10

In a comic strip, Superman extends his arm and brings a large speeding truck to a stop before it reaches his torso (Fig. 3.19). To see if this is consistent with the principles of physics, assume the truck is moving at 30 m s^{-1},

Figure 3.19. Superman attempting to stop a speeding truck. Since the coefficient of friction for his feet on the road is at most 1, he cannot exert a horizontal force greater than his weight without slipping due to the equally large reaction force exerted on him by the truck. Despite his superstrength, his effort to stop the truck within an arm's length will fail badly.

its mass M is 50,000 kg, and Superman's mass m is 100 kg. If the force he exerts is limited by the frictional force between his feet and the ground, and $\mu_s = \mu_k = 1$, what is the minimum distance over which he can stop the truck?

The maximum force he can exert is $F = \mu_s N = \mu_s mg$, so

$$F = (1)(100 \text{ kg})(9.8 \text{ m s}^{-2}) = 980 \text{ N}$$

Thus the acceleration of the truck is at most

$$a = \frac{F}{M} = \frac{980 \text{ N}}{50,000 \text{ kg}} = 0.0196 \text{ m s}^{-2}$$

Using $v^2 = v_0^2 + 2a\,\Delta x$ with $v = 0$, the stopping distance is

$$\Delta x = \frac{-v_0^2}{2a} = -\frac{(30 \text{ m s}^{-1})^2}{2(0.0196 \text{ m s}^{-2})}$$

$$= -23,000 \text{ m} = -23 \text{ km}$$

Because Superman's maximum force is limited here to his weight, it takes him 23 km to stop the truck! (The minus sign arises because Δx and a are in opposite directions.)

The next example shows how the coefficient of static friction may be measured in an introductory physics laboratory.

Example 3.11

A block is at rest on an inclined plane (Fig. 3.20). The coefficient of static friction is μ_s. What is the maximum possible angle of inclination $\theta(\text{max})$ of the surface for which the block will remain at rest?

In this problem it is convenient to choose the coordinate axes as shown. By definition, the normal force $\mathbf{N}$ is perpendicular to the surface, or in the $-x$ direction. Similarly, the frictional force $\mathbf{f}_s$ is parallel to the surface or along the y direction. From Fig. 3.20c, the components of the weight $\mathbf{w}$ are

$$w_x = w \cos \theta, \qquad w_y = -w \sin \theta$$

When the block remains at rest, the first law requires that the x and y components of the forces cancel. Thus

$$f_s = w \sin \theta$$
$$N = w \cos \theta$$

Taking the ratio of these equations, the weight w cancels, leaving

$$\frac{f_s}{N} = \frac{\sin \theta}{\cos \theta} = \tan \theta$$

When the block is about to slip, $f_s = f_s(\text{max}) = \mu_s N$ and $\theta = \theta(\text{max})$. Thus we find

$$\mu_s = \tan \theta(\text{max})$$

An arrangement of this type provides a simple way to measure the coefficient of static friction. The angle is gradually increased until the block will not stay in place. For example, if $\theta(\text{max}) = 37°$, then $\mu_s = \tan 37° = 0.754$.

SUMMARY

The mass of an object is defined as the ratio of its weight w to the gravitational acceleration g. It can also be defined as the net force on the object divided by the acceleration produced. The density of an object is its mass to volume ratio, $\rho = m/V$. Its relative density or specific gravity is the ratio of the density to that of water at $0°$ C and atmospheric pressure.

Newton's three laws of motion allow us to predict the motion of an object from the forces acting upon it. The first law states that in an inertial coordinate frame, the object remains at rest or in motion with constant velocity unless a net force is applied. When an object has no net force on it even though two or more forces act upon it, the object is said to be in equilibrium. The type of equilibrium—stable, unstable, or neutral—is determined by noting whether the

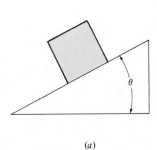

(a)

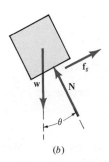

(b)

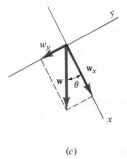

(c)

Figure 3.20. (a) A block on an inclined plane. (b) The forces on the block. (c) The components of the weight $\mathbf{w}$.

object tends to return to its original state of rest or uniform motion when it is disturbed slightly.

The second law states that the net force **F** needed to produce an acceleration **a** satisfies

$$\mathbf{F} = m\mathbf{a}$$

Here m is the mass or inertia of the object.

The third law states that if object A exerts a force B, then B exerts an equal and opposite force on A. Since the forces act on different objects, their effects do not cancel.

All objects exert gravitational forces on each other. The forces between two spheres or point particles are proportional to the products of their masses and inversely proportional to the square of their separation. The mass of an object is the same everywhere in the universe; its weight $\mathbf{w} = m\mathbf{g}$ depends on the acceleration due to gravity at that object's location. The perceived or effective weight also depends on the acceleration, and is given by

$$\mathbf{w}^e = m(\mathbf{g} - \mathbf{a}).$$

A person in free fall has zero effective weight.

When the force applied to an object resting on a surface exceeds the maximum static friction force $\mu_s N$, it starts to move. The force of sliding or kinetic friction $\mu_k N$ is usually smaller than the maximum static friction force. The coefficients μ_s and μ_k depend on the surfaces and are usually less than 1. The frictional forces between two surfaces are independent of their contact area.

Checklist
Define or explain:

force	tension
contact force	Newton's third law of
weight	motion
newton	action-reaction forces
gravitational mass	Newton's second law of
kilogram	motion
density	inertial mass
relative density	law of universal
specific gravity	gravitation
Newton's first law of	gravitational constant
motion	inverse square law
inertial coordinate	effective weight
system	friction force
equilibrium: stable,	coefficients of static and
unstable, neutral	kinetic or sliding
normal force	friction

REVIEW QUESTIONS

Q3-1   Forces exerted only when two objects are touching are called _____.

Q3-2   The weight of an object is the _____ force on that object.

Q3-3   The mass of an object is its weight divided by the _____.

Q3-4   The density is the ratio of the mass of an object to its _____.

Q3-5   A material with a specific gravity of 1 has the same density as _____.

Q3-6   According to Newton's first law, an object in motion with a constant velocity relative to an inertial frame has _____ acting on it.

Q3-7   If an object returns to its original resting place when moved slightly, it is in _____.

Q3-8   If object A exerts a force on object B, B exerts _____ on A.

Q3-9   The acceleration of an object is equal to the net force acting on the object divided by its _____.

Q3-10   The universal law of gravitation is referred to as an inverse square law because the gravitational force varies as _____.

Q3-11   The gravitational mass of an object equals its _____.

Q3-12   The _____ of an object is the same everywhere; its _____ depends on where the object is located in the universe.

Q3-13   The effective weight of an object is zero when it is in _____.

Q3-14   The maximum frictional force between two given surfaces is independent of the _____ and proportional to the _____.

Q3-15   The force needed to keep an object sliding is smaller than that required to _____.

Q3-16   The coefficient of static friction is usually less than _____, and greater than the coefficient of _____.

EXERCISES

Section 3.1 | Force, Weight, and Gravitational Mass

3-1   Find the direction and magnitude of the net force on the object in Fig. 3.21.

3-2   Find the direction and magnitude of the net force on the object in Fig. 3.22.

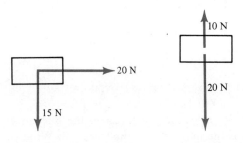

Figure 3.21. Exercise 3-1.　　**Figure 3.22.** Exercise 3-2.

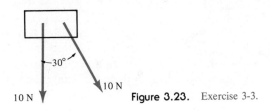

Figure 3.23. Exercise 3-3.

3-3 Find the direction and magnitude of the net force on the object in Fig. 3.23.

3-4 A man weighs 980 N. What is his mass in kilograms?

3-5 A woman has a mass of 50 kg. What is her weight in newtons?

3-6 What is the weight of a 1 kilogram steak?

3-7 Find the weight of 500 grams of candy in (a) newtons; (b) pounds.

3-8 A candy bar weighs 1 ounce (1 oz) (16 oz = 1 lb). Find its mass in kilograms.

3-9 A woman weighs 120 lb. What is her mass in kilograms?

3-10 A large oil tanker weighs 200,000 tons, where 1 ton = 2000 lb. What is its mass in kilograms?

Section 3.2 | Density (See Table 3.2 for required densities.)

3-11 What is the mass of a litre of whole blood? (1 litre = 10^{-3} m^3.)

3-12 Hydrogen atoms are the most common type of matter in many regions of interstellar space. A hydrogen atom has a mass of 1.67×10^{-27} kg. If there is an average of one hydrogen atom in each cubic centimetre of an interstellar "gas cloud," what is the density of hydrogen in S.I. units?

3-13 (a) Using the solar and terrestrial data tabulated on the inside back cover of this book, calculate the average density of the sun. (b) Is this density consistent with that given for the sun in Table 3.2? Explain.

3-14 A cylindrical iron rod has a radius of 1 cm and a length of 20 cm. What is its mass?

3-15 The nucleus of a uranium atom is approximately described as a sphere of radius 8.7×10^{-15} m and of mass 3.5×10^{-25} kg. (a) What is its average density? (b) What is its specific gravity?

3-16 Neutron stars are a late stage in stellar evolution. A typical neutron star has a radius of 10^4 m and a mass of 2×10^{30} kg. (a) What is its average density? (b) Find the ratio of this density to the density of lead.

3-17 What is the specific gravity of water at a temperature of 0° C and a pressure of 50 atmospheres?

3-18 A gold foil has a thickness of 10 micrometres (1 μm = 10^{-6} m). What is the mass of a square of side 10 cm?

3-19 In the petroleum industry, a barrel is defined to be 42 gallons, where 1 gallon = 3.786 litres = 3.786×10^{-3} m^3. Find the mass in kilograms of one barrel of oil with a relative density of 0.8.

3-20 Find the density of gasoline if 5 kg has a volume of 7.35×10^{-3} m^3.

3-21 Battery acid has a density of 1290 kg m^{-3} and contains 35 percent sulfuric acid by weight. What is the mass of the sulfuric acid in 1 litre? (1 litre = 10^{-3} m^3 of battery acid.)

3-22 (a) Calculate the percentage change in the density of air when it is heated at atmospheric pressure from 0° C to 100° C. (b) Calculate the corresponding change for water.

3-23 (a) What is the specific gravity of lead? (b) What is the mass of a cube of lead whose side is 10 cm?

Section 3.3 | Newton's First Law and Section 3.4 | Equilibrium

3-24 For which of the following observers does Newton's first law of motion hold true in the form stated in this chapter? (a) A person on a plane moving at constant speed in a constant direction. (b) A parachutist who has just stepped out of an airplane. (c) A parachutist who has reached terminal velocity and is falling at constant speed. (d) The pilot of a plane that is taking off from a runway. Explain your reasoning.

3-25 A person is in a car traveling around a circular track. Do the observations of this person agree with Newton's first law as stated in this chapter? Does your answer depend on whether the speed is constant? Explain.

3-26 A car is moving at a constant speed on a straight road, which is at an angle of 10° to the horizontal direction. (a) Is it in equilibrium? (b) What forces act on the car?

3-27 A 2000-kg airplane is in level flight at constant velocity. (a) What is the net force on the plane? (b) What is the upward lift force on the plane due to the air?

3-28 A round stone resting in the back of a pickup truck begins to roll toward the back as the truck begins moving. The driver is at rest with respect to the truck and concludes that a force is causing the stone to move toward the rear of the truck. (a) Is he correct? (b) What forces are acting on the stone? (c) Does the stone move forward or backward with respect to the ground?

3-29 In Fig. 3.4, the cables are at an angle of 30° to the horizontal direction. How large are the forces F_1 and F_2 they exert on the traffic light if its weight is w?

3-30 (a) A pencil is placed on its side on a table. If its cross section is hexagonal (six-sided), what type of equilibrium is the pencil in? (b) What type of equilibrium is it if the pencil has a circular cross section? (c) Suppose the pencil is balanced on its point. What type of equilibrium is this?

3-31 Which type of equilibrium is illustrated by the traffic light in Fig. 3.4? Explain what will happen if it is slightly displaced horizontally or vertically.

3-32 A package of emergency supplies is parachuted from a plane. The air resistance force increases roughly as the velocity squared, so the package rapidly reaches a constant, maximum velocity directed straight down. (a) Once it reaches this velocity, is this package in equilibrium? (b) What will happen to its state of motion if a brief gust of wind pushes it sideways? (c) What will happen to its state of motion if there is a sudden brief downdraft of air?

3-33 A car goes around a circular track at a constant speed. Is it in equilibrium? Explain.

Section 3.5 | Newton's Third Law

3-34 A boat in a flowing river is tied by a rope to a post on a dock. (a) Draw a diagram showing all of the forces acting on the boat. (b) Identify the horizontal forces acting on the boat and the associated

reaction forces. (c) Identify the vertical forces acting on the boat and the associated reaction forces. (d) What is the net force on the boat?

3-35 A large airplane is pulled at constant velocity relative to the runway by a truck. The two are connected by an iron bar. (a) What are the forces on the airplane? (b) What are the forces on the truck? (c) What is the net force on the airplane? (d) What is the net force on the truck? (e) What is the net force on the iron bar? (f) Identify the action-reaction forces acting on the airplane, bar, and truck.

3-36 An airplane is flying horizontally with a constant velocity. The propellers are pushing backward on the air. (a) Is the airplane in equilibrium? (b) What forces are acting in the horizontal plane?

3-37 (a) A girl holds a ball motionless in her hand. Identify the forces acting on the ball and their reactions. (b) She throws the ball into the air. What are the forces on the ball while it is in the air? What are the reactions to these forces?

3-38 A car coasts to a stop on a flat, straight road. (a) What are the forces acting on the car? (b) What are the reactions to those forces?

Section 3.6 | Newton's Second Law

3-39 The takeoff acceleration of a woman jumping vertically is 20 m s^{-2}. Her mass is 50 kg. (a) How large a force does the ground exert on the woman? (b) What is the ratio of this force to her weight?

3-40 What acceleration is produced when a 100-N net force is applied to a 10-kg rock?

3-41 What net force is needed to give a 1000-kg car an acceleration of 3 m s^{-2}?

3-42 A baseball of mass 0.15 kg is struck by a bat with a force of 5000 N. What is the acceleration of the ball?

Section 3.8 | Some Examples of Newton's Laws

3-43 An elevator of mass 900 kg accelerates upward at 3 m s^{-2}. What is the tension in the cable where it is attached to the elevator?

3-44 A horse can exert a horizontal force of 3.5×10^4 N on a rope that passes over a pulley to lift loads vertically. What is the acceleration of a load of weight (a) 3.5×10^4 N; (b) 3×10^4 N? (Neglect the masses of the rope and pulley.)

3-45 An elevator cable that is light in weight compared to the elevator car can support a weight of 10,000 N. If the elevator and occupants weigh

8000 N, what is the maximum possible vertical acceleration of the elevator?

3-46 A 60-kilogram man hangs from a light cable suspended from a helicopter. Find the tension in the cable if the acceleration is (a) 5 m s^{-2} upward; (b) 5 m s^{-2} downward.

3-47 A human femur will fracture if the compressional force is 2×10^5 N. A person of mass 60 kg lands on one leg, so that there is a compressional force on the femur. (a) What acceleration will produce fracture? (b) How many times the acceleration of gravity is this?

3-48 A 55-kg woman wishes to slide down a stationary rope that will support a force of 400 N. What is the minimum acceleration of the woman if she safely slides down the rope?

3-49 An engine with a mass of 4×10^4 kg pulls a train with a mass of 2×10^5 kg on a level track with an acceleration of 0.5 m s^{-2}. What would the acceleration be if the train had a mass of 10^5 kg?

3-50 In a collision, an automobile of mass 1000 kg stops with constant acceleration in 2 m from an initial speed of 20 m s^{-1}. (a) What is the acceleration of the car? (b) What is the net force on the car during the collision?

3-51 A tennis ball of mass 0.058 kg initially at rest is served at a velocity of 45 m s^{-1}. If the racket is in contact with the ball for 0.004 s, what is the net force on the ball during the serve? (Assume the acceleration is constant.)

Section 3.9 | Gravitational Forces

3-52 The moon is 3.9×10^5 km from the center of the earth. The mass of the moon is 7.3×10^{22} kg, and the mass of the earth is 6.0×10^{24} kg. How far from the earth's center are the gravitational forces on an object due to the earth and moon equal and opposite? (Assume the object is on the line connecting the earth and moon.)

3-53 The mass of the sun is 2.0×10^{30} kg, and the distance from the moon to the sun is 1.5×10^8 km. Using the data of the preceding example, find the ratio of the forces exerted by the earth and the sun on the moon.

3-54 When a rocket ship is at a distance R_E from the surface of the earth, the earth's gravitational attraction on the ship is 144,000 N. What is the earth's gravitational attraction when the ship is at a distance $3R_E$ from the surface? (R_E is the radius of the earth.)

Section 3.10 | Weight

3-55 The acceleration of gravity on the surface of Mars is 3.62 m s^{-2}. How much would a person who weighs 800 N on earth weigh on Mars?

3-56 The mass of Mars is 6.42×10^{23} kg and the acceleration of gravity on its surface is 3.62 m s^{-2}. What is the radius of Mars?

3-57 The acceleration of gravity at the surface of a planet is half that on the surface of the earth. If the radius of the planet is half the radius of the earth, how is its mass related to the mass of the earth?

3-58 Planet Y has a radius one-third times that of the earth, and a mass $(1/3)^3 = 1/27$ times that of the earth. What is the weight of an astronaut with a mass of 70 kg?

3-59 If the earth is assumed to be a uniform sphere of radius 6.38×10^6 m, compute its mass from G and g.

3-60 An airline stewardess has a mass of 50 kg. (a) What is her weight on the ground? (b) By what fraction does her weight change when she is in a plane 6.38 km above the ground? (The radius of the earth is 6380 km.)

Section 3.11 | Effective Weight

3-61 A fighter plane dives straight down with an acceleration of $3g$. What is the magnitude and direction of the pilot's effective weight if his weight is w?

3-62 A racing car accelerates with an acceleration equal to g on a straight flat track. If the driver has a mass of 60 kg, what is the magnitude and direction of her effective weight?

3-63 A car initially moving on a straight, flat road at 30 m s^{-1} comes to a stop in 10 s. (a) Assuming the acceleration is constant, how large is it? (b) The driver has a mass m. What is the magnitude and direction of his effective weight as the car slows down?

3-64 An astronaut of mass m is in a spaceship that takes off straight up from the surface of the earth. Its acceleration is kept equal to 9.8 m s^{-2}. (a) What is the effective weight of the astronaut just after takeoff? (b) What is his effective weight when the spaceship is at a distance from the surface of the earth equal to the radius of the earth?

Section 3.12 | Friction

3-65 A horse weighing 7500 N is able to exert a horizontal force of 6500 N on a load. What is the coefficient of static friction between the horse's feet and the ground? (Assume the force exerted by the horse is limited by its tendency to slip.)

3-66   A refrigerator weighs 1000 N. A horizontal force of 200 N is applied, but the refrigerator does not move. (a) What is the frictional force? (b) What can we conclude about the coefficient of static friction?

3-67   A box weighing 100 N is at rest on a horizontal floor. The coefficient of static friction is 0.3. What is the minimum force needed to start the box in motion?

3-68   A box weighing 100 N is pushed on a horizontal floor. The coefficient of sliding friction is 0.2. What acceleration will result if a horizontal force of 40 N is applied?

3-69   On racing cars, surfaces called *spoilers* are sometimes placed so that there will be a downward force due to the air rushing over the car. What is their purpose?

3-70   A refrigerator of mass 120 kg is at rest on a kitchen floor. ($\mu_s = 0.4$ and $\mu_k = 0.2$.) (a) If nobody touches the refrigerator, what is the frictional force exerted on the refrigerator by the floor? (b) A boy of mass 40 kg leans on the refrigerator, exerting a horizontal force on it equal to half his weight. What is the frictional force exerted by the floor on the refrigerator?

3-71   A sled weighing 1000 N is pulled along flat, snow-covered ground. The coefficient of static friction is 0.3, and the coefficient of sliding friction is 0.15. Find the force needed to (a) start the sled moving, and (b) keep it moving at constant velocity.

3-72   A stone boat of total weight 60,000 N is used in a horse-pulling contest. The coefficient of static friction between the stone boat and the earth is 0.6, and the coefficient of sliding friction is 0.4. (a) What force must a pair of horses exert to start the stone boat moving? (b) What force must the horses exert to keep the stone boat moving at a constant velocity?

3-73   How can the adjustable inclined plane in Fig. 3.20 be used to measure the coefficient of kinetic friction for an object on the plane?

PROBLEMS

3-74   A planet of radius R is made up of a core of radius $R/2$ and density ρ and an outer shell of den-

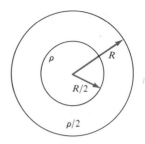

Figure 3.24.   Problem 3-74.

sity $\rho/2$ (Fig. 3.24). What is the average density of the planet as a whole?

3-75   A 1000-kg automobile is traveling at 15 m s^{-1} and skids to a stop with constant acceleration in 100 m. What is the frictional force on the car?

3-76   A 0.5-kg ball is initially at rest. If a 10-N force is applied for 2 s, what is the final velocity of the ball?

3-77   A hockey player who weighs 800 N comes to rest from 10 m s^{-1} in 1 s. (a) What is his mass? (b) What is his average acceleration? (c) What force is required to provide this acceleration?

3-78   A runner's foot strikes the ground with a velocity of 10 m s^{-1} downward. If the effective mass of the foot and leg that is brought to rest is 9 kg, what is the force on the foot as it comes to rest with constant acceleration in (a) 0.03 m on a soft surface; (b) 0.005 m on a hard surface?

3-79   A woman of mass 55 kg steps off a rock, striking the ground at 5 m s^{-1}. (a) If she lands on her feet with her body held rigid, she stops in 0.15 m. What average force is exerted upward on the woman during the impact? (b) If she flexes her legs and body during impact, she stops in 0.5 m. What average force does she now experience during impact?

3-80   A boy is fishing with a line that will sustain a maximum force of 40 N. If he hooks a 3-kg fish, which can exert a force of 60 N for several seconds, what is the minimum acceleration with which the line must be played out during that time interval?

3-81   A subway train has three cars, each weighing 1.2×10^5 N. The frictional force on each car is 10^3 N, and the first car, acting as an engine, exerts a horizontal force of 4.8×10^4 N on the rails. (a) What is the acceleration of the train? (b) What is the tension in the coupling between the first and second cars? (c) What is the tension between the second and third cars?

3-82 If a 60-kg woman walks up a 45° hill at a constant velocity, what is the force she must exert parallel to the ground if (a) she walks straight up the hill; (b) she takes a zigzag path up the hill so the effective angle of ascent is 30°?

3-83 A 60-kg man pulls horizontally on a rope. The frictional force on his feet is 600 N. The rope passes over a pulley and is attached to a post of mass 25 kg set in the ground. If the frictional force opposing the vertical motion of the post is 300 N, (a) what is the vertical acceleration of the post? (Assume the man moves as the post does.) (b) If the frictional force is still 300 N, what is the maximum mass of a post that the man can pull from the ground?

3-84 A wagon weighing 5×10^3 N is pulled along a muddy, horizontal road by a 500-kg horse. The coefficient of friction between the wagon wheels and the road is 0.2. (a) If the wagon is pulled at constant speed, what force must the horse exert on the ground to pull the wagon? (b) If the wagon is accelerated from rest to a velocity of 5 m s^{-1} in 5 s, what force must the horse exert on the ground?

3-85 The radius of the planet Venus is 6.1×10^3 km and that of the earth is 6.4×10^3 km. The mass of Venus is 82 percent of the earth's mass. What is the acceleration of gravity on the surface of Venus?

3-86 Two lead spheres of radius 0.1 m are in contact. (a) What is the mass of each sphere? (b) What is the gravitational force between them?

3-87 Neutron stars with a density comparable to that of atomic nuclei, 10^{17} kg m^{-3}, are believed to exist. Suppose two spheres of radius 0.01 m of such a density were somehow placed 1 m apart on the earth. (a) What would be the weight of each sphere? (b) What would be the gravitational attraction between them?

3-88 A 60-kilogram man wishes to run on ice. The coefficient of static friction between his shoes and the ice is 0.1. What is his maximum possible acceleration?

3-89 A girl of mass 40 kg skis down a slope, which is at an angle of 37° with the horizontal. (Neglect air resistance.) If the coefficient of kinetic friction between her skis and the snow is 0.1, what is her acceleration?

3-90 In Fig. 3.25, the strings and pulleys are

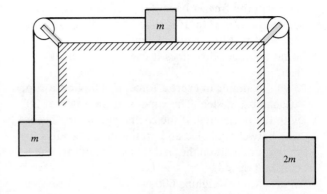

Figure 3.25. Problems 3-90 and 3-91.

massless, and there is no friction. Find (a) the tensions in the strings, and (b) the acceleration of the system.

3-91 Repeat the preceding problem if the coefficient of kinetic friction between the block on the surface and the surface is 0.1.

3-92 In Fig. 3.26 the string and pulley are massless, and there is no friction. Find (a) the tension in the string, and (b) the acceleration.

3-93 In the preceding problem, $m_1 = 2$ kg and $m_2 = 3$ kg. Find (a) the tension in the string, and (b) the acceleration. (c) If the system is released from rest, what are its velocity and position after 0.5 seconds?

*3-94 Two people wish to push a food freezer weighing 2000 N up a ramp inclined at an angle of 37° with the horizontal. The coefficient of sliding friction between the freezer and ramp is 0.5. (a) What minimum force must the people exert to slide the freezer up the ramp? (b) What accelera-

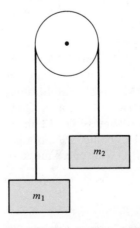

Figure 3.26. Problems 3-92 and 3-93.

tion will the freezer have if it is released and slides down the ramp? (c) If it slides 4 m down the ramp and strikes a heavy object, coming to rest in 0.5 m, what average force does it exert on the heavy object?

*3-95 A man can exert a force of 700 N on a rope attached to a sled. The rope is at an angle of 30° with the horizontal. If the coefficient of kinetic friction between the sled and ground is 0.4, what is the maximum load on the sled that the man can pull at constant speed?

*3-96 A box weighing 600 N is at rest on a ramp at an angle of 37° with the horizontal. The coefficient of static friction between the box and ramp is 0.8. Find the minimum force required to move the box down the ramp if the force is applied (a) parallel to the ramp; (b) horizontally.

*3-97 In a ski jump, the slope is initially at an angle of 45° to the horizontal. If the coefficient of kinetic friction between the skis and the snow is 0.1, find (a) a skier's acceleration, and (b) the velocity reached after 40 m on the ramp.

3-98 If there is no friction, what is the acceleration of the block in Fig. 3.27?

*3-99 If the coefficient of sliding friction is 0.2, what is the acceleration of the block in Fig. 3.27?

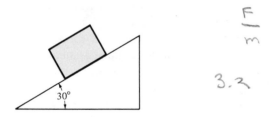

Figure 3.27. Problems 3-98 and 3-99.

ANSWERS TO REVIEW QUESTIONS

Q3-1, contact forces; **Q3-2,** gravitational; **Q3-3,** gravitational acceleration; **Q3-4,** volume; **Q3-5,** water at 0° C; **Q3-6,** no net force; **Q3-7,** stable equilibrium; **Q3-8,** an equal but opposite (reaction) force; **Q3-9,** mass; **Q3-10,** $1/r^2$; **Q3-11,** inertial mass; **Q3-12,** mass, weight; **Q3-13,** free fall; **Q3-14,** contact area, normal force; **Q3-15,** start it moving; **Q3-16,** one, kinetic or sliding friction.

Additional Reading

E.N. daCosta Andrade, *Sir Isaac Newton: His Life and Work,* Science Study Series, Doubleday and Co., Garden City, New York, 1958.

Richard S. Westfall, *Never at Rest,* Cambridge University Press, New York, 1981. A biography of Newton.

Herbert Butterfield, *The Origins of Modern Science,* The Macmillan Co., New York, 1960. Chapters 8 and 10 discuss Newton's work.

Robert W. Zimmerman, The Measurement of Mass, *The Physics Teacher,* vol. 21, 1983, p. 354.

William J. Kaufman III, Listening for the Whisper of Gravity Waves, *Science 80,* May/June 1980, p. 64.

J.W. Beams, Finding a Better Value for G, *Physics Today,* vol. 24, May 1971, p. 34.

R. McNeill Alexander, *Animal Mechanics,* University of Washington Press, Seattle, 1968. Pages 55–66 discuss friction in animals.

E.H. Freitag, The Friction of Solids, *Contemporary Physics,* vol. 2, 1961, p. 198.

Ernest Rabinowitz, Friction (Resource Letter F-1), *American Journal of Physics,* vol. 31, 1963, p. 897. A bibliography.

John R. Cameron and James R. Skofronick, *Medical Physics,* John Wiley and Sons, Inc., New York, 1978. Friction in the body is discussed on page 27 and on pages 56–58.

Scientific American articles:

I. Bernard Cohen, Newton, December 1955, p. 73.

D. Sciama, Inertia, February 1957, p. 99.

Michael McCloskey, Intuitive Physics, April 1983, p. 122. Common misconceptions about the laws of motion.

Carl Gans, How Snakes Move, June 1970, p. 82.

W.A. Heiskane, The Earth's Gravity, September 1955, p. 164.

George Gamow, Gravity, March 1961, p. 94.

R.H. Dicke, The Eötvös Experiment, December 1961, p. 84.

Clifford C. Will, Gravitational Theory, November 1974, p. 24.

Stillman Drake, Newton's Apple and Galileo's Dialogue, August 1980, p. 150.

I. Bernard Cohen, Newton's Discovery of Gravity, March 1981, p. 166.

Daniel Z. Freedman, Supergravity and the Unification of the Laws of Physics, February 1978, p. 126.

T.C. Van Flandern, Is Gravity Getting Weaker, February 1976, p. 44.

Bryce S. DeWitt, Quantum Gravity, December 1983, p. 112.

F. Palmer, Friction, February 1956, p. 54.

Ernest Rabinowitz, Stick and Slip, May 1956, p. 109.

CHAPTER 4

STATICS

Statics is the study of the forces acting on an object that is in equilibrium and at rest. Even though no motion occurs, a number of interesting questions arise concerning these forces that can be answered with the aid of Newton's laws. For example, one can find the forces acting on various parts of engineering structures, such as bridges or buildings, or of biological structures, such as jaws, limbs, or backbones. Statics can be used to understand the force multiplication or *mechanical advantage* obtained with simple machines, such as the many levers found in the human body. It also addresses problems of balance and stability for objects as well as for animals. The range of questions answerable with statics makes it invaluable in fields as diverse as engineering, comparative anatomy, physical therapy, and orthodonture.

Our discussion of statics is based on an idealized object, the *rigid body:* an object extended in space which does not change its size or shape when subjected to a force. Real objects are made up of a large number of particles (atoms and molecules) held together by forces acting among the particles, and they may vibrate or bend when they are subjected to forces. However, objects such as baseballs, bones, and steel beams are often rigid enough so that these deformations are negligibly small.

A rigid body will be in equilibrium if two conditions are satisfied. The condition that the net force is zero is sufficient to ensure that a point particle at rest remains at rest; for a rigid body, this condition means that the body as a whole will not accelerate, or that it is in *translational equilibrium*. However, a rigid body will start to rotate if forces act so that there is a net turning effect, or *torque*. Thus the absence of a net torque is the second condition necessary for the equilibrium of a rigid body.

Applications of statics to questions of stability and balance also require the concept of the *center of gravity*. This is the point at which the weight of a rigid body may be considered to be concentrated.

4.1 | TORQUES

Suppose an object is subjected to two equal but opposite forces. The net force is zero, so the object is in translational equilibrium. Nevertheless, it may not be in rotational equilibrium. As an example, consider a lunch counter stool with a freely rotating seat (Fig. 4.1). If a child applies equal but opposite forces $\mathbf{F}_1$ and $\mathbf{F}_2 = -\mathbf{F}_1$ to opposite sides of the seat, it will

Figure 4.1. Equal but opposite forces applied to opposite sides of the seat will cause it to start rotating. Hence, it is not in equilibrium, even though the net force is zero.

67

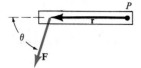

Figure 4.2. The torque on the rigid body about point P has a magnitude $\tau = rF \sin \theta$.

begin to rotate. Thus the seat does not remain at rest even though $\mathbf{F} = \mathbf{F}_1 + \mathbf{F}_2 = 0$, and there is no net force.

Clearly, in addition to $\mathbf{F} = 0$, we require another equilibrium condition to exclude the possibility of rotational motion. The quantity that indicates the ability of a force to cause rotation is called the *torque*. *A rigid body is in rotational equilibrium when there is no net torque acting on it.* The torque τ depends on the force $\mathbf{F}$, the distance $\mathbf{r}$ from a point on the axis of rotation to the point where the force acts on the object, and the angle θ between $\mathbf{r}$ and $\mathbf{F}$ (Fig. 4.2). (τ is

the Greek letter tau.) Deferring the specification of the direction for several paragraphs, the magnitude of the torque about point P is

$$\tau = rF \sin \theta \qquad (4.1)$$

The dimensions of a torque are force times length, so the S.I. torque unit is a newton metre (N m).

This expression for the torque can be illustrated in various ways. For example, suppose we need to unscrew a large nut that is rusted into place (Fig. 4.3). To maximize the torque, we use the longest wrench available and exert as large a force as possible. When $\theta = 90°$, $\sin \theta = 1$, which is the largest value of $\sin \theta$. Consequently, we should pull at right angles to the wrench. Note that when θ is $0°$ or $180°$, $\sin \theta$ is zero, and there is no torque.

Another example is provided by the screen door in Fig. 4.4. A spring opposes the rotation of the door and permits it to open through an angle ϕ that increases with the torque. We observe that the torque is

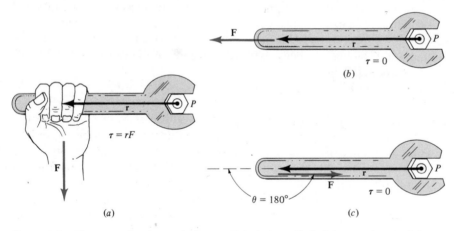

Figure 4.3. The torque has a magnitude $\tau = rF \sin \theta$. Accordingly it is a maximum when $\mathbf{r}$ and $\mathbf{F}$ are at right angles as in (a). The torque is zero when $\mathbf{r}$ and $\mathbf{F}$ are parallel ($\theta = 0°$) as in (b) or opposite ($\theta = 180°$) as in (c).

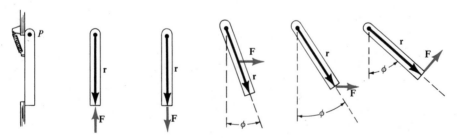

Figure 4.4. A top view of a screen door hinged at P. A spring resists opening of the door, so that the deflection angle ϕ increases with the applied torque. When a force $\mathbf{F}$ is applied opposite or parallel to $\mathbf{r}$, no torque results. The greatest torque occurs when $\mathbf{F}$ is perpendicular to $\mathbf{r}$ and $\mathbf{r}$ is as large as possible.

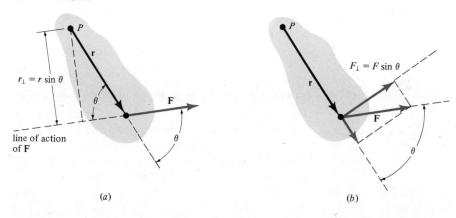

(a)

(b)

Figure 4.5. The magnitude of the torque about P is $\tau = rF \sin \theta$. This can be rewritten as (a) $\tau = r_\perp F$, where $r_\perp = r \sin \theta$ is the lever arm. (b) Also, $\tau = rF_\perp$, where $F_\perp = F \sin \theta$.

greatest when the force is applied as far as possible from the hinge and at right angles to the door.

The magnitude of the torque $\tau = rF \sin \theta$, can be rewritten as (Fig. 4.5a)

$$\tau = r_\perp F \qquad (4.2)$$

Here the subscript $_\perp$ means perpendicular. That is, the magnitude of the torque is the product of the force and the perpendicular distance to the *line of action* of the force. The perpendicular distance or *lever arm* $r_\perp = r \sin \theta$ depends both on the distance from P to the point where the force acts and on the angle; it is greatest when $\theta = 90°$ and $\sin \theta = 1$. The torque can also be written (Fig. 4.5b) in terms of the force component $F_\perp = F \sin \theta$ perpendicular to $\mathbf{r}$:

$$\tau = rF_\perp \qquad (4.3)$$

Direction of the Torque | Specifying the direction of the torque is simplest for an object such as a wrench or door constrained to move about a given axis. Then only torques due to forces acting perpendicular to that axis have to be considered. A force (or a component of a force) parallel to the axis will have no effect on the state of rotational motion, because the hinges or other contraints will exert compensating torques.

In this kind of situation, we can draw two dimensional pictures, as in Figs. 4.3 and 4.4, with the forces and the distance vectors $\mathbf{r}$ from a point P on the axis lying in a plane perpendicular to the axis of rotation. The torques that tend to produce a *counterclockwise* rotation are then said to be vectors directed along the axis outward from the page, and are conventionally taken to be *positive* (Fig. 4.6). In the figure, $\boldsymbol{\tau}$ is shown

as a circled dot representing the tip of an arrow. Similarly, torques causing *clockwise* rotations are directed along the axis into the page and are conventionally taken to be *negative*. Here $\boldsymbol{\tau}$ is shown as a circled cross symbolizing the tail of an arrow.

A more general definition of the direction of the torque is needed for objects, such as balls or gymnasts, that are free to rotate about any axis. This definition, which reduces to that given previously for objects rotating about fixed axes, is given in terms of the *vector product* or *cross product* of two vectors.

The vector product of $\mathbf{A}$ and $\mathbf{B}$ is a vector $\mathbf{C}$, which is perpendicular to both $\mathbf{A}$ and $\mathbf{B}$ and denoted by

$$\mathbf{C} = \mathbf{A} \times \mathbf{B} \qquad (4.4)$$

The magnitude of $\mathbf{C}$ is (see Fig. 4.7)

$$C = AB \sin \theta \qquad (4.5)$$

where θ is the angle between $\mathbf{A}$ and $\mathbf{B}$.

Since $\mathbf{C}$ is perpendicular to both $\mathbf{A}$ and $\mathbf{B}$, it points along the direction perpendicular to the plane de-

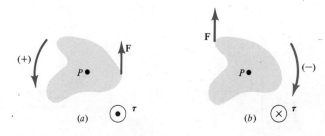

(a)

(b)

Figure 4.6. (a) Counterclockwise torques are taken to be positive. The torque vector is out of the page and is symbolized by a dot representing the tip of an arrow. (b) Clockwise torques are taken to be negative, or into the page. The cross represents the tail of an arrow.

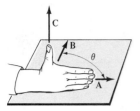

Figure 4.7. $C = A \times B$ has a magnitude $AB \sin \theta$. Its direction is given by the right-hand rule.

fined by **A** and **B**. You can determine whether **C** points toward or away from this plane with the aid of the *right-hand rule* as follows:

1 Place the two vectors tail to tail, and put your right hand at their intersection.
2 Point the fingers of your right hand along the first vector (**A** in Fig. 4.7).
3 Orient your arm so that you can bend your hand at the wrist and rotate your palm forward, through an angle of less than 180°, until your fingers point toward **B**.
4 Your thumb now points in the same direction as **C** = **A** × **B**.

From the definition of the cross product, it follows that **A** × **B** and **B** × **A** are in opposite directions, or **A** × **B** = −**B** × **A**. Thus the order of the factors in a cross product is significant; this product does not obey the *commutative principle*. This contrasts with ordinary algebra, where $x \cdot y = y \cdot x$ holds, and with vector addition, where **A** + **B** = **B** + **A**. These operations are commutative.

In terms of this notation the torque τ can be written as

$$\tau = r \times F \qquad (4.6)$$

We note that the magnitude of τ is $\tau = rF \sin \theta$ as before. The direction of τ is given by the right-hand rule and indicates the axis about which rotation will tend to occur.

To illustrate the right-hand rule, suppose **r** is in the $+x$ direction and **F** is in the $+y$ direction (Fig. 4.8*a*). Using the right-hand rule, we point the fingers of our right hand in the $+x$ direction. When our palm faces the $+y$ direction, and our thumb is out of the page, we can rotate our fingers 90° toward the $+y$ direction. Thus $\tau = r \times F$ *is out* of the page.

Suppose instead that **r** is in the $-x$ direction and **F** is in the $+y$ direction (Fig. 4.8*b*). Pointing our fingers toward $-x$ with the palm toward $+y$, we can rotate 90° from **r** to **F**. In this case our thumbs and $\tau = r \times F$ are into the page. Note that these results are equivalent to those in Fig. 4.6.

Couples | A pair of forces with equal magnitudes but opposite directions acting along different lines of action is called a *couple*. The pair of forces applied to the lunch counter seat in Fig. 4.1 is an example of a couple. Couples do not exert a net force on an object even though they do exert a net torque. They have the interesting property that the net torque is independent of the choice of the point P from which distances are measured. This is seen explicitly in the following example.

Example 4.1

Two forces with equal magnitudes but opposite directions act on an object with different lines of action (Fig. 4.9). Find the net torque on the object resulting from these forces.

If we compute the torque about the point P in the figure, the torque resulting from the force at x_2 is $\tau_2 = -x_2 F$. (The minus sign indicates a clockwise torque.) The torque resulting from the force at x_1 is $\tau_1 = x_1 F$. The net torque is

$$\tau = \tau_1 + \tau_2 = x_1 F - x_2 F$$
$$= (x_1 - x_2)F = -lF$$

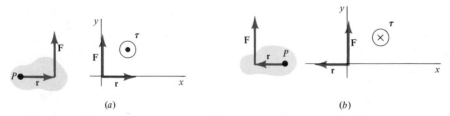

(a)   (b)

Figure 4.8. (*a*) When torques are computed about point P, $\tau = r \times F$ is out of the page. (*b*) τ is into the page.

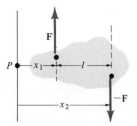

Figure 4.9. The torque due to a couple is the same about every point.

The minus sign means that the net torque tends to cause clockwise rotation, and is directed into the page. Note that only the distance l between the lines of action of the forces appears in the result, so that the torque is independent of the location of the point P.

In the next section we see how torques are used to determine whether a system is in rotational equilibrium.

4.2 | EQUILIBRIUM OF RIGID BODIES

From the discussion in the preceding section, we see that there are two conditions for the equilibrium of a rigid body:

1 *The net force on the object must be zero,*

$$\mathbf{F} = 0 \qquad (4.7)$$

2 *The net torque on the object computed about any convenient point must be zero,*

$$\boldsymbol{\tau} = 0 \qquad (4.8)$$

These two conditions ensure that a rigid body will be in both translational and rotational equilibrium.

They have been motivated earlier by qualitative or intuitive arguments, but they can be derived by applying Newton's laws of motion to the forces acting on a rigid body and to the forces among its constituent particles.

A familiar application of these conditions is provided by two children balancing on a seesaw (Fig. 4.10). If their weights are unequal, they soon find that the heavier child must be closer to the pivot. For example, if one child is twice as heavy as the other, he or she must sit half as far from the pivot. Let us see how this comes about from the equilibrium conditions.

Example 4.2

Two children of weights w_1 and w_2 are balanced on a board pivoted about its center (Fig. 4.10). (a) What is the ratio of their distances x_2/x_1 from the pivot? (b) If $w_1 = 200$ N, $w_2 = 400$ N, and $x_1 = 1$ m, what is x_2? (For simplicity, we assume the board to be weightless; this will not affect the result.)

(a) According to the first equilibrium condition, the force $\mathbf{N}$ exerted by the support must balance their weights so that the net force is zero:

$$N - w_1 - w_2 = 0, \qquad N = w_1 + w_2$$

This gives us no information about their positions. However, we have not yet used the condition that the net torque is zero. We will compute torques about the pivot point, P. Then the lever arm is zero for $\mathbf{N}$, since its line of action passes through the pivot, and it has no torque. The torques resulting from the weights are $\tau_1 = x_1 w_1$ and $\tau_2 = -x_2 w_2$. Thus, $\tau = \tau_1 + \tau_2 = 0$ requires

$$x_1 w_1 - x_2 w_2 = 0$$

or

$$\frac{x_2}{x_1} = \frac{w_1}{w_2}$$

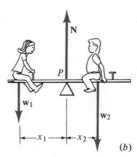

(a)   *(b)*   *(c)*

Figure 4.10. *(a)*, *(b)* When two children balance on a see-saw, the heavier child must be closer to the pivot. *(c)* Another equilibrium configuration.

This is the condition relating the positions x_1 and x_2 when the seesaw is balanced. Either position may be chosen arbitrarily; the other is then determined by this condition.

(b) If $w_1 = 200$ N, $w_2 = 400$ N, and $x_1 = 1$ m, then we must have

$$x_2 = x_1 \frac{w_1}{w_2} = (1 \text{ m}) \frac{(200 \text{ N})}{(400 \text{ N})} = 0.5 \text{ m}$$

This is consistent with our initial statement that if one child is twice as heavy as the other, he or she must sit half as far from the pivot.

In this example we chose to compute the torques about the pivot point. However, the equilibrium condition says that the torques calculated about *any* point must add up to zero. In the next example we will solve the same problem, computing torques about a different point. We will find that the result is the same.

Example 4.3

Again find x_2/x_1 for the seesaw of the preceding example, calculating torques about the point P_1, where the child of weight w_1 is seated.

To proceed, we redraw the force diagram of Fig. 4.10 as in Fig. 4.11. Computing torques about P_1, N and $\mathbf{w}_2$ produce torques x_1N and $-(x_1 + x_2)w_2$, respectively; $\mathbf{w}_1$ produces no torque, since its lever arm is zero. In equilibrium the sum of these torques must be zero, so

$$-(x_1 + x_2)w_2 + x_1N = 0 \qquad \text{(i)}$$

But the forces must also add up to zero, so

$$N - w_1 - w_2 = 0$$

or

$$N = w_1 + w_2$$

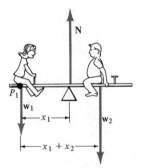

Figure 4.11. Calculation of the torques about P_1.

Substituting for N in Eq. i, we find

$$-(x_1 + x_2)w_2 + x_1(w_1 + w_2) = 0$$

Cancelling some terms, we obtain our earlier result

$$x_2 w_2 = x_1 w_1 \qquad \text{or} \qquad \frac{x_2}{x_1} = \frac{w_1}{w_2}$$

This example illustrates that we can often simplify our calculations with a clever choice of the point about which we compute the torques. For example, using the pivot eliminated the unknown normal force **N** from the torque equation. The same final results are always obtained no matter what point is used to calculate torques.

We now use the equilibrium conditions to find the forces on a human forearm.

Example 4.4

A model for the forearm in the position shown in Fig. 4.12 is a pivoted bar supported by a cable. The weight **w** of the forearm is 12 N and can be treated as concentrated at the point shown. Find the tension **T** exerted by the biceps muscle and the force **E** exerted by the elbow joint.

The tension **T** and the weight **w** have no horizontal components. Since the net horizontal force must be zero, the force **E** exerted by the joint cannot have a horizontal component. We assume tentatively that **E** is directed downward; a negative result will indicate it points the opposite way.

Applying the condition $\mathbf{F} = 0$,

$$T - E - w = 0$$

This contains both unknowns, T and E. Calculating torques about the pivot, **E** produces no torque, **w** produces a torque $-(0.15 \text{ m})w$, and **T** produces a torque $(0.05 \text{ m})T$. Thus, $\tau = 0$ becomes

$$-0.15w + 0.05T = 0$$

or

$$T = 3w = 3(12 \text{ N}) = 36 \text{ N}$$

The first equation then gives

$$E = T - w = 36 \text{ N} - 12 \text{ N} = 24 \text{ N}$$

E is positive, so it is directed downward as initially assumed.

Note that both the tension **T** exerted by the muscle and the force **E** exerted by the elbow are considerably larger than the weight they are supporting. This is because, relative to the joint, the lever arm for the weight is larger than that for the muscle. When a weight is held in the hand, it is even further from the joint, and the addi-

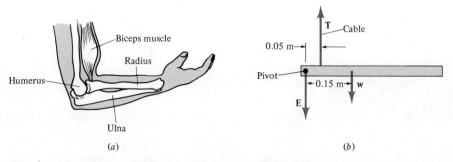

Figure 4.12. (*a*) The forearm is supported by the biceps muscle and pivoted at the elbow. (*b*) The forearm can be considered as a pivoted bar supported by a cable. The pivot represents the elbow joint, and the cable represents the biceps. (Adapted from Williams and Lissner.)

tional forces the muscle and joint must provide are correspondingly larger.

4.3 | THE CENTER OF GRAVITY

The torque about any point produced by the weight of an object is equal to that due to a concentrated object of the same weight placed at a point called the *center of gravity* (C.G.) (Fig. 4.13). This fact simplifies

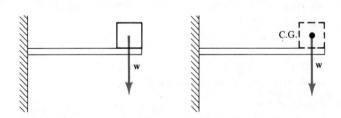

Figure 4.13. The torque produced by the weight of a rigid body is equal to that due to a concentrated object of the same weight placed at the center of gravity.

the mechanics of both static and moving objects and was implicitly assumed in Example 4.4 when we treated the weight of the forearm as concentrated at a point. The centers of gravity of uniformly dense symmetric objects are at their geometric centers (Fig. 4.14). For less symmetric objects the C.G. can be calculated mathematically or located experimentally.

A suspended object always hangs so that its center of gravity is directly below the point of suspension, since in this position the torque resulting from the weight about that point is zero (Fig. 4.15). This observation provides a way of locating the C.G. experimentally. If an object is suspended from a point P_1, it comes to rest with the C.G. on the vertical line through P_1. Suspending the object from a second point P_2, the C.G. lies on the vertical line through P_2. The C.G. is then at the intersection of the two lines (Fig. 4.16).

We now show how to find the center of gravity mathematically, starting with the simplest possible system: two point weights on a weightless rod (Fig. 4.17). We take the C.G. to be a distance x_1 from w_1

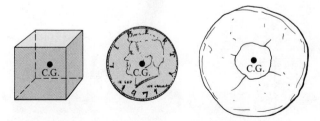

Figure 4.14. The center of gravity of a uniformly dense symmetric object is at its geometric center. Note that the C.G. of the doughnut is at the center of the hole and hence not in the object itself.

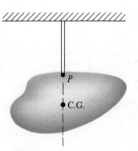

Figure 4.15. An object will hang so that its center of gravity is below the point of suspension, *P*.

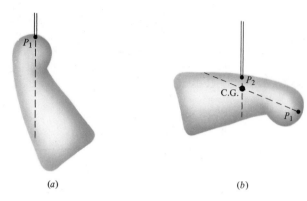

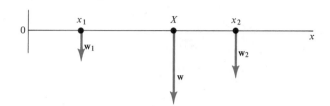

Figure 4.18. The center of gravity of two point weights on a weightless rod is at X.

A single weight w at X gives a torque $\tau = -Xw$. Equating the two expressions for τ, we find the center of gravity is at

$$X = \frac{x_1 w_1 + x_2 w_2}{w}$$

Suppose, for example, that $w_1 = w_2$. Then $w = w_1 + w_2 = 2w_1$, and

$$X = \frac{x_1 w_1 + x_2 w_1}{2w_1} = \frac{x_1 + x_2}{2}$$

Thus the center of gravity X is midway between the two weights, as was found before.

If there are more than two weights, the center of gravity is found in the same way. The result is

$$X = \frac{x_1 w_1 + x_2 w_2 + x_3 w_3 + \cdots}{w} \qquad (4.9)$$

where

$$w = w_1 + w_2 + w_3 + \ldots \qquad (4.10)$$

The following example illustrates how this result is used when there are three weights.

Example 4.5

A weightless plank 4 m long has one concrete block at the left end, another at the center, and two blocks at the right end (Fig. 4.19). Where is the center of gravity?

Figure 4.16. (*a*) The center of gravity lies on the vertical line through P_1. (*b*) The C.G. also lies on the vertical line through P_2, so it is located at the intersection of the two lines.

and x_2 from w_2. Then the net torque about the C.G. due to the two weights must be equal but opposite. This is similar to the seesaw of Example 4.2, and we can use the balance condition obtained there, $x_2/x_1 = w_1/w_2$. For example, if $w_1 = w_2$, then $x_1 = x_2$; the C.G. is at the midpoint, as is always true for a symmetric object. If $w_2 = 2w_1$, then $x_2 = x_1/2$; the C.G. is closer to the heavier weight, as we might have expected.

Another way of finding the C.G. of two weights leads to a formula that can immediately be adapted to any number of weights. Figure 4.18 shows the same two weights on a weightless rod located on an x axis. The C.G. is at an unknown point X. From the definition, a weight $w = w_1 + w_2$ concentrated at X will produce a torque equal to the sum of the torques due to w_1 and w_2. The individual torques about the origin are $\tau_1 = -x_1 w_1$ and $\tau_2 = -x_2 w_2$. Thus the total torque is

$$\tau = \tau_1 + \tau_2 = -x_1 w_1 - x_2 w_2$$

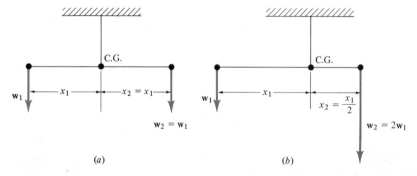

Figure 4.17. The center of gravity of two point weights. (*a*) Equal weights. (*b*) Unequal weights.

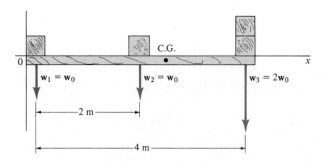

Figure 4.19. The center of gravity is between the center of the plank and the heavier end.

Let us call the weight of a block w_0; the numerical value is not needed since it will cancel out. We can choose our origin arbitrarily, so we will take it at the left end. Then the total weight is $w = w_1 + w_2 + w_3 = 4w_0$, and Eq. 4.9 becomes

$$X = \frac{x_1 w_1 + x_2 w_2 + x_3 w_3}{w}$$

$$= \frac{0 + (2\text{ m})w_0 + (4\text{ m})(2w_0)}{4w_0} = 2.5\text{ m}$$

Thus the C.G. is between the center of the plank and the heavier end.

The equation for the center of gravity, Eq. 4.9, contains weights in both the numerator and denominator. If we substitute $w = mg$ for each weight, the factors of g cancel out. Then X involves the masses instead of the weights, and it is called the *center of mass* (C.M.). There is no difference between the C.G. and C.M. as long as **g** has the same direction and magnitude for each weight.

Although we have only discussed weights placed at several points on a straight line, the procedure for

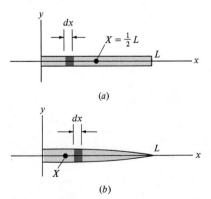

(a)

(b)

Figure 4.20. (a) The weight of a segment of width dx is $\lambda_0 dx$. The center of gravity of a uniformly dense rod is at its midpoint. (b) The center of gravity is nearer the heavier end.

finding the center of gravity for more complicated arrangements is basically the same. If weights are at several points in a plane, then the C.G. is at a point (X, Y) in the plane. Equation 4.9 is then used to find X, and an analogous equation involving the y coordinates of the weights is used to find Y.

When the weight is distributed continuously, rather than being concentrated at several discrete points, the sum in Eq. 4.9 must be replaced by an integral

$$X = \frac{1}{w} \int x\, dw \tag{4.11}$$

Here w is the total weight

$$w = \int dw \tag{4.12}$$

These equations are used in the next example.

Example 4.6

(a) Find the center of gravity X of a thin metal rod of length L with a uniform weight per unit length λ_0 (Fig. 4.20a). (b) Find the center of gravity for the rod in Fig. 4.20b that is shaped so that its weight per unit length λ varies according to the formula $\lambda = \lambda_0(1 - x/L)$.

(a) Since this rod is symmetric about the point $x = L/2$, its C.G. is clearly located there. Nevertheless, we will use Eqs. 4.11 and 4.12 to derive this result to illustrate the procedure in the simplest type of situation.

We imagine cutting the rod into small segments of length dx. The product of dx and the weight per unit length λ_0 is the weight of the segment,

$$dw = \lambda_0 dx$$

Thus the total weight of the rod is the integral

$$w = \int_0^L \lambda_0 dx = \lambda_0 x \Big|_0^L = \lambda_0 L$$

The total weight is just the weight per unit length times the length.

Now each small segment contributes $x\,dw$ to the numerator of Eq. 4.11. Thus, the center of gravity is located at

$$X = \frac{1}{w} \int_0^L x\, dw = \frac{1}{\lambda_0 L} \int_0^L \lambda_0 x\, dx$$

$$= \frac{1}{L} \int_0^L x\, dx = \frac{1}{2L} x^2 \Big|_0^L$$

$$= L^2/2L = L/2$$

The C.G. is at the midpoint, as anticipated.

(b) Again we imagine the cutting the rod into small segments of length dx. The weight of a segment is $dw = \lambda dx = \lambda_0(1 - x/L)dx$, and the total weight of the

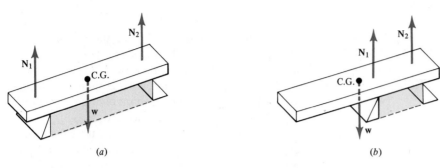

Figure 4.21. The beam is (*a*) in equilibrium; (*b*) not in equilibrium. The base area defined by the supports is shown in color.

rod is

$$x = \int_0^L \lambda_0 (1 - x/L) dx = \lambda_0 \left[x - x^2/2L \right]_0^L$$

$$= \lambda_0 (L - L^2/2L) = \lambda_0 L/2$$

Then the center of gravity is located at

$$X = \frac{1}{w} \int_0^L x \, dw = \frac{1}{\lambda_0 L/2} \int_0^L x \lambda_0 (1 - x/L) \, dx$$

$$= \frac{1}{L/2} \int_0^L (x - x^2/L) dx$$

$$= \frac{(x^2/2 - x^3/3L)}{L/2} \bigg|_0^L = \frac{L}{3}$$

The center of gravity is at $X = L/3$. It is to the left of the midpoint $L/2$ of the rod because the left side is heavier.

4.4 | STABILITY AND BALANCE

The number and position of an animal's legs have apparently been partially determined by its requirements for stability and balance. The basic idea is illustrated by the beam in Fig. 4.21. If its center of gravity is between the two supports, the torques about the C.G. due to N_1 and N_2 are opposite and cancel, so the beam is in equilibrium. However, when the center of gravity is to the left of the supports, the torques about the C.G. due to N_1 and N_2 are both positive. Since the net torque is not zero, the beam topples over. Thus, an object is balanced only when its *center of gravity lies above the base area* defined by its supports.

An animal standing on four legs is analogous to a table. A table placed on a surface that is gradually tilted will topple over when the center of gravity is no longer over the portion of the surface defined by the tips of the four legs (Fig. 4.22). The shorter the legs for a given tabletop, the larger the angle θ at which this occurs and the greater its stability; a low table is more stable than a high one. Similarly, the centers of gravity of automobiles, ships, and even vases must be kept low for good stability to be achieved. Thus, we

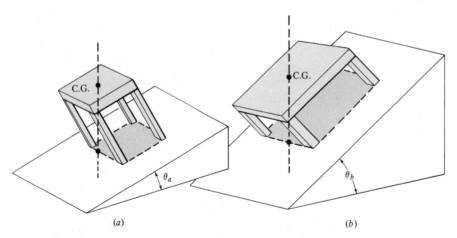

Figure 4.22. A four-legged table will tip when a vertical line from the center of mass passes just outside the base area formed by the legs. Since $\theta_a < \theta_b$, a long-legged table (*a*) is less stable than a short-legged one (*b*).

can see that rats and squirrels, which have relatively short legs, are adapted well for living on steep slopes or on tree branches. The long-legged horse and antelope are proportioned for efficient running on nearly flat terrain.

If a quadruped lifts one leg, it will remain balanced if its center of gravity is over the triangular base area formed by its remaining three legs. By moving its legs in the correct sequence, it also can walk slowly while always keeping three feet on the ground and the center of mass over the triangle they define (Fig. 4.23). This sequence is right front, left rear, left front, right rear, and it is used by all four-legged animals and by human infants. Of course, humans, birds, and some animals can balance on one or two feet, but their large feet make it relatively easy to do this.

When a quadruped runs rapidly, it can happen that only one or two legs are on the ground at one time. The tendency to tip diagonally forward or to roll sideways is quickly countered when the other feet are set down. Thus, brief periods of instability are required for fast motion of quadrupeds and bipeds (Fig. 4.24).

Three feet on the ground is the minimum for stability for animals with small feet, so insects that have six legs can move three at a time and still be stable at all times. Since they have very small masses, even a slight gust of wind would tip them over if they had periods of instability. The need to be stable in a moderate wind also explains why their legs do not point nearly vertically as in mammals, but are instead splayed outward.

The fact that animals usually have the minimum number of legs consistent with stability is apparently related to considerations of strength and weight. A single leg weighing as much as two thinner legs is better able to withstand torques tending to bend the leg. Accordingly, the portion of the body weight composing the legs is minimized by keeping the number of legs as small as possible.

4.5 | LEVERS; MECHANICAL ADVANTAGE

Levers, pulley systems, and screw jacks are all examples of *machines*. In each case, a force $\mathbf{F}_a$ is applied and a load force $\mathbf{F}_L$ is balanced. The *mechanical advantage* (M.A.) of the machine is defined as the ratio of the magnitudes of these forces,

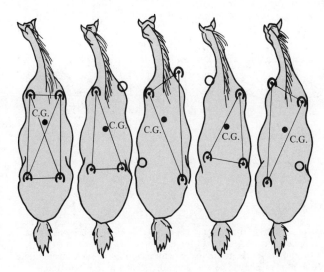

Figure 4.23. The diagram of a quadruped walking as seen from above. The open circle represents the foot that is off the ground. Notice that the center of gravity is always within the triangle formed by the three feet on the ground.

$$\text{mechanical advantage} = \text{M.A.} = \frac{F_L}{F_a} \quad (4.13)$$

Levers | A lever in its simplest form is a rigid bar used with a fulcrum (Fig. 4.25). Three classes of levers are defined according to the relative positions of $\mathbf{F}_L$, $\mathbf{F}_a$, and the fulcrum. The following example illustrates the effects of a lever.

Example 4.7

Suppose the load $\mathbf{F}_L$ on a class I lever (Fig. 4.25*a*) has a magnitude of 2000 N. A person exerts a force $F_a = 500$ N to balance the load. (a) What is the ratio of the distances x_a and x_L? (b) What is the mechanical advantage of this lever?

(a) To find x_a/x_L, we compute the torques about the fulcrum. The torque due to $\mathbf{F}_a$ is $\tau_a = -x_a F_a$, and the

Figure 4.24. A sprinter, at the start of and even during a race, has his center of mass well forward of his feet as shown in this drawing. This means that he is in a very unstable position. He maintains his balance by bringing his legs forward just in time to keep from falling. This extreme position aids the athlete in exerting a larger force on the ground to increase his acceleration.

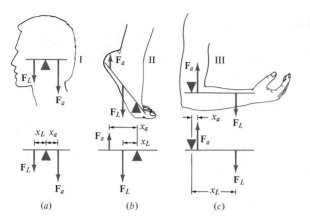

Figure 4.25. (a), (b), and (c) show the relative positions of the applied force $\mathbf{F}_a$, the load force $\mathbf{F}_L$, and the fulcrum for class I, II, and III levers, respectively, with examples of such levers in the human body. The load force $\mathbf{F}_L$ shown here is equal and opposite to the force produced by the lever on the load, $^-\mathbf{F}_L$.

torque due to $\mathbf{F}_L$ is $\tau_L = x_L F_L$. For balance, these must sum to zero, so

$$x_L F_L - x_a F_a = 0$$

and

$$\frac{x_a}{x_L} = \frac{F_L}{F_a} = \frac{2000\ \text{N}}{500\ \text{N}} = 4$$

(b) The mechanical advantage of the lever used in this way is then

$$\text{M.A.} = \frac{F_L}{F_a} = \frac{x_a}{x_L} = 4$$

For all classes of levers the mechanical advantage can be expressed as a ratio of distances from the fulcrum, as in the preceding example. If the forces are at right angles to the lever, in equilibrium the ratio of the magnitudes of the load force to the applied force is

$$\frac{F_L}{F_a} = \frac{x_a}{x_L}$$

Hence, for all classes of levers,

$$\text{M.A.} = \frac{x_a}{x_L} \qquad \text{(forces} \perp \text{lever)} \qquad (4.14)$$

With the forces at right angles to the lever the M.A. of class III levers is always less than 1, and the M.A. of class II levers is always greater than 1. Class I levers can have an M.A. larger or smaller than 1.

For all levers the M.A. given in Eq. 4.13 is an ideal value. Real machines always have frictional forces present that reduce the actual mechanical advantage below the ideal value.

4.6 | MUSCLES

Many examples of levers are found in the bodies of animals. Muscles provide the forces for using these levers.

A muscle is composed of thousands of long, thin fibers. When a muscle is stimulated by an electrical pulse from the nervous system, it contracts briefly or *twitches,* thereby exerting a force. A series of pulses sent to a muscle causes a series of twitches in the fibers. The twitches are close together in time but occur at different times in different parts of the muscle, so the apparent result is a smooth contraction of the muscle. If the frequency of the twitches is increased, the tension in the muscle increases up to a state of maximum tension. Further nerve impulse rate increases cause no further increase in tension.

The maximum tension of a muscle is proportional to its cross-sectional area at the widest point. This maximum tension also depends on the length of the muscle, which can be varied. The greatest tension can be achieved when the muscle is only slightly elongated relative to its resting or undisturbed state and is about 30 to 40 newtons per square centimetre of cross-sectional area. The maximum tension possible drops rapidly if the muscle is substantially elongated or shortened. To see an example of this, bend your wrist forward as far as you can and try to make a tight fist. Most people either fail to close their fingers completely or else do so with a rather weak grip.

4.7 | LEVERS IN THE BODY

The evolutionary development of the limbs and other skeletal structures in animals has been strongly affected by the animals' needs. We recall that when the forces are perpendicular to a lever, its mechanical advantage is $F_L/F_a = x_a/x_L$. Thus, short limbs with small values of x_L will have relatively large mechanical advantages and be able to exert large forces (Fig. 4.26). However, the distance that the end of a limb moves is proportional to its length x_L, so rapid motion requires a long limb. Consequently, a compromise must be made between strength and speed of movement. For example, the foreleg of a fast-moving horse has a mechanical advantage of 0.08. The armadillo, which is a burrowing animal, has a foreleg with an M.A. of about 0.25. Hence, it cannot move as quickly, but it has the strength needed for burrowing.

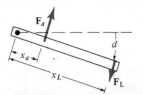

Figure 4.26. A limb being flexed may be represented by a pivoted bar supported by a cable. The cable represents a muscle and the pivot represents a joint. The M.A. of the limb is x_a/x_L, while the distance d that the end of the limb moves is proportional to x_L.

The Spinal Column

The human spinal column is made up to 24 *vertebrae* separated by fluid-filled *disks*. When a person bends, the spine is effectively a lever with a very small M.A. Hence, bending over to pick up even a light object produces a very large force on the *lumbrosacral* disk, which separates the last vertebra from the *sacrum*, the bone supporting the spine (Fig. 4.27). If weakened, this disk can rupture or be deformed, causing pressure on nearby nerves and severe pain.

To understand why this force is so large, we can use a model that treats the spine as a pivoted rod. The pivot corresponds to the sacrum and exerts a force **R** (Fig. 4.28). The various muscles of the back are equivalent to a single muscle producing a force **T** as shown. When the back is horizontal, the angle α is 12°. **w** is the weight of the torso, head, and arms, about 65 percent of the total body weight.

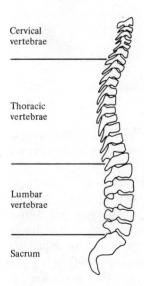

Figure 4.27. The anatomy of the spinal column.

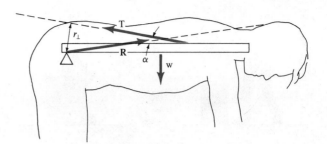

Figure 4.28. Force diagram for the spine of a person bending over with the back horizontal.

Notice that because α is small, the line of action of **T** passes close to the pivot, so its lever arm $r_\perp$ is small. However, the weight **w** acts at right angles to the spine, and its lever arm is much longer. Hence for their torques to balance, the muscle force **T** must be much greater than the weight. Because **T** is large, its horizontal component is also large. In equilibrium, the force **R** due to the sacrum must have an equal but opposite horizontal component, so this force due to the sacrum is also much larger than the weight.

If this calculation is carried out in detail, the numbers obtained are impressively large. For a man weighing 750 N (a mass of 77 kg), **T** and **R** are each close to 2200 N! If the man is also lifting a 175-N (18 kg) child, so that there is an extra 175-N weight at the right end of the bar in Fig. 4.28, **T** and **R** are each about 3300 N! Such forces in the muscles and on the disk are potentially quite hazardous.

Since even bending over without lifting a load puts great stress on the spine, it should be avoided. If, instead, one flexes the knees but keeps the back vertical, then the centers of gravity of all the weights lie almost directly above the sacrum. Consequently their torques about the sacrum are small, and the muscles need not exert any appreciable force. The force on the disk is then approximately the total weight supported. For the 750-N man, this weight is about 490 N for the body alone and 665 N with a 175-N load. This is a far safer way to lift even a light object (Fig. 4.29).

SUMMARY

The quantity that indicates the ability of a force to cause rotation is the torque τ. If a force **F** acts at a distance **r** from a point P, the torque about that point has a magnitude $rF\sin\theta$, where θ is the angle between **r** and **F**. The torque is a vector quantity and is directed perpendicular to the plane of **r** and **F**. In the vector or cross-product notation, $\tau = \mathbf{r} \times \mathbf{F}$, the direc-

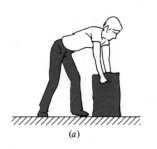 

(a)   (b)

Figure 4.29. (a) Incorrect and (b) correct ways to lift a weight.

tion of τ is then given by a right-hand rule. Alternatively, if the motion is in a specific plane, counterclockwise torques are conventionally taken as positive and clockwise torques as negative.

For a rigid body to be in equilibrium, two vector equations must be satisfied:

$$\mathbf{F} = 0$$
$$\tau = 0$$

The first of these is the condition that the net force is zero and ensures that there is no change in the translational motion. The second states that the net torque about any conveniently chosen point is zero. If it is satisfied, the torque about all points is zero, and the rotational motion remains the same.

The entire weight of an object can be considered to be concentrated at the center of gravity. This fact facilitates the understanding of stability and balance. For point weights on the x axis, the C.G. is at

$$X = (x_1 w_1 + x_2 w_2 + \ldots)/w$$

The mechanical advantage of a lever or other simple machine is defined as the ratio M.A. $= F_L/F_a$ of the load force $\mathbf{F}_L$ balanced by an applied force $\mathbf{F}_a$. In the bodies of animals, limbs with small mechanical advantages are suited to fast motion, while those with large mechanical advantages are able to exert large forces.

Checklist

Define or explain:

rigid body
torque
vector product or cross
   product
couple
equilibrium conditions

center of gravity
center of mass
base area
machines
mechanical advantage

REVIEW QUESTIONS

Q4-1 A rigid body does not change its _____ or _____ when subjected to a force.

Q4-2 The quantity that indicates the ability of a force to cause rotation is called the _____.

Q4-3 The perpendicular distance from the axis of rotation to the line of action of a force is the _____.

Q4-4 The greatest torque is obtained when a force is applied _____ to a wrench.

Q4-5 The vector product of two vectors points _____ to the plane of those vectors.

Q4-6 According to our sign convention, clockwise torques are _____ and counterclockwise torques are _____.

Q4-7 A pair of forces with equal magnitudes but opposite directions is called a _____.

Q4-8 For a rigid body to be in translational equilibrium, the _____ on it must be zero.

Q4-9 For a rigid body to be in rotational equilibrium, the _____ on it must be zero.

Q4-10 The rotational equilibrium condition may be applied about _____.

Q4-11 The weight of an object is effectively concentrated at its _____.

Q4-12 An object is balanced when its C.G. lies above its _____.

Q4-13 The M.A. is the ratio of the _____ to the _____.

Q4-14 When forces are applied perpendicular to a lever, its M.A. is the distance from the fulcrum to the _____ divided by the distance to the _____.

EXERCISES

Section 4.1 | Torques

4-1 Find the magnitude and sign of the torque due to each of the weights in Fig. 4.30 relative to point P.

4-2 Find the magnitude and sign of the torque due to each of the weights in Fig. 4.30 relative to point Q.

4-3 In Fig. 4.31, consider the vector products $\mathbf{A} \times \mathbf{A}$, $\mathbf{A} \times \mathbf{B}$, $\mathbf{A} \times \mathbf{C}$, $\mathbf{A} \times \mathbf{D}$, and $\mathbf{A} \times \mathbf{E}$. (a) Which of these vector products are zero? (b) Which are directed into the page? (c) Which are directed out of the page? (d) Which are equal to each other in magnitude and direction?

4-4 In Fig. 4.31, what are the directions of (a) $\mathbf{B} \times \mathbf{C}$; (b) $\mathbf{C} \times \mathbf{B}$; (c) $\mathbf{B} \times \mathbf{E}$?

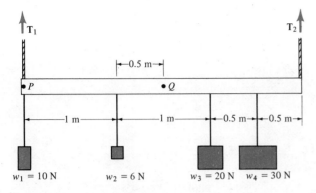

Figure 4.30. Exercises 4-1, 4-2, and 4-9.

4-5 In Fig. 4.32, how large are the lever arms for the torques about point P due to $\mathbf{F}_1$ and $\mathbf{F}_2$?

4-6 In Fig. 4.32, find the torques due to $\mathbf{F}_1$ and $\mathbf{F}_2$ relative to point P.

4-7 A cyclist applies a downward force $\mathbf{F}$ of magnitude 100 N to the pedal of her bicycle (Fig. 4.33). (a) Find the magnitude and direction of the torques in each position shown. (b) In which position is the torque a maximum?

Section 4.2 | Equilibrium of Rigid Bodies

4-8 The bar in Fig. 4.32 is pivoted at point P. Will it tend to start rotating if it is initially at rest?

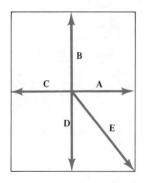

Figure 4.31. The vectors are drawn from the center of a rectangle. Exercises 4-3 and 4-4.

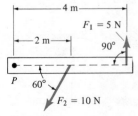

Figure 4.32. Exercises 4-5, 4-6, and 4-8.

Explain, and indicate in which direction it would rotate if your answer is yes.

4-9 A weightless bar supported by two vertical ropes has four weights hung from it (Fig. 4.30). Find the tensions $\mathbf{T}_1$ and $\mathbf{T}_2$ in the ropes.

4-10 Two children balance on a weightless seesaw. One weights 160 N and is seated 1.5 m from the fulcrum. The second is seated 2 m on the other side of the fulcrum. What is the weight of the second child?

4-11 Find the forces $\mathbf{F}_1$ and $\mathbf{F}_2$ on the tooth in Fig. 4.34. (In orthodontics, forces applied to the teeth lead to forces on the supporting bones. Gradually the bone tissue breaks down and permits the tooth to rotate or translate. New bone tissue grows in the space left behind. The forces must be small enough to avoid damaging the root of the tooth.)

4-12 Figure 4.35 shows the forearm considered in Example 4.4 when the person is holding a 12-N weight ($\mathbf{w}_1$) in the hand ($\mathbf{w}$ is the weight of the forearm). (a) Find the force $\mathbf{T}$ exerted by the biceps muscle and the force $\mathbf{E}$ exerted by the elbow joint. (b) In Example 4.4, with $\mathbf{w}_1 = 0$, we found $T = 36$ N and $E = 24$ N. Why are these forces more than twice as large here?

4-13 Children with weights w_1 and w_2 balance on a seesaw. The weight w of the seesaw can be considered to act at its center of gravity, which is directly over the pivot. In terms of w, w_1, and w_2, (a) find the force exerted by the pivot, and (b) find the ratio x_2/x_1 of the children's distances from the pivot.

4-14 A heavy load of wet laundry is hung from a clothesline. Is the line more likely to break if it is stretched tightly or allowed to sag considerably? Explain.

Section 4.3 | The Center of Gravity

4-15 Three weights are positioned on a weightless rod, as shown in Fig. 4.36. Where is their center of gravity?

4-16 Two weights are hung from the ends of a horizontal metre stick. If the weight at $x = 0$ is 10 N and the center of gravity is at $x = 0.8$ m, what is the weight at $x = 1$ m? (Neglect the weight of the stick.)

4-17 A woman's forearm has a mass of 1.1 kg and her upper arm has a mass of 1.3 kg. When her arm is held straight out, the C.G. of the forearm is 0.3 m from the shoulder joint, and the C.G. of the upper

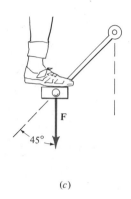

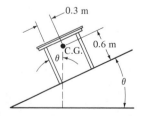

(a)   (b)   (c)

Figure 4.33. Exercise 4-7.

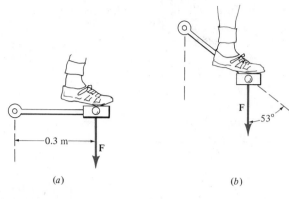

Figure 4.34. Exercise 4-11.

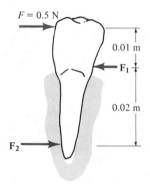

Biceps muscle

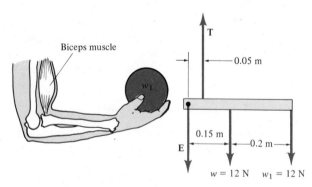

Figure 4.35. Exercises 4-12 and 4-30.

Figure 4.36. Exercise 4-15.

arm is 0.07 m from the shoulder joint. What is the position of the center of gravity of the entire arm with respect to the shoulder joint?

4-18 An 80-kilogram hiker carries a 20-kg pack. The center of gravity of the hiker is 1.1 m above the ground when he is not wearing the pack. The C.G. of the pack is 1.3 m from the ground when it is worn. How far above the ground is the C.G. of the hiker and the pack?

4-19 The axles of a car are 3 m apart. The front wheels support a total weight of 9000 N and the rear wheels support 7000 N. How far is the center of gravity from the front axle?

4-20 Using the data on the inside rear cover, find the position of the center of mass of the earth-moon system.

Section 4.4 | Stability and Balance

4-21 At what angle θ will the table in Fig. 4.37 tip over?

4-22 Ships returning to their home ports without a cargo are sometimes loaded with rocks in their holds or water in their tanks. Why is this done?

4-23 An amusing toy consists of a figure holding a curved pole with weights at either end (Fig. 4.38). It is stable in the position shown and will not fall when pushed gently. Explain why? (*Hint:* Where is the center of gravity?)

Figure 4.37. Exercise 4-21.   **Figure 4.38.** Exercise 4-23.

4-24 A steel girder with a mass of 1000 kg and a length of 10 m rests on a concrete slab, with 4 m overhanging the edge. How far can a 100-kg man walk on the girder?

Section 4.5 | Levers; Mechanical Advantage

4-25 A man places a 2-m-long bar under a boulder weighing 4500 N. He uses a fulcrum 0.2 m from the point where the bar touches the rock (Fig. 4.39). What force **F** must he exert to lift the rock?

4-26 An oar is held 0.4 m from the oarlock (Fig. 4.40). If it contacts the water at an average of 1.4 m from the oarlock, what is its mechanical advantage?

4-27 Figure 4.41 shows a pair of tweezers. What is its mechanical advantage?

4-28 Figure 4.42 shows a pair of pliers. (a) What is its mechanical advantage? (b) If a force $F = 10$ N is applied, what force is exerted on the object?

4-29 Give examples of class I levers with M.A.s equal to 1, less than 1, and greater than 1.

Section 4.7 | Levers in the Body

4-30 Figure 4.35 shows the forearm represented as a pivoted bar. T is the force exerted by the biceps muscle. (a) What class of lever does this represent? (b) What is the mechanical advantage of the forearm for supporting its own weight, w? (c) What is its M.A. for supporting a load w_1 held in the hand? (d) If the muscle contracts 1 centimetre, how far will the load in the hand move?

4-31 The head pivots about the atlanto-occipital joint (Fig. 4.43). The splenius muscles attached behind the joint support the head. (a) What class of lever does this represent? (b) The anterior muscles produce forward motions of the head. What class of levers does their action represent? (c) Which

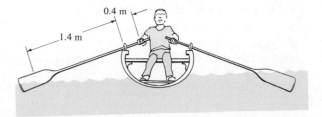

Figure 4.40. Exercise 4-26.

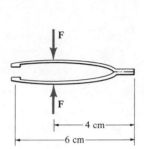 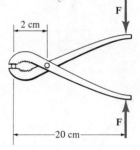

Figure 4.41. Exercise 4-27.   **Figure 4.42.** Exercise 4-28.

muscles have the larger mechanical advantage? Speculate on the reasons for this.

PROBLEMS

4-32 Vector **A** points north, and vector **C** = **A** × **B** points straight up. What can you say about (a) the vertical component of **B**; (b) the component of **B** toward the east; (c) the component of **B** toward the north?

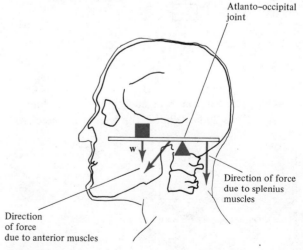

Figure 4.43. Muscles moving and supporting the head. Exercise 4-31.

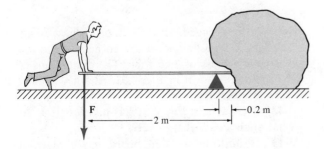

Figure 4.39. Exercise 4-25.

4-33 Find the tension in the ropes in Fig. 4.44.

4-34 In Fig. 4.45, an object is supported by a hinged, weightless rod and a cable. Find the tension in the cable and the force exerted by the hinge.

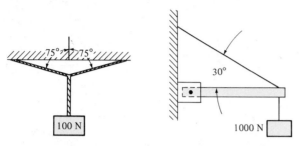

Figure 4.44. Problem 4-33.   **Figure 4.45.** Problem 4-34.

4-35 In Fig. 4.46, an object is supported by a hinged, weightless rod and cable as shown. If $w = 1000$ N, find (a) the tension in the cable, and (b) the force on the rod at the hinge.

4-36 In Fig. 4.46, the hinged rod and cable are weightless. The cable will break when the tension exceeds 2000 N. What is the maximum weight w that can be supported?

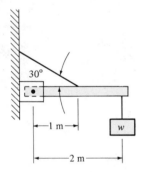

Figure 4.46. Problems 4-35 and 4-36.

*4-37 The tension T at each end of the chain in Fig. 4.47 is 20 N. What is the weight of the chain?

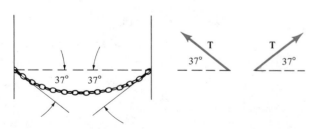

Figure 4.47. Problem 4-37.

4-38 A horse stands with its left forefoot off the ground (Fig. 4.48). The left rear and right front legs each support 1500 N of its weight, which is 5000 N. (a) What force is exerted by the right hind leg? (b) Find the position (X, Y) of the center of gravity.

4-39 What is the position of the center of mass of the three masses of Fig. 4.49?

*4-40 The man in Fig. 4.50 has a mass of 100 kg. His arms are held straight out to the side, and in one hand he holds a mass M. (a) Find the horizontal and vertical positions of the C.G. of the man plus the mass M. (Choose the origin to be midway between his feet). (b) What is the maximum mass the man can hold without falling over?

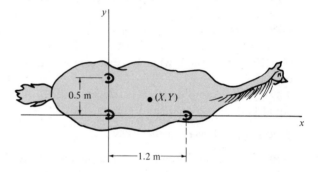

Figure 4.48. A horse viewed from above standing on three legs. Problem 4-38.

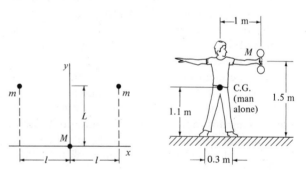

Figure 4.49. Problem 4-39.   **Figure 4.50.** Problem 4-40.

*4-41 A four-legged table has a top of mass 20 kg. The legs are positioned at the corners of the top and have a mass of 2 kg each. The dimensions of the table are shown in Fig. 4.51. At what angle θ will the table tip over?

4-42 The table in Fig. 4.51 has massless legs. At what angle θ will it tip over?

4-43 A uniform wooden board has a mass of 20 kg and a length of 2 m. A circular hole is cut out

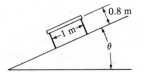

Figure 4.51. Problems 4-41 and 4-42.

with its center 0.5 m from one end. If the C.G. is now 0.9 m from the opposite end, what is the mass of the wood removed?

*4-44 The ammonia (NH_3) molecule is a pyramid, with the three hydrogen (H) atoms forming the base, and the nitrogen (N) atom at the apex. The centers of the hydrogen atoms are separated by 16.3 nm (1 nm $= 10^{-9}$ m), and the nitrogen atom is 3.8 nm above the center of the base. Where is the center of mass of the molecule relative to the nitrogen atom? (The mass of a nitrogen atom is 14 times that of a hydrogen atom.)

*4-45 The deltoid muscle raises the upper arm to a horizontal position (Fig. 4.52). (a) Find the tension T exerted by the muscle and the components R_x and R_y of the force exerted by the shoulder joint. (b) What is the mechanical advantage of the muscle for lifting the arm?

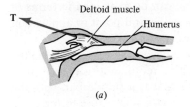

(a)

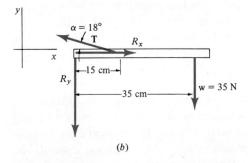

(b)

Figure 4.52. Problem 4-45. (Adapted from Williams and Lissner.)

*4-46 A vase 0.4 m tall has its center of gravity 0.15 m from the bottom, which is a circle of radius 0.05 m (Fig. 4.53). How far can the top of the vase be pushed to the side without toppling it?

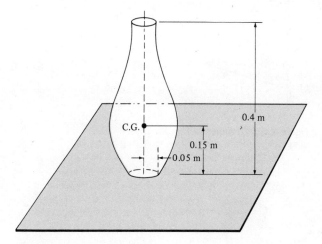

Figure 4.53. Problem 4-46.

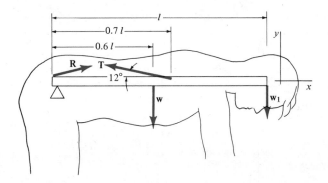

Figure 4.54. Problem 4-49.

4-47 Show that the M.A. of a class III lever is always less than 1, assuming the forces are perpendicular to the lever.

4-48 Show that the M.A. of a class II lever is always greater than 1, assuming the forces are perpendicular to the lever.

*4-49 In Fig. 4.54, the weight of the upper body is $w = 490$ N. Find the force T exerted by the spinal muscles and the components R_x and R_y of the force R exerted by the pivot (sacrum) if the weight w_1 is (a) zero; (b) 175 N.

4-50 Show that 1 N m = 1 kg m^2 s^{-2}.

^{c}4-51 A rod is shaped so that its weight per unit length λ is $\lambda_0(1 - x/2L)$ for $0 \leq x \leq L$. Find its center of gravity.

^{c}4-52 Find the center of gravity of the triangular metal plate in Fig. 4.55.

^{c}4-53 A solid cone of uniform density has a height h (Fig. 4.56). (a) Show that its weight per unit

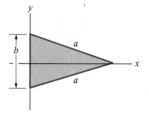

Figure 4.55.   Problem 4-52.

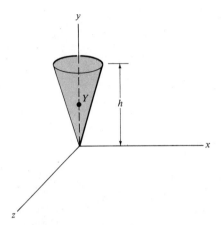

Figure 4.56.   Problem 4-53.

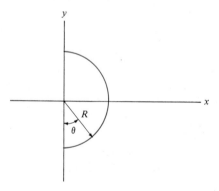

Figure 4.57.   Problem 4-54.

length along the y direction varies as y^2. (b) Find its center of gravity, Y.

^{c}**4-54**   A wire is bent into a semicircle of radius R (Fig. 4.57). Find its center of gravity. [*Hint:* Write $\int x\, dw$ in terms of the angle θ.]

ANSWERS TO REVIEW QUESTIONS

Q4-1, size or shape; **Q4-2,** torque; **Q4-3,** lever arm; **Q4-4,** perpendicular; **Q4-5,** perpendicular; **Q4-6,** negative, positive; **Q4-7,** couple; **Q4-8,** net force; **Q4-9,** net torque; **Q4-10,** any convenient point; **Q4-11,** center of gravity; **Q4-12,** base; **Q4-13,** load force, applied force; **Q4-14,** point where applied force acts, load.

SUPPLEMENTARY TOPICS
4.8 | THE JAWS OF ANIMALS

Mechanics enables us to understand why many anatomical structures have evolved to their present state, since the mechanical functions of bones, muscles, and joints largely determine their sizes and shapes. This is illustrated nicely by the development of the lower jaws of mammals.

It is often advantageous for an animal to be able to bite very hard. The biting force depends on the magnitude, direction, and point of application of the forces exerted by the muscles closing the jaw. This leads to certain optimal shapes and sizes of jaws. In addition, the bones of the jaw joint connecting the upper and lower jaws must be strong enough to prevent fractures and dislocations. From fossil records, we known that mammals evolved from mammal-like reptiles. As this occurred, the muscles attached to the lower jaw progressively *increased* in size, while the bones forming the jaw joint steadily *decreased* in size. This apparent paradox can be explained in terms of the changes in the direction and point of application of the muscular forces.

Figure 4.58 shows the basic differences between the lower jaws of a primitive reptile and a typical present-day mammal. The first is a simple bar, with the muscles pulling upward at a point close to the joint. The mammalian jaw has a large bump or projection called the *coronoid process*. Attached to this is the *temporalis* muscle, which pulls *backward* as well as upward (force **T** in Fig. 4.58). The *masseter* and *pterygoideus* muscles pull *forward* as well as upward (force **M**).

A primitive reptile biting with an upward force $-\mathbf{B}$ on food between its back teeth experiences an equal but opposite reaction force $\mathbf{B}$ downward on the jaw. Since the muscular force $\mathbf{M}$ is applied close to the joint, static equilibrium can be achieved only if a large downward force $\mathbf{R}$ is exerted by the joint.

Calculating torques about the point 0, the net torque is zero if

$$x_B B - x_R R = 0 \qquad \text{or} \qquad R = \frac{x_B}{x_R} B$$

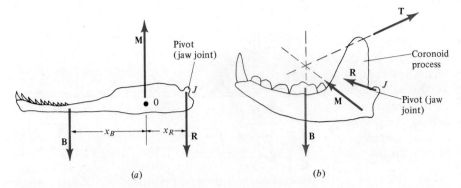

Figure 4.58. (a) Lower jaw of a primitive reptile. **M** is the force due to the muscle, **B** is the reaction force from the object being bitten, and **R** is the force due to the jaw joint at *J*. (b) A mammalian jaw. Muscle forces are shown as **T** and **M**. As explained in the text, the force **R** due to the jaw joint can be zero if the lines of action of the three forces **T**, **B**, and **M** intersect as shown here.

Since the net force on the jaw must be zero, $M - B - R = 0$, and the required muscle force is

$$M = B + R = B\left(1 + \frac{x_B}{x_R}\right)$$

For example, if $x_B = 2x_R$ and $B = 1$ N, then $R = 2$ N, and $M = 3$ N. Thus the force B on the food is smaller than the forces M and R exerted by the muscle and the joint, respectively. Clearly the strength of the joint is a limiting factor in how hard the reptile can bite and in how large a muscle it can safely employ.

In the mammalian jaw, the force **M** is applied further from the joint and another muscular force, **T**, is also present (Fig. 4.59). If the lines of action of **T**, **M**, and **B** all cross at a point, their torques about this point are zero. Then the second equilibrium condition, $\tau = 0$, requires that the line of action of **R** pass through this point as well. Furthermore, when the

forces also satisfy $\mathbf{T} + \mathbf{M} + \mathbf{B} = 0$, *no force **R** need be supplied by the joint to satisfy the requirement* $\mathbf{F} = 0$. If $\mathbf{T} + \mathbf{M} + \mathbf{B}$ is not zero, or if their lines of action do not exactly meet at one point, a force, **R**, will have to be supplied by the joint, but **R** will still be much smaller than in the reptile. Thus, a smaller joint structure is adequate, and the strength of the joint does not limit the size of the muscle the animal can have.

Example 4.8

To illustrate the superiority of the mammalian jaw, suppose that the muscle forces **T** and **M** of Fig. 4.58 are both at $\theta = 45°$ to the horizontal. If there is to be no force **R** supplied by the joint, how is **M** related to **T**, and how large a force, **B**, is exerted on the food? (Assume the lines of action of **B**, **T**, and **M** all cross at a point, so the second equilibrium condition $\tau = 0$ is satisfied.)

Since the forces have both x and y components, $\mathbf{F} = 0$ can be used in component form as $F_x = 0$ and $F_y = 0$. With $F_x = 0$, we obtain

$$T\cos\theta - M\cos\theta = 0$$

so $M = T$. With $F_y = 0$,

$$T\sin\theta + M\sin\theta - B = 0$$

Using $M = T$ and $\sin\theta = \sin 45° = \sqrt{2}/2$,

$$B = (T + M)\sin\theta = 2T\sin\theta = \sqrt{2}T$$

Thus the force B exerted by the jaw on the food is greater than either of the muscular forces T and M, and the force due to the joint is zero. (This conclusion is also evident from the lengths of the vectors in Fig. 4.59*b*.) By contrast, we found that in the reptile, B is smaller than the muscular force or the joint force.

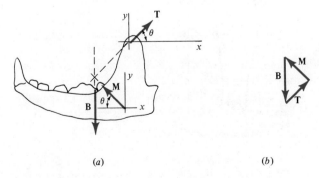

Figure 4.59. Forces on the mammalian jaw when no force is supplied by the jaw joint.

Just as we have compared mammalian and more primitive jaws, one can also compare different mammals. Carnivores use their powerful front teeth to seize or tear their prey, while herbivores grind their food sideways between their molars. The weight of a carnivore's temporalis muscle is one-half to two-thirds of the total weight of the jaw-closing muscles. However, in herbivores, this muscle weighs only about one-tenth of the total. It is left as an exercise to show why this is the proper adaptation for the needs of these two types of animals.

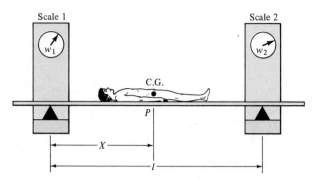

Figure 4.60. A method for finding the center of gravity of a human.

4.9 | CENTER OF GRAVITY OF HUMANS

Information about the center of gravity of humans is useful in many applications. The center of gravity of a freely falling object follows the same trajectory as a simple particle, even though the object may be rotating or changing its shape. This simplifies the analysis of jumping, gymnastics, and other athletic activities. In physical therapy, an amputee fitted with an artificial limb that is lighter than the natural limb has his or her center of gravity shifted. This must be taken into account in planning the person's rehabilitation.

In Section 4.3, we described how the center of gravity can be located by suspending an object from two different points. Another technique is better suited for live humans and animals (Fig. 4.60). A board of length l is supported at its ends by knife edges resting on scales adjusted to read zero with the board alone. When the person lies on the board, the scales read w_1 and w_2, respectively.

The condition $\tau = 0$ for the net torque on the board can be used to find X. Calculating torques about point P,

$$-Xw_1 + (l - X)w_2 = 0$$

or

$$X = \frac{lw_2}{w_1 + w_2} \tag{4.15}$$

The measurement is repeated twice more with the subject first standing up and then turning 90°. In this way all three coordinates of the center of gravity are determined.

The detailed measurement of the masses, sizes, and centers of gravity of body segments is difficult, and the results vary from individual to individual. Data for a typical man are given in Figs. 4.61 and 4.62 and Table 4.1 and are used in some of the exercises.

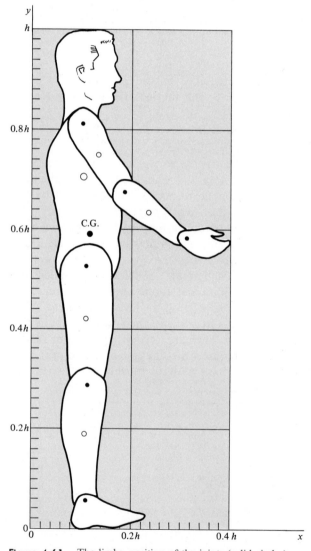

Figure 4.61. The limbs, position of the joints (solid circles), and the position of the centers of gravity (open circles) of several body segments for a typical man. (Adapted from Williams and Lissner.)

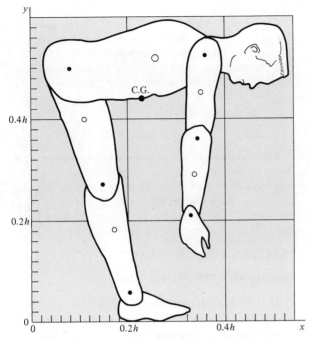

Figure 4.62. The man shown in Fig. 4.61 is now bending over so that his back is nearly horizontal. Note that his center of gravity is still over his feet. (Adapted from Williams and Lissner.)

4.10 | PULLEY SYSTEMS

Pulleys, like levers, are simple machines that are used in many situations. A single pulley is used to change the direction of a force, while combinations of pulleys can be used to reduce the force needed to lift a heavy load.

If friction in the bearings is negligible, the equilibrium tension in the rope or cable is the same on either side of the pulley. This property is used in discussing some typical pulley arrangements in the following

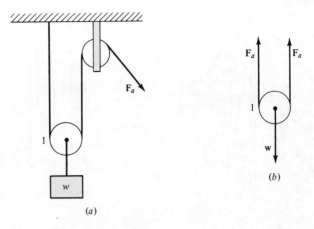

Figure 4.63. The tension in the rope is the same throughout the system, so the two forces on either side of pulley 1 are equal. Example 4.9.

examples. We assume in these examples that the friction is negligible and that the pulleys and ropes are massless.

Example 4.9

What applied force $\mathbf{F}_a$ is necessary to lift the weight $\mathbf{w}$ in Fig. 4.63?

The forces on pulley 1 are shown in Fig. 4.63b. The rope is continuous, and the tension on both sides of the pulleys is the same. If the weight is being raised at constant speed, the system is in equilibrium. Hence, $2F_a - w = 0$, and $F_a = w/2$. The force required is only half the weight, and the mechanical advantage is

$$\text{M.A.} = \frac{w}{F_a} = 2$$

Example 4.10

What applied force $\mathbf{F}_a$ is necessary to raise the weight $\mathbf{w}$ with the pulley system of Fig. 4.64? What is the mechanical advantage of this system?

TABLE 4.1

Masses and centers of gravity of body segments of the man in Figs. 4.61 and 4.62. His total mass is m and his height is h. For example, if his mass is 700 kg, then the mass of his trunk and head is $0.593m = 0.593(70 \text{ kg}) = 41.5 \text{ kg}$.

Segment	Mass	Center of Gravity Position For Segment			
		Fig. 4.57		Fig. 4.58	
		x	y	x	y
Trunk and head	$0.593m$	$0.10\,h$	$0.70\,h$	$0.26\,h$	$0.52\,h$
Upper arms	$0.053m$	$0.14\,h$	$0.75\,h$	$0.35\,h$	$0.45\,h$
Forearms and hands	$0.043m$	$0.24\,h$	$0.64\,h$	$0.34\,h$	$0.29\,h$
Upper legs	$0.193m$	$0.12\,h$	$0.42\,h$	$0.11\,h$	$0.40\,h$
Lower legs and feet	$0.118m$	$0.10\,h$	$0.19\,h$	$0.17\,h$	$0.18\,h$

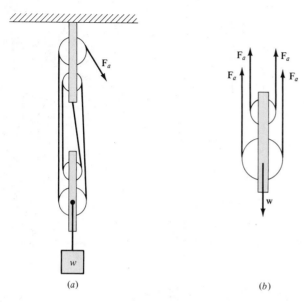

Figure **4.64.** Example 4.10.

Again the tension in each vertical segment of rope is the same, so $4F_a - w = 0$, and $F_a = w/4$. The mechanical advantage for this system is

$$\text{M.A.} = \frac{w}{F_a} = 4$$

From the preceding two examples, we can infer a rule for the mechanical advantage of pulley systems used to lift weights. *The mechanical advantage of the system is equal to the number of parallel ropes supporting the pulley to which the load is attached:* 2 in Fig. 4.63 and 4 in Fig. 4.64. Note that this rule does *not* apply when, as in the next example, the forces applied to the load are not all parallel.

Example 4.11

Leg traction is applied to a patient's leg as shown in Fig. 4.65. What horizontal force is exerted on the leg?

The sum of the forces on each pulley is zero, since the pulleys are at rest. From Fig. 4.65*b*, the horizontal forces that act on the pulley attached to the foot satisfy

$$2w \cos \theta - F_L = 0$$

or

$$F_L = 2w \cos \theta$$

This force can be changed by altering either w or θ. Since $\cos \theta$ varies from 1 to 0 as θ goes from 0° to 90°, any force from zero to $2w$ can be obtained by choosing the correct angle. When θ is large, $\cos \theta$ is small. Then the weight w and the tension in the rope are much larger than the force F_L exerted on the foot.

EXERCISES ON SUPPLEMENTARY TOPICS

Section 4.8 | The Jaws of Animals

4-55 A snake exerts a muscle force $M = 5$ N (see Fig. 4.58*a*). M acts at a distance of 0.03 m from the joint, and the resulting bite force is 2 N. Find (a) the distance from the joint to the line of action of the bite force, and (b) the force exerted by the jaw joint.

4-56 (a) In a typical herbivore, the maximum magnitude of the force T is one-tenth the maximum magnitude of the force M in Fig. 4.66. Assuming that there is no force at the joint, would you expect the animal to exert the largest biting force near the front or back of the jaw? (b) In a carnivore the maximum value of T is about twice that of M. Would you expect the maximum biting force to be exerted further from or closer to the jaw joint than in the herbivore? Explain.

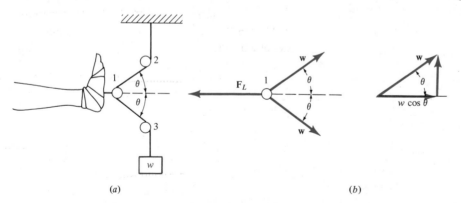

Figure **4.65.** (*a*) A system of pulleys used to apply a force in leg traction. The magnitude of this force can be adjusted by changing the angle θ. Pulley 1 is attached to the foot, and pulleys 2 and 3 are mounted on a rigid frame that is now shown. (*b*) The forces on pulley 1.

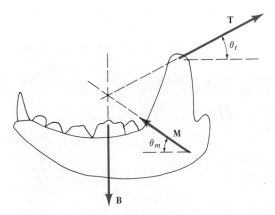

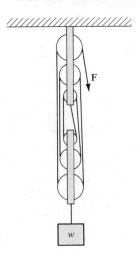

Figure 4.66. Exercise 4-54; Problems 4-61 and 4-62.

Section 4.9 | Center of Gravity of Humans

4-57 A board 4 m long is used to find the C.G. as in Fig. 4.60. When a person is on the board, the scale readings are $w_1 = 200$ N and $w_2 = 600$ N. What is the position of the C.G. of the person?

4-58 Using the data of Table 4.1, find the C.G. of the man in Fig. 4.61.

4-59 Using the data of Table 4.1, find the C.G. of the man in Fig. 4.62.

Section 4.10 | Pulley Systems

4-60 What force F must be applied in Fig. 4.67 to lift the load?

4-61 In Fig. 4.65, a 50-N force is to be applied to the leg. If a 10-kg mass is hung from the cable, what angle θ should be used?

4-62 Suppose the rope in Fig. 4.64 is pulled at 0.25 m s^{-1}. How fast will the load rise?

Figure 4.67. Exercise 4-58.

PROBLEMS ON SUPPLEMENTARY TOPICS

4-63 A mammal bites so that the muscle force **M** of Fig. 4.66 has a magnitude of 30 N. What is the force **B** of the bite? (Assume $\theta_t = \theta_m = 45°$.)

*4-64 In a particular carnivore the magnitude of the force T is 1.3 times the magnitude of M (Fig. 4.66). There is no force at the jaw joint. If $\theta_m = 60°$, find (a) θ_t, and (b) the ratio B/M.

Additional Reading

John M. Cooper and Ruth B. Glassow, *Kinesiology,* 4th ed., The C.V. Mosby Co., St. Louis, 1976. Balance and body levers in humans.

Signi Brunnstrom, *Clinical Kinesiology,* 3rd ed., F.A. Davis Co., Philadelphia, 1972. Balance and body levers in humans.

Katherine F. Wells and Janet Wessel, *Kinesiology,* 5th ed., W.B. Saunders Co., Philadelphia, 1971. Balance, body levers in humans.

Marian Williams and Herbert R. Lissner, *Biomechanics of Human Motion,* W.B. Saunders Co., Philadelphia, 1962. Physical therapy applications.

John J. Hay, *The Biomechanics of Sports Techniques,* Prentice-Hall, Inc., Englewood Cliffs, N.J., 1973.

Sir James Gray, *How Animals Move,* Cambridge University Press, Cambridge, 1953. Chapter 3 discusses balance and stability.

R.A.R. Tricker and B.J.K. Tricker, *The Science of Movement,* Mills and Boon Ltd., London, 1967. Chapters 3, 4, 15, and 16 discuss balance and stability.

R.C. Thurow, *Edgewise Orthodontics,* 3rd ed., The C.V. Mosby Co., St. Louis, 1972. Applications of statics.

R. McNeill Alexander, *Animal Mechanics,* University of Washington Press, Seattle, 1968. Pages 5–13 discuss the jaws of animals.

A.W. Crompton, On the Lower Jaw of Diarthrognathus and the Origin of the Mammalian Lower Jaw, *Proceedings of the Zoological Society of London,* vol. 140, 1963, pp. 697–753.

Scientific American articles:

Milton Hildebrand, How Animals Run, May 1960, p. 148.

Jearl Walker, In Judo and Aikido Application of the Physics of Forces Makes the Weak Equal to the Strong, *The Amateur Scientist,* July 1980, p. 150.

Marc H. Raibert and Ivan E. Sutherland, Machines That Walk, January 1983, p. 44.

CHAPTER 5

CIRCULAR MOTION

Newton's laws of motion permit us to find the motion of any object if we know the forces acting upon the object and its initial position and velocity. In the preceding chapter, we considered objects that are in equilibrium and remain at rest. Earlier, we discussed the uniformly accelerated motion that occurs when the net force is constant in magnitude and direction. In this chapter, we consider another kind of motion frequently encountered, motion in a circular path.

When an object moves in a circular path at a constant speed, its acceleration is directed toward the center of the circle. The force required to produce this acceleration can be provided in various ways. Friction provides the required force for a car on a flat, circular track; gravity for an artificial satellite orbiting the earth; and electrical forces for an electron orbiting an atomic nucleus. In this chapter, we discuss these and other examples of circular motion.

Closely related to the motion of an object in a circular path is the rotation of a rigid body about an axis. Examples are easy to find: lawn mower blades, automobile wheels, the earth spinning on its axis. *Each point* in such an object is undergoing circular motion. Again, we will see that Newton's laws permit us to predict and analyze such motions.

5.1 | CENTRIPETAL ACCELERATION

Consider a car going around a cricular track at a constant speed (Fig. 5.1*a*). The car is said to be undergoing *uniform circular motion*. Because the magnitude of the velocity is constant, the car has no acceleration

component along the direction of motion; its *tangential acceleration* is zero. However, the direction of the velocity vector is changing, so there is an acceleration component along the direction at right angles to the velocity. We now develop a formula for this *centripetal acceleration*. We will see examples of its use in this and later sections.

In Fig. 5.1*b*, **v** is the velocity at some instant, and **v** + Δ**v** is the velocity a short time interval Δ*t* later. The velocity change Δ**v** that occurs in the short interval Δ*t* points toward the center of the circle. The instantaneous acceleration is the ratio Δ**v**/Δ*t* evaluated in the limit as Δ*t* approaches zero. Thus the acceleration **a**$_r$ points along Δ**v** or toward the center of the circle. The subscript r reminds us that the acceleration is *radially inward*. **a**$_r$ is also called the *centripetal acceleration* because it is directed toward the *center*.

The magnitude of **a**$_r$ can be found with the aid of the triangles in Fig. 5.1*b*. The motion is along the circumference of the circle, so the velocity is always tangential to the circle or perpendicular to the radius. Thus **v** is perpendicular to **r**. A short time Δ*t* later, the velocity vector **v** + Δ**v** is perpendicular to **r** + Δ**r**. Since they remain perpendicular, the velocity vector and position vector rotate through the same angle. Hence the angle θ in the two triangles is the same. Since the triangles are isosceles, their other two angles are each (180° − θ)/2, and all three angles in the two triangles are equal. Thus the triangles are similar, and their sides are proportional, or

$$\frac{1}{v}|\Delta \mathbf{v}| = \frac{1}{r}|\Delta \mathbf{r}|$$

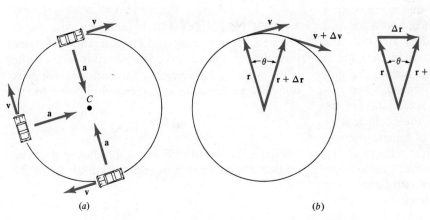

Figure 5.1. (a) A car going around a circular track at a constant speed has an acceleration directed toward the center C of the circle. Its tangential acceleration component is zero since the speed is constant. (b) The velocity vector $\mathbf{v}$ and the position vector $\mathbf{r}$ rotate at the same rate, so the angle θ is the same in both triangles. Thus the triangles are similar, and $|\Delta\mathbf{v}|/v = |\Delta\mathbf{r}|/r$.

Now if we divide by Δt, we have

$$\frac{1}{v}\left|\frac{\Delta\mathbf{v}}{\Delta t}\right| = \frac{1}{r}\left|\frac{\Delta\mathbf{r}}{\Delta t}\right|$$

Taking the limit $\Delta t \to 0$ gives

$$\frac{a_r}{v} = \frac{v}{r}$$

Hence the centripetal acceleration has a magnitude

$$a_r = \frac{v^2}{r} \qquad (5.1)$$

The acceleration varies inversely with the radius; the smaller the circle, the greater the acceleration. It also varies as v^2; the centripetal acceleration needed to negotiate a curve increases rapidly with the speed.

The following example illustrates the use of this radial acceleration formula.

Example 5.1

A car goes around a flat, circular track of radius 200 m at a constant speed of 30 m s^{-1}. What is its acceleration?

Since the speed is constant, there is no tangential acceleration. However, because the velocity vector is changing direction, there is an acceleration toward the center of the circle. This acceleration has a magnitude of

$$a_r = \frac{v^2}{r} = \frac{(30 \text{ m s}^{-1})^2}{200 \text{ m}} = 4.5 \text{ m s}^{-2}$$

In the absence of any net force, an object moves in a straight line at a constant speed in accordance with Newton's first law. If the object moves in a circular path, it has a radial acceleration $a_r = v^2/r$, so some force must be producing this acceleration. From the second law, the force must be equal to the mass times the acceleration. *Thus the net force F needed to produce an acceleration a_r is*

$$F = ma_r = \frac{mv^2}{r} \qquad (5.2)$$

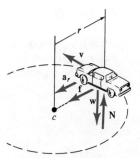

Figure 5.2. Forces on a car on a flat circular track. Not shown are the forces due to the air and the road along the direction of motion. A frictional force, $\mathbf{f}$, is directed toward the center of the circle.

In the case of a car on a flat track, the centripetal acceleration results from a frictional force exerted on the tires by the road (Fig. 5.2). Other forces also act on the car: its weight, directed downwards; an equal but opposite upward normal force due to the road; a backward force due to air resistance; and a forward force exerted by the road on the tires. However, the only force that can produce a radial acceleration is the frictional force acting perpendicular to the motion. Accordingly, the maximum frictional force possible sets a limit to the centripetal acceleration of the car. This limit is determined by the coefficient of *static* friction, not the coefficient of kinetic friction, since the tire is momentarily at rest relative to the road at the point of contact. If the driver tries to negotiate a curve with a small radius of curvature at too high a speed, the maximum frictional force will be exceeded and the car will skid. Once a skid begins, the car is sliding rather than rolling, so the frictional force is then determined by the coefficient of kinetic friction and is reduced in magnitude. This reduction in the frictional force makes it difficult to regain control of the vehicle and can lead to an accident. The force needed to produce a centripetal acceleration is found in the next example.

Example 5.2

The car in the preceding example travels on a flat, circular track of radius 200 m at 30 m s^{-1} and has a centripetal acceleration $a_r = 4.5$ m s^{-2}. (a) If the mass of the car is 1000 kg, what frictional force is required to provide the acceleration? (b) If the coefficient of static friction μ_s is 0.8, what is the maximum speed at which the car can circle the track?

(a) Since the mass is 1000 kg and the acceleration is 4.5 m s^{-2}, Newton's second law gives

$$F = ma_r = (1000 \text{ kg})(4.5 \text{ m s}^{-2}) = 4500 \text{ N}$$

This is the frictional force required to circle the track at 30 m s^{-1}.

(b) From Fig. 5.2, the normal force **N** is equal in magnitude to the weight **w** = m**g**. Thus the maximum frictional force possible is $\mu_s N = \mu_s mg$, and the maximum velocity satisfies

$$\frac{mv^2}{r} = \mu_s mg$$

or

$$v = \sqrt{\mu_s rg}$$

Note that the mass has cancelled, so the maximum velocity on this track is the same for any car as long as the coefficient of static friction is the same. Substituting the numerical values,

$$v = [(0.8)(200 \text{ m})(9.8 \text{ m s}^{-2})]^{1/2} = 39.6 \text{ m s}^{-1}$$

If the driver attempts to exceed 39.6 m s^{-1}, the car will not be able to continue on the circular course, and it will skid as in Fig. 5.3.

Vector Notation
It is useful to write the expression for the centripetal acceleration in vector form. In Fig. 5.4, **r** is a vector directed from the center of the circle, and $\hat{\mathbf{r}} = \mathbf{r}/r$ is a unit vector in that direction. The acceleration $\mathbf{a}_r$ is directed opposite to **r**, so

$$\mathbf{a}_r = -\frac{v^2}{r}\hat{\mathbf{r}} \qquad (5.3)$$

Similarly the net force $\mathbf{F} = m\mathbf{a}_r$ producing the acceleration can be written in vector form as

$$\mathbf{F} = -\frac{mv^2}{r}\hat{\mathbf{r}} \qquad (5.4)$$

The equation $\mathbf{a}_r = -(v^2/r)\hat{\mathbf{r}}$ is also applicable in some cases other than uniform circular motion. For example, when a car moves on a circular track with a variable speed, its acceleration can be resolved into two components. One is parallel or tangential to the track, and the other is perpendicular to the track. The tangential component $\mathbf{a}_T$ is equal in magnitude to the rate of change of the speed, dv/dt (Fig. 5.5). However, its acceleration component perpendicular to the motion is still v^2/r. Also, if the car travels on a road that is not circular, any small segment of the road can be considered as part of a circle. The radius of that circle

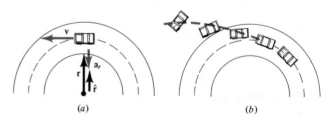

Figure 5.3. (a) A car on a flat circular track of radius r requires an acceleration $a_r = v^2/r$. (b) If the road cannot supply a frictional force mv^2/r, then the car will tend to move more nearly in a straight line and will skid.

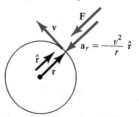

Figure 5.4. The vector $\hat{\mathbf{r}} = \mathbf{r}/r$ is parallel to **r** and has unit magnitude. $\mathbf{a}_r$ and $\mathbf{F} = m\mathbf{a}_r$ are directed opposite to **r** and to $\hat{\mathbf{r}}$.

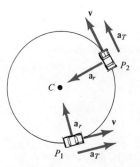

Figure 5.5. A car going around a circular track with an increasing speed. It has a centripetal acceleration $a_r = v^2/r$ and a tangential acceleration a_τ equal to the rate of change of the speed. (Note that both $\mathbf{v}$ and $\mathbf{a}_\tau$ are larger at P_2 than at P_1.)

is called the *radius of curvature* at the point P (Fig. 5.6). Again v^2/r is the acceleration component perpendicular to the motion.

5.2 | EXAMPLES OF CIRCULAR MOTION

In the preceding section, we considered in some detail the example of a car on a flat, circular track. We now discuss some other examples of circular motion.

Banked Turns | Good highways usually have banked or slanted curves, so that the normal force exerted by the road on the car has a horizontal component. This horizontal component can provide part of or all the force needed to produce the centripetal acceleration, reducing the role of the frictional force. The road is then much safer, especially under slippery conditions.

To examine this idea in detail, we consider a car on a banked curve. We suppose that the driver wants to

drive at the correct speed v, so that no frictional force is required to provide the centripetal acceleration. The problem, then, is to find what that speed should be.

Figure 5.7 shows the relevant forces acting on the car: its weight $\mathbf{w} = m\mathbf{g}$ and the normal force $\mathbf{N}$. (The forces along the direction of motion play no role here.) Since the horizontal component of $\mathbf{N}$ must provide all the centripetal acceleration,

$$N \sin \theta = \frac{mv^2}{r}$$

There is no vertical acceleration component, so the net vertical force is zero, and

$$N \cos \theta = mg$$

Dividing the first equation by the second, the masses cancel out, leaving

$$\frac{\sin \theta}{\cos \theta} = \frac{v^2}{rg}$$

Since $\sin \theta / \cos \theta = \tan \theta$, this gives

$$v^2 = rg \tan \theta \qquad (5.5)$$

This equation gives the speed v at which the car can negotiate a curve banked at an angle θ without any assistance from frictional forces. At any other speed, a frictional force will be needed to supplement the horizontal component of the normal force. On a very slippery day, a car going faster will tend to slide out of the curve and a slow car will tend to slide toward the inside of the curve. The use of this result is illustrated by the following example.

Example 5.3
A curve of radius 900 m is banked, so no friction is required at a speed of 30 m s^{-1}. What is the banking angle θ?

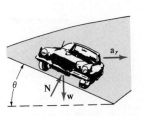

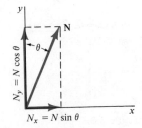

Figure 5.6. At P the road has a radius of curvature r. The centripetal acceleration at P is $a_r = v^2/r$.

Figure 5.7. A car on a banked curve is moving at the correct speed v so that the horizontal component N_x of the normal force provides the centripetal acceleration and no frictional force is required.

Solving Eq. 5.5 for $\tan\theta$,

$$\tan\theta = \frac{v^2}{rg} = \frac{(30 \text{ m s}^{-1})^2}{(900 \text{ m})(9.8 \text{ m s}^{-2})} = 0.102$$

$$\theta = 6°$$

As we have seen, on a banked road a component of the normal forces provides part of or all the force accelerating a car toward the center of the curve. Banking also supplies the entire force needed for a bird or an airplane to turn. A bird is held aloft by aerodynamic lift forces perpendicular to the surface of its wings. When one wing or the tip of one wing is rotated about its long axis, the lift forces become unbalanced, and the bird tilts or banks. In airplanes, banking is accomplished with ailerons, which are movable surfaces on the trailing edges of the wings. Banking causes the aerodynamic lift force to have a horizontal component and to make the bird or airplane turn (Fig. 5.8).

This is exactly the same situation as in the banked curve, and Eq. 5.5 can therefore be used to relate the banking angle, the speed, and the radius of the turn. For example, birds such as swallows, which can maneuver very rapidly, experience accelerations of a few g's.

Motion in a Vertical Circle | Swinging a water-filled bucket in a vertical circle provides a simple but dramatic example of circular motion. Now we see why spilling can be avoided if the pail is swung fast enough.

Figure 5.9 shows a bucket of total mass m swinging in a circle of radius R. At the top of the circle (point

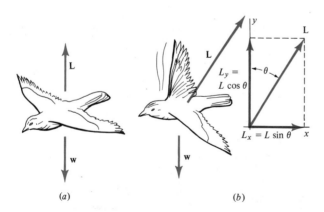

Figure 5.8. (a) Forces on a bird include its weight **w** and a lift force **L**. (The forward thrust and backward drag forces are not shown.) (b) When the bird banks, the lift force **L** has a horizontal component.

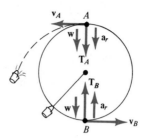

Figure 5.9. Forces acting on a pail swung in a vertical circle. If the rope breaks at the top of the circle, the pail follows the dashed path.

A), the tension $\mathbf{T}_A$ in the rope is parallel to the weight $\mathbf{w} = m\mathbf{g}$, so $\mathbf{F} = m\mathbf{a}_r$ gives

$$T_A + mg = \frac{mv_A^2}{R} \qquad T_A = \frac{mv_A^2}{R} - mg$$

Thus the tension is less than mv_A^2/R by an amount equal to the weight. If v_A^2/R is less than g, then T is negative, corresponding to a force opposite to the assumed direction. However, a rope can only pull, not push, so this cannot happen. Consequently, if the pail is swung too slowly, the pail and its contents will drop below the circular path; to avoid wet clothing, we must have $T_A \geq 0$ at the top of the circle, or

$$\frac{v_A^2}{R} \geq g$$

For example, if $R = 1$ m, v_A must satisfy

$$v_A \geq \sqrt{gR} = \sqrt{(9.8 \text{ m s}^{-2})(1.0 \text{ m})} = 3.13 \text{ m s}^{-1}$$

It is natural to ask why the pail and the water do not fall when they reach the top of the circle. The answer is *that they do fall, but not fast enough*. One way to see this is to consider what would happen if the rope broke as the pail reached the top of the circle. It would then follow the dashed curve in Fig. 5.9 and fall less rapidly then when the rope is intact. Thus the rope forces the pail and its contents to fall even faster than they do under the influence of gravity alone.

At the bottom of the circle (point B), the tension in the rope $\mathbf{T}_B$ and the acceleration are opposite to the weight, so

$$T_B - mg = \frac{mv_B^2}{R} \qquad T_B = \frac{mv_B^2}{R} + mg$$

Thus the tension here is mv_B^2/R *plus* the weight. Even though we discussed what would happen if the rope broke at the top of the circle, it is more likely to break at the bottom. This is true because the tension is *less* than $mv_A{}^2/R$ at the top and *more* than $mv_B{}^2/R$ at the bottom. Also, the gravitational force accelerates the pail as it falls, so the speed is greater at the bottom than at the top, which further increases the tension at the bottom.

Effective Weight |

We noted in Chapter Three that our perceived or effective weight is determined by the forces exerted by the floor or chair supporting us. The effective weight is zero for a person in free fall, and more generally is given by

$$\mathbf{w}^e = m\mathbf{g} - m\mathbf{a} \tag{5.6}$$

Rather large accelerations are encountered in some carnival rides (Fig. 5.10), and the effective weight may be quite different from $m\mathbf{g}$.

Example 5.4

On a carnival ride, a girl of mass m stands in a cylindrical cage of radius R. The cage is spun about its cylindrical axis so that her speed is v (Fig. 5.11). What is her effective weight?

From Eq. 5.6, $\mathbf{w}^e = m\mathbf{g} - m\mathbf{a}$. Taking horizontal and vertical components,

$$w_x{}^e = \frac{mv^2}{R} \qquad w_y{}^e = -mg$$

Using the Pythagorean relation (Fig. 5.11c),

$$w^e = m\sqrt{g^2 + \left(\frac{v^2}{R}\right)^2}$$

Thus the girl's effective weight is greater than $m\mathbf{g}$.

The sedimentation of material in a liquid occurs because the weight of the material exceeds that of an

Figure 5.10. Amusement park rides can have large centripetal accelerations. Riders experience large effective weights. (M. Sternheim.)

equal volume of liquid. The sedimentation rate is proportional to the acceleration of gravity g, which can in effect be increased by spinning the material

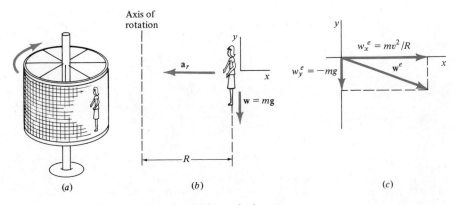

Figure 5.11. Effective weight of a girl in a spinning cage.

rapidly in a centrifuge. Ultracentrifuges used in research laboratories achieve accelerations of up to 500,000 g.

5.3 | ANGULAR VARIABLES

Suppose a runner has gone 100 m along a circular track of circumference 400 m. If we say that she has gone one-fourth of the way or 90° around the track, we are describing her motion in terms of the change in her *angular position*. The rate at which the angular position changes is the *angular velocity,* and its rate of change is the *angular acceleration*. These angular variables are especially useful in discussing the rotation of a rigid body about a fixed axis. For example, on the colored spoke of the wheel in Fig. 5.12, every point moves through 360° in one full rotation, even though points near the rim have traveled further than points near the axis.

Angles can be measured either in degrees or in *radians* (rad). Referring to Fig. 5.12, we define the *angular position θ in radians by*

$$\theta = \frac{s}{r} \tag{5.7}$$

This result for θ is the same for any point on the colored spoke of the wheel. The value of s for a point a distance r from the rotation axis is $r\theta$. For a complete rotation of the wheel, s is equal to the circumference $2\pi r$, and $\theta = 2\pi$ rad. Since a full circle is 360°, 2π rad = 360° and 1 rad = 360°/2π = 57.3° (Fig. 5.13).

The definition $\theta = s/r$ implies that θ is dimensionless; it is the ratio of two lengths. Thus the radian is not a unit in quite the same sense as the metre or kilogram. We use radians as a reminder of how the angle is defined.

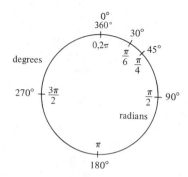

Figure 5.13. Typical angles in degrees and radians. An object that rotates through one and one-half revolutions is said to have rotated through $2\pi + \pi = 3\pi$ radians. 1 radian = 57.3°.

Angular Velocity | The magnitude of the angular velocity $\boldsymbol{\omega}$ (omega) is equal to the rate of change of the angle. The average angular speed is defined by $\bar{\omega} = \Delta\theta/\Delta t$, where $\Delta\theta$ is the change in angle in the time Δt. As Δt becomes arbitrarily small, the angular speed at a given instant in radians per second is

$$\omega = \frac{d\theta}{dt} \tag{5.8}$$

If θ is in radians and t is in seconds, then the units of ω are radians per second. For example, if a runner goes once around a circular track in 50 s, her average angular velocity is $(2\pi$ rad$)/(50$ s$) = 0.126$ rad s^{-1}.

The vector $\boldsymbol{\omega}$ is conventionally taken to be directed *along the axis of rotation*. It is directed out of the page, parallel to the rotation axis, if the rotation is counterclockwise (Fig. 5.14a). If the rotation is clockwise, as in Fig. 5.14b, $\boldsymbol{\omega}$ is directed into the page. (These conventions are similar to those given for the directions of torques in Chapter Four.) One way of identifying the direction of $\boldsymbol{\omega}$ is to curl the fingers of the right hand around the rotation axis in the direc-

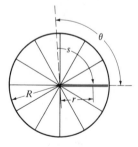

Figure 5.12. A wheel of radius R rotates about its axle. The spoke shown in color was initially vertical and the wheel has rotated through an angle θ. The point at a distance r from the center has moved a distance $s = r\theta$ along a circular path.

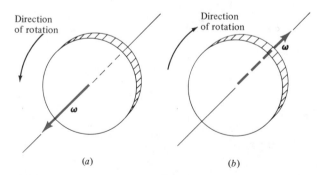

Figure 5.14. (a) For counterclockwise rotations $\boldsymbol{\omega}$ is directed out of the page. (b) $\boldsymbol{\omega}$ is directed into the page for clockwise rotations.

tion of rotation. The right thumb then points in the direction of ω (Fig. 5.15). The following example illustrates these ideas.

Example 5.5

A disk rotates clockwise with a constant angular velocity at two revolutions per second (Fig. 5.14b). (a) What is the direction and magnitude of the angular velocity? (b) What is the rotation angle after 4 s?

(a) The disk rotates twice or through 4π rad each second, so its angular velocity is

$$\omega = \frac{\Delta\theta}{\Delta t} = \frac{4\pi \text{ rad}}{1 \text{ s}} = 4\pi \text{ rad s}^{-1}$$

The direction of ω is into the page since the motion is clockwise.

(b) In 4 s, the disk rotates through an angle

$$\Delta\theta = \omega\Delta t = (4\pi \text{ rad s}^{-1})(4 \text{ s}) = 16\pi \text{ rad}$$

The speed $v = ds/dt$ of a point on a rotating object can be related to ω. By definition, $\theta = s/r$, so $\omega = d\theta/dt = (ds/dt)(1/r) = v/r$, and

$$v = r\omega \tag{5.9}$$

In S.I. units, r has units of metres and ω is in radians per second (rad s^{-1}), but the product $r\omega$ has units of rad m s^{-1} = m s^{-1}; the radians are dropped in the final result.

The speed of points on a rotating object is proportional to the distance r from the rotation axis. The high speeds sometimes encountered when objects rotate are illustrated in the following example.

Example 5.6

The maximum speed of the blades on rotary lawn mowers is limited to reduce the hazard from flying stones and other debris. A currently available model has a rotation rate of 3700 revolutions per minute and a blade 0.25 m in radius. What is the speed at the tip of the blade?

To use $v = r\omega$, we need to find ω. Converting 3700 rev min^{-1} to rad s^{-1},

$$\omega = 3700 \; \frac{\text{rev}}{\text{min}} \left(\frac{2\pi \text{ rad}}{\text{rev}}\right) \left(\frac{1 \text{ min}}{60 \text{ s}}\right) = 387 \text{ rad s}^{-1}$$

Then the velocity at the blade tip is $v = r\omega = (0.25 \text{ m})(387 \text{ rad s}^{-1}) = 97 \text{ m s}^{-1}$, which is nearly 350 km h^{-1}.

Angular Acceleration | The rate of change of the angular velocity is the angular acceleration α (alpha),

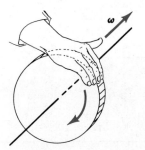

Figure 5.15. Curling the fingers of the right hand in the direction of rotation, the thumb points perpendicular to the disk in the direction of ω.

$$\alpha = \frac{d\omega}{dt} \tag{5.10}$$

The S.I. units of α are radians per second per second. If the orientation of the rotation axis does not change, α points along the axis and is either parallel or opposite to ω. For example, if the disk of Fig. 5.16 is increasing its rotation rate, α and ω are parallel. If the disk is slowing down, α is opposite to ω.

If the angular velocity of the object is changing, so it has an angular acceleration, then points on the object will experience tangential accelerations (Fig. 5.16). The magnitude of the tangential acceleration of a point is $a_T = dv/dt$, so from $v = r\omega$, we have $dv/dt = r\, d\omega/dt$, and

$$a_T = r\alpha \tag{5.11}$$

Points on a rotating object also experience a radial or centripetal acceleration of magnitude v^2/r. Since $v = \omega r$, the centripetal acceleration can be written in angular variables as

$$\mathbf{a}_r = -\omega^2 r\hat{\mathbf{r}} \tag{5.12}$$

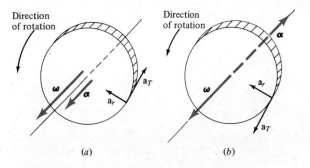

Figure 5.16. In (a) the angular velocity of the disk is increasing, so α and ω are parallel. In (b) the angular velocity is decreasing so α and ω are antiparallel. The tangential and radial accelerations $\mathbf{a}_T$ and $\mathbf{a}_r$ of a point on the disk are also shown.

TABLE 5.1

Translational and rotational motion comparisons

Quantity	Translation	Rotation	Relationship
Position, displacement	x, s	θ	$s = r\theta$
Velocity	**v**	ω	$v = r\omega$
Acceleration	$\mathbf{a} = \mathbf{a}_r + \mathbf{a}_T$	$\boldsymbol{\alpha}$	$a_T = r\alpha$
			$a_r = \omega^2 r$

If the forces among the particles of a rapidly spinning object such as a centrifuge or a star are not large enough to provide this centripetal acceleration, it will break up when a critical angular velocity ω_c is reached. For example, consider the rotating star in Fig. 5.17 that has a mass M and radius R. A particle of mass m on its equator has an acceleration $\omega^2 R$. The gravitational force GmM/R^2 due to the star must be at least equal to ma_r. Hence, the star is just barely able to hold together for an angular velocity ω_c satisfying $m\omega_c^2 R = GmM/R^2$, or

$$\omega_c^2 = GM/R^3$$

The density of the star is its mass-to-volume ratio, $\rho = M/\left(\frac{4}{3}\pi R^3\right)$. Hence we find

$$\omega_c^2 = \frac{4}{3}\pi G\rho \qquad (5.13)$$

This result implies that the maximum angular velocity of a star is determined by its average density. A star with the density of the sun cannot rotate faster than about once every 3 hours; the sun actually rotates about once in 27 days. *Pulsars,* however, are stars which rotate as fast as every 0.03 seconds. According to Eq. 5.13, they must have enormous densities. Pulsars are believed to be neutron stars, with densities comparable to those of atomic nuclei, and greater than that of the sun by a factor of 10^{12}.

Finding the Motion | The description of rigid body rotation can be rather complicated if the angular acceleration and velocity are not parallel. This situation arises, for example, in Chapter Seven when we discuss the motion of a gyroscope or the precession of the equinoxes of the Earth. However, when an object is constrained to rotate about an axis fixed in space, the angular variables θ, ω, and α are related to each other in exactly the same way as are the variables x, v, and a for motion along a straight line (see Table 5.1). Consequently, given the angular acceleration and the initial angular position and velocity, we can find the subsequent motion, just as in Chapter One. For example, if the angular acceleration is constant, the derivation of the equations of motion will proceed precisely as before, with obvious changes in notation. The results are listed in Table 5.2 for convenient reference, and are illustrated by the next example.

Example 5.7

In Example 5.6, a lawn mower blade rotates at 387 rad s^{-1}. If the blade comes to a stop with a constant acceleration in 3 s, find the number of turns it makes as it slows down.

The angle it moves through is

$$\Delta\theta = \frac{1}{2}(\omega_0 + \omega)\,\Delta t = \frac{1}{2}(387 \text{ rad s}^{-1} + 0)(3 \text{ s})$$

$$= 581 \text{ rad}$$

The number of turns is $581/2\pi = 92.4$

5.4 | TORQUE, ANGULAR ACCELERATION, AND THE MOMENT OF INERTIA

We saw in Chapter Four that when the net torque is zero on a rigid body constrained to move about a fixed axis, its rotational motion remains constant.

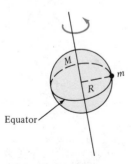

Figure 5.17. A star spinning on its axis. Particles at the equator move the fastest, and have the greatest centripetal acceleration.

TABLE 5.2

Equations for constant angular acceleration α along the axis of rotation and their translational motion analogs. In using these equations, one direction along the rotation axis is defined as positive and the other as negative; θ, ω, and α can each be positive or negative

Constant Linear Acceleration a	Constant Angular Acceleration α
$v = v_0 + a\,\Delta t$	$\omega = \omega_0 + \alpha\,\Delta t$
$\Delta x = v_0\,\Delta t + \frac{1}{2}a\,(\Delta t)^2$	$\Delta \theta = \omega_0\,\Delta t + \frac{1}{2}\alpha\,(\Delta t)^2$
$\bar{v} = \frac{1}{2}(v_0 + v)$	$\bar{\omega} = \frac{1}{2}(\omega_0 + \omega)$
$\Delta x = \frac{1}{2}(v_0 + v)\,\Delta t$	$\Delta \theta = \frac{1}{2}(\omega_0 + \omega)\,\Delta t$
$v^2 = v_0^2 + 2a\,\Delta x$	$\omega^2 = \omega_0^2 + 2\alpha\,\Delta \theta$

When there is a net torque, the object experiences an angular acceleration proportional to that torque. We can identify the proportionality factor by applying Newton's second law of motion to a simple object.

Figure 5.18 shows a point mass m at the end of a string swinging on a frictionless plane in a horizontal circle. The vertical forces on the mass are its weight $\mathbf{w}$ and a normal force $\mathbf{N}$. These forces are equal in magnitude and their torques about the center C cancel. The tension in the string, which produces the centripetal acceleration, $\mathbf{a}_r$, is directed toward the center; since its line of action passes through C, its torque about that point is zero. Only the force $\mathbf{F}_a$ applied at right angles to the string produces a torque about C. From Newton's second law, $\mathbf{F}_a = m\mathbf{a}_T$, and by Eq. 5.11, $a_T = r\alpha$. Thus the torque due to $\mathbf{F}_a$ can be written as $\tau = rF_a = r(ma_T) = rm(r\alpha) = (mr^2)\alpha$. The quantity mr^2 is the *moment of inertia* I of the point mass. In vector notation, our result is

$$\boldsymbol{\tau} = I\boldsymbol{\alpha} \qquad (5.14)$$

This equation is similar in form to Newton's second law, $\mathbf{F} = m\mathbf{a}$, relating the net force and the accelera-

tion; Eq. 5.14 relates the net torque and the angular acceleration. The moment of inertia is analogous to the mass, and indicates the inertia of the object to changes in its rotational motion. Although we have obtained Eq. 5.14 for a particularly simple object, it applies quite generally to the motion of a rigid body about a fixed axis if the moment of inertia is calculated as described in the following subsection.

Moment of Inertia | Despite the fact that $\tau = I\alpha$ is similar in form to $\mathbf{F} = m\mathbf{a}$, it is important to realize that both the torque τ and the moment of inertia I depend on the position of the axis of rotation. We will also find that I depends on the shape and mass of the rotating object.

To calculate the moment of inertia of a complex object, we must mentally separate the object into N small pieces of mass m_1, m_2, $\ldots$, m_N. Then each piece is a distance r_1, r_2, $\ldots$, r_N from the axis of rotation. The moment of inertia of the first piece is $m_1 r_1^2$, that of the second is $m_2 r_2^2$, and so on. The net moment of inertia is the sum of all such terms

$$I = m_1 r_1^2 + m_2 r_2^2 + \ldots + m_N r_N^2$$
$$= \sum_{i=1}^{N} m_i r_i^2 \qquad (5.15)$$

From this we see that the moment of inertia is large when the pieces are far from the rotation axis. In the limit where the masses are arbitrarily small, the sum becomes an integral, and the moment of inertia is given by

$$I = \int r^2 dm \qquad (5.16)$$

The following examples illustrate the calculation of the moment of inertia, and Table 5.3 gives the moments of inertia for several shapes.

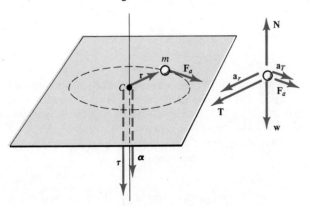

Figure 5.18. The force $\mathbf{F}_a$ exerts a torque on m about the point C, resulting in a tangential acceleration $\mathbf{a}_T$.

TABLE 5.3

Moments of inertia. The mass of each object is taken to be *m*. The dotted line is the rotation axis

	$I = \frac{1}{2}mR^2$	Uniform disk or cylinder of radius R
	$I = \frac{1}{12}ml^2$	Thin rod of length l with the rotation axis through its center
	$I = \frac{1}{3}ml^2$	Thin rod of length l with the rotation axis through one end
	$I = mR^2$	Thin ring or cylindrical shell of radius R
	$I = \frac{2}{5}mR^2$	Sphere of radius R
	$I = \frac{2}{3}mR^2$	Spherical shell of radius R

Example 5.8

Two equal point masses m_0 are at the ends of a massless thin bar of length l (Fig. 5.19). Find the moment of inertia for an axis perpendicular to the bar through (a) the center, and (b) an end.

(a) For an axis through the center, each mass is a distance $l/2$ from the axis. Summing the mr^2 terms for each of the two masses,

$$I = m_0 \left(\frac{l}{2}\right)^2 + m_0 \left(\frac{l}{2}\right)^2 = \frac{m_0 l^2}{2}$$

(b) For an axis through an end, the mass at that end has $r = 0$, while the other mass is at a distance l, so

$$I = 0 + m_0 l^2 = m_0 l^2$$

Thus we see that the moment of inertia depends on the position of the rotation axis.

Example 5.9

Find the moment of inertia of a bicycle wheel of radius R if the mass m is concentrated in the rim and tire. (The spokes are considered massless.)

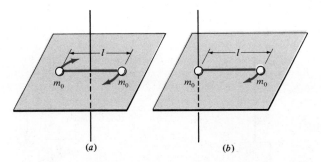

Figure 5.19. Two point masses on a massless bar with axis of rotation (a) through the center; (b) through an end.

If we break up the tire into small pieces, each one is the same distance R from the axle. Thus the moment of inertia is just the total mass m times R^2

$$I = mR^2$$

This arrangement yields the maximum moment of inertia for a wheel of a given mass and radius. It is this large moment of inertia that gives a bicycle some of its stability. Any torques caused, for example, by a rough road will only produce small angular accelerations, since $\alpha = \tau/I$ and I is large.

Example 5.10

Find the moment of inertia of a thin rod of length l and mass m about an axis through its center (Fig. 5.20).

The moment of inertia is found by breaking the rod into many small masses and integrating their contributions to Eq. 5.16. The mass per unit length of the rod is m/l, and the mass of a small segment of length dx is $dm = mdx/l$. Its contribution to I is then $x^2dm = x^2mdx/l$. Integrating over the length of the rod, we find the moment of inertia

$$I = \frac{m}{l} \int_{-l/2}^{l/2} x^2 \, dx = \frac{m}{l} \left[\frac{x^3}{3} \right]_{-l/2}^{l/2}$$

$$= \frac{m}{3l} [(l/2)^3 - (-l/2)^3] = \frac{ml^2}{12}$$

Parallel Axis Theorem | Given the moment of inertia about an axis passing through the center of mass, we can immediately write down the moment of inertia about any axis parallel to that axis. To see this, consider the point masses in Fig. 5.21. The center of mass is at a distance X from the y axis, which is the axis of rotation. Mass m_i is located at a distance x_i from the axis of rotation or at a distance $r_i = x_1 - X$ from the C.M. The contribution of mass m_i to the moment of inertia is then

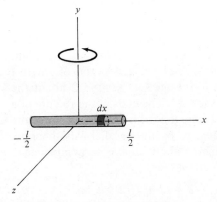

Figure 5.20. The segment of length dx has a mass $dm = mdx/l$, and contributes $x^2dm = x^2mdx/l$ to the moment of inertia.

$$m_i x_i^2 = m_i(r_i + X)^2$$
$$= m_i(X^2 + r_i^2 + 2r_iX)$$

Summing over the masses, the first term is $(\Sigma m_i)X^2 = mX^2$, where m is the total mass of the system. The second term is $\Sigma m_i r_i^2 = I_{C.M.}$, the moment of inertia relative to the C.M. The last term, $2(\Sigma m_i r_i)X$, is proportional to the net torque $\Sigma w_i r_i = \Sigma(m_i g)r_i$ about the C.M. due to the weights of the particles. Since the C.M. (or C.G.) was defined in Chapter Four as the point where the weight is effectively concentrated, the net torque about the C.M. is zero, and the last term vanishes. Thus the moment of inertia about the y axis is given by the *parallel axis theorem*,

$$I = I_{C.M.} + mX^2 \quad \text{(parallel axis theorem)} \quad (5.17)$$

The moment of inertia about any axis is determined by the moment of inertia about the C.M., $I_{C.M.}$, the

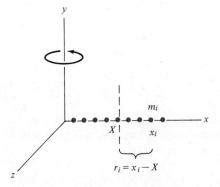

Figure 5.21. The dashed axis passes through the C.M. (point X), and the moment of inertia about this axis is $I_{C.M.}$. The moment of inertia about a parallel axis (the y axis) is related to $I_{C.M.}$ by the parallel axis theorem.

total mass of the system, m, and the distance from the axis to the C.M., X.

A simple illustration of this theorem is provided by Example 5.8. This example showed that two point masses m_0 at the ends of a massless rod of length l have a moment of inertia $I_{C.M.} = m_0 l^2/2$ about the center, which is the C.M. The ends of the rod are at a distance $X = l/2$ from the C.M. The total mass of the system is $m = 2m_0$, so by the parallel axis theorem, the moment of inertia about an axis perpendicular to the rod through an end is

$$I = m_0 l^2/2 + (2m_0)(l/2)^2 = m_0 l^2$$

This is the result found in Example 5.8 by directly summing the $m_i r_i^2$ terms.

Now let us apply the parallel axis theorem to a more complex situation. Consider the third object in Table 5.3, a thin rod of length l rotating about an axis through its end. Its C.M. is the center of the rod. We found in Example 5.10 that $I_{C.M.} = ml^2/12$. With $X = l/2$, the parallel axis theorem gives the moment of inertia about an axis through the end of the rod,

$$I = ml^2/12 + m(l/2)^2 = ml^2/3$$

In this case, I can also be obtained readily by integration, but this is not always true.

Radius of Gyration | For irregular objects such as a bone or forearm, it is usually necessary to determine the moment of inertia experimentally. Such experimental results are often expressed by giving the mass m of the object and *radius of gyration k* defined by

$$I = mk^2 \quad \text{or} \quad k = \sqrt{\frac{I}{m}} \quad (5.18)$$

A point mass m at a distance k from the rotation axis would have the same moment of inertia as the actual object. For example, the radius of gyration of the sphere in Table 5.3 is found from

$$k = \sqrt{\frac{I}{m}} = \sqrt{\frac{\frac{2}{5}mR^2}{m}} = R\sqrt{\frac{2}{5}} = 0.63R$$

The following example shows how $\tau = I\alpha$ can be used in a problem involving both rotation and translation.

Example 5.11

A wheel is on a horizontal axis of radius $r = 0.01$ m supported by frictionless bearings (Fig. 5.22). A 5-kg

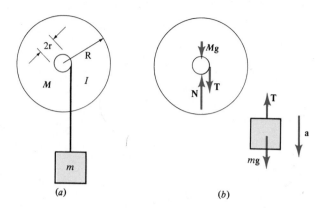

Figure 5.22. (*a*) A wheel and axle mounted on frictionless bearings have a block of mass m hanging from a string wrapped around the axle. (*b*) The forces on the axle and on the falling mass. **T** is the tension on the string, and **N** is the force exerted by the bearings.

block attached to a string is wrapped around the axle and released from rest. It falls with an acceleration of 0.02 m s^{-2}. (a) What is the tension in the string? (b) What is the moment of inertia of the wheel and axle?

(a) Using Newton's second law for the falling mass, $mg - T = ma$. Solving for T,

$$T = m(g - a) = (5 \text{ kg})(9.8 - 0.02) \text{ m s}^{-2} = 48.9 \text{ N}$$

(b) The lines of action of the bearing force **N** and weight $M\mathbf{g}$ pass through the rotation axis, so they give zero torque; the magnitude of the torque due to **T** is rT. The acceleration a of the block is also the tangential acceleration of a point on the edge of the axle. With $a = r\alpha$, $\tau = I\alpha$ becomes $rT = Ia/r$. Solving for I,

$$I = \frac{r^2 T}{a} = \frac{(0.01 \text{ m})^2(48.9 \text{ N})}{0.02 \text{ m s}^{-2}} = 0.245 \text{ kg m}^2$$

This example illustrates one method of determining I experimentally. By measuring the time for the mass to fall we can find the acceleration and obtain I. The procedure is often used in the introductory physics laboratory.

5.5 | ELECTRIC CHARGES; FUNDAMENTAL FORCES

We have seen that gravitational forces are responsible for keeping the planets in their nearly circular orbits around the sun. *Electric* forces, which are in some ways very similar to gravitational forces and in other ways quite different, are responsible for keeping electrons in their orbits in atoms. Although we will defer a full discussion of electric forces until Chapter Sixteen, we introduce them here and discuss a simple model of an atom.

The gravitational force is a fundamental force of nature. Related to this force is the concept of mass, which plays a basic role in the description of the physical world. An equally basic concept is that of *charge.* Just as two masses produce forces on each other, so also do two charges produce forces on one another. However, an important difference is that there are two kinds of charges, described in mathematical terms as positive and negative. Although two masses always attract each other, like charges (both positive or both negative) repel each other, while unlike charges attract each other. A further complication is that charges may produce two types of forces. Charges at rest exert forces on each other called *electric forces,* but charges in motion exert additional forces on each other called *magnetic forces.*

Charge and the Structure of Matter

While it requires sophisticated mathematics to fully describe atoms and molecules, a simple atomic model (Fig. 5.23) gives some insight into charges, their properties, and the role they play in nature. There are slightly more than 100 naturally occurring or artificially produced atomic species or *elements.* The typical atomic radius is 10^{-10} m. The model pictures atoms as rather like a solar system, with negatively charged *electrons* orbiting a dense, positively charged *nucleus.* The attractive electric force between the negative electrons and the positive nucleus holds the atom together, just as the gravitational attraction holds the solar system together. When an atom has its full complement of electrons, it is electrically neutral, that is, its total charge is zero. If one or more electrons are added or taken away the atom is said to be *ionized.*

The constituents of the massive nucleus were not completely identified until the 1930s. These constituents, collectively called *nucleons,* are *neutrons* that are uncharged and *protons* that have a positive charge. The proton charge is denoted by e and is exactly equal in magnitude to the electron charge, $-e$. A neutron or proton is approximately 1800 times as massive as an electron.

A typical nuclear radius is 10^{-14} m, some 10,000 times smaller than that of the atom. One may well ask how a nucleus can have many positively charged protons so close together if they repel each other. The explanation depends on the existence of still another

Figure 5.23. A simple model of an atom. Negatively charged electrons orbit a small, massive nucleus under the influence of electrical forces. The nucleus, which is not shown to scale, contains two kinds of nucleons: positively charged protons and neutral neutrons.

fundamental force of nature that does not depend upon charge. If two protons approach each other, they electrically repel each other more and more strongly until they are close enough for the *strong nuclear force* to take effect. This force overwhelms the electric repulsion and holds the protons together.

Atoms combine in various ways, all of which depend on the electric forces produced by charges, to form molecules and macroscopic objects that, in their normal state, are electrically neutral. Thus, common objects are not ordinarily charged, but their existence and solidity depend on charges being present in their microscopic structure.

Two remarkable facts concerning charges have emerged from physics research up to the present. Using particle accelerators, complex nuclear reactions have been produced in which particles are created or destroyed. In this way, many short-lived particles have been discovered that are not ordinarily present in matter. Every one of these particles has an electric charge that is an exact integer multiple of the electronic charge: $0, \pm e, \pm 2e, \ldots$. In recent years, evidence has accumulated that nucleons are not the most elementary objects, and that they too are composite structures made up of particles called *quarks.* Quarks appear to have charges that are a fraction of the electronic charge. However, no particles having such fractional charges have ever been observed directly. Thus, it currently seems that isolated charges cannot exist in units other than that found on the electron.

The second observation is that when particles are created or destroyed, the net charge remains constant. This is referred to as *conservation of charge.* For example, if a positive charge is destroyed, a negative charge is also destroyed, leaving the net charge unchanged. The reasons for these characteristics of charge are not known.

The Four Fundamental Forces | In addition
to electric forces, charges exert magnetic forces on each other, if they are in motion. Since objects at rest in one reference frame are in motion in another, it is clear that electric and magnetic forces are intimately related, and may be regarded as types of *electromagnetic forces.* Electromagnetic and gravitational forces are two of the fundamental forces in nature, and Newton's laws are applied to both in the same way.

There are a total of four known fundamental forces, the other two being the strong nuclear force just mentioned and the so-called *weak force,* which is responsible for the fact that some nuclei are radioactive and decay or change into other nuclear species. It is now thought that the weak and electromagnetic forces may be intrinsically related, so there are at most three truly fundamental forces. In order of decreasing strength, the fundamental forces among subatomic particles are the strong, electromagnetic, weak, and gravitational.

Unlike electromagnetic and gravitational forces, the strong and weak forces act only at distances comparable to nuclear dimensions. At these very short distances, Newton's laws cannot be applied without modifications. Nevertheless, many of the concepts developed in our discussions of Newton's laws are still applicable.

5.6 | COULOMB'S LAW

The earliest published studies of the detailed properties of the forces between charges were made in 1784 by Charles Augustin de Coulomb (1736–1806). He found that the electric force, like the gravitational force, is inversely proportional to the distance squared. However, because there are two kinds of charge, the force can be either attractive or repulsive.

Two point charges, q_1 and q_2, exert equal and opposite electric forces on each other. If the distance to q_1 from q_2 is r, then the *force experienced by q_1 due to q_2 is given by* Coulomb's law,

$$\mathbf{F}_{12} = \frac{kq_1q_2}{r^2}\hat{\mathbf{r}} \qquad (5.19)$$

Here $\hat{\mathbf{r}}$ is a unit vector pointing from q_2 to q_1, and k is an experimentally determined constant.

The S.I. unit of charge is the *coulomb* (C). In S.I. units, it is found that

$$k = 9.0 \times 10^9 \text{ N m}^2 \text{ C}^{-2} \qquad (5.20)$$

and that the magnitude of the charge on a proton or electron is

$$e = 1.60 \times 10^{-19} \text{ C}$$

The coulomb is a large unit of charge. For example, according to Coulomb's law, the force between a pair of 1 C charges 1 m apart is

$$F = (9 \times 10^9 \text{ N m}^2 \text{ C}^{-2})\frac{(1 \text{ C})(1 \text{ C})}{(1 \text{ m})^2} = 9 \times 10^9 \text{ N}$$

This is about 1 million tons! Consequently isolated charges as large as a coulomb are rarely encountered.

The direction of the electric force depends on the relative signs of the charges q_1 and q_2 (Fig. 5.24). If the two charges have the same sign (both positive or both negative), then q_1q_2 is positive, so that $\mathbf{F}_{12}$ is directed along $\hat{\mathbf{r}}$, and the charges repel each other. If the charges have opposite signs, with one positive and the other negative, then q_1q_2 is negative; $\mathbf{F}_{12}$ is then directed along $-\hat{\mathbf{r}}$, and the charges attract. In each case, Newton's third law holds true: the force $\mathbf{F}_{21}$ on q_2 due to q_1 equals $-\mathbf{F}_{12}$.

The gravitational force law (see Chapter Three) can be written in a vector form similar to Eq. 5.19 for the electric force. The gravitational force on a particle of mass m_1 due to a particle of mass m_2 is

$$\mathbf{F}_{12} = -\frac{Gm_1m_2}{r^2}\hat{\mathbf{r}} \qquad (5.21)$$

Again $\hat{\mathbf{r}}$ is a unit vector pointing from particle 2 toward particle 1; the minus sign indicates the force is toward particle 2, or attractive. Since both the gravitational and electric forces vary as $1/r^2$, their ratio for a pair of particles of given mass and charge is independent of the separation.

Figure 5.24. (*a*) Like charges, both positive or both negative, repel each other. (*b*) Unlike charges, one positive and one negative, attract each other.

CHARLES AUGUSTIN de COULOMB
(1736–1806)

HENRY CAVENDISH
(1731–1810)

(The Bettmann Archive.)

(Radio Times Hulton Picture Library.)

Although we know it today as Coulomb's law, the expression for the electrostatic force between two charges was first discovered by Henry Cavendish. This fact is not surprising when one learns about Cavendish's bizarre personality.

Charles Augustin de Coulomb was a French nobleman who began a career as a military engineer and gradually became interested in scientific research. At the beginning of the French Revolution, he wisely retired to the safety of a provincial town to concentrate on his experiments. In 1777, he competed for a prize offered by the French Academy of Sciences for the improvement of magnetic compasses. He found that if he suspended a compass needle from a fine hair or thread, the torque exerted on the needle was proportional to the angle through which it had rotated. This *torsion balance* principle made it possible for him to measure electrostatic forces accurately and to obtain the force law. His results were published in the normal fashion, and he received the credit for this important discovery.

Unknown to Coulomb, his English contemporary, Cavendish, had already done the same kinds of electrostatic experiments with a torsion balance. Cavendish was an extraordinarily eccentric person: shy, absentminded, and a recluse to the point of insisting on dying alone. He never completed his studies at Cambridge because he could not bear to face his professors at the required examinations. He avoided all people, especially women, as much as possible.

Cavendish came from a wealthy family and never cared for nor needed to worry about money. Devoting himself to scientific research for his entire life, he made many major discoveries. However, he published very few of them and had no concern about receiving credit for his advances. In particular, his electrical researches anticipated many discoveries of the following decades but remained unknown until James Clark Maxwell examined Cavendish's notes decades after his death. Cavendish did publish some of his early work on the properties of hydrogen gas, but his experiments in which he discovered the inert gas now called argon were ignored until they were repeated a century later.

The most important experiment performed by Cavendish was, in effect, a measurement of the mass of the earth. From Newton's law of universal gravitation, it follows that the gravitational acceleration at the surface of the earth is $g = GM_E/R_E{}^2$, where G is the gravitational costant and M_E and R_E are the earth's mass and radius, respectively. Since g and R_E are easily measured, a determination of either G or M_E determines the other quantity. Cavendish used a torsion balance to measure the small gravitational force exerted on two small spheres at the ends of a pivoted rod by two other large spheres. This gave him a value for G and therefore for the mass of the earth.

The next two examples illustrate some features of the simplest atom, hydrogen, which has a single electron orbiting a proton.

Example 5.12

In a simple model of the hydrogen atom, an electron moves around a proton in a circular orbit of radius 5.29×10^{-11} m. The proton mass is $M = 1.67 \times 10^{-27}$ kg and the electron mass is $m = 9.11 \times 10^{-31}$ kg (Fig. 5.25). What are the electric and gravitational forces exerted on the electron by the proton?

Since the proton and electron have opposite charges, $+e$ and $-e$, the electric force is attractive, and its magnitude is

$$F = \frac{ke^2}{r^2}$$

$$= (9 \times 10^9 \text{ N m}^2 \text{ C}^{-2}) \frac{(1.6 \times 10^{-19} \text{ C})^2}{(5.29 \times 10^{-11} \text{ m})^2}$$

$$= 8.23 \times 10^{-8} \text{ N}$$

The gravitational force is also attractive; from Eq. 5.21, its magnitude is

$$F_G = \frac{GmM}{r^2}$$

$$= (6.67 \times 10^{-11} \text{ N m}^2 \text{ kg}^{-2})$$

$$\times \frac{(1.67 \times 10^{-27} \text{ kg})(9.11 \times 10^{-31} \text{ kg})}{(5.29 \times 10^{-11} \text{ m})^2}$$

$$= 3.63 \times 10^{-47} \text{ N}$$

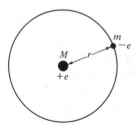

Figure 5.25. A simple model of the hydrogen atom. A negative electron moves in a circular orbit around a massive positive proton.

Comparison of these two forces shows that in this situation the electric force is about 10^{39} times stronger than the gravitational force. Thus, in atomic physics the gravitational force can be completely ignored.

Example 5.13

Using the numbers given in the preceding example, find the speed of the electron in the hydrogen atom.

The centripetal acceleration of the electron results from the electric force, so the speed v satisfies

$$\frac{mv^2}{r} = \frac{ke^2}{r^2}$$

or

$$v = \sqrt{\frac{ke^2}{mr}}$$

Substituting the numerical values from the preceding example,

$$v = \left[\frac{(9 \times 10^9 \text{ N m}^2 \text{ C}^{-2})(1.6 \times 10^{-19} \text{ C})^2}{(9.11 \times 10^{-31} \text{ kg})(5.29 \times 10^{-11} \text{ m})} \right]^{1/2}$$

$$= 2.19 \times 10^6 \text{ m s}^{-1}$$

This is a large velocity; it is about 1 percent of the speed of light, 3×10^8 m s^{-1}.

We will see later that this simple model of the hydrogen atom and its generalization for atoms with many electrons correctly predicts some, but not all, of the observed properties of atoms. The failure of Newtonian mechanics to account for a variety of atomic phenomena led to revolutionary changes early in this century in our view of the world at atomic and subatomic levels.

SUMMARY

An object undergoing circular motion has a centripetal acceleration $\mathbf{a}_r = -(v^2/r)\hat{\mathbf{r}}$ directed toward the center of the circle. If its speed is changing, it also has a tangential acceleration $\mathbf{a}_T$ along its direction of motion.

The centripetal acceleration is produced by a net force equal to $m\mathbf{a}_r$ acting on the object. On a flat curve in a road, the net force is provided entirely by friction, but part of all of the required force can be provided by the normal force in a banked road. When an object swings rapidly in a vertical circle, a downward force in addition to its weight is necessary to keep the object in its path.

The angular position, velocity, and acceleration conveniently describe the rotational motion of a rigid body about a fixed axis. The definitions and relationships are analogous to those for translational motion along a straight line, and are summarized in Tables 5.1 and 5.2.

When a rigid body constrained to move about a fixed axis has a net torque about a point on that axis, it has an angular acceleration found from

$$\tau = I\alpha$$

The moment of inertia I is mr^2 for a point mass, so it is large when the mass is far from the axis of rotation. For a complex object, we divided the object into many small parts and sum the individual mr^2 contributions.

Atoms consist of negatively charged electrons orbiting around a dense, positively charged nucleus. Charges at rest exert electric forces on each other, while moving charges also exert magnetic forces. The known fundamental forces are the strong, electromagnetic, weak, and gravitational forces, in order of decreasing strength.

Coulomb's law states that the electric force between two charges is proportional to the product of the charges and inversely proportional to their separation squared:

$$F_{12} = \frac{kq_1q_2}{r^2}$$

The force is attractive for unlike charges and repulsive for like charges. Electric forces keep electrons in atomic orbits.

Checklist
Define or explain:

uniform circular motion
centripetal acceleration
tangential acceleration
radius of curvature
banked curve
radian
angular position,
   velocity, acceleration
torque
moment of inertia

electric force
nucleus
atomic model
nucleons, protons,
   neutrons
electronic charge
quarks
conservation of charge
magnetic force
electromagnetic forces

radius of gyration
electric charge

strong, weak force
Coulomb's law

REVIEW QUESTIONS

Q5-1 Motion in a circular path with a constant speed is _____.

Q5-2 When an object moves along a circular path with a constant speed, it has an acceleration directed toward the _____.

Q5-3 An object undergoing uniform circular motion must have a net force on it directed toward _____.

Q5-4 Skids occur when the maximum frictional force of a flat road on a car is less than _____.

Q5-5 If an object moving along a circular path has its speed changing, it has a _____ equal to the rate of change of the speed.

Q5-6 A complete circle is _____ radians.

Q5-7 The angular velocity is the rate of change of the _____, and points along the _____.

Q5-8 The angular acceleration is the rate of change of the _____.

Q5-9 The angular acceleration of a rigid body about a fixed axis is proportional to the _____.

Q5-10 The moment of inertia of a point mass depends on its mass and on the _____.

Q5-11 For a given object, the moment of inertia depends on the location of the _____.

Q5-12 Atoms and molecules are held together by _____.

Q5-13 Nuclei are held together by _____.

Q5-14 Like charges _____, unlike charges _____.

Q5-15 A nucleus is made up of two types of _____, _____ and _____.

Q5-16 A nucleus is about _____ times smaller than an atom.

Q5-17 Nucleons are thought to be made up of _____.

Q5-18 The four fundamental forces are the _____.

Q5-19 The electric force between two charges is inversely proportional to the _____.

EXERCISES

Section 5.1 | Centripetal Acceleration

5-1 A woman runs on a circular track of radius 100 m at a speed of 8 m s^{-1}. What is her acceleration?

5-2 A boy rides a bicycle at 10 m s^{-1} on a flat curve of radius 200 m. (a) What is his acceleration? (b) If the boy and the bike have a total mass of 70 kg, how large a force is needed to provide this acceleration?

5-3 A racing car rounds a turn at 60 m s^{-1}. If the force needed to provide the centripetal acceleration is equal to the weight of the car, what is the radius of the turn?

5-4 A man sits without a seat belt in a car. He tends to slide to the left as the car makes a right turn. Is there a force pushing the man to the left? Explain.

5-5 The speeds of centrifuges are limited in part by the strength of the materials used in their construction. A centrifuge spins a 10 g = 10^{-2} kg sample at a radius of 0.05 m at 60,000 revolutions per minute. (a) What force must the centrifuge exert on the sample? (b) What is the mass of a sample at rest with a weight equal to this force?

5-6 A child sits 4 m from the center of a merry-go-round, which turns completely each 10 s. What is the child's acceleration?

5-7 A jet fighter plane flying at 500 m s^{-1} pulls out of a dive on a circular path. What is the radius of the path if the pilot is subjected to an upward acceleration of 5g?

5-8 The radius of the earth's orbit about the sun is 1.5 × 10^8 km, and its period is 365 days. What is the centripetal acceleration of the earth?

5-9 A trained pilot can pull out of a dive on a circular path with an upward acceleration of 5.5g at the bottom of the path. An untrained pilot can perform the same maneuver at the same speed but only with an acceleration of 3g. What is the ratio of the minimum radii of the paths in which the two pilots can fly?

5-10 A centrifuge used for testing human tolerance to acceleration has a gondola at a distance of 16 m from the vertical axis of rotation. What speed is required to produce a horizontal acceleration of 11g?

5-11 A car is traveling around a flat curve of radius 0.25 km. The coefficient of static friction between the tires and the road is 0.4. At what speed will the car begin to skid?

5-12 A woman of mass 60 kg runs around a flat, circular track of radius 200 m at 6 m s^{-1}. (a) What is her acceleration? (b) What force produces this acceleration? (c) How large is this force?

5-13 Show that v^2/r has the dimensions of an acceleration.

Section 5.2 | Examples of Circular Motion

5-14 At what angle should the track of Exercise 5-12 be banked so that there is no frictional force required?

5-15 A racing car travels around a curve of radius 1300 m. If the frictional force is zero and the speed is 63 m s^{-1}, at what angle is the curve banked?

5-16 A curved track has a radius of 336 m and is banked at 35°. At what speed is the frictional force zero?

5-17 Why might it be unsafe to drive on steeply banked curves at low speeds under very slippery conditions?

5-18 A swallow flies in a horizontal arc of radius 15 m at 18 m s^{-1}. (a) What is its acceleration? (b) What is the banking angle?

5-19 A curve is banked so that no friction is required at 60 km h^{-1}. If a car is moving at 40 km h^{-1} on the curve, in which direction is a frictional force exerted by the road?

5-20 An airplane climbs and then turns downward in a circular arc of radius R. If its speed is 400 m s^{-1}, at what radius will the pilot experience weightlessness at the top of the arc?

5-21 An ultracentrifuge spins samples so that their effective weight is 10^5 times their normal weight. If the sample is at a radius of 0.05 m, how many revolutions per minute does the machine make?

5-22 A bird of weight w flies at 15 m s^{-1} in a horizontal circle of radius 15 m. What is the bird's effective weight?

5-23 A Ferris wheel of radius 16 m rotates in a vertical circle uniformly once every 20 s. (a) What is the centripetal acceleration? (b) What is the effective weight of a 45-kg rider at the highest point of the ride? (c) What is the rider's effective weight at the lowest point of the ride?

Section 5.3 | Angular Variables

5-24 A point on a bicycle wheel moves a distance $s = 1$ m. If this point is 0.4 m from the wheel axis, through what angle has the wheel rotated in (a) radians, and (b) degrees?

5-25 The following angles are given in radians. Find the corresponding angles in degrees, and sketch these angular coordinates on circles as in

Fig. 5.13. (a) $\theta = \pi/3$ rad; (b) $\theta = 3\pi/4$ rad; (c) $\theta = 9\pi/4$ rad.

5-26 Figure 5.26 shows a right triangle with two of the angles given. (a) Find these two angles in radians. (b) The sum of the internal angles of any triangle is 180°. Find the third angle in Fig. 5.26 in radians.

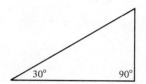

Figure 5.26. Exercise 5-26.

5-27 Figure 5.27 shows a sphere rotating about a vertical axis. What is the direction of the angular velocity in (*a*) and in (*b*)?

5-28 If the sphere in Fig. 5.27a rotates at a constant rate, what is the direction and magnitude of ω if the angular displacement is (a) π rad in 0.4 s; (b) 270° in 0.6 s?

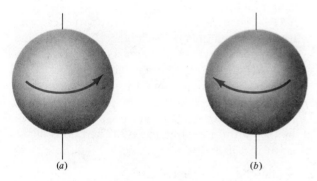

(*a*)   (*b*)

Figure 5.27. The colored arrows indicate the direction of rotation of the sphere about a vertical axis. Exercises 5-27 and 5-28.

5-29 A bicycle rider moves past us from left to right. If her speed is 5 m s^{-1}, (a) what is the direction and magnitude of the angular velocity of one of the wheels, which is 0.4 m in radius? (b) The rider is increasing her speed as she passes, and her acceleration is 1 m s^{-2}. What is the angular acceleration of a wheel?

5-30 (a) Find the radial acceleration at the edge of a phonograph record of radius 0.15 m rotating at 78 rev min^{-1}. (b) The record comes to a stop with a uniform angular acceleration in 2 s. Find the tan-

gential acceleration at the edge and the angular acceleration.

5-31 An ultracentrifuge produces a radial acceleration of 300,000 times the acceleration of gravity at a distance of 0.05 m from the rotation axis. What is the angular velocity in radians per second and revolutions per minute?

5-32 A Ferris wheel at an amusement park rotates in a vertical circle once every 20 s. Its radius is 10 m. (a) What is its angular velocity in radians per second? (b) What is the radial acceleration of a passenger?

5-33 Pulsars are astronomical objects that rotate as fast as once every 0.03 s. Using the data in Table 3.2, determine whether they are (a) ordinary stars like our sun; (b) white dwarf stars; (c) neutron stars.

5-34 Assume a wheel has an initial angular velocity of $\omega_0 = 10$ rad s^{-1}. The angular acceleration is 2.5 rad s^{-2} directed opposite to ω_0. (a) How long does it take for the wheel to stop turning? (b) Through what angle has the wheel turned in this time?

5-35 A car accelerates uniformly from rest to 20 m s^{-1} in 15 s. The wheels have a radius of 0.3 m. (a) What is the final angular velocity of the wheels? (b) What is the angular acceleration of the wheels? (c) What is their angular displacement during the 15-s interval?

5-36 The flywheel of an automobile engine is turning at 700 revolutions per minute. The accelerator is depressed, and in 6 s the speed is 3500 rev min^{-1}. (a) Find the initial and final angular velocities in rad s^{-1}. (b) Find the average angular acceleration. (c) Assuming constant angular acceleration, find the angular displacement during the 6 second acceleration period. (d) Find the tangential acceleration of a point on the flywheel of the engine that is 0.2 m from the axis.

Section 5.4 | Torque, Angular Motion, and the Moment of Inertia

5-37 A bicycle wheel has a mass of 2 kg and a radius of 0.35 m. What is its moment of inertia?

5-38 Two wheels of mass m each have a radius R. Wheel A is a uniform disk, while wheel B has nearly all its mass at the rim. Find the ratio of the moments of inertia, I_B/I_A.

5-39 Find the radius of gyration of a rod of length l pivoted about an axis through its center.

5-40   What is the radius of gyration of a spherical shell of radius R rotating about an axis through its center?

5-41   A grinding wheel, a disk of uniform thickness, has a radius of 0.08 m and a mass of 2 kg. (a) What is its moment of inertia? (b) How large a torque is needed to accelerate it from rest to 120 rad s^{-1} in 8s?

5-42   Assuming that the earth is a uniform sphere, find its moment of inertia about an axis through its center. (The mean radius of the earth is 6.38×10^6 m. Its total mass is 5.98×10^{24} kg.)

5-43   Two masses m_1 and m_2 hang on a pulley of mass M (Fig. 5.28). The pulley is a solid cylinder of radius R and rotates without friction. What is the tangential acceleration of the wheel if $M = m_2$ and $m_1 = \frac{1}{2}m_2$?

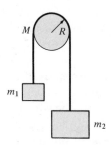

Figure 5.28.   Exercises 5-43 and 5-44.

5-44   Two masses m_1 and m_2 hang on a frictionless pulley of mass M (Fig. 5.28). If all the mass of the pulley can be considered to be at its rim, and $M = 2m_2 = 3m_1$, find the acceleration of the masses m_1 and m_2.

5-45   Using the parallel axis theorem, find the moment of inertia of a rod of mass m and length l about an axis perpendicular to the rod at a distance $l/4$ from an end.

^{c}**5-46**   Solve the preceding exercise using integration rather than the parallel axis theorem.

5-47   A thin ring of radius R and mass m lies on a horizontal plane. Find its moment of inertia for rotation about a vertical axis passing through the edge of the ring.

^{c}**5-48**   (a) Show that a cone of mass m and length l (Fig. 5.29) has a mass per unit length which is proportional to $(l - x)^2$. (b) Find its moment of inertia for rotation about the y axis.

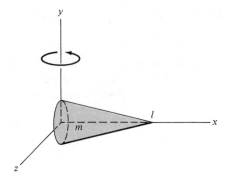

Figure 5.29.   Exercise 5-48.

Section 5.5 | Electric Charges; Fundamental Forces

5-49   Objects can be given a net charge by rubbing, as is observed by people walking on woolen rugs in dry weather. How many electrons must be transferred to give an object a net charge of $+10^{-6}$ C? Must electrons be added or taken away?

5-50   A gram of hydrogen contains about 6×10^{23} atomic electrons. What fraction of the electrons must be removed to give the sample a net charge of 10^{-3} C?

5-51   Two identical charges 0.1 m apart exert electrical forces of 10 N on each other. (a) What is the magnitude of one of the charges? (b) Find the ratio of this charge to that on an electron.

5-52   Cosmic ray protons strike the upper atmosphere at an average rate of 1500 protons per square metre each second. How much charge is received by the earth in a 24-hour day? (The radius of the earth is 6.38×10^6 m.)

Section 5.6 | Coulomb's Law

5-53   A kilogram of molecular hydrogen contains 3.01×10^{26} molecules, each consisting of two hydrogen atoms. (a) What is the total charge on the electrons in 1 kg of hydrogen? (b) What is the total charge on the protons? (c) If all the electrons were removed and placed 1 m away from the protons, what would be the electrical force between them?

5-54   In a nucleus the minimum separation between two protons is about 10^{-15} m. (a) Find the electrical force between two protons at this distance. (b) Find the ratio of this force to the force between a proton and an electron separated by 10^{-10} m.

5-55   A salt crystal is made up of Na$^+$ ions, which

are lacking one electron, and Cl⁻ ions, which have one extra electron. What is the force between a Na⁺ ion and a Cl⁻ ion separated by 5×10^{-10} m?

5-56 In a simple model of the normal hydrogen atom, the radius of the circular electronic orbit is 5.29×10^{-11} m and the speed of the electron is 2.19×10^6 m s⁻¹. Find (a) the acceleration of the electron, and (b) the number of orbits completed per second.

PROBLEMS

5-57 A car travels on a curve of radius 100 m banked at 20° at a speed such that no friction is needed. (a) What is the speed of the car? (b) Find the ratio of the normal force to the weight.

*5-58 A car of weight w travels on a curve of radius 200 m banked at 10°. (a) At what speed is no friction required? (b) What frictional force is required if the car travels 5 m s⁻¹ faster than this speed?

*5-59 A curve of radius 300 m is banked at an angle of 10°. (a) At what speed is no friction required? (b) If the coefficient of friction is 0.8, what are the maximum and minimum speeds at which the curve can be traveled?

5-60 A bird of mass 0.3 kg flies in a horizontal curve of radius 20 metres at 15 m s⁻¹. (a) What is the banking angle? (b) What is the lift force exerted by the air on the bird?

5-61 The earth's radius is 6.38×10^6 m and it rotates on its axis once every 24 hours. (a) What is the centripetal acceleration at the equator? (b) If a man weighs 700 N at the North Pole, what is his effective weight at the equator? (c) The earth is actually not quite spherical; it is flattened slightly at the poles and broadened at the equator. What qualitative effect does this have on your answer to part (b)?

5-62 A 2-kg rock is tied to a light rope 1 metre long and is swung in a horizontal circle. The rope is at an angle of 30° to the horizontal. (a) What is the tension in the rope? (b) What is the speed of the rock?

5-63 An airplane moving at 400 m s⁻¹ can safely be subjected to a $8g$ acceleration when banking. How long will it take for the plane to turn 180°?

5-64 On an Olympic bobsled course, a sled goes through a horizontal turn at 120 km h⁻¹, subjecting the crew to an effective weight 5 times their actual weight. What is the radius of the turn?

5-65 An amusement park ride rotates its occupants in a vertical circle at a constant speed. At the top of the circle, a rider has an effective weight that is directed upward and has a magnitude of twice his actual weight, w. (a) What is his effective weight at the bottom of the circle? (b) How large is his effective weight when he is halfway to the top of the circle?

5-66 A thin washer of mass m is made by drilling a hole of radius $0.4R$ in the center of a circular disk of radius R. Find its moment of inertia.

*5-67 A wheel of radius R has a thickness a from $r = 0$ to $r = R/2$, and a thickness $2a$ from $r = R/2$ to $r = R$. If its density is ρ, what is its moment of inertia?

5-68 A person stands on the earth halfway between the equator and the North Pole. Find his speed in m s⁻¹ due to the daily rotation of the earth. (The radius of the earth is 6.38×10^6 m.)

5-69 A grinding wheel is a uniformly thick disk of mass 3 kg and radius 10 cm. It is initially moving at 2400 revolutions per minute. A tool is pressed against the wheel with a normal force of 10 N. If the coefficient of friction is 0.7 and no other torques act on the wheel, how long will it take for the wheel to come to rest?

5-70 A block of mass 10 kg rests on a horizontal surface. The coefficient of sliding friction is 0.1. A massless horizontal string attached to this block passes over a frictionless pulley and is tied to a hanging block of mass 200 kg. When released, the system moves 2 m in 1 s. What is the mass of the pulley if it is a solid cylinder?

5-71 Blocks of mass 10 kg and 30 kg hang on either side of a pulley on a massless string. The pulley has a mass of 3 kg, a radius of 0.1 m, and a radius of gyration of 0.08 m. If the system has an acceleration of 3 m s⁻², what torque is exerted by the frictional forces in the bearings of the pulley?

5-72 Two lead balls of mass 5 kg have a distance of 1 m between their centers. (a) A lead atom has a mass of 3.44×10^{-25} kg. How many atoms are there in each ball? (b) If each atom has 82 electrons, what fraction of the electrons must be transferred from one ball to the other so that their gravitational and elecrical attrctions are equal? (Neglect the mass of the transferred electrons.)

5-73 Two identical postive charges q are separated by a distance $2a$. (a) If a third positive charge Q is placed midway between them, what is the net force on Q? (b) If the charge Q is instead placed a distance $2a$ from each of the other charges, as in Fig. 5.30, find the magnitude and direction of the net force on it.

5-74 Positive chages q and $2q$ are a distance a apart. Where can we place a third charge so that the forces due to these two charges will cancel?

c5-75 A square plate of side b has a mass m. (a) Find its moment of inertia about an axis perpendicular to the plate and passing through its center. (b) Find the moment of inertia about an axis perpendicular to the plate and passing through a corner.

c5-76 Derive the formula in Table 5.3 for the moment of inertia of a thin circular disk. [*Hint:* Break up the disk into thin rings.]

*c5-77 Using the formula in Table 5.3 for the moment of inertia of a disk, derive the formula for the moment of inertia of a sphere rotating about an axis through its center. [*Hint:* Slice the sphere into thin disks.]

5-78 Two identical solid spheres have radii R and masses m. They are attached to the ends of a bar of mass $m/5$ and length $5R$ to form a dumbbell. Find the moment of inertia of the dumbbell about an axis perpendicular to the bar through its center.

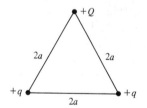

Figure 5.30.   Problem 5-73 (b).

ANSWERS TO REVIEW QUESTIONS

Q5-1, uniform circular motion; **Q5-2,** center of the circle; **Q5-3,** the center of the circle; **Q5-4,** mv^2/r; **Q5-5,** tangential acceleration; **Q5-6,** 2π; **Q5-7,** angular position, axis of rotation; **Q5-8,** angular velocity; **Q5-9,** net torque about that axis; **Q5-10,** position of the rotation axis; **Q5-11,** axis of rotation; **Q5-12,** electric forces; **Q5-13,** strong nuclear forces; **Q5-14,** repel, attract; **Q5-15,** nucleons, protons, neutrons; **Q5-16,** 10,000; **Q5-17,** quarks; **Q5-18,** strong, electromagnetic, weak, gravitational; **Q5-19,** distance squared.

SUPPLEMENTARY TOPICS

5.7 | SATELLITES; TIDES

Newton's great triumph was his demonstration that, with the laws of motion and the universal law of gravitation, the motion of the planets about the sun and of the moon about the earth could be understood in detail. He was also able to use these laws to give a qualitative explanation of the tides.

We can see what is involved in the motion of satellites by considering an artificial satellite in a circular earth orbit. Like a water pail swung in a vertical circle, the satellite has an acceleration due to gravity toward the earth. It is falling just fast enough to stay in that orbit (Fig. 5.31). We can find a formula relating the orbital radius r to the *period T*, the time needed for one full orbit.

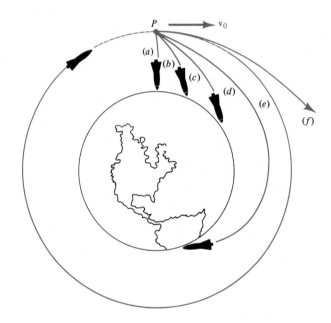

Figure 5.31.   Rockets placed at P with their engines off experience a gravitational acceleration toward the earth. Trajectory (*a*) corresponds to a rocket with no initial velocity; it falls straight down. As the rocket is given an increasingly large initial velocity $\mathbf{v}_0$ in the direction shown, the trajectory changes as indicated by paths (*b*), (*c*), and so on. Trajectory (*e*) is a closed circular orbit, corresponding to an artificial earth satellite. If the initial velocity is slightly larger than in (*e*), the rocket will move in a closed elliptical orbit. If the initial velocity is sufficiently large, the rocket will escape from the earth (*f*). Note that the rocket is always falling, just as a thrown rock falls. In a circular orbit, it moves parallel to the surface just fast enough so that its fall carries it around the earth.

If the satellite mass is m and the earth mass is M_E, the gravitational force on the satellite has a magnitude GmM_E/r^2. Since $F = ma_r$, we have

$$\frac{GmM_E}{r^2} = \frac{mv^2}{r}$$

In the period T the satellite travels a distance $2\pi r$, so its speed is $v = 2\pi r/T$, and the equation above can be rewritten as

$$\frac{GmM_E}{r^2} = \frac{m}{r}\left(\frac{2\pi r}{T}\right)^2$$

Solving for T^2, we find the relation between T and r,

$$T^2 = Cr^3 \qquad (5.22)$$

The constant C is

$$C = \frac{4\pi^2}{M_E G} \qquad (5.23)$$

Notice that C is independent of the satellite mass m. Hence the motion of the moon and of all artificial earth satellites will satisfy $T^2 = Cr^3$ with the *same* value of C.

The relation $T^2 = Cr^3$ also holds true for the planets in their nearly circular orbits about the sun. (In this case M_E in the constant C must be replaced by the solar mass, and r must be replaced by the average orbital radius.) This was one of the three laws of planetary motion discovered by Kepler in the early part of the seventeenth century from the careful analysis of observations made by earlier workers. Newton showed that all three laws could be derived using the gravitational force law and the equations of motion.

In the next example, we calculate the orbital radius of a communications satellite that has a period of exactly 1 day and consequently is always above the same point on the earth.

Example 5.14

The moon has a period $T_m = 27.3$ days and an orbital radius $r_m = 3.84 \times 10^5$ km. What is the orbital radius r_s of a satellite that has a period $T_s = 1$ day?

We can apply $T^2 = Cr^3$ to both the artificial satellite and the moon:

$$T_s^2 = Cr_s^3$$
$$T_m^2 = Cr_m^3$$

If we take the ratio of these equations, C cancels, leaving

$$\frac{T_s^2}{T_m^2} = \frac{r_s^3}{r_m^3}$$

or

$$r_s = r_m\left[\frac{T_s}{T_m}\right]^{2/3} = (3.84 \times 10^5 \text{ km})\left[\frac{1 \text{ day}}{27.3 \text{ days}}\right]^{2/3}$$
$$= 4.24 \times 10^4 \text{ km}$$

The motion of the earth's natural satellite, the moon, provided another test of Newton's ideas. Its acceleration $a_m = v^2/r$ can be calculated from its period and its distance from the earth. Using the $1/r^2$ dependence of the gravitational force and the value of g at the earth's surface, one can also predict the gravitational acceleration g' at the radius of the moon's orbit. Newton found that a_m and g' agreed well. (See Problem 5-86.)

Tides |

Newton gave the first explanation of the intervals between successive high tides, and he also accounted for their typical height. His explanation involves a rather subtle interplay between gravitational forces and circular motion.

Initially, let us imagine that the earth and the moon are isolated from the sun, and are at rest, except for the earth's daily rotation. The gravitational force exerted by the moon on the water covering most of the earth pulls it toward the side facing the moon, producing a bulge (Fig. 5.32*a*). As the earth turns, a land mass will encounter the bulge once a day, producing a daily high tide and a low tide 12 hours later.

However, high and low tides occur approximately *twice* each day. Newton realized that this happens because the moon and the earth are *both* moving in nearly circular paths about their common center of mass under the influence of their gravitational attractions. (This effect is not too obvious, since the earth is about 80 times as massive as the moon, and therefore moves in a much smaller circle.) The water furthest from the center of mass has the greatest centripetal acceleration $\omega^2 r$, but the pull of the moon is weakest there. Therefore the effective weight of the water on that side is reduced, and a second bulge develops (Fig. 5.32*b*).

This reasoning suggests high tides every 12 hours. Actually, they occur every 12 hours and 25 minutes, because the moon changes its position as the earth rotates. The gravitational pull of the sun also contributes to the tides, but this effect is less than half that due to the moon. When the moon and the sun are in

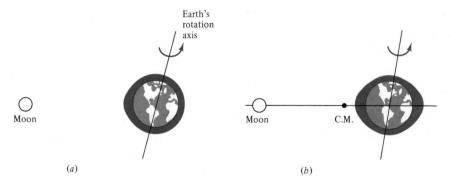

Figure 5.32. (a) The moon and the earth, not to scale. If both were at rest except for the daily rotation of the earth, water would be pulled toward the side of the earth facing the moon. (b) The moon and earth move in nearly circular orbits (perpendicular to the page) around their common center of mass, C.M. Water furthest from the C.M. has the greatest centripetal acceleration but the weakest attraction to the moon. The effective weight of that water is reduced.

line at full moon or new moon, larger than average *spring* tides occur. When their directions appear to be at right angles (quarter or three-quarter moon), smaller than average *neap* tides occur.

Newton's arguments can be made quantitative, and predict that the water will rise and fall about 0.5 m on an earth fully covered by water. This is approximately what is observed far away from land. However, near the continents, the effects of varying depths, and of bays in which the water tends to resonate much like water sloshing in a bathtub, can lead to much larger tides. Quantitative understanding of the tides remains a difficult and not fully solved problem.

5.8 | PHYSIOLOGICAL EFFECTS OF ACCELERATION

The advent of high-speed aviation and, more recently, of space travel has spurred considerable research into the physiological reactions of humans to acceleration. Some representative accelerations and their time durations are given in Table 5.4.

Most of us have experienced moderate vertical accelerations in high-speed elevators. The effects we feel are related to the fact that most, but not all, of the body is fairly rigid. The blood flows in distensible vessels, so when the body is accelerated upward, the blood accumulates in the lower body. When the acceleration is downwards, the blood volume increases in the upper body. Also, the internal organs of the body are not rigidly held in place, and there may be unpleasant feelings produced by the displacement of these organs during acceleration.

The ability of a person to withstand an acceleration depends on both the magnitude and the duration of the acceleration. Because of the inertia of the blood and distensible organs, the effects on them of moderate accelerations (a few g's) are unimportant if the acceleration lasts for only a small fraction of a second. The limit of tolerance is then some tens of g's and is set by the structural strength of the vertebrae. As the time duration increases, so also do the dangers.

Extensive studies have been made of the circulatory disturbances of pilots subjected to accelerations lasting a few seconds or longer. Pilots in an airplane pulling out of a dive for several seconds may experience two successive types of *blackouts*. A visual blackout occurs first at about $3g$ as a result of reduced blood pressure in the retina, which is very sensitive to oxygen deprivation. The blood pressure drops because the heart has difficulties in pumping the blood with its increased effective weight. By modifying the pilot's position, training him to tense his abdominal muscles, and equipping him with a flight suit that reduces blood pooling in the lower body, the visual blackout threshold can be increased to about $5g$. In addition, the reduced supply of blood to the brain leads to a complete blackout or unconsciousness at about $6g$. Since many high-performance planes can sustain about $9g$ when pulling out of a dive, the limits of human tolerance may easily be exceeded.

When a plane climbs and then circles downward, the plane and pilot are both more susceptible than they are at the bottom of a dive. An engorgement of the vessels in the head causes a reduction of heart

TABLE 5.4

The approximate duration and magnitude of some brief accelerations in multiples of the gravitational acceleration, $g = 9.8 \text{ m s}^{-2}$

Type of Acceleration	Acceleration in Multiples of g	Duration (seconds)
Elevators		
Fast service	0.1–0.2	1–5
Comfort limit	0.3	
Emergency stop	2.5	
Automobiles		
Comfortable stop	0.25	5–8
Very unpleasant	0.45	3–5
Maximum possible	0.70	3
Crash (possibly survivable)	20–100	0.1
Aircraft		
Normal takeoff	0.5	10–20
Catapult takeoff	2.5–6	1.5
Crash landing (possibly survivable)	20–100	
Seat ejection	10–15	0.25
Humans		
Parachute opening	8–33	0.2–0.5
Parachute landing	3–4	0.1–0.2
Fall into firemen's net	20	0.1

Adapted fom D.E. Goldman and H.E. von Gierke in Harris & Crede (eds.), *Shock and Vibration Handbook,* McGraw-Hill, New York, 1961, Chapter 44.

activity and hence a reduction of the oxygen supplied to the retina and brain. Planes are usually not designed to withstand as large a stress in this maneuver as in a dive, so they also will be structurally sound only under lower g stresses. An illustration of the upward effective weight attained during some maneuvers is provided by the next example.

Example 5.15

An airplane of mass m is flown at a speed of 300 m s^{-1} in a vertical circle (Fig. 5.33). What is the minimum radius r of the circle so that the upward effecive weight does not exceed $3mg$?

Since the effective weight $\mathbf{w}^e$ is now upward and equal to $3mg$ in magnitude, $\mathbf{w}^e = 3mg$. According to Eq. 5.6, $\mathbf{w}^e = m\mathbf{g} - m\mathbf{a}$, so

$$-3mg = mg - ma$$

or $\mathbf{a} = 4\mathbf{g}$. The acceleration here is the centripetal acceleration, so we have

$$\frac{v^2}{r} = 4g$$

Thus the radius that will lead to an upward effective weight of $3mg$ is

$$r = \frac{v^2}{4g} = \frac{(300 \text{ m s}^{-1})^2}{4(9.8 \text{ m s}^{-2})} = 2300 \text{ m}$$

A smaller radius will lead to a larger effective weight.

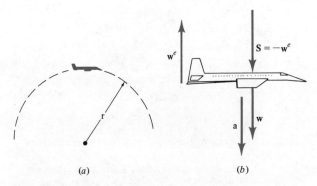

Figure 5.33. $\mathbf{S}$ is the downward force exerted by the air on the airplane.

5.9 | SENSORY PERCEPTION OF ANGULAR MOTION

If a person is blindfolded and put on a chair that can rotate smoothly, the following results are observed.

1   If the chair is turned 90° clockwise and stopped, the person can identify the direction of rotation and the fact that it has stopped. Reversal of the process yields the same results.

2   If the chair is rapidly accelerated and the rotation is continued, the person can correctly identify the direction of rotation for about 20 s. After 20 s the answers are uncertain.

3   If after about 30 s the chair is slowed to a lower angular velocity, the person will generally indicate that motion has stopped.

4   If the chair is then stopped, the person usually perceives rotation in the opposite direction.

To interpret these results we must examine the structure of the inner ear. The inner ear is composed of two portions (Fig. 5.34). One portion, the *cochlea*, contains the auditory response elements. The other portion if comprised of three *semicircular canals*, which primarily detect motion of the head and have little auditory function.

The structure of one of the semicircular canals is shown in Fig. 5.35. The canal contains a watery fluid called the *endolymph* and has a projection somewhat like a swinging door, called the *cupola*, which senses relative motion of the fluid. To understand the operation of the canal, consider what happens to a bucket of water suspended from a rope when it is set into rotation (Fig. 5.36). The water is initially at rest and it

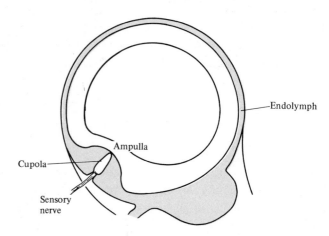

Figure 5.35.   A semicircular canal of a human. The endolymph is a fluid that can flow around the canal. The ampulla is a swelling into which the cupola projects. The cupola can block the entire canal, but it is elastic and will bend when the fluid moves. When the cupola is pushed aside by the flowing fluid; the sensory nerve detects this motion and the information is transmitted to the brain.

remains at rest when the bucket starts spinning, since the frictional force between the wall of the bucket and the water is small. However, the torque resulting from this force gradually increases the angular velocity of the water, and after some seconds the water is rotating with the same angular velocity as the bucket. If the bucket is now stopped, the water continues to rotate for some time. The *relative* motion of the water and bucket is now opposite to that when the bucket was first set into motion.

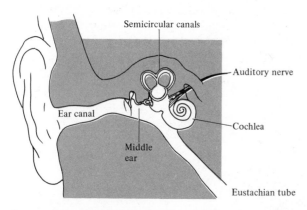

Figure 5.34.   A schematic view of the right ear of a human. The three mutually perpendicular semicircular canals function as indicators of rotations about three perpendicular axes in space.

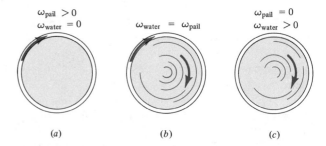

Figure 5.36.   (*a*) A bucket of water is spun, with the water initially at rest. (Small scraps of paper on the surface make its motion more readily visible.) The water remains nearly at rest for a few seconds, and the pail is moving *clockwise* relative to the water. (*b*) After a while the frictional torques have increased the angular velocity of the water and it has the same angular velocity as the bucket. (*c*) When the bucket is stopped the water continues to move for some time until slowed by the frictional torques. Relative to the water, the pail is now moving *counterclockwise*.

In the inner ear, it takes a second for the endo-lymph to start moving at about the speed of the canal, so initially the cupola is deflected by its motion relative to the fluid. This deflection is sensed by the nerve, and information is transmitted to the brain. Once the canal, fluid, and cupola are all rotating at the same rate, the cupola starts to return to its normal position. However, the elastic or spring force on the cupola is so weak that it takes about 20 s for this return to be completed. The perception of rotation persists during this 20-s period, but afterward the ear can no longer detect rotation.

If the rotation rate is now reduced, the canal slows while the fluid continues to move briefly at the higher speed. Thus the cupola is deflected in the opposite direction, and the subject perceives a change in the direction of rotation. If this slower rotation continues for about 20 s, the cupola once more returns to its normal position, and the subject again perceives no rotation. Finally, if the rotation is stopped, the cupola is again deflected, and the subject perceives a rotation opposite to the original direction.

When a subject is put through the same sequence but is not blindfolded, the eyes do not normally stare directly outward at passing objects. Instead, the eyes focus briefly on a given object, then fix on another object, and so on. This is called the *nystagmus reflex*. The flicking of the eyes will continue as long as the sensory nerve detects a bending of the cupola. When the cupola resumes its normal position, the reflex action of the eyes stops and the subject, if untrained, sees objects streaking by. Then balance is lost.

When the rotation is stopped, the cupola is again deflected and the nystagmus reflex acts, this time in the opposite direction. Objects at rest appear to be moving steadily and then moving back again. The postural muscles are then trying to respond to nonexistent motion, resulting in staggering.

EXERCISES ON SUPPLEMENTARY TOPICS

Section 5.7 | Satellites; Tides

5-79 An artificial earth satellite moves in a circular orbit with a radius one-fourth that of the moon's orbit. What is its period?

5-80 An artificial satellite is to be placed in an orbit around the sun so that its period is 8 earth years. By definition, the radius of the earth's orbit is 1 astronomical unit (A.U.). What is the radius of the satellite's orbit in A.U.?

5-81 The distance from the sun to the earth is 1 astronomical unit. What is the length of a "year" on a planet 9 A.U. from the sun?

5-82 The average distance from Mars to the sun is 1.524 times the distance from the Earth to the sun. How long does it take for Mars to go around the sun?

5-83 Using the data on the inside back cover, find the time an artificial satellite would need to go once around the sun in an orbit whose radius is twice that of the sun.

5-84 The period of the moon's orbit around the earth is 27.3 days, and its mean distance is 3.84×10^8 m. From these data, determine the mass of the earth.

5-85 Newton's theory of the tides predicts that the difference in the height of the oceans at high and low tides should be $h = 3GMR_E{}^2/2gr^3$, where G is the gravitational constant, M is the mass of the moon, R_E is the average radius of the earth, g is the acceleration due to gravity at the surface of the earth, and r is the earth-moon distance. Using the numerical values listed on the inside back cover, evaluate h.

PROBLEMS ON SUPPLEMENTARY TOPICS

5-86 The orbital radius of the moon is 3.84×10^5 km and its period is 27.3 days. (a) Find its acceleration, a_m. (b) The gravitational acceleration at the earth's surface, which is 6380 km from its center, is $g = 9.81$ m s^{-2}. Using the $1/r^2$ dependence of the gravitational force law, what gravitational acceleration g' would one expect at the radius of the moon's orbit? (c) Compare g' and a_m. (This was one of Newton's original tests of the universal law of gravitation.)

5-87 Suppose the gravitational force were proportional to $1/r^3$ instead of $1/r^2$. What would be the relation between the period of a planet and its orbital radius? Compare this result with the observed relation, $T^2 = Cr^3$.

Additional Reading

D.E. Goldman and H.E. von Gierke, in Cyril M. Harris and Charles E. Crede (eds.), *Shock and Vibration Handbook*, McGraw-Hill Book Co., New York, 1961, Chapter 44, Effects of Shock and Vibration on Humans.

Otto Glasser (ed.), *Medical Physics,* vol. I, The Year Book Publishers, Inc., Chicago, 1944, p. 22. Effects of acceleration.

J.L.E. Dreyer, *Tycho Brahe: A Picture of Scientific Life and Work in the Sixteenth Century,* Dover Publication Inc., New York, 1963. Brahe and his observations of planetary motions.

Judah Levine, The Earth Tides, *The Physics Teacher,* vol. 20, 1982, p. 588.

Scientific American articles:

J.W. Beams, Ultrahigh-Speed Rotation, April 1961, p. 134.

Terence A. Rogers, The Physiological Effects of Acceleration, February 1962, p. 60.

Gerald Feinberg, Ordinary Matter, May 1967, p. 126.

Thomas C. Van Vlandern, Is Gravity Getting Weaker? February 1976, p. 44.

Richard P. Post and Stephen F. Post, Flywheels, December 1973, p. 17.

Curtis Wilson, How Did Kepler Discover His First Two Laws?, March 1972, p. 92.

UNIT TWO

ADDITIONAL TOPICS IN MECHANICS

(Alan Carey/The Image Works).

The principles described in Unit One form the basis for analyzing all mechanics problems. In particular, Newton's laws of motion tell us how to use the forces on objects to predict their motion. There are, however, large classes of problems for which it is either difficult to determine all the forces when a skier descends an uneven slope, or two objects collide, or a skater begins a rapid spin. Nevertheless, there are convenient ways based on *conservation laws* to handle many such problems.

In the first chapter of this unit, we define *work* and *energy,* and show that the work done on an object equals the change in its energy. When no work is done, the energy remains constant, and it is said to be *conserved.* In such a situation, if we know the initial energy of an object—or a system of objects—we can immediately make some statements about its state of motion at a later time, even though we may lack detailed information about the forces that are present.

In Chapter Seven, we discuss *linear* and *angular momentum.* When no net force acts on a system, its linear momentum is conserved, and when no net torque acts, its angular momentum is conserved. These conservation laws make it possible to partially or fully analyze some rather difficult looking problems. For example, linear momentum is conserved in collisions between two objects—two cars, a bat and a ball, two astronomical bodies—where the external forces due to other objects are either absent or negligible in comparison to the very large forces the colliding objects exert on each other. Similarly, the fact that the angular momentum of a skater, a diver, or a gymnast remains constant during certain maneuvers helps to explain what they can and cannot accomplish.

Unlike the idealized rigid bodies used to illustrate the principles of mechanics, real objects may deform significantly or break if they are subjected to large forces or torques. In Chapter Eight, we see how the strength of an object depends on its size, shape, and composition. Construction materials, bones, and trees provide obvious examples of the importance of these questions.

The final chapter of this unit applies Newton's laws of motion to mechanical oscillations and vibrations. From a knowledge of the forces, we can tell whether vibrational motion may occur, and predict its frequency. The general features of this kind of motion are usually similar, whether we are dealing with molecules in a solid oscillating about their equilibrium positions, human vocal cords vibrating to produce sound, or a spider web vibrating as an insect struggles.

This unit concludes our formal development of mechanics, but it is by no means the end of its usefulness to us. Throughout this text, we make explicit or implicit use of the principles of mechanics. Since these first two units are based almost entirely on Newton's laws, the general range of application of mechanics is as broad as the various forces that appear in nature. Even in our discussion of atoms and molecules, where Newton's laws break down, many of the principles we have discussed will survive. In particular, the conservation laws will remain as universally valid concepts.

CHAPTER 6

WORK, ENERGY, AND POWER

Work, energy, and *power* are words with a variety of meanings in everyday usage. However, to the scientist these terms have very specific definitions. In this chapter, we consider these definitions and the relationships between work and various kinds of energy for mechanical systems. Although these relationships are obtained from Newton's laws, they can often be readily used when the forces are unknown or when the systems are so complex as to make the direct use of Newton's laws impracticably difficult.

This chapter also provides us with our first look at a *conservation law.* We find that under certain conditions the mechanical energy of a system is *constant* and is then said to be *conserved.* This fact provides a very powerful tool for understanding and solving certain mechanical problems.

However, it is now known that a much broader energy conservation law is valid in nature. If we calculate or measure the *total energy*—mechanical, electrical, thermal, and so on—then this total energy is constant, even though any one type need not be conserved. What is observed in nature is an interchange of energy from one form to another, with the *sum* remaining constant. This total energy conservation law became fully understood when Einstein showed that matter and energy are two forms of the same quantity. He saw that not only may energy be converted from one form to another but it may also be converted to matter and vice versa.

Energy is a concept that plays a key role in an enormous range of applications. Biological processes, the weather, the evolution of astronomical systems, chemical reactions—these are all constrained by the fact of energy conservation and the limitations on how energy may be used and transformed. Some of these limitations are considered in Chapter Eleven, when we study thermal energy, the energy associated with the motions of the molecules in an object. Also, we have become increasingly aware of the importance of energy in our technological society. We will see that the basic principles developed in this chapter can help us to understand the possibilities and difficulties associated with alternatives to our dwindling fossil fuel resources.

6.1 | WORK

In this section, we define the work done on an object by a force and show how it is calculated. It becomes clear in the following sections that the concept of work plays a fundamental role in the analysis of many mechanical problems.

Suppose that an object is displaced a distance **s**, and that a force **F** acting on the object has a constant component F_s along **s** (Fig. 6.1). Then the *work done by the force* is defined as the product of the force component and the displacement,

$$W = F_s s \qquad (6.1)$$

If **F** is at an angle θ to **s**, as in Fig. 6.2, then $F_s = F \cos \theta$, and the work can be written as

$$W = Fs \cos \theta \qquad (6.2)$$

The S.I. unit of work is the *joule* (J). Since work has dimensions of force times distance, a joule is a newton-metre.

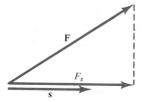
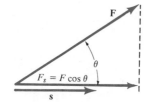
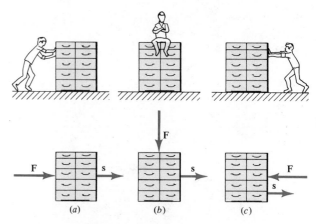

Figure 6.1. The work done by the force **F** during a displacement **s** is $W = F_s s$.

Figure 6.2. The component of **F** along **s** is $F_s = F\cos\theta$. The work is then $W = Fs\cos\theta$.

Figure 6.3. A force is applied (a) parallel to the motion, (b) perpendicular to the motion, and (c) opposite to the motion. In each case the work done by the force **F** is different. In part (c) we may imagine that the dresser is being slowed and brought to rest. In (a) the man does work on the dresser; in (b) he does no work; and in (c) the dresser does work on the man.

The definition of work can be stated conveniently in terms of a product of two vectors called the *scalar* or *dot* product. Suppose we have two vectors **A** and **B**, with an angle θ between them. The scalar product **A · B** is the product of their magnitudes and of $\cos\theta$,

$$\mathbf{A} \cdot \mathbf{B} = AB\cos\theta \qquad \text{scalar product} \qquad (6.3)$$

Note that **A · B** is an ordinary number or scalar, unlike the cross product **A × B** which is a vector. Thus $W = F_s s$ can be rewritten as

$$W = \mathbf{F} \cdot \mathbf{s} \qquad (6.4)$$

Notice that our definition of work differs in some ways from the nontechnical meaning of the word. According to Eq. 6.2, we do twice as much work in pushing an object along the floor if we double the force or the distance it moves. This is consistent with the everyday notion of work. However, this is not so if we stand in one spot supporting a heavy weight. We would consider that we are working rather hard, but since there is no displacement, we conclude that no work is done on the weight.

However, work is being done in the body as nerve impulses repeatedly trigger muscle fiber contractions. Unlike a bone or a steel post, a muscle fiber cannot sustain a static load. Instead, it must repeatedly relax and contract, doing work in each contraction. We are unaware of this process because of the large number of fibers and the rapidity of the contractions.

In the following examples we show how the work done by a force acting on an object is computed.

Example 6.1

A 600-N force is applied by a man to a dresser that moves 2 m. Find the work done if the force and displacement are (a) parallel; (b) at right angles; (c) oppositely directed (Fig. 6.3).

(a) When **F** and **s** are parallel, $\cos\theta = \cos 0° = 1$, and

$$W = \mathbf{F} \cdot \mathbf{s} = Fs\cos\theta = (600\text{ N})(2\text{ m})(1) = 1200\text{ J}$$

The man does 1200 J of work on the dresser. Since **F** is parallel to **s**, $F_s = F$, and we obtain the same result using $W = F_s s$.

(b) When **F** is perpendicular to **s**, $\cos\theta = \cos 90° = 0$, and $W = 0$. No work is done when the force is at right angles to the displacement, since $F_s = 0$.

(c) When **F** and **S** are opposite, $\cos\theta = \cos 180° = -1$, and

$$W = \mathbf{F} \cdot \mathbf{s} = Fs\cos\theta = (600\text{ N})(2\text{ m})(-1) = -1200\text{ J}$$

In this case the work done by the force is negative, so *the object is doing work on the man.* Note that here **F** is opposite to **s**, so $F_s = -F$.

Example 6.2

_____ ulls a barge along a canal with a rope in which the tension is 1000 N (Fig. 6.4). The rope is at an angle of 10° with the towpath and the direction of the barge. (a) How much work is done by the horse in pulling the barge 100 m upstream at a constant velocity? (b) What is the net force on the barge?

(a) The work done by the constant force **T** in moving the barge a distance **s** is given by $W = \mathbf{T} \cdot \mathbf{s} = Ts\cos\theta$,

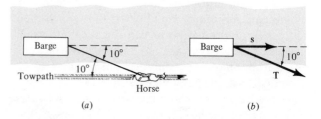

Figure 6.4. (a) A horse pulling a barge at constant velocity. (b) The tension in the rope is **T**.

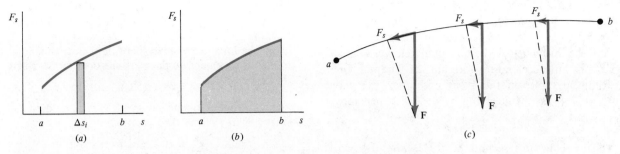

Figure 6.5. (a) The work $\Delta W_i = F_s \Delta s_i$ in a small displacement Δs_i is approximately equal to the area of the rectangle. (b) The work done when the object is moved from a to b is the area under the curve. (c) The curved path followed by a projectile. The force **F** due to gravity is constant, but the component F_s varies because the displacements change direction.

where θ is the angle between **T** and **s**. Using $\cos 10° = 0.985$,

$$W = (1000 \text{ N})(100 \text{ m})(0.985) = 9.85 \times 10^4 \text{ J}$$

(b) Since the barge moves at a constant velocity, the sum of all the forces on it must be zero. There must be another force acting that is not shown in Fig. 6.4, a force exerted on the barge by the water that is equal in magnitude and opposite to **T**.

In this example the net work done by all the forces acting on the barge is zero because the net force is zero. The work done by the force due to the water is -9.85×10^4 J. In other words, the barge does 9.85×10^4 J of work on the water.

In our equation defining work, $W = F_s s = \mathbf{F} \cdot \mathbf{s}$, we assumed that the force component F_s was constant over the displacement. In many situations, this is, at best, only approximately true. If the force varies in magnitude or direction relative to the displacement, we must consider many small displacements. In each of these F_s is approximately constant. Then $\Delta W_i = \mathbf{F} \cdot \Delta \mathbf{s}_i$ is the work done in one small displacement, and the total work done is the sum $\Sigma \Delta W_i$. In the limit as the displacements approach zero, the sum becomes an integral. Thus the work done when the object is moved from point a to point b is the integral

$$W = \int_a^b \mathbf{F} \cdot \mathbf{ds} = \int_a^b F_s \, ds \qquad (6.5)$$

This integral has a direct interpretation in terms of a graph of F_s versus s. The work $\Delta W_i = F_s \Delta s_i$ done in a small displacement Δs_i is approximately equal to the area of the colored rectangular strip (Fig. 6.5a). The work done when the object is moved from a to b is the area under the curve between these two points (Fig. 6.5b).

Equation 6.5 can be used to calculate the work W even if as in Fig. 6.5c the path is not a straight line.

The dot product $\mathbf{F} \cdot \mathbf{ds} = F_s ds = F ds \cos \theta$ must be calculated everywhere along the path in order to evaluate the integral. Technically, the integral in Eq. 6.5 is then a *line integral*.

One other point should be made about computing the total or net work done on an object. If several forces act on the object, we can find their vector sum as usual to find the net force, and then calculate the work done by this net force. Alternatively, we can calculate the work done by each of the forces, and sum these scalar quantities. Either procedure gives the same result.

6.2 | KINETIC ENERGY

The *kinetic energy* of an object is a measure of the work an object can do by virtue of its motion. As we show in the following, the translational kinetic energy of an object of mass m and velocity $\mathbf{v}$ is $\frac{1}{2}mv^2$.

The work done on an object and its kinetic energy obey the following fundamental principle:

The final kinetic energy of an object is equal to its initial kinetic energy plus the total work done on it by all the forces acting upon it.

This work-energy principle is derived at the end of this section quite generally from Newton's laws of motion. However, it is instructive to see first how it comes about in a simple situation.

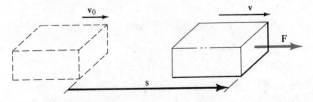

Figure 6.6. A force **F** does work on an object as it moves a distance **s**. The velocity changes from $\mathbf{v}_0$ to $\mathbf{v}$.

Consider an object of mass m subjected to a constant force $\mathbf{F}$ (Fig. 6.6). The object moves a distance $\mathbf{s}$ parallel to $\mathbf{F}$. Since its acceleration $\mathbf{a} = \mathbf{F}/m$ is constant, the initial velocity $\mathbf{v}_0$ and final velocity $\mathbf{v}$ satisfy the constant acceleration formulas of Chapter One. From the table on the front endpapers we have $v^2 = v_0^2 + 2as$. Multiplying by $m/2$, this becomes

$$\tfrac{1}{2}mv^2 = \tfrac{1}{2}mv_0^2 + mas \qquad (6.6)$$

Using Newton's second law, $\mathbf{F} = m\mathbf{a}$, the work done by the force $\mathbf{F}$ is $W = Fs = mas$, which is the last term in Eq. 6.6. The final kinetic energy K and the initial kinetic energy K_0 are defined by

$$K = \tfrac{1}{2}mv^2 \quad \text{and} \quad K_0 = \tfrac{1}{2}mv_0^2 \qquad (6.7)$$

Hence we can rewrite Eq. 6.6 as

$$K = K_0 + W \qquad (6.8)$$

Therefore, the final kinetic energy is equal to the initial kinetic energy of the object plus the work done on it. Note that work and kinetic energy have the same dimensions and units.

One consequence of Eq. 6.8 is that if work is done on the object, its kinetic energy increases. Conversely, if the object does work on an external agency, its kinetic energy decreases. This is illustrated by the work that is done on a person who slows down or stops a moving object. The following examples should serve to clarify these ideas.

Example 6.3

A woman pushes a toy car, initially at rest, toward a child by exerting a constant horizontal force $\mathbf{F}$ of magnitude 5 N through a distance of 1 m (Fig 6.7a). (a) How much work is done on the car? (b) What is its final kinetic energy? (c) If the car has a mass of 0.1 kg, what is its final speed? (Assume no work is done by frictional forces.)

(a) The force the woman exerts on the car is parallel to the displacement, so the work she does on the car is

$$W = \mathbf{F} \cdot \mathbf{s} = Fs = (5 \text{ N})(1 \text{ m}) = 5 \text{ J}$$

(b) The initial kinetic energy K_0 is zero, so the final kinetic energy of the car is

$$K = K_0 + W = 0 + (5 \text{ J}) = 5 \text{ J}$$

(c) The final kinetic energy is $K = \tfrac{1}{2}mv^2$, so

$$v = \sqrt{\frac{2K}{m}} = \sqrt{\frac{(2)(5 \text{ J})}{0.1 \text{ kg}}} = 10 \text{ m s}^{-1}$$

Example 6.4

In the preceding example the woman releases the toy car with a kinetic energy of 5 J. It moves across the floor and reaches the child who stops the car by exerting a constant force $\mathbf{F}'$ opposite to its motion. The car stops in 0.25 m. Find $\mathbf{F}'$ (Fig. 6.7b) if no work is done on the car by frictional forces.

While the car is moving toward the child, no work is done on the car, and its kinetic energy remains 5 J until it reaches the child. The initial kinetic energy K_0 is 5 J, and the final kinetic energy K is zero, since the car comes to rest, so

$$W = K - K_0 = 0 - 5 \text{ J} = -5 \text{ J}$$

Since $\mathbf{F}'$ is opposite to $\mathbf{s}'$, the work done is $W = -F's'$. Thus

$$F' = -\frac{W}{s'} = -\frac{(-5 \text{ J})}{(0.25 \text{ m})} = 20 \text{ N}$$

The negative sign for the work done, $W = -5$ J, indicates that *the car does work on the child*. These two examples show that positive work done on an object gives it kinetic energy, which is then available to do work, in this case on the child.

We now derive the work-energy principle without making the constant force assumption. Suppose a small displacement $d\mathbf{s}$ of an object is parallel to the force $\mathbf{F}$ acting on it, so $\mathbf{F} \cdot d\mathbf{s} = Fds$. According to the second law of motion, $F = ma$. Using the chain rule for differentiation, we can rewrite the acceleration a as

$$a = \frac{dv}{dt} = \frac{dv}{ds}\frac{ds}{dt} = \frac{dv}{ds}v = \frac{1}{2}\frac{d}{ds}(v^2)$$

Then the work can be written as

$$W = \int_a^b F ds = \int_a^b mads = \int_a^b \frac{m}{2}\frac{d}{ds}(v^2)ds$$
$$= \frac{m}{2}v_b^2 - \frac{m}{2}v_a^2$$

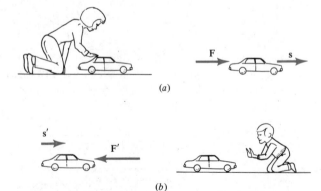

Figure 6.7. (a) A woman pushes a toy car to the right. She exerts a horizontal force $\mathbf{F}$ on the car, which is parallel to its displacement $\mathbf{s}$. (b) A child stopping the car exerts a force $\mathbf{F}'$ on the car. $\mathbf{F}'$ is opposite to the displacement $\mathbf{s}'$ of the car as it is brought to rest.

Thus the work done equals the change in kinetic energy when **F** is parallel to the displacement but variable in magnitude.

If the force is not parallel to the displacement, only its component F_s along the motion affects the speed. The component perpendicular to v changes its direction, but not its magnitude. Hence the derivation still applies if F is replaced by F_s.

6.3 | POTENTIAL ENERGY AND CONSERVATIVE FORCES

The work-energy relation of the preceding section includes the work done by *all* the forces acting on the object. However, it is useful when gravitational forces do work to treat them separately, and to refer to the other forces acting on the object as *applied forces,* for want of a better name. The work done by gravity on the object can be taken care of very conveniently if we introduce another form of energy called *potential energy.* Forces that can be dealt with in this way are called *conservative forces.* The condition a force must satisfy to be conservative is discussed at the end of this section.

To motivate the introduction of the gravitational potential energy, consider a ball thrown straight up. Its velocity decreases steadily as it rises. From the point of view of this chapter, the gravitational force $m\mathbf{g}$ is doing negative work, since it is opposite to the displacement, and the kinetic energy is diminishing correspondingly. Once the ball starts to come down, the gravitational force does an equally large amount of positive work, and the kinetic energy returns to its initial value once the ball returns to its starting point.

Alternatively, we can think of the rising ball losing kinetic energy and gaining *potential energy.* This potential energy is converted back into kinetic energy when the ball falls. In general, potential energy is

energy associated with the position or configuration of a mechanical system. The potential energy can, at least in principle, be converted into kinetic energy or used to do work.

We now put these ideas into quantitative form. In Fig. 6.8a, a ball rises from an initial height h_0 to a height h. The gravitational force $m\mathbf{g}$ is opposite in direction to the displacement $s = (h - h_0)$, so the work done is *negative:*

$$W(\text{grav}) = -mg(h - h_0)$$

However, according to our discussion the potential energy increases in this situation, and the change in potential energy $\Delta \mathcal{U} = \mathcal{U} - \mathcal{U}_0$ is *positive.* The magnitude of $\Delta \mathcal{U}$ is defined to be equal to the magnitude of $W(\text{grav})$, so we write

$$\mathcal{U} - \mathcal{U}_0 = -W(\text{grav}) \qquad (6.9)$$

The minus sign takes care of the differences in sign. Using our expression for $W(\text{grav})$, Eq. (6.9) becomes

$$\mathcal{U} - \mathcal{U}_0 = mg(h - h_0) \qquad (6.10)$$

This result for the *potential energy change* involves a difference of two terms on each side, and suggests that we define the *potential energies themselves* at h and h_0, respectively, by

$$\mathcal{U} = mgh \qquad \text{and} \qquad \mathcal{U}_0 = mgh_0 \qquad (6.11)$$

As we see shortly, these definitions do not uniquely specify the potential energies, but they are nevertheless very useful.

By the work-energy principle, $K = K_0 + W(\text{grav}) = K_0 - (\mathcal{U} - \mathcal{U}_0)$, so we have the important result

$$K + \mathcal{U} = K_0 + \mathcal{U}_0 \qquad (W_a = 0) \qquad (6.12)$$

The notation $W_a = 0$ reminds us that the work done by applied forces is zero; only the gravitational force is doing work here. The sum of the kinetic energy and

(a) (b)

Figure 6.8. (a) A ball rises a distance $s = h - h_0$. The gravitational force does work $W(\text{grav}) = -mg(h - h_0)$. (b) A block slides up an inclined plane. The component of **w** along **s** is $w_s = -mg \cos \theta$, so $W(\text{grav}) = -mgs \cos \theta$. Since $s \cos \theta = h - h_0$, we again find $W(\text{grav}) = -mg(h - h_0)$.

the potential energy is called the *total mechanical energy*,

$$E = K + \mathcal{U} \qquad (6.13)$$

Thus Eq. 6.12 means that *when there is no work done by applied forces, the total mechanical energy is constant or conserved.*

If applied forces also do work, Eq. 6.12 must be generalized to include this work, W_a. Then we have

$$K + \mathcal{U} = K_0 + \mathcal{U}_0 + W_a \qquad (6.14)$$

or

$$E = E_0 + W_a \qquad (6.15)$$

The final mechanical energy $E = K + \mathcal{U}$ is equal to the initial mechanical energy $E_0 = K_0 + \mathcal{U}_0$ plus the work done by the applied forces.

Several points should be noted in interpreting these results. First, we considered a ball moving straight upward. However, even when an object follows a more complex path, the work done by the gravitational force is still $-mg(h - h_0)$, so *the potential energy change $\Delta \mathcal{U} = \mathcal{U} - \mathcal{U}_0$ depends only on the difference in heights.* This is verified, for example, in Fig. 6.8b for an object moving up an inclined plane.

Second, suppose we measure our heights from a different level so that h and h_0 change by the same amount. Then $\Delta \mathcal{U} = mg(h - h_0)$ remains the same, even though $\mathcal{U}$ and $\mathcal{U}_0$ are changed. Accordingly, we can measure heights from any convenient reference level: the ground, the top of a building, and so on.

Third, if an object goes high enough, the gravitational force is not constant, and mgh is no longer the appropriate form for the potential energy. Nevertheless, the potential energy change is still defined as the negative of the work done by the gravitational force, as in Eq. 6.9. A similar definition of $\Delta \mathcal{U}$ also applies for other conservative forces. These points are considered later in this chapter.

Finally, we must remember that $E = E_0 + W_a$ is just another form of our original work-energy principle, $K = K_0 + W$, which involves the work done by *all* the forces. In $E = E_0 + W_a$, the work done by the gravitational forces is still present, but it is now separated from the work W_a done by the other "applied" forces and taken into account through the potential energy.

The following example shows how energy conservation can be used to solve an otherwise complex problem.

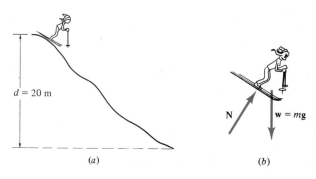

Figure 6.9. The forces on the skier are a normal force **N** and the weight **w** = mg.

Example 6.5

A woman skis from rest down a hill 20 m high (Fig. 6.9). If friction is negligible, what is her speed at the bottom of the slope?

The forces acting on the skier are her weight and a normal force due to the ground. The effects of the weight (the gravitational force) are contained in the potential energy, and the normal force does no work because it is perpendicular to the displacement. Thus, no work is done by applied forces, and the total energy $E = K + \mathcal{U}$ is constant.

Since we can choose the reference level for measuring potential energy as we wish, we choose the bottom of the slope as the level at which $\mathcal{U} = 0$. The kinetic energy at the top is $K_0 = 0$, since she starts from rest; her potential energy there is $\mathcal{U}_0 = mgd$. Her final kinetic energy at the bottom of the hill is $K = \frac{1}{2}mv^2$, and her final potential energy is $\mathcal{U} = 0$. Thus $K + \mathcal{U} = K_0 + \mathcal{U}_0$ becomes

$$\tfrac{1}{2}mv^2 + 0 = 0 + mgd$$

Her speed v at the bottom of the slope is then

$$v = \sqrt{2gd} = \sqrt{2(9.8 \text{ m s}^{-2})(20 \text{ m})}$$
$$= 19.8 \text{ m s}^{-1}$$

This example shows the advantage of the energy conservation approach to solving mechanics problems. We cannot use $\mathbf{F} = m\mathbf{a}$ directly unless we know the exact shape of the slope so that we can compute the force. Even with this information the calculation would be difficult. Energy conservation immediately tells us the velocity at any height.

Conservative Forces |
The gravitational force has the interesting property that when an object moves from one point to another, the work done by the force does not depend on the choice of path. For example, in Fig. 6.10 the work done by gravity when an object moves from B to C is $-mg(h - h_0)$. No

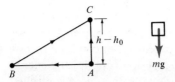

Figure 6.10. The work done by gravity is the same for the paths *ABC* and *AC*. Whenever the work done by a force is the same for all paths, the force is said to be conservative, and its effects can be included in the potential energy.

work is done by gravity when the object moves horizontally from *A* to *B*, so the total work done by gravity along the path *ABC* is $-mg(h - h_0)$. When the block is moved vertically from *A* to *C*, the work done by gravity again is $-mg(h - h_0)$. Therefore the work is the same for both paths.

Any force that has the property that the work it does is the same for all paths between any two given points is said to be a conservative force. This property makes it meaningful to associate a potential energy with a position. Gravitational, electrical, and spring forces are examples of conservative forces; friction and many other forces are not conservative. The effects of *any* conservative force can always be taken into account by introducing a suitable potential energy term.

6.4 | DISSIPATIVE FORCES

We have seen that the work done by conservative forces can be handled conveniently by the introduction of the potential energy concept. This is not true for frictional forces, and they must be treated as applied forces.

Frictional forces are not conservative, since the work done by friction depends on the path. Also, friction always opposes the motion of an object, so it always does negative work. The energy expended by an object against the frictional forces is usually converted into thermal energy and hence is lost as mechanical energy, as in the next example.

Example 6.6

Suppose that as in the preceding example, a woman skis down a 20-m-high hill. However, this time frictional forces are not negligible, so her speed at the bottom of the hill is only 10 m s^{-1}. How much work is done by frictional forces if her mass is 50 kg?

Again we choose the bottom of the hill as our reference level for calculating potential energies, which means her final potential energy is $\mathcal{U} = 0$. Also, her initial kinetic energy is still $K_0 = 0$. Using $E = E_0 + W_a$, we have in this case

$$\tfrac{1}{2}mv^2 + 0 = 0 + mgd + W_a$$
$$W_a = \tfrac{1}{2}mv^2 - mgd$$
$$= \tfrac{1}{2}(50 \text{ kg})(10 \text{ m s}^{-1})^2$$
$$\quad - (50 \text{ kg})(9.8 \text{ m s}^{-2})(20 \text{ m})$$
$$= -7300 \text{ J}$$

As anticipated, the work done by the applied force is negative, since frictional forces always oppose the motion. The skier has done 7300 J of work against friction, and this mechanical energy has been converted into thermal energy.

The frictional force on a moving object was written earlier as the coefficient of kinetic friction times the normal force. (Since work is done only on moving objects, it is the coefficient of kinetic friction that must always be used. When an object rolls, the point of contact between the object and the supporting surface is instantaneously at rest. In this special case the frictional force does no work.) The frictional force is always directed opposite to the motion, so if an object moves a distance *s* against a frictional force $\mu_k N$, the mechanical energy dissipated is

$$W_a = -\mu_k Ns \qquad (6.16)$$

We illustrate this in the following example.

Example 6.7

A skier reaches the flat ground at the bottom of the slope with a speed of 19.8 m s^{-1} and then, by turning her skis sideways, quickly comes to a stop. If the coefficient of kinetic friction is 2.5, how far will she skid before coming to a halt?

Since the ground is level, there is no potential energy change and all her kinetic energy, $\tfrac{1}{2}mv^2$, must be dissipated. The normal force is equal and opposite to her weight, so the work done by the frictional force over a distance *s* is $W_a = -\mu_k mgs$. Thus $E = E_0 + W_a$ becomes

$$0 = \tfrac{1}{2}mv^2 - \mu_k mgs$$

and solving for *s*,

$$s = \frac{v^2}{2g\mu_k} = \frac{(19.8 \text{ m s}^{-1})^2}{2(9.8 \text{ m s}^{-2})(2.5)} = 8 \text{ m}$$

The coefficient of friction in this example is large because the skier is actually deforming and moving snow as she stops. Her mechanical energy is then dissipated partially as thermal energy and partially as work done in disturbing the snow.

6.5 | OBSERVATIONS ON WORK AND ENERGY

In this section we discuss the general result,

$$K + \mathcal{U} = K_0 + \mathcal{U}_0 + W_a$$

or

$$E = E_0 + W_a$$

The total mechanical energy of an object was defined as $E = K + \mathcal{U}$, where the kinetic energy is $K = \frac{1}{2}mv^2$, and the potential energy $\mathcal{U}$ may be characterized as the energy by virtue of position.

As we saw, *if no work is done by applied forces, the total mechanical energy E is constant.* Then

$$K + \mathcal{U} = K_0 + \mathcal{U}_0 \qquad (W_a = 0)$$

This result is referred to as *mechanical energy conservation.* Under these circumstances, the total mechanical energy or the sum of potential and kinetic energies remains constant, although either may change at the expense of the other.

In many situations, dissipative forces that convert mechanical energy into other forms of energy are present. The heat and noise generated by a saw or drill are examples of this. *Heat* represents energy transferred to the random motion of the molecules of a substance, increasing their average velocity or their thermal energy. As we see in Chapter Ten, increasing the average molecular energy is equivalent to raising the temperature.

Thermal energy can also be converted to mechanical energy, as in a steam engine, where hot steam expands and does work. The limitations inherent in such processes are discussed in Chapter Eleven.

Forms of Energy | Energy exists in many forms

in addition to mechanical and thermal energy. A hot object transmits energy to its surroundings not only by direct contact but also by the emission of electromagnetic waves that travel with the speed of light. The chemical bonds in molecules arise from electric forces. These chemical bonds can be broken or altered, causing the chemical energy to be released. For example, when a fossil fuel and oxygen are brought together at an elevated temperature, chemical changes occur that release energy. Similarly, the body uses food to synthesize molecules that later break down to provide energy as needed.

The energy associated with the forces that bind atomic nuclei together is very large. Some of this is released in a nuclear weapon or power reactor when uranium nuclei *fission* into smaller nuclei. Energy is also released when hydrogen nuclei combine or *fuse* to form a larger nucleus. Fusion supplies the energy in stars and in hydrogen bombs. In a more controlled form, fusion may supply much of our future energy if present research programs are successful. Table 6.1 lists the energies associated with various phenomena.

TABLE 6.1

Approximate energy in joules associated with various events and phenomena

Description	Energy
Big Bang	10^{68}
Radio energy emitted by the Galaxy during its lifetime	10^{55}
Rotational energy of the Milky Way	10^{52}
Energy released in a supernova explosion	10^{44}
Oceans' hydrogen in fusion	10^{34}
Rotational energy of the earth	10^{29}
Annual solar energy incident on the earth	5×10^{24}
Annual wind energy dissipated near earth's surface	10^{22}
Annual global energy usage by humans	3×10^{20}
Annual energy dissipated by the tides	10^{20}
Annual U.S. energy usage	8×10^{19}
Energy release during Krakatoan eruption of 1883	10^{18}
Energy release of 15-megaton fusion bomb	10^{17}
Annual electrical output of large generating plant	10^{16}
Thunderstorm	10^{15}
Energy released in burning 1000 kg of coal	3×10^{10}
Kinetic energy of a large jet aircraft	10^{9}
Energy released in burning 1 litre of gasoline	3×10^{7}
Daily food intake of a human adult	10^{7}
Kinetic energy of a home run in baseball	10^{3}
Work done by a human heart per beat	0.5
Turning this page	10^{-3}
Flea hop	10^{-7}
Discharge of a single neuron	10^{-10}
Typical energy of a proton in a nucleus	10^{-13}
Typical energy of an electron in an atom	10^{-18}
Energy to break one bond in DNA	10^{-20}

Conservation of Total Energy | We see that energy occurs in many forms. It has been found experimentally that energy can be changed into different forms, but it is never created or destroyed. *This is the principle of conservation of total energy.* Historically, failures of energy conservation have led to the identification of new forms of energy. Since energy occurs in so many forms, energy conservation is an underlying concept that unifies all of science. Although mechanical energy conservation occurs only under special circumstances, the total energy is always conserved. In this chapter, our discussion is mainly limited to mechanical energy, but the concepts developed here will appear often in the remainder of the book.

6.6 | SOLVING PROBLEMS USING WORK AND ENERGY

We have used simple examples to infer a general result

$$E = E_0 + W_a$$

In this section we present examples that illustrate the great variety of situations in which work and energy are useful concepts.

A systematic procedure is used in solving problems using work and energy. The procedure includes the following steps:

1 We draw a force diagram showing all the forces acting on the object of interest. This step may be done mentally in simple cases, but it is very helpful in complex situations to actually draw the diagram.

2 We identify the conservative forces—such as the gravitational force—that may be included in the potential energy, as well as the applied forces that do work.

3 We calculate the work done by the applied forces and the mechanical energy at two points in the motion. These quantities are then related by $E = E_0 + W_a$.

The following examples make use of this procedure.

Example 6.8

Two identical balls of mass m are thrown from a window a height h above the ground. The initial speed of

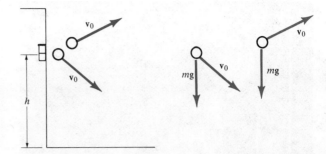

Figure 6.11. Two identical balls thrown in different directions with the same initial speed. The force on each is mg downward.

each ball is v_0 but they are thrown in different directions (Fig. 6.11). What is the speed of each ball as it strikes the ground? (Neglect air resistance.)

Each ball is subject only to the gravitational force mg after it is released, so its total mechanical energy is constant. For either ball, choosing the ground as the reference level, the initial potential energy is $\mathcal{U}_0 = mgh$ and the final potential energy is $\mathcal{U} = 0$. Also, both balls have the same initial kinetic energy, $K_0 = \frac{1}{2}mv_0^2$. Thus, $K + \mathcal{U} = K_0 + \mathcal{U}_0$ tells us that both final kinetic energies are the same, and the balls land with the same speed! With $K = \frac{1}{2}mv^2$, that speed satisfies

$$\tfrac{1}{2}mv^2 = \tfrac{1}{2}mv_0^2 + mgh$$

or

$$v = \sqrt{v_0^2 + 2gh}$$

Thus, while the two balls do not strike the ground at the same time, they do have the same speed when they hit.

Example 6.9

In the pole vault, an athlete uses a pole to convert the kinetic energy of running into potential energy when the pole is vertical (Fig. 6.12). A good sprinter runs at a speed of 10 m s^{-1}. Disregarding the additional height the athlete gains by using his arms to raise his center of gravity well above the position of his hands on the pole, how high can the athlete raise his center of gravity?

Initially, his C.G. is about 1 m above the ground. Just before the athlete begins to use the pole, $\mathcal{U}_0 = 0$ and $K_0 = \frac{1}{2}mv^2$, where $v = 10$ m s^{-1}. At the top of an ideal jump, $v = 0$, so $K = 0$ and $\mathcal{U} = mgh$, where h is the height of his center of gravity above its initial position. Since only the force of gravity acts on the airborne jumper, $E = E_0$ and

$$0 + mgh = \tfrac{1}{2}mv^2 + 0$$

Then

$$h = \frac{v^2}{2g} = \frac{(10 \text{ m s}^{-1})^2}{2(9.8 \text{ m s}^{-2})} = 5.1 \text{ m}$$

Figure 6.12. Photo of a pole vaulter at equally spaced time intervals. Note that the vaulter is nearly at rest at the top of the jump where all his kinetic energy has been converted to potential energy. (Dr. Harold Edgerton, MIT, Cambridge, Mass.)

His center of gravity is then about 6.1 m above the ground. The present world pole vault record is near 5.8 m.

Example 6.10

The tides are used to generate electrical energy at a dam across the mouth of the Rance River in France. At this location the tidal rise—the difference in height between high and low tides—averages 8.5 m. The river basin is closed off after it fills at high tide, and at low tide approximately 6 hours later the water is allowed to fall through turbines that drive the electrical generators (Fig. 6.13). The area of the basin is 23 km^2 = 23 × 10^6 m^2. How much work does the falling water do, assuming that its initial and final kinetic energies are negligible?

Since the kinetic energies are both zero, we have $\mathcal{U} = \mathcal{U}_0 + W_a$ or $W_a = \mathcal{U} - \mathcal{U}_0$. To calculate this potential energy difference, we need the mass of the water and the distance it falls. The mass is the product of the density ($\rho = 10^3$ kg m^{-3}) and the volume, which is the area A times the depth d. Hence the mass is $m = \rho Ad$. The water at the top of the basin falls the full 8.5 m, but as the level drops, the water falls a shorter distance; on

the average, the drop is half the full depth. Thus the work done by the applied forces is

$$W_a = \mathcal{U} - \mathcal{U}_0 = -mgh = -(\rho Ad)g\left(\frac{d}{2}\right) = -\tfrac{1}{2}\rho Agd^2$$
$$= -\tfrac{1}{2}(1000 \text{ kg m}^{-3})(23 \times 10^6 \text{ m}^2)(9.8 \text{ m s}^{-2})(8.5 \text{ m})^2$$
$$= -8.14 \times 10^{12} \text{ J}$$

The work is negative because the water does work *on* the turbines. Some of this work is lost as thermal energy, but most of it is converted into electrical energy. This facility can supply enough electricity for a few hundred thousand people. Much larger power plants have been proposed for parts of the Bay of Fundy. These could supply a large part of the electrical energy needed in eastern Canada and in the New England region of the United States.

6.7 | GRAVITATIONAL POTENTIAL ENERGY

We saw in Section 6.3 that the gravitational force on an object near the earth's surface can be taken into account by defining its potential energy to be *mgh*. We assumed that the gravitational force *m***g** was constant, and used the property that the work it did depended only on the change in height and not on the path taken. Once we move a distance that is a significant fraction of the earth's radius R_E, it is no longer correct to treat the gravitational force as constant. Nevertheless, it can be shown that the work it does still depends only on the initial and final heights. In more formal terms, the gravitational force is still a conservative force. Thus the potential energy concept can still be used to take account of the gravitational force, although the formula for $\mathcal{U}$ must be modified.

We saw in Chapter Five that the attractive gravitational force between two points or spherical masses M and m can be written in vector form as

$$\mathbf{F} = -\frac{GMm}{r^2}\hat{\mathbf{r}} \qquad (6.17)$$

When the separation r becomes very large, this force becomes negligibly small, which suggests defining the potential energy of the two masses to be zero when r is infinite. With this choice of the reference level, two very distant objects—which exert no forces on each other—also have no potential energy. As the objects approach, the gravitational force does work on them, increasing the kinetic energy and decreasing the potential energy. (Recall $\mathcal{U}$ *increases* when we raise an object, and *decreases* when we lower it.) Since $\mathcal{U}$ has

Figure 6.13. The tidal power plant at the mouth of the Rance River in France. Water enters the river basin through the turbines at high tide, and leaves the basin through the turbines at low tide. The turbines can also use electricity from other installations to pump extra water in each direction to increase the subsequent "head," the distance the water will fall. (French Embassy Press and Information Center.)

been chosen to be zero at $r = \infty$, it becomes negative as r decreases.

To find the magnitude of $\mathfrak{U}$, we must apply the definition, $\mathfrak{U} - \mathfrak{U}_0 = -W(\text{grav})$; the change in potential energy equals the negative of the work done by gravity. Suppose M is stationary at the origin. We move m so that its displacement $d\mathbf{r}$ is parallel to the vector $\mathbf{r}$ from M to m (Fig. 6.14). Then, using Eq. 6.17 for $\mathbf{F}$,

$$-dW(\text{grav}) = -\mathbf{F} \cdot d\mathbf{r} = \frac{-(-GMm\hat{\mathbf{r}})}{r^2} \cdot d\mathbf{r}$$

$$= \frac{GMm\,dr}{r^2}$$

Integrating from r_0 to r_1, we obtain $-W(\text{grav})$, and the potential energy change is

$$\mathfrak{U} - \mathfrak{U}_0 = \int_{r_0}^{r_i} GMm \frac{dr}{r^2} = GMm \left[-\frac{1}{r} \right]_{r_0}^{r_1}$$

$$= \frac{-GMm}{r_1} + \frac{GMm}{r_0}$$

This is the difference between the potential energies at separations of r_1 and r_0. Although we assumed a displacement along $\mathbf{r}$, the result is quite general, since no additional work is needed to move m along a circular path centered at M.

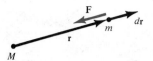

Figure 6.14. Since the gravitational force **F** is opposite to $d\mathbf{r}$, **F** · $d\mathbf{r}$ is negative. When m moves away from M, the gravitational force does negative work, and the potential energy increases.

With the choice of reference level $\mathfrak{U}_0 = 0$ at $r_0 = \infty$, the potential energy at a distance $r = r_1$ is

$$\mathfrak{U} = \frac{-GMm}{r} \qquad (6.18)$$

Note that this result can be obtained quickly but incorrectly by arguing "the work equals the force GMm/r^2 times the distance r." Since the masses move from a separation of infinity to one of r, which is quite different from moving a distance r, this approach is not correct.

At the end of this section we show that Eq. 6.18 is equivalent to the simpler expression $\mathfrak{U} = mgh$ when an object remains near the surface of the earth. However, it must be used whenever the height changes by a significant fraction of the earth's radius, as in the following applications to satellites.

Energy of a Satellite

The total mechanical energy of a satellite in a circular orbit can be calculated with the aid of Eq. 6.18. Applying Newton's second law **F** = m**a** to the circular motion of a mass m under the gavitational attraction of the earth, we have

$$\frac{GM_{\mathrm{E}}m}{r^2} = \frac{mv^2}{r}$$

or

$$mv^2 = \frac{GM_{\mathrm{E}}m}{r}$$

Thus the kinetic energy can be written as

$$K = \frac{1}{2}mv^2 = \frac{GM_{\mathrm{E}}m}{2r} = -\frac{1}{2}\mathfrak{U}$$

Since $\mathfrak{U}$ is negative, K is positive and half as large as $\mathfrak{U}$ in magnitude. The total energy is

$$E = K + \mathfrak{U} = -\tfrac{1}{2}\mathfrak{U} + \mathfrak{U} = \tfrac{1}{2}\mathfrak{U}$$

or

$$E = -\frac{GM_{\mathrm{E}}m}{2r} \qquad (6.19)$$

Example 6.11

How much work must be done to lift an artificial satellite of mass m from the surface of the earth and put it in a circular orbit with a radius equal to twice the earth's radius?

Initially, $K_0 = 0$ and $\mathfrak{U}_0 = -GM_{\mathrm{E}}m/R_{\mathrm{E}}$. When the satellite is in orbit, according to Eq. 6.19 its total mechanical energy is $E = -GM_{\mathrm{E}}m/4R_{\mathrm{E}}$. Since $E = E_0 + W_a$, we have

$$W_a = E - E_0 = -\frac{GM_{\mathrm{E}}m}{4R_{\mathrm{E}}} - \frac{(-GM_{\mathrm{E}}m)}{R_{\mathrm{E}}} = \frac{3GM_{\mathrm{E}}m}{4R_{\mathrm{E}}}$$

Escape Velocity

The escape velocity is the minimum initial velocity v_0 required for a projectile fired vertically at the earth's surface to escape its gravitational force (Fig. 6.15). At the surface of the earth the velocity is v_0, so

$$E_0 = K_0 + \mathfrak{U}_0 = \frac{1}{2}mv_0^2 - \frac{GM_{\mathrm{E}}m}{R_{\mathrm{E}}}$$

If the projectile is to escape permanently from the earth, it will eventually reach a very large value of r, so that $\mathfrak{U} = 0$. If it has the minimum energy needed to do this, its velocity and kinetic energy will also be zero at this distance. Thus the minimum total energy needed to escape the earth is $E = K + \mathfrak{U} = 0$. Since the mechanical energy is conserved

$$\frac{1}{2}mv_0^2 - \frac{GM_{\mathrm{E}}m}{R_{\mathrm{E}}} = 0$$

Then

$$v_0 = \sqrt{\frac{2GM_{\mathrm{E}}}{R_{\mathrm{E}}}}$$

The weight of an object of mass m at the surface of the earth is $mg = GM_{\mathrm{E}}m/R_{\mathrm{E}}^2$. Using this, we have

$$v_0 = \sqrt{\frac{2GM_{\mathrm{E}}}{R_{\mathrm{E}}}} = \sqrt{2gR_{\mathrm{E}}} \qquad (6.20)$$

This is the minimum velocity necessary at the earth's surface to escape from the earth.

The escape velocity for any planet can be found from its gravitational acceleration and radius. For the earth, $g = 9.8$ m s^{-2} and $R_{\mathrm{E}} = 6.4 \times 10^6$ m, so the escape velocity is

$$v_0 = \sqrt{2(9.8 \text{ m s}^{-2})(6.38 \times 10^6 \text{ m})}$$
$$= 1.12 \times 10^4 \text{ m s}^{-1}$$

Figure 6.15. A rocket fired from the earth's surface must have a velocity at least as great as the escape velocity v_0 in order to fully escape the earth's gravitational attraction. A velocity less than v_0 is sufficient if the engines remain on until a higher altitude is reached. (NASA.)

Objects Near The Surface of The Earth

Earlier in this chapter, we used $\mathcal{U} = mgh$ for the gravitational potential energy, while we have used $\mathcal{U} = -GM_{E}m/r$ here. It is left as a problem to show that if an object is at a small height $h \ll R_{E}$ above the earth's surface, the potential energy is

$$\mathcal{U} = -\frac{GM_{E}m}{(R_{E} + h)} \simeq -\frac{GM_{E}m}{R_{E}}\left(1 - \frac{h}{R_{E}}\right)$$

(The symbol $\simeq$ means approximately equal.)

Using $g = GM_{E}/R_{E}^2$, this becomes

$$\mathcal{U} = -mgR_{E} + mgh$$

Since we always discuss potential energy differences,

the constant term $-mgR_{E}$ drops out of all calculations. Thus, as long as we are near the earth's surface, we can use $\mathcal{U} = mgh$.

6.8 | ELECTRICAL POTENTIAL ENERGY

The gravitational force is not the only conservative force that we encounter. The electric force is also conservative. In Chapter Five, we saw that the force on a charge q due to a second charge Q a distance r away is

$$\mathbf{F} = k\frac{qQ}{r^2}\hat{\mathbf{r}} \qquad (6.21)$$

Here $k = 9 \times 10^9$ N m² C⁻². Except for a minus sign and the fact that qQ can be positive or negative, this is identical in mathematical form to the gravitational force law. Thus we can immediately say that the *electrical force is conservative.* If we replace $-GMm$ by kqQ in the gravitational potential energy formula, $\mathcal{U} = -GMm/r$, we obtain an expression for the potential energy of the two charges q and Q,

$$\mathcal{U} = \frac{kqQ}{r} \qquad (6.22)$$

Because the electric force is conservative the change in potential energy if the charge q is moved from r_1 to r_2 is the same *no matter what path is taken.*

Since the electrical and potential energies are identical in mathematical form, the results obtained in the preceding section can be immediately applied to electrical forces, as we do in the next example.

Example 6.12

In a hydrogen atom the electron orbits about a proton under the influence of the electric force. If the orbit is a circle with radius $a_0 = 5.3 \times 10^{-11}$ m, what is the energy of the electron? (The electron and proton charges are $-e = -1.60 \times 10^{-19}$ C and $+e = +1.60 \times 10^{-19}$ C, respectively.)

In the preceding section, the energy of a satellite orbiting under gravitational forces was calculated to be $E = -GM_{E}m/2r$. The result holds true for this problem also if we replace $GM_{E}m$ by $(-kqQ)$. Then $E = +kqQ/2r$. With $q = -e$, $Q = e$, and $r = a_0$, the energy of the electron is

$$E = -\frac{1}{2}\frac{ke^2}{a_0}$$

$$= -\frac{1}{2}\frac{(9 \times 10^9 \text{ N m}^2 \text{ C}^{-2})(1.6 \times 10^{-19} \text{ C})^2}{5.3 \times 10^{-11} \text{ m}}$$

$$= -2.17 \times 10^{-18} \text{ J}$$

To remove the electron from the atom, we must supply 2.17×10^{-18} J in some way.

In atomic physics, a joule is a very large energy unit. A more convenient unit is the *electron volt* (eV), where

$$1 \text{ eV} = 1.60 \times 10^{-19} \text{ J}$$

To illustrate, in the previous example the hydrogen atom has an energy of magnitude 2.17×10^{-18} J, or

$$(2.17 \times 10^{-18} \text{ J}) \frac{(1 \text{ eV})}{(1.6 \times 10^{-19} \text{ J})} = 13.6 \text{ eV}$$

6.9 | POWER

We are often less concerned with the net amount of work done or energy transferred than with the rate at which this occurs. For example, we may remove snow from a driveway either with a hand shovel or with a snowplow. The same work is done, but the plow does it much faster. We say that the plow is more powerful. In this section, we describe the relationship between work and power.

When an amount of work ΔW is done in a time Δt, the *average power* is defined as the average rate of doing work,

$$\bar{\mathcal{P}} = \frac{\Delta W}{\Delta t} \tag{6.23}$$

The instantaneous power $\mathcal{P}$ is found by considering the limit as Δt approaches zero,

$$\mathcal{P} = \frac{dW}{dt} \tag{6.24}$$

From this definition, we see that the S.I. power unit is a joule per second, which is called a *watt* (W). A watt is rather a small unit for many purposes. It takes about 9 kilowatts (1 kW $= 10^3$ W) to overcome the dissipative forces acting on a 2000-kg car moving at a constant 65 km h^{-1}. A medium-sized electric power plant might generate 200 megawatts (1 MW $= 10^6$ W), and a rather large one might generate 1 gigawatt (1 GW $= 10^9$ W).

Energy is often sold by electrical utilities by the kilowatt hour (kW h). This is 1 kilowatt of power for 1 hour. In terms of S.I. units,

$$1 \text{ kW h} = (10^3 \text{ W})(3600 \text{ s}) = 3.6 \times 10^6 \text{ J}$$

In the following examples we find the power output of a man and of a wind-powered electrical generator.

Example 6.13
A 70-kg man runs up a flight of stairs 3 m high in 2 s. (a) How much work does he do against gravitational forces? (b) What is his average power output?

(a) The work done, ΔW, is equal to his change in potential energy, mgh. Thus

$$\Delta W = mgh = (70 \text{ kg})(9.8 \text{ m s}^{-2})(3 \text{ m})$$
$$= 2060 \text{ J}$$

(b) His average power is the work done divided by the time,

$$\bar{\mathcal{P}} = \frac{\Delta W}{\Delta t} = \frac{2060 \text{ J}}{2 \text{ s}} = 1030 \text{ W}$$

This is a high power output for a human.

Example 6.14
The blades of a wind-powered electrical generator sweep out a circle of area A. (a) If the wind moves at a velocity v perpendicular to the circle, what is the mass of the air passing through it in time t? (b) What is the kinetic energy of that air? (c) Suppose the machine converts 30 percent of the wind's energy into electrical energy, and that $A = 30$ m^2 and $v = 10$ m s^{-1} (36 km h^{-1}). What is the electrical power output? (The density of air at 20° C is 1.2 kg m^{-3}.)

(a) In time t, the wind moves a distance vt. Hence all the air in a cylinder of cross-sectional area A, length vt, and volume $V = Avt$ will cross the circle. Multiplying V by the density of the air ρ, the mass is

$$m = \rho A v t$$

(b) The kinetic energy of the air crossing in time t is

$$K = \tfrac{1}{2}mv^2 = \tfrac{1}{2}(\rho A v t)v^2 = \tfrac{1}{2}\rho A v^3 t$$

(c) The power available from the wind is its kinetic energy divided by the time it takes to cross the circle defined by the blades. Thirty percent is converted to electrical power, so

$$\mathcal{P} = \frac{(0.30)K}{t} = \frac{(0.30)(\tfrac{1}{2}\rho A v^3 t)}{t} = 0.15\rho A v^3$$
$$= (0.15)(1.2 \text{ kg m}^{-3})(30 \text{ m}^2)(10 \text{ m s}^{-1})^3$$
$$= 5400 \text{ W} = 5.4 \text{ kW}$$

This is enough electrical power for about five typical American homes. Note, however, we need a fairly high wind speed of 36 km h^{-1}. $\mathcal{P}$ varies as v^3, so if the wind speed were to decrease 50 percent, the power output would become only $(5.4/2^3)$ kW $= 0.675$ kW.

Another expression for the power is often useful. The work done by a force **F** acting through a small displacement **Δs** in a short time Δt is $\Delta W = F_s\, \Delta s$. Dividing by Δt gives the power,

$$\overline{\mathcal{P}} = F_s \frac{\Delta s}{\Delta t}$$

Since ds/dt is the velocity, the power is also given by,

$$\mathcal{P} = F_s v = \mathbf{F} \cdot \mathbf{v} \qquad (6.25)$$

Thus the power is the force component F_s times the velocity. This result is used in the next example.

Example 6.15

A 250-kg piano is raised by a hoist at a constant velocity of 0.1 m s^{-1}. What is the power expended by the hoist?

Since the force $m\mathbf{g}$ is parallel to the velocity,

$$\mathcal{P} = Fv = (250 \text{ kg})(9.8 \text{ m s}^{-1})(0.1 \text{ m s}^{-1})$$
$$= 245 \text{ W}$$

6.10 | WORK AND ENERGY FOR ROTATIONAL MOTION

Objects can have kinetic energy associated with rotational motion as well as with translation. We now obtain expressions for the work, power, and kinetic energy appropriate for objects rotating about a fixed axis.

Consider a wheel of radius r rotating about its axis. When it rotates through an angle θ, a point on the rim moves a distance $s = r\theta$. A force F acting tangential to the wheel during this displacement will do work $Fs = Fr\theta$. Since Fr is the torque τ due to this force, the work done can be written as

$$W = \tau\theta \qquad (6.26)$$

The power is defined as $\Delta W/\Delta t$. With this result for W, $\mathcal{P} = \tau\Delta\theta/\Delta t$. Since $\Delta\theta/\Delta t = \omega$, we have

$$\mathcal{P} = \tau\omega \qquad (6.27)$$

When an object moves in a circular path, $v = \omega r$, and the kinetic energy $\frac{1}{2}mv^2$ can be rewritten as $\frac{1}{2}mr^2\omega^2$. If a rigid body rotates about a fixed axis, each of its constituent particles moves in a circle with the same angular velocity. Adding up the individual kinetic energies therefore amounts to summing up the mr^2 factors, or to finding the moment of inertia.

Hence the kinetic energy is

$$K = \tfrac{1}{2}I\omega^2 \qquad (6.28)$$

All of these results are analogs of the corresponding formulas for translation, with the natural replacements $s \to \theta$, $v \to \omega$, $F \to \tau$, and $m \to I$. If rotating objects are present, the work and energy terms in $E = E_0 + W_a$ must take this into account, as in the next example.

Example 6.16

A 20-kg bucket is held above a well by a massless rope wound about a windlass (Fig. 6.16). The windlass is a cylinder with a radius of 0.2 m and a moment of inertia of 0.2 kg m^2. If the bucket is released from rest, what is its speed just before it hits the water 10 m below? (Assume there is no friction or air resistance.)

With the reference height for the potential energy at the water, $\mathcal{U}_0 = mgh$ and $\mathcal{U} = 0$. The kinetic energy K_0 at the top is zero. When the bucket hits the water, if its speed is v, the angular velocity of the windlass is $\omega = v/r$, and their combined kinetic energy is $K = \frac{1}{2}mv^2 + \frac{1}{2}I\omega^2 = \frac{1}{2}mv^2 + \frac{1}{2}Iv^2/r^2$. Since the mechanical energy is conserved, $E = E_0$, and

$$\frac{1}{2}mv^2 + \frac{1}{2}I\frac{v^2}{r^2} = mgh$$

Solving for v, we find

$$v = \left[\frac{2mgh}{m + I/r^2}\right]^{1/2}$$
$$= \left[\frac{(2)(20 \text{ kg})(9.8 \text{ m s}^{-2})(10 \text{ m})}{(20 \text{ kg}) + (0.2 \text{ kg m}^2)/(0.2 \text{ m})^2}\right]^{1/2}$$
$$= 12.5 \text{ m s}^{-1}$$

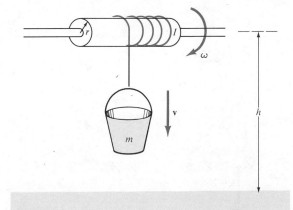

Figure 6.16. When the bucket falls, gravitational potential energy is converted into kinetic energy of the bucket and windlass.

In the previous example, if the bucket were not attached to the windlass, it would have acquired a greater speed. Some of its potential energy was converted into kinetic energy of the windlass. Similarly, if an object rolls down a hill, some of the potential energy is converted into rotational kinetic energy, and it reaches the bottom of the hill with a lower speed than an object sliding frictionlessly down the hill.

6.11 | JUMPING; SCALING LAWS IN PHYSIOLOGY

In Chapter One, we related the heights reached by jumping animals to the velocities and accelerations involved. Here we describe a procedure invented by Galileo called *scaling*, which we use in this section to draw some general conclusions about jumping. We begin by reanalyzing the vertical jump of a human in terms of work and energy.

Figure 6.17 shows a man jumping. Initially he crouches, lowering his center of gravity a distance d called the acceleration distance. We shall measure potential energy from this reference level. As he accelerates and straightens, he does work to raise his potential and kinetic energies. At the instant of takeoff, his potential energy is $\mathcal{U}_0 = mgd$. If his upward velocity at this instant is v_0, his kinetic energy is $\frac{1}{2}mv_0^2$. Thus, in reaching the erect takeoff position, he has done work W_a where

$$W_a = mgd + \tfrac{1}{2}mv_0^2 \qquad (6.29)$$

From takeoff to the top of the jump, the only force acting on the man is his weight. Thus, during the airborne part of the jump, his mechanical energy is constant. At the top of the jump, $K = 0$ and $\mathcal{U} = mg(h + d)$. This must equal the energy at takeoff, so

$$W_a = mgd + \tfrac{1}{2}mv_0^2 = mg(h + d) \qquad (6.30)$$

Thus the total energy that the human must supply to make the jump is $W_a = mg(h + d)$. In our subsequent discussion, we neglect d compared to h and use

$$W_a \simeq mgh$$

This is only a fair assumption for humans, since $d/h \simeq \frac{1}{2}$, but it is a good approximation for smaller animals. Notice that Eq. 6.30 also gives us the takeoff velocity

$$v_0 = \sqrt{2gh}$$

Having found the total energy in terms of the height reached in the jump, we might hope to determine the energy expended by various muscles and compare these numbers for various animals. Unfortunately, the large number of bones and muscles involved makes such an analysis very complicated, a difficulty characteristic of many biological problems. However, we can make some progress in comparing the jumping abilities of different animals with the aid of a procedure called *scaling*. Scaling assumes that

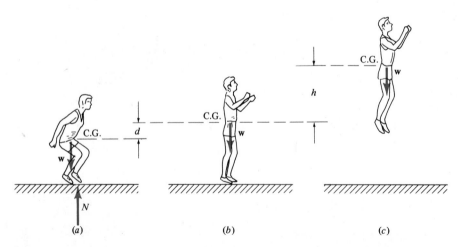

Figure 6.17. (*a*) A man preparing to jump. The forces on the man are his weight **w** down and the normal force **N** of the ground acting upward. (*b*) The man at takeoff. (*c*) During the airborne phase the only force on the man is **w**. The total change in the position of his center of gravity during the jump is $d + h$.

the basic characteristics of a biological system change or *scale* with its overall size in a simple but plausible way. This procedure, which can be used in many physiological problems, leads to predicted *scaling laws* for the jumping ability versus the size of an animal, which can be compared to the data.

The simplest scaling model assumes that the mass m of an animal is proportional to its volume and that this volume is proportional to the cube of some characteristic length l. Thus the scaling assumption is $m = cl^3$, where c is a constant. For example, the ratio of the characteristic lengths of a 0.02-kg mouse and a 700-kg cow would be

$$\frac{l_{\text{cow}}}{l_{\text{mouse}}} = \left(\frac{m_{\text{cow}}}{m_{\text{mouse}}}\right)^{1/3} = \left(\frac{700 \text{ kg}}{0.02 \text{ kg}}\right)^{1/3} = 32.7$$

This may, for example, represent the relative lengths of the legs.

Using this length, we note that the volume of an animal or any of its organs is proportional to l^3, the surface area of the body and the cross-sectional areas of the muscles are proportional to l^2, and the length of the limbs is proportional to l.

Comparing the jumping ability of animals we notice the interesting fact that the heights of the jumps are not so very different for animals that are extremely different in size. Thus the rat kangaroo, about the size of a rabbit, jumps nearly as high as a large kangaroo. Locusts and the much smaller flea jump to roughly the same height. We may ask, what characteristics do they share that would explain this result?

Two possibilities come immediately to mind.

1 The energy supplied per unit mass of muscle is the same for all animals.
2 The power supplied per unit mass of muscle is the same for all animals.

We examine the first assumption here, leaving the second as a problem (Problem 6-110). The first assumption says that an animal should be able to do an amount of work proportional to its mass, m. But we saw that the work done during a jump of height h is mgh. Thus $mgh \propto m$, and h does not depend on m or l.

This prediction, that the height reached is independent of the size of an animal, is roughly correct experimentally. By contrast, the predictions obtained with the second hypothesis disagree with the data. Thus, we may tentatively conclude that the *energy*

expended per unit mass is about the same for all animals of the same general type.

Another conclusion follows by noting that the takeoff velocity $v_0 = \sqrt{2gh}$ is roughly independent of size. The acceleration distance d is proportional to the characteristic length l, so the takeoff time $t = d/\bar{v} = d/(\frac{1}{2}v_0)$ also scales as l. The power expended per unit mass is the energy expended per unit mass divided by this time. Since the energy expended per unit mass is independent of l, the power must vary as $1/l$. *This predicts that larger animals will expend their energy at a slower rate.*

Comparisons between mammals and insects are complicated by the different ways they use their leg muscles. Mammals use muscle contractions directly, but insects employ a catapult arrangement. For example, a flea has an elastic material called *resilin* in its knee joint. The flea gradually bends its leg, stretching the resilin, and locks the knee in position. In a jump, the knee is unlocked, and the resilin contracts rapidly, straightening the leg. Thus, insects employ stored *elastic* potential energy and use their muscles somewhat indirectly.

The discussion of scaling laws as applied to jumping shows how interesting qualitative insights can sometimes be obtained in complex biological systems. The model we have used assumed that all body dimensions scale in exactly the same way with the mass of the animal. However, it is possible to make other scaling assumptions. Another approach to scaling is described in Chapter Eight.

SUMMARY

Many problems can be most easily solved using the concepts of work and energy. The relationship between these quantities is contained in the two most important statements of this chapter.

1 The final kinetic energy of an object is equal to its initial kinetic energy plus the work W done by all the forces acting on it:

$$K = K_0 + W$$

In this equation, $W = F_s s$, where s is the displacement of the object, and F_s is the component of the net force along the direction of the displacement. $K_0 = \frac{1}{2}mv_0^2$ and $K = \frac{1}{2}mv^2$ are the kinetic energies of the object before and after the displacement, respectively.

2 If conservative forces act on an object, their effects can be taken into account by including

appropriate terms in the potential energy, $\mathcal{U}$. The other forces acting on the object are called applied forces, and they do work W_a. Then the general work-energy relationship just given is rewritten in the more useful form

$$K + \mathcal{U} = K_0 + \mathcal{U}_0 + W_a$$

A force is conservative if the work it does as an object moves from one location to another is independent of the path. Gravitational and electrical forces are conservative.

When there is no work done by applied forces, the total mechanical energy $E = K + \mathcal{U}$ is conserved. The kinetic and potential energies may each change, but their sum must remain constant. When dissipative forces are present, mechanical energy is converted into thermal energy. Energy is never destroyed or created—it is only transformed from one form to another.

Near the earth's surface, the gravitational force is approximately constant, and the potential energy is $\mathcal{U} = mgh$. The more general form for the gravitational potential energy is $\mathcal{U} = -GMm/r$. The electrical potential energy of two point charges is $\mathcal{U} = kqQ/r$.

The rate dW/dt at which work is done or energy is transferred is the power. An equivalent expression is $\mathcal{P} = F_s v$.

For an object rotating about a fixed axis, the formulas for the work, power, and kinetic energy are obtained by replacing s, v, F, and m by their analogs θ, ω, τ, and I, respectively.

Checklist

Define or explain:

work	escape velocity
joule	gravitational potential
scalar product	energy
kinetic energy	electrical potential
potential energy	energy
conservative force	electron volt
total mechanical energy	power
dissipative force	watt
applied force	kilowatt hour
mechanical energy	rotational kinetic energy
conservation	scaling

REVIEW QUESTIONS

Q6-1 The work done by a force is positive when **F** is _____ to **s**, negative when **F** is _____ to **s**, and zero when **F** is _____ to **s**.

Q6-2 The S.I. unit for work is the _____.

Q6-3 The kinetic energy of an object is a measure of its ability to _____.

Q6-4 The translational kinetic energy of an object of mass m and velocity v is _____.

Q6-5 The final kinetic energy of an object equals its _____ plus the total work done by _____.

Q6-6 Forces that can be included in the potential energy are said to be _____.

Q6-7 The total mechanical energy is the _____ plus the _____.

Q6-8 Mechanical energy is conserved when _____.

Q6-9 Potential energy is energy associated with the _____.

Q6-10 The place where the potential energy is zero is _____.

Q6-11 In general, the increase in the potential energy is equal to the negative of the _____.

Q6-12 Dissipative forces usually convert mechanical energy into _____.

Q6-13 The work done by frictional forces is always _____.

Q6-14 The gravitational potential energy formula $\mathcal{U} = mgh$ holds for objects _____.

Q6-15 The minimum velocity for an object fired upward at the earth's surface to permanently leave the earth is called the _____.

Q6-16 Power is the rate at which _____.

Q6-17 The S.I. power unit is the _____.

Q6-18 A kilowatt hour is a unit of _____.

Q6-19 _____ assumes that the basic characteristics of a biological system change with its overall size in a direct way.

EXERCISES

Section 6.1 | Work

6-1 A child pulls a toy car with a 10-N force at a 20° angle to the horizontal (Fig. 6.18). If the car moves a distance of 6 m, how much work does the child do?

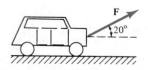

Figure 6.18. Exercise 6.1.

6-2   A woman pushes horizontally on a chair with a force of 300 N. Compute the work she does on the chair if (a) it moves 2 m parallel to the force; (b) it moves 1 m opposite to the force; (c) the chair does not move.

6-3   A woman exerts a constant horizontal force of 200 N on a child and a tricycle. The tricycle moves 2 m, and the work done by the woman is 100 J. What is the angle between the force and the displacement of the tricycle?

6-4   A motorcycle comes to a skidding stop in 5 m. During stopping, the force on the cycle due to the road is 200 N and is directly opposed to the motion. (a) How much work does the road do on the cycle? (b) How much work does the cycle do on the road?

6-5   A girl pulls a box weighing 40 N a distance of 10 m across the floor at constant speed. How much work does she do if the coefficient of sliding friction is 0.2?

6-6   A box of mass 10 kg falls straight down a distance of 2 m. How much work is done by gravitational forces?

6-7   A car of mass 1300 kg rolls down a hill a distance of 100 m. The road is at an angle of 10° to the horizontal direction. How much work is done on the car by gravitational forces?

Section 6.2 | Kinetic Energy

6-8   A 1000-kg automobile has a velocity of 40 km h^{-1}. What is its kinetic energy?

6-9   What is the kinetic energy of a 0.25-kg stone moving at 10 m s^{-1}?

6-10   A baseball of mass 0.15 kg is thrown at 30 m s^{-1}. (a) What is its kinetic energy? (b) If it is thrown by a man who exerts a constant force over a distance of 1.5 m, what force does he exert?

6-11   A 100-kg man is in a car traveling at 20 m s^{-1}. (a) Find his kinetic energy. (b) The car strikes a concrete wall and comes to rest after the front of the car has collapsed 1 m. The man is wearing a seat belt and harness. What is the average force exerted by the belt and harness during the crash?

6-12   Show that kinetic energy has dimensions of force times distance.

6-13   A 200-kg swordfish swimming at 5 m s^{-1} rams a wooden yacht which is at rest. Its sword penetrates the yacht, and the fish is stopped in 1 m. (a) What is the initial kinetic energy of the fish? (b) How much work is done by the fish?

6-14   Fishing line is usually sold as being capable of withstanding a certain force. What strength line is necessary to handle a 10-kg salmon swimming at 3 m s^{-1} if it is stopped in 0.2 m?

6-15   A baseball is thrown from center field to second base, and its velocity diminishes from 20 m s^{-1} to 15 m s^{-1}. If its mass is 0.15 kg, how much energy was lost due to air resistance? (Assume the initial and final heights are the same.)

6-16   A ball of mass 0.2 kg falls straight down a distance of 10 m. (a) How much work is done by gravitational forces on the ball? (b) If it was initially at rest, what is its speed after falling the 10 m?

6-17   A golf club strikes a ball lying on the grass. It remains in contact for a distance of 2 cm. If the ball acquires a speed of 60 m s^{-1}, and its mass is 0.047 kg, what is the average force exerted by the club?

Section 6.3 | Potential Energy and Conservative Forces

6-18   A swimmer dives off a board into a pool, swims to the edge, and climbs back onto the board. Identify and discuss the kinds of forces present and the work they do.

6-19   Mountain roads usually wind back and forth up the mountain side rather than going straight up. Why?

6-20   A child on a swing reaches a maximum height of 2 m above her lowest position. What is the speed of the swing at the lowest point? (Neglect frictional forces.)

6-21   A beer can is dropped from a window 30 m above the ground. How fast will it be moving just before it lands? (Neglect air resistance.)

6-22   A car moving at 40 m s^{-1} hits a concrete wall. From what height would a car have to be dropped onto concrete to achieve the same result?

6-23   A boy sits on a Ferris wheel. How much work is done on him by gravitational forces when the wheel makes one complete turn?

6-24   A 6 kg salmon swims a distance of 5 m up a fish ladder at constant velocity. It experiences a 1.3-N dissipative force from the water. The fish rises a distance of 0.5 m while swimming up the ladder (Fig. 6.19). (a) How much work must the fish do to overcome the dissipative force? (b) What is the change in the fish's potential energy? (c) How much total work does the fish do in swimming up the ladder?

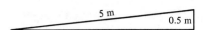

Figure 6.19. Exercise 6-24.

6-25   A baseball thrown straight up reaches a height of 50 m. What was its initial velocity? (Neglect air resistance.)   3 1 . 3

6-26   Water flows over a dam and falls a distance h in a smooth stream, as shown in Fig. 6.20. Assume that the water has zero velocity at the top of the falls. What is the velocity at the base of the falls?

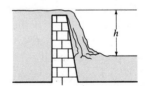

Figure 6.20. Water going over a dam. It has zero velocity at the top and drops a distance h. Exercise 6-26.

6-27   A car goes off a cliff with a speed of 30 m s^{-1}. How fast will it be moving after falling 20 m?

6-28   How much work must a pump do to deliver 100 kg of water from a 300-m-deep well to the surface of the earth? (Assume the kinetic energy change is negligible.)

6-29   An elevator and its contents have a mass of 2000 kg. It is counterweighted by a piece of metal of mass 1700 kg that falls as the elevator rises. How much work must the motor do against gravitational forces to raise the elevator 30 m?

Section 6.4 | Dissipative Forces

6-30   A hockey puck with an initial velocity of 4 m s^{-1} slides on ice. The coefficient of kinetic friction is 0.1. How far will the puck slide before stopping?

6-31   A box with an initial speed of 2 m s^{-1} slides to rest on a horizontal floor in 0.5 m. What is the coefficient of kinetic friction?

6-32   At some amusement parks, one can slide down a ramp, as shown in Fig. 6.21. (a) What is the velocity at the bottom? (Assume the ramp is frictionless.) (b) What distance l is necessary to stop if the coefficient of friction at the bottom is 0.5?

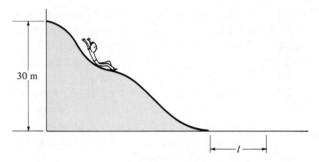

Figure 6.21. Profile of the amusement park slide. l is the stopped distance. Exercise 6-32.

6-33   A sled slides 100 m down a hill that slopes at an angle of 30° with the horizontal direction. The sled attains a final velocity of 20 m s^{-1} at the base of the slope. What fraction of the energy was lost due to friction?

Section 6.5 | Observations on Work and Energy and Section 6.6 | Solving Problems Using Work and Energy

6-34   A 50-kg woman climbs a mountain 3000 m high. (a) How much work does she do against gravitational forces? (b) A kilogram of fat supplies about 3.8×10^7 J of energy. If she converts fat into mechanical energy with a 20 percent efficiency rate, how much fat will she consume in the climb?

6-35   A dieter lifts a 10-kg mass a distance of 0.5 m 1000 times. (a) How much work does he do against gravitational forces? (Assume the potential energy lost each time he lowers the mass is dissipated.) (b) Fat supplies 3.8×10^7 J of energy per kilogram, which is converted to mechanical energy with a 20 percent efficiency rate. How much fat will the dieter use up?

6-36   A man lifts a 20-kg mass 0.5 m vertically. (a) How much work does he do against gravitational forces? (b) When he lowers the mass to its original height, how much work is done on the man? (c) What happens to this energy?

6-37   A reservoir has a volume of 10^7 m^3. If the water in the reservoir falls an average distance of 30 m, and 80 percent of the potential energy lost is actually converted into electrical energy by turbines, how much electrical energy is generated? (The density of water is 1000 kg m^{-3}.)

6-38   Water passing through a rapids has a speed of 3 m s^{-1} as it enters the rapids, and a speed of 15 m s^{-1} as it leaves. The elevation of the river changes along the rapids from 200 m above sea level to 180 m. What fraction of the potential energy lost by the water is dissipated?

6-39 As it leaves Lake Ontario, the St. Lawrence River has a flow of 6800 m³ s⁻¹. Lake Ontario is 75 m above sea level. Ignoring any water entering the river further downstream, what is the maximum energy that could, in principle, be extracted by electric generating plants every 24 hours? (The density of water is 1000 kg m⁻³.)

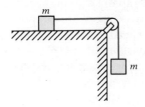

Figure 6.22. Exercises 6-40, 6-41, and 6-79.

6-40 In Fig. 6.22, the pulley and strings are massless and there is no friction. If the system is released from rest, what is its speed after the masses have moved a distance d?

6-41 In Fig. 6.22, the pulley and strings are massless. The coefficient of kinetic friction between the block and the plane is 0.2, and $m = 5$ kg. (a) How much work is done against friction as the system moves 3 m? (b) If it is initially at rest, what is its speed after it has moved 3 m?

6-42 A man lifts a box of mass 20 kg to a height of 1.6 m and throws it at a speed of 6 m s⁻¹. How much work does he do on the box?

6-43 A 40-kg girl is pulled to the top of a ski slope by a rope tow. The top of the tow is higher than the bottom by 100 m. She is initially at rest, and she gets off the tow moving at 1 m s⁻¹. If the friction is negligible, how much work does the tow do on the girl?

Section 6.7 | Gravitational Potential Energy
(Use the solar and terrestrial data on the inside back cover.)

6-44 A spaceship of mass m is in circular orbit around the earth. Its distance from the earth's surface is equal to the radius of the earth. How much energy is required to double its distance from the surface of the earth?

6-45 Suppose a spaceship of mass m is in a cir-

cular orbit around the sun at the same distance as the earth. (a) In terms of the solar mass M_S and the earth-sun distance R, how much energy E_S is required for the spaceship to escape from the solar system? (b) How much energy E_E was needed for this spaceship to escape the earth? (c) Find the numerical value of the ratio E_S/E_E.

6-46 Two identical space stations collide head on and come to rest at a distance above the earth equal to twice its radius. At what speed will the fragments reach the earth's surface, neglecting the dissipative forces?

6-47 What is the change in their gravitational potential energy when two balls of mass 10 kg with their centers initially 10 m apart are moved so that they are only 0.1 m apart?

6-48 How much energy would be required for the moon to "escape" from the earth?

6-49 On planet X, the acceleration due to gravity is four times that on earth, and the radius of the planet is twice that of the earth. What is the escape velocity?

6-50 The average center-to-center distance from the earth to the moon is 3.84×10^5 km. What is the smallest distance from the center of the earth at which the gravitational potential energies of the earth and moon are equal? The mass of the moon is approximately $1.2 \times 10^{-2} M_E$.

6-51 What is the escape velocity from the moon?

Section 6.8 | Electrical Potential Energy

6-52 A proton is fired directly toward a second proton, which is held stationary. The velocity of the incident proton is 2×10^6 m s⁻¹. How close will the protons be when they both are at rest?

6-53 An electron is fired directly at a proton that is held stationary. The initial velocity of the electron is 10^7 m s⁻¹. What is its speed when it is 10^{-11} m from the proton?

6-54 According to Bohr's theory of the hydrogen atom, the only possible radii for the electron are a_0, $2^2 a_0$, $3^2 a_0$, . . . , where $a_0 = 5.3 \times 10^{-11}$ m. How much energy in electron volts must be supplied to make the transition from the orbit of radius a_0 to the orbit with radius $n^2 a_0$, $n = 2, 3, . . .$? (This energy can be supplied by electromagnetic radiation.)

6-55 In a uranium atom, the nucleus contains 92 protons. The electrons move in orbits with varying radii, but the closest electrons have a radius $\frac{1}{92}$ times that of the radius of the electron in the hydrogen atom. Find the ratio of the energy needed to remove one of these closest electrons from a uranium atom to that needed to remove the electron from the hydrogen atom.

6-56   Find the ratio of the electrical and gravitational potential energies of the electron and proton in a hydrogen atom.

6-57   An electron in a cathode-ray tube in a television set is accelerated by electric forces from rest to 8×10^7 m s^{-1}. What is the change in its electrical potential energy in (a) joules; (b) electron volts?

6-58   In an X-ray machine, electrons are accelerated by electrical forces. If they are initially at rest and they lose 50,000 eV of electrical potential energy, what is their speed?

Section 6.9 | Power

6-59   Two teams of students pull a rope in a tug of war. Team A is gaining over B, since the rope is moving in its direction at a constant rate of 0.01 m s^{-1}. The tension in the rope is 4000 N. How much power is expended by team A?

6-60   If a man can do mechanical work at a rate of 8 W kg^{-1} of body mass during sustained activity, how fast can he run up stairs?

6-61   A girl of mass 40 kg climbs a rope 8 m long at constant speed in 15 s. What power does she expend against gravitational forces during the climb?

6-62   An elevator motor produces 2000 W of power. How fast can it lift a 1000-kg load?

6-63   Cruising speeds for fish 0.3 m long are about 0.35 m s^{-1}. The average power expended is about 4.5 W kg^{-1} of body mass. Assume the fish has a mass of 0.4 kg. (a) What is the average power expended by the cruising fish? (b) What is the average force exerted by the fish on the water? (c) How much work does the fish do on the water in 10 minutes?

6-64   A motor operating a water pump can produce 1000 W of power. If the kinetic energy change is negligible, how many kilograms of water per second can it pump from a well 20 m deep?

6-65   How much does it cost to keep a 100-W bulb running per 24-hour day if electrical energy costs 10 cents per kilowatt-hour (kWh)?

6-66   A man and a bicycle have a total mass of 100 kg. He is cycling at a constant 8 m s^{-1} up a hill that makes a 4° angle with the horizontal direction. What power must the rider expend against gravitational forces?

6-67   A bicyclist moving on flat ground at a constant speed of 5 m s^{-1} expends 100 W against dissipative forces. (a) If the dissipative forces are independent of the velocity, what power will she expend at a constant speed of 10 m s^{-1}? (b) The portion of the dissipative forces arising from air resistance actually increases rapidly with the speed. If we suppose the dissipative forces are proportional to the square of the speed, what power will she have to expend at a constant 10 m s^{-1}?

6-68   A 2000-kg car accelerates from rest to 30 m s^{-1} in 10 s. What average power is required?

6-69   A chair lift delivers two skiers to the top of a 500-m-tall hill every 12 s. The average mass of a skier plus equipment is 80 kg. Assuming dissipative forces are negligible, find the power expended by the chair lift motor.

6-70   Direct sunlight reaches a horizontal surface at an average rate of 200 watts per square metre of surface area, averaging over day and night, times of the year, and cloudy and clear weather. Suppose 10 percent of this solar energy could be converted into electrical power. How large an area in square kilometres would be required to replace a large nuclear power plant that supplies 10^9 W?

6-71   An average American family of 4 persons uses about 8 kW of power in all forms in its home. (a) Direct solar energy is incident on a horizontal surface at an average rate of 200 W per square metre. If 20 percent of this energy is collected and used, how large an area is needed to supply the 8 kW? (b) Compare this area to that of the roof on a typical single-family home.

6-72   Solar energy is received at a rate of 350 W per square metre of the earth's surface, averaging over latitude and time of day and year. (The energy is received no matter what the weather.) About 2 percent of that energy is converted into wind energy. (a) Find the ratio of the wind power generated by the sun over the globe to the total power used by humans, about 10^{13} W. (The radius of the earth is 6.38×10^6 m.) (b) It has been suggested that a maximum of about 3 percent of the wind power could in principle be extracted for human activities. Would this be adequate to supply our global energy needs?

Section 6.10 | Work and Energy for Rotational Motion (See Table 5.3 for moments of inertia.)

6-73 A tricycle wheel of moment of inertia 0.04 kg m^2 is rotating once per second. What is its kinetic energy?

6-74 Using the astronomical data on the inside back cover, find the kinetic energy associated with the earth's daily rotation. (Assume the earth is a uniform sphere.)

6-75 A 3-kg cylinder of radius 0.2 m is rotating about its axis at 40 rad s^{-1}. (a) Find the kinetic energy if the cylinder is solid. (b) Find the kinetic energy if the cylinder is a thin shell.

6-76 Refinements in materials technology have opened the possibility of using the energy stored in flywheels to operate vehicles. (a) If a flywheel is a solid cylinder of mass 1000 kg and radius 1 m, what is its kinetic energy when it rotates at 1000 rad s^{-1}? (b) How long can it supply energy to a vehicle at an average power of 20 kW?

6-77 A lawn mower blade has a mass of 3 kg and a radius of gyration of 0.1 m. It is rotating at 300 rad s^{-1}. (a) What is its kinetic energy? (b) When the engine is stopped, the blade comes to rest after 100 full rotations. What is the average torque acting on the blade? (c) What has happened to the kinetic energy?

6-78 An electric drill is rotating at 200 rad s^{-1} as the bit bores through a block of wood. If the drill motor is expending 400 W of power, what torque is the wood exerting on the drill bit?

6-79 In Fig. 6.22, the pulley is a uniform cylindrical disk of mass m and radius r. The strings are massless and there is no friction. If the system is initially at rest, find the speed of the blocks after they have moved a distance d.

6-80 A bicycle wheel has a mass of 4 kg and a radius of 0.35 m. All its mass may be considered to be concentrated at the edge. When the bicycle is lifted off the ground and the wheel is spun at 5 rev s^{-1}, it comes to rest after 20 full revolutions. (a) What is the average torque acting on the wheel? (b) If this torque is due entirely to friction in the bearings, and this force acts opposite to the motion at a distance of 1 cm from the axis, how large is it?

Section 6.11 | Jumping; Scaling Laws in Physiology

6-81 Using the data in Table 1.4, calculate the work and power per unit mass in jumping for (a) a kangaroo; and (b) a locust.

6-82 Which of the animals listed in Table 1.4 does the most work per unit mass in jumping vertically? Which does the least?

6-83 Show that if the rate at which oxygen is absorbed by an animal and supplied to its tissues varies as the surface area of the arteries, then its power output per unit mass should vary as l^{-1}. What does this suggest about the pulse rate?

6-84 How does the rate of heat loss per unit body mass in a cold environment vary with the characteristic length l of an animal? (The rate of heat loss is proportional to the surface area.)

PROBLEMS

6-85 A horizontal force **F** applied to a block of mass m pushes it at constant speed a distance l along a plane at an angle θ to the horizontal (Fig. 6.23). If the plane is frictionless, how much work is done by the force **F**?

6-86 The force needed to compress a certain spring a distance s from its normal length when undisturbed is $F = ks$, where $k = 300$ N m^{-1}. (a) Draw a graph of F versus s. (b) How much work must be done to compress the spring from its normal length of 0.2 m less? (c) How much more work must be done to compress it an additional 0.2 m?

*6-87 If a salmon encounters a falls when swimming upstream, it will attempt to ascend the falls in one of two ways. If it can swim fast enough, it will swim up the falls. If it cannot, it will jump from the base of the falls into the falls at a height where the water is moving slowly enough so that it can swim up to the top.

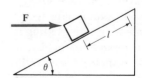

Figure 6.23. Problem 6-85.

Assume that the salmon has a maximum swimming speed of 5 m s^{-1} in still water and that the water at the top of the falls and in the pool at the base of the falls is at rest. (a) What is the maximum height of a falls that the salmon can swim up without jumping? (b) If the falls is 1 m high, what is the velocity of the fish with respect to the ground when it begins to swim upward at the bottom of the falling stream? (c) If the falls is 2 m high, what is the minimum height in the stream to which the fish must jump in order to swim up the rest of the way? (d) In order to jump the distance necessary in part (c) , what must the fish's initial velocity be when it leaves the water?

*6-88   On a certain river, a falls 6 m high is encountered by the salmon of Problem 6-87. To facilitate the progress of the salmon upstream, a fish ladder is constructed. It is a series of sloping flow passages with connecting resting pools where the water velocty is virtually zero. If we refer to each pool with its flow passage as a step, what is the minimum number of steps necessary to allow the salmon to ascend the falls without jumping?

6-89   An upper limit for the amount of energy that can be extracted from water power in the United States can readily be calculated. The average depth of the annual rainfall is about 0.75 m, and the area of the United States is 8×10^6 km^2. (a) What is the mass of the annual rainfall? (b) Averaging over mountains, plains, and coastal areas, the average elevation is about 500 m. If all the rainwater eventually reached the oceans, how much potential energy would be dissipated? (b) Actually, two thirds of the water evaporates and returns to the atmosphere. Suppose all the remaining water were used to generate electric power. What would be the average power produced, assuming no energy was dissipated as heat? (The installed hydroelectric capacity in the United States is about 65 GW.)

6-90   Large fossil fuel or nuclear electrical power plants operate most economically when run at full capacity 24 hours a day. Excess electrical energy can be "stored" and recovered by pumping water into a reservoir on top of a mountain and letting it fall through turbines during periods of peak demand. At such a facility at Northfield Mountain in Massachusetts, the water falls an average distance of 250 m. The reservoir has a surface area of 1.3 km^2 and a depth averaging 10 m. (a) How much energy is available each time the reservoir is emptied? (b) If it is emptied over a period of 10 hours, and 80 percent of the energy is converted into electrical energy, what is the power generated?

6-91   In parts of the Bay of Fundy, the height difference between high and low tide can reach the huge value of 17 m. The average height difference in the bay is about 4 m. The bay is about 300 km long and 65 km wide. (a) How much gravitational potential energy is lost when the bay empties? (b) Suppose half this energy could be transformed into electrical power each time the bay emptied and also each time it filled. There are approximately two high and two low tides daily. What would be the average electrical power generated? (c) Find the ratio of this power to the electrical power currently used in Canada, about 2×10^{10} W.

6-92   Water on the sides of the earth facing and opposite the moon is higher than the water in between. The rotation of the earth under these bulges causes high and low tides approximately twice each day. (a) Show that when water in an area A of the ocean falls a distance d from high to low tide, the decrease in gravitational potential energy is $\rho g A d^2 / 2$, where ρ is the density of the water. (b) Assuming the earth is fully covered with water, show that the global total of the daily decreases in potential energy is $4\pi R_E^2 \rho g d^2$, where R_E is the radius of the earth. (Some of this energy is dissipated as thermal energy each day, and is replaced by kinetic energy from the rotational motion of the earth. In the process, the rotation slows slightly, so that the day lengthens by 1.5×10^{-3} seconds per century.)

6-93   In the preceding problem, a formula is given for the daily potential energy changes associated with the tides. (a) Calculate this energy if the average height change d from high to low tide is 0.5 m. (The density of water is 1000 kg m^{-3}, and the radius of the earth is 6.38×10^6 m.) (b) What is the average power corresponding to this energy? (c) What is the ratio of this power to the total power used by people worldwide, about 10^{13} W?

6-94   The Horseshoe Falls on the Niagara River in Canada is about 50 m high and 800 m wide. The water is moving at 10 m s^{-1} and has a depth of 1 m as it goes over the falls. (a) What volume of water goes over the falls every second? (b) What is the change in potential energy of this volume of water? (One cubic metre of water has a mass of 10^3 kg.) (c) If this potential energy could be converted directly into electrical energy, how much electrical power would be produced? (d) The total electrical power-generating capacity in the United States is about 5×10^{11} W. What percentage of this power could be produced by harnessing the Horseshoe Falls with 80 percent efficiency?

6-95   In order for a 2000-kg car to maintain a constant speed of 65 km h^{-1}, work must be done against dissipative forces at a rate of 9 kW. (a) How large are the dissipative forces? (b) The efficiency of a gasoline engine is only 20 percent, some power is lost in the transmission and drive train, and additional power is required to operate the lights, generator, water pump, and other accessories. Hence only 12.5 percent of the energy obtained by burning the gasoline is actually used to keep the car in motion. How far can the car travel at this speed on a litre of gasoline, which contains 3.4×10^7 J of chemical energy?

6-96   (a) What is the kinetic energy of a 2000-kg car traveling at 65 km h^{-1}? (b) If this speed is reached in 10 s, what average power is required? (c) Find the ratio of this power to the 9 kW required to maintain a steady 65 km h^{-1}. (d) Using the data in the preceding problem, calculate the amount of gasoline in litres needed to bring a car from rest to 65 km h^{-1}.

6-97   When a car goes uphill, extra power must be supplied in addition to that needed to maintain a constant speed on flat ground. For the car moving at 65 km h^{-1} in Problem 6-95, at what angle is the road to the horizontal if the power requirements are double that on flat ground?

6-98   Some of the power required to keep a car moving is dissipated by air resistance, and some by the work done in deforming the tires as they roll along the road (road resistance). At 65 km h^{-1}, these are about equal. (a) The air resistance varies approximately as the square of the velocity, while the road resistance forces are nearly independent of the velocity. By what factor will the power expenditure increase if the speed of the car is doubled? (b) By what factor will the number of kilometres driven per litre be reduced?

*6-99   The potential energy of an object of mass m at a height h above the earth's surface is $\mathfrak{U} = -GM_\mathrm{E}m/(R_\mathrm{E} + h)$. R_E is the earth's radius and M_E its mass. Using Eq. B.15 in Appendix B, show for $h \ll R_\mathrm{E}$ that

$$\mathfrak{U} \simeq -\frac{GM_\mathrm{E}m}{R_\mathrm{E}}\left(1 - \frac{h}{R_\mathrm{E}}\right).$$

*6-100   Show that for a rocket fired in the vicinity of the earth, the minimum velocity to escape the solar system is $v_s = \sqrt{2}C/T$, where C is the circumference of the earth's orbit and T is 1 year, the period of the earth's rotation about the sun. (Neglect the energy needed to escape the earth.)

6-101   Using the formula in the preceding problem, compare the escape velocities needed to escape the earth and to escape the solar system.

6-102   Assuming all planets have the same density or mass-to-volume ratio, show how the escape velocity varies with the radius of the planet.

6-103   At one time it was thought that the sun's energy came from gravitational sources. If the sun were originally larger and it had contracted to its present size, gravitational potential energy would have been transformed into heat. To estimate the energy involved, suppose the sun originally had two distant parts, each of mass $M_\mathrm{S}/2$, where M_S is the mass of the sun. (a) If the pieces were brought from infinity to an average separation of R_S, show that the energy released would be $GM_\mathrm{S}{}^2/4R_\mathrm{S}$. ($R_\mathrm{S}$ is the radius of the sun.) (b) Calculate that energy using the data on the inside rear cover of this textbook. (c) The sun is believed to have radiated energy at its present rate, 3.8×10^{26} W, for 5×10^9 years. How long could it radiate at that rate, based on the estimate calculated in part (b)? (The short time the energy would last was once used to set a limit on the age of the solar system. This limit served as an argument against Darwin's theory of evolution, since it required a much longer time scale. The sun's main source of energy is nuclear fusion.)

6-104   An orbiting space station at a distance above the earth's surface equal to its radius ejects an instrument package with an initial velocity of 5000 m s^{-1} relative to the earth. What is its velocity when it reaches the atmosphere just a few kilometres above the earth?

*6-105   The energy dissipated in the oceans by the tides is replenished via frictional forces from the earth's rotational kinetic energy. Suppose we could somehow extract the present global needs of 10^{13} W from the tides. (This is 100 times as much power as could realistically be obtained from the tides.) If this were an additional drain on the earth's kinetic energy, by how much would the day lengthen after a century? (Assume the earth is a uniform sphere and use the solar and terrestrial data on the inside rear cover.)

*6-106   The day lengthens by 1.5×10^{-3} s each century due to tidal friction. At what rate is power being dissipated by the tides? (Assume the earth is a uniform sphere and use the solar and terrestrial data on the inside rear cover.)

6-107   The muscle force is proportional to the cross-sectional area of the muscle. Show that if the speed of contraction is constant, the power per kilogram is proportional to l^{-1}.

*6-108   Show that if the power output of an animal of length l varies as l^2, the speed with which it can run uphill varies as l^{-1}.

6-109   Show that if the consumption of oxygen for a mammal of characteristic length l varies as l^2, the time it can remain under water without breathing varies as l.

6-110   Show that if the power expended per unit mass of muscle were the same for all animals of the same type, the height jumped would vary as $l^{2/3}$, where l is the characteristic size of the animal.

^{c}6-111   A car starts from rest with a constant acceleration a. (a) After a time t has elapsed, how much work has been done on the car? (b) At time t, what is the instantaneous power delivered to the car?

^{c}6-112   A car of mass m starts from rest and accelerates so that the instantaneous power delivered to the car has the constant magnitude $\mathcal{P}_0$. (a) After a time t has elapsed, how much work has been done on the car? (b) At time t, what is the instantaneous velocity of the car?

ANSWERS TO REVIEW QUESTIONS

Q6-1, parallel, opposite, perpendicular; **Q6-2,** joule; **Q6-3,** do work; **Q6-4,** $\frac{1}{2}mv^2$; **Q6-5,** initial kinetic energy, all the forces acting on it; **Q6-6,** conservative; **Q6-7,** kinetic energy, potential energy; **Q6-8,** applied forces do no work; **Q6-9,** position or configuration; **Q6-10,** arbitrary; **Q6-11,** work done by conservative forces; **Q6-12,** thermal energy; **Q6-13,** negative;

Q6-14, near the surface of the earth; **Q6-15,** escape velocity; **Q6-16,** work is done; **Q6-17,** watt; **Q6-18,** energy; **Q6-19,** scaling.

SUPPLEMENTARY TOPICS
6.12 | RUNNING

In our discussion of jumping, we assumed that all the work done by the muscles is converted into mechanical energy and that no energy is dissipated. This cannot be true for an animal running at constant speed on flat ground. In this case the mechanical energy remains constant. However, in each stride, energy is expended by the muscles in accelerating the legs and in lifting the center of gravity of the body. This energy is dissipated as the legs slow down and the center of gravity falls. A variety of dissipative forces are involved, making any model that relies on a direct analysis based on Newton's laws very difficult to use for running.

Another kind of model proposed recently by a mathematician, J.B. Keller, sidesteps the details of the forces. The runner is assumed to exert a net force f which varies with the time but never exceeds a maximum force, F_{max}. The power expended by the runner is the force times the velocity, fv. A dissipative force, D, opposes the motion and is assumed to be proportional to the velocity, so $D = cv$. The runner is able to draw on an initial reserve of stored energy E_0, but once that is exhausted the speed is limited by the rate σ (assumed constant) at which additional energy is supplied by metabolic processes.

Given these assumptions, the mathematical problem is to find the force f that the runner must exert to run a given race as fast as possible. The four physiological parameters of this model—the quantities F_{max}, c, E_0, and σ—are chosen by comparing the calculated times with the men's world records for several distances. If the model then predicts many other records correctly, it can be considered successful and its physiological parameters may be of use in studying other activities.

Keller's solution to this problem states that for races shorter than 291 m, a runner should exert the maximum force F_{max} throughout the race. In longer events, he should accelerate as fast as possible for a second or two and then run at constant speed until he slows slightly in the last few steps. These results are

TABLE 6.2

Physiological parameters in S.I. units for the Keller model of running for a man of mass 80 kg. These constants are all proportional to the body mass. For example, for a runner of mass 100 kg, the parameters should be multiplied by 100/80 = 1.25

Maximum force	$F_{max} = 976$ N
Coefficient of the dissipative force	$c = 89.7$ N s m^{-1}
Stored metabolic energy	$E_0 = 193,000$ J
Rate of metabolic energy conversion	$\sigma = 3330$ W

consistent with the usual advice to maintain a steady pace. However, runners often finish long races with a burst of speed rather than by slowing down. The competitive aspects of racing may be the reason for this.

Table 6.2 gives the four physiological parameters found by comparison with the men's world records. Figure 6.24 shows that the predicted average speed-versus-distance curve agrees very well with the data at all distances up to about 10,000 m. The striking initial rise and later decrease in this speed as the distance increases is correctly predicted.

The use of the physiological parameters is illustrated by the following example.

Example 6.17

The world record for the 5000-m run is 796.6 s, corresponding to an average velocity of 6.28 m s^{-1}. Find the

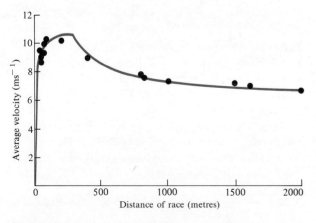

Figure 6.24. The average velocity for running races of various lengths. The dots represent average speeds for men's world records. The curve is the theoretical prediction using the values of the four parameters in Keller's model which give the best overall fit to the world records. (From Joseph B. Keller, *Physics Today.* © American Institute of Physics.)

power dissipated at this speed by a runner with a mass of 80 kg, and compare this with the rate at which additional energy is supplied by metabolic processes.

The power dissipated is $\mathcal{P}_d = Dv = (cv)v$. From Table 6.2, $c = 89.7$ N m^{-1}, and

$$\mathcal{P}_d = cv^2 = (89.7 \text{ N s m}^{-1})(6.28 \text{ m s}^{-1})^2$$
$$= 3538 \text{ W}$$

Energy is supplied at a rate of $\sigma = 3330$ W, so the runner is expending stored energy. The difference between the power dissipated and the power supplied is 208 W. In 797 s, this corresponds to an energy expenditure of $E = (208 \text{ W})(797 \text{ s}) = 166,000$ J. This is close to the total stored energy $E_0 = 193,000$ J, indicating as expected that the runner has used nearly all his stored energy.

It remains to be seen whether this particular model correctly describes the most important features of running. In any event, it shows how a simple model incorporating energy considerations can illuminate a very complex problem.

EXERCISES ON SUPPLEMENTARY TOPICS

Section 6.12 | Running

6-113 (a) According to Keller's model, how large is the dissipative force on an 80-kg man running at 8 m s^{-1}? (b) What is his power expenditure?

6-114 Metabolizing a gram of fat yields about 8000 J of mechanical energy. If all his energy is supplied from metabolizing fat, how much does an 80-kg runner consume at 7 m s^{-1} according to Keller's model?

PROBLEMS ON SUPPLEMENTARY TOPICS

6-115 In Keller's model, the power expenditure of a runner equals the power input when $v^2 = \sigma/c$. Calculate the velocity at which this occurs, and compare it to the world record average velocity for a 5000-m race, which is 6.28 m s^{-1}.

6-116 (a) Calculate the maximum force a 60-kg runner can exert. (b) In a 2000-m race, the world record average speed is 6.64 m s^{-1}. What is the dissipative force? (c) Explain the difference.

6-117 (a) Suppose a runner exerted the maximum possible force and did not have any additional energy supply. How far could he run using his stored energy? (b) What period of time is needed for the metabolic processes to supply an amount of energy equal to the stored reserves?

Additional Reading

A.Z. Hendel, Solar Escape, *The American Journal of Physics,* vol. 51, 1983, p. 746. Velocity for a rocket fired from the earth to escape the solar system. A problem treated incorrectly in many textbooks.

Elmer L. Offenbacher, Physics and the Vertical Jump, *American Journal of Physics,* vol. 38, 1970, p. 829.

Sir James Gray, *How Animals Move,* Cambridge University Press, Cambridge, 1953. Pages 69–80 discuss jumping.

R. McNeill Alexander, *Animal Mechanics,* University of Washington Press, Seattle, 1968. Pages 28–33 discuss jumping.

E.F. Adolph, Quantitative Relations in the Physiological Constitutions of Mammals, *Science,* vol. 109, 1949, p. 579. Scaling in biology.

Walter R. Stahl, Similarity and Dimensional Methods in Biology. *Science,* vol. 137, 1962, p. 205.

A. Gold, Energy Expenditure in Animal Locomotion, *Science,* vol. 181, 1973, p. 275.

John Maynard Smith, *Mathematical Ideas in Biology,* Cambridge University Press, Cambridge, 1968. Scaling, running.

Joseph B. Keller, A Theory of Competitive Running, *Physics Today,* vol. 26, September 1973, p. 42.

Knut Schmidt-Nielson, Locomotion: Energy Cost of Swimming, Flying, and Running, *Science,* vol. 177, 1972, p. 222.

Knut Schmidt-Nielsen, *How Animals Work,* Cambridge University Press, London, 1972. Applications of physical principles to physiology, scaling.

Galileo Galilei, *Dialogues Concerning Two New Sciences,* translated by Henry Crew and Alfonso de Salvio. The Macmillan Co., New York, 1914; Dover Publications Inc., New York 1954. The first discussion of scaling in biology.

David Pilbeam and Stephen Jay Gould, Size and Scaling in Human Evolution, *Science,* vol. 186, 1974, p. 892.

V.A. Tucker, The Energetic Cost of Moving About, *American Scientist,* vol. 63, 1975, p. 413. Advantages of flying, swimming, bicycling compared to walking and running.

The additional reading in Chapter Eleven includes many references to energy-related topics. *Science* and *Scientific American* print many up-to-date articles on alternative energy sources and related topics. Some articles relating to the topics mentioned in this chapter are:

George F.D. Duff, Tidal Power in the Bay of Fundy, *Technology Review,* November, 1978, p. 34.

M.R. Gustavson, Limits to Wind Power Utilization, *Science,* vol. 204, April 6, 1979, p. 13.

John D. Isaacs and Walter R. Schmitt, Ocean Energy: Forms and Prospects, *Science,* vol. 207, January 18, 1980, p. 265.

G. Waring, Energy and the Automobile, *The Physics Teacher,* vol. 18, 1980, p. 494.

J. Kelly Beatty, Solar Satellites, *Science 80,* December 1980, p. 28.

Bernard L. Cohen, Cost per Million BTU of Solar Heat, Insulation, and Conventional Fuels, *The American Journal of Physics,* vol. 52, 1984, p. 614.

The following books present detailed discussions of the consequences of finite energy resources, environmental effects, and the science and technology of alternative energy sources.

Robert H. Romer, *Energy, An Introduction to Physics,* W. H. Freeman and Co., San Francisco, 1976. A basic physics text with many energy applications.

Joseph Priest, *Energy for a Technological Society,* Second Edition, Addison-Wesley Publishing Co., Reading, 1979. Paperback.

John M. Fowler, *Energy and the Environment,* McGraw-Hill Book Co., New York, 1975. Paperback.

Energy, Readings from Scientific American, W.H. Freeman and Co., San Francisco, 1979. Paperback.

Delbert W. Devins, *Energy: Its Physical Impact on the Environment,* J. Wiley & Sons, New York, 1982.

Scientific American articles:

Milton Hildebrand, How Animals Run, May 1960, p. 149.

Henry W. Ryder, Harry J. Carr, and Paul Herget, Future Performances in Footracing, June 1976, p. 109.

T.J. Dawson, Kangaroos, August 1977, p. 78. Energetics of hopping.

Charles L. Gray, Jr., and Frank von Hippel, The Fuel Economy of Light Vehicles, May 1981, p. 48.

Jearl Walker, Thinking About Physics While Being Scared to Death (on a Falling Roller Coaster), *The Amateur Scientist,* October 1983, p. 162.

CHAPTER 7

LINEAR AND ANGULAR MOMENTUM

We have seen how energy conservation can be used in many situations to analyze the motion of both simple and complex systems. In this chapter we introduce two other quantities that are conserved under certain conditions: *linear momentum* and *angular momentum*. Like energy conservation, the conservation of linear and of angular momentum are concepts of widespread applicability, and they are useful even in the subatomic world where Newton's laws do not hold.

Many common events can be regarded as collisions even though we may not think of them as such. A baseball player hitting or catching a ball, a blow struck in a boxing match, and the action of seat belts on passengers during an automobile accident are all examples of collisions. Although collisions are not unusual in our experience, they are difficult to analyze using the methods we have described so far in this textbook. In particular, since the forces acting during a collision are often difficult to determine, Newton's second law is hard to use directly. Furthermore, in many collisions, mechanical energy is not conserved, and the methods of energy conservation described in the previous chapter are also frustrated.

Linear momentum, which is often referred to more concisely as *momentum* when there is no chance of confusion with the angular momentum, is especially useful in dealing with collisions. The momentum of an object is the product of its mass and velocity. When two objects collide for a brief time, the momentum of each changes, but the *total* momentum of the system often remains constant, either exactly or to a very good approximation.

The change in the linear momentum of an object is very closely related to the forces on the object. Because of this, we can sometimes find the average forces acting during a very complex event by measuring the momentum changes. The forces acting when momentum changes occur are contained in a quantity called the *impulse*.

The angular momentum of a rigid body rotating about a fixed axis is its moment of inertia times its angular velocity. Just as linear momentum conservation can be used to analyze translational motion, angular momentum conservation provides insight into complex rotational motion problems. For example, spinning figure skaters, tumbling gymnasts, and falling cats all have a constant angular momentum. This fact is important in understanding their motion.

7.1 | IMPULSE AND LINEAR MOMENTUM

With the aid of an example, we now define *impulse* and *linear momentum* and find the relationship between them. Figure 7.1 shows a man pushing a dresser with casters. We will suppose that the frictional forces on the dresser are negligible. With Newton's laws, we can find a relation between the forces acting on the dresser, its velocity change, and the time the force acts.

The force **F** acts on the dresser for a time Δt. During this period the velocity of the dresser changes from **v** to **v**′. By definition, the average acceleration $\bar{\mathbf{a}}$ is the velocity change divided by the time, $\bar{\mathbf{a}} = (\mathbf{v}' - \mathbf{v})/\Delta t$. However, from Newton's second

153

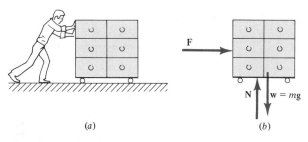

(a)

(b)

Figure 7.1. (*a*) A man pushes a dresser, which moves frictionlessly on casters. (*b*) The forces on the dresser are **F**, the weight **w** = *m***g**, and the normal force **N**. The net force is equal to **F**.

(a)   (b)   (c)

Figure 7.2. A passenger in a vehicle during a collision: (*a*) without seat belts; (*b*) with a lap belt only; (*c*) with a lap belt and shoulder harness.

law, the average acceleration and net force are related by $\overline{\mathbf{F}} = m\overline{\mathbf{a}}$, so

$$\overline{\mathbf{F}} = m\left(\frac{\mathbf{v}' - \mathbf{v}}{\Delta t}\right)$$

or

$$\overline{\mathbf{F}}\,\Delta t = m\mathbf{v}' - m\mathbf{v} \qquad (7.1)$$

The product of the average force and the time $\overline{F}\Delta t$ is called the *impulse*. An equivalent expression is $\int F\,dt$, where the integral is over the time Δt the force acts. The product $m\mathbf{v}$ is the *linear momentum*,

$$\mathbf{p} = m\mathbf{v} \qquad (7.2)$$

We will refer to **p** more simply as the *momentum* only when there is no angular motion and hence no chance of confusion with the angular momentum. From its definition, the S.I. momentum unit is the kilogram-metre per second (kg m s^{-1}). Equation 7.1 means that *the impulse equals the change in momentum,*

$$\overline{\mathbf{F}}\,\Delta t = \mathbf{p}' - \mathbf{p} \qquad (7.3)$$

The following example illustrates how this result is used.

Example 7.1

A baseball, initially at rest, is struck with a bat. The velocity of the 0.15-kg ball just after it is hit is 40 m s^{-1}. If the impact time is 10^{-3} s, what is the average force on the ball?

The initial momentum **p** of the ball is zero, since it starts from rest; the final momentum is $\mathbf{p}' = m\mathbf{v}'$. Thus, from $\overline{\mathbf{F}}\,\Delta t = \mathbf{p}' - \mathbf{p}$, the average force $\overline{F}$ on the ball is

$$\overline{F} = \frac{mv'}{\Delta t} = \frac{(0.15\text{ kg})(40\text{ m s}^{-1})}{(10^{-3}\text{ s})} = 6000\text{ N}$$

Highway Safety | Automobile accidents are complex events involving many variables. However, many questions about the safety of passengers in-

volved in such accidents can be explored using the concepts of impulse and momentum.

To visualize what happens in an accident, suppose that a car hits a stone wall or a tree head-on at an appreciable speed. The front end of the car will collapse, and the passenger compartment will come to rest in a metre or so. The time required for this to occur is typically a few tenths of a second. A passenger who is wearing a seat belt or shoulder harness will therefore come to rest in a few tenths of a second. However, an unrestrained passenger will slide forward at roughly the original speed of the car, striking the windshield or dashboard (Fig. 7.2). Such a passenger will come to rest in a very short time and experience much larger forces during the impact. The force may also be applied over a very small area, increasing the severity of the injuries. The following example illustrates the difference between the two situations.

Example 7.2

A car traveling at 10 m s^{-1} (36 km h^{-1}) collides with a tree. (a) An unrestrained passenger strikes the windshield headfirst and comes to rest in 0.002 s. The contact area between the head and windshield is 6×10^{-4} m^2, and the mass of the head is 5 kg. Find the average force and the force per unit area exerted on the head. (b) A passenger of mass 70 kg wearing a shoulder harness comes to rest in 0.2 s. The area of the harness in contact with the passenger is 0.1 m^2. Find the average force and the average force per unit area.

(a) Again we use $\overline{\mathbf{F}}\,\Delta t = \mathbf{p}' - \mathbf{p}$. The final momentum is zero, since the windshield is stationary, and the initial momentum is the head mass times the velocity. Thus the magnitude of the average force is

$$\overline{F} = \frac{p}{\Delta t} = \frac{(5\text{ kg})(10\text{ m s}^{-1})}{0.002\text{ s}} = 25{,}000\text{ N}$$

The average force per unit area is

$$\frac{\overline{F}}{A} = \frac{25{,}000\text{ N}}{6 \times 10^{-4}\text{ m}} = 4.16 \times 10^7\text{ N m}^{-2}$$

This is a very large force per unit area and is certain to cause serious injury.

(b) The average force is found from the change in momentum of the entire body as the speed of the car changes from 10 m s^{-1} to zero. Thus the magnitude of $\bar{F}$ is

$$\bar{F} = \frac{p}{\Delta t} = \frac{(70 \text{ kg})(10 \text{ m s}^{-1})}{0.2 \text{ s}} = 3500 \text{ N}$$

This is much smaller than the force exerted on the head of the unrestrained passenger in part (a). The average force per unit area is

$$\frac{\bar{F}}{A} = \frac{3500 \text{ N}}{0.1 \text{ m}^2} = 3.5 \times 10^4 \text{ N m}^{-2}$$

Since this is smaller than the force per unit area on the unrestrained passenger by a factor 1200, the chance of serious injury is much smaller for this passenger.

Improving the safety of passengers in automobile collisions involves increasing the time over which the momentum changes or increasing the area over which the decelerating forces are applied. In the first category are improved passenger restraints and front-end designs that permit more gradual slowing of the passenger compartment. The removal of sharp objects and the use of inflatable air bags are examples of attempts to increase the contact area.

7.2 | MOMENTUM CONSERVATION

The concept of momentum is most useful when two or more objects act on one another. For example, in Fig. 7.3 we show two objects that collide. Friction is negligible, and each object experiences a gravitational force and a normal force. During the collision, they also exert forces on each other. These forces, $\mathbf{F}_{12}$ and $\mathbf{F}_{21}$, are an action-reaction pair, so they are equal and opposite and $\mathbf{F}_{12} + \mathbf{F}_{21} = 0$.

If the collision time is Δt, the change in momentum of each object can be computed from the average forces $\bar{\mathbf{F}}_{12}$ and $\bar{\mathbf{F}}_{21}$. For m_1, $\bar{\mathbf{F}}_{12} \Delta t = \mathbf{p}'_1 - \mathbf{p}_1$, and for m_2, $\bar{\mathbf{F}}_{21} \Delta t = \mathbf{p}'_2 - \mathbf{p}_2$. But $\bar{\mathbf{F}}_{12} + \bar{\mathbf{F}}_{21} = 0$, so if we add these two equations, we find

$$\mathbf{p}'_1 - \mathbf{p}_1 + \mathbf{p}'_2 - \mathbf{p}_2 = 0$$

or

$$\mathbf{p}_1 + \mathbf{p}_2 = \mathbf{p}'_1 + \mathbf{p}'_2 \qquad (7.4)$$

This equation means that the net momentum of the two objects before and after the collision is the same, or the total momentum of the system is conserved.

This is important because it means that we can relate the velocities of the objects before and after the collision *without knowing anything about the forces between them during the collision*. The fact that these forces are always equal and opposite is sufficient to

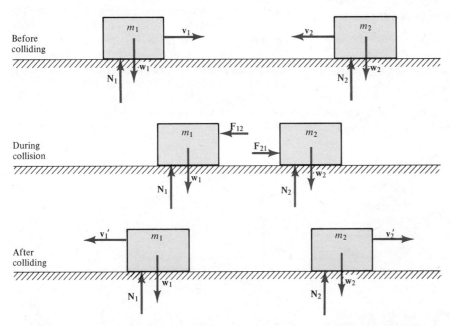

Figure 7.3. Two objects slide frictionlessly towards each other. Their initial momenta are $\mathbf{p}_1 = m_1\mathbf{v}_1$ and $\mathbf{p}_2 = m\mathbf{v}_2$. During the collision the objects exert equal and opposite forces on each other. After the collision the momenta of the two objects are $\mathbf{p}'_1 = m_1\mathbf{v}'_1$ and $\mathbf{p}'_2 = m_2\mathbf{v}'_2$.

obtain the result. The following example illustrates this idea.

Example 7.3

A neutron moving at 2700 m s^{-1} collides head-on with a nitrogen nucleus at rest and is absorbed. The neutron and nitrogen masses are $m = 1.67 \times 10^{-27}$ kg and $M = 23.0 \times 10^{-27}$ kg, respectively. What is the final velocity of the combined object? (Fig. 7.4.)

The only forces affecting the motion during the collision are those between the neutron and nucleus. Thus the net momentum is constant. Before the collision, $p = mv$ and afterwards $p' = (m + M)v'$, where v' is the final velocity. Then $mv = (m + M)v'$, so

$$v' = \frac{mv}{(m + M)} = \frac{(1.67 \times 10^{-27}\text{ kg})(2700\text{ m s}^{-1})}{(1.67 + 23.0) \times 10^{-27}\text{ kg}}$$
$$= 183\text{ m s}^{-1}$$

In order to know when momentum is conserved, we must understand the role of *internal* and *external* forces. In Fig. 7.3b, the normal and gravitational forces are external forces. Since there is no vertical motion, $\mathbf{N}_1 + \mathbf{w}_1 = 0$ and $\mathbf{N}_2 + \mathbf{w}_2 = 0$, and there is no net impulse as a result of these forces. The force $\mathbf{F}_{12}$ is an external force on m_1, and $\mathbf{F}_{21}$ is an external force on m_2. However, if m_1 and m_2 are considered as *one system*, we can say that $\mathbf{F}_{12}$ and $\mathbf{F}_{21}$ are internal forces of the system. Thus, in Fig. 7.3, there is *no net external force acting on the system*. The momentum of m_1 alone is not conserved nor is that of m_2. However, the net momentum of the system is conserved.

Whenever there is no net external force acting on a system, its momentum is conserved. Momentum is therefore always conserved for an isolated system, one subjected only to internal forces.

In many situations of interest to us, the momentum is only approximately conserved because there are net external forces acting. However, the impulses resulting from these external forces are often small enough to be neglected. For example, when two cars collide, they exert huge forces on each other for a very short time, and the impulse on each is substantial. An external force such as a frictional force exerted by the road may also be present, but its magnitude is not large enough to contribute any appreciable impulse in the brief time of the collision. Hence, only the internal forces between the cars are really important, and momentum conservation can be used to analyze the collision.

Momentum is also conserved to a very good approximation when an object explodes, since no external forces contribute an appreciable impulse during the brief time of the explosion. For example, when a rifle is fired, it must recoil backward to conserve momentum when the bullet and gases resulting from the gunpowder explosion move forward. A situation of this type is discussed in the next example.

Example 7.4

A cannon is mounted inside a railroad car, which is initially at rest but can move frictionlessly (Fig. 7.5). It fires a cannonball of mass $m = 5$ kg with a horizontal velocity $v = 15$ m s^{-1} relative to the ground at the opposite wall. The total mass of the cannon and railroad car is $M = 15,000$ kg. (Assume that the mass of the exhaust gases is negligible.) (a) What is the velocity $\mathbf{V}$ of the car while the cannonball is in flight? (b) If the cannonball becomes embedded in the wall, what is the velocity of the car and ball after impact?

(a) When fired, the cannon exerts a force to the right on the ball. The ball exerts an equal and opposite force on the cannon, so the car and cannon recoil to the left. The net momentum is conserved because there is no external frictional force. The momentum before firing is zero, so after firing the momentum of the ball to the right must be equal in magnitude to that of the car and cannon to the left. Thus $mv = MV$ and

$$V = \frac{mv}{M} = \frac{(5\text{ kg})(15\text{ m s}^{-1})}{(15,000\text{ kg})}$$
$$= 5 \times 10^{-3}\text{ m s}^{-1}$$

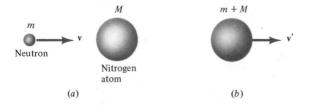

Figure 7.4. (a) A neutron collides with a stationary nitrogen atom. (b) The neutron is absorbed and a single object of mass $m + M$ is formed.

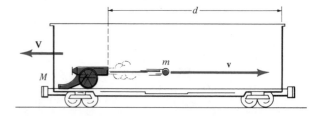

Figure 7.5. The cannon fires a ball toward the right. The cannon is mounted firmly to the floor of the railroad car.

The recoil speed of the car and cannon is very small because of their large mass.

(b) As the ball becomes embedded in the wall, it exerts a force on the wall to the right in Fig. 7.5. The wall, in turn, exerts a force to the left on the ball. The ball and car both stop moving when this happens, since the net momentum is still zero. Meanwhile the car will have rolled to the left as the ball traveled to the right.

It is important to realize that momentum is a *vector quantity,* and that if the total momentum of a system is constant, each *component* must be constant. This is illustrated by the next example.

Example 7.5

A car of mass $m = 1000$ kg moving at 30 m s^{-1} collides with a car of mass $M = 2000$ kg traveling at 20 m s^{-1} in the opposite direction. Immediately after the collision, the 1000-kg car moves at right angles to its original direction at 15 m s^{-1}. Find the velocity of the 2000-kg car right after the collision (Fig. 7.6).

Let us take the x and y axes as in Fig. 7.6. Then the x component of the total momentum of the two cars is conserved, so

$$mv_x + MV_x = mv'_x + MV'_x$$

Since $v'_x = 0$, we can solve for V'_x and substitute:

$$V'_x = \frac{mv_x + MV_x}{M} = \frac{mv_x}{M} + V_x$$

$$= \frac{1000 \text{ kg}}{2000 \text{ kg}} (30 \text{ m s}^{-1}) + (-20 \text{ m s}^{-1}) = -5 \text{ m s}^{-1}$$

The initial momentum components along the y direction were zero, so $mv'_y + MV'_y = 0$, and

$$V'_y = -\frac{mv'_y}{M} = -\frac{1000 \text{ kg}}{2000 \text{ kg}} (-15 \text{ m s}^{-1}) = 7.5 \text{ m s}^{-1}$$

We see from these examples that momentum conservation is a very powerful tool for analyzing many kinds of problems. However, like energy conservation, it does have limits to what it can tell us, and we must usually have some additional information about the final state of the system. Thus in Example 7.3, we know the neutron and nucleus move together as a single object, and in Example 7.4, we know the velocity of the cannonball. Momentum conservation by itself tells us only that the total momentum of a system is conserved in the absence of a net external force, and not how this momentum is shared among the objects in a system.

Ballistocardiography | Ballistocardiography is an application of the principle of momentum conservation to a medical problem, the study of heart functions and abnormalities. When the left ventricle of the heart contracts, it pumps blood into the aorta. The aorta has flexible walls and expands, allowing a net displacement of blood toward the head. A person lying on a table supported almost frictionlessly by air jets will recoil in the opposite direction. The mass of the person is very large compared to that of the displaced blood, so the recoil velocity is small, typically 1 millimetre per second or less. The recoil velocities change steadily, averaging out to zero over a full heart cycle. A record of these motions is called a *ballistocardiogram*. Attempts to use such measurements for diagnostic purposes have had only limited success.

7.3 | MOTION OF THE CENTER OF MASS

The motion of two colliding objects or of the pieces of an exploding artillery shell may be quite complicated and hard to predict. Nevertheless, the motion of the center of mass (C.M.) of such a system is unaffected by the internal forces and is determined entirely by the external forces acting on the system. For example, when the neutron in Example 7.3 is absorbed by the nucleus, the velocity of the center of mass of the two objects is the same before and after the collision. Similarly, when the cannonball is fired in Example 7.4, the ball moves further to the right than does the car to the left, and the C.M. of the ball and car remains at rest. In both cases, there is no net external force, so the C.M. velocity V is constant.

This important result follows from the definition of the center of mass. For example, suppose there is a mass m_1 at x_1 and a mass m_2 at x_2; their total mass is

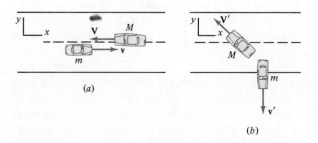

Figure 7.6. The x and y components of the total momentum of the two cars are each conserved in the collision.

$M = m_1 + m_2$. Then we saw in Chapter Four that their C.M. is located at

$$X = \frac{m_1 x_1 + m_2 x_2}{M}$$

In a time interval Δt, x_1 changes by $v_1 \Delta t$, x_2 by $v_2 \Delta t$, and X by $V \Delta t$. These changes are related by

$$V \Delta t = \frac{m_1 v_1 \Delta t + m_2 v_2 \Delta t}{M}$$

Dividing by Δt and using the definition of the momentum, we have in vector notation an expression for the C.M. velocity,

$$\mathbf{V} = \frac{\mathbf{p}_1 + \mathbf{p}_2}{M} \qquad (7.5)$$

From this result, we see that when there is no net external force and the total momentum $\mathbf{p}_1 + \mathbf{p}_2$ is constant, the velocity of the center of mass remains constant. When a net external force $\mathbf{F}_{ext}$ does act on a system, the center of mass has an acceleration equal to $\mathbf{F}_{ext}/M$. Thus the center of mass moves under the influence of $\mathbf{F}_{ext}$ in exactly the same fashion as a single particle of mass M. This principle is illustrated by the following example.

Example 7.6

An artillery shell explodes in the air. Neglecting air resistance, what can be said about the subsequent motion of its fragments?

Before the explosion, the shell moves in the trajectory described in our discussion of projectiles in Chapter Two. After the explosion the fragments fly off in many directions, and we cannot determine anything about their individual paths without more information. However, their center of mass is unaffected by the internal forces of the explosion, so it continues to follow the orig-

Figure 7.8.   The diver's center of mass moves in a projectile trajectory after he leaves the board.

inal trajectory (Fig. 7.7). Similarly, divers or gymnasts may execute complicated maneuvers, but their centers of mass follow simple trajectories (Fig. 7.8).

7.4 | ELASTIC AND INELASTIC COLLISIONS

In a brief collision, the total momentum is conserved, either exactly or to a very good approximation. However, mechanical energy may or may not be conserved in such situations. For example, when we drop a lively rubber ball onto a concrete floor, it rises again to nearly its original height; little mechanical energy is lost as it strikes the floor. By contrast, if we drop a ball of putty, it remains where it lands. All its kinetic energy is lost as heat or as work done in deforming the putty.

A collision that conserves mechanical energy is said to be *elastic*. If mechanical energy is not conserved, the collision is called *inelastic*. In a *completely inelastic* collision, *relative* motion ceases; the objects join and move as a single object, as when the bumpers of two vehicles lock (Fig. 7.9a). This type of collision dissipates the maximum amount of kinetic energy

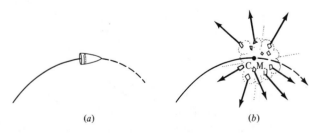

(a)   (b)

Figure 7.7.   (a) The trajectory of an artillery shell. (b) The motion of the center of mass of a system is determined solely by the external forces. Therefore the center of mass of the shell continues to follow the original trajectory after the shell explodes. None of the fragments need follow this path.

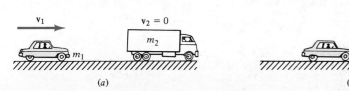

Figure 7.9. (*a*) A car strikes a stationary truck. (*b*) In a completely inelastic collision, the two vehicles lock together and move as a single unit.

possible within the constraints of momentum conservation.

If we know the extent to which mechanical energy is conserved in a given collision, we can use this information together with momentum conservation in analyzing the motion. We will illustrate this idea for the two extreme cases of elastic and completely inelastic collisions. Similar calculations can also be done for situations that are intermediate between these extremes, but they are more complicated.

Completely Inelastic Collisions | We have
already examined one completely inelastic collision in Example 7.3, where a neutron was captured by a nitrogen nucleus. We now study such collisions in more detail.

In a completely inelastic collision, the amount of mechanical energy lost to heat and deformation work depends on the relative masses of the two objects. For example, suppose a car of mass m_1 and velocity v_1 strikes a stationary truck of mass m_2 and they move together afterward (Fig. 7.9). The ratio of the final and initial kinetic energies is

$$\frac{K'}{K_1} = \frac{\frac{1}{2}(m_1 + m_2)v'^2}{\frac{1}{2}m_1v_1^2}$$

Momentum conservation requires $m_1v_1 = (m_1 + m_2)v'$, or $v' = m_1v_1/(m_1 + m_2)$. With this expression for v', we find

$$K' = \left(\frac{m_1}{m_1 + m_2}\right)K_1 \qquad \text{(totally inelastic (7.6) collision)}$$

This result implies that the final kinetic energy is small when the moving mass m_1 is small compared to the stationary mass m_2; most of the kinetic energy is lost in the collision. This result is illustrated by the next example.

Example 7.7

(a) A 1000-kg car traveling at 10 m s^{-1} hits a stopped truck of mass 9000kg, and the vehicles lock together

(Fig. 7.9). How much kinetic energy is lost? (b) If instead the truck is initially moving at 10 m s^{-1} and the car is stationary, how much kinetic energy is dissipated?

(a) The kinetic energy of the car before the collision is $K_1 = \frac{1}{2}m_1v_1^2 = \frac{1}{2}(1000 \text{ kg})(10 \text{ m s}^{-1})^2 = 5 \times 10^4$ J. After the collision, the total kinetic energy of the vehicles is

$$K' = \left(\frac{m_1}{m_1 + m_2}\right)K_1$$

$$= \frac{(1000 \text{ kg})(5 \times 10^4 \text{ J})}{(1000 + 9000) \text{ kg}} = 5 \times 10^3 \text{ J}$$

Thus the kinetic energy dissipated is

$$K_1 - K' = (5 \times 10^4 \text{ J}) - (5 \times 10^3 \text{ J}) = 4.5 \times 10^4 \text{ J}$$

Ninety percent of the kinetic energy is dissipated.

(b) Now $K_1 = \frac{1}{2}(9000 \text{ kg})(10 \text{ m s}^{-1})^2 = 4.5 \times 10^5$ J, and

$$K' = \frac{(9000 \text{ kg})(4.5 \times 10^5 \text{ J})}{(9000 + 1000)\text{kg}} = 4.05 \times 10^5 \text{ J}$$

Hence $K_1 - K' = (4.5 \times 10^5 \text{ J}) - (4.05 \times 10^5 \text{ J}) = 4.5 \times 10^4$ J. Since this equals the kinetic energy dissipated in part (a), the collisions are equally destructive. However, here only 10 percent of the kinetic energy is dissipated because the more massive truck has much more kinetic energy than the car has at the same speed.

Karate provides an interesting application of inelastic collisions. A karate fighter attempts to disable an opponent by transforming kinetic energy into deformation work in a vulnerable area. Since the fraction of the kinetic energy transformed is greatest when the moving mass is small, the karate fighter tries to deliver a large amount of kinetic energy with a relatively small part of his or her body, such as an arm. An arm strike (Fig. 7.10) is aimed so that the fist makes contact at the instant of most rapid motion, which occurs when the arm is about 70 percent extended (Fig. 7.11). Stepping forward increases the velocity and hence the kinetic energy at impact. Karate experts seldom follow through after a blow. Contact made during the follow-through of a wide swing

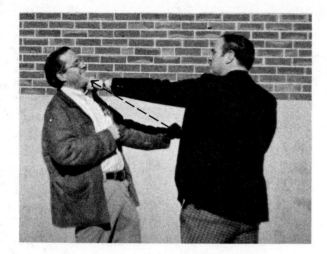

Figure 7.10. A forward punch in karate. The dashed line shows the path of the fist.

involves lower speeds and larger body masses; there is less deformation work and a risk of a loss of balance.

Elastic Collisions
After two objects collide elastically, they may move in various directions. If we know the direction of either object, we can use energy and momentum conservation to calculate both final velocity vectors. The procedure is straightforward in principle, but the results are complicated in appearance and are not very easy to interpret, except in special cases.

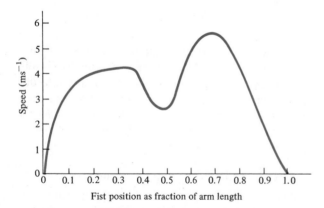

Figure 7.11. The speed of the fist versus the fractional arm extension in a forward karate punch. The data are taken from the study of high-speed motion pictures. (From J.D. Walker, *American Journal of Physics*, vol. 43, October 1975, p. 845.)

Figure 7.12. (*a*) Before an elastic head-on collision of a moving and a stationary object. (*b*) Motion after the collision.

Fortunately, a number of general features can be seen in one such case: a head-on elastic collision of a moving object and a stationary object, with both objects moving afterward either parallel or opposite to the original direction of motion (Fig. 7.12). Applying the momentum conservation to this situation,

$$m_1 v_1 = m_1 v_1' + m_2 v_2' \qquad (7.7)$$

Since the collision is elastic, $K_1 = K_1' + K_2'$, or

$$\tfrac{1}{2} m_1 v_1^2 = \tfrac{1}{2} m_1 v_1'^2 + \tfrac{1}{2} m_2 v_2'^2 \qquad (7.8)$$

If we solve Eq. 7.7 for v_1' and substitute this result in Eq. 7.8, we can find v_2' after several lines of algebra. We then obtain for the kinetic energies

$$K_1' = \frac{(m_1 - m_2)^2}{(m_1 + m_2)^2} K_1 \qquad (7.9)$$

$$K_2' = \frac{4 m_1 m_2}{(m_1 + m_2)^2} K_1 \qquad (7.10)$$

These results hold for any masses m_1 and m_2. When $m_1 = m_2$, we get $K_1' = 0$ and $K_2' = K_1$. This result is familiar to billiard players: a cue ball comes to rest when it hits a stationary ball head-on, and all the kinetic energy is transferred to the struck ball. By contrast, when the masses are very different, little energy is transferred no matter which of the objects is initially in motion. This is true for any elastic collision, not just the special head-on case considered here. The next example illustrates one elastic head-on collision.

Example 7.8

A neutron of mass m and velocity v_1 has an elastic head-on collision with a carbon nucleus of mass $12m$. (a) What fraction of the neutron's kinetic energy is transferred to the carbon nucleus? (b) What are the velocities of the neutron and the carbon nucleus after the collision?

(a) Using Eq. 7.10 with $m_1 = m$ and $m_2 = 12m$,

$$K_2' = \frac{4 m_1 m_2}{(m_1 + m_2)^2} K_1 = \frac{4m(12m)}{(m + 12m)^2} K_1$$

$$K_2'/K_1 = 48/13^2 = 0.284$$

The carbon nucleus has acquired 28.4 percent of the kinetic energy, so the neutron still retains 71.6 percent of its initial energy.

(b) From part (a),

$$\frac{K_2'}{K_1} = \frac{\frac{1}{2}m_2 v_2'^2}{\frac{1}{2}m_1 v_1^2} = \frac{48}{13^2}$$

Substituting $m_1 = m$ and $m_2 = 12m$, we find

$$v_2'^2 = \frac{4v_1^2}{13^2}$$

and

$$v_2' = \frac{2}{13} v_1$$

We choose the positive square root for v_2', since the nucleus moves forward when struck by the neutron, not backward or opposite to v_1. Then using $m_1 v_1 = m_1 v_1' + m_2 v_2'$, we find

$$v_1' = v_1 - \frac{m_2}{m_1} v_2'$$

$$= v_1 - \frac{12m}{m} \left(\frac{2v_1}{13} \right) = -\frac{11}{13} v_1$$

The neutron has reversed its direction, and its speed is $(11/13)v_1$. This reversal always happens in a head-on collision when the moving object is less massive than the staionary target.

In nuclear reactors, a neutron from the fission of a uranium nucleus must be slowed by a *moderator* before it can be captured by another uranium nucleus and cause this nucleus to fission as the next step in a *chain reaction*. Carbon in the form of graphite is sometimes used as a moderator, but water is much more common. Each water molecule contains two hydrogen nuclei (protons) plus an oxygen nucleus. Since the proton mass differs from the neutron mass by only 0.1 percent, protons are ideal for slowing neutrons.

7.5 | ANGULAR MOMENTUM OF A RIGID BODY

We have seen that when there is no net external force acting on an object or a system of objects, the linear momentum is conserved. Similarly, when there is no net torque due to external forces, the *angular momentum* is conserved.

This important result can be obtained readily from Newton's second law when it is written, as in Chapter Five, in the form convenient for rotational motion,

$$\tau = I\alpha \tag{7.11}$$

As before, τ is the torque, I is the moment of inertia, and α is the angular acceleration. If the angular velocity of an object rotating about a fixed axis changes from ω to ω' in a time Δt, the angular acceleration is $\alpha = (\omega' - \omega)/\Delta t$. Substituting this expression for α in Eq. 7.11 and multiplying by Δt, we have

$$\tau \Delta t = I\omega' - I\omega \tag{7.12}$$

We define the *angular momentum* by

$$\mathbf{L} = I\omega \tag{7.13}$$

Then Eq. 7.12 becomes

$$\tau \Delta t = \mathbf{L}' - \mathbf{L} \tag{7.14}$$

This result states that the *angular impulse* $\tau \Delta t$ equals the change in the angular momentum, $\mathbf{L}' - \mathbf{L}$. When the net torque on an object is zero, its angular momentum remains constant, or is conserved. Similarly, if there is no net external torque on a system of objects, its total angular momentum is conserved.

These results look simple. In fact, they could have been guessed immediately from the analogous equations for the linear momentum, with the replacements $\mathbf{F} \to \tau$, $m \to I$, $\mathbf{v} \to \omega$. Nevertheless, angular momentum conservation is a distinct and very important principle, with applications as diverse as atomic physics, figure skating, and astronomy.

Angular momentum conservation can be illustrated quite readily. Suppose you spin on a stool that has well-lubricated bearings. The friction is small, as is the torque tending to change the angular momentum. If you spin with your arms and legs folded and then extend them, your moment of inertia increases. Since the angular momentum $\mathbf{L} = I\omega$ is nearly constant, the angular velocity ω decreases. Conversely, if you initially spin with your limbs extended and then bring them closer to the rotation axis, your angular velocity increases. This is precisely what a figure skater does: she starts a spin with her limbs extended and then increases her angular velocity by pulling her limbs closer to her body (Fig. 7.13).

In both of these examples the angular momentum is conserved, but there is a change in the rotational kinetic energy. When the limbs are extended or retracted, there are equal and opposite forces acting on

Figure 7.13.   A figure skater can increase her angular speed by folding her arms. (Photos by David Leonardi.)

the body and limbs. These two forces produce no net torque, but they each do work. Thus, when we retract our limbs to spin faster, we do work, increasing our kinetic energy. This energy change is calculated in the following example.

Example 7.9

A figure skater begins to spin at 3π rad s^{-1} with her arms extended. (a) If her moment of inertia with arms folded is 60 percent of that with arms extended, what is her angular velocity when she folds her arms? (b) What is the fractional change in her kinetic energy?

(a) If we assume that the ice is nearly frictionless, angular momentum is conserved: $I'\omega' = I\omega$. With $I' = 0.6I$ and $\omega = 3\pi$ rad s^{-1}, we have

$$\omega' = \frac{I\omega}{I'} = \frac{3\pi \text{ rad s}^{-1}}{0.6} = 5\pi \text{ rad s}^{-1}$$

(b) Her initial and final kinetic energies are $K = \frac{1}{2}I\omega^2$ and $K' = \frac{1}{2}I'\omega'^2$. To find the fractional change, we compute

$$\frac{\Delta K}{K} = \frac{K' - K}{K}$$

$$= \frac{\frac{1}{2}(0.6I)(5\pi \text{ rad s}^{-1})^2 - \frac{1}{2}I(3\pi \text{ rad s}^{-1})^2}{\frac{1}{2}I(3\pi \text{ rad s}^{-1})^2}$$

$$= \frac{2}{3}$$

The skater increases her kinetic energy by 67 percent when she folds her arms. This is equal to the work she has to do to bring her arms in.

Conservation of angular momentum does *not* mean that the angular position of an object must remain the same in the absence of external torques. One example of this is given in Fig. 7.14, where a man rotates his swivel chair by executing a suitable sequence of moments.

A more interesting example of this principle is provided by a falling cat, which can always land on its feet if the fall is long enough. Looking at Fig. 7.15, in (b) the cat prepares to turn about a horizontal axis by drawing the front paws close to this axis, diminishing the moment of inertia of the fore part. The hind legs are left extended, so the rear part has a larger moment of inertia. In (c) the cat rotates the fore part in one direction and the rear part in the opposite direction as required by angular momentum conservation. In (d) the fore legs are extended and the hind legs drawn in, so the rear portion has the smaller moment of inertia. Thus in (e) the cat rotates the hind portion with relatively little counterrotation of the fore part.

Like figure skaters, divers—as well as astronauts floating in spaceships and gymnasts—can twist or

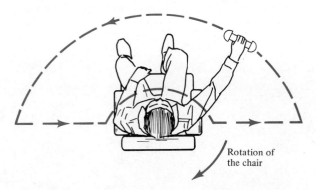

Figure 7.14. Top view of a person sitting in a swivel chair that turns easily. If an arm is moved repeatedly along the dotted path, the swivel chair gradually rotates clockwise. The chair is in motion only when the hand is moving along the circular parts of the path, since the hand's angular momentum is zero when moving along a line through the axis of rotation. Holding a heavy weight in the hand makes the chair rotate faster.

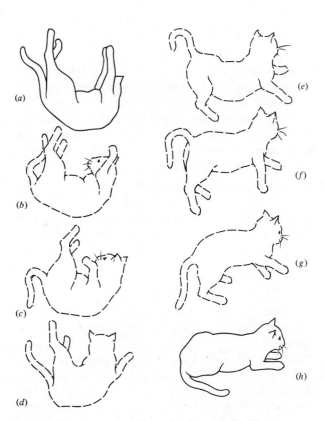

Figure 7.15. The sequence of motions performed by a cat falling from rest when initially upside down. The sketches are from photos at 1/20 second intervals. The motions are described in the text. (Adapted from Tricker and Tricker.)

spin. This is a rotation about an axis passing through their centers of mass and extending from the head to the toes. In addition, they can somersault, or rotate about an axis through their C.M. from one side to the other, and they can change their moments of inertia about *either* axis. Once in the air, a diver is not subjected to any torques about the C.M., so his or her *total* angular momentum is constant. However, there can be an interchange between the angular momenta associated with twisting and somersaulting. An understanding of angular momentum conservation has aided athletes and coaches in developing increasingly complex maneuvers.

7.6 | ANGULAR MOMENTUM OF A PARTICLE

The angular momentum of a particle (Fig. 7.16a) is defined to be the vector cross product of its distance **r** from a specified point and its linear momentum **p** = m**v**:

$$\mathbf{L} = \mathbf{r} \times \mathbf{p} \qquad (7.15)$$

The magnitude of **L** is $rp \sin \theta$. When **r** and **p** are perpendicular, $\sin \theta = 1$, and $L = rp = rmv$. Its direction is given by the right-hand rule and is perpendicular to **r** and **p**.

This definition of **L** is equivalent to that given earlier, $\mathbf{L} = I\boldsymbol{\omega}$, if we consider a system of particles rotating about a common center (Fig. 7.16b). If a particle of mass m moves in a circle of radius r with a velocity $v = r\omega$, then its angular momentum has a magnitude $rmv = rm(r\omega) = (mr^2)\omega$ and is directed along $\boldsymbol{\omega}$. Summing over all the particles, ω is the same for each and we get the total of the mr^2 terms. Since this total is the moment of inertia I, the total angular momentum is $I\omega$, as expected.

The basic result of the preceding section, $\tau \Delta t = \mathbf{L}' - \mathbf{L}$, applies not only to a rigid body but to a single particle as well. Hence, when there is no net torque on a particle about a given point, its angular momentum relative to that point is conserved. For example, from the discussion in Section 7.3, the motion of the earth about the sun can be found by considering the earth as a single particle located at its C.M. The line of action of the gravitational force exerted by the sun passes through the center of the sun, so there is no torque about that point. Consequently, the angular momentum of the earth relative to the center of the sun is constant. The earth's orbit is almost circular, but not quite; it is actually an ellipse,

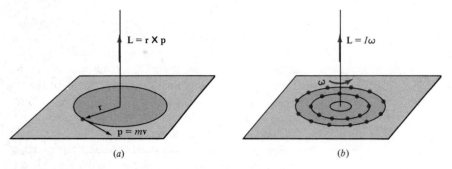

Figure 7.16. (a) The angular momentum of a particle is $\mathbf{L} = \mathbf{r} \times \mathbf{p}$. (b) The total angular momentum of a system of particles rotating about a common center is the sum of the individual angular momenta. This sum is equivalent to $\mathbf{L} = I\omega$.

with the earth-sun distance varying by about 3 percent between its extremes. We see that although r, v, and θ may change, the product $L = mrv \sin \theta$ must remain constant throughout the motion (Fig. 7.16).

The total angular momentum of the earth is composed of two parts. Its *orbital angular momentum* is associated with the annual revolution of the earth about the sun, and its *spin angular momentum* arises from the daily rotation of the earth about its axis. Similarly, an electron in an atom has both an orbital angular momentum, due to its motion around the nucleus, and a spin angular momentum. In the absence of any torques, these angular momenta are conserved. Angular momentum considerations play an important role in the description of atomic properties.

SUMMARY

If a net average force $\overline{\mathbf{F}}$ acts for a time Δt on an object, then

$$\overline{\mathbf{F}} \, \Delta t = \mathbf{p}' - \mathbf{p}$$

The product of the average force and the time during which it acts is called the impulse. The linear momentum $\mathbf{p}$ of an object with mass m and velocity $\mathbf{v}$ is $\mathbf{p} = m\mathbf{v}$. The impulse equals the change in the momentum.

When two objects collide, each suffers a change in momentum. However, the total momentum is constant or conserved if no net external force acts on the two objects. Consequently, the total momentum of an isolated system is always conserved. Momentum conservation is particularly useful in collision problems, even though energy may or may not be conserved.

When there is no net external force acting on an object or a system of objects, the total momentum is constant and the velocity of the center of mass remains constant. When there is a net external force, the center of mass moves under its influence in ex-

actly the same way as a single particle with a mass equal to the total mass of the object or objects.

Collisions that conserve mechanical energy are called elastic; those which do not are called inelastic. In a completely inelastic collision, all relative motion ceases and the objects subsequently move together. When an object with a small mass collides completely inelastically with an object with a large mass that is at rest, most of the mechanical energy is dissipated as heat or deformation work. A much smaller fraction of the mechanical energy is dissipated when the larger object is in motion and the smaller is initially at rest. In elastic collisions, the greatest kinetic energy transfer occurs when the two masses are equal.

The angular momentum of a rigid body rotating about a fixed axis is $\mathbf{L} = I\omega$. The angular momentum of a particle is $\mathbf{L} = \mathbf{r} \times \mathbf{p}$. The two definitions are equivalent for a system of particles rotating about a common center. If a net torque $\boldsymbol{\tau}$ acts for a time Δt, then

$$\boldsymbol{\tau} \, \Delta t = \mathbf{L}' - \mathbf{L}$$

The angular impulse equals the change in angular momentum. If the net torque on a system is zero, its angular momentum is conserved.

Checklist
Define or explain:

impulse	angular momentum of a
momentum	rigid body
collision	angular momentum
momentum conservation	conservation
ballistocardiography	external torques
center of mass motion	angular momentum of a
elastic collision	particle
inelastic collision	orbital angular
completely inelastic	momentum
collision	spin angular momentum
angular impulse	

REVIEW QUESTIONS

Q7-1 The impulse is the product of the _____ and the _____.

Q7-2 The impulse equals the _____.

Q7-3 Momentum is the product of the _____ and _____ of an object.

Q7-4 Momentum is usually conserved in brief collisions because the impulse due to _____ is negligible.

Q7-5 The total momentum of a system is constant if the _____.

Q7-6 If the momentum of a system is conserved, each _____ of the momentum vector is constant.

Q7-7 The center of mass of a system moves like a particle with a mass equal to _____ subjected to the _____.

Q7-8 A completely inelastic collision dissipates the _____ mechanical energy consistent with momentum conservation.

Q7-9 In an elastic collision, _____ mechanical energy is dissipated.

Q7-10 In inelastic collisions, mechanical energy is transformed into _____ and _____.

Q7-11 The angular momentum of a rigid body rotating about a fixed axis is _____.

Q7-12 The angular momentum of a particle relative to a point is _____.

Q7-13 The angular momentum of a system is constant when there is no _____.

EXERCISES

Section 7.1 | Impulse and Linear Momentum

7-1 A golf ball initially at rest is struck with an average force of 2600 N for 1.25×10^{-3} s. What is the final velocity of the 0.047-kg ball?

7-2 If the momentum of a car increases by 9×10^4 kg m s^{-1} in 12 s, what is the average force accelerating the car?

7-3 (a) What is the change in momentum of a car that experiences a force of 6000 N for 6 s? (b) If the mass of the car is 1000 kg, what is its change in velocity?

7-4 A pilot ejected from a plane suffers an average acceleration of 12 g's for a period of 0.25 s. (a) If the mass of the pilot is 70 kg, what is the average force he experiences? (b) Find the magnitude of the change in his velocity.

7-5 A batter hits a baseball of mass m, traveling at a velocity $\mathbf{v}$, directly back toward the pitcher at the same speed. If the ball and bat are in contact for a time Δt, what is the average force exerted by the bat?

7-6 A rubber ball of mass m is thrown against a wall at a speed v and rebounds with the same speed in the opposite direction. (a) How large is the impulse acting on the ball? (b) A ball of putty with the same mass and speed is thrown against the wall and sticks to it. How large is the impulse acting on the ball?

Section 7.2 | Momentum Conservation

7-7 A propeller-driven airplane pushes air toward the rear of the plane as it moves forward. Describe the force on the plane in terms of the change in momentum of the air.

7-8 A man sits on a chair fixed to a cart that is initially at rest on frictionless rails. (a) If he throws a sandbag off the side of the cart, will the cart move? Explain. (b) If he throws a sandbag off the back of the cart, will the cart move? Explain.

7-9 A person seated at the stern of a sailboat attempts to move forward by blowing on the sails. Explain what will happen.

7-10 A rocket of mass M has a velocity v_0 at its maximum height. It explodes into two pieces just as it reaches this height. One piece of mass m_1 *stops and then falls* vertically to the earth. What is the velocity of the other piece of mass m_2 just after the rocket separates?

7-11 A cart of mass m on a frictionless air track has an initial velocity $\mathbf{v}_0$ toward a cart of mass M that is at rest. What are the final velocities of the two carts after they collide if the carts lock together?

7-12 A 1-g = 10^{-3}-kg bullet has a horizontal velocity of 200 m s^{-1}. It strikes and becomes embedded in a 1-kg block of wood that rests on a frictionless table top. What is the velocity of the block and bullet after impact?

7-13 A truck of mass 4500 kg moving at 10 m s^{-1} strikes the back of a car that is at rest. The car and occupants have a mass of 950 kg. (a) What is the speed of the car just after impact if the car and truck lock together? (b) If the impact lasts 0.3 s, what is the average force on a 60-kg passenger in the car?

7-14 The engine of a motorboat fails, the boat coming to rest in still water with its bow pointed toward the shore 5 m away. Its disgusted operator hurls a six-pack of beverages horizontally from the stern at a speed of 12 m s^{-1} relative to the water. The mass of the boat plus operator is 240 kg, and that of the six-pack is 3 kg. (a) What is the recoil velocity of the boat? (b) How long would it take the boat to reach the shore at this velocity? (c) Will it actually reach the shore in this time? Explain.

7-15 When the left ventricle of the heart contracts, there is a net displacement of blood toward the head. Suppose a person lies on a horizontal table which can move frictionlessly and that is initially at rest. In a contraction lasting 0.2 s, 0.8 kg of blood is pumped 0.1 m. The mass of the person plus table is 80 kg. What is the velocity of the person and table at the end of the contraction?

7-16 Is momentum conserved when a ball of putty hits a floor? Explain.

7-17 An object collides with a stationary object of the same mass. After the collision, the first object is deflected by an angle θ from its original direction, and the two objects are observed to have the same speed. Find the angle ϕ between the direction of the second object and the original direction of the first object.

7-18 A 2000-kg car and a 1000-kg car traveling at 40 m s^{-1} in opposite directions collide head-on and lock together. What are their speed and direction right after the collision?

7-19 Suppose a large meteor with a mass of 10^{10} kg and a velocity of 2×10^4 m s^{-1} were to hit the earth and disintegrate. (a) With what velocity would the earth recoil? (Use the solar and terrestrial data on the inside back cover.) (b) What fraction of the earth's orbital velocity about the sun does this recoil velocity represent?

7-20 A car traveling north at 30 m s^{-1} has a mass of 1500 kg. It hits a car of mass 1000 kg and they both come immediately to rest. What was the speed and direction of the other car just before the collision?

7-21 A car is traveling at a velocity **v**. It is struck from behind by a car of the same mass with a velocity 2**v**. If the cars lock together, what will their velocity be after the collision?

Section 7.3 | Motion of the Center of Mass

7-22 A man with a mass of 70 kg is seated at the center of a stationary canoe of mass 30 kg. If he moves to a seat 2 m forward in the canoe, how far will the canoe move?

7-23 The moon does not rotate about the earth;

rather, they both rotate about a common point. (a) What is that point called? (b) What is the radius of the circular path of the earth about that point? (Use the solar and terrestrial data on the inside back cover.)

7-24 An 80-kg man stands on a 120-kg raft that is initially at rest. If the man starts to walk relative to the raft at 1.5 m s^{-1}, how fast will the raft move?

Section 7.4 | Elastic and Inelastic Collisions

7-25 A 1000-kg car moving at 20 m s^{-1} hits a 2000-kg parked car head-on. How much mechanical energy is dissipated if the collision is (a) elastic; (b) completely inelastic?

7-26 A ship of mass m and speed v strikes a stationary iceberg of mass 10 m. Find the resulting velocity of the iceberg if (a) the collision is completely inelastic; (b) the collision is elastic and the ship bounces off with its direction reversed.

7-27 One of the disadvantages of using the protons in water molecules to slow neutrons in a reactor is that occasionally a proton will capture a neutron when they collide, forming a *deuteron, a heavy hydrogen* or *deuterium* nucleus. If the proton is originally at rest when it captures a neutron, what is the ratio of the kinetic energy of the resulting deuteron to that of the neutron? (Take the proton and neutron masses to be half that of the deuteron.)

7-28 The molecules in heavy water contain an oxygen atom plus two atoms of *heavy hydrogen* or *deuterium*. A deuterium nucleus has a mass twice that of a neutron. (a) What fraction of its kinetic energy is transferred when a neutron strikes a stationary deuterium nucleus head-on and collides elastically? (b) What is the ratio of the initial and final neutron velocities in such a collision? (Heavy water is used as a moderator in several nuclear power plants that have been built in Canada.)

Section 7.5 | Angular Momentum of a Rigid Body

7-29 A bicycle wheel has a radius of 0.36 m, and the bicycle is moving at 6 m s^{-1}. The mass of the wheel is 2 kg. (a) What is the angular velocity of the wheel? (b) Assuming the mass of the wheel is entirely at its edge, find its angular momentum.

7-30 An acrobat holds a long pole and walks on a tightrope. (a) Using angular momentum arguments, explain which way he should tip the pole if he starts tipping toward the right side. (b) What effect will weights at the ends of the pole have?

7-31 A tricycle wheel has a moment of inertia of 0.04 kg m². If the wheel rotates once per second, what is its angular momentum?

7-32 A 3-kg cylinder of radius 0.2 m is rotating about its axis at 40 rad s⁻¹. Find the angular momentum if the cylinder is (a) solid; (b) a thin shell.

7-33 Why does a girl walking on the top of a fence keep her arms outstretched?

7-34 A car is at rest with the engine idling. If the accelerator is suddenly depressed, the left side of the car drops slightly and the right side rises. Which way is the crankshaft turning?

7-35 Why must a single-engine plane be trimmed for normal horizontal flight by raising one wing flap or aileron and lowering the opposite one? Is this necessary in planes with two engines?

7-36 A grinding wheel has a moment of inertia of 0.5 kg m² and an angular velocity of 120 rad s⁻¹. (a) What is its angular momentum? (b) A tool is pressed against the wheel, and it comes to a stop in 10 s. What is the average torque due to the tool? (Assume no other torques act on the wheel.)

7-37 A flywheel of mass 500 kg and radius of gyration of 0.5 m rotates at 1000 rad s⁻¹. How long can it exert a torque of 250 N m on a shaft?

7-38 A diver leaves a springboard with his body extended, somersaulting at 3 rad s⁻¹. If he bends into a tuck position, his moment of inertia decreases by a factor of 5. (a) By what factor will his angular velocity change? (b) By what factor will his rotational kinetic energy change?

7-39 A woman with her arms outstretched has a moment of inertia of 2 kg m². She rotates on a stool at an angular velocity of 6 rad s⁻¹. At a certain instant she grabs a 3 kg object in each hand, holding them 0.8 m from the axis of rotation. If the objects are initially at rest, what is her angular velocity?

Section 7.6 | Angular Momentum of a Particle

7-40 Calculate (a) the orbital and (b) the spin angular momenta of the earth. (Use the solar and terrestrial data on the inside back cover and assume the earth is a uniformly dense sphere.)

7-41 An artificial satellite moves in an elliptical orbit with the earth at one focal point. (Fig. 7.17) At point A its speed is v and its distance from the center of the earth is r. At point B its distance from the center of the earth is $2r$. What is its speed?

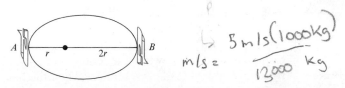

Figure 7.17. Exercise 7-41.

7-42 In Bohr's model of the hydrogen atom, the electron in its smallest circular orbit has an angular momentum of 1.055×10^{-34} kg m² s⁻¹. The radius of the orbit is 5.29×10^{-11} m, and the mass of the electron is 9.11×10^{-31} kg. (a) What is the speed v of the electron? (b) Find the ratio v/c, where $c = 3.00 \times 10^8$ m s⁻¹ is the speed of light.

PROBLEMS

7-43 A 1000-kg car and a 2000-kg truck are each moving at 20 m s⁻¹ when they collide head-on. Find their final velocities just after the collision, (a) if the collision is elastic; (b) if they remain locked together.

7-44 A man is sitting on a sled on an ice pond so that the frictional forces are zero. He has a machine gun that fires bullets of mass 1.3×10^{-2} kg with a muzzle velocity of 800 m s⁻¹. (a) What is the momentum of each bullet? (b) What is the average force on the man, per bullet, if that force is experienced for 0.2 s? (c) What is the speed of the man on the sled after he fires 100 bullets? Assume that the mass of the man, sled, and gun is 90 kg, and neglect the loss of mass because of the release of 100 bullets.

7-45 An empty railroad gondola car rolls initially at 5 m s⁻¹ and then collects 1000 kg of water in a storm in which the rain falls vertically. If there is no friction and the empty car has a mass of 12,000 kg, what is its final speed?

7-46 A freight car with a mass of 30,000 kg rolls at 5 m s⁻¹ under a grain elevator where 10,000 kg of grain is dropped vertically into the car. (a) Assuming frictional forces between the car and rails are negligible, what is the final velocity of the car? (b) How much kinetic energy is lost by the car, and how is this energy transformed?

*7-47 A billiard ball strikes an identical billiard ball initially at rest and is deflected 45° from its original direction. Show that if the collision is elastic, the other ball must move at 90° to the first and with the same speed.

168

7-48 (a) If the distance d in Fig. 7.5 is 6 m, how far does the cannonball of Example 7.4 travel before striking the wall? (b) How far do the car and cannon move? (c) Show that the center of gravity of the entire system remains stationary when the ball is fired.

*7-49 A bullet of mass 10^{-2} kg and initial horizontal velocity of 250 m s^{-1} strikes and is embedded in a 1-kg wooden block. The wooden block hangs on the end of a long string. (a) What is the velocity of the block and bullet after impact? (b) How high will the block and bullet swing upward?

*7-50 Explain how the toy pictured in Fig. 7.18 operates. The toy is composed of steel balls hung on strings. When one ball is initially raised and released, the last ball at the other end swings up. When two balls are initially displaced, two on the other end swing upward.

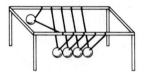

Figure 7.18. When the raised ball is released it swings downward and collides elastically with the second ball. After a sequence of collisions, the right-hand ball swings upward while the others remain at rest. Problem 7-50.

7-51 A vehicle with velocity v_0 collides with (a) a stationary vehicle of the same mass, (b) a solid wall, and (c) a vehicle of the same mass moving in the opposite direction with speed v_0. Which is (are) the most severe as far as the effect on passengers? Explain.

7-52 A carbon-14 nucleus emits a beta particle (electron) and a neutrino (ν) and is transformed into a nitrogen-14 nucleus. The beta particle and nitrogen nucleus can be observed because they leave a trail of ions in a detector. The neutrino, however, is very difficult to detect directly because it rarely interacts with the atoms it passes. Suppose in a particular decay event the beta particle has a momentum **p** and the nitrogen nucleus has a momentum of magnitude $\frac{4}{3}p$ at an angle of 90° to **p**. What is the magnitude and direction of the momentum of the neutrino?

7-53 Two objects with masses m_1 and m_2 are at either end of a spring. If the objects are pulled apart and then released from rest, what is the ratio K_1/K_2 of their kinetic energies at any instant?

7-54 A plutonium-239 nucleus decays at rest into an alpha particle (helium nucleus) plus a uranium-235 nucleus. The kinetic energy of the alpha particle is found to be 5.06 MeV, where 1 MeV = 10^6 eV = 1.60×10^{-13} J. The mass of a uranium nucleus is $\frac{235}{4}$ times that of an alpha particle; the mass of an alpha particle is 6.64×10^{-27} kg. (a) What is the velocity of the alpha particle? (b) What is the velocity of the uranium nucleus? (c) What is the kinetic energy of the uranium nucleus in MeV?

7-55 A nucleus at rest decays into an alpha particle of mass m and a nucleus of mass M. What fraction of the total kinetic energy of the system is carried by the nucleus of mass M?

7-56 A spaceship initially of mass M and velocity **V** fires a projectile of mass m and velocity **v**. (Both velocities are relative to the earth.) Find the magnitude of the resulting velocity **V'** of the spaceship if (a) **v** is parallel to **V**; (b) **v** is opposite to **V**; (c) **v** is perpendicular to **V**.

7-57 A 1000-kg car traveling at 20 m s^{-1} toward the north collides with a 10,000-kg truck heading south at the same speed. Immediately after the collision the car is moving east at 20 m s^{-1}. (a) How fast is the truck moving and in what direction immediately after the collision? (b) How much mechanical energy is dissipated in the collision?

7-58 On a moving bicycle (Fig. 7.19), the angular momentum of the spinning wheels is directed to the rider's left. If the bicycle is ridden "no-handed," what is the effect of the rider leaning to her left? Explain.

7-59 A girl of mass 50 kg standing at the center of a turntable is spun at 1.5 revolutions per second. In each hand she holds a 6-kg mass, and initially these

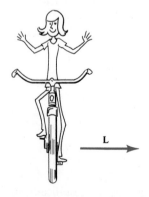

Figure 7.19. Problem 7-58.

are close to her body. Estimate her rotational velocity once she extends her arms outward. (Neglect the moment of inertia of the turntable.)

7-60 An insect of mass 10^{-3} kg walks around the edge of a phonograph turntable of mass 0.5 kg and radius 0.15 m. If the insect comes back to the spot on the record it started from, how far has the record turned? (Neglect friction in the bearings, and treat the turntable as a solid cylinder.)

7-61 If the earth's polar ice caps melt, what will happen to the length of the day? Explain.

*7-62 In the Bohr model of the hydrogen atom, the only possible electronic circular orbits are those for which the angular momentum is $nh/2\pi$, where n is an integer and h is Planck's constant. (a) What relation must the radius and linear momentum satisfy in this model? (b) We saw in Chapter Six that the kinetic energy of an electron in an orbit of radius r is $\frac{1}{2}mv^2 = ke^2/2r$. Show that the possible radii of the orbits are $n^2h^2/(4\pi^2kme^2)$. (c) What are the kinetic and total energies for these radii?

*7-63 For some purposes, a diatomic molecule can be thought of as a dumbbell, with two masses at the ends of a rigid rod and a moment of inertia I. According to Bohr's model for such a molecule, such a dumbbell can only have an angular momentum that is an integer multiple of $h/2\pi$, where h is a number called Planck's constant; the angular momentum is said to be *quantized*. What are the possible kinetic energies of such a dumbbell?

7-64 A single-engine airplane is viewed from the front. The propeller begins to rotate counterclockwise. (a) Are the forces due to the ground on the two wheels, one mounted under each wing, the same? If not, which is larger, that on the left wheel or the right? (b) Once the propeller is turning with a constant angular velocity, which wheel experiences the largest force due to the earth?

7-65 A man seated on a stool that rotates freely holds a 5-kg mass in each hand. With his hands in his lap, the moment of inertia about a vertical axis is 20 kg m². With his arms outstretched, the total moment of inertia is 35 kg m². Initially the man has an angular velocity of 3 rad s⁻¹ with his arms in his lap. (a) What is the angular velocity of the man with his arms extended? (b) If the man drops the weights while his arms are extended, what is his angular velocity?

7-66 A high diver of height h wishes to do a forward double flip dive in a tuck position, with her arms around her folded legs. She leaves the board in an erect position with an angular velocity of 2.0 rad s⁻¹. Her radius of gyration is 0.25h when she is erect and is 0.1h in the tuck position. (a) What is her angular velocity in the tuck position? (b) What is the minimum time necessary for two complete turns? (c) If the diver initially has no vertical translational velocity, what is the minimum height of the diving board above the water if the dive is to be successful? (d) Does a tall person have any advantage in performing this dive?

7-67 Figure 7.20 shows a helicopter in flight. The large propeller which provides the lift force rotates clockwise when viewed from above. Explain, using angular momentum and torque, the function of the small propeller on the tail of the aircraft. This small propeller spins about a horizontal axis.

7-68 A star is initially similar to our sun, with a radius of 7×10^8 m and a period of rotation about its axis of 27 days. It evolves eventually into a neutron star with a radius of only 10^4 m and a period of 0.1 s. Assuming that the mass stays the same, calculate the ratio of its initial and final (a) angular momentum; (b) kinetic energy. (The angular momentum decreases, indicating that some angular momentum must have been carried off by material escaping from the star. The kinetic energy increases as a result of the conversion of gravitational potential energy into other forms of energy as the star collapses.)

*7-69 One of Kepler's laws of planetary motion states that a line joining a planet to the sun sweeps out equal areas in equal times. Show that this law follows from the fact that the angular momentum of the planet is conserved.

*7-70 A long wooden beam of mass m and length $2a$ can pivot in a horizontal plane about its midpoint. A bullet of mass $m/60$ is fired horizontally at a velocity v into the beam at right angles to its length at the end. What is the resulting angular velocity of the beam?

ANSWERS TO REVIEW QUESTIONS

Q7-1, force, elapsed time; **Q7-2,** momentum change; **Q7-3,** mass, velocity; **Q7-4,** external forces; **Q7-5,** net external force is zero; **Q7-6,** component; **Q7-7,** that of the system, net external force; **Q7-8,** maximum; **Q7-9,** no; **Q7-10,** heat, deformation work; **Q-11,** $I\omega$; **Q7-12,** $\mathbf{r} \times \mathbf{p}$; **Q7-13,** net external torque.

Figure 7.20.   Problem 7-67. (Courtesy of Sikorsky Aircraft, Division of United Technologies.)

SUPPLEMENTARY TOPICS

7.7 | MOMENTUM AND THE USE OF THE BODY

In many athletic activities, one tries to maximize the momentum transfer. For example, in boxing, punches thrown by a simple extension of the arm are not nearly as effective in imparting momentum to an opponent as those thrown in conjunction with body movement. However, in karate, large momentum transfers are often attained by high-speed motion of the limbs instead of by entire body movements.

Contact sports are not the only area where momentum transfer is important. For example, the primary aim of a shot-putter is to transform the low-velocity motion of the mass of the entire body into high-velocity motion of the smaller ball (Fig. 7.21). Momentum also plays an important role in activities where a ball is struck.

Table 7.1 lists the typical velocities and times measured for good athletes in various sports using balls.

The striker is the instrument used to hit the ball, such as a baseball bat, tennis racket, or foot. To obtain useful information from these data using impulse and momentum, we must recognize that the mass of the striker, and hence its momentum, is somewhat ambiguous. Using tennis as an example, the striker is a racket of mass 0.4 kg. However, in using the racket, the arm and part of the body can be considered as part of the striker; the racket acts as an extension of the body. The effective striker mass depends on what part of the body is used and on how it is used. When a person swings a tennis racket primarily through wrist action, the effective striker mass is small, and the swing cannot be carried through firmly.

It must also be remembered that the person involved in striking the ball will usually be in contact with the ground. This means that the ball and striker cannot truly be regarded as parts of an isolated system with no external forces. When a tennis racket hits a ball, by Newton's third law the ball exerts a force on the racket, the racket exerts a force on the body, and

Figure 7.21. A shot-putter in action. Ideally, the momentum of the body parts is very small just after release. The momentum of the body is transferred to the ball.

the body exerts a force on the ground. Therefore, if a racket imparts momentum to a ball, momentum is also transferred to the earth.

To avoid the difficult details of this situation, it is convenient to define the *effective mass* of the striker. That is, we pretend the striker and ball are two parts of a system on which no external forces act, so the total momentum of the striker and ball can be considered constant. This effective mass is not just a trick to allow us to compute a meaningless number. A detailed study of the motions of an athlete can aid in understanding how to increase the effective mass of the striker and thereby achieve higher ball velocities.

We denote the mass of the ball by m and its final velocity by $\mathbf{v'}$; its initial velocity is zero. The effective mass of the striker is denoted by M and its initial and final velocities by $\mathbf{V}$ and $\mathbf{V'}$. If the motion is in a straight line and momentum is conserved,

$$MV = mv' + MV' \qquad (7.16)$$

Assuming m, v', V, and V' are known, we can solve this equation for the effective mass of the striker,

$$M = \frac{mv'}{V - V'} \qquad (7.17)$$

Notice that this effective mass can be increased by performing the motion in such a way as to minimize the change in velocity of the striker during the impact (Fig. 7.22).

Example 7.10

(a) Using the data of Table 7.1 for a tennis serve, calculate the effective mass of the striker. (b) What is the average force on the tennis ball during impact?

(a) With the data of Table 7.1, the effective mass of the striker is

$$M = \frac{mv'}{V - V'} = \frac{(0.058 \text{ kg})(51 \text{ m s}^{-1})}{(38 - 33) \text{ m s}^{-1}} = 0.59 \text{ kg}$$

This is greater than the mass of the racket alone, which is 0.4 kg.

(b) During impact the average force on the ball $\bar{F}$ is found from $\bar{F}\Delta t = m(v' - v)$. With $v = 0$ and the impact time of 4×10^{-3} s,

$$\bar{F} = \frac{m(v' - v)}{\Delta t} = \frac{(0.058 \text{ kg})(51 \text{ m s}^{-1})}{4 \times 10^{-3} \text{ s}} = 740 \text{ N}$$

TABLE 7.1

Masses of balls, velocities of strikers before and after impact, and the impact times during which the balls and strikers are in contact

Ball	Ball Mass (kg)	Ball Velocity (m s^{-1})		Striker Velocity (m s^{-1})		Impact Time (s)
		Before	After	Before	After	
Baseball (hit from rest)	0.15	0	39	31	27	1.35×10^{-3}
Football (punt)	0.42	0	28	18	12	8×10^{-3}
Golf ball (driver)	0.047	0	69	51	35	1.25×10^{-3}
Handball (serve)	0.061	0	23	19	14	1.35×10^{-2}
Soccer ball (kick)	0.43	0	26	18	13	8×10^{-3}
Squash ball (serve)	0.032	0	49	44	34	3×10^{-3}
Softball (hit from rest)	0.17	0	35	32	22	3×10^{-3}
Tennis ball (serve)	0.058	0	51	38	33	4×10^{-3}

Figure 7.22. A batter has his feet firmly set and his body is moving toward the ball on impact. This increases the effective striking mass of the bat.

7.8 | GYROSCOPIC MOTION

The relationship between the torque and angular momentum leads to some startling motions. The behavior of a gyroscope is an example of this (Fig. 7.23). A gyroscope is constructed so that a spinning wheel is supported beneath its center of gravity and is isolated from any external torques. Thus the angular momentum of the spinning wheel continues to point in one direction, even if the frame is tilted or rotated.

Gyroscopes of very sophisticated design have virtually no friction and are used in inertial guidance systems for planes, rockets, and ships. Three such gyroscopes can be used to indicate three fixed directions in space. From measurements of a vehicle's acceleration relative to these axes, the velocity and position changes of the vehicle can be computed electronically.

Figure 7.24 shows a toy top on a pedestal. The wheel spins on an axle, which is supported at one end. If the top is placed as shown and released, it will fall if the wheel is not spinning. However, if the wheel is spinning rapidly, the top does not fall. Instead it *precesses:* the axis slowly turns in the horizontal plane. To analyze this motion, we must see how the torque affects the angular momentum of the wheel.

The gravitational force **w** exerts a torque on the top about the end resting on the pedestal. If the wheel is not spinning, the initial angular momentum $\mathbf{L}_i$ is zero. The torque produces an angular momentum $\mathbf{L}_f$ in a time Δt; $\tau \Delta t = \mathbf{L}_f$. Since the torque $\tau = \mathbf{r} \times \mathbf{w}$ and the resulting angular momentum are directed perpendicular to **r** and **w**, the top rotates clockwise and falls.

When the wheel is spinning, the situation changes. Now there is an initial angular momentum $\mathbf{L}_i$ along the top axis. Also, τ is perpendicular to $\mathbf{L}_i$, and the final angular momentum after a short time is $\mathbf{L}_f = \mathbf{L}_i + \tau \Delta t$. Since $\tau \Delta t$ is perpendicular to $\mathbf{L}_i$, τ acts to change the direction but not the magnitude of the angular momentum. Thus $\mathbf{L}_f$ and $\mathbf{L}_i$ have the same magnitudes but different directions (Fig. 7.24c). The torque causes the angular momentum and the top to precess in a circle.

Tops are fascinating devices to experiment with, and many different kinds of motion can be produced. When the end of the axis away from the pivot is released with exactly the right initial horizontal velocity, the precessional motion of the axis previously described occurs. If, instead, the end of the axis is simply dropped, the top initially begins to fall, and it then gradually acquires the sideways precessional velocity. The result is a vertical bobbing motion called *nutation,* which occurs along with the horizontal precession. If the top is spinning fast enough, the

Figure 7.23. A toy gyroscope. Its three pairs of pivots (1, 2, 3) isolate it from any external torques. (Courtesy of Sperry Division, Sperry Rand Corporation.)

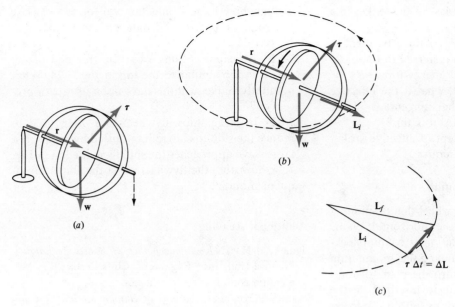

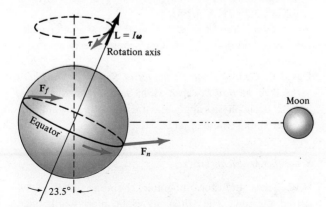

Figure 7.24. (a) The torque produced by **w** causes the top to rotate clockwise and fall. The wheel is not spinning. (b) and (c) When the wheel is spinning, the torque causes the angular momentum to precess in a horizontal plane. The end of the top moves along the dashed circle shown.

magnitude of the vertical or nutational displacement is small initially and rapidly diminishes as a result of frictional effects. Only the precessional motion is observed after a brief interval has elapsed.

Precession of the Equinoxes |

The earth rotates about its axis in a period of one day. The axis of rotation precesses very slowly resulting in what is called the precession of the equinoxes.

When the earth was formed, somewhat more mass

accumulated along its equator than at the poles because of the centripetal acceleration. The moon, and to some extent the sun, exert unequal forces on the opposite sides of the earth (Fig. 7.25). The net gravitational torque on a spherical planet is zero, so the net torque on the earth results entirely from the unequal forces on the *equatorial bulge* and is quite small.

Because the torque is small, the precession is very slow, with one complete cycle lasting 25,800 years. One effect of this precession is to cause a slow change in the time of year when a star is seen in a given position. Over the last 2000 years this time has shifted about 1 month. A second result is the gradual shifting of the seasons. After 12,900 years winter will occur in the position of the earth's orbit where summer occurs now.

EXERCISES ON SUPPLEMENTARY TOPICS

Section 7.7 | Momentum and the Use of the Body

7-71 What is the average force on the foot in a football punt?

7-72 Using the data of Table 7.1, find the average force on the hand of a handball player when serving.

7-73 Discuss, in terms of impulse and momentum, the difference between boxing with bare knuckles and with gloves.

Figure 7.25. The earth's axis of rotation is tilted as shown. The gravitational forces exerted on the equatorial bulge are F_n and F_f. F_n is larger than F_f because the near side of the earth is closer to the moon. The precession is along the dotted curve.

7-74 What is the effective mass of the striker in a soccer kick?

7-75 Use the data in Table 7.1 for a baseball to find (a) the kinetic energy transferred to the baseball, and (b) the mechanical energy lost.

7-76 (a) in a football punt, what is the effective mass of the striker? (b) What percentage of the body weight of an 80-kg person is this?

7-77 Discuss how the concept of effective striker mass applies to boxing techniques.

Section 7.8 | Gyroscopic Motion

7-78 A boy holds a bicycle wheel that is spinning rapidly. The wheel is in the horizontal plane. Viewed from above, the rotation is counterclockwise. (a) Which way does the angular momentum point? (b) The boy tries to turn the plane of rotation of the wheel by pushing to his right on the upper end of the axle and to his left on the lower end. What is the direction of the torque? (c) What happens to the wheel?

7-79 "A top will fall to the floor if released when it is not spinning but will precess if it is spinning." Since this statement does not state any minimum rotational velocity, it suggests that an arbitrarily small rotational speed is sufficient to prevent the top from falling. Explain how nutation resolves this apparent paradox.

PROBLEMS ON SUPPLEMENTARY TOPICS

7-80 (a) If a torque applied to the earth reduced the length of the day by 1 hour, how large an angular impulse would be required? (b) If this torque arose from a pair of forces (a couple) applied along the equator on opposite sides of the earth acting for 1 hour, how large would the two forces be? (Use the solar and terrestrial data on the inside back cover.)

*7-81 A weight is hung from the free end of the axis of a top similar to the top in Fig. 7.24. What qualitative effect will this have upon its rate of precession?

7-82 In large ships, flywheels are often used to reduce the side-to-side rolling caused by waves. If a large wave approaches the ship from the left in Fig. 7.26, how does the flywheel affect the ship's subsequent motion?

Additional Reading

James G. Hay, *The Biomechanics of Sports Techniques,* Prentice-Hall, Inc., Englewood Cliffs, N.J., 1973. Momentum conservation in athletics.

Stanley Plagenhoef, *Patterns of Human Motion: A Cinematographic Analysis,* Prentice-Hall, Inc., Englewood Cliffs, N.J., 1971. Momentum conservation in athletics.

Jearl D. Walker, Karate Strikes, *American Journal of Physics,* vol. 43, October 1975, p. 845.

F.I. Ordway, Principles of Rocket Engines, *Sky and Telescope,* vol. 14, 1954, p. 48.

R.A.R. Tricker and B.J.K. Tricker, *The Science of Movement,* Mills and Boon Ltd., London, 1966. Applications of angular momentum to balance and motion.

R.L. Page, The Mechanics of Swimming and Diving, *The Physics Teacher,* vol. 14, 1976, p. 72.

Chris D. Zafiratos, An Alternative Treatment of Gyroscopic Behavior, *The Physics Teacher,* vol. 20, 1982, p. 34.

Cliff Frohlich, Do Springboard Divers Violate Angular Momentum Conservation?, *American Journal of Physics,* vol. 47, July 1979, p. 583.

James Gray, *How Animals Move,* Cambridge University Press, Cambridge, 1953. Gyroscopic effects in insect motion.

David F. Griffing, *The Dynamics of Sports—Why That's the Way the Ball Bounces,* Mohican, Loudonville, Ohio, 1982. Basic physics applied to track and field, swimming, water skiing, football, etc.

Scientific American articles:

H.W. Lewis, Ballistocardiography, February 1958, p. 89.

Alfred Gessow, The Changing Helicopter, April 1967, p. 38.

James E. McDonald, The Coriolis Effect, May 1952, p. 72.

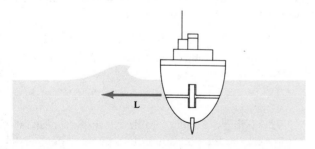

Figure 7.26. A flywheel mounted on a horizontal axle viewed from the rear of the ship. The rotation of the wheel is such that its angular momentum is directed as shown. Problem 7-82.

Cornelius T. Leondes, Inertial Navigation for Aircraft, March 1970, p. 80.

Cliff Frohlich, The Physics of Somersaulting and Twisting, March 1980, p. 154.

Jearl Walker, The Mysterious "Rattleback": a Stone That Spins in One Direction and Then Reverses, The Amateur Scientist, October 1979, p. 172.

Jearl Walker, The Physics of Spinning Tops, Including Some Far-out Ones, The Amateur Scientist, March 1981, p. 182.

Jearl Walker, Delights of the "Wobbler," a Coin or a Cylinder That Precesses as it Spins, *The Amateur Scientist*, October 1982, p. 184.

Jearl Walker, The Essence of Ballet Maneuvers Is Physics, *The Amateur Scientist*, June 1982, p. 146.

Jearl Walker, The Physics of the Follow, the Draw, and the Massé (in Billiards and Pool), *The Amateur Scientist,* July 1983, p. 124.

CHAPTER 8

ELASTIC PROPERTIES OF MATERIALS

We have discussed the motion of objects using the implicit assumption that these objects never changed in size or shape. However, an object made from any real material will always be deformed at least slightly and may even break when forces or torques are applied. For example, a steel or wooden beam will bend when a weight is hung from it, and a bone will twist and perhaps fracture when subjected to a torque.

Although materials are held together by complicated electric and magnetic forces among the molecules, the effects of these forces can be categorized quite adequately using a few measured quantities. With these quantities one can determine the size and shape of a steel beam needed to safely support a given load or the torque a particular bone can withstand without breaking.

The first portion of this chapter is devoted to the description of the *stresses* that produce deformations or *strains* in a material. We find that the strain depends on the way in which the stress is applied. Using experimentally determined parameters, we can then discuss the strength of materials and the optimal design of objects. We also apply our results to find the relation between the lengths and radii of columns. This is used to discuss tree sizes and to develop a very interesting scaling hypothesis relating the structure and function of animals.

8.1 | GENERAL ASPECTS OF STRESS AND STRAIN

If a certain force stretches a rubber band a given distance, then a force twice as great is needed to produce the same elongation in two such bands, or equivalently, in a *single* band of twice the cross-sectional area. Thus the deformations of materials are determined by the *force per unit area, and not by the total force*. Because of this it is useful to define the *stress σ* in a bar of cross-sectional area A (Fig. 8.1a) subjected to a force **F** as the ratio of the force to the area,

$$\sigma = \frac{F}{A} \tag{8.1}$$

The stress is opposed by the intermolecular forces within the material. For example, if the leg of a table is supporting a 100-N weight, then the intermolecular

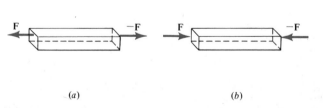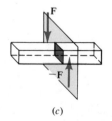

Figure 8.1. A bar subjected to (*a*) tension; (*b*) compression; (*c*) shear forces.

forces must exert an upward 100-N force on the top layer of molecules in the leg.

Three kinds of stress are commonly defined (Fig. 8.1). *Tension stress* is the force per unit area producing elongation of an object. *Compression stress* acts to compress an object. *Shear stress* corresponds to the application of scissorlike forces.

The change in the length of the bar under tension or compression stress is proportional to its length. For example, if a bar of length l subjected to a tension force **F** stretches a distance Δl, then each half of the bar stretches $\frac{1}{2}\Delta l$. The *strain ε* is the *fractional change in length* (Fig. 8.2),

$$\varepsilon = \frac{\Delta l}{l} \qquad (8.2)$$

From this definition, we see that ε is dimensionless and does not depend on the length of the bar. There are three kinds of strains: tension, compression, and shear. Any deformation of an object can be considered as a combination of these three strains. Equation

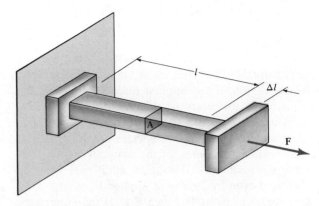

Figure 8.3. An experiment to measure the relationship between stress and strain.

8.2 defines both tension and compression strains. We describe shear strains later in the chapter.

The relation between the stress and the strain for a material under tension can be found experimentally. A bar clamped tightly at each end is gradually stretched, and the applied force F needed to do this is recorded at intervals (Fig. 8.3). The fractional change in length is then the strain, and the force per unit area is the stress. Typical results obtained in this way are shown in Fig. 8.4. Analogous graphs for compression and shear stresses can also be obtained.

For small values of the strain, the stress-strain graph (Fig. 8.4) is a straight line; the stress σ is linearly proportional to the strain ε. This is called the *linear region* for a material. Beyond the *linear limit A,* the stress is no longer linearly proportional to the

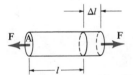

Figure 8.2. A bar of length l and area A subjected to a tension stress has its length increased by Δl. The ratio $\Delta l/l$ is the strain, ε.

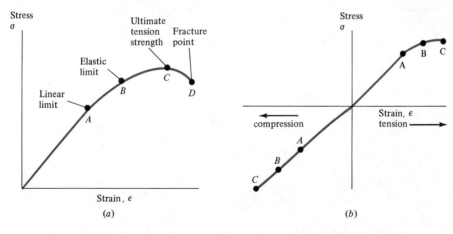

Figure 8.4. (*a*) The stress associated with a given tension strain for a ductile metal. (*b*) Stress-strain graph for a brittle material such as bone. The fracture point D (not shown) is now very close to ultimate strength point C. Note that the slopes in the linear portion of the curve are not the same for tension and compression.

strain. However, from *A* to the *elastic limit* or *yield point B,* the object still returns to its original dimensions when the applied force *F* is removed. The deformation up to *B* is said to be elastic. If the applied force is further increased, the strain increases rapidly. In this region, if the applied force is removed, the object does not return completely to its original dimensions; it retains a permanent deformation. The highest point *C* on the stress-strain graph is the *ultimate tension strength σ_t* of the material or its maximum stress. Beyond this point, additional strain is produced even by a reduced applied force, and *fracture* occurs at point *D*. From *B* to *D* the material is said to undergo *plastic deformation*. If the ultimate tension strength and fracture points *C* and *D* are close together, as in Fig. 8.4*b*, the material is *brittle;* if they are far apart, as in Fig. 8.4*a*, the material is said to be *ductile.*

When discussing the properties of materials, it must be remembered that all materials exhibit the phenomenon of *fatigue*. After many cycles of applying and removing a load, their ultimate strength gradually diminishes, and the material finally fails, even under weak stresses. For example, a paper clip bent back and forth several times eventually breaks easily. The effects of fatigue must be considered in situations as diverse as designing a bridge and developing pins to insert into bone fractures.

The reasons for material fatigue are not completely understood, but it is believed that after repeated deformations the internal molecular structure of the material is changed. These changes result in decreased intermolecular forces and hence in a reduction in strength of the material.

8.2 | YOUNG'S MODULUS

The elastic deformations of a solid are related to the associated stresses by quantities called *elastic moduli*. In the linear region of the stress-strain graph for tension or compression, the slope equals the stress-to-strain ratio and is called the *Young's modulus E* of the material:

$$E = \frac{\sigma}{\varepsilon} \qquad (8.3)$$

For homogeneous materials such as steel, the Young's moduli for compression and tension are usually equal. For inhomogeneous materials such as concrete or bone, the moduli for compression and tension are different. Table 8.1 lists representative Young's moduli, ultimate tension strengths σ_t, and ultimate compression strengths σ_c for various materials.

The following example illustrates the relationships we have described.

Example 8.1

(a) If the minimum cross-sectional area of the femur of a human adult is 6×10^{-4} m^2, what is the compressional load at which fracture occurs? (The femur is the

TABLE 8.1

Young's moduli and ultimate strengths for representative materials. All quantities have units of N m^{-2}

Material	Young's Modulus, E	Ultimate Tension Strength, σ_t	Ultimate Compression Strength, σ_c
Aluminum	7×10^{10}	2×10^8	
Steel	20×10^{10}	5×10^8	
Brick	2×10^{10}	4×10^7	
Glass	7×10^{10}	5×10^7	11×10^8
Bone (along axis)			
Tension	1.6×10^{10}	12×10^7	
Compression	0.9×10^{10}		17×10^7
Hardwood	10^{10}		10^8
Tendon	2×10^7		
Rubber	10^6		
Blood vessels	2×10^5		

main bone in the upper leg.) (b) Assuming the stress-strain relationship is linear until fracture, find the strain at which the fracture occurs.

(a) From Table 8.1, the ultimate compression strength σ_c for bone is 17×10^7 N m^{-2}. This is the force per unit area that will lead to fracture, and the total force is found by multiplying by the cross-sectional area of the bone. Thus

$$F = \sigma_c A = (17 \times 10^7 \text{ N m}^{-2})(6 \times 10^{-4} \text{ m}^2)$$
$$= 1.02 \times 10^5 \text{ N}$$

This force is large; it is about 15 times the weight of a 70-kg person. However, it is readily exceeded if one falls several metres and lands in a rigid position.

(b) Using the definition of Young's modulus, $E = \sigma/\varepsilon$, with $E = 0.9 \times 10^{10}$ N m^{-2} from Table 8.1,

$$\varepsilon = \frac{\sigma}{E} = \frac{17 \times 10^7 \text{ N m}^{-2}}{0.9 \times 10^{10} \text{ N m}^{-2}} = 0.0189$$

Thus the bone is reduced in length by 1.89 percent at the load that will cause fracture. The experimental value of the strain at fracture is slightly larger, since the stress-strain curve is not linear (Fig. 8.4b). It flattens out as shown in Fig. 8.4b as the ultimate strength is approached.

The linear stress-strain region of Fig. 8.4 is also called the *Hooke's law* region. In this region, since the stress is linearly related to the strain, the force is linearly related to the elongation. This can be seen using the definition of Young's modulus rewritten as $\sigma = E\varepsilon$. With the definitions of the stress $\sigma = F/A$ and the strain $\varepsilon = \Delta l/l$, this becomes

$$\frac{F}{A} = E \frac{\Delta l}{l}$$

Thus, in tension or compression the force on an object is proportional to its elongation,

$$F = k\,\Delta l \tag{8.4}$$

k is called the *spring constant,* and

$$k = \frac{EA}{l} \tag{8.5}$$

Equation 8.4 is called *Hooke's law*. As long as an object under stress is in the linear region, Hooke's law is valid. For example, coil springs, leaf springs, and rubber bands obey this relation if the deformations are not too large. The spring constant k is large for strong springs. From the definition of k, we see that increasing the cross-sectional area and decreasing the length both serve to strengthen the spring properties of the object. Spring forces will be discussed further in Chapter Nine.

8.3 | BENDING STRENGTH

Almost all mechanical structures ranging from beams to tree trunks and human limbs are subjected to various kinds of stresses. When the stress is a simple compression or tension stress, the shape of the object is unimportant, since the deformation depends only on the cross-sectional area. However, the ability of an object to resist bending or to bend without breaking depends not only on the composition but also on the shape of the object. For example, a hollow tube made from a given amount of material is stronger than a solid rod of the same length constructed from an equal amount of material. Similarly, there is a definite relationship between the lengths and radii of tree trunks and of animal limbs imposed by their shape and composition. In this section, we see how humans and nature design structures for both strength and lightness.

Figure 8.5 shows a bar of length l and rectangular cross section with sides a and b. Placed on two supports, it bends somewhat under its own weight. When we look at the left half of the bar (Fig. 8.5c), we notice

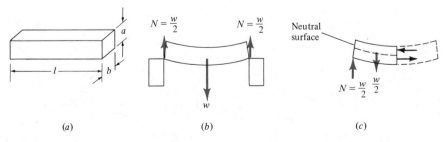

| (a) | (b) | (c) |

Figure 8.5. (a) and (b) A rectangular bar will bend under its own weight when supported at two points. (c) The left half of the bar experiences forces from one support and its weight. It also experiences forces from the other half of the bar (shown in black).

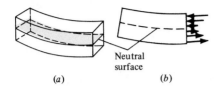

Figure 8.6. (a) The shaded plane, called the neutral surface, suffers no change in dimension as the bar bends. (b) Detailed internal forces acting on the left half of the bar. The pairs of forces at the upper and lower edges produce the largest torques because they are large and far apart.

that the vertical force from the left support and the weight of this half are equal and opposite. However, these forces have different lines of action, so they form a couple that rotates the half-bar clockwise. Since the bar is in equilibrium, the right half must exert forces that produce an equal but opposite torque.

From Fig. 8.5c we note that the upper part of the bar is being compressed and the lower part is under tension. The *neutral surface* suffers no change in length. This means that the strength of the bar depends on the elastic properties of the bar.

The upper and lower surfaces of the bar are distorted most, so the largest internal forces will appear at these surfaces (Fig. 8.6b). These forces produce a torque opposing that of the weight and support. The further from the neutral surface they act, the greater will be their contribution to the torque. Thus, with thick bars we can obtain large torques with relatively small internal forces, making it possible to support large loads.

To make this idea quantitative, suppose the bar is bent with a radius of curvature R (Fig. 8.7). We mentally divide the cross section into strips like the one shown at a distance x below the neutral surface. At this strip, the bar is stretched a distance Δl, so its total length is $l + \Delta l$. In radians, the angle θ formed by the bar is the arc length l at the neutral surface divided by the radius R, $\theta = l/R$. θ is also equal to $(l + \Delta l)/(R + x)$. Equating the two expressions for θ, we find after some algebra that $\Delta l/l = x/R$. Thus the strain in the strip at x is

$$\varepsilon = \frac{\Delta l}{l} = \frac{x}{R}$$

The strip has an area $\Delta A = b\Delta x$. Since the stress σ is $E\varepsilon$, where E is Young's modulus, the force on the strip is

$$\Delta F = \sigma \Delta A = E\varepsilon \Delta A = \frac{Ex\Delta A}{R}$$

The torque due to this force about the neutral surface is $x\Delta F = Ex^2\Delta A/R$.

If we sum the torques due to all the strips, the total internal torque is the integral

$$\tau = \int \frac{Ex^2}{R}\, dA$$

Since E and R are constants, they can be taken out of the integral, leaving

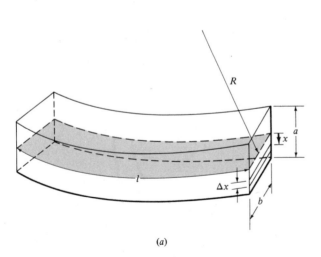

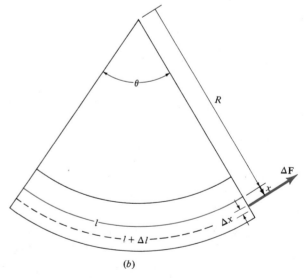

Figure 8.7. (a) A beam of length l is bent with a radius of curvature R. (b) At a distance x below the neutral surface, the bar is stretched so that its length is $l + \Delta l$.

TABLE 8.2

Area moments of inertia for vertical loading

Cross section	I_A
Rectangle:	$I_A = \dfrac{a^3 b}{12}$
Solid cylinder:	$I_A = \dfrac{\pi r^4}{4}$
Hollow cylinder:	$I_A = \dfrac{\pi (a^4 - b^4)}{4}$
I beam: Each section has a thickness t: a is the distance between the midpoints of the members	$I_A = \dfrac{a^2 b t}{2} + \dfrac{a^3 t}{12} \qquad (t \ll a, b)$

$$\tau = \frac{E I_A}{R} \qquad (8.6)$$

The quantity I_A is called the *area moment of inertia*, and it is defined by

$$I_A = \int x^2 \, dA \qquad (8.7)$$

The integral is over the cross section of the bar. We calculate I_A for a rectangular bar in the next example, and apply the result in the following example to show that thick boards resist bending better than thin ones. Additional area moments of inertia are listed in Table 8.2.

Example 8.2

Find the area moment of inertia for the bar in Fig. 8.7.

The strip shown has an area $\Delta A = b \Delta x$, and x goes from $-a/2$ to $+a/2$. Thus the area moment of inertia is

$$I_A = \int_{-a/2}^{a/2} b x^2 \, dx = b \left[\frac{x^3}{3} \right]_{-a/2}^{a/2} = \frac{a^3 b}{12}$$

Note that I_A increases rapidly as a is increased, since a enters to the third power.

Example 8.3

Two identical wooden 2 cm $\times$ 6 cm boards are supported at each end (Fig. 8.8). Each supports its own

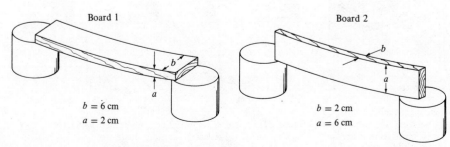

Board 1

Board 2

$b = 6$ cm
$a = 2$ cm

$b = 2$ cm
$a = 6$ cm

Figure 8.8.   Identical boards resting flat and on edge.

weight, but one is resting with its wide side down and the other with its narrow side down. Which board bends most and what is the ratio of the radii of curvature for the two boards?

Since each board supports its own weight, the internal torque of each must be the same. Thus, from Eq. 8.6, $I_{A1}/R_1 = I_{A2}/R_2$, where R_1 and R_2 are the radii of curvature of the two boards. Using $I_A = a^3b/12$ for each of the two rectangular boards, we see that $I_{A1} = (2\ \text{cm})^3(6\ \text{cm})/12 = 4\ \text{cm}^4$ and $I_{A2} = (6\ \text{cm})^3(2\ \text{cm})/12 = 36\ \text{cm}^4$. Then

$$\frac{4\ \text{cm}^4}{R_1} = \frac{36\ \text{cm}^4}{R_2}$$

or

$$\frac{R_2}{R_1} = 9$$

The radius of curvature of the board with its narrow side down is nine times that of the other. Since a large radius of curvature implies little bending, board 2 does not bend as much as board 1. Consequently, board 2 is less likely to break when a heavy weight is placed upon it.

These results suggest that to construct strong, light structural members, most of the material should be located as far as practical from the neutral surface. A horizontal I beam (Fig. 8.9) is better able to withstand bending torques due to vertical forces than a beam with a square cross section made from the same amount of material. By contrast the two beams can support the same applied *compressional force,* since their cross-sectional areas are the same. Similarly, a hollow tube has more bending strength than a solid rod of the same length and weight. This is why metal chair and table legs are usually hollow (Fig. 8.10). They can better withstand forces applied in any direction perpendicular to their length, since on the average the metal is further from the neutral surface.

Figure 8.10. The legs of both chairs have the same length and are constructed from the same amount of material. Those of (*a*) are solid cylinders, while those of (*b*) are hollow cylinders. The hollow cylinders are better able to withstand bending torques.

From this discussion, it would seem advantageous to make structural members with very large diameters and very thin walls. However, a limit to how far one may go in this direction is imposed by the tendency of thin-walled structures to *buckle* under compressional stresses. Figure 8.11 shows an experiment one can readily perform to illustrate this point. One sheet of notebook paper is rolled into a cylinder a single layer thick and fastened with tape. A second sheet is rolled into a cylinder an inch or so in diameter and taped. When the cylinders are stood on end and this textbook is placed on top, the larger cylinder immediately buckles and collapses, while the smaller one is able to support the load. Thus the thinner walls of the larger cylinder are unable to withstand a force applied approximately along its axis. We will consider buckling in more detail in the next section.

Nature has made extensive use of the principle that hollow structures are stiffer than solid ones of the same cross-sectional area. Bones are generally hol-

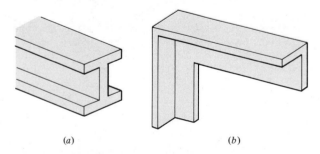

Figure 8.9. (*a*) An I beam is constructed so that most of the material is near the outer surfaces during bending due to a vertical force. (*b*) Brackets with an *L*-shaped cross section are designed for the same purpose. This bracket could be used as a wall-mounted support for a shelf.

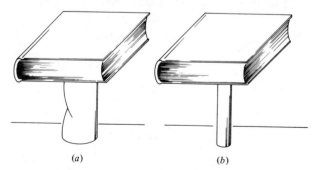

Figure 8.11. (*a*) A large radius, thin walled paper tube buckles easily. (*b*) A narrower tube made from an identical sheet of paper with thicker walls will not buckle under the same load.

low. For example, in the human femur, the ratio of the inner and outer radii is about 0.5, and the cross-sectional area is only 78 percent of that of a solid bone with the same bending strength. Smaller mammals and birds have bones with relatively thinner walls. For example, the ratio of the inner to outer radii for the humerus of a swan is 0.9, and the cross-sectional area is 38 percent of that of a solid bone with the same strength. The danger of collapse by buckling in this thin-walled bone is reduced by thin, reinforcing struts of bone placed across the interior of the humerus.

8.4 | BUCKLING STRENGTH AND STRUCTURAL DESIGN IN NATURE

In nature the failure of structural members usually results from large torques of various types rather than from simple compression or tension stresses. For example, except when someone falls from a great height, fractures of limb bones are usually the result of bending or twisting. We noted in the preceding section that a thin tube will readily buckle if a force is applied along its axis. More generally, any beam or column may buckle under such a force. Here we discuss the buckling strength of a cylindrical column and illustrate our results by seeing how nature seems to have used buckling strength as a criterion in designing the approximately cylindrical trunks of trees.

To understand how buckling occurs, consider the long, thin, cylindrical column in Fig. 8.12. It is held almost but not exactly vertically, so that its center of gravity is not quite over the center of the base, point P. The weight therefore exerts a torque about P that causes the column to bend. If the material is strong enough, the bending stops when the torques due to the internal forces in the material become large

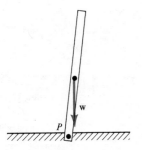

Figure 8.12. A cylindrical column tilted so that its weight is not directly over the center of its base, point P.

enough to balance the torque due to the weight. However, if the column is very tall and thin, as it bends the torques due to the weight will grow faster than will the torques due to the internal forces. The column will then buckle and collapse.

More generally, any vertical column supporting a load or even just its own weight will eventually buckle if its radius is held fixed and its height is increased. This *critical height* is determined by the Young's modulus of the material. This is because Young's modulus determines the internal forces for a given deformation.

For a solid cylinder of radius r supporting only its own weight, the critical height is shown in Section 8.7 to be

$$l_{cr} = cr^{2/3} \qquad (8.8)$$

Here c is a constant that depends on the weight per unit volume and the Young's modulus of the material composing the column. This result implies, for example, that if a column is just barely stable against buckling, doubling its radius does not permit doubling its height. This is illustrated in the following example.

Example 8.4

Two columns are made of the same material. One has a radius r_1 and the other has a radius of $2r_1$. If both columns can just support their own weight without buckling, what is the ratio of their lengths?

The length of the column of radius r_1 is $l_1 = cr_1^{2/3}$. The other column has a length $l_2 = c(2r_1)^{2/3}$. The ratio l_2/l_1 is

$$\frac{l_2}{l_1} = \frac{(2r_1)^{2/3}}{r_1^{2/3}} = (2)^{2/3} = 1.59$$

Thus the column with twice the radius can only be about 1.6 times as long.

The Height of Trees | The result $l_{cr} = cr^{2/3}$ is quite general. For example, with suitable values of the constant c, it holds true for tapered columns, hollow columns, and columns supporting loads. It must also hold true for trees; that is, the maximum height a tree can have and be stable against buckling must vary as $r^{2/3}$. Whether buckling is, in fact, the limiting factor in determining the height of trees can be investigated by comparing their measured heights and radii (Fig. 8.13). While there is considerable scatter in the data, the results support the idea that buckling strength is the key factor in determining the proportions of trees. We see in Section 8.6 that there is evi-

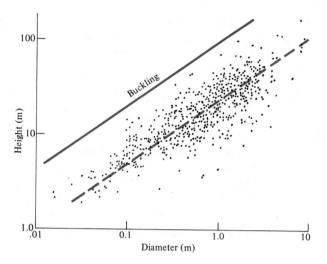

Figure 8.13. Data for North American trees. The dashed line is that of $l = cr^{2/3}$, with $c = 34.9$ chosen to fit the data. The solid line is the theoretical result for a tapered column that is just on the verge of buckling. Presumably no data points appear above this line since such trees would buckle under their own weight. (From T. McMahon, *Science,* vol. 179, pp. 1201–1204. March 23, 1973. Copyright 1973 by the American Association for the Advancement of Science.

dence that buckling strength also determines the body proportions of animals.

8.5 | SHEARING AND TWISTING TORQUES

Thus far, we have only considered compression and tension stress. However, it is also common to have forces acting on an object that cause shearing or twisting. In this section, we describe shear forces qualitatively and twisting torques more analytically.

A simple example of shearing stresses and strains that you can easily try is provided by placing a book on a table and exerting equally large forces in opposite directions on its covers (Fig. 8.14). Each page moves slightly relative to the next one, and the shape of the book changes even though its height h and width w stay nearly the same.

In Fig. 8.14, the book is deformed through an angle α. The upper cover moves a distance δ relative to the lower one. The *shear stress* on the upper cover is

$$\sigma_s = \frac{F}{A} \tag{8.9}$$

The *shear strain* is

$$\varepsilon_s = \frac{\delta}{h} = \tan\alpha \tag{8.10}$$

The ratio of these quantities defines the *shear modulus,*

$$G = \frac{\sigma_s}{\varepsilon_s} \tag{8.11}$$

Table 8.3 gives the shear moduli for some materials. Usually the shear modulus is between one-third and one-half of Young's modulus for the material.

Even though an object is subjected only to shearing forces, compression or tension stresses may result on various planes in the object. For example, the cube in Fig. 8.15 is in translational and rotational equilib-

TABLE 8.3

Shear moduli for some materials in N m⁻²

Material	Shear Modulus, G
Aluminum	2.4×10^{10}
Bones (long)	10^{10}
Copper	4.2×10^{10}
Glass	2.3×10^{10}
Hardwood	10^{10}
Steel	8.4×10^{10}
Tungsten	11.4×10^{10}

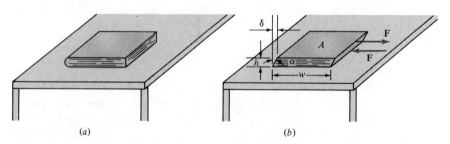

Figure 8.14. A book subjected to shearing forces changes its shape. The upper cover is displaced a distance δ relative to the lower, and the back of the book makes an angle α with the vertical direction.

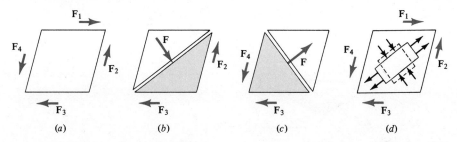

Figure 8.15. (*a*) A cube subjected to four shearing forces of equal magnitude. The cube deforms slightly, but it is in static equilibrium. (*b*) The parts formed by an imaginary cut along a diagonal plane are subjected to compressive stresses. (*c*) The other diagonal plane forms parts that experience tension stresses. (*d*) The dashed lines show a square on the side of the cube before the shearing forces are applied. A square drawn with this orientation is deformed into a rectangle by the shearing forces, even though the cube face itself is deformed from a square into an equilateral parallelogram or rhombus.

rium under the action of four shearing forces of equal magnitude acting along the faces. The cube is deformed slightly by these forces as shown. If we examine an imaginary cut along a diagonal plane as in Fig. 8.15*b*, then the shaded part of the cube must experience a compressive force **F** in the direction shown, since this part is in equilibrium. Similarly, for the plane in Fig. 8.15*c*, the shaded part experiences a tension force. The net result is that the square shown on the side of the cube in Fig. 8.15*d* becomes a rectangle when the cube is deformed by shearing forces.

From this discussion it is clear that stress and strain are complex quantities except in rather special cases. As we saw, externally applied shear forces result in both compressive and tension forces inside the material. Full specification of the stress must include the three components of the force on each of three perpendicular planes at each point in the object. Such complete analyses of stress-strain relations are beyond the scope of this textbook.

We now briefly discuss the effect of a torque directed along the axis of a cylinder arising from twisting forces. Such torques occur when a skier's leg is twisted in a fall or when power is transmitted by a rotating shaft. Figure 8.16 shows a cylinder fixed at one end. A couple is applied at the free end, so that there is a torque directed along the axis. If the resulting deformation is not too large, it is found that a plane drawn along the axis of the cylinder becomes twisted, as in Fig. 8.16*b*. The angle of twist increases linearly with the distance from the fixed end, so that the radial lines remain straight. Lines originally drawn along the outside of the cylinder parallel to the axis become slightly curved.

To find the relation between the toruqe τ and the deformation or twisting angle α, we divide the cylinder into thin concentric cylindrical layers, seen at the end as narrow concentric rings (Fig. 8.16*c*). The layers are twisted by shear forces, so their stress and strain are related by the shear modulus: $\sigma_s = G\varepsilon_s$. A ring of radius r has a width Δr and an area $\Delta A = 2\pi r \Delta r$. The ring rotates through an angle α and a distance $r\alpha$. With the replacements $\delta = r\alpha$ and $h = l$, the shear strain of the ring is $\varepsilon_s = \delta/h = r\alpha/l$. The corresponding shear stress, the force per unit area needed to cause this deformation, is $\sigma_s = G\varepsilon_s = G(r\alpha/l)$. The torque about the axis on this ring is then $\Delta\tau = r\sigma_s\Delta A = r(Gr\alpha/l)(2\pi r\Delta r)$. Summing over the rings gives the total torque needed to cause this deformation,

$$\tau = \int_0^a \frac{G\alpha}{l} 2\pi r^3 \, dr = \frac{G\alpha\pi a^4}{2l}$$

We can rewrite this result as

$$\tau = GI_p \frac{\alpha}{l} \tag{8.12}$$

This is similar in form to the result for bending torques, but I_p is the *polar moment of inertia*. For a cylinder of radius a,

$$I_p = \frac{\pi a^4}{2} \quad \text{(solid cylinder)} \tag{8.13}$$

Since the polar moment of inertia increases with the fourth power of the radius, doubling the radius of a cylinder increases its resistance to twisting by a factor of $2^4 = 16$.

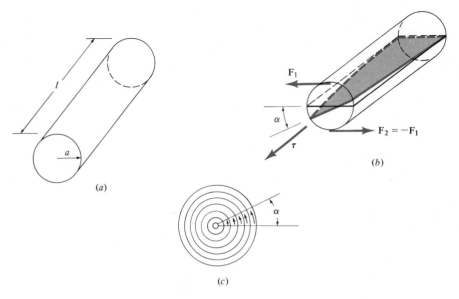

(a)

(b)

(c)

Figure 8.16 (a) A cylinder of length *l* and radius *a*. (b) The far end of the cylinder is held fixed by a constraint (not shown), and the near end is subjected to forces producing a torque along the axis. The resulting twist of the planes in the cylinder increases linearly with the distance from the fixed end if the applied torque is not too large. (c) The near end. Adjacent cylindrical layers deform more as one goes further from the cylinder axis. This is a shearing deformation.

If an object is subjected to an increasingly large twisting torque, it will eventually fracture. The torques and corresponding angles at which this happens for several human limb bones are listed in Table 8.4. Under the right conditions, a relatively small force can lead to a fracture. For example, a ski tip is about a metre from the heel of a skier's boot, so a 100-N force at the tip produces a torque of 100 N m. According to the table, this is sufficient to break the tibia, the larger of the two bones in the lower leg.

The toe of the boot is about 0.3 m from the heel. If the foot and ski do not rotate, an applied torque of 100 N m requires that the skier exert a force of 100 N m/0.3 m = 330 N at the toe (Fig. 8.17). If the

toe release is set to open at a lower force level, the risk of injury is reduced.

Twisting fractures in bones or metal cylinders normally are not breaks at right angles to the axis, but instead are spiral fractures (Fig. 8.18). The explanation for this phenomenon is suggested by our discussion of the stresses and strains in a cube subjected only to shearing forces. We saw that for some surfaces in the cube, there was pure tension or compression stress, and a square drawn appropriately on a face of the cube was deformed into a rectangle. Similarly,

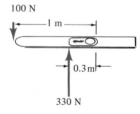

Figure 8.17. An overhead view of a skier's boot and ski. If the ski and foot do not rotate, the net torque about the heel must be zero. Thus a 100-N force applied to the ski results in a 330-N force exerted on the ski by the toe.

TABLE 8.4

The breaking torque and breaking angle for twisted bones in humans

Bone	Breaking Torque (N m)	Breaking Angle of Twist
Leg		
Femur	140	1.5°
Tibia	100	3.4°
Fibula	12	35.7°
Arm		
Humerus	60	5.9°
Radius	20	15.4°
Ulna	20	15.2°

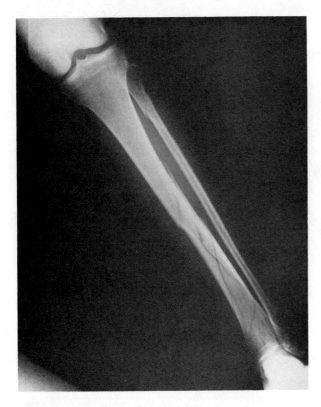

Figure 8.18. A spiral fracture of a tibia. (Courtesy of Dr. Robert C. Runyon.)

when a rod is twisted, planes drawn at 45° angles to the axis experience pure tension or compression. Fracture occurs when the smaller of the ultimate strengths for tension and compression is exceeded on one of these planes (Fig. 8.19).

SUMMARY

Objects subjected to forces or torques change their shape and may break. The fractional change in size or shape is the strain ε, and the force per unit area producing the deformation is the stress σ. For small applied forces or torques, the stress and strain in a material are usually linearly related. The proportionality constant relating stress and strain in the linear region is Young's modulus for compression or tension,

$$E = \frac{\sigma}{\varepsilon}$$

In the presence of a shear stress, the relationship is

$$G = \frac{\sigma_s}{\varepsilon_s}$$

where G is the shear modulus.

When a beam bends with a radius of curvature R, the torque due to the internal force is

$$\tau = E \frac{I_A}{R}$$

I_A is the area moment of inertia. If the area moment of inertia of the beam is large, the radius of curvature for a given torque is also large, and the beam does not bend much.

A column of radius r supporting its own weight has a maximum or critical length

$$l_{cr} = c r^{2/3}$$

c depends on Young's modulus for the material and on the weight per unit volume. If the length exceeds l_{cr}, the column will buckle. This criterion appears to be a limiting factor in the growth of trees.

A twisting torque τ applied to a uniform cylinder of length l is related to the angle α through which the cylinder twists by

$$\tau = G I_p \frac{\alpha}{l}$$

I_p is the polar moment of inertia and depends on the radius of the cylinder.

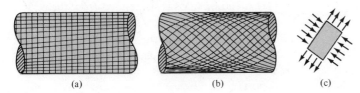

(a) (b) (c)

Figure 8.19. (a) When a cylinder is twisted, lines parallel to the axis are twisted, but lines in planes perpendicular to the axis are unaffected. (b) When the cylinder is twisted, one set of parallel lines drawn at 45° to the axis is pushed together and the other set is stretched out. (c) A single square from part (b becomes deformed into a rectangle. The stress is pure tension along the long sides and pure compression along the short sides.

REVIEW QUESTIONS

Q8-1 If a force F is applied to a bar of cross section A, the stress is _____.

Q8-2 The strain in an object subjected to a stress is the _____.

Q8-3 The three types of stress are _____, _____, and _____.

Q8-4 In the linear region, the _____ and _____ are linearly proportional.

Q8-5 Up to the _____, an object returns to its original length when the stress is removed.

Q8-6 If a material is easy to compress, it has a small _____.

Q8-7 If the force needed to stretch an object is proportional to the elongation, the object obeys _____.

Q8-8 A bar with a large area moment of inertia is _____ to bend than one with a small area moment of inertia.

Q8-9 Buckling of a column refers to collapse under forces approximately along its _____.

Q8-10 The shear modulus is the ratio of the _____ to _____.

Q8-11 Twisting a cylinder produces _____ stresses.

Checklist

Define or explain:

tension stress and strain	Hooke's law
compression stress and strain	spring constant
	neutral surface
linear limit	internal torque
elastic limit	area moment of inertia
ultimate tension strength	buckling
	critical height
brittle	shear stress and strain
ductile	shear modulus
fatigue	polar moment of inertia
Young's modulus	

EXERCISES

Section 8.1 | General Aspects of Stress and Strain

8-1 A tension force of 100 N is applied to the 0.1-m² face of a bar. What is the stress in the bar?

8-2 A tension stress of 2×10^6 N m⁻² is applied to a bar 0.05 m² in cross section. What is the applied force?

8-3 A 0.4-m pipe under compressional stress changes length by 0.005 m. What is the strain in the pipe?

8-4 The largest tension strain that can occur before fracture in aluminum is 0.003. What is the maximum change in length of a 1-m aluminum pipe?

8-5 A man's leg can be thought of as a shaft of bone 1.2 m long. If the strain is 1.3×10^{-4} when the leg supports his weight, by how much is his leg shortened?

8-6 A rubber rod of length 0.5 m and radius 10^{-3} m stretches 0.1 m when a 140 N force is applied. How large a force is needed to stretch a rubber rod 0.1 m if its length is 0.5 m and its radius is 2×10^{-3} m?

8-7 An automobile jack supports half the weight of a 1500-kg vehicle. If the stress is not to exceed 10^8 N m⁻², and the jack has a solid circular cross section, what is its minimum radius?

8-8 A steel wire 10 m long has a radius of 1 mm = 10^{-3} m. Its linear limit is 2.5×10^8 N m⁻², and its ultimate tension strength is 5×10^8 N m⁻². The wire is attached at one end and hangs vertically with a weight at its lower end. (a) If the wire is just at its linear limit, how large is the weight? (b) What is the largest load the wire can support?

8-9 A steel cable with a diameter of 3 cm = 3×10^{-2} m supports a chairlift at a ski area. If the maximum stress is not to exceed 10^8 N m⁻², what is the greatest load the cable can support?

Section 8.2 | Young's Modulus

8-10 An aluminum wire is 20 m long and has a radius of 2 mm = 2×10^{-3} m. The linear limit for aluminum is 0.6×10^8 N m⁻². (a) How large a tension force must be applied to stretch the wire to its linear limit? (b) How much will the wire stretch when this force is applied?

8-11 A 100-kg mass is suspended from the end of a vertical, 2-m-long steel post with a cross-sectional area of 0.1 m². (a) Find the stress and strain in the post. (b) How much does the post stretch? (c) What is the maximum mass that can be suspended from this post?

8-12 A hardwood post with dimensions 10 cm by 15 cm by 3 m supports a load of 1000 N along its length. (a) Find the stress and strain in the post. (b) What is its change in length?

8-13 If the minimum cross-sectional area of a human femur is 6.45×10^{-4} m^2, what is the tension load at which fracture occurs?

8-14 A sheet of glass has an area of 0.5 m^2 and is 0.005 m thick. (a) If it is placed horizontally, what is the uniformly distributed load at which fracture occurs? (b) What is the change in thickness if half that load is applied?

8-15 A vertical steel post is 3 m long and has a radius of 0.1 m. It is supporting a load of 10^5 N. (a) Find the stress and strain in the post. (b) Find the change in length.

8-16 The average cross-sectional area of a woman's femur is 10^{-3} m^2, and it is 0.4 m long. The woman weighs 750 N. (a) What is the length change of this bone when it supports half of the weight of the woman? (b) Assuming the stress-strain relationship is linear until fracture, what is the change in length just prior to fracture? (c) Is the answer to part (b) an overestimate or underestimate?

8-17 What is the spring constant of a human femur under compression of average cross-sectional area 10^{-3} m^2 and length 0.4 m?

Section 8.3 | Bending Strength

8-18 A cylindrical rubber rod is 0.5 m long and has a radius of 0.005 m. (a) What is its area moment of inertia? (b) What torque is exerted by the internal elastic forces on the ends of a rod when it is bent into a circle?

8-19 A cylindrical steel rod 2 m long has a radius of 0.01 m. If it is loaded so that it bends elastically with a 20-m radius of curvature, what is the torque due to this load?

8-20 A board is 1 cm by 6 cm in cross section. (a) Compute the area moments of inertia for loads parallel to the longer dimension and for loads parallel to the shorter dimension. (b) What is the ratio of the radii of curvature for the two deflections if equal loads are applied to the board in both orientations?

8-21 Two boards have the same length. (a) Board A has a 4-cm-by-4-cm cross section. What is its area moment of inertia for forces perpendicular to one of its sides? (b) Board B has a 2-cm-by-8-cm cross section. Find its two area moments of inertia for forces perpendicular to its shorter and longer sides. (c) Which board would be stronger if forces were always applied along a given direction perpendicular to its length? (d) Which board would be the best choice if the forces were to be applied in various directions perpendicular to its length?

8-22 A 10-m-long hollow steel cylinder is secured to a concrete base and used as a flagpole. Its inner and outer radii are 7 and 8 cm, respectively. (a) What is its area moment of inertia? (b) If the wind exerts a horizontal force at the top of 10^3 N, what is the radius of curvature of the flagpole?

8-23 In Fig. 8.14, two equal but opposite forces of magnitude **F** are shown acting on a book that remains at rest although it is deformed. (a) Is the book in equilibrium? (b) What are the other two forces acting on the book? (c) Do the other forces have the same line of action? Explain.

Section 8.4 | Buckling Strength and Structural Design in Nature

8-24 In a monument, a column is just strong enough to withstand buckling under its own weight. The column is 10 m tall and 0.1 m in radius. If a similar column is to be 40 m tall, what is its minimum radius?

8-25 A tall, slender column or tower is less likely to buckle if it is supported by guy wires attached to its top and to the ground some distance from the bottom. Why is a relatively small amount of material in the form of guy wires more effective than adding a similar amount to the structure itself?

8-26 A tree is just stable against buckling. If it grows until its height is doubled, and again it is just stable against buckling, by what factor does its cross-sectional area at the base change?

8-27 Using $l = cr^{2/3}$ with the experimental value $c = 34.9$ m$^{1/3}$, find the height of a tree of trunk radius $\frac{1}{8}$ m. Comment on whether your answer seems reasonable.

8-28 Give an argument as to why the data of Fig. 8.13 do not fall on a single straight line.

8-29 A uniform column will buckle under its own weight when

$$l = \left(\frac{2E}{w_0}\right)^{1/3} r^{2/3}$$

For hardwood $E = 10^{10}$ N m^{-2} and $w_0 = 5900$ N m^{-3}. Compare the length of a uniform column of wood of radius $\frac{1}{8}$ m with that of a typical tree of the same trunk radius for which $l = cr^{2/3}$ and $c = 34.9$ m$^{1/3}$.

Section 8.5 | Shearing and Twisting Torques

8-30   Two bones of equal radius are subjected to equal twisting torques. If one is longer than the other, which will fracture first?

8-31   A steel bar is clamped in a vise, so that a cube with sides 0.01 m protrudes above the jaws of the vise. (a) If a force of 100 N is applied along the top face of the cube, what are the stress and strain? (b) What is the horizontal displacement of the top face?

8-32   A 75 kg bicyclist puts all her weight on one pedal. (Fig. 8.20) The diameter of the central shaft of the pedal is 1.5 cm. (a) Find the shear stress on the pedal shaft. (b) Find the ratio of this stress to the maximum shear stress, 10^8 N m^{-2}.

8-33   Two metal plates overlap slightly and are attached by a row of 10 rivets. Each rivet has a radius of 3 mm $= 3 \times 10^{-3}$ m. If the shearing stress on the rivets is not to exceed 10^8 N m^{-2}, what is the greatest force that can be applied to the ends of the plates?

8-34   A brake block on a bicycle has a shear modulus of 10^7 N m^{-2}. When the brake is applied, the block exerts a 100 N force on the wheel rim. The surface in contact with the rim is 1 cm by 5 cm, and the block is 0.8 cm thick. (a) What is the shear stress on the block? (b) By what distance is the surface in contact displaced?

8-35   A truck pulls a travel trailer with a mass of 2000 kg. They are connected by a hitch whose weakest link is a steel pin 2 cm in diameter that slides through a hole across the shaft of the hitch. The maximum shearing stress for the steel is 10^8 N m^{-2}. (a) During a panic stop, the hitch must support a load equal to 20 percent of the weight of the trailer. What is the shearing stress on the pin? (b) Find the ratio of this stress to the maximum shearing stress.

8-36   A steel rod is 0.4 m long and has a radius of 0.5 cm. (a) Find its polar moment of inertia.

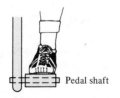

Figure 8.20   Exercise 8-32.

(b) One end is held fixed and the other is twisted. How large a torque must be applied to twist the end 0.1 radians (5.7°)? (c) This torque is applied by a wrench. If the force applied normal to the wrench is 100 N, how long is the wrench?

PROBLEMS

8-37   Three vertical 10-m-long steel tubes support a water tank. The inner and outer radii of the tubes are 15 and 17 cm, respectively. The tank is designed so that the stress on the tubes will not exceed 10^8 N m^{-2}. (a) Find the maximum volume of water the tank can hold, assuming the weight of the tank is negligible compared to that of the water. (The density of water is 1000 kg m^{-3}.) (b) By how much are the tubes shortened when they support the maximum load?

8-38   Suppose the I beam shown in Table 8.2 is rotated 90° about a horizontal axis, so that the "I" becomes an "H". Its area moment of inertia for supporting a vertical load is then $b^3t/6$, assuming $t \ll a, b$. For an I beam with $a = b$, find the ratio of this area moment of inertia to that of the I beam in its original orientation.

8-39   An amusement park ride whirls a car and its passengers with a total mass of 700 kg in a vertical circle of radius 8 m. At the bottom of the circle the car is moving at 12 m s^{-1}. The car is at the end of a steel arm. (a) If the maximum stress is to be 1 percent of the ultimate tension strength of the arm, what is its cross-sectional area? (b) What is the maximum elongation of the arm due to the moving car?

8-40   A freight elevator and its contents have a mass of 10,000 kg and are at rest. The steel cable supporting it has a stress equal to 10 percent of its ultimate tension strength. (a) What is the radius of the cable? (b) Find the fractional change in length $\Delta l/l$ of the cable when the motor is turned on and it accelerates the elevator upward at 2 m s^{-2}.

8-41   Two stone columns are barely stable against buckling under their own weight. If one is twice as tall as the other, find the ratio of the weight of the taller column to that of the other.

8-42   Estimate the twisting torques on the femur of a football player when he makes a sharp pivot on one foot. Does your result have any bearing on the types of cleats and playing surfaces that should be used?

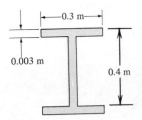

Figure 8.21. Problem 8-43.

8-43 The I beam in Fig. 8.21 is constructed from plates 0.003 m thick and is 0.3 by 0.4 m. (a) Calculate its area moment of inertia for loads applied vertically. (b) Calculate the area moment of inertia for a beam with square cross section having the same weight per unit length.

8-44 A 100-N bar is 5 m long. What is the magnitude of the torque on one half due to forces from the other half when it is supported horizontally? (*Hint:* See Fig. 8.5c and assume that the bar bends very little.)

*8-45 Two cylinders are constructed, one solid with radius r and one hollow with radii $a = 2r$ and $b = 3r/2$. If both cylinders are subjected to the same load, perpendicular to their long axes, what is the ratio of their radii of curvature?

8-46 A rod of radius a is replaced by a hollow tube of the same length with inner radius a. (a) If the tube is to have the same area moment of inertia as the rod, what must its outer radius be? (b) What is the ratio of the weights of the tube and rod?

8-47 The inner radius of a bone is half the outer radius. What would the outer radius of the bone be if, for a given torque, it were to twist through the same angle as a steel rod of the same length with a 1-cm radius? (The polar moment of inertia of a hollow cylinder is $\pi(a^4 - b^4)/2$, where a and b are the outer and inner radii, respectively.)

8-48 A steel shaft connecting an electric motor to a machine rotates at 1800 revolutions per minute. The shaft is 0.4 m long, has a radius of 1 cm, and delivers 2 kW of power. What is the twisting angle at the end of the shaft?

c8-49 Show that the area moment of inertia of a solid cylinder of radius r is $\pi r^4/4$. [*Hint:* Work in polar coordinates.]

c8-50 Show that the area moment of inertia of a hollow cylinder with outer radius a and inner radius b is $\pi(a^4 - b^4)/4$. [*Hint:* Work in polar coordinates.]

c8-51 Show that the polar moment of inertia of a hollow cylinder with outer radius a and inner radius b is $\pi(a^4 - b^4)/2$.

*c8-52 Derive the formula in Table 8.2 for the area moment of inertia of an I beam.

ANSWERS TO REVIEW QUESTIONS

Q8-1, F/A; **Q8-2,** fractional deformation; **Q8-3,** tension, compression, shear; **Q8-4,** stress and strain; **Q8-5,** elastic limit; **Q8-6,** Young's modulus; **Q8-7,** Hooke's law; **Q8-8,** harder; **Q8-9,** axis; **Q8-10,** shear stress, shear strain; **Q8-11,** shear.

SUPPLEMENTARY TOPICS

8.6 | STRUCTURE AND FUNCTION

The buckling strength criterion was employed in Section 8.4 to relate structure and size in trees. It can also be used to relate structure and physiological function in animals with a different form of the scaling concept introduced in Chapter Six. We will see that this different scaling approach leads to much better agreement with observations.

The simple version of scaling adopted earlier assumes that all of an animal's body dimensions scale with a single characteristic length l. Thus the body volume, and hence the mass m, varies as l^3; the length varies as $l \propto m^{1/3}$. Any area A then varies as l^2 or as $m^{2/3} = m^{0.67}$, so it is predicted that the body surface area will vary as $m^{0.67}$. It is also predicted that the *metabolic rate*—the rate at which food energy is used in the body—will vary in the same way, since the rate of oxygen absorption must vary as the surface area of the lungs, and the rate of heat loss must vary as the body surface area.

The actual scaling laws found by experiment are not always in exact agreement with the predictions of this simple model (Fig. 8.22). For example, one failure of the simple scaling procedure was observed by Kleiber in 1932. He found that the rate of heat production in mammals ranging in size from mice to elephants is not proportional to $m^{0.67}$ but to $m^{0.75}$. This difference is small, but it is sufficient to raise doubts about the validity of the model.

Recently McMahon suggested that one should take account of the fact that most body segments are cylindrical and are perhaps built to withstand buckling. If this is true, the length l and radius r of each body segment are related as we found in Section 8.4,

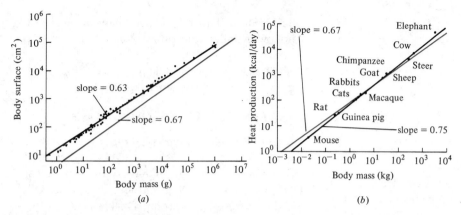

Figure 8.22. (a) The body surface area of mammals of various sizes. (b) The metabolic rates of various mammals versus mass. The black lines are the predictions of the buckling strength model; the colored lines are those of the simple scaling model. (From T. McMahon *Science*, vol. 179, pp. 1201–1204, March 23, 1973. Copyright 1973 by the American Association for the Advancement of Science.)

$l \propto r^{2/3}$. Because the volume of a cylindrical body segment is $\pi r^2 l$, the mass m should be proportional to $r^2 l$. Using $r \propto l^{3/2}$ we find $m \propto r^2 l = (l^{3/2})^2 l = l^4$. This is significantly different from the other scaling assumption, $m \propto l^3$.

In this model, since $m \propto l^4$, lengths should scale as $l \propto m^{1/4}$. Also because $l \propto r^{2/3}$, $r^{2/3} \propto m^{1/4}$ or $r \propto m^{3/8}$. Thus, our scaling assumption is that lengths and radii of body segments scale with mass as

$$l \propto m^{1/4}, \qquad r \propto m^{3/8} \qquad (8.14)$$

We can use these results to find how the body surface area of mammals with cylindrical body segments scales with their mass.

To find the surface area, we note that most cylindrical body segments are connected to other segments at either one or both ends. The surface area of one segment will then be proportional to the area of the sides of the cylinder. $A_{\text{surf}} = 2\pi r l$. Using the relations of Eq. 8.14, $A_{\text{surf}} \propto r l$, so

$$A_{\text{surf}} \propto m^{3/8} m^{1/4} = m^{5/8}$$

$\frac{5}{8} = 0.625$ is very close to 0.63 and fits the surface area data in Fig. 8.22a quite well.

To find the metabolic rate, we calculate not the surface area but the power used to flex a muscle. This power $\mathcal{P}$ is the force exerted, F, times the velocity of muscle contraction, v. Since all mammalian muscles are found to exert the same force per unit area $\sigma = F/A$, we write the force as $F = \sigma A$. A is the cross-sectional area of the muscle. Thus,

$$\mathcal{P} = Fv = \sigma A v$$

For voluntary muscle fiber, it has been found experimentally that the velocity of muscle contraction is also the same for all mammals. Thus σ and v do not depend on the mass and $\mathcal{P} \propto A$. The cross-sectional area A depends on the square of the radius of the muscle, $A \propto r^2$, so

$$\mathcal{P} \propto A \propto r^2 \propto m^{0.75}$$

This is the result shown in Fig. 8.22b, if we assume that the power expended and the heat production scale in the same way. The rate of work done by the heart muscle should also vary as $m^{0.75}$ and, because metabolic processes use oxygen absorbed through the lung wall, the lung area should depend on $m^{0.75}$. Both of these results have been well established experimentally.

From these results, we can also find how pulse rates scale. The metabolic rate, and hence the oxygen demand of the body, is proportional to $m^{0.75}$. The volume of blood pumped per heartbeat is proportional to the heart volume or to m. The blood pumped per second will vary as mf, where f is the heart rate. Then $mf \propto m^{0.75}$ or $f \propto m^{-0.25}$. This means that larger animals should have lower pulse rates. Again, this has been observed experimentally.

To summarize, the assumption that the cylindrical segments of animals have their shape determined by the buckling criterion gives rise to scaling laws that are in good agreement with a variety of experiments. Even if further study should show that this model is not adequate to explain other kinds of data, it does illustrate clearly how physical principles can influence the relationship between size and function in biological systems.

8.7 | DERIVATION OF $l_{cr} = cr^{2/3}$

We consider the particular example of a uniform column of radius r and length l which bends under its own weight with a radius of curvature R (Fig. 8.23). The torque due to the weight w is $\tau = wd$ and must be counteracted by a torque acting at the base of the column equal to $\tau = EI_A/R$. For a cylinder, $I_A = \pi r^4/4$. Thus, when the column is just at the buckling point,

$$\frac{Emr^4}{4R} = wd \qquad (8.15)$$

We now need to find expressions for w and d.

If the weight per unit volume is w_0, the total weight is w_0 times the volume or $w = w_0(\pi r^2 l)$. The distance d can be found from the colored triangle in Fig. 8.23. The sides of the triangle have lengths R, $R - d$, and h. If the radius of curvature is large compared to l, then $h \simeq \frac{1}{2}l$. Using the Pythagorean theorem

$$(R - d)^2 + \left(\frac{l}{2}\right)^2 = R^2$$

Squaring and neglecting the term d^2, we find $d = l^2/8R$.

Using our results for w and d in Eq. 8.15,

$$\frac{E\pi r^4}{4R} = w_0(\pi r^2 l)\frac{l^2}{8R}$$

or

$$l = \left(\frac{2E}{w_0}\right)^{1/3} r^{2/3}$$

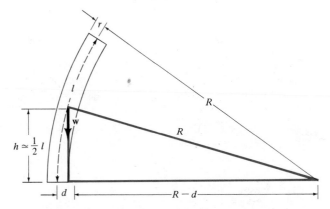

Figure 8.23. A column of length l and radius r, supported at its base, bends with a radius of curvature R.

This is the critical length l_{cr} of Eq. 8.8 for a uniform column. Here, $c = (2E/w_0)^{1/3}$, where E is Young's modulus and w_0 is the weight per unit volume of the column.

EXERCISES ON SUPPLEMENTARY TOPICS

Section 8.6 | Structure and Function

8-53 If the properties of animals did not depend on buckling strength but on compressional strength, how would the cross-sectional area of the legs of animals depend on body weight?

8-54 How does the time an animal can stay under water scale with body mass? Use the scaling hypothesis of Section 8.6. (Assume the oxygen used varies with volume.)

8-55 How does the heat loss to the environment scale with body mass? (Use the scaling hypothesis of Section 8.6.) Are your results consistent with the fact that small animals are not generally found in arctic areas?

8-56 If all mammals lived the same number of heartbeats, which animals would live longest?

PROBLEM ON SUPPLEMENTARY TOPICS

8-57 Find how the speed with which an animal can run uphill varies with its mass, assuming (a) the simplest scaling law; (b) the buckling strength criterion.

8-58 Assume that as a human grows from an infant to an adult, the bones grow according to the relationship $l \propto r^{2/3}$. (a) If the breaking angle during twisting is always the same, are adults or infants more susceptible to fractures due to twisting forces? (b) How does the twisting torque scale with body mass?

8-59 If the proportions of animal body segments are determined by the structural buckling strength criterion, how does the height jumped vary with the mass? Compare your result with that of Chapter Six.

8-60 The column of Fig. 8.23 will fracture when the stress at the outer (and inner) edges reaches the maximum, σ_t. Show that this occurs when $\sigma_i = Er/R$.

8-61 Using the results of Problem 8-60, find the maximum radius of curvature for a hardwood tree with $r = 0.25$ m. Use $E = 10^{10}$ N m^{-2} and $\sigma_t = 10^8$ N m^{-2}.

8-62 What is the radius of a cylindrical steel rod that fractures when the radius of curvature is 4 m? (Use the results of Problem 8-60).

Additional Reading

Harold M. Frost, *An Introduction to Biomechanics,* Charles C Thomas, Springfield, Ill., 1967. Brief introduction to the properties of biological materials.

Hiroshi Yamada, in F. Gaynor Evans (ed.), *Strength of Biological Materials,* Williams and Wilkins Co., Baltimore, 1970. Results of extensive research.

T. McMahon, Size and Shape in Biology, *Science,* vol. 179, 1973, p. 1201. Buckling strength and scaling.

R. McNeill Alexander, *Animal Mechanics,* University of Washington Press, Seattle, 1968. Chapters 3 and 4 discuss elastic properties, strength.

Francis Gaynor Evans, *Stress and Strain in Bones,* Charles C Thomas, Springfield, Ill., 1957.

Haywood Blum, Physics and the Art of Kicking and Punching, *American Journal of Physics,* vol. 45, 1977, p. 61. Karate and the strength of materials.

John R. Cameron and James G. Skofronick, *Medical Physics,* John Wiley and Sons, Inc., New York, 1978. Chapter 3 on composition and strength of bones.

Horace Freeland Judson, "Take That, King Richard!", *Science 80,* July/August, p. 44.

Scientific American articles:

John J. Gilman, Fracture in Solids, February 1960, p. 94.

Carl W. Condit, The Wind Bracing of Buildings, February 1974, p. 93.

Francis P. Bundy, Superhard Materials, August 1974, p. 62.

Thomas A. McMahon, The Mechanical Design of Trees, July 1975, p. 92.

Jearl Walker, Strange to Relate, Smokestacks and Pencil Points Break in the Same Way, The Amateur Scientist, February 1979, p. 158.

CHAPTER 9

VIBRATIONAL MOTION

When an object moves back and forth repeatedly over the same path, it is said to be oscillating or vibrating. Some familiar examples are a child on a swing, a clock pendulum, and a violin string. Oscillations also play an important role in many physical phenomena outside the field of mechanics. For example, the molecules of solids vibrate about their equilibrium positions and the currents in electrical circuits can reverse directions or oscillate. There are also many biological examples of oscillations, such as the production of sound by the human vocal cords and the motions of insect wings.

Although the physical nature of vibrating systems can vary greatly, the equations describing small oscillations of an object about an equilibrium position often relate its acceleration, velocity, and displacement in exactly the same special way. A vibration of this kind is called *simple harmonic motion,* and its mathematical description is always the same except for differences in the symbols used.

Simple harmonic motion is characterized by several quantities. The *amplitude* is the maximum displacement of the oscillating object from equilibrium. A complete oscillation back and forth, which returns the system to its original state, is called a *cycle.* The *period T* is the time needed for one full cycle. The *frequency f* of the oscillation is the number of cycles in a unit time. If the period is $\frac{1}{2}$ s, there are two full oscillations per second. In general, the period and the frequency are related by $f = 1/T$. Frequencies are measured in cycles per second or *Hertz* (Hz).

In this chapter, we first find the general relations describing simple harmonic motion and discuss several examples of this motion. We then discuss the effects of a frictional force that dissipates energy and of external forces that vary in time and provide energy. We find that there is a special frequency of the external force for which the oscillations have the greatest amplitude. Examples of this *resonance* phenomenon occur in molecular, mechanical, biological, and other systems.

9.1 | SIMPLE HARMONIC MOTION

In Chapter Eight, we found that many objects, when stretched or compressed, exert a force opposing this action that is directly proportional to the distance they are stretched or compressed. Simple coil springs have this property. When a mass at the end of a coil spring is displaced from its equilibrium position and released, the resulting oscillatory motion is referred to as *simple harmonic motion*. This means the position, velocity, and acceleration are related in a specific way that we now determine.

When a coil spring is stretched by the application of a force, the resulting elongation x and the applied force F are proportional

$$F = kx \qquad (9.1)$$

The proportionality factor k is called the *spring constant*.

The spring exerts a restoring force that is opposite in direction

$$F_r = -kx \qquad (9.2)$$

The minus sign indicates that the restoring force is

195

always opposite to the displacement, or toward the equilibrium point.

In Fig. 9.1, a mass resting on a frictionless table is attached to a spring. Suppose now we pull the mass away from its equilibrium point and release it. Then it will move under the influence of the restoring force $-kx$ in accordance with Newton's second law of motion, $F = ma$. Thus the motion will satisfy $-kx = ma$ or

$$a = -\frac{kx}{m} \qquad \text{(simple harmonic motion)} \quad (9.3)$$

This equation states that the acceleration a is proportional to the magnitude of the displacement x from the equilibrium position and is opposite in direction. Whenever such a relationship holds, the object undergoes a specific type of motion called *simple harmonic motion*. Equation 9.3 can be considered a definition of simple harmonic motion (SHM).

To find the explicit description of this motion, we recall that $v = dx/dt$, and that $a = dv/dt = d^2x/dt^2$. Hence $a = -kx/m$ implies that the second derivative of x is proportional to $-x$. Two functions that have this property are sines and cosines. For example, we can try a solution of the form

$$x = A \cos \omega t \qquad (9.4)$$

Here A and ω are constants that we determine shortly. Then, using Eq. B.26 in Appendix B, $(d/dt) \cos \omega t = -\omega \sin \omega t$, and

$$v = \frac{dx}{dt} = -A\omega \sin \omega t \qquad (9.5)$$

Similarly, $(d/dt) \sin \omega t = \omega \cos \omega t$, and

$$a = \frac{dv}{dt} = -A\omega^2 \cos \omega t \qquad (9.6)$$

Thus we have found a solution of the SHM equation of motion, $a = -kx/m$, if the constant ω satisfies $\omega^2 = k/m$ or

$$\omega = \sqrt{k/m} \qquad (9.7)$$

If we had chosen $x = B \sin \omega t$, we would again have had a satisfactory solution, but with peak values of x and v at different times from those of a cosine solution. A linear combination of a sine and a cosine is the most general solution.

Now let's see what the constants A and ω mean. The cosine varies from -1 to $+1$, so x varies from $-A$ to $+A$ (Fig. 9.2). Thus A is the *amplitude*, the maximum displacement in either direction from the equilibrium point. The cosine function repeats itself when its argument ωt goes through 360° or 2π radians, or after a time T that satisfies $\omega T = 2\pi$. The time $T = 2\pi/\omega$ required for one full cycle is called the *period*. Its reciprocal, $f = 1/T$, the number of oscillations per second, is the *characteristic frequency* of the oscillator. Thus

$$f = \frac{1}{T} = \frac{\omega}{2\pi} = \frac{1}{2\pi}\sqrt{\frac{k}{m}} \qquad (9.8)$$

This equation shows that if the mass is increased, the frequency decreases, and the spring oscillates less

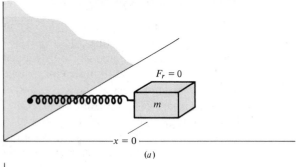

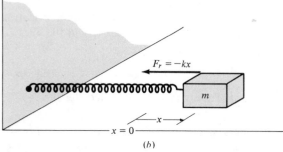

Figure 9.1. (a) A mass on a frictionless horizontal table. (b) When the spring is stretched a distance x, it exerts a restoring force $F_r = -kx$ opposite to the displacement.

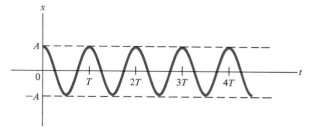

Figure 9.2. A plot of $x = A \cos \omega t$ versus t. The cosine varies from -1 to $+1$, so x varies from $-A$ to $+A$. Since the cosine repeats when its argument increases by 2π, the time for one full oscillation is $T = 2\pi/\omega$.

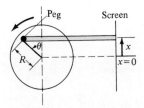

Figure 9.3. A peg at distance R from the center of a wheel that is rotating with a constant angular velocity ω. The peg is illuminated by a lamp at the left, and it casts a shadow on the screen at the right. If $\theta = \omega t$, the position of the shadow is $x = R \cos \omega t$.

rapidly. If the spring is made stiffer so that k is increased, the spring oscillates more rapidly. The quantity $\omega = 2\pi f$ is measured in radians per second and is called the *angular frequency*. The oscillator is not undergoing angular motion, so the name is somewhat inappropriate. However, an object undergoing circular motion at a constant angular velocity ω in the x-y plane has x and y coordinates that undergo SHM with a frequency $f = \omega/2\pi$ (Fig. 9.3).

In summary, SHM occurs whenever the acceleration is related to the displacement by $a = -kx/m$, or with $\omega^2 = k/m$, by

$$a = -\omega^2 x = -(2\pi f)^2 x \qquad (9.9)$$

The following example shows how the spring constant and frequency of a mass and spring may be found.

Example 9.1

The object in Fig. 9.1 has a mass of 0.1 kg and is on a frictionless table. If a 5-N force is applied, the spring is stretched 0.2 m. (a) What is the spring constant? (b) Find the characteristic frequency and period of oscillation when the mass is set in motion.

(a) The force applied and the elongation are related by $F = kx$, so

$$k = \frac{F}{x} = \frac{5 \text{ N}}{0.2 \text{ m}} = 25 \text{ N m}^{-1}$$

(b) The characteristic frequency is

$$f = \frac{1}{2\pi} \sqrt{\frac{k}{m}} = \frac{1}{2\pi} \sqrt{\frac{25 \text{ N m}^{-1}}{0.1 \text{ kg}}} = 2.52 \text{ Hz}$$

The period is $T = 1/f = 0.397$ s.

9.2 | THE WEIGHT ON A SPRING

It is much easier in practice to illustrate oscillatory motion with an object hanging on a spring than it is to arrange for nearly frictionless horizontal motion. We see now that despite the complication of the weight of the object, the motion is essentially the same.

In Fig. 9.4, a weight $w = mg$ is hung on a spring. In equilibrium, the spring will be stretched a distance d such that its restoring force $F_r = -kd$ balances the weight mg, or $-kd + mg = 0$. If the weight is displaced downward an *additional* distance x, then the

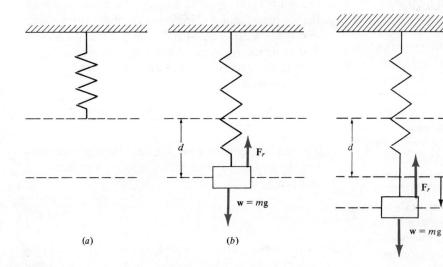

(a) (b)

Figure 9.4. (a) A spring with constant k. (b) A weight w is hung from the spring. It is in equilibrium and remains at rest when $kd = mg$. (c) The weight is displaced a distance x from the equilibrium position. The force F_r due to the spring is $-kd(d + x)$.

restoring force is $-k(d + x)$. Applying $F = ma$, we have

$$-k(d + x) + mg = ma$$

Since $-kd + mg = 0$, this reduces to $-kx = ma$, exactly the same result we just obtained for the mass on the frictionless horizontal table.

Thus the hanging weight also executes simple harmonic motion with a frequency given by Eq. 9.8. The only difference is that here it oscillates about an equilibrium point displaced a distance d from the end of the unstretched spring. *The frequencies and periods are the same for vertical and horizontal motions.*

Example 9.2

In the preceding example, a 0.1-kg mass moved horizontally on a frictionless table at the end of a spring with spring constant 25 N m⁻¹. Suppose instead the mass is hung from the spring as in Fig. 9.4. (a) How far is the spring stretched when the mass is in equilibrium? (b) Find the characteristic frequency when the mass is set into motion.

(a) In equilibrium, the restoring force balances the weight, or $-kd + mg = 0$. Solving for d,

$$d = \frac{mg}{k} = \frac{(0.1 \text{ kg})(9.8 \text{ m s}^{-2})}{25 \text{ N m}^{-1}} = 0.0392 \text{ m}$$

(b) Since the frequency is the same whether the motion is horizontal or vertical, the frequency has the same value as in Example 9.1, or 2.52 Hz.

9.3 | THE PHYSICAL PENDULUM

An object swinging back and forth also undergoes simple harmonic motion. Such an object is called a *physical pendulum*. We will find that the physical pendulum serves as a useful model for the analysis of certain kinds of body motions. A *simple pendulum* is an idealized example of a physical pendulum, consisting of a massless rod with a point mass at its end. Results for the simple pendulum can be obtained directly from those for the physical pendulum.

We consider an object of arbitrary shape that can swing frictionlessly about an axis A (Fig. 9.5). The object has mass m and moment of inertia I about A. The distance from A to the center of mass C.G. is d. The object is initially displaced and released. Figure 9.5 shows the object when the displacement is θ, which is taken to be positive when the C.G. is to the right of the vertical line through A. Since the motion is not linear but angular, we must compute torques and angles rather than forces and distances.

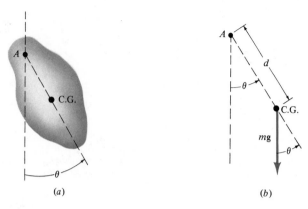

Figure 9.5. (a) An object of mass m and moment of inertia I with respect to the axis at A. (b) A schematic diagram showing the force $\mathbf{w} = m\mathbf{g}$. The distance from A to the center of gravity (C.G.) is d. At equilibrium the C.G. is on the vertical dotted line beneath A.

The torque is $\tau = rF \sin \theta$, where r is the length of the line from A to C.G., and $F \sin \theta$ is the component of the force perpendicular to this line. From the figure we see that $r = d$ and $F = mg$. The torque is then $\tau = -mgd \sin \theta$, where the minus sign indicates that the torque is clockwise. Using the angular form of Newton's second law, $\tau = I\alpha$, we have

$$-mgd \sin \theta = I\alpha$$

or

$$\alpha = -\frac{mgd}{I} \sin \theta$$

This is not the equation for harmonic motion even though α is the angular acceleration associated with the displacement θ. However, at small angles $\sin \theta$ and θ are almost equal if θ is measured in radians. For example, $\sin \theta$ and θ differ by only about 1 percent at 15° and by about 5 percent at 30°. Thus for *small angles* we have

$$\alpha = -\frac{mgd}{I} \theta$$

This has the correct form for harmonic angular motion, $\alpha = -(2\pi f)^2 \theta$, if we identify the characteristic frequency as

$$f = \frac{1}{T} = \frac{1}{2\pi} \sqrt{\frac{mgd}{I}} \qquad (9.10)$$

A metrestick swinging about one end is an example of a physical pendulum.

Example 9.3

A metrestick is suspended from one end and released. Its initial angle with the vertical direction is

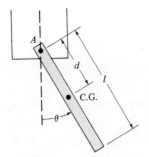

Figure 9.6. A metrestick pivoted at one end.

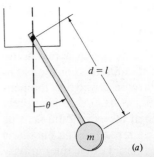

Figure 9.7. (*a*) A simple pendulum is a point mass at the end of a mass rod. (*b*) The period of a simple pendulum is independent of its mass.

$\pi/24$ rad $= 7.5°$ (Fig. 9.6). When does the metrestick first become vertical?

According to Table 5.3, the moment of inertia I for a rod of mass m and length l rotating about an axis through its end is $\frac{1}{3} ml^2$; the distance d from the C.G. to the axis of rotation is $\frac{1}{2} l$. With Eq. 9.10,

$$f = \frac{1}{2\pi} \sqrt{\frac{mgd}{I}} = \frac{1}{2\pi} \sqrt{\frac{mgl/2}{ml^2/3}} = \frac{1}{2\pi} \sqrt{\frac{3g}{2l}}$$

$$= \frac{1}{2\pi} \sqrt{\frac{3(9.8 \text{ m s}^{-2})}{2(1 \text{ m})}} = 0.61 \text{ Hz}$$

Notice that the frequency does not depend on the mass. The metrestick is vertical after one-quarter of a period or $t = T/4 = 1/4f$. Thus

$$t = \frac{1}{4f} = \frac{1}{4(0.61 \text{ s}^{-1})} = 0.41 \text{ s}$$

Simple Pendulum

The simple pendulum is an idealized system that is also sometimes a good approximation to a real system. It consists of a point mass m at one end of a massless rod of length l pivoted about the other end. Its moment of inertia is ml^2 and $d = l$ (Fig. 9.7).

Using the results for the physical pendulum, the frequency of the simple pendulum is

$$f = \frac{1}{T} = \frac{1}{2\pi} \sqrt{\frac{mgl}{ml^2}} = \frac{1}{2\pi} \sqrt{\frac{g}{l}} \qquad (9.11)$$

The following example describes a simple pendulum composed of a rock on a string.

Example 9.4

A small rock swings at the end of a string 1 m long. What is its characteristic frequency?

Using the preceding results,

$$f = \frac{1}{2\pi} \sqrt{\frac{g}{l}} = \frac{1}{2\pi} \sqrt{\frac{9.8 \text{ m s}^{-2}}{1 \text{ m}}} = 0.5 \text{ Hz}$$

Note that this frequency is less than that of the metre-

stick in Example 9.3. This is because most of the metrestick mass is closer to the axis of rotation.

The result, $f = (1/2\pi) \sqrt{g/l}$ for a simple pendulum, is particularly simple and interesting. Since g is nearly constant over the earth's surface, f depends only on the length of the simple pendulum. The mass is not important except that it must be much larger than that of the rod or string, so that we may assume $I = ml^2$. One way of observing this is to notice that the period of oscillation of different people on identical swings is nearly independent of their weight. The dependence of the period on length may also be verified roughly by observing the way the period of a weight tied to a string varies as the length of the string is changed.

The implications of the variation of period with length are often important in our daily activities. Normal walking speeds are very nearly those determined by considering the legs as swinging rigid rods. In running, the leg is bent during its forward motion. This reduces its length and hence its period. The bent leg moves forward more quickly than a rigid leg, reducing the effort required. Furthermore, the arms are moved in opposition to the legs to provide balance but are kept bent to reduce their period and the effort required to move them back and forth.

9.4 | ENERGY IN SIMPLE HARMONIC MOTION

In the preceding section, we discussed a pendulum executing simple harmonic motion. It is convenient to define the potential energy to be zero when the pendulum is at its lowest point, so its energy is entirely kinetic at this position. When the pendulum is at its highest point the velocity is zero, and the energy is entirely potential. Since no forces other than gravity do work, the total mechanical energy is conserved, and the maximum potential energy is equal to the maximum kinetic energy. As the pendulum swings back and forth, there is a continual conversion of kinetic to potential energy and then back to kinetic energy (Fig. 9.8).

Similarly, when a mass oscillates on a spring, the total energy is constant, and there is a continual interchange of potential and kinetic energy. For a mass on a spring sliding on a frictionless table, we choose the origin at the equilibrium position where the spring force is zero. Again, it is convenient to define potential energy to be zero at the equilibrium point. As the mass passes through $x = 0$, its energy is entirely kinetic. At the points of maximum displacement, the mass is momentarily at rest, and the kinetic energy is

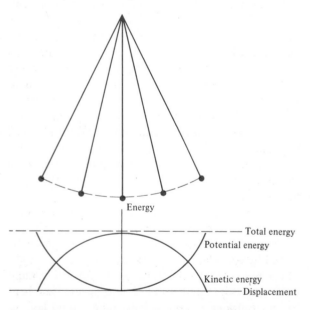

Figure 9.8. As a pendulum swings further from its equilibrium point, the potential energy increases and the kinetic energy decreases. Their sum, which is the total energy, remains constant throughout the motion.

zero. All the energy is then potential energy stored in the spring.

The potential energy at a displacement x is equal to the work that must be done against the restoring force to stretch the spring to that extent. We saw in Section 6.1 that when an object is displaced, the work done by a force $\mathbf{F}$ is $\int \mathbf{F} \cdot d\mathbf{s}$. The force needed to stretch a spring is $F = kx$, so the work done in stretching the spring from 0 to x is

$$W = \int_0^x F \, dx = \int_0^x kx \, dx = \tfrac{1}{2}kx^2$$

(This result can also be obtained by using geometric arguments to find the area under the F-x graph.) Hence the potential energy at x is

$$\mathcal{U} = \tfrac{1}{2}kx^2 \qquad (9.12)$$

The following example illustrates this result.

Example 9.5

A mass of 2 kg on a spring is extended 0.3 m from the equilibrium position and released from rest. The spring constant is 65 N m^{-1}. (a) What is the initial potential energy of the spring? (b) What is the maximum speed of the mass after it is released? (c) Find the speed when the displacement is 0.2 m.

(a) Initially the displacement is 0.3 m, so

$$\mathcal{U}_0 = \tfrac{1}{2}kx^2 = \tfrac{1}{2}(65 \text{ N m}^{-1})(0.3 \text{ m})^2 = 2.92 \text{ J}$$

(b) The energy is totally kinetic when the spring and mass pass through the unstretched position, $x = 0$. The kinetic energy there is $\tfrac{1}{2}mv^2$ and is equal to the initial potential energy. Thus

$$\tfrac{1}{2}mv^2 = \mathcal{U}_0$$

or

$$v = \sqrt{\frac{2\mathcal{U}_0}{m}} = \sqrt{\frac{2(2.92 \text{ J})}{(2 \text{ kg})}} = 1.71 \text{ m s}^{-1}$$

(c) When $x = 0.2$ m, the system has both potential and kinetic energies that are nonzero. The potential energy can readily be calculated from $\tfrac{1}{2}kx^2$. Since the total energy is conserved, it is equal to the initial potential energy $\mathcal{U}_0$ found in (a). Hence we can get the kinetic energy and determine the speed:

$$\tfrac{1}{2}mv^2 + \tfrac{1}{2}kx^2 = \mathcal{U}_0$$

$$v = \sqrt{\frac{2}{m}\left(\mathcal{U}_0 - \frac{1}{2}kx^2\right)}$$

$$= \sqrt{\frac{2}{(2 \text{ kg})}\left[2.92 \text{ J} - \frac{1}{2}(65 \text{ N m}^{-1})(0.2 \text{ m})^2\right]}$$

$$= 1.27 \text{ m s}^{-1}$$

9.5 | DAMPED OSCILLATIONS

Most real situations involving vibrational motion cannot be described precisely by the equations of simple harmonic motion because of the presence of dissipative forces such as friction or air resistance. For example, a child on a swing, a clock pendulum, and a violin string all gradually come to rest unless energy is supplied to replace the losses. In this section, we briefly describe the reduction in amplitude or *damping* caused by dissipative forces. In the next section, we discuss what happens when energy is fed into a system by an external source.

Dissipative forces typically depend on the velocity, but the exact dependence varies and can be quite complicated for some systems. Often it is assumed that the dissipative force F_d is linearly proportional to v, or $F_d = -\gamma v$. (The qualitative behavior of an oscillator does not greatly depend on the exact form of the force law.) Here γ is the *damping constant,* and the minus sign indicates that the damping force opposes the motion.

The effect of including the damping force in the equation of motion for a weight on a spring is illus-

trated in Fig. 9.9. If γ is zero, oscillations continue with the same amplitude indefinitely. When a small amount of damping is present, the oscillations steadily decrease in amplitude until they are negligibly small. If γ is larger, then the oscillations diminish faster. Oscillations cannot occur at all when γ is very large; if displaced, the weight returns to its equilibrium position without oscillating.

The theory of damped oscillations has some interesting biological applications. For example, it provides a way of measuring the friction present in the joints of mammalian limbs. As we noted in Chapter Three, this friction is quite small because of the lubricating effects of synovial fluid. If you sit so that your lower leg can swing freely from the knee and set it in motion, it will gradually come to rest if no muscular forces are exerted. Measurement of the rate at which the amplitude diminishes leads to information about the frictional forces. In a normal knee, there is very little friction, and the oscillations damp out slowly, much as in Fig. 9.9b.

An example of strongly damped motion is provided by the cupola in the inner ear of vertebrates. We learned in Chapter Five that this is a swinging door type of structure that is easily displaced from its equilibrium position by the relative motion of the endolymph fluid and the semicircular canal. Here the spring constant k is very small, and the damping due to the fluid is comparatively large. As a result, the cupola returns very slowly over a period of about 20 s to its equilibrium position, and it does not oscillate at all. This is similar to the situation shown in Fig. 9.9d.

Another structure in the ear of vertebrates, the *otolith,* provides an example of an intermediate amount of damping. All vertebrates have two or three otoliths, which provide information about acceleration and tilting of the animal. The otolith is made of calcium carbonate and is about three times as dense as water. It is connected by springlike tissue to a cavity filled with watery fluid (Fig. 9.10a). When the head tilts or accelerates, the otolith moves relative to the fluid, and this motion is detected by sensory nerves. Although the otolith does go past the new equilibrium position when the head tilts or accelerates, it oscillates only a few times before coming to rest. This is the kind of situation shown in Fig. 9.9c. Thus, evolution has arranged that, in a very short time, information about the amount of tilt or acceleration will be available to the brain without significant ambiguity due to continued oscillations.

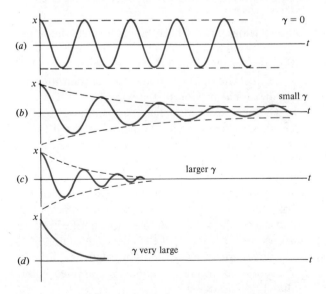

Figure 9.9. The displacement of a weight on a spring versus time. (a) No dissipative forces. The displacement is alternately positive and negative with constant amplitude. (b) With a dissipative or damping force, the amplitude gradually diminishes. (c) With more damping, the amplitude diminishes more rapidly. (d) If the damping is very large, no oscillations occur.

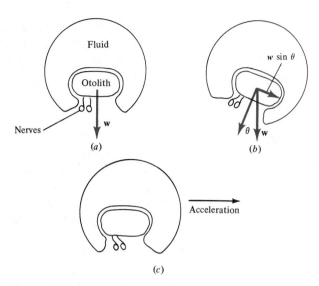

Figure 9.10.   (a) The otolith is denser than the surrounding fluid. Tilting (b) or accelerating (c) causes the otolith to move. Nerve endings adjacent to the otolith sense its motion. Note that in (b) and (c) the side portions of the fluid filled chamber change size. (Adapted from Alexander.)

9.6 | FORCED OSCILLATIONS AND RESONANCE

We saw in the preceding section that unless energy is supplied, the amplitude of an oscillator usually decreases in time because of frictional effects. To overcome such losses, clock pendulums are driven by coiled springs, and children on swings pump their feet. When energy is fed into a vibrating system, the system is said to be undergoing *forced oscillations*.

A singer holding a note of a certain frequency can set up vibrations in a glass. If the singer persists, the energy absorbed by the glass may cause large enough vibrations so that the glass will shatter (Fig. 9.11). This is only true of well-made glassware. In poorer quality glasses, which are less homogeneous in composition, various portions of the glass have different characteristic frequencies, and no single frequency will cause shattering.

Soldiers crossing bridges break step because the regular march step may be at just the right frequency to set the bridge vibrating and perhaps cause its ultimate destruction. A spectacular example of a bridge being set into motion and collapsing is that of the Tacoma Narrows Bridge in Washington. The wind set the bridge oscillating with a steadily increasing amplitude (Fig. 9.12).

Figure 9.11.   A trained singer can break a glass by sustaining the right note. (Ella FItzgerald in an ad for Memorex tape. Courtesy Memorex Corporation.)

Insect wings may vibrate at up to 120 times per second, although only about three nerve impulses per second trigger the wing muscles. The nerve impulses arrive at just the right frequency to maintain the natural vibrational motion of the wing.

In all these examples, there are both dissipative forces that reduce the vibrations and external forces that supply energy. Depending on the physical circumstances, there may be a balance of these two energies, so that the amplitude of the motion is constant, as with the clock or the insect wing. Sometimes energy enters the system faster than it is dissipated, and disaster follows, as with the glass and the Tacoma Narrows Bridge. Finally, if the energy does not enter the system at very nearly the right frequency, little or no vibration occurs, since the energy supplied is immediately dissipated.

As we have seen, energy is most effectively supplied to an oscillator when the external force acts at the correct frequency, which is usually close to the frequency of the oscillator with no external force. This phenomenon is called *resonance,* and the opti-

Figure 9.12. Winds caused the Tacoma Narrows Bridge to collapse after several hours of increasing vibrations. After this disaster, which occurred in 1940, some similar bridges were substantially modified. (Wide World Photos.)

mum frequency is referred to as the *resonant* frequency (Fig. 9.13). A child pumping a swing or a parent pushing from behind learns to apply forces at just the right interval to attain the maximum amplitude. Similarly, people trying to push a car struck in snow or mud are most successful when they allow the car to rock back and forth and time their pushes appropriately.

A spectacular example of resonance is provided by the enormous tides in Canada's Bay of Fundy. The tidal rise in the ocean averages about 0.3 metres, but at the head of the Bay it averages 11 metres. One

reason for this is that the characteristic frequency of oscillation of the water as it sloshes back and forth in the Bay is about 13 hours, just slightly more than the 12.4 hours between successive high tides. Since the driving force—the ocean tide—has a frequency close to the characteristic frequency, large resonant amplitudes result. Proposals have been made to dam up part of the flow and use it to drive electrical generators. It is expected that the dams would in effect shorten the Bay and decrease the period. In that event, the frequencies would be even closer, and the tidal rise might increase further! (Fig. 9.14)

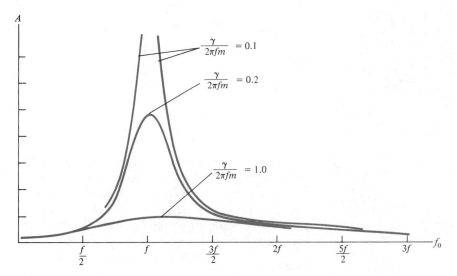

Figure 9.13. The amplitude of an oscillator versus the frequency f_0 of the external force driving the oscillator. The amplitude has its greatest value at the resonant frequency, which is very close to the characteristic frequency of the oscillator. Notice that as the damping constant γ is reduced, the amplitude at resonance becomes larger.

(a)

(b)

Figure 9.14. The Hopewell Rocks along the New Brunswick coast at (a) low tide; (b) high tide. At the head of the Bay of Fundy, the tidal rise averages 11 metres. The characteristic frequency of oscillation of the water sloshing in the bay is close to that of the ocean tides, resulting in large resonant amplitudes. (New Brunswick Department of Tourism.)

SUMMARY

Any motion in which the displacement x and acceleration a are related by

$$a = -\omega^2 x = -(2\pi f)^2 x$$

is simple harmonic motion; f is the natural or characteristic frequency of the motion. The period T is the time for one full cycle, and $f = 1/T$.

A mass attached to a spring undergoes simple harmonic motion. If the mass is m and the spring constant is k, the frequency of the motion is

$$f = \frac{1}{2\pi}\sqrt{\frac{k}{m}}$$

A common example of simple harmonic motion is a physical pendulum that swings back and forth at relatively small angles. If the pendulum has a mass m and moment of inertia I, and d is the distance from the rotation axis to the center of gravity, the characteristic frequency of the motion is

$$f = \frac{1}{2\pi}\sqrt{\frac{mgd}{I}}$$

For a simple pendulum with all the mass concentrated at a distance l from the pivot,

$$f = \frac{1}{2\pi}\sqrt{\frac{g}{l}}$$

The total mechanical energy of an object undergoing simple harmonic motion is constant. There is a repeated interchange between the kinetic and potential energy portions. The potential energy is

$$\mathfrak{U} = \frac{1}{2}kx^2$$

If dissipative forces are present, the energy of an oscillating system is not constant, and the motion is said to be damped. The rate at which the motion dies out is proportional to the magnitude of the dissipative force.

If an external driving force is also present, the motion is again quite similar. However, the amplitude of the motion depends on the frequency of the external force. When the frequency of the external force is equal to the resonant frequency of the oscillator, the amplitude is largest, and the system is at resonance.

Checklist

Define or explain:

simple harmonic motion   cycle
amplitude   frequency

hertz
period
spring motion
spring constant
restoring force
physical pendulum
simple pendulum

potential energy of a
    spring
damping
forced oscillations
resonance
resonant frequency

REVIEW QUESTIONS

Q9-1   The time needed for one complete oscillation is the _____; its reciprocal is the _____.

Q9-2   The shadow of a peg on a rotating wheel executes _____.

Q9-3   An object will undergo simple harmonic motion if its acceleration is proportional to _____.

Q9-4   The force needed to change the length of a spring is proportional to the _____.

Q9-5   Comparing an object hanging from a spring with a similar system moving horizontally without friction, the characteristic frequencies of oscillation are _____.

Q9-6   The characteristic frequency of oscillation of a mass on a spring decreases if the mass is _____, and increases if the spring constant is _____.

Q9-7   If the length of a simple pendulum is increased, its period _____.

Q9-8   The kinetic energy of a simple harmonic oscillator is greatest when the displacement equals _____, and least when the displacement equals _____.

Q9-9   Oscillations die out quickly if the _____ is large.

Q9-10   When an external force acts on an oscillator, the amplitude is greatest at the _____.

EXERCISES

Section 9.1 | Simple Harmonic Motion; An Experiment

9-1   An object undergoing simple harmonic motion has its greatest displacement, 0.2 m, at $t = 0$. Its characteristic frequency is 8 Hz. (a) Find the earliest times at which the displacements will be 0.1 m, 0 m, −0.1 m, and −0.2 m. (b) Find the velocities at those times.

9-2   An object undergoing simple harmonic motion with an amplitude of 0.5 m and a period of 2 s has a velocity of 1.11 m s^{-1}. What is its displacement?

9-3   An object undergoing simple harmonic motion with a frequency of 10 Hz has a maximum velocity of 3 m s^{-1}. What is the amplitude of the motion?

9-4   At what displacement of an object undergoing simple harmonic motion is the magnitude greatest for (a) the velocity; (b) the acceleration?

9-5   The characteristic frequency of a mass on a spring is 5 Hz. What is the acceleration of the mass when the displacement is 0.15 m?

9-6   The period of a 0.75-kg mass on a spring is 1.5 s. What is the spring constant?

9-7   By what factor must the mass of an object attached to a spring be increased to double the period of oscillation?

Section 9.2 | The Weight on a Spring

9-8   Masses m and M are suspended from two identical springs of spring constant k. When set in motion, the characteristic frequency of M is three times that of m. What is the ratio M/m?

9-9   When a 3-kg mass is hung from a spring, it oscillates once every 4 s. What is the spring constant?

9-10   When a 30-N force is applied to a spring, it stretches 0.2 m. (a) If a 5-kg mass is hung from the spring and remains at rest, how much is the string stretched from its original length? (b) What is the period of oscillation of the mass and spring?

9-11   When a passenger with a mass of 80 kg enters a car, the springs are compressed by his weight a distance of 1.2 cm. If the total mass supported by the springs (including the passenger) is 900 kg, find the characteristic frequency of oscillation of the car and passenger.

9-12   An object of mass 10 kg hanging on a spring has a characteristic frequency of 2 Hz. How much will the length of the spring change when the object is detached?

Section 9.3 | The Physical Pendulum

9-13   A small weight swings at the end of a string. If the period is 1 s, how long is the string?

9-14   A simple pendulum has a 1.5-s period on Earth. When it is set swinging at the surface of another planet, the period is found to be 0.75 s. What is the acceleration of gravity on this planet?

9-15   A steel ball on the end of a cable is used in demolition work. The period of swing is found to be 7 s. What is the length of the cable? (Neglect the mass of the cable.)

9-16 (a) Estimate the moment of inertia of your lower leg and foot when pivoted at the knee. (b) What is your estimate of the characteristic frequency of the lower leg and foot when pivoted at the knee? (c) How does your estimate in part (b) compare to your observations?

9-17 A uniform rod is suspended at one end. The characteristic period of its swing is 2 s. What is the length of the rod?

9-18 What is the percentage error in using $\sin \theta = \theta$, where θ is in radians, at (a) 10°; (b) 20°; (c) 30°; (d) 40°?

9-19 By what factor must the length of a simple pendulum be changed to double the period of oscillation?

9-20 The gravitational acceleration g increases by 0.44 percent when one goes from the equator to Greenland. If a pendulum has a period of 1 s at the equator, what is its period in Greenland?

9-21 A wrench is allowed to pivot about a hole through one end. Its radius of gyration is 0.15 m and its center of gravity is 0.1 m from the end. If it swings as a physical pendulum, what is its frequency?

Section 9.4 | Energy in Simple Harmonic Motion

9-22 A 0.5-kg mass on a spring has a period of 0.3 s. The amplitude of the motion is 0.1 m. (a) What is the spring constant? (b) What is the potential energy stored in the spring at maximum displacement? (c) What is the maximum speed of the mass?

9-23 A 0.05-kg mass is hung from a massless rubber band and stretches it 0.1 m. (a) What is the spring constant of the rubber band? (b) What is the characteristic frequency of oscillation of the system? (c) What is the period of the oscillation? (d) If the mass is pulled 0.05 m below the equilibrium position and released, what will be the energy associated with the oscillations?

9-24 A 5-kg mass is attached to a spring with a spring constant of 100 N m^{-1}. If it oscillates with a maximum velocity of 4 m s^{-1}, what is the amplitude of the motion?

9-25 A simple harmonic oscillator has an amplitude of 0.1 m. At what displacement will its kinetic and potential energies be equal?

9-26 When the displacement of a simple harmonic oscillator is half its amplitude, what fraction of the total energy is kinetic energy?

9-27 A 10-kg mass is attached to a spring with spring constant 50 N m^{-1}. It is pulled 0.2 m from its equilibrium position and released. Find its velocity (a) at the equilibrium position; (b) at a displacement of -0.1 m.

Section 9.5 | Damped Oscillations

9-28 The otolith in a fish has a mass of $0.022 \text{ g} = 2.2 \times 10^{-5}$ kg, and the effective spring constant is 3 N m^{-1}. (a) What is the characteristic frequency of the otolith? (b) Is the frequency consistent with the idea that the otolith should respond rapidly to accelerations or changes in orientation?

9-29 The characteristic frequency of a man's lower leg and foot, when swung at the knee, is 1.3 Hz. The motion is damped out after six swings. (a) What is the period of the motion? (b) How long does the leg swing?

9-30 The otolith of a fish has a mass of $0.1 \text{ g} = 10^{-4}$ kg and a spring constant at 3 N m^{-1}. (a) Find the characteristic frequency. (b) Assuming the damped otolith comes to rest in a time equal to twice its natural period, how long is the otolith in motion after the fish's head is tilted?

PROBLEMS

9-31 During the motion of a weight on a spring, the position, velocity and acceleration are given by

$$x = x_0 \cos (\omega t)$$
$$v = -\omega x_0 \sin(\omega t)$$
$$a = -\omega^2 x_0 \cos(\omega t)$$

Evaluate x, v, and a for (a) $t = 0$; and (b) $t = \pi/\omega$. (c) In words, describe how x, v, and a change with time between $t = 0$ and $t = \pi/\omega$.

9-32 Sketch the x-t, v-t, and a-t graphs for an object undergoing simple harmonic motion for one full cycle.

9-33 Equations 9.4, 9.5, and 9.6 were obtained assuming the clock was started at $t = 0$ when the displacement of the spring had its greatest value. Find the analogous equations for the case when the velocity has its greatest value at $t = 0$.

9-34 An object attached to a spring undergoes

simple harmonic motion. Its maximum velocity is 3 m s^{-1}, and its maximum displacement is 0.4 m. (a) What is the displacement when $v = 3$ m s^{-1}? (b) What is the displacement when $v = 1.5$ m s^{-1}?

9-35 Using sense organs in their legs, spiders can detect vibrations in their webs when their prey is captured. When trapped in one web, a 1-$g = 10^{-3}$-kg insect causes the web to vibrate at 15 Hz. (a) What is the spring constant of the web? (b) What would be the frequency of a 4-kg insect caught in the web?

9-36 Young's modulus for bone is $E = 1.6 \times 10^{10}$ N m^{-2}. The tibia is 0.2 m long and has an average cross-sectional area of 0.02 m^2. (a) What is the spring constant of the bone? (b) A man weighs 750 N. How much is the bone compressed if it supports half his weight? (c) If the bone oscillates lengthwise with half the body weight on it, what is the characteristic frequency of oscillation?

9-37 An instructor wishes to prepare a class demonstration of simple harmonic motion. She has only a 2-kg mass available but has a selection of springs. (a) What spring constant should she choose to have a period of 2 s? (b) She selects the spring by measuring the extension of the available springs with the 2-kg mass. What extension is she seeking?

9-38 The characteristic frequency of the simple pendulum of a clock is 0.7 Hz. The pendulum is 0.5 m long. How should the length be changed to change the frequency to 0.8 Hz?

9-39 The human leg can be approximated by a cylinder. (a) Estimate the characteristic frequency of your legs when swung from the hip with the knee locked. (b) If normal walking were performed with the legs swinging at their natural frequency, how far could you walk in 1 hour?

9-40 (a) If one assumes that the arm is a uniform rod, estimate the characteristic frequency of the extended arm. (b) Estimate the frequency if the forearm is at a right angle with the upper arm as in running.

9-41 A ball on a 30-m cable is used for demolition work. If the maximum angular displacement is 20°, what is the velocity of the ball at its lowest point? (Neglect the mass of the cable.)

9-42 A 1-m-long rod is suspended about a pivot at one end and has a small sphere attached at the other. The rod and sphere have the same masses. What is the frequency of oscillation?

9-43 A 50-kg boy rides on a Pogo stick, a pole with a spring on its bottom end. He jumps into the air 0.3 m and lands on the ground, compressing the spring 0.05 m. (Neglect the mass of the stick.) (a) How much energy is stored in the spring? (b) What is its spring constant? (c) What is the characteristic frequency of oscillation?

9-44 A weight hung from a spring stretches it a distance d when it is in equilibrium. Show that when the weight is set into motion, its characteristic frequency is

$$f = \frac{1}{2\pi} \sqrt{\frac{g}{d}}$$

9-45 A toy gun uses a spring to fire darts tipped with suction cups. The darts have a mass of 0.03 kg, and the spring constant is 200 N m^{-1}. If the spring is compressed 0.1 m and released, and all the stored energy is transferred to a dart, how high will it rise when fired straight up?

9-46 Show that the potential energy of a simple pendulum displaced from equilibrium by a small angle θ is

$$\mathcal{U} = \frac{1}{2} mg\ell\theta^2$$

9-47 Using the result of the preceding problem, find the maximum angular velocity of a simple pendulum if its amplitude of oscillation is θ_0.

*9-48 Derive a formula for the potential energy of a physical pendulum displaced from equilibrium by a small angle, θ.

9-49 Sketch the graph of $\mathcal{U}$ and K versus time for an object undergoing simple harmonic motion for one full cycle.

9-50 The otolith of a fish has a mass of 0.3 g $= 3 \times 10^{-4}$ kg and an effective spring constant $k = 3$ N m^{-1}. The damping constant $\gamma = 1.5 \times 10^{-2}$ N m^{-1} s. (a) What is the characteristic frequency of the otolith? (b) What is the ratio $\gamma/(2\pi fm)$? (c) Does the otolith have a fairly sharp resonance?

^{c}9-51 (a) Show that $x = A \cos (\omega t + \alpha)$ is a solution of the SHM equation $a = -kx/m$. (b) What is the significance of α?

9-52 Show that the area under the F-x curve (Section 9.4) is $kx^2/2$ using geometric arguments.

9-53 Show from the motion of the peg in Fig. 9.3 that its shadow has (a) a velocity $v = -\omega R \sin \omega t$; (b) an acceleration $a = -\omega^2 R \cos \omega t = -\omega^2 x$.

9-54 Using the series $\sin\theta = \theta - \theta^3/6 + \cdots$, estimate the angle in radians and in degrees at which the approximation $\sin\theta = \theta$ is in error by 10 percent.

ANSWERS TO REVIEW QUESTIONS

Q9-1, period, frequency; **Q9-2,** simple harmonic motion; **Q9-3,** the negative of the displacement; **Q9-4,** elongation; **Q9-5,** equal; **Q9-6,** increased, increased; **Q9-7,** increases; **Q9-8,** zero, the amplitude; **Q9-9,** damping constant; **Q9-10,** resonant frequency.

SUPPLEMENTARY TOPICS

9.7 | THE EFFECTS OF VIBRATION ON HUMANS

People have always experienced vibrational motion in activities such as walking or running. However, they are now routinely subjected to vibrational motion when they travel in cars or airplanes, operate tractors, use power tools, or work near machinery. The effects of vibration may range from mild annoyances to injury or death, depending on the amplitude, frequency, and duration of the vibration. Considerable research has been directed at measuring these effects, so that safer and more comfortable vehicles and machines can be designed.

Discussions of the effects of vibration normally use the peak acceleration $a_{\max}$ rather than the amplitude as a variable. According to Eq. 9.9, the acceleration is related to the displacement by $a = -(2\pi f)^2 x$. When x is equal to the amplitude A or to $-A$, the magnitude of the acceleration is

$$a_{\max} = (2\pi f)^2 A \qquad (9.13)$$

Laboratory experiments are usually done with platforms or seats that vibrate at a single frequency, although practical vibrational motion situations often involve a combination of oscillations with many different frequencies. Figure 9.15 shows the reactions to vertical oscillations for subjects sitting on a hard seat. The greatest sensitivity occurs at 6 to 7 Hz; larger accelerations can be tolerated at higher or lower frequencies.

These ideas are illustrated in the following example.

Example 9.6

What peak acceleration is perceived as alarming at 6 Hz? What is the corresponding vibrational amplitude?

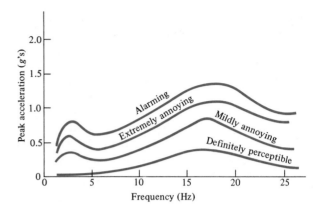

Figure 9.15. The average reactions of men to simple harmonic motion. The subjects are on hard seats that vibrate vertically. (Adapted from S. Lippert, *Vibration Research.*)

From Fig. 9.15, the peak acceleration at 6 Hz that is perceived as alarming is about 0.65 g, where g is the acceleration due to gravity. With Eq. 9.13, the amplitude is

$$A = \frac{a_{\max}}{(2\pi f)^2} = \frac{(0.65)(9.8 \text{ m s}^{-2})}{[2\pi(6 \text{ s}^{-1})]^2}$$
$$= 4.48 \times 10^{-3} \text{ m}$$

Mathematical models can correlate the observations of the response of the body to vibrations. (Fig. 9.16). At frequencies up to 2 Hz, the body reacts as though it were a single mass attached to a spring with some damping present. Above 2 Hz, there is relative motion of the body parts, and a much more complex model is needed. The body as a whole has a resonance at about 6 Hz, but the abdominal mass resonates at 3 Hz, the pelvis at 5 and 9 Hz, the head relative to the shoulders at 20 Hz, and the eyeballs in their sockets at 35 and 75 Hz (Fig. 9.17). Vibration well below the "alarming" level causes physiological changes in the circulatory and nervous systems and impairs coordination, vision, and speech. High levels of vibration can cause serious or even fatal lung, heart, intestinal, and brain damage.

Above 20 Hz it is relatively easy to protect people from excessive vibration using cushioned seats and various simple suspension arrangements. However, tractors and trucks have most of their vibration at 1 to 7 Hz, and peak accelerations up to $1g$ sometimes occur. Springs with low resonant frequencies and hydraulic shock absorbers can be used to construct seats that reduce these vibrations to acceptable levels.

It is useful to know that when a person's legs are slightly bent, considerably less vibration is transmit-

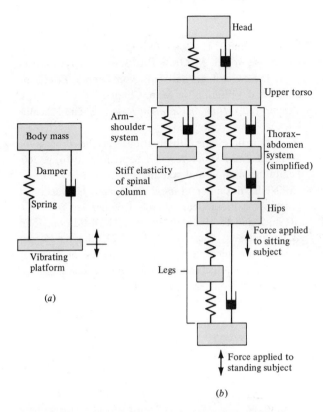

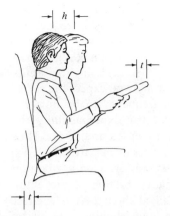

Figure 9.17. The amplitude of the motion of various body parts can exceed that of the original vibration. This often happens when the frequency is close to the resonant frequency of a particular part. Here a subject is seated on a test platform built to resemble an automobile. The figure shows the maximum displacements of the car and the subject. Note that the subject's head moves a distance h which is larger than the distance t moved by the test car.

Figure 9.16. (a) A model that treats the body as a single mass describes the motion adequately for frequencies up to 2 Hz. The device resembling a piston in a fluid symbolizes the damping. (b) A more elaborate model that takes into account the relative motion of the body parts. (Adapted from R.R. Coermann et. al., *Aerospace Med.*, vol. 31, p. 443, 1960.)

ted to the upper body than when the person stands or sits erect. Operators of farm machinery sometimes stand in this way to reduce the vibration to a level lower than that they experience when sitting. Similarly, if one rides a bicycle on a rough road, the discomfort can be reduced by supporting part or all of the weight on the pedals rather than on the seat.

EXERCISES ON SUPPLEMENTARY TOPICS

Section 9.7 | The Effects of Vibration on Humans

9-55 When a man is placed on a shaking platform, his head is observed to oscillate relative to the shoulders with an amplitude that is greatest at 20 Hz. Estimate the spring constant for this motion.

9-56 A tractor seat is mounted on springs. When a 70-kg adult sits on the seat, the characteristic frequency is 7 Hz. What is the characteristic frequency when a child of mass 25 kg sits on the seat? (Neglect the mass of the seat itself.)

9-57 A 50-kilogram woman sits on a spring-mounted seat. The spring is compressed 5×10^{-3} m. (Neglect the mass of the seat itself.) (a) What is the spring constant? (b) What is the characteristic frequency of the motion? (c) Would this be an appropriate size spring for use in a vehicle?

9-58 What is the spring constant of a tractor seat if the resonant frequency is 2 Hz with a 70-kg person seated on it? (Neglect the mass of the seat.)

PROBLEMS ON SUPPLEMENTARY TOPICS

9-59 A human can endure vibration at 4 Hz with a maximum acceleration of $4g$ for a brief period. What is the maximum displacement of the body under these conditions?

9-60 Since various parts of the bodies of animals have characteristic frequencies of vibration, we can think of these parts as being connected by springs (Fig. 9.16b). The spring is formed by the flexible connections of these parts of the body. In Chapter Eight, we learned that the spring constant is proportional to the cross-sectional area divided by the length of the spring material. (a) Using the scaling hypothesis of Section 8.6, $l \propto r^{2/3}$, show that the spring constant $k \propto m^{1/2}$, where m is the body mass. (b) Show that the characteristic frequencies should scale as $f \propto m^{-1/4}$.

9-61 The abdomen and thorax of a 60-kg human has a resonance at about 3 Hz. (a) Using the results of Problem 9-60, what would you expect the corresponding characteristic frequency to be in a 20-g $= 2 \times 10^{-2}$ kg mouse? (b) Experimentally the frequency in mice is between 18 and 25 Hz. How does this compare to the result in part (a)?

9-62 When a person stands erect on a vibration table, there is a resonant vibration of the entire body, including the head, at about 2 Hz. The effect on the upper body is only slightly reduced by standing with the legs slightly bent. When the frequency is increased, the legs become much better absorbers. The amplitude at the head is only about 30 percent of that of the table at 5 Hz. Explain this in terms of the high-frequency portion of Fig. 9.13.

Additional Reading

C.M. Harris and C.E. Crede (eds.), *Shock and Vibration Handbook,* vol. 3, McGraw-Hill Book Co., New York, 1961, D.E. Goldman and H.E. von Gierke, Effects of Shock and Vibration on Man, Chapter 44.

S. Lippert (ed.), *Human Vibration Research,* The Macmillan Co., New York, 1963.

C.H. Bachman, Some Observations on the Process of Walking, *The Physics Teacher,* September 1976, p. 360. Pendulum motion and walking.

R. McNeill Alexander, *Animal Mechanics,* University of Washington Press, Seattle, 1968. Chapter 7.

R. Resnick and D. Halliday, *Physics.* Third Edition, John Wiley and Sons, Inc., New York, 1977. Chapter 15 presents a more complete mathematical discussion of vibrational motion, including a quantitative discussion of forced oscillations and resonance.

UNIT THREE

HEAT

(Peter Menzel/Stock Boston).

THE study of heat and the thermal properties of matter is actually a study of energy and energy transfer. Heat phenomena can be interpreted on a molecular basis; for example, a warm substance has a greater degree of molecular motion than a cold one. The temperature can then be thought of as a measure of the kinetic energy of this motion.

In the first chapter of this unit (Chapter Ten), we develop the molecular basis for thermal phenomena and temperature. This is accomplished by considering a simplified model of real gases, the *ideal gas*. This model is a good approximation to a real gas at low pressures and densities. Surprisingly enough, the ideal gas model also gives a very good description of the phenomena of diffusion and osmosis.

Long before the relationship between molecular energies and thermal phenomena was understood or even recognized, a great deal of effort went into the study of what is now called *thermodynamics*. Recognizing that the thermal properties of materials could be characterized by such general quantities as the pressure, volume, and temperature, it was found that very general and amazingly important predictions could be made about how systems exchange thermal energy and do work. At the basis of these studies are the first and second laws of thermodynamics, which are covered in Chapter Eleven. As an example of the generality of thermodynamics, we show how Carnot, in 1824, predicted the maximum possible efficiency for a heat engine. Such engines are used in automobiles and in the production of electricity in power plants. Carnot's result shows that the optimal operation of such engines depends only on the operating temperatures. Thermodynamics has proved to be a powerful and useful approach to many problems. This is particularly evident to students of chemistry, since thermodynamics is frequently used to study chemical reactions.

In the final chapter of this unit, we examine the thermal properties of various kinds of materials. We consider thermal expansion, the absorption of heat, and phase changes such as melting or vaporization. We also discuss the three types of heat transport—conduction, convection, and radiation. The phenomena discussed in this chapter are all well understood at the molecular level. However, we do not concentrate on this molecular description but describe, in terms of the temperature, the transfer of energy from one object to another and its effects.

CHAPTER 10

TEMPERATURE AND THE BEHAVIOR OF GASES

The concept of temperature plays an important role in the physical and biological sciences. As we learn in this chapter, this is because the temperature of an object is directly related to the average kinetic energy of the atoms and molecules composing the object. Since natural processes often involve energy changes, the temperature plays the role of a label for these changes.

In our everyday experience, this same idea holds true. Our perception of hot and cold is actually a measure of how rapidly energy exchange occurs between objects. Touching something hot results in a rapid and sometimes damaging transfer of energy into our bodies.

In this chapter, we first discuss several temperature scales. The measurement of temperature is based on the variation of physical properties of materials with temperature. *Dilute gases*—those with average intermolecular separations that are very large compared to the molecular dimensions, so that molecular forces are unimportant—are sometimes used to measure temperature. This is because their pressure and temperature are related accurately by a simple expression called the *ideal gas law*. The ideal gas law can be derived by applying Newton's laws of motion to a model that depicts gas molecules as noninteracting particles. This analysis also makes clear the relationship between temperature and energy and provides a framework for discussing the processes of diffusion and osmosis.

10.1 | TEMPERATURE SCALES

Many physical quantities always have the same value at a given temperature. For example, the length of a rod varies with the temperature, but it has the same value every time it is placed in a container of ice and water. Because of the reproducibility of experiments of this nature, these properties may be used to define a temperature scale.

One common thermometer uses the volume of a fixed mass of mercury to indicate the temperature. A fine glass tube is attached to a larger bulb. The bulb and part of the tube are filled with mercury, and the rest of the tube is evacuated and sealed. As the temperature increases, the volume of mercury increases faster than that of the bulb, so the mercury rises in the tube.

To calibrate the thermometer, one usually chooses two reference temperatures and divides the interval between them into some number of equal steps. Thus we might take the freezing and boiling points of water at normal atmospheric pressure as our reference temperatures and divide the interval between them into 100 equal steps. We would then have the *Celsius* (centigrade) temperature scale if we set the freezing temperature equal to 0° C and the boiling temperature equal to 100° C. This scale is in everyday use in most of the world and is widely used in scientific work. The *Fahrenheit* (° F) scale used in the United States was originally defined by setting the lowest temperature obtained with a prescribed ice-water–salt mixture equal to 0° F and the temperature of the human body to 96° F. Because of the variability of the body temperature, this scale was later redefined, so that water freezes at 32° F and boils at 212° F. The relationship between the Celsius temperature T_C and the Fahrenheit temperature T_F is given exactly by the equation

$$T_C = \tfrac{5}{9}(T_F - 32° \text{ F}) \qquad (10.1)$$

For example, normal body temperature is 98.6° F. On the Celsius scale, this is

$$\begin{aligned}
T_C &= \tfrac{5}{9}(T_F - 32° \text{ F}) \\
&= \tfrac{5}{9}(98.6° \text{ F} - 32° \text{ F}) \\
&= 37.0° \text{ C.}
\end{aligned}$$

The procedure for defining a temperature scale does not clarify the meaning of temperature and depends on the choice of materials. For example, suppose the length of a steel rod were chosen to measure the temperature, with the same calibration procedure used again. Although the steel and mercury thermometers would necessarily agree at the two reference points, they might differ slightly at intermediate temperatures. In the following sections, we will see how these difficulties can be avoided by using the properties of gases to define a temperature scale.

Later in this chapter we will also introduce a third temperature scale, the Kelvin or absolute scale. This scale is intimately related to the molecular motion and hence is used extensively in scientific work.

10.2 | MOLECULAR MASSES

The kinetic energy of a molecule depends on its mass, and the net kinetic energy of a collection of molecules also depends on the mass of the molecules. As a result, the temperature and pressure of a collection of molecules depend on the mass of the molecules. Before discussing these relationships, we describe the *molecular mass* and the *gram-mole*.

Even a relatively small sample of a pure gas contains a large number of molecules. Some gases such as helium (He) and argon (A) are *monatomic;* that is, their molecules are single atoms. Molecules of *polyatomic* gases such as oxygen (O_2), nitrogen (N_2), and ammonia (NH_3) contain two or more atoms.

The masses of atoms and molecules are tabulated using a scale defined so that the mass of a ^{12}C atom is exactly 12 atomic mass units (u). It has been determined that

$$1 \text{ u} = 1.660 \times 10^{-27} \text{ kg} \qquad (10.2)$$

In addition to ^{12}C, naturally occurring carbon contains a small amount of ^{13}C, which has one more neutron in its nucleus. The average mass of natural carbon is therefore 12.011 u.

Atomic masses for the elements are tabulated in Appendix A. The *molecular mass* (often imprecisely termed the molecular weight) of a molecule is the sum of the masses of its constituent atoms in atomic mass units. Molecular masses are calculated in the following example.

Example 10.1

Find the molecular masses of carbon dioxide (CO_2) and molecular hydrogen (H_2). The atomic masses of H, O, and C are 1.008 u, 15.999 u, and 12.011 u respectively.

$$\begin{aligned}
M(CO_2) &= M(C) + 2\,M(O) \\
&= 12.011 \text{ u} + 2(15.999 \text{ u}) \\
&= 44.009 \text{ u}
\end{aligned}$$

$$M(H_2) = 2\,M(H) = 2(1.008 \text{ u}) = 2.016 \text{ u}$$

A *gram-mole* or more simply, a *mole*, of a substance is an amount whose mass in grams is *numerically equal* to the molecular mass in atomic mass units. Thus, 1 mole of CO_2 has a mass of 44.009 g, and 20.16 g of H_2 is 10 moles. Because of its definition, 1 mole of CO_2 contains exactly the same number of molecules as does 1 mole of H_2 or 1 mole of any other substance. The number of molecules in 1 mole is called *Avogadro's number* N_A, where

$$N_A = 6.02 \times 10^{23} \text{ molecules mole}^{-1} \qquad (10.3)$$

More generally, if the particles of a substance are ions or atoms rather than molecules, or a mixture of several kinds of particles, a mole still consists of N_A particles.

10.3 | PRESSURE

When we refer to the pressure of a gas on its container walls, or to the pressure exerted on a liquid to produce flow, we usually think of a force. In fact, the pressure of a fluid (a liquid or gas) is intimately related to but not the same as a force. The pressure is related to the magnitude of the forces a sample exerts in all directions on its surroundings. The surroundings may be the remainder of the sample or the walls of a container.

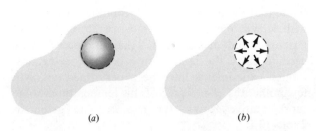

Figure 10.1. (*a*) A cross section of a segment of fluid in equilibrium with a spherical portion. (*b*) Some of the forces necessary to maintain equilibrium when the portion is removed.

To illustrate the first case, consider a fluid in equilibrium as in Fig. 10.1*a*. Suppose we wished to remove a spherical portion of the fluid in such a way as to leave the remaining fluid undisturbed. Then we would have to apply, in some manner, forces such as those shown in Fig. 10.1*b*. *These forces would replace the actual effect of the removed fluid* and would be perpendicular or normal to the spherical surface at each point.

The sum of the *magnitudes* of the normal forces divided by the surface area is the *average pressure* $\bar{P}$ on the surface of the sphere:

$$\bar{P} = \frac{\text{magnitude of normal forces on surface}}{\text{surface area}}$$

$$= \frac{F_N}{A} \qquad (10.4)$$

To obtain the *pressure P at a point,* we shrink the imaginary sphere until its radius and area are arbitrarily small.

The pressure exerted by a gas on its container walls is similarly defined. Figure 10.2 shows a gas in a cylinder fitted with a piston of area A. The magnitude of the force F the gas exerts on the piston divided by A is the *pressure P*:

$$P = \frac{F}{A} \qquad (10.5)$$

The gas exerts a force per unit area equal to P in magnitude and normal to the walls of its container at

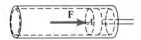

Figure 10.2. The pressure on the piston is equal to the force **F** exerted on it by the gas divided by the area A of the piston.

every point. Methods for measuring pressures are discussed in Chapter Thirteen.

The S.I. unit of pressure is the *pascal;* $1 \text{ Pa} = 1 \text{ N m}^{-2}$. Normal atmospheric pressure is

$$
\begin{aligned}
1 \text{ atmosphere} = 1 \text{ atm} &= 1.013 \times 10^5 \text{ Pa} \\
&= 1.013 \text{ bars} \\
&= 760 \text{ torr} \\
&= 760 \text{ mm Hg}
\end{aligned}
$$

The *bar* and *millibar* are used extensively in meteorology. The *torr* or the *millimetre of mercury* (mm Hg) is used in medicine and physiology. Normal atmospheric pressure will support a column of mercury of height 760 mm = 0.76 m.

The relationship between force and pressure is used in the following example.

Example 10.2

A gas at a pressure of 10 atm is in a cubical container of side 0.1 m. If the pressure outside is atmospheric pressure, what is the net force on one wall of the container?

The force due to the gas inside is

$$
\begin{aligned}
F_i = P_i A &= (10 \text{ atm})(1.013 \times 10^5 \text{ Pa atm}^{-1})(0.1 \text{ m})^2 \\
&= 1.013 \times 10^4 \text{ N}
\end{aligned}
$$

The force on the outside due to the atmosphere is

$$
\begin{aligned}
F_a = P_a A &= (1.013 \times 10^5 \text{ Pa})(0.1 \text{ m})^2 \\
&= 0.1013 \times 10^4 \text{ N}
\end{aligned}
$$

The net outward force is the difference

$$
\begin{aligned}
F_i - F_a &= 1.013 \times 10^4 \text{ N} - 0.1013 \times 10^4 \text{ N} \\
&= 0.912 \times 10^4 \text{ N}
\end{aligned}
$$

In this example the force on the wall is proportional to the difference between the internal pressure P_i and the atmospheric pressure P_a. The difference $P_i - P_a$ is called the *gauge pressure* of the gas. For example, the gauge on a tire pump registers the gauge pressure. In this textbook, the term "pressure" without any modifier always means the *absolute pressure,* not the gauge pressure.

10.4 | THE IDEAL GAS LAW

The concept of an ideal gas was developed in the mid-nineteenth century as the simplest model for which the then new methods of *kinetic theory* could be applied. Fortuitously, the results of these model calculations were in complete agreement with the observed behavior of *dilute* real gases, that it, gases at low pressures and densities.

In the ideal gas model, one assumes that the molecules are point particles and that they never collide or interact in any way with each other. Each molecule travels in a straight line until it strikes and recoils from the container walls. With these assumptions and Newton's laws of motion (see Section 10.9) one can predict how the pressure P, volume V, and temperature T_C of such a collection of molecules are related. The calculations predict that the product PV should be related to the Celsius temperature T_C as

$$PV = aT_C + b \qquad (10.6)$$

Here a and b are constants that are determined experimentally.

Equation 10.6 predicts that the product of P and V should vary linearly with temperature. Figure 10.3 shows the straight-line graph of $PV = aT_C + b$ for P constant and with a and b chosen to fit the data. Notice that there is a minimum temperature here; any attempt to lower the temperature below $T_C = -273.15°$ C would require the volume of the gas to become negative, a clear impossibility. The existence of this minimum or *absolute zero* of temperature makes it convenient to define an *absolute or Kelvin temperature,* T, measured from absolute zero. Thus,

$$T = T_C + 273.15 \qquad (10.7)$$

The increments on this scale are the same as those of the Celsius scale ($1°$ C $= 1$ K), but $T = 0$ K (zero degrees Kelvin) represents a minimum temperature.

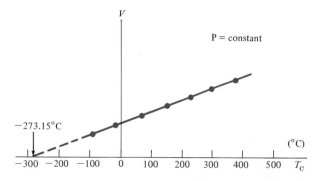

Figure 10.3. The volume of a gas decreases uniformly as the temperature decreases. The solid line is drawn through actual data points and the dashed line is the extension down to zero volume. This dashed line intersects the temperature axis at $-273.15°$ C.

(Note that the degree symbol ° is not usually written with Kelvin temperatures or temperature changes.) Extensive study has confirmed that no process can be used to carry the temperature below $T = 0$ K, although experiments have been performed within a few millionths of a degree of this temperature. Some representative Kelvin temperatures are given in Table 10.1.

Using the Kelvin temperature and the slope of the graph (Fig. 10.3) to find a and b, the *ideal gas law* $PV = aT_C + b$ can be rewritten as

$$PV = nRT \qquad (10.8)$$

Here n is the amount of the gas present in moles and R, called the *universal gas constant,* is

$$R = 8.314 \text{ J mole}^{-1} \text{ K}^{-1}$$
$$= 0.08207 \text{ litre atm mole}^{-1} \text{ K}^{-1}$$
$$(1 \text{ litre} = 10^{-3} \text{ m}^3 = 10^3 \text{ cm}^3)$$

Defining a temperature scale using the ideal gas law has the advantage of not depending on the properties of any one material such as mercury or steel. Gas thermometers are maintained at standards laboratories and are used to calibrate more convenient thermometers, such as mercury thermometers. Another important advantage of defining temperature in this way is that it enables us to relate temperature to molecular quantities, as shown in the next section.

TABLE 10.1

Kelvin temperatures at which representative physical and biological phenomena occur

Description	T
Absolute zero	0
Melting point of nitrogen	67
Gasoline freezes	123
Dry ice (CO_2 freezes)	195
Water freezes	273.15
Hibernating squirrel	275
Body temperature in man	310
Body temperature in birds	315
Water boils	373.15
Fireplace fire	1,100
Gold melts	1,336
Gas flame (stove)	1,900
Surface of sun	6,000
Center of earth	16,000
Center of sun	10^7

Standard conditions for a gas are defined to be $P = 1$ atm and $T_C = 0°$ C. The following example illustrates an important property of ideal gases at standard conditions.

Example 10.3

What is the volume of one mole of an ideal gas at standard conditions?

From Eq. 10.7, $T = T_C + 273.15 = 273.15$ K. The ideal gas law then gives

$$V = \frac{nRT}{P}$$

$$= \frac{(1 \text{ mole})}{1 \text{ atm}} (0.08207 \text{ litre atm mole}^{-1} \text{ K}^{-1})(273.15 \text{ K})$$

$$= 22.4 \text{ litres}$$

Thus under standard conditions, a mole of an ideal gas occupies 22.4 litres.

Often in experiments or applications, either the pressure, or the volume, or the temperature of a given amount of a gas is held constant. If the temperature is constant, the ideal gas law predicts (Fig. 10.4)

$$PV = \text{constant} \qquad (T \text{ constant}) \qquad (10.9)$$

This is *Boyle's law,* discovered experimentally by Robert Boyle (1627–1691) long before the ideal gas model was developed.

If the pressure is constant, the ideal gas law predicts

$$\frac{V}{T} = \text{constant} \qquad (P \text{ constant}) \qquad (10.10)$$

This is *Charles' law;* it was discovered by Jacques Charles (1746–1823) and Joseph Gay-Lussac (1778–1850). It states that a proportional increase in volume accompanies an increase in the temperature at constant pressure.

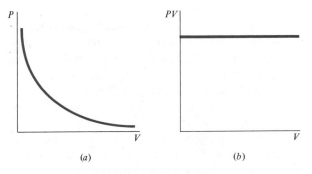

Figure 10.4. Boyle's law, for a gas at fixed temperature, (a) P is inversely proportional to V, or (b) PV is constant.

In summary, all real gases satisfy the ideal gas law to a high degree of accuracy if they are sufficiently dilute. The ideal gas law also provides a good approximation to the behavior of real gases at moderate pressures and temperatures.

10.5 | GAS MIXTURES

Very often, particularly in considering life supporting processes, we must deal with mixtures of gases. For example, each mole of dry air contains 0.78 mole of nitrogen (N_2), 0.21 mole of oxygen (O_2), 0.009 mole of argon, 0.0004 mole of carbon dioxide, and traces of several other gases. These proportions are nearly constant up to an altitude of 80 km.

Discussions of dilute gas mixtures are simplified by the fact that each constituent behaves as though the others were not present. This follows from the ideal gas model in which we assume that molecules do not interact with one another. Thus, suppose there are $n(O_2)$ moles of oxygen and $n(N_2)$ moles of nitrogen in a volume V of air at a temperature T. The *partial pressures* of oxygen and nitrogen, $P(O_2)$ and $P(N_2)$, will *each* satisfy an ideal gas law,

$$P(O_2)V = n(O_2)RT$$
$$P(N_2)V = n(N_2)RT \qquad (10.11)$$

The total pressure P (ignoring the small amount of other gases present) of the air is the sum $P = P(O_2) + P(N_2)$. The number of moles of air is $n = n(O_2) + n(N_2)$, so by adding Eqs. (10.11) we recover $PV = nRT$ for the air.

The partial pressure of oxygen is easy to calculate if the total air pressure is known, as shown in the following example.

Example 10.4

What are the partial pressures of oxygen at sea level and at an altitude of 7000 m, where the air pressure is 0.45 atm?

Dividing $P(O_2)V = n(O_2)RT$ by $PV = nRT$, we have at sea level, where $P = 1$ atm,

$$\frac{P(O_2)}{P} = \frac{n(O_2)}{n} = \frac{0.21 \text{ mole}}{1 \text{ mole}} = 0.21$$

Thus $P(O_2) = 0.21$ atm. The ratio is the same at 7000 m, but $P = 0.45$ atm, so $P(O_2) = (0.21)(0.45 \text{ atm}) = 0.096$ atm, less than half the pressure at sea level.

The amount of a gas present in our bodies is directly proportional to the partial pressure of the gas in the air we breathe. Thus, any time the air pressure changes, so does the oxygen and nitrogen content of our bodies. This is of great importance to divers.

As a diver descends, the water pressure increases rapidly. The pressure of the air the diver inhales also increases, since the pressures inside and outside the body must be kept equal. For example, the pressure 10.3 m below the surface of a lake is 2 atm, and the partial pressures of oxygen and nitrogen in a diver's lungs must be twice normal. This increased nitrogen pressure can lead to problems because nitrogen is much more soluble in the blood stream and body tissues than oxygen.

As the diver in this example breathes the high-pressure air, the amount of nitrogen in the body tissue and blood gradually increases to a level twice normal. If the diver now ascends too rapidly to the surface, the external partial pressure of nitrogen drops and the excess nitrogen in the body comes out of solution. Since it cannot escape rapidly, it forms bubbles in tissues and in the blood stream, causing the severe symptoms called "bends." A slow ascent and gradual decompression avoid this problem.

10.6 | TEMPERATURE AND MOLECULAR ENERGIES

The ideal gas law, $PV = nRT$, was originally obtained from several types of experiments. However, it is also possible to construct a theoretical model of a gas that yields the ideal gas law. A by-product of this model is the direct identification of the average kinetic energy of the gas molecules with the Kelvin temperature.

In the ideal gas model, we regard the molecules as particles that never collide with each other but that do collide with the container walls. These collisions are assumed to be elastic, so the molecules lose no energy, but they do change direction. The change in direction involves a change in momentum of the molecules, and this means that there is a reaction force on the container walls. The average force per unit area exerted by the molecules on the walls is the pressure of the gas.

In Section 10.9, we use this model to derive the expression for the pressure on the walls due to the colliding molecules. The result of that calculation is that the product of the pressure and volume, PV, is

related to the mean or average kinetic energy of the molecules, $(K)_{ave}$, by

$$PV = \tfrac{2}{3}nN_A(K)_{ave} \qquad (10.12)$$

Here, n is the number of moles of gas present, N_A is Avogadro's number, m is the mass of a molecule, and

$$(K)_{ave} = \frac{m}{2}(v^2)_{ave} \qquad (10.13)$$

The quantity $(v^2)_{ave}$ is called the mean square speed, and it represents the average value of v^2.

If we compare these results of the model with the ideal gas law, $PV = nRT$, we have

$$nRT = \tfrac{2}{3}nN_A(K)_{ave}$$

Thus

$$(K)_{ave} = \frac{3}{2}\left(\frac{R}{N_A}\right)T = \frac{3}{2}k_BT \qquad (10.14)$$

The ratio $k_B = (R/N_A)$ is called *Boltzmann's constant* and has the value

$$k_B = 1.38 \times 10^{-23}\,\text{J K}^{-1}$$

This is an extremely important result. It gives us a direct interpretation of the absolute temperature in terms of the average kinetic energy of the molecules in a gas.

The thermal energy k_BT is a ubiquitous factor in the natural sciences. By knowing the temperature we have a direct measure of the energy available for initiating chemical, physical, and biological processes.

We can also use Eqs. 10.13 and 10.14 to identify what is called the root mean square (rms) speed for the molecules at a given temperature. With the rms speed defined by $(v_{rms})^2 = (v^2)_{ave}$, we have

$$v_{rms} = \sqrt{(v^2)_{ave}} = \sqrt{\frac{2(K)_{ave}}{m}} = \sqrt{\frac{3k_BT}{m}} \qquad (10.15)$$

Individual molecules of a gas can have much larger or smaller speeds. However, as shown in Fig. 10.5, there is a speed at which the largest number will be found for a given temperature. The peaks in the curves occur at speeds that are slightly below v_{rms}. Thus, in an approximate sense we can think of v_{rms} as a typical speed of a molecule in the gas.

These relationships allow us to compute the average molecular energy and the rms speed of molecules of any ideal gas at a given temperature.

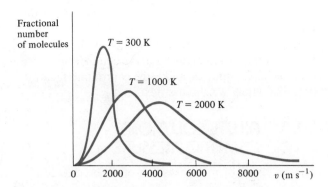

Fractional number of molecules

$T = 300$ K

$T = 1000$ K

$T = 2000$ K

v (m s^{-1})

Figure 10.5. The distribution of molecular speeds of diatomic hydrogen gas H_2. At higher temperatures the rms speed is greater and there is a greater spread.

Example 10.5

(a) What is the average kinetic energy of a hydrogen molecule at 27° C = 300 K? (b) What is the rms speed?

(a) From $(K)_{ave} = \frac{3}{2}k_B T$,

$$(K)_{ave} = \frac{3}{2}k_B T$$
$$= \frac{3}{2}(1.38 \times 10^{-23} \text{ J K}^{-1})(300 \text{ K})$$
$$= 6.21 \times 10^{-21} \text{ J}$$

(b) In Ex. 10.1, we found that the molecular mass of H_2 is

$$2.016 \text{ u} = (2.016 \text{ u})(1.66 \times 10^{-27} \text{ kg u}^{-1})$$
$$= 3.35 \times 10^{-27} \text{ kg}$$

Thus, using $v_{rms} = \sqrt{2(K)_{ave}/m}$,

$$v_{rms} = \sqrt{\frac{2(6.21 \times 10^{-21} \text{ J})}{(3.35 \times 10^{-27} \text{ kg})}} = 1930 \text{ m s}^{-1}$$

We have calculated $(K)_{ave}$ and v_{rms} for the molecule H_2 in Example 10.5. Our calculations are correct for the *translational* motion, but such a molecule can also *rotate* and *vibrate,* and each of these motions has kinetic energy. In the equations we have developed $(K)_{ave}$ should be interpreted as the translational kinetic energy only.

Two processes that depend on the thermal energy and are very important in biological systems are *osmosis* and *diffusion*. Both may be understood from the kinetic theory of an ideal gas and are described in the next two sections.

10.7 | DIFFUSION

When perfume is sprayed into still air, the scent eventually spreads to all parts of a room. Similarly, a drop

A

Figure 10.6. A container of air has some helium atoms released into it at A. The helium atoms spread out, and at some instant, we draw an imaginary hemispherical surface approximately separating the regions of high and low concentrations of helium.

of dye placed in a solvent will gradually spread throughout the container, even though we are careful not to disturb or stir the liquid. The rather slow process by which the molecules spread out evenly is called *diffusion*.

We may visualize the diffusion process for a typical gas by referring to Fig. 10.6. A small amount of helium gas is released at point A in an air-filled container. At a given moment, we draw an imaginary hemispherical surface so that most of the helium atoms are inside the surface, but a few are outside. These helium atoms are in constant random motion, bouncing off air molecules, other helium atoms, and the walls. A certain number of helium atoms will cross the surface from the inside outward and some from the outside inward. Because most of the helium is inside, more helium atoms will pass outward than inward, increasing the number outside. We see that there is a net drift of helium atoms away from A into the remainder of the container. *This flow from higher to lower concentrations is diffusion.*

The average distance the helium atoms diffuse increases with the time, but in a somewhat surprising fashion. Consider the path of a typical atom as illustrated in Fig. 10.7. The atom makes many collisions

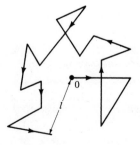

Figure 10.7. The random motion of an atom. On the average the distance l is proportional to the square root of the number of straight-line steps.

and changes direction repeatedly. Often the effect of one straight-line section or *step* is nearly cancelled by a later step. On the average, the distance l from the initial point 0 increases, but much more slowly than the number of steps, N_s. When statistical methods are applied to such a path, it is found that l increases as the square root of the number of steps, or $l^2 \propto N_s$. Since the number of steps is proportional to the time, l^2 is proportional to t. This result for the case of *random motion* is quite different from the relationship $l \propto t$ that holds true for straight-line motion.

Conventionally, instead of $l^2 \propto t$, one writes an equation for the *mean squared displacement* x^2_{rms} in one direction,

$$x^2_{\text{rms}} = 2Dt \qquad (10.16)$$

where D is called the *diffusion constant*. Similar equations hold for y^2_{rms} an z^2_{rms}. The value of D depends on the nature of the diffusing atom or molecule and the choice of the solvent or medium (Table 10.2). The following numerical example illustrates how slow the diffusion process is in a typical situation.

Example 10.6

How long will it take a hemoglobin molecule to diffuse an rms distance of 1 cm = 10^{-2} m along the x direction in water?

Using Eq. 10.16 with $D = 6.9 \times 10^{-11}$ m^2 s^{-1} and $x_{\text{rms}} = 10^{-2}$ m, we have

$$t = \frac{x^2_{\text{rms}}}{2D} = \frac{(10^{-2} \text{ m})^2}{2(6.9 \times 10^{-11} \text{ m}^2 \text{ s}^{-1})} = 7.24 \times 10^5 \text{ s}$$

This is 201 hours or about 8.4 days!

Even though diffusion is a slow process, it is the primary mechanism used by the body in absorbing and distributing the substances required by living cells. The release of the by-products of cellular function, such as carbon dioxide, also proceeds by diffusion.

TABLE 10.2

Representative values of the diffusion constant D at 20°C = 293 K

Molecule	Solvent	D(m^2 s^{-1})
Hydrogen (H$_2$)	Air	6.4×10^{-5}
Oxygen (O$_2$)	Air	1.8×10^{-5}
Oxygen (O$_2$)	Water	1.0×10^{-9}
Glucose (C$_6$H$_{12}$O$_6$)	Water	6.7×10^{-10}
Hemoglobin	Water	6.9×10^{-11}
DNA	Water	1.3×10^{-12}

It is the slow process of diffusion that determines the rate at which nitrogen is absorbed and released during diving, as described in Section 10.5. Diffusion is also an important factor in nerve conduction, as we shall see in Chapter 18, and is the controlling factor in osmosis, which we discuss next.

10.8 | DILUTE SOLUTIONS; OSMOTIC PRESSURE

The behavior of real gases can often be closely approximated by the ideal gas law, which is derived with a model that depicts the gas molecules as small, noninteracting particles. Somewhat surprisingly, the *osmotic pressure of dilute solutions* can also be accurately predicted using a form of the ideal gas law.

Figure 10.8 illustrates the meaning of osmotic pressure. The outer vessel contains water, and the inner one is filled initially to the same height with a solution of water and sugar. Water molecules can pass freely through the membrane separating the two vessels, but the larger sugar molecules cannot. Because the membrane is *permeable* to water molecules and *impermeable* to sugar molecules, it is called *semipermeable*.

Since the concentration of water molecules is greater in the outer vessel, there is a net diffusion of water molecules into the inner vessel raising the level of the solution until an equilibrium level is reached (Fig. 10.8b). At this point equal numbers of water molecules cross the membrane in each direction. The additional pressure in the solution just above the

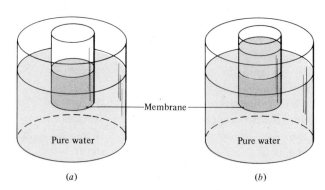

Figure 10.8. (a) Initially the levels of the water outside the container and the sugar solution inside are the same. (b) The membrane at the bottom of the inner vessel is not permeable to sugar molecules. Water molecules enter the container until equilibrium is reached, with the level of the solution inside higher than that of the pure water.

membrane due to the weight of the raised column of solution is called the *osmotic pressure, π*. In other words, the osmotic pressure is the extra pressure that must be applied to stop the flow of water into the solution.

The osmotic pressure can be predicted on the basis of a surprisingly simple model. We associate a pressure P_w^0 with the pure water in the outer vessel and a pressure P_w^i with the water inside. The sugar inside has a pressure P_s, so the total pressure inside is $P_w^i + P_s$. In equilibrium (Fig. 10.8b), the pressure difference across the membrane is the osmotic pressure,

$$\pi = (P_w^i + P_s) - P_w^0 \tag{10.17}$$

If we assume that the net flow of water across the membrane stops when the water pressures are the same, and that the sugar obeys the ideal gas law, we find

$$\pi = P_s = \frac{nRT}{V} \tag{10.18}$$

where n is the number of moles of sugar, and V is the volume of the fluid in the inner vessel.

This prediction is in good agreement with experiment. It can be interpreted as meaning that, in the dilute solution, the sugar molecules behave like an ideal gas. It appears that because the sugar molecules seldom encounter one another, although they are almost constantly in collision with water molecules, they behave like noninteracting particles. This is just the basic assumption of the ideal gas model.

Sometimes it is convenient to rewrite the equation for the osmotic pressure in terms of the *concentration* of the dissolved material or *solute*. If n moles of solute are in a volume V, the concentration is

$$c = \frac{n}{V} \tag{10.19}$$

The S.I. units of c are moles per cubic metre. The osmotic pressure is then given by

$$\pi = cRT \tag{10.20}$$

The following example illustrates the use of this result and shows the role that osmosis plays in maple trees.

Example 10.7

Sap rises in the early spring in maple trees as a result of the osmotic pressure difference between the sugar solution (sap) within the tree and the water in the ground around the roots. The sap contains 1 percent by weight of sucrose ($C_{12}H_{22}O_{11}$) in water. Assuming the temperature is $27°$ C find (a) the concentration in moles m^{-3}; (b) the osmotic pressure; and (c) the height the sap rises.

(a) The molecular mass of sucrose is $[12(12\ u) + 22(1\ u) + 11(16\ u)] = 342\ u$, so a mole has a mass of 342 g. One cubic metre of solution has a mass nearly equal to that of 1 m^3 of water, or 10^3 kilograms. The mass of the sucrose is 1 percent of that, or 10 kg = 10^4 g. Thus 1 m^3 of solution contains 10^4 g/(342 g mole^{-1}) = 29.2 moles of sucrose. Its concentration is 29.2 moles m^{-3}.

(b) The osmotic pressure is

$$\begin{aligned} \pi &= cRT \\ &= (29.2 \text{ moles m}^{-3})(8.314 \text{ J mole}^{-1} \text{ K}^{-1})(300 \text{ K}) \\ &= 7.28 \times 10^4 \text{ Pa} \end{aligned}$$

(c) The weight of the column of sap divided by the area A of its base equals the osmotic pressure. The weight can be written as $w = mg = \rho Vg$, where ρ is the density of sap and is well approximated by that of water, 1000 kg m^{-3}. If the height of the tree is h, $V = Ah$, and

$$\pi = \frac{w}{A} = \frac{\rho Ahg}{A} = \rho hg$$

Then

$$h = \frac{\pi}{\rho g} = \frac{7.28 \times 10^4 \text{ Pa}}{(1000 \text{ kg m}^{-3})(9.8 \text{ m s}^{-2})} = 7.43 \text{ m}$$

The concentration of sugar in maple trees can be higher than 1 percent, and it may be that osmosis is largely responsible for the spring flow of sap in maples. In most other trees the concentration of solutes that will not pass through the root wall is much smaller. In these cases, osmotic pressure is insufficient to account for the movement of fluids in trees. We discuss another mechanism for this fluid transport in Chapter Fifteen.

Osmosis is extremely important in understanding a wide variety of biological processes. Both plant and animal tissues are composed of cells containing complex solutions. Solutes that can pass through the cell membrane are present along with impermeable solutes. The fluids surrounding the cells are also complex solutions but with different compositions. In equilibrium the total osmotic pressures due to impermeable molecules and ions must be the same inside and outside the cell; otherwise, the difference in osmotic pressures will cause water to enter or leave the cell, along with dissolved permeable materials.

To illustrate this, suppose a person drinks a large quantity of water. Water enters the blood, thereby reducing the concentration of solutes relative to the body tissue. Consequently, the body tissues take on water. The flow of water into the kidneys is also increased because of the increased osmotic pressure difference. The kidneys then excrete more dilute urine until the blood concentration returns to equilibrium values. By contrast, a person with a high fever may lose a great deal of water from the tissues and hence from the blood until it is impossible for the kidneys to absorb water and dissolved permeable solutes.

Fluids given to a patient intravenously are usually adjusted so their concentrations of impermeable solutes and osmotic pressures balance those of the tissues. Such solutions are called *isotonic*. If a cell is placed in a solution with a lower concentration of impermeable solutes, there will be a tendency for water to enter the cell. A red blood cell has relatively rigid walls. When it is placed in pure water, an influx of water molecules occurs, which raises the internal pressure since the cell cannot expand appreciably. Equilibrium would occur at about 8 atm, but the cells will usually rupture before this pressure is reached. Cells that can change volume readily will expand or contract when placed in solutions that are not isotonic.

Osmotic Pressure and Energy | Recently there has been a revived interest in the osmotic pressure difference between fresh water and sea water. Where a large river empties into the sea, the large osmotic pressure difference combined with the enormous amounts of water involved suggest that energy may be extracted in large amounts.

In its simplest form, we may envision an apparatus such as that of Fig. 10.8 where the pure water is river water and the solution is sea water. As the fresh water flows through the membrane, the solution can be drawn off at the top of the cylinder so the flow never stops. The concentration of the solution can be maintained since the volume of fresh water is small compared to the volume of sea water. Using a turbine, energy could then be extracted from the flowing water. Recalling from Chapter Six that power is force times velocity, $\mathcal{P} = Fv$, and writing the force as the pressure times the area, we have the power and pressure related by

$$\mathcal{P} = PAv \qquad (10.21)$$

Here P is the osmotic pressure, A the membrane area, and v the velocity of flow. In the following example

we estimate the power available where a major river empties into the ocean.

Example 10.8

Sea water contains approximately 1000 moles m^{-3} of salt ions. The mouth of a typical major river has a cross-sectional area of 900 m^2 and an average flow velocity of 0.5 m s^{-1}. (a) What is the osmotic pressure at the fresh-water–sea-water junction? (b) How much power is available? (Assume a temperature of 300 K.)

(a) The osmotic pressure is found from

$$\pi = cRT$$
$$= (1000 \text{ moles } m^{-3})(8.314 \text{ J moles}^{-1} \text{ K}^{-1})(300 \text{ K})$$
$$= 2.5 \times 10^6 \text{ Pa}$$

This is almost 25 atmospheres!

(b) Using Eq. 10.21,

$$\mathcal{P} = PAv = (2.5 \times 10^6 \text{ Pa})(900 \text{ m}^2)(0.5 \text{ m s}^{-1})$$
$$= 1.125 \times 10^9 \text{ W} = 1125 \text{ MW}$$

This is roughly the output of a very large fossil-fuel or nuclear power plant.

In practice, harnessing this energy is not easy. For example, the difficulties in constructing a membrane of appropriate size and its maintenance under actual conditions could preclude its use. However, this energy may still be available using vapor pressure differences between sea and fresh water, which does not involve a membrane.

SUMMARY

Temperature is a measure of the average translational kinetic energy of the molecules in a substance. The temperature scale that most directly characterizes the energy is the Kelvin scale, which starts at absolute zero.

Atomic and molecular masses are measured in terms of atomic mass units, where the mass of a ^{12}C atom is defined to be exactly 12 u. A mole of any substance is an amount whose mass in grams is numerically equal to its molecular mass in atomic mass units, and it contains a number of molecules equal to N_A, Avogadro's number.

The pressure of a gas or liquid is the force it exerts per unit area on its surroundings or on the walls of its container.

The ideal gas model assumes that the gas molecules do not interact with one another. The model predicts that the pressure, volume, amount of gas, and temperature are related by

$$PV = nRT$$

The ideal gas law holds very well for dilute real gases.

When a dilute real gas is composed of several types of molecules, the net pressure is just the sum of the ideal gas pressures for each type of molecule. The ideal gas model also leads to the identification of the Kelvin temperature with the average kinetic energy per molecule,

$$(K)_{ave} = \tfrac{3}{2}k_B T$$

Molecules diffuse slowly from regions of high to low concentrations. The root mean squared diffusion distance in one direction, x_{rms}, is related to the time by

$$x_{rms}^2 = 2Dt$$

Diffusion also plays a role in the osmotic flow of fluids through semipermeable membranes. The permeable fluid will flow between regions of different concentration as if there were a pressure difference across the membrane equal to the osmotic pressure,

$$\pi = cRT$$

where c is the concentration of the impermeable solute.

Checklist

Define or explain:

Celsius (centigrade)	Boltzmann's constant
Fahrenheit	root mean square speed
molecular mass	diffusion
mole	diffusion constant
Avogadro's number	osmosis
pressure	osmotic pressure
gauge, absolute pressure	random motion
dilute gas	semipermeable
ideal gas	membrane
Kelvin temperature	concentration
absolute zero	solute
partial pressure	isotonic

REVIEW QUESTIONS

Q10-1 For use as a thermometer, a substance must have some characteristic property which changes with _____.

Q10-2 The Fahrenheit temperature scale is based on defining _____ fixed and reproducible temperatures.

Q10-3 The atomic mass unit is chosen so the mass of a ^{12}C atom is exactly _____.

Q10-4 The molecular mass is the sum of the _____.

Q10-5 The S.I. unit of pressure is the pascal, which is 1 _____ per _____.

Q10-6 Gas station attendents check tires for the correct _____ pressure.

Q10-7 A collection of noninteracting molecules and/or atoms is _____.

Q10-8 Under some conditions, _____ gases behave like _____ gases.

Q10-9 For constant pressure, absolute zero is the extrapolated temperature at which the _____ becomes zero.

Q10-10 Boyle's and Charles' laws are special cases of the _____ law.

Q10-11 For a gas with several types of molecules, the total gas pressure is the _____ of the partial pressures.

Q10-12 The amount of dissolved gas in a liquid in contact with that gas is directly proportional to the _____ of the gas.

Q10-13 The Boltzmann constant k_B is the proportionality constant relating the _____ and the _____ of gas molecules.

Q10-14 The root mean square speed for the molecules of a gas is proportional to the _____ of the Kelvin temperature.

Q10-15 Diffusion occurs from regions of _____ concentration to regions of _____ concentration.

Q10-16 A _____ membrane is one through which only certain types of molecules can move.

Q10-17 The osmotic pressure equation $\pi = cRT$ makes it appear as if the molecules of the solute behaved like an _____.

Q10-18 A solution with the same concentrations of impermeable solutes on both sides of a semipermeable membrane is _____.

EXERCISES

Section 10.1 | Temperature Scales

10-1 What temperature on the Celsius scale corresponds to 100° on the Fahrenheit scale?

10-2 What temperature on the Fahrenheit scale corresponds to 50° on the Celsius scale?

10-3 At what temperature are the readings on a Fahrenheit and a Celsius thermometer the same?

10-4 What temperature on the Celsius scale corresponds to 105° F, a dangerously high temperature in humans?

Section 10.2 | Molecular Masses

Refer to Appendix A for atomic masses where needed.

10-5   Find the molecular mass of HCl.

10-6   In a crude experiment the molecular mass of NH_3 is found to be 17.5 u. If the accepted value is used for the hydrogen mass, what is the atomic mass of nitrogen from this experiment?

10-7   What is the mass of two moles of H_2?

10-8   What is the mass of 0.3 mole of ammonia, NH_3? (The molecular mass of NH_3 is 17.03 u.)

10-9   How many molecules are in 3 moles of sucrose?

10-10   How many molecules are in 0.7 mole of mercury?

10-11   How many moles of carbon tetrachloride (CCl_4) will contain 9.5×10^{23} molecules?

10-12   What is the mass of 6.02×10^{23} atoms of magnesium?

10-13   What is the mass of 17.4×10^{23} molecules of carbon dioxide? (The molecular mass of CO_2 is 44.0 u.)

10-14   How many molecules are there in 18 g of hydrogen chloride, HCl? (The molecular mass of HCl is 34.46 u.)

Section 10.3 | Pressure

10-15   What is the force exerted by the atmosphere on a field 50 m by 100 m?

10-16   What is the force due to the atmosphere on one side of a door of area 2 m²?

10-17   If the difference in pressure on the two sides of a closed door of area 2 m² is 0.01 atm, what is the net force on the door? Do you think you could open it by hand?

10-18   What pressure on the 12-m² base of an automobile of mass 1300 kg would be necessary to support it?

10-19   Estimate the pressure you exert on the ground (a) when standing erect on both feet; (b) when lying down.

Section 10.4 | The Ideal Gas Law

10-20   A gas occupies 5 m³ at a pressure of 1 atm. What is the pressure when the volume is 1.5 m³ if the temperature is held constant?

10-21   The pressure of a gas changes from 1.5 atm to 0.3 atm at constant temperature. What is the ratio of the final and initial volumes?

10-22   The volume of a gas doubles at constant pressure. What is the final temperature if the gas was initially at 30° C?

10-23   If the temperature of a gas is raised from 0° C to 100° C at constant pressure, by what factor is its volume changed?

10-24   What is the volume of 3 moles of an ideal gas at $P = 2$ atm and $T = 300$ K?

10-25   What is the temperature of 1 mole of an ideal gas with a pressure of 0.3 atm in a volume of 3 m³?

10-26   One mole of an ideal gas occupies 2.24×10^{-2} m³ at standard conditions. What will the pressure be if the volume is increased to 1 m³ at constant temperature?

Section 10.5 | Gas Mixtures

10-27   A tank of gas at a pressure of 3 atm is composed of 0.35 mole of oxygen and 0.65 mole of helium. What is the partial pressure of oxygen?

10-28   Normal air contains 0.21 mole of oxygen per mole of air. It is desired to change this proportion so that at an altitude where the air pressure is 0.40 atm, the oxygen partial pressure is the same as at sea level. How many moles of oxygen per mole of air are needed?

10-29   If each increase in altitude of 1000 m resulted in an air pressure drop of 0.078 atm, above what altitude would even the use of pure oxygen result in a below normal oxygen intake?

10-30   A human will suffer from oxygen toxicity when the partial pressure of oxygen reaches about 0.8 atm. If the pressure increases 1 atm for each 10.3 m below the water surface, at what depth would breathing air lead to oxygen toxicity?

Section 10.6 | Temperature and Molecular Energies

10-31   If the average kinetic energy of a gas is doubled and the volume remains constant, what is the change in pressure?

10-32   What is the ratio of the rms speeds of hydrogen gas (H_2) and oxygen gas (O_2) if both gases are at the same temperature?

10-33   Natural uranium is composed of 99.3 percent ^{238}U, with a mass of 238 u, and 0.7 percent ^{235}U, with a mass of 235 u. It is the ^{235}U that is mainly used in reactors and weapons. The two isotopes are separated using diffusion processes. This method utilizes the fact that the rms speed of the gas UF_6 is different for the two isotopes. What is the ratio of their speeds at $37°$ C?

10-34 Suppose that all the translational molecular kinetic energy of 1 mole of an ideal gas at a temperature of 300 K could be used to raise a 1-kg mass. How high would the block be raised?

10-35 The rms speed of the molecules in an ideal gas with molecular mass 32.0 u is 400 m s^{-1}. (a) What is the average translational kinetic energy? (b) What is the temperature of the gas?

Section 10.7 | Diffusion

10-36 What average distance will O_2 molecules diffuse in air in 1 hour? Assume $T_C = 20°$ C.

10-37 How long will it take glucose molecules to diffuse an average distance of $1 \text{ mm} = 10^{-3}$ m in water at $20°$ C?

10-38 If a solute diffuses through water an average distance of 10^{-2} m in 6 hours, what is its diffusion constant?

Section 10.8 | Dilute Solutions; Osmotic Pressure

10-39 What is the osmotic pressure of a brine solution with an ion concentration of 1500 moles m^{-3} separated by a semipermeable membrane from pure water? Assume $T_C = 27°$ C.

10-40 A sugar solution in an apparatus such as that of Figure 10.8 will support a water column 14 m high. What is the sugar concentration? Assume $T_C = 27°$ C.

10-41 What concentration difference of impermeable solutes across a cell membrane would result in an osmotic pressure of 5 atm? Assume $T_C = 37°$ C.

PROBLEMS

10-42 How many molecules are in 500 g of sucrose, $C_{12}H_{22}O_{11}$?

10-43 What is the molecular mass of tributyrin (fat), $C_3H_5O_3(OC_4H_7)_3$?

10-44 How many molecules are in 100 g of ethyl alcohol, C_2H_5OH?

10-45 A cylinder contains 0.02 m^3 of oxygen at a temperature of $25°$ C and a pressure of 15 atm. (a) What volume does this gas occupy at $25°$ C and a pressure of 1 atm? (b) A man is breathing pure oxygen through a face mask at a rate of 0.008 $m^3 \text{ min}^{-1}$ at atmospheric pressure. How long will the cylinder of gas last?

10-46 A diver, originally 20 m below the surface, ascends, expelling air as she rises to keep her lung volume constant. The air bubbles rise faster than she does. If her lung volume is 2.4 litres, what is the total volume of expelled air in the bubbles at the water surface? (The pressure changes by 1 atm for each 10.3 m of depth change in water.)

10-47 Mixtures of oxygen and helium can be tolerated by divers. What proportion should be oxygen if the diver works 50 m below the surface and the partial pressure of oxygen should be 0.3 atm? (The pressure changes by 1 atm for each 10.3 m of depth change in water.)

10-48 The rms speed for an ideal monatomic gas at 300 K is 299 m s^{-1}. What is the atomic mass of the atoms? Identify the gas from the periodic table in Appendix A.

10-49 In Example 10.5, the rms speed of hydrogen molecules (H_2) at $27°$ C (300 K) was calculated to be 1930 m s^{-1}. At what temperature would hydrogen molecules have an rms speed of $1.1 \times 10^4 \text{ ms}^{-1}$, which is sufficient to escape from the earth?

*10-50 The walls of blood capillaries are impermeable to proteins. The two major protein groups in blood plasma are

Protein Group	Concentration	Ave. Molecular Mass
Albumin	0.045 g m^{-3}	69,000 u
Globulin	0.025 g m^{-3}	140,000 u

(a) Calculate the concentration of each protein group in moles m^{-3}. (b) Calculate the osmotic pressure of blood plasma at a temperature of 310 K due to the proteins dissolved in it. (c) How tall a column of water can this pressure support?

10-51 The alveoli of the lungs are small air sacs about 10^{-4} m in radius. The membrane walls of the sacs that separate the air space from the blood capillaries are about 0.25×10^{-4} m thick. The capillaries themselves have a radius of about 5×10^{-6} m. (a) Assuming that O_2 diffuses through the wall and blood as it does through water, what average time is required for O_2 to diffuse from the center of an alveolus to the center of a capillary? (b) How does this compare with the time, 0.1 s, that the blood is in transit around an alveolus?

10-52 The osmotic pressure of ocean water is 22 atm at 300 K. What is the salt concentration? (*Hint:* NaCl dissociates in water into Na^+ and Cl^- ions.)

*10-53 A solution of sodium chloride (NaCl) in water of 160 moles m^{-3} concentration is isotonic with blood cells. What is the osmotic pressure in atmospheres in the cells if the temperature is 300 K? (*Hint:* NaCl dissociates in water into Na$^+$ and Cl$^-$ ions.)

*10-54 The osmotic pressure of a red blood cell is 8 atms. The cell is placed in a water solution containing 100 moles m^{-3} of a solute to which the cell membrane is impermeable. Will the cell tend to expand, contract, or remain the same size? Explain. (Assume $T = 300$ K.)

10-55 How much energy must be added to 1 mole of an ideal monatomic gas to raise its temperature 1 K? (The volume is kept fixed.)

*10-56 The earth's atmosphere contains only small traces of hydrogen (H$_2$) and helium (He) gases, even though these were once present. What happened to these gases? (*Hint:* How do the rms speeds for H$_2$ and He compare with the rms speeds for N$_2$ and O$_2$?)

ANSWERS TO REVIEW QUESTIONS

Q10-1, temperature; **Q10-2**, two; **Q10-3**, 12; **Q10-4**, atomic masses; **Q10-5**, newton, square metre; **Q10-6**, gauge; **Q10-7**, an ideal gas; **Q10-8**, real, ideal; **Q10-9**, volume of an ideal gas; **Q10-10**, ideal gas law; **Q10-11**, sum; **Q10-12**, partial pressure; **Q10-13**, temperature, average kinetic energy; **Q10-14**, square root; **Q10-15**, high, low; **Q10-16**, semipermeable; **Q10-17**, ideal gas; **Q10-18**, isotonic.

SUPPLEMENTARY TOPICS

10.9 | MODEL DERIVATION OF THE IDEAL GAS LAW

One of a large number of molecules that collide only with the walls of a container is shown in Fig. 10.9. The volume of the container is $V = Al$, and there are nN_A molecules inside.

Each time a molecule hits the piston, its x component of velocity reverses itself (Fig. 10.10). Since the y component of the velocity is unchanged, the momentum change in one collision is $2mv_x$.

We can find the force on the piston due to one molecule from Newton's second law in the form $F\,\Delta t = \Delta p$, where Δp is the momentum change and Δt the time interval. The time interval Δt between impacts with the pistons is the time for the molecule to

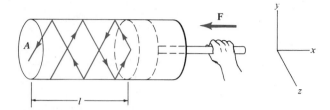

Figure 10.9. A single molecule moving in a container of area A and length l. The molecule collides elastically with the walls and with the piston. The piston must be held in place with a force **F** to offset the force due to the molecule.

travel the length of the container and back, so $2l = v_x\,\Delta t$. Thus the force is

$$F_1 = \frac{\Delta p}{\Delta t} = \frac{2mv_x}{\left(\dfrac{2l}{v_x}\right)} = \frac{mv_x^2}{l}$$

For the nN_A molecules actually present, the total force on the piston is nN_A times the average value of mv_x^2/l,

$$F = \frac{nN_A}{l}(mv_x^2)_{\text{ave}}$$

The pressure is the force per unit area. Using $V = Al$,

$$P = \frac{F}{A} = \frac{nN_A}{V}(mv_x^2)_{\text{ave}}$$

Since m is the same for each molecule, we need to compute $(v_x^2)_{\text{ave}}$. This average can be more conven-

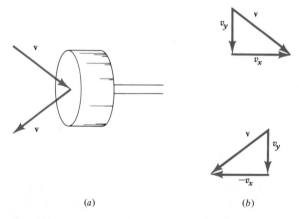

(a)   (b)

Figure 10.10. (a) When a molecule collides with a piston, its speed does not change, but the molecule does change directions. (b) v_y is the same before and after the collision, but v_x is reversed.

iently written by noticing that $(v_x{}^2)_{ave}$ should be just the same as $(v_y{}^2)_{ave}$ and $(v_z{}^2)_{ave}$. Since v^2, the total velocity squared, is the sum of the squares of its components, we can use $(v_x{}^2)_{ave} = \frac{1}{3}(v^2)_{ave}$ and then write

$$PV = \frac{2nN_A}{3}\left(\frac{mv^2}{2}\right)_{ave}$$
$$= \tfrac{2}{3}nN_A(K)_{ave} \qquad (10.12)$$

This is the result we quoted earlier.

Additional Reading

R.E. Wilson, Standards of Temperature, *Physics Today,* vol. 6, January 1953, p. 10.

Leslie A. Guildner, The Measurement of Thermodynamic Temperature, *Physics Today,* December 1982, p. 24.

Robert H. Romer, Temperature Scales: Celsius, Fahrenheit, Kelvin, Reámur, and Rømer, *The Physics Teacher,* vol. 20, 1982, p. 450.

D.K.C. MacDonald, *Near Zero: The Physics of Low Temperatures,* Science Study Series, Doubleday and Co., Garden City, N.Y., 1961. Paperback.

Mark W. Zemansky, *Temperatures Very High and Very Low,* Momentum Series, D. Van Nostrand and Co., Princeton, N.J., 1964. Paperback.

C. Barber Jorgensen and Erik Skadhauge, *Osmotic and Volume Regulation* (Proceedings of a symposium, Copenhagen, June 1977). Academic Press, New York, 1978.

Ultrahigh Pressure: New Highs Spur Pursuit of Exotic Goals, *Science,* vol. 201, 1978, p. 429.

T.H. Maugh II, Birds Fly, Why Can't I? *Science,* vol. 203, 1979, p. 1230.

Gerald L. Wich and John D. Isaacs, Salt Domes: Is There More Energy Available from Their Salt than from Their Oil? *Science,* vol. 199, 1978, p. 1436.

W. Gary Williams, Mineral Salt: A Costly Source of Energy? *Science,* vol. 203, 1979, p. 376.

Mark Olsson, Gerald L. Wich, and John D. Isaacs, Salinity Gradient Power: Utilizing Vapor Pressure Differences, *Science,* vol. 206, 1979, p. 452.

Eric Perlman, Walking on Thin Air, *Science 80,* July/August 1980, p. 89.

Scientific American articles:

Marie Boas Hall, Robert Boyle, August 1967, p. 84.

Arthur K. Solomon, The State of Water in Red Cells, February 1971, p. 88.

Reuben Hersh and Richard J. Griego, Brownian Motion and Potential Theory, March 1969, p. 66.

Holger W. Iannosch and Carl O. Wirsen, Microbial Life in the Deep Sea, June 1977, p. 42.

Warren G. Proctor, Negative Absolute Temperatures, August 1978, p. 90.

CHAPTER 11

THERMODYNAMICS

Thermodynamics is the study of the transformation of energy from one form to another and from one system to another. While it was originally developed to explain and quantify the relationship between mechanical and *thermal energy*—the energy associated with the motions of the atoms and molecules within a substance—its scope is actually much broader. The first law of thermodynamics is a universal energy-conservation law and the second law provides information about how processes can and will occur.

Virtually all discussions of thermodynamics contain the word "heat." Historically, heat was thought to be a property of an object and could be transferred from object to object as a sort of fluid, called "caloric." While this caloric theory has long since been discarded, the use of the words heat and heat flow has persisted and is often confusing.

We shall use these terms in a very precise way. If energy is transferred from one substance or object to another because of a temperature difference between the source and destination points, we will refer to this transfer as *heat flow*. The *amount* of energy transferred is the *heat*.

Thermodynamics provides the broad framework for finding the relationships between the macroscopic properties of systems, such as the pressure, volume, and temperature. Its utility derives from the predictions that it allows us to make without reference to the detailed microscopic properties of the system. In some cases, we can even make numerical predictions without any reference to or knowledge of the materials involved.

The first law of thermodynamics is a generalization of the fundamental result of Chapter Six. This result states that the work done on a system equals its change in *internal* energy plus the energy that *leaves* the system.

The second law of thermodynamics can be phrased in several ways. In one form it is a statement that a quantity called the *entropy* tends to increase in all real processes. The entropy change of a system is related to the heat entering or leaving a system divided by the absolute temperature of the system. Microscopic theories of matter have shown that the entropy of a system is also closely related to the randomness or disorder of its constituents. We use the second law to obtain the ultimate limits of efficiency of thermal energy conversion processes. Such processes are used in both fossil and nuclear electrical power-generating plants.

11.1 | MECHANICAL WORK

Work can be done on or by a system in many ways. A gas may be compressed or allowed to expand against a piston. A liquid may be stirred, and a solid may be pounded with a hammer. Electric charges may be brought near a material, so that the electric forces alter the arrangements of charges inside the material. Thus the kinds of work that can be done are as varied as the forces that can be exerted on a system. In this section, we obtain an expression for the work done by a substance or system when its volume changes.

When a substance or system expands or contracts, the work W done by the system can be related to the volume change of the material. It is conventional in thermodynamics to take W as positive when work is done *by* the system. Note by contrast that in Chapter Six we took W to be positive when work was done *on* an object.

It is most convenient to develop our ideas using the example of a gas. Figure 11.1 shows a gas at a pressure P in a closed cylinder. A movable piston of cross-sectional area A forms one end of the enclosure. The gas exerts a force $F = PA$ on the piston. When the piston moves a small distance Δx parallel to the force, the work done by the gas is $\Delta W = F\Delta x = PA\Delta x$. Since $\Delta V = A\Delta x$ is the change in volume of the gas, the work done by the gas is

$$\Delta W = P\Delta V \tag{11.1}$$

A large displacement can be broken up into a series of small displacements Δx_j, with volume changes $\Delta V_j = A\Delta x_j$ and work $\Delta W_j = P\Delta V_j$. If we sum over these, in the limit as ΔV_j approaches zero, the work done by the system when its volume changes from V_1 to V_2 is the integral

$$W = \int_{V_1}^{V_2} P\,dV \tag{11.2}$$

The work done by the system is equal to the area under the PV curve (Fig. 11.2). This result is valid for a gas in a container of any shape and also for volume changes in liquids and solids.

Equation 11.2 has a simple form if the work is done in an *isobaric* process, that is, at constant pressure. If the initial and final volumes of the system are denoted by V_i and V_f, then the work done is

$$W = P(V_f - V_i) \tag{11.3}$$

For an isobaric process, the work done by the system

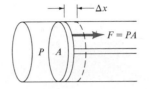

Figure 11.1. The work done by a gas or a piston during a small displacement Δx is $\Delta W = F\Delta x = P\Delta V$.

is positive if $(V_f - V_i)$ is positive; work is done on the system if $(V_f - V_i)$ is negative.

The following example describes an isobaric process.

Example 11.1

A gas at a pressure of 2 atm = 2.02×10^5 Pa is heated and is allowed to expand against a frictionless piston at constant pressure. If the volume change is 0.5 m³, how much work is done by the gas?

Using Eq. 11.3,

$$W = P(V_f - V_i) = (2.02 \times 10^5 \text{ Pa})(0.5 \text{ m}^3)$$
$$= 1.01 \times 10^5 \text{ J}$$

Another illustration of the formula for the work is provided by the expansion of an ideal gas kept at a constant temperature, an *isothermal* process. From the ideal gas law, $PV = nRT$, we have $P = nRT/V$. Hence the work done by the gas is

$$W = \int_{V_1}^{V_2} P\,dV = \int_{V_1}^{V_2} nRT\frac{dV}{V}$$
$$= nRT\ln V \Big|_{V_1}^{V_2} = nRT\ln\frac{V_2}{V_1}$$

Here ln is the natural logarithm (see Appendix B.10). Since $P_1V_1 = P_2V_2$ when T is constant, $V_2/V_1 = P_1/P_2$. Thus the work done by an ideal gas in an isothermal expansion can also be written as

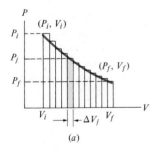

(a)

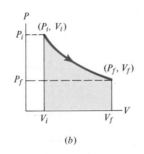

(b)

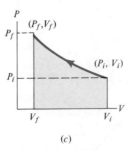

(c)

Figure 11.2. (a) The work done by the system is the sum of the areas of the small segments. In each segment, P_j is assumed consant, and for the colored segment shown, $\Delta W_j = P_j\Delta V_j$. (b) The total work is exactly equal to the total colored area under the P versus V curve. W is positive in this example. (c) A process in which work is done on the system, so W is negative.

$$W = nRT\ln(P_1/P_2) \quad \text{(ideal gas, constant } T) \quad (11.4)$$

If the gas expands, the pressure drops. Then $P_1/P_2 > 1$, and $\ln(P_1/P_2)$ is positive. As we would expect, the gas does positive work as it expands.

11.2 | THE FIRST LAW OF THERMODYNAMICS

The first law of thermodynamics relates the heat transferred to a system to the work done by the system and to the changes in the thermal energy of the system, its *internal energy U.*

The internal energy of a system depends in general on the pressure and the absolute temperature. However, in ideal gases U depends only on the temperature. For example, in our discussion of a monatomic ideal gas we found that the average translational kinetic energy of a single molecule was $3k_BT/2$. The internal energy of such a gas is the total kinetic energy which, for N molecules, is

$$U = \tfrac{3}{2}Nk_BT \qquad \text{(ideal gas)} \qquad (11.5)$$

More generally, the internal energy of a substance includes the kinetic energies associated with translational, rotational, and vibrational motions of the particles. It also includes the potential energy due to the interactions of the particles with one another. Just as in our discussion of potential energy in Chapter Six, the internal energy is defined with respect to some reference configuration. Usually this choice is of no

practical importance because only changes in internal energy affect the properties of the system.

The heat Q added to or taken from a system is the amount of thermal energy transferred due to a temperature difference. For example, we may say that heat flows from a wood stove to the air and to objects in a room because they have a lower temperature than the stove.

We now discuss the first law of thermodynamics, using a gas as an example. Consider a container of gas fitted with a movable piston (Fig. 11.3). If we add heat Q to the system, but do not allow the piston to move, the temperature and hence the internal energy U of the gas will increase. We can also change the internal energy by doing work on the gas. Thus, if we insulate the container walls and push the piston in, we compress the gas. The work done on the system equals the change in internal energy (Fig. 11.3b), since no heat enters or leaves the gas.

More generally we may add heat Q to the gas and have the gas do work W. The difference is the change in internal energy of the gas, $\Delta U = U_f - U_i$. U_f and U_i are the final and initial energies of the gas, respectively. This is the *first law of thermodynamics*

$$U_f - U_i = Q - W \qquad (11.6)$$

Q is positive if heat is *added* to the system, and W is positive if work is *done* by the system. Although we have used a gas in this example, this result is generally true for all systems and does not depend on the presence of a gas.

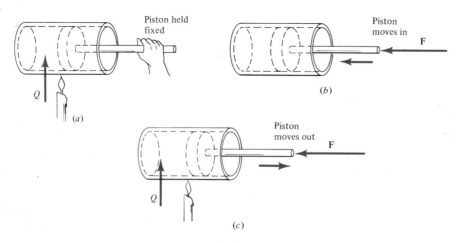

Figure 11.3. (a) With the piston held fixed, the heat added, Q, equals the internal energy increase ΔU. (b) The walls are insulated. The force **F** does work on the piston and hence on the gas. This work is equal to the internal energy increase. (c) Heat Q is added to the gas, and it expands against the external force **F**. The gas does work W. The difference between Q and W is the change in internal energy of the gas ΔU.

Two statements about the physical world are contained in the first law. First, heat and work are to be treated on an equal basis. Second, because the same change in internal energy can be obtained by adding heat, or having work done on the system, or by a combination of these processes, *the internal energy change is independent of how the change was achieved.* The difference between the initial and final internal energies of the system must depend *only* on its initial and final *states,* that is, on the change in quantities such as the temperature, pressure, and volume. This idea is illustrated in Fig. 11.4. Suppose a system undergoes two processes, represented by curves (1) and (2), in which the initial and final pressures and volumes are (P_i, V_i) and (P_f, V_f). Along path (1) the area beneath the curve is larger than that beneath curve (2). Hence, more work is done by the system in process (1) than in (2). From the first law, since $U_f - U_i$ is the same for both processes, more heat must be added to the system in process (1) than in (2) to achieve the same final state.

Two idealized types of processes will be of particular importance in our later discussion of heat cycles. One is an *isothermal* or constant temperature process. In practice, such processes are difficult to achieve, but they may be visualized for an ideal gas. The internal energy of an ideal gas depends only on the temperature. We can picture adding heat very slowly to such a gas and allowing it to expand, doing work. If the process is slow, the temperature, and hence the internal energy, will remain constant. Thus, for an isothermal process in an ideal gas, $Q = W$.

A second, more easily approximated process is an *adiabatic* process. This is one in which no heat enters or leaves the system. Thus, $U_f - U_i = -W$ for an adiabatic process. This condition is common because we can insulate systems to minimize heat transfer.

Alternatively, processes may occur so rapidly that no heat flow occurs. If we pinch the hose on a bicycle pump, pushing the plunger once results in a sudden increase in the internal energy of the air. It will take a few seconds before heat flow from the inside to outside of the pump will be noticed. This flow results from the fact that the air inside is at a higher temperature.

From our presentation of the concepts of heat and temperature, the first law seems an almost obvious statement of the conservation of energy. Nevertheless, the first law forms one of the most fundamental cornerstones of thermodynamics. Historically, the first law was by no means obvious. Until the work of Mayer, Joule, and Helmholtz in the 1840s, heat was regarded as a material substance inside an object. This substance, called caloric, could flow from one object to another. Many observations were satisfactorily explained with this idea. Mayer was the first to suggest that heat and internal energy were closely related. Joule then showed that one could produce as much heat as one wished by doing work. The idea of a fixed amount of caloric being contained in a substance died with these developments. Heat, work, and internal energy were then identified as different manifestations of the same quantity, energy.

11.3 | THE SECOND LAW OF THERMODYNAMICS

The first law of thermodynamics is useful in understanding the flow of energy during a given process. However, it yields no information about which energy-conserving processes are possible nor does it allow us to predict what state a system will be in under a given set of conditions. The second law can be used to answer some of these questions.

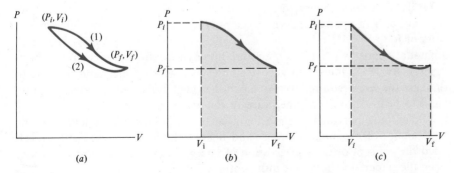

Figure 11.4. (a) A system can change from a state (P_i, V_i) to state (P_f, V_f) by many paths, two of which are shown. (b) The work done by the system is the area beneath the PV curve and is greater for process (1) than for process (2) shown in (c).

JULIUS ROBERT MAYER (1814–1878)

JAMES PRESCOTT JOULE (1818–1889)

HERMANN VON HELMHOLTZ (1821–1894)

(Brown Brothers)

(Radio Times Hulton Picture Library)

(Culver Pictures)

Until nearly the middle of the nineteenth century, heat was generally regarded as a fluid called *caloric*. This weightless fluid, which could flow from one object to another but could neither be created nor destroyed, provided a rather complete explanation of many experiments of this period. The three men pictured above are now linked historically with the demise of the caloric theory and with extending the conservation of energy principle to include thermal phenomena.

Mayer was the first to suggest that the energies associated with gravity, electricity, heat, and motion are intimately connected; when one kind of energy disappears, an equivalent amount of other types appears. However, Mayer and his ideas both suffered from neglect. Educated in Germany as a physician, Mayer became interested in physics and published his observations in 1842. His article was written in a metaphysical style and was not viewed as very convincing. Part of the resistance to Mayer's ideas arose because it was not even clear that he understood Newton's laws, much less the concepts involved in a theory of energy conservation.

While the caloric theory was dying even before Mayer's work, the real acceptance of heat as another form of energy came as a result of the independent work of Joule and Helmholtz in 1847. Joule was an English brewery owner and amateur scientist who had built a reputation as a careful and ingenious experimenter. By 1847, Joule had very accurately shown that mechanical work, in this case the work required to turn a paddle wheel immersed in water, is equivalent to heat, since the temperature of the water increased.

While Joule provided the most convincing experimental evidence for the equivalence of mechanical energy and heat, Helmholtz systematically developed the concept of the conservation of energy in one of the most important scientific papers of the nineteenth century.

Helmholtz was also a physician, serving as a surgeon in the Prussian army.

In 1849, he became a professor of physiology in Konigsberg. His work on energy conservation, stimulated by his observations on muscle motion, was only one example of his remarkable ability to understand the physics behind biological systems. Helmholtz invented the ophthalmoscope for peering into the eye and also developed the ophthalmometer for measuring the curvature of the eye. In addition, he revived a theory of color vision credited to Young and expanded on it. Helmholtz's study of the ear included the role played by the bones of the middle ear and the cochlea and went on into the difficult but important subject of the quality of sounds. His book, *Sensations of Tone,* is a cornerstone of physiological acoustics. He was the first to measure the speed of a nerve impulse. In his later life, his interests and research led directly to the experimental discovery of electromagnetic waves by one of his students.

While Joule and Helmholtz pursued successful scientific careers, Mayer did not fare so well. Mayer was deeply affected by his lack of recognition and attempted suicide in 1849. After a period of mental illness from which he never fully recovered, he lived out his life in obscurity.

For example, suppose fuel is burned and the heat produced is supplied to a steam engine. The first law requires that the work done by the engine plus the heat rejected by it to the surroundings must equal the heat added, since the internal energy of the engine does not change. However, the first law gives no clue as to the ratio of the work done to the heat supplied, that is, the *efficiency* of the engine. The second law makes it possible to calculate the efficiency for an idealized engine and to set limits on the efficiencies of real engines.

A second example of its use is in chemical reactions. When a reaction occurs the first law allows one to predict how much energy is absorbed or liberated. However, the second law allows one to predict, for given conditions of temperature and pressure, what the equilibrium state of this system will be.

In this section, we discuss the microscopic and macroscopic forms of the second law. The efficiency of heat engines is discussed in the next section. Applications to chemical systems are treated in detail in chemistry and biochemistry texts.

Microscopic Form of the Second Law |

The microscopic form of the second law is a statement about the probable behavior of a large number of molecules or other particles. It states that systems tend to evolve from highly ordered, relatively improbable configurations to more disordered, more statistically probable configurations. Equivalently,

systems tend toward states of maximum molecular disorder or chaos. For example, Fig. 11.5 shows two ways in which the molecules of a gas may move. Both pictures show a gas with the same internal energy, but Fig. 11.5a represents a highly ordered situation unlike Fig. 11.5b. The second law says that a situation *similar* to the disordered state is *more probable* than one similar to the ordered state. Thus, it is possible to imagine a great many pictures similar to 11.5b, but only a few like Fig. 11.5a. Poker players will recognize an equivalent idea: there are only a few hands with a royal flush, but many hands with no value at all, so the odds strongly favor hands of the latter type. Similarly, from this point of view, parking in a tight space at a curb is more difficult than moving onto the street, because there are many more moving configurations than parked ones.

Macroscopic Form of the Second Law |

The second law was first formulated as a statement about large or macroscopic systems. This form is eas-

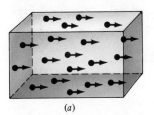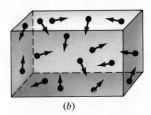

(a)   (b)

Figure 11.5. Two patterns of molecular motion in a gas. (a) Highly ordered. (b) Much less ordered or more chaotic.

ier to use for many applications, although its physical interpretation is perhaps more subtle than that of the molecular chaos previously discussed. The two forms have been shown to be equivalent using microscopic statistical theories.

The macroscopic form of the second law is a statement that a quantity called the *entropy* tends to assume a maximum value. Like the internal energy, the entropy of a system depends only on its state and not on how that state is achieved.

The definition of entropy involves the concept of *reversible* and *irreversible* processes. A reversible process is one in which the system can be returned to its original state with no net change in either the system or its surroundings. For example, if no friction, turbulence, or other dissipative effect is present, the adiabatic expansion of a gas is reversible (Fig. 11.6). This is because an adiabatic compression can return the system to its original state. The work done on the gas during the compression is equal to that done by the gas during the expansion; the net work done by the gas and its surroundings is zero.

No known natural process is reversible. When heat is transferred between objects at different temperatures, the heat can be returned to the higher temperature object, but to do so requires that the surroundings do work, as in a refrigerator. Thus the surroundings must be modified to return the system to its original state. Reversible processes, like frictionless systems in mechanics, are an idealization only approximately realized in real systems.

We can now define the entropy of a system. Suppose a small quantity of heat ΔQ is added to a system at a Kelvin temperature T during a reversible proc-

ess. We use ΔQ to emphasize the small changes involved. The entropy change of this system is then defined by

$$\Delta S = \frac{\Delta Q}{T} \qquad \text{(reversible process)} \qquad (11.7)$$

If a large quantity of heat is transferred, it can be divided into many small amounts ΔQ_i such that the temperature T_i is nearly constant during the transfer of ΔQ_i. Then, in a reversible process, the total change in entropy is found by summing the small entropy changes $\Delta Q_i/T_i$. Note that when heat *leaves* a system, ΔQ is *negative* and so is the associated entropy change of the system. For an irreversible process, the entropy change of an isolated system can be evaluated by considering reversible processes that would bring the system to the same final state.

We can now give the macroscopic form of the second law. For any process, *the total entropy of a system plus its surroundings may never decrease:*

$$\Delta S(\text{total}) \geq 0 \qquad (11.8)$$

The total entropy change is zero for a reversible process and positive for an irreversible process. This is called the *second law of thermodynamics*. Microscopically this is equivalent to saying that the molecular disorder of a system and its surroundings remains constant if the process is reversible and increases if it is not.

It is possible to derive Eq. 11.8 starting from either of two experimental observations. One is that heat never flows spontaneously from a colder to a hotter object. This is called the Clausius form of the second law. The second observation is that it is impossible to extract heat from an object and convert it *entirely* into work. This is the Kelvin form of the second law. The path from either of these statements of the second law to the entropy or molecular chaos forms is complex and beyond the scope of this textbook. We can, however, see that Eq. 11.8 is indeed satisfied in some simple processes. The following examples also show how one computes entropy changes in reversible and irreversible processes.

Example 11.2

Find the entropy changes of the system and of its surroundings for a reversible adiabatic process.

In an adiabatic process the heat entering a system is zero. Since the process is reversible, by Eq. 11.7 the entropy change of the system is zero. Similarly, since the surroundings transfer no heat, their entropy change is

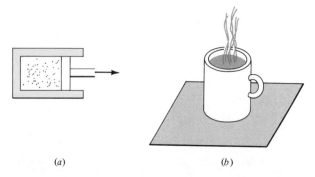

(a)   (b)

Figure 11.6. In a reversible process the system and surroundings can both be restored to their original states. (*a*) A frictionless adiabatic expansion is reversible. (*b*) Heat transfer between objects at different temperatures is irreversible.

also zero. Thus $\Delta S(\text{total}) = 0$ as required for reversible processes by the second law.

Example 11.3

If 3.33×10^5 J of heat is removed from a kilogram of liquid water at $0°$ C, it will turn to ice. Suppose heat is reversibly withdrawn from 10^{-2} kg of liquid water at $0°$ C, until it is entirely converted to an ice cube at the same temperature. (a) What is the entropy change of the water? (b) What is the net entropy change of the water and surroundings?

(a) The heat withdrawn is $(3.33 \times 10^5 \text{ J kg}^{-1}) \times (10^{-2} \text{ kg}) = 3.33 \times 10^3$ J, so the entropy change is

$$\Delta S = \frac{\Delta Q}{T} = \frac{-3.33 \times 10^3 \text{ J}}{273 \text{ K}} = -12.2 \text{ J K}^{-1}.$$

The minus sign appears because heat is withdrawn from the water, and its entropy decreases.

(b) The process occurs reversibly, so the net entropy change of the water plus surroundings is zero. Since the entropy of the water decreases, the entropy of the surroundings must *increase* by the same amount, $\Delta S = 12.2 \text{ J K}^{-1}$.

In this example we can see the relationship between entropy and order. When the water solidifies to an ordered solid phase, the entropy decreases. Since the net change in entropy is zero, the entropy of the surroundings increases. Conversely, when the ice melts, its entropy increases. If the melting occurs reversibly, the entropy of the surroundings decreases.

Example 11.4

Two large objects are isolated from their surroundings. They are at temperatures T_1 and T_2 with $T_2 > T_1$ and are placed in thermal contact. A small quantity of heat Q is transferred, leaving their temperatures nearly unchanged. Find the entropy changes.

The entropy change of the surroundings is zero, since the system is isolated. This is an irreversible process, so we calculate the entropy changes for reversible paths leading to the same final states. We may, for example, consider the case where heat Q is reversibly removed from the object at T_2 by placing a cylinder of gas in contact with it and allowing the gas to expand isothermally. This gives an entropy change $\Delta S_2 = -Q/T_2$. Similarly, a second cylinder can be placed in contact with the cooler object and heat transferred reversibly. The entropy change of the cooler object is then $\Delta S_1 = Q/T_1$. Thus the total entropy change of the two objects is

$$\Delta S(\text{total}) = -\frac{Q}{T_2} + \frac{Q}{T_1} = Q\left(\frac{1}{T_1} - \frac{1}{T_2}\right)$$
$$= Q\frac{(T_2 - T_1)}{T_1 T_2}$$

Since T_2 is greater than T_1, this is positive as predicted for an irreversible process.

Although the second law of thermodynamics in the form of Eq. 11.8 states that no *net* decrease in entropy can occur, it is not true that the entropy of a *system* cannot be reduced. When water is frozen, the entropy of the water is lowered. However, if careful consideration is taken of the surroundings, it is found that the *total entropy of the system plus its surroundings remains constant or increases.*

11.4 | THE CARNOT THEOREM AND CONVERSION OF ENERGY

If one applies the second law of thermodynamics to an idealized reversible heat engine, one finds that the efficiency of the engine in converting heat into mechanical work is always less than 100 percent. Real heat engines, such as automobile engines and electric power-generating plants, always have some friction or turbulence and are therefore irreversible. Their efficiencies are necessarily still lower. This remarkable property of heat engines was discovered by Sadi Carnot (1796–1832) in 1824 and is called *Carnot's theorem.*

Carnot's theorem shows that the conversion of thermal energy into other forms of energy is qualitatively different from other kinds of energy conversion. For example, a swinging pendulum can completely convert mechanical potential energy into mechanical kinetic energy and back again. The kinetic energy of moving water in a turbine can be transformed into electrical energy by a generator; the efficiency is limited only by friction and turbulence and can be quite high in practice. Electrical energy can also be converted into mechanical energy by a motor; again there is no theoretical limit on the efficiency, and the actual efficiencies can be 90 percent or better. Also, chemical energy can be converted into electrical energy with very high efficiency, as in a fuel cell where hydrogen and oxygen combine and produce an electrical current. Electrical energy can also be transformed into chemical energy without any fundamental limitation. In the human body the conversion of chemical energy from food into mechanical work is usually done at an efficiency of about 30 percent at most. This limit on the efficiency does not arise from the second law. It occurs because the body loses energy in converting foods into compounds that the body can utilize effectively at the cellular level.

Any heat engine can be thought of as acting in a *cycle*. First, a substance at a high temperature is permitted to do work and also to transfer heat to its surroundings, then the cycle is completed by returning the substance to its original state while transferring enough heat to it to replace the lost energy.

Carnot's theorem is proved by considering a particular reversible heat engine, called a *Carnot engine* or a *Carnot cycle*. This cycle consists of four reversible processes, which are illustrated in Fig. 11.7 for the special case of an ideal gas. However, the *calculated efficiency is independent of the material* used in the engine; we only use the ideal gas because it is easy to discuss.

The path from a to b is an isothermal expansion in which heat Q_2 is absorbed from a reservoir at T_2. The path bc is an adiabatic expansion; no heat is transferred. The path cd is an isothermal compression in which heat Q_1 leaves the system at a lower temperature T_1. Finally the material is returned to its original state by an adiabatic compression da; again, no heat is transferred.

In expanding, the gas does work equal to the area under the curve abc (Fig. 11.7b). When it is compressed, work must be done on the gas equal to the smaller area under the curve cda (Fig. 11.7c). The net work W done by the gas over one full cycle therefore equals the shaded area enclosed by the path (Fig. 11.7a). Since the gas at the end of the cycle is in its original state, the net change in internal energy is zero. The first law therefore requires that the net work done by the gas must equal the heat absorbed minus the heat rejected, or

$$W = Q_2 - Q_1 \qquad (11.9)$$

This is shown symbolically in Fig. 11.8.

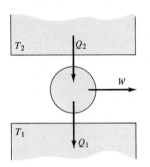

Figure 11.8. The work done in a complete cycle equals the heat absorbed at T_2 less the heat rejected at T_1.

Since the Carnot cycle is reversible, the total entropy change of the system is zero. For the reservoir at T_2 the entropy change is $\Delta S_2 = -Q_2/T_2$. The entropy change of the lower temperature reservoir is $\Delta S_1 = Q_1/T_1$. The gas returns to its original state, so its entropy change is zero. Thus

$$\Delta S(\text{total}) = \Delta S_1 + \Delta S_2 = -\frac{Q_2}{T_2} + \frac{Q_1}{T_1} = 0$$

or

$$\frac{Q_1}{Q_2} = \frac{T_1}{T_2}$$

The efficiency is defined as the work done divided by the heat absorbed, $\varrho = W/Q_2$. Using $W = Q_2 - Q_1$, we find

$$e = \frac{W}{Q_2} = 1 - \frac{Q_1}{Q_2} = 1 - \frac{T_1}{T_2} \qquad (11.10)$$

This ratio of the work done to the heat supplied is the *efficiency* of the Carnot engine. According to Eq. 11.10, the efficiency is always less than 1 unless the

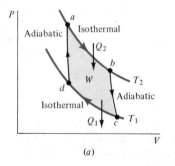
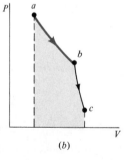
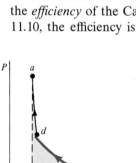

Figure 11.7. (*a*) The Carnot cycle for an ideal gas. (*b*) The work done by the gas during the expansion equals the colored area. (*c*) The work done on the gas during the compression equals the colored area.

colder reservoir is at absolute zero, which is, in fact, unattainable even in principle.

The following example shows the efficiency of a Carnot engine operating between two easily achieved temperatures.

Example 11.5

A Carnot engine operates between 100° C and 0° C. What is its efficiency?

Converting the temperatures to the Kelvin scale, $T_1 = 273$ K and $T_2 = 373$ K. Then

$$e = 1 - \frac{T_1}{T_2} = 1 - \frac{273}{373} = 0.268$$

Only 26.8 percent of the heat supplied is converted into mechanical work; the remainder is rejected into the colder reservoir.

11.5 | IMPLICATIONS OF CARNOT'S THEOREM

As we noted earlier, the efficiency of a Carnot engine is independent of the substance used. This is clear from the derivation, because the calculation of the efficiency used only the fact that the cycle consisted of two isothermal and two adiabatic processes and did not rely on any properties of the material. This means that one cannot hope to improve the efficiency of an engine by an ingenious choice of materials if that engine is already fairly close to the Carnot efficiency for the temperatures of its cycle. Carnot also showed that *no heat engine operated cyclically between two temperature reservoirs is more efficient than that described by the Carnot cycle.*

Real engines always have some losses resulting from friction and turbulence and must be less efficient than Carnot engines operating between the same temperatures. To maximize the efficiency, T_2/T_1 is made as large as practicable. High-performance auto engines have large ratios of maximum to minimum cylinder volumes or compression ratios in order to achieve correspondingly large gas temperature ratios in the cylinders.

The temperatures and pressures of the steam boilers in electric power-generating plants are also very high. Modern fossil fuel electric-generating plants are about 40 percent efficient; the excess heat lost when the steam condenses is given off into a lake or river or transferred to the atmosphere with cooling towers. Present commercial nuclear power plants are at best 34 percent efficient, so the thermal pollution prob-

lems they pose are somewhat larger. These efficiencies should be compared with the ideal Carnot efficiencies of 52 percent and 44 percent, respectively, for the two types of plants.

The lower efficiency of the nuclear plants is related to limitations on the high temperatures allowable for the uranium in the reactor. The uranium oxide in the center of the fuel rods must be kept well below melting temperature. This limits the temperature at the outside of the fuel rods. In all power plants the maximum temperature is limited by design and materials problems, and the low-temperature exhaust temperature is determined by the available cooling water in lakes and rivers. The components of a major steam-generating plant are shown in Fig. 11.9.

Example 11.6

A nuclear power plant generates 500 MW = 5×10^8 W at 34 percent efficiency. The waste heat goes into a river, such as the Connecticut River, with an average flow of 3×10^4 kg s^{-1}. How much does the water temperature rise? (It requires 4.18×10^3 J to raise the temperature of 1 kg of water 1 K.)

If the plant operates at 34 percent efficiency, it must have a total heat input of 1.47×10^9 W. Of this, 66 percent or 9.7×10^8 W must go into the river. In 1 second the heat $\Delta Q = 9.7 \times 10^8$ J will raise the water temperature by ΔT. With $m = 3 \times 10^4$ kg,

$$\Delta T = \frac{\Delta Q}{(4.18 \times 10^3 \text{ J kg}^{-1} \text{ K}^{-1})m}$$

$$= \frac{9.7 \times 10^8 \text{ J}}{(4.18 \times 10^3 \text{ J kg}^{-1} \text{ K}^{-1})(3 \times 10^4 \text{ kg})}$$

$$= 7.7 \text{ K}$$

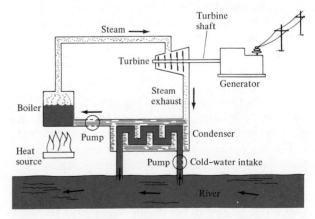

Figure 11.9. The major components of a steam electric power generating plant. The heat source may be coal or oil or a uranium-fueled reactor.

This temperature rise could cause serious damage to the organisms in the river. In a case such as this, cooling towers are used to dissipate heat to the atmosphere rather than the river. A 40 percent efficient fossil fuel plant of the same output power would raise the river temperature by almost 6 K.

The expression $e = 1 - T_1/T_2$ for the efficiency of a Carnot engine provides a useful way of interpreting the second law. If heat is transferred from a hotter object to a colder one, reducing their temperature difference, then the efficiency of a Carnot engine operating between these two heat reservoirs decreases. This means that less work can be done with a given heat input and that the energy stored in these objects is therefore less available. For example, with tanks of water at 0° C and 100° C, we can operate an engine; this cannot be done if we mix these tanks together so that they are at a common temperature. The statement that systems tend toward states of maximum entropy therefore implies that *energy tends to become less and less available for doing mechanical work.*

Ocean Thermal Differences | The intensifying search for alternative energy sources often involves consideration of the limits imposed by the Carnot theorem. One source of power now under investigation employs the temperature differences between the ocean surface and water layers deep below. The warm surface water evaporates a working fluid such as ammonia. The gas drives a turbine, much as steam does in a conventional power plant, and is then condensed using the colder water from lower levels. The temperature differences range from 18° C to 25° C in tropical waters. Assuming a surface temperature of 30° C or 303 K, the ideal Carnot efficiency of the scheme is at best

$$e = 1 - \frac{T_1}{T_2} = 1 - \frac{278}{303} = 0.0825$$

Since the theoretical Carnot efficiency is only 8.25% and there are other unavoidable losses, a 3 percent efficiency in producing electrical power is considered somewhat optimistic. For such a plant to produce 500 MW of electrical power requires pumping approximately 200,000 litres of water per second through the heat exchangers. In a small-scale 1979 test of this scheme in the Pacific Ocean near Hawaii, the pumps used 80 percent of the electrical power generated; eventually, this might be reduced to 30 percent. By contrast, conventional power plants use

about 1 percent of their output on site. Thus, while the supply of thermal energy in the oceans is enormous, the problems inherent in using it are very large.

11.6 | REFRIGERATORS AND HEAT PUMPS

Refrigerators and heat pumps are devices that remove heat from a low-temperature region or reservoir and release it into a higher temperature reservoir. The refrigerator removes heat from the cooling compartment and freezer section and releases it into the room. The heat pump, usually operating between the inside and outside of a building, can be used to cool the inside during hot weather and to heat the inside during cold weather.

If one of these devices removes heat Q_1 from a low-temperature reservoir at T_1 and releases heat Q_2 into a high-temperature reservoir at T_2, from the first law of thermodynamics the work done *by* the system is

$$W = Q_1 - Q_2$$

The internal energy change is zero because the device acts cyclically, returning the system to the same thermodynamic state repeatedly.

The *coefficient of performance* (C.P.), which gives a useful measure of effectiveness, is defined differently for the refrigerator and heat pump. For the refrigerator, C.P.$_R$ is defined as the ratio of the heat absorbed at the *lower temperature* to the work done *on* the system. The work done on the system is $-W = Q_2 - Q_1$, so

$$\text{C.P.}_R = \frac{Q_1}{-W} = \frac{Q_1}{Q_2 - Q_1} \qquad (11.11)$$

A typical refrigerator has C.P.$_R \simeq 5$.

For the heat pump, C.P.$_{HP}$ is defined as the ratio of the heat released at the *higher temperature* to the work done *on* the system.

Thus

$$\text{C.P.}_{HP} = \frac{Q_2}{-W} = \frac{Q_2}{Q_2 - Q_1} \qquad (11.12)$$

Commercially available heat pumps have C.P.$_{HP}$ in the range of 2 to 4.

We can find the theoretical limits on the performance of ideal refrigerators and heat pumps just as we did for the Carnot engine. An ideal device operates reversibly so the total entropy change of the system plus surroundings per cycle is zero. Adding the entropy changes of the low- and high-temperature reservoirs,

$$\Delta S = \frac{-Q_1}{T_1} + \frac{Q_2}{T_2} = 0$$

Hence $Q_1/T_1 = Q_2/T_2$. Using this result, the work done on the system per cycle is $-W = Q_2 - Q_1$ or

$$-W = Q_2 \left(1 - \frac{T_1}{T_2}\right) = Q_1 \left(\frac{T_2}{T_1} - 1\right)$$

The corresponding ideal coefficients of performance are

$$\text{C.P.}_R = \frac{T_1}{T_2 - T_1} \qquad \text{(ideal)} \qquad (11.13)$$

and

$$\text{C.P.}_{HP} = \frac{T_2}{T_2 - T_1} \qquad \text{(ideal)} \qquad (11.14)$$

Usually the amount of heat that must be transferred by a refrigerator or heat pump is proportional to the temperature difference, $T_2 - T_1$. However, we see that as the temperature difference *increases,* the coefficient of performance for an ideal refrigerator or heat pump *decreases.* Just when the outside temperature is low, and the need for heat transfer into the house is large, the coefficient of performance of a heat pump may not be much greater than 1. Also, heat pumps tend to condense moisture and freeze up in cold weather. Thus, their main use so far has been in areas with mild winters, such as the southern parts of the United States.

Example 11.7

A commercial heat pump has $\text{C.P.}_{HP} = 3$ when the indoor temperature is $20°$ C and the outdoor temperature is $6°$ C. (a) What is the ideal value of C.P._{HP}? (b) How much work is required to operate the commercial heat pump if 3×10^6 J of heat must be transferred to the room each hour?

(a) The indoor and outdoor temperatures are 293 K and 279 K, respectively. For an ideal heat pump,

$$\text{C.P.}_{HP} = \frac{T_2}{T_2 - T_1} = \frac{293 \text{ K}}{293 \text{ K} - 279 \text{ K}} = 20.9$$

(b) The work done on the heat pump in an hour is

$$-W = \frac{Q_2}{\text{C.P.}_{HP}} = \frac{3 \times 10^6 \text{ J}}{3} = 10^6 \text{ J}$$

Thus energy must be provided to operate the heat pump at the rate of 10^6 J per hour or 278 W. In one hour, the heat transferred from the outdoors to the room is $Q_1 = Q_2 + W = 3 \times 10^6 \text{ J} - 10^6 \text{ J} = 2 \times 10^6 \text{ J}$.

SUMMARY

Thermodynamics is the study of the transformation of energy from one form to another. In this chapter, we were specifically concerned with internal energy, heat, and mechanical work. The first law is a relationship between these quantities: the internal energy change of a system is equal to the heat added to the system less the work done by the system,

$$\Delta U = U_f - U_i = Q - W$$

The work done by a system at a pressure P when a small volume change ΔV occurs is

$$\Delta W = P \Delta V$$

The second law of thermodynamics is expressed in terms of the entropy. In its microscopic form, it is a statement that systems tend to evolve toward increasing disorder. The macroscopic statement of the second law, while it is somewhat subtle in its content, can be used to make very general statements about processes and their efficiencies.

Macroscopically, if a small amount of heat ΔQ is added to a system in a reversible process, the entropy change of the system is

$$\Delta S = \frac{\Delta Q}{T}$$

The second law states that the total entropy of a system and its surroundings will never decrease.

Using the first and second laws of thermodynamics, Carnot showed that the maximum efficiency of a heat engine operating with high and low temperatures of T_2 and T_1, respectively, is

$$e = 1 - \frac{T_1}{T_2}$$

Refrigerators and heat pumps can also be described and analyzed in terms of the first and second laws. The maximum possible coefficients of performance for these devices are

$$\text{C.P.}_R = \frac{T_1}{T_2 - T_1} \qquad \text{(ideal)}$$

and

$$\text{C.P.}_{HP} = \frac{T_2}{T_2 - T_1} \qquad \text{(ideal)}$$

Checklist

Define or explain:

work

internal energy

first law of
  thermodynamics

adiabatic process

isothermal process

disordered state

reversible and
  irreversible processes

entropy

second law of
  thermodynamics

Carnot's theorem

Carnot efficiency

refrigerator

heat pump

coefficient of
  performance

REVIEW QUESTIONS

Q11-1   When a gas at a constant pressure P expands by an amount ΔV, the work done by the system is _____.

Q11-2   What is meant by the word heat?

Q11-3   What is the internal energy?

Q11-4   If heat is added to a system and some work is done by the system, what is the difference between these quantities?

Q11-5   The _____ is a measure of the disorder in a system.

Q11-6   If a small amount of heat ΔQ enters a system at an absolute temperature T, what is the entropy change of the system if the process is reversible?

Q11-7   The second law of thermodynamics states that the entropy of a system plus its surroundings may never _____.

Q11-8   If a heat engine takes in heat at a higher temperature T_2 and loses heat at a lower temperature T_1, what is its maximum possible efficiency?

Q11-9   How is the coefficient of performance of a refrigerator defined?

EXERCISES

Section 11.1 | Mechanical Work

11-1   In Fig. 11.10, how much work is done by the system in the process (a) along path (2) from A to B? (b) If the system is returned from B to A along the same path, how much work is done by the system?

11-2   A gas does work in an isobaric process at $P = 10^5$ Pa. How much work is done by the gas if
(a) $V_i = 10^{-2}$ m^3,   and $V_f = 2.24 \times 10^{-2}$ m^3;
(b) $V_i = 2 \times 10^{-2}$ m^3, and $V_f = 0.5 \times 10^{-2}$ m^3?

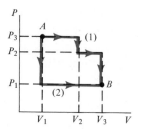

Figure 11.10.   Exercises 11-1 and 11-4.

11-3   A 10-N force directed to the left is applied to the piston of Fig. 11.1. If the piston moves 0.14 m, how much work is done on the gas?

Section 11.2 | The First Law of Thermodynamics

11-4   In Fig. 11.10, a system undergoes a process from point A to point B. How much work is done

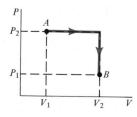

Figure 11.11. $P_1 = 1$ atm, $P_2 = 3$ atm, $V_1 = 0.02$ m^3, and $V_2 = 0.10$ m^3. Exercise 11-5.

by the system on (a) path (1), and (b) path (2)? (c) If process (1) is carried out adiabatically, what is the change in the internal energy?

11-5   In undergoing the process from A to B in Fig. 11.11, the increase in internal energy of a substance is 3×10^5 joules. How much heat is absorbed by the system?

11-6   Find the change in internal energy of the system when (a) a system absorbs 2000 J of heat and produces 500 J of work; (b) a system absorbs 1100 J of heat and 400 J of work is done on it.

11-7   An electric heater supplies heat to a gas at a rate of 100 W. If the expanding gas does 75 joules of work in each second, at what rate is the internal energy increasing?

11-8   If the gas of Exercise 11-7 has a constant pressure of 1 atm, by how much does it expand in 10 s?

11-9   The gas of Figure 11.1 is heated. Under which condition will the internal energy increase most rapidly, when the piston is held fixed or when the piston is allowed to move to the right? Explain your reasoning.

Section 11.3 | The Second Law of Thermodynamics

11-10 A new deck of playing cards has cards of each suit together and ordered numerically. The deck is then shuffled. Has its entropy changed? If so, why?

11-11 The second law of thermodynamics states that the disorder of the universe is always either constant or increasing. (a) How is this compatible with the fact that plants and animals develop into highly ordered systems? (b) How can the second law of thermodynamics be compatible with the theory of evolution, which states that more highly ordered complex living things evolve from simpler species?

11-12 A system at a constant temperature of 300 K absorbs 10^4 J of heat, and no work is done. (a) What is the change in entropy of the system? (b) What is the change in its internal energy?

11-13 A helium-filled balloon is punctured, and the gas spreads uniformly throughout a room. (a) Is the process reversible? Explain. (b) Has the entropy of the helium and the air in the room increased or decreased? Explain.

11-14 In terms of the microscopic form of the second law of thermodynamics, explain why repeatedly tossing a coin should yield heads about as often as tails.

11-15 A coin is flipped 6 times. (a) How many ways can these flips result in all tails, one head and five tails, two heads and four tails . . . to five heads and one tail, and to all heads? (b) From your results, what is the most probable result of the six flips? Explain.

Section 11.4 | The Carnot Theorem and Conversion of Energy

11-16 What is the maximum efficiency of a heat engine working between 100° C and 400° C?

11-17 An internal combustion engine using natural gas and air as a working substance has a temperature in the spark-ignited firing chamber of 2150 K and an exhaust temperature of 900 K. The difference between the heat supplied and the work

done by the engine in each second is 4.6×10^6 J. (a) What is the ideal Carnot efficiency of this engine? (b) How much work is actually done per second if the heat input is 7.9×10^6 W? (c) What is the actual efficiency of this engine?

11-18 If a heat engine operates at 40 percent efficiency and it absorbs 10^4 W from the high-temperature reservoir, at what rate does it do mechanical work?

Section 11.5 | Implications of Carnot's Theorem

11-19 If fossil fuel and nuclear generating plants operate with efficiencies of 40 percent and 30 percent efficiency, respectively, and the low-temperature reservoir is at 300 K for both, what is the minimum temperature of the steam produced by the fuel in each case?

11-20 If the power plant of example 11.6 were a fossil plant operating at 40 percent efficiency, what would the temperature rise of the water be as a result of the waste heat?

Section 11.6 | Refrigerators and Heat Pumps

11-21 A refrigerator takes heat from the freezer unit at −3° C at a rate of 100 W. The heat is released to a room at 26° C. (a) What is the maximum possible coefficient of performance of the refrigerator? (b) If the actual C.P.$_R$ is 4, how much power is needed to maintain the freezer unit at −3° C?

11-22 A heat pump is used to provide heat to the inside of a house at a rate of 5000 W. If energy is provided to operate the heat pump at a rate of 2000 W, what is the coefficient of performance of the heat pump?

PROBLEMS

11-23 A system goes from state a to c via path abc in Fig. 11.12. 10^5 J of heat flow into the system, and 4×10^4 J of work are done. (a) How much heat flows into the system along path adc if 10^4 J of work are done? (b) If the system returns to state a from state c via the zigzag path, the work is 2×10^4 J. How much heat enters or leaves the system? (c) If $U_a = 10^4$ J and $U_d = 5 \times 10^4$ J, what is the heat absorbed along the paths ad and dc?

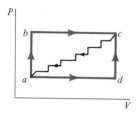

Figure 11.12. Problem 11-23.

11-24 Eight litres of air at room temperature and atmospheric pressure are compressed isothermally to a volume of 3 litres. The air then expands adiabatically to a volume of 8 litres. Show the process on a *P-V* diagram.

11-25 An engine operates on the cycle shown in Fig. 11.13, a temperature-entropy diagram. What is the efficiency of the engine?

*11-26 Draw a schematic *T-S* diagram for a Carnot engine operating between temperatures T_2 and T_1, where $T_2 > T_1$.

11-27 Two dice, each having the numbers one through six on its sides, are thrown together. (a) In how many ways can the two dice be thrown so that the "up" numbers total 2, 3, 4, . . . , 12? (b) Which total is most probable? (c) What is the implication of part (b) in terms of entropy and disorder?

11-28 A heat pump is used to heat a building when the outdoor temperature is 0° C and the indoor temperature is 25° C. The coefficient of performance for the pump under these conditions is 3.2. (a) If the heat pump delivers heat inside at a rate of 5×10^6 J per hour, at what rate must work be done to run the heat pump? (b) How much elec-

trical energy would be used to heat the building directly? (c) A litre of oil provides 3.7×10^7 J of energy when burned. How many litres of oil per hour must be burned at 80% efficiency to provide the heat needed for the building? (d) If the oil is burned to produce electric power at 40% efficiency and used to run the heat pump, how much oil per hour must be burned?

11-29 About 2×10^5 J of heat must be removed from a tray of water to produce ice cubes. (a) How long will it take to remove this much heat in a 200-W input Carnot refrigerator operating between inner and outer temperatures of 270 K and 310 K, respectively? (b) Is this time consistent with your estimate of how long it takes to freeze an ice tray? Explain the difference, if any.

11-30 The motor in an ideal (Carnot) refrigerator delivers 200 W of useful power. The freezing compartment of the refrigerator is at $T_1 = 270$ K, and the room air is at $T_2 = 300$ K. Find the maximum amount of heat that can be removed from the refrigerator in 1 minute.

11-31 A nuclear boiling-water reactor heats steam to 285° C, and the cooling water is at 40° C. The actual operating efficiency of the power plant is 34 percent. (a) What is the ideal efficiency of the plant? (b) What is the ratio of the power actually lost to that lost in the ideal situation?

11-32 Using the temperature difference between surface and deep ocean water to produce electricity involves transferring the heat from the higher temperature water to a working substance and finally to the cooling water. To increase the temperature of a kilogram of water by 1 K requires 4.169×10^3 J. In order to produce 500 MW (500×10^6 W) of electricity at 4 percent efficiency, a mass *m* of cooling water must have its temperature raised from 12° C to 30° C each second. What is *m*?

11-33 The temperature difference between the surface and deep water behind large dams could be used to produce power in the same way as described for ocean water. If the deep water is assumed to always be at 5° C, what is the ideal efficiency for utilizing this energy source in January and in July if the surface temperature of the water is 8° C and 23° C, respectively, in these two months?

c11-34 For real gases, an improvement over the ideal gas equation of state is given by the *van der Waal's equation,*

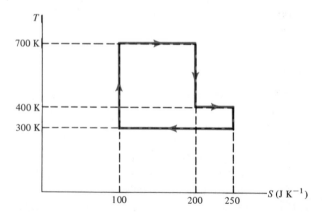

Figure 11.13. Problem 11-25.

$$(P + a/V^2)(V - b) = RT$$

Here a and b are constants. Show that the work done when a gas expands from initial volume V_0 to a volume V at a constant temperature T is $W = a/V - a/V_0 + RT \ln [(V - b)/(V_0 - b)]$.

^c**11-35** In an adiabatic expansion of an ideal gas, the pressure and volume are related by $PV^\gamma = K$, where K and γ are constants. Show that when the pressure and volume change in an adiabatic expansion from P_0, V_0 to P, V, the work done by the gas is

$$W = (PV - P_0V_0)/(1 - \gamma)$$

ANSWERS TO REVIEW QUESTIONS

Q11-1, $P \Delta V$; **Q11-2,** the thermal energy transferred; **Q11-3,** the energy of a system due to the translational, rotational, and vibrational kinetic energy of its molecules plus that due to molecular interactions; **Q11-4,** the change in internal energy; **Q11-5,** entropy; **Q11-6,** $\Delta Q/T$; **Q11-7,** decrease; **Q11-8,** $1 - T_1/T_2$; **Q11-9,** the ratio of the heat absorbed at the low temperature to the work done on the system.

SUPPLEMENTARY TOPICS

11.7 | HUMAN METABOLISM

All living things require energy to sustain the life process. Green plants obtain their energy directly from the sun through the process of photosynthesis. Plants such as mushrooms, which do not utilize photosynthesis, as well as animals require food capable of providing chemical energy. In all cases, living plants and animals operate within the constraints described by thermodynamics.

The first law of thermodynamics provides a convenient scheme for cataloging the factors that enter the complex subject of human metabolism. Suppose that in a time Δt, a person does an amount of mechanical work ΔW. This can be directly measured in activities such as bicycling, shoveling snow, or pushing a cart. Heat will usually leave the body, so ΔQ will be negative. It can be measured by finding the amount of heat that must be removed to keep the temperature constant in a room in which the person is working. From the first law, the internal energy change ΔU will be given by $\Delta U = \Delta Q - \Delta W$. Dividing by Δt, this becomes a relation among the rates of change,

$$\frac{\Delta U}{\Delta t} = \frac{\Delta Q}{\Delta t} - \frac{\Delta W}{\Delta t} \qquad (11.15)$$

The rate of change of internal energy can be measured accurately by observing the rate at which a person uses oxygen in converting food into energy and waste materials. For example, a mole (180 g) of glucose, which is a typical carbohydrate, combines with 134.4 litres of oxygen gas in a series of steps to form carbon dioxide and water. In the process, 2870 kJ of energy are released. The *energy equivalent of oxygen* is defined as the ratio of the energy released to the oxygen consumed. For glucose, this ratio is 2870 kJ/134.4 litres = 21.4 kJ litre^{-1}. The *energy content per unit mass* is defined as the energy released divided by the mass. For glucose, this ratio is 2870 kJ/180 g = 15.9 kJ g^{-1}.

Table 11.1 lists the average energy content per unit mass and energy equivalent of oxygen for the carbohydrates, proteins, and fats usually consumed. The calorific equivalent of oxygen of all of these is the same to within a few percent. Thus, an average value of 20.2 kJ litre^{-1} is used in converting measured oxygen consumption rates to rates of internal energy change. For example, if a person consumes oxygen at the high rate of 100 litres h^{-1}, the rate of internal energy change is (100 litres h^{-1})(20.2 kJ litre^{-1}) = 2020 kJ h^{-1} = 561 W.

Basal Metabolic Rate | All animals, including humans, use internal energy even while sleeping. The rate of energy consumption while resting but awake is called the basal metabolic rate. It is about 1.2 W kg^{-1} for an average 20-year-old man and 1.1 W kg^{-1} for a woman of the same age. In the units often used in

TABLE 11.1

The average energy content per unit mass of food and the energy equivalent of oxygen for a typical diet

Food	Energy Content per Unit Mass (kJ g^{-1})	Energy Equivalent of Oxygen (kJ litre^{-1})
Carbohydrate	17.2	21.1
Protein	17.6	18.7
Fat	38.9	19.8
Ethanol	29.7	20.3
Standard average		20.2

discussions of nutrition, kilocalories, this corresponds to about 1700 kcal per day and 1400 kcal per day for a 70-kg man and a 60-kg woman, respectively. Much of the energy consumed by a resting person is converted directly into heat. The remainder is used to do work in the body and then converted into heat.

Food materials are not used directly in the body. Instead they are converted into materials such as ATP (adenosine triphosphate), which can be used in the tissues. About 55 percent of the internal energy is lost as heat in this conversion. The remaining 45 percent is available to do internal work in the body organs or to enable the skeletal muscles to contract and do work on an external object.

When a person is performing an activity such as running up stairs or doing housework, the metabolic rate increases (Table 11.2). Part of the increased conversion of internal energy is needed to provide the mechanical work done by the person. The remainder results from the increased internal demands of the body. For example, in shoveling, the metabolic rate is about eight times the basic metabolic rate, but little mechanical work is actually done. The metabolic energy is used primarily by the skeletal muscles in changing and maintaining the body position.

The following example illustrates some of the ideas we have developed to this point.

Example 11.8

(a) How much internal energy is used by a 65-kg man when bicycling for 4 hours? (b) If this energy is obtained

by the metabolism of body fat, how much fat is used in this period?

(a) From Table 11.2, the metabolic rate while cycling is 7.6 W kg^{-1}. A 65-kg man then uses energy at a rate of (7.6 W kg^{-1})(65 kg) = 494 W. Four hours is 1.44×10^4 s, so the net energy usage is

$$-\Delta U = (494\text{ W})(1.44 \times 10^4\text{ s})$$
$$= 7.1 \times 10^6\text{ J}$$
$$= 7100\text{ kJ}$$

(b) The energy equivalent of fat is 38.9 kJ g^{-1} so the mass of fat needed to supply the necessary energy is

$$\text{Mass of fat} = \frac{(7100\text{ kJ})}{(38.9\text{ kJ g}^{-1})}$$
$$= 180\text{ g} = 0.18\text{ kg}$$

To appreciate this result, it is useful to compare it with the energy equivalent of the food required by a fairly sedentary man during a 24-hour day, which is 10,500 kJ or 2500 kcal. Hence the exercise of bicycling for 4 hours will use up about two-thirds of the daily energy required by a sedentary man. This indicates that limiting food intake is a more practical way to decrease body weight than is exercise for most people.

The Efficiency of Food Utilization | The efficiency of a human using the chemical energy in food to do useful work can be defined in several ways. The usual convention is to compare the measured rate at which mechanical work is done with the actual metabolic rate during the activity minus the basal metabolic rate. The efficiency e in percent is then

$$e = \frac{100\dfrac{\Delta W}{\Delta t}}{\left|\dfrac{\Delta U}{\Delta t} - \dfrac{\Delta U}{\Delta t}_{\text{basal}}\right|}\% \qquad (11.16)$$

The denominator is the magnitude of the difference in actual and basal metabolic rates. The efficiency would be 100 percent if all the additional energy were converted to mechanical work. Table 11.3 lists some measured efficiencies.

The following example illustrates the calculation of efficiency for hiking up a mountain.

Example 11.9

A 20-year-old woman of mass 50 kg climbs a mountain 1000 m high in 4 hours. Her metabolic rate per unit mass during this activity is 7 W kg^{-1}. (a) What is the difference between her actual and basal metabolic rate? (b) How much work is done in the climb? (c) What is her efficiency?

TABLE 11.2

Approximate metabolic rate per unit mass of a 20-year-old man during various activities

Activity	$-\dfrac{1}{m}\dfrac{\Delta U}{\Delta t}$ (W kg^{-1})
Sleeping	1.1
Lying awake	1.2
Sitting upright	1.5
Standing	2.6
Walking	4.3
Shivering	Up to 7.6
Bicycling	7.6
Shoveling	9.2
Swimming	11.0
Lumbering	11.0
Skiing	15.0
Running	18.0

TABLE 11.3

Maximum efficiencies for physical work

Activity	Efficiency in Percent
Shoveling in stooped posture	3
Weight lifting	9
Turning a heavy wheel	13
Climbing ladders	19
Climbing stairs	23
Cycling	25
Climbing hills with a 5° slope	30

Adapted from E. Grandjean, *Fitting the Task to the Man; An Ergonomic Approach,* Taylor and Francis, London, 1969.

(a) Since the woman's basal rate is 1.1 W kg^{-1}, the difference per unit mass is $(7 - 1.1)$W kg$^{-1} =$ 5.9 W kg^{-1}. The total difference in rates is this times her mass or

$$\left|\frac{\Delta U}{\Delta t} - \frac{\Delta U}{\Delta t}_{\text{basal}}\right| = (50 \text{ kg})(5.9 \text{ W kg}^{-1})$$
$$= 295 \text{ W}$$

(b) The work done during the climb is equal to the change in the woman's potential energy so

$$\Delta W = mgh = (50 \text{ kg})(9.8 \text{ m s}^{-2})(1000 \text{ m})$$
$$= 4.9 \times 10^5 \text{ J}$$

The rate at which she does work in 4 hours = 1.44×10^4 s is

$$\frac{\Delta W}{\Delta t} = \frac{4.9 \times 10^5 \text{ J}}{(1.44 \times 10^4 \text{ s})} = 34 \text{ W}$$

(c) Her efficiency is found from the definition

$$e = \frac{100 \dfrac{\Delta W}{\Delta t}}{\left|\dfrac{\Delta U}{\Delta t} - \dfrac{\Delta U}{\Delta t}_{\text{basal}}\right|} = \frac{100(34 \text{ W})}{295 \text{ W}} = 11.5\%$$

Thus the woman is utilizing food energy for mechanical work at 11.5 percent efficiency. In general the efficiencies in human activities are below 30 percent.

It is important to notice that the rate at which work can be done depends on how long the activity is continued. A person in good condition can produce a power of nearly 21 W kg^{-1} during bicycle racing, but only for about 5 or 6 seconds. When doing work for a period of as long as 5 hours, the maximum metabolic rate is 6 or 7 W kg^{-1}. For a person doing physical labor the annual average metabolic rate should be 4 W kg^{-1} or less to be safe.

EXERCISES ON SUPPLEMENTARY TOPICS

Section 11.7 | Human Metabolism

11-36 A woman of mass 55 kg produces heat at a rate of 1.1 W kg^{-1} while lying at rest on a hot day. If her body temperature is constant, (a) what is the rate at which her internal energy changes? (b) How much internal energy will she use up in 8 hours? (c) If all of this energy is produced by metabolizing carbohydrates, what mass of carbohydrate is consumed?

11-37 A woman with a normal diet uses internal energy at a rate of 3 W kg^{-1} and has a mass of 50 kg. (a) What is her rate of oxygen consumption? (b) How much oxygen does she use in 8 hours?

11-38 A 60-kg man shovels dirt with an efficiency of 3 percent, and his metabolic rate is 8 W kg^{-1}. (a) What is his power output? (b) How much work does he do in 1 hour? (c) How much waste heat does his body produce in 1 hour?

11-39 A resting 45-kg woman has a normal basal metabolic rate. (a) What volume of oxygen does she use in 1 hour? (b) If she walks for 1 hour and has a metabolic rate of 4.3 W kg^{-1}, how much oxygen does she consume?

11-40 If a 70-kg, 20-year-old man uses 1 litre of oxygen per minute, (a) what is his metabolic rate? (b) If he could do work at 100 percent efficiency, what would his mechanical power output be?

11-41 Experimentally it was found by J.P. Joule in 1846 that in 24 hours a horse was capable of doing work equivalent to raising a weight of 10^8 N to a height of 0.3 m. The hay and corn consumed during this period were equivalent to a reserve of internal energy of $1.2 - 10^8$ J. (a) What fraction of this internal energy was used by the horse to do mechanical work? (b) If the horse's actual efficiency was 30 percent, what is the basal metabolic rate of the horse?

PROBLEMS ON SUPPLEMENTARY TOPICS

11-42 The basal metabolic rate for most humans steadily decreases after age 20, decreasing by about 20 percent by age 70. (a) Will an older person become cold on a winter day more quickly than a younger person? Why? (b) If a 70-year-old man does work at the same rate as a younger man and both work with the same efficiency, which man will have the higher metabolic rate? Explain.

11-43 If a 70-kg person's diet yields a food energy equivalent of 1.25×10^7 J, how much work can that person do at 15 percent efficiency with a metabolic rate of 250 W before all of the food energy is used up?

11-44 If a 90-kg man exercises with a metabolic rate of 7.5 W kg^{-1}, how long must he continue to use up 1 kg of fat?

11-45 A 70-kg sprinter does work at a rate of 820 W during a bicycle dash lasting 11 s. If the efficiency is 20 percent and only carbohydrates are consumed, what mass of carbohydrates is used?

11-46 A 45-kg woman runs up a flight of stairs 5 m high in 3 s. (a) What is her mechanical power output? (b) If the woman's basal metabolic rate is 1 W kg^{-1} and she works with 10 percent efficiency, what is her metabolic rate when ascending the stairs? (c) What is the total amount of oxygen consumed by the woman during the climb?

11-47 If the woman of Problem 11-46 descends the stairs, her change in potential energy is negative. Is any metabolic energy required in this process? Explain.

11-48 In Chapter Eight the buckling strength scaling model led to the conclusion that the metabolic rate of an animal of mass m should vary as $m^{0.75}$. This also means that the metabolic rate per unit mass varies as $m^{0.75}/m = m^{-0.25}$. The basal metabolic rate of a 60-kg man is 1.2 W kg^{-1}. What is the basal metabolic rate per unit mass of a 960-kg horse?

11-49 Using the scaling law described in Problem 11-48, what is the expected basal metabolic rate of a 6400-kg elephant if that of a 0.04-kg mouse is 0.3 W?

11-50 A hummingbird requires an expenditure of 0.06 W to hover. The measured oxygen consumption rates for a hummingbird at rest and hovering are 5×10^{-6} litre s^{-1} and 35×10^{-6} litre s^{-1}, respectively. What is the efficiency of the hummingbird while hovering?

11-51 A dieter consumes 10,500 kJ or 2500 kcal per day and uses 12,600 kJ per day. If the deficit is made up by the use of stored body fat, in how many days will the dieter lose 1 kg?

Additional Reading

W.F. Magie, *A Source Book in Physics,* McGraw-Hill Book Co., New York, 1935, pp. 196–211. Excerpts from papers by Mayer and Joule.

V.V. Raman, Where Credit is Due—The Energy Conservation Principle, *Physics Teacher,* vol. 13, 1975, p. 80.

T.C. Ruch and H.D. Patton (eds.), *Physiology and Biophysics,* vol. 3, W.B. Saunders Co., Philadelphia, 1973. Chapter 5 by Arthur C. Brown discusses human metabolism.

George B. Benedek and Felix M.H. Villars, *Physics: With Illustrations from Medicine and Biology,* vol. I, Addison-Wesley Publishing Co., Reading, Mass., 1973, pp. 5–115. Human metabolism.

Knut Schmidt-Nielsen, *Animal Physiology,* 3rd ed., Prentice-Hall, Inc., Englewood Cliffs, N.J., 1970. Human and animal metabolism.

I. Prigogine, G. Nicolis, and A. Babloyantz, Thermodynamics of Evolution, *Physics Today,* November 1972, p. 23; December 1972, p. 38. Difficult but interesting articles.

Beverly Karplus Hartline, Tapping Sun-Warmed Ocean Warmed Water for Power, *Science,* vol. 209, 1980, p. 794. The first demonstration that power can actually be generated from ocean thermal differences.

The Science-and-Art-of Keeping Warm, *Natural History,* vol. 90, October 1981. A special issue about how people, plants, and animals survive cold weather.

John Tierney, Perpetual Commotion, *Science 83,* May, p. 30. Thermodynamics and perpetual motion.

Ronald Giedd, Real Otto and Diesel Engine Cycles, *The Physics Teacher,* vol. 21, 1983, p. 29.

Scientific American articles:

Mitchell Wilson, Count Rumford, October 1960, p. 158.

J.W.L. Köhler, The Stirling Refrigeration Cycle, April 1965, p. 119.

Graham Walker, The Stirling Engine, August 1973, p. 80.

C.M. Summers, The Conversion of Energy, September 1971, p. 148.

Lynwood Bryant, Rudolf Diesel and His Rational Engine, August 1969, p. 108.

Freeman Dyson, Energy in the Universe, September 1971, p. 50.

W. Ehrenberg, Maxwell's Demon, November 1967, p. 103.

Stanley W. Angrist, Perpetual Motion Machines, January 1968, p. 114.

O.V. Lounasmaa, New Methods for Approaching Absolute Zero, December 1969, p. 26.

John R. Clark, Thermal Pollution and Aquatic Life, March 1969, p. 18.

James B. Kelley, Heat, Cold and Clothing, February 1956, p. 107.

R.E. Newell, The Circulation of the Upper Atmosphere, March 1964, p. 62.

C.B. Chapman and J.H. Mitchell, The Physiology of Exercise, May 1965, p. 88.

J.R. Brett, The Swimming Energetics of Salmon, August 1965, p. 80.

John W. Kanwisher and Sam H. Ridgway, The Physiological Ecology of Whales and Porpoises, June 1983, p. 110.

Vance A. Tucker, The Energetics of Bird Flight, May 1969, p. 70.

David M. Gates, The Flow of Energy in the Biosphere, September 1971, p. 88.

Bernd Heinrich, The Energetics of the Bumblebee, April 1973, p. 97.

Eugene S. Ferguson, The Measurement of the Man-Day, October 1971, p. 96.

Rudolfo Margaria, The Sources of Muscular Energy, March 1972, p. 84.

Salvador E. Luria, Colicins and the Energetics of Cell Membranes, December 1975, p. 30.

Jules Janick, Carl H. Noller, and Charles L. Rhykerd, The Cycles of Plant and Animal Nutrition, September 1976, p. 74.

S.S. Wilson, Sadi Carnot, August 1981, p. 134.

The September 1970 issue of *Scientific American* deals with *The Biosphere,* and has several articles relating to energy cycles and life.

CHAPTER 12

THERMAL PROPERTIES OF MATTER

In Chapter Ten, we described the relationship between the temperature of a substance and the average energy of its molecules, using an ideal gas as an example. The relationship connecting this internal energy, the thermal energy transferred, and the work done by the system was discussed in Chapter Eleven. Most of these discussions make little reference to specific materials. Here we focus on how several properties of materials vary with temperature and how materials transfer thermal energy. In each case, account is taken of the nature of the material involved.

Because the properties of matter depend on temperature, the exchange of thermal energy is extremely important. Since biological process can only function properly over a small temperature range, both people and nature have devised ways of either limiting or improving the means by which this energy is transferred. For example, we insulate our homes, and our bodies respond to high temperatures by perspiring.

We begin our study of the thermal properties of matter with the thermal expansion of solids and liquids. We then discuss how the temperature of an object changes as thermal energy or *heat* is added or withdrawn. This leads us to a discussion of the transfer of thermal energy, and we use our ideas to study temperature control in warm-blooded animals.

12.1 | THERMAL EXPANSION

When a substance is heated, its volume usually increases, and each dimension increases correspondingly. This increase in size can be understood in terms of the increased kinetic energy of the atoms or molecules. The additional kinetic energy results in each molecule colliding more forcefully with its neighbors. The molecules effectively push each other further apart and the material expands.

At the macroscopic level, we can find a convenient relation between the change in length of an object and the temperature change. Suppose the original length of the object in Fig. 12.1 is l, and a small increase in length Δl occurs when the temperature increases by a small amount ΔT. If we divide the object into two equal parts, each part will have length $l/2$ and will expand by $\Delta l/2$. Consequently, the change in length Δl is directly proportional to the length l. In addition, we find from experiment that if we double the temperature change by raising the temperature by $2\,\Delta T$, the expansion also doubles. A single equation expressing both proportionalities is

$$\Delta l = \alpha l\,\Delta T \qquad (12.1)$$

The constant α is the *coefficient of linear expansion*. It is a property of a given material and depends somewhat on the temperature. Typical values of α are given in Table 12.1. We see that α has the dimensions of inverse temperature, so its units are K^{-1}. *Since only the change in temperature is important, we can measure ΔT in Kelvins or in Celsius degrees.* We use this fact throughout this chapter.

Since α depends somewhat on the temperature, the proportionality of the length and temperature changes is exact only for infinitesimal changes. Thus Eq. 12.1 should be replaced by

$$dl = \alpha l\,dT$$

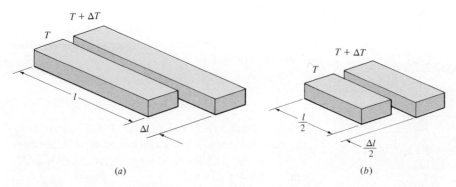

Figure 12.1. The expansion of a bar is proportional to its length. When the bar is halved in length, the expansion is also halved.

However, Eq. 12.1 is often quite accurate if an average value of α is used even when ΔT is 100 K or more. We assume the validity of this approximation in our examples and problems.

The following example illustrates the importance of thermal expansion.

Example 12.1

The roadbed of the Golden Gate Bridge is 1280 m long. During a certain year the temperature varies from $-12°$ C to $38°$ C. What is the difference in the lengths at those temperatures if the road is supported by steel girders? For steel, $\alpha = 1.27 \times 10^{-5}$ K^{-1}.

With $\Delta T = 38°$ C $- (-12°$ C$) = 50°$ C $= 50$ K,

$$\Delta l = \alpha l \, \Delta T$$
$$= (1.27 \times 10^{-5} \text{ K}^{-1})(1280 \text{ m})(50 \text{ K})$$
$$= 0.81 \text{ m}.$$

This substantial change in the roadbed length must be allowed for in the design of the bridge. If the structure could not alter its length with changes in temperature,

huge forces would be developed and severe damage would result.

Two interesting applications of thermal expansion are the *thermostat* and the *ultramicrotome*. A thermostat has two metal strips with different coefficients of linear expansion attached to each other (Fig. 12.2). When they are heated the unequal expansion causes the strips to bend; if they bend far enough, they open or close a switch that may, for example, control a heating system or air conditioning.

The *ultramicrotome* is an apparatus designed to make very thin tissue slices for use in microscopes. A sample, mounted on a rotating metal arm, passes a knife edge (Fig. 12.3). If the metal arm is heated at a constant rate, it will expand uniformly, and a thin specimen slice will be cut on each turn. The metal arm can be extended as slowly as 1 micrometre (10^{-6} m) per minute.

Area and Volume Expansion | When an object is heated, all of its dimensions increase (Fig.

TABLE 12.1

Coefficient of linear expansion for various materials

Material	Temperature (°C)	α (K^{-1})
Aluminum	-23	2.21×10^{-5}
	20	2.30×10^{-5}
	77	2.41×10^{-5}
	527	3.35×10^{-5}
Diamond	20	1.00×10^{-6}
Celluloid	50	1.09×10^{-4}
Glass (most types)	50	$8.3 \ \times 10^{-6}$
Glass (Pyrex)	50	$3.2 \ \times 10^{-6}$
Ice	-5	5.07×10^{-5}
Steel	20	1.27×10^{-5}
Platinum	20	$8.9 \ \times 10^{-6}$

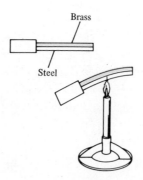

Figure 12.2. Brass and steel strips attached to each other expand at different rates when heated. ($\alpha_{\text{brass}} > \alpha_{\text{steel}}$).

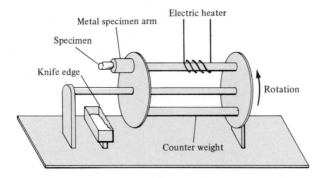

Figure 12.3.   A schematic view of an ultramicrotome. The mechanism for turning the holder is not shown.

12.4). Thus the lengths of the sides increase, the areas of surfaces increase, and the volume increases.

Consider what happens to a square surface of a cube with an initial area $A = l^2$. If a temperature increase dT changes l by $dl = \alpha l dT$, then the area expands by

$$dA = \frac{dA}{dl}dl = 2ldl$$

$$= 2l(\alpha l dT) = 2\alpha l^2 dT$$

Using $A = l^2$, the change in area is $dA = 2\alpha A dT$. For a finite temperature change ΔT, the change in area is approximately

$$\Delta A = 2\alpha A\, \Delta T \qquad (12.2)$$

Thus the area increases at twice the rate of a linear dimension. Note that although we obtained Eq. 12.2 by considering a square area, it is correct for any shape. We use this result in the following example.

Example 12.2

A circular steel disk has a circular hole through its center. If the disk is heated from 10° C to 100° C, what is the fractional increase in the area of the hole?

When the disk is heated, all of its dimensions increase. The area of the hole increases also, just as the area inside

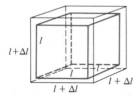

Figure 12.4.   When the temperature is increased, each dimension of a cube changes from l to $l + \Delta l$.

a penciled circle on a solid disk increases. The area of the hole increases as if it were filled with steel. Using Eq. 12.2 and α from Table 12.1, the fractional increase in the area of the hole is

$$\frac{\Delta A}{A} = 2\alpha\, \Delta T = 2(1.27 \times 10^{-5}\ \mathrm{K^{-1}})(90\ \mathrm{K})$$

$$= 2.29 \times 10^{-3}$$

The rate at which the volume $V = l^3$ expands as the cube is heated is also determined by α. A temperature change dT causes a length change $dl = \alpha l dT$, and a volume change

$$dV = \frac{dV}{dl}dl = 3l^2 dl$$

$$= 3l^2(\alpha l dT) = 3\alpha l^3 dT$$

$$= 3\alpha V dT$$

We define the *coefficient of volume expansion* β by

$$dV = \beta V dT$$

Comparing this with the equation above, we have

$$\beta = 3\alpha$$

Thus, the volume increases three times as fast as the length. Again, for most applications it is adequate to use the approximation

$$\Delta V = \beta V \Delta T \qquad (12.3)$$

WATER | In discussing thermal expansion, special mention must be made of liquid water because it is one of the very few substances with a negative coefficient of volume expansion at some temperatures. Figure 12.5 shows both β and the density (mass per unit volume) of water plotted against the temperature. β varies as the temperature changes and even reverses sign at 3.98° C. Thus, as T rises from 0° C, water *contracts* up to 3.98° C and then expands as the temperature increases further; water has its greatest mass per unit volume at 3.98° C.

This characteristic of water is extremely important for aquatic life. As the air temperature decreases in early winter, the surface water of lakes cools. When this surface water reaches 3.98° C, it sinks to the bottom; the warmer, less dense water from beneath floats to the surface. The cool descending water carries oxygen with it. Once the entire lake has undergone this mixing and has reached 3.98° C, further cooling occurs at the surface and ice forms. The lower

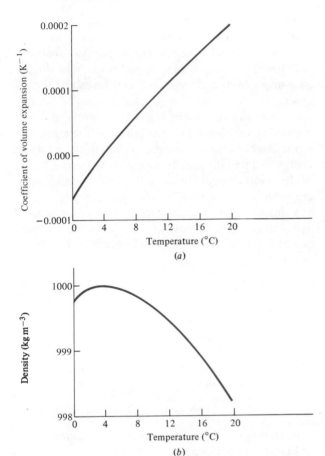

Figure 12.5. (a) Coefficient of volume expansion for water. (b) The density (mass per unit volume) of water.

density ice floats, so that lakes freeze from the surface downward. Aquatic life survives the winter in the freshly oxygenated water beneath the ice.

12.2 | HEAT CAPACITY

When an object at one temperature is placed near or in contact with another object at a higher temperature, energy is transferred to the cooler object, and its temperature rises. The ratio of the amount of energy transferred to the temperature change is called the *heat capacity.*

When energy transfer occurs because of a temperature difference, we say that thermal energy or *heat* is being transferred. Alternatively, energy may be transferred to a substance by doing work on the substance as, for example, by stirring a liquid or compressing a gas.

Suppose a small amount of heat ΔQ is transferred to n moles of a substance. We know from the first law of thermodynamics that the internal energy will increase or the substance will do work, or possibly both will happen. The *molar heat capacity C* is defined as

$$C = \lim_{\Delta T \to 0} \frac{1}{n} \frac{\Delta Q}{\Delta T} = \frac{1}{n} \frac{dQ}{dT} \qquad (12.4)$$

That is, C is the ratio of the heat added per mole to the rise in temperature. Substances such as water, which have a high molar heat capacity, experience relatively small temperature changes when a given amount of heat is transferred.

In practice the heat capacity is usually measured under one of two special conditions. If heat is added with the *volume* of the substance kept *constant,* no work is done. Then the heat dQ added is equal to the internal energy change, dU, and the molar heat capacity at constant volume is

$$C_V = \frac{1}{n} \frac{dU}{dT} \qquad \text{(constant volume)} \qquad (12.5)$$

More often, the heat capacity is measured at *constant pressure.* Here, when heat is added, there is an increase in internal energy and also some work is done by the substance. The heat capacity at constant volume is easiest to calculate theoretically, while the heat capacity at constant pressure is easiest to measure. However, for the special case of an ideal monatomic gas, we can easily derive both results.

We found in Chapter Ten that the average kinetic energy per molecule in an ideal monatomic gas is

$$(K)_{\text{ave}} = \frac{3}{2} \frac{R}{N_A} T$$

R is the gas constant, and N_A is Avogadro's number. For n moles or nN_A molecules, the internal energy U is the sum of the kinetic energies of the molecules $U = nN_A(K)_{\text{ave}} = 3nRT/2$. If the gas does not change its volume as the heat is added, and the temperature increases by dT, the change in internal energy is $dU = 3nR\,dT/2$. Hence the molar heat capacity at constant volume is

$$C_V = \frac{1}{n} \frac{dU}{dT} = \frac{3R}{2} \qquad \text{(ideal monatomic gas)} \qquad (12.6)$$

If the pressure, not the volume, is held constant when heat dQ is added to a substance, the internal energy increases and some work is done; $dQ = dU + W$. Using the ideal gas law, $PV = nRT$, the work done at constant pressure when the temperature changes by dT is $W = P\,dV = nR\,dT$. Thus, $dQ = dU + nR\,dT$, and the molar heat capacity at constant pressure is

$$C_P = \frac{1}{n}\frac{dQ}{dT} = \frac{1}{n}\frac{dU}{dT} + R$$

But $dU/n\,dT$ is C_V, so

$$C_P = C_V + R = \frac{5R}{2} \quad \text{(ideal monatomic gas)} \quad (12.7)$$

This result is in very good agreement with experiments on real monatomic gases at low to moderate densities. The relationship $C_P = C_V + R$ also agrees well with measurements on real polyatomic gases.

While we have been able to calculate the heat capacity of a monatomic gas from first principles, the problem of calculating the energy required to raise the temperature of a polyatomic gas, a liquid, or a solid is more difficult. For example, we noted that for polyatomic molecules the rotational and vibrational energies must be included in the internal energies. Modern theory predicts that at low temperatures the molar heat capacity of any gas results entirely from the translational motion, and C_V is $\frac{3}{2}R$. At intermediate temperatures rotations also contribute and C_V is $\frac{5}{2}R$ for a diatomic gas. At high temperatures, vibrational, rotational, and translational motions all contribute to the specific heat, and C_V is $\frac{7}{2}R$ for a diatomic gas. The molar heat capacity of hydrogen gas H_2 versus temperature is shown in Fig. 12.6, where the three contributions are clearly seen.

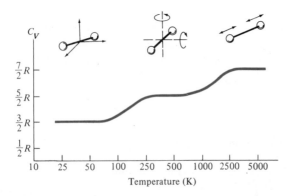

Figure 12.6. Molar heat capacity at constant volume of hydrogen gas H_2. At low temperatures the heat capacity is due to the translational kinetic energy. At higher temperatures the rotational kinetic energy must be included, and the vibrational kinetic energy is important at still higher temperatures. Note that the temperature scale is logarithmic.

As a gas is cooled and becomes a liquid, further difficulties in making accurate calculations are encountered. For example, the heat capacity of steam is only half as large as that of liquid water. The difference arises from a change in the effects of the forces among molecules. In the gas, the molecules are far apart, and the forces are negligible. In a liquid, the molecules are closer to each other, so their mutual attraction becomes significant, adding to the internal energy and making specific heat calculations very difficult. Oddly enough, in the solid form, specific heats are again relatively easy to calculate.

A quantity closely related to the molar heat capacity is the *specific heat capacity, c*. This is the heat required for a unit temperature change in a unit mass of a substance. It is related to C by

$$c = \frac{C}{M} \quad (12.8)$$

Here M is the mass of 1 mole. For example, helium gas has a molecular mass of 4 u; 1 mole is 4 g = 4×10^{-3} kg, and $C_V = 12.47$ J mole^{-1} K^{-1}. Thus,

$$c_V = \frac{12.47 \text{ J mole}^{-1}\text{ K}^{-1}}{4 \times 10^{-3}\text{ kg mole}^{-1}}$$
$$= 3.12 \times 10^3 \text{ J kg}^{-1}\text{ K}^{-1}$$

In terms of c, the heat required for a temperature change ΔT in a mass m is

$$\Delta Q = mc\,\Delta T \quad (12.9)$$

Since c depends on the temperature, this equation is exact only for very small ΔT. However, like the thermal expansion formula, it is usually adequate for large values of ΔT if the average specific heat capacity is used. Specific heat capacities of representative substances are given in Table 12.2.

A simple but effective way of measuring heat capacities at constant pressure uses a *calorimeter* (Fig. 12.7). Heat ΔQ is supplied by an electrical heater to a well-insulated calorimeter that holds the sample, and a thermometer measures the temperature rise ΔT. A sample with mass m and specific heat capacity c absorbs an amount of heat equal to $mc\,\Delta T$. The container also absorbs some heat; if its mass is m_c and its specific heat capacity is c_c, this heat equals $m_c c_c\,\Delta T$. Adding the heat absorbed by the sample and the container, we have

$$\Delta Q = mc\,\Delta T + m_c c_c\,\Delta T \quad (12.10)$$

This result may be used to find the specific heat capacity as illustrated in the next example.

TABLE 12.2

Specific heat capacities at constant pressure at 25°C, except where noted, of various substances in kJ kg^{-1} K^{-1}

Substance	Specific Heat Capacity, c_P
Aluminum	0.898
Steel	0.447
Diamond	0.518
Lead	0.130
Copper	0.385
Helium (gas)	5.180
Hydrogen (H_2) (gas)	14.250
Iron	0.443
Nitrogen (N_2) (gas)	1.040
Oxygen (O_2) (gas)	0.915
Water (liquid)	4.169
Ice ($-10°$ to $0°C$)	2.089
Steam ($100°$ to $200°C$)	1.963

Example 12.3

There are 0.1 kg of carbon in a calorimeter at 15° C. The container has a mass of 0.02 kg and is made of aluminum. The addition of 0.892 kJ of heat energy brings the temperature to 28° C. What is the specific heat capacity of carbon? Assume the specific heat capacity of aluminum in this temperature range is 0.9 kJ kg^{-1} K^{-1}.

Substituting in Eq. 12.10, with $\Delta T = 28° C - 15° C = 13$ K.

$$c = \frac{\Delta Q - m_c c_c \, \Delta T}{m \, \Delta T}$$

$$= \frac{0.892 \text{ kJ} - (0.02 \text{ kg})(0.9 \text{ kJ kg}^{-1} \text{ K}^{-1})(13 \text{ K})}{(0.10 \text{ kg})(13 \text{ K})}$$

$$= 0.506 \text{ kJ kg}^{-1} \text{ K}^{-1}$$

When the specific heat capacities are known, the same ideas can be used to predict the final temperature, as shown in the next example.

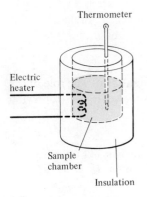

Figure 12.7. A calorimeter.

Example 12.4

A copper pipe of mass 0.5 kg is originally at 20° C. If its ends are capped after 0.6 kg of water at 98° C is poured into it, what is the final temperature of the pipe? (Assume the pipe is insulated so no heat is lost to the surroundings.)

Thermal energy is transferred from the water to the pipe until both are at the same temperature, T_f. The heat transferred to the pipe is its mass times its specific heat capacity times its temperature change, which is $T_f - 20°$ C. The heat lost by the water is its mass times specific heat capacity times its temperature change, $98°$ C $- T_f$. No heat enters or leaves through the insulation. Thus, using Table 12.2,

$$(0.5 \text{ kg})(0.385 \text{ kJ kg}^{-1} \text{ K}^{-1})(T_f - 20° \text{ C})$$
$$= (0.6 \text{ kg})(4.169 \text{ kJ kg}^{-1} \text{ K}^{-1})(98° \text{ C} - T_f)$$

Multiplying this out and solving for T_f, we find

$$T_f = 92.43° \text{ C}$$

In our discussion of heat capacities based on the ideal gas model, the relationship between heat and energy is quite clear. Historically, scientists studied heat long before any molecular theory was developed. Since it was not realized that heat is just another form of energy, a separate set of units was developed to measure heat. The gram-calorie (cal) was defined as the heat required to heat 1 g of water from 14.5° C to 15.5° C. Later, it was discovered that the energy represented by 1 calorie is equivalent to 4.18 J. The first suggestion of this equivalence was made in 1842 by Julius Robert Mayer (1814–1878), a German physician, on the basis of physiological observations. Mayer also suggested that energy can be neither created nor destroyed. His work was largely ignored, and it was not until the work of James Prescott Joule (1818–1889) several years later that the equivalence of heat and energy was accepted. The principle of the conservation of energy was also independently proposed by Hermann von Helmholtz (1821–1894).

Although 125 years have elapsed in which scientists could have adopted a common energy unit for both heat and work, it is still common to measure mechanical and electrical energies in joules and heat energies in calories. To make matters worse, American engineers frequently use the British thermal unit (BTU), defined as the heat required to heat 1 lb of water from 63° F to 64° F, and the kilowatt-hour (kWh), which is equal to 3.6×10^6 J. Further, nutritionists measure the energy supplied by food in kilocalories (kcal), which are 10^3 times as large as the gram-calorie. Conversion factors for these energy units may be found on the inside of the front cover of this book.

12.3 | PHASE CHANGES

Most substances can exist in solid, liquid, or gas *phases*. For example, water may be ice, liquid, or steam. A transition from one of these phases to another is called a *phase change*. Many other kinds of phase changes occur in nature. A solid may change from one crystalline structure to another; at low temperatures, a material may become magnetic or lose its electrical resistance. All these changes take place very abruptly at a sharply defined temperature.

The temperature at which a phase change occurs usually depends on additional variables, such as pressure. This is illustrated by the *phase diagram* for water (Fig. 12.8). At the temperature and pressure corresponding to point *A*, water can exist only as ice. If we keep the pressure fixed and add heat, the temperature rises until point *B* is reached. Now as more heat is added the temperature does not rise. Instead, the ice gradually melts into water, and the *temperature remains constant until all the ice is melted* (Fig. 12.9). As more heat is added, the temperature of the liquid again steadily increases until point *D* is reached. *Here again the temperature remains constant* until all the liquid has been converted into water vapor. Additional heat will now increase the temperature of the gas.

If we repeat this experiment at a lower pressure, the phase changes occur at *B'* and *D'*, which corre-

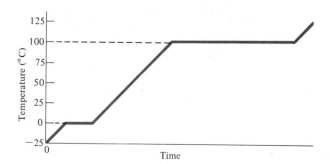

Figure 12.9. The temperature of a sample of water versus time. Heat is added at a uniform rate and the pressure is held constant at one atmosphere.

spond to different temperatures than do *B* and *D*. If we repeat the experiment once more at an even lower pressure, we find that ice *sublimes* at *E* directly into vapor without passing through the liquid phase.

Sublimation has become an important part of preserving food. Food is frozen and placed in a low-pressure enclosure. As heat is added, the ice sublimates and the vapor is drawn off. This freeze-drying process does not damage the food and preserves its shape and taste. The food can later be reconstituted by adding water.

Two points with special significance are noted on the diagram. At the *triple point*, liquid, solid, and vapor may all exist together. At the *critical point*, a kilogram of liquid and a kilogram of gas have the same volume, and the distinction between the two phases vanishes. Then if the pressure and temperature are adjusted so that a sample passes from point *F* to *F'* along the path shown in Fig. 12.8, the change of phase from liquid to vapor is never observed.

The energy absorbed or liberated in a phase change is called the *latent heat*. At atmospheric pressure, the *latent heat of fusion* L_f needed to melt ice is 333 kJ kg^{-1}. The *latent heat of vaporization* L_v needed to boil water at atmospheric pressure is 2255 kJ kg^{-1}. Some other latent heats are given in Table 12.3. The heat ΔQ needed to change the phase of a mass m is

$$\Delta Q = Lm \qquad (12.11)$$

where L is the appropriate latent heat. The role of the latent heat of fusion is illustrated in the following example.

Example 12.5

How much heat is required to melt 5 kg of ice at 0° C?

Since the latent heat of fusion L_f is 333 kJ kg^{-1}, $Q = Lm$ gives

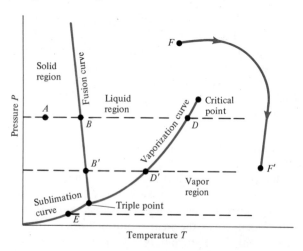

Figure 12.8. Phase diagram for water. The lines indicate the temperature and pressures at which phase changes occur.

TABLE 12.3
Latent heats at atmospheric pressure

Substance	Melting Point (°C)	Latent Heat of Fusion (kJ kg^{-1})	Boiling Point (°C)	Latent Heat of Vaporization (kJ kg^{-1})
Helium			−268.9	21
Nitrogen	−209.9	25.5	−195.8	201
Ethyl alcohol	−114	104	78	854
Mercury	−39	11.8	357	272
Water	0	333	100	2255
Silver	96	88.3	2193	2335
Lead	327	24.5	1620	912
Gold	1063	64.4	2660	1580

$$Q = L_f m = (333 \text{ kJ kg}^{-1})(5 \text{ kg}) = 1665 \text{ kJ}$$

The latent heat and specific heat capacity are both important in determining the equilibrium state of a system. This is shown in the following example.

Example 12.6

If 20 kg of water at 95° C is mixed with 5 kg of ice at 0° C, what is the final temperature of the mixture?

In this example it is useful to determine first whether enough thermal energy is available in the water to melt the ice. If not, the equilibrium temperature will be 0° C with only a portion of the ice melting. From the previous example, we know that 1665 kJ is required to melt 5 kg of ice. If the water is cooled to 0° C, it gives up

$$\Delta Q = mc_P \Delta T = (20 \text{ kg})(4.169 \text{ kJ kg}^{-1} \text{ K}^{-1})(95 \text{ K})$$
$$= 7921 \text{ kJ}$$

This is more than enough to melt the ice so the final temperature will be above 0° C.

If the final temperature is T_f, the heat transferred to the ice is $1665 \text{ kJ} + m_i c_P (T_f - 0° \text{ C})$. That given up by the water is $mc_P(95° \text{ C} - T_f)$, so equating these

$$1665 \text{ kJ} + m_i c_P(T_f - 0° \text{ C}) = mc_P(95° \text{ C} - T_f)$$

Substituting $m_i = 5 \text{ kg}$, $m = 20 \text{ kg}$, and $c_P = 4.169 \text{ kJ kg}^{-1} \text{ K}^{-1}$, multiplying, and solving for T_f, we find $T_f = 60.0° \text{ C}$.

The following example shows how to find the mass of ice melted when the final temperature is 0° C.

Example 12.7

A 0.6-kg pitcher of tea at 50° C is cooled with 0.4 kg of ice cubes at 0° C. What is the equilibrium condition if no heat is lost to the surroundings?

The heat needed to melt all of the ice is $\Delta Q = L_f m_i = (333 \text{ kJ kg}^{-1})(0.4 \text{ kg}) = 133.2 \text{ kJ}$. The heat given up by the tea if it is cooled to 0° C is $\Delta Q = (0.6 \text{ kg})(4.169 \text{ kJ kg}^{-1} \text{ K}^{-1})(50 \text{ K}) = 12.51 \text{ kJ}$, not enough to melt all the ice. Thus some of the ice will not melt and the final temperature will be 0° C.

To find the mass of the melted ice, we equate the heat loss of the cooled tea to the heat needed to melt a mass m_1 of ice, $125.1 \text{ kJ} = m_1 L_f$. Thus, with $L_f = 333 \text{ kJ kg}^{-1}$, the mass of ice melted is

$$m_1 = \frac{125.1 \text{ kJ}}{333 \text{ kJ kg}^{-1}} = 0.376 \text{ kg}$$

12.4 | HEAT CONDUCTION

Heat transfer always occurs from regions of higher to regions of lower temperature, so that two objects isolated from their surroundings gradually approach a common temperature. In this section, we discuss heat *conduction* between objects in contact. Heat transfer by *convection* and *radiation* is considered in the following two sections.

When two objects at temperatures T_1 and T_2 are connected by a rod, their temperature difference $\Delta T = T_2 - T_1$ will steadily diminish (Fig. 12.10). If we cut the rod lengthwise so that each section has a cross section of $A/2$, then half the heat flow will be conducted in each section. Thus the *rate* at which heat flows from the hotter to the colder object must be proportional to the cross-sectional area A.

The rate of heat flow also depends on ΔT and the length l. If we double ΔT and double the length at the same time, the heat flow is unchanged. If the temperature difference is doubled and the length is unchanged, the heat flow doubles. The same thing happens if ΔT is unchanged and the length is halved. Thus the heat flow must depend on the ratio $\Delta T/l$,

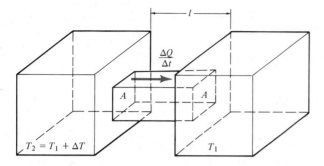

Figure 12.10. A rod of cross-sectional area A and length Δl conducts heat from the higher to the lower temperature object.

which is called the *temperature gradient*. We can summarize these observations algebraically in an equation for the *heat flow* $H = \Delta Q/\Delta t$,

$$H = \kappa A \frac{\Delta T}{l} \qquad (12.12)$$

Here κ is a proportionality constant called the *thermal conductivity*. Equation 12.12 is exact when ΔT is very small. However, the discussion of heat flow becomes more complicated if κ varies with temperature or when the geometry is not so simple. In this chapter, we assume that κ is constant in the temperature range of interest.

Representative values of κ are given in Table 12.4. Typically, metals are good thermal conductors. Their conductivities are greater by factors of 10^3 or 10^4 than those of thermal insulators such as asbestos or rock wool.

One of the most important insulators is air. Insulation in homes and in the material for warm clothing utilize this fact. The fibers of the material trap air in the material, and this air acts as an insulator. Storm or double-pane windows use air trapped between two glass panes to reduce conductive heat losses.

Body tissue is also a good insulator. When the environment is warm, the interior body temperature is quite uniform (Fig. 12.11a). Because body tissues are poor conductors, the inner core of the body can be kept warm in a cold environment (Fig. 12.11b).

Example 12.8

A person walking at a modest speed generates heat at a rate of 280 W. If the surface area of the body is 1.5 m² and if the heat is assumed to be generated 0.03 m below the skin, what temperature difference between the skin and interior of the body would exist if the heat were conducted to the surface? Assume the thermal conductivity is the same as that for animal muscle, 0.2 W m⁻¹ K⁻¹.

Despite the dissimilarity between humans and the rod in Fig. 12.10, we may still apply Eq. 12.12 to a small section of tissue. Summing over the sections is approximately equivalent to using the total body surface area for the area A.

Solving $H = \kappa A \Delta T/l$ for ΔT,

$$\Delta T = \frac{lH}{\kappa A} = \frac{(0.03 \text{ m})(280 \text{ W})}{(0.2 \text{ W m}^{-1} \text{ K}^{-1})(1.5 \text{ m}^2)}$$

$$= 28 \text{ K}$$

TABLE 12.4

Thermal conductivities in W m⁻¹ K⁻¹

Substance	Thermal Conductivity, κ
Silver	420
Copper	400
Aluminum	240
Steel	79
Ice	1.7
Glass, concrete	0.8
Water	0.59
Animal muscle, fat	0.2
Wood, asbestos	0.08
Felt, rock wool	0.04
Air	0.024
Down	0.019

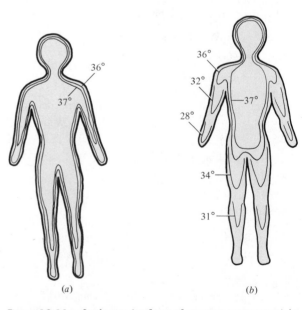

(a)   (b)

Figure 12.11. Isotherms (surfaces of constant temperature) in the body for (a) a warm environment and (b) a cold environment. (Adapted from Aschoff and Wever, *Naturwissenschaften*, vol. 45, p. 477, 1958.)

Since the actual temperature difference is only a few degrees (Fig. 12.11), we can conclude that heat is not removed from the body by conduction through tissue from the interior to the exterior of the body. In fact, the flow of warm blood from the interior of the body to the cooler exterior is the major factor in body heat transport.

Example 12.9

A copper hot water pipe is 2 m long and 0.004 m thick, and has a surface area of 0.12 m². If the water is at 80° C and the temperature in the room is 15° C, at what rate is heat conducted through the pipe wall?

If we assume provisionally that the outside surface of the pipe is at 15° C, then Eq. 12.12 yields

$$H = \kappa A \frac{\Delta T}{l} = (400 \text{ W m}^{-1} \text{ K}^{-1})(0.12 \text{ m}^2) \frac{(65 \text{ K})}{(0.004 \text{ m})}$$

$$= 780{,}000 \text{ W}$$

This is a huge heat loss rate, much larger than is actually observed. As we will find, the air cannot carry off heat at this rate, nor can radiation, and the outer surface of the pipe is actually at a temperature much higher than 15° C. The resulting smaller temperature difference reduces the rate of conduction. This also corresponds to our experience, since if we touch a heating pipe the outer surface feels very hot.

The quantity l/κ in Eq. 12.12 is sometimes called the "R-value." Using this definition, the heat flow can be written as

$$H = \frac{A \Delta T}{R} \tag{12.13}$$

For example, a 3-cm-thick section of rock wool has

$$R = \frac{l}{\kappa} = \frac{(0.03 \text{ m})}{(0.04 \text{ W m}^{-1} K^{-1})} = 0.75 \text{ m}^2 \text{ K W}^{-1}$$

It can be shown that R-values are additive (Problem 12-68). For example, if the thickness of a wall is doubled, the R-value is doubled, and the heat loss through the wall is halved. If a second layer of attic insulation is placed over the original insulation, the net R-value is the sum of the two R-values. Recommended R-values have steadily increased as energy costs have risen.

12.5 | HEAT TRANSFER BY CONVECTION

Although some heat is transported by conduction in liquids and gases, a much larger quantity may be carried by the motion of the fluid itself. This is called

convection. In Fig. 12.12a, the liquid near the heat source is heated and expands slightly, becoming lighter than the overlying cooler fluid. It then rises and is replaced by cooler, heavier fluid. When the warmer fluid arrives at the cooler region of the container, it cools, contracts, and begins to sink again. Had the container been heated from the top, convection would not have occurred, and the bulk of the fluid would have been heated by the much slower conduction process.

A hot water or steam radiator provides another illustration of convection (Fig. 12.12b). Air near the radiator is heated and rises, while air near the outside walls and the windows is cooled and sinks. This establishes the flow pattern shown.

There are many difficulties in developing a quantitative theory of convection. For example, a given surface loses heat slower when it is vertical than when it is horizontal. Despite such difficulties, we can make some progress using an approximate formula. In still air the rate of convective heat transfer for a surface area A is given approximately by the empirical formula,

$$H = qA \Delta T \tag{12.14}$$

Here ΔT is the temperature difference between the surface and the air distant from the surface. The *convective heat transfer constant* q depends on the shape and orientation of the surface and to some extent on ΔT. For a naked human, we use the average value $q = 7.1 \text{ W m}^{-2} \text{ K}^{-1}$. Heat loss by convection is important for humans, as shown in the following example.

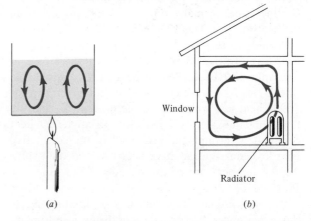

Window

Radiator

(a)

(b)

Figure 12.12. Paths of convective flow in (a) a container of liquid; (b) in a room heated by a radiator.

Example 12.10

In a warm room, a naked resting person has a skin temperature of 33° C. If the room temperature is 29° C and the body surface area is 1.5 m², what is the rate of heat loss due to convection?

Using $q = 7.1$ W m⁻² K⁻¹, we have

$$H = qA \,\Delta T$$
$$= (7.1 \text{ W m}^{-2} \text{ K}^{-1})(1.5 \text{ m}^2)(33° \text{ C} - 29° \text{ C})$$
$$= 43 \text{ W}$$

A resting person in this situation will generate heat at about twice this rate. Thus, under these moderate conditions, convection provides the mechanism for about 50 percent of the body's heat loss. If there is a breeze or if the room temperature is lower, convective heat losses will increase correspondingly.

We can return to the question of the water-filled pipe to gain some insight into the interplay between conduction and convection. In Example 12.9 we found that conduction from a heating pipe would be enormous if the outside of the pipe was at room temperature. Suppose we assume instead that the outside of the pipe is at the water temperature, 80° C. The maximum convective heat loss would then be, using a typical value of $q = 9.5$ W m⁻² K⁻¹ for the pipe,

$$H = qH \,\Delta T = (9.5 \text{ W m}^{-2} \text{ K}^{-1})(0.12 \text{ m}^2)(65 \text{ K})$$
$$= 74.1 \text{ W}$$

This is smaller than the conductive heat loss estimated in Example 12.9 by a factor of more than 10⁴.

Radiation, discussed in the next section, transfers a comparable amount of heat (Exercise 12-39). Thus we can conclude that radiation and convection are limiting factors in the heat loss from any metal hot water or steam heating element. In most baseboard heating systems, many thin metal fins project from the heating pipe. This effectively increases the surface area and hence the rate of convective and radiative heat loss. On the other hand, the section of pipe in Example 12.9 is still losing heat at an appreciable rate. If this pipe is used to transport hot water, considerable heat loss will occur in the process. To analyze this problem exactly, it is necessary to equate the heat conduction through the pipe wall to the convective heat loss plus the radiative heat loss and then solve for the temperature of the outside of the pipe (see Problem 12-67).

Convection plays a significant role in many everyday experiences. In a warm room the air a centimetre or so away from a cold windowpane feels rather cold.

The air in contact with the glass has the same temperature as the pane; the air temperature rises perceptibly for some distance moving away from the window (Fig. 12.13). A similar gradual decrease in air temperature occurs outside if there is no wind. However, because the conduction of heat across the glass is very efficient, only a small temperature difference exists across the window.

If there is a wind, the nearly stagnant layer of warm air just outside the window is removed more rapidly than when convective forces alone are acting. This results in a lower temperature at the outside window surface and hence increases the rate of heat loss from the window. This is equivalent to having a lower outside temperature and is referred to in weather reports as the *wind chill factor*. The *effective temperature* decreases rapidly as the wind velocity increases, as shown in Table 12.5. Skiers and snowmobilers who create their own wind must be conscious of the potential hazards. Exposed flesh can freeze in about a minute at an effective temperature of −30° C, which can, for example, be achieved by moving at 40 km h⁻¹ in air at −10° C. Effective temperatures below −60° C are extremely dangerous, since freezing can occur in seconds.

Convection plays a major role in the determination of weather patterns. Moist warm air masses heated over bodies of water are relatively light and tend to rise. As they rise into regions of lower pressure, they expand, doing work in the process. According to the first law of thermodynamics, if an air mass does work when no heat enters it, the internal energy must decrease. Thus the temperature of the air mass drops, causing some of the moisture to condense into clouds

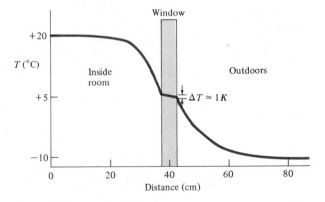

Figure 12.13. Air and glass temperatures near a window on a cold windless day. For clarity, the window thickness has been exaggerated.

TABLE 12.5

Effect of moving air on temperature

Wind Speed (km h⁻¹)	Actual Temperature, °C								
Calm	−10	−15	−20	−25	−30	−35	−40	−45	−50
	Effective Temperature, °C								
10	−15	−20	−25	−30	−35	−40	−45	−50	−55
20	−20	−25	−35	−40	−45	−50	−55	−60	−65
30	−25	−30	−40	−45	−50	−60	−65	−70	−75
40	−30	−35	−45	−50	−60	−65	−70	−75	−80
50	−35	−40	−50	−55	−65	−70	−75	−80	−85

▢ —Moderate danger   ▨ —Great danger

and release its heat of vaporization to the air mass. This slows the cooling of the air mass, permitting it to rise still further. Under some conditions this process will continue until large thunderclouds form and violent showers result.

12.6 | RADIATION

Conduction and convection require the presence of some material, be it solid, liquid, or gas. However, we know that heat can also be transmitted through a vacuum, since the sun's energy traverses millions of kilometres of space before reaching the earth. The process by which this occurs is called *radiation*. Radiant heat transfer also occurs in transparent media.

The term "radiation," as used in this chapter, is another word for electromagnetic waves. These are waves that are of electric and magnetic origin and carry energy. In a hot object the atomic charges oscillate rapidly, sending out energy as electromagnetic waves, somewhat like ripples on a pond. These waves travel with the speed of light, $c = 3 \times 10^8$ m s⁻¹. Visible light, radio waves, and X rays are all examples of electromagnetic waves (Fig. 12.14). The energy carried by the waves depends on the motion of the charges and hence on the temperature. These electromagnetic waves are discussed further in later chapters.

A wave is characterized by its *wavelength* λ and *frequency f*. The wavelength is the distance between successive wave crests; the frequency is the number

of crests passing a given point each second and is equal to the frequency of vibration of the charge producing the electromagnetic wave. The distance between crests, λ, times *f*, the number of crests per second passing a point, must equal the velocity of the wave, so

$$f\lambda = c \qquad (12.15)$$

For example, red light has a wavelength of about 7×10^{-7} m, which corresponds to a frequency

$$f = \frac{c}{\lambda} = \frac{3 \times 10^8 \text{ m s}^{-1}}{7 \times 10^{-7} \text{ m}} = 4.2 \times 10^{14} \text{ Hz}$$

In principle, every object at a nonzero temperature emits some radiation at all wavelengths. However,

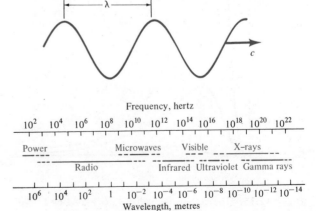

Figure 12.14. Frequencies and wavelengths of various kinds of electromagnetic waves. The scales are logarithmic.

the amount of energy radiated at each wavelength depends on the temperature (Fig. 12.15). An object at 800° C looks red since it emits some radiation in the longest wavelength or red portion of the visible spectrum but very little in the blue portion. An object heated to 3000° C looks white because it is emitting substantial amounts of radiation throughout the visible light range. Similarly, very hot stars appear relatively blue, while cooler ones are somewhat red.

The wavelength at which the radiation is most intense is given by the *Wien displacement law,*

$$\lambda = \frac{B}{T} \tag{12.16}$$

The constant B has a numerical value of 2.898×10^{-3} m K.

The temperature of the sun determines the wavelength at which most of its radiation reaches us. This is shown in the following example.

Example 12.11

The sun has a surface temperature of 6000 K. What is the wavelength of maximum radiation?

From the Wien displacement law,

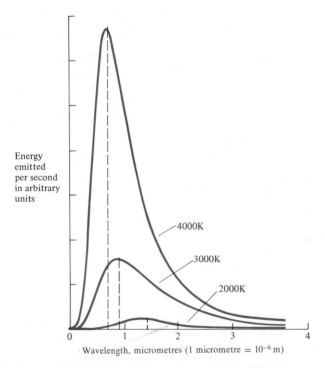

Figure 12.15. The heat energy radiated per second by 1 m² of surface. As the temperature increases. λ_{max} shifts to shorter wavelengths (vertical dashed lines), and the total energy radiated increases rapidly.

$$\lambda_{max} = \frac{2.898 \times 10^{-3} \text{ m K}}{6000 \text{ K}} = 4.83 \times 10^{-7} \text{ m}$$

Thus the maximum solar radiation is in the visible portion of the spectrum as noted earlier. It is no accident that the eyes of animals are most sensitive very near the wavelength at which there is the most radiation available for seeing.

This variation of λ_{max} with temperature has many extremely important consequences. One is commonly called the *greenhouse effect.* In a greenhouse, the incident radiation from the sun is most intense at $\lambda_{max} = 4.83 \times 10^{-7}$ m and readily passes through glass. Inside, the radiation is absorbed by objects that in turn reemit radiation. However, since the average temperature in the greenhouse is approximately 300 K, the wavelength of greatest radiation intensity is much longer,

$$\lambda_{max} = \frac{2.898 \times 10^{-3} \text{ m K}}{300 \text{ K}} = 96.6 \times 10^{-7} \text{ m}$$

This is in the infrared part of the radiation spectrum and *will not* readily pass through glass. Since the incident radiation enters the greenhouse and the radiation from the objects within is trapped inside, the greenhouse warms up. Of course, the outside of the glass radiates energy, but this balances the input from the sun only after the greenhouse temperature has risen considerably. The glass also prevents convective losses by stopping the upward flow of warm air.

In a sense, the earth is a huge greenhouse, with the atmosphere replacing the glass. The water vapor and carbon dioxide in the air are good absorbers of infrared radiation. Thus, sunlight passes through the atmosphere more readily than infrared, and the temperature at the surface is higher and more stable than it would be without an atmosphere. Burning fossil fuels increases the carbon dioxide level. This could significantly increase the average temperature of the earth and lead to major climatic changes. This is an important and complex problem, and it is the subject of considerable research.

In Fig. 12.15, we note that the area beneath the curves increases very rapidly as the temperature increases. The area represents the energy emitted per second. There is a simple formula for this emission rate. The rate at which energy radiates from a surface of area A at temperature T was given by Josef Stefan (1835–1893) in 1879 and is called *Stefan's law,*

$$H = e\sigma AT^4 \tag{12.17}$$

Here
$$\sigma = 5.67 \times 10^{-8} \text{ W m}^{-2} \text{ K}^{-4}$$

is *Stefan's constant*. The quantity e is called the *emissivity*, its value depends on the surface, but is always between 0 and 1.

Before illustrating Eq. 12.17 with an example, we remark on several features of radiating objects in general and Stefan's law in particular.

Notice first that Stefan's law contains only the absolute temperature of the emitting object, not a temperature difference as found for conduction and convection. At first glance Stefan's law thus apparently means that an object will radiate energy until its temperature reaches absolute zero. In fact, the object does radiate energy to its surroundings, *but it also absorbs energy from those same surroundings.* Consider an object at a temperature T_2 that is in a large container kept at a temperature T_1. If initially $T_2 > T_1$, one finds that the temperature of the object decreases until the object and the container are both at a common temperature T_1; afterward, no further temperature change occurs. At this stage the object is emitting energy at a rate given by $e\sigma A T_1^4$, and it is absorbing radiated energy from the container at the same rate, $e\sigma A T_1^4$.

A single result contains all of the information about this process. At a time that the object's temperature T is between T_2 and T_1, its net rate of heat loss is

$$H_{\text{net}} = H_{\text{out}} - H_{\text{in}} = e\sigma A(T^4 - T_1^4) \quad (12.18)$$

When $T = T_1$, the net heat leaving the object becomes zero. Equation 12.18 is indeed correct, but we have not explained why the same value of e, the emissivity, can be used for emission and absorption.

The value of the emissivity depends on the surface of the object. A shiny surface has a small value of e; a black surface has e near 1. Because the object reaches equilibrium at the temperature of the walls, it must absorb energy at the same rate as it is emitted. Thus, a good emitter must also be a good absorber. Conversely, a good reflector is a poor emitter; therefore, small values of e correspond to surfaces that emit radiation poorly and reflect radiation well. Large values of e describe surfaces with the opposite characteristics.

A perfect absorber (emitter) is one for which $e = 1$ and all incident radiation is absorbed; none is reflected. Any object that absorbs all incident radiation appears black (unless it is hot enough to radiate visibly); hence, a perfect absorber (and emitter) is called a *blackbody*. A blackbody has $e = 1$.

The emissivity usually varies somewhat with the wavelength. Sunlight is most intense in the visible light region, which corresponds approximately to wavelengths from 4×10^{-7} m to 7×10^{-7} m. The darkest human skin has an emissivity of 0.82 for visible light and the lightest skin has $e = 0.65$. The radiation from objects at or near typical room or body temperatures is predominantly in the longer infrared wavelengths. All human skin has an emissivity of nearly 1 at these wavelengths.

Radiation losses are important for people. This is illustrated in the following example.

Example 12.12

The person in Example 12.10 had a skin temperature of $33° \text{ C} = 306 \text{ K}$ and was in a room where the walls were at $29° \text{ C} = 302 \text{ K}$. If the emissivity is 1 and the body surface area is 1.5 m^2, what is the rate of heat loss due to radiation?

We must consider two competing processes. First, the person at a temperature $T_2 = 33° \text{ C} = 306 \text{ K}$ will radiate heat at a rate

$$\begin{aligned} H_{\text{out}} &= e\sigma A T_2^4 \\ &= (1)(5.67 \times 10^{-8} \text{ W m}^{-2} \text{ K}^{-4})(1.5 \text{ m}^2)(306 \text{ K})^4 \\ &= 746 \text{ W} \end{aligned}$$

This is roughly six times the typical heat output of a human; a person losing this much heat would quickly freeze. However, heat is also received from the surroundings. If these are at temperature $T_1 = 29° \text{ C} = 302 \text{ K}$, heat will be received at the rate

$$\begin{aligned} H_{\text{in}} &= e\sigma A T_1^4 \\ &= (1)(5.67 \times 10^{-8} \text{ W m}^{-2} \text{ K}^{-4})(1.5 \text{ m}^2)(302 \text{ K})^4 \\ &= 707 \text{ W} \end{aligned}$$

Thus the net rate of heat loss is $(746 \text{ W} - 707 \text{ W}) = 39 \text{ W}$. This is roughly equal to the convective loss rate calculated in Example 12.10.

In almost all radiation problems, one must compute the difference between the rates at which energy is emitted at two temperatures. Thus if one temperature is $T + \Delta T$ and the other is T, the difference is $\Delta H = e\sigma A[(T + \Delta T)^4 - T^4]$. In many such problems ΔT is very much less than T, and it is a good approximation to write $(T + \Delta T)^4 - T^4 \simeq 4T^3 \Delta T$. Then

$$\Delta H = 4e\sigma A T^3 \Delta T \quad (\Delta T \ll T) \quad (12.19)$$

The fractional error introduced by the approximation leading to this result is of the order $\Delta T/T$. In the preceding example, $\Delta T/T = 0.013$, so the error would be about 1.3 percent if the approximate formula was used. The following example illustrates a situation where this approximate result can be used accurately.

Example 12.13

Comparing two patches of skin of area A on a person's chest, the radiation rate is found to differ by 1 percent. What is the difference in skin temperature?

Suppose the lower temperature patch is at $T = 37° C = 310 K$, and the temperature of the other is $T + \Delta T$. Then the difference divided by the heat loss at 310 K is

$$\frac{\Delta H}{H} \simeq \frac{4e\sigma AT^3 \Delta T}{e\sigma AT^4} = \frac{4 \Delta T}{T} = 0.01$$

Solving for ΔT,

$$\Delta T = 0.01 \frac{T}{4} = (0.01) \frac{(310 \text{ K})}{4} = 0.775 \text{ K}$$

Thus, a temperature change of less than 1 K alters the radiation rate by 1 percent.

The preceding example illustrates the physical principle underlying *thermography,* a technique with uses in many fields. In medical applications, special infrared detectors are used to produce a photographic record of the infrared radiation emitted by a patient (Fig. 12.16). In photos of this type, areas with higher temperatures appear darker; differences as small as 0.1° C are detectable. The photographs show, for example, how circulation is reduced by smoking. Thermography is sometimes used in preliminary screenings to identify patients who may have breast cancers, thyroid tumors, or other diseases.

It is also possible to measure accurately the weak radiation emitted by the body at microwave wavelengths, which are longer than those of infrared radiation. Recent experiments show that such measurements may help doctors to detect tumors up to 10 cm

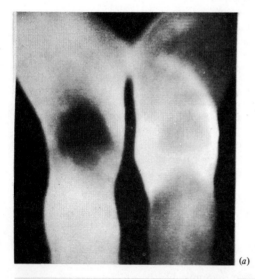

(a)

(b)

Figure 12.16. Thermograms showing warmer areas as darker. (a) Joint inflammation accompanying rheumatoid arthritis. (b) Increased peripheral circulation is shown in a foot after treatment (right). (Courtesy of AGA Medical Division.)

below the surface. By contrast, infrared radiation is absorbed more readily in body tissues, so thermography is sensitive only to tumors closer to the skin. Both techniques are noninvasive and intrinsically safe, since the patient is not subjected to any external source of radiation or other probe.

Wien's and Stefan's laws together can be used to explain many phenomena of our everyday experience. Consider, for example, why clear nights are cooler than cloudy nights. On a clear night the earth radiates energy into space at a rate proportional to the fourth power of its temperature, about 300 K. The incoming radiation from space is very small because its average temperature is near absolute zero. On the other hand, with cloud cover, the earth radiates at 300 K, but the radiation is absorbed in the clouds, which radiate energy back to earth. Again the radiation is trapped, rather like the greenhouse effect. The cooling that occurs on clear nights is often referred to as radiational cooling.

SUMMARY

Many properties of materials depend on temperature. The temperature is a measure of the different states of motion of the molecules. For example, most objects will experience a fractional change in length $\Delta l/l$ proportional to the temperature change ΔT,

$$\frac{\Delta l}{l} = \alpha \, \Delta T$$

Similarly, fractional changes in the area and volume of uniform materials are given by

$$\frac{\Delta A}{A} = 2\alpha \, \Delta T$$

and

$$\frac{\Delta V}{V} = \beta \, \Delta T = 3\alpha \, \Delta T$$

where α is the linear expansion coefficient, and β is the volume expansion coefficient.

When heat is added to an object, its temperature will change or it will undergo a phase change. If n moles of a substance change temperature by dT when heat dQ is transferred, the molar heat capacity is given by

$$C = \frac{1}{n} \frac{dQ}{dT}$$

The specific heat capacity is the molar heat capacity divided by the mass of one mole of the substance,

$c = C/M$. The heat ΔQ entering or leaving an object of mass m when the temperature changes by ΔT is

$$\Delta Q = mc \, \Delta T.$$

In practice, molar and specific heat capacities are measured or calculated either at constant pressure or at constant volume.

If heat ΔQ is added to a substance but no temperature change occurs, then a mass m of the substance undergoes a phase change, where

$$\Delta Q = Lm$$

Here L is the latent heat for the phase change.

Heat is transferred between objects at different temperatures by conduction, convection, and radiation. If a sample of material of cross-sectional area A has a temperature difference ΔT along its length, the rate of heat transfer from the high- to low-temperature end by conduction is

$$H = \kappa A \frac{\Delta T}{l}$$

where κ is the thermal conductivity of the material. The factor l/κ is called the R-value and is a measure of the resistance of the material to the flow of heat.

An object may transfer heat by convection. If the object has a surface area A and is at a temperature ΔT above that of a fluid that can circulate, the heat loss rate is given approximately by

$$H = qA \, \Delta T$$

Convective heat transfer increases dramatically if the fluid is forced to move. For example, convective losses are much greater on windy days than on calm days.

Heat transfer by radiation occurs through the emission and absorption of electromagnetic radiation. An object at a temperature T will emit most of its radiation at wavelengths near

$$\lambda = \frac{B}{T}$$

Here B has a value 2.898×10^{-3} m K. Knowledge of this wavelength is very useful since materials transmit, absorb, and reflect radiation differently at different wavelengths. The total rate at which energy is emitted depends on the fourth power of the temperature,

$$H = e\sigma A T^4$$

Here A is the surface area, e the emissivity of the surface, and $\sigma = 5.67 \times 10^{-8}$ W m^{-2} K^{-4} is Stefan's constant. The heat lost by radiation is the difference between that emitted by the object and that absorbed from the object's surroundings.

Checklist

Define or explain:

linear expansion	conduction
thermostat	thermal conductivity
volume expansion	R-value
internal energy	convection
molar heat capacity	radiation
specific heat capacity	frequency
calorimeter	wavelength
phase change	emissivity
sublimation	blackbody
triple point	Wien displacement law
critical point	Stefan's law
latent heat	

REVIEW QUESTIONS

Q12-1   The fractional change in length of an object is proportional to the temperature change. The proportionality constant is called the _____.

Q12-2   The change in length of a heated object is proportional to α, the coefficient of linear expansion. The fractional area change is proportional to _____, and the fractional volume change is proportional to _____.

Q12-3   The ratio of the heat added to one mole of a substance to the temperature change is called the _____.

Q12-4   The specific heat capacity is defined as the molar heat capacity divided by the _____ of the substance.

Q12-5   The heat capacity is usually measured or calculated under the conditions of constant _____ or constant _____.

Q12-6   The latent heat of vaporization of a substance is the heat required to change 1 kg of a substance from _____ to _____.

Q12-7   The transfer of heat from one place to another by the actual movement of material is called _____.

Q12-8   Heat conduction refers to the transfer of thermal energy between objects in _____.

Q12-9   The wavelength at which the radiation from an object is most intense increases as the temperature _____.

Q12-10   Stefan's law describes the fact that the rate of heat loss through radiation is proportional to the _____ power of the temperature.

EXERCISES

Section 12.1 | Thermal Expansion

12-1   A steel railroad track is 20 m long at 20° C. How much longer is it at 40° C?

12-2   A steel railroad track is 30 m long at 0° C. How much shorter is it at $-20°$ C?

12-3   An aluminum metrestick is exactly 1 m long at 20° C. How much shorter is it at 0° C? Use $\alpha = 2.30 \times 10^{-5}$ K^{-1}.

12-4   For Pyrex glass, β is about one-third that of ordinary glass. What does this imply about thermal stresses?

12-5   Why does heating a jar lid make it easier to open?

12-6   By how much does the area of a rectangular steel plate 0.5 m by 2.5 m change when it is heated from 0° C to 40° C?

12-7   A common lecture demonstration uses a steel ball that will not fit through a steel ring unless the ring is heated. If the diameter of the ball is 3 cm = 0.03 m at 20° C, what is the inner diameter of the ring at 20° C if the ball slips through when the ring reaches 250° C?

12-8   A container of water is filled to the top. The temperature increases 8 K, but no water spills. What was the original temperature?

Section 12.2 | Heat Capacity

12-9   How many kilojoules are needed to heat 0.15 kg of helium gas from 20° C to 80° C at constant pressure?

12-10   A snowball is dropped from a rooftop 20 m above the ground. If its temperature is initially $-10°$ C, and if all of its kinetic energy is converted to internal energy, what is its final temperature?

12-11   A container with a mass of 0.6 kg is at a temperature of 20° C. When 2.5 kg of boiling water is poured into it, the final temperature is 90° C. What is the specific heat capacity of the container?

12-12   A calorimeter of mass 0.4 kg and a specific heat capacity of 0.63 kJ kg^{-1} K^{-1} contains a sample with a mass of 0.55 kg. If 2.45 kJ of energy is supplied electrically and the temperature rises 4° C, what is the specific heat capacity of the sample?

12-13   A 5-kg meteorite hits the ground at 2000 m s^{-1}. How much thermal energy is liberated if all its kinetic energy is converted to thermal energy?

12-14 The temperature of 15 kg of water increases at 0.003° C per second. At what rate is the internal energy of the water increasing? (Neglect the work done by the water.)

Section 12.3 | Phase Changes

12-15 How much heat is needed to melt a 10-kg block of ice that is initially at $-10°$ C?

12-16 How much heat is required to heat 1 kg of water from 20° C at atmospheric pressure to boiling and to convert it entirely into steam?

12-17 If 0.15 kg of ice at 0° C is added to 0.25 kg of water at 20° C, (a) does all the ice melt? (b) What is the final temperature?

12-18 An ice cube at 0° C has a mass of 0.01 kg. It melts in 5 minutes. At what rate is its internal energy increasing?

Section 12.4 | Heat Conduction

12-19 A picnic jug contains 1.3 kg of water and 0.6 kg of ice. If 35.6 W of heat enters through the insulation, how long does it take for all the ice to melt?

12-20 A cabin wall is made of wood 0.05 m thick and has an area of 12 m². If the outside surface of the wall is at 0° C and the inside surface is at 20° C, at what rate is heat lost through the wall?

12-21 At what rate will heat be conducted by a copper rod 4 m long with a cross-sectional area of 0.015 m² if one end is at 250° C and the other is at 40° C?

12-22 At what rate will heat be conducted across a wooden wall of area 25 m² and thickness 0.1 m if the temperature inside is 20° C and the temperature outside is $-10°$ C?

12-23 At what rate is heat lost by conduction through a rectangular pane of glass 0.003 m thick with sides 0.1 m and 0.2 m? Assume the inner surface is at 10° C and the outer surface is at 0° C.

12-24 (a) What is the R-value of a 1-m² slab of glass 0.5 cm = 0.005 m thick? (b) How much heat per hour passes through this slab if the temperature difference is 10° C between the two faces of the glass?

12-25 (a) What is the R-value for a 1-cm thickness of fiberglass with $\kappa = 0.038$ W m⁻¹ K⁻¹? (b) What is R for a 15-cm thickness of fiberglass?

12-26 The R-value of a 1-cm = 0.01-m thickness of a certain material is 0.2 m² K W⁻¹. If 50 W of thermal energy passes through a 1-m² area of the material, what is the temperature difference between the faces?

12-27 Equation 12.12 does not apply to thick insulation covering a pipe, since it holds only when the cross section is at least approximately uniform. However, one can still see the effect of insulating a water pipe qualitatively by considering a sheet of copper 0.002 m thick insulated with a felt layer 0.02 m thick. If water in contact with the copper is at 80° C, and the outer surface of the felt is at 15° C, what is the temperature of the other side of the copper sheet?

Section 12.5 | Heat Transfer by Convection

12-28 How much energy per second will a naked person with surface area 1.4 m² lose by convection in air at 0° C? Assume the average q is 7.1 W m⁻² K⁻¹ and that the skin temperature is 30° C.

12-29 A naked person with a surface area of 1.8 m², a skin temperature of 31° C, and an average q of 7.1 W m⁻² K⁻¹ loses 126 W by convection. What is the air temperature?

12-30 A windowpane is at 10° C and has an area of 1.2 m². If the outside air temperature is 0° C, at what rate is energy lost by convection? The q for the window is 4 W m⁻² K⁻¹.

12-31 When the wind is a factor, Eq. 12.14 can be used with the temperature far from the surface taken from the wind chill chart, Table 12.5. Suppose the temperature on the outside surface of a window is 10° C and the air temperature far from the window is $-10°$ C. Find the ratio of the convective heat loss when the wind speed is 20 km h⁻¹ to the heat loss when there is no wind.

Section 12.6 | Radiation

12-32 What is the wavelength of maximum radiation intensity for a surface at 37° C?

12-33 What is the wavelength of maximum radiation intensity for a surface at 1000° C?

12-34 What is the wavelength of maximum radiation intensity for a surface at 2000° C?

12-35 Calculate the error made if Eq. 12.19 rather than Eq. 12.18 were used if $T + \Delta T = 319$ K and $T = 270$ K.

12-36 Estimate the error made if Eq. 12.19 rather than Eq. 12.18 were used if $T + \Delta T = 297$ K and $T = 293$ K.

12-37 A naked person with surface area 1.8 m²

and skin temperature 33° C is in a room at 10° C. (Assume $e = 1$.) (a) At what rate does the person radiate energy? (b) What is the person's net rate of energy loss due to radiation?

12-38 An object is at 300 K, and an identical object is at 900 K. What is the ratio of the energies radiated? (Assume the emissivity is the same for both objects.)

12-39 A 2-m length of copper pipe containing hot water has its outer surface at 80° C. If the surroundings are at 20° C, at what rate does the pipe lose thermal energy due to radiation? (The surface area of the pipe is 0.12 m² and $e = 1$.)

12-40 A black-surfaced road at a temperature of 320 K receives radiant energy from the sun at a rate of 700 W m⁻². What is the net rate at which a square metre of road surface absorbs thermal energy?

PROBLEMS

*12-41 The pendulum of a large clock consists of a thin steel rod with a heavy weight at the end. At 20° C the rod is 1.22 m long and the clock keeps accurate time. (a) By how much does the length change if the temperature rises to 40° C? (b) Does the clock run fast or slow? (c) What is the fractional change in the period of the pendulum? (*Hint:* See Chapter Nine.) (d) What is the error of the clock in seconds per day?

12-42 (a) A car with a full 40-litre steel gasoline tank is parked in the sun. If its temperature rises 30° C, how much gasoline spills out? (For gasoline, $\beta = 9.50 \times 10^{-5}$ K⁻¹ and for steel, $\beta = 3.81 \times 10^{-5}$ K⁻¹.) (b) At what time of day is gasoline cheapest?

12-43 (a) Show that the coefficient of volume expansion β for an ideal gas is $1/T$, where T is the Kelvin temperature. (Assume that the pressure remains constant.) (b) What is β at 20° C?

12-44 Assuming that no heat is transferred to or from the surroundings, how much does the temperature of a river rise in going over a 30-m falls? (Assume the velocity of the water is the same above and below the falls.)

12-45 A tightly sealed small house has a total floor area of 90 m² and ceilings 2.5 m high. If the inside temperature is 21° C and the outside air is at −10° C, what is the rate of energy loss if all the air in the house is exchanged with outside air every 3

hours? (The specific heat of air at constant pressure is 1.0 kJ kg⁻¹ K⁻¹; the density of air at 20° C is 1.20 kg m⁻³.)

*12-46 A steel rod is initially at 20° C and has a length of 2 m. It has a cross-sectional area of 0.001 m². (a) If it is heated to 120° C, by how much does its length increase? (b) How large a force must be applied to its ends to restore the original length? (Young's modulus for steel is 2×10^{11} N m⁻².)

12-47 A 75-kg man utilizes energy at the rate of 10,000 kJ per day. Suppose that 10 percent of this energy is used for work and 90 percent is waste heat. If his body had no way of releasing that heat, by how much would his temperature rise per hour on the average? (The specific heat of animal tissue is approximately equal to that of water.)

12-48 How fast must a lead bullet travel so that it melts when fired into a thick slab of wood, assuming that no heat is lost to the wood? (Assume initially that $T_C = 20°$ C.)

12-49 Using the data in Fig. 12.6, find the energy required to heat, at constant volume, 1 kg of hydrogen gas from (a) 30 K to 40 K, and from (b) 260 K to 270 K.

12-50 A steel-wool pad is rubbed on a steel frying pan with a force component of 10 N along the direction of motion. Each stroke is 0.1 m long, and there are 0.8 strokes per second. (a) At what rate is thermal energy generated? (b) If no heat is lost to the surroundings and the combined mass of the pan plus pad is 1.25 kg, by how much will their temperature rise after 1 minute?

12-51 How much heat is needed to melt a mole of solid N_2 initially at its melting point?

*12-52 The prevailing winds in the western United States are from the west. In many areas the region east of mountain ranges is hot and dry. Suggest a reason for this. (*Hint:* Typically, there is relatively great precipitation over the mountains. What effect does this have on the temperature of the air?)

*12-53 A pot has a copper bottom layer 0.005 m thick and an inner steel layer 0.002 m thick. The inside of the pot is at 100° C and the outside of the bottom is at 103° C. (a) What is the temperature of the copper-steel junction? (b) At what rate is heat conducted if the area of the bottom is 0.04 m²?

*12-54 (a) At what rate is heat conducted across a wooden wall of area 20 m² and thickness 0.03 m if the temperature difference is 40° C? (b) How much heat is conducted per second if there is a layer of rock wool 0.04 m thick on one side of the wall?

12-55 A pot with a steel bottom 0.01 m thick rests on a hot stove. The area of the bottom is 0.1 m². The water inside the pot is at 100° C, and 0.05 kg are evaporated every 3 minutes. Find the temperature of the lower surface of the pot, which is in contact with the stove. (Assume that no heat is lost to the room.)

12-56 A picnic chest has dimensions 0.5 × 0.3 × 0.35 m. It is insulated with a 0.02-m-thick layer of a material with conductivity of 0.04 W m⁻¹ K⁻¹. (a) If the temperature difference across the layer is 35° C, at what rate will heat be conducted? (b) How many kilograms of ice will melt each hour in the chest?

12-57 A homeowner turns down the thermostat from 23° C to 18° C. If the average outside temperature is 0° C, estimate the percentage reduction in fuel consumption.

12-58 The average outside temperature is 0° C; inside a house the temperature is 23° C. To reduce fuel bills by 10 percent, what thermostat setting should be chosen?

12-59 A surface at 20° C faces a cloudless night-time sky. (a) What is the rate of radiation per square metre of the surface? (Assume $e = 1$.) (b) The amount of radiation from the sky is very small by comparison. Why?

12-60 The surface of the sun is at about 6000 K. (a) How much power does it radiate per square metre of surface area? (b) How much power reaches the upper atmosphere of the earth per square metre? (The diameter of the sun is 1.39×10^6 km, and the mean radius of the earth's orbit is 1.49×10^8 km.)

12-61 A man having surface area 1.8 m² wears a garment 0.01 m thick with a thermal conductivity of 0.04 W m⁻¹ K⁻¹. (a) If his skin temperature is 34° C and the outside of his garment is at −10° C, what is his rate of heat loss? (b) Is he adequately dressed? Explain.

12-62 A girl having a surface area of 1.2 m² wears a down jacket and pants 0.03 m thick. If her skin temperature is at 34° C and she can safely lose 85 W by conduction, what is the lowest temperature for which the clothing is adequate? (Assume that the outside of the clothing is at the same temperature as the air.)

12-63 A thin sheet of ice with an area of 0.15 m² on each side hangs from a roof. The ice is at 0° C

and the surrounding air is at 10° C. The net rate at which radiant energy is transferred to the ice from the sunlight and from the surroundings is 20 W; $q = 9.5$ W m⁻² K⁻¹. What mass of ice melts in 1 minute?

12-64 It is desired to insulate an attic floor to an R-value of 3 m² K W⁻¹. If rock wool is used, how thick must it be?

12-65 (a) If the R-value of the insulation in a 40-m² ceiling is 3.8 m² K W⁻¹, and the temperature difference across it is 10° C, how much energy is lost through the ceiling in 7 hours due to conduction? (b) A litre of heating oil provides 3.846×10^7 J of energy. At 39.6 cents per litre, what is the cost of this energy per hour?

12-66 What is the cost per hour of replacing the heat lost through a single-pane window 0.003 metres thick and 1 m² in area if the inner and outer surface temperature differ by 10° C? (Assume heating oil provides 3.846×10^7 J of heat per litre and costs 39.6 cents per litre.)

12-67 A copper heating pipe 2 m long of outer radius 0.01 m has walls 0.002 m thick. It contains water at 80° C and is in a room with still air at 20° C. (Neglect radiation effects.) (a) Using $q = 9.5$ W m⁻² K⁻¹, what is the temperature of the outer pipe surface? (b) At what rate is energy lost through the pipe walls?

*12-68 Show that when slabs with R-values R_1 and R_2 are in contact, the effective R-value is $R_1 + R_2$. (*Hint:* If the temperatures on either side are T_1 and T_2, respectively, find the temperature between the two slabs.)

ANSWERS TO REVIEW QUESTIONS

Q12-1, coefficient of linear expansion; **Q12-2**, 2α, $\beta = 3\alpha$; **Q12-3**, molar heat capacity; **Q12-4**, mass of one mole; **Q12-5**, pressure, volume; **Q12-6**, liquid, vapor, or gas; **Q12-7**, convection; **Q12-8**, contact; **Q12-9**, decreases; **Q12-10**, fourth.

SUPPLEMENTARY TOPICS

12.7 | TEMPERATURE REGULATION IN WARM-BLOODED ANIMALS

Most biological processes are temperature-dependent, so the body temperature of an animal must be kept within a narrow range. Warm-blooded animals, such as birds and mammals, control their own temperatures by regulating the heat loss from their bodies. By contrast, cold-blooded animals are dependent on their environment for the maintenance of their body temperature. For example, snakes are often found sunning themselves on sun-warmed rocks. Many insects must beat their wings before takeoff to raise the temperature of their flight muscles. In this section, we are concerned only with the ways in which warm-blooded animals regulate their temperatures on a short-term basis.

In addition to the mechanisms used daily, biological adaptations to changing climatic conditions also occur on seasonal and evolutionary time scales. The heavy coats of animals, bird migration, and hibernation are all seasonal adaptations. On a longer time scale, nature has favored the development of larger animals in the coldest climates. Such animals have large ratios of volume to surface area, and the corresponding heat production to heat loss ratios are high.

In warm-blooded animals, the main object of temperature regulation is to keep vital organs and muscles at a nearly ideal temperature. Since heat loss occurs either at the body surface or through water evaporation from the lungs, the object is to adjust the flow of heat from the organs and muscles to the lung and body surfaces. The blood plays an important role in this process, since it carries heat from the internal to the external portions of the body. Also, the body's thermostat, the *hypothalamus*, which is located in the brain, uses the blood temperature to monitor the system. The hypothalamus acts to keep itself at a nearly constant temperature. In doing so, the temperature of other internal organs may vary much more widely. This is analogous to a house with a single thermostat in one room. The temperature in that room will be nearly uniform, while the temperature in other rooms may fluctuate considerably.

The source of body heat is the chemical metabolism of food. A resting 70-kg human generates about 80 W; during heavy exercise the rate may be up to 20 times greater. Depending on the air temperature and the clothing worn, the heat generated may be needed to overcome convective and radiative losses, or it may be a waste product to be disposed of by the body.

A warm-blooded animal has a number of mechanisms to use in controlling its temperature. To raise its temperature, the body reduces the blood flow through the capillaries nearest the skin surface. Flesh is a poor conductor of heat, so this is effective in reducing heat losses. Also, body hair can be fluffed up to increase insulation. (Even humans have a vestige of this manifested as goose bumps; the body is attempting to fluff almost nonexistent hair.) Finally, heat production may be increased by shivering.

The body is cooled by convection and radiation from the skin and by the evaporation of sweat from the skin and of water from the lungs. If the interior temperature begins to increase, the body first increases blood flow near the skin surface to increase convective and radiative losses and then, if necessary, uses the evaporative loss mechanism. Humans and horses, among other animals, sweat from glands located over much of the body and thus benefit from evaporation over a large surface area. As much as 1.5 kg of sweat per hour may be evaporated from the human body. The latent heat of vaporization for sweat at 37° C is about equal to that of water at the same temperature, 2427 kJ kg^{-1}. Note that this is larger than the 2255 kJ kg^{-1} necessary to vaporize water at 100° C. Under many conditions the evaporation of sweat is the primary cooling mechanism used by the body.

Furry animals such as dogs do not sweat but take major advantage of evaporative cooling in the lungs. They exhale large volumes of air by panting. Panting itself generates heat but fortunately less than is lost by evaporation.

The various contributions to the rate of heat loss and production for a typical human adult are approximately as follows. If the area A is given in square metres, the skin temperature T_s and air temperature T_a are given in degrees Celsius, and the rate r of sweating is in units of kilograms per hour, then

H_m = rate heat is generated by metabolism:
   80 to 1600 W

H_c = rate heat is lost by convection (still air):
   $D_c A(T_s - T_a)$

H_r = rate heat is lost by radiation: $D_r A(T_s - T_a)$

H_s = rate heat is lost by evaporation of sweat: $D_s r$

H_l = rate heat is lost by evaporation from the lungs: D_l

The D's are constants,

$$D_c = 7.1 \text{ W m}^{-2} \text{ K}^{-1}$$
$$D_r = 6.5 \text{ W m}^{-2} \text{ K}^{-1}$$
$$D_s = 674 \text{ W h kg}^{-1}$$
$$D_l = 10.5 \text{ W}$$

The value of H_l is given for normal breathing rates; it

increases in proportion to the rate of breathing. Typically, H_l is a small fraction of the heat loss, and we neglect its variation.

If the body temperature is held constant, then the heat losses are related by

$$H_m = H_c + H_r + H_s + H_l \qquad (12.20)$$

The following example illustrates the importance of sweating during moderate levels of activity.

Example 12.14

A person generates heat at a rate of 230 W. How much sweat is produced per hour if the surface area is 1 m²? Assume the body temperature is 37° C and that of the air is 28° C.

Using the results given above,

$$\begin{aligned}
H_s &= H_m - H_c - H_r - H_l \\
&= 230 \text{ W} - (7.1)(1.0)(9) \text{ W} \\
&\quad - (6.5)(1.0)(9) \text{ W} - 10.5 \text{ W} \\
&= 97.1 \text{ W}
\end{aligned}$$

Since the rate at which heat is removed by sweating is $D_s r$, the sweat required is

$$r = \frac{H_s}{D_s} = \frac{97.1 \text{ W}}{674 \text{ W h kg}^{-1}} = 0.14 \text{ kg h}^{-1}$$

12.8 | ADIABATIC EXPANSION OF AN IDEAL GAS

The ratio of the specific heats, $\gamma = C_P/C_V$, determines the behavior of an ideal gas when its volume changes with no heat entering or leaving the gas. We show now that this follows directly from the first law and the equation of state.

Consider a small adiabatic volume change dV. Since $Q = 0$, from the first law $dU = -W = -PdV$. From the definition of C_V, $dU = nC_V dT$. Thus we have $nC_V dT = -PdV$ or

$$dT = -PdV/nC_V$$

Also, from the equation of state, $PV = nRT$, it follows that $PdV + VdP = nRdT$, or

$$dT = (PdV + VdP)/nR$$

Equating the two expressions for dT, we find

$$(C_V + R)PdV + C_V VdP = 0$$

Now $C_V + R = C_P$ for an ideal gas, so dividing by $C_V PV$ and using the definition $C_P/C_V = \gamma$,

$$\frac{\gamma dV}{V} + \frac{dP}{P} = 0$$

Integrating, using $\int dx/x = \ln x$, we get $\gamma \ln V + \ln P = $ constant, or $\ln(PV^\gamma) = $ constant. Therefore,

$$PV^\gamma = \text{constant (adiabatic process in ideal gas)} \qquad (12.21)$$

Note that for an ideal monatomic gas, $C_V = \frac{3}{2}R$, so $C_P = \frac{5}{2}R$ and $\gamma = \frac{5}{3}$.

A PV plot shows curves corresponding to a constant quantity of an ideal gas (Fig. 12.17). The adiabatic curves drop more steeply than the isothermal curves as V increases. Hence a given adiabatic curve such as A_3 crosses successive isothermals corresponding to decreasing temperatures as the volume increases, as seen at points a and b in Fig. 12.17. This is what we should expect. As the gas expands without any heat transfer, it does work and loses internal energy. Hence its temperature must decrease.

The use of this result is illustrated by the next example.

Example 12.15

A monatomic ideal gas initially at 0° C and a pressure of 1 atm is suddenly compressed to half its original volume. Find its new pressure and temperature.

Since the compression is very rapid, we can assume that there is not significant heat flow, and that the proc-

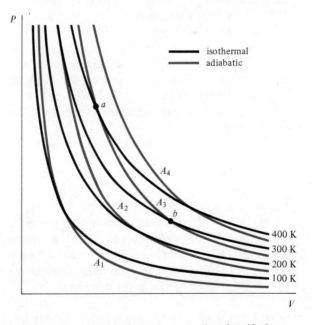

Figure 12.17. PV plots for a given quantity of an ideal gas. Black curves show isothermal volume changes, $PV = nRT$. Colored curves show adiabatic processes, $PV^\gamma = $ constant, for $\gamma = 1.40$.

ess is adiabatic. For an ideal monatomic gas, $\gamma = 5/3 = 1.67$. Thus when the pressure and volume change from P_0, V_0 to P, V, the pressure and volume satisfy $PV^\gamma = $ constant, or

$$PV^{1.67} = P_0 V_0^{1.67}$$
$$P = P_0(V_0/V)^{1.67} = (1 \text{ atm})(2)^{1.67}$$
$$= 3.18 \text{ atm}$$

The pressure is more than three times its original value. Note that in an isothermal compression, PV remains constant, so P doubles when V is halved.

To find the new temperature, we use the ideal gas law, $PV = nRT$, or

$$T = PV/nR = (3.18 \, P_0)(0.5 \, V_0)/nR$$
$$= 1.59 \, P_0 V_0/nR = 1.59 \, T_0$$

The original temperature is $0°$ C or $T_0 = 273$ K. Thus the final temperature is $1.59(273 \text{ K}) = 434$ K or $161°$ C. The temperature rises rapidly when the gas is compressed.

EXERCISES ON SUPPLEMENTARY TOPICS

Section 12.8 | Adiabatic Expansion of an Ideal Gas

12-69 Suppose the gas in Example 12.15 is cooled back to $0°$ C at constant pressure after the compression. If the volume before the gas was compressed was V_0, what is its final volume?

12-70 An ideal diatomic gas with $\gamma = 1.4$ is at a temperature of 300 K and a pressure of 1 atm. It is suddenly compressed so that its volume is a third of its original volume. (a) Find its new pressure and temperature. (b) The gas is now cooled to 300 K at constant volume. What is its final pressure?

12-71 Show that in an adiabatic expansion of an ideal gas, $TV^{\gamma-1}$ is constant.

12-72 Show that in an adiabatic expansion of an ideal gas, $TP^{(1-\gamma)/\gamma}$ is constant.

12-73 An ideal gas expands adiabatically so that its temperature changes from T_1 to T_2. Show that the work it does is given by $C_V(T_1 - T_2)$.

12-74 For a monatomic ideal gas, $C_V = 3/2$, while for a diatomic ideal gas $C_V = 5/2$. An ideal gas at 300 K is compressed adiabatically to one-fourth its original volume. Find its final temperature if it is (a) monatomic; (b) diatomic.

PROBLEMS ON SUPPLEMENTARY TOPICS

12-75 A person produces thermal energy at a rate of 175 W. If all this heat is dissipated by evapora-

tion of sweat, how much sweat per hour is required?

12-76 A naked person with a surface area of 1.5 m^2 and a skin temperature of $40°$ C is in a sauna at $85°$ C. (a) At what rate does the person absorb energy by radiation from the walls, assuming emissivity 1? (b) At what rate does the person radiate energy to the surroundings? (c) How much sweat must be evaporated per hour, assuming that no energy is transferred by convection? (Neglect the metabolic heat production.)

Additional Reading

T.C. Ruch and H.D. Patton (ed.), *Physiology and Biophysics,* vol. 3, W.B. Saunders Co., Philadelphia, 1973. Chapter 5 discusses temperature regulation in humans.

John R. Cameron and James G. Skofronick, *Medical Physics,* John Wiley and Sons, Inc., New York, 1978. Chapters 4 and 5.

J. Hansen, et al., Climate Impact of Increasing Atmospheric Carbon Dioxide, *Science,* vol. 213, 1981, p. 957.

Phillip B. Allen, Conduction of Heat, *The Physics Teacher,* vol. 21, 1983, p. 569.

Albert A. Bartlett and Thomas J. Brown, Death in a Hot Tub: The Physics of Heat Stroke, *The American Journal of Physics,* vol. 51, 1983, p. 127.

Scientific American articles:

Robert L. Sproull, The Conduction of Heat in Solids, December 1962, p. 92.

David M. Gates, Heat Transfer in Plants, December 1965, p. 76.

G. Yale Eastman, The Heat Pipe, May 1968, p. 38.

G. Neugebauer and Eric E. Becklin, The Brightest Infrared Sources, April 1973, p. 28.

Jacob Gershon-Cohen, Medical Thermography, February 1967, p. 94.

David Turnbull, The Undercooling of Liquids, January 1965, p. 38.

Laurence Irving, Adaptations to Cold, January 1966, p. 94.

T.H. Benzinger, The Human Thermostat, January 1961, p. 134.

Bernd Heinrich and George A. Bartholomew, Temperature Control in Flying Moths, June 1972, p. 70.

Francis G. Carey, Fishes with Warm Bodies, February 1973, p. 36.

Manuel G. Velarde and Christiane Normand, Convection, July 1980, p. 92.

H. Craig Heller, Larry I. Crawshaw, and Harold T. Hammel, The Thermostat of Vertebrate Animals, August 1978, p. 102.

Knut Schmidt-Nielsen, Countercurrent Systems in Animals, May 1981, p. 118.

Bernd Heinrich, The Regulation of Temperature in the Honeybee Swarm, June 1981, p. 146.

Eric A. Newman and Peter H. Hartline, The Infrared "Vision" of Snakes, March 1982, p. 116.

Roger Revelle, Carbon Dioxide and World Climate, August 1982, p. 35. The greenhouse effect.

Jearl Walker, Gismos That Apply Non-Obvious Physical Principles to the Enjoyment of Cooking, The Amateur Scientist, June 1984, p. 146.

Jearl Walker, What Happens When Water Boils Is a Lot More Complicated Than You Might Think, The Amateur Scientist, December, 1982, p. 12.

Jearl Walker, Thermal Oscillators: Systems That Seesaw, Buzz Or Howl Under the Influence of Heat, The Amateur Scientist, February 1983, p. 146.

John T. Snow, The Tornado, April 1984, p. 86.

UNIT FOUR

FLUIDS

Flow of fluid around a moving array of cylinders. (Onera photo by Henre Werle).

FLUIDS play a unique role in our lives and in our study of science, mainly because of their characteristic of flowing and the fact that they conform to the shape of their container. In this context, we can regard both liquids and gases as fluids.

Animals transport nutrients and remove wastes via the fluids of their circulatory systems. Similarly, the transport of materials in plants occurs in fluids. The flight of birds and planes involves fluid motion, as do the weather, waves, and ocean currents.

All these phenomena can be described by applying the principles of mechanics to fluids. However, because fluids do not remain in a fixed shape and because they may be compressed, a complete analysis can become very complicated. In order to simplify matters, in this unit we assume that fluids are *incompressible;* that is, they remain constant in density. For most liquids, this is a good first approximation. For gases, we must realize that our methods will only apply to those situations where the pressure and temperature variations are small.

In the first chapter of this unit (Chapter Thirteen), we also assume that there are no frictional forces between segments of a fluid moving with respect to one another. This is fine for fluid at rest and for some applications when fluids move. However, we see in Chapter Fourteen that these frictional or *viscous* effects are often important. In Chapter Fifteen, we discuss some important properties arising from the intermolecular forces in fluids.

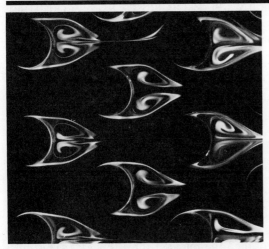

CHAPTER 13

THE MECHANICS OF NONVISCOUS FLUIDS

In this chapter, we discuss fluids at rest and non-viscous (frictionless) fluids in motion. We first develop an understanding of why an object may either sink or float in a fluid at rest. We then develop *Bernoulli's equation*, which puts work and energy concepts into a form suitable for fluids. This is the central point of this chapter and can be used, for example, to understand why fluids in connected containers tend to have the same surface levels and how fluids flow from one place to another.

An important condition on this discussion is the assumption that the fluid is incompressible: a given mass of fluid always occupies the same volume though its shape may change. This condition is described mathematically by the *equation of continuity*, which simply states that the amount of fluid entering a tube must equal the amount leaving. This equation plays an important part in our discussions.

In mechanics, we found that if we could identify the forces acting on an object or objects, we could then predict the subsequent motion or describe the state of equilibrium. For our discussion of fluids, we use the same philosophy, but we need to take account of the fact that a given mass of fluid does not have a fixed shape. This complication is handled by using the *density* and *pressure* instead of the mass and force concepts we used previously.

The definitions of the density (the mass per unit volume) and the pressure (the force per unit area) in Sections 3.2 and 10.3, respectively, should be reviewed in preparation for this Chapter. Typical densities of fluids are given in Table 13.1.

13.1 | ARCHIMEDES' PRINCIPLE

An object floating or submerged in a fluid experiences an upward or *buoyant* force due to the fluid. To understand this force, B, consider a segment of fluid

TABLE 13.1

Densities of some fluids at atmospheric pressure
(At 0°C, 1 ft³ of water weighs 62.5 lbs; 1 cm³ of water has a mass of 1 gram.)

Fluid	Density (kg m^{-3})	Temperature (°C)
Hydrogen (H$_2$)	0.0899	0
Helium (He)	0.178	0
Nitrogen (N$_2$)	1.25	0
Carbon dioxide (CO$_2$)	1.98	0
Oxygen (O$_2$)	1.43	0
Air	1.29	0
	1.20	20
	0.95	100
Water, pure	1000	0
	958	100
Sea water	1025	15
Alcohol, ethyl	791	20
Chloroform	1490	20
Ether	736	0
Linseed oil	930	0
Glycerin	1260	0
Mercury	13600	0
Whole blood	1059.5	25
Blood plasma	1026.9	25

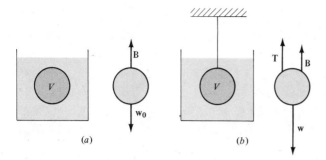

Figure 13.1. (a) An imaginary segment of fluid with the forces acting on it. (b) An object suspended by a string in a fluid and the forces acting on it.

of volume V and density ρ_0. The fluid segment has a mass equal to $\rho_0 V$ and a weight $w_0 = \rho_0 g V$, the mass times the acceleration of gravity (Fig. 13.1a). The segment is in equilibrium with the surrounding fluid, so the buoyant force must be equal and opposite to the weight. Thus, $B = w_0$, or

$$B = \rho_0 g V$$

The buoyant force is the force exerted by the remainder of the fluid to keep the segment at rest.

Now suppose that the imaginary segment is replaced by an object of volume V suspended on a string. The density ρ of the object is greater than that of the fluid. The forces on the object are its weight $w = \rho g V$, the tension in the string T, and the buoyant force B. The fluid does not distinguish between the object and the fluid it replaces, so again $B = \rho_0 g V$. Since the object is in equilibrium, $T = w - B$, or

$$T = (\rho - \rho_0) g V \qquad (13.1)$$

The tension in the string is reduced by the weight of the displaced fluid. The principle that *the buoyant force on the object is equal to the weight of the displaced fluid* was deduced by Archimedes (287–212 B.C.) and is called *Archimedes' principle.* Archimedes' principle provides a convenient way to determine densities. This is illustrated in the next example.

Example 13.1

A piece of metal of unknown volume V is suspended from a string. Before submersion the tension in the string is 10 N. When the metal is submerged in water, the tension is 8 N. What is the density ρ of the metal?

Before submersion the tension is $T_i = \rho g V$. After submersion the tension is $T_f = (\rho - \rho_0) g V$, where the density of water is $\rho_0 = 10^3$ kg m^{-3}. Dividing the second

equation by the first to eliminate V gives

$$\frac{T_f}{T_i} = \frac{\rho - \rho_0}{\rho}$$

Solving for ρ,

$$\rho = \frac{\rho_0 T_i}{T_i - T_f}$$

$$= \frac{(1000 \text{ kg m}^{-3})(10 \text{ N})}{(10 \text{ N} - 8 \text{ N})}$$

$$= 5000 \text{ kg m}^{-3}$$

An object less dense than a fluid will float partially submerged. If a portion V_s of its volume V is submerged, then the buoyant force is $\rho_0 g V_s$. This must equal the weight $\rho g V$ of the object, so $\rho_0 g V_s = \rho g V$, or

$$\frac{\rho}{\rho_0} = \frac{V_s}{V}$$

Thus the ratio of the densities is equal to the fraction of the volume submerged. This is illustrated in the following example of an iceberg.

Example 13.2

The density of ice is 920 kg m^{-3} while that of sea water is 1025 kg m^{-3}. What fraction of an iceberg is submerged?

The fraction is

$$\frac{V_s}{V} = \frac{\rho}{\rho_0} = \frac{920 \text{ kg m}^{-3}}{1025 \text{ kg m}^{-3}} = 0.898$$

Almost 90 percent of the iceberg is submerged.

13.2 | THE EQUATION OF CONTINUITY; STREAMLINE FLOW

Consider a situation where an incompressible fluid completely fills a channel such as a pipe or an artery. Then if more fluid enters one end of the channel, an equal amount must leave the other end. This principle, which can be put into various mathematical forms, is called the *equation of continuity.* It will be quite useful in many of our discussions.

Suppose that a certain volume of the incompressible fluid enters the channel each second in units of m^3 s^{-1}. We will call this the *flow rate* Q_1. If the fluid enters one end at a rate Q_1, it must leave the other end at a rate Q_2, which is the same rate. Thus the equation of continuity can be written as

$$Q_1 = Q_2 \qquad (13.2)$$

ARCHIMEDES
(287–212 B.C.)

(The Bettmann Archive.)

Archimedes was a mathematical and scientific genius whose equal did not appear until Newton, 2000 years later. Born the son of an astronomer, his relationship with King Hiero II of Syracuse afforded him both independent means and the attention of a man of power.

Archimedes is best remembered for his ingenious machines, which captured the popular imagination, played a role in the wartime defense of Syracuse from the Romans, and in the end resulted in larger-than-life stories about him. For example, the seige of Syracuse by the Romans during the Carthaginian war lasted 3 years, largely because of the defenses developed by Archimedes. It is said that he developed giant reflecting mirrors that focused sunlight on the Roman ships, setting them on fire. Giant cranes were also used to lift and upset ships. The war ended badly for Archimedes, as he is said to have been stabbed to death by one of the conquering Roman soldiers. At the time, he was doing geometry in the sand and refused to be disturbed.

Many of the stories about Archimedes' inventions are distorted because he himself considered them beneath the dignity of the pure scientist and never left any written record of them. His written testaments include treatises on geometry, where he approached the foundations of calculus, equilibrium and the center of gravity, and hydrostatics.

The principle of hydrostatics that carries his name grew from a problem concerning the gold content of a crown made for Hiero. Archimedes was asked to determine whether the crown was pure gold or was adulterated with silver. After much thought, he realized while at the public baths that he could measure the volume of the crown very accurately by placing it in water and measuring the amount displaced. He then compared the weight of the crown with the weight of an equal volume of gold. This discovery apparently had at least two immediate results. The idea so excited him that Archimedes dashed home naked shouting "Eureka, eureka," and the goldsmith was executed because the crown was not pure.

Archimedes fully understood and demonstrated the importance of the lever. He is said to have arranged a lever and pulley system with which the king himself pulled a large ship onto the shore. His work on the lever and on the importance of the center of gravity laid much of the groundwork for modern mechanics.

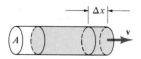

Figure 13.2. The flow rate in a tube is $Q = Av$.

For example, if $1 \ \text{m}^3 \ \text{s}^{-1}$ enters, then a $1 \ \text{m}^3 \ \text{s}^{-1}$ must leave.

This can be put into a more useful form if all the fluid in the channel is moving with a uniform velocity v. Consider a section of a tube with a constant cross-sectional area A (Fig. 13.2). In a time Δt, the fluid moves a distance $\Delta x = v \, \Delta t$, and the volume of fluid leaving the tube is $\Delta V = A \, \Delta x = Av \, \Delta t$. Alternatively, ΔV equals the flow rate Q times the time interval Δt, or $\Delta V = Q \, \Delta t$. Comparing these expressions for ΔV, we see that

$$Q = Av \qquad (13.3)$$

The flow rate equals the cross-sectional area of the channel times the velocity of the fluid.

For a channel whose cross section changes from A_1 to A_2, this result together with $Q_1 = Q_2$ gives another form of the equation of continuity,

$$A_1 v_1 = A_2 v_2 \qquad (13.4)$$

The product of the cross-sectional area and the velocity of the fluid is constant. If at some point, A decreases, v must increase. For example, if the area is halved, then the velocity doubles.

Usually the flow velocity is not uniform in a channel. For example, in the next chapter we will encounter situations where the fluid near the walls of the channel is moving at a lower speed than the fluid near its center. The equation of continuity still holds for such cases if it is written in terms of the average flow velocity $\bar{v}$. The flow rate is $Q = A\bar{v}$, and at two points in a channel, $A_1 \bar{v}_1 = A_2 \bar{v}_2$.

Streamline Flow | A type of fluid flow, called *streamline flow,* is most readily discussed quantitatively and is important in many applications. It is best defined by imagining a simple experiment. Suppose we use a very fine eyedropper to inject some ink into a fluid, without interrupting the flow (Fig. 13.3a). If the ink line formed in this way does not disperse or mix but remains narrow and well defined, the flow is called *streamline.* If several eyedroppers injected ink side by side, the ink pattern might look like Fig.

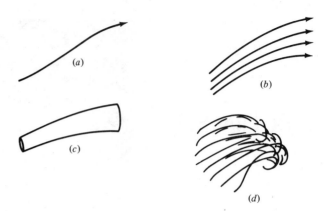

Figure 13.3. (a) A single streamline. (b) A group of adjacent streamlines. (c) A flow tube. The walls of the tube are composed of streamlines. (d) Turbulent flow.

13.3b. Finally, we can imagine a *tube* of streamlines, as in Fig. 13.3c. The fluid in this tube is undergoing streamline flow. If, instead, the ink lines swirl and mix, the flow is said to be *turbulent* (Fig. 13.3d).

The tube is a useful concept because, by definition, the streamlines do not cross one another; no fluid flows into or out of the sides of the tube. Accordingly, the equation of continuity can be applied to this flow tube, and *the product Av is the same at all points in the flow tube.* This property of streamline flow is used in the next section.

13.3 | BERNOULLI'S EQUATION

We now consider Bernoulli's equation.* It states the consequences of the principle that the work done on a fluid as it flows from one place to another is equal to the change in its mechanical energy. Bernoulli's equation can be used under the following conditions:

1 The fluid is *incompressible;* its density remains constant.
2 The fluid does not have appreciable frictional effects; it is *nonviscous.* Consequently, no mechanical energy is lost due to friction.

*This important equation is usually said to have first been derived in the 1730's by Daniel Bernoulli (1700–1783). However, it may actually have been obtained somewhat earlier by his father Johann Bernoulli (1667–1748). In any case, the brilliant mathematician Leonhard Euler (1707–1783) provided the first completely rigorous derivation some years later. For a discussion of an unusual family squabble and episode in the history of science, see the preface by Hunter Rouse to the English translation of *Hydrodynamics* by Daniel Bernoulli and *Hydraulics* by Johann Bernoulli, Dover Publications, Inc., New York, 1968.

3 The flow is *streamline, not turbulent*.

4 The velocity of the fluid at any point does not change during the period of observation. (This is called the *steady-state* assumption.)

In this section, we see how Bernoulli's equation is obtained by using the relation between work and mechanical energy. Its applications are discussed in the following sections.

Consider the fluid in a section of a flow tube with a constant cross section A (Fig. 13.4*a*). According to the equation of continuity, the product Av remains constant. Thus the velocity v does not change as the fluid moves through the tube, and its kinetic energy remains the same. However, the potential energy changes as the fluid rises.

The net force on the fluid in the tube due to the surrounding fluid is the cross-sectional area A times the difference in pressures on the ends, or $(P_a - P_b)A$. If the fluid in the section moves a short distance Δx, then the work done on it is the product of the force and the displacement $(P_a - P_b)A \, \Delta x$. Since $A \, \Delta x$ is the volume ΔV of the fluid leaving the section (Fig. 13.4*b*), the work done on the fluid is

$$W = (P_a - P_b) \, \Delta V$$

This work done on the fluid must equal the increase $\Delta \mathcal{U}$ in its potential energy. $\Delta \mathcal{U}$ can be calculated if we note that the fluid leaving the section has a mass $\rho \, \Delta V$ and a potential energy $(\rho \, \Delta V)gy_b$, while the fluid entering at the bottom of the section

has a potential energy $(\rho \, \Delta V)gy_a$. Thus $\Delta \mathcal{U} = \rho g \, \Delta V(y_b - y_a)$. Equating this to W, we have

$$P_a - P_b = \rho g(y_b - y_a)$$

or

$$P_a + \rho g y_a = P_b + \rho g y_b \qquad (v = \text{constant}) \quad (13.5)$$

Thus the pressure P plus the *potential energy per unit volume* $\rho g y$ of the fluid is the same everywhere in the flow tube if the velocity remains constant.

More generally, if the cross-sectional area of the flow tube changes, the fluid velocity v and *kinetic energy per unit volume* $\frac{1}{2}\rho v^2$ will also change. The work done on the fluid must then be set equal to the change in the potential plus kinetic energy of the fluid. The result is *Bernoulli's equation*,

$$P_a + \rho g y_a + \tfrac{1}{2}\rho v_a{}^2 = P_b + \rho g y_b + \tfrac{1}{2}\rho v_b{}^2 \quad (13.6)$$

The pressure plus the total mechanical energy per unit volume, $P + \rho g y + \frac{1}{2}\rho v^2$, is the same everywhere in a flow tube.

Bernoulli's equation is the main result of this chapter. As is clear from its derivation, it is a restatement of the relationship between work and energy in a form suitable for use in the physics of fluids. Examples of its uses are given in the following sections.

13.4 | STATIC CONSEQUENCES OF BERNOULLI'S EQUATION

We first examine the implications of Bernoulli's equation when the fluid is at rest, so $v = 0$ and $P + \rho g y$ is constant.

Fluid at Rest in a Container | Suppose fluid is at rest in a container shaped like the one in Fig. 13.5. We can find the pressure at a point B in terms of the pressure at the surface and the depth. To do this, we calculate $P + \rho g y$ at points A and B, choosing the

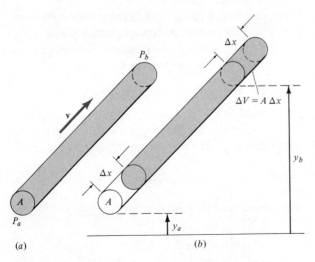

(a)   (b)

Figure 13.4. (*a*) The fluid in a section of flow tube with a constant cross section has the same velocity everywhere. (*b*) The fluid has moved a short distance Δx.

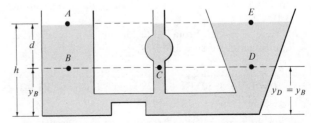

Figure 13.5. Fluid at rest in a container. As is explained in the text, the level of the surface of the fluid in each of the segments is the same.

y axis so that $y = 0$ at the bottom of the container. At A the pressure is atmospheric pressure, P_{atm}; at B the pressure is P_B. Thus, $P_{atm} + \rho gh = P_B + \rho gy_B$ or, with $h - y_B = d$,

$$P_B = P_{atm} + \rho gd \qquad (13.7)$$

This result shows that the pressure at a depth d in a fluid at rest is equal to the surface pressure plus the potential energy density change ρgd corresponding to this depth. Equation 13.7 can also be interpreted as a statement of the force on a unit area a distance d below the surface of the liquid. The force per unit area P_B is the sum of two terms: P_{atm}, the pressure due to the atmosphere, and ρgd, the pressure due to the weight of the liquid above the point B.

Calculating $P + \rho gy$ at points B and D gives $P_B + \rho gy_B = P_D + \rho gy_D$. Or, since $y_B = y_D$,

$$P_B = P_D$$

Thus the pressure at the same depth at two places in a fluid at rest is the same. In particular, since points A and E are both at atmospheric pressure, the surface of the liquid is at the same height at both points. Thus the surfaces of liquids at rest in connected containers of any shape must be at the same height if they are open to the atmosphere.

These ideas are illustrated in the next two examples.

Example 13.3

What is the pressure on a swimmer 5 m below the surface of a lake?

Using $d = 5$ m and $\rho = 1000$ kg m^{-3}, we find

$$
\begin{aligned}
P_B &= P_{atm} + \rho gd \\
&= 1.013 \times 10^5 \text{ Pa} + (1000 \text{ kg m}^{-3})(9.8 \text{ m s}^{-2})(5 \text{ m}) \\
&= 1.50 \times 10^5 \text{ Pa}
\end{aligned}
$$

Example 13.4

The pressure 1 m above a floor is measured to be normal atmospheric pressure, 1.013×10^5 Pa. How much greater is the pressure at the floor if the temperature is 0° C?

Here, $d = 1$ m. From Table 13.1 the density of air at atmospheric pressure and 0° C is 1.29 kg m^{-3}. Thus

$$
\begin{aligned}
P_B &= P_{atm} + \rho gd \\
&= 1.013 \times 10^5 \text{ Pa} + (1.29 \text{ kg m}^{-3})(9.8 \text{ m s}^{-2})(1 \text{ m}) \\
&= (1.013 \times 10^5 + 12.6) \text{ Pa}
\end{aligned}
$$

Thus the pressure at the floor is greater by 12.6 Pa or by about 1 part in 10^4, which is negligible for most purposes. The pressure change is small, unlike in the preceding example, because the density of the air is very small compared to that of a typical liquid. The proper-

ties of gases were discussed in Chapter Ten, with the assumption that the pressure in a gas-filled container is the same everywhere. This is usually a good approximation. However, if the pressure of a gas is measured at two very different heights, a considerable difference will be observed. For example, the atmospheric pressure at Aspen, Colorado, which is 2500 m above sea level, is about 80 percent of the sea level pressure.

The Manometer | The open-tube manometer is a U-shaped tube used for measuring gas pressures. It contains a liquid that may be mercury or, for measurements of low pressures, water or oil. One end of the tube is open to the atmosphere, and the other end is in contact with the gas in which the pressure is to be measured (Fig. 13.6). The manometer can also be used to measure pressures in a liquid, provided that the liquid does not mix with the manometer fluid.

Measuring heights from the base of the U tube, $P + \rho gy$ is $P + \rho gy_1$ at the left surface of the column and $P_{atm} + \rho gy_2$ at the right surface. Equating these,

$$P + \rho gy_1 = P_{atm} + \rho gy_2$$

or

$$
\begin{aligned}
P &= P_{atm} + \rho g(y_2 - y_1) \\
&= P_{atm} + \rho gh \qquad (13.8)
\end{aligned}
$$

Thus, a measurement of the height difference h of the two columns determines the gas pressure P. In a blood pressure gauge, the sphygmomanometer, the pressure measured is that in an air sack wrapped around the upper arm. We discuss such measurements in Section 13.6.

The pressure P in Eq. 13.8 is the *absolute pressure*. The difference between this and atmospheric pressure, $P - P_{atm}$, is the *gauge pressure*. The gauge pressure is then exactly equal to ρgh.

Blood Pressure Measurements by Cannulation | In many experiments with anesthetized animals, the blood pressure in an artery or vein is measured by the direct insertion into the vessel

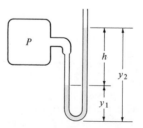

Figure 13.6. The open-tube manometer.

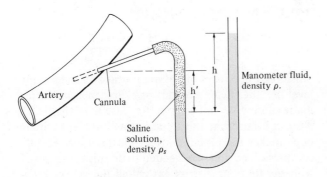

Figure 13.7. Measurement of blood pressure by cannulation.

of a *cannula*, which is a small glass or plastic tube containing saline solution plus an anticlotting agent. The saline solution, in turn, is in contact with the fluid in a manometer. It is necessary to have the surface of contact between the saline solution and the manometer fluid either at the same level as the insertion point of the cannula or to correct for the height difference (Fig. 13.7). Calculating $P + \rho gy$ at suitable points (Problem 13-45), it follows that the blood pressure P_B is given by

$$P_B = P_{atm} + \rho gh - \rho_s gh' \qquad (13.9)$$

Mercury is commonly used as the manometer fluid in measurements of arterial pressures. However, pressures in the veins are relatively low, and the use of mercury as a manometer fluid would give poor accuracy because h would be very small. Consequently the saline solution is used as the manometer fluid.

Physiologists often use electromanometers in which the manometer fluid pushes against a membrane instead of rising up a tube. The flexing of the membrane is proportional to the pressure, and the extent that the membrane is flexed is translated into an electrical signal. This signal drives a recorder that automatically gives a continuous record of the blood pressure.

13.5 | THE ROLE OF GRAVITY IN THE CIRCULATION

When animals evolved to the point where they spent a significant amount of time standing upright, a surprising number of changes in the circulatory system were required. Of particular importance is the venous system used to return blood from the lower extremeties to the heart. Humans have adapted to the problems of moving blood upward a large distance against the force of gravity. Animals that have not, such as

snakes, eels, and even rabbits, will die if held head upwards; the blood remains in the lower extremeties, and the heart receives no blood from the venous system.

Figure 13.8 shows what is observed if a person's large arteries are cannulated. In the reclining position, the pressures everywhere are almost the same. The small pressure drop between the heart and the feet or brain is due to the viscous forces. However, the pressures at the three points are quite different in the standing person, reflecting the large difference in their heights.

Since the viscous effects are small, we can use Bernoulli's equation, $P + \rho gh + \frac{1}{2}\rho v^2 =$ constant, to analyze this situation. The velocities in the three arteries are small and roughly equal, so the $\frac{1}{2}\rho v^2$ term can be ignored. Hence the gauge pressures at the heart P_H, at the foot P_F, and at the brain P_B are related by

$$P_F = P_H + \rho gh_H = P_B + \rho gh_B \qquad (13.10)$$

where ρ is the density of blood.

In discussions of the circulatory system, it is convenient to measure pressures in *kilopascals*, kPa, where $1\,\text{kPa} = 10^3$ Pa is a multiple of the basic S.I. pressure unit. Many discussions of the circulatory system use an older unit, the *torr*; 1 torr = 0.1333 kPa. In Eq. 13.10, typical values for adults are $h_H \doteq 1.3$ m and $h_B = 1.7$ m. With $\rho = 1.0595 \times 10^3$ kg m^{-3}, we find

$$\begin{aligned} P_F - P_H &= \rho gh_H \\ &= (1.0595 \times 10^3 \text{ kg m}^{-3})(9.8 \text{ m s}^{-2})(1.3 \text{ m}) \\ &= 1.35 \times 10^4 \text{ Pa} = 13.5 \text{ kPa} \end{aligned}$$

P_H is typically 13.3 kPa, so $P_F \simeq 26.8$ kPa. In a similar way we find $P_B = 9.3$ kPa. This explains why the

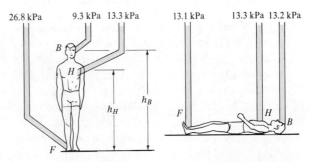

Figure 13.8. Schematic view of the results of cannulation of arteries in various parts of the human body when standing and reclining. The pressures shown are averaged over the heart cycle.

pressures in the lower and upper parts of the body are very different when the person is standing, although they are about equal when reclining.

This situation poses several problems. The most important are the tendency for blood to drain out of the venous side of the upper body back to the heart and the difficulty of lifting blood from the lower extremeties up to the heart. To retard drainage from the venous side of the upper body, particularly from the brain where constant volume and flow rate are extremely important, the muscles surrounding the veins contract and cause constriction. In the lower extremeties, because the veins have a much larger capacity for passive expansion and blood storage than do arteries, the problem is to pump the blood "uphill." The veins in the extremeties contain valves that open when blood flows toward the heart and close if the blood moves away from the heart. Blood is returned to the heart, at least partially, by the pumping action associated with breathing and by the flexing of skeletal muscle, as in walking. These muscle contractions squeeze the veins, and the valves ensure that the resultant blood flow is toward the heart. The importance of this is illustrated by the fact that a soldier who is required to stand at strict attention may faint because of insufficient venous return. Once horizontal, the pressures are equalized, and the soldier regains consciousness.

Effects of Acceleration |
When a person in an erect position experiences an upward acceleration **a**, the effective weight becomes $m(g + a)$. Applying Bernoulli's equation to the brain and heart with g replaced by $(g + a)$,

$$P_B + \rho(g + a)h_B = P_H + \rho(g + a)h_B$$

or

$$P_B = P_H - \rho(g + a)(h_B - h_H) \qquad (13.11)$$

Thus the blood pressure in the brain will be reduced even further. It has been found that if a is two or three times g, a human will lose consciousness because of the collapse of arteries in the brain. This factor limits the speed with which a pilot can pull out of a dive (Chapter Five). A related experience is the feeling of light-headedness that sometimes occurs when one suddenly stands up. Since muscular movement is required to activate the venous return mechanism, blood will tend to collect in the lower veins until normal activity is resumed.

13.6 | BLOOD PRESSURE MEASUREMENTS USING THE SPHYGMOMANOMETER

Since the upper arm of a human is at about the same level as the heart, blood pressure measurements made there give values close to those near the heart. Also, the fact that the upper arm contains a single bone makes the brachial artery located there easy to compress. The pressure needed to do this is measured with the familiar instrument called the *sphygmomanometer,* which is convenient and painless (Fig. 13.9).

During a complete heart pumping cycle, the pressure in the heart and circulatory system goes through both a maximum (as the blood is pumped from the heart) and a minimum (as the heart relaxes and fills with blood returned from the veins). The sphygmomanometer is used to measure these extreme pressures. Its use relies on the fact that blood flow in the arteries is not always streamline. When the arteries are constricted and the blood flow rate is large, the flow becomes turbulent. This turbulent flow is noisy and can be heard with a stethoscope.

In the sphygmomanometer the gauge pressure in an air sack wrapped around the upper arm is measured using a manometer or a dial pressure gauge. The pressure in the sack is first increased until the brachial artery is closed entirely. The pressure in the sack is then slowly reduced, while a stethoscope is used to listen for noises in the brachial artery below the sack. When the pressure is slightly below the *systolic* (peak) pressure produced by the heart, the artery will open briefly. Because it is only partially opened, the flow

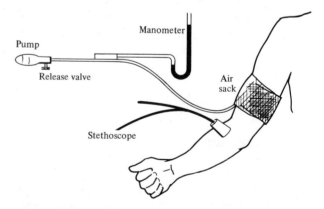

Figure 13.9. The use of the sphygmomanometer for measuring blood pressures.

velocity is high and turbulent and therefore noisy. This resulting noise is heard as a tapping sound.

When the pressure in the sack is lowered further, the artery remains open during longer portions of the heart cycle but is still closed during the *diastolic* (minimum) pressure portion of the cycle. Hence, sounds are heard, but they are interrupted by periods of silence. When the pressure in the sack reaches the diastolic pressure, the artery remains open during the entire heart cycle. At this pressure the flow is still turbulent and noisy (particularly at diastolic pressure), but the sound is continuous. Thus, both the systolic and diastolic pressures can be measured without the cannulation technique.

Blood pressures are usually presented as systolic/diastolic ratios. Typical readings for a resting healthy adult are about 120/80 in torr and 16/11 in kPa. The borderline for high blood pressure (hypertension) is usually defined to be 140/90 in torr and 19/12 in kPa. Pressures appreciably above that level require medical attention, because prolonged high blood pressure can lead to serious damage of the heart or other organs before a person is aware of any problem. Increasing emphasis has been placed in recent years on mass screenings to discover people with undetected high blood pressure.

13.7 | DYNAMIC CONSEQUENCES OF BERNOULLI'S EQUATION

Sometimes the velocity terms in Bernoulli's equation, Eq. 13.6, are comparable to or larger than the gravitational terms. In this section and the next, we discuss examples where these dynamic terms are important.

A very simple demonstration of the effect of flow rate on pressure can be performed by the reader. Tear a piece of paper in half and, holding the halves side by side about 2 cm apart, as in Fig. 13.10, blow be-

tween them. Since the flow is horizontal, the terms in Bernoulli's equation containing y are equal and cancel. If the pressure and velocity between the sheets are P_B and v_B, respectively, and those outside the sheets are P_0 and v_0, then

$$P_B + \tfrac{1}{2}\rho v_B{}^2 = P_0 + \tfrac{1}{2}\rho v_0{}^2$$

If we rearrange these terms, we find

$$P_0 - P_B = \tfrac{1}{2}\rho(v_B{}^2 - v_0{}^2)$$

Now v_B is larger than v_0, so the right-hand side is positive and P_0 must be greater than P_B. This pressure difference results in the sheets moving toward one another, as you will observe if you try it.

The fact that the pressure drops when the velocity increases for a fluid moving at a constant height is a consequence of energy conservation. The kinetic energy can increase only if work is done. That means a net force must act on a segment of the moving fluid, and that the pressure must be lower at the end where the velocity is higher.

This pressure drop associated with increasing fluid velocities has many everyday implications. For example, you may have been in a car that was pulled toward a large truck as it passed you on a highway. The air rushes around the vehicles and the portion passing between them is forced through a small area. From the continuity equation, Eq. 13.4, this air moves faster as a result. In the region of high velocity, the pressure is reduced and the vehicles are pulled toward each other. From the point of view of the drivers, it is rather like blowing between sheets of paper.

Part of the problem with windows blowing out in Boston's Hancock Building was due to the relatively low pressure outside (compared to inside), which occurred when the wind was blowing (Fig. 13.11).

In the following section we describe two quantitative uses of Bernoulli's equation.

13.8 | FLOW METERS

In this section we apply the complete Bernoulli equation to two kinds of flow meters. With minor modifications, they can be used to measure the flow in blood vessels, the airspeed of planes, and many other flow rates.

The Venturi Tube | In a *venturi tube,* fluid flows through different cross-sectional areas in different portions of the tube. When the tube narrows, the

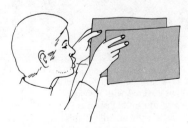

Figure 13.10. A simple test of Bernoulli's equation. When you blow between two pieces of paper, they move closer together.

Figure 13.11.   Hancock Building in Boston, Massachusetts. In 1973, a prominent feature of the Boston skyline was this unoccupied skyscraper with its windows covered with plywood (white rectangles) because of their unfortunate tendency to blow out during high winds. The pressure differences predicted by Bernoulli's equation provide a partial explanation of this phenomenon. (The Boston Globe.)

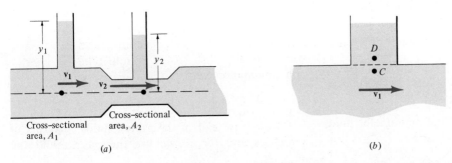

Figure 13.12. (a) Venturi tube. (b) Enlarged view of the region where column 1 connects to the flow tube.

fluid velocity increases. Then, according to Bernoulli's equation, the pressure drops. Measurment of this pressure change determines the fluid velocity. The pressure measurement can be made with narrow vertical columns inserted into the main tube (Fig. 13.12a) or with electrical sensors.

Before using Bernoulli's equation, we must clear up a difficulty that arises because the liquid in the columns is at rest while the liquid in the tube is moving. Bernoulli's equation cannot be applied directly to relate the pressures at points C and D in Fig. 13.12b because the fluid at the two points is not in the same streamline. However, if the pressures were unequal, fluid would flow from one point to the other. Since this does not occur, $P_C = P_D$. Thus the pressure in the columns is the same as the pressure in the streamline.

Bernoulli's equation requires that $P + \rho g y + \frac{1}{2}\rho v^2$ is the same everywhere in a flow tube. Applying Bernoulli's equation to points at the same height in the flow stream just below the columns,

$$P_1 + \tfrac{1}{2}\rho v_1^2 = P_2 + \tfrac{1}{2}\rho v_2^2$$

From the continuity equation, $A_1 v_1 = A_2 v_2$, or

$$v_2 = \frac{A_1}{A_2} v_1 \qquad (13.12)$$

Using this expression for v_2, the preceding equation can be written as

$$P_1 - P_2 = \frac{1}{2}\rho v_1^2 \left[\frac{A_1^2}{A_2^2} - 1 \right] \qquad (13.13)$$

Thus, a measurement of $P_1 - P_2$ and knowledge of the areas determines v_1; v_2 can also be found using Eq. 13.12.

The following example shows how this flow meter can be used to measure the velocity of the blood in an artery.

Example 13.5

The flow of blood through a large artery in a dog is diverted through a venturi flow meter. The wider part of the flow meter has an area $A_1 = 0.08$ cm^2, which equals the cross-sectional area of the artery. The narrower part of the flow meter has an area $A_2 = 0.04$ cm^2. The pressure drop in the flow meter is 25 Pa. What is the velocity v_1 of the blood in the artery?

The ratio of the areas A_1/A_2 is dimensionless and has the value $0.08/0.04 = 2$. From Table 13.1, the density of whole blood is 1059.5 kg m^{-3}. Dropping the units, Eq. 13.13 becomes

$$25 = \tfrac{1}{2}(1059.5)v_1^2(2^2 - 1)$$

Solving for v_1,

$$v_1 = \sqrt{\frac{(2)(25)}{(1059.5)(2^2 - 1)}} = 0.125 \text{ m s}^{-1}$$

The Prandtl Tube | Figure 13.13 shows a *Prandtl tube* inserted in a flow stream. It interrupts the flow pattern very little except at point A, where

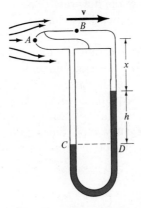

Figure 13.13. A Prandtl tube in a constant velocity flow stream. The right arm of the U-tube connects to the chamber opening at B. The left connects to the opening at A, where the fluid has zero velocity.

the fluid has zero velocity. At point B the velocity is assumed to be the streamline flow velocity v. From Bernoulli's equation, neglecting the small difference in heights at A and B,

$$P_A - P_B = \frac{1}{2}\rho v^2$$

where ρ is the fluid density. If the manometer fluid has density ρ_m, then $P_C = P_D$ gives

$$P_A + \rho g(h + x) = P_B + \rho g x + \rho_m g h$$

or

$$P_A - P_B = (\rho_m - \rho)g h$$

Comparing the two expressions for $P_A - P_B$,

$$\tfrac{1}{2}\rho v^2 = (\rho_m - \rho)g h \qquad (13.14)$$

Thus, a manometer reading gives a direct measurement of the flow velocity. As in the venturi tube, electromanometers may be employed instead of the open-tube manometer.

SUMMARY

The fundamental variables used in describing fluids are the density and pressure. The density is the mass per unit volume. The pressure is the force per unit area that a segment of fluid exerts on adjacent segments.

The definition of the pressure leads directly to Archimedes' principle: the buoyant force on an object in a fluid is equal to the weight of the fluid displaced.

If a fluid may be treated as incompressible, the flow satisfies the continuity equation. This equation states that when fluid flows in a tube, the flow rate Q must be constant even if the tube dimensions change, so the product Av is constant.

With the additional assumption that frictional or viscous forces are unimportant, we can equate the work done on a fluid to its change in mechanical energy. The result is Bernoulli's equation, which states that $P + \rho g y + \frac{1}{2}\rho v^2$ is constant during steady-state flow everywhere in a streamline flow tube. This equation can be applied to fluids in motion or at rest.

Checklist

Define or explain:

buoyant force	equation of continuity
Archimedes' principle	streamline flow
flow rate	turbulent flow
flow tube	absolute pressure
Bernoulli's equation	gauge pressure
manometer	venturi tube
cannulation	Prandtl tube
sphygmomanometer	

REVIEW QUESTIONS

Q13-1   Instead of describing fluids in terms of masses and forces, we use the _____ and the _____.

Q13-2   What two forces act on a segment of fluid at rest to keep it in equilibrium?

Q13-3   The equation of continuity can be used to explain why a stream of fluid moves faster in _____ portions of the stream than in _____ portions.

Q13-4   If a stream of fluid has its cross-sectional area halved in a certain region, its average velocity is _____.

Q13-5   In Bernoulli's equation, the terms $\rho g y + \frac{1}{2}\rho v^2$ represent the _____.

Q13-6   In static fluids, the pressure difference between two points in the fluid is determined by the fluid density, g, and the _____.

Q13-7   A person suffering from "light headedness" can be relieved by _____.

Q13-8   When using the sphygmomanometer, why is the blood pressure usually measured in the upper arm?

Q13-9   In fluid motion, regions of high average velocity tend to have _____ pressures.

Q13-10   In a pipe with a constriction in it, will the fluid pressure in the constriction be higher or lower than in the wider part of the pipe?

EXERCISES

Section 13.1 | Archimedes' Principle

13-1   A 75-kg man just floats in fresh water with virtually all of his body below the surface. What is his volume?

13-2   An object weighs 100 N in air and 75 N in water. What is the relative density of the object?

13-3   A tank of water on a scale weighs 200 N. A 65 N salmon is placed in the tank and swims around. What is the reading on the scale?

13-4   A balloon has a capacity of 0.1 m³. What weight can it lift when it is filled with helium? Use the densities of helium and air at 0° C of Table 13.1.

13-5 A log of mass 40 kg is dropped into a river at 0° C. If the relative density of the log is 0.8, what will be the volume of the log above the surface?

Section 13.2 | The Equation of Continuity; Streamline Flow

13-6 The radius of a water pipe decreases from 0.2 to 0.1 m. If the average velocity in the wider portion is 3 m s^{-1}, find the average velocity in the narrower region.

13-7 A garden hose with a cross-sectional area of 2 cm^2 has a flow of 200 cm^3 s^{-1}. What is the average velocity of the water?

13-8 A blood vessel of radius r splits into four vessels, each with radius $r/3$. If the average velocity in the larger vessel is v, find the average velocity in each of the smaller vessels.

Section 13.3 | Bernoulli's Equation

13-9 Can Bernoulli's equation be used to describe the flow of water through a rapids in a stream? Explain.

13-10 A baseball is thrown by a pitcher and curves as it approaches the batter. Can Bernoulli's equation be applied to this problem using a reference frame (a) moving with the ball; (b) fixed with respect to the ground? Explain.

13-11 If water moves through pipes of constant cross section from the basement to the first floor, will the pressure stay the same? Explain.

Section 13.4 | Static Consequences of Bernoulli's Equation

13-12 Deep-sea photographs have been made at depths of 8000 m. (a) What is the pressure at this depth? (b) What is the force on the camera window if it measures 0.1 by 0.15 m?

13-13 What is the pressure difference between the heart and brain of a giraffe if the brain is 2 m above the heart? (Assume that the velocity of the blood is the same in both locations.)

13-14 Estimate the drop in the pressure of the atmosphere as one goes from sea level to the top of a hill 500 m high if the temperature is 0° C.

13-15 How high can water rise in the pipes of a building if the gauge pressure at the ground floor is 2 × 10^5 Pa?

13-16 A submarine dives to a depth of 100 m in sea water. What gauge pressure is necessary to expel water from its ballast tanks?

13-17 A certain pressure can support a column of pure water 0.7 m high. The same pressure will support a column of saline solution 0.6 m high. What is the density of the saline solution?

13-18 A large artery is cannulated, and a saline solution of density 1300 kg m^{-3} is used as the manometer fluid. What is the blood pressure (gauge pressure) if the height difference in the manometer tubes is 0.67 m?

Section 13.5 | The Role of Gravity in the Circulation

13-19 When a man stands, his brain is 0.5 m above his heart. If he bends so that his brain is 0.4 m below his heart, by how much does the blood pressure in his brain change?

13-20 Explain why we are uncomfortable when we bend with our heads lower than our heart.

13-21 A pilot is accelerating downward at four times the acceleration of gravity during a maneuver. If he is erect, what is the blood pressure in his brain?

13-22 A jet pilot pulls his plane out of a dive in such a way that his upward acceleration is 3 g's. What would you predict as the blood pressure in the brain?

13-23 At what acceleration would you expect the blood pressure in the brain to drop to zero for an erect person? (Assume there are no body mechanisms operating to compensate for such conditions.)

13-24 (a) If an elevator accelerates upward at 9.8 m s^{-2}, what is the average blood pressure in the brain? What is the average blood pressure in the feet? Assume the person is standing. (b) If the elevator accelerates downward at 9.8 m s^{-2}, what is the average blood pressure in the brain and feet?

Section 13.6 | Blood Pressure Measurements Using the Sphygmomanometer

13-25 If a sphygmomanometer were used to measure the blood pressure in the leg of a man sitting at rest, would the results give the pressure at the heart? Explain.

13-26 Suppose a man ran into a doctor's office and immediately had his blood pressure measured. Would a reading above the normal 120/80 in torr (or 16/11 in kPa) indicate he is suffering from high blood pressure? Explain.

Section 13.8 | Flow Meters

13-27   A venturi tube has a radius of 1 cm in its narrower portion and 2 cm in its wider sections. The velocity of water in the wider sections is 0.1 m s^{-1}. Find (a) the pressure drop, and (b) the velocity in the narrower portion.

13-28   A Prandtl tube flow meter is used to measure the aortic blood velocity of a dog. If water is the manometer fluid, what is the height difference in the manometer tube when the blood velocity is 0.1 m s^{-1}?

13-29   Suppose a venturi tube is equipped with vertical columns, as in Fig. 13.12. Show that

$$g(y_1 - y_2) = \frac{1}{2} v_1^2 \left(\frac{A_1^2}{A_2^2} - 1 \right)$$

13-30   In a standard classroom demonstration, a Ping-Pong ball will stay in an upside-down funnel when a stream of air passes through the funnel from the small end (Fig. 13.14). Explain briefly why the ball does not fall.

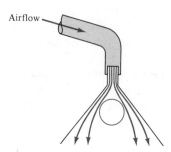

Figure 13.14.   A ping-pong ball stays in an inverted funnel when air is blown through it. Exercise 13-30.

PROBLEMS

13-31   A block of oak weighs 90 N in air. A lead weight weighs 130 N when immersed in water. When attached together, they weigh 100 N in water. What is the density of the wood?

13-32   Find the initial acceleration of an iron ball with a density 8 times that of water when placed in (a) water; (b) mercury. (c) What is the direction of the acceleration in each case?

13-33   A man rows out to the center of a lake with a number of large rocks in the boat. He then throws them overboard. When he finishes, has the lake level risen, fallen, or remained the same? Explain.

13-34   An ice cube floats in a glass filled to the top with water. What happens to the water level as the ice cube melts? Explain.

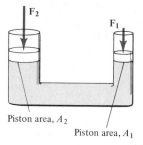

Figure 13.15.   A hydraulic jack. Problems 13-35 and 13-36.

13-35   A *hydraulic jack* is constructed as shown in Fig. 13.15. The cross-sectional areas of the pistons are A_1 and A_2. Show that in equilibrium the forces F_1 and F_2 are related by

$$\frac{F_1}{A_1} = \frac{F_2}{A_2}$$

13-36   A hydraulic lift similar to that in Fig. 13.15 with pistons of cross-sectional areas 1500 cm^2 and 75 cm^2 is used to lift a dentist's chair of weight 1500 N. (a) What force is necessary on the small piston to raise the chair? (b) What distance must the small piston be moved to raise the chair 0.1 m?

13-37   A tank contains oxygen gas at 0° C. The pressure at the bottom of the tank is 100 atm. If the tank is 1 m tall, what is the pressure at the top? (Assume the average density of the oxygen is 143 kg m^{-3}.)

13-38   A *barometer* is constructed by filling a long tube with mercury and then inverting it, open side down, in a container of mercury exposed to the atmosphere (Fig. 13.16). (a) Show that atmospheric pressure P_{atm} is equal to $\rho g h$, where ρ is the density of mercury. (b) What is the height h when $P_{atm} = 1.013 \times 10^5$ Pa?

13-39   During a whole-blood transfusion, the needle is inserted in a vein where the pressure is

Figure 13.16.   A mercury barometer. Problem 13-38.

2000 Pa. At what height must the blood container be placed, relative to the vein, so that the blood just enters the vein?

13-40 In Section 13.5, we assumed that the blood in the various large arteries had the same velocities. Suppose the velocity were 0.2 m s^{-1} in the artery near the heart and 0.01 m s^{-1} in the artery near the foot. How large a percentage error would that introduce in $P_F - P_H$?

*13-41 A *siphon tube* of cross-sectional area 3×10^{-4} m^2 is used to drain a tank of water. The tube is initially filled with water, and with the ends closed, one end is placed in the tank 0.25 m below the water surface. The other end is outside the tank and at a distance of 0.5 m below the immersed end. (a) What is the velocity of the water flowing out of the tube shortly after the ends are opened? (b) Is the flow continuous? (c) What is the velocity of the water when the water surface in the tank is 0.1 m above the immersed end?

*13-42 A siphon tube is filled with gasoline and closed at each end. One end is inserted in a gasoline tank 0.3 m below the surface of the gasoline. The other end is placed 0.2 m below the end in the tank and both ends of the tube are opened. The tube has an inner cross-sectional area of 4×10^{-4} m^2. The density of gasoline is 680 kg m^{-3}. (a) What is the velocity of the gasoline in the tube shortly after it is opened? (b) What is the corresponding rate of flow?

13-43 An artery or vein may become partially blocked or *occluded,* when some material reduces the vessel radius over a small portion of its length. (a) Does the velocity in the occluded region change? (b) Does the pressure in the occluded region change? (c) Describe the possible sequence of events in the flexible vessel if the pressure inside the vessel becomes very small compared to the pressure outside.

13-44 A sealed bottle has water in it, and two tubes enter the bottle through the seal, as shown in Fig. 13.17. Tube B is attached to a pump at one end, and the other end is above the water surface. If tube A is immersed 0.15 m below the water surface and the other end is open to the air, what is the minimum pressure obtained inside the bottle? What will occur when this pressure is reached? (Bottles of this type are used for draining unwanted fluids from the body, such as those that collect in and around the lungs in some diseases.)

13-45 The blood pressure in an artery is meas-

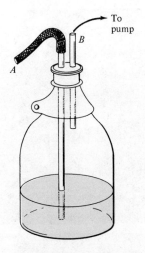

Figure 13.17. Problem 13-44.

ured by cannulation (Fig. 13.7). The cannula contains saline solution of density ρ_s, and the manometer fluid has density ρ. Show that the blood pressure is given by Eq. 13.9.

ANSWERS TO REVIEW QUESTIONS

Q13-1, density, pressure; **Q13-2,** weight, buoyant force; **Q13-3,** narrower, wider; **Q13-4,** doubled; **Q13-5,** mechanical energy per unit volume; **Q13-6,** height difference; **Q13-7,** lowering their head; **Q13-8,** it is close to the pressure in the heart; **Q13-9,** low; **Q13-10,** lower.

SUPPLEMENTARY TOPICS

13.9 | THE FLIGHT OF ANIMALS AND AIRPLANES

A full discussion of flight involves a combination of elaborate mathematical theories and practical experimentation because the flow patterns about wings are very complex. However, with the aid of Bernoulli's equation, we can derive some qualitative results applicable to airplanes and to birds, insects, and other flying animals.

If we watch a plane from the ground, the air at any point is disturbed briefly as the plane passes and then returns to its original state. Since the fluid motion is not constant in time, we may not apply Bernoulli's equation to our observations. However, to an observer on the plane, the fluid flow is steady in time. *Thus, Bernoulli's equation holds true provided we apply it in a coordinate frame at rest relative to the plane.*

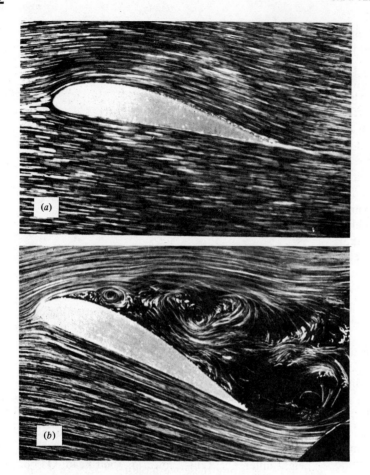

Figure 13.18. (*a*) Streamline flow about a wing. (*b*) Turbulence, loss of lift, and stalling result from large angles of attack. (F. Homann, *Forschung auf dem Gebiete des Ingenieurwesens,* 1936, courtesy VDI Verlag.)

Fig. 13.18*a* shows the flow of air of initial velocity v and density ρ about a stationary airplane wing. The streamlines above the wing are closer together than those below the wing. According to the continuity equation, this indicates that the velocity v_a of the air above the wing is greater than that below, v_b (Fig. 13.19). Since Bernoulli's equation says that

$P + \rho g y + \frac{1}{2}\rho v^2$ is constant in a streamline, the pressure P_a above the wing must be less than the pressure P_b below. (We can neglect the $\rho g y$ terms, since the wing thickness is small.) Then the upward lift force F_L on a wing area A is

$$F_L = (P_b - P_a)A = A\frac{\rho}{2}(v_a{}^2 - v_b{}^2)$$

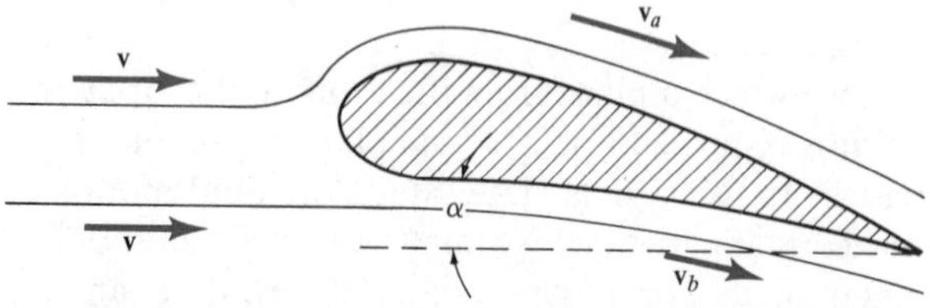

This expression is not yet useful, since we do not know the velocities v_a and v_b, and there is no simple way to predict their values. However, we would expect that both v_a and v_b are proportional to the initial air velocity v. Hence $(v_a{}^2 - v_b{}^2)$ is proportional to v^2, and the lift force can be rewritten as

$$F_L = AC_L\frac{\rho}{2}v^2 \tag{13.15}$$

Figure 13.19. The air above the wing moves at a speed v_a, which is greater than the speed below the wing v_b. Both v_a and v_b are proportional to v.

Figure 13.20. The angle of attack is increased just before the jet takes off. (Courtesy Boeing.)

The proportionality factor C_L is called the *lift coefficient*. In a few idealized cases, it can be predicted theoretically using advanced mathematical techniques, but it is usually measured experimentally. The lift coefficient depends in a complicated way on the shape of the wing and on its *angle of attack α*, the angle between the wing and the direction of the airflow. When the angle of attack is small, the lift is approximately proportional to the angle of attack. However, if α is increased sufficiently, turbulence sets in, the lift diminishes, and a stall may result (Fig. 13.18b).

Equation 13.15 for the lift force is all we need to discuss flight qualitatively. For flight at constant altitude the lift force must equal the weight. Jet aircraft have high flight speeds, so the area A and the angle α can be made small, thereby reducing dissipative drag forces. However, this also has the effect of making the takeoff velocities quite high. In order to take off at moderate speeds, jets must make an abrupt increase in the angle of attack after reaching a critical speed (Fig. 13.20). This increases the lift coefficient and permits takeoff. Propeller planes, which fly at lower speeds, have proportionately larger wing areas as well as larger angles of attack. This makes the lift coefficient large enough to permit takeoff at lower speeds.

Because the wings of flying animals provide both propulsion and lift, their design and motion are complex. The lift force is still determined by $F_L = AC_L\rho v^2/2$ but A, C_L, and v vary during different phases of the wing motion (Fig. 13.21).

Scaling and Flight

Small birds, such as sparrows, and larger ones, such as ducks, take off and land quite differently. We can understand this by once again using the scaling model. We use the simplest form of scaling, which assumes that the volume or weight of a bird varies as the cube of a characteristic length l. (The more complicated scaling procedure of Chapter Eight, which employs the buckling strength criterion $l \propto r^{2/3}$, leads to results very similar to those we obtain here.)

The cross-sectional area A of a wing varies as l^2, so the lift force $F_L = AC_L\rho v^2/2$ varies as Av^2 or l^2v^2. In level flight, F_L must equal the weight of the bird, $w \propto l^3$. Equating these we have $l^2v^2 \propto l^3$, or

$$v \propto l^{1/2} \qquad (13.16)$$

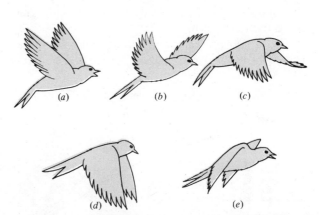

Figure 13.21. During level flight the downstroke provides the greatest propulsive and lift force. The front of the wing is lower than the trailing edge during the sequence (a) through (d). During the upstroke the wing is pulled forward and up, nearly parallel to the plane of the wing. Little propulsive or lift force is obtained, but little flying speed is lost, and little energy is required to execute the motion. (Adapted from Alexander, *Animal Mechanics.*)

Figure 13.22. A large bird builds up speed before taking off. (Horst Schafer/Photo Trends.)

This result means that a large bird has a greater minimum flying speed than a small bird. A small bird can jump into the air and achieve flying speed with a beat or two of its wings. A large bird first builds up speed by taxiing along the ground or water (Fig. 13.22) or by descending from a high perch.

It is interesting to compare two birds of greatly different sizes. The minimum flying speed for a swift is about 21 km h^{-1}. An ostrich has a characteristic length about 25 times that of the swift, so from $v \propto l^{1/2}$, its minimum flying speed is $\sqrt{25} = 5$ times that of the swift or 105 km h^{-1}. It is small wonder that the ostrich cannot fly.

The largest present-day flying animal is the albatross, which has a wingspan of 3.3 m. Larger flying

reptiles (pterosaurs) existed during the age of dinosaurs. The largest discovered to date had a wingspan of 15.5 m! It is speculated that such large creatures could not take off except by scaling cliffs and soaring off them like gliders. Their continued stay aloft would then depend on their ability to locate rising air currents or columns. (Fig. 13.23).

EXERCISES ON SUPPLEMENTARY TOPICS

Section 13.9 | The Flight of Animals and Airplanes

13-46 An airplane with a mass of 9000 kg must reach 120 m s^{-1} to take off. If the plane carries an additional 7000 kg, what is its minimum takeoff speed?

13-47 Why do very large airplanes usually require long runways?

13-48 The minimum flying speed of a 0.05-kg swift is 6 m s^{-1}. What is the minimum flying speed of a 3.2-kg goose?

13-49 The swift has a wingspread of about 0.25 m, while the pterosaur had a wingspread of about 16 m. The swift has a minimum flying speed of 6 m s^{-1}. (a) Using scaling, estimate the minimum flying speed of the pterosaur. (b) Comment on the validity of the scaling model for this case.

PROBLEMS ON SUPPLEMENTARY TOPICS

13-50 The power required for a hummingbird (or a helicopter) to hover is

$$\mathscr{P}_H = \sqrt{\frac{w^3}{2\rho A}}$$

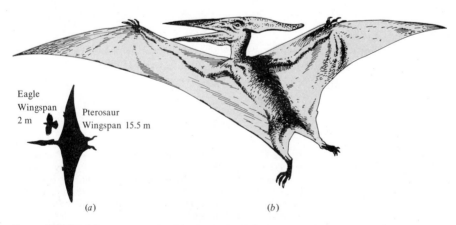

Eagle
Wingspan
2 m

Pterosaur
Wingspan 15.5 m

(a)   (b)

Figure 13.23. The pterosaur. (a) The size scale. (b) An artist's reconstruction of the pterosaur hunting its prey.

where w is the weight of the bird, A is the area swept out by its wings as they swing back and forth, and ρ is the density of air. (a) Show that the right-hand side has the dimensions of power. (b) Can another expression of the form $w^a \rho^b A^c$ be found that has dimensions of power?

13-51 Using the formula for the power expended in hovering flight in Problem 13-50, show how this power scales with the characteristic length of a bird.

13-52 A hovering hummingbird of mass 3×10^{-3} kg has wings that sweep out an area of 3×10^{-3} m². (a) Using the formula in Problem 13-50, find the power expended at 20° C. (b) The bird's muscles have a mass of 0.75×10^{-3} kg. Compare their power output with the 80 W kg^{-1} maximum output of human muscle.

13-53 Apply the scaling law $l^3 \propto r^2$ to the flight of birds. How is Eq. 13.16 modified?

Additional Reading

R. McNeill Alexander, *Animal Mechanics,* University of Washington Press, Seattle, 1968. Chapters 5 and 6.

R.A.R. Tricker and B.J.K. Tricker, *The Science of Movement,* Mills and Boon Ltd., London, 1967.

Arthur C. Guyton, *Circulatory Physiology: Cardiac Output and Its Regulation,* W.B. Saunders Co., Philadelphia and London, 1963. Chapter 6 discusses flow meters and their use.

John R. Cameron and James G. Skofronick, *Medical Physics,* John Wiley and Sons, Inc., New York, 1978. Chapter 6 discusses pressures in the body.

Alan P. Lightman, If Birds Can Fly Why Can't I?, *Science 83,* October, p. 22. Scaling and flight.

Lloyd Hunter, The Art and Physics of Soaring, *Physics Today,* April 1984, p. 34.

William F. Allman, Pitching Rainbows, *Science 82,* October, p. 32. The physics of the curve ball in baseball.

Scientific American articles:

Wallace O. Fenn, The Mechanism of Breathing, January 1960, p. 138.

J.V. Warren, The Physiology of the Giraffe, November 1974, p. 96.

Stanley J. Dudrick and Jonathan E. Rhoads, Total Intravenous Feeding, May 1972, p. 73.

D. James Baker, Jr., Models of Oceanic Circulation, January 1970, p. 114.

Suk Ki Hong and Hermann Rahn, The Diving Women of Korea and Japan, May 1967, p. 34.

Eric Denton, The Buoyancy of Marine Animals, July 1960, p. 118.

John P. Campbell, Vertical-Takeoff Aircraft, August 1960, p. 41.

John H. Storer, Bird Aerodynamics, April 1952, p. 24.

Carl Welty, Birds as Flying Machines, March 1955, p. 88.

Clarence D. Cone, Jr., The Soaring Flight of Birds, April 1962, p. 130.

David S. Smith, The Flight Muscles of Insects, June 1965, p. 76.

Alfred Gessow, The Changing Helicopter, April 1967, p. 38.

Felix Hess, The Aerodynamics of Bommerangs, November 1968, p. 124.

Vance A. Tucker, The Energetics of Bird Flight, May 1969, p. 70.

C.J. Pennynick, The Soaring Flight of Vultures, December 1973, p. 102.

Torkel Weis-Fogh, Universal Mechanisms for the Generation of Lift in Flying Animals, November 1975, p. 80.

Wann Langston, Jr., Pterosaurs, February 1981, p. 122.

Jearl Walker, Boomerangs! How to Make Them and Also How They Fly, The Amateur Scientist, March 1979, p. 162.

Jearl Walker, More on Boomerangs, Including Their Connection With the Dimpled Golf Ball, The Amateur Scientist, April 1979, p. 180.

Jearl Walker, Introducing the Musha, the Double Lozenge, and a Number of Other Kites to Build and Fly, The Amateur Scientist, February 1978, p. 156.

Ronald E. Rosenzweig, Magnetic Fluids, October 1982, p. 136. The strange properties of a liquid with magnetic particles in suspension.

Michael A. Markowski, Ultralight Airplanes, July 1982, p. 62.

CHAPTER 14

VISCOUS FLUID FLOW

Real fluids in motion always exhibit some effects of frictional or viscous forces. Whenever the work done against these dissipative forces is comparable to the total work done on the fluid or to its mechanical energy change, Bernoulli's equation cannot be used. Bernoulli's equation always applies for fluids at rest, since viscous forces then have no effect, but an estimate of the viscous forces must be made for moving fluids. For example, Bernoulli's equation can adequately describe the flow of blood in the large main arteries of a mammal, but not in the narrower blood vessels.

We begin this chapter by defining the *viscosity* of a fluid. We then examine the effects of viscous forces on the flow of a fluid in a tube. Viscosity is also responsible for the drag force experienced by a small object moving slowly through a fluid. Consequently viscous forces determine the velocities of molecules and small particles in solution in a centrifuge.

14.1 | VISCOSITY

Viscosity is readily defined by considering a simple experiment. Figure 14.1 shows two flat plates separated by a thin fluid layer. If the lower plate is held fixed, a force is required to move the upper plate at a constant speed. This force is needed to overcome the viscous forces due to the liquid and is greater for a highly viscous fluid, such as molasses, than for a less viscous fluid, such as water.

The force F is observed to be proportional to the area of the plates A and to the velocity of the upper plate Δv, and inversely proportional to the plate separation Δy:

$$F = \eta A \frac{\Delta v}{\Delta y} \qquad (14.1)$$

When the layers lack the symmetry of Fig. 14.1, the ratio $\Delta v/\Delta y$ for the *velocity gradient* must be replaced by dv/dy, and Eq. 14.1 becomes

$$F = \eta A \frac{dv}{dy} \qquad (14.2)$$

The proportionality constant η (eta) is called the *viscosity*. From this equation the dimensions, denoted with brackets [ ], of the viscosity are

$$[\eta] = \left[\frac{F/A}{\Delta v/\Delta y}\right] = \left[\frac{MLT^{-2}/L^2}{LT^{-1}/L}\right]$$
$$= [ML^{-1}T^{-1}] \qquad (14.3)$$

Here M, L, and T stand for mass, length and time,

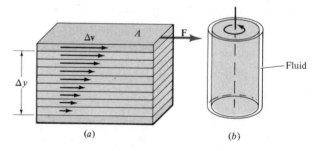

Figure 14.1. (*a*) Apparatus for measuring viscosity. (*b*) To prevent fluid leaking out, an actual measurement of viscosity uses rotating and fixed concentric cylinders.

TABLE 14.1

Typical values of viscosity in units of pascal-seconds (Pa s)

Temperature, °C	Castor Oil	Water	Air	Normal Blood*	Blood Plasma*
0	5.3	1.792×10^{-3}	1.71×10^{-5}		
20	0.986	1.005×10^{-3}	1.81×10^{-5}	3.015×10^{-3}	1.810×10^{-3}
37	—	0.6947×10^{-3}	1.87×10^{-5}	2.084×10^{-3}	1.257×10^{-3}
40	0.231	0.656×10^{-3}	1.90×10^{-5}		
60	0.080	0.469×10^{-3}	2.00×10^{-5}		
80	0.030	0.357×10^{-3}	2.09×10^{-5}		
100	0.017	0.284×10^{-3}	2.18×10^{-5}		

* The relative viscosities (η/η_{water}) of blood and of plasma remain nearly constant for temperatures between 0°C and 37°C.

respectively. The S.I. unit of viscosity is $1 \text{ kg m}^{-1}\text{ s}^{-1} = 1 \text{ Pa s}$.

Table 14.1 lists some typical viscosities. Usually as the temperature decreases, liquids become more viscous, as is readily observed in engine oil, honey, and other viscous fluids. By contrast, gases usually become less viscous as the temperature is lowered.

Because viscous forces are usually small, fluids are often used as lubricants to reduce friction. This is illustrated in the following example.

Example 14.1

An air track used in physics lecture demonstrations supports a cart that rides on a thin cushion of air $1 \text{ mm} = 10^{-3} \text{ m}$ thick and 0.04 m² in area. If the viscosity of the air is 1.8×10^{-5} Pa s, find the force required to move the cart at a constant speed of 0.2 m s⁻¹.

Figure 14.1a can be used to model this apparatus, with the upper surface representing the moving cart. With Eq. 14.1, the force required is

$$F = \eta A \frac{\Delta v}{\Delta y}$$

$$= (1.8 \times 10^{-5} \text{ Pa s})(0.04 \text{ m}^2)\frac{(0.2 \text{ m s}^{-1})}{(10^{-3} \text{ m})}$$

$$= 1.44 \times 10^{-4} \text{ N}$$

This is a very small force and is consistent with the observation that an air track is nearly frictionless.

Laminar Flow

In the type of flow illustrated by Fig. 14.1, because of the intermolecular forces, each layer moves at almost the same rate as the adjacent layer; the velocity changes continuously. Thus the fluid in contact with the moving plate has the same velocity as the plate, as indicated by the arrows. The fluid layer just below moves slightly more slowly, and each successive layer lags a bit more. The layer next to the stationary plate is at rest. This layered structure or *laminar flow* is the kind of streamline flow characteristic of viscous fluids at low velocities. When the fluid velocity is increased sufficiently, the flow changes its character and becomes turbulent.

Turbulent flow is frequently undesirable because it dissipates more mechanical energy than does laminar flow. Airplanes and cars are often designed so that the flow of air around them is as streamlined as possible. Also, nature has arranged that the flow in blood vessels is normally laminar rather than turbulent. These ideas are discussed further in later sections.

14.2 | LAMINAR FLOW IN A TUBE

Many interesting applications of the physics of fluids involve laminar flow in cylindrical tubes such as copper pipes or human arteries. In this section, we consider a formula for the flow rate called *Poiseuille's law*. It was first discovered experimentally by a physician, Jean Louis Marie Poiseuille (1799–1869), who was investigating the flow in blood vessels. Poiseuille's law relates the flow rate to the viscosity of the fluid, the pressure drop, and the radius and length of the tube. We use it extensively in the next section in discussing the flow in blood vessels.

Consider a fluid moving through a tube slowly enough so that there is no turbulence, and the flow is laminar. Just as in the case of the two flat surfaces discussed in the preceding section, the fluid in contact with the wall of the tube clings to it and is at rest. The thin, cylindrical layer of fluid adjacent to this station-

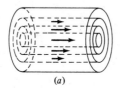
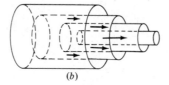

Figure 14.2. Laminar flow in a tube. The fluid in contact with the wall is at rest and successive thin cylindrical layers move with increasing speeds. The fluid at the center has the maximum velocity. If initially the fluid in the tube is as in (a), then a short time later the layers will have moved as in (b).

ary layer moves very slowly, and successive thin layers move at increasing velocities (Fig. 14.2). Hence the fluid at the center has the maximum velocity, v_{max}.

In a horizontal tube with a constant cross section, the equation of continuity implies that the average velocity $\bar{v}$ remains the same, since $Q = A\bar{v}$ must be constant. Nevertheless, the pressure drops as the fluid moves along the tube. This is because work is done against the viscous forces. If the cross section varies or if the tube is not horizontal, additional pressure changes arise in accordance with Bernoulli's equation (Fig. 14.3).

We can find the explicit velocity variation with the distance r from the axis of the tube by relating the viscous force and the pressure drop. Consider a tube with constant radius R and length l (Fig. 14.4a). The pressure drop between its ends is $\Delta P = P_1 - P_2$. If we consider a cylinder of fluid of radius r', the net driving force on it is the pressure drop times the cross-sectional area, $\Delta P\pi r'^2$. The surface area around the cylinder is $2\pi r'l$, so the viscous force on it is $-\eta A\,dv'/dy = -\eta(2\pi r'l)dv'/dr'$. Equating this viscous force to $\Delta P\pi r'^2$, we find

$$dv' = \frac{-\Delta Pr'\,dr'}{2\eta l}$$

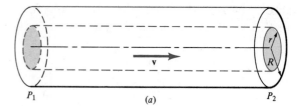
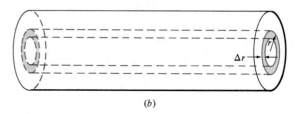

Figure 14.4. (a) In equilibrium, the net force on the ends of the fluid in the cylinder must balance the viscous force opposing its motion. (b) The total flow rate is found by summing the flow rates in thin cylindrical shells.

If we integrate the right side from r to R, the velocity goes from v to 0, since the fluid at the wall of the tube is at rest. Then we have

$$\int_v^0 dv' = \frac{-\Delta P}{2\eta l}\int_r^R r'\,dr' \quad \text{or} \quad v'\Big|_v^0 = \frac{-\Delta P}{2\eta l}\frac{r'^2}{2}\Big|_r^R$$

Thus

$$v = \frac{\Delta P}{4\eta l}(R^2 - r^2) \tag{14.4}$$

This formula for the velocity variation with r is used in sketching the velocity vectors shown in Fig. 14.2. It contains a great deal of information. We look first at the maximum velocity, that as expected occurs at $r = 0$. Its value is

$$v_{max} = \frac{\Delta PR^2}{4\eta l} \tag{14.5}$$

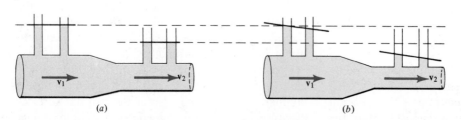

Figure 14.3. Pressures in a horizontal tube containing a moving fluid. (a) Ideal nonviscous fluid. In accordance with Bernoulli's equation, the pressure drops when the tube narrows and the average velocity increases. (b) Viscous fluid. Additional pressure drops occur due to the work done against viscous or frictional forces.

Next we calculate the flow rate Q through the tube. A thin cylindrical shell of thickness dr' has a cross-sectional area $2\pi r'dr'$ and a flow rate $dQ' = v'dA' = 2\pi r'v'dr'$ (Fig. 14.4b). Using Eq. 14.4 for the velocity and integrating from $r' = 0$ to $r' = R$, we obtain the flow rate (see Problem 14-34)

$$Q = \frac{\Delta P\pi R^4}{8\eta l} \quad \text{(Poiseuille's law)} \quad (14.6)$$

This is *Poiseuille's law*. It indicates that high viscosity leads to low flow rates, which is reasonable. The flow rate is proportional to the pressure gradient, $\Delta P/l$. Note the analogy with the case of heat flow discussed in Chapter Twelve; the heat flow rate is proportional to the temperature gradient, $\Delta T/l$. Also, the flow rate is proportional to R^4, which is somewhat surprising; the rate is extremely dependent on the radius of the tube. This implies, for example, that in blood vessels, moderate adjustments in the radius can produce large changes in the flow rates. Thus, increasing R to $1.19R$ means a factor of $(1.19)^4 = 2$ increase in the flow rate.

Finally, we recall that the average velocity is related to the flow rate by $Q = A\bar{v} = \pi R^2\bar{v}$, so with the formulas above

$$\bar{v} = \frac{\Delta PR^2}{8\eta l} = \frac{v_{max}}{2} \quad (14.7)$$

The average velocity is half the maximum velocity, which occurs at the center of the tube. Hence, a flow meter placed in the center measures twice the average velocity.

We can illustrate these results by considering the blood flow in an artery.

Example 14.2

A large artery in a dog has an inner radius of 4×10^{-3} m. Blood flows through the artery at the rate of $1\ cm^3\ s^{-1} = 10^{-6}\ m^3\ s^{-1}$. Find (a) the average and maximum velocities of the blood; (b) the pressure drop in a 0.1-m-long segment of the artery.

(a) The average velocity is

$$\bar{v} = \frac{Q}{A} = \frac{Q}{\pi R^2}$$

$$= \frac{10^{-6}\ m^3\ s^{-1}}{\pi(4 \times 10^{-3}\ m)^2}$$

$$= 1.99 \times 10^{-2}\ m\ s^{-1}$$

The maximum velocity occurs at the center of the artery and is

$$v_{max} = 2\bar{v} = 2(1.99 \times 10^{-2}\ m\ s^{-1})$$
$$= 3.98 \times 10^{-2}\ m\ s^{-1}$$

(b) From Table 14.1, $\eta = 2.084 \times 10^{-3}$ Pa s. Thus the pressure drop is found from $\bar{v} = \Delta PR^2/8\eta l$, or

$$\Delta P = \frac{8\eta l\bar{v}}{R^2}$$

$$= \frac{8(2.084 \times 10^{-3}\ Pa\ s)}{(4 \times 10^{-3}\ m)^2}(0.1\ m)(1.99 \times 10^{-2}\ m\ s^{-1})$$

$$= 2.07\ Pa$$

Power Dissipation | We can readily compute the power dissipated in a tube by viscous forces. That power is, of course, equal to the power that must be supplied to maintain the flow.

Referring again to Fig. 14.4a, the net force on a segment is the pressure drop in the segment times the cross-sectional area, $F = (P_1 - P_2)A = \Delta PA$. The average power required to maintain the flow is this force times the average velocity, $\mathcal{P} = F\bar{v} = \Delta PA\bar{v}$. Notice that $A\bar{v}$ is the flow rate Q; hence, the power required to maintain the flow is

$$\mathcal{P} = \Delta P\,Q \quad (14.8)$$

For the special case where the flow is through a cylindrical tube of radius R, $A = \pi R^2$, and Eq. 14.8 becomes

$$\mathcal{P} = \Delta P(\pi R^2)\bar{v} \quad \text{(cylindrical tube)} \quad (14.9)$$

This result is particularly suitable for flow in blood vessels, as we show in the following example.

Example 14.3

What is the power required to maintain the blood flow in the dog's artery described in Example 14.2?

Using the results of Example 14.2 in Eq. 14.9, we find the power is

$$\mathcal{P} = \Delta P(\pi R^2)\bar{v}$$
$$= (2.07\ Pa)\pi(4 \times 10^{-3}\ m)^2(1.99 \times 10^{-2}\ m\ s^{-1})$$
$$= 2.07 \times 10^{-6}\ W$$

The metabolic rate of a dog is 10 W or greater, so the power expended in pumping blood through the large arteries is negligible. We see in Section 14.4 that most of the pressure drop and loss of energy due to viscous forces occurs in the small arterial subbranches and in the capillaries.

14.3 | TURBULENT FLOW

Poiseuille's laws hold true only for laminar flow. However, often the flow is not laminar but turbulent;

it is similar to the wake of a fast boat, with swirls and eddies. The ink lines we used to describe streamline flow in Chapter Thirteen would become mixed or blurred. Some examples of turbulent flow are shown in Fig. 14.5.

We noted earlier that the mechanical energy dissipated is typically much larger in turbulent flow than in laminar flow. Hence, it is often desirable to ensure that the flow does not become turbulent.

It is much harder to analyze turbulent flow than laminar flow. For example, Poiseuille's law for the laminar flow rate in a tube has no analog for turbulent flow. In practice turbulent flow is treated using a variety of empirical rules and relationships developed from extensive experimental studies.

In order to determine whether the flow is laminar and thus whether Poiseuille's law can be applied, we can make use of one of these empirical rules. It states that the value of a dimensionless quantity called the *Reynolds number* N_R determines whether the flow is turbulent or laminar. Consider a fluid of viscosity η and density ρ. If it is flowing in a tube of radius R and

has an average velocity $\bar{v}$, then the Reynolds number is defined by

$$N_R = \frac{2\rho\bar{v}R}{\eta} \quad \text{(flow in a tube of radius R)} \quad (14.10)$$

In *tubes*, it is found experimentally that if

$$
\begin{aligned}
N_R &< 2000 &&\text{flow is laminar} \\
N_R &> 3000 &&\text{flow is turbulent} \quad (14.11) \\
2000 &< N_R < 3000 &&\text{flow is unstable (may change from laminar to turbulent or vice versa)}
\end{aligned}
$$

Thus whether the flow is laminar or turbulent is determined by a particular combination of variables; doubling the tube radius and halving the average velocity will leave the character of the flow unchanged.

The Reynolds number also indicates whether the flow around an obstacle, such as a ship hull or an airplane wing, will be turbulent or laminar. In general, the Reynolds number at which turbulence sets in depends significantly on the shape of the obstacle and is *not* given by Eqs. 14.10 or 14.11.

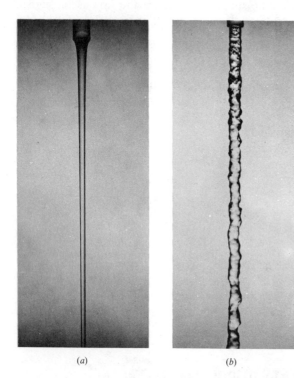

(a)   (b)   (c)

Figure 14.5. (a) Laminar flow and (b) turbulent flow of water. (c) First laminar, then turbulent flow of cigarette smoke. (From Sears, Zemansky and Young, *University Physics,* fifth edition, Copyright © 1976, Addison-Wesley, Reading, Mass.)

In the preceding section, we computed the flow rate in an artery using Poiseuille's law, which is valid only if the flow is laminar. We now compute the Reynolds number for that case to see whether the flow is laminar or turbulent.

Example 14.4

In Example 14.2, the radius of the artery is 4×10^{-3} m, the average velocity of the blood is 1.99×10^{-2} m s^{-1}, and the viscosity is 2.084×10^{-3} Pa s. Also, the density of the blood from Table 13.1 is 1.0595×10^3 kg m^{-3}. Find the Reynolds number and determine whether the flow is laminar.

The Reynolds number is

$$N_R = \frac{2\rho\bar{v}R}{\eta}$$

$$= \frac{2(1.0595 \times 10^3 \text{ kg m}^{-3})}{2.084 \times 10^{-3} \text{ Pa s}}$$

$$\times (1.99 \times 10^{-2} \text{ m s}^{-1})(4 \times 10^{-3} \text{ m})$$

$$= 80.9$$

This is much less than 2000, so the flow is laminar.

In turbulent flow, some energy is dissipated as sound and some as heat. The noise associated with turbulent flow in arteries facilitates blood pressure measurements as we described in Chapter Thirteen, and it makes possible the detection of some heart abnormalities.

14.4 | FLOW IN THE CIRCULATORY SYSTEM

We now apply some of these ideas to the flow of blood in blood vessels. To do this, it is useful to summarize briefly some aspects of the cardiovascular system.

The Blood | The circulatory system transports the substances required by the body and the waste products of metabolism. In order to perform a large number of functions, the blood contains many different constituents, including red blood cells, white blood cells, platelets, and proteins. However, for our purposes, it is sufficient to treat the blood as a uniform fluid, with viscosity $\eta = 2.084 \times 10^{-3}$ Pa s and density $\rho = 1.0595 \times 10^3$ kg m^{-3} at normal body temperature. Only more subtle physical properties of the circulatory system require a fuller description.

The Cardiovascular System | The cardiovascular system includes the heart and an extensive system of *arteries, vascular beds* containing *capillaries,* and *veins* (Fig. 14.6). The arteries carry blood to the organs, muscles, and skin, and the veins transport the return flow. Each large artery branches to form several smaller arteries, which in turn branch further. The blood ultimately reaches the vascular beds, where materials are exchanged with the surrounding tissues. The branching process is then reversed in the venous system, culminating in the *vena cava,* which returns blood to the heart. Table 14.2 gives the number and dimensions of several types of blood vessels in one vascular bed. Table 14.3 lists the properties of the human cardiovascular system.

A particularly interesting component of the cardiovascular system is the *arteriovenous anastomosis* (AVA) or *shunt.* Figure 14.6 shows only one of the many AVAs present in the body. These shunts are particularly important, since the surrounding smooth muscle tissue can adjust the vessels' diameter. In the interior of the body, they help to adjust the blood flow to various organs as conditions change. Smaller shunts in the skin are open if the body needs to release heat or to increase the skin temperature. Under more moderate conditions, they are closed in order to reduce the load on the heart and direct the major portion of the blood supply to other parts of the body.

Flow Resistance | The *flow resistance R_f* is defined, in general, as the ratio of the pressure drop to the flow rate,

$$R_f = \frac{\Delta P}{Q} \tag{14.12}$$

When the flow is laminar, we can compare this with Eq. 14.5. Then we see

$$R_f = \frac{8\eta l}{\pi R^4} \quad \text{(laminar flow)} \tag{14.13}$$

The flow resistance is seldom computed for physiological systems because of the complexity of the systems. Instead, it is usually arrived at by measuring ΔP and Q and hence determining R_f. Note that $R_f = \Delta P/Q$ defines a flow resistance whether the flow is laminar or not. However, Eq. 14.13 is valid only for laminar flow.

From its definition, it follows that the units of flow resistance are those of pressure divided by a volume

TABLE 14.2

Detailed structure of the mesenteric (intestinal) vascular bed of a small dog. This is one of many such vascular beds in the body

Structure	Number, N	Inner Radius, R (m)	Total Inner Cross-Sectional Area, $N\pi R^2$ (m²)	Length, l (m)	Equivalent Flow Resistance R_{f1}/N (kPa s m⁻³)
Mesenteric artery	1	1.5×10^{-3}	7.0×10^{-6}	6.0×10^{-2}	6.67×10^3
Main branches	15	5.0×10^{-4}	1.2×10^{-5}	4.5×10^{-2}	2.55×10^5
Secondary branches	45	3.0×10^{-4}	1.3×10^{-5}	3.91×10^{-2}	5.69×10^5
Tertiary branches	1,900	7.0×10^{-5}	2.9×10^{-5}	1.42×10^{-2}	1.65×10^6
Terminal arteries	26,600	2.5×10^{-5}	5.2×10^{-5}	1.1×10^{-3}	5.61×10^5
Terminal branches	328,500	1.5×10^{-5}	2.32×10^{-4}	1.5×10^{-3}	4.79×10^5
Arterioles	1,050,000	1.0×10^{-5}	3.3×10^{-4}	2.0×10^{-3}	1.01×10^6
Capillaries	47,300,000	4.0×10^{-6}	2.378×10^{-3}	1.0×10^{-3}	4.38×10^5
Venules	2,100,000	1.5×10^{-5}	1.484×10^{-3}	1.0×10^{-3}	4.93×10^4
Terminal branches	160,000	3.7×10^{-5}	6.73×10^{-4}	2.4×10^{-3}	4.27×10^4
Terminal veins	18,000	6.5×10^{-5}	2.39×10^{-4}	1.5×10^{-3}	2.53×10^4
Tertiary veins	1,900	1.4×10^{-4}	1.17×10^{-4}	1.42×10^{-2}	1.03×10^5
Secondary veins	60	8.0×10^{-4}	1.47×10^{-4}	4.19×10^{-2}	9.33×10^2
Mesenteric vein	1	3.0×10^{-3}	2.8×10^{-5}	6.0×10^{-2}	4.0×10^2

TABLE 14.3

Properties of the human cardiovascular system for a typical adult. All pressures listed are gauge pressures (1 atm = 1.013×10^5 Pa = 101.3 kPa)

Mean pressure in large arteries	12.8 kPa
Mean pressure in large veins	1.07 kPa
Volume of blood (70-kg man)	5.2 litres = 5.2×10^{-3} m³
Time required for complete circulation (resting)	54 seconds
Heart flow rate (resting)	9.7×10^{-5} m³ s⁻¹
Viscosity of blood (37°C)	2.084×10^{-3} Pa s
Density of blood (37°C)	1.0595×10^3 kg m⁻³

per unit time. The basic S.I. unit is the pascal-second per cubic metre; we will use kilopascal-seconds per cubic metre (kPa s m⁻³). In texts and literature on physiology pressures are usually measured in torr and lengths in centimetres, so the unit of flow resistance is

$$1 \text{ torr s cm}^{-3} = 1.333 \times 10^5 \text{ kPa s m}^{-3}$$

In the units we are using it is necessary to use Q in cubic metres per second (m³ s⁻¹) and ΔP in kilopascals (kPa).

From the following example, we see that the flow resistance in a large artery is small. This explains why the pressure drop in such arteries is small.

Example 14.5

The aorta of an average adult human has a radius of 1.3×10^{-2} m. What are the resistance and pressure drop over a 0.2-m distance, assuming a flow rate of 10^{-4} m³ s⁻¹?

From Table 14.3, $\eta = 2.084 \times 10^{-3}$ Pa s, so the flow resistance of the aorta is

$$R_f = \frac{8\eta l}{\pi R^4} = \frac{8}{\pi} \frac{(2.084 \times 10^{-3} \text{ Pa s})(0.2 \text{ m})}{(1.3 \times 10^{-2} \text{ m})^4}$$

$$= 3.72 \times 10^4 \text{ Pa s m}^{-3}$$
$$= 37.2 \text{ kPa s m}^{-3}$$

The pressure drop over the 0.2-m distance is then

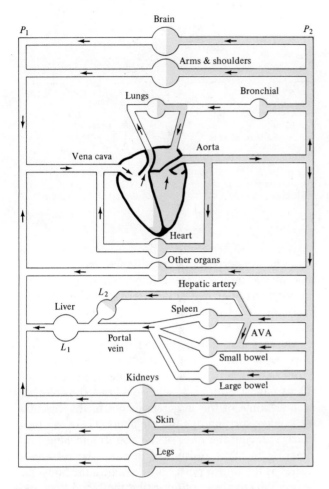

P_1

P_2

Brain

Arms & shoulders

Lungs

Bronchial

Aorta

Vena cava

Heart

Other organs

Hepatic artery

L_2

Liver

Spleen

L_1

Portal vein

AVA

Small bowel

Kidneys

Large bowel

Skin

Legs

Figure 14.6. Schematic diagram of the mammalian circulatory system. The circles represent vascular beds. The small arrows represent the direction of blood flow. The colored shaded areas are those with oxygen-rich blood. The arterial system is on the right at pressure P_2, and the venous system on the left is at pressure P_1.

$$\Delta P = R_f Q$$
$$= (37.2 \text{ kPa s m}^{-3})(10^{-4} \text{ m}^3 \text{ s}^{-1})$$
$$= 0.00372 \text{ kPa}$$

This is very small compared to the total pressure drop in the system, which is about 13.3 kPa. Most of the flow resistances and pressure drops occur in the smaller arteries and vascular beds of the body (Table 14.4).

When several blood vessels with flow resistances $R_{f1}, R_{f2}, \ldots$, are connected at each end, they have a common pressure drop ΔP, and they are said to be in *parallel*. The total flow Q is divided among them, with

$Q_1 = \Delta P/R_{f1}$ through R_{f1}, $Q_2 = \Delta P/R_{f2}$ through R_{f2}, and so on. Adding up these flow rates

$$Q = \Delta P \left[\frac{1}{R_{f1}} + \frac{1}{R_{f2}} + \frac{1}{R_{f3}} + \cdots \right]$$

If we were to replace these several resistances with a single equivalent resistance R_p, we would have $Q = \Delta P/R_p$. Hence for several resistances in parallel

$$\frac{1}{R_p} = \frac{1}{R_{f1}} + \frac{1}{R_{f2}} + \frac{1}{R_{f3}} + \cdots \quad (14.14)$$

When there are N equal parallel resistances, the equivalent resistance is

$$R_p = R_{f1}/N \quad (14.15)$$

The flow resistance of a collection of arteries, such as the mesenteric bed of the dog, can be measured or calculated. The calculation can be done by considering each category of artery separately.

Example 14.6

From Table 14.2, the radius of a single capillary is 4×10^{-6} m, and its length is 10^{-3} m. What is the net resistance of the 4.73×10^7 capillaries in the mesenteric

TABLE 14.4

Approximate flow rates and resistances for the resting, reclining adult. The total flow rate from the aorta is 9.7×10^{-5} m^3 s^{-1} and the average pressure drop across the beds is 11.7 kPa (Refer to Fig 14.6 for the relationships of the various beds.)

Vascular Bed	Flow Rate (m^3 s^{-1})	Flow Resistance (kPa s m^{-3})
Brain	12.5×10^{-6}	9.3×10^5
Arm and shoulders	6.8×10^{-6}	1.7×10^6
Lungs	97.0×10^{-6}	9.0×10^3
Bronchial	1.0×10^{-6}	1.2×10^7
Heart	4.2×10^{-6}	2.8×10^6
Other organs	10.0×10^{-6}	1.2×10^6
Liver, L_1	25.0×10^{-6}	4.0×10^4
Liver, L_2	5.0×10^{-6}	2.2×10^6
Spleen	8.3×10^{-6}	1.3×10^6
Small bowel	8.8×10^{-6}	1.2×10^6
Large bowel	2.9×10^{-6}	3.7×10^6
Kidneys	18.3×10^{-6}	6.4×10^5
Skin	5.5×10^{-6}	2.1×10^6
Legs	13.7×10^{-6}	8.5×10^5

vascular bed of a dog if they are assumed to be in parallel?

With $\eta = 2.084 \times 10^{-3}$ Pa s, the resistance of one capillary is

$$R_{f1} = \frac{8\eta l}{\pi R^4}$$

$$= \frac{8(2.084 \times 10^{-3} \text{ Pa s})(10^{-3} \text{ m})}{\pi(4 \times 10^{-6} \text{ m})^4}$$

$$= 2.073 \times 10^{16} \text{ Pa s m}^{-3}$$

$$= 2.073 \times 10^{13} \text{ kPa s m}^{-3}$$

There are $N = 4.73 \times 10^7$ capillaries in parallel, so their effective resistance is

$$R_f = \frac{R_{f1}}{N} = \frac{2.073 \times 10^{13} \text{ kPa s m}^{-3}}{4.73 \times 10^7}$$

$$= 4.38 \times 10^5 \text{ kPa s m}^{-3}$$

The effective resistance of all the capillaries in one vascular bed was calculated in the preceding example by assuming that all the capillaries are in parallel, which is only approximately true. We also assumed that Poiseuille's law holds true for the flow in calculating the resistance of a single capillary. Again, this is an approximation, since the radii of the capillaries are comparable to the sizes of some of the constituents of the blood. Nevertheless, the results of such calculations are in rough agreement with the values found by measuring the pressure drop and flow rate and using $R_f = \Delta P/Q$.

Similar calculations for other portions of the vascular bed yield the results in the last column of Table 14.2. These results can be used to obtain the resistance of several sections of the vascular bed or of the entire bed. Suppose we know the resistances of N sections, each of which leads into the next. The total pressure drop is $\Delta P = \Delta P_{f1} + \Delta P_{f2} + \cdots + \Delta P_{fN}$. Each pressure drop, for example $\Delta P_1 = QR_{f1}$, is the total flow rate Q times the resistance of that section. Adding all the pressure changes yields $\Delta P = Q(R_{f1} + R_{f2} + \cdots + R_{fN})$. Thus, the effective flow resistance R_s of these sections, which are said to be in *series*, is the sum of the resistances

$$R_s = R_{f1} + R_{f2} + \cdots + R_{fN} \qquad (14.16)$$

If we add all the resistances of the arterial sections of the vascular bed of Table 14.2, we find $R_s = 45.3 \times 10^5$ k Pa s m^{-3}. The same procedure can be used for the venous portion of the bed, and these results are summarized in Fig. 14.7. The arterial portion of this vascular bed has 87 percent of the total

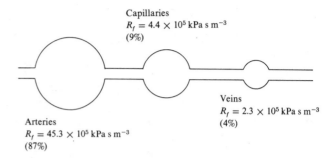

Figure 14.7. Simplified flow resistance diagram for the mesenteric vascular bed of a dog. The total equivalent resistance is $(45.3 + 4.4 + 2.3) \times 10^5 = 52.0 \times 10^5$ kPa s m^{-3}

resistance and hence of the pressure drop. The capillaries have 9 percent of the resistance, and the venous portion has only 4 percent. Similar results are found in other vascular beds; the arteries have the largest resistance, the capillaries a relatively small resistance, and the veins least of all.

The fact that most of the pressure drop and resistance occur in the arterial systems has important consequences for how the body regulates the circulation. The smallest arterial subdivisions, the arterioles, as well as some of the larger branches in the vascular beds, have surrounding muscle fibers. These can contract, reducing the radii of the vessels and hence increasing the flow resistance. Because the resistance varies as the fourth power of the radius, the body has a very effective way of adjusting the blood flow and responding to changing requirements.

Since the blood in the arteries is at a high pressure, severing an artery can lead to a serious loss of blood. The risk of this occurring is reduced by the fact that many arteries are deep within the body. When there is a drop in arterial pressure as a result of hemorrhaging, the body responds by constricting the vessels in many of the vascular beds. This temporarily conserves the reduced blood supply for use in the heart and the brain. However, the buildup of metabolic waste products and the shortage of oxygen eventually cause the vessels in the vascular beds to dilate. This lowers the blood pressure further, and the body is then in an unstable and poorly functioning condition referred to as shock.

SUMMARY

Viscous or frictional forces in fluids must be considered whenever the work done against them is appreciable. Viscous forces are proportional to the viscosity of the fluid and to the velocity difference between parts of a fluid,

$$F = \eta A \frac{\Delta v}{\Delta y}$$

If the velocity is low enough to avoid turbulence, the flow of a viscous fluid is laminar in structure, with a smooth variation of the velocity from layer to layer.

In a tube of radius R, the laminar flow rate is proportional to the pressure gradient and to R^4:

$$Q = \frac{\Delta P \pi R^4}{8\eta l}$$

The velocity of the fluid is a maximum at the center of the tube and decreases gradually in successive lamina to zero at the walls.

Turbulence sets in when the fluid velocity is increased sufficiently. The Reynolds number, a dimensionless combination of variables, can be used to predict whether the flow will be laminar or turbulent.

In many applications, such as the cardiovascular system, it is convenient to define the flow resistance, R_f, as

$$R_f = \frac{\Delta P}{Q}$$

For laminar flow

$$R_f = \frac{8\eta l}{\pi R^4} \quad \text{(laminar flow)}$$

When there are several flow resistances in parallel, the equivalent flow resistance is given by

$$\frac{1}{R_p} = \frac{1}{R_{f1}} + \frac{1}{R_{f2}} + \frac{1}{R_{f3}} + \cdots$$

When there are N equal parallel resistances, the equivalent flow resistance is

$$R_p = R_{f1}/N$$

If N resistances are in series, the effective resistance is

$$R_s = R_{f1} + R_{f2} + \cdots + R_{fN}$$

Checklist

Define or explain:

viscous force
viscosity
velocity gradient

laminar flow
pressure gradient
Poiseuille's law

vascular bed
flow resistance
parallel

series
Reynolds number

REVIEW QUESTIONS

Q14-1 The viscosity characterizes the force between moving _____ of fluid.

Q14-2 Is there a viscous force between a moving fluid and the container walls?

Q14-3 Streamline flow of a viscous fluid is called _____ flow.

Q14-4 Dimensional analysis can sometimes be used to obtain equations for various quantities. Can we also obtain the numerical coefficient in this way?

Q14-5 Poiseuille's law describes the _____ of fluid through a channel.

Q14-6 The Reynolds number is used to categorize the type of fluid flow. Roughly speaking, low Reynolds numbers correspond to _____ flow.

Q14-7 Each category of arterial sections in a vascular bed consists of a number of nearly identical arteries in _____ with each other.

Q14-8 The arterial sections of a vascular bed are in _____ with each other.

Q14-9 Is the flow resistance defined only for laminar flow?

Q14-10 If a fluid is forced through a channel, which type of flow requires the least energy to sustain, laminar or turbulent flow?

EXERCISES

Use 1.2 kg m^{-3} for the density of air and 1000 kg m^{-3} for the density of water.

Section 14.1 | Viscosity

14-1 In Fig. 14.1, at what point in the fluid is its velocity half that of the upper plate? Explain.

14-2 Why do automobile manufacturers recommend the use of "multiviscosity" engine oil in cold weather?

14-3 An experimental car rides on an air cushion 0.06 m thick with an area of 15 m^2. How much power is expended against viscous forces at a speed of 20 m s^{-1} if the viscosity of the air is 1.8×10^{-5} Pa s?

14-4 The air track cart in Example 14.1 has a mass of 0.5 kg. If its initial speed is 1 m s^{-1}, estimate the change in its speed after it has traveled 2 m.

Section 14.2 | Laminar Flow in a Tube

14-5 A blood vessel of radius 10^{-3} m has a pressure gradient $\Delta P/l$ of 600 Pa m^{-1}. (Assume laminary flow.) (a) What is the flow rate of blood at $37°$ C in the vessel? (b) What is the maximum velocity of the blood in the vessel?

14-6 An artery has an inner radius of 2×10^{-3} m. If the temperature is $37°$ C, the average velocity of the blood is 0.03 m s^{-1} and the flow is laminar, find (a) the maximum velocity; (b) the flow rate; and (c) the pressure drop in 0.05 m, if the artery is horizontal.

14-7 The pressure drop along a length of a horizontal artery is 100 Pa. The radius of the artery is 0.01 m, and the flow is laminar. (a) What is the net force on the blood in this portion of the artery? (b) If the average velocity of the blood is 1.5×10^{-2} m s^{-1}, find the power expended in maintaining the flow.

14-8 The radius of an artery is increased by a factor of 1.5. (a) If the pressure drop remains the same, what happens to the flow rate? (b) If the flow rate stays the same, what happens to the pressure drop? (Assume laminar flow.)

14-9 A hypodermic needle of length 0.02 m and inner radius 3×10^{-4} m is used to force water at $20°$ C into air at a flow rate of 10^{-7} m^3 s^{-1}. (a) What is the average velocity of the water in the needle? (Assume laminar flow.) (b) What is the pressure drop necessary to achieve the flow rate?

Section 14.3 | Turbulent Flow

14-10 The average velocity of water at $20°$ C in a tube of radius 0.1 m is 0.2 m s^{-1}. (a) Is the flow laminar or turbulent? (b) What is the flow rate?

14-11 A water pipe of radius 0.02 m delivers water at a flow rate of 0.01 m^3 s^{-1} at $20°$ C. (a) What is the average velocity of the water? (b) Is the flow laminar or turbulent? (c) Is there enough information to determine the maximum velocity of the water in the pipe?

14-12 Calculate the Reynolds number for Exercise 14.9. Is the flow laminar as assumed?

14-13 (a) What is the greatest average velocity of blood flow at $37°$ C in an artery of radius 2×10^{-3} m if the flow is to remain laminar? (b) What is the corresponding flow rate Q?

14-14 (a) Draw an imaginary plane at right angles to the laminar flow in a cylinder. Is the pressure the same everywhere on this plane? Explain. (b) Does the pressure vary on this plane if the flow becomes turbulent? Why?

Section 14.4 | Flow in the Circulatory System

14-15 A blood vessel of radius R branches into several vessels of smaller radius r. If the average fluid velocity in the smaller vessel is half that in the large vessel, how many vessels of radius r must there be?

14-16 Which of the various portions of the arterial system in the vascular bed of Table 14.2 will have (a) the largest pressure drop; (b) the smallest?

14-17 A small artery has a length of 0.11 cm = 1.1×10^{-3} m and a radius of 2.5×10^{-5} m. (a) Calculate its resistance. (b) If the pressure drop across the artery is 1.3 kPa, what is the flow rate?

14-18 A glass tube has a radius of 10^{-3} m and a length of 0.1 m. (a) What is its resistance to the flow of a liquid with a viscosity of 10^{-3} Pa s? (b) If the pressure drop along the tube is 10^3 Pa, what is the flow rate?

PROBLEMS

14-19 A glass plate 0.25 m^2 in area is pulled at 0.1 m s^{-1} across a larger glass plate that is at rest. What force is necessary to pull the upper plate if the space between them is 0.003 m thick and filled with (a) water with $\eta = 1.005 \times 10^{-3}$ Pa s; (b) oil with $\eta = 0.01$ Pa s? (c) Why might oil be preferred to water as a lubricant?

14-20 The average flow rate of blood in the aorta is 4.20×10^{-6} m^3 s^{-1}. The aorta is 1.3×10^{-2} m in radius. (a) What is the average blood velocity in the aorta? (b) What is the pressure drop along 0.1 m of the aorta? (c) What is the power required to pump blood through this portion of the aorta?

14-21 Water at $20°$ C flows in a pipe of radius 0.02 m. For each metre of length, there is a pressure drop of 100 Pa, and a power dissipation of 0.8 W. (a) Find the flow rate. (b) Find the average velocity. (c) Is the flow laminar? Explain.

14-22 A fluid is pumped through a pipe at a flow rate of 0.01 m^3 s^{-1}. The power loss is 1.5 W per metre of pipe length. What is the flow resistance per metre?

14-23 What is the power loss in a pipe carrying

fluid at $0.15 \, \text{m}^3 \, \text{s}^{-1}$ if the flow resistance is $4 \times 10^4 \, \text{Pa s m}^{-3}$?

14-24 (a) Estimate the power lost due to viscous effects in a gas pump hose if the viscosity of gasoline is $0.8 \times 10^{-3} \, \text{Pa s}$. (b) Will the actual power dissipated be considerably larger? Explain. (The density of gasoline is $670 \, \text{kg m}^{-3}$.)

14-25 When fluid is pumped through a pipe of radius R at a flow rate Q, the viscous power loss is $\mathcal{P}_0$. What is the power loss at the same flow rate if 10 pipes of the same length but radius $R/8$ are used?

14-26 Two pipes, each of radius R and length L, feed into a single pipe of radius $1.5 \, R$ and length $1.5 \, L$. What is the total flow resistance of this configuration for a fluid of viscosity η? (Assume laminar flow.)

14-27 (a) Using the resistances in Table 14.4, compute the pressure drops across the legs and lungs. (b) Explain the differences. What does this imply about the right ventricle, which pumps blood to the lungs?

14-28 Which of the vascular beds in Table 14.4, excluding the lungs, dissipates (a) the most power; (b) the least?

14-29 The pressure drop along a tube is increased gradually, and the corresponding flow rates are measured. (a) How would one get the flow resistance from these data? (b) What would happen to the flow resistance once the pressure became large enough so the flow became turbulent? Explain.

14-30 In Table 14.2, it is noted that the total cross-sectional area of all the capillaries in a particular vascular bed is $2.378 \times 10^{-3} \, \text{m}^2$, while the corresponding area for the main branches is only $1.2 \times 10^{-5} \, \text{m}^2$. How can this be reconciled with the fact that the main branches have a much smaller resistance?

14-31 The pressure drop along a copper tube is $10^3 \, \text{Pa}$, and the flow rate of a liquid in this tube is $0.01 \, \text{m}^3 \, \text{s}^{-1}$. (a) Find the flow resistance. (b) How much power is required to maintain this flow? (c) Does this calculation assume laminar flow? Explain.

14-32 When a calf is at rest, its heart pumps blood at a rate of $6 \times 10^{-5} \, \text{m}^3 \, \text{s}^{-1}$. The pressure drop from the arterial to venous systems is 12 kPa. (a) What is the flow resistance of its circulatory system? (b) How much work does the heart do in pumping the blood? (c) An experimental artificial

heart powered by an electrical pump is implanted in place of the animal's heart. If the pump has an efficiency of 50 percent, how much electrical power does it require?

14-33 (a) Calculate the flow resistance of a typical human capillary $2 \times 10^{-6} \, \text{m}$ in radius, and $10^{-3} \, \text{m}$ long. (b) Estimate the number of capillaries in a human using this result, given that the net flow rate through the aorta is $9.7 \times 10^{-5} \, \text{m}^3 \, \text{s}^{-1}$ and that the pressure drop from the arterial system to the venous system is 11.6 k Pa. Assume all the capillaries are in parallel, and that 9 percent of the pressure drop occurs in the capillaries.

^{c}**14-34** Show that integrating $dQ' = v'dA' = 2\pi r v' dr'$ leads to Eq. 14.6.

^{c}**14-35** (a) When a fluid in a tube undergoes laminar flow as described by Eq. 14.4, what is the flow rate for the portion of the fluid from $r = 0$ to $r = R/2$? (b) What is the ratio of this flow rate to the total flow rate through the tube? (c) What is the corresponding ratio for the cross-sectional areas?

ANSWERS TO REVIEW QUESTIONS

Q14-1, layers; **Q14-2,** yes; **Q14-3,** laminar; **Q14-4,** no; **Q14-5,** flow rate; **Q14-6,** laminar; **Q14-7,** parallel; **Q14-8,** series; **Q14-9,** no; **Q14-10,** laminar flow.

SUPPLEMENTARY TOPICS

14.5 | VISCOUS DRAG FORCES

Anyone who has held a hand out of a car window has discovered that an object moving through a fluid experiences a force that increases rapidly with the velocity. At very low velocities, this *drag force* results primarily from viscous forces that are proportional to the velocity v. At slightly higher but still relatively low speeds, the object accelerates the fluid moving around it, causing the force to vary approximately as v^2. In this section, we consider *low-speed* or *viscous drag forces*. "High-speed" drag is discussed in the next section.

Viscous drag forces result because the layer of fluid adjacent to an object is at rest with respect to the object. When the object moves through the fluid, this layer experiences a frictional force from the more rapidly moving layer next to it. Successive adjacent layers near the object produce frictional forces on each other, and the net result is to retard the motion of the object through the fluid.

An exact derivation of the viscous drag force is very difficult and has been carried out only for the simplest cases. However, given the observed direct proportionality at low speeds of the force and the velocity, we can obtain the general low-velocity form of the force by using a method called *dimensional analysis*.

The basic condition under which the drag force has the form we will obtain is that the object moves very slowly through the fluid and the Reynolds number is very small. Figure 14.8 shows a spherical object of radius R moving with a small velocity v through a fluid of viscosity η and density ρ_0. Apart from dimensionless factors, the only physical quantities that could enter an expression for the drag force F_d are R, η, ρ_0, and v.

To construct from these variables a force proportional to v, we assume that the formula involves a factor of v and unknown powers of R, η, and ρ_0

$$F_d = \phi v R^a \eta^b \rho_0{}^c \qquad (14.17)$$

Here ϕ is a dimensionless numerical factor, and the powers a, b, and c must be chosen so that F_d has the correct dimensions.

The dimensions of η were given in Eq. 14.3; ρ_0 is a mass per unit volume, and $[F] = [ma]$. Thus the physical quantities in Eq. 14.17 have the following dimensions:

$$[F] = [ML^1T^{-2}], \qquad [v] = [LT^{-1}], \qquad [R] = [L]$$
$$[\rho_0] = [ML^{-3}], \qquad [\eta] = [ML^{-1}T^{-1}]$$

From Eq. 14.17, we require

$$[ML^1T^{-2}] = [LT^{-1}][L]^a[ML^{-1}T^{-1}]^b[ML^{-3}]^c$$

We must choose a, b, and c so that the units on both sides of the equation match. To get T^{-2} on the right, we must make $b = 1$. Then to obtain M^1, we assign $c = 0$. Finally, to obtain L^1, we choose $a = 1$. Thus the viscous drag on an object must have the form

$$F_d = \phi v R \eta \qquad (14.18)$$

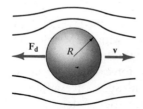

Figure 14.8. $\mathbf{v}$ is the velocity of the spherical object, and $\mathbf{F}_d$ is the viscous drag force. Also shown are the laminar flow lines.

Note that the density ρ_0 does not enter the result. This is not too surprising, since the fundamental viscous force law $F = \eta A\,\Delta v/\Delta y$ does not depend upon the fluid density.

This result is valid whenever the velocity is small enough. In terms of the Reynolds number for a sphere of radius R, the requirement is

$$N_R = \frac{\rho_0 v R}{\eta} < 1 \qquad \text{(sphere of radius } R) \qquad (14.19)$$

If $N_R \gtrsim 1$, the drag force becomes proportional to v^2. ($\gtrsim$ means "greater than or approximately equal to.") This occurs at velocities well below the onset of turbulence and is due, as noted earlier, to the kinetic energy imparted to the fluid.

For a sphere, ϕ is known to be exactly 6π, so

$$F_d = 6\pi R v \eta \qquad \text{(sphere)} \qquad (14.20)$$

This is called *Stokes' law*. For complex shapes, Eq. 14.18 can still be used if ϕ is determined by experiment. R must be interpreted as some characteristic length of the object, such as the average radius of a red blood cell.

It is sometimes possible to get information about the size or shape of a small object if the viscous drag force on the object is determined from its motion. This idea underlies some of the applications of the centrifuge, which are discussed in Section 14.7.

To illustrate the use of Stokes' law, we obtain the maximum or terminal velocity v_T for a small sphere of radius R and density ρ falling through a fluid of viscosity η and density ρ_0. Terminal velocity is reached when the drag force $\mathbf{F}_d$ exactly balances the weight $\mathbf{w}$ and the buoyant force $\mathbf{B}$ (Fig. 14.9).

The volume of the sphere is $V = \frac{4}{3}\pi R^3$, and its weight is $w = \rho g V$. The upward buoyant force is the weight of the fluid displaced $B = \rho_0 g V$. From Stokes' law, the viscous drag force at the terminal velocity v_T is $F_d = 6\pi R v_T \eta$. Thus terminal velocity is reached when $F_d = w - B$, or

$$6\pi R v_T \eta = \tfrac{4}{3}\pi R^3 \rho g - \tfrac{4}{3}\pi R^3 \rho_0 g$$

Solving for v_T, the terminal velocity of the sphere is

$$v_T = \frac{2}{9}\frac{R^2}{\eta}g(\rho - \rho_0) \qquad (14.21)$$

If v_T and R are measured, this equation provides a way of finding the viscosity of the fluid. The terminal velocity of a dust particle is found in the next example.

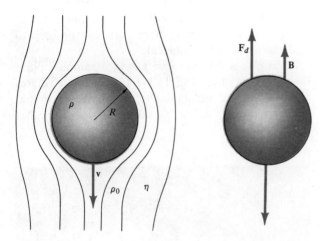

Figure 14.9. A sphere falling through a viscous fluid. Terminal velocity is reached when $F_d = w - B$.

Example 14.7

(a) What is the terminal velocity in air at 20° C of a spherical dust particle of radius 10^{-5} m and density 2×10^3 kg m^{-3}? (b) What is the Reynolds number at terminal velocity? (c) Find the drag force at terminal velocity.

(a) At 20° C, the density of air is 1.20 kg m^{-3}, which is negligible compared to that of the particle. Using $\eta = 1.81 \times 10^{-5}$ Pa s from Table 14.1 and assuming the velocity is low enough that Stokes' law applies,

$$v_T = \frac{2}{9} \frac{R^2}{\eta} g(\rho - \rho_0)$$

$$= \frac{2}{9} \frac{(10^{-5} \text{ m})^2}{(1.81 \times 10^{-5} \text{ Pa s})}$$

$$\times (9.8 \text{ m s}^{-2})(2 \times 10^3 \text{ kg m}^{-3})$$

$$= 2.41 \times 10^{-2} \text{ m s}^{-1}$$

(b) The Reynolds number is

$$N_R = \frac{\rho_0 v R}{\eta}$$

$$= \frac{(1.22 \text{ kg m}^{-3})(2.41 \times 10^{-2} \text{ m s}^{-1})(10^{-5} \text{ m})}{(1.81 \times 10^{-5} \text{ Pa s})}$$

$$= 0.0162$$

This is much less than 1, so Stokes' law and Eq. 14.21 for v_T apply.

(c) The drag force is

$$F_d = 6\pi R v \eta$$
$$= 6\pi(10^{-5} \text{ m})(2.41 \times 10^{-2} \text{ m s}^{-1})(1.81 \times 10^{-5} \text{ Pa s})$$
$$= 8.23 \times 10^{-11} \text{ N}$$

Equation 14.21 for the terminal velocity is valid only for very small objects, such as dust particles in air or macromolecules in solution. For larger objects the terminal velocity predicted by this equation corresponds to a Reynolds number much larger than 1, so that Stokes' law is not applicable. However, Eq. 14.21 is useful for finding the motion of molecules or small particles in a centrifuge, as is discussed in Section 14.7.

In this section, we obtained the viscous drag force using dimensional analysis. This method is very useful in many scientific problems where the exact theory is difficult to apply or is not even known, and we use it again in this chapter. However, it has limitations. For example, if several lengths enter a problem, it may not be possible to determine what combinations enter the formula, since $(l_1/l_2)^3$, $(l_1/l_2)^{1/2}$, and e^{-l_1/l_2} are all dimensionless quantities. In general, we must have only three dimensionally independent variables so that unique combinations can be formed that determine the units of mass, time, and length. Another drawback is that dimensionless constant factors such as ϕ occasionally turn out to be numerically very large or small.

14.6 | "HIGH-SPEED" DRAG FORCES

At the beginning of the preceding section, we noted that as the speed of an object moving in a fluid increases so that $N_R = \rho_0 v R/\eta \gtrsim 1$, the drag force is no longer proportional to v. Instead, it is approximately proportional to v^2. The term "high speed" is somewhat misleading, since the v^2 dependence is observed whenever N_R is appreciably more than 1, and this happens at surprisingly low speeds. For example, for a sphere of radius 1 cm in air, N_R is 1 at $v = 1.5$ mm^{-1}. Thus, high-speed drag applies to essentially all problems involving the motion of macroscopic objects.

The formula for the high-speed drag force on a sphere in a fluid can again be found using dimensional analysis. The force is assumed to be proportional to v^2 and to unknown powers of the radius of the sphere R, the viscosity of the fluid η, and the density of the fluid ρ_0. The calculation is left as a problem. The result is that F_d is proportional to $R^2\rho_0 v^2$, or that

$$F_d = C_D A \frac{\rho_0 v^2}{2} \qquad (14.22)$$

Here factors are grouped for convenience: $A = \pi R^2$ is the cross-sectional area of the moving sphere, $\rho_0 v^2/2$

is the kinetic energy per unit volume of fluid with velocity v, and C_D is the *drag coefficient* that must be obtained from measurements. The viscosity η is absent, which is consistent with the idea that the drag results from accelerating the fluid moving around the sphere and not from viscous effects.

The result just given holds true for objects of any shape. Figure 14.10 gives experimental values of C_D versus the Reynolds number for a long cylinder moving perpendicular to its axis. We see that C_D is nearly constant; it varies between 0.3 and 3.0 as N_R varies from 10 to 10^6. This occurs despite the fact that as N_R increases, the motion of the fluid around the cylinder changes from laminar to turbulent when $N_R \simeq 100$.

Equation 14.22 can be used to obtain the terminal velocity of a macroscopic falling object. The result for an object of cross-sectional area A, length L, and mass ρAL, is

$$v_T = \sqrt{\frac{(\rho - \rho_0)}{\rho_0} \frac{2gL}{C_D}} \qquad (14.23)$$

Note that the cross-sectional area does not appear in the result. This derivation and numerical examples are left for the exercises and problems.

14.7 | CENTRIFUGATION

Centrifuges use the very large accelerations experienced by rapidly rotating objects to perform many tasks in biological and medical laboratories. For example, centrifuges can separate large and small particles and molecules. The speed of a molecule in a centrifuge is determined by the viscous drag force and

the molecular mass. Accordingly, a measurement of this speed, together with other data, can be used to find the mass of the molecule.

A centrifuge consists basically of a sample holder that can be rotated (Fig. 14.11). If the sample rotates at a radius r and an angular velocity ω, its centripetal acceleration is $a_r = \omega^2 r$. We saw in Chapter Five that

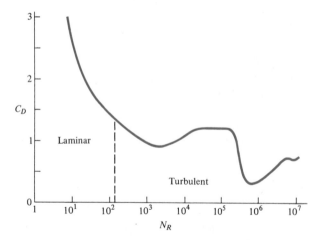

Figure 14.10. Drag coefficients versus Reynolds number for a long cylinder moving perpendicular to its axis. Here $N_R = \rho_0 v R / \eta$, where R is the radius of the cylinder.

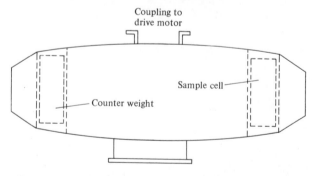

Figure 14.11. A refrigerated centrifuge. Rotational speeds of up to 6000 revolutions per minute corresponding to 6835 g are achieved with this unit. Other types of centrifuges achieve speeds of 60,000 revolutions per minute and 500,000 g. (Photo courtesy of Beckman Instruments, Inc.)

the effective weight of a rotating object of mass m is $\mathbf{w}^e = m(\mathbf{g} - \mathbf{a}_r)$. Since a_r may be up to $500{,}000g$ and is always much greater than g, the effective weight is given to a good approximation by $\mathbf{w}^e = -m\mathbf{a}_r$. Thus the sample will behave as though it were on a planet where $w^e = mg^e$ and the gravitational acceleration has a magnitude $g^e = \omega^2 r$. For example, molecules denser than a solvent will *sediment* or settle to the bottom of the holder at a much greater terminal or *sedimentation* velocity than in a stationary solution.

The sedimentation velocity v_s of particles experiencing a "gravitational acceleration" g^e is determined by their effective weight w^e, the buoyant force B due to the fluid, and the low speed or viscous drag force F_d (Fig. 14.12). If the particles are spherical, Eq. 14.21, which was obtained from Stokes' law, can be used with g replaced by g^e to find this velocity. However, another expression for v_s is often more useful.

To obtain this expression we use the general form for the low-speed drag force, Eq. 14.18, with R representing an average dimension of the particle,

$$F_d = \phi R v \eta$$

If the particles have mass m and volume V, their density is $\rho = m/V$, and their effective weight is $w^e = mg^e$. If the density of the fluid is ρ_0, the buoyant force is $B = \rho_0 g^e V = (\rho_0/\rho)mg^e$. The forces are balanced at a velocity v_s if $F_d = w^e - B$. Thus

$$\phi R v_s \eta = mg^e - \frac{\rho_0}{\rho} mg^e$$

or

$$v_s = \frac{mg^e}{\phi R \eta}\left(1 - \frac{\rho_0}{\rho}\right) \qquad (14.24)$$

This is the sedimentation velocity for molecules in a solution if the acceleration is $g^e = \omega^2 r$.

This result can be used in various ways. If the quantities on the right side are known, the sedimenta-

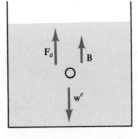

Figure 14.12. Forces on a particle in a centrifuge. The particle moves downward with a sedimentation velocity $\mathbf{v}_s$.

tion velocity can be calculated, and one can determine how long it will take for the material to collect at the bottom of the sample holder (Fig. 14.13). Alternatively, the sedimentation velocity can be measured by shining light through the spinning sample and exploiting the fact that proteins and many other materials absorb light at specific wavelengths. This velocity measurement then provides information about the other variables in Eq. 14.24. Also, the product ϕR can be inferred from diffusion experiments, and the densities are readily measured, so sedimentation velocities can provide a determination of molecular masses. If several molecules of different sizes are present, v_s will be different for each. This permits various components of a mixture to be identified.

Example 14.8

Hemoglobin has a density of 1.35×10^3 kg m^{-3}, and a molecular mass of 68,000 u. The factor ϕR for hemoglobin in water is 9.46×10^{-8} m. If it is in a centrifuge with a centripetal acceleration of $10^5 g$, find its sedimentation velocity in water at $37°$ C.

In kilograms, the mass of the molecule is (68,000 u) $\times$ (1.66 $\times$ 10^{-27} kg u^{-1}) = 1.129 $\times$ 10^{-22} kg. The density of water is 10^3 kg m^{-3}. Thus the sedimentation velocity is

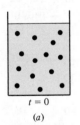 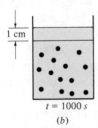 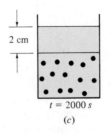 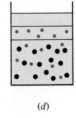

$t = 0$	$t = 1000\ s$	$t = 2000\ s$	
(a)	(b)	(c)	(d)

Figure 14.13. (a), (b), and (c). Molecules sedimenting in a centrifuge at a velocity $v_s = 10^{-3}$ cm s^{-1}. A sharp boundary between the solution and pure solvent moves downward with velocity v_s. (d) In a mixture of two kinds of solutes, the larger molecules (black dots) will sediment faster than the smaller ones (colored dots).

$$v_s = \frac{mg^e}{\phi R\eta}\left(1 - \frac{\rho_0}{\rho}\right)$$

$$= \frac{(1.129 \times 10^{-22}\text{ kg})(10^5)(9.8\text{ m s}^{-2})}{(9.46 \times 10^{-8}\text{ m})(0.695 \times 10^{-3}\text{ Pa s})}\left(1 - \frac{1}{1.35}\right)$$

$$= 4.37 \times 10^{-7}\text{ m s}^{-1}$$

At this velocity the molecules will travel 3.7 cm in a 24-hour period. We can see why it is quite common to run a centifuge for many hours.

Example 14.9

A protein of density 1.3×10^3 kg m^{-3} has a sedimentation velocity of 10^{-6} m s^{-1} in a centrifuge with acceleration 10^6 m s^{-2}. At this sedimentation velocity the drag force is 2.07×10^{-16} N. Find the molecular mass of the protein.

Solving Eq. 14.24 for the molecular mass gives

$$m = \frac{\phi R\eta v_s}{g^e}\frac{1}{1 - \rho_0/\rho}$$

Since $\phi R\eta v_s$ is the drag force, and $\rho_0 = 10^3$ kg m^{-3},

$$m = \frac{(2.07 \times 10^{-16}\text{ N})}{(10^6\text{ m s}^{-2})}\frac{1}{1 - 1/1.3}$$

$$= (7.98 \times 10^{-22}\text{ kg})\left(\frac{1\text{ u}}{1.66 \times 10^{-27}\text{ kg}}\right) = 481{,}000\text{ u}$$

EXERCISES ON SUPPLEMENTARY TOPICS

Section 14.5 | Viscous Drag Forces

14-36 A spherical blood cell of radius 5×10^{-6} m and density 1.3×10^3 kg m^{-3} is in water at 37° C. What is its terminal velocity? (Assume Stokes' law holds true.)

14-37 A large spherical molecule has a radius of 2×10^{-8} m and a density of 1.5×10^3 kg m^{-3}. (a) What is its terminal velocity when falling in water at 20°C? (b) What is the maximum velocity at which Stokes' law holds true?

14-38 For spherical dust particles of density 3×10^3 kg m^{-3}, find the maximum radius for which Stokes' law can be used to find the terminal velocity (a) in air at 20° C; and (b) in water at 20° C.

14-39 The terminal velocity of a spherical oil droplet falling in air at 20° C is 2×10^{-7} m s^{-1}. What is the radius of the spherical droplet if its density is 930 kg m^{-3}? (Assume that Stokes' law holds true.)

Section 14.6 | "High-Speed" Drag Forces

14-40 A spherical pebble of radius 0.04 m and density 3×10^3 kg m^{-3} is dropped into a pond.

Estimate its terminal velocity, assuming the drag coefficient is 1.

14-41 Estimate the force a person experiences standing in a wind blowing at 20 m s^{-1}, assuming the drag coefficient is 1.

14-42 A fish swimming in water at a velocity of 0.3 m s^{-1} experiences a drag force proportional to the velocity squared. The cross-sectional area presented to the water is 2.2×10^{-3} m^2. (a) What is the drag force on the fish if the drag coefficient C_D is 1? (b) What is the power expended by the fish against the drag force?

14-43 A baseball has a mass of 0.149 kg and a Ping-Pong ball has a mass of 3×10^{-3} kg. Their radii are 0.037 m and 0.018 m, respectively. Both are thrown in air with initial horizontal velocities of 15 m s^{-1}. (a) What is the ratio of drag forces on the two balls? (b) What is the initial acceleration of each ball due to the drag force, assuming $C_D = 1$?

14-44 A baseball has a mass of 0.149 kg and a radius of 0.037 m. (a) Calculate its terminal velocity in air at 0° C using Eq. 14.21. (b) Calculate the terminal velocity using Eq. 14.23, assuming $C_D = 1$. (c) Explain the significance of the difference in the results in parts (a) and (b).

Section 14.7 | Centrifugation

14-45 A protein molecule has a mass of 10^5 u and a density of 1.35×10^3 kg m^{-3}. It is in a centrifuge with an acceleration of $2 \times 10^5 g$. If it is in water, find (a) its effective weight; and (b) the buoyant force.

14-46 The sample in a centrifuge is 0.1 m from the axis of rotation. If its centripetal acceleration is $200{,}000 g$, how many times per minute does the centrifuge rotate?

14-47 Bushy stunt virus has a molecular mass of 1.06×10^7 u and a density of 1.35×10^3 kg m^{-3}. The factor ϕR in water is 3.58×10^{-7} m at 20° C. How much time is required for it to sediment 10^{-2} m if the acceleration in the centrifuge is $10^5 g$?

14-48 A red blood cell can be approximated by a sphere of radius 2×10^{-6} m and density 1.3×10^3 kg m^{-3}. How long will it take for it to sediment 1 cm $= 10^{-2}$ m in blood at 37° C (a) under the earth's gravitational acceleration; (b) in a centrifuge with an acceleration of $10^5 g$? (The density of blood is 1.0595×10^3 kg m^{-3}.)

14-49 Tobacco mosaic virus has a density of 1370 kg m^{-3} and $\phi R = 1.16 \times 10^{-6}$ m at 37° C. If the acceleration in a centrifuge is $2 \times 10^5 g$, the sedimentation velocity in water at 300 K is observed to be 3.7×10^{-5} m s^{-1}. What is the molecular mass of the tobacco mosaic virus?

PROBLEMS ON SUPPLEMENTARY TOPICS

14-50 Red blood cells are usually not spherical in shape. What is the product ϕR in Eq. 14.18 for a cell of mass 10^{-12} kg if the terminal velocity in water at 37° C is 10^{-5} m s^{-1}? (The density of the cell is 1.3×10^3 kg m^{-3}.)

14-51 Estimate the maximum velocity at which Stokes' law applies for a small fish in water.

14-52 Estimate the terminal velocity of an 800-N man falling through air (a) if he falls in a tuck position; (b) if he spreads his body and arms out. (Assume the drag coefficient C_D is 1.)

14-53 Assuming that the flow rate Q is proportional to the pressure gradient $\Delta P/l$, use dimensional analysis to derive the powers in Eq. 14.6.

14-54 Derive Eq. 14.22 using dimensional analysis.

14-55 When the drag force is proportional to the velocity squared, show that the terminal velocity of an object of density ρ falling through a fluid of density ρ_0 is given by Eq. 14.23.

14-56 For a spherical molecule of radius R in a fluid with viscosity η, the drag force is given both by $F_d = 6\pi\eta R v$ (Stokes' law) and by $F_d = \phi R v \eta$. What is the product ϕR for a spherical molecule with radius 10^{-8} m?

14-57 Nucleohistone has a mass of 2.1×10^6 u and a density of 1520 kg m^{-3}. (a) What is its volume? (b) If this molecule is spherical, what is its radius? (c) Assuming the molecule is spherical, use Stokes' law to find the product ϕR. (d) The product ϕR is actually 4.33×10^{-7} m. Is this molecule spherical? Explain.

14-58 (a) What is the acceleration of a large air bubble in water at 20° C when it reaches half of its terminal velocity? (The density of water at 20° C is 990 kg m^{-3}, and the density of air is 1.20 kg m^{-3}.) (b) The acceleration in part (a) turns out to be a very large number. What does this imply about the time required to approach terminal velocity?

14-59 Figure 14.14 shows a viscous fluid trapped between the two gas bubbles as the gas and fluid move through a tube. Putting yourself in a frame of reference moving with the gas, draw arrows in the fluid showing its velocity at various points in the tube. Using your results, suppose the gas bubbles are actually red blood cells, the fluid blood plasma,

Figure 14.14. Problem 14-59.

and the tube a capillary. What does the velocity pattern you've drawn suggest about the role of the plasma in aiding diffusion of gases between the capillary and the red blood cells?

14-60 Some animals have very streamlined shapes and relatively low drag coefficients. For example, a dolphin has a drag coefficient C_D of 0.055. (a) If the dolphin presents an area of 0.11 m^2 to the water when swimming at 8.3 m s^{-1}, what is the drag force on it? (b) For the dolphin, about 15 percent of the 90-kg body mass is swimming muscle. Estimate the power output per kilogram of muscle for the dolphin. (c) Find the ratio of this result to the maximum power output of human muscles, 40 W kg^{-1}. (The large value of this ratio has led a nuamber of researchers to study whether dolphin muscles are more efficient or if the flow along the body is less turbulent than expected.)

14-61 Assume that two swimmers, A and B, each swim 110 m in 80 s. A does so by swimming each 25 m stretch in 20 s. B swims the 4 segments in 15, 15, 25, and 25 s, respectively. Which swimming does more work against drag forces?

14-62 An air bubble of radius 0.5 mm = 5×10^{-4} m rises in a liquid of density 900 kg m^{-3} and viscosity 8.37×10^{-3} Pa s. Find the terminal velocity of the bubble and justify your answer by checking the Reynolds number. Assume the air density is 1.2 kg m^{-3}. (Assume $C_D = 1$.)

Additional Reading

Richard B. Setlow and Ernest C. Pollard, *Molecular Biophysics,* Addison-Wesley Publishing Co., Reading, Mass., 1962. Chapter 4 discusses centrifugation.

R. McNeill Alexander, *Animal Mechanics,* University of Washington Press, Seattle, 1968. Chapters 5 and 6.

E.M. Purcell, Life at Low Reynolds Numbers. *American Journal of Physics,* vol. 45, 1977, p. 3.

Alan C. Burton, *Physiology and Biophysics of the Circulation,* Yearbook Medical Publishers, Inc., Chicago, 1965.

Vernon B. Mountcastle (ed.), *Medical Physiology,* Vol. 1, 12th ed., The C.V. Mosby Co., St. Louis, 1968.

Arthur C. Guyton, *Circulatory Physiology: Cardiac Output and Its Regulation,* W.B. Saunders Co., Philadelphia, 1963.

R.L. Whitmore, *Rheology of the Circulation,* Pergamon Press, Oxford and New York, 1968.

Gerald Aiello, Pierre LaFrance, Rogers C. Ritter, and James S. Trefil, The Urinary Drop Spectrometer, *Physics Today,* vol. 27, September 1974, p. 23.

J. Richard Shanebrook, Fluid Mechanics of Coronary Artery Bypass Surgery, *American Journal of Physics,* vol. 45, 1977, p. 677.

James A. Lock, The Physics of Air Resistance, *The Physics Teacher,* vol. 20, 1982, p. 158.

Richard A. Kerr, How Does Fluid Flow Become Turbulent?, *Science,* vol. 221, 1983, p. 140.

M. Van Dyke, *Album of Fluid Motion,* Parabolic, Stanford, CA, 1982. 300 photographs of many phenomena.

Cliff Frohlich, Aerodynamic Drag Crisis and Its Possible Effect on the Flight of Baseballs, *The American Journal of Physics,* vol. 51, 1984, p. 325.

Herman Erlichson, Maximum Projectile Range with Drag and Lift, with Particular Application to Golf, *The American Journal of Physics,* vol. 51, 1983, p. 347.

Scientific American articles:

Jesse W. Beams, Ultrahigh-Speed Rotation, April 1961, p. 134.

F.H. Harlow and J.E. Fromm, Computer Experiments in Fluid Dynamics, March 1965, p. 104.

R.W. Stewart, The Atmosphere and the Ocean, September 1969, p. 76.

D. James Baker, Jr., Models of Oceanic Circulation, January 1970, p. 114.

R.C. Chanaud, Aerodynamic Whistles, January 1970, p. 40.

Victor P. Starr and Norman E. Gant, Negative Viscosity, July 1970, p. 72.

Norman C. Chigier, Vortexes in Aircraft Wakes, March 1974, p. 76.

Kay Johansen, Aneurysms, July 1982, p. 110.

J. Edwin Wood, The Venous System, January 1968, p. 86.

James V. Warren, The Physiology of the Giraffe, November 1974, p. 96.

Donald R. Olander, The Gas Centrifuge, August 1978, p. 37.

E. Eugene Larrabee, The Screw Propeller, July 1980, p. 134.

Jearl Walker, Serious Fun with Polyox, Silly Putty, Slime, and Other Non-Newtonian Fluids, The Amateur Scientist, November 1978, p. 186.

Jearl Walker, Delights of Forming Water Into Sheets and Bells With Knives, Spoons, and Other Objects, August 1979, p. 188.

Jearl Walker, Easy Ways to Make Holograms and View Fluid Flow, and More About Funny Fluids, The Amateur Scientist, February 1980, p. 158.

Jearl Walker, The Charm of Hydraulic Jumps, Starting With Those Observed in the Kitchen Sink, The Amateur Scientist, April 1981, p. 176.

Albert C. Gross, Chester R. Kyle, and Douglas J. Malewicki, The Aerodynamics of Human-Powered Land Vehicles, December 1983, p. 142.

M.A.R. Koehl, The Interaction of Moving Water and Sessile Organisms, December, 1982, p. 124. Effects of drag forces on shallow water organisms.

Paul W. Webb, Form and Function in Fish Swimming, July 1984, p. 72.

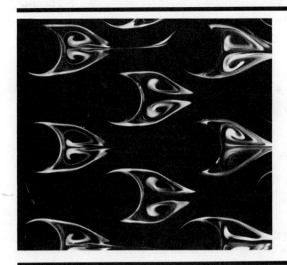

CHAPTER 15

COHESIVE FORCES IN LIQUIDS

When we consider pulling on a substance until it breaks, we usually think of solids. However, liquids also have a strong tendency to remain intact. For example, if pure water with no air dissolved in it is pressed between two smooth plates, enormous forces are necessary to separate the plates.

As in solids, the cohesiveness of liquids results from the attractions among the molecules. Because of these attractions, liquids have well-defined surfaces which, like stretched membranes or rubber sheets, tend to have a minimum surface area. Ripples in a quiet, smooth pool are suppressed because they require an increase in surface area. Aquatic insects are able to move on water surfaces because their weight is opposed by the resistance of the surface to deformation.

In addition to the attractive forces among themselves, the molecules of a liquid experience attractive or repulsive interactions with the molecules of other substances. Thus, water rises near a vertical glass surface, while the opposite is true for mercury.

The cohesive properties of liquids can be changed by the addition of small amounts of other substances. For example, molecules in oils are *hydrophobic* (water-hating), and oils will not dissolve in pure water. Molecules in soaps and detergents have both hydrophobic and *hydrophilic* (water-loving) portions. The hydrophilic part attaches itself to the water surface, and the hydrophobic portion surrounds oil or grease. This aids in the solution and removal of oil and grease.

In this chapter, we consider some effects of the cohesiveness of liquids. Topics discussed include the rise of liquids in narrow tubes or *capillaries,* the formation of bubbles, and the rise of sap in trees.

15.1 | SURFACE TENSION

One way to observe the effects of surface tension is to dip the apparatus shown in Fig. 15.1 into a liquid. This consists of a U-shaped wire, a slide wire of weight w_1, and a suspended weight w_2. A thin film of liquid fills the enclosed area between the wires. If the total weight $w = w_1 + w_2$ is properly chosen, the two surfaces of the film exert a force equal and opposite to the weight and the slide wire remains stationary. Thus, in this situation the force F due to the surface tension is equal in magnitude to w.

The *surface tension* γ (gamma) is defined as the force per unit length exerted by *one* surface. Then, if the straight wire of Fig. 15.1 has a length l, the net force upward due to the two surfaces if $F = 2\gamma l$. Thus the surface tension is

$$\gamma = \frac{F}{2l} \qquad (15.1)$$

Representative surface tensions are given in Table 15.1.

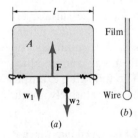

Figure 15.1. (*a*) A liquid film filling the area A exerts a force F, which balances $w = w_1 + w_2$. (*b*) A cross-sectional view of the sliding straight wire and the film attached to it. The film has two surfaces.

315

TABLE 15.1

The surface tension for representative liquids in contact with air

Liquid	Surface Tension ($N m^{-1}$)	Temperature ($0°C$)
Ethyl alcohol	2.23×10^{-2}	20
Olive oil	3.20×10^{-2}	20
Glycerine	6.31×10^{-2}	20
Water	7.56×10^{-2}	0
	7.28×10^{-2}	20
	6.62×10^{-2}	60
	5.89×10^{-2}	100
Mercury	0.465	20
Silver	0.800	970
Gold	1.000	1070
Copper	1.100	1130
Oxygen	1.57×10^{-2}	-193
Neon	5.15×10^{-3}	-247

One may ask why the forces in Fig. 15.1 are attributed to the surfaces. The point is that the liquid film acts somewhat differently than a rubber sheet. If the apparatus is in equilibrium and the straight wire is moved up or down, it is in equilibrium again at each position, even though the film thickness has changed. Thus the thickness of the film is not important. If the film is stretched, molecules from the bulk of the fluid move to the surface, and the surface area is increased. Since the force remains the same while the bulk thickness varies, the force must be attributed to the surface of the film.

We can understand the reason for this behavior with the aid of a microscopic picture of the boundary between a liquid and its vapor (Fig. 15.2). A molecule in the interior of the liquid experiences attractive forces due to all its neighbors, which lowers its potential energy. A molecule at the surface does not have as many neighbors, and its potential energy is not so low. Consequently, the molecules at the surface arrange themselves so that they have as many neighbors as possible. In this process, they minimize the surface area and the potential energy and produce the observed surface tension. The following example shows that these surface forces are relatively small.

Example 15.1

The U-shaped loop in Fig. 15.1 is dipped into water at 20° C. The slide wire is 0.1 m long and has a mass $m_1 = 1$ g $= 10^{-3}$ kg. (a) How large is the surface tension

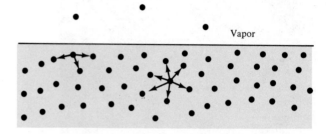

Figure 15.2. The molecules of a substance in the liquid and the coexistant vapor phase. The attractive forces on molecules near the surface and in the interior are denoted by arrows.

force? (b) If the wire is in equilibrium, how large is the mass m_2 suspended from the wire?

(a) According to Table 15.1, $\gamma = 7.28 \times 10^{-2}$ N m^{-1}. Thus the force due to the two surfaces of the film of water is

$$F = 2\gamma l = 2(7.28 \times 10^{-2} \text{ N m}^{-1})(0.1 \text{ m})$$
$$= 1.46 \times 10^{-2} \text{ N}$$

(b) In equilibrium the force F exerted by the film must equal the total weight $m_1 g + m_2 g$. Solving for m_2,

$$m_2 = \frac{F}{g} - m_1 = \frac{1.46 \times 10^{-2} \text{ N}}{9.8 \text{ m s}^{-2}} - 10^{-3} \text{ kg}$$
$$= 0.49 \times 10^{-3} \text{ kg} = 0.49 \text{ g}$$

Thus the surface forces support the weight of a total mass, $m_1 + m_2$, which is approximately 1.5 g.

15.2 | CONTACT ANGLES AND CAPILLARITY

The surface of a liquid in contact with a solid surface forms an angle with respect to the solid surface (Fig. 15.3). The *contact angle* θ is determined by the competition between the liquid-liquid molecular forces and the liquid-solid forces and depends on the particular solid and liquid involved (Table 15.2). It

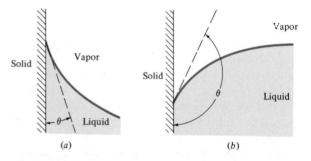

Figure 15.3. The contact angle θ for two liquids in contact with a solid surface. (a) $\theta < 90°$; (b) $\theta > 90°$.

TABLE 15.2

The contact angles for several liquid-solid interfaces

Interface	Contact Angle
Water—clean glass	0°
Ethyl alcohol—clean glass	0°
Mercury—glass	140°
Water—silver	90°
Water—paraffin	107°
Methylene iodide—Pyrex glass	30°

also depends on how smooth and clean the solid surface is.

If θ is less than 90°, the liquid surface appears as in Fig. 15.3a and will rise in a narrow tube. If θ is greater than 90°, the liquid appears as in Fig. 15.3b and will be depressed in a tube. If $\theta = 90°$, the liquid will neither rise nor fall. The rise or fall of a liquid in a narrow tube or capillary is referred to as *capillarity* or *capillary action*.

We now look at how the contact angle determines the height of a liquid in a column. Figure 15.4 shows a liquid of density ρ in a tube of radius r. The contact angle is less than 90°. The net vertical force F_{up} is the vertical component of the surface tension times the length l of the liquid surface in contact with the tube; l equals the circumference $2\pi r$. Thus

$$F_{up} = 2\pi r \gamma \cos \theta \qquad (15.2)$$

The volume of the column of liquid up to the bottom of the curved liquid surface or *meniscus* is $V = \pi r^2 h$, and its weight is $w = \rho g V = \rho g \pi r^2 h$. The liquid rises until $F_{up} = w$, or

$$2\pi r \gamma \cos \theta = \rho g \pi r^2 h$$

This gives, for the height of the liquid in a capillary,

$$h = \frac{2\gamma \cos \theta}{\rho g r} \qquad (15.3)$$

This result has several notable features. If $\theta = 90°$, then $h = 0$, and the fluid neither rises nor is depressed. If θ is greater than 90°, $\cos \theta$ is negative and so is h. This means that the liquid is depressed. The height h is proportional to γ; the greater the surface tension, the greater the capillary effect. Conversely, h depends on $1/r$, so the effect is most dramatic when r is small.

Although Eq. 15.3 was derived for a circular cross section, capillary rise occurs in tubelike structures or connected channels of any geometry. The height still depends on $\gamma/\rho g r$, where r is some typical dimension, although the overall numerical coefficient changes. For example, water will be easily absorbed by fine, porous fabric if the contact angle is less than 90°. We apply this result to trees in the next example.

Example 15.2

The sap in trees, which consists mainly of water in summer, rises in a system of capillaries of radius $r = 2.5 \times 10^{-5}$ m. The contact angle is 0°. The density of water is 10^3 kg m^{-3}. What is the maximum height to which water can rise in a tree at 20° C?

From Table 15.1, $\gamma = 7.28 \times 10^{-2}$ N m^{-1}. With $\cos \theta = 1$, the height the sap rises is

$$h = \frac{2\gamma \cos \theta}{\rho g r}$$
$$= \frac{2(7.28 \times 10^{-2}\ \text{N m}^{-1})}{(10^3\ \text{kg m}^{-3})(9.8\ \text{m s}^{-2})(2.5 \times 10^{-5}\ \text{m})}$$
$$= 0.594\ \text{m}$$

Since trees grow to heights of many metres, capillary action cannot account for the supply of water to the top of a tree.

Figure 15.5a shows the shape of a drop of water on paraffin. A wetting agent, a molecule with hydrophobic and hydrophilic portions, would cause the droplet

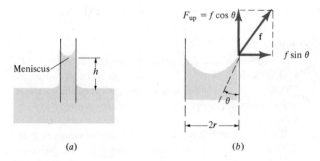

(a)   (b)

Figure 15.4. Liquid in a tube of radius r with a contact angle θ rises to a height h. The force on a small segment of the liquid in contact with the tube wall is **f**.

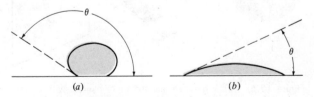

(a)   (b)

Figure 15.5. Water droplets on paraffin. (a) Pure water. (b) Wetting agent added.

on the paraffin to appear as shown in Fig. 15.5*b*. The hydrophilic part of the molecule attaches itself to the water surface, and the hydrophobic part stays away from the water surface but is attracted to the paraffin surface. Waterproofing agents have the opposite effect. They cause the angle of contact to be increased, and droplets are less likely to penetrate narrow fabric pores.

15.3 | LAPLACE'S LAW

Laplace's law relates the pressure difference across a closed elastic membrane or liquid film to the tension in the membrane or film. Its specific form depends on the shape of the closed surface.

We begin by considering a spherical membrane or rubber balloon filled with a fluid. The membrane wall exerts a force per unit length or wall tension γ (Fig. 15.6); this force per unit length depends on the thickness of the wall and therefore is associated with the membrane as a whole, *not* with each of the two surfaces as in the case of a liquid film. The pressure inside is P_i and that outside is P_0. The lungs and the ventricles of the heart may be approximated by such a model.

The total force to the left on the hemisphere of Fig. 15.6*b* due to the wall tension is the product of γ and the circumference $2\pi r$ of the hemisphere, or $2\pi r\gamma$. The forces due to the pressures are perpendicular to the surface at each point. All the components cancel except those to the right, which add up to the pressure

difference $P_i - P_0$ times the *projected area* πr^2. The two forces must balance in equilibrium, so

$$2\pi r\gamma = (P_i - P_0)\pi r^2$$

or

$$P_i - P_0 = \frac{2\gamma}{r} \quad \text{(spherical membrane)} \quad (15.4)$$

This is called *Laplace's law for a spherical membrane.* It is named after Marquis Pierre Simon de Laplace (1749–1827), a French physicist and mathematician.

Laplace's law implies that it requires a greater pressure difference to sustain a small sphere than a larger one. It holds true not only for the model discussed, a membrane or a balloon, but also for a spherical liquid drop. Both situations are illustrated by the next two examples.

Example 15.3

A rubber balloon is inflated to a radius of 0.1 m. The pressure inside is 1.001×10^5 Pa, and the pressure outside is 10^5 Pa. What is the wall tension γ?

Solving Laplace's law for γ,

$$\gamma = \frac{r}{2}(P_i - P_0) = \left(\frac{0.1\,\text{m}}{2}\right)(1.001 \times 10^5\,\text{Pa} - 10^5\,\text{Pa})$$

$$= (0.05\,\text{m})(10^2\,\text{Pa}) = 5\,\text{N m}^{-1}$$

This is the tension in the wall when the balloon is inflated to a radius of 0.1 m. Since the balloon is elastic, the wall tension will increase if the radius is increased.

When a liquid is in equilibrium with its own vapor, the pressure of the gas phase is called the *vapor pressure.* It may also be thought of as the pressure necessary to keep more of the liquid from evaporating; it balances a pressure difference $P_i - P_0$ across the liquid-vapor surface. Thus, a liquid drop in equilibrium with its vapor will have a pressure difference $P_i - P_0$ equal to the vapor pressure. The droplet size arising from this condition is found in the next example.

Example 15.4

From Table 15.1 the surface tension of water at 20° C is 7.28×10^{-2} N m^{-1}. The vapor pressure of water at 20° C is 2.33×10^3 Pa. What is the radius of the smallest spherical water droplet which can form without evaporating?

The pressure difference $P_i - P_0$ must be no more than 2.33×10^3 Pa. Solving Laplace's law for r, we have

$$r = \frac{2\gamma}{P_i - P_0} = \frac{2(7.28 \times 10^{-2}\,\text{N m}^{-1})}{2.33 \times 10^3\,\text{Pa}}$$

$$= 6.25 \times 10^{-5}\,\text{m}$$

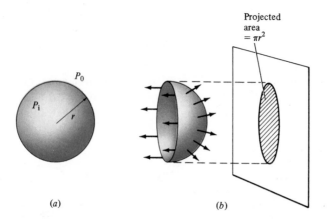

Figure 15.6. (*a*) A spherical membrane or balloon with inner and outer pressures P_i and P_o. (*b*) The sphere is cut in half by an imaginary plane. The arrows to the left represent the forces exerted by the wall. The arrows normal to the surface represent the forces due to the pressure difference.

In a soap bubble the film has two surfaces, each of which exerts a tension force. Hence if the derivation is repeated, there is an additional factor of 2, and we find

$$P_i - P_0 = \frac{4\gamma}{r} \qquad \text{(spherical bubble)} \quad (15.5)$$

Also, for a cylindrical membrane of radius r, Laplace's law is

$$P_i - P_0 = \frac{\gamma}{r} \qquad \text{(cylindrical tube)} \quad (15.6)$$

Again we observe the inverse dependence on the radius. In physiology, $P_i - P_0$ is called the *transmural* pressure, the pressure difference across the walls of a blood vessel. This is quite different from the driving pressure that forces fluid through the vessel.

Equations 15.4, 15.5, and 15.6 can be deceptive in their apparent simplicity. Implicit in each result is the fact that the pressure and wall tension may depend on the radius. Suppose that we use a soap bubble as an example. Then the surface tension γ is constant as the radius changes and, because there are two surfaces, Eq. 15.5 applies. Suppose the outside pressure is reduced gradually to zero. Then $P_i - P_0$ is replaced by P_i, and Eq. 16.5 becomes

$$P_i = \frac{4\gamma}{r}$$

Now P_i is larger than $P_i - P_0$, so the left-hand side of this equation is larger than before. Since γ is constant, the right-hand side can only increase if r decreases! But this is contrary to our intuition; we know the soap bubble will become larger. The solution is that the inside pressure P_i does not remain constant as P_0 decreases. The following example illustrates this.

Example 15.5

The pressure outside a soap bubble of radius r is initially half the pressure inside, so $P_0 = \frac{1}{2}P_i$. The outside pressure is then reduced to $P_0' = 0$. Find the new pressure inside P_i' and the new radius r', assuming the surface tension γ and temperature remain constant.

Initially $P_0 = \frac{1}{2}P_i$ and $P_i - P_0 = \frac{1}{2}P_i$. Thus $P_i - P_0 = 4\gamma/r$ becomes

$$\frac{1}{2}P_i = \frac{4\gamma}{r} \qquad (i)$$

When the outside pressure is reduced to zero, $P_i' - P_0' = P_i'$ and

$$P_i' = \frac{4\gamma}{r'} \qquad (ii)$$

Dividing Eq. (ii) by (i), Laplace's law requires for equilibrium

$$\frac{P_i'}{P_i} = \frac{r}{2r'}$$

Since the temperature remains constant, the ideal gas law gives $P_i V = P_i' V'$. The volume of a sphere is $4\pi r^3/3$, so we find a second relationship between P_i and P_i',

$$\frac{P_i'}{P_i} = \frac{V}{V'} = \frac{r^3}{r'^3}$$

Equating the two expressions for P_i'/P_i, we find $r'^2 = 2r^2$, and $r' = 1.41r$. Thus the radius increases as expected.

15.4 | SURFACTANT IN THE LUNGS

The small air sacs of the lungs, the *alveoli*, expand and contract an average of 15,000 times per day in a normal adult (Fig. 15.7). It is across the membrane of the alveoli that oxygen and carbon dioxide are transported. The tension in the walls results from both the membrane tissue and a liquid on the walls containing a long lipoprotein, *surfactant*. We will see that the surfactant gives the membrane the elasticity needed to make the required adjustments in the wall tension.

The alveoli may be approximated by small spheres with tiny entrances. Thus we may use Laplace's law, Eq. 15.4, rewritten as

$$r(P_i - P_0) = r\,\Delta P = 2\gamma \qquad (15.7)$$

For a spherical membrane the product of the radius and the pressure difference must be equal to 2γ for the membrane to be in equilibrium. In the alveoli, there is a special problem. During exhalation the pleural pressure P_0 increases, so ΔP decreases. At the same time, muscle contraction reduces the radii of the

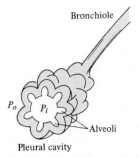

Figure 15.7. A cross section exposing alveoli at the end of a bronchiole, the smallest branch of a bronchial tube. The pressure inside the alveoli is P_i; the pressure in the fluid of the pleural cavity is P_o. The pleural cavity encloses the entire lung.

alveoli. If both r and ΔP decrease and γ is constant, the equilibrium condition, Eq. 15.7, cannot be satisfied, and the alveoli will collapse, because the inward force due to the walls exceeds the outward force due to the pressure difference. During inhalation the pleural pressure P_0 decreases and the radius increases. Again, if γ is constant, Eq. 15.7 is not satisfied, and the alveoli would tend to increase in size and rupture; the force due to the pressure difference would exceed the force due to the wall.

Nature solves this problem by having the surfactant present. Its long molecules prefer to lie nearly beside each other, making the membrane highly elastic. During inhalation, as the radius increases the molecules are pulled apart and the wall tension increases. Thus as $r\,\Delta P$ increases, so does γ, and Eq. 15.7 is satisfied. During exhalation the molecules slide back together, and the wall tension decreases along with $r\,\Delta P$. Thus the surfactant serves to change the wall tension so that equilibrium is maintained. Insufficient surfactant in the lungs is the cause of death of many newborn infants.

15.5 | THE HEART AS A PUMP

As our last example of how the physics of fluids can be used in understanding the circulatory system, we examine the mechanical properties of the heart. In particular, we calculate the work done by the heart in pumping blood and compare the result with the total metabolic energy expended.

In Section 14.2, we found that the power necessary to overcome viscous forces and maintain fluid flow is $\mathcal{P} = \Delta P\,Q$. We can use this result for the heart.

In a normal resting adult the heart pumps $9.7 \times 10^{-5}\,\mathrm{m^3}$ of blood per second. The pressure drop from the arterial to the venous systems is 11.7 kPa. Hence the power expended by the heart in overcoming viscous forces is

$$\begin{aligned}
\mathcal{P} &= Q\,\Delta P \\
&= (9.7 \times 10^{-5}\,\mathrm{m^3\,s^{-1}})(11.7\,\mathrm{kPa}) \\
&= 1.1 \times 10^{-3}\,\mathrm{kW} = 1.1\,\mathrm{W}
\end{aligned}$$

Note that $1\,\mathrm{Pa\,m^3\,s^{-1}} = 1\,\mathrm{watt} = 1\,\mathrm{W}$.

The total metabolic energy used by the heart can be estimated by measuring its oxygen consumption. For a 70-kg resting man, it is found that the rate of energy usage by the entire heart is about 5.5 W. Of this, about 2.5 W are available to do useful work; the remainder is converted directly to heat during metabolism. We have just seen that 1.1 W are used to do

mechanical work pumping blood. A significant fraction of the remaining 1.4 W is used to maintain the wall tension of the heart.

We can gain some insight into this wall tension by using a model that depicts the heart chamber that pumps blood to the body, the left ventricle, as a sphere. When the ventricle contracts, its wall must be under active tension to generate the increased blood pressure. The force per unit length or tension γ exerted by the wall was found in the preceding chapter to be related to the pressure difference $\Delta P = P_i - P_0$ and the radius r by Laplace's law, $\Delta P = 2\gamma/r$ or $\gamma = r\,\Delta P/2$. The total force produced along an imaginary cut in the wall (Fig. 15.6b) is then

$$F = 2\pi r\gamma = \pi r^2\,\Delta P$$

The energy required by the heart muscle is related to this force, but the exact relationship is complex. However, we can see qualitatively what may happen. If a person suffers from high blood pressure, ΔP is large, and the power, $Q\,\Delta P$, the heart must expend to pump blood is larger than normal. The energy necessary to maintain the wall tension is also greater than normal. After months or even years, this high pressure may lead to enlargement of the heart. Since the wall force increases with r^2, even more energy will be required, and the heart's oxygen requirements will be correspondingly larger. Congestive heart failure occurs when the oxygen supply is insufficient to meet the increased demands of the heart muscle.

15.6 | THE RISE OF SAP IN TREES; NEGATIVE PRESSURES

Sap is water plus the products of photosynthesis, including sugar. The fact that sap can reach the top of Douglas fir trees, 60 m or more high, has long been a puzzle. We have already seen that osmotic pressure (Chapter Ten) and capillary action (Example 15.2) cannot explain this phenomenon.

Negative pressures arising from the cohesive forces in water appear to provide the answer to this puzzle. These cohesive forces of water can be measured with an apparatus such as that of Fig. 15.8. When the piston is pulled upward the water expands very slightly, exerting a downward force on the piston. The force per unit area on the piston is found experimentally to be between 25 and 300 atm just before the column separates from the piston. This pressure is called negative pressure because the water is pulling *inward* on the piston instead of pushing *outward*. Because of the

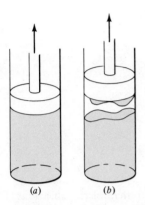

(a)   (b)

Figure 15.8. (a) Pure water, containing no dissolved gases, fills the volume beneath a piston in a chamber. The piston is pulled upward; the water pulls downward. (b) The water finally separates from the piston leaving water and its vapor.

attractive forces among the molecules, the water and piston behave much like a solid rod that resists tension forces applied to its ends.

In trees the sap moves through the *xylem,* which forms channels with radii from 2.5×10^{-5} to 2.5×10^{-4} m. The xylem channels are filled with water up to the leaves. As water is evaporated from the leaf, the water column moves upward to keep itself intact. We can determine whether the experimentally observed maximum values of the negative pressure are consistent with the rise of water in tall trees.

If we suppose that the pressure at the base of a tree trunk is atmospheric pressure P_A, then the pressure at a height h above the ground is

$$P_h = P_A - \rho g h \qquad (15.8)$$

If h is large enough, P_h can become negative. For a tree 60 m in height, $P_h = -4.8$ atm. This is well within the range of the observed negative pressures in water. Thus, it is believed that the negative pressure arising from the cohesive forces among the water molecules is responsible for the rise of sap to the tops of trees.

SUMMARY

Molecules at the surface of a liquid are subject to forces that attract them to the bulk of the liquid. Because of these forces the molecules of the liquid tend to stick together, producing the surface tension. The surface tension is defined as the force exerted by the surface of the liquid per unit length and is denoted by γ.

When a liquid is in contact with a solid, there is a competition between the liquid-liquid molecular forces and the liquid-solid molecular forces for those molecules near the region of contact. The results of this competition determine the shape of the liquid surface and the contact angle between the solid and liquid surfaces. If the liquid-liquid forces dominate, the liquid surface is depressed when a tube is lowered into the liquid. The liquid rises in the tube if the liquid-solid forces dominate.

Laplace's law describes the relation between the pressure difference across the surface of a closed membrane and the wall tension in the membrane. It is particularly useful in understanding the relation between the size of the membrane and the pressure as illustrated by the discussions of surfactant in the lungs and the heart's pumping capacity.

Checklist

Define or explain:

surface tension	meniscus
contact angle	Laplace's law
capillarity	alveoli
hydrophobic	ventricle
hydrophilic	negative pressure

REVIEW QUESTIONS

Q15-1 Is the surface tension γ defined as a force or as a force per unit length?

Q15-2 True or false? The molecules in the bulk of a fluid play absolutely no role in the surface tension.

Q15-3 The _____ between liquid and solid surfaces depends on the molecular forces between solid-liquid and liquid-liquid molecules.

Q15-4 If the contact angle at a liquid-solid interface is near zero, the liquid will _____ in a narrow tube made of the solid.

Q15-5 Wetting agents result in _____ contact angles between the solid and liquid.

Q15-6 A general feature of Laplace's law for the membrane shapes considered in the text is that it requires a _____ pressure difference to sustain a small radius surface than it does to sustain a large one.

Q15-7 The presence of surfactant in the lungs ensures that the wall tension of the alveoli varies with the _____.

Q15-8 The wall force in the heart, required to

maintain blood flow, depends on the blood pressure and on the _____ of the heart.

Q15-9 The phenomenon described as _____ in trees is due to the cohesiveness, or stick togetherness, of a pure liquid.

EXERCISES

Section 15.1 | Surface Tension

15-1 The surface tension of a liquid is measured using the apparatus of Fig. 15.1. If the straight wire is 0.05 m long and the weight of the wire plus attached weights is 2×10^{-3} N, what is the surface tension of the liquid?

15-2 If the surface tension of a liquid is 3×10^{-2} N m^{-1} and the length of the wire is 0.2 m, what total weight will the liquid support in the apparatus of Fig. 15.1?

15-3 Figure 15.9 shows a loop of thread attached to a wire ring that has been dipped into a soap solution. Why does the loop assume a circular shape when the film inside is punctured?

 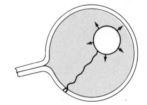

Figure 15.9.   Exercise 15-3.

Section 15.2 | Contact Angles and Capillarity

15-4 How high will water at 0° C rise in a tube of radius 10^{-3} m if it is made (a) of glass; (b) of paraffin?

15-5 How far will the meniscus of mercury be depressed in a glass tube of radius 10^{-4} m?

15-6 A glass tube of radius 1 mm = 10^{-3} m is used to construct a mercury barometer (Fig. 15.10). (a) How far will the surface be shifted by capillary effects? Is it raised or lowered? (b) If the height of the mercury column is 0.76 m, what fractional error is introduced if the capillary effects are ignored?

15-7 Normal atmospheric pressure supports a column of mercury 0.76 m high. A mercury barometer consists of a glass tube filled with mercury (Fig. 15.10). If capillary effects are to be less than 0.01 percent of the height of the column, what is the minimum radius of the tube?

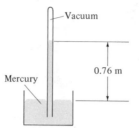

Figure 15.10.   Exercises 15-6 and 15-7.

15-8 How high will ethyl alcohol rise in a glass tube 0.04 mm = 4×10^{-5} m in diameter? (The density of ethyl alcohol is 791 kg m^{-3}.)

Section 15.3 | Laplace's Law, and
Section 15.4 | Surfactant in the Lungs

15-9 A soap bubble has a radius of 0.05 m. If the pressure difference between the inside and outside is 2 Pa, what is the surface tension of the soap film?

15-10 A typical alveolus has a radius of 10^{-4} m. The wall tension is 0.05 N m^{-1}. (a) What is the pressure difference between the inside and outside of the alveolus? (b) There is an outward normal force everywhere on the alveolar surface. Find the sum of the magnitudes of these forces. (The surface area of a sphere is $4\pi r^2$.)

15-11 At the end of exhalation the radius of the alveoli is 0.5×10^{-4} m. The gauge pressure inside the alveoli is -400 Pa, and in the pleural cavity it is -534 Pa. What is the wall tension in the alveoli? How does this compare with the wall tension of 0.05 N m^{-1} in the absence of surfactant?

15-12 Why can arteries with small radii have thinner walls than arteries with larger radii when the pressures are the same?

15-13 A spherical soap bubble has a radius of 2 cm. Its surface tension is 0.02 N m^{-1}. How much larger is the pressure inside the bubble than the pressure outside.

15-14 Derive Eq. 15.5.

15-15 Derive Eq. 15.6.

15-16 What is the pressure difference between the inside and outside of a droplet of glycerine of radius 10^{-4} m?

Section 15.5 | The Heart as a Pump

15-17 Using the assumption that a ventricle may be approximated by a sphere, by what factor must the wall force increase if the ventricle radius increases by 10 percent?

15-18 A man exercising vigorously is using metabolic energy at a rate 10 times as large as when he is at rest. His arterial blood pressure rises by 50 percent above the resting level. Assuming the rate of blood flow is proportional to the metabolic rate, estimate (a) the blood flow rate for this person; (b) the work done by the heart against viscous forces in pumping the blood. (In a typical resting human, the blood flow rate is 9.7×10^{-5} m³ s⁻¹.)

Section 15.6 | The Rise of Sap in Trees; Negative Pressures

15-19 If the pressure of sap in a tree is negative, will fluid flow out of a fine tube inserted into the xylem?

15-20 The pressure of sap in the xylem is measured at two points 1 m apart in height. What pressure difference is observed?

15-21 What is the negative pressure at the base of a tree 20 m high?

15-22 Show that if the pressure is measured in atmospheres and h in metres, Eq. 15.8 can be written as $P_h = 1 - (0.0967 \text{ m}^{-1})\, h$.

PROBLEMS

15-23 A thin wire loop of weight w and radius r floats on a liquid with surface tension γ (Fig. 15.11). What force F must be exerted to remove the loop?

15-24 Each leg of a six-legged insect standing on water at 20° C produces a depression of radius $r = 10^{-3}$ m (Fig. 15.12). The angle ϕ is 30°.

(a) What is the surface tension force acting upward on each leg? (b) What is the weight of the insect?

15-25 A narrow cylindrical tube of radius r is inserted slightly into a liquid of surface tension γ (Fig. 15.13). If one blows into the top so that a hemispherical bubble forms at the lower end, show that the pressure in the tube P is given by $P = P_A + 2\gamma/r$, where P_A is the atmospheric pressure.

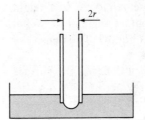

Figure 15.13. Problem 15-25.

15-26 Two bubbles with radii r and R are at either end of a tube that is closed off by a clamp (Fig. 15.14). If the clamp is released, what happens to the two bubbles?

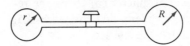

Figure 15.14. Problem 15-26.

15-27 A soap bubble is formed by dipping one end of a tube of radius r in a soap solution and blowing on the other end until the bubble is hemispherical and has the same radius as the tube. Show that the pressure in the tube P is related to atmospheric pressure P_A by

$$P = P_A + \frac{4\gamma}{r}$$

*15-28 A capillary tube is constructed as shown in Fig. 15.15. The lower segment of length l has radius $2r$ and the upper segment has radius r. How high in the upper segment will a liquid with surface tension γ (where $\gamma > 2rl\rho g$) climb if the contact angle is 0°?

15-29 When a glass tube is dipped into water at 20° C, the water rises to a height of 0.2 m. When the tube is dipped into a liquid of density

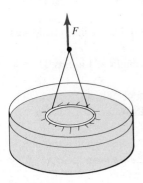

Figure 15.11. Problem 15-23.

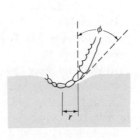

Figure 15.12. An insect leg supported by water. Problem 15-24.

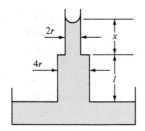

Figure 15.15.   Problem 15-28.

700 kg m⁻³, the liquid is observed to rise to a height of 0.15 m and to make a contact angle of 0°. What is the surface tension of that liquid? (The density of water at 20° C is 990 kg m⁻³)

15-30   An alternative way of defining the surface tension is the potential energy per unit area, or the energy that must be supplied to stretch a surface divided by the increase in area. (a) In Fig. 15.1, how much work must be done on the two surfaces of the film to lower the sliding wire a distance d? (b) What is the increase in area of the two surfaces? (c) Show that the ratio of the work done to the increase in area equals the surface tension γ.

15-31   A soap bubble is blown up from a radius of 3 cm to a radius of 5 cm. The surface tension of the film is 2.5×10^{-2} N m⁻¹. (a) How much work must be done against the atmosphere? (b) Using the result of the preceding problem, find the work that must be done to stretch the surface.

15-32   A wire with a radius of 0.05 mm is dipped vertically into water at 20° C. The water makes a contact angle of 0° with the wire. What length of wire is under the water when the buoyant force exactly balances the downward force exerted by the surface of the water on the wire?

15-33   (a) To what height will water rise in a vertical glass capillary tube of radius 3×10^{-5} m? (b) If the tube is only 0.2 m long, will the water overflow the top of the tube? (Assume that the top edges of the tube are smoothly rounded, and that the water is at 0° C.)

15-34   Capillary rise or fall is not limited to cylindrical tubes. Show that when two flat plates separated by a distance, d, are placed vertically in a fluid, the fluid rises a height $h = 2\gamma \cos\theta/\rho g d$ between the plates. (γ is the surface tension of the fluid, ρ is its density, and θ is the contact angle between the fluid and the plates.)

15-35   Use the result of the previous problem to find the height that water at 20° C rises between

two flat glass plates separated by 0.3 mm = 3×10^{-4} m.

15-36   A person suffering from high blood pressure has an average difference between the arterial and venous systems of 20 kPa instead of the normal 11.7 kPa. Assume that the energy expended in maintaining the tension in the heart wall is proportional to the force it exerts. (a) If the heart is not enlarged, by what factor are its metabolic requirements increased? (b) If the heart is enlarged in radius by 5 percent, estimate the additional metabolic demands this implies.

ANSWERS TO REVIEW QUESTIONS

Q15-1, force per unit length; **Q15-2,** false; **Q15-3,** contact angle; **Q15-4,** rise; **Q15-5,** small; **Q15-6,** larger; **Q15-7,** radii of the alveoli; **Q15-8,** radius; **Q15-9,** negative pressure.

Additional Reading

Martin H. Zimmerman and C.L. Brown. *Trees: Structure and Function,* Springer-Verlag, New York, 1971.

Alan T. Hayward, Negative Pressure in Liquids: Can It Be Harnessed to Serve Man? *American Scientist,* vol. 59, 1971, p. 434.

P.F. Scholander, Tensile Water, *American Scientist,* vol. 60, 1972, p. 584.

R. McNeill Alexander, *Animal Mechanics,* University of Washington Press, Seattle, 1968. Chapters 5 and 6.

Scientific American articles:

James E. McDonald, The Shape of Raindrops, February 1954, p. 64.

V.A. Greulach, The Rise of Water in Plants, October 1952, p. 78.

John A. Clements, Surface Tension in the Lungs, December 1962, p. 120.

Martin H. Zimmermann, How Sap Moves in Trees, March 1963, p. 132.

Robert E. Apfel, The Tensile Strength of Liquids, December 1972, p. 58.

Jearl Walker, Funny Things Happen When Drops of Oil or Other Substances Are Placed in Water, *The Amateur Scientist,* December 1983, p. 164.

Jearl Walker, What Causes the "Tears" That Form on the Inside of a Glass of Wine?, *The Amateur Scientist,* May 1983, p. 162. Surface tension effects.

See also selected references on the heart in Chapter 14.

UNIT FIVE

ELECTRICITY
AND
MAGNETISM

(Eric Roth/The Picture Cube).

Although electric and magnetic forces have been observed for many centuries, most of the laws governing these forces were first discovered in the nineteenth century. To this day the study of the theoretical implications of these remarkable laws and the development of their practical applications remain fields of very active research.

We have several reasons for studying the properties of electric and magnetic forces in some detail. These forces represent fundamental natural laws, and they are responsible for the structure and indeed the existence of atoms, molecules, and bulk matter. Consequently, any serious attempt to understand the properties of atoms or aggregates of atoms depends on a knowledge of these forces. At the macroscopic level, electric charge flows or *currents* are present in systems as different as the circuits in television sets and the nerve fibers of animals. Since electric currents exert forces, do work, transmit information, and produce elecromagnetic waves, they are of major importance.

The first of the five chapters in this unit deals with the basic properties of electric forces, and the next two consider electric currents in circuits and nerves, respectively. The fourth chapter explores magnetic forces, and the final chapter discusses the surprising interrelations between electric and magnetic phenomena.

CHAPTER 16

ELECTRIC FORCES, FIELDS, AND POTENTIALS

We have already touched lightly upon electric charges and forces (Chapter Five) and on the electric potential energy (Chapter Six). In this chapter, we explore these and related concepts in more depth, so that we have a better understanding of how charges interact. We frequently employ the ideas developed here in later chapters.

16.1 | ELECTRIC FORCES

Coulomb's law states that the force between two electric charges is proportional to the product of the charges and inversely proportional to their separation squared. If a charge q is at a distance r from a second charge Q, then the force on q (Fig. 16.1) is

$$\mathbf{F} = \frac{kqQ}{r^2}\hat{\mathbf{r}} \qquad (16.1)$$

Here, $\hat{\mathbf{r}}$ is a unit vector directed toward q. In S.I. units, the charge is measured in coulombs, and k has the experimentally determined value

$$k = 9.0 \times 10^9 \,\text{N m}^2\,\text{C}^{-2} \qquad (16.2)$$

The force is attractive if q and Q have opposite signs. It is repulsive if they have like signs, both positive or both negative.

When a charge q is near two or more additional charges, the net force on q is the vector sum of the forces due to each of the other charges. The following example illustrates this point.

Example 16.1

A positive charge q is near a positive charge Q and a negative charge $-Q$ (Fig. 16.2). (a) Find the magnitude and direction of the force on q. (b) If $q = 10^{-6}$ C, $Q = 2 \times 10^{-6}$ C, and $a = 1$ m, find the force on q.

(a) Probably the easiest way to find the directions of the forces on q is to remember that like charges repel and opposites attract. Since q is positive, it is repelled by the positive charge Q. Thus the force $\mathbf{F}_+$ on q due to Q is directed away from Q, or in the $\hat{\mathbf{y}}$ direction. Similarly, q is attracted toward $-Q$, so the force $\mathbf{F}_-$ due to $-Q$ also points upward or in the $\hat{\mathbf{y}}$ direction.

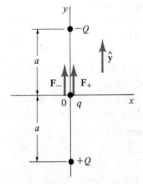

Figure 16.2. Both q and Q are positive. The force $\mathbf{F}_+$ on q due to the charge Q is repulsive, so it is directed upward in the direction of $\hat{\mathbf{y}}$. The force $\mathbf{F}_-$ on q due to $-Q$ is attractive and is also directed upward.

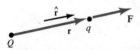

Figure 16.1. The force $\mathbf{F}$ between the two charges has a magnitude kqQ/r^2. When the charges have like signs, the force on q is repulsive or away from Q.

These directions can also be found by keeping track of the signs in Coulomb's law, Eq. 16.1. The distance from Q to q is a, and the unit vector $\hat{\mathbf{r}}$ directed from Q toward q is $\hat{\mathbf{y}}$. Thus the force on q due to this charge is

$$\mathbf{F}_+ = \frac{kqQ}{a^2}\hat{\mathbf{y}}$$

Since both q and Q are positive, the force is parallel to $\hat{\mathbf{y}}$, or upward. Similarly, the distance from $-Q$ to q is a, and here $\hat{\mathbf{r}}$ is along the $-y$ direction. Thus $\hat{\mathbf{r}} = -\mathbf{y}$, and the force on q due to $-Q$ is

$$\mathbf{F}_- = \frac{kq(-Q)}{a^2}(-\hat{\mathbf{y}}) = \frac{kqQ}{a^2}\hat{\mathbf{y}}$$

Again the force is parallel to $\hat{\mathbf{y}}$. Since $\mathbf{F}_+$ and $\mathbf{F}_-$ are parallel, the net force on q is

$$\mathbf{F} = \mathbf{F}_+ + \mathbf{F}_- = \frac{2kqQ}{a^2}\hat{\mathbf{y}}$$

(b) Substituting the numerical values given, the net force on q is

$$\begin{aligned}
\mathbf{F} &= \frac{2kqQ}{a^2}\hat{\mathbf{y}} \\
&= \frac{2(9 \times 10^9 \text{ N m}^2 \text{ C}^{-2})(10^{-6} \text{ C})(2 \times 10^{-6} \text{ C})}{(1 \text{ m})^2}\hat{\mathbf{y}} \\
&= 3.6 \times 10^{-2}\,\hat{\mathbf{y}} \text{ N}
\end{aligned}$$

The force is directed upward.

16.2 | THE ELECTRIC FIELD

We have seen that when two or more charges exert forces on a given charge q, the net force is the vector sum of the forces. When there are many charges, it is often more convenient to do this sum indirectly by introducing a quantity called the *electric field*. The electric field is also useful because it characterizes the effects of the other charges without explicit reference to the charge q.

Before considering electric fields, it is helpful to look at some examples of objects subjected to gravitational forces. We experience the gravitational attraction of the earth: the earth pulls us toward its center with a force described by the universal law of gravitation, $F = GmM/r^2$. Alternatively, we may say that the earth produces a *gravitational field* in its vicinity, and that field in turn exerts a force on us. Similarly, a spaceship sent to Mars experiences gravitational forces due to the sun, earth, Mars, and the other planets. The net gravitational force on the spaceship at

any point will be determined by the net gravitational field there. This field is produced by the sun, earth, and so on (Fig. 16.3). The gravitational force on the spaceship is proportional to its mass: the larger the ship, the larger the gravitational force. It is also proportional to the strength of the gravitational field, which is determined by the masses and locations of the sun and planets.

For the applications considered in this text, the concept of a gravitational field is not particularly useful, and we will not pursue it in detail. However, we will find electric and magnetic fields quite important.

When we have one or more electric charges, we may say that they produce an electric field in their vicinity. If another charge q is present, it experiences a force proportional to the electric field $\mathbf{E}$ and to q itself:

$$\mathbf{F} = q\mathbf{E} \tag{16.3}$$

Since the force $\mathbf{F}$ is a vector, the electric field $\mathbf{E}$ must also be a vector. If q is a positive charge, the force due to the electric field is parallel to the field. If q is negative, $\mathbf{F}$ is proportional to $-\mathbf{E}$, so the force is opposite to $\mathbf{E}$. *Positive charges experience forces parallel to the field, and negative charges experience forces opposite to the field.*

From Eq. 16.3 it follows that the units of the electric field are those of force divided by charge. Thus, in S.I. units, the electric field has units of newtons per coulomb ($N\,C^{-1}$).

Figure 16.3. A spaceship encounters gravitational forces due to the sun and the planets. Equivalently, it encounters a force proportional to its mass and to the net gravitational field produced by the sun and planets.

The expression for the field due to a single point charge can be deduced from Coulomb's law. As we saw, the force on a charge q due to a charge Q at a distance r is

$$\mathbf{F} = \frac{kqQ}{r^2}\hat{\mathbf{r}}$$

Here $\hat{\mathbf{r}}$ is a unit vector directed from the charge Q toward the point P where q is located (Fig. 16.4). The electric force on q can also be written as $\mathbf{F} = q\mathbf{E}$. Thus, it follows that at point P the field due to Q is $\mathbf{E} = \mathbf{F}/q$ or

$$\mathbf{E} = \frac{kQ}{r^2}\hat{\mathbf{r}} \qquad (16.4)$$

If Q is positive, $\mathbf{E}$ points along $\hat{\mathbf{r}}$ or away from Q; if Q is negative, $\mathbf{E}$ points along $-\hat{\mathbf{r}}$ or toward Q. *The electric field due to a charge points away from the charge if it is positive, and toward it if the charge is negative.*

When there are several charges $Q_1, Q_2, \ldots$ at various positions, the electric field $\mathbf{E}$ at a point P is the vector sum of the individual electric fields $\mathbf{E}_1, \mathbf{E}_2, \ldots$ due to all the charges. If a charge q is placed at P, then the force on it is again given by $\mathbf{F} = q\mathbf{E}$. This is equivalent to the statement that the total force on q is the sum of the forces due to the individual charges.

Once we have measured the electric field at a point, we can immediately find the force on any charge placed there. It is not necessary to know the magnitude or location of the charges producing the field. For example, a Na^+ ion has a charge $q = e = 1.6 \times 10^{-19}$ C. If the field in a cell membrane is 10^6 N C^{-1}, the force on the ion has a magnitude $F = qE =$

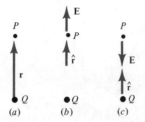

Figure 16.4. (a) A point P is at a distance $\mathbf{r}$ from a charge Q. (b) If the charge Q is positive, the electric field $\mathbf{E}$ at P is away from Q or along $\hat{\mathbf{r}}$. (c) If Q is negative, the field is along $-\hat{\mathbf{r}}$.

$(1.6 \times 10^{-19}\text{ C})(10^6\text{ N C}^{-1}) = 1.6 \times 10^{-13}$ N and is directed along the electric field.

The following example demonstrates the calculation of the electric field due to two point charges and the force exerted by this field on another charge.

Example 16.2

The charges Q and $-Q$ of the preceding example are shown again in Fig. 16.6. (a) If $Q = 2 \times 10^{-6}$ C and $a = 1$ m, find the electric field at the origin. (b) Find the force on the charge $q = 10^{-6}$ C placed at the origin.

(a) As before, the distance from the charge Q to the origin is a, and $\hat{\mathbf{r}} = \hat{\mathbf{y}}$. Using Eq. 16.4, the field at the origin due to this charge is

$$\mathbf{E}_+ = \frac{kQ}{a^2}\hat{\mathbf{y}}$$

The field points upward, *away from the positive charge.* Also, the distance from the charge $-Q$ to the origin is a, and $\hat{\mathbf{r}} = -\hat{\mathbf{y}}$, so its field there is

$$\mathbf{E}_- = \frac{k(-Q)}{a^2}(-\hat{\mathbf{y}}) = \frac{kQ}{a^2}\hat{\mathbf{y}}$$

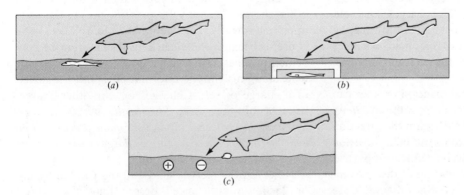

Figure 16.5. Sharks are sensitive to the minute electric fields produced by charges in a body. (a) The shark attacks a fish hidden beneath the sand. (b) A chamber blocks all but electrical stimuli, and the shark still attacks. (c) An artificially produced electric field elicits the same response. Here the shark is ignoring an obvious piece of food to follow the electrical stimulus.

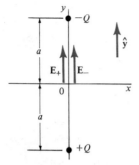

Figure 16.6. The electric field due to a positive charge is directed away from the charge, while the field due to a negative charge is directed toward the charge. Hence, both contributions to the total field at the origin point along the $+y$ direction.

This field points upward as well; *it is pointed toward the negative charge.* The net electric field at the origin is the sum,

$$\mathbf{E} = \mathbf{E}_+ + \mathbf{E}_- = \frac{2\,kQ}{a^2}\,\widehat{\mathbf{y}}$$

$$= \frac{2(9 \times 10^9 \text{ N m}^2 \text{ C}^{-2})(2 \times 10^{-6} \text{ C})}{(1 \text{ m})^2}\,\widehat{\mathbf{y}}$$

$$= 3.6 \times 10^4\,\widehat{\mathbf{y}}\text{ N C}^{-1}$$

The net field is directed upward.

(b) The force on a charge $q = 10^{-6}$ C at the origin is

$$\mathbf{F} = q\mathbf{E} = (10^{-6}\text{ C})(3.6 \times 10^4)\widehat{\mathbf{y}}\text{ N C}^{-1}$$
$$= 3.6 \times 10^{-2}\,\widehat{\mathbf{y}}\text{ N}$$

As expected, this is identical to the force calculated in the previous example by the direct use of Coulomb's law.

Electric Field Diagrams

Figure 16.7*a* and *b* shows the electric field vectors calculated from Eq. 16.4 for several locations near point charges. The field points away from the positive charge and toward the negative charge. In both cases the field becomes weaker as the distance from the charge increases, since it varies as $1/r^2$.

Another kind of electric field diagram or map can be drawn by joining the vectors with continuous *lines*, as in Fig. 16.7*c* and *d*. In such a diagram the direction of the lines at any point indicates the field direction there. Also, since the lines become further apart as the field becomes weaker, the spacing of the lines indicates the magnitude of the field. Such diagrams are especially useful in understanding the electric fields due to complex arrangements of charges.

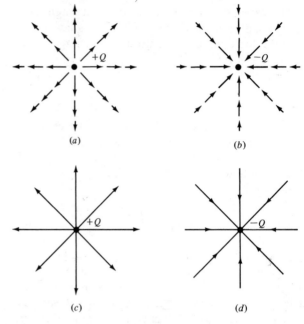

Figure 16.7. (*a*) and (*b*) Electric field vectors at locations near charges that are (*a*) positive, (*b*) negative. The corresponding electric field lines are shown in (*c*) and (*d*).

We can actually make an electric field visible by exploiting the fact that oblong objects, such as grass seeds, suspended in a liquid tend to line up along the field lines (Fig. 16.8). The pattern formed near a single point charge is strikingly similar to that in Fig. 16.7. We will see corresponding similarities shortly for more complex charge arrangements.

Now let's discuss the basic concept of an electric field more fully. When there is a charge at rest at some point, we say that it produces an electric field in its vicinity. This field in turn exerts a force on any other charge that may be present. So far this field approach is really equivalent to Coulomb's law, but it is useful because the electric field can often be conveniently measured or calculated. Also, the electric field diagrams convey information about the effects of a charge or set of charges on *any* other charges brought into their field. *Specifically, positive charges will experience forces along (or, more precisely, tangent) to the field lines. Negative charges will experience forces in the opposite direction.*

However, the electric field and the Coulomb's law or "action-at-a-distance" descriptions of electrical forces are *not* equivalent for charges in motion. Electric fields are not merely a bookkeeping device for

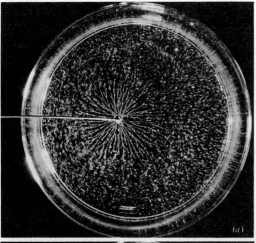

keeping track of electric forces. Suppose a charge Q is moved, so its field changes. The effect of relocating Q is not felt immediately by a nearby charge q, because the change in the field does not propagate instantaneously. Instead, it travels at the speed of light, 3×10^8 m s^{-1}. The force on q changes only when the changed field has reached it. Coulomb's law implies an instantaneous change in this force, contrary to what is observed.

To make this point more concrete, consider a television transmitter. Charges oscillating back and forth in its antenna produce an *electromagnetic wave* that carries the programming in coded form. In the

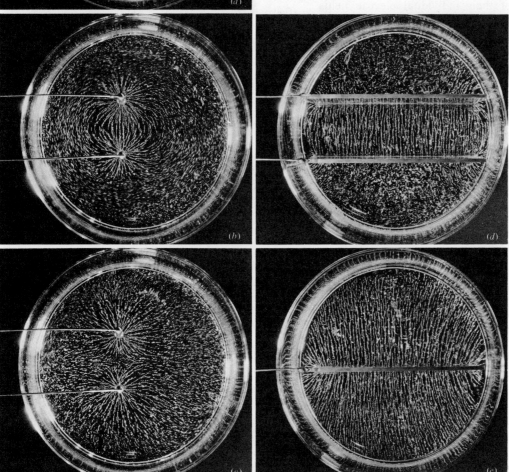

Figure 16.8. Photographs of patterns formed by grass seeds in a liquid near various charge arrangements. (*a*) A point charge. (*b*) Two point charges, opposite signs. (*c*) Two point charges, like signs. (*d*) Two charged metal plates, opposite signs. (*e*) One charged metal plate. (Reproduced by permission of the publisher from *PSSC Physics,* fourth edition, 1976, by D.C. Heath and Company, Lexington, Massachusetts.)

antenna of a distant receiver, electric charges respond to the wave produced earlier at the station as it passes by. Similarly, electromagnetic waves from distant stars reach us and have an effect many years after being emitted by the stars. Thus a full description of electrical forces requires a careful study of the electric field, and involves physical principles not contained in Coulomb's law. These ideas will be explored further in Chapter Twenty.

16.3 | THE ELECTRIC FIELD DUE TO ARRANGEMENTS OF CHARGES

The total electric field due to two or more charges is the vector sum of their individual electric fields. When there is a continuous distribution of charges, say along a wire or on the surface of a conducting plane, this sum becomes an integral. If the individual fields point in different directions, cancellations will occur in the sum or integral. As a result, the field may change with distance in a way that is quite different from the $1/r^2$ variation of the individual point charge fields. We will see, for example, that the field may vary as $1/r^3$ or as $1/r$, or it may even be constant.

The Field of an Electric Dipole | A pair of charges with equal magnitudes and opposite signs, $+q$ and $-q$, is called an *electric dipole*. We see later in this chapter that dipoles are useful in characterizing atoms and molecules and in discussing the effect of an electric field on an insulator.

Consider the electric dipole in Fig. 16.9. At the point P, the fields $\mathbf{E}_+$ due to $+q$ and $\mathbf{E}_-$ due to $-q$ are

$$\mathbf{E}_+ = \frac{kq}{(r-a)^2}\widehat{\mathbf{y}}, \quad \mathbf{E}_- = \frac{-kq}{(r+a)^2}\widehat{\mathbf{y}}$$

The total field at P is

$$\mathbf{E} = \mathbf{E}_+ + \mathbf{E}_- = kq\left[\frac{1}{(r-a)^2} - \frac{1}{(r+a)^2}\right]\widehat{\mathbf{y}}$$

If r is large compared to a, the total field $\mathbf{E}$ is quite small, since the fields due to the two charges nearly cancel. Rewriting $\mathbf{E}$ by putting both terms over a common denominator, some terms cancel in the numerator, and we find

$$\mathbf{E} = \frac{4kqar}{(r^2-a^2)^2}\widehat{\mathbf{y}}$$

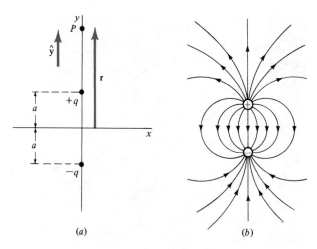

Figure 16.9. (*a*) A dipole is a pair of equal but opposite charges. (*b*) The field due to a dipole. Notice that the lines go from the positive to the negative charge. Note also the similarity with Fig. 16.8*b*.

When r is much greater than a, we can neglect a in the denominator. Then

$$\mathbf{E} = \frac{4kqa}{r^3}\widehat{\mathbf{y}}$$

Thus the field diminishes as $1/r^3$ at large distances, which is a faster rate of decrease than the $1/r^2$ associated with a single charge. If we repeat this calculation for a distant point on the x axis or anywhere else, the overall numerical factor and the direction of the field will vary, but the field will again diminish as $1/r^3$.

The Field of a Long Charged Wire | A very long, uniformly charged straight wire provides one of the simplest illustrations of a continuous distribution of electric charges. Such a wire might be found in an electrical device or in an antenna. Although any real wire has a finite length, for simplicity we consider the idealized case of an infinitely long wire (Fig. 16.10a). The result will be accurate for a real wire at points that are located much closer to the wire than to its ends (Problem 16-16). The basic procedure employed here applies also to other problems involving continuous charge distributions.

We will specify the charge on the wire in terms of its *charge per unit length*, λ (lambda). Suppose a length L of the wire has a charge Q. Then the charge per unit length is

$$\lambda = Q/L \qquad (16.5)$$

For a segment of the wire of length dx, the charge is λdx.

In Fig. 16.10a, the distance to an arbitrary point P is R, where

$$R^2 = r^2 + x^2 \qquad R = (r^2 + x^2)^{1/2}$$

Thus the magnitude of the field due to the charge λdx is

$$dE = \frac{k\lambda dx}{R^2} = \frac{k\lambda dx}{r^2 + x^2}$$

The total field due to a charge distribution is found by adding up the fields due to all the charges. Here this means integrating the fields due to all the segments. To do this, we first find the horizontal and vertical components of **dE,**

$$dE_x = dE \cos\theta \qquad dE_y = dE \sin\theta$$

Now we can see immediately that the net horizontal field component E_x is zero. This is because the segments located symmetrically on either side of the origin have canceling dE_x contributions (Fig. 16.10b). Thus we need only consider E_y. Since

$$\sin\theta = r/R = r/(r^2 + x^2)^{1/2}$$

$dE_y = dE \sin\theta$ becomes

$$dE_y = \frac{k\lambda dx}{(r^2 + x^2)} \frac{r}{(r^2 + x^2)^{1/2}}$$

$$= \frac{kr\lambda dx}{(r^2 + x^2)^{3/2}}$$

Summing over all the segments means integrating over the length of the wire, or from $x = -\infty$ to $+\infty$. Since E_y is the net field E, we have

$$E = E_y = kr\lambda \int_{-\infty}^{+\infty} \frac{dx}{(r^2 + x^2)^{3/2}}$$

Using Eq. B.46 in Appendix B,

$$r^2 \int dx/(r^2 + x^2)^{3/2} = x/(r^2 + x^2)^{1/2}$$

Thus

$$E = \frac{k\lambda}{r} \left[\frac{x}{(r^2 + x^2)^{1/2}} \right]_{-\infty}^{+\infty} = \frac{k\lambda[1 - (-1)]}{r}$$

and

$$E = \frac{2k\lambda}{r} \qquad \text{(long straight wire)} \qquad (16.6)$$

This is the magnitude of the field **E** at a distance r

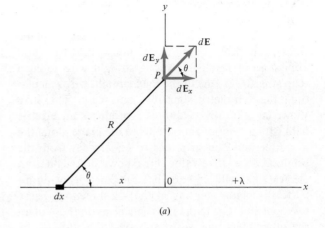

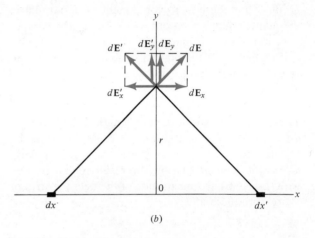

Figure 16.10. (a) The field due to one segment of an infinitely long uniformly charged wire. (b) Two equally long segments dx and dx' are located symmetrically about the origin produce fields with horizontal components $d\mathbf{E}_x$, that are equal in magnitude but opposite in direction.

from a long straight wire. The field is in the plane perpendicular to the wire, and directed radially outward, assuming the charge per unit length λ is positive. If λ is negative, the field points radially inward toward the wire.

The Field of a Uniformly Charged Plane | Continuous charge distributions are found on things as different as the surfaces of metallic objects and the membranes of plant and animal cells. The simplest example of finding the field due to the charge on a surface is provided by a uniformly charged plane. Again, we consider an idealized case, an infinitely large plane. The results hold for a finite

plane at points much closer to the plane than to its edges.

The electric field due to a uniformly charged plane can be found by cutting it into many thin strips and applying our result for a long straight wire to each strip (Fig. 16.11). As in the long straight wire calculation, the symmetry simplifies our work. The strip shown at a positive value of z produces an electric field with a component parallel to the plane along the $-z$ direction. The strip located equally far from the origin at a negative value of z (shown dotted) has an electric field with an equally large component parallel to the plane pointing in the $+z$ direction. These two contributions to the field cancel, as do those of all the other pairs of strips drawn on either side of the origin. Thus there is no component of $\mathbf{E}$ parallel to the plane, and we need only evaluate E_y, the component normal to the plane.

If an area A of the plane has a charge Q, then the *charge per unit area* σ (sigma) is defined as

$$\sigma = Q/A \qquad (16.7)$$

Then a strip of width dz and length L has an area $dA = L\,dz$, and a charge $\sigma\,dA = \sigma L\,dz$. Dividing by the length L gives its charge per unit length, $d\lambda = \sigma\,dA/L = \sigma\,dz$. Using the result for an infinite wire (Eq. 16.6), the magnitude of the field the strip or "wire" produces at P is

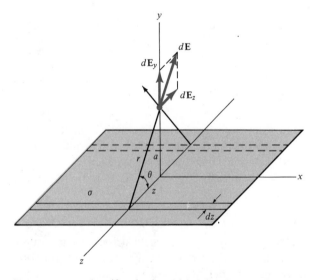

Figure 16.11. A uniformly charged infinite plane can be broken up into many narrow strips. The strips on opposite sides of the origin produce fields with components parallel to the plane that cancel. The resulting field is normal to the plane and is uniform.

$$dE = \frac{2k\,d\lambda}{r} = \frac{2k\sigma\,dz}{r}$$

The vertical component of $d\mathbf{E}$ is $dE_y = dE \sin\theta$, where $\sin\theta = a/r$. Thus, with $r^2 = a^2 + z^2$

$$dE_y = \frac{2k\sigma\,dz}{r}\frac{a}{r} = \frac{2k\sigma a\,dz}{a^2 + z^2}$$

We can now sum over all the strips by integrating over z. Replacing dE_y by dE and using Eq. B.44 to evaluate the integral

$$E = 2k\sigma a \int_{-\infty}^{+\infty} \frac{dz}{(a^2 + z^2)}$$

$$= 2k\sigma a \left[\frac{1}{a}\tan^{-1}\frac{z}{a}\right]_{-\infty}^{+\infty} = 2k\sigma\left[\frac{\pi}{2} - \frac{(-\pi)}{2}\right] = 2k\sigma\pi$$

In vector form, we can write this result as

$$\mathbf{E} = 2\pi k\sigma\hat{\mathbf{n}} \quad \text{(uniformly charged plane)} \quad (16.8)$$

$\hat{\mathbf{n}}$ is a unit vector normal to the plane.

The interesting feature of this result is that the distance from P to the plane has disappeared in the final result for the field due to the uniformly charged infinite plane. The field everywhere is perpendicular to the plane and constant in magnitude, or uniform.

With two parallel planes having charges per unit area that are equal in magnitude but opposite in sign, the fields add between them and cancel elsewhere (Fig. 16.12). Hence, between the planes

$$\mathbf{E} = 4\pi k\sigma\hat{\mathbf{n}} \quad \text{(oppositely charged planes)} \quad (16.9)$$

The field is directed from the positive toward the negative plane.

This type of uniform field can be produced to a good approximation with two metal plates having dimensions large compared to their separation. When a metal plate has a net charge, the mutual repulsion of the charges causes them to be almost uniformly distributed. The field between two such oppositely charged plates is nearly uniform. A charged particle, such as an electron, moving through the uniform field experiences a constant acceleration. We see later how this is put to use in an oscilloscope (Section 16.7) and in other kinds of apparatus employing charged particle beams (Chapter Nineteen). The electric fields in a cell membrane are also approximately uniform, as seen in the next example.

Example 16.3

A thin, flat membrane separates a layer of positive ions outside a cell from a layer of negative ions inside

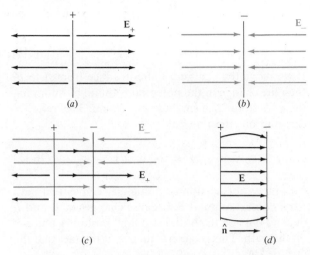

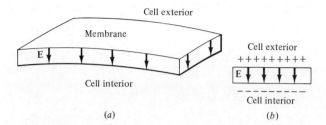

Figure 16.12. (a) The field of a positively charged plane. (b) The field of a negatively charged plane shown in color for clarity. (c) The field lines due to the two planes when they are close together. (d) The total field due to the two charged planes. Note the similarity with Fig. 16.8d. The field is not quite uniform at the edges. The effects of these *fringing fields* at the edges may often be neglected.

(Fig. 16.13). If the electric field due to these charges is 10^7 N C^{-1}, find the charge per unit area in the layers on either side of the membrane.

Since the membrane is flat, the charges form uniformly charged planes on either side. Using $E = 4\pi k\sigma$, we have

$$\sigma = \frac{E}{4\pi k} = \frac{10^7 \text{ N C}^{-1}}{4\pi(9 \times 10^9 \text{ N m}^2 \text{ C}^{-1})}$$

$$= 88.4 \times 10^{-6} \text{ C m}^{-2} = 88.4 \, \mu\text{C m}^{-2}$$

using 1 microcoulomb = $1 \, \mu\text{C} = 10^{-6}$ C. Thus 1 m^2 of membrane will have a net charge of $+88.4 \, \mu$C along the outside surface and a net charge of $-88.4 \, \mu$C on the inside.

16.4 | THE ELECTRIC POTENTIAL

In Chapter Six, we noted that the electric forces among charges at rest are conservative, so that their effects can be included in the potential energy of a system. Here we will introduce a related concept, the *electrical potential,* which is the potential energy per unit charge. Like the electric field, it permits us to characterize the effects of one or more charges without specifying the magnitude or sign of a charge located at the position of interest.

Suppose that at a certain position a charge q has an electrical potential energy $\mathfrak{U}$. *Then the electric poten-*

Figure 16.13. (a) A portion of a cell membrane seen in perspective. (b) A cross-sectional view showing the charge layers.

tial V at that position is defined to be the potential energy divided by the charge,

$$V = \frac{\mathfrak{U}}{q} \tag{16.10}$$

The unit of potential is the volt (V), where from this definition 1 volt = 1 joule per coulomb. (The standard abbreviation V for the volt should not be confused with the symbol V for the potential.) Colloquially, potential differences are often referred to as *voltages.* We see in the next chapter that it is the potentials rather than the fields that are most useful in discussing electric circuits.

We saw in Chapter Six that many problems in mechanics can be solved quite readily if the potential energies at two points are known. Similarly, given the potential difference between two points, we can say many things about the motion of charged particles without using detailed information about the electrical forces or fields. This is illustrated by the next example.

Example 16.4

In the cathode-ray tube of an oscilloscope or a television picture tube, electrons are accelerated from rest through a potential difference of +20,000 V. What is their velocity? (The electron mass is 9.11×10^{-31} kg, and the charge is $-e = -1.6 \times 10^{-19}$ C.)

From the definition of the electric potential, the change in potential energy is $\Delta\mathfrak{U} = q \Delta V = (-e) \Delta V$, where $\Delta V = 20,000$ V. Then, from energy conservation,

$$\tfrac{1}{2}mv^2 = e \Delta V$$

Hence,

$$v = \sqrt{\frac{2e \Delta V}{m}}$$

$$= \sqrt{\frac{2(1.60 \times 10^{-19} \text{ C})(20,000 \text{ V})}{9.11 \times 10^{-31} \text{ kg}}}$$

$$= 8.38 \times 10^7 \text{ m s}^{-1}$$

Often, energies of electrons or other atomic particles are most conveniently expressed in units of *electron volts* (eV). An electron volt is the kinetic energy acquired when a charge e is accelerated by a potential difference of one volt,

$$1 \text{ eV} = (1.60 \times 10^{-19} \text{ C})(1 \text{ V})$$
$$= 1.60 \times 10^{-19} \text{ J}$$

For example, when an electron is accelerated by a 20,000-V potential difference, it acquires a kinetic energy of 20,000 eV, or

$$(20,000 \text{ eV}) \frac{(1.60 \times 10^{-19} \text{ J})}{1 \text{ eV}} = 3.2 \times 10^{-15} \text{ J}$$

We make extensive use of the electron volt and multiples of this unit in our discussions of atomic and molecular phenomena in later chapters.

Relation Between Electric Fields and Potentials

If we know the electric field in a region, we can use it to calculate the potential differences between various points. We show how this is done in general and illustrate the result for a uniform field, a point charge, and a uniformly charged wire.

Consider a positive charge q in an electric field **E** (Fig. 16.14). We suppose a force **F** equal in magnitude but opposite in direction to the electric force $q\mathbf{E}$ is applied, so that the charge moves at a constant velocity over a short distance $d\mathbf{l}$. Then the work done (Chapter Six) by the applied force **F** is $\mathbf{F} \cdot d\mathbf{l} = -q\mathbf{E} \cdot d\mathbf{l}$. Since the kinetic energy remains constant, this work must equal the change in the potential energy of the charge

$$d\mathcal{U} = -q\mathbf{E} \cdot d\mathbf{l}$$

Dividing by q, we obtain the change in the electric potential,

$$dV = -\mathbf{E} \cdot d\mathbf{l} \tag{16.11}$$

If we move the charge a finite distance from A to B, then the change in the potential is found by summing the dV's from each small displacement. This requires carrying out the integral

$$\Delta V = \int_A^B \mathbf{E} \cdot d\mathbf{l} \tag{16.12}$$

This integral is evaluated along any convenient path from point A to point B, because changes in $\mathcal{U}$ and V depend only on the initial and final positions, and not on the path. This is a consequence of the fact that the electric force is a conservative force.

Equation 16.12 is the general formula for the difference in the potential between two points in an electric field. A charge q displaced from A to B would have a potential energy change $\Delta\mathcal{U} = q\Delta V$. We now illustrate the use of this result.

Uniform Field

We saw in the preceding section that the field is uniform in the region between two closely spaced large metal plates with opposite charges (Fig. 16.15). If we move in a direction opposite to a uniform field, then the scalar product simplifies to

$$\mathbf{E} \cdot d\mathbf{l} = Edl \cos 180° = -Edl$$

E is constant, so it can be taken out of the integral, and Eq. 16.12 becomes

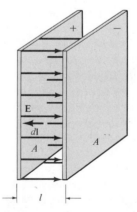

Figure 16.15. The potential difference between the parallel plates is $\Delta V = El$. The positively charged plate is at the higher potential, since work must be done against the field to move a positive charge $+q$ from the negative plate to the positive plate.

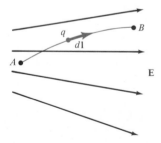

Figure 16.14. When a charge q is moved a distance $d\mathbf{l}$, its electrical potential energy changes by $-q\mathbf{E} \cdot d\mathbf{l}$. For a positive charge, the increase is greatest when $d\mathbf{l}$ is opposite to **E**.

$$\Delta V = E \int_A^B dl = El \quad \text{(uniform field)} \quad (16.13)$$

The field between two plates with charges per unit area $+\sigma$ and $-\sigma$ is $E = 4\pi k\sigma$. Thus, for plates a distance l apart, the potential difference is

$$\Delta V = 4\pi k\sigma l \quad \text{(oppositely charged plates)} \quad (16.14)$$

The positive plate is at the higher potential. This result is illustrated by the following example.

Example 16.5

Two oppositely charged parallel plates have an area of 1 m² and are separated by 0.01 m. The potential difference between the plates is 100 V. Find (a) the field between the plates, and (b) the magnitude of the charge on a plate.

(a) Since the field is uniform, $\Delta V = El$ can be used, and

$$E = \frac{\Delta V}{l} = \frac{100 \text{ V}}{0.01 \text{ m}} = 10^4 \text{ V m}^{-1}$$

The field is directed from the positive plate at the higher potential toward the negative plate.

(b) Using Eq. 16.14 and the definition $\sigma = Q/A$, the charge on a plate has a magnitude

$$Q = \sigma A = \frac{A \Delta V}{4\pi kl} = \frac{(1 \text{ m}^2)(100 \text{ V})}{4\pi (9 \times 10^9 \text{ N m}^2\text{C}^{-2})(10^{-2} \text{ m})}$$
$$= 8.84 \times 10^{-8} \text{ C}$$

Notice that a small charge is sufficient to produce a 100-V potential difference.

Potential energy of any kind depends only on the position of an object and not on how it arrived there. Stated somewhat differently, the work done by the conservative electric force is independent of the path followed between the initial and final positions. Therefore, we may obtain the difference in potential between two points using any convenient path to find the work per unit charge done against the electric field. For example, if we wish to know the difference in potential between points A and C in Fig. 16.16, we can choose the path ABC. AB is a displacement opposite to the field, so $\Delta V = El$. BC is perpendicular to the field, so no work is done, and the potential is the same at B as at C. Using the dashed path AC, we would find the same potential difference. However, the calculation would be slightly more complicated.

Point Charge

Finding the potential of a point charge $+Q$ is a bit more complicated since $E = kQ/r^2$ is not constant. If we move radially outward, or away from the charge, then the displacement is parallel to the field (Fig. 16.17a). Thus writing $d\mathbf{l} = d\mathbf{r}$,

$$\mathbf{E} \cdot d\mathbf{l} = Edr = \frac{kQ}{r^2} dr$$

Suppose we move a finite distance from point A along this line at a distance r_1 from Q to point B at a distance r_2. Then Eq. 16.12 gives, with $\int dr/r^2 = -1/r$,

$$\Delta V = V(r_2) - V(r_1) = -\int_{r_1}^{r_2} Edr$$
$$= -\int_{r_1}^{r_2} \frac{kQ dr}{r^2} = \frac{kQ}{r_2} - \frac{kQ}{r_1}$$

Since only potential differences can be measured, we can define the potential to be zero at any convenient point. It is customary to choose the potential at

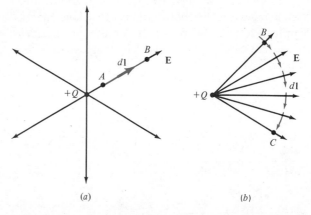

(a)   (b)

Figure 16.17. (a) Moving a charge from A to B can be broken up into infinitesimal steps dl that are parallel to the field **E** for the case shown. (b) If we move a charge from B to a point C equally far from the charge Q, the displacement can be made in steps perpendicular to the field, so no work is required. Thus the potential at B and at C is the same, and V depends only on the distance from the charge.

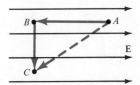

Figure 16.16. The work done by the conservative electric force is the same for the paths ABC and AC.

$r = \infty$ to be zero. With this assignment, putting $r_2 = \infty$ and $r_1 = r$ leads to

$$V(r) = \frac{kQ}{r} \qquad \text{(point charge)} \qquad (16.15)$$

$V(r)$ is the amount of work we must do per unit charge to move a second charge from infinity to a point a distance r from Q. To restate this, suppose there is a charge q at a distance r. Then from the definition of the potential, $V = \mathfrak{U}/q$, the potential energy of the two charges (q and Q) is $\mathfrak{U} = qV = kqQ/r$. This is the formula stated in Chapter Six for the potential energy of two point charges.

One other point should be noted. We assumed that the points A and B were on the same radial line directed outward from Q. However, the potential is the same everywhere on the sphere of radius r_2 centered at Q, because no work is required to move at right angles to the field (Fig. 16.17b). We return to this point in the next section.

When we want the potential due to several charges, or to several charge distributions, two equivalent procedures can be used. We can calculate the total electric field of the system and then use the defining equation (16.12) to obtain the potential. This is effectively how we found the potential difference between two plates. Alternatively, the potentials due to the various charges can be added. Potentials are scalars, ordinary positive and negative numbers, so adding them is a simpler task than the vector summation required when combining electric fields. Again, cancellations between the contributions of various charges can lead to varied dependences on the distance. We see such a cancellation in the following example of a dipole.

Example 16.6

For the dipole in Fig. 16.18, find the electric potential at (a) point P_1 on the y axis, and (b) at point P_2 on the x

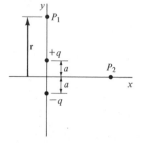

Figure 16.18. Example 16.6.

axis. (c) How much work is required to move a charge q from infinitely far away to point P_1 if $r = 3a$? (d) How much work is needed to move q from infinity to P_2?

(a) At P_1, the potentials due to the two charges are

$$V_+ = \frac{kq}{r - a}, \qquad V_- = \frac{-kq}{r + a}$$

The total electric potential is then

$$V = V_+ + V_- = kq\left[\frac{1}{r - a} - \frac{1}{r + a}\right] = \frac{2kqa}{r^2 - a^2}$$

If r is large compared to a, then a can be neglected in the denominator, so

$$V = \frac{2kqa}{r^2}$$

Thus even though the potential due to a single point charge varies as $1/r$, this combination of two point charges of equal magnitude and opposite sign has a potential that diminishes as $1/r^2$.

(b) The two charges are equally distant from *any* point on the x axis. Therefore, at P_2 the two potentials will have the same magnitude but opposite signs, and their sum will be zero. The net potential of the dipole is zero everywhere on the x axis.

(c) Using the result for V in part (a) with $r = 3a$, the potential at P_1 is

$$V = \frac{2kqa}{(3a)^2 - a^2} = \frac{2kqa}{8a^2} = \frac{kq}{4a}$$

Thus the potential energy of the system consisting of the dipole and the charge q is $\mathfrak{U} = qV = kq^2/4a$. This is the amount of work needed to move q from infinity to P_1.

(d) Since $V = 0$ at point P_2, $\mathfrak{U} = qV = 0$, and no work is needed to move the charge from infinity to P_2.

Long Straight Wire | The calculation of the potential of a uniformly charged infinitely long straight wire is very similar to that for a point charge. Here the field is radially outward in the plane perpendicular to a wire carrying a positive charge per unit length λ. If we go radially outward from the wire, then $d\mathbf{l} = dr$ is parallel to $\mathbf{E}$, and $\mathbf{E} \cdot d\mathbf{l} = E dr$ (Fig. 16.19). Suppose point A is at a distance r_1 from the wire, and B is at a distance r_2. Then, with $E = 2k\lambda/r$,

$$\Delta V = V(r_2) - V(r_1) = -\int_{r_1}^{r_2} E \, dr$$

$$= -2k\lambda \int_{r_1}^{r_2} \frac{dr}{r} = -2k\lambda \ln r \,\Big|_{r_1}^{r_2}$$

and

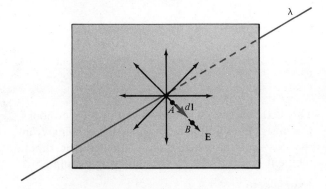

Figure 16.19. The field of an infinitely long straight wire carrying a uniform positive charge is directed radially outward. A displacement $d\mathbf{l} = d\mathbf{r}$ is parallel to the field.

$$V(r_2) - V(r_1) = -2k\lambda \ln \frac{r_2}{r_1} \quad \begin{array}{l}(\text{long} \\ \text{straight wire})\end{array} \qquad (16.16)$$

Here ln is the natural logarithm (see Appendix B.10). Again, as in the case of a point charge, the potential difference depends only on the distance from the wire, and not on the direction.

Note that when $r_2/r_1 > 1$, the logarithm is positive. This means that the potential decreases as we go away from a positively charged wire, as would be expected. However, here we cannot conveniently assign the potential a value of zero at infinity, since $\ln (r_2/r_1)$ is infinite when $r_2 = \infty$. This behavior is related to the artificial nature of an infinitely long charged wire, and disappears when a finite wire is considered.

16.5 | EQUIPOTENTIAL SURFACES

At any point on the surface of an imaginary sphere of radius r centered on a point charge, Q, the potential has the same value, $V = kQ/r$. A surface on which the potential is the same everywhere is called an *equipotential surface*. Thus, for a point charge the equipotential surfaces are concentric spheres (Fig. 16.20). The equipotential surfaces for a uniform electric field are planes normal to the field.

When a charge moves at right angles to the electric field, no work is done against electrical forces, so its potential energy remains constant. For this reason, *the equipotential surfaces are always perpendicular to the electric field lines.* Charges may move along an equipotential surface with no change in potential energy.

Conductors and Insulators | Most materials may be considered either electrical *conductors* or *insulators*. A conductor is a material, such as a metal or an ionic solution, in which some charges are relatively free to move about. In an insulator, such as paper or glass, all the charges are relatively immobile.

Conductors have the important property of being equipotential objects when there are no charges in motion. To see this, we note that if all the charges that can move are at rest, the electric field must be zero everywhere in the conductor. Alternatively, only if the field is zero everywhere will the charges stop mov-

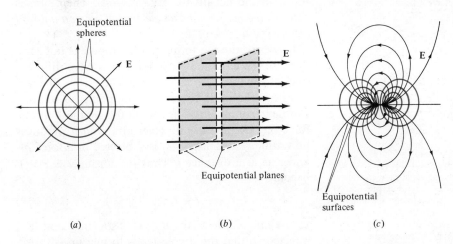

(a) (b) (c)

Figure 16.20. Equipotential surfaces for (*a*) a point charge; (*b*) a uniform electric field (*c*) a dipole. Note that the equipotential surfaces and the electric field lines are always mutually perpendicular.

ing. Consequently the potential is the same at all points.

The effect of a conductor on an electric field is illustrated in Fig. 16.21.

16.6 | ELECTRIC DIPOLES

Earlier we defined an electric dipole as a pair of equal and opposite charges, and we briefly discussed the field and potential due to a dipole. Here we consider a dipole placed in a uniform electric field due to other charges.

Atoms and molecules provide many illustrations of electric dipoles. For example, a water molecule has an excess of negative charge near its oxygen atom and a similar positive excess near the hydrogen atoms. Accordingly, it behaves like a small electric dipole, and many of its physical and chemical properties are related to its dipole character. Also, atoms and molecules can develop *induced* dipoles in response to an electric field. This is one way in which insulators are affected by electric fields.

A dipole is characterized by its *electric dipole moment* (Fig. 16.22). If l is the distance vector from $-q$ to $+q$, then the electric dipole moment is defined by

$$\mathbf{p} = q\mathbf{l} \qquad (16.17)$$

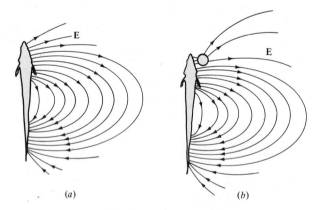

(a)   (b)

Figure 16.21. A number of fish use electric fields for detection and communication. (*a*) A charge separation in the body of the fish produces electric fields of a dipole type. The strength of the field is monitored by receptors along the body of the fish. (*b*) The field is distorted by the presence of a conducting object. The field lines intersect the surface of the object perpendicular to the surface, which is an equipotential. The resulting distortion of the field is apparent at the fish's body and the receptors sense the change. Thus the fish effectively "sees" using electric fields.

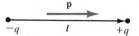

Figure 16.22. An electric dipole consists of equal and opposite charges, $+q$ and $-q$. The electric dipole moment is $\mathbf{p} = q\mathbf{l}$.

This is the definition usually given in physics books. Chemistry books often define $\mathbf{p}$ so that it points from $+q$ to $-q$. The calculation of the electric dipole moment is illustrated by the following example.

Example 16.7

In a hydrogen atom the electron and proton are separated by 5.29×10^{-11} m. (a) Find the electric dipole moment at one instant in time. (b) The electron moves in a circular path about the proton. Find the average of the electric dipole moment vector over one full orbit.

(a) The proton charge is $e = 1.60 \times 10^{-19}$ C, and the electronic charge is $-e$. The dipole moment points toward the positively charged proton and has a magnitude of

$$p = ql = (1.60 \times 10^{-19} \text{ C})(5.29 \times 10^{-11} \text{ m})$$
$$= 8.46 \times 10^{-30} \text{ C m}$$

(b) If the electric dipole initially points in one direction, half an orbit later it will point in the opposite direction. The average of two vectors equal in magnitude but opposite in direction is zero. Thus the average of the electric dipole moment vector over a complete circular orbit must be zero. For this reason the hydrogen atom has no permanent electric dipole moment.

Figure 16.23 shows an electric dipole in a uniform field $\mathbf{E}$. The force on the positive charge is $q\mathbf{E}$, and the force on the negative charge is $-q\mathbf{E}$. Their sum is zero, so *the net force on an electric dipole in a uniform electric field is zero.*

However, the net torque on the dipole is not zero, because the equal but opposite forces have different lines of action and form a couple. The torque due to a couple is the same relative to any point, so we may choose to compute torques about the charge $-q$. The force on $-q$ then has no lever arm, and it produces no torque. The charge $+q$ is subjected to a force $q\mathbf{E}$ and acts at a distance l. Thus the torque $\boldsymbol{\tau} = \mathbf{r} \times \mathbf{F}$ becomes

$$\boldsymbol{\tau} = \mathbf{l} \times (q\mathbf{E}) = \mathbf{p} \times \mathbf{E} \qquad (16.18)$$

The magnitude of the torque is $pE \sin \theta$, and it is directed so that the dipole tends to line up with the field. When the dipole is oriented as in Fig. 16.23a, the torque is directed into the page, tending to rotate

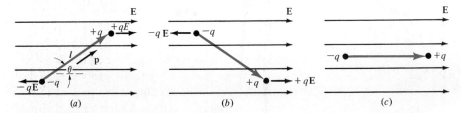

Figure 16.23. The torque on the dipole tends to produce rotations that are (a) clockwise; (b) counterclockwise. (c) The dipole is in equilibrium.

the dipole clockwise. The torque is directed out of the page in Fig. 16.23b and tends to rotate the dipole counterclockwise. The torque is zero, and the dipole is in stable equilibrium when it is directed along the field (Fig. 16.23c).

Conveniently the potential energy $\mathcal{U}$ of a dipole in a uniform field is taken to be zero when $\theta = 90°$; the dipole is then perpendicular to the field (Fig. 16.24a). The energy is a minimum when the dipole is parallel to the field. Thus, when the dipole is in the position shown in Fig. 16.24b, its potential energy must be negative. We can find an expression for $\mathcal{U}$ by imagining that the dipole is rotated from $\theta = 90°$, where $\mathcal{U} = 0$, to θ. If the rotation is performed with $-q$ held in place, the energy of the negative charge remains unchanged. The displacement of the positive charge along the field is $l \cos \theta$, so the work done against the electrical force qE is $-qE(l \cos \theta)$. Equating this work to the change in potential energy, we find

$$\mathcal{U} = -qEl \cos \theta = -pE \cos \theta$$

or $\qquad \mathcal{U} = -\mathbf{p} \cdot \mathbf{E} \qquad (16.19)$

These ideas are illustrated by the following example.

Example 16.8

An atom with an electric dipole moment of 8.46×10^{-30} C m is in a uniform electric field of 10^4 N C^{-1}. If the angle between $\mathbf{p}$ and $\mathbf{E}$ is 30°, find

(a) the magnitude of the torque, and (b) the potential energy

(a) When $\theta = 30°$, $\sin \theta = 0.5$, and the torque has a magnitude

$$\begin{aligned}\tau &= pE \sin \theta \\ &= (8.46 \times 10^{-30}\ \text{C m})(10^4\ \text{N C}^{-1})(0.5) \\ &= 4.23 \times 10^{-26}\ \text{N m}\end{aligned}$$

(b) At $\theta = 30°$, $\cos \theta = 0.866$, and the potential energy is

$$\begin{aligned}\mathcal{U} &= -\mathbf{p} \cdot \mathbf{E} = -pE \cos \theta \\ &= -(8.46 \times 10^{-30}\ \text{C m})(10^4\ \text{N C}^{-1})(0.866) \\ &= -7.33 \times 10^{-26}\ \text{J}\end{aligned}$$

16.7 | THE OSCILLOSCOPE

Except for some simple meters, probably no scientific instrument is used as widely as the *oscilloscope*. The electrical circuitry contained in an oscilloscope is quite elaborate, and a sophisticated model has so many control knobs and switches that it takes some time to become acquainted with its operation. Nevertheless, the basic principles of its major component, the *cathode-ray tube*, can be understood fully with the ideas developed in this chapter.

In the first part of the tube, which is the electron gun, a constant potential difference accelerates electrons emitted from a hot filament (Fig. 16.25). If there are no other forces on the electron, they travel along a straight line and strike the center of the fluorescent screen, producing a bright spot. Horizontal deflections can be produced with horizontal electric fields between the parallel plates marked 1 in Fig. 16.25. These plates have a separation l. If they are at a potential difference V_1, there is a uniform horizontal electric field between them $E_1 = V_1/l$. This field accelerates the electrons, which then strike the screen at a horizontal distance from the center, which is proportional to V_1. If V_1 is gradually increased, the spot gradually moves or *sweeps* across the screen; when V_1

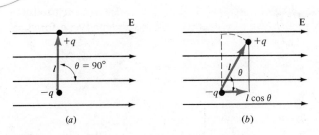

Figure 16.24. The energy of the dipole $\mathcal{U}$ is zero in (a); negative in (b).

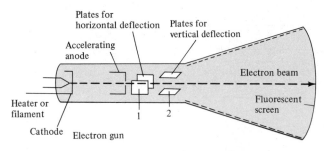

Figure 16.25. A cathode-ray tube. Electrons are accelerated from the negative cathode toward the positive accelerating anode and then pass between the two pairs of deflecton plates. When they strike the fluorescent screen, light is emitted.

returns to its original value, the spot returns to its starting point. If this sweep is repeated at a high enough frequency, the persistence of the image on the screen and in the eye conceals the motion, and a straight line is seen (Fig. 16.26b).

In the same way, a potential difference V_2 applied to the plates marked 2 in Fig. 16.25 will cause vertical deflections. For example, a constant V_2 will shift the line as in Fig. 16.26c. Usually an unknown potential difference V_2, varying at some frequency f, produces the vertical deflection. The horizontal sweep frequency is adjusted until it is equal to f (or f divided by some integer), so every time the vertical signal V_2 repeats itself the horizontal sweep is at the same point in its cycle. Thus the beam repeatedly hits the same points on the screen, and a stable pattern is seen (Fig. 16.26d, e). Accurate measurements of both the fre-

quency and the magnitude of the unknown potential difference can readily be made in this way.

Since almost any type of information can be converted into electric potential differences, oscilloscopes are used in laboratories of virtually every kind. They are extremely valuable in qualitative and quantitative studies, not only of electrical variables but also of mechanical, acoustical, and other quantities. Microphones and television cameras are examples of devices that convert energy from other forms into varying electrical potentials. In general, devices that convert energy from one form to another are called *transducers*.

16.8 | CAPACITANCE

Suppose that two conductors are initially electrically neutral and that we then take small amounts of charge from one and place them on the other. As this process continues, a potential difference develops. The ratio of the amount of charge transferred to the potential difference resulting is defined to be the *capacitance* of the two conductors and turns out to be independent of the charge transferred. An arrangement of two conductors separated by a vacuum or an insulator is called a *capacitor*.

As our description suggests, the capacitance is a measure of the amount of charge separation that can be maintained at a given potential difference. The energy required to separate the charges is stored in

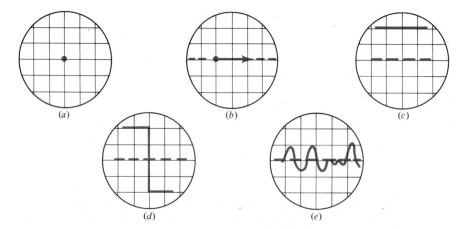

Figure 16.26. (a) If there are no deflecting fields, a bright spot is seen at the center of the oscilloscope screen. (b) Gradually changing the horizontal potential difference causes the beam spot to move steadily across the screen. If this sweep is repeated rapidly enough, a continuous line is perceived. (c) A constant vertical field displaces the line. (d) If the vertical field quickly reverses after each half sweep, this pattern is seen. (e) Complex variations of the vertical field with time can be displayed if they repeat at the sweep frequency.

the capacitor. Hence the capacitance is also a measure of the ability to store energy.

Examples of capacitance can be found in nature. For instance, cell membranes separate thin layers of ions in the fluids inside and outside a cell. Hence the membrane and the adjacent fluids are considered to have capacitance.

Capacitors are widely used in electrical circuits. They are used in radio and television tuners, in automobile ignition systems, and in the starting circuits of electric motors. Capacitors also influence the way currents change in time. In nerve cells the rate of transmission of a nerve pulse depends on the membrane capacitance. Some of these applications are discussed in later chapters.

If two conductors have equal and opposite charges $\pm Q$ and a corresponding potential difference V, the ratio Q/V is usually found to be a constant independent of Q. The ratio is the capacitance,

$$C = \frac{Q}{V} \qquad (16.20)$$

Note that V is defined to be the potential of the positive plate less that of the negative plate, so C is positive.

The unit of capacitance is the *farad* (F); $1\,F = 1\,C\,V^{-1}$. Because the coulomb is a large unit, the farad is also large, and most capacitors have small values in terms of the farad. Hence we define the microfarad and picofarad by

$$1\,\mu F = 10^{-6}\,F, \qquad 1\,pF = 10^{-12}\,F$$

The Parallel Plate Capacitor | The simplest capacitor is composed of two parallel plates in a vacuum. The plates have surface area A, charges $\pm Q$, and a separation l (Fig. 16.27). From Eq. 16.14, $V = 4\pi k\sigma l = 4\pi kQl/A$. Thus the capacitance $C = Q/V$ is

$$C = \frac{A}{4\pi kl} \qquad (16.21)$$

It is conventional to define a new constant, ε_0, as

$$\varepsilon_0 = \frac{1}{4\pi k} = 8.85 \times 10^{-12}\,C^2\,N^{-1}\,m^{-2}$$

Our final expression for the capacitance of a parallel plate capacitor is then

$$C = \frac{\varepsilon_0 A}{l} \qquad (16.22)$$

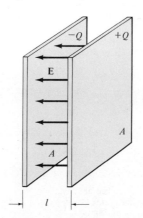

Figure 16.27. A parallel plate capacitor.

Note that C depends only on the geometry of the arrangement and not on the charge Q. The capacitance increases with the area of the plates and decreases with their separation. This is also true for capacitors with more complicated shapes. Numerical examples of parallel plate capacitors are given in the next section.

16.9 | EFFECTS OF DIELECTRICS

If an insulator or *dielectric* is introduced between the plates of a parallel plate capacitor with fixed charges, the capacitance is increased. This occurs because the electric field due to the charges on the plates distorts the charge distributions of the molecules in the dielectric, giving each molecule a small *induced* electric dipole moment. These dipoles, in turn, reduce the overall electric field and hence the potential difference between the plates.

To study this in detail, consider a neutral molecule that normally has its centers of positive and negative charge coincident so that it has no permanent dipole moment. When an external electric field is applied, it separates the charge centers by a distance proportional to the strength of the field (Fig. 16.28). Thus, each molecule acquires an *induced* electric dipole moment.

Inside the dielectric the effects of the displaced positive and negative charges cancel. However, at the left surface, there is an excess of positive charge, and at the right surface, there is an excess of negative charge. This produces an electric field $\mathbf{E}'$ directed opposite to $\mathbf{E}$. Thus the actual electric field between the plates is reduced to an effective value,

$$E_{\text{eff}} = E - E'$$

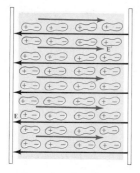

Figure 16.28. The molecules in a dielectric are distorted by the electric field **E** between the parallel plates. The charge on the plates is the same before and after the dielectric is introduced.

Since the separation of the charges increases with **E**, the field **E′** they produce also increases with **E**. Generally, except for very strong fields, **E′** is proportional to **E**. Thus we can write

$$\mathbf{E}_{\text{eff}} = \frac{1}{K}\mathbf{E} \qquad (16.23)$$

The *dielectric constant K* is a dimensionless number that indicates the reduction of the field due to the dielectric. K is one for a vacuum, and greater than one for a dielectric.

Table 16.1 lists representative dielectric constants. In some substances, such as water, the molecules have permanent electric dipole moments that tend to line up with the field. The degree of alignment increases with the field, but decreases as the temperature rises due to thermal disordering. Thus K depends on temperature.

TABLE 16.1

Dielectric constants of several common insulators. Note that for air we can usually use $K = 1$, the value for a vacuum

Material	Temperature, °C	Dielectric Constant, K
Air (dry, at 1 atm)	20	1.00059
Glass	25	5–10
Water	25	78
	80	61
Plastics	20	3–20
Titanium dioxide	20	100
Axon membrane (unmyelinated)	37	8
Paper	20	3.5

The potential difference between the plates is $V = E_{\text{eff}}l = El/K$. Since K is greater than one for a dielectric, V is reduced when a dielectric is inserted. Consequently, the *capacitance* $C = Q/V$ *increases by the factor K*. An increase in C occurs for any capacitor, not just a parallel plate arrangement. In a parallel plate capacitor, the capacitance changes from $\varepsilon_0 A/l$ to

$$C = \frac{K\varepsilon_0 A}{l} \qquad (16.24)$$

The following examples illustrate the effects of dielectrics on the capacitance.

Example 16.9

A capacitor is made of two foils, each of surface area 1 m², separated by paper 0.05 mm = 5×10^{-5} m thick. What is its capacitance?

According to Table 16.1, $K = 3.5$ for paper. Thus the capacitance is

$$
\begin{aligned}
C &= \frac{K\varepsilon_0 A}{l} \\
&= \frac{(3.5)(8.85 \times 10^{-12}\,\text{C}^2\,\text{N}^{-1}\,\text{m}^{-2})(1\,\text{m}^2)}{(5 \times 10^{-5}\,\text{m})} \\
&= 6.19 \times 10^{-7}\,\text{F} = 0.619\,\mu\text{F}
\end{aligned}
$$

Example 16.10

The ions inside and outside a cell are separated by a flat membrane 10^{-8} m thick with a dielectric constant $K = 8$. Find the capacitance of 1 cm² of membrane (Fig. 16.29).

For an area $A = 1$ cm² = 10^{-4} m², the capacitance is

$$C = \frac{K\varepsilon_0 A}{l} = \frac{8(8.85 \times 10^{-12}\,\text{C}^2\,\text{N}^{-1}\,\text{m}^{-2})(10^{-4}\,\text{m}^2)}{10^{-8}\,\text{m}}$$

$$= 7.08 \times 10^{-7}\,\text{F} = 0.708\,\mu\text{F}$$

When the electric field in a dielectric becomes sufficiently strong, large numbers of free electrons and ions are produced, and the material becomes an excellent conductor. This *breakdown* occurs at a critical electric field called the *dielectric strength*. Lightning is

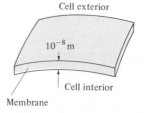

Figure 16.29. A small section of a cell membrane.

a spectacular example of this phenomenon in which the air becomes a conductor (Fig. 16.30). Commercial capacitors are marked with the maximum voltage that can be applied without risk of exceeding the dielectric strength and damaging the capacitor or other parts of a circuit.

16.10 | ENERGY STORED IN A CAPACITOR

A charged capacitor stores electrical energy. If its plates are connected by conducting wire, electrons will move in the wire from the negative plate to the positive plate. This charge flow or *current* continues until the plates are neutralized and can be used to operate an electronic flash gun or trigger an artificial heart pacemaker. Thus, stored electrical energy an be transformed into other forms of energy.

The energy stored in a capacitor may initially be supplied by a battery, which maintains a potential difference between its two terminals. When the plates of an uncharged capacitor are connected by two wires to the terminals, electrons flow in the wires from one plate through the battery to the other until the potential difference across the capacitor reaches a maximum value. We can find the energy stored in the capacitor by calculating the work that must be done by the battery in building up this charge from zero to the final value Q.

To avoid minus signs, we suppose that positive charges are gradually transferred from one plate to the other. (The final result is independent of the sign of the moving charges.) As one plate acquires an increasingly large positive charge, and the other a corresponding negative charge, the potential difference between them increases. If at some instant the charges are $+q$ and $-q$, then from the definition of capacitance, $C = q/v$, we have $v = q/C$. Transferring a small additional amount of charge therefore requires that work dW be done by the battery, where

$$dW = v\,dq = \frac{q}{C}\,dq$$

Since the electric force is conservative, the work done by the battery dW must equal the increase $d\mathfrak{U}$ in the stored energy of the capacitor, so

$$d\mathfrak{U} = \frac{q}{C}\,dq$$

Figure 16.30. When low-lying clouds accumulate a charge, there is a potential difference between the cloud and the ground. If this potential difference is large enough, the electric field exceeds the dielectric strength of the air. The air then ionizes and becomes a good electrical conductor. (Wide World Photos.)

To find the total energy stored in the capacitor, we integrate from 0 to the final charge Q:

$$\mathfrak{U} = \int_0^Q \frac{q}{C}\,dq = \tfrac{1}{2}\frac{Q^2}{C}$$

Using $C = Q/V$, the energy can be expressed in three equivalent forms,

$$\mathfrak{U} = \tfrac{1}{2}QV = \tfrac{1}{2}\frac{Q^2}{C} = \tfrac{1}{2}CV^2 \qquad (16.25)$$

The perhaps surprising factor of $\tfrac{1}{2}$ reflects the fact that, on the average, during the charging process the potential is just half its final value.

The energy stored in a cell membrane is calculated in the following example.

Example 16.11

One square centimetre of membrane has a capacitance of 7.08×10^{-7} F. If the potential difference across the membrane is 0.1 V, find the electrical energy stored in 1 cm² of membrane.

Since we know C and V, we write the energy as $\tfrac{1}{2}CV^2$. Then

$$\mathfrak{U} = \tfrac{1}{2}CV^2 = \tfrac{1}{2}(7.08 \times 10^{-7}\text{ F})(0.1\text{ V})^2$$
$$= 3.54 \times 10^{-9}\text{ J}$$

SUMMARY

A charge Q exerts an electrical force on a second charge q, which is given by Coulomb's law

$$\mathbf{F} = \frac{kqQ}{r^2}\hat{\mathbf{r}} \quad \text{(Coulomb's law)}$$

Alternatively, we can say that the electric field due to a point charge Q is

$$\mathbf{E} = \frac{kQ}{r^2}\hat{\mathbf{r}} \quad \text{(point charge } Q)$$

The force on q is the product of its charge and the electric field at its location, $\mathbf{F} = q\mathbf{E}$.

The field due to a system of charges is found by summing or integrating their individual fields. The resulting field may vary with position in many ways. The field of an infinitely long uniformly charged straight wire is radial and has a magnitude

$$E = \frac{2k\lambda}{r} \quad \text{(long straight wire)}$$

A uniformly charged infinite plane produces a uniform field. A pair of oppositely charged plates has a field between them which is normal to the plates and has a magnitude

$$E = 4\pi k\sigma \quad \text{(parallel plates)}$$

Outside the plates the field is zero.

The electrical potential is the potential energy of a charge divided by that charge,

$$V = \frac{\mathcal{U}}{q}$$

Moving a distance $d\mathbf{l}$ in an electric field $\mathbf{E}$ changes the potential by

$$dV = -\mathbf{E} \cdot d\mathbf{l}$$

Summing or integrating over such small displacements, the potential change is

$$\Delta V = -\int \mathbf{E} \cdot d\mathbf{l}$$

This equation can be used to find formulas for some simple situations:

$$V = \frac{kQ}{r} \quad \text{(point charge)}$$

$$\Delta V = 4\pi k\sigma l \quad \text{(parallel plates)}$$

$$\Delta V = -2k\lambda \ln \frac{r_2}{r_1} \quad \text{(long straight wire)}$$

Two charges $+q$ and $-q$ separated by a distance l have a dipole moment $\mathbf{p} = ql$. In a uniform electric field, there is no net force on a dipole, but there is a torque tending to align it with the field. The torque on the dipole is

$$\boldsymbol{\tau} = \mathbf{p} \times \mathbf{E}$$

and its potential energy is

$$\mathcal{U} = -pE \cos \theta$$

A capacitor is a pair of conductors separated by a vacuum or an insulator. If the conductors are given equal but opposite charges, the ratio $C = Q/V$ of the charge to the resulting potential difference is the capacitance. Other arrangements of conductors and insulators, such as those in living cells, can also be regarded as having capacitance.

When an insulator is placed between the plates of a charged capacitor, electric dipoles are induced in the material. The dipole fields oppose the applied field and reduce the potential difference between the plates, thereby increasing the capacitance. Capacitors store electrical energy according to the relationship $\mathcal{U} = \frac{1}{2}QV$.

Checklist

Define or explain:

Coulomb's law	insulator
electric field	cathode-ray tube
electric field lines	capacitance
electric dipole	dielectric
electric potential	induced dipole moment
voltage	dielectric constant
equipotential surface	dielectric strength
conductor	

REVIEW QUESTIONS

Q16-1 The force between two charges of opposite sign is _____.

Q16-2 The force on a charge q in an electric field $\mathbf{E}$ is _____.

Q16-3 The electric force on a positive charge is _____ to the field; the electric force on a negative charge is _____ to the field.

Q16-4 The electric field due to a positive charge points _____ the charge; the field due to a negative charge points _____ the charge.

Q16-5 The spacing of electric field lines indicates the _____, and their direction gives the _____.

Q16-6 The field between two oppositely charged metal plates is nearly _____.

Q16-7 The change in potential energy of a charge q moved through a potential difference ΔV is _____.

Q16-8 The equipotential surfaces near a point charge are _____.

Q16-9 Two charges $+q$ and $-q$ a distance l apart have a dipole moment of magnitude _____ directed toward _____.

Q16-10 The potential energy of a dipole is least when it points _____.

Q16-11 If a capacitor has a capacitance C and its plates have charges $\pm Q$, the potential difference across it is _____.

Q16-12 Inserting a dielectric between two charged plates _____ the field, _____ the potential difference, and _____ the capacitance.

Q16-13 If the voltage across a capacitor is doubled, the energy stored changes by a factor of _____.

EXERCISES

Several of the following exercises involve atomic quantities. The magnitude of the charge on an electron or a proton is $e = 1.60 \times 10^{-19}$ C, the electron mass is 9.11×10^{-31} kg, and the proton mass is 1.673×10^{-27} kg.

Section 16-1 | Electric Forces

16-1 Find the magnitude and direction of the force on the charge Q in Fig. 16.31.

16-2 If $Q = 10^{-6}$ C and $b = 0.1$ m, what is the magnitude and direction of the force on the charge $2Q$ in Fig. 16.31?

16-3 An additional charge Q is placed at the origin in Fig. 16.31. What is the force on this charge?

16-4 Find the magnitude and direction of the force on the charge $-Q$ in Fig. 16.32.

16-5 In a NaCl molecule, a Na^+ ion with charge e is 2.3×10^{-10} m from a Cl^- ion with charge $-e$. What is the magnitude of the force between them?

16-6 A cell membrane 10^{-8} m thick has positive ions on one side and negative ions on the other. What is the force between two ions with charges $+e$ and $-e$ at this separation?

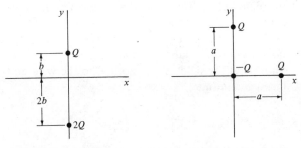

Figure 16.31. Exercises 16-1, 16-2, 16-3, 16-11, and 16-12.

Figure 16.32. Exercise 16-4.

Section 16.2 | The Electric Field and Section 16.3 | The Electric Field Due to Arrangements of Charges

16-7 A uranium nucleus has a charge of $92e$. (a) What is the direction and magnitude of the electric field due to the nucleus at a distance of 10^{-10} m from the nucleus? (b) What is the direction and magnitude of the force on an electron at this distance?

16-8 What is the force on an electron in a field of 10^5 N C^{-1}?

16-9 An electron is accelerated at 10^8 m s^{-2} by an electric field. What is the direction and magnitude of the field?

16-10 Find the magnitude and direction of the electric field 0.1 m from a charge of -10^{-4} C.

16-11 In Fig. 16.31, find the electric field at the origin.

16-12 In Fig. 16.31, find the electric field at $x = 0$, $y = -b$.

16-13 The electric field near a uniformly charged circular plate of area 0.1 m^2 is directed toward the plate and has a magnitude of 10^4 N C^{-1}. Find the charge on the plate.

16-14 Two square plates of side 0.1 m have equal and opposite charges of $\pm 10^{-6}$ C, which are uniformly distributed. The plates are separated by 10^{-2} m. (a) What is the magnitude and direction of the electric field? (b) What is the magnitude and direction of the force on an electron placed in this field? (c) How much work must be done against the field to move an electron from the positive plate to the negative plate?

16-15 A 10-m straight wire has a total charge of 10^{-5} C. (a) Find the charge per unit length, assuming the charge is uniformly distributed. (b) What is the field 0.1 m from the wire at a point near its center?

^{c}16-16   (a) Suppose that in Fig. 16.10 the wire is not infinite but rather extends from $-L$ to $+L$. Show that the field at P is normal to the wire and has a magnitude

$$E = \frac{2k\lambda L}{r(r^2 + L^2)^{1/2}}$$

(b) Show that this formula reduces to Eq. 16.6 in the limit of an infinitely long wire.

16-17   The fractional error Δ in a formula for the field $\mathbf{E}$ is defined as

$$\Delta = |E(\text{exact}) - E(\text{approx})|/E(\text{exact})$$

(a) If $r = L/10$, what is the fractional error Δ produced by using the formula for the field due to an infinitely long straight wire instead of the equation in the preceding exercise? (b) Find the corresponding fractional error when $r = L$.

16-18   Show that the electric field at the center of a uniformly charged circular ring is zero.

Section 16.4 | The Electric Potential

16-19   A carbon nucleus has a charge of $+6e$. At a distance of 10^{-10} m from a carbon nucleus, find (a) the electric potential; (b) the potential energy of an electron in electron volts and joules.

16-20   In Fig. 16-33, what is the potential at (a) the origin; (b) $x = 3a$, $y = 0$?

16-21   At what points on the x axis in Fig. 16.33 is the potential zero?

16-22   Two uniformly charged metal plates separated by 0.04 m produce a uniform field between them of 10^4 N C^{-1}. Find (a) the charge per unit area Q/A on the plates, and (b) the potential difference between the plates.

16-23   At the center of the square in Fig. 16.34, find (a) the electric field, and (b) the potential.

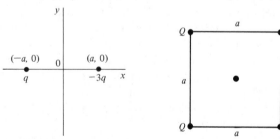

Figure 16.33. Exercises 16-20, and 16-21.

Figure 16.34. Exercise 16-23.

16-24   An alpha particle is a helium nucleus with a charge of $2e$ and a mass of 6.64×10^{-27} kg. Suppose an alpha particle is accelerated from rest to a speed of 10^7 m s^{-1} by electrical forces in the first stage of a particle accelerator. What potential difference is required to accomplish this?

16-25   An electron and a proton are separately placed at rest midway between two oppositely charged metal plates. (a) Which way will the electron accelerate? (b) Which way will the proton accelerate? (c) Which particle, if either, will acquire more kinetic energy just before striking a plate? (d) What is the ratio of their velocities just before they strike the plates?

16-26   In Bohr's model of the hydrogen atom, the electron moves in a circle of radius 5.29×10^{-11} m. Find (a) the electric potential due to the proton at that circle; (b) the potential energy of the electron in electron volts and joules.

16-27   If the potential is constant in a certain region, what can be said about the electric field in that region?

16-28   If the electric field points in the $+x$ direction in some region, in what direction is the most rapid increase in the electric potential observed?

16-29   An infinitely long straight wire has a charge per unit length of 10^{-4} C m^{-1}. (a) If a proton moves from a distance of 0.01 m from the wire to a distance of 0.5 m, by how much will its kinetic energy increase? (b) If the proton was initially at rest 0.01 m from the wire, what is its speed when it is 0.5 m away?

16-30   Charged particles such as protons are accelerated to high velocities and allowed to collide with atomic nuclei in order to probe their internal structure. The electric potential outside of a nucleus with Z protons is equal to that of a point charge Ze. (a) A lead nucleus ($Z = 82$) is approximately described as a sphere of radius 7×10^{-15} m. How much kinetic energy in MeV (1 MeV = 10^6 electron volts) must a proton have initially to overcome the electrical repulsion and reach its surface? (b) What is the corresponding initial velocity?

Section 16.6 | Electric Dipoles

16-31   The NH_3 molecule has a permanent electric dipole moment of 5.0×10^{-30} C m. If this arises from net charges of $+e$ and $-e$ in two regions of the molecule, what is their separation?

16-32 Although a neutron has no net charge, it has a charge distribution consisting of spherical layers of positive and negative charge. Its electric dipole moment has not been detected; if it has one, the moment must be less than 10^{-45} C m according to recent experiments. The diameter of a neutron is about 10^{-15} m. How large must equal and opposite charges at a distance of 10^{-15} m be to produce an electrical dipole moment of 10^{-45} C m? Express your result in multiples of e.

16-33 What energy is required to "flip" an electric dipole **p** from its position parallel to an electric field **E** to the antiparallel position?

16-34 An electric dipole consists of two charges of $\pm 10^{-4}$ C separated by 10^{-5} m. (a) What is the magnitude of the electric dipole moment? (b) If the dipole is in a field of 10^3 N C^{-1}, find its minimum and maximum potential energies.

16-35 An electric dipole consists of charges $\pm e$ separated by 10^{-10} m. It is in a field of 10^6 N C^{-1}. Find the magnitude of the torque on the dipole when it is (a) parallel to the field; (b) at right angles to the field; (c) opposite to the field.

Section 16.8 | Capacitance

16-36 What is the charge on a 100-μF capacitor when its potential difference is 1000 V?

16-37 A capacitor has a potential difference of 100 V when its plates have charges of magnitude 10^{-5} C. What is its capacitance?

16-38 Two square metal plates with sides of length 0.1 m are separated in vacuum by 10^{-3} m. Find their capacitance.

Section 16.9 | Effects of Dielectrics

16-39 A capacitor made of thin foils of aluminum separated by paper 10^{-4} m thick is found to have a capacitance of 1 μF. What is the area of the foil?

16-40 Metal plates are placed on either side of a sheet of glass, and their capacitance is measured to be 10^{-5} F. If they are kept at the same separation in air, their capacitance is found to be 2×10^{-6} F. What is the dielectric constant of the glass?

16-41 A lightning bolt travels 500 m from a cloud to a mountain top. What is the potential difference between the cloud and the peak? (Assume that the electric field is uniform and that the air breaks down and becomes a conductor when the field reaches 8×10^5 V m^{-1}.)

16-42 A parallel plate capacitor has a capacitance of 2 μF when the plates are separated by a vacuum. The plates are 10^{-3} m apart, and they are connected to a battery that maintains a 50 V potential difference between them. (a) What is the charge on the plates? (b) What is the electric field between the plates? (c) If a slab with dielectric constant equal to 5 is inserted between the plates, what is the new charge? (d) What is the electric field with the slab in place?

16-43 A capacitor is made of metal foils 5×10^{-5} m thick separated by paper 10^{-4} m thick. (a) How large an area of foil is needed to make a 0.1-μF capacitor? (b) If the capacitor is tightly rolled into a cylinder 10 cm long, what is its radius?

Section 16.10 | Energy Stored in a Capacitor

16-44 A 50-μF capacitor in an electronic flash gun supplies an average power of 10^4 W for 2×10^{-3} seconds. (a) To what potential difference must the capacitor initially be charged? (b) What is its initial charge?

16-45 An electronic flash gun uses energy stored in a capacitor. (a) How large a capacitor is needed to provide 30 J of energy if it is charged to 1000 V? (b) How much charge is on its plates?

16-46 Can you run a car on energy stored in a capacitor? Make a rough estimate to support your answer.

16-47 Two parallel metal plates of area 0.1 m^2 are separated in air by 0.01 m. Their potential difference is 1000 V. Find (a) the charge on the plates, and (b) the stored energy. (c) A slab of dielectric with $K = 10$ is inserted so that it fills the space between the plates. The charge on the plates remains the same. Find the new potential difference and stored energy. (d) Explain the change, if any, in the stored energy.

PROBLEMS

16-48 An electron is accelerated by a constant electric field from rest to a velocity of 10^6 m s^{-1}. If the accelerating region is 0.2 m long, what is the magnitude of the electric field?

16-49 In his oil-drop measurements of the electronic charge (1900–1913), R. A. Millikan suspended small charged oil droplets by adjusting a vertical electric field to balance their weight. The size of a droplet was measured by observing its terminal velocity in the air with the field turned off.

Millikan found that the droplets always had charges that were integer multiples of a basic unit, which he identified with the electronic charge e. (a) If a droplet had a radius of 1.2×10^{-6} m, and the field required to balance it was 1.26×10^5 N C^{-1}, what was the charge on the droplet? (The density of the oil was 851 kg m^{-3}.) (b) What is the ratio of this charge to the magnitude of the charge e on the electron? (c) If the field was produced by a pair of metal plates 0.015 m apart, what potential difference was required to produce this field?

16-50   A charge of -10^{-6} C is placed at the origin of an x-y plane. Find the magnitude and direction of the electric field at (a) $x = 1$ m, $y = 0$; (b) $x = 0$, $y = -2$ m; (c) $x = 2$m, $y = 2$ m.

16-51   A circular metal plate of radius 0.2 m has 10^{10} excess electrons uniformly distributed over its surfaces. What is the magnitude and direction of the field just outside the plate near its center?

16-52   Estimate the magnitude and direction of the field at a distance of 10 m from the center of the plate in the preceding problem.

16-53   An electron is projected with a velocity v_0 into a uniform electric field (Fig. 16.35). (Neglect the gravitational force on the electron.) (a) Find the direction and magnitude of its acceleration. (b) How long will it be in the field? (c) How far will it be deflected vertically as it leaves the field? (d) Find the angle between its velocity as it leaves the field and its original direction.

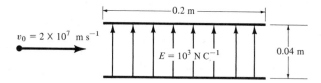

$v_0 = 2 \times 10^7$ m s^{-1}

$E = 10^3$ N C^{-1}

0.2 m

0.04 m

Figure 16.35.   Problem 16-53.

16-54   A charge q is placed at the origin and a charge $2q$ is placed at $x = a$, $y = 0$. Find the potential at $x = a$, $y = a$.

16-55   A particle of mass m and positive charge q is projected with a velocty $\mathbf{v}$ into a region where a uniform electric field $\mathbf{E}$ is opposite to $\mathbf{v}$. How far will the particle travel before it comes momentarily to rest?

16-56   A thin ring of radius a has a total charge Q distributed uniformly around it (Fig. 16.36). Show

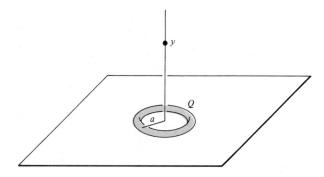

Figure 16.36.   Problems 16-56, 16-61.

that the electric field on the axis at a distance y from the center of the ring is directed along the axis and has a magnitude

$$E = \frac{kQy}{(y^2 + a^2)^{3/2}}$$

ᶜ**16-57**   A disk of radius R has a uniform charge per unit area σ (Fig. 16.37). (a) Show that the electric field on the axis at a distance y from the center of the disk has a magnitude

$$E = 2\pi k\sigma \left[1 - \frac{y}{(y^2 + R^2)^{1/2}} \right]$$

[*Hint:* Use the result of the preceding problem for the field of a ring.] (b) Show that this formula reduces to Eq. 16.8, the expression for the field due to an infinite plane, when the disk becomes very large compared to y ($R \gg y$). (c) Show that in the limit $y \gg R$, the formula reduces to the usual field for a point charge.

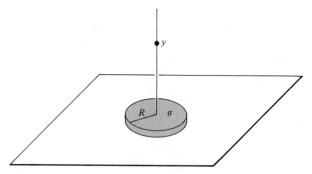

Figure 16.37.   Problems 16-57, 16-62.

ᶜ**16-58**   A wire is bent into a semicircle of radius R and has a uniform charge per unit length λ (Fig. 16.38). Find the electric field at its center. [*Hint:*

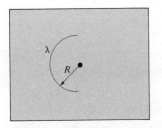

Figure 16.38. Problem 16-58.

Write the length of a segment as $R\,d\theta$ and integrate over θ.]

16-59 (a) Find the magnitude and direction of the electric field at a point on the x axis for the dipole in Fig. 16.9. (b) For $x \gg a$, find the ratio of the magnitude of this field to that of the field at a point equally far from the origin on the y axis.

16-60 When an electron is 2 cm from a long wire carrying a uniform positive charge, its speed is 10^4 m s^{-1}. When it is 1 cm from the wire, its speed is 2×10^4 m s^{-1}. At what distance from the wire will its speed be 4×10^4 m s^{-1}?

16-61 A uniformly charged ring with a radius a has a total charge Q (Fig. 16.36). Show that the potential at a point on its axis a distance y from its center is

$$V = \frac{kQ}{(y^2 + a^2)^{1/2}}$$

^c**16-62** A uniformly charged disk with a radius R has a total charge Q. (Fig. 16.37). Show that the potential at a point on its axis a distance y from its center is

$$V = \frac{2kQ}{R^2}[(R^2 + y^2)^{1/2} - y]$$

[*Hint:* Divide the disk into rings and use the result of the preceding problem.]

16-63 Show that the expression for the potential along the axis of a uniformly charged disk in the preceding problem reduces to the potential for a point charge in the limit $y \gg R$.

16-64 If an atom is placed in an electric field $\mathbf{E}$, its charge distribution will be distorted so that an induced electric dipole moment $\mathbf{p} = \alpha\mathbf{E}$ is produced, where α is the *polarizability* of the atom. (a) An atom of polarizability α is a distance r from an ion with charge $+e$, where r is large compared to the size of the atom. What is the induced dipole mo-

ment? (b) What is the potential energy of the atom and ion?

*16-65 Two dipoles $\mathbf{p}_1$ and $\mathbf{p}_2$ are a distance R apart, which is large compared to their charge separation. Find the total electric energy of the dipoles when they are oriented as in Fig. 16.39a and b. (A dipole at the origin directed along the y axis has an electric field at a distant point on the x axis that is $\mathbf{E} = -kp\hat{\mathbf{y}}/x^3$, and on the y axis the field is $\mathbf{E} = 2kp\hat{\mathbf{y}}/y^3$.)

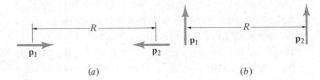

Figure 16.39. Problem 16-65.

16-66 The electric dipole moment of a water molecule is 6.13×10^{-30} C m. (a) If this dipole moment is due to a pair of point charges $\pm e$, how far apart must they be? Find the ratio of this distance to the radius of a hydrogen atom, 5.29×10^{-11} m. (b) If the dipole is parallel to an electric field of 10^6 V m^{-1}, how much energy is needed in joules and in electron volts to "flip" the dipole so that it is opposite to the field? (c) At room temperature, the average kinetic energy of a molecule is about 0.04 eV. What implication does this have for the orientation of water molecules in the relatively strong field of 10^6 V m^{-1}?

*16-67 Figure 16.40 shows a model of a water molecule. Each hydrogen atom has a net positive charge q, and the oxygen atom has a net charge $-2q$. The distance l is 9.65×10^{-11} m. The positive charge on a hydrogen atom and half the negative charge on the oxygen atom form a dipole. The total molecular electric dipole moment $\mathbf{p}$ is the vector

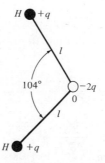

Figure 16.40. Problem 16-67.

sum of the two H—O dipole moments. (a) What is the direction of **p**? (b) If $p = 6.0 \times 10^{-30}$ C m, what is the charge q in multiples of the proton charge e?·

16-68 A certain nerve fiber (axon) is a cylinder 10^{-4} m in diameter and 0.1 m long. Its interior has a potential 0.09 V below that of the surrounding fluid; it is separated from that fluid by a thin membrane. Na⁺ ions are transported by a chemical reaction out of the fiber at the rate of 3×10^{-11} moles per second per cm² of membrane. (a) How many coulombs of charge per hour are transported out of the fiber? (b) How much work per hour must be done against the electrical forces?

*16-69 The two capacitors in Fig. 16.41 are said to be connected in *parallel*. They are connected to a battery so the potential difference across each is V. Show that a single capacitor C_p will store the same amount of charge if $C_p = C_1 + C_2$. (C_p is called the *equivalent capacitance*.)

16-70 In Fig. 16.41, $C_1 = 2\,\mu\text{F}$ and $C_2 = 4\,\mu\text{F}$. Using the result of the preceding problem, find the equivalent single capacitance that would have the same charge when connected to the battery.

16-71 Using the result of Problem 16-69, show how an equivalent capacitance of $10\,\mu\text{F}$ can be assembled from a supply of 2-μF capacitors.

*16-72 The capacitors in Fig. 16.42 are said to be in *series*. When they are connected to the battery as shown, the charge Q on each is the same. Show that a single capacitor C_s with charge Q will have a potential difference $V = V_1 + V_2$ if

$$\frac{1}{C_s} = \frac{1}{C_1} + \frac{1}{C_2}$$

(C_s is called the equivalent capacitance.)

16-73 In Fig. 16.42, $C_1 = 2\,\mu\text{F}$ and $C_2 = 4\,\mu\text{F}$. Using the result of the preceding problem, find the equivalent single capacitance that would maintain the same potential difference $V = V_1 + V_2$ with the same charge Q.

ANSWERS TO REVIEW QUESTIONS

Q16-1, attractive; **Q16-2** qE; **Q16-3**, parallel, opposite; **Q16-4**, away from, toward; **Q16-5**, magnitude of the electric field, direction of the electric field; **Q16-6**, uniform; **Q16-7**, $\Delta\mathfrak{U} = q\,\Delta V$; **Q16-8**, concentric spheres; **Q16-9**, ql, $+q$; **Q16-10**, along the field; **Q16-11**, Q/C; **Q16-12**, decreases, decreases, increases; **Q16-13**, 4.

SUPPLEMENTARY TOPICS

16.11 | GAUSS' LAW

In the main portion of this chapter, we computed the electric field for various arrangements of charges. The methods we used, based directly on Coulomb's law, work well for some problems. However, there are many other simple looking arrangements for which the Coulomb's law calculations turn out to be very complex. We see in this section that Gauss' law offers a useful alternative approach to finding the electric fields for certain important types of charge distributions.

For many situations, Gauss' law is just another formulation of the information contained in Coulomb's law. However, in the presence of time-varying magnetic fields, Gauss' law correctly describes the entire electric field, while Coulomb's law does not. We will not prove nor use this fact. Instead we will use Gauss' law as a tool to investigate some charge configurations that we cannot deal with so readily in any other way.

Our description of Gauss' law has a very strong geometric basis, as does our use of it. We will obtain the integral form of Gauss' law with the aid of some geometric ideas and the concept of electric field lines. There is also a differential form of Gauss' law that we will not discuss. It is equivalent in content to the integral form, and it is useful in more advanced applications.

Our analysis relies on the fact that electric field lines begin at positive charges and end at negative charges. Conventionally, the number of electric field lines drawn is proportional to the number and size of the charges producing the field. Thus if we have one positive charge Q and two negative charges of

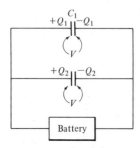

Figure 16.41. Problems 16-69 and 16-70.

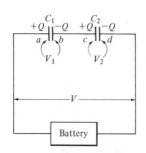

Figure 16.42. Problems 16-72 and 16-73.

- $Q/2$ each, then all the electric field lines begin at the lone positive charge and half end at each of the negative charges (Fig. 16.43a).

In Fig. 16.43b, we have focused on the positive charge Q and have drawn three *closed surfaces* around it. A closed surface is one where we cannot pass from the inside to the outside without passing through the surface. For example, a complete sphere is a closed surface, but one with a hole in it is not. Notice in Fig. 16.43b that the net number of electric field lines passing through each of the three closed surfaces is the same. Since the number of such lines is proportional to the positive charge shown, we con-

clude that the number of lines leaving a closed surface is proportional to the total amount of charge inside that surface.

This general statement can be verified by trying several surfaces other than those shown. For example, in Fig. 16.43c, the surface S4 has as many lines leaving it as entering it. The net number of lines passing through the surface (the number leaving minus the number entering) is zero, and so is the charge enclosed by it. S5 in Fig. 16.43c has half as many net lines passing through it as does S3 in Fig. 16.43b, and it encloses a total charge half as large, namely $Q - Q/2 = Q/2$.

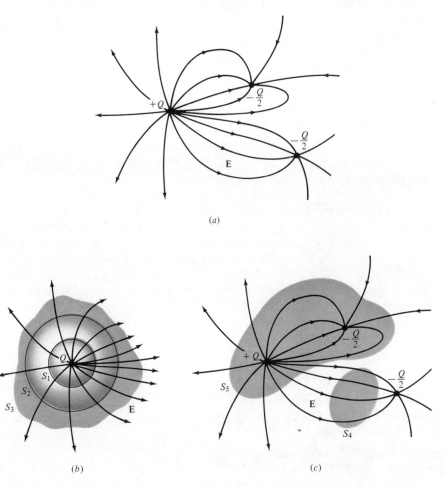

Figure 16.43. Approximate field line pattern for charges $+Q$, $-Q/2$, and $-Q/2$. (a) All the field lines start at $+Q$. Half end at each negative charge. (b) The same number of lines passes through the spheres S_1 and S_2 with different radii and also through the irregular surface S_3. (c) As many lines enter as leave S_4, so the net number leaving is zero. There is no net charge inside S_4. Half as many lines leave S_5 as leave the surfaces in (b); S_5 encloses half as much charge.

We can now make these observations more quantitative. The number of electric field lines crossing a surface ΔA depends on how $\mathbf{E}$ is oriented relative to ΔA. For example, in Fig. 16.44a, several lines cross ΔA, but none do so in Fig. 16.44b. The area that is effective in intercepting the field lines is the projection of ΔA onto the plane perpendicular to the field lines. In Fig. 16.44c, θ is the angle between $\mathbf{E}$ and the unit vector $\hat{\mathbf{n}}$ drawn normal to the area ΔA. The projection of ΔA on the plane normal to $\mathbf{E}$ is $\Delta A \cos \theta$. Note, for example, in Fig. 16.44a, the normal to the plane is parallel to $\mathbf{E}$, so that $\cos \theta = \cos 0° = 1$, and $\Delta A \cos \theta = \Delta A$ there. Similarly, the normal is perpendicular to $\mathbf{E}$, and $\Delta A \cos \theta = \Delta A \cos 90° = 0$ in Fig. 16.44b.

We see then that the number of lines crossing the surface is proportional to the magnitude of the field E, to the area ΔA, and to $\cos \theta$. Thus it is proportional to $E \Delta A \cos \theta$. We can define an *area vector* by $\Delta \mathbf{A} = \hat{\mathbf{n}} \Delta A$; $\Delta \mathbf{A}$ is equal in magnitude to the area and directed normal to it. Then the number of lines is proportional to the scalar product $\mathbf{E} \cdot \Delta \mathbf{A}$, since

$$\mathbf{E} \cdot \Delta \mathbf{A} = \mathbf{E} \cdot \hat{\mathbf{n}} \, \Delta A = E \, \Delta A \cos \theta$$

$\mathbf{E} \cdot \Delta \mathbf{A}$ is called the *electric flux* (Fig. 16.44d).

The flux though a surface is a quantity that provides a numerical way of discussing the pictorial concept of the number of field lines crossing the surface. If the field varies, then the surface is broken up into infinitesimal areas dA such that $\mathbf{E}$ is nearly constant in each. The total flux is then found by summing or integrating the products $\mathbf{E} \cdot d\mathbf{A}$.

Gauss' law is just the mathematical form of the statement that the number of lines leaving a closed surface indicates the amount of charge it contains. It says that the net electric flux leaving a closed surface S is proportional to the total charge $\Sigma \, Q_i$ within the surface:

$$\oint_S \mathbf{E} \cdot d\mathbf{A} = 4\pi k \Sigma \, Q_i \qquad \text{(Gauss' law)} \quad (16.26)$$

We will see shortly that the factor $4\pi k$ is the correct proportionality constant. The circle on the integral sign reminds us that the surface S is closed. The total charge $\Sigma \, Q_i$ is the algebraic sum of all the charges within S, and at each point on S the outward normal direction is used to define $\hat{\mathbf{n}}$.

Most of the utility of Gauss' law as stated in Eq. 16.26 derives from the fact that the closed surface S can be any surface one chooses. Some choices provide useless or trivial information, while others allow one to reach conclusions or calculate fields in cases where

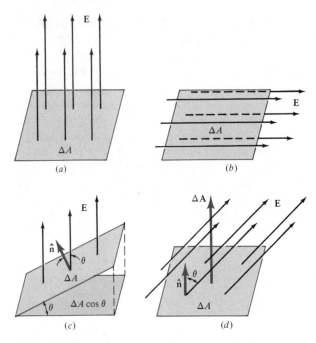

Figure 16.44. (a) The number of field lines passing through A is a maximum when $\mathbf{E}$ is perpendicular to the surface. (b) No field lines pass through the surface when $\mathbf{E}$ is parallel to it. (c) The projection of ΔA on the plane perpendicular to $\mathbf{E}$ determines how many lines cross ΔA. $\hat{\mathbf{n}}$ is a unit vector normal or perpendicular to the surface, and θ is the angle between $\mathbf{E}$ and $\hat{\mathbf{n}}$. (d) $\Delta \mathbf{A}$ is a vector of magnitude ΔA directed along $\hat{\mathbf{n}}$. The electric flux is $\mathbf{E} \cdot \Delta \mathbf{A} = \mathbf{E} \cdot \hat{\mathbf{n}} \Delta A = E \Delta A \cos \theta$.

the use of Coulomb's law is very difficult or not helpful. It is this judicious choice of surfaces to use in Gauss' law that the reader should focus on in the remainder of this section and the next.

We can demonstrate the equivalence of Gauss' law and Coulomb's law and the correctness of the $4\pi k$ factor by considering the field of a point positive charge $+Q$ (Fig. 16.45). This demonstration also illustrates how we exploit geometry in using Gauss' law.

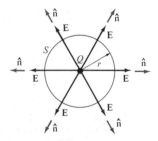

Figure 16.45. By symmetry, the field of a point charge must be radial, and its magnitude is the same at all points a distance r from the charge.

By symmetry, since no direction is different from any other, and because the lines must start at the postive charge, the field of the point charge $+Q$ can only be radial. Then, everywhere on the surface of a sphere S centered at Q, $\mathbf{E}$ must be perpendicular to the sphere, that is, along its normal $\hat{\mathbf{n}}$. Thus $\cos\theta = 1$, and $\mathbf{E} \cdot d\mathbf{A} = E\,dA$. Also, the same symmetry requires that $\mathbf{E}$ have the same magnitude *everywhere on this sphere;* otherwise, all directions would not be the same. Since E is constant, we can take it out of the integral sign. The total charge inside the sphere is $+Q$, so Gauss' law reduces to

$$E \oint_S dA = 4\pi k Q$$

Now the integral $\int dA$ over the surface of a sphere gives us its area, $4\pi r^2$. Thus this equation becomes $E(4\pi r^2) = 4\pi k Q$, or

$$E = \frac{kQ}{r^2}$$

If there is a second charge q in this field, it will experience a force $F = qE = kqQ/r^2$. This is precisely Coulomb's law, Eq. 16.1.

Besides showing the equivalence of Gauss' law and Coulomb's law, this calculation illustrates how to find electric fields using Gauss' law. If the geometric symmetry of a situation can be used to define surfaces called *Gaussian surfaces* on which $\mathbf{E}$ is constant in magnitude, E can be taken out of the integral. The integral then reduces to a surface area. However, when there isn't enough symmetry to do this, even though Gauss' law as stated in Eq. 16.26 is still correct, it is not sufficient to determine $\mathbf{E}$. The equivalent form of Gauss' law involving derivatives of the field components can then be used, but such calculations are beyond the level of this text.

The use of Gauss' law to solve problems involving symmetrical charge distributions is further illustrated in the next two examples. In the first we rederive a result found earlier by using Coulomb's law but with a good deal more effort, while in the second we obtain a new result. Notice carefully that the key point in both examples is that we are able to exploit Gauss' law because the symmetry enable us to find Gaussian surfaces on which the electric field has a constant magnitude.

Example 16.12

Use Gauss' law to find the field of a uniformly charged infinite plane with a charge per unit area $+\sigma$.

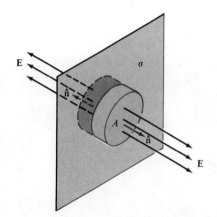

Figure 16.46. Part of a uniformly charged infinite plane. By symmetry, the field is perpendicular to the plane. Thus $\mathbf{E} \cdot d\mathbf{A} = \mathbf{E} \cdot \hat{\mathbf{n}} dA$ is zero except on the two ends of the Gaussian cylinder.

The field lines start at the charges and must be perpendicular to the charged plane, since any other direction would not be consistent with the symmetry (Fig. 16.46). We draw a Gaussian surface in the form of a circular cylinder, so that $\mathbf{E}$ is parallel to its sides and normal to both of its ends. Then the Gauss' law integral can be divided into three parts: the integral over the sides of the cylinder, and the integrals over each of the two ends. The field is constant on each of these three surfaces. At each end it is has some constant value E, and it is perpendicular to the area. Hence $\int \mathbf{E} \cdot d\mathbf{A} = E\int dA = EA$ on each of the ends; the integrals total $2EA$ for the two ends. On the sides, $\mathbf{E}$ is parallel to $d\mathbf{A}$, so $\mathbf{E} \cdot d\mathbf{A} = 0$, and the integral over the sides is zero. The total charge inside the cylinder is $Q = \sigma A$. Thus Gauss' law becomes

$$2EA = 4\pi k\sigma A \qquad \text{or} \qquad E = 2\pi k\sigma$$

This agrees with Eq. 16.8, which was obtained from Coulomb's law after a complicated integration.

Example 16.13

Find the electric field (a) outside and (b) inside a uniformly charged sphere of radius R and total charge Q.

(a) To find the field outside the sphere, we choose as our Gaussian surface a sphere S of radius $r > R$ (Fig. 16.47a). As in the case of a point charge, the symmetry demands that $\mathbf{E}$ be radial and constant in magnitude on the Gaussian surface. The charge inside S is the total charge on the sphere, Q. Thus Gauss' law gives

$$E \oint_S dA = 4\pi k Q$$

Since the surface area of the sphere is $4\pi r^2$, we find

$$E = \frac{kQ}{r^2}$$

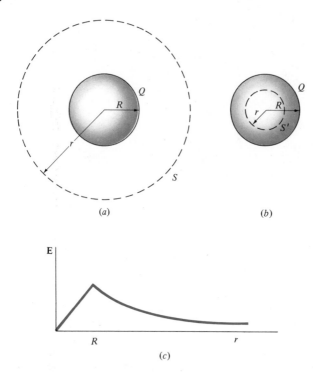

(c)

Figure 16.47. (*a*) A Gaussian sphere larger than the uniformly charged sphere encloses the full charge *Q*. (*b*) A sphere of radius *r* < *R* contains only part of the charge. (*c*) The field increases linearly inside the charged sphere. Outside it is identical to the field of a point charge.

The field outside the sphere is identical to that of a point charge. This would also be the case if the sphere carried a uniformly distributed surface charge or any other spherically symmetric charge distribution. Thus, from the field outside the charge, we cannot tell whether its source is a point charge or a more complicated spherically symmetric distribution. Similarly, the field *outside* a cylindrically symmetric distribution is identical to that of a uniformly charged long straight wire.

(b) To find the field inside the sphere, we choose a spherical Gaussian surface *S'* with *r* < *R* (Fig. 16.47*b*). The portion of the charge inside *S'* is the total charge *Q* times the ratio of the volumes of the Gaussian sphere and the entire charged sphere of radius *R*. Since the volume of a sphere of radius *a* is $4\pi a^3/3$, this volume ratio is r^3/R^3. Applying the same symmetry argument to the left side of Gauss' law as in part (a), we find

$$E(4\pi r^2) = 4\pi kQ\,\frac{r^3}{R^3}$$

or

$$E = \frac{kQr}{R^3}$$

Notice that here *E* increases linearly as *r* increases. At the surface of the sphere, *r* = *R*, *E* has its largest magnitude. The formulas found in parts (a) and (b) give the same value for *E* at this point (Fig. 16.47*c*).

The results obtained in this example with the aid of Gauss' law can also be found from Coulomb's law, but the calculation is much more difficult.

16.12 | GAUSS' LAW AND CONDUCTORS

Gauss' law enables us to derive some very important properties of conductors when there are no charges in motion. Specifically, it tells us where the net charge must reside, and it allows us to determine the field near the surface.

We noted in Section 16.5 that inside a conductor the electric field must be zero when there are no charges in motion. Thus if we consider a solid conductor, the field must vanish everywhere within it (Fig. 16.48*a*). Thus, if we draw any closed surface *S* inside the conductor, **E** · *d***A** is zero everywhere on *S* because **E** is zero, and the integral in Gauss' law is

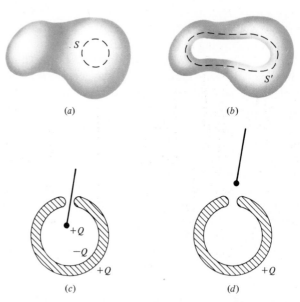

Figure 16.48. A conductor with no charges in motion must have *E* = 0 inside of it. (*a*) The net charge inside any surface *S* is zero, so the charge is zero everywhere in the conductor. (*b*) *S'* includes the cavity and contains no net charge, so there is no charge on the interior of the cavity. (*c*) Introducing a charge +*Q* through a small hole in the cavity causes a charge −*Q* to be induced on the interior of the conductor. This leaves a charge +*Q* on its outside, since the conductor has no net charge. (*d*) If the object is touched to the inner surface, the charges +*Q* and −*Q* cancel. All the charge is then on the outer surface of the conductor.

zero. Accordingly, the net charge contained within S is also zero. Since this is true for any closed surface within the metal, it must be true that there is no net charge anyplace inside the metal. *Any net charge at rest on any solid conductor must therefore reside on its surface.*

Suppose now that the conductor is hollow rather than solid. Again the field within the conductor itself is zero, so that $\mathbf{E} \cdot d\mathbf{A}$ vanishes everywhere on the Gaussian surface S' that encloses the cavity. Thus there can be no net charge within S' (Fig. 16.48b).

A simple experiment using a hollow conductor shows some interesting consequences of this fact. When we introduce an object carrying a charge $+Q$ through a small opening into a cavity, an opposite charge $-Q$ must appear on the inner surface of the conductor to balance or neutralize it. (This must happen or the field in the conductor would be nonzero, since the Gaussian surface surrounding the cavity would contain a net, nonzero charge.) If the conductor as a whole was initially electrically neutral, it follows that a charge $+Q$ will now be present on its outer surface (Fig. 16.48c). If we finally touch the object carrying the charge $+Q$ to the inside of the conductor, the charges $+Q$ and $-Q$ exactly neutralize. When the object is withdrawn, all the charge $+Q$ will reside on the outside of the conductor (Fig. 16.48d).

Field Near a Conductor |
The field at the surface of a conductor must be perpendicular to the surface when all the charges are at rest. This follows because if the field had a component parallel to the surface, electrons would tend to move. The field is zero inside the conductor. Thus if we draw a small, flat Gaussian cylinder as in Fig. 16.49, the field is zero on its inner end, normal to its outer end, and parallel to its sides. The charge contained in the cylinder is σA, so Gauss' law becomes

$$EA = 4\pi k \sigma A$$

$$E = 4\pi k \sigma \quad \text{(field just outside} \quad (16.27) \\ \text{conductor)}$$

This is the field just outside a charged conductor of any shape. It is exactly *double* the field derived in Section 16.3 and again in Example 16.12 for a uniformly charged plane. The difference is that here all the field lines go through a single end of the Gaussian cylinder, rather than just half of the lines. The field is zero inside the conductor, and twice as large outside.

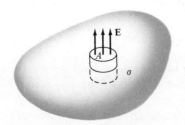

Figure 16.49. The field at the surface of a conductor is perpendicular to the surface on the outside and zero inside. Shown is a tiny cylindrical Gaussian surface. The cylinder is supposed to be small enough so that no matter what the actual shape of the surface, the field is perpendicular to the outer surface and parallel to the sides.

EXERCISES ON SUPPLEMENTARY TOPICS

Section 16.11 | Gauss' Law

16-74 A hollow metallic sphere has an outer radius b and an inner radius a. The total charge on the sphere is Q. Find the electric field for (a) $r > b$; (b) $a < r < b$; (c) $r < a$.

16-75 Two concentric thin conducting spherical shells have radii a and $2a$. The inner sphere has a charge $-Q$ and the outer has a charge $+Q$. Find the magnitude and direction of the field for (a) $r > 2a$; (b) $a < r < 2a$; (c) $r < a$.

16-76 According to the modern view of the atom, its positive charge is contained in a small nucleus. A lead nucleus contains 82 protons and is a sphere of radius 7×10^{-15} m. (a) What is the electric field at the surface of the nucleus? (b) The early Thomson model assumed the positive charge was spread over the entire atom, or over a sphere of radius 10^{-10} m. Ignoring the effects of the electrons, what would the field be at the surface of the atom according to this model?

16-77 A sphere of radius 0.1 m is uniformly charged. Its total charge is 10^{-4} C. (a) Where does the electric field have its greatest magnitude? (b) How large is that maximum field? (c) At what two distances from the center is the field half its maximum magnitude?

16-78 A long thin cylindrical metallic shell has a radius R. It has a charge per unit area σ. (a) What is the charge per unit length on the cylinder? (b) What is the electric field outside the cylinder? (c) What is the field inside the cylinder?

16-79 A cube of side a has a cube of side $a/2$ centered within it. The inner cube has a total charge Q that is uniformly distributed over its surface. (a) For the surface of the outer cube, find

$$\oint_S \mathbf{E} \cdot d\mathbf{A}$$

(b) Is this sufficient information to find the electric field at points on the surface of the outer cube? Explain.

16-80 A thin spherical shell of radius r_1 has a net charge per unit area σ_1. A larger concentric spherical shell of radius r_2 has a net charge per unit area σ_2. If the field outside the larger shell is zero, what is the ratio of σ_2 to σ_1?

Section 16.12 | Gauss' Law and Conductors

16-81 Show that Eq. 16.27 for the field just outside a charged conductor agrees with the formula for the field of a charged conducting sphere, kQ/r^2.

16-82 A hollow metal sphere has radii a and $0.8a$ and has no net charge. A charge Q is introduced through a small hole and placed at the center of the sphere. Find the charge per unit area on (a) the inner surface of the sphere; (b) the outer surface. (c) If Q were placed not on the center but elsewhere within the cavity, what can be said about the induced charge distribution on each surface?

16-83 When a conducting wire is connected to the terminals of a battery, there is a current in the wire. Do the remarks in Section 16.12 concerning the fields and charge distributions in conductors apply? Explain.

PROBLEMS ON SUPPLEMENTARY TOPICS

^c**16-84** A thin metallic sphere has an outer radius R_1 and a total charge Q. A larger concentric metallic sphere has an inner radius R_2 and a total charge $-Q$. (a) Show that the magnitude of the potential difference between them is

$$V = kQ \left[\frac{1}{R_1} - \frac{1}{R_2} \right]$$

(b) Find the capacitance of the two spheres.

^c**16-85** Find the potential at the center of a uniformly charged solid sphere of radius R and total charge Q.

^c**16-86** A thin metallic cylinder has an outer radius R_1 and a charge per unit length λ. A larger concentric metallic cylinder has an inner radius R_2 and a charge per unit length $-\lambda$. (a) Show that the magnitude of the potential difference between them is

$V = 2k\lambda \ln(R_2/R_1)$. (b) Find the capacitance per unit length of the cylinders, $C = \lambda/V$.

^c**16-87** An infinitely long uniformly charged cylinder has a charge per unit length λ and a radius R. (a) Find the electric field outside the cylinder. (b) Show that inside the cylinder $E = 2k\lambda r/R^2$. [*Hint:* Use a concentric cylinder as a Gaussian surface.]

* ^c**16-88** A sphere of radius R has a charge distribution that is linear in the distance from the center. This means that its charge per unit volume ρ is given by $\rho = Cr$. (a) Find the total charge inside the sphere. (b) Find the electric field inside the sphere.

Additional Reading

R.E. Orville, The Lightning Discharge, *Physics Teacher,* vol. 14, 1976, p. 7.

R.M. Alexander, *Functional Design in Fishes,* Hutchinson University Library, London, 1967. Chapter 6 describes the electrical senses of fish.

Robert Burton, *Animal Senses,* Taplinger Publishing Company, Inc., New York, 1970.

R.H. Bullock, Seeing the World Through a New Sense: Electroreception in Fish, *American Scientist,* May/June 1973, p. 316.

Carl D. Hopkins, Electric Communication in Fish, *American Scientist,* July/August 1974, p. 426.

R.T. Cox, Electric Fish, *American Journal of Physics,* vol. 11, 1943, p. 13.

Russell K. Hobbie, The Electrocardiogram as an Example of Electrostatics, *American Journal of Physics,* vol. 41, 1973, p. 824.

Scientific American articles:

George Shiers, Ferdinand Brown and the Cathode Ray Tube, March 1974, p. 92.

H. Kondo, Michael Faraday, October 1953, p. 90.

A.D. Moore, Electrostatics, March 1972, p. 46.

Herbert A. Pohl, Nonuniform Electric Fields, December 1960, p. 106.

C. Andrave Bassett, Electrical Effects in Bone, October 1965, p. 18.

L.B. Loeb, The Mechanism of Lightning, January 1949, p. 22.

A.A. Few, Thunder, July 1975, p. 80.

J.E. McDonald, The Earth's Electricity, April 1953, p. 32.

Harry Grundfest, Electric Fishes, October 1960, p. 115.

H.W. Lissman, Electric Location by Fishes, March 1963, p. 50.

H.B. Steinback, Animal Electricity, February 1950, p. 40.

CHAPTER 17

DIRECT CURRENTS

Most applications of electricity and magnetism involve moving charges or *electrical currents* in conductors. Direct currents (dc) are produced when a conducting path exists between the terminals of a battery or a dc generator. These devices tend to maintain a constant potential difference between their terminals and convert other kinds of energy, such as chemical or mechanical energy, into electrical energy. An alternating current (ac) is produced by an ac generator, which has a terminal potential difference that alternates in sign at some characteristic frequency. Many of the ideas we consider in this chapter can be applied immediately or with minor changes to alternating currents as well as to direct currents.

We begin this chapter by defining *electric current* and *resistance*. We then consider sources of energy and the transformation of energy in circuits. In the remaining sections, we discuss methods of analyzing complex circuits, electrical meters, the charging of a capacitor through a resistor, and electrical safety.

17.1 | ELECTRIC CURRENT

The *electric current* in a wire is the rate at which charge moves in the wire. For example, in Fig. 17.1,

charges move through a conducting wire under the influence of an applied electric field. If a net charge ΔQ crosses the shaded cross-sectional area in a time Δt, the *average current* is

$$\bar{I} = \frac{\Delta Q}{\Delta t} \qquad (17.1)$$

and the *instantaneous current* is

$$I = \frac{dQ}{dt} \qquad (17.2)$$

The S.I. current unit is the *ampere* (A). Often it is convenient to use the *milliampere* (mA); $1 \text{ mA} = 10^{-3}$ A. From the definition, it follows that an ampere is a coulomb per second. (The ampere is defined by the magnetic force between two currents under specified conditions. This, in turn, defines the coulomb. See Section 19.8.) The definition of current is used in the following example taken from electrochemistry.

Example 17.1
An electrochemical cell consists of two silver electrodes placed in an aqueous solution of silver nitrate. A constant 0.5-A current is passed through the cell for 1 hour. (a) Find the total charge transported through the cell in coulombs and in multiples of the electronic charge. (b) Each electron reaching the cell discharges one positively charged silver ion, which is then deposited on the negative electrode (cathode). What is the total mass of the deposited silver? (The atomic mass of silver is 107.9 u.)

(a) Since the current is constant,

$$\Delta Q = I \Delta t = (0.5 \text{ A})(1 \text{ h})$$
$$= (0.5 \text{ C s}^{-1})(3600 \text{ s}) = 1800 \text{ C}$$

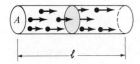

Figure 17.1. A segment of a conducting wire. Charges enter through the left end and leave through the right end.

The ratio of ΔQ to the electronic charge is

$$N = \frac{\Delta Q}{e} = \frac{1800 \text{ C}}{1.60 \times 10^{-19} \text{ C}} = 1.13 \times 10^{22}$$

This is the number of silver ions transported through the cell and deposited in one hour.

(b) The mass of the deposited silver is the number of atoms N times the mass of an atom. Using $1 \text{ u} = 1.66 \times 10^{-27} \text{ kg}$,

$$m = (1.13 \times 10^{22})(107.9 \text{ u})(1.66 \times 10^{-27} \text{ kg u}^{-1})$$
$$= 2.02 \times 10^{-3} \text{ kg}$$

Conventionally the current in a conductor is assumed to be in the direction of motion of positive charges. However, in metallic conductors the moving charges are electrons. In metals, some of the electrons become detached, leaving behind positively charged ions. The heavy ions form a regularly spaced crystalline lattice and vibrate about their equilibrium positions with an energy and an amplitude that increases with the temperature. The detached *conduction electrons* move randomly among the ions. In the absence of an applied electric field, the average charge flow in any direction is zero. When an electric field is applied, the electrons acquire an average *drift velocity* opposite to the field, and there is a net current.

We can relate the current in a wire to the density of conduction electrons and their drift velocity v. If there are n electrons per unit volume, then the total number of electrons in a volume V is nV. The segment of wire in Fig. 17.1 has length ℓ and cross-sectional area A, so its volume is $V = \ell A$. Hence there are $n\ell A$ electrons in the wire, with a total charge of magnitude $en\ell A$. The time needed for all of them to pass through the end of the segment shown is $\Delta t = \ell/v$, so the magnitude of the current is

$$I = \frac{\Delta Q}{\Delta t} = \frac{en\ell A}{\ell/v} = enAv \qquad (17.3)$$

The current is the product of the electronic charge, the density of conduction electrons, the area, and the average drift velocity. Note that a positive charge flow in one direction is equivalent to a negative flow in the opposite direction. The following example shows that the drift velocity in a typical metal is surprisingly small.

Example 17.2

Number 12 copper wire is often used to wire household electrical outlets. Its radius is 1 mm = 10^{-3} m. If it carries a current of 10 A, what is the drift velocity of the electrons? (Metallic copper has one conduction electron per atom, the atomic mass of copper is 64 u, and the density of copper is 8900 kg m^{-3}.)

From Eq. 17.3, the drift velocity is $v = I/neA$. We are given the current I, and we can immediately find the area A from the radius of the wire, leaving only the electron density to be found. Since copper has one conduction electron per atom, n equals the number of atoms per unit volume. The number of atoms per unit volume times the mass of one atom M equals the mass of a unit volume of copper, which is its density d. (We will use d here for density rather than ρ as in Chapter Three to avoid confusion with the resistivity ρ defined later.) Thus $nM = d$, or

$$n = \frac{d}{M} = \frac{8900 \text{ kg m}^{-3}}{(64 \text{ u})(1.66 \times 10^{-27} \text{ kg u}^{-1})}$$
$$= 8.38 \times 10^{28} \text{ m}^{-3}$$

The drift velocity is then

$$v = \frac{I}{neA} = \frac{I}{ne\pi r^2}$$
$$= \frac{(10 \text{ A})}{(8.38 \times 10^{28} \text{ m}^{-3})(1.6 \times 10^{-19} \text{ C})\pi(10^{-3} \text{ m})^2}$$
$$= 2.37 \times 10^{-4} \text{ m s}^{-1}$$

Thus the electrons move very slowly, contrary to what one might have supposed; it takes about 4200 seconds or over an hour for them to move a metre! The average electron thermal velocity at room temperature is about 10^5 m s^{-1}, which is about 10^{10} times the drift velocity. On the other hand, changes in the electric field travel in a wire at nearly the speed of light. This is analogous to the situation in a fluid such as air or water, where the effects of pressure changes travel much faster than the fluid itself.

17.2 | RESISTANCE

In Chapter Fourteen, we defined the resistance of a section of pipe to the flow of a fluid as the pressure difference divided by the flow rate. Similarly, the *electrical resistance* R of a conductor is the potential difference V between its ends divided by the current I,

$$R = \frac{V}{I} \qquad (17.4)$$

The S.I. unit for resistance is the *ohm;* an ohm is a volt per ampere.

For many materials, the potential difference and the current are directly proportional, so the resistance

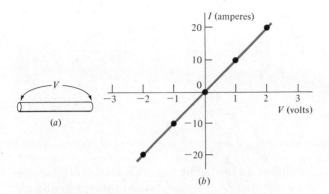

Figure 17.2. (a) A copper wire has a variable potential difference V between its ends. (b) The current varies linearly with the potential difference in the wire, so it is an ohmic conductor.

is a constant independent of the current. This is illustrated in Fig. 17.2 for a length of copper wire with its ends at an adjustable potential difference V. The current increases linearly with V and reverses direction when V is reversed, so the ratio $R = V/I$ is constant. Materials with a constant resistance are said to obey *Ohm's law* and are called *ohmic conductors*.

The resistance of some conductors varies with the magnitude or direction of the applied potential difference. The operation of many electronic devices, such as vacuum tubes and transistors, is based on their *nonohmic* character (Fig. 17.3). The calculation of the resistance from the definition $R = V/I$ for an ohmic conductor is illustrated by the following example.

Example 17.3

Find the resistance of the wire in Fig. 17.2.

The potential difference and current are proportional, so the resistance is a constant. When the current is 10 A, the potential difference is 1 V, and

$$R = \frac{V}{I} = \frac{1\ V}{10\ A} = 0.1\ \text{ohm}$$

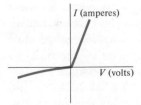

Figure 17.3. The current versus potential difference graph for a rectifying transistor. An applied voltage, which produces a large current in one direction, produces only a small current in the opposite direction. This is an example of a conductor that does not satisfy Ohm's law.

The resistance of a conductor depends on its size, shape, and composition. We can understand the size and shape dependence with the following argument. If we put two identical wires side by side, the current doubles and hence the resistance is halved. Thus, R must vary inversely with the cross-sectional area A. Now if we halve the length ℓ, the potential change and hence the resistance are also halved, so R must be proportional to ℓ. Thus, we can write the resistance in terms of geometric factors and a constant as

$$R = \frac{\rho \ell}{A} \qquad (17.5)$$

The proportionality constant ρ (rho) depends only on the properties of the material and is called the *resistivity*. The S.I. unit for resistivity is the ohm-metre. Typical resistivities are listed in Table 17.1.

The *conductivity* σ (sigma), defined by

$$\sigma = \frac{1}{\rho} \qquad (17.6)$$

is sometimes used instead of the resistivity to characterize conductors. The S.I. unit of conductivity is the ohm^{-1} m^{-1}. The calculation of the resistance of a wire is illustrated by the following example.

Example 17.4

Find the room temperature resistance of a copper wire 100 m long with a radius of 1 mm = 10^{-3} m.

According to Table 17.1, the resistivity of copper at room temperature (20° C) is 1.72×10^{-8} ohm m. Hence the resistance of this wire is

$$R = \frac{\rho \ell}{A} = \frac{\rho \ell}{\pi r^2}$$

$$= \frac{(1.72 \times 10^{-8}\ \text{ohm m})(100\ \text{m})}{\pi (10^{-3}\ \text{m})^2}$$

$$= 0.547\ \text{ohm}$$

TABLE 17.1

Resistivities at 20°C in ohm-metres

	Substance	Resistivity (ohm-metres)
Conductors	Silver	1.47×10^{-8}
	Copper	1.72×10^{-8}
	Aluminum	2.63×10^{-8}
Semiconductors	Germanium	0.60
	Silicon	2300
Insulators	Sulfur	10^{15}
	Glass	10^{10}–10^{14}
Ionic conductors	Body fluids	approx. 0.15

17.3 | ENERGY SOURCES IN CIRCUITS

A nonelectrical source of energy, such as a battery or a generator, is needed to maintain a continuous current in a closed conducting path or *circuit*. Such an energy source is called a *source of EMF*, which is short for the archaic term electromotive force. EMF, as we define it below, has the dimensions of potential, not force.

To clarify the role of an EMF in a circuit, consider the analogous situation of water moving under the influence of gravitational forces. Figure 17.4 shows an artificial stream and waterfall constructed for decorative purposes in a garden. Water drops over the falls a vertical distance h, so that the gravitational potential energy of a mass m decreases by mgh. This energy is converted into kinetic energy as the water accelerates and is then transformed into heat $Q = mgh$ as the water splashes among the rocks in the falls. If no water is supplied to the beginning of the stream, the flow stops almost immediately. However, water is returned to the stream by a pump, which does work $W = mgh$ on the water against the gravitational force. The conservative gravitational force does positive work equal to mgh on the water as it falls and negative work $-mgh$ as it rises. These exactly cancel, so the conservative gravitational force does no net work when the water goes completely around the system. The energy W supplied by the pump exactly equals the heat energy Q generated in the falls. Note that if we introduced a waterwheel connected to a

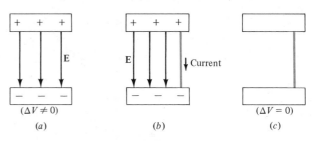

Figure 17.5. The electrical analog to a falls without a pump.

machine into the falling stream, then some of the energy supplied by the pump would be used to do mechanical work rather than being converted into heat.

Figure 17.5 shows an electrical system that is analogous to a waterfall *without* a pump. Two conductors have equal and opposite charges, so there is a potential difference between them. When they are connected by a metal wire, there is a brief charge flow, which stops when the charges are neutralized and the potential difference is reduced to zero.

To maintain a steady current, a *pump* or source of EMF must supply energy. In Fig. 17.6, a battery is connected to a resistor by wires that are perfect conductors. As a charge goes through the battery, it is *pumped* by nonelectrical forces to a position of higher potential energy. The energy required for this is supplied by chemical reactions occurring in the battery. In the resistor, the charge moves in the direction of the electric force, and the kinetic energy it acquires is transformed into heat. The conservative electric field does positive work in the resistor, negative work in the battery, and zero net work when a charge goes completely around the circuit. The net effect of the various energy transfers is that stored chemical energy in the battery is transformed into heat in the resistor. This is completely analogous to the energy supplied by the pump, which is converted into heat in the waterfall.

Two other aspects of the analogy should be noted. The resistor can be replaced in the circuit by some other *load*, such as an electric motor. Then the chemical energy of the battery is converted into mechanical energy, just as the energy supplied by the pump is used to do mechanical work if a waterwheel is introduced. Also, just as the pump supplied energy but did not alter the original supply of water, the battery cannot produce or destroy charge. Whatever current enters the battery in Fig. 17.6 must leave it, and the current in the resistor and in the connecting wires must also be the same. *A battery or generator converts*

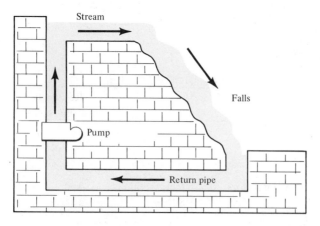

Figure 17.4. The work done by the pump equals the energy dissipated as heat in the falls. The conservative gravitational force does no net work on water that goes completely around the system.

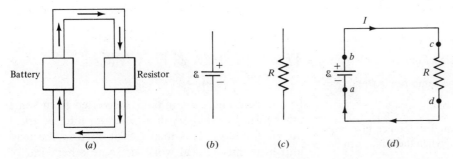

Figure 17.6. (a) A battery connected by perfectly conducting wires to a resistor. (b) The symbol for an EMF used in circuit diagrams. The longer line and + symbol indicate the higher potential terminal. (c) The symbol for a resistance. (d) The circuit of (a) redrawn using the conventional symbols.

some other kind of energy into electrical energy, but it is not a source of charge.

The EMF ε of a battery or generator is defined as the *work done per unit charge by the nonelectrical forces.* Its S.I. unit is a joule per coulomb or volt, which is also the unit of the electric potential.

For the circuit of Fig. 17.6, we can relate the current I to the battery EMF ε and to the resistance R by using the potential changes observed as we go completely around the circuit. This is equivalent to finding the potential energy changes for a unit positive charge. Proceeding clockwise, these changes are:

From *a* to *b*: *V increases* by $\varepsilon(\Delta V = \varepsilon)$; the potential increase in the battery is equal to the EMF.
From *b* to *c*: *V is constant* $(\Delta V = 0)$; the potential difference is *IR*, the current times the resistance, and the resistance is zero for a perfect conductor.
From *c* to *d*: *V decreases* by *IR* $(\Delta V = -IR)$.
From *d* to *a*: *V is constant* $(\Delta V = 0)$; again, no potential change occurs in a perfect conductor.

When a charge goes around a closed path and returns to the starting point, its potential energy must return to its original value, since the conservative electrical forces do no net work. Thus the sum of the potential changes must be zero, or $\varepsilon - IR = 0$. Equivalently, the potential increase in the battery must equal the decrease in the resistor:

$$\varepsilon = IR \qquad (17.7)$$

This result is illustrated by the following example.

Example 17.5

A dry cell with an EMF of 6 V is connected to a light bulb with resistance 4 ohms. Find the current.

With $\varepsilon = IR$, the current is

$$I = \frac{\varepsilon}{R} = \frac{6 \text{ V}}{4 \text{ ohms}} = 1.5 \text{ A}$$

Real batteries and generators usually have various dissipative effects associated with them that can be thought of as *internal resistances*. These cause the terminal voltages to be different from the EMFs and can result in substantial heat being produced. For example, the electrolytic fluid in storage batteries will boil if the current is large enough. These effects are illustrated in the next two examples.

Example 17.6

A battery with an EMF of 6 V and an internal resistance $r = 2$ ohms is connected to an $R = 4$-ohm lightbulb (Fig. 17.7). Find (a) the current, and (b) the terminal potential difference of the battery.

(a) The potential increase due to the EMF must equal the total of the potential decreases in the resistances, so

$$\varepsilon = IR + Ir$$

and

$$I = \frac{\varepsilon}{R + r} = \frac{6 \text{ V}}{(2 + 4) \text{ ohms}} = 1 \text{ A}$$

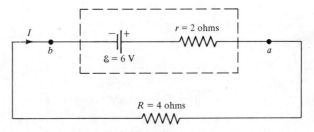

Figure 17.7. The terminal voltage of the battery is less than the EMF here because of the internal resistance.

(b) The terminal voltage is the potential difference V_{ab} between points a and b. It equals the EMF minus the potential drop associated with the internal resistance, so

$$V_{ab} = \mathcal{E} - Ir = 6\text{ V} - (1\text{ A})(2\text{ ohms}) = 4\text{ V}$$

Alternatively, we can obtain the same result from

$$V_{ab} = IR = (1\text{ A})(4\text{ ohms}) = 4\text{ V}$$

We see that the terminal potential difference of a real battery is different from its EMF when there is a current.

Example 17.7

Two batteries are connected to a resistor as shown in Fig. 17.8. Find (a) the current, and (b) the terminal potential difference of each battery.

(a) Since the batteries are connected so that their polarities are opposite, the direction of the current is determined by the larger EMF and is counterclockwise in Fig. 17.8. As we follow the current through the larger EMF, the potential *increases* by $\mathcal{E}_2$, and in the other EMF the potential *decreases* by $\mathcal{E}_1$. The current is found from

$$\mathcal{E}_2 - \mathcal{E}_1 = I(r_1 + r_2 + R)$$

Thus

$$I = \frac{\mathcal{E}_2 - \mathcal{E}_1}{r_1 + r_2 + R} = \frac{(18 - 6)\text{ V}}{(2 + 1 + 3)\text{ ohms}} = 2\text{ A}$$

(b) The potential differences across the batteries are

$$V_{ab} = \mathcal{E}_2 - Ir_2 = 18\text{ V} - (2\text{ A})(1\text{ ohm}) = 16\text{ V},$$
$$V_{dc} = \mathcal{E}_1 + Ir_1 = 6\text{ V} + (2\text{ A})(2\text{ ohms}) = 10\text{ V}$$

Note that the terminal potential difference V_{dc} is greater than the EMF in this case. In calculating the potential difference between points c and d, we are proceeding opposite to the current. Thus, we see a potential *rise* in the resistance as well as in the EMF. The terminal potential difference exceeds the EMF only when the cur-

rent is driven "backwards" through the battery by another larger EMF.

17.4 | POWER IN ELECTRICAL CIRCUITS

In a circuit, energy is initially converted from some other form by a battery or a generator into electrical potential energy. It is then transformed in the load into heat, mechanical work, or some other kind of energy. In this section, we calculate the power or rate of energy conversion for various parts of a circuit.

Figure 17.9 shows a *circuit element* (one part of a circuit) with a potential difference V across its terminals and a current I through it. If the element is a battery, V is the terminal voltage; if it is a resistance R, then V is equal to IR in magnitude. In a time Δt, a charge $\Delta Q = I\,\Delta t$ passes through the element. Its change in potential energy is $V\,\Delta Q = VI\,\Delta t$, which must equal the work ΔW done by the element on the charge. Dividing this work by the time Δt and taking the limit as Δt approaches zero, the power in any circuit element is

$$\mathcal{P} = \frac{dW}{dt} = IV \qquad \text{(any circuit element)} \quad (17.8)$$

If the current is in the direction of increasing potential, the circuit element supplies energy. When the current is in the direction of decreasing V, as in a resistor or a motor, the element receives or dissipates energy. For an EMF, the potential difference is $\mathcal{E}$, so

$$\mathcal{P} = I\mathcal{E} \qquad \text{(EMF)} \qquad (17.9)$$

For a resistor, $V = IR$, and

$$\mathcal{P} = I^2R \qquad \text{(resistor)} \qquad (17.10)$$

In any electrical circuit, the power supplied is always equal to the power dissipated, unless a capacitor or other device that can store energy is present. These ideas are illustrated by the following example.

Example 17.8

Calculate the power supplied to or by each element in the circuit of Fig. 17.8.

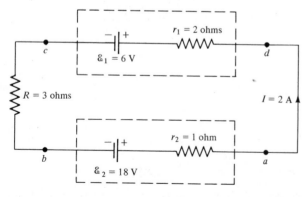

Figure 17.8. Examples 17.7 and 17.8.

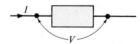

Figure 17.9. A circuit element is one part, such as a battery or resistor, of a complete circuit.

In the larger battery, the EMF supplies power $I\mathcal{E}_2$, and the internal resistance dissipates heat at the rate I^2r_2. Thus the net power *supplied* by this battery is

$$\mathcal{P}_2 = I\mathcal{E}_2 - I^2r_2$$
$$= (2\text{ A})(18\text{ V}) - (2\text{ A})^2(1\text{ ohm})$$
$$= 32\text{ W}$$

The other battery has *work done on it* by the current, since the charge is moving from the high potential terminal to the low potential side. Thus the power *absorbed* by this battery is

$$\mathcal{P}_1 = I\mathcal{E}_1 + I^2r_1$$
$$= (2\text{ A})(6\text{ V}) + (2\text{ A})^2(2\text{ ohms})$$
$$= 20\text{ W}$$

Finally the resistor R generates heat at the rate

$$\mathcal{P}_R = I^2R = (2\text{ A})^2(3\text{ ohms}) = 12\text{ W}$$

Notice that the power supplied by the larger battery equals the power dissipated in R plus the power absorbed by the smaller battery. This is just the energy conservation law in another form.

Electric utilities charge their customers for the electric energy used, which is the product of the electric power and the time. Electric energy is usually sold by the kilowatt-hour (kW h). This is a kilowatt of power for one hour, so

$$1\text{ kW h} = (10^3\text{ W})(3600\text{ s})$$
$$= 3.6 \times 10^6\text{ J} \qquad (17.11)$$

Most household electric power in North America is equivalent to 120 V dc, but is actually alternating current at a frequency of 60 Hz. As we see in Chapter Twenty, for resistive current and power calculations, an *effective* voltage of 120 V ac is the same as 120 V dc. The calculation of the cost of operating a typical electrical device is illustrated by the following example.

Example 17.9

For a household 60-W light bulb operated at 120 V, find (a) the current; (b) the resistance; and (c) the 24-hour operating cost, if energy costs 10 cents per kilowatt-hour.

(a) Since the power $\mathcal{P}$ in any circuit element equals VI,

$$I = \frac{\mathcal{P}}{V} = \frac{60\text{ W}}{120\text{ V}} = 0.5\text{ A}$$

(b) The resistance is

$$R = \frac{V}{I} = \frac{120\text{ V}}{0.5\text{ A}} = 240\text{ ohms}$$

(c) The bulb uses power at the rate of 60 W or 0.060 kW. Thus the cost of operating it for 24 hours is

$$(0.06\text{ kW})(24\text{ h})[10\text{¢ (kW h)}^{-1}] = 14.4\text{¢}$$

17.5 | SERIES AND PARALLEL RESISTORS; KIRCHHOFF'S RULES

Looking inside a television set or stereo one quickly discovers that circuits are often quite complex. Nevertheless, the most complex dc circuit can be analyzed to find the currents and voltages at each point using two basic rules formulated by Kirchhoff. These rules can also be generalized to deal with ac circuits. Here we show how they can be used to simplify the discussion of circuits containing certain combinations of resistors.

The first rule is already familiar, and follows from the conservative nature of electrical forces:

1 *The sum of the potential changes around any closed path is zero.*

The second rule is a restatement of another basic idea, charge conservation. Suppose a current divides into two or more currents at a point where a number of wires come together. No charge can be lost or created nor can it collect at that point. Consequently,

2 *The current entering any point must equal the current leaving.*

We can use Kirchhoff's rules to discuss series and parallel combinations of resistors. Two or more resistors are in *series* if the same current goes through each (Fig. 17.10). According to the first rule, the sum of the potential changes around the circuit is zero, or

$$\mathcal{E} - IR_1 - IR_2 - IR_3 = 0$$

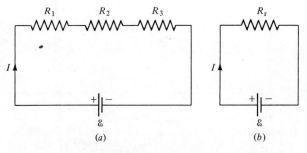

Figure 17.10. (*a*) Three resistors in series. (*b*) The equivalent resistance R_s leads to the same current I.

Thus

$$I = \frac{\mathcal{E}}{R_1 + R_2 + R_3}$$

A single *equivalent resistance* R_s connected to the battery will produce the same current if $I = \mathcal{E}/R_s$, or

$$R_s = R_1 + R_2 + R_3 + \cdots \qquad (17.12)$$

Two or more resistors are in *parallel* if they have a common potential difference (Fig. 17.11.) According to the second rule, the current entering point a must equal that leaving, or

$$I = I_1 + I_2 + I_3$$

Applying $I = \mathcal{E}/R$ to each resistor,

$$I_1 = \mathcal{E}/R_1, \qquad I_2 = \mathcal{E}/R_2, \qquad \text{and} \qquad I_3 = \mathcal{E}/R_3$$

Hence

$$I = \frac{\mathcal{E}}{R_1} + \frac{\mathcal{E}}{R_2} + \frac{\mathcal{E}}{R_3}$$

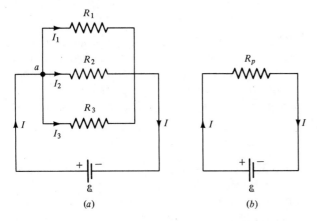

Figure 17.11. (*a*) Three resistors in parallel. (*b*) The equivalent single resistance R_p produces the same current I.

A single equivalent resistance R_p will produce the same current I if

$$I = \frac{\mathcal{E}}{R_p}$$

Comparing, we see that

$$\frac{1}{R_p} = \frac{1}{R_1} + \frac{1}{R_2} + \frac{1}{R_3} + \cdots \qquad (17.13)$$

The following example illustrates how the series and parallel formulas may sometimes be used together in analyzing a complex arrangement of resistors.

Example 17.10

(a) Find the equivalent resistance of the resistors in Fig. 17.12*a*. (b) Find the current in each resistor.

(a) The current I splits at point a into two currents, I_1 and I_2. I_1 goes through each of the 1-ohm resistors, so they are in series. Their equivalent resistance is then

$$R_s = 1 \text{ ohm} + 1 \text{ ohm} + 1 \text{ ohm} = 3 \text{ ohms}$$

This reduces the circuit to two parallel 3-ohm resistors (Fig. 17.12*b*). The effective resistance of the entire system is found from

$$\frac{1}{R_p} = \frac{1}{3 \text{ ohms}} + \frac{1}{3 \text{ ohms}} = \frac{2}{3 \text{ ohms}}$$

Thus

$$R_p = 1.5 \text{ ohms}$$

as shown in Fig. 17.12*c*.

(b) The current I in the battery is found using the equivalent resistance of the system, 1.5 ohms. Thus

$$I = \frac{\mathcal{E}}{R} = \frac{6 \text{ V}}{1.5 \text{ ohms}} = 4 \text{ A}$$

In Fig. 17.12*b*, the voltage drop across each resistor must be the same. Since their resistances are equal, the cur-

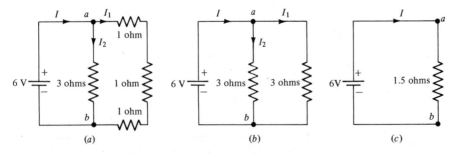

Figure 17.12. (*a*) The complete network. (*b*) The simplified equivalent network obtained using the series resistance formula. (*c*) The final equivalent network obtained with the parallel resistance formula.

rent in each must also be the same. Hence, each current is half the total, or

$$I_1 = I_2 = \tfrac{1}{2}I = \tfrac{1}{2}(4 \text{ A}) = 2 \text{ A}$$

The current in each resistor is 2 A.

Some circuits cannot be analyzed by use of the series and parallel formulas, and Kirchhoff's rules must be applied directly. We show how this is done in Section 17.12.

17.6 | VOLTMETERS AND AMMETERS

The most basic dc measuring instruments are the *voltmeter* and *ammeter,* which measure potential differences and currents, respectively. Each of these contains a *galvanometer,* which consists of a coil of wire suspended near a magnet. The coil is attached to a spring, which opposes rotational motion. When there is a current in the coil, the magnetic force on the moving charges causes the coil to rotate by an amount proportional to the current. Typically, a few milliamperes will cause full-scale deflection, and the resistance of the coil is about 10 to 100 ohms.

Figure 17.13 illustrates how a voltmeter and an ammeter are used. To measure the potential difference across a circuit element, the *voltmeter is connected in parallel* with that element. According to the parallel resistance formula, the voltmeter resistance must be large compared to that of the element to avoid large changes in the current in the circuit. To measure the current in an element, the *ammeter is inserted in series* with the element. Consequently, from the series resistance formula, the ammeter resistance must be small to minimize current changes.

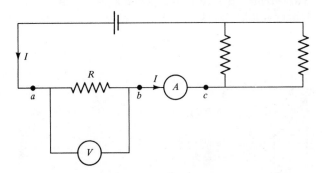

Figure 17.13. The voltmeter (V) is placed in parallel with the resistor R and indicates the potential difference across R. The ammeter (A) is in series with R and gives the current through R.

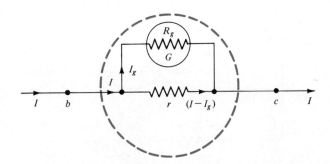

Figure 17.14. An ammeter is a galvanometer coil in parallel with a small resistance r. The two points b and c label points in the circuit of Fig. 17.13. I is the current being measured.

To construct an ammeter, we connect a small resistance r in parallel with a galvanometer coil (Fig. 17.14). If the circuit has a current I, only a small part of the current, I_g, goes through the relatively high resistance galvanometer coil. The remaining current, $I - I_g$, goes through r. Since the potential drop across the galvanometer and the parallel resistor must be the same, it follows that

$$I_g R_g = (I - I_g)r \qquad (17.14)$$

The choice of r is determined by the desired range of the ammeter. This is illustrated in the following example.

Example 17.11

A galvanometer with a resistance of 100 ohms gives a full-scale deflection for a current of 1 mA. (a) How large a parallel resistor is needed to convert it into an ammeter with a 20-A range? (b) What is the ammeter resistance?

(a) Using Eq. 17.14,

$$r = \frac{R_g I_g}{(I - I_g)} = \frac{(100 \text{ ohms})(0.001 \text{ A})}{(20 \text{ A} - 0.001 \text{ A})}$$

Since we are working to three significant figures, the 0.001 A in the denominator can be neglected. Thus

$$r \simeq \frac{(100 \text{ ohms})(0.001 \text{ A})}{20 \text{ A}} = 0.005 \text{ ohm}$$

($\simeq$ means approximately equal.)

(b) Since r is much smaller than R_g, the resistance of the ammeter is very close to r:

$$\frac{1}{R_A} = \frac{1}{r} + \frac{1}{R_g}$$

$$= \frac{1}{0.005 \text{ ohm}} + \frac{1}{100 \text{ ohms}} \simeq \frac{1}{0.005 \text{ ohm}}$$

and $R_A = 0.005$ ohm.

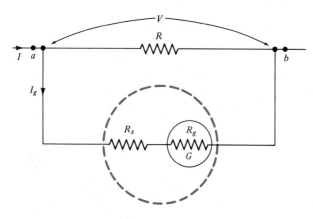

Figure 17.15.   A voltmeter is a galvanometer in series with a large resistance R_s. Points a and b label points in the circuit of Fig. 17.13. V is the potential difference between a and b that is being measured.

To construct a voltmeter, a large resistance R_s is placed in series with the galvanometer coil (Fig. 17.15). The potential difference V across the voltmeter is

$$V = I_g(R_g + R_s)$$

so

$$R_s = \frac{V}{I_g} - R_g \qquad (17.15)$$

Again the desired range of the meter determines R_s, as is seen in the next example.

Example 17.12

   The galvanometer of Example 17.11 has a resistance of 100 ohms and gives a full-scale deflection for a current of 1 mA. (a) How can it be converted into a voltmeter with a 100-V range? (b) What is the voltmeter resistance?

(a) The series resistance needed is

$$R_s = \frac{V}{I_g} - R_g = \frac{100 \text{ V}}{0.001 \text{ A}} - 100 \text{ ohms}$$
$$= (100{,}000 - 100) \text{ ohms} = 99{,}900 \text{ ohms}$$

(b) The total voltmeter resistance is

$$R_V = R_s + R_g = 99{,}900 \text{ ohms} + 100 \text{ ohms}$$
$$= 100{,}000 \text{ ohms}$$

17.7 | CIRCUITS CONTAINING RESISTANCE AND CAPACITANCE

If we connect an uncharged capacitor to a battery, charge moves from one capacitor plate to the other through the battery and the connecting wires. This current stops once the potential difference across the capacitor equals the EMF of the battery. Similarly, if the plates of a charged capacitor are connected by a wire, there will be a current in the wire until the capacitor is fully discharged. These short-lived currents are referred to as *transients*. The constant currents we have been discussing until now are referred to as *steady* currents.

   We first consider the transient currents associated with charging a capacitor. Fig. 17.16*a* shows a circuit containing an initially uncharged capacitance C, a resistance R, and an EMF $\mathcal{E}$. If at time $t = 0$ the switch is closed, the charge q on the capacitor and the current i in the circuit vary with time as in Fig. 17.16*b* and *c*. As we show below, the time required for either q or i to reach a specific fraction of its final value is proportional to the time constant T given by

$$T = RC \qquad (17.16)$$

The greater the resistance or the capacitance, the longer it takes to charge the capacitor. This is physi-

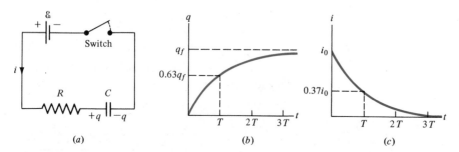

(a)	(b)	(c)

Figure 17.16.   (*a*) A circuit containing a resistance, a capacitance, and an EMF. (*b*) and (*c*) If the capacitor is initially uncharged and the switch is closed at time $t = 0$, the charge q and current i vary with the time t as shown. The time constant T is equal to RC.

cally reasonably, since a larger capacitance requires a larger final charge, and a larger resistance leads to smaller charging currents.

To study this in more detail, we calculate the potential differences across the circuit elements during the charging process. Suppose at time t the charge on the capacitor is q and the current in the circuit is i. Proceeding counterclockwise, there is a potential rise, $\mathcal{E}$, in the EMF, and a drop, iR, in the resistor. For the capacitor, $C = q/V$, so the potential drop is q/C. The sum of these changes is zero. Thus

$$\mathcal{E} - iR - \frac{q}{C} = 0$$

or

$$i = \frac{\mathcal{E}}{R} - \frac{q}{RC} \qquad (17.17)$$

This equation directly gives the current i at $t = 0$ and at $t = \infty$. At $t = 0$, when the switch has just been closed, the charge q on the capacitor is its initial value, zero. Thus the current i_0 at $t = 0$ is

$$i_0 = \frac{\mathcal{E}}{R} \qquad (t = 0) \qquad (17.18)$$

This equals the constant current that would occur if only the resistance and the EMF were connected in series. If we wait a long time, the capacitor will be fully charged, and its potential difference must equal the EMF $\mathcal{E}$. The charge q is then

$$q_f = \mathcal{E}C \qquad (t = \infty) \qquad (17.19)$$

If we substitute q_f for q in Eq. 17.17, we find that the current is zero. As expected, once the capacitor is fully charged, the current ceases.

We now find the charge and current as functions of time. First, using $i = dq/dt$, we rewrite Eq. 17.17 as

$$RC\frac{dq}{dt} - \mathcal{E}C + q = 0 \qquad (17.20)$$

The simplest way to solve this equation is to write down the solution for q and verify that it is correct. Thus we take

$$q(t) = \mathcal{E}C(1 - e^{-t/RC})$$

Here e is the base of natural logarithms, and has the value 2.718. . . (See Appendix B for a discussion of the exponential function.) Note that at $t = 0$, $q = 0$, and that when $t = \infty$, $q = \mathcal{E}C$, as required. Now according to Eq. B.25 from Appendix B,

$(d/dt)e^{-at} = -ae^{-at}$. Thus, differentiating the charge, we obtain the current

$$i = \frac{dq}{dt} = \mathcal{E}C\frac{1}{RC}e^{-t/RC} = \frac{\mathcal{E}}{R}e^{-t/RC}$$

Now if we substitute these expressions for q and dq/dt into the left side of Eq. 17.20, we obtain

$$\mathcal{E}Ce^{-t/RC} - \mathcal{E}C + \mathcal{E}(1 - Ce^{-t/RC})$$

Since these add up to zero, we have found a correct solution.

Using $i_0 = \mathcal{E}/R$, $q_f = \mathcal{E}C$, and $T = RC$, our formulas for q and i can be rewritten as

$$\begin{aligned} q(t) &= q_f(1 - e^{-t/T}) \\ i(t) &= i_0 e^{-t/T} \end{aligned} \qquad (17.21)$$

Note that when $t = T$, $e^{-t/T} = e^{-1} = 1/2.718 \ldots = 0.37. \ldots$ Thus, after one time constant has elapsed, the current has dropped to about 37 percent of its initial value, i_0. Also, since

$$(1 - e^{-1}) = 1 - 0.37\ldots = 0.63\ldots$$

after a time T has elapsed, $q = 0.63q_f$. That means the capacitor charge is equal to approximately 63 percent of its final charge q_f. Alternatively, we can say it is within 37 percent of its final value. When the elapsed time is equal to twice the time constant, $e^{-t/T} = e^{-2} = 0.14. \ldots$ At this time the charge is $(1 - 0.14)q_f = 0.86q_f$, and the current is $0.14i_0$. After a time equal to several time constants, the charge q and current i are very close to their final values, as is seen in Fig. 17.16. The calculation of some of these quantities is illustrated by the following example.

Example 17.13

In Fig. 17.16, $C = 2\,\mu\text{F}$, $R = 1000$ ohms, and $\mathcal{E} = 6$ V. Find (a) the final charge on the capacitor, (b) the initial current, and (c) the time constant.

(a) The final charge is

$$q_f = \mathcal{E}C = (6\text{ V})(2 \times 10^{-6}\text{ F}) = 1.2 \times 10^{-5}\text{ C}$$

(b) The initial current is

$$i_0 = \frac{\mathcal{E}}{R} = \frac{6\text{ V}}{1000\text{ ohms}} = 6 \times 10^{-3}\text{ A}$$

(c) The time constant is

$$T = RC = (1000\text{ ohms})(2 \times 10^{-6}\text{ F}) = 2 \times 10^{-3}\text{ s}$$

We have seen that the time constant $T = RC$ determines the rate at which the charge on a capacitor increases. The time constant also determines the rate at which the charge *decreases*. For example, in Fig. 17.17 a capacitor discharges through a resistance R when the switch is closed. If initially it has a charge q_0, after a time t the charge q is reduced to

$$q = q_0 e^{-t/T} \qquad (17.22)$$

The initial potential difference across the capacitor is $V_0 = q_0/C$, and the initial current is directed away from the positive plate and has a magnitude of $i_0 = V_0/R$. The current after a time t is

$$i = i_0 e^{-t/T} \qquad (17.23)$$

Thus, both the charge on the capacitor and the current in the circuit steadily diminish, reaching very small values after a few time constants. Verifying that these are the correct expressions for q and i is left as a problem.

The artificial pacemaker provides an example of a circuit in which a capacitor is repeatedly charged and discharged.

Example 17.14

Each cycle in the human heart begins with an electrical *pacemaker* pulse from a group of nerve fibers. Some heart patients are now being helped by surgically implanted *artificial pacemakers,* which are battery-powered circuits that pulse if the person's pacemaker fails to do so. One model has pulses triggered 75 times per minute by a 0.4-μF capacitor, which rapidly charges through a very small resistance r and then slowly discharges through a large resistance R (Fig. 17.18). When the charge drops to $e^{-1} = 0.37$ times its initial value, transistors deliver a short pulse to the heart and then recharge the capacitor almost immediately through r. (a) Find the time constant of the discharging RC circuit, neglecting the small time needed to recharge the capacitor through r. (b) Find the resistance R.

(a) There are 75 pulses per minute, $75/60 = 1.25$ per second, or one pulse each $1/1.25 = 0.8$ seconds. This equals the time for the capacitor charge to drop to e^{-1}

times its initial value, since the recharging time is negligible. Such a decrease in the charge requires one time constant to elapse, so $T = 0.8$ seconds.

(b) From $T = RC$, using $C = 0.4\,\mu$F,

$$R = \frac{T}{C} = \frac{0.8\text{ s}}{0.4 \times 10^{-6}\text{ F}} = 2 \times 10^6 \text{ ohms}$$

The artificial pacemaker in this example stimulates the heart at a fixed frequency. More advanced demand pacemakers trigger only if the patient's natural pacemaker fails to operate within a specified time.

17.8 | ELECTRICAL SAFETY

Electrical equipment must be designed and used with care to avoid possible fire and electrocution hazards. In this section, we consider these hazards and the means of reducing them.

Electric power in American homes and offices is usually supplied as 120-V, 60-Hz alternating current. The outlets in each building are connected in parallel to two power lines from a neighborhood substation (Fig. 17.19). One line is connected to the earth, which is a good conductor. This line is said to be *grounded*. The other line is *hot* or *live;* its potential alternates relative to that of the grounded line, which is considered to be at zero potential.

Each circuit in a building is protected by a *fuse* or a *circuit breaker*. If the current exceeds a conservatively chosen maximum, the fuse melts or the breaker trips, opening the circuit. This reduces the danger of a fire starting because of the heat generated by excessive currents. One cause of excess currents is an *overload:* too many electrical devices in operation on one circuit at the same time. Another cause is a *short circuit:* a very low resistance path between the two power lines leading to a very large current. This sometimes occurs when electrical insulation becomes defective as a result of wear or gradual deterioration.

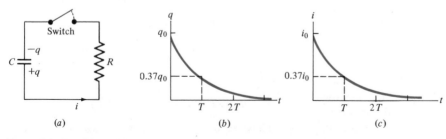

Figure 17.17.   (*a*) An initially charged capacitor discharging through a resistor. (*b*) The charge on the capacitor versus time. (*c*) The current in the circuit versus time.

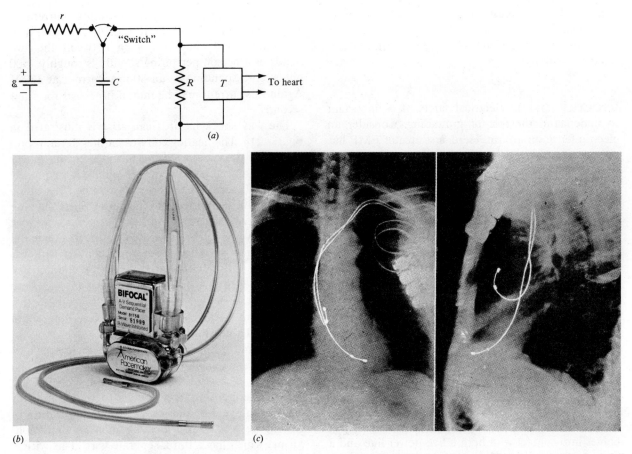

Figure 17.18. (a) The basic pacemaker timing circuit. The capacitor C charges quickly through the small resistance r. The "switch," which is actually a transistor, then changes position, and the capacitor slowly discharges through the large resistance R. When the voltage across R reaches a preset level, the triggering circuit T sends a pulse to the heart. (b) If a patient's heart fails to initiate a pulse, this artificial pacemaker provides both atrial and ventricular stimulation. Its height is 6 cm, its mass is 150 grams, and it contains four lithium-silver chromate cells that may supply power for over four years. (c) The implanted pacemaker and leads. ((b) and (c) courtesy of American Pacemaker Corporation.)

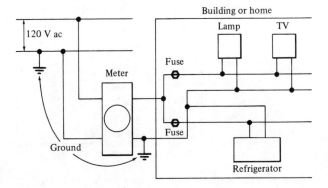

Figure 17.19. Each circuit in every house in a neighborhood is connected in parallel to the power lines. Each circuit contains a fuse or circuit breaker. The meter is normally placed in series with the lines that enter the building.

Electrocution is a serious hazard, since relatively small currents through the human torso can cause injury or death. The average adult can detect a current as small as 1 mA $= 10^{-3}$ A; a direct current produces slight warmth, and a 60-Hz alternating current causes a tingling sensation. Someone reacting involuntarily to a small, unexpected current might drop a hot pan or fall from a ladder, so the maximum allowable leakage currents are set at less than 1 mA. Considerable muscular reaction and pain occur at a few milliamperes. Between 10 and 20 mA will paralyze some muscles and prevent a person from releasing a conductor; about 18 mA contracts the chest muscles and causes breathing to stop. Unconsciousness and death will occur in minutes unless the current stops. One hundred milliamperes for a few seconds causes ventricular fibrillation, rapid uncoordinated movements of the heart muscle that prevent it from pumping blood. Fibrillation rarely stops spontaneously, so death usually results unless prompt medical treatment is available.

Grounding

In electrical safety, it is important to understand the role of grounding. Consider an electric fence on a farm. Here, a source of EMF has either one of its terminals attached to a wire that is totally insulated from electrical contact with anything else. The other terminal is attached to a metal rod driven into the earth and is said to be grounded (Fig. 17.20). If an animal standing on the ground touches the wire, it will complete a conducting path from the EMF through the wire and the ground back to the EMF. Thus, a current will pass through the animal, and it will experience a shock. On the other hand, if a bird sits on the wire, its body is at the potential of the wire. It does not receive a shock, since it does not form part of a closed conducting path.

The point of this example is that electric power lines usually have one wire grounded. If anything is simultaneously in contact with the ground and with a part of the circuit not at ground potential, charge will flow through that object.

Electrical accidents usually occur when a person comes into contact with both a live power line and a good path to ground. The current through the body is determined by its electrical resistance, which is highly variable. Dry skin has a resistance of 10^5 ohms per square centimetre or greater, but the resistance may decrease by a factor of 100 when the skin is wet. A person standing in water in a bathtub is in excellent electrical contact with the ground via the water pipes. If this person grasps a defective electrical appliance, the resistance of the body can be as small as 500 ohms. At 120 V the resulting current is $I = V/R = 120$ V/500 ohms $= 0.240$ A $= 240$ mA, which is lethal. The resistance between the two hands of a person perspiring slightly is roughly 1500 ohms, corresponding to an 80-mA current at 120 V. Again this current may be fatal if it persists for a few seconds.

The way shock hazards can arise is illustrated in Fig. 17.21a. The heating coil in a clothes dryer, represented by the resistor R, is connected to the power line with a common two-prong plug. Consequently the ground and hot lines are determined by which way the plug is inserted into the outlet. Since the plug can usually be inserted in either of two ways, no wire in the dryer may be permitted to touch the metal cabinet. If a wire does come into contact with the cabinet, it will be either at ground or at the hot line voltage, depending on how the plug is inserted into the outlet. A person simultaneously touching a live cabinet and a water faucet or a concrete floor could receive a lethal shock.

To minimize this hazard, electrical devices such as laundry appliances or power tools should be grounded. This is sometimes done by connecting their metal housings or cabinets to water pipes or other grounded conductors. A more reliable way to ground devices is provided by modern household wiring, which has a third, separate wire. This wire is grounded and is connected to the third prong of the electrical outlets. Thus, when an electrical device with a three-prong plug is plugged into such an outlet, its housing is automatically grounded (Fig. 17.21b). When the housing is grounded, accidental contact between the live wire and the housing will result in a short circuit, and the fuse or circuit breaker will open the circuit. Once this happens, there is no danger of a shock.

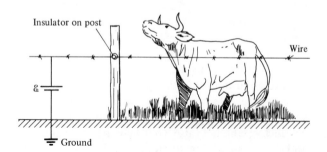

Figure 17.20. A person or animal touching the fence completes a path to ground and receives a mild shock.

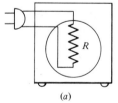 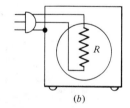

Figure 17.21. An electric clothes dryer. (a) In a two-wire circuit the cabinet is not grounded. A defect in the wiring can therefore make the cabinet "hot." (b) In a three-wire circuit the cabinet is connected directly to ground.

Ground-Fault-Interrupters | If the failure of

some component in an electrical appliance or instrument leads to large enough currents leaking to ground, the fuse or circuit breaker opens the circuit, eliminating any shock hazard. However, if an electrical device fails in such a way that smaller currents leak to ground, then a person touching a component not quite at ground potential may receive a severe shock.

A device called a *ground-fault-interrupter* is now sometimes installed in the ground wire of three-wire circuits. If the current in the ground wire exceeds some very small preset limit, say 5 mA, this device opens the circuit. The majority of the electrocutions that occur in American homes could probably be prevented if ground-fault-interrupters were in general use.

All-Insulated Construction | In recent years,

many power tools and appliances have appeared with *all-insulated* or *double-insulated* construction. They are designed so that, in the event of an insulation failure, no exposed metal part will become electrically live. These devices do not require grounding and represent a useful advance in electrical safety.

Hospitals | The electrical shock hazards in hospi-

tals are generally similar to those in homes and offices, although the use of electrical devices near grounded metal beds and in the presence of water or other liquids does pose potential problems. However, a patient with catheters, probes, and needles inserted in the body is highly vulnerable to very small electrical currents, since 0.02 mA applied directly to the heart is sufficient to produce ventricular fibrillation. Since deaths due to such small currents are indistinguishable from those due to natural causes, there is some dispute over how often they occur. The special hazards associated with patients connected directly to electrical circuits have led to recommendations that their beds be specially insulated so that they are not grounded and that no other objects within reach of the patient be grounded. Then, like the bird on the fence wire, these patients cannot complete a conducting path to ground. Chapter Twenty discusses how devices called transformers can also be used to reduce shock hazards.

SUMMARY

If an amount of charge ΔQ passes through a cross section of a wire in a time Δt, the average current is

$$\bar{I} = \frac{\Delta Q}{\Delta t}$$

The instantaneous current is $I = dq/dt$.

In a metallic conductor, conduction electrons can move readily among the positive lattice ions. When there is an electric field in a wire, the electrons acquire a small net drift velocity v in the direction of the electric force. If there are n conduction electrons per unit volume, and the cross-sectional area of the wire is A, the magnitude of the current is

$$I = enAv$$

The ratio of the potential drop to the current in a wire is its resistance,

$$R = \frac{V}{I}$$

Materials in which the resistance is independent of the current are said to obey Ohm's law. The resistance is determined by the geometry and by the resistivity ρ of the material according to

$$R = \rho \frac{\ell}{A}$$

In an electric circuit a continuous current can exist only if energy is supplied by an EMF. The power supplied by or to any circuit element is VI, where I is the current in the element and V is the potential change across the element. This power is I^2R for a resistor and $I\mathcal{E}$ for an EMF.

Kirchhoff's rules state that the sum of the potential changes around any closed path is zero, and that the current entering any point must equal the current leaving. These rules can be used to analyze any dc circuit, and lead to formulas for series and parallel resistors,

$$R_s = R_1 + R_2 + R_3 + \cdots$$
$$\frac{1}{R_p} = \frac{1}{R_1} + \frac{1}{R_2} + \frac{1}{R_3} + \cdots$$

A galvanometer is a device for measuring currents. A voltmeter is made by adding a large series resistance to a galvanometer; it measures the voltage across a circuit element when it is placed in parallel with that element. An ammeter is constructed by placing a small resistance in parallel with a galvanometer. Placed in series in a circuit, the ammeter measures the current through that part of the circuit.

When the plates of a capacitance C are connected to an EMF through a resistance R, the rate at which its charge changes is determined by the time constant

$$T = RC$$

When a time equal to T has elapsed, an initially uncharged capacitor is within 37 percent of its final charge. Similarly, an initially charged capacitor being discharged through a resistance will lose all but 37 percent of its charge in one time constant.

Checklist

Define or explain:

electric current	series, parallel resistors
ampere	voltmeter
conduction electrons	ammeter
drift velocity	galvanometer
resistance	transient, steady
Ohm's law	currents
resistivity	time constant
conductivity	overload
EMF	short circuit
kilowatt-hour	ground
Kirchhoff's rules	

REVIEW QUESTIONS

Q17-1 If 3 C of charge pass a point in a wire in 10 s, the current is _____.

Q17-2 The resistance of a wire is the ratio of the voltage difference across the wire to the _____.

Q17-3 In an ohmic conductor, the resistance is independent of the _____.

Q17-4 If a material is a good conductor, its resistivity is _____.

Q17-5 A battery converts _____ energy into _____ energy.

Q17-6 Electrical energy is sold by the _____.

Q17-7 The power supplied to or by a circuit element is the product of the _____ and _____.

Q17-8 If two identical resistances R are connected in series, the equivalent resistance is _____; if they are connected in parallel, the equivalent resistance is _____.

Q17-9 A voltmeter is used in _____ with the circuit element whose potential difference is being measured.

Q17-10 An ammeter must have a _____ resistance to avoid disturbing the circuit it is placed in.

Q17-11 When a resistor and capacitor are connected in series to a battery, the current has its greatest value at _____ and drops to $1/e = 0.37 \ldots$ times its peak value after a time _____.

Q17-12 Appliances should be _____ to minimize the risk of _____.

EXERCISES

Resistivities needed in some of the exercises are given in Table 17.1.

Section 17.1 | Electric Current

17-1 A proton accelerator has a beam with a current of 10^{-6} A. How many protons are accelerated per second?

17-2 In an electrochemical experiment, a current of 0.5 A passes through a cell for 1 hour. If two electrons are needed to discharge an ion, how many ions are discharged?

17-3 Silver has 5.8×10^{28} free electrons per cubic metre. If the current in a silver wire is 10 A, and the wire has a radius of 10^{-3} m, what is the drift velocity of the electrons?

17-4 In a hydrogen discharge tube positive hydrogen ions (protons) and negative electrons travel in opposite directions under the influence of an electric field. Find the direction and magnitude in amperes of the current if in 1 second 8×10^{18} electrons and 3×10^{18} protons reach the electrodes.

17-5 Two copper plates are immersed in a copper sulfate solution and connected to a battery. If there is a 0.4-A current for 1 hour, how much copper is deposited on the plates? (The atomic mass of copper is 63.6 u, and two electrons are needed to discharge one copper ion.)

17-6 The conduction electrons in a metal can be considered for some purposes as an ideal gas. (a) What is the average thermal kinetic energy $\frac{3}{2}k_B T$ of an electron at 300 K? (b) What is the corresponding velocity? (c) Calculate the ratio of this velocity to the drift velocity found in Example 17.2.

Section 17.2 | Resistance

17-7 A wire has a 10-V potential difference across its ends when it carries a 4-A current. What is its resistance?

17-8 A 10-A current in a wire results in a potential difference of 2 V across its ends. If it is an ohmic conductor, what current will produce a 6-V potential difference?

17-9 The thinnest copper wire normally manufactured has a radius of 4×10^{-5} m. Find the resistance of a 10-m-long segment.

17-10 A nerve fiber (axon) may be approximated as a long cylinder. If its diameter is 10^{-5} m and its resistivity is 2 ohm m, what is the resistance of a 0.3-m-long fiber?

17-11 A 2-m-long copper wire has a resistance of 0.01 ohms. Find its radius.

17-12 The voltage difference across a resistor is varied, and the current is measured. The results are $\Delta V = 2$ V, $I = 0.4$ A; 4 V, 0.8 A; 6 V, 1.2 A; 8 V, 1.6 A. (a) What resistance is implied by the measurement at 2 V? (b) What resistance is implied by the measurement at 8 V? (c) Is the resistor an ohmic conductor? Explain.

Section 17.3 | Energy Sources in Circuits

17-13 The heating element in a hot water heater has a current of 20 A when connected to a 230-V line. What is its resistance?

17-14 An automobile lamp draws 1.2 A when connected to a 12-V battery. What is its resistance?

17-15 An electric iron draws 6 A when plugged into a 120-V outlet. What is its resistance?

17-16 A 120-V power line contains a fuse that will open when the current exceeds 15 A. What is the minimum resistance of an appliance operated by this circuit?

17-17 A 12-V battery is connected to a 2-ohm resistor. (a) What is the current? (b) How much charge is transported through the circuit in 10 seconds? (c) How much work is done on the charge by electric fields in the battery? (d) How much work is done on the charge by electric fields in the resistor? (e) What is the total work done by the electric fields on the charge? (f) How much energy is converted into heat? (g) What is the source of that energy?

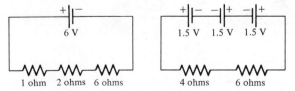

Figure 17.22. Exercises 17-18 and 17-20.

Figure 17.23. Exercise 17-19.

17-18 For the circuit in Fig. 17.22, find (a) the current, and (b) the potential difference across each resistance.

17-19 (a) Find the current in the circuit of Fig. 17.23. (b) Find the potential difference across each circuit element.

Section 17.4 | Power in Electrical Circuits

17-20 Find the power supplied to or by each circuit element in Fig. 17.22.

17-21 (a) A 12-V automobile storage battery has an internal resistance of 0.004 ohms. Find the current and power dissipation when the terminals are connected by a conductor of negligible resistance. (b) The battery delivers 80 A to the starter motor. What resistance would draw the same current? (c) How much power is supplied to the motor? What happens to that power? (d) How much power is dissipated in the battery?

17-22 A toaster uses 1500 W when plugged into a 120-V line. It takes 1 minute to toast a slice of bread. If electrical energy costs 6 cents per kilowatt-hour, how much does it cost to do this?

17-23 (a) What is the resistance of a 100-W light bulb designed for use in 120-V circuits? (b) What current will it draw?

17-24 A 20-A circuit in a home is wired with No. 12 copper wire. A single No. 12 wire 1 m long has a resistance of 5.2×10^{-3} ohms. If a two-wire cable carries 20 A for 30 m, find (a) the potential drop along each wire, and (b) the total power dissipated.

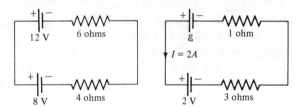

Figure 17.24. Exercise 17-25. **Figure 17.25.** Exercise 17-26.

17-25 For the circuit in Fig. 17.24, find (a) the current, (b) the power supplied by each battery, and (c) the power dissipated by each resistor.

17-26 (a) In Fig. 17.25, what is the EMF $\mathcal{E}$? (b) Find the power supplied to each of the two resistors. (c) Find the power supplied to or by each EMF. Compare the results with those in part (b).

17-27 A room air conditioner uses 900 W of electrical power. If it operates an average of 12 hours

378

CHAPTER 17 DIRECT CURRENTS

per day, find the daily operating cost at 6 cents per kilowatt-hour.

17-28 A heater uses 1400 W of power when connected to a 120-V line. (a) What is its resistance? (b) What current does it draw? (c) If the line voltage is reduced to 112 V, how much power does the heater use? (Assume that the resistance remains constant.)

Section 17.5 | Series and Parallel Resistors; Kirchhoff's Rules

17-29 A string of 25 Christmas tree lights wired in series uses 500 W when connected to a 120-V line. (a) What is the current in the lights? (b) What is the resistance of a single bulb?

17-30 Some strings of Christmas lights are connected in series, and others are connected in parallel. Which is less affected by the burning out of one light bulb? Explain your reasoning.

17-31 An electrically heated home is supplied with 200 A at 230 V. How many 100-W appliances could be operated simultaneously in this home?

17-32 An electric heater uses 1800 W when connected to a 120-V source. (a) What is the resistance of the heater? (b) The heater is plugged into a 10-m-long extension cord made of No. 18 wire, which is intended only for lamps and small wattage appliances. One metre of a single No. 18 wire has a resistance of 0.021 ohms. How much power is dissipated in the cord?

17-33 Six 60-W light bulbs are used in parallel in a lighting fixture connected to a 120-V power line. (a) What is the resistance of one bulb? (b) What is the effective resistance of the six bulbs?

17-34 Two light bulbs use 100 W each when separately connected to a 120-V line. (a) If they are connected in series to this line, how much power will they draw? (Assume that the resistance of a bulb does not change when it is used in this way.) (b) Will the bulbs be brighter or dimmer? Explain.

17-35 Find the equivalent resistance for the network in Fig. 17.26.

17-36 Find the equivalent resistance for the network in Fig. 17.27.

17-37 For the circuit in Fig. 17.28, find (a) the current through the 2-ohm resistance; (b) the voltage difference across the 3-ohm resistance; (c) the current through the 3-ohm resistance.

17-38 Find the current through the battery in Fig. 17.29.

Section 17.6 | Voltmeters and Ammeters

17-39 A galvanometer has a resistance of 10 ohms and gives full-scale deflection with a current of 10^{-3} A. How can one convert it into a voltmeter with a 0.1-V range?

17-40 A galvanometer has an internal resistance of 100 ohms and has full-scale deflection for a current of 10^{-5} A. (a) Design a voltmeter with full-scale deflection for 10 V. (b) Design an ammeter with full-scale deflection for 10 A.

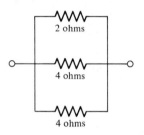

Figure 17.26. Exercise 17-35.

Figure 17.27. Exercise 17-36.

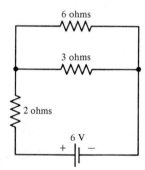

Figure 17.28. Exercise 17-37.

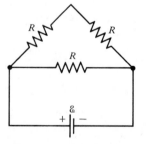

Figure 17.29. Exercise 17-38.

Section 17.7 | Circuits Containing Resistance and Capacitance

17-41 A 1-μF capacitor is connected with copper wires to a 12-V battery. The initial current is 120 A. (a) What is the total resistance of the wires and the battery? (b) What is the time constant for the circuit?

17-42 A 1000-ohm resistor and a 10^{-5}-F capacitor are connected in series to a 100-V EMF. (a) Find the time constant. (b) What is the final charge on the capacitor?

17-43 A 10^{-4}-F capacitor is discharged through a 100-ohm resistor. How long does it take for the charge on the capacitor to drop to $1/e^2$ times its original value?

17-44 Verify that the product RC has the dimensions of a time. (*Hint:* Use the defining relations for R and C.)

17-45 A 500-ohm resistor is connected in series with a 6-V battery and a 10-μF capacitor. (a) What is the initial current? (b) How long does it take for the current to drop to $1/e$ times its initial value?

Section 17.8 | Electrical Safety

17-46 A bird stands on one foot on a high-voltage power line. What happens when it puts the other foot down on the same line? What would happen if the bird put this foot instead on a grounded conductor?

17-47 (a) If a person with wet hands grasps two conductors and has a resistance of 1000 ohms, how large a potential difference is needed to produce a 10 mA current that may freeze the hands to the conductors? (b) How large a potential difference is needed to produce a 100-mA current that will cause ventricular fibrillation in a second or so?

17-48 (a) Is one likely to get a dangerous electrical shock from a 12-V automobile battery? (b) What other possible hazards are associated with such batteries?

17-49 An electric shaver plugged into an electric outlet is accidentally dropped into a sink full of water. Is it safe to reach into the water and remove the shaver? Does it matter whether it is turned on? Explain.

17-50 Why is it especially important to properly ground appliances when used outdoors or in basements?

PROBLEMS

17-51 In a determination of Avogadro's number, a constant 4-A current is passed for 30 minutes through a cell consisting of two silver electrodes in a silver nitrate solution. (a) How many coulombs of charge are transported through the system? (b) How many electrons are transported? (c) If 7.84 g of silver are found to be deposited on the plates, what value of Avogadro's number is obtained? (Each electron discharges one ion, and the atomic mass of silver is 107.9 u.)

17-52 The resistivity of body fluids is about 0.15 ohm m. Estimate the resistance of a finger end to end, ignoring the resistance of the skin.

17-53 (a) What is the room temperature resistance of a 1-m-long aluminum wire of radius 0.002 m? (b) What is the radius of a 1-m-long copper wire with the same resistance? (c) Compare the weights of the two wires. (The density of copper is 8900 kg m^{-3}, and the density of aluminum is 2700 kg m^{-3}.)

17-54 When the terminals of a dry cell are connected by a wire, the current is 2.2 A and the terminal voltage is 1.4 V. When the circuit is opened, the terminal voltage is 1.52 V. Find the internal resistance and the EMF, neglecting any errors due to the meters.

17-55 An electric dryer connected to a 230-V power line draws 20 A. (a) How much power does it use? (b) If it takes 2600 J to evaporate 1 g of water, how long will it take to dry a load of wet laundry containing 4 kg of water? (Assume no heat is lost to the surroundings.)

17-56 A motor is driven by a 12-V battery. This load is equivalent to a 0.2-ohm resistor. (a) What is the current? (b) What is the power supplied to the motor? (c) If the motor operates at 80 percent efficiency, at what rate can it lift a 100-N weight?

17-57 In a student laboratory experiment to measure the mechanical equivalent of heat, a resistor immersed in 0.7 kg of water carries a current of 4.2 A and a potential difference of 12 V. (a) How much energy is supplied in 5 minutes? (b) If no heat is lost to the surroundings, what is the temperature change of the water?

17-58 A calorimeter is used to measure the latent heat of fusion of a material. The calorimeter has a 50-ohm resistor connected to a 120-V dc power line. Once the sample reaches the melting point, it takes 2 minutes to melt 0.3 kg. What is the latent heat of fusion of the material?

17-59 A 12-V storage battery with an internal resistance of 0.003 ohms is charged by a generator at a rate of 20 A. (a) Which terminal of the battery is connected to the high potential side of the generator? (b) How much power is supplied by the generator? (c) What is the terminal voltage of the battery while it is being charged?

17-60 Find the equivalent resistance for the network in Fig. 17.30.

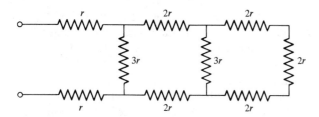

Figure 17.30.   Problem 17-60.

17-61 If an EMF $\mathcal{E}$ is connected to the terminals of the network shown in Fig. 17.31, what is the current in (a) the EMF; (b) the lower $\frac{3}{2}r$ resistor?

17-62 A large supply of 20-ohm resistors is available. What are the simplest ways to combine several such resistors so as to produce an equivalent resistance of (a) 60 ohms; (b) 70 ohms; (c) 75 ohms?

17-63 The capacitor in an electronic flash gun has a capacitance of 100 μF and is charged so that its potential difference is 1000 V. (a) What is the charge on the capacitor plates? (b) The capacitor is discharged through the flashbulb, and 0.001 sec-

onds after the switch is closed, the charge remaining is 0.37 times the initial charge. What is the resistance of the circuit? (c) What is the current after 0.001 seconds?

*17-64 If an uncharged capacitance C is connected in series with a resistance R to an EMF, how long does it take for the charge to reach 99 percent of its final value?

*17-65 A capacitance C and a resistance R are connected in series to a *square-wave generator*. This instrument has a constant terminal potential difference $\mathcal{E}_0$ for a time $T_0/2$ and then quickly reverses polarity so its terminal potential difference is $-\mathcal{E}_0$ for an equal time. (a) Sketch a graph of the terminal potential difference of the square-wave generator versus time for a time interval equal to $3T_0$. (b) Sketch graphs of the potential differences across the capacitor and the resistor if $T_0 \gg RC$. (c) Repeat part (b) assuming $T_0 = RC$.

17-66 Figure 17.32 shows a *Wheatstone bridge*, an instrument used to measure resistance. R_1 and R_2 are 10 ohms, and R_3 is adjusted until the galvanometer current I_g is zero. If R_3 is then 98.2 ohms, find the unknown resistance R_x.

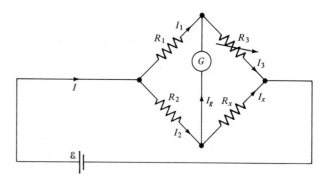

Figure 17.32.   A Wheatstone bridge. Problem 17-66.

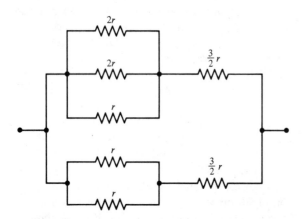

Figure 17.31.   Problem 17-61.

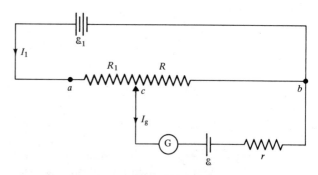

Figure 17.33.   A potentiometer. Problem 17-67.

*17-67 Figure 17.33 shows a *potentiometer,* which is used to measure the EMF of a battery without drawing any current from it. The contact c is moved along the resistive wire ab until the galvanometer current is zero. When this happens, how is the unknown EMF $\mathcal{E}$ related to $\mathcal{E}_1$, R, and R_1?

c17-68 Show that Eqs. 17.22 and 17.23 for q and i satisfy Eq. 17.17 for the case where $E = 0$ and $q = q_0$ at time $t = 0$.

17-69 When a capacitor is discharged, how many time constants are required for its stored energy to decrease to half its initial value?

ANSWERS TO REVIEW QUESTIONS

Q17-1, 0.3 A; **Q17-2,** current; **Q17-3,** current; **Q17-4,** small; **Q17-5,** chemical, electrical; **Q17-6,** kilowatt-hour; **Q17-7,** current, potential difference; **Q17-8,** 2R, $R/2$; **Q17-9,** parallel; **Q17-10,** small; **Q17-11,** $t = 0$, $t = RC$; **Q17-12,** grounded, electrical shock.

SUPPLEMENTARY TOPICS

17.9 | ATOMIC THEORY OF RESISTANCE

We now discuss a simple atomic model that gives some insight into the physical origin of electrical resistance. The model assumes that the conduction electrons in a metal move freely among the positive ions with an average *thermal velocity u* determined by the temperature. Their directions are frequently changed by collisions with the ions. The average distance between collisions is called the *mean free path* λ (lambda) and plays a key role in determining the resistance.

In the absence of an applied electric field, the electrons move randomly, and there is no net charge flow in any direction. When an electric field **E** is maintained in the metal, the negatively charged electrons experience a force opposite to **E**, and they acquire an average drift velocity along the force. Although this drift velocity v is very small compared to the thermal velocity u, it is responsible for the current (Fig. 17.34).

Suppose an electric potential difference V is maintained between the ends of a wire of length ℓ. If the field is uniform, then $E = V/\ell$. The force on an electron has a magnitude $eE = eV/\ell$, and the acceleration is

$$a = \frac{eV}{m\ell}$$

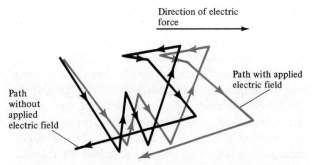

Figure 17.34. A typical random electron path with and without an applied electric field. An electric force produces a small net drift velocity in the direction of the electric force. Note how the field steadily deflects the electron toward the right.

Every time an electron collides with an ion, it is deflected randomly and loses its tendency to drift with the electric force. Its next collision will occur after a time t that satisfies $ut = \lambda$ or $t = \lambda/u$. Between the collisions, it will acquire a velocity at. The average drift velocity v is half this, so

$$v = \frac{1}{2} \frac{eV}{m\ell} \frac{\lambda}{u}$$

If the wire has a cross-sectional area A and there are n conduction electrons per unit volume, then the net current is

$$I = envA = \frac{ne^2\lambda}{2mu} \frac{A}{\ell} V$$

This result states that the current and the potential difference are proportional. Thus the model has led to Ohm's law, which is found experimentally to correctly describe the behavior of most metals. Now using $R = \rho\ell/A$, the current $I = V/R$ is

$$I = \frac{V}{\rho\ell/A} = \frac{VA}{\rho\ell}$$

Comparing the two previous equations for I, we find that in this atomic model the resistivity is given by

$$\rho = \frac{2\,mu}{ne^2\lambda} \qquad (17.24)$$

Only the factors n and λ in this formula for the resistivity depend on the choice of the material, since the electron charge e and mass m are constants, and the thermal speed u is determined by the temperature. For a metal, the number of conduction electrons is typically one or two per atom. Estimating the mean free path λ is a little more difficult. Classical mechanics suggests that λ is comparable to the interatomic spacing, but this value leads to overestimating the room temperature resistivity by a factor of 100 or more. Modern atomic theory leads to the remarkable but experimentally correct result that the electrons travel completely unimpeded in the crystal lattice until they encounter a deviation from the regular lattice structure (Fig. 17.35). Such an imperfection can be an impurity or a lattice ion located some distance from its equilibrium position because of its thermal vibrational motion. At very low temperatures, the amplitude of the lattice ion vibrations is very small, so in very pure metals the mean free path of an electron may be thousands of interatomic spacings.

The resistivities of electrical insulators are larger than those of conductors by a factor of up to 10^{22}. This is a huge number, much larger than the variation of 10^4 between good thermal conductors and thermal insulators. Materials that are good electrical conductors are usually good thermal conductors, because the free electrons transport both charge and thermal energy. Insulators have very few free electrons, so they do not readily conduct either electricity or heat. Some materials called *semiconductors* have resistivities that are intermediate between those of conductors and insulators. They have only a few charge carriers, which can be either electrons or missing electrons (holes). These materials are employed in the manufacture of transistors and other solid-state electronic components.

As the temperature rises, the average electron velocity increases. Also, the mean free path decreases, since the lattice ion vibrational amplitude increases. Thus, Eq. 17.24 predicts that the resistivity will also increase, as is observed experimentally in metals (Fig. 17.36a). However, the resistivity of a semiconductor decreases as the temperature rises (Fig. 17.36b), because the number of charge carriers rises rapidly with the temperature, more than offsetting the changes in the path length λ and velocity u in Eq. 17.24.

At temperatures a few degrees above absolute zero, some materials become superconducting: their resistances apparently drop to zero (Fig. 17.36c). Once currents are started in a superconducting loop, they persist for years; no battery is needed to maintain the currents. Discovered by Heike Kamerlingh Onnes (1853–1926) in 1911, this spectacular effect has been studied extensively by physicists and has a number of useful applications. Powerful magnets built with superconducting coils are used in many laboratories, and superconducting electric power transmission lines with no power loss due to heating may soon be practical. The major difficulty in all applications is that the materials must be kept at very low temperatures, since no material has yet been found that is superconducting much above 15 K.

17.10 | APPLICATIONS OF RESISTANCE MEASUREMENTS

Resistance measurements often provide very useful information. The resistance of some semiconductors varies as much as 5 percent if the temperature changes by 1° C. A small semiconducting bead or *thermistor* can rapidly detect temperature changes as

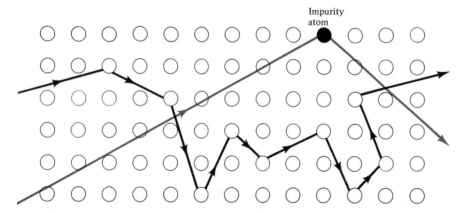

Figure 17.35. Classical mechanics suggests that the mean free path for an electron in a crystal lattice is comparable to the interatomic spacing (black path). According to modern atomic theory, collisions occur only when an electron encounters an imperfection in the lattice, such as the impurity atom in the top row (colored path). The modern theory is in agreement with measured resistivities.

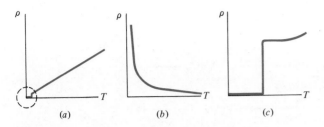

Figure 17.36. Resistivity versus temperature for (a) a conductor; (b) a semiconductor. (c) At low enough temperatures a conductor will become superconducting. This graph is an expanded view of the circled part of (a).

small as 10^{-3} degrees over large temperature ranges (Fig. 17.37).

A *strain gauge* exploits the dependence of the resistance of a wire on its geometry. A small wire grid is bonded to a flexible backing material, which is attached to a flexible diaphragm; the entire unit is covered with rubber. If the diaphragm flexes, the wire grid is distorted, and its resistance changes. This strain gauge and its attached wires can be swallowed by a patient to measure pressures within the digestive tract or inserted with a catheter into veins or arteries to measure blood pressures.

Resistance measurements are useful in chemistry and biochemistry laboratories. The resistivity of a conducting or *electrolytic* solution is measured with a *conductivity cell,* which consists of two metal plates immersed in the solution. The cell is calibrated by

using it to measure the resistance of a solution of known resistivity. For example, strongly acidic or basic solutions have many free ions and therefore have a low resistivity or high conductivity. If one gradually adds measured amounts of a basic solution to an acidic solution of unknown concentration, the conductivity of the sample drops sharply as the point of exact acid-base neutralization is approached, and the number of free ions becomes very small. This procedure is called *titration* and can be used to find the concentration of the acid (Fig. 17.38).

Resistance measurements are also used to study changes in living cells. A decrease in the cell resistance means that ions can more easily pass through the cell membrane. Therefore the changes in the resistance that occur when various materials are added to the cellular environment give clues to their effects on the membrane.

17.11 | ELECTROPHORESIS

Electrophoresis is an efficient technique for separating and analyzing the mixtures of proteins found in human blood and in other biological materials. It is based on the fact that the drift velocities of molecules in an electric field depend on their masses. When a solution is placed in an electric field, large protein molecules with a net charge of a few times the electronic charge but with masses of thousands of atomic mass units experience small accelerations. Thus, they drift much more slowly than do small ions, such as Na^+ or Cl^-.

In an electrophoretic technique commonly used for medical diagnostic studies, one end of a moistened filter paper strip is placed in the protein solution. A potential difference is applied across the ends of the strip, and the protein molecules of various sizes migrate at different rates along the strip. If the process is stopped after a while, the various proteins have traveled different distances and are separated into several

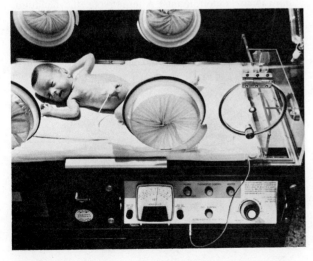

Figure 17.37. The resistance of the thermistor taped to the abdomen of the infant changes rapidly with the temperature. This provides information to an electronic circuit that controls the incubator temperature, compensating for the baby's erratic temperature control system. (Courtesy of Air Shields, Inc.)

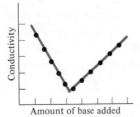

Figure 17.38. The conductivity of a solution is smallest at the point of exact acid-base neutralization. (The conductivity σ is the inverse of the resistivity ρ.)

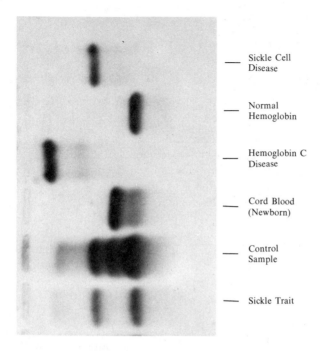

Figure 17.39. Electrophoretic separation of hemoglobin samples can be used to help diagnose many disorders. (Courtesy Helena Laboratories.)

components (Fig. 17.39). Comparison with standard electrophoresis patterns then indicates whether certain abnormalities are present. More elaborate electrophoretic techniques can separate as many as 40 proteins in human blood plasma.

17.12 | KIRCHHOFF'S RULES IN COMPLEX CIRCUITS

We noted in Section 17.5 that Kirchhoff's rules permit us to analyze any dc circuit, including circuits too complex for the parallel and series formulas to apply. The procedure is simple in principle, but one has to be careful to get all the signs right in constructing the equations, and the algebra rapidly gets messy.

Recall the two rules: (1) the sum of the potential drops around any closed path is zero; (2) the current entering any point must equal the current leaving. We apply these in the following example.

Example 17.15

Find the currents in the circuit in Fig. 17.40.

We start by applying the second rule to point d, where I_1 and I_3 enter and I_2 leaves. According to the rule,

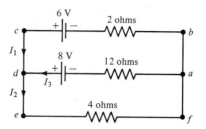

Figure 17.40. Example 17-15.

$$I_1 + I_3 = I_2 \tag{i}$$

We can also apply this rule at point a, but the same equation results with an overall minus sign. *In general, if there are n points where currents branch, the second rule can give us $n - 1$ independent equations.*

We now apply the first rule to the loops $abcda$ and $adefa$. Dropping the units, we obtain

$$-2I_1 + 6 - 8 + 12I_3 = 0 \tag{ii}$$

$$-12I_3 + 8 - 4I_2 = 0 \tag{iii}$$

We can also apply the first rule to the larger loop $abcdefa$; this yields the same equation we get by adding Eqs. ii and iii, so nothing new is obtained. *In general, once the first rule has been applied to each small loop, we have extracted all the information it can provide.*

In order to solve for the three unknown currents, we must have three independent equations; we have now found three such equations. Solving three linear equations for three unknowns is straightforward if you know how to use determinants. Alternatively, the equations can be combined two at a time to eliminate one unknown, reducing the problem to two equations in two unknowns. These are then solved as in Appendix B.4.

The set of equations we have here is relatively simple to solve. From Eq. ii, we find immediately

$$I_1 = 6I_3 - 1 \tag{iv}$$

Also, from Eq. iii, we have

$$I_2 = 2 - 3I_3 \tag{v}$$

If we substitute these expressions for I_1 and I_2 into Eq. i, we get

$$(6I_3 - 1) + I_3 = (2 - 3I_3)$$

Solving for I_3,

$$I_3 = 0.3 \text{ A}$$

Substituting this result into Eqs. iv and v, we obtain

$$I_1 = 0.8 \text{ A}, \qquad I_2 = 1.1 \text{ A}$$

EXERCISES ON SUPPLEMENTARY TOPICS

Section 17.9 | Atomic Theory of Resistance

17-70 At room temperature the average random thermal speed of a conduction electron in copper is about 10^5 m s^{-1}, and there are approximately 10^{29} conduction electrons per cubic metre. (a) Using the measured resistivity, what is the mean free path of an electron in the copper? (b) Find the ratio of this distance to the interatomic spacing, 4.2×10^{-10} m.

Section 17.10 | Applications of Resistance Measurements

17-71 The two parallel electrodes in a conductivity cell each have an area of 1.4×10^{-4} m^2, and they are separated by 4×10^{-2} m. The cell has a resistance of 370 ohms when filled with a particular electrolytic solution. Find the resistivity of the solution.

17-72 A conductivity cell is calibrated by filling it with a solution of known conductivity σ and measuring its resistance R. The *cell constant k* is defined as $R\sigma$. (a) When a cell is filled with an aqueous solution of potassium chloride with conductivity 0.277 ohm^{-1} m^{-1}, the resistance is 190 ohms. What is the cell constant? (b) If the area of each plate is 1.2×10^{-4} m^2, what is their average separation?

17-73 The resistance R and resistivity ρ of a sample in a conductivity cell are related by $\rho = R/k$, where k is the *cell constant*. The cell constant for a particular cell is 42 m^{-1}. (a) When the cell is filled with a potassium sulfate solution, the resistance is 570 ohms. What is the resistivity of the solution? (b) What is the conductivity of the solution?

17-74 A conductivity cell has a resistance of 156 ohms when filled with a potassium chloride solution with conductivity 0.277 ohm^{-1} m^{-1}. Its resistance is 755 ohms when it is filled with a sodium chloride solution. Find the conductivity of the sodium chloride solution.

PROBLEMS ON SUPPLEMENTARY TOPICS

17-75 In Fig. 17.41, find (a) I_2; (b) r; (c) $\mathcal{E}$.

17-76 Find I_1, I_2, and I_3 in Fig. 17.42.

17-77 Two batteries with EMFs of 6 V and internal resistances of 0.5 ohm are connected in parallel to a 10-ohm resistor. How much power is dissipated in the resistor?

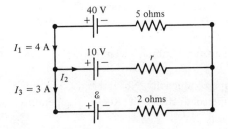

Figure 17.41. Problem 17-75.

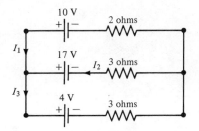

Figure 17.42. Problem 17-76.

17-78 When the direction of a current is guessed incorrectly, its value turns out to be negative; this tells us that the current is in the opposite direction. Repeat Example 17.15 assuming I_3 is in the opposite direction (to the right), and show that the results have the same physical meaning.

Additional Reading

O.H. Smith, An Inexpensive High Resistance Voltmeter, *American Journal of Physics,* vol. 19, 1951, p. 224.

Charles F. Dalziel, Electric Shock Hazard, *IEEE Spectrum,* vol. 7, February 1972, p. 41.

G.D. Friedlander, Electricity in Hospitals, Elimination of Lethal Hazards, *IEEE Spectrum,* vol. 8, September 1971, p. 40.

Brian B. Schwartz and Simon Foner, Large Scale Applications of Superconductivity, *Physics Today,* vol. 30, July 1977, p. 34.

B.N. Turman, Is a Swimmer Safe in a Lightning Storm?, *The Physics Teacher,* May 1980, p. 388.

Scientific American articles:

Giorgio de Sautillana, Alessandro Volta, January 1965, p. 82.

Henry Ehrenreich, The Electrical Properties of Materials, September 1967, p. 194.

Werner Rieder, Circuit Breakers, January 1971, p. 76.

L.O. Barthold and H.G. Pfeiffer, High Voltage Transmission, May 1964, p. 3.

Curtis A. Williams, Jr., Immunoelectrophoresis, March 1960, p. 130.

R.T. Mattias, Superconductivity, November 1957, p. 92.

W.A. Little, Superconductivity at Room Temperature, February 1965, p. 21.

N.B. Brandt and N.I. Ginzberg, Superconductivity at High Pressure, April 1971, p. 83.

T.H. Geballe, New Superconductors, November 1971, p. 22.

Donald P. Snowden, Superconductors for Power Transmission, April 1972, p. 84.

T.H. Geballe and J.K. Hulm, Superconductors in Electric-Power Technology, November 1980, p. 138.

Juri Matisoo, The Superconducting Computer, May 1980, p. 50.

CHAPTER 18

NERVE CONDUCTION

In the preceding chapter, we applied the concepts of electric potential, current, resistance, and capacitance to electric circuits and instruments. These concepts can also be applied to the biological phenomenon of nerve conduction.

Information is transmitted in the human body by electrical pulses in nerve fibers called *axons*. These pulses differ greatly from pulses in copper telephone wires because an axon is a complex structure in which biochemical processes play an important role. An axon has a very high resistance and is poorly insulated from its surroundings, so after a very short distance nerve pulses become very weak and must be amplified. By comparison, telephone messages need to be amplified only after traveling many kilometres. Also, the velocity of a nerve pulse is only about one-millionth that of a pulse in a wire; the latter travels with nearly the speed of light.

Our discussion in this chapter centers on understanding two electrical potentials observed in nerves. In its undisturbed or *resting state*, the interior of an axon is at a lower potential than the surrounding *interstitial fluid*. This is the *resting potential*. When a nerve is suitably stimulated, a current pulse travels along the axon. The associated transient potential change is termed the *action potential*.

18.1 | THE STRUCTURE OF NERVE CELLS

Like other cells, a nerve cell is separated from its surroundings by a membrane that restricts the flow of materials. However, it is atypical in shape (Fig. 18.1). Protuberances called *dendrites,* as well as a long thin structure, the *axon,* are attached to the central core of the cell, the *cell body.* Axons are typically 1 to 20 micrometres in diameter (1 micrometre = 1 μm = 10^{-6} m) and may be quite long. For example, nerves controlling the muscles have their cell bodies in the spinal column. Since some axons reach as far as the foot, a human axon may be 1 m long. The dendrites are typically shorter and thinner, but like the axon, they may have several branches. One nerve cell can influence another at points called *synapses* where dendrites make functional contact.

Wrapped around some axons of higher animals are *Schwann cells,* which form a multilayered *myelin sheath,* reducing the membrane capacitance and increasing its electrical resistance. This sheath allows a nerve pulse to travel further without amplification, reducing the metabolic energy required by the nerve cell. Each Schwann cell is about 1 mm = 10^{-3} m long, but the distance between successive Schwann cells is only about 1 μm = 10^{-6} m. In these short spaces between successive cells, called the *nodes of Ranvier,* the axon is in close contact with the surrounding intersitial fluid. We will see that it is at the nodes that the amplification of nerve pulses occurs in a myelinated nerve. Thus, a myelinated axon resembles an intercontinental submarine cable, with periodic amplifiers used to prevent the signal from becoming too weak. By contrast, signals in unmyelinated axons become weak in a very short distance and require virtually continuous amplification.

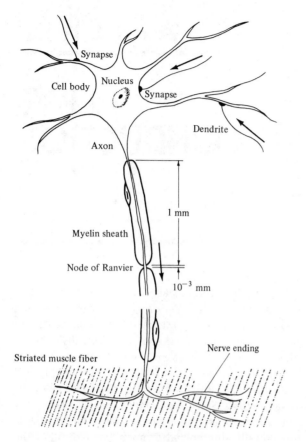

Figure 18.1. A typical nerve. Arrows indicate the direction of nerve impulse conduction.

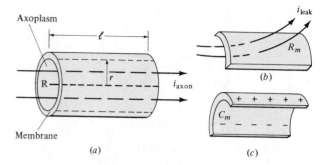

Figure 18.2. (a) The conducting axoplasm has a resistance R to a current i_{axon} along the axon. (b) The resistance of a unit area of membrane to a leakage current i_{leak} is R_m. (c) Charges can accumulate on the two sides of the membrane. The capacitance of a unit area is C_m.

18.2 | THE RESISTANCE AND CAPACITANCE OF AN AXON

We can understand many of the electrical properties of an axon with the aid of a model that resembles an electrical cable covered with defective insulation so that current leaks to the surroundings in many places. More specifically, we assume the axon consists of a cylindrical membrane containing a conducting fluid, the *axoplasm* (Fig. 18.2). The current can travel along the axon in this fluid and can also leak out through the membrane.

The electrical properties of the axon are determined by several quantities. The resistance R of a length of the axon to a current i_{axon} along the axon is proportional to the axoplasm resistivity, ρ_a. The resistance of a unit area of membrane to a leakage current i_{leak} is labeled R_m. The membrane also has capacitance, since charges of opposite signs accumulate

on the two sides of the membrane. The charge per unit area divided by the resulting potential difference is the capacitance of a unit area, C_m.

From Eq. 17.5, the resistance of a wire of length ℓ, cross-sectional area $A = \pi r^2$, and resistivity ρ_a is $R = \rho_a \ell/A$. Using the typical axon parameters given in Table 18.1, a 1-cm $= 10^{-2}$-m length of axon has a resistance

$$R = \frac{\rho_a \ell}{\pi r^2} = \frac{(2 \text{ ohm m})(0.01 \text{ m})}{\pi(5 \times 10^{-6} \text{ m})^2}$$
$$= 2.5 \times 10^8 \text{ ohms} \qquad (18.1)$$

This is a huge resistance! It equals that of 70,000 kilometres of No. 40 copper wire, the thinnest wire normally manufactured, with a diameter of 0.08 mm. Nature has paradoxically devised an efficient communication system using a "wire" that would usually be considered a good insulator.

The membrane is quite thin, so a small section appears nearly flat. This permits us to use the parallel plate capacitance formula, Eq. 16.19, which states that the capacitance is proportional to the plate area, A. A length ℓ of membrane has a surface area $A = 2\pi r\ell$. Since the capacitance per unit area is C_m, the capacitance of the length ℓ of axon is

$$C = C_m(2\pi r\ell) \qquad (18.2)$$

Using the data of Table 18.1, a 1-cm length of an unmyelinated axon of radius 5 μm has a capacitance of 3.1×10^{-9} F. A similar myelinated axon segment has a much larger distance between the interstitial fluid and the axoplasm. Its capacitance is therefore smaller by a factor of about 200, because the capaci-

TABLE 18.1

Values of axon parameters used in examples. Measured values vary somewhat with the type of axon. (The resistivity of the interstitial fluid is small and can be neglected)

Quantity	Myelinated Axon	Unmyelinated Axon
Axoplasm resistivity, ρ_a	2 ohm m	2 ohm m
Capacitance per unit area of membrane, C_m	5×10^{-5} F m^{-2}	10^{-2} F m^{-2}
Resistance of a unit area of membrane, R_m	40 ohm m^2	0.2 ohm m^2
Radius, r	$5\,\mu m = 5 \times 10^{-6}$ m	$5\,\mu m = 5 \times 10^{-6}$ m

tance of any capacitor diminishes as its two conductors are separated.

Since a membrane is not a perfect insulator, charge leaks from the axoplasm through the membrane into the interstitial fluid. The resistance of a conductor is inversely proportional to its cross-sectional area. If the resistance to leakage currents through a unit surface area of membrane is R_m, then a portion of the membrane with surface area A has a resistance $R' = R_m/A$. For a length ℓ of axon, the surface area of the membrane is $2\pi r\ell$, and the membrane leakage resistance is

$$R' = \frac{R_m}{2\pi r\ell} \qquad (18.3)$$

Using the values in Table 18.1, a 1-cm length of unmyelinated axon has a leakage resistance $R' = 6.4 \times 10^5$ ohms, which is less than 1 percent of the axoplasm resistance R. Hence, most of a current entering an axon segment leaks out through the walls in much less than 1 cm.

According to our model, the axoplasm resistance R is proportional to the length ℓ of the axon segment, and the leakage resistance R' is proportional to $1/\ell$. Thus, there is some distance λ (lambda) for which the resistances R and R' are equal. Using Eqs. 18.1 and 18.3 for R and R' respectively, λ must satisfy

$$\frac{\rho_a\lambda}{\pi r^2} = \frac{R_m}{2\pi r\lambda}$$

or

$$\lambda = \sqrt{\frac{R_m r}{2\rho_a}} \qquad (18.4)$$

The distance λ, called the *space parameter,* indicates how far a current travels before most of it has leaked out through the membrane. The values in Table 18.1

give $\lambda = 0.05$ cm for a typical unmyelinated axon, and 0.7 cm for a myelinated axon. Thus, a current pulse can travel much farther without amplification in a myelinated nerve.

18.3 | IONIC CONCENTRATIONS AND THE RESTING POTENTIAL

So far, we have considered the capacitance and the resistances associated with an axon. In this section, we consider some additional information that is needed to understand the behavior of an axon in its undisturbed or resting state: the differences in the ionic concentrations and potentials inside and outside the axon.

Figure 18.3 lists the concentrations of various ions inside, c_i, and outside, c_o, a resting axon. Of the ions that can pass through the membrane, sodium (Na$^+$) and chlorine (Cl$^-$) are much more numerous outside, while potassium (K$^+$) has a larger concentration in-

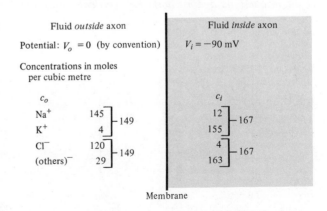

Figure 18.3. Concentrations and potentials inside and outside a typical mammalian axon in the resting state.

side. Conventionally, the electric potential V_o in the fluid outside the cell is taken to be zero. The potential inside the axon is found to be 90 mV lower, so $V_i = -90$ mV. A number of important consequences follow from these observations, as we see shortly.

According to Fig. 18.3, there are equal numbers of positive and negative ions inside and outside the cell. However, there is a potential difference across the membrane, so there must be small net charges $\pm Q$ on either side of the membrane. These can be calculated from $Q = CV$, using the capacitance found earlier and the measured resting potential. The excess of negative ions over positive ions inside the cell turns out to be only about 1/100,000 times the number of negative ions in the cell. An equal number of excess positive ions is present in the interstitial fluid. The excess ions form thin charge layers on either side of the membrane.

Let us now consider the effects arising from the fact that the concentration of Na$^+$ is much higher outside the cell than inside. We saw in Chapter Ten that because of their random thermal motion, particles tend to diffuse from regions where their concentrations are high to regions where they are lower. The permeability of a cell membrane is a measure of the ease with which a given molecule or ion can pass through the membrane. Sodium ions will diffuse *into* the cell at a rate proportional to the Na$^+$ concentration difference $(c_o - c_i)$ and to the permeability of the membrane to Na$^+$ (Fig. 18.4). The resting potential is negative, so the electric field is directed into the cell, and it drives additional positive Na$^+$ ions through the membrane *into* the cell. Since the Na$^+$ concentration remains much larger outside, *sodium must be continually brought back out against both the electrical forces and the effects of the concentration differences.* This return process requires an ongoing expenditure of metabolic energy.

The discussion of the Cl$^-$ and K$^+$ flows is more complicated. There are more Cl$^-$ ions outside, so diffusion produces a net flow into the cell. However, the electrical force on negative ions is opposite to the field, so the electric field draws some Cl$^-$ ions outward. Conversely, K$^+$ ions diffuse outward because of the concentration difference, but they also flow inward under electrical forces. Thus Cl$^-$ and K$^+$ each have flows in both directions across the membrane (Fig. 18.4). In a resting axon, the effect of the K$^+$ concentration difference exceeds that of the potential difference, and there is a net outward movement of K$^+$ ions. As in the case of sodium, there must be a mechanism that returns potassium into the cell and maintains the imbalanced concentrations. There is no net movement of Cl$^-$ ions, since the effect of the Cl$^-$ concentration difference is exactly balanced by the effect of the resting potential difference.

Nernst Equation

We can determine whether an ion is in equilibrium by calculating a theoretical resting potential for which there would be no net flow of that ion across the cell membrane. At this *equilibrium potential difference* across the cell membrane, the flows due to the concentration and potential differences are exactly balanced.

The equilibrium potential difference for an ion can be found from the *Nernst equation*. Although its derivation involves mathematics beyond the level of this textbook, it is based on a model that treats ions in a dilute solution as an ideal gas. A potential difference across a membrane will produce a concentration im-

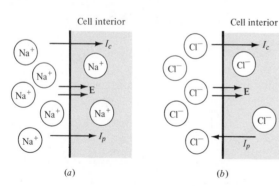
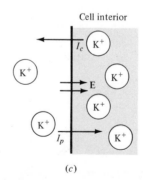

(a)   (b)   (c)

Figure 18.4. Ionic currents I_c due to concentration differences and I_p due to potential differences. (a) Both Na$^+$ flows are into the cell. (b) The flows exactly balance for Cl$^-$. (c) For K$^+$, I_c is slightly larger than I_p, and there is a slight outward flow.

balance on the two sides of the membrane. Equilibrium occurs when the potential energy of one ion with charge q, $q(V_i - V_o)$, is equal to the work necessary to transfer it to the region of higher concentration. According to the model, this work is $k_B T \ln(c_o/c_i)$, where k_B is the Boltzmann constant, T is the Kelvin temperature, and $\ln(c_o/c_i)$ is the natural logarithm (to the base e) of the concentration ratio. (Logarithms are reviewed in Appendix B.10.) Hence the ion is in equilibrium if the concentrations satisfy the Nernst equation,

$$q(V_i - V_o) = k_B T \ln \frac{c_o}{c_i} \qquad (18.5)$$

Example 18.1 illustrates the use of the Nernst equation by considering the K^+ concentrations.

Example 18.1

Compare the equilibrium potential difference for K^+ with the observed resting potential, -90 mV. (Use the data in Fig. 18.3 and assume a temperature of $37°$ C $= 310$ K.)

The charge e on a K^+ ion is $q = e = 1.60 \times 10^{-19}$ C, and $k_B = 1.38 \times 10^{-23}$ J K^{-1}. From Fig. 18.3, $\ln c_o/c_i = \ln 4/155 = -\ln 155/4$. Hence the Nernst equation gives the equilibrium potential difference.

$$
\begin{aligned}
V_i - V_o &= \frac{k_B T}{q} \ln \frac{c_o}{c_i} \\
&= \frac{(1.38 \times 10^{-23} \text{ J K}^{-1})(310 \text{ K})}{(1.60 \times 10^{-19} \text{ C})} \left(-\ln \frac{155}{4} \right) \\
&= -98 \text{ mV}
\end{aligned}
$$

This is slightly larger in magnitude than the resting potential, -90 mV. Therefore, in the resting axon the inward flow due to the potential difference is not quite as large as the outward flow due to the concentration difference. If $(V_i - V_o)$ were -98 mV, the two flows would balance exactly.

If we apply the Nernst equation to Cl^-, we find that the equilibrium potential difference is -90 mV. This is equal to the resting potential, so the Cl^- flows are balanced as noted before. The equilibrium potential difference for Na^+ is $+66$ mV, which is opposite in sign to the resting potential. In other words, if enough positive charge entered the axon to reverse its polarity and change the axon potential to $+66$ mV, the outward flow of Na^+ due to the potential difference would balance the inward flow due to the concentration difference. One way to visualize this situation is to think of the concentration difference as producing the same inward flow as a 66-mV battery. When the

charge inside the axon builds up enough so that it produces a 66-mV axon potential, there will be no net flow of Na^+ ions.

The Sodium-Potassium Pump | The net flows of Na^+ into the cell and of K^+ out of the cell due to diffusion and to the electrical forces are referred to as *passive flows*, because energy does not have to be supplied for them to occur. Some as yet undetermined process returns the Na^+ and K^+ across the membrane and maintains the nonequilibrium ionic concentrations. This energy expending process is called the *active Na-K transport* or the *Na-K pump*.

In the resting state, the axon membrane is about 100 times more permeable to K^+ than to Na^+. This means that equal concentration or potential differences would produce much larger flows of potassium ions than of sodium ions. However, the net passive sodium and potassium flows are about equal, because the resting potential is much closer to the equilibrium potential for K^+ than for Na^+. It is thought that the pump transports one K^+ ion into the cell for each Na^+ ion it removes.

The Na-K pump is responsible for the establishment and maintenance of the resting potential and the concentration imbalances. We can see this explicitly if we suppose that initially the potentials V_i inside the cell and V_o outside are both zero and that the concentrations of Na^+, K^+, and Cl^- are the same on both sides of the membrane. In a short time the pump transports a few K^+ ions into the cell and as many Na^+ ions out of the cell. Since the permeability of the membrane to K^+ ions is relatively large, some K^+ ions diffuse back out of the cell. A smaller number of Na^+ ions diffuse into the cell, since the membrane is less permeable to Na^+ ions. Thus a net positive charge has developed outside the cell, even though the pump has not transferred any net charge across the membrane. This positive charge attracts negative ions from inside the cell. The only negative ion inside the cell that can readily pass through the membrane is Cl^-, so some Cl^- ions move out of the cell, reducing but not quite eliminating the excess of positive ions outside the cell. Since the total system is electrically neutral, there is then an equal slight excess of negative ions inside the cell. The excess charges on either side of the membrane produce a potential difference across it, and now V_i is negative if V_o is defined to be zero.

As the pumping continues, the outside Na^+ and

Cl⁻ concentrations, the inside K⁺ concentration, and the magnitude of V_i all gradually increase. Eventually, equilibrium is reached and the concentrations stop changing when the passive Na⁺ and K⁺ flows due to diffusion and to the electrical forces exactly balance the active transport due to the pump. Because the membrane is much more permeable to K⁺ ions than to Na⁺ ions, the resting potential $V_i = -90$ mV, at which equilibrium occurs, is much closer to the -98-mV K⁺ equilibrium than to the $+66$-mV Na⁺ equilibrium potential.

It is important to note that if the membrane were suddenly to become much more permeable to Na⁺, the balance would be upset briefly by an increased flow of Na⁺ ions into the axon. A small increase in the number of sodium ions entering the cell would substantially modify V_i without significantly altering the ionic concentrations. We see in Section 18.5 that such a change of permeability is believed to occur when the nerve is stimulated and an action potential is produced.

In summary the observed resting state potential and ionic concentrations imply a passive flow of Na⁺ ions into the axon and of K⁺ ions out of the axon due to diffusion and electrical forces. The continued existence of these imbalances is due to the Na-K pump, which actively transports Na⁺ out of the cell and K⁺ into the cell, expending metabolic energy in the process. The magnitude and sign of the potential of the cell, in turn, is largely determined by the ratio of the permeabilities of the membrane for K⁺ and Na⁺.

18.4 | THE RESPONSE TO WEAK STIMULI

Having considered the resting state of the axon, we now examine the response of an axon to a weak stimulus. In most experiments, the stimuli are electrical, since these are easily controlled and do not injure the cell if they are sufficiently mild. For an electrical stimulus smaller than a critical threshold value, the response of the axon is similar to that of an analog network of resistors and capacitors. Specifically, if a weak stimulus is applied at some point on an axon, no significant axon potential changes occur beyond a few millimetres from that point. By contrast, a stimulus *above* the threshold level produces a current pulse that travels the length of the axon without attenuation. This current pulse and the associated action potential are discussed in the next section.

We can develop the analog circuit for the axon by dividing the axon into many short segments. The interstitial fluid surrounding the axon has very little resistance and may be represented by a perfect conductor. Each axon segment has a resistance R to a current i_{axon} along its length. The membrane has a resistance R' to a leakage current i_{leak} plus a capacitance C (Fig. 18.5a,b). A series of several segments is then analo-

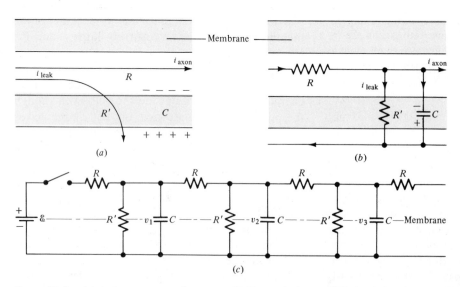

Figure 18.5. (*a*) A short segment of an axon. (*b*) The equivalent analog circuit for the segment. (*c*) The analog circuit for several axon segments. The EMF $\mathcal{E}$ represents the stimulus.

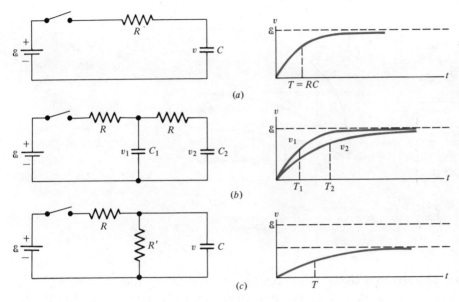

(a)

(b)

(c)

Figure 18.6. Circuit diagrams and capacitor potential differences versus time. The capacitors are initially uncharged, and the switch is closed at $t = 0$. (a) Simplest RC circuit. (b) v_2 rises more slowly than v_1, but ultimately reaches the same final value, the battery EMF $\mathcal{E}$. (c) If a leakage resistor is placed in parallel with the capacitor, the final potential difference v is less than $\mathcal{E}$.

gous to the complex network of resistors and capacitors in Fig. 18.5c. The EMF shown there represents an applied stimulus.

The behavior of this complex axon analog circuit is more readily understood if we first consider the simplest RC network in Fig. 18.6a. Suppose that the initial charge q on C is zero, and the switch is closed at $t = 0$. The charge q and the potential difference $v = q/C$ will gradually increase. As we saw in Chapter Seventeen, the time needed to reach a characteristic fraction of the final values of q and v is determined by the time constant, $T = RC$.

When there are two resistors and two capacitors, as in Fig. 18.6b, the charging process is more complicated. The potential difference v_2 across C_2 must increase more slowly than the potential difference v_1 across C_1, since the path from the battery to C_2 and back has resistance $2R$. As more and more RC pairs are added, the potential difference across each added capacitor rises ever more slowly. Consequently, in the axon analog circuit (Fig. 18.5c), v_2 will change more slowly than v_1, v_3 still more slowly, and so on.

The effect of a leakage resistor can be seen in Fig. 18.6c. There is always some current in the conducting path through R and R'. Thus, there is a corresponding potential drop across R, and the final potential differ-

ence across the capacitor is less than the EMF. In the axon analog circuit, the final potential differences steadily diminish as we move to the right because of the current lost through the leakage resistors R'.

To summarize, when the switch is closed or a "stimulus" is applied in the axon analog circuit, the potential differences across the capacitors gradually change. As one goes further from the stimulus, the changes occur more slowly, and their final magnitudes diminish (Fig. 18.7).

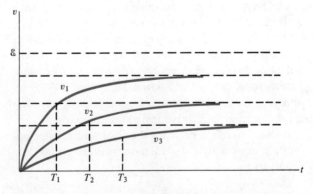

Figure 18.7. The capacitor potential differences for the axon analog circuit of Fig. 18.5c, assuming the switch is closed at $t = 0$. Each potential difference rises more slowly than the preceding one and approaches a successively smaller final value.

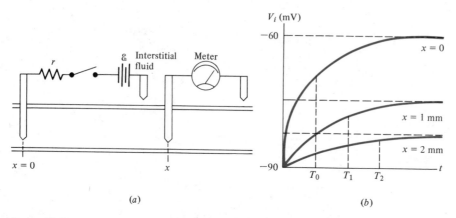

Figure 18.8. (a) Apparatus for exchanging the axon potential at $x = 0$ and observing the resulting change at position x. The "meter" is usually a cathode-ray oscilloscope. (b) The axon potential versus time at $x = 0$, 1, and 2 mm. Note the similarities with Fig. 18.7.

Very similar behavior is observed when an unmyelinated axon is stimulated weakly, as in Fig. 18.8. A probe connected to a battery is inserted at $x = 0$, and gradually the axon potential V_i at that point changes from -90 to -60 mV. The time required for this change to occur is determined by the membrane capacitance and the external series resistance r. At other values of x, the potentials change more slowly, reaching a final potential between -90 and -60 mV. As in the axon analog circuit, the time needed to change the potential appreciably increases with the distance x from the stimulus, reflecting the time needed to alter the charges on the membrane. The final magnitude of the potential changes diminishes as x increases, because of the leakage of current through the membrane. Thus the effects of a weak stimulus propagate rather slowly and become negligible after a few millimetres.

We have used the axon analog circuit to make qualitative predictions of the response of an axon to a weak stimulus. The analog circuit can also yield quantitative predictions relating the final axon potential at a distance x from the stimulus to the space parameter λ defined in Section 18.2. This is done by applying Ohm's law to the currents leaking through the membrane and traveling along the axon.

If the difference between the final and resting potentials at $x = 0$ is V_d, then the difference at x is found to be

$$V(x) = V_d e^{-x/\lambda} \qquad (18.6)$$

If λ is 0.05 cm, a typical value for an unmyelinated axon, then $V(x)$ diminishes to $V_d e^{-1} = 0.37 V_d$ after

0.05 cm, to $v_d e^{-2} = 0.135\, V_d$ after 0.1 cm, and so on. This predicted dependence on x agrees with the experimental data.

18.5 | THE ACTION POTENTIAL

We have seen that when a weak electrical stimulus is applied to an axon, the potential changes are proportional to the stimulus. The situation is very different if a battery such as that shown in Fig. 18.8 briefly increases the potential at $x = 0$ to a value just above the action potential threshold, which is typically at -50 mV. Shortly after this stimulus is applied, the axon potential at x suddenly increases and becomes positive, reaching a value as high as $+50$ mV for some axons (Fig. 18.9a). The potential then gradually returns to its resting value. For a particular axon, the shape and peak size of the action potential curve are *independent of the strength of the initial above-threshold stimulus* or the distance x from the stimulus except very near $x = 0$. Thus the action potential is not proportional to the stimulus. Instead, it is a transient *all-or-nothing* response.

In an unmyelinated axon, the action potential is accompanied by dramatic changes in the permeability of the membrane to Na$^+$ and K$^+$ (Fig. 18.9b). When the axon potential V_i in an unmyelinated axon rises above the action potential threshold at some point, the Na$^+$ permeability there suddenly increases by a factor of more than 1000. This causes a rapid influx of positive sodium ions, which changes the sign of V_i from negative to positive. After about 0.3 ms, the potential approaches the Na$^+$ equilibrium poten-

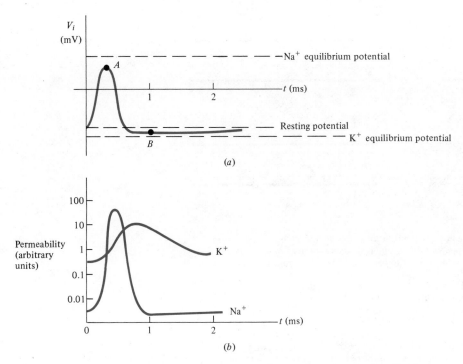

Figure 18.9. (a) The axon potential at a point some distance from the stimulus. (b) The associated changes in the membrane permeabilities. Note that the vertical scale is logarithmic.

tial (point A in Fig. 18.9a) as determined from the Nernst equation, and the sodium influx diminishes. Also the sodium permeability begins to decrease toward its normal low level. Meanwhile the potassium permeability has gradually risen by about a factor of 30. Consequently, potassium ions now begin to flow rapidly out of the cell, and V_i again becomes negative. V_i actually reaches a value below the resting potential (point B in Fig. 18.9a), close to the K^+ equilibrium potential which, as we saw before, is slightly more negative than the resting potential. This return over about 1 ms to a potential close to the resting potential is due to the changes in the K^+ permeability and *not* to the effects of the Na-K pump, which acts much more slowly. The pump does gradually reestablish the resting Na^+ and K^+ concentrations that were altered slightly during the action potential pulse; this process takes about 50 ms.

We now see how this mechanism amplifies a pulse and permits an action potential to travel the length of an axon without attenuation. Figure 18.10 shows an unmyelinated axon segment that has been excited at one end, so that V_i is positive at this point. Positive ions move toward this end on the outside of the mem-

brane and away from it inside. This decreases the charge on the adjacent portion of the membrane, so that the axoplasm potential there becomes less negative and rises to the action potential threshold. This triggers an increased sodium permeability, leading to a sodium influx and an action potential at the adjacent portion of the membrane. In this way the action potential proceeds from point to point along the entire length of the axon.

The term "amplification" is applied to the process that occurs each time an action potential is generated at some point along an axon, because it is here that energy is expended. The Na-K pump is continually maintaining both the resting potential and the ionic imbalances of Na^+ and K^+ ions across the membrane. In so doing, considerable electric potential energy is stored in the membrane, much like water is stored above a dam. An action potential occurs because of the increases that occur in the permeability of the membrane to Na^+ and K^+ ions. The increases of permeability are analogous to opening a floodgate in the dam. There is a sudden flow of ions through the cell membrane due to the large concentration differences, and it is this flow of ions that provides the current of

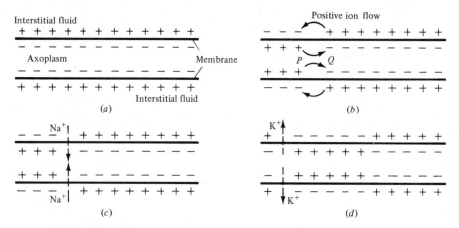

Figure 18.10. Propagation of an action potential in an *unmyelinated* axon. (*a*) The resting state. (*b*) An action potential pulse has traveled through the axon segment to point *P*. Positive ions move away from *P* toward the right inside the axon and toward *P* from the right outside. The charges on either side of the adjacent portion of the membrane at *Q* gradually diminish in magnitude, and the potential there increases toward the action potential threshold. (*c*) Once that threshold is reached, the membrane allows an influx of sodium ions (dotted arrows). The axon potential at this point now rises rapidly and becomes positive. (*d*) The action potential pulse has traveled further along the axon, and the left end of the segment has returned to a negative potential close to the resting potential. This quick return to a negative axon potential is due to an outward flow of potassium ions (dotted arrows) caused by an increase in the K^+ permeability. The resting state potential and ionic concentrations are restored over a much longer interval by the active Na-K pump.

an action potential. The Na-K pump then acts analogously to a pump at the dam, which redeposits the water above the dam for the next flow. Like the pump at the dam, the Na-K pump requires much more time to restore the ions than elapses during an action potential.

Myelinated Axon | In a myelinated nerve, relatively few ions pass through the myelin sheath except at the nodes of Ranvier, which are about 1 mm apart (Fig. 18.11). At the nodes the membrane responds to an above-threshold stimulus much as in an unmyelinated axon: the Na^+ permeability increases rapidly, producing an influx of Na^+ and the charac-

teristic action potential pulse. This, in turn, produces flows of positive ions away from the node inside the axon and toward the node outside. Some of the axon current leaks through the membrane, but most of it reaches the next node, since the 1-mm internodal distance is small compared to the space parameter λ, which is equal to several millimetres in a typical myelinated axon. The axon current reduces the charges on the membrane and increases the potential at the next node to the action potential threshold level. Consequently, this next node is triggered, and the action potential travels along the axon.

Because the amplification and the associated ionic transfers occur only at the nodes, less metabolic energy is needed to restore a myelinated axon to its resting state after an action potential pulse than is required for an unmyelinated axon. The propagation of an action potential is also faster in a myelinated axon. Thus, such axons are better suited to transmit rapidly the large amounts of information required by more advanced animals and represent an important evolutionary step.

The velocity of propagation of an action potential between two nodes in a myelinated axon can be predicted with the simple axon model because amplifica-

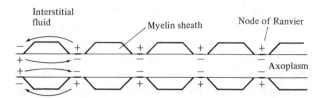

Figure 18.11. Propagation of an action potential in a *myelinated* axon. An above-threshold axon potential at one node triggers an influx of sodium and an action potential pulse, which in turn triggers the next node.

tion occurs only at the nodes. Although a rigorous derivation involves advanced mathematics, we can quickly obtain the correct formula for the velocity using plausible physical arguments. This formula is interesting because it indicates some compromises that have been made in the process of evolution.

The action potential velocity v is the distance X between two nodes divided by the time T needed to reduce the charge on the membrane and increase the potential at the second node to the threshold level. This time must be comparable to the time constant RC of the series circuit containing the axoplasm resistance R and membrane capacitance C. R is the resistance from the first node to $X/2$, or to the midpoint of the "capacitor" being charged. Using $\ell = X/2$ in Eq. 18.1 for R and $\ell = X$ in Eq. 18.2 for C gives, for an axon of radius r,

$$T = RC = \left(\frac{\rho_a X}{2\pi r^2}\right)(2\pi r X C_m) = \rho_a C_m \frac{X^2}{r}$$

and

$$v = \frac{X}{T} = \frac{r}{\rho_a C_m X} = 10r \text{ m s}^{-1} \mu\text{m}^{-1} \quad (18.7)$$

This last expression is obtained using $X = 1$ mm and the values of ρ_a and C_m in Fig. 18.3. Measured values of v in metres per second are usually between $12r$ and $17r$, so the simple axon model gives reasonable results.

Equation 18.7 shows that nerve pulses travel rather slowly. For example, if $r = 5$ μm, v is 50 m s^{-1}, and it takes 0.08 s to send messages from a toe stepping on a sharp object to the brain and back again. Increasing the radius of an axon increases the velocity, but it also increases the metabolic energy requirements, so that a compromise must be made. Clearly, long axons that transmit information requiring rapid responses must have large radii, but other axons need not. Accordingly, many, but by no means all, long axons have large radii. For example, about 60 percent of the axons going to muscles are 6 to 10.5 μm in radius. By contrast, in parts of the brain where the nerves are very short, 90 percent of the axons are 2 μm or less in radius.

A subtle process is also involved in choosing the optimum internode spacing, since the velocity varies as $1/X$. If X is increased, the velocity decreases, which is undesirable, but the metabolic energy requirements are also reduced. Thus again, a compromise between high velocity and low energy usage must be made. In nature, X varies over the relatively small range of 1 to 2 mm, suggesting that deviations from this range render the axon less useful in carrying out its function.

18.6 | ELECTROENCEPHALOGRAPH AND ELECTROCARDIOGRAPH

Direct observations of nerve pulses require inserting a probe into a nerve and are impractical for routine medical diagnostic studies. Fortunately, electrodes placed on the skin can pick up small signals related to large-scale electrical activity within the body. This makes possible the *electroencephalograph* (EEG) and *electrocardiograph* (EKG), instruments that are useful aids in studying brain and heart disorders, respectively.

Electrical signals arising within the body reach the surface because the resistance of the interstitial fluid is not quite zero. If many axons are simultaneously active, then the potential of the body nearby changes relative to the potential elsewhere in the body. Although the changes observed at the surface are at most 50 μV or about 0.1 percent of the full action potential changes, they can be amplified and measured or recorded.

An *electroencephalograph* is shown in Fig. 18.12, along with EEGs typical of various states of awareness. Electrodes are placed at standardized locations on the head, and the signals are recorded. Deviations from the norm can aid in the diagnosis of epilepsy, tumors, accidental brain damage, and other disorders.

An *electrocardiograph* records electrical activity associated with the heart. The usual positions for the electrodes are shown in Fig. 18.13 along with typical EKG patterns.

SUMMARY

An axon resembles a poorly insulated electrical cable. Many of its electrical properties can be understood using this model or an equivalent RC network.

A resting axon has a potential below that of the surrounding interstitial fluid. The resting potential and nonequilibrium concentrations of Na$^+$ and K$^+$ ions are maintained by the active Na-K pump mechanism, which offsets passive ionic flows due to diffusion and to electric forces.

Weak electrical stimuli produce proportional axon responses that are similar to those of the analog RC

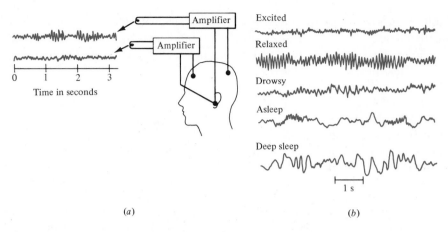

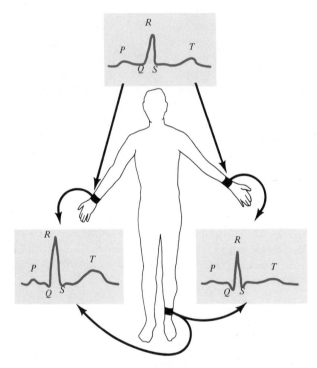

Figure 18.12. Electroencephalograph. (*a*) An electrode attached to the ear provides the reference potential. The two recordings are typical of the response from different areas. (*b*) EEG patterns typical of different states of awareness of a patient.

Figure 18.13. The standard limb leads of the electrocardiogram and normal patterns recorded from these leads. The *P* peak corresponds to the initial electrical pulse that triggers a heart cycle. The interval from *P* to *Q* lasts about 0.12 to 0.20 seconds in a normal person and corresponds to the spread of the pulse over the atria (the upper heart chambers.) The *QRS* interval lasts about 0.06 to 0.10 seconds and corresponds to the spread of the pulse over the ventricles (lower chambers). The *T* peak is associated with the return of the heart to its normal unstimulated state after a pulse.

circuit and diminish rapidly with distance. A stimulus that raises the axon potential above a critical threshold triggers a current pulse and an associated action potential. These travel without attenuation along the entire length of the axon, amplified by processes associated with changes in the ionic permeabilities of the membrane.

The velocity of nerve pulses in a myelinated axon is proportional to the axon radius and inversely proportional to the distance between the nodes of Ranvier where amplification occurs. Thus there is a delicate evolutionary compromise between speed and economy of energy usage.

Checklist
Define or explain:

axon	Nernst equation
myelin sheath	Na-K pump
nodes of Ranvier	axon analog circuit
leakage resistance	action potential
space parameter	amplification
equilibrium potential	EEG
difference	EKG

REVIEW QUESTIONS

Q18-1 Compared to a telephone cable, an axon has much _____ resistance, is much more _____ insulated, and needs much _____ frequent amplification.

Q18-2 A myelinated axon is surrounded by _____.

Q18-3   The distance a current can travel without amplification is characterized by the _____.

Q18-4   In the resting state of an axon, there are far more _____ ions on the outside of the membrane than inside, and far more _____ ions on the inside than outside.

Q18-5   The Nernst equation relates the potential difference across a membrane to the equilibrium ratio of _____.

Q18-6   When an axon is subjected to a weak stimulus, the potential differences change at a _____ rate and with a _____ final magnitude as one goes further from the stimulus.

Q18-7   The action potential involves a sudden increase in the membrane permeability to _____ ions.

Q18-8   In a myelinated axon, amplification occurs at the _____.

Q18-9   If the radius of a myelinated axon is doubled, the velocity of the action potential changes by a factor of _____.

Q18-10   Axons with larger radii require more _____ than those with smaller radii.

EXERCISES

When numerical values are required that are not given in the exercise, the data in Table 18.1 and Fig. 18.3 should be used.

Section 18.1 | The Structure of Nerve Cells

18-1   Estimate the number of axons in the human spinal column if the average axon is 10 μm in diameter and the spinal column is 1 cm in diameter.

18-2   An axon 1 m long has nodes of Ranvier every 10^{-3} m. How many times is a nerve pulse amplified when it is transmitted along this axon?

Section 18.2 | The Resistance and Capacitance of an Axon

18-3   Calculate the axoplasm resistance of a 1-cm long segment of unmyelinated axon of radius 2 μm.

18-4   Calculate the axoplasm resistance of a 1-cm long segment of myelinated axon of radius 2 μm.

18-5   An unmyelinated segment of axon has a radius of 2 μm and a length of 1 cm. Find its (a) membrane capacitance, and (b) membrane leakage resistance.

18-6   A myelinated segment of axon has a radius of 2 μm and a length of 1 cm. Find its (a) membrane capacitance; (b) membrane leakage resistance.

18-7   An axon membrane is 7.5×10^{-9} m thick. (a) In the resting state the axon potential is -90 mV. What is the direction and magnitude of the electric field in the membrane? (b) If the membrane has a capacitance of 0.01 F m^{-2}, what is its dielectric constant?

18-8   Find the space parameter for an axon of radius 0.2 μm if it is (a) myelinated; (b) unmyelinated.

18-9   A membrane has a potential difference of 90 mV across it. How much charge per square metre is located on either side if it is (a) myelinated; (b) unmyelinated.

18-10   What is the space parameter for an axon of radius 3 μm if it is (a) unmyelinated; (b) myelinated?

18-11   A myelinated axon has a space parameter of 1 cm. Find its radius.

18-12   A myelinated axon and an unmyelinated axon have the same space parameter. Find the ratio of their radii.

18-13   Find the radius of an unmyelinated axon with a space parameter of 2×10^{-4} m.

Section 18.3 | Ionic Concentrations and the Resting Potential

18-14   How many potassium ions are there inside a 1-cm-long segment of an axon of radius 2 μm?

18-15   How many sodium ions are there in a 1-cm length of an axon of radius 6 μm?

18-16   The concentration of an ion is 100 moles m^{-3}. How many ions are there in 1 m^3?

18-17   Find the equilibrium potential at 37° C for an ion with charge $+e$ if its concentration outside the axon is 160 moles m^{-3} and inside is 10 moles m^{-3}.

18-18   The equilibrium potential at 37° C for Cl$^-$ ions in a particular axon is -80 mV. If the Cl$^-$ concentration outside the cell is 110 moles m^{-3}, what is the concentration inside?

18-19   The active sodium transport due to the Na-K pump occurs at the rate of 3×10^{-7} moles m^{-2} s^{-1}. For 1 m^2 of membrane, find (a) the current in amperes due to the sodium ions; (b) the power expended against electrical forces if the resting potential is -90 mV.

18-20 The K^+ concentration inside an axon is 165 moles m^{-3} and outside it is 8 moles m^{-3}. (a) What is the equilibrium potential at 37° C? (b) In which directions are the potassium flows due to diffusion and to the electric field if the axon potential is -90 mV? Which flow is greater?

Section 18.4 | The Response to Weak Stimuli

18-21 Estimate the space parameter for the axon in Fig. 18.8*b*.

18-22 A myelinated axon of radius 3 μm is stimulated so that the potential at $x = 0$ is -100 mV, which is 10 mV below the resting potential. (a) What is the space parameter? (b) Sketch the axon potential-versus-distance graph when the potential has achieved its final steady-state value.

18-23 A myelinated nerve with a space parameter of 0.5 cm is disturbed at a point so that its potential is raised from the resting value of -90 mV to -80 mV. Find the steady-state potential at distances from this point of (a) 0.5 cm; (b) 1 cm.

Section 18.5 | The Action Potential

18-24 A probe with a resistance of 10^7 ohms connects an electrical circuit to an axon. Estimate the time constant for charging a 1-cm long axon segment assuming $C_m = 10^{-2}$ F m^{-2}, $\rho_a = 0$, $R_m = \infty$, and a radius of 5 μm.

18-25 Find the velocity of propagation for an action potential and the time required for it to travel 2 m in a myelinated nerve (a) with a radius of 1 μm; (b) with a radius of 20 μm.

18-26 A nerve pulse can travel the length of a 0.5-m-long myelinated axon in 0.05 s. What is its radius?

PROBLEMS

Use the data in Table 18.1 and Fig. 18.3 when numerical values are required that are not given in the problem.

18-27 A square metre of axon membrane has a resistance of 0.2 ohms. The membrane is 7.5×10^{-9} m thick. (a) What is the resistivity of the membrane? (b) Suppose the membrane resistance is due to fluid-filled cylindrical pores through the membrane. The pores have a radius of 3.5×10^{-10} m and a length equal to the membrane thickness, 7.5×10^{-9} m. The fluid in the pores has a resistivity of 0.15 ohm m, and the remainder of the membrane is assumed to be a perfect insulator. How many pores must there be in 1 m^2 to account for the observed resistance? (c) If the pores are in a square pattern, how far apart are they?

*18-28 If an ion in an electric field E has a drift velocity v, then the *mobility* of the ion is $\mu = v/E$. (a) Show that the resistivity of a fluid containing a single type of ion is $(qn\mu)^{-1}$, where q is the magnitude of the charge on an ion, and n is the number of ions per unit volume. (b) Show that if there are m different kinds of ions in a solution, the resistivity is $(q_1n_1\mu_1 + q_2n_2\mu_2 + \cdots + q_mn_m\mu_m)^{-1}$.

18-29 Using the relation between mobility and resistivity given in the preceding problem and the ionic concentrations in Fig. 18.3, estimate the resistivities of axoplasm and interstitial fluid. Explain any difference between the calculated result for ρ_a and the value in Table 18.1. (The mobilities of Na^+, K^+, and Cl^- are 5.20×10^{-8}, 7.64×10^{-8}, and 7.91×10^{-8} m^2 V^{-1} s^{-1}, respectively. Neglect the effects of the "other" ions.)

18-30 A 1-cm-long segment of myelinated axon has a radius of 10^{-5} m. Its capacitance is 6×10^{-9} F, and the resting potential is -90 mV. (a) The resting potential is due to an excess of positive ions along the membrane outside the axon and an equal negative excess inside. How much excess charge is there on either side? (b) If these excesses are due to singly charged ions, how many excess ions are there on either side? (c) Find the ratio of this number of ions to the total number of negative ions inside the segment.

18-31 A 1-cm-long segment of unmyelinated axon has a capacitance of 3×10^{-9} F. (a) If the axoplasm potential changes from -90 to $+40$ mV, by how much do the excess charges on either side of the membrane change? (b) If this change is due to an influx of sodium ions, how many sodium ions enter the axon?

18-32 A 1-cm-long segment of unmyelinated axon has a radius of 5×10^{-6} m and a capacitance of 3×10^{-9} F. (a) If the axoplasm potential changes from $+40$ to -96 mV, by how much do the excess charges on either side of the membrane change? (b) If this change is due to an outflow of potassium ions, how many ions must leave the axon segment? (c) If the original potassium concentration inside the axon is 155 moles m^{-3}, what fraction of the potassium ions leave the axon?

18-33 The axon analog network in Fig. 18.5*c* can

be made more complete by adding EMFs that simulate the resting potential. Show how this may be done.

18-34 The *Goldman equation*

$$V_i - V_o = \frac{k_B T}{e} \ln \left[\frac{c_{oNa} + c_{oK}(P_K/P_{Na})}{c_{iNa} + c_{iK}(P_K/P_{Na})} \right]$$

gives the net equilibrium potential difference for an axon, where P_K/P_{Na} is the ratio of the membrane permeabilities for potassium and sodium, the c's are the ionic concentrations, and e is the proton charge, Assume $T = 310$ K. (a) If the resting potential is -90 mV, what is the permeability ratio? (b) An above-threshold stimulus alters the permeabilities so the potential jumps to $+40$ mV. What is the new ratio? (c) Explain to what extent it is correct to use the same concentrations in parts (*a*) and (*b*).

*18-35 Draw a rough graph of the equilibrium potential versus the permeability ratio using the Goldman equation given in Problem 18-34.

*18-36 Using the Goldman equation in Problem 18-34, find the resting potential of an axon if the concentrations are as in Fig. 18.3 and the potassium to sodium permeability ratio is 100.

18-37 The total influx of Na^+ during the transmission of a single action potential pulse is 4×10^{-8} moles m^{-2} of active axon surface. (a) How many pulses may be transmitted in an unmyelinated nerve of radius 5 μm before the Na^+ concentration within the axon is increased by 10 percent? (Neglect the effects of the Na-K pump.) (b) How many pulses in a myelinated axon of radius 5 μm are required to increase the Na^+ concentration by 10 percent? (Assume the active surface is only at the nodes, which are 1 μm long and 1 mm apart. Again neglect the effects of the Na-K pump.)

ANSWERS TO REVIEW QUESTIONS

Q18-1, more, poorly, more; **Q18-2,** Schwann cells; **Q18-3,** space parameter; **Q18-4,** sodium, potassium; **Q18-5,** ionic concentrations; **Q18-6,** slower, smaller; **Q18-7,** sodium; **Q18-8,** nodes of Ranvier; **Q18-9,** two; **Q18-10,** metabolic energy.

Additional Reading

Bernard Katz, *Nerve, Muscle, and Synapse,* McGraw-Hill Book Co., New York, 1966. Paperback.

Arthur C. Guyton, *Textbook of Medical Physiology,* 4th ed., W.B. Saunders Co., Philadelphia, 1971, Chapters 4 and 5. More advanced but quite readable.

Theodore C. Ruch, Harry D. Patton, J. Walter Woodbury, and Arnold L. Towe, *Neurophysiology,* W.B. Saunders Co., Philadelphia, 1965. A very complete discussion is given in Chapters 1 and 2.

D.V. Aidley, *Physiology of Excitable Cells,* Cambridge University Press, Cambridge, 1971.

John R. Cameron and James G. Skofronick, *Medical Physics,* John Wiley and Sons, Inc. New York, 1978. Chapter 9 is an introduction to electrocardiograms, electroencephalograms, and other aspects of electricity in the body.

Russell K. Hobbie, *Intermediate Physics for Medicine and Biology,* John Wiley and Sons, Inc., New York, 1978. Chapter 6 is a more mathematical introduction to the electrical properties of nerves than that presented in this textbook.

Scientific American articles:

C.L. Stong, How to Make an Electrocardiogram of a Water Flea and Investigate other Bioelectric Effects, The Amateur Scientist, January 1962, p. 145. Description of a demonstration apparatus for nerve and muscle potentials in aquatic animals.

Arthur K. Solomon, Pumps in the Living Cell, August 1962, p. 100.

Peter F. Baker, The Nerve Axon, March 1966, p. 74.

Sir John Eccles, The Synapse, January 1965, p. 56.

Bernard Katz, The Nerve Impulse, November 1952, p. 55.

Werner R. Lowenstein, Biological Transducers, August 1960, p. 98.

Graham Hoyle, How is Muscle Turned On and Off?, April 1970, p. 85.

H.E. Huxley, The Contraction of Muscle, November 1958, p. 66.

Allen M. Scher, The Electrocardiogram, November 1961, p. 132.

Bruce I.H. Scott, Electricity in Plants, October 1962, p. 107.

Keir Pearson, The Control of Walking, December 1976, p. 72.

A.J. Hudspeth, The Hair Cells of the Inner Ear, January 1983, p. 54. Mechanical forces are converted into electrical signals to the brain.

CHAPTER 19

MAGNETISM

Most of us have had the opportunity to observe some of the fascinating properties of permanent magnets. The north pole of one magnet attracts the south pole and repels the north pole of another magnet. Either pole of a magnet attracts unmagnetized iron or steel objects, and when such objects are in contact with a magnet, they can attract other iron or steel objects.

It has been known since antiquity that *lodestones,* pieces of the mineral *magnetite* (iron oxide), are able to attract iron objects. Also, the fact that the earth is a large magnet was long known from its orienting effect on compass needles. However, it was only in 1820 that Hans Christian Oersted (1777–1851) discovered that an electric current in a wire can deflect a compass needle. Electric currents in wires, as well as charges moving in a vacuum, produce magnetic effects indistinguishable from those due to permanent magnets. Since currents exert forces on magnets, we would expect from Newton's third law relating action and reaction forces that magnets also exert forces on currents. Such forces are indeed observed.

In discussing electric forces, we found it advantageous to avoid the direct use of Coulomb's law for the force between two charges. Instead, we said that one charge produces an electric field **E**, which in turn exerts a force on a second charge. Similarly, we will consider that a magnet or moving charge produces a magnetic field **B**. The magnetic field exerts a force on another magnet or moving charge.

We begin this chapter by describing magnetic fields. We then discuss the forces on moving charges and currents, along with some applications. Finally we show how the fields due to currents can be calculated.

19.1 | MAGNETIC FIELDS

The direction and magnitude of the magnetic field **B** produced by a magnet can be determined with the aid of a compass, a permanently magnetized steel needle pivoted at its center. The equilibrium orientation of the needle gives the direction of the field, and the torque tending to align the needle is proportional to its magnitude (Fig. 19.1).

Magnetic fields are represented by diagrams or maps similar to those used for electric fields. Again the lines indicate the field direction, and the field is strongest where the lines are closest together (Figs. 19.2 and 19.3). *Notice that by convention the magnetic*

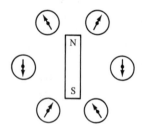

Figure 19.1. The equilibrium orientation of a compass needle indicates the direction of the magnetic field.

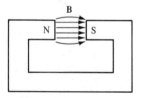

Figure 19.2. If the pole faces of a magnet are closely spaced, the field between them is nearly uniform.

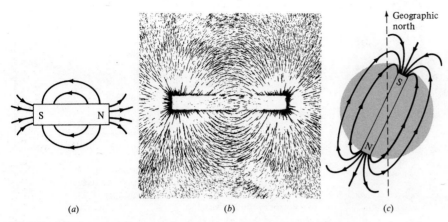

Figure 19.3. (*a*) Magnetic field of a bar magnet. (*b*) The pattern formed by iron filings near a bar magnet. (*c*) The magnetic field of the earth resembles that of a bar magnet. The *magnetic* south pole is located near the *geographic* north pole; it attracts the north pole of a compass needle. [Part (*b*) reproduced by permission of the publisher from *PSSC Physics,* fourth edition, 1976, by D.C. Heath and Company, Lexington, Massachusetts.]

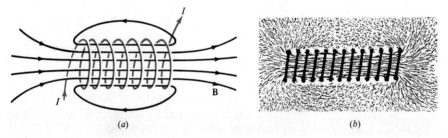

Figure 19.4. (*a*) Magnetic field lines and (*b*) iron filing patterns due to a current in a long coil of wire. Note the similarities to the field lines and patterns for the long bar magnet in Fig. 19.3. [Part (*b*) reproduced by permission of the publisher from *PSSC Physics,* fourth edition, 1976, by D.C. Heath and Company, Lexington, Massachusetts.]

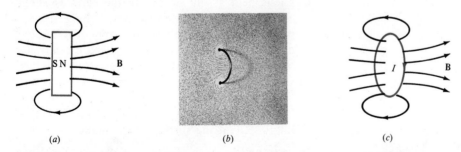

Figure 19.5. (*a*) Magnetic field lines for a short permanent bar magnet. (*b*) Iron filing patterns, and (*c*) magnetic field lines due to a single current turn. [Part (*b*) reproduced by permission of the publisher from *PSSC Physics,* fourth edition, 1976, by D.C. Heath and Company, Lexington, Massachusetts.]

field lines are always directed from the north pole to the south pole outside a magnet. The field lines around a magnet can be made visible by sprinkling iron filings on a sheet of paper. The filings become magnetized and tend to align themselves with the field (Fig. 19.3).

As we noted earlier, electric currents also produce magnetic fields (Figs. 19.4, 19.5, and 19.6). The field due to a current in a long coil of wire looks much like that of a long bar magnet, and the field of a single current turn is similar to that of a short bar magnet. A current in a long straight wire produces circular magnetic field lines.

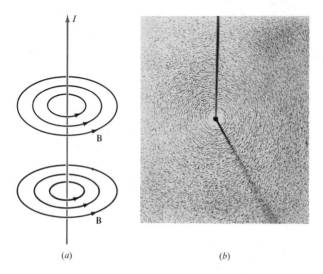

Figure 19.6. (*a*) Magnetic field lines and (*b*) iron filing patterns due to a current in a long straight wire. [Part (*b*) reproduced by permission of the publisher from *PSSC Physics,* fourth edition, 1976, by D.C. Heath and Company, Lexington, Massachusetts.]

A convenient rule aids in remembering the observed relationship between the directions of the magnetic field and the current. We may use the long straight wire as an example. If the right thumb is placed along the wire in the direction of the current, the fingers curl in the direction of the field (Fig. 19.7*a*). Similarly, for the circular current loop in Fig. 19.7*b*, this rule indicates that inside the loop, **B** points out of the page; outside the loop, **B** points into the page.

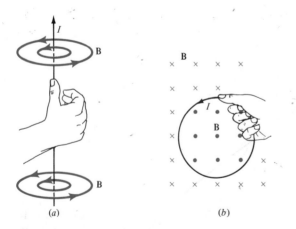

Figure 19.7. (*a*) With the right thumb along the direction of the current, the fingers curl in the direction of the magnetic field. (*b*) Inside the loop, **B** points out of the page; outside the loop, **B** points into the page.

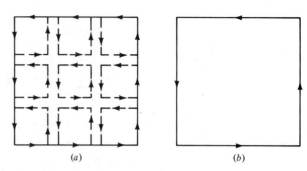

Figure 19.8. (*a*) The microscopic currents in a permanent magnet. (*b*) Since the effects of the opposing pairs of currents cancel in the interior, the net field produced is the same as for a current around the edge.

The close similarity between the magnetic fields of currents and permanent magnets led André Marie Ampère (1775–1836) to realize that microscopic currents in atoms or molecules are responsible for the fields of permanent magnets. Within a permanent magnet, the atoms are aligned so that their currents are in the same direction (Fig. 19.8*a*). In the interior of the material the currents occur in opposing pairs, so their fields exactly cancel. The net field is then due to the currents around the edge of the material and is identical to that produced by an electrical current in a wire with the same shape (Fig. 19.8*b*).

The S.I. magnetic field unit, the *tesla* (T), is a rather large unit. The largest fields produced in the laboratory for an extended time are about 10 T; fields up to about 100 T can be produced very briefly. The earth's magnetic field at the surface of the earth is about 10^{-4} T. One centimetre away from a long straight wire carrying the relatively large current of 100 A, the field is only 2×10^{-3} T. The small electrical currents associated with the human nervous system produce very weak magnetic fields. These fields are about 10^{-11} T near the chest (Fig. 19.9). The magnetic field due to a current pulse in a single axon (nerve fiber) is about 10^{-10} T at the surface of the nerve.

Magnetic Navigation in Animals | A wide

variety of animals use the earth's magnetic field as a navigational aid. On cloudy days when they cannot use the sun to navigate, pigeons become disoriented if small magnets are attached to their heads. European robins kept in cages during the migration season orient themselves according to the magnetic field in the cage. Honey bees display several behavioral patterns correlated with the direction of the local magnetic

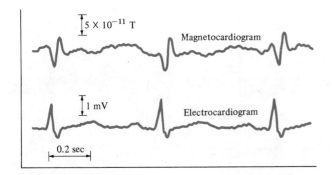

Figure 19.9. Magnetic fields near the human chest vary in time much like the electrical potential differences measured with an electrocardiograph. Possible diagnostic uses of magnetic field measurements near the heart and brain are under investigation. (From Alexander Kolin, *Physics Today*, November 1968, p. 39 © American Institute of Physics.)

field. Mud bacteria normally swim downward, seeking the soft mud that is their habitat; when magnets are used to cancel out and reverse the magnetic field of the earth, they swim upward. Also, mud bacteria from the southern hemisphere—where the vertical component of the magnetic field is opposite to that in the northern hemisphere—swim upward when brought to the United States, while mud bacteria from Brazil—where the field is horizontal—are equally likely to swim in either direction relative to the magnetic field. One scientist has even reported that humans display some homing ability; blindfolded people transported several kilometres from home in vehicles could make moderately accurate estimates of the direction toward home. This ability was impaired when the subjects wore magnets. However, other investigators have been unable to observe such effects.

In homing pigeons, robins, honey bees, and mud bacteria, studies have located bits of permanently magnetized magnetite. These species possess small permanent magnets that apparently behave like compass needles and experience torques when placed in a magnetic field (Fig. 19.10). Just how this information is sensed and processed is a tantalizing question currently under investigation by many scientists.

19.2 | THE MAGNETIC FORCE ON A MOVING CHARGE

Electric charges in motion near a magnet experience forces that can readily be demonstrated. For example, a magnet brought near a cathode-ray tube will

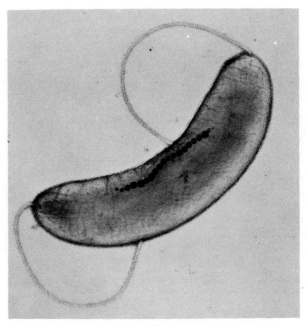

Figure 19.10. Electron microscope photograph of a *magnetotactic* bacterium, one that is sensitive to magnetic fields. The dark chain of particles consists of magnetite crystals that serve as a compass, tending to align the bacterium along the earth's magnetic field. The existence of magnetotactic bacteria was discovered in 1975 by Richard P. Blakemore, then a graduate student at the University of Massachusetts in Amherst. Many different types of magnetotactic bacteria have since been found. (Photo courtesy of Richard P. Blakemore and Nancy Blakemore.)

deflect a beam of electrons and alter the places where they strike a fluorescent screen (Fig. 19.11).

The magnetic force law is more complicated in form than is the electric force law, $\mathbf{F} = q\mathbf{E}$. A charge q moving with velocity $\mathbf{v}$ in a magnetic field $\mathbf{B}$ experiences a force perpendicular to $\mathbf{v}$ and to $\mathbf{B}$. The magnitude of this force is the product of q, v, and the com-

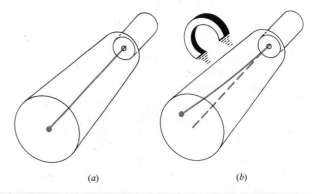

(a)   (b)

Figure 19.11. (a) Electrons in a cathode-ray tube strike a fluorescent screen. (b) The electrons are deflected by a magnet.

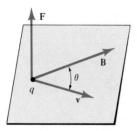

Figure 19.12. The magnetic force **F** on a positive charge q has a magnitude $qvB \sin \theta$ and is directed as shown.

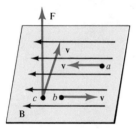

Figure 19.13. Positively charged particles in a magnetic field experience no force if the velocity is parallel (a) or opposite (b) to the field. The force has its maximum magnitude qvB when **v** is perpendicular to the field (c).

ponent of **B** perpendicular to **v**. Using the cross product notation introduced in Chapter Four, the magnetic force **F** is given (Fig. 19.12) by

$$\mathbf{F} = q\mathbf{v} \times \mathbf{B} \qquad (19.1)$$

The magnitude of this force is

$$F = |q|vB \sin \theta \qquad (19.2)$$

where θ is the angle between **v** and **B**. The force is greatest when $\sin \theta = 1$ or when the velocity is at right angles to the field. It is zero when $\sin \theta = 0$ or when the velocity is parallel or opposite to **B** (Fig. 19.13). Note that a charge at rest experiences no force; magnetic fields exert forces only on moving charges. The following example illustrates the calculation of the magnetic force on a moving charge.

Example 19.1

At Boston, Massachusetts, the magnetic field due to the earth is at 17° to the vertical direction and has a magnitude of 5.8×10^{-5} T (Fig. 19.14). (a) Find the force **F** on an electron moving straight down at 10^5 m s^{-1}. (b) Find the ratio of F to the weight mg.

(a) With $q = -e = -1.6 \times 10^{-19}$ C and $\sin 17° = 0.292$, the magnitude of the magnetic force is

$F = evB \sin \theta$
$= (1.6 \times 10^{-19} \text{ C})(10^5 \text{ m s}^{-1})(5.8 \times 10^{-5} \text{ T})(0.292)$
$= 2.71 \times 10^{-19}$ N

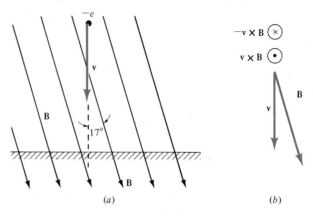

Figure 19.14. (a) The magnetic field of the earth varies in strength and direction. At Boston, it is at 17° to the vertical. (b) **v** $\times$ **B** is out of the page. The force on an electron is **F** $= -e\mathbf{v} \times \mathbf{B}$ and is opposite to **v** $\times$ **B** or into the page.

Since $\mathbf{F} = -e\mathbf{v} \times \mathbf{B}$, the force is opposite to **v** $\times$ **B** or into the page in Fig. 19.14.

(b) The mass of an electron is 9.1×10^{-31} kg, so

$$\frac{F}{mg} = \frac{2.71 \times 10^{-19} \text{ N}}{(9.1 \times 10^{-31} \text{ kg})(9.8 \text{ m s}^{-2})}$$
$$= 3.04 \times 10^{10}$$

The force on the electron due to the relatively weak magnetic field of the earth is larger than its weight by a factor of more than 10^{10}. For this reason the motion of charged cosmic ray particles entering the upper atmosphere is determined primarily by the earth's magnetic field and not by gravitational forces.

The magnetic force $q\mathbf{v} \times \mathbf{B}$ is perpendicular to the velocity, so it can change the direction in which a particle is moving but not its speed. Magnetic forces never change the kinetic energy, since they are always perpendicular to the motion and hence do no work.

19.3 | ELECTROMAGNETIC FLOW METERS

Electromagnetic flow meters, which employ the force on charges moving in a magnetic field, provide an excellent way to measure blood flow rates in patients undergoing heart or arterial surgery. Unlike the flow meters discussed in Chapter Thirteen, they do not require the insertion of a probe into a vessel and may be used even if the flow is not laminar but turbulent.

Figure 19.15a illustrates the principle of this device. A magnetic field is applied perpendicular to the motion of the blood. This produces magnetic forces to the left on positive ions in the blood and to the right

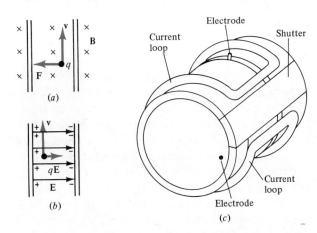

(a)

(b)

(c)

Figure 19.15. The electromagnetic flow meter as used for blood flow measurements. (a) When the magnetic field is into the page, the magnetic force on a positive ion is toward the left. (b) Charges accumulate at the edges, producing an electric field, **E**, which exerts a force toward the right on a positive ion. In equilibrium the electric force balances the magnetic force. (c) In one apparatus the artery is inserted into a metal sleeve by opening a shutter. The magnetic field is provided by a single current loop. The voltage between the electrodes is proportional to the flow rate.

on negative ions. Thus, charges accumulate at the edges of the vessel. Equilibrium occurs when the electric field due to these accumulations produces a force qE on an ion of charge q, which exactly balances the magnetic force qvB, that is, when $E = vB$ (Fig. 19.15b). The potential difference associated with the electric field is then proportional to the average velocity v of the blood and can be measured with a sensitive voltmeter.

In a version of this instrument used for arteries 4 mm or larger in diameter, a closely fitting metal sleeve is placed around the artery to assure that its diameter remains consant (Fig. 19.15c). A current in a single loop of wire provides a small magnetic field that is rapidly alternated in direction to avoid *polarization,* the formation of an insulating gas layer. Voltmeter leads are attached to electrodes on opposite sides of the sleeve. The meter is calibrated by making measurements on an artery with a known flow rate.

19.4 | THE MAGNETIC FORCE ON A CURRENT-CARRYING WIRE

We can find the expression for the magnetic force on a current-carrying wire directly from the force $\mathbf{F} = q\mathbf{v} \times \mathbf{B}$ on a moving charge. Consider a straight segment of wire of length ℓ, which is part of a closed

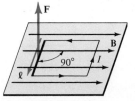

Figure 19.16. When the segment of wire carrying the current is perpendicular to the field, the force has a magnitude $F = I\ell B$.

circuit carrying a current I (Fig. 19.16). If the moving charges in the wire have a velocity v, they travel the length of the segment in a time $t = \ell/v$. The total moving charge q is the product of the current and the time,

$$q = It = \frac{I\ell}{v} \tag{19.3}$$

If the magnetic field **B** is perpendicular to the wire, then the magnetic force on the wire is

$$F = qvB = \left(\frac{I\ell}{v}\right)vB$$

or

$$F = I\ell B \quad (\ell \text{ perpendicular to } \mathbf{B}) \tag{19.4}$$

If the wire and the field are not perpendicular, $\mathbf{F} = q\mathbf{v} \times \mathbf{B}$ leads to

$$\mathbf{F} = I\boldsymbol{\ell} \times \mathbf{B} \tag{19.5}$$

where $\boldsymbol{\ell}$ is directed along the wire in the direction of the current (Fig. 19.17). If the wire is not straight, or if **B** varies with position, the wire may be regarded as having many small segments so that Eq. 19.5 holds true for each. The total force on the wire is then the vector sum of the forces on each segment.

In the next example, we see that the force due to the earth's magnetic field on a current-carrying wire is relatively small.

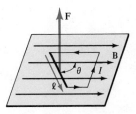

Figure 19.17. If the segment is at an angle θ to the field, the magnitude of the force is $F = I\ell B \sin \theta$.

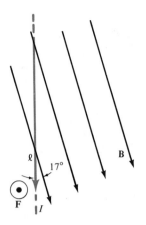

Figure 19.18. The force on the current-carrying wire is directed out of the page.

Example 19.2

The magnetic field of the earth at Boston, Massachusetts, is at 17° to the vertical and has a magnitude of 5.8×10^{-5} T (Fig. 19.18). A vertical wire carries a current of 10 A. Find the force on a 2-m length of the wire.

The magnitude of the force is

$$F = I\ell B \sin \theta$$
$$= (10 \text{ A})(2 \text{ m})(5.8 \times 10^{-5} \text{ T})(\sin 17°)$$
$$= 3.39 \times 10^{-4} \text{ N}$$

With the right-hand rule for cross products, the force is out of the page in Fig. 19.18.

19.5 | MAGNETIC DIPOLES

In Chapter Sixteen, we discussed the electric dipole, two neighboring electric charges $+q$ and $-q$. A loop of wire carrying a current behaves in a magnetic field much as an electric dipole does in an electric field and

is called a *magnetic dipole*. Magnetic dipoles associated with orbiting and spinning charges are important in discussions of the magnetic properties of atoms and molecules, including many molecules of importance in biochemistry.

We saw in Chapter Sixteen that an electric dipole in a uniform electric field experiences no net force, but it does experience a torque. We now show that a magnetic dipole in a uniform magnetic field also experiences no net force but experiences a torque. The rectangular current loop in Fig. 19.19 is an example of a magnetic dipole. The uniform magnetic field is parallel to sides 1 and 3, so no forces are exerted on them. Using $\mathbf{F} = I\boldsymbol{\ell} \times \mathbf{B}$, the forces on sides 2 and 4 are

$$F_2 = IbB \text{ (downward)}, \qquad F_4 = IbB \text{ (upward)}$$

These are equal but opposite, so the net force on the loop is zero. However, the net torque on the loop is not zero, since $\mathbf{F}_2$ and $\mathbf{F}_4$ have different lines of action and form a couple.

We found in Chapter Four that the torque due to a couple is independent of the position of the axis of rotation. With the convenient choice of an axis along side 4, the torque due to $\mathbf{F}_4$ is zero. The lever arm for $\mathbf{F}_2$ is a, so its torque, $\boldsymbol{\tau} = \mathbf{r} \times \mathbf{F}$, is equal in magnitude to $\tau = a(IbB)$. Since $ab = A$ is the area of the loop, $\tau = IAB$ and is directed as shown.

If the loop is oriented as in Fig. 19.20, then the torque has a magnitude of $aF \sin \theta = IAB \sin \theta$. This can be rewritten in vector form as

$$\boldsymbol{\tau} = IA\hat{\mathbf{n}} \times \mathbf{B}$$

Here $\hat{\mathbf{n}}$ is a unit vector perpendicular or *normal* to the plane of the loop; its direction can be found using the

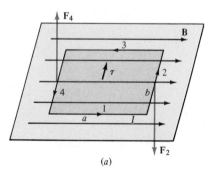

(a)

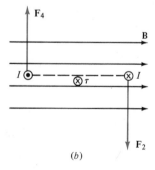

(b)

Figure 19.19. A magnetic dipole in a uniform magnetic field experiences no net force, but it does experience a torque. (a) A current loop with its plane parallel to the field. (b) The same loop seen in cross section. The dashed line indicates the plane of the loop.

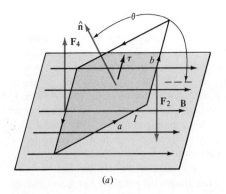

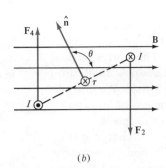

(a)   (b)

Figure 19.20. Two views of a loop tilted relative to the field. The unit vector $\hat{n}$ makes an angle θ with the magnetic field direction. The torque on the loop is $\tau = IA\hat{n} \times B = \mu \times B$.

right-hand rule as in Fig. 19.21. The *magnetic dipole moment* μ (mu) is defined by

$$\mu = IA\hat{n} \qquad (19.6)$$

With this definition, the torque on a magnetic dipole in a uniform magnetic field is

$$\tau = \mu \times B \qquad (19.7)$$

This result is correct for a current loop of *any* shape. It is identical in form to the torque $p \times E$ on an electric dipole in a uniform electric field.

The torque due to the magnetic field tends to align the dipole so that μ is parallel to B. In that position, $\mu \times B = 0$, and the dipole is in stable equilibrium; a slight displacement of the dipole results in a torque tending to decrease the displacement. The torque also vanishes when μ is opposite to B, but that is an unstable equilibrium situation; small displacements produce torques tending to increase the displacements.

The potential energy $\mathcal{U}$ of an electric dipole moment p at an angle θ to a uniform electric field E is

$-pE \cos \theta = -p \cdot E$. Similarly, a magnetic dipole moment μ at an angle θ to a uniform magnetic field has a potential energy

$$\mathcal{U} = -\mu B \cos \theta = -\mu \cdot B \qquad (19.8)$$

The potential energy has its minimum value, $-\mu B$, at $\cos \theta = 1$ or $\theta = 0°$, which is its stable equilibrium orientation. It has the greatest value, $+\mu B$, when $\cos \theta = -1$ or $\theta = 180°$.

These results are illustrated in the following example.

Example 19.3

A circular turn of wire of radius 0.5 m carries a current of 4 A. (a) What is its magnetic moment? (b) If the normal to the loop is at 90° to a field of 2 T, what is the magnitude of the torque? (c) What is its potential energy at this angle?

(a) The area of the loop is πr^2, so its magnetic moment is

$$\mu = IA = I\pi r^2 = (4 \text{ A})\pi(0.5 \text{ m})^2 = 3.14 \text{ A m}^2$$

(b) With $\sin 90° = 1$,

$$\tau = \mu B \sin \theta = (3.14 \text{ A m}^2)(2 \text{ T})(1) = 6.28 \text{ N m}$$

The direction of the torque is shown in Fig. 19.22.

(c) Since $\cos 90° = 0$, the potential energy of the loop is $\mathcal{U} = -\mu B \cos \theta = 0$.

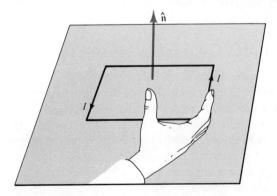

Figure 19.21. The normal direction $\hat{n}$ indicated by the right thumb is upward when the fingers point along the current direction.

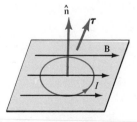

Figure 19.22. Example 19.3.

Magnetic Moment of an Orbiting Charge

When an electron moves around an atomic nucleus, there is a flow of charge or a current around the nucleus. The corresponding magnetic dipole moment determines how an electron interacts with an applied magnetic field. We now obtain a formula for the dipole moment of an orbiting charge, which we will use in later chapters.

Suppose a particle of mass m and charge q (Fig. 19.23) moves in a circular orbit. To find the magnetic moment $\mu = IA$, we need the current I and the area A. The time t required for the charge to complete one orbit of radius r at a speed v satisfies $vt = 2\pi r$ or $t = 2\pi r/v$. The magnitude of I is then

$$I = \frac{q}{t} = \frac{q}{2\pi r/v} = \frac{qv}{2\pi r}$$

Since the area A of the orbit is πr^2, the magnetic moment is

$$\mu = IA = \frac{qv}{2\pi r}\pi r^2 = \frac{qrv}{2}$$

This result can be put into another form involving the angular momentum of the particle. We saw in Chapter Seven that the angular momentum of a particle is $\mathbf{L} = \mathbf{r} \times \mathbf{p}$, and that when the linear momentum $\mathbf{p} = m\mathbf{v}$ is perpendicular to the distance vector $\mathbf{r}$, the magnitude of the angular momentum is $L = mvr$. Hence $rv = L/m$ and the magnetic dipole moment can be rewritten in vector form as

$$\boldsymbol{\mu} = \frac{q}{2m}\mathbf{L} \qquad (19.9)$$

The magnetic moment of a charged particle is proportional to its angular momentum; it is parallel to the angular momentum for a positive charge and opposite for a negative charge (Fig. 19.23). This result is illustrated by the next example.

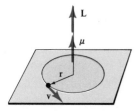

Figure 19.23.   The magnetic moment due to an orbiting positive charge is parallel to its angular momentum.

Example 19.4

In the Bohr model of the hydrogen atom, the electron moves in circular orbits for which the angular momentum $L = nh/2\pi$, where n is an integer, and h is Planck's constant; $h = 6.63 \times 10^{-34}$ J s. (a) Find the magnetic moment due to the orbital motion in the lowest ($n = 1$) orbit. (b) Suppose an electron in the $n = 1$ orbit has its moment parallel to a magnetic field of magnitude 10 T. How much energy in electron volts must be supplied to reverse the orientation of the moment?

(a) With $n = 1$, the angular momentum is $L = h/2\pi$. Since $q = -e$, the magnetic moment has a magnitude

$$\mu = \frac{e}{2m}L = \frac{eh}{4\pi m}$$

Substituting $e = 1.6 \times 10^{-19}$ C and $m = 9.11 \times 10^{-31}$ kg,

$$\mu = \frac{(1.60 \times 10^{-19}\text{ C})(6.63 \times 10^{-34}\text{ J s})}{4\pi(9.11 \times 10^{-31}\text{ kg})}$$
$$= 9.26 \times 10^{-24}\text{ A m}^2$$

(b) Since the potential energy of the magnetic dipole is $\mathcal{U} = -\mu B \cos\theta$, the potential energy is $-\mu B$ when $\boldsymbol{\mu}$ is along the field and $+\mu B$ when $\boldsymbol{\mu}$ is opposite to the field. Hence the energy needed to reverse the direction of the dipole is

$$2\mu B = 2(9.26 \times 10^{-24}\text{ A m}^2)(10\text{ T})$$
$$= 1.85 \times 10^{-22}\text{ J}\left(\frac{1\text{ eV}}{1.60 \times 10^{-19}\text{ J}}\right)$$
$$= 1.16 \times 10^{-3}\text{ eV}$$

We will use the relation between the magnetic dipole moment and the angular momentum of a particle in atomic and molecular applications in Chapters Twenty-eight and Twenty-nine. There we will see that the energy needed to reorient the dipole moment as in the previous example can be supplied by a time-varying magnetic field.

19.6 | MOTORS; GALVANOMETERS

The magnetic force on a current-carrying wire has many useful applications. Two of the most important ones are electrical motors and galvanometers.

Figure 19.24 illustrates the principle of the dc motor. A loop connected to an EMF is mounted so that it can rotate in a magnetic field that is produced either by a permanent magnet or by an electromagnet. When the normal to the loop becomes parallel to the field, the *split ring* or *commutator* reverses the current direction. As a result the torque on the loop is always in the same direction. This causes the loop to

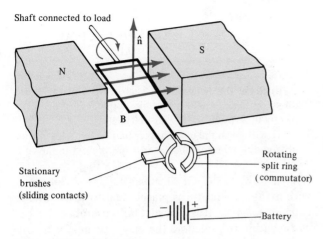

Figure 19.24. Elements of a dc motor.

rotate and enables it to do work on a load. Practical dc motors have many turns with each successive one rotated slightly, so that the net torque is nearly constant as the motor turns. Alternating current motors of various types are also widely used. In these no commutator is needed, since an alternating current is supplied to the electromagnet in such a way that the field reverses along with the current in the coil.

The pivoted coil galvanometer used in voltmeters and ammeters has a coil suspended in a radial magnetic field (Fig. 19.25). Hairsprings provide restoring torques proportional to the angular displacement. When there is a current in the coil, the torque due to the magnetic field causes it to rotate until equilibrium is established. A pointer is used to indicate the deflection of the coil, and a calibrated scale gives the corresponding current.

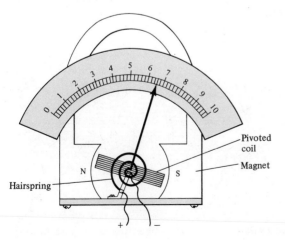

Figure 19.25. A galvanometer.

19.7 | MAGNETIC FIELDS PRODUCED BY CURRENTS

We saw in Chapter Sixteen how to calculate the electric field due to an arrangement of charges. We now discuss the analogous procedure for finding the magnetic field due to a current.

The basic law for the magnetic field is an expression for the field due to a short current segment. It is named the *Biot-Savart law,* after the two physicists who deduced it from experimental studies soon after Oersted discovered that currents produce magnetic fields. Consider a short segment of length $d\ell$ in a closed circuit with a current I (Fig. 19.26). At point P, a distance $\mathbf{r}$ from the segment, the magnetic field due to the segment is

$$d\mathbf{B} = k'I\frac{d\boldsymbol{\ell} \times \hat{\mathbf{r}}}{r^2} \qquad (19.10)$$

Here $\hat{\mathbf{r}}$ is a unit vector directed along $\mathbf{r}$. The net magnetic field $\mathbf{B}$ at P is the sum of the fields due to all the elements in the complete circuit. In S.I. units, the proportionality constant k' has the exact value

$$k' = 10^{-7}\,\text{T m A}^{-1} \qquad (19.11)$$
$$= 10^{-7}\,\text{N A}^{-2}$$

Like Coulomb's law for the electric field of a point charge, the Biot-Savart law is an inverse square law. However, it is more complex, since it involves a vector cross product. Also, we can never directly observe the field $d\mathbf{B}$ at P due to the single element in Fig. 19.26, because only the net field $\mathbf{B}$ due to the complete circuit can be measured.

Magnetic Field of a Circular Current Loop | To illustrate the use of the Bio-Savart law, we calculate the field at the center of a circular wire loop of radius a carrying a current I (Fig. 19.27). We

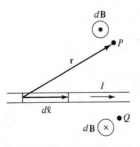

Figure 19.26 The field $d\mathbf{B}$ due to the current through $d\ell$ is out of the page at P and into the page at Q.

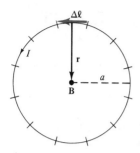

Figure 19.27. A circular current loop is divided into many small segments to calculate the magnetic field. The field is directed out of the page at the center of the circle.

divide the loop into many small segments. For any one segment $d\ell$, the distance r to the center is a. Since $d\ell$ is perpendicular to the unit vector $\hat{\mathbf{r}}$, $|d\ell \times \hat{\mathbf{r}}| = d\ell$; $d\ell \times \hat{\mathbf{r}}$ is directed out of the page. Thus the field due to the one segment is directed out of the page and has a magnitude

$$dB = \frac{k'I\,d\ell}{a^2}$$

Because $d\mathbf{B}$ is out of the page for all the $d\ell$'s, summing (or integrating) over the segments just replaces $d\ell$ by the circumference, $2\pi a$, and the net field has a magnitude

$$B = \frac{2\pi k'I}{a} \qquad (19.12)$$

Thus at the center of the circle $\mathbf{B}$ varies inversely with the first power of the radius, even though the fundamental Biot-Savart law varies as $1/r^2$. (The complete field was shown in Fig. 19.5.)

A single loop carrying a relatively large current produces a fairly weak field. For this reason, many turns of wire are needed to produce even modest sized fields. This is illustrated by the following example.

Example 19.5

A coil consists of 1000 circular turns of thin wire with an average radius of 0.1 m (Fig. 19.28). If the current in the coil is 10 A, find the magnetic field at its center due to (a) one turn; (b) the entire coil.

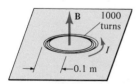

Figure 19.28. Example 19.5.

(a) For any one turn, the field has a magnitude

$$B = \frac{2\pi k'I}{a} = \frac{2\pi(10^{-7}\text{ T m A}^{-1})(10\text{ A})}{0.1\text{ m}}$$
$$= 6.28 \times 10^{-5}\text{ T}$$

(b) Since the field due to each turn is in the same direction, the net field is 1000 times the field of one turn or $(1000) \times (6.28 \times 10^{-5}\text{ T}) = 0.0628$ T.

We will see in the next chapter that larger fields can be obtained if materials such as iron are present in the coil.

Magnetic Field of a Solenoid | A long coil

with many circular turns placed next to each other is called a *solenoid*. It is a particularly useful device because within the solenoid the magnetic field is nearly uniform (Fig. 19.29). A solenoid with length ℓ and N turns has $n = N/\ell$ turns per unit length. The field inside the solenoid is

$$B = 4\pi k'In \qquad \text{(solenoid)} \qquad (19.13)$$

This formula is exact for an infinite solenoid, and it is a good approximation near the center of a long, thin solenoid. It will be derived in Section 19.12 in the Supplementary Topics.

Magnetic Field of a Long Straight Wire | A current in an infinitely long straight wire

is an idealized situation which cannot really occur. Its field, however, is a good approximation to the actual field at points near a long straight wire which are far from its ends and from other currents.

For a long straight wire along the x axis (Fig. 19.30), we replace $d\ell$ by dx in the Biot-Savart law. Point P is at a perpendicular distance r to the wire,

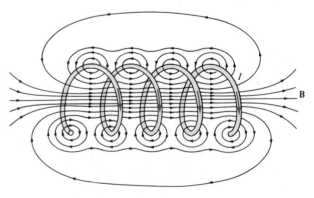

Figure 19.29. The field of a loosely wound solenoid. Inside the solenoid the field is nearly uniform.

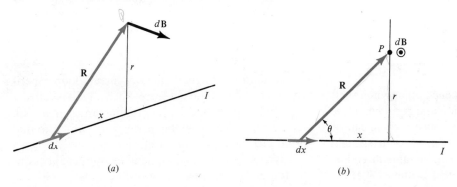

Figure 19.30. (a) The field at P due to one segment of a current in an infinitely long wire. (b) The fields at P due to all the segments point out of the page.

and it is a distance R from dx. The field due to this segment, $d\mathbf{B}$, is out of the page no matter where dx is on the axis. Thus summing (integrating) the $d\mathbf{B}$ vectors means adding their magnitudes.

Since $d\ell$ is at an angle θ with respect to $\dot{\mathbf{R}}$, the magnitude of $d\mathbf{B}$ is

$$dB = \frac{k'I dx \sin\theta}{R^2}$$

Now $\sin\theta = r/R$, and $R^2 = r^2 + x^2$, so integrating from $-\infty$ to $+\infty$,

$$B = \int_{-\infty}^{+\infty} \frac{k'Ir\,dx}{(r^2 + x^2)^{3/2}} = k'Ir \frac{1}{r^2}\left[\frac{x}{(r^2 + x^2)^{1/2}}\right]_{-\infty}^{+\infty}$$

$$B = \frac{2k'I}{r} \qquad \text{(long straight wire)} \qquad (19.14)$$

The lines of $\mathbf{B}$ form circles about the wire as shown before in Fig. 19.6. The following example illustrates how the field due to two wires is found.

Example 19.6

Two parallel long straight wires a distance d apart each carry a current I in the same direction (Fig. 19.31a). Find the net magnetic field (a) midway between the wires; (b) at points P and Q in Fig. 19.31a.

(a) According to the right-hand rule, the field $\mathbf{B}_l$ due to the left-hand wire points into the page in the region between the wires. Similarly, the field $\mathbf{B}_r$ due to the right-hand wire points out of the page in this region. At the midpoint, the distance to either wire is the same, so $\mathbf{B}_l$ and $\mathbf{B}_r$ are equal in magnitude. Since they are oppositely directed, their vector sum is zero.

(b) At point P, the fields due to both wires point out of the page. Using $B = 2k'I/r$, we have

$$B_l = \frac{2k'I}{d} \qquad B_r = \frac{2k'I}{2d}$$

Hence

$$B = B_l + B_r = \frac{2k'I}{d} + \frac{2k'I}{2d} = \frac{3k'I}{d}$$

At point Q, the field has the same magnitude, but it points into the page.

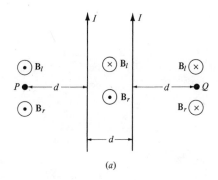

(a)

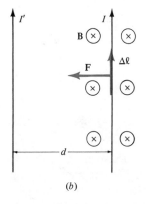

(b)

Figure 19.31. (a) The fields due to the two currents are opposite in the region between the two wires and in the same directions elsewhere, in the plane of the wires. (b) Two currents in the same direction attract each other. $\mathbf{B}$ is the field due to I'.

19.8 | THE FORCE BETWEEN TWO PARALLEL CURRENT-CARRYING WIRES

Two long, parallel wires attract each other when their currents are in the same direction and repel when the currents are opposite. The force between them can be found using $B = 2k'I/r$ for the field of a straight wire. In Fig. 19.31b, I and I' are in the same direction, and their separation is d. The field $\mathbf{B}$ due to I' at the segment $\Delta \boldsymbol{\ell}$ is into the page, and its magnitude is $2k'I'/d$. The force $\mathbf{F}$ on the segment is $I\,\Delta \boldsymbol{\ell} \times \mathbf{B}$. By the right-hand rule for cross products, $\mathbf{F}$ is directed to the left or toward the other wire, so the force is attractive as anticipated. Since $\Delta \boldsymbol{\ell}$ is perpendicular to $\mathbf{B}$,

$$F = I\,\Delta\ell B = I\,\Delta\ell\,\frac{2k'I'}{d}$$

The *force per unit length* of wire is

$$\frac{F}{\Delta\ell} = \frac{2k'II'}{d} \qquad (19.15)$$

This result is illustrated by the following example.

Example 19.7

Two parallel wires carry currents of 10 A and are separated by 1 mm = 10^{-3} m. What is the force on a 2-m-long portion of a wire?

The force per unit length is

$$\frac{F}{\Delta\ell} = \frac{2k'II'}{d} = \frac{2(10^{-7}\ \text{T m A}^{-1})(10\ \text{A})(10\ \text{A})}{10^{-3}\ \text{m}}$$

$$= 0.02\ \text{N m}^{-1}$$

Thus the force on a 1-m segment of the wire is 0.02 N. The force on a 2-m segment is twice this force, or 0.04 N.

A standard arrangement of this type is used to define the ampere as the current that gives rise to a stated force between the two wires. The unit of charge, the coulomb, is defined as the charge carried by an ampere of current in 1 second.

SUMMARY

Magnetic fields are established by permanent magnets and by moving charges. These fields, in turn, exert forces on other permanent magnets and on moving charges. The magnetic force on a charge q moving with velocity $\mathbf{v}$ in a magnetic field $\mathbf{B}$ is

$$\mathbf{F} = q\mathbf{v} \times \mathbf{B}$$

From the expression for the magnetic force on a moving charge it follows that the magnetic force on a

wire segment of length $\boldsymbol{\ell}$ carrying a current I is

$$\mathbf{F} = I\,\boldsymbol{\ell} \times \mathbf{B}$$

In a uniform magnetic field, a current loop experiences a torque but no net force. Such a loop is called a magnetic dipole, and its dipole moment is $\boldsymbol{\mu} = IA\hat{\mathbf{n}}$. Its behavior in a magnetic field is very similar to that of an electric dipole in an electric field.

Given a current, the magnetic field it produces can be found by applying the Biot-Savart law,

$$d\mathbf{B} = k'I\,\frac{d\boldsymbol{\ell} \times \hat{\mathbf{r}}}{r^2}$$

The constant k' has a value of 10^{-7} T m A^{-1}. At the center of a circular loop,

$$B = \frac{2\pi k'I}{a}$$

In a solenoid with n turns per unit length,

$$B = 4\pi k'In$$

Near a long, straight wire,

$$B = \frac{2k'I}{r}$$

Two long, parallel wires attract each other when their currents are in the same direction and repel when they are opposite. The force per unit length has a magnitude

$$\frac{F}{\Delta\ell} = \frac{2k'II'}{d}$$

Checklist

Define or explain:

magnetic field	magnetic dipole
magnetic force on a	motor
moving charge	galvanometer
magnetic force on a	Biot-Savart law
current-carrying wire	solenoid

REVIEW QUESTIONS

Q19-1 The magnetic field is largest where the field lines are _____.

Q19-2 The force on a charge moving in a magnetic field is largest when the velocity is _____ to the magnetic field.

Q19-3 If a positive charge experiences a magnetic force straight up, a negative charge moving in the same direction will experience a magnetic force _____.

Q19-4 The magnetic force on a current-carrying wire is zero when the magnetic field is _____ or _____ to the current.

Q19-5 Magnetic dipoles tend to align themselves _____ to the magnetic field.

Q19-6 The net force on a magnetic dipole is _____ when it is in a uniform magnetic field.

Q19-7 In a galvanometer, the deflection is proportional to the _____ and _____.

Q19-8 The magnitude of the magnetic field due to a short segment of a circuit varies as the _____ of the distance.

Q19-9 A nearly uniform magnetic field can be found inside a _____.

Q19-10 The field due to a long straight wire varies _____ with the distance.

EXERCISES

Section 19.1 | Magnetic Fields

19-1 Consider the magnetic field in Fig. 19.32 at points P_1, P_2, P_3, P_4. (a) At which point is the field largest? (b) At which point is it smallest?

19-2 In the magnetic field shown in Fig. 19.32, near which point is the field most nearly uniform?

Section 19.2 | The Magnetic Force on a Moving Charge

19-3 An electron passing through an apparatus is undeflected. Does that imply that the apparatus has no magnetic field? Explain.

19-4 A particle with charge 10^{-6} C is moving in the $+y$ direction at 10^4 m s^{-1} at right angles to a magnetic field of 2 T (Fig. 19.33). (a) Find the magnitude and direction of the force on the particle. (b) What would the force be if the velocity were in the $-y$ direction?

19-5 A particle of charge $+q$ and mass m moves with a velocity **v** in a magnetic field **B** (Fig. 19.33).

Find the magnitude and direction of the acceleration if **v** is (a) in the $+x$ direction; (b) $-x$ direction; (c) $+y$ direction; (d) $-y$ direction; (e) out of the page; (f) into the page.

19-6 A negatively charged particle moves in a uniform magnetic field (Fig. 19.33). Find the direction of the force on the particle if it is moving (a) in the $+x$ direction; (b) $+y$ direction; (c) $-y$ direction; (d) out of the page; (e) into the page.

19-7 An object of mass 0.01 kg moves at 100 m s^{-1} at an angle of 30° to a magnetic field of 10^{-2} T (Fig. 19.34). If its charge is -10^{-3} C, find the magnitude and direction of (a) the magnetic force; (b) the acceleration.

19-8 The velocity of a typical hydrogen ion (proton) at room temperature is about 3000 m s^{-1}. (a) If the magnetic field of the earth is 10^{-4} T, what is the maximum magnetic force on the hydrogen ion? (b) Find the ratio of this force to the electric force between two protons 10^{-10} m apart, a typical interatomic distance. (c) Is the earth's magnetic field likely to affect most biochemical processes? Explain.

Section 19.4 | The Magnetic Force on a Current-Carrying Wire

19-9 A 0.5-m-long segment of a wire carrying a 20-A current experiences a force of 5 N when it is perpendicular to a magnetic field. What is the magnitude of the field?

19-10 A horizontal wire experiences a 10-N force directed straight up when it carries a 5-A current toward the right. What will the force be if the current is reversed and its magnitude is doubled?

19-11 A straight, 2-m-long segment of a circuit is at 30° to a magnetic field. (a) If the field is 3 T and the current is 10 A, find the magnitude of the force on the segment. (b) Indicate the force direction with the aid of a sketch.

19-12 A power line carries a current of 1000 A. The earth's magnetic field makes an angle of 73° with the line and has a magnitude of 7×10^{-5} T. Find the magnitude of the magnetic force on a 100-m-long section of the line.

19-13 (a) Find the magnitude and direction of the force on each of the three straight segments in Fig. 19.35. (b) What is the net force on the circuit?

Section 19.5 | Magnetic Dipoles

19-14 What is the torque on the circuit in Fig. 19.35?

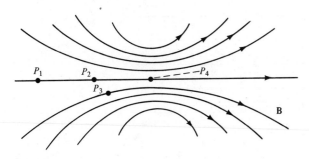

Figure 19.32. Exercises 19-1 and 19-2.

P_1 P_2 P_3 P_4 B

19-15   A square wire loop with sides 0.1 m and a current of 10 A is placed in a magnetic field of magnitude 0.1 T. (a) What is magnetic dipole moment of the loop? (b) What is the largest torque the loop can experience in this field? (c) For what orientation of the loop will this torque occur?

19-16   What orientation of a magnetic dipole in a magnetic field will result in (a) zero energy; (b) maximum torque?

19-17   A square current loop is placed in a magnetic field directed into the page (Fig. 19.36). The magnitude of **B** increases as x increases. Is there a net force on the loop? If there is, what is its direction? Explain.

19-18   The z component of the magnetic dipole moment due to the spin of an electron about its axis is $\mu_z = \pm eh/4\pi m$, where $h = 6.63 \times 10^{-34}$ J s is Planck's constant. (a) If μ_z is parallel to a magnetic field of 10 T, how much energy in electron volts must be supplied to reverse its direction so that it is opposite to the field? (b) Find the ratio of this energy to the 13.6 eV needed to remove an electron from a normal hydrogen atom.

Section 19.7 | Magnetic Fields Produced by Currents

19-19   A circular current loop of radius 0.2 m has a resistance of 100 ohms and is connected to a 12-V battery. What is the magnetic field at the center of the loop?

19-20   The magnetic field at the center of a circular current loop is 0.05 T. If the radius of the loop is 1.2 m, find the current.

19-21   Fifty circular turns of wire are wound closely together, so that all of them have a radius of close to 0.05 m. If the wire carries a current of 2 A, what is the magnitude of the magnetic field at its center?

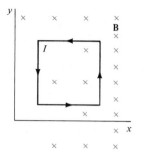

Figure 19.36.   Exercise 19-17.

19-22   It is desired to have a field inside a solenoid of magnitude 0.1 T. How many turns per unit length are required if the current is 10 A?

19-23   A thousand turns of wire are wound on a thin tube 0.4 m long. If the current in the wire is 2 A, find the field in the tube.

19-24   The magnetic field 0.1 m from a long, straight wire is 10^{-4} T. How large is the current in the wire?

19-25   A long, straight wire carries a current of 4 A. Find the magnitude and direction of the field at a distance of (a) 0.1 m, and (b) 1 m.

19-26   Show that the two sets of units used for k' in Eq. 19.11 are equivalent.

19-27   The magnetic field due to a current pulse in a single long, straight axon (nerve fiber) is found to be 1.2×10^{-10} T at a distance of 1.3 mm = 0.0013 m from the axon. How large is the current in the axon?

Section 19.8 | The Force Between Two Parallel Current-Carrying Wires

19-28   Two long, straight parallel wires 0.1 m apart each carry a current of 10 A. Find the direction and magnitude of the force on a 0.5-m-long

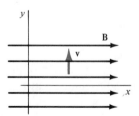

Figure 19.33.   Exercises 19-4, 19-5, 19-6.

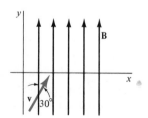

Figure 19.34.   Exercise 19-7.

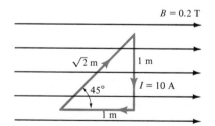

Figure 19.35.   Exercises 19-13 and 19-14.

segment of one of the wires if the currents are (a) in the same direction; (b) in opposite directions.

19-29 A 10-m-long segment of wire carries a 5-A current and has a weight of 0.5 N. It is constrained so that it can move vertically but not horizontally above another wire carrying a current of 10 A in the opposite direction. (a) At what separation would the weight of the wire be supported by the magnetic forces? (b) If the wires each have a radius of 1 mm = 10^{-3} m, will they be in contact?

19-30 Three parallel wires lie in a plane. The separation between adjacent wires is 0.1 m, and each wire carries a 10-A current in the same direction. Find the magnitude of the net force per unit length on (a) the central wire; (b) one of the outer wires.

19-31 Two parallel wires attract each other with a force of 10^{-4} N per metre of length. The wires are 0.01 m apart. The current in one wire is 20 A. What is the magnitude and relative direction of the current in the other wire?

PROBLEMS

19-32 An electron is moving at 10^5 m s^{-1} directly toward a long, straight wire that carries a 50-A current. (a) Find the force on the electron when it is 0.5 m from the wire. (Using a sketch, indicate the direction of the force.) (b) Find the resulting acceleration.

19-33 In Fig. 19.37, electrons in a conductor are moving into the page with a drift velocity $\mathbf{v}_d$. A magnetic field $\mathbf{B}$ is perpendicular to $\mathbf{v}_d$. (a) Find the magnitude and the direction of the magnetic force on the electrons. (b) Find the direction and magnitude of the electric field $\mathbf{E}$ that would exert an equal but opposite force on the electrons. (c) What potential difference would have to be applied across the conductor to produce this electric field? (d) If no external electric field is applied, then electrons will tend to move toward one side of the conductor until a constant potential difference is estab-

lished across the conductor. How large is this potential difference? (The establishment of a potential difference across a conductor in a magnetic field is called the *Hall effect*.) (e) Find the Hall potential difference if $v_d = 10^{-3}$ m s^{-1}, $B = 2$ T, and $a = 10^{-2}$ m. (f) What is the direction of the current? (g) In some conductors, the charge carriers are missing electrons, or "holes," which behave as positive charges. Will the Hall potential difference be the same as when the carriers are negative? Explain.

19-34 It has been proposed to build huge magnets using superconducting wires to store energy for periods of peak electrical power demands. In one design, turns of radius 100 m carry 150,000 A. The average field due to this current is 5 T. The magnet is placed in a tunnel cut into bedrock to obtain the necessary structural support, since each portion of the wire is subjected to large forces due to the field. (a) If the magnetic field is parallel to the coil axis and perpendicular to the wires, what is the force on 1 m of wire? (b) Show that the force is radially outward. (c) If there are 10 turns per metre of length in the coil, what is the average outward pressure on the bedrock?

19-35 A wire with a resistance of 10 ohms is bent into a rectangle 0.5 m by 0.8 m and connected to a 6-V battery. (a) What is the dipole moment of the loop? (b) The loop is placed in a magnetic field of 0.5 T. What is the maximum torque on the loop? (c) How is it oriented when the torque is a minimum?

19-36 Suppose that the wire in the preceding problem is bent to obtain the maximum magnetic dipole moment. (a) What is the shape of the wire? (b) What is the dipole moment?

19-37 Find the net force on the loop in Fig. 19.38.

*19-38 In Fig. 19.39, $I = I'$. Show that the field between the wires at a distance x from the origin is

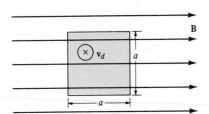

Figure 19.37. Problem 19-33.

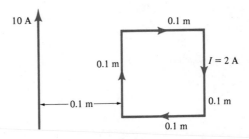

Figure 19.38. Problem 19-37.

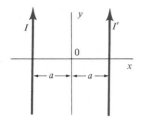

Figure 19.39.   Problems 19-38 and 19-39.

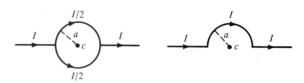

Figure 19.41.   Problem 19-44. **Figure 19.42.**   Problem 19-45.

$4k'Ix/(a^2 - x^2)$ if $|x| < a$. (Positive B values are out of the page, and negative values are into the page.)

19-39 In Fig. 19.39, $I = 10$ A, $I' = 5$ A, and $a = 0.1$ m. Find the magnitude and direction of **B** at (a) $x = 0$; (b) $x = 0.2$ m; (c) $x = -0.2$ m. (d) Where is the field zero?

19-40 The coil of a galvanometer has 100 turns, each with an area of 10^{-3} m^2. The magnetic field is 0.02 T and is arranged so that it is always in the plane of the coil. A current of 1 mA $= 10^{-3}$ A produces a rotation of $\theta = 10°$. If the torque due to the hairspring equals $k\theta$, find the spring constant k in newton-metres per radian.

19-41 Five hundred metres of wire are wound on a tube 0.5 m long and 0.2 m in circumference. How large a current is needed to produce a field inside the tube of 0.1 T?

*19-42 (a) Show that the magnetic field along the axis of a circular loop of radius a carrying a current I at a distance y from its center is equal in magnitude to $2\pi Ik'a^2/(a^2 + y^2)^{3/2}$ (Fig. 19.40). (b) Find an approximate expression for the field in terms of the dipole moment and y when y is large compared to a. (c) Compare this result to the corresponding formula for the electric field due to an electric dipole (Chapter Sixteen).

*19-43 According to the Bohr model, in the normal hydrogen atom the electron orbits the proton

in a circle of radius 5.1×10^{-11} m at a frequency of 6.8×10^{15} Hz. (a) What is the current due to the orbital motion of the electron? (b) What is the magnetic field at the proton due to this current?

*19-44 A current splits into two equal currents in Fig. 19.41. What is the magnetic field at the center c of the circle?

*19-45 What is the magnetic field at the center c of the semicircle in Fig. 19.42?

19-46 In a laboratory, the magnetic field of the earth has a magnitude of 6×10^{-5} T and is 20° to the vertical; its vertical component is downward. It is desired to reverse this field in an experiment to study the magnetic senses of bacteria. (a) If a solenoid with 1000 turns per metre is used, how large a current is required? (b) How should the solenoid and current be oriented?

ANSWERS TO REVIEW QUESTIONS

Q19-1, closest; **Q19-2,** perpendicular; **Q19-3,** straight down; **Q19-4,** parallel, opposite; **Q19-5,** parallel; **Q19-6,** zero; **Q19-7,** current, magnetic field; **Q19-8,** inverse square; **Q19-9,** solenoid; **Q19-10,** inversely.

SUPPLEMENTARY TOPICS

19.9 | MEASUREMENT OF CHARGE TO MASS RATIOS

In this and the following two sections, we consider some applications of the magnetic force on a moving charge. Here we discuss an arrangement of electric and magnetic fields that can be used to measure the charge to mass ratio of a charged particle.

Figure 19.43 shows a particle of mass m and charge q accelerated from rest by a known potential difference V. Since the sum of the kinetic and potential energies must remain constant, the resulting velocity **v** satisfies $\frac{1}{2}mv^2 = qV$, and the charge to mass ratio is

$$\frac{q}{m} = \frac{v^2}{2V} \qquad (19.16)$$

Thus the charge to mass ratio q/m of the particle can be found if the velocity is measured. This is ac-

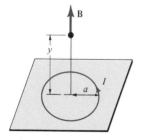

Figure 19.40.   Problem 19-42.

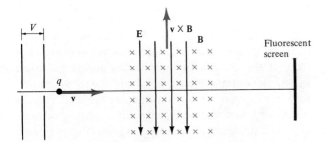

Figure 19.43. The charged particle is accelerated from rest to a velocity **v** by a potential difference *V*. The magnetic field **B** is directed into the page. The vector **v** ✕ **B** is opposite to **E** so the net force **F** = *q***E** + *q***v** ✕ **B** is zero for some value of *E/B*.

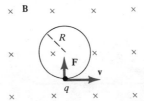

Figure 19.44. A charged particle moves with velocity **v** perpendicular to a uniform magnetic field **B** directed into the page. Since the force and the acceleration are always perpendicular to **v** and are constant in magnitude, the particle moves in a circular orbit with constant speed.

complished with the aid of the *crossed* or perpendicular electric and magnetic fields shown in the center of Fig. 19.43. The magnetic force on the particle is *q***v** ✕ **B**, and the electric force in *q***E**. Since **v** ✕ **B** points upward or opposite to **E**, these forces exactly balance when *qvB* = *qE* or

$$v = \frac{E}{B} \qquad (19.17)$$

This means the velocity may be determined by adjusting the crossed fields until the beam spot on the fluorescent screen is undeflected from its position when there are no crossed fields.

J.J. Thomson (1856–1940) used an arrangement of this kind in 1897 to measure the charge to mass ratio for the negative cathode rays in a cathode-ray tube. He concluded that these cathode rays were negatively charged particles (electrons). He also found that their velocity in his apparatus was about one-tenth of the velocity of light, which was much greater than any velocity previously observed.

One other feature of this apparatus is significant. The crossed **E** and **B** fields permit charged particles to travel through them without deflection whenever the condition *v = E/B* is satisfied, no matter what the charge or mass of the particles. If a beam of several kinds of charged particles with various velocities passes through such a crossed field region, the particles emerging undeflected all have the same velocity. Thus, perpendicular electric and magnetic fields act as a *velocity selector*. One use of such an arrangement is seen in the next section.

19.10 | MASS SPECTROMETERS

The *mass spectrometer* was originally developed as a nuclear physics research tool. Today, mass spectrom-

eters are widely used in many kinds of laboratories to measure and identify minute quantities of various substances.

The principle of the mass spectrometer can be understood by considering a particle with positive charge *q* and mass *m* moving perpendicular to a uniform magnetic field **B** (Fig. 19.44). The magnetic force **F** = *q***v** ✕ **B** is perpendicular to the velocity **v**, so it changes direction but not magnitude. The magnitude of the force, *qvB*, does not change, because **v** and **B** remain constant in magnitude and perpendicular to each other.

Since the acceleration **a** = **F***/m* is constant in magnitude and always perpendicular to **v**, the particle moves in a circular orbit with constant speed. The centripetal acceleration $a_r = v^2/R$ is produced by the magnetic force. Thus the radius *R* of the orbit must satisfy $qvB = mv^2/R$, and

$$R = \frac{mv}{qB} \qquad (19.18)$$

The use of this result is illustrated by the following example.

Example 19.8

How large a magnetic field is needed to cause an O_2^+ ion to move in a circular orbit of radius 2 m at 10^6 m s⁻¹? (The mass of the O_2^+ ion is approximately 32 u, where 1 u = 1.66×10^{-27} kg.)

The charge on the ion is *e* = 1.60×10^{-19} C. Using *R = mv/qB*, the magnetic field required is

$$B = \frac{mv}{qR}$$

$$= \frac{(32 \times 1.66 \times 10^{-27} \text{ kg})(10^6 \text{ m s}^{-1})}{(1.60 \times 10^{-19} \text{ C})(2 \text{ m})}$$

$$= 0.167 \text{ T}$$

Figure 19.45 shows the major parts of the mass spectrometer. In the ion source, molecules are ionized

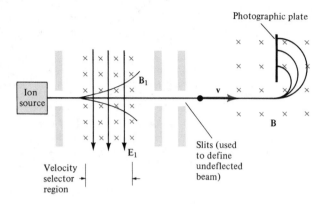

Figure 19.45. A mass spectrometer. Crossed fields E_1 and B_1 select only those particles that have the same velocity. The radius of the circular path in the **B** field then determines m/q.

by bombarding them with electrons, and the ions are extracted by an electric field. Crossed electric and magnetic fields in the velocity selector only permit ions with a velocity $v = E_1/B_1$ to proceed without deflection through the slits into the uniform magnetic field **B**. There the ions move in circular paths of radius $R = mv/qB$ until they strike the photographic plate at a distance proportional to their mass, since the radius of curvature increases with the mass. This allows one to separate ions with the same charges but different masses.

Historically the mass spectrometer made possible the systematic study of *isotopes*. Isotopes are forms of an element with very nearly the same chemical properties but different numbers of neutrons in their nuclei and hence different atomic masses. Later the mass spectrometer was also used to separate the fissionable uranium 235 isotopes from the more abundant ^{238}U in the World War II development of nuclear weapons.

The ability of the mass spectrometer to distinguish among nuclear isotopes makes it invaluable in situations involving *stable* rather than *radioactive* isotopes. The latter spontaneously transform into other nuclear species, emitting ionizing radiation in the process. If a radioactive material is administered to a plant or animal, its motion within the organism may be observed by detecting the ionization due to the radiation. Such *radioactive tracers* are widely used in biological research and medical diagnosis. However, some biologically important elements, such as nitrogen and oxygen, lack suitable radioactive isotopes, although they do have stable isotopes. For example, oxygen nor-

mally has 99.756 percent oxygen 16, which is a nucleus containing 8 protons and 8 neutrons; 0.039 percent oxygen 17, which has one more neutron; and 0.205 percent oxygen 18, which has yet another neutron. If the oxygen in material administered to an organism contains extra amounts of the rare isotopes, samples can get taken from the organism at various places or times and analyzed with a mass spectrometer. The presence of the rare oxygen isotopes signals the arrival of the administered substance or its metabolic derivatives. Thus the mass spectrometer makes possible the use of stable isotopes as tracers. It is also used to identify the ratios of the abundances of stable isotopes in geological samples to aid in determining their source or age.

Because the masses of different isotopes of an element differ by a considerable percentage in the lighter elements, small but sometimes significant variations occur in the rates of chemical reactions, evaporation, and so on. As a result measurable variations arise in the ratios of the hydrogen, carbon, and oxygen isotopes, among others, in minerals, bodies of water, and organisms. A vast amount of contemporary research in biology, planetary science, oceanography, and archaeology is made possible by using mass spectrometers to study these minute isotopic ratio variations. For example, investigators traced the source of the carbon used by plankton in a lake, and fragments of several Greek columns bearing inscriptions were sorted out in this way.

Mass spectrometers are sometimes employed in situations where one is not concerned about the isotopic composition. They have been used to analyze the respiratory gases of patients with various disorders, and to study the composition of the gas surrounding plants during photosynthesis. Also, they are used in the petroleum industry to distinguish among complex compounds having identical chemical compositions but different molecular configurations, and a mass spectrometer was one of the instruments carried by the space ships that landed on Mars and Venus.

19.11 | CYCLOTRONS

The *cyclotron,* invented by Ernest O. Lawrence (1901–1958) in 1930, was the first machine developed to accelerate charged particles to high velocities by causing them to pass repeatedly through the same accelerating region. Its operation depends on the remarkable fact that the period or the time required for a charged particle to complete one circular orbit in a uniform magnetic field **B** is independent of the speed of the particle v.

For a particle of charge q and mass m, the radius R of the orbit was shown in the preceding section to be $R = mv/qB$. The period T satisfies $vT = 2\pi R$, or

$$T = \frac{2\pi R}{v} = \frac{2\pi m}{qB} \qquad (19.19)$$

Thus, increasing the velocity increases the radius of the orbit but has no effect on the period T or the orbital frequency $f = 1/T$.

A cyclotron consists of two evacuated hollow metal *dees* in a uniform magnetic field perpendicular to their plane (Fig. 19.46). Protons or other positive ions are injected near the center. An electric generator reverses the potential difference between the dees at the orbital frequency of the ions, so they are accelerated each time they pass through the gap between the dees. This increases their velocity and consequently their orbital radius, $R = mv/qB$, but does not alter their period.

The operation of the cyclotron depends on the fact that the period is independent of the velocity. However, it is found that the inertial mass of a particle increases rapidly as its velocity approaches the speed of light. This was originally predicted by Einstein's

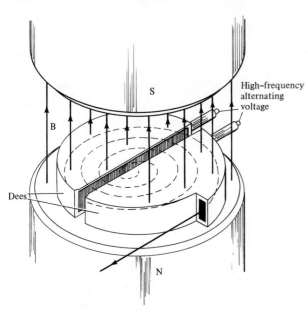

Figure 19.46. The principle of the cyclotron. A magnetic field perpendicular to the plane of the dees maintains the ions in circular orbits. The orbital radius increases as the ions are accelerated by the potential difference between the dees, which reverses at the orbital frequency of the ions.

special theory of relativity in 1905 and has been well confirmed. This mass increase causes the period to increase and sets a limit to the velocity and kinetic energy attainable with a cyclotron. Higher energies can be achieved in more complex accelerators that gradually vary the magnetic field or the generator frequency as the ions accelerate.

Originally developed for nuclear physics research, cyclotrons have largely been replaced in that area by newer machines. However, they are sometimes used today in hospitals to bombard various targets, inducing nuclear reactions that produce medically useful radioactive materials. Most radioactive materials are supplied to hospitals by nuclear reactor facilities, but some are so short-lived that they cannot be transported, and others can only be made in cyclotrons.

19.12 | AMPERE'S LAW

The Biot–Savart law permits us to find the magnetic field by summing the fields due to small current segments. A different looking but equivalent way of finding magnetic fields is provided by Ampere's law.

Like Gauss' law in electrostatics, Ampere's law involves integrals that we can evaluate when there is sufficient geometric symmetry to determine the field direction. Also, it has a differential form that is useful in more advanced applications. Both forms of Ampere's law contain precisely the same physical information as the Biot–Savart law.

Although Ampere's law can be derived quite generally from the Biot–Savart law, we obtain it here for the special case of a current in a long straight wire. Using the Biot–Savart law, we showed that

$$B = \frac{2k'I}{r} \qquad (19.14)$$

The lines of **B** form concentric circles about the wire (Fig. 19.47a). If we take the circular path C in Fig. 19.47b about the wire, at each point B is parallel to the path. Consider one small segment of the path, $d\ell$. Since **B** is parallel to $d\ell$, $\mathbf{B} \cdot d\ell = B\,d\ell$. According to Eq. 19.14, B depends only on r, so it has the same magnitude for any segment on the path. Thus, if we sum up the $\mathbf{B} \cdot d\ell$ products around the circle, we find

$$\oint_C \mathbf{B} \cdot d\ell = B \oint_C d\ell$$

The circle on the integral sign indicates a closed path C. Since $\oint d\ell$ is the circumference of the circle,

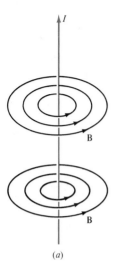

(a)

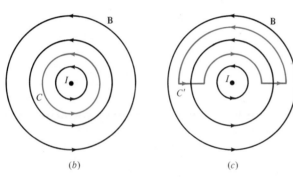

(b)   (c)

Figure 19.47. (a) A current in a long straight wire, perpendicular to the page, has magnetic field lines that are concentric circles. (b) The current crossing the surface within the circle C is I. (c) There is no current crossing the surface within C'.

$2\pi r$, with Eq. 19.14 we have

$$\oint_C \mathbf{B} \cdot d\boldsymbol{\ell} = \frac{2k'I}{r} 2\pi r = 4\pi k'I$$

Notice that the factors of r have canceled; the result is independent of the radius of the circle. This cancellation would also have occurred if we had integrated over a semicircle or some other fraction of a circle.

It is left as a problem to show that the integral is zero for a path such as C' in Fig. 19.47c that does not encircle the wire. Therefore, our result, which is known as Ampere's law, is

$$\oint_C \mathbf{B} \cdot d\boldsymbol{\ell} = 4\pi k'I \quad \text{(Ampere's law)} \quad (19.20)$$

Here I is the current passing through the surface enclosed by the closed path C. When there are two or more currents, I is their algebraic sum. If, as in Fig. 19.47a, you can put your fingers of your right hand around the curve, and your thumb is approximately in the direction of the current, the current is positive. Your thumb will be opposite to a negative current.

Despite the lack of generality of our derivation, Ampere's law is, in fact, equivalent to the Biot–Savart law for any path and current distribution. However, when there are time-varying electric fields present, neither law as presented here correctly predicts the entire magnetic field. This will be discussed further in the next chapter.

The fact that Ampere's law holds for any closed path we wish to choose is what makes it useful in finding magnetic fields. Much as with Gauss' law, the judicious choice of integration paths and the exploi-

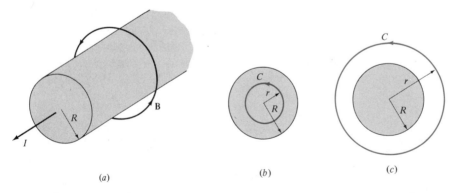

(a)   (b)   (c)

Figure 19.48. (a) An infinitely long cylinder with a current I uniformly distributed over its cross section. The symmetry indicates that the lines of B are circles centered at the axis. (b) A circular path within the cylinder is parallel to B and encloses part of the current. (c) A path outside the wire encloses all the current.

tation of the geometric symmetries are the keys to its applications.

We now illustrate the use of Ampere's law by considering a current in a long cylindrical conductor and in a solenoid. In both cases the direct application of the Biot–Savart law would involve very difficult calculations.

Field of a Long Cylindrical Conductor | In

Fig. 19.48a, a current I is distributed uniformly across a long cylindrical conductor of radius R. In the simpler case of a long straight wire, the Biot–Savart law showed the field lines are circles centered at the wire. The same symmetry is present here; nothing changes when we move about the wire at a fixed radius. We still have a current with symmetry about the axis, so we again expect to find circular field lines. Thus, our task is to determine how the magnitude of **B** depends on r.

First consider a circular path C inside the wire, $r < R$ (Fig. 19.48b). The current in Ampere's law is the total current I times the fraction of the cross-sectional area of the wire enclosed by C, or

$$i = \frac{I\pi r^2}{\pi R^2} = \frac{Ir^2}{R^2}$$

B is tangent to the path C, so $\mathbf{B} \cdot d\boldsymbol{\ell} = B\,d\ell$. Since **B** is constant in magnitude for a fixed r

$$\oint_C \mathbf{B} \cdot d\boldsymbol{\ell} = B \oint_C d\ell$$

The integral is just the circumference, $2\pi r$. Thus Ampere's law gives

$$B(2\pi r) = 4\pi k'I(r^2/R^2)$$
$$B = \frac{2k'Ir}{R^2} \qquad r < R \qquad (19.21)$$

The field inside the wire increases in magnitude linearly with r.

To find the field outside the wire, consider a circle with radius $r > R$ (Fig. 19.48c). Then the entire current I passes through the circle. The left side of Ampere's law is again $B(2\pi r)$. Hence we have

$$B(2\pi r) = 4\pi k'I$$
$$B = \frac{2k'I}{r} \qquad r > R \qquad (19.22)$$

This result for the field is identical to that of a long straight wire (Eq. 19.14). Measurements of the field outside the wire give no clue as to whether its source is a thin wire, a current uniformly distributed in a wire of finite radius, or any other current distribution with circular symmetry about the axis. Note that the results for $r < R$ and for $r > R$ agree at $r = R$, as expected.

Field of a Solenoid | We now derive Eq. 19.13

for the field of a solenoid. As the turns of a solenoid come closer together, and the number of turns becomes very large, the cancellations among the separate turns produce a field that is along the axis within the solenoid and very small outside it (Fig. 19.29). The field lines outside become further and further apart as the solenoid is increased in size. In the limit of an infinitely long solenoid, the symmetry of the current leads to the field in Fig. 19.49. In this idealized case, **B** is uniform and directed along the axis inside the coil, and it is zero outside the coil.

Suppose now we apply Ampere's law to the rectangular path C. The integral splits into four parts, corresponding to the sides of the rectangle. The field is zero along the top side, so this integral is zero. Also, **B** is perpendicular to the short sides, so $\mathbf{B} \cdot d\boldsymbol{\ell}$ is zero everywhere on these sides. Finally, along the bottom side, **B** is parallel to $d\boldsymbol{\ell}$, so $\mathbf{B} \cdot d\boldsymbol{\ell} = B\,d\ell$. Also, B is constant along this side, so it can be taken out of the integral. Hence, if the length of the rectangle is h, we find

$$\oint_C \mathbf{B} \cdot d\boldsymbol{\ell} = B \oint d\ell = Bh$$

If a length L of the solenoid has N turns, there are $n = N/L$ turns per unit length. The number of turns crossing the rectangle is nh, and the total current through it is nhI. Hence Ampere's law becomes

$$Bh = 4\pi k'nhI$$
$$B = 4\pi k'nI$$

This is Eq. 19.13.

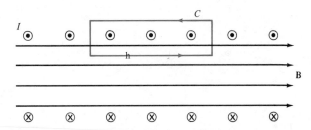

Figure 19.49. Part of an infinitely long solenoid. The field is parallel to the axis.

EXERCISES ON SUPPLEMENTARY TOPICS

Section 19.10 | Mass Spectrometers

19-47 A beam of particles with charge $+e$ moves in a circle of radius 3 m in a magnetic field of magnitude 0.2 T. (a) What is the momentum of the particles? (b) If the particles are protons, what is their velocity? (The proton mass is 1.67×10^{-27} kg.)

19-48 An electron moves at a speed of 10^7 m s^{-1} in a circle of radius 2 m in a magnetic field. (a) How large is the field? (b) How large is the acceleration? (c) Draw a sketch showing the directions of the electron path and the field.

19-49 Protons with a kinetic energy of 5 MeV $= 5 \times 10^6$ eV move along a circular path in a magnetic field. Alpha particles, which are helium nuclei, have twice the charge and about four times the mass of a proton. How much energy must they have to move along the same path?

19-50 A beam of ions passes through a velocity selector in a mass spectrometer that has an electric field of 1.4×10^5 N C^{-1}. (a) If the ions emerging from the selector have a velocity of 2×10^5 m s^{-1}, what is the magnetic field in the selector? (b) The magnetic field in the bending region is 1 T. What is the radius of the path followed by a He$^+$ ion with charge e and mass 6.68×10^{-27} kg?

19-51 Charged particles passing through a bubble chamber leave visible tracks that consist of very small hydrogen gas bubbles in liquid hydrogen that is almost at the boiling point. In Fig. 19.50, the magnetic field is directed into the page, and the tracks are in the plane of the page and moving in the directions indicated by the arrows. (a) Which of the tracks C, D, and E correspond to positively charged particles? (b) If all three particles have the same mass and their charges are equal in magnitude, which is moving the fastest? (c) If all three particles are moving with the same speed, which has the greatest mass?

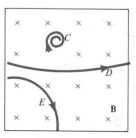

Figure 19.50.　Exercise 19-51.

Section 19.11 | Cyclotrons

19-52 Helium nuclei with mass 6.68×10^{-27} kg and charge $2e$ are accelerated in a cyclotron. The period of their orbit is 10^{-7} s. (a) What is the magnetic field? (b) If the maximum orbital radius is 2 m, what is the maximum velocity attained?

PROBLEMS ON SUPPLEMENTARY TOPICS

19-53 A mass spectrometer has an electric field of 10^5 N C^{-1} and a magnetic field of 0.6 T in its velocity selector, and a magnetic field of 0.8 T in its bending region. (a) What is the velocity of the ions passing through the velocity selector? (b) Find the spatial separation of singly ionized neon 20 and neon 22 isotopes with charges $+e$ after they have been bent through a half circle. (Neon 20 has a mass of approximately 20 u, and neon 22 has a mass of approximately 22 u. 1 u $= 1.66 \times 10^{-27}$ kg.)

19-54 A cyclotron accelerates protons of mass 1.67×10^{-27} kg to a velocity 3×10^7 m s^{-1}, a tenth that of light. The magnetic field is 1.5 T. Find the maximum orbital radius and the orbital frequency.

*19-55 An electron is projected with velocity **v** at an angle of 30° into a uniform magnetic field **B**. (a) Find the magnitude and direction of the acceleration. (b) Find the component of the velocity along the magnetic field. What is its rate of change? (c) Find the component of the velocity perpendicular to the field. Show that this component rotates about the field and find the associated radius. (d) How far does the electron move along the field direction during one full rotation? (e) Sketch the trajectory of the electron.

*19-56 (a) Electrons in a television picture tube are accelerated from rest by a potential difference of 2×10^4 V. What is their speed? (b) If the earth's magnetic field is at right angles to the electron beam, and its magnitude is 6×10^{-5} T, what is the magnetic force on an electron? (c) Estimate the displacement resulting from this force when the beam travels 0.15 m through the tube.

c19-57 A solenoid of finite length bent into a circular or doughnut shape is a *toroid* (Fig. 19.51). If it has N turns and a current I in its windings, using Ampere's law show that B is (a) zero for $r < a$; (b) $2k'IN/r$ for $a < r < b$; (c) zero for $r > b$.

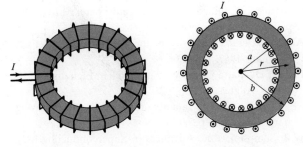

Figure **19.51.** A toroid. Problem 19-57.

Additional Reading

Francis Bitter, *Magnets: The Education of a Physicist,* Science Study Series, Doubleday and Co., Garden City, N.Y., 1959. Paperback.

Louis W. McKeehan, *Magnets,* Momentum Series, D. Van Nostrand and Co., Princeton, N.J., 1967. Paperback.

Alexander Kolin, Magnetic Fields in Biology, *Physics Today,* November 1968, p. 39.

David Cohen, Magnetic Fields of the Human Body, *Physics Today,* vol. 29, August 1975, p. 34.

Richard Blakemore, Magnetotactic Bacteria, *Science,* vol. 190, 1975, p. 377.

D. Brenner, S.J. Williamson, and L. Kaufman, Visually Evoked Magnetic Fields of the Human Brain, *Science,* vol. 190, 1975, p. 480.

Wolfgang Wiltschko and Roswitha Wiltschko, Magnetic Compass of European Robins, *Science,* vol. 176, 1972, p. 62.

Robert P. Green, Orientation of Homing Pigeons Altered by a Change in the Direction of an Applied Magnetic Field, *Science,* vol. 184, 1974, p. 180.

Charles Walcott, The Homing of Pigeons, *American Scientist,* vol. 62, 1974, p. 542.

J.P. Wikswo, Jr., J.P. Barach, and J.A. Freeman, Magnetic Field of a Nerve Impulse: First Measurements, *Science,* vol. 208, 1980, p. 53.

R.B. Frankel, R.P. Blakemore, and R.S. Wolfe, Magnetite in Freshwater Magnetotactic Bacteria, *Science,* vol. 203, 1979, p. 1355.

R.B. Frankel, R.P. Blakemore, F.F. Torres de Araujo, D.M.S. Esquival, and J. Danon, Magnetotactic Bacteria at the Geomagnetic Equator, *Science,* vol. 212, 1981, p. 1269.

James L. Gould, J.L. Kirschvink, and K.S. Deffeyes, Bees Have Magnetic Remanance, *Science,* vol. 201, 1978, p. 1026.

Charles Wolcott, James L. Gould, and J.L. Kirschvink, Pigeons Have Magnets, *Science,* vol. 205, 1979, p. 1027.

James L. Gould and Kenneth P. Able, Human Homing: An Elusive Phenomenon, *Science,* vol. 212, 1981, p. 1961.

R.R. Baker, Goal Orientation by Blindfolded Humans After Long-Distance Displacement: Possible Involvement of a Magnetic Sense, *Science,* vol. 210, 1980, p. 555.

Samuel Epstein, Peter Thompson, and Crayton J. Yapp, Oxygen and Hydrogen Isotopic Ratios in Plant Cellulose, *Science,* vol. 198, 1977, p. 1209.

Norman Herz and David B. Wenner, Assembly of Greek

ᶜ**19-58** Two long parallel plates of width a are separated by a distance small compared to a (Fig. 19.52). The plates carry uniformly distributed currents in opposite directions along their length. The total current in either plate is I. Show that the field between them is given approximately by $4\pi k'I/a$.

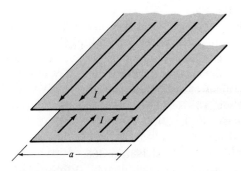

Figure **19.52.** Problem 19-58.

ᶜ**19-59** A coaxial cable has a solid central conductor of radius a separated by insulators from an outer concentric tube with inner radius b and outer radius c. The two conductors carry uniformly distributed currents in opposite directions parallel to their axis. The total current in either conductor is I. Find B in the regions (a) $r < a$; (b) $a < r < b$; (c) $b < r < c$; (d) $r > c$.

ᶜ**19-60** A long hollow tube has an inner radius b and an outer radius c. It carries a uniformly distributed current parallel to its axis. The total current is I. Find the field in the regions (a) $r < b$; (b) $b < r < c$; (c) $r > c$.

ᶜ**19-61** Show that the integral of $\mathbf{B} \cdot d\boldsymbol{\ell}$ around path C' in Fig. 19.47c is zero along the radial lines, positive on the outer semicircle, negative on the inner semicircle, and zero for the complete path.

Marble Inscriptions by Isotopic Methods, *Science,* vol. 199, 1978, p. 1070.

Greg Rau, Carbon-13 Depletion in a Subalpine Lake; Carbon Flow Implications, *Science,* vol. 201, 1978, pl 901.

Richard A. Kerr, Isotopic Anomalies of Meteorites: Complications Multiply, *Science,* vol. 202, 1978, p. 203.

Roger Lewin, Isotopes Give Clues to Past Diets, *Science,* vol. 220, 1983, p. 1369.

Woodfin V. Ligon, Jr., Molecular Analysis by Mass Spectrometry, *Science,* vol. 205, 1979, p. 151.

W.H. Berger and J.S. Killingley, Migrations of California Gray Whales Tracked by Oxygen-18 Variations in their Epizoic Barnacles, *Science,* vol. 207, 1980, p. 759.

F.W. McLafferty, Tandem Mass Spectrometry, *Science,* vol. 214, 1981, p. 280.

R. Thompson, *et al.,* Environmental Applications of Magnetic Measurements, *Science,* vol. 207, 1980, p. 481.

Scientific American articles:

M. Wilson, Joseph Henry, July 1954, p. 72.

Joseph J. Becker, Permanent Magnets, December 1970, p. 92.

H.P. Furth, Strong Magnetic Fields, February 1958, p. 28.

Henry H. Kolm and Arthur J. Freeman, Intense Magnetic Fields, April 1965, p. 66.

Francis Bitter, Ultrastrong Magnetic Fields, July 1965, p. 64.

Alfred O.C. Nier, The Mass Spectrometer, March 1953, p. 68.

A.B. Benfield, The Earth's Magnetism, June 1950, p. 20.

S.K. Runcorn, The Earth's Magnetism, September 1955, p. 152.

W.M. Elsasser, The Earth on a Dynamo, May 1958, p. 44.

Charles R. Carrigan and David Gubbins, The Source of the Earth's Magnetic Field, February 1979, p. 118.

S.W. Angrist, Galvanomagnetism and Thermomagnetism, December 1961, p. 124.

Raymond Wolfe, Magnetothermoelectricity, June 1964, p. 70.

Laurence L. Cahill, Jr., The Magnetosphere, March 1965, p. 58.

Allen Cox *et al.,* Reversals of the Earth's Magnetic Field, February 1967, p. 44.

Palmer Dyal and Curtis W. Parkin, The Magnetism of the Moon, August 1971, p. 62.

J.E. Kurszler and M. Tanenbaum, Superconducting Magnets, June 1962, p. 60.

W.B. Sampson *et al.,* Advances in Superconducting Magnets, March 1967, p. 114.

John H. Reynolds, The Age of the Solar System, November 1960, p. 171.

Kenneth W. Ford, Magnetic Monopoles, December 1963, p. 122.

Henry Koln, John Oberteuffer, and David Kelland, High-Gradient Magnetic Separation, November 1975, p. 46.

R.R. Wilson, The Batavia Accelerator, February 1974, p. 72.

William T. Keeton, The Mystery of Pigeon Homing, December 1974, p. 96.

Richard P. Blakemore and Richard B. Frankel, Magnetic Navigation in Bacteria, December 1981, p. 58.

R.K. O'Nions, P.J. Hamilton, and N.M. Evensen, The Chemical Evolution of the Earth's Mantle, May 1980, p. 120. Applications of mass spectrometers.

Robert R. Wilson, The Next Generation of Particle Accelerators, January 1980, p. 42.

Richard A. Carrigan, Jr. and Peter W. Trower, Superheavy Magnetic Monopoles, April 1982, p. 106.

Ronald E. Rosenzweig, Magnetic Fluids, October 1982, p. 136. The strange properties of a liquid with magnetic particles in suspension.

Richard A. Mewalt, Edward C. Stone, and Mark E. Wiedenbeck, Samples of the Milky Way, December 1982, p. 108. Mass spectrometers and cosmic rays.

Jearl Walker, Motors in Which Magnets Attract Other Magnets in Apparent Perpetual Motion, *The Amateur Scientist,* May 1983, p. 162.

CHAPTER 20

INDUCED CURRENTS AND FIELDS

We have seen that charged objects at rest may experience electrical forces and that moving charged objects may experience electrical and also magnetic forces. However, we have not yet encountered any situations in which the electrical and magnetic forces are related; electricity and magnetism have appeared to be essentially distinct phenomena. We see in this chapter that when electric and magnetic fields are changing in time, they are, in fact, related to each other in a remarkable fashion. A changing magnetic field produces an *induced* electric field, and a changing electric field produces an *induced* magnetic field. One major consequence of these effects is the existence of *electromagnetic waves,* which travel with the speed of light.

20.1 | FARADAY'S LAW

As we saw in the preceding chapter, Oersted discovered in 1820 that electric currents produce magnetic fields. Michael Faraday (1791–1867), a self-educated English physicist and chemist, guessed that magnetic fields might also produce electric currents. He found that this did not happen if a wire loop and a magnet were both stationary. However, he discovered in 1831 that when the magnetic field through a loop is changing, a current is *induced* in the loop.

This phenomenon, called Faraday's law, can be illustrated with a coil of wire connected to a battery and a second separate loop of wire attached to an ammeter (Fig. 20.1). When the switch is closed, the ammeter indicates that a current occurs momentarily in the second loop; when it is reopened, a current in

the opposite direction is briefly observed. Moving the loop further away from the coil or reducing its area by bending the wires produces a current in the same direction, as does opening the switch. Rotating the loop so that its plane is not horizontal also induces a current.

Faraday's studies led him to observe that the single common feature in these and related phenomena was the occurrence of a change in the "amount" of the magnetic field passing through the loop in which the current is induced. Suppose, for example, that we count the number of magnetic field lines that pass

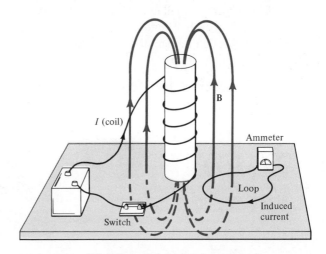

Figure 20.1. When the switch is closed, a current is temporarily produced in the loop in the indicated direction; a current in the opposite direction occurs when the switch is opened. If the switch is left closed so that a constant magnetic field is present, moving the loop or changing its size or orientation will also produce a current.

427

through the loop of Fig. 20.1. Then, a current will be induced in the loop when any change occurs in the number of lines intersecting the loop area. Such a change occurs in each of the situations we just described.

To make this idea more precise, we define the *magnetic flux*. We will see shortly that the magnetic flux is proportional to the number of lines that pass through an area. Suppose a loop of area A has its normal vector $\hat{\mathbf{n}}$ at an angle θ to a uniform magnetic field $\mathbf{B}$ (Fig. 20.2). The vector area $\mathbf{A}$ is defined then as $\mathbf{A} = A\hat{\mathbf{n}}$; the magnitude of $\mathbf{A}$ is equal to the area, and it is directed perpendicular to the surface. The magnetic flux Φ (capital phi) through the loop is a scalar quantity defined by

$$\Phi = BA \cos \theta = B_n A = \mathbf{B} \cdot \mathbf{A} \qquad (20.1)$$

Thus the magnetic flux is the product of the area and the component B_n of the magnetic field normal to the loop. Notice that the flux is proportional to the area, since a larger area has more lines passing through it. The $\cos \theta$ factor indicates that the maximum number of lines of $\mathbf{B}$ will pass through the area when $\hat{\mathbf{n}}$ and $\mathbf{B}$ are parallel, and this number decreases until no lines pass through the loop when $\theta = 90°$. We will see shortly that it is the changes in the magnetic flux that are most directly related to the induced currents.

The definition we have given above for the magnetic flux holds true only when the magnetic field is constant over the loop. If $\mathbf{B}$ varies, the loop must be subdivided into areas small enough so that the field is nearly constant in each. The total flux is then the sum of the fluxes through the small areas, or the integral over the area of the loop

$$\Phi = \int \mathbf{B} \cdot d\mathbf{A} \qquad (20.2)$$

In S.I. units, magnetic flux is measured in *webers* (Wb). Since the field $\mathbf{B}$ is measured in teslas and the area A in square metres, $\Phi = \mathbf{B} \cdot \mathbf{A}$ implies

$$1 \text{ weber} = 1 \text{ tesla(metre)}^2 = 1 \text{ T m}^2 \qquad (20.2)$$

Magnetic fields are sometimes given in webers per square metre (Wb m^{-2}) instead of in teslas.

The calculation of the flux is illustrated by the following example.

Example 20.1

The loop in Fig. 20.3 has an area of 0.1 m². The magnetic field is perpendicular to the plane of the loop and has a constant magnitude of 0.2 T. Find the magnetic flux through the loop.

The normal vector $\hat{\mathbf{n}}$ must be perpendicular to the loop, but it may be chosen to be either into or out of the page in Fig. 20.3. If we choose $\hat{\mathbf{n}}$ to be into the page, then by the convention discussed in the preceding chapter, clockwise currents will be positive, and counterclockwise currents will be negative. (See Section 19.5.) Since $\mathbf{B}$ is parallel to $\hat{\mathbf{n}}$, $B_n = B$, and

$$\begin{aligned} \Phi = \mathbf{B} \cdot \mathbf{A} &= BA \\ &= (0.2 \text{ T})(0.1 \text{ m}^2) = 0.02 \text{ Wb} \end{aligned}$$

Faraday's law can now be written in terms of the magnetic flux. Suppose that the magnetic flux through a loop is changed by an amount $\Delta\Phi$ in a time Δt.

Faraday's law states that the average EMF induced in the circuit equals the average rate of change of the flux:

$$\bar{\mathcal{E}} = -\frac{\Delta\Phi}{\Delta t} \qquad (20.3a)$$

The instantaneous EMF is

$$\mathcal{E} = -\frac{d\Phi}{dt} \qquad (20.3b)$$

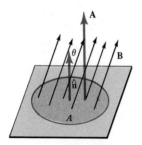

Figure 20.2. $\mathbf{A} = A\hat{\mathbf{n}}$ is a vector equal in magnitude to the area of the loop and directed parallel to the normal vector $\hat{\mathbf{n}}$. The magnetic flux through the loop is $\Phi = BA \cos \theta = B_n A = \mathbf{B} \cdot \mathbf{A}$.

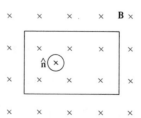

Figure 20.3. $\hat{\mathbf{n}}$ is chosen to be directed into the page. (With this choice, clockwise currents are positive, counterclockwise currents are negative.) Hence $\mathbf{B}$ is parallel to $\hat{\mathbf{n}}$ and $B_n = B$.

The minus signs indicate the direction of the induced EMF and current, as can be seen in the following examples. *Note that Faraday's Law contains only the rate of change of the flux and does not depend on how that rate is achieved.* This is consistent with the observation that moving the loop away in Fig. 20.1, reducing the loop area, or rotating the loop, all produce the same qualitative effect as reducing the field; all of these result in a reduction in the flux. Calculation of the induced EMF and the resulting current is illustrated by the next example.

Example 20.2

A wire loop of area 0.1 m² has a resistance of 10 ohms (Fig. 20.4). A magnetic field **B** normal to the loop initially has a magnitude of 0.2 T and is reduced to zero at a uniform rate in 10^{-4} s. Find the induced EMF and the resulting current.

Since the flux changes at a uniform rate, the average EMF and the instantaneous EMF are the same. Let us take $\hat{n}$ parallel to the field or into the page, so that clockwise currents are positive. The initial flux is $\Phi_i = BA$ and the final flux Φ_f is zero, so $\Delta\Phi = \Phi_f - \Phi_i = -BA$. Thus

$$\mathcal{E} = \bar{\mathcal{E}} = -\frac{\Delta\Phi}{\Delta t} = -\frac{(-BA)}{\Delta t} = \frac{BA}{\Delta t}$$

$$= \frac{(0.1 \text{ m}^2)(0.2 \text{ T})}{10^{-4} \text{ s}} = 200 \text{ V}$$

and

$$I = \frac{\mathcal{E}}{R} = \frac{200 \text{ V}}{10 \text{ ohms}} = 20 \text{ A}$$

Since I is positive, the current is clockwise.

Note that $\hat{n}$ can be chosen to be out of the page, so that counterclockwise currents are positive. With this choice of $\hat{n}$, $\Delta\Phi$ is positive, and the induced EMF and the current are negative. Since with this convention a negative current is clockwise, we have the same result as before.

It is important to notice that the induced current will also produce a magnetic field. In the previous example, the induced current is clockwise. The direction of the field can be found with the right-hand rule given in the last chapter (see Fig. 19.7). Applying the right-hand rule, we place our thumb along the current; our fingers show that the current produces a field into the page within the loop. This means that the field inside the loop due to the induced current is in the same direction as the initial field and *tends to slow the reduction in the field and in the flux.* Similarly, if the original field is increasing, the field produced by the induced current is opposite to the original field, *tending to retard its increase.* For example, if in Fig. 20.4, **B** is into the page and increasing, then the induced current will produce a field out of the page, so the current must be counterclockwise. *The observation that the field due to the induced current always opposes the change in flux is called Lenz's law.*

In solving problems, it is often easier to use Lenz's law to find the direction of the induced current than to carefully keep track of the minus signs in applying Faraday's law. However, we must remember that it is the *change* in flux, not the flux, that is opposed. This is seen again in the next example.

Example 20.3

According to Faraday's law, an EMF is induced when the flux through a loop is changed by varying its area. To illustrate this, consider a metal bar of length l sliding with a velocity **v** along two conducting rails that form a closed circuit (Fig. 20.5). A uniform magnetic field is directed into the page. What is the induced EMF?

We choose the normal vector $\hat{n}$ into the page or parallel to **B**; with this choice, clockwise currents are positive. If the area of the circuit is A, then the flux is $\Phi = BA$. A short time Δt later, the bar will have moved a distance $\Delta x = v\,\Delta t$. Hence the area will have increased by $\Delta A = l\,\Delta x = lv\,\Delta t$, and the flux will have increased by

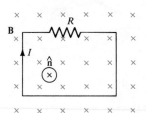

Figure 20.4. Example 20.2.

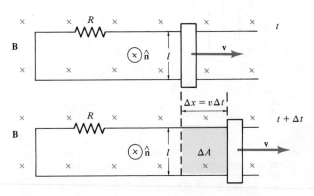

Figure 20.5. Example 20.3.

$\Delta\Phi = B\Delta A = Blv\Delta t$. The instantaneous induced EMF is therefore

$$\mathcal{E} = -\lim_{\Delta t \to 0} \frac{\Delta\Phi}{\Delta t} = -Blv$$

The minus sign indicates that the induced current, $I = \mathcal{E}/R$, will be negative, or counterclockwise. Using the right-hand rule, we see that the field within the loop due to this current will be out of the page, or opposite to **B**. This opposes the increase in the flux that occurs as the area increases, and again our result is in accordance with Lenz's law.

20.2 | EDDY CURRENTS

So far, we have considered examples of currents induced in circuits with a single current path. However, whenever the magnetic flux through a conducting object changes, EMFs and currents are induced in that object. Such *eddy currents* occur, for example, in electrical machinery. As we see in Section 20.6, iron has the property of greatly increasing an applied magnetic field. For this reason, motors and generators usually have large amounts of iron in magnetic fields. Since the flux through the iron is continually changing, eddy currents are produced that generate heat and reduce the efficiency of the machine. This effect is minimized by using thin iron sheets separated by insulating coatings rather than large solid pieces. This increases the resistance of the iron and decreases the eddy currents and hence the power loss (Fig. 20.6).

Eddy currents also arise when the flux through a conductor changes because of its motion. For example, if a metal pendulum swings between the poles of a strong magnet, it rapidly comes to rest (Fig. 20.7). This happens because the changing flux induces a current, and the magnetic force on that current is opposite to the motion of the pendulum. The kinetic energy lost by the pendulum is dissipated as heat by the eddy currents.

Eddy currents can also be induced in nonmetallic conductors, such as biological tissues. For example, a field of about 1 T alternating at 60 Hz will induce a large enough current in the retina of a human eye to produce a sensation of intense brightness.

20.3 | ELECTRIC GENERATORS

Faraday's law is the basic physical principle underlying electric power generators. To understand their operation, consider a wire loop of area A rotated by some power source at an angular velocity ω in a magnetic field (Fig. 20.8a). Its ends slide along two fixed rings. If at some instant the angle between the normal vector $\hat{\mathbf{n}}$ and the field **B** is $\theta = \omega t$, then the flux through the loop is

$$\Phi = \mathbf{B} \cdot \mathbf{A} = BA \cos \omega t \qquad (20.4)$$

Since the flux is continually changing, there is an EMF induced in the loop, which is the generator voltage. According to Eq. B.27 in Appendix B,

$$\frac{d}{dt}(\cos at) = -a \sin at$$

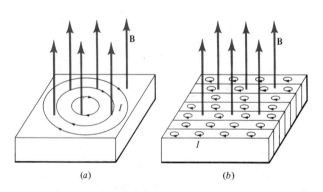

Figure 20.6. (*a*) An increasing magnetic field will induce eddy currents in a conducting object. (*b*) If the object is made of thin conducting sheets separated by insulating layers, the induced currents are confined to the individual sheets. The average current is smaller than in (*a*) and less power is dissipated as heat.

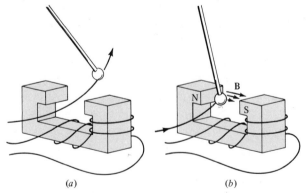

Figure 20.7. (*a*) If the magnet is turned off, the metal pendulum swings freely between the poles. (*b*) When the magnet power is on, the pendulum slows abruptly as it enters the magnetic field due to the magnetic forces on the eddy currents induced in the pendulum.

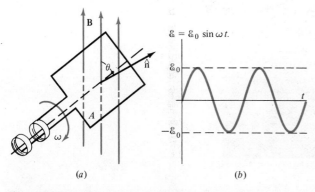

Figure 20.8. (a) A simplified ac generator consists of a coil rotating in a magnetic field. (b) The potential difference between the rings.

Thus $\mathcal{E} = -d\Phi/dt = -BA(-\omega \sin \omega t)$, or

$$\mathcal{E} = \mathcal{E}_0 \sin \omega t \qquad (20.5)$$

where the *amplitude* $\mathcal{E}_0$ is

$$\mathcal{E}_0 = \omega BA \qquad (20.6)$$

Since $\sin \omega t$ varies between $+1$ and -1, the potential difference between the two ends of the loop varies between $+\mathcal{E}_0$ and $-\mathcal{E}_0$ with a frequency $f = \omega/2\pi$ (Fig. 20.8b). Hence this rotating loop is an alternating current (ac) generator.

Direct current or dc generators supply an EMF that does not reverse polarity. The two rings of the ac generator are replaced by a single *split ring* or *commutator* (Fig. 20.9a). This reverses the connectors every half-turn and produces the EMF plotted in Fig. 20.9b. A practical dc generator has many loops each

at slightly different angles, so the resulting total EMF is nearly constant (Fig. 20.9c).

20.4 | TRANSFORMERS

The fact that an ac voltage can efficiently be increased or decreased with a *transformer* is of great practical importance to the transport and distribution of electrical power. To see why, we recall that the rate at which electrical energy is lost as heat in a power line of resistance R carrying a current i is i^2R. Since the power supplied by a pair of lines with a potential difference v is iv, choosing the largest feasible voltage and a correspondingly small current minimizes the heat losses. For example, power is sent hundreds of kilometres from hydroelectric plants to cities at some hundreds of thousands of volts. The power lines used are of moderate size. By contrast, if power were transported over such distances at a low voltage, huge amounts of copper would be needed to build lines with acceptable losses. However, safety and insulation limitations require that electricity be supplied to consumers at low voltages. This requirement is satisfied by using transformers to reduce the voltage in several steps in the local transmission system.

Figure 20.10 illustrates the principle of a transformer. *Primary* and *secondary* coils with N_1 and N_2 turns, respectively, are wound around an iron core. A variable current i_1 in the primary produces a magnetic field through both coils. The iron serves to increase the field and the flux. The induced EMF in any one turn of either coil is $-\Delta\Phi/\Delta t$; the total induced EMF in a coil is this times the number of turns. Since

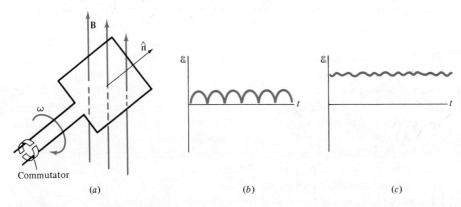

Figure 20.9. (a) A split ring or commutator reverses the connections each half turn. (b) The terminal voltage measured at the split ring. (c) The total EMF due to a large number of turns, each at slightly different angles.

MICHAEL FARADAY
(1791–1867)

(The Royal Institution, London.)

Michael Faraday discovered many of the fundamental laws of physics and chemistry, despite the fact that he had virtually no formal education. The son of an English blacksmith, he was apprenticed at the age of 14 to a bookseller and bookbinder. He read every book on science in the bookshop and attended lectures given at the Royal Institute by various scientists, including Sir Humphrey Davy, the discoverer of 12 chemical elements. In 1812, he applied to Davy for a job, citing his interest in science and showing Davy the extensive lecture notes he had taken. Davy hired Faraday to assist with his research and lecture demonstrations.

Within a few years, Faraday began to do original research on his own, submitting two papers on chemistry to the Royal Society in 1820. In that same year, Oersted discovered that a current in a wire will deflect a compass needle. Faraday repeated Oersted's experiments and found that a magnet also exerts a force on a wire carrying an electric current. Soon afterward, he also showed how to liquefy chlorine, and he isolated benzene, a compound now widely used in chemical products.

Faraday's important discoveries brought him considerable fame, much to the discomfort of Davy, who felt he should have shared the credit for some of the advances. Davy preferred to regard Faraday as a technical assistant and even forced him to serve as a valet on an extended tour of European research centers. Despite Davy's objections, Faraday was elected to the Royal Society in 1824 and was made director of the laboratory at the Royal Institute in 1825.

After Faraday discovered, in 1831, that a changing magnetic field can induce a current, he performed a series of experiments that showed clearly that the induced EMF is equal to the rate of change of magnetic flux. Also, generalizing from the patterns formed by iron filings around magnets, he invented the concepts of magnetic and electric field lines. Faraday knew little mathematics and found this concrete approach to electricity and magnetism much more useful than equations giving the forces between charges or currents. He also suggested that the propagation of light through space consisted of vibrations of these lines. His concepts of electric and magnetic fields were put into a mathematical form a generation later by Maxwell, who showed that light is, in fact, an oscillatory electromagnetic disturbance.

Faraday made many other notable contributions. He devised the first electrical generator, which consisted of a copper disk rotating between the poles of a magnet. He discovered the correct laws of electrochemistry after proving that earlier theories disagreed with experiments. He studied optical phenomena and found that when light passes through a medium, a magnetic field will rotate the direction of the oscillating electric field. Ignoring scorn from his contemporaries, he attempted unsuccessfully in laboratory experiments to find a link between gravitation and electromagnetism. Such a link was observed 70 years later in a test of Einstein's general theory of relativity, when light rays passing near the sun were found to be deflected.

Despite his achievements, Faraday remained a modest and humble person. He declined to be knighted or to receive honorary degrees and only reluctantly accepted a small pension on his retirement in 1858.

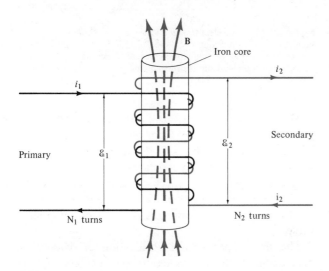

Figure 20.10. A transformer consists of a primary coil and a secondary coil wound around a core that is usually made of iron to increase the field and the flux.

$\mathcal{E}_1 = -N_1(\Delta\Phi/\Delta t)$ and $\mathcal{E}_2 = -N_2(\Delta\Phi/\Delta t)$, it follows that

$$\frac{\mathcal{E}_2}{\mathcal{E}_1} = \frac{N_2}{N_1} \qquad (20.7)$$

In the next example we see that if the secondary coil has more turns than the primary, the EMF will be increased by the transformer.

Example 20.4

The primary coil of a transformer has 100 turns and is connected to a 120-V ac power line. If the secondary coil has 1000 turns, what is its terminal voltage?

According to Eq. 20.7,

$$\mathcal{E}_2 = \frac{N_2}{N_1}\mathcal{E}_1 = \left(\frac{1000}{100}\right) 120 \text{ V}$$
$$= 1200 \text{ V}$$

Transformers are useful in increasing electrical safety. Since the primary and secondary coils are not directly connected to each other, the secondary is electrically "*isolated*," and its potential relative to ground is independent of that in the primary. If a person simultaneously touches a wire in the secondary circuit and a ground, this will ground the wire, but no current will pass through the person. This is especially useful in reducing the risk associated with high voltages. Also, the extreme vulnerability of patients wired into EKGs, pacemakers, and other devices has led to the proposal that hospital units serving such people should have all their instrumentation supplied with power via carefully designed isolation transformers.

20.5 | INDUCED FIELDS AND ELECTROMAGNETIC WAVES

We have seen that a varying magnetic field induces an EMF in a wire loop. This EMF, in turn, does work on the charges in the conductor and sustains a current. Alternatively, we may say that the *changing magnetic field* **B** *produces an induced electric field* **E** in the loop, and this electric field exerts an electric force on the charges.

Since an induced electric field *does net work* on a charge as it goes around a closed path, *an induced electric field is not conservative.* In this respect, an induced electric field due to a changing magnetic field is quite different from the conservative electrostatic field produced by electrical charges, which does no net work around a closed path.

It is not necessary to have a conductor present in the changing magnetic field in order to observe the effects of the induced electric fields. One example of this provided by the *betatron,* which accelerates electrons to energies of about 100 MeV = 10^8 eV. If the electrons are allowed to strike a target, they produce highly penetrating X rays that can be used for cancer therapy. In a betatron, a magnetic field maintains the electrons in circular orbits, just as in the cyclotron

(Chapter Nineteen). However, the magnetic field is not kept constant. Instead, it is gradually increased with time, and the resulting induced electric field accelerates the electrons. With proper shaping of the magnetic field, the radius of the electronic orbits remains constant as the electronic speed and the field increase. Thus the electrons can be confined to a narrow doughnut-shaped tube instead of requiring a large chamber as in a cyclotron.

With the introduction of the idea of an induced electric field resulting from a changing magnetic field, there is a fundamental connection between electrical and magnetic phenomena. However, the situation is not symmetric unless we suppose that a changing electric field can also induce a magnetic field.

Maxwell's Hypothesis and Electromagnetic Waves

James Clerk Maxwell (1831–1879) made the brilliant hypothesis in 1864 that *changing electric fields do indeed induce magnetic fields*. He was led to this idea by considering the relationships between the basic laws of electromagnetism, which had been discovered decades before. These were:

1 Coulomb's law for the force between two charges, or equivalently, for the electric field due to a point charge.
2 The Biot-Savart law for the magnetic field due to a current.
3 Faraday's law, which states that a changing magnetic field induces an electric field.
4 The conservation of electric charge.

Maxwell showed that these laws were not mathematically consistent when the electric fields were changing in time. However, if one hypothesized that a changing electric field could induce a magnetic field, the inconsistency was removed.

Although it was based on purely theoretical reasoning, Maxwell's hypothesis led immediately to the prediction that *electromagnetic waves* can be produced by oscillating charges or currents. For example, suppose a current oscillates (reverses direction) at some frequency. The current produces a magnetic field nearby that also oscillates at this frequency. According to Faraday's law, the magnetic field induces an oscillatory electric field in its vicinity. Maxwell's hypothesis provides the critical next step: *the changing electric field induces a magnetic field*. This magnetic field in turn induces an electric field, and so on (Fig. 20.11).

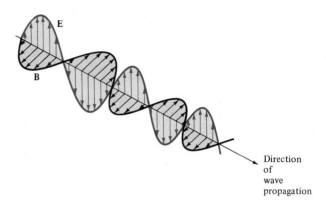

Figure 20.11.   An electromagnetic wave. An oscillating magnetic field induces an oscillating electric field, which in turn induces an oscillating magnetic field, and so on. This disturbance travels at the speed of light.

Maxwell showed that this electromagnetic disturbance would travel away from the oscillating current with a velocity equal to $(k/k')^{1/2} = 3.00 \times 10^8$ m s^{-1}, where k and k' are the constants in the electric and magnetic force laws, respectively. Since this velocity was equal to the velocity of light to the accuracy then known, he inferred that light is an electromagnetic wave.

The first experimental verification of Maxwell's prediction of electromagnetic waves came nearly a quarter-century after he had published his work and several years after his death. In 1887, Heinrich Hertz (1857–1894) constructed two circuits that tended to oscillate at the same frequency. He found that if one circuit was connected to an EMF so that it had a current, the other circuit located some distance away would also have a current.

The "technological fallout" from this demonstration of electromagnetic waves quickly followed. In 1890, Guglielmo Marconi (1874–1937) became interested in wireless telegraphy, and by 1901 he succeeded in transmitting signals across the Atlantic. Today, we enjoy the benefits of radio, television, and other communications systems based on electromagnetic waves produced by oscillating currents.

The wavelength of the radiation produced by electric circuits ranges from millimetres to kilometres or longer, while visible light has a wavelength of 4×10^{-7} to 7×10^{-7} m. It has been verified to a very high accuracy that all the waves produced by oscillating currents have the same velocity as that of visible light. Thus, Maxwell's remarkable theoretical predictions have been fully verified.

20.6 | MAGNETIC MATERIALS

We saw in Chapter Sixteen that the introduction of a dielectric into an applied electric field **E** reduces the field to **E**/K, where K is the dielectric constant. Similarly, the introduction of materials into an applied magnetic field **B** changes that field to K_m**B**, where K_m is the *magnetic constant.*

There are three major kinds of magnetic materials. In *diamagnetic* materials, K_m is slightly less than 1. In *paramagnetic* materials, K_m is slightly greater than 1. Finally, in *ferromagnetic* materials, K_m varies with the applied field and with the way the material has been treated, but it is typically very large compared to 1.

When a diamagnetic material is placed in a changing applied magnetic field, changes are induced in the electron currents in the atoms. When the applied field is increasing, the field due to the induced currents opposes the applied field, in accordance with Lenz's law, so the net field is smaller than the applied field. Diamagnetism is a weak effect; typically, K_m is less than 1 by 0.001 to 0.01 percent in diamagnetic materials. Diamagnetism is present in all materials, although it is sometimes masked by other effects.

In many materials, when there is no applied field, the magnetic effects associated with the orbital and spin motions of the atomic electrons cancel exactly. However, in some materials, which are called paramagnetic, there is a permanent residual magnetic dipole moment associated with the individual atoms. In the absence of an applied field, thermal agitation will cause these dipoles to be randomly aligned, and no macroscopic magnetic field will be observed. However, when such a paramagnetic substance is placed in a magnetic field, the dipoles tend to align themselves with their moments along the field, increasing the total field inside the material. Since the effects of the permanent dipoles are usually larger than those of induced currents, paramagnetism, when present, will mask diamagnetic effects. Paramagnetic materials typically have values of K_m that are larger than 1 by about 0.01 percent.

Five elements (Fe, Co, Ni, Gd, and Dy) are ferromagnetic, as are many alloys. Ferromagnets are characterized by strong interactions between neighboring atomic dipoles that are sufficiently strong to cause a spontaneous alignment of the dipoles, even in the absence of an applied magnetic field.

Although the dipoles align themselves without an applied field, the fields produced outside the material are often very small. This is associated with the be-

havior of *magnetic domains,* regions in which all of the dipoles have the same alignment. If the directions of the domains are sufficiently varied, their net field will be small. These domains are rather stable, but when an external magnetic field is applied, the domains most nearly parallel with the field grow in size at the expense of others. The domains may also rotate into alignment with the field. In both cases the ordering of the dipoles may enhance the field by a factor of 1000 or more. For this reason, ferromagnetic materials play an important role in transformers and other devices requiring large magnetic fields. When the applied field is removed the domains in a ferromagnet may retain some order. In this case the sample is a permanent magnet (Fig. 20.12).

20.7 | INDUCTANCE

A current in a circuit always produces a magnetic field and a magnetic flux through that circuit. If this current changes, so does the flux, and this causes a *self-induced* EMF $\mathcal{E}$ in the circuit. According to Lenz's law, the induced current opposes the change in the flux and therefore also opposes the change in the current. *Thus self-induced EMFs tend to prevent rapid current changes in circuits.*

One familiar example of the effect of self-induced EMFs in limiting current changes is the flash seen

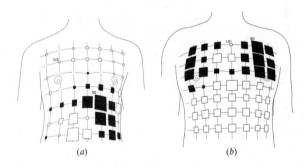

Figure 20.12. Ferromagnetic particles in the body can be detected by magnetizing them with a strong magnetic field and observing the fields they produce afterwards. (*a*) This man has eaten beans from a can, and his stomach contains about 10^{-4} grams of iron oxide. The areas of the squares are proportional to the field at their centers; the largest and smallest nonzero values are indicated in units of 10^{-11} T. Solid squares indicate fields in one direction, and open squares show fields in the opposite direciton. (*b*) A similar plot for a man with 5×10^{-4} g of iron oxide in his lungs inhaled while doing arc welding. Since asbestos has small amounts of iron oxide adhering to its fibers, measurements of this type can also be used to determine the amount of the dangerous asbestos dust inhaled by workers. (D. Cohen, *Physics Today,* August, 1975, p. 41. © American Institute of Physics.)

when a high-current appliance, such as a toaster, is on and its plug is pulled out of the electric outlet. This flash or arc occurs because the sudden current change induces a large EMF, which produces an electric field strong enough to ionize the air and permit current to pass through it briefly.

The self-induced EMF in a circuit can be related as follows to the rate of change of the current in that circuit. The EMF is proportional to the rate of change of flux. The flux, in turn, is proportional to the field and therefore to the current. Thus the EMF is proportional to the rate of change of current, and the average induced EMF is

$$\overline{\mathcal{E}} = -L\frac{\Delta i}{\Delta t} \qquad (20.8a)$$

The constant L is called the *inductance* of the circuit. The minus sign indicates that the induced EMF opposes the current change. The instantaneous EMF is

$$\mathcal{E} = -L\frac{di}{dt} \qquad (20.8b)$$

The inductance depends on the geometry of the circuit and on the magnetic materials present in its vicinity. In S.I. units, L is measured in *henries* (H). This unit is named after Joseph Henry (1797–1878), an American contemporary of Faraday, who independently discovered induced EMFs.

We can calculate L if the flux Φ through the circuit is known. In a coil with a flux Φ through each of N turns, the total average induced EMF is $-N\,\Delta\Phi/\Delta t$. Comparing this with Eq. 20.8, we have

$$Li = N\Phi \qquad (20.9)$$

Consider, for example, a long thin coil with n turns per unit length, length ℓ, and cross-sectional area A wound on a core with magnetic constant K_m (Fig. 20.13). From Chapter Nineteen, the field in the coil is

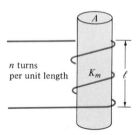

Figure 20.13. A solenoid wound on a core with magnetic constant K_m.

approximately $B = 4\pi K_m k'in$, where we have added a factor K_m to the earlier formula to account for the core. The inductance is then $L = N\Phi/i = (n\ell)(BA)/i$ or

$$L = 4\pi K_m k'n^2 A\ell \quad \text{(long thin coil)} \quad (20.10)$$

Note that the inductance is proportional to K_m and to the coil volume $A\ell$. Such a coil is referred to as an *inductor*. Its effect is seen in the next example.

Example 20.5

A coil with 1000 turns has a cross section of $1\text{ cm}^2 = 10^{-4}\text{ m}^2$ and a length of 0.1 m. (a) If K_m is 1, find the inductance. (b) The current increases from 0 to 1 A in 10^{-3} s. Find the average EMF in the coil (c) Suppose the current in the coil varies as $i = i_0 \cos \omega t$, where $i_0 = 10$ A, and $\omega = 1000$ rad s^{-1}. Find the induced EMF as a function of time.

(a) Using Eq. 22.10, the inductance is

$$L = 4\pi K_m k'n^2 A\ell$$

$$= 4\pi(1)(10^{-7}\text{ T m A}^{-1})\left(\frac{1000}{0.1\text{ m}}\right)^2(10^{-4}\text{ m}^2)(0.1\text{ m})$$

$$= 1.26 \times 10^{-3}\text{ H} = 1.26\text{ mH}$$

Note that this relatively large coil has an inductance of only a small fraction of a henry. The henry is a comparitively large unit.

(b) The average EMF induced in the coil is

$$\overline{\mathcal{E}} = -L\frac{\Delta i}{\Delta t} = -(1.26 \times 10^{-3}\text{ H})\frac{1\text{ A}}{10^{-3}\text{ s}} = -1.26\text{ V}$$

The minus sign indicates that the EMF opposes the current change.

(c) The instantaneous induced EMF is

$$\mathcal{E} = -L\frac{di}{dt} = -Li_0(-\omega)\sin \omega t = Li_0 \omega \sin \omega t$$

$$= (1.26 \times 10^{-3}\text{ H})(10\text{ A})(1000\text{ s}^{-1})\sin(1000t\text{ s}^{-1})$$

$$= 12.6\sin(1000t\text{ s}^{-1})\text{ V}$$

The EMF varies from -12.6 V to $+12.6$ V during each cycle.

A current in one circuit can produce a magnetic flux through another circuit. If that is the case, then a change in the current will induce an EMF in the second circuit. The ratio of the EMF induced in the second circuit to the rate of change of the current in the first is the *mutual inductance* of the two circuits (See Problem 20-40). This phenomenon provides a way to transfer power and information from one circuit to another. It also represents a potential problem in complex electronic systems, since unwanted signals from one part of the device can be picked up elsewhere. These effects are minimized by the careful placement of components. Also, magnetic fields will not penetrate far into metals except at very low frequencies, so metallic shielding is sometimes used.

20.8 | ENERGY STORED IN AN INDUCTOR

Just as a capacitor can store electrical energy, so also can an inductor store magnetic energy. To calculate this energy, we note that the power that must be *supplied* to change a current i in an inductor at a rate di/dt is $\mathcal{P} = -\mathcal{E}i = L(di/dt)i$. In a time dt, the work done *against* the induced EMF $\mathcal{E}$ is

$$dW = \mathcal{P}dt = \left(Li\frac{di}{dt}\right)dt = Li\,di$$

Now the energy stored in an inductor with current I equals the work done in increasing the current from 0 to I. Thus

$$\mathcal{U} = \int_0^I Li\,di = \tfrac{1}{2}L\left[i^2\right]_0^I$$

or

$$\mathcal{U} = \tfrac{1}{2}LI^2 \qquad (20.11)$$

For example, if a 10^{-2}-H inductor carries a current of 1 A, the energy stored is $\tfrac{1}{2}LI^2 = \tfrac{1}{2}(10^{-2}\text{ H})(1\text{ A})^2 = 5 \times 10^{-3}$ J.

The quantitative effects of inductances on circuits are considered in the supplementary topics at the end of this chapter.

SUMMARY

In a uniform magnetic field, the magnetic flux through a loop of area A is $\Phi = B_n A$, where B_n is the normal component of the field. Whenever the flux is changing, Faraday's law states that there is an induced EMF in the loop equal to the time rate of change of the flux:

$$\mathcal{E} = -\frac{d\Phi}{dt}$$

Lenz's law states that the current due to an induced EMF always has a magnetic field that tends to retard the change in the flux. The induced EMF is determined only by the flux change; it does not matter whether it is the magnetic field or the area of the loop in the field that is changing.

When the magnetic flux through a conducting object changes, eddy currents are induced. The changing magnetic flux through the coil of a generator is the source of its EMF. In a transformer, a changing flux due to the current in the primary coil induces an EMF in the secondary coil. The ratio of the EMFs equals the ratio of the number of turns,

$$\frac{\mathcal{E}_1}{\mathcal{E}_2} = \frac{N_1}{N_2}$$

A changing magnetic field will induce an EMF even if no conducting loop is present, as is illustrated by the betatron. This is equivalent to stating that the changing magnetic field induces a nonconservative electric field. Since changing electric fields also induce magnetic fields, oscillating charges or currents produce electromagnetic waves that travel with the speed of light.

The self-induced EMF in a circuit slows current changes. It is proportional to the inductance and to the rate of change of the current,

$$\mathcal{E} = -L\frac{di}{dt}$$

The energy stored in an inductor is

$$\mathcal{U} = \tfrac{1}{2}LI^2$$

Checklist

Define or explain:

induced current	induced electric field
Faraday's law	induced magnetic field
magnetic flux	electromagnetic waves
weber	diamagnetism
Lenz's law	paramagnetism
eddy currents	ferromagnetism
amplitude	magnetic domains
generator	inductance
transformer	henry
primary, secondary coils	

REVIEW QUESTIONS

Q20-1 The flux through a loop is the product of its area and the _____.
Q20-2 The induced EMF is equal to the _____.
Q20-3 Magnetic flux is measured in _____.
Q20-4 The induced current produces a magnetic field that tends to _____ the change in flux.

Q20-5   In the design of machinery, eddy currents are minimized in order to reduce _____.

Q20-6   In a dc generator, the connections are reversed every _____.

Q20-7   If the secondary coil in a transformer has more turns than the primary coil, the EMF will be _____ in the secondary.

Q20-8   Maxwell proposed that a changing _____ field induces a _____ field.

Q20-9   Maxwell predicted the existence of _____ that travel at the speed of _____.

Q20-10   Diamagnetic materials slightly _____ magnetic fields, while paramagnetic materials slightly _____ magnetic fields.

Q20-11   The strong magnetic effects in ferromagnetic materials arise from the alignment of _____.

Q20-12   Self-induced EMFs tend to prevent _____ in circuits.

Q20-13   If the current in an inductor is doubled, the stored energy increases by a factor of _____.

EXERCISES

Section 20.1 | Faraday's Law

20-1   In Fig. 20.14 the magnetic field is uniform in the region shown and is into the page. A wire loop is in the plane of the page. State whether there is an induced current when the loop is moved from (a) P_1 to P_2; (b) P_2 to P_3; (c) P_3 to P_1. Give reasons for your answers.

20-2   The wire loop in Fig. 20.14 is moved in the plane of the page from position P_1 to P_2. Find the direction of the induced current, if any.

20-3   A copper loop and a rubber loop of the same size and shape are placed in a uniform magnetic field perpendicular to the plane of the loops, which is gradually increased. Discuss (a) The EMFs induced in the two loops; (b) the currents induced in the two loops.

20-4   In Fig. 20.15, (a) what is the direction of the field inside the rectangular wire loop due to the current I in the long straight wire? (b) If I is increasing, what is the direction of the current induced in the loop? (c) If I is constant and the loop is moved toward the right, what is the direction of the induced current?

20-5   A wire loop of radius 0.1 m has a 2-ohm resistance. A uniform magnetic field perpendicular to the plane of the loop increases from zero to 2 T in 0.01 s. Find the average induced current.

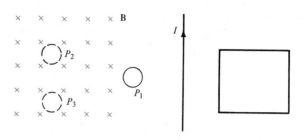

Figure 20.14.   Exercises   **Figure 20.15.**   Exercise 20-4.
20-1 and 20-2.

20-6   A circular wire loop has a resistance of 10 ohms and a radius of 0.1 m. Its normal is parallel to a uniform magnetic field of magnitude 3 T. If the loop is rotated so that in 0.3 s its normal is perpendicular to the field, find (a) the average induced EMF; (b) the average induced current.

20-7   A loop of area 0.1 m² is at right angles to a uniform magnetic field. The field alternates direction at 60 Hz and has a peak magnitude of 2 T. (a) What is the average induced EMF during one-half cycle when the field varies from normal to the loop to zero and to normal again, but in the opposite direction? (b) What is the average EMF over a full cycle?

20-8   In Fig. 20.5, the resistance of the loop is 100 ohms, the velocity of the moving conductor is 0.7 m s⁻¹, its length is 0.05 m, and B is 0.5 T. Find the induced current.

Section 20.2 | Eddy Currents

20-9   In Fig. 20.16, the field through a conductor is upward and increasing in magnitude. What is the direction of the eddy currents?

20-10   Suppose the field in Fig. 20.16 is constant and the conductor is gradually removed to the left. What is the direction of the eddy currents induced?

Section 20.3 | Electric Generators

20-11   Find the average EMF induced during the one-quarter cycle it takes for the normal vector in Fig. 20.8 to rotate from parallel to the field to perpendicular to the field. (Express the result in terms of A, B, and ω.)

20-12   A square wire loop of resistance 10 ohms with sides 0.2 m rotates 100 times per second about a horizontal axis (Fig. 20.17). The magnetic field is vertical and has a magnitude of 0.5 T. Find the amplitude of the induced current.

20-13   When a generator turns at 60 Hz, the am-

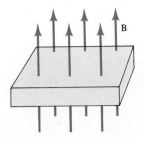

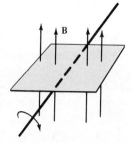

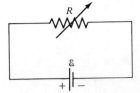

Figure 20.16. Exercises 20-9 and 20-10.　　**Figure 20.17.** Exercise 20-12.

Figure 20.18. Exercise 20-21.

plitude of the induced EMF is 50 V. What is its amplitude when it turns at 180 Hz if the magnetic field remains the same?

Section 20.4 | Transformers

20-14 The tube of an X-ray machine requires an ac voltage of 50,000 V. If this is obtained from a 120-V line, what is the ratio N_2/N_1 of the number of turns in the secondary and primary coils of the transformer?

20-15 In a transformer used with a set of toy trains, the 120-V ac line voltage is stepped down to 9 V. What is the ratio N_2/N_1 of the turns?

20-16 A transformer has 100 turns in one coil and 500 in the other. It is connected to a 120-V ac power line. What is the voltage across the secondary if (a) the 100-turn coil is connected to the power line; (b) the 500-turn coil is connected to the power line?

Section 20.5 | Induced Fields and Electromagnetic Waves

20-17 Find the ratio of $(k/k')^{\frac{1}{2}}$ to the speed of light in a vacuum, c. (Using the best experimental values of the constants involved gives agreement to better than one part per million.)

Section 20.6 | Magnetic Materials

20-18 The field in a solenoid is 0.5 T when it is in a vacuum. A rod is placed inside the solenoid, and the field drops to 0.498 T. (a) What is the magnetic constant of the material? (b) What kind of magnetic material is the rod made of?

Section 20.7 | Inductance

20-19 Why is it important that the materials used in or near electrical switches and outlets be fireproof?

20-20 How can one make a variable inductance?

20-21 The variable resistance in the left-hand loop of Fig. 20.18 is increased. Find the direction of the current induced in the (a) left-hand loop, and (b) right-hand loop.

20-22 The current in a 0.2-H inductor is increased from zero to 10 A in 0.1 s. Find the induced EMF.

20-23 Figure 20.19 shows the current in a solenoid during some time interval. Sketch how the voltage drop across the solenoid varies during that interval.

20-24 A solenoid with 1000 turns is wound on a core of radius 0.01 m and length 0.2 m. The magnetic constant of the core is 1000. Find the inductance of the coil.

20-25 Show that 1 henry = 1 ohm-second = 1 weber-ampere^{-1}.

Section 20.8 | Energy Stored in an Inductor

20-26 When the current in a coil is increased uniformly from zero to 10 A in 2 s, the self-induced EMF is 8 V. (a) How large is the inductance of the coil? (b) How much energy does the coil store when the current is 10 A?

20-27 An inductance stores 4 J of energy when the current is 5 A. How large is it?

20-28 A coil stores 50 J of energy when the current is 10 A. How much current is required for it to store 450 J?

20-29 Show that 1 henry = 1 joule-ampere^{-2}.

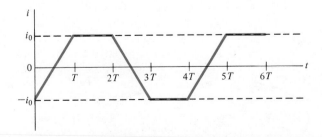

Figure 20.19. Exercise 20-23.

PROBLEMS

20-30 Suppose the opposite to Lenz's law were true, so that induced currents would tend to increase changes in flux instead of opposing them. Would this be consistent with the energy conservation principle? Explain.

20-31 (a) In Fig. 20.5, what is the magnetic force on a charge q in the moving conductor? (b) If this charge moves the length l of the moving conductor, how much work W is done on it? (c) The EMF is defined as the work done per unit charge, W/q. Compare this result with the EMF obtained in Example 20.3.

20-32 A wire loop of resistance R and area A has its normal along the direction of a uniform magnetic field, **B**. The loop is then flipped over in a time Δt so that its normal is opposite to the field. (a) Calculate the average induced EMF. (b) Find the average induced current. (c) If the field is out of the page, what is the direction of the induced current? (d) What is the total charge transported through the circuit in the time interval Δt? (This charge turns out to be independent of Δt and proportional to the field. Therefore, it is possible to determine magnetic fields by using such a coil and measuring the total charge.)

20-33 A bar magnet is dropped through a wire loop (Fig. 20.20). Find the direction of the induced current when the magnet is (a) above the loop; (b) halfway through the loop; (c) below the loop. Draw a rough graph of the induced current versus time, assuming the velocity of the magnet is constant.

20-34 In Fig. 20.21, a metal pendulum is moving into a uniform magnetic field. (a) Are the induced eddy currents generally clockwise or counterclockwise? (b) What is the direction of the magnetic force on the pendulum? (Explain your answers in both parts.)

20-35 When a motor coil turns, the magnetic field in the motor induces a *back EMF* in its coil that opposes the current. (a) How does the back EMF depend upon the speed of the motor? (b) How does the EMF affect the current? (c) When a motor is connected to a large load, its speed decreases. What effect will this have on the current in the motor? (d) Why is there a danger that an excessive load will burn out a motor?

20-36 The ratio of the power transmitted into the secondary of a transformer to that supplied to the primary is usually over 90 percent, but it is never 100 percent. Suggest some reasons for the power losses.

20-37 Explain why the magnetic constant of a diamagnetic material does not change as the temperature is increased. Would a paramagnetic material have the same temperature dependence? Explain.

20-38 It has been proposed to build very large magnets using superconducting wires that have no resistance and therefore dissipate no heat. In one proposed design, 1000 turns of radius 100 m form a coil of height 100 m in rock ($K_m = 1$). (a) Using the formula for the inductance of a solenoid, estimate the inductance of this coil. (This is only a rough estimate since this coil is not long in relation to its width). (b) Estimate the energy stored if the current is 1.5×10^5 A. (c) A very large nuclear reactor or fossil fuel electric power plant produces 10^9 W of power. For how long can the superconducting magnet supply energy at this rate if its current is initially 1.5×10^5 A?

20-39 A solenoid carrying a current I has N turns, area A, length l, and magnetic constant K_m. (a) What is the magnetic field B inside the solenoid? (b) How much energy is stored per unit volume of the solenoid? (c) The energy stored per unit volume in a magnetic field is a constant times B^2/K_m. Use your result in part (b) to find that constant.

*20-40 A change in the current i_1 in a loop (1) will induce an EMF $\mathcal{E}_2$ in a second loop (2) nearby. The *mutual inductance M* is defined by $\mathcal{E}_2 = -M di_1/dt$. Two solenoids with N_1 and N_2 turns, respectively, are wound on a common core of area A, length l, and magnetic constant K_m. Show that the mutual inductance is

$$M = 4\pi K_m k' \frac{N_1 N_2}{l} A.$$

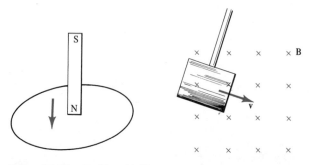

Figure 20.20. Problem 20-33. **Figure 20.21.** Problem 20-34.

*20-41 Show that two inductances L_1 and L_2 connected in series have an equivalent inductance equal to $L_1 + L_2$.

*20-42 Show that two inductances L_1 and L_2 in parallel are equivalent to a single inductance L, where

$$\frac{1}{L} = \frac{1}{L_1} + \frac{1}{L_2}$$

20-43 Using the results of Problems 20-41 and 20-42, find the equivalent inductance of a 2-H and a 4-H inductor connected (a) in parallel; (b) in series.

*c20-44 In Fig. 20.22, the current in the long straight wire is i. (a) What is the flux due to this current through a strip of width dr in the rectangular loop if the strip is at a distance r from the wire? (b) What is the total flux due to i through this loop? (c) If i is changing at a known rate di/dt, what is the EMF induced in the loop? (d) The rectangular loop has a resistance R. What is the current i_1 induced in this loop? (e) If i is increasing, what is the direction of i_1?

c20-45 The current through a coil with inductance L is $i_0 e^{-t/T}$. (a) Find the EMF induced in the coil at time t. (b) Find the energy stored in the coil at t.

ANSWERS TO REVIEW QUESTIONS

Q20-1, normal component of the magnetic field; **Q20-2,** rate of change of magnetic flux; **Q20-3,** webers; **Q20-4,** oppose; **Q20-5,** power losses; **Q20-6,** half cycle; **Q20-7,** greater; **Q20-8,** electric, magnetic; **Q20-9,** electromagnetic waves, light; **Q20-10,** decrease, increase; **Q20-11,** magnetic domains; **Q20-12,** rapid current changes; **Q20-13,** 4.

SUPPLEMENTARY TOPICS

20.9 | *RL* CIRCUITS

We saw in Section 20.7 that any circuit always has some inductance. The inductance retards current changes, but it has no effect on a current that is constant. We now examine this in more detail for a series circuit consisting of a resistance R, an EMF $\mathcal{E}$ provided by a battery, and an inductance L (Fig. 20.23).

When the switch is closed in the circuit, the inductance prevents a sudden change in the current. Instead, the current gradually increases from zero toward the current predicted by Ohm's law,

$$i_f = \frac{\mathcal{E}}{R} \tag{20.12}$$

The *time constant* T_L, which characterizes this buildup, is proportional to L, since the greater the inductance the longer it takes to achieve the final current. If the resistance R is large, then the magnitude of the overall current change is small, and the final current may be achieved more rapidly. Thus, we also expect T_L to be proportional to $1/R$, and the time constant is

$$T_L = \frac{L}{R} \tag{20.13}$$

To see how the current changes in detail, we suppose that in Fig. 20.23 the switch is closed at $t = 0$. Since the current was zero before the switch was closed, it is zero immediately after because the inductance prevents sudden current changes. At a later time t, there is a current i and a corresponding potential drop iR across the resistor. Since the current is increasing, there is an induced EMF in the inductance equal in magnitude to $L di/dt$. The voltage drops by this amount as we go through the inductance with the current. Also, the voltage rises by $\mathcal{E}$ in the EMF.

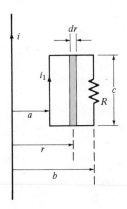

Figure 20.22. Problem 20-44.

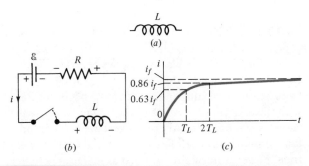

Figure 20.23 (*a*) The symbol for an inductor. (*b*) An *RL* series circuit. The switch is closed at $t = 0$. (*c*) The current in the circuit versus time.

Setting the algebraic sum of the voltage changes around the circuit to zero,

$$-iR - L\frac{di}{dt} + \mathcal{E} = 0 \qquad (20.14)$$

The solution to this equation is

$$i = i_f(1 - e^{-t/T_L}) \qquad (20.15)$$

Here the final current $i_f = \mathcal{E}/r$ and the time constant $T_L = L/R$ are as defined above. Note that at $t = 0$, $i = 0$ as required. Also, when $t/T_L \gg 1$, i approaches i_f and di/dt approaches zero, as we anticipated. It is left as an exercise to verify by differentiating that Eq. 20.15 does satisfy Eq. 20.14.

It is clear from the equation for i that the time constant T_L plays a role in the RL series circuit that closely parallels that of the time constant RC in an RC series circuit. When $t = T_L$, the current differs from its final value i_f by $i_f e^{-1} = 0.37 i_f$; when $t = 2T_L$, the current differs from i_f by $0.14 i_f$, and so on. These results are illustrated by the following example.

Example 20.6

A coil with a resistance of 10 ohms and an inductance of 0.1 H is connected to a 12-V battery. (a) Find the time constant. (b) What is the final current? (c) Find the current after 0.03 seconds.

(a) The time constant is

$$T_L = \frac{L}{R} = \frac{0.1\ \text{H}}{10\ \text{ohms}} = 0.01\ \text{s}$$

(b) From Ohm's law, the final current is

$$i_f = \frac{\mathcal{E}}{R} = \frac{12\ \text{V}}{10\ \text{ohms}} = 1.2\ \text{A}$$

(c) Since 0.03 s is equal to $3T_L$, the current at that time is

$$i = i_f(1 - e^{-t/T_L}) = (1.2\ \text{A})(1 - e^{-3}) = 1.14\ \text{A}$$

The current is within a few percent of its final value.

20.10 | EFFECTIVE OR ROOT MEAN SQUARE ALTERNATING CURRENTS AND VOLTAGES

In the remainder of this chapter, we discuss some basic features of alternating currents, begining here with the concept of *effective* or *root mean square (rms) currents* and *voltages*. Alternating currents are important in many applications. As we noted earlier, electrical power is normally supplied as 60 Hz ac. Higher frequency currents are present in the circuits used to produce radio waves with an antenna or sound waves with a loudspeaker. Combinations of currents with many frequencies can also occur. For example, to reproduce a musical sound or a human voice, a loudspeaker is supplied with a complex current that is a sum of many small currents with various frequencies. Such complex currents can be discussed by analyzing each small current separately.

In discussing ac circuits, it is customary to use effective or rms currents and voltages, since these are convenient for power calculations. Alternating current ammeters and voltmeters are normally calibrated to read these quantities, rather than the amplitudes or peak values.

To define the effective current, consider a circuit connected to an ac generator. The current varies with the generator frequency $f = \omega/2\pi$ (Section 20.3) and has the form $i = i_0 \sin \omega t$, where i_0 is the amplitude of the current. Since $\sin \omega t$ is positive for one half-cycle and negative for the other, the current averages to zero over a full cycle. Nevertheless, the current still has some effects; for example, it heats a resistance R at an instantaneous rate $\mathcal{P} = i^2 R$ (Fig. 20.24). Since i^2 is never negative, its average cannot be zero. According to Eq. B.13 in Appendix B, the average of $\sin^2 \omega t$ over a full cycle is $1/2$. Hence the average value of $i^2 = i_0^2 \sin^2 \omega t$ is $i_0^2/2$, and the average power dissipated in a resistor is

$$\bar{\mathcal{P}} = \frac{i_0^2}{2} R = i_e^2 R \qquad (20.16)$$

Here the *effective* or *root mean square (rms) current* i_e is defined by

$$i_e = \frac{i_0}{\sqrt{2}} \qquad (20.17)$$

Thus, an effective current of 1 ampere in an ac circuit produces exactly as much heat in a resistor as a 1-ampere dc current. Notice that both i_0 and i_e are defined as positive quantities.

Effective ac voltages are defined in an analogous fashion. If the voltage amplitude is v_0, the *effective or rms voltage* is

$$v_e = \frac{v_0}{\sqrt{2}} \qquad (20.18)$$

Again v_0 and v_e are always positive.

The effective voltage across a resistor v_e^R can be related to the current using Ohm's law, $v = iR$. At

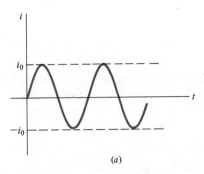
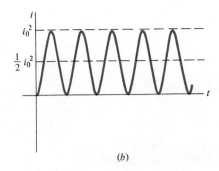

Figure 20.24. (a) An alternating current $i = i_0 \sin \omega t$ has an average value of zero. (b) The average of $\sin^2 \omega t$ over a full cycle is $\frac{1}{2}$, so the average value of i^2 is $i_0^2/2$.

any instant, the voltage and current are related by

$$v^R = (i_0 \sin \omega t)R \qquad (20.19)$$

Squaring and averaging, we find $(v_e^R)^2 = \frac{1}{2}i_0^2R^2$, or

$$v_e^R = i_e R \qquad (20.20)$$

The average power dissipated in a resistor can then be expressed as

$$\bar{\mathcal{P}} = \frac{(v_e^R)^2}{R} = i_e v_e^R \qquad (20.21)$$

Thus with effective quantities, we have exactly the same formulas as for the power dissipation in a dc circuit. These quantities are used in the following example.

Example 20.7

A broiler draws 1000 W from a 120-V rms power line. Find (a) the amplitude of the power line voltage; (b) the rms current; (c) the amplitude of the current.

(a) Since the rms voltage is 120 V, $v_e = v_0/\sqrt{2}$ gives for the amplitude or peak voltage

$$v_0 = \sqrt{2}v_e = 1.414(120 \text{ V}) = 170 \text{ V}$$

(b) Using $\bar{\mathcal{P}} = i_e v_e^R$, with v_e^R equal to the rms line voltage, the rms current is

$$i_e = \frac{\bar{\mathcal{P}}}{v_e^R} = \frac{1000 \text{ W}}{120 \text{ V}} = 8.33 \text{ A}$$

(c) From $i_e = i_0/\sqrt{2}$, the current amplitude is

$$i_0 = \sqrt{2}i_e = 1.414(8.33 \text{ A}) = 11.8 \text{ A}$$

20.11 | REACTANCE

In an ac circuit, the rms current and the potential drop across a resistor are determined by the resistance R. The analogous quantities for inductance and capacitance are frequency dependent quantities called *reactances*. They can be used to determine the effective and instantaneous potential differences in an ac circuit.

Inductive Reactance | Suppose the current in an inductance is $i = i_0 \sin \omega t$. Then the induced EMF is $\mathcal{E} = -L\, di/dt$, and the voltage drop is $-\mathcal{E}$, or

$$v^L = L\frac{di}{dt} = Li_0 \frac{d}{dt} \sin \omega t$$

Now $(d/dt)\sin \omega t = \omega \cos \omega t$. Thus, defining the *inductive reactance* X_L by

$$X_L = \omega L \qquad (20.22)$$

we have $v^L = Li_0(\omega \cos \omega t)$ or

$$v^L = i_0 X_L \cos \omega t \qquad (20.23)$$

This is the instantaneous potential drop across the inductor. To find the effective or rms voltage, we must square this equation and average over a full cycle. By definition, the square of the left side averages to $(v_e^L)^2$. Now $\cos^2 \omega t$, like $\sin^2 \omega t$, averages to 1/2. Hence we find $(v_e^L)^2 = i_0^2 X_L^2/2 = i_e^2 X_L^2$, and the rms voltage drop across an inductance carrying an effective current i_e is

$$v_e^L = i_e X_L \qquad (20.24)$$

This is similar in form to the voltage drop across a resistor, $v_e^R = i_e R$. The resistance R determines the effective voltage across a resistor, and the inductive reactance X_L determines the effective voltage across an inductor. Resistors, inductors, and capacitors are referred to as *impedances*. From the definition of the inductive reactance, $X_L = \omega L$, we see that for $\omega = 0$, a constant current, an inductor does not impede the current at all; there is no voltage drop across the inductor. On the other hand, as ω becomes large, the inductive reactance X_L becomes large. An inductor opposes changes in the current, and the more rapidly one attempts to change the current in an inductor, the more it opposes that change.

The reactance of an inductor is calculated and used in the following example.

Example 20.8

A 5-mH $= 5 \times 10^{-3}$-H inductor is connected to a 120-V rms 60-Hz power line. Find (a) the inductive reactance, and (b) the rms current.

(a) Since the frequency f is 60 Hz, the inductive reactance is

$$X_L = \omega L = 2\pi f L = 2\pi(60 \text{ s}^{-1})(5 \times 10^{-3} \text{ H})$$
$$= 1.88 \text{ ohms}$$

(b) Since the rms voltage across the inductor is 120 V, $v_e^L = i_e X_L$ gives

$$i_e = \frac{v_e^L}{X_L} = \frac{120 \text{ V}}{1.88 \text{ ohms}} = 63.8 \text{ A}$$

Capacitive Reactance | The charge on a capacitor is related to the current entering it by $i = dq/dt$. With $\int \sin \omega t \, dt = (-1/\omega) \cos \omega t$, we have

$$q = \int i \, dt = \int i_0 \sin \omega t$$
$$= -\frac{i_0}{\omega} \cos \omega t$$

Hence the voltage drop across the capacitor at time t is

$$v^C = \frac{q}{C} = -\frac{i_0}{\omega C} \cos \omega t$$

If we define the *capacitive reactance* X_C by

$$X_C = 1/\omega C \qquad (20.25)$$

we have the instantaneous potential drop across the capacitor

$$v^C = -i_0 X_C \cos \omega t \qquad (20.26)$$

As with the inductive reactance, squaring and averaging gives the rms voltage drop across the capacitor

$$v_e^C = i_e X_C \qquad (20.27)$$

Notice that the capacitive reactance $X_C = 1/\omega C$ *decreases* with the frequency. This happens because the voltage drop q/C is proportional to the charge, and the charge has less and less time to built up as the frequency increases. Since the inductive reactance

increases with the frequency, circuits containing both capacitors and inductors can exhibit complicated frequency variations.

If a resistor, inductor, and capacitor are connected in series to a generator which supplies a current $i_0 \sin \omega t$, their voltage maximum and minima occur at different times (Fig. 20.25). The potential drops across the three elements vary with time as

$$v^R = i_0 R \sin \omega t \qquad (20.19)$$
$$v^L = i_0 X_L \cos \omega t \qquad (20.23)$$
$$v^C = -i_0 X_C \cos \omega t \qquad (20.26)$$

These potential differences and the current are plotted in Fig. 20.25c. The drop v^R across the resistor is always proportional to the instantaneous current, but this is not true for the other two voltages. For example, at $t = 0$, the current is zero, but its rate of change is a maximum. Since the induced EMF is proportional to di/dt, v^L has its maximum positive value. Also, at $t = 0$, the current has just completed a negative half cycle. Hence the capacitor charge q and its

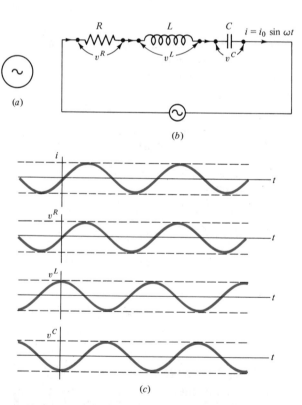

Figure 20.25. (a) Symbol for an ac generator. (b) An *RLC* series circuit. (c) The potential drops across each of the elements and the current versus time.

potential drop $v^C = q/C$ have their most negative values. These concepts are illustrated by the next example.

Example 20.9

The current in Fig. 20.26 has an amplitude of 0.5 A and a frequency $f = \omega/2\pi = 159$ Hz, so $\omega = 1000\ \text{s}^{-1}$. (a) Find the instantaneous voltage drop across each circuit element at $t = 0$. (b) Find the potential difference v between points a and b at an arbitrary time t, and the effective voltage between these points. (c) Find the effective voltages across the reactances, X_L and X_C.

(a) At $t = 0$, $\sin \omega t = \sin 0 = 0$, and $\cos \omega t = \cos 0 = 1$. With $X_L = X_C = 20$ ohms, the voltage drops are

$$v^R = i_0 R \sin \omega t = 0,$$

$$v^L = i_0 X_L \cos \omega t = (0.5\ \text{A})(20\ \text{ohms})(1) = 10\ \text{V}$$

$$v^C = -i_0 X_C \cos \omega t = -(0.5\ \text{A})(20\ \text{ohms})(1) = -10\ \text{V}$$

Thus at this instant, if we proceed along the circuit in the direction of positive currents, we see no voltage drop across the resistor, a 10-V drop across the inductor, and a 10-V *rise* across the capacitor.

(b) Since X_L and X_C are equal, $v^L = -v^C$ at all times. Thus the sum $v^R + v^L + v^C$ of the voltage drop across the three circuit elements equals v^R, and with $R = 10$ ohms, the total drop from point a to point b is

$$\begin{aligned} v = v^R &= i_0 R \sin \omega t \\ &= (0.5\ \text{A})(10\ \text{ohms}) \sin (1000\ t) \\ &= 5 \sin (1000\ t)\ V \end{aligned}$$

Since the voltage amplitude is 5 V, the effective voltage is

$$v_e = v_e^R = \frac{v_0}{\sqrt{2}} = \frac{(5\ \text{V})}{1.414} = 3.54\ \text{V}$$

(c) The potential differences across the inductor and the capacitor each have an amplitude $v_0 = i_0 X_L = i_0 X_C = (0.5\ \text{A})(20\ \text{ohms}) = 10\ \text{V}$, so

$$v_e^L = v_e^C = \frac{v_0}{\sqrt{2}} = \frac{(10\ \text{V})}{1.414} = 7.07\ \text{V}$$

From this example, we can see that it is *not* correct to proceed in analogy with dc circuits and simply add rms ac voltages. Alternating current voltmeters con-

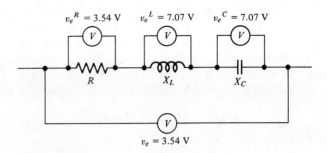

$v_e^R = 3.54$ V $v_e^L = 7.07$ V $v_e^C = 7.07$ V

$v_e = 3.54$ V

Figure 20.27. Effective voltages are *not* additive.

nected to the terminals of the resistor, inductor, and capacitor will give rms voltages of 3.54, 7.07, and 7.07 V, respectively. These add up to 17.68 V, but the rms voltage across the three elements according to part (b) is 3.54 V (Fig. 20.27). The procedure for combining ac voltages is considered in the next section.

20.12 | IMPEDANCE

The current in a network of resistors connected to a dc EMF can be found by calculating the equivalent resistance of the network and using Ohm's law. Similarly, the current in a network of resistors, inductors, and capacitors can be found by calculating the equivalent *impedance Z* of the network and using an analog to Ohm's law.

As we saw in the preceding section, the instantaneous potential drop v from point a to point b in Fig. 20.28 is the sum of the individual voltage drops

$$\begin{aligned} v &= v^R + v^L + v^C \\ &= i_0 R \sin \omega t + i_0 X_L \cos \omega t - i_0 X_C \cos \omega t \\ &= i_0(R \sin \omega t + X \cos \omega t) \end{aligned}$$

Here the *net reactance* is

$$X = X_L - X_C = \omega L - 1/\omega C \qquad (20.28)$$

If a circuit contains an inductor but no capacitor, $X = X_L$; similarly, if it has only a capacitor, $X = -X_C$.

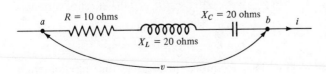

Figure 20.26. Example 20.9.

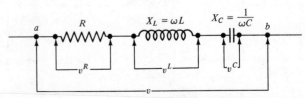

Figure 20.28. Reactances and voltages in an *RLC* circuit.

To find the effective voltage squared, v_e^2, we square this equation and average over a full cycle. This leads to terms proportional to $\sin^2 \omega t$, $\cos^2 \omega t$, and $(\sin \omega t)(\cos \omega t)$. Both $\sin^2 \omega t$ and $\cos^2 \omega t$ average to $\frac{1}{2}$, while $(\sin \omega t)(\cos \omega t)$ averages to zero. (See Appendix B, Eqs. B.13 and B.14.) Thus we find

$$v_e^2 = (i_0^2/2)(R^2 + X^2)$$

The impedance Z is defined as

$$Z = (R^2 + X^2)^{1/2} \qquad (20.29)$$

Equivalently, with $X = X_L - X_C$,

$$Z = [R^2 + (X_L - X_C)^2]^{1/2} \qquad (20.30)$$

With this definition and $i_e^2 = i_0^2/2$, v_e satisfies

$$v_e = i_e Z \qquad (20.31)$$

This equation is very similar to Ohm's law, $v = iR$. However, it relates effective or rms voltages and currents, rather than instantaneous quantities. Also, the impedance, unlike the resistance, depends on the frequency. These results are used in the next example.

Example 20.10

An RLC series circuit is connected to an ac generator (Fig. 20.29). If $\omega = 1000$ rad s^{-1}, find (a) the impedance, and (b) the effective current.

(a) The net reactance is

$$X = X_L - X_C = \omega L - \frac{1}{\omega C}$$

$$= (1000 \text{ rad s}^{-1})(0.4 \text{ H}) - \frac{1}{(1000 \text{ rad s}^{-1})(10^{-5} \text{ C})}$$

$$= 400 \text{ ohms} - 100 \text{ ohms} = 300 \text{ ohms}$$

Hence the impedance is

$$Z = (R^2 + X^2)^{1/2}$$
$$= [(400 \text{ ohms})^2 + (300 \text{ ohms})^2]^{1/2}$$
$$= 500 \text{ ohms}$$

(b) The effective voltage across the RLC series combination must equal the effective generator voltage $\mathcal{E}_e$, so $v_e = i_e Z$ gives

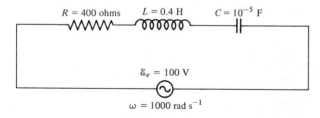

Figure 20.29. Examples 20.10, 20.11, and 20.12.

$$i_e = \frac{\mathcal{E}_e}{Z} = \frac{100 \text{ V}}{500 \text{ ohms}} = 0.2 \text{ A}$$

In the preceding section, we also saw that the sum of the rms ac voltages is *not* equal to the rms potential difference v_e across two or more circuit elements. The correct procedure for combining these voltages can be found from Eq. 20.30 for the impedance. If we multiply this by i_e, then we have

$$i_e Z = [(i_e R)^2 + (i_e X_L - i_e X_C)^2]^{1/2}$$

Now, $i_e Z = v_e$, $i_e R = v_e^R$, $i_e X_L = v_e^L$,   and $i_e X_C = v_e^C$, so

$$v_e = [(v_e^R)^2 + (v_e^L - v_e^C)^2]^{1/2} \qquad (20.32)$$

These ideas are illustrated by the next two examples.

Example 20.11

(a) Find the rms voltage drops across each of the elements in the preceding example, where $i_e = 0.2$ A, $X_L = 400$ ohms, $X_C = 100$ ohms, and $R = 400$ ohms. (b) Calculate the rms voltage drop across the three elements, and show it is equal to the generator rms EMF, 100 V.

(a) The rms drops across the three elements are

$$v_e^R = i_e R = (0.2 \text{ A})(400 \text{ ohms}) = 80 \text{ V}$$
$$v_e^L = i_e X_L = (0.2 \text{ A})(400 \text{ ohms}) = 80 \text{ V}$$
$$v_e^C = i_e X_C = (0.2 \text{ A})(100 \text{ ohms}) = 20 \text{ V}$$

(b) Using Eq. 20.32,

$$v_e = [(80 \text{ V})^2 + (80 \text{ V} - 20 \text{ V})^2]^{1/2}$$
$$= [10{,}000 \text{ V}^2]^{1/2} = 100 \text{ V}$$

Example 20.12

Suppose the frequency of the generator in Fig. 20.29 is varied. (a) At what value of $\omega = 2\pi f$ will the impedance be smallest? (b) Find the effective current at that value of ω.

(a) Since $Z = [R^2 + (X_L - X_C)^2]^{1/2}$, its minimum value occurs when $X_L - X_C = \omega L - 1/\omega C = 0$, or when $\omega L = 1/\omega C$. Thus $\omega^2 = 1/LC$, or

$$\omega = \sqrt{\frac{1}{LC}}$$

$$= \sqrt{\frac{1}{(0.4 \text{ H})(10^{-5} \text{ F})}} = 500 \text{ rad s}^{-1}$$

(b) Since X is zero, $Z = R$, and the effective current is

$$i_e = \frac{v_e}{Z} = \frac{v_e}{R} = \frac{100 \text{ V}}{400 \text{ ohms}} = 0.25 \text{ A}$$

This last example illustrates the phenomenon called *resonance*. At a frequency below or above the resonance frequency, the impedance is greater, so the current for a given applied generator EMF is smaller (Fig. 20.30). This is very similar to the behavior of a mechanical oscillator. We found in Chapter Nine that an oscillating object will have its greatest amplitude if the driving force is applied at the characteristic frequency of the oscillator. The similarity in the behavior of the *RLC* circuit and the mechanical oscillator reflects the fact that the two systems satisfy the same equation of motion, except for a relabeling of the variables.

Resonant ac circuits have many applications. For example, in a television or radio antenna, currents are produced by the electromagnetic waves broadcast at different frequencies by many stations. A tuner in the receiver contains an inductor and a variable capacitor adjusted so that the resonant frequency corresponds to the desired station. Since the resistance of the tuner is small, the impedance rises rapidly as the frequency varies from the resonant frequency, and therefore only one station will normally be received.

20.13 | POWER IN ALTERNATING CURRENT CIRCUITS

The equation for the power dissipated in an ac circuit differs somewhat from the dc formula. During half of each cycle, the capacitor charge is increasing, so the capacitor is absorbing energy. However, this stored electrical energy is released when the capacitor discharges during the other half cycle. Similarly, the inductor temporarily stores and releases magnetic energy. Neither of these circuit elements absorbs any *net* energy over a full cycle. However, as we have seen in Section 20.10, a resistor dissipates energy as heat at the average rate $\overline{\mathcal{P}} = i_e^2 R$. Since $v_e = i_e Z$, this can be rewritten as $\overline{\mathcal{P}} = i_e(v_e/Z)R$, or

$$\overline{\mathcal{P}} = i_e v_e \left(\frac{R}{Z}\right) \qquad (20.33)$$

The power dissipation in a dc circuit is iv, so this result has an extra factor (R/Z) called the *power factor*. It is always less than 1 if there is any reactance. If R is small compared to Z, then very little power is dissipated by the circuit, even though i_e is large; energy is absorbed and released by the inductor and the capacitor, but little heating occurs in the resistor. The power factor is less than 1 for ac motors, as in the next example.

Example 20.13

An ac motor connected to a 120-V rms power line is equivalent in the circuit to a 80-ohm resistor and a 60-ohm inductive reactance in series. Find (a) the effective current, and (b) the power supplied to the motor.

(a) To find the effective current, we first calculate the impedance:

$$Z = [R^2 + X^2]^{1/2}$$
$$= [(80 \text{ ohms})^2 + (60 \text{ ohms})^2]^{1/2}$$
$$= 100 \text{ ohms}$$

Then

$$i_e = \frac{v_e}{Z} = \frac{120 \text{ V}}{100 \text{ ohms}} = 1.2 \text{ A}$$

(b) The power supplied to the motor is

$$\overline{\mathcal{P}} = i_e v_e \left(\frac{R}{Z}\right) = (1.2 \text{ A})(120 \text{ V})\left(\frac{80 \text{ ohms}}{100 \text{ ohms}}\right)$$
$$= 115 \text{ W}$$

20.14 | IMPEDANCE MATCHING

It is often desirable to arrange that the maximum possible power transfer occurs between a power source and a load. These could be a battery and a resistor, respectively, as in the example we discuss shortly. They could also be an amplifier and a loudspeaker or probes connected to a person and an EKG recorder. *In general, the maximum power transfer occurs when the impedances of a source and a load are matched.*

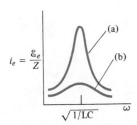

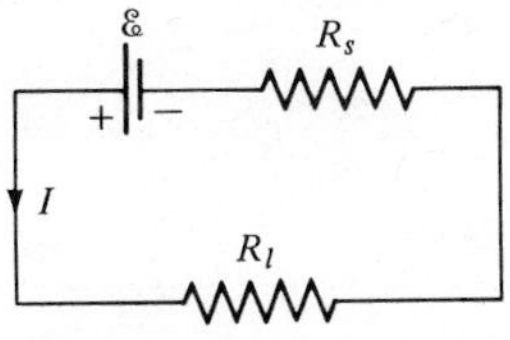

Figure 20.30. Effective current versus frequency for two circuits with the same EMF and resonant frequency. The narrow curve (*a*) corresponds to a circuit with a small resistance. The broad flat curve (*b*) corresponds to a circuit with a large resistance.

Figure 20.31. The power transfer to R_l is greatest if the resistances are matched ($R_l = R_s$).

We begin by considering the dc case. Figure 20.31 shows a battery with EMF $\mathcal{E}$ and internal resistance R_s connected to a load that is a resistance R_l. With Ohm's law, the current is

$$I = \frac{\mathcal{E}}{(R_s + R_l)}$$

The power supplied to the load is then

$$\mathcal{P} = I^2 R_l = \frac{R_l}{(R_l + R_s)^2}\mathcal{E}^2$$

To see how this varies with the ratio $x = R_l/R_s$ of the load and source resistances, we divide out a factor of R_s^2 in the denominator and rewrite $\mathcal{P}$ as

$$\mathcal{P} = \frac{\mathcal{E}^2}{R_S}\frac{x}{(1 + x)^2} \qquad (20.34)$$

The factor $\mathcal{E}^2/R_S$ is the power that would be dissipated in the battery if the load resistance R_l were equal to zero, or if the battery were short-circuited. The second factor is proportional to x, when x is small compared to 1 (Fig. 20.32). It is left as a problem to show that its maximum occurs when $x = 1$, or $R_l = R_S$.

Thus the maximum dc power transfer occurs when the load and source resistances are equal.

Note that when $x = 1$, the power supplied to R_l is $(1/4)(\mathcal{E}^2/R_s)$, or one-fourth the power dissipated in a short-circuited source.

Now let us consider, instead, an ac generator with effective EMF $\mathcal{E}_e$, internal resistance R_s, and internal reactance X_s connected to a load with resistance R_l and net reactance X_l (Fig. 20.33). Again we want to know what load will receive the maximum power transfer. If the previous analysis is repeated using the formulas appropriate to alternating currents, the result is that the maximum power transfer occurs if

$$R_l = R_s \qquad \text{and} \qquad X_l = -X_s \qquad (20.35)$$

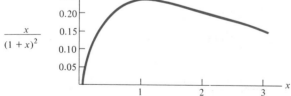

Figure 20.32. The peak power transfer to the load is at $x = R_l/R_s = 1$.

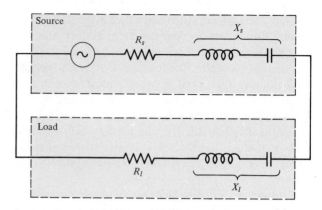

Figure 20.33. The power transfer to the load is greatest when the impedances are matched, ($R_l = R_s$, $X_l = X_s$.)

This means that the maximum power transfer in an ac circuit occurs when the source and load resistances are equal and the reactances cancel, so that the total reactance $X_l + X_s$ of the circuit is zero.

We recall that the net reactance is $X = X_L - X_C$, which may be positive or negative. Thus, for example, if the source has a resistance of 50 ohms and an inductive reactance of 100 ohms, then to be matched the load must have a resistance of 50 ohms and a capacitive reactance of 100 ohms.

Impedance matching is important whenever "black boxes" are assembled in a laboratory. Even though we can often ignore the internal details of complex electronics instruments, we must be sure that the input and output electrical impedances of two interconnected instruments are not badly mismatched. Impedance matching also is important in applications other than electrical circuits. For example, if sound is to be transferred efficiently without reflections from one medium to another, then the *acoustical impedances* must be matched. Thus, impedance matching is a concept with wide applicability.

EXERCISES ON SUPPLEMENTARY TOPICS

Section 20.9 | RL Circuits

20-46 Using the definitions of the inductance and resistance, show that L/R has the units of time.

20-47 A 10-ohm resistor and a 0.2-H inductor are connected in series to a 12-V battery. (a) What is the final current? (b) What is the time constant? (c) What is the current after one time constant? (d) What is the current after 4 seconds?

20-48 A coil of wire connected to a 6-V battery

has a current of 0.063 A after 0.01 second and a current of 0.1 A after a few seconds. (a) What is its resistance? (b) What is its time constant? (c) What is its inductance?

20-49 An *RL* circuit has a time constant of 4 s and a final current of 8 A. The inductance is 0.1 H. How much energy is stored in the inductor 8 s after the switch is closed?

20-50 A coil with a resistance of 10 ohms and an inductance of 0.5 H is attached to a 12 V battery. (a) What is the final current? (b) How much energy is stored in the coil when this current is reached? (c) At what rate is power dissipated in the coil due to its resistance once the final current is reached?

ᶜ20-51 Verify by differentiating that Eq. 20.15 for *i* is a correct solution of Eq. 20.14.

ᶜ20-52 Suppose the switch in Fig. 20.34 has been closed in position 1 for a long time, so that $i = \mathcal{E}/R$. At $t = 0$, the switch is suddenly shifted to position 2. (a) Show that the equation satisfied by the current for $t > 0$ is $-iR - L\,di/dt = 0$. (b) Show that for $t > 0$, the current is

$$i = (\mathcal{E}/R)e^{-t/T_L}$$

(c) Show that this formula for the current is also correct at $t = 0$.

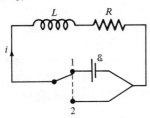

Figure 20.34. Exercise 20-52.

Section 20.10 | Effective or Root Mean Square Alternating Current and Voltages

20-53 A fuse is designed to open if the current exceeds 20 A rms. What is the peak current allowed if this fuse is used in an ac circuit?

20-54 A light bulb draws 60 W from a 120-V rms 60-Hz line. (a) What is the rms current? (b) What is the peak current? (c) How many times per second does this peak current occur?

20-55 A light bulb draws 1.67 A from a 120-V dc line. (a) How much power does the light bulb use? (b) If it is used with a 120-V rms ac line, what is the rms current? (c) What is the peak current when it is used with the ac line?

20-56 An electric heater is designed to use 1500 W of power when connected to a 120-V rms ac line. It is accidentally connected to a 240-V rms ac line. Assuming its resistance does not change and that it does not burn out, how much power will it use?

20-57 Homes are often supplied with 240-V rms alternating current to provide the power for high-wattage appliances or electric heat. The lines coming into the home include a ground and two "hot" lines, which have voltages equal in magnitude but opposite in sign at all times. Explain how 120-V rms alternating current can be obtained from these lines.

20-58 Number 14 wire is rated to safely carry 15 A rms. How much power can this wire supply at (a) 120-V rms alternating current; (b) 240-V rms alternating current.

20-59 The EMF of a generator is given by $\mathcal{E} = (100\ V)\sin(1000s^{-1})t$. (a) What is the frequency *f* of the generator? (b) What is its rms voltage?

20-60 A generator rotates at 400 Hz and has an rms EMF of 1000 V. Find the formula for the EMF versus time.

Section 20.11 | Reactance

20-61 A capacitor has a reactance of 10 ohms at 400 Hz. (a) Find the capacitance. (b) What is the reactance at 60 Hz? (c) What will be the rms voltage drop across the capacitor if it is connected to a 240-V 60-Hz line?

20-62 At 60 Hz, find the reactance of a (a) 0.05-H inductor; (b) a 1-$\mu F = 10^{-6}$-F capacitor. (c) At what frequency will the inductive and capacitive reactances be equal?

20-63 At 1000 Hz, an inductor and a capacitor have equal reactances. What is the ratio of the capacitance reactance to the inductive reactance at 100 Hz?

20-64 A 10^{-4}-F capacitor is connected to a 120 V 60-Hz ac line. Find (a) the rms current; (b) the peak current.

20-65 When a 0.4-H inductor is connected to a 120-V 60-Hz ac line, what are (a) the rms current; (b) the peak current?

20-66 A 2-H inductor is connected to a 120-V rms 60-Hz power line. (a) Find its reactance. (b) Find the rms potential difference across the inductor.

(c) What is the peak voltage drop across the inductor?

Section 20.12 | Impedance

20-67 A 100-ohm resistance and a 0.2-H inductance are connected in series to a 240-V rms 60-Hz generator. (a) Find the reactance of the inductor. (b) Find the impedance of the circuit. (c) Find the rms current.

20-68 For the circuit in the preceding exercise, draw sketches showing how the following quantities vary with time: (a) the current; (b) the potential difference across the resistor; (c) the potential difference across the inductor.

20-69 A 100-ohm resistor, a 10^{-4}-F capacitor, and a 0.1-H inductor are connected in series to a 120-V rms generator. If the generator frequency is 60 Hz, find (a) the impedance, and (b) the effective current.

20-70 A 30-ohm resistor and a capacitor with a reactance of 40 ohms are connected in series to a 50-V rms ac generator. (a) What is the rms current? (b) What is the peak current?

20-71 A tuning circuit in a radio transmitter has a 10^{-6}-H inductance in series with a 10^{-12}-F capacitance. Find (a) the frequency of the waves transmitted; (b) their wavelength.

20-72 A tuning circuit in a radio has a coil of inductance 2×10^{-5} H in series with a variable capacitor. What is the capacitance if the radio is tuned to receive a station at $f = \omega/2\pi = 10^6$ Hz?

20-73 Alternating current voltmeters are connected across the terminals of an inductance, a capacitance, and a resistance connected in series. Their readings are 100 V, 300 V, and 150 V, respectively. What is the rms voltage across the three circuit elements?

Section 20.13 | Power in Alternating Current Circuits

20-74 An air conditioner connected to a 120-V rms ac line is equivalent to a 10-ohm resistance and a 1-ohm inductive reactance in series. (a) What is its impedance? (b) What is the power supplied to the air conditioner?

20-75 A 5-ohm resistor, a 10-ohm inductive reactance, and a 22-ohm capacitive reactance are connected in series to a 120-V rms ac generator. (a) Find the impedance. (b) How much power is dissipated in the resistor?

20-76 A 5-ohm resistor, a 0.01-H inductor, and a 10^{-4}-F capacitor are connected in series to a power supply with a variable frequency. (a) At what value of ω is the current the greatest? (b) What is the corresponding frequency f? (c) How much power is dissipated at this frequency if the effective EMF of the power supply is 10 V?

20-77 For the circuit in the preceding exercise, find (a) the impedance at $\omega = 2000$ rad s^{-1}; (b) the power supplied to the resistor if the effective EMF is 10 V.

Section 20.14 | Impedance Matching

20-78 It is more economical to extract energy from a battery by using a load resistance that is larger than the internal source resistance rather than a load resistance equal to the source resistance. Explain why.

20-79 Why is impedance matching *not* desirable when using a voltmeter to measure a potential difference?

20-80 A 12-V battery has an internal resistance of 0.005 ohms. A resistance R is connected across its terminals. (a) If $R = 0.005$ ohms, find the ratio of the power dissipated in the internal resistance to that dissipated in R. (b) Find this ratio when $R = 0.05$ ohms.

20-81 A dc generator has an internal resistance of 10^{-4} ohms. It is desired that no more than 1 percent of the total power generated to be dissipated in the generator itself. What is the minimum resistance one can connect to the generator?

PROBLEMS ON SUPPLEMENTARY TOPICS

20-82 A 10-ohm resistor and a 0.1-H inductor are connected in series to a battery. (a) What is the time constant? (b) How long will it take for the current to reach 99 percent of its final value?

20-83 A coil with a resistance R and inductance L is connected to a battery with an EMF $\mathcal{E}$ and a negligible internal resistance. Show that when the current has reached its final value, the energy stored in the coil is $\frac{1}{2}\mathcal{E}^2 T_L/R$.

20-84 (a) Show that the initial rate of change of the current in an RL circuit when it is connected to a battery is $i_f R/L$. (b) If the current continued to increase at this initial rate, how long would it take to reach i_f?

20-85 The current in an RL circuit reaches half its

final value in 4 s. What is the time constant of the circuit?

20-86 Suppose a light bulb and a variable inductance are connected in series to an ac power line. (a) How would the brightness of the bulb vary as the inductance is increased? (b) One can also vary the brightness by replacing the inductor with a variable resistance. Why is that less satisfactory?

*20-87 A tuning circuit in a radio receiver consists of a coil with an inductance of 10^{-5} H and a variable capacitor. (a) What is the capacitance if the receiver is tuned to $f_0 = 1.4 \times 10^6$ Hz? (b) Find the reactance at a frequency 1 percent above f_0. (c) At this higher frequency the impedance is larger than at f_0 by a factor of 4. Find the resistance of the circuit.

c20-88 Show by differentiating Eq. 20.34 and setting $d\mathcal{P}/dx = 0$ that the maximum power transfer occurs in a dc circuit when the load and source resistances are equal.

*20-89 Verify that if the resistance and reactance of a load are adjusted as in Eq. 20-35, then the maximum power is transferred to the load.

20-90 A battery with EMF $\mathcal{E}$ and internal resistance R_s is connected to a resistance R_l. Compare the power dissipated in the battery and in the load if (a) $R_l = R_s$; (b) R_l is much larger than R_s.

Additional Reading

E. Schlömann, Recovery of Nonmagnetic Metals from Municipal Wastes, *Physics Teacher*, vol. 14, 1976, p. 116. The process uses eddy currents.

Kurt S. Lion, Elements of Electrical and Electronic Instrumentation, McGraw-Hill Book Co., New York, 1975.

E.B. Forsyth, The Brookhaven Superconducting Power Transmission Line, *The Physics Teacher*, vol. 21, 1983, p. 285.

Scientific American articles:

Herbert Kondo, Michael Faraday, October 1953, p. 90.

J.R. Newman, James Clerk Maxwell, June 1955, p. 58.

P. Morrison and E. Morrison, Heinrich Hertz, December 1957, p. 98.

H.L. Sharlin, From Faraday to the Dynamo, May 1961, p. 107.

G. Shiers, The Induction Coil, May 1971, p. 80.

James R. Heirtzler, The Longest Electromagnetic Waves, March 1962, p. 128.

E.N. Parker, Magnetic Fields in the Cosmos, August 1983, p. 44.

UNIT SIX

WAVE MOTION

(Jeff Albertson/Stock Boston).

MOST of the wave motions with which we are familiar involve a large-scale coordinated disturbance of many particles or objects. While the individual particles do not move far, the disturbance may travel great distances, carrying with it energy and momentum. The motions of the particles vary with the type of wave. For example, in a water wave, the water molecules move in small, approximately circular paths; in a sound wave, molecules move back and forth; and in a wave on a string, the parts of the string move up and down.

Light waves are also coordinated disturbances involving changing electric and magnetic fields. Here, no particles move, but the waves nevertheless carry energy and momentum, and the mathematical description of these waves is nearly identical to the mechanical examples given above.

All waves have a number of characteristics in common. Many of these are described in the first chapter of this unit (Chapter Twenty-one) where we mainly use strings and springs for illustration. The special features of sound waves are described in Chapter Twenty-two and the properties of light are covered in Chapters Twenty-three and Twenty-four.

The importance of waves in physics derives from the transmission of energy and momentum from one place to another or from a source to a detector. These wave motions are described using the ideas of mechanics or electromagnetic theory. Their properties follow from the ideas presented earlier in this book and merely represent cooperative or coordinated motions in systems we have already studied.

A severe jolt to physics in the early twentieth century was the realization that the sharp distinction between wave and individual particle motions was not valid when studying atoms and molecules. In the years since, a dramatic revolution in our description of nature has taken place. Modern quantum physics, which was developed to describe the microscopic properties of molecules, atoms, and nuclei, has forced us to regard particles as having some wave properties and to regard waves as though they are composed of particles or *quanta*. This wave-particle duality of our present description of nature may be due to our lack of a true understanding of nature or of our inability to express our understanding clearly. Nevertheless, with the language we do have, we must, for example, regard an electron as an entity with both particle and wavelike characteristics.

This revolution of modern physics does not mean that all of our studies have been for naught. It simply means that when very small dimensions are involved, we must further refine our understanding. The remaining units of this book are devoted to those refinements.

CHAPTER 21

THE DESCRIPTION OF WAVE MOTION

In this chapter, concepts common to all kinds of wave phenomena are developed. We primarily use the readily visualized waves along strings and springs as examples and analogies for other types of wave motion that are more difficult to observe.

When a stone is dropped into a lake or when a string is wiggled briefly at an end, a single wave *pulse* travels away from the disturbance. However, waves also occur in a regular continuing series; they are then said to be *periodic*. For example, a vibrating tuning fork produces alternate compressions and rarefactions of the air nearby. These disturbances, which are perceived as sound, occur at the frequency of the tuning fork.

When a string is disturbed, its particles move at right angles to the string. Waves with displacements perpendicular to the direction of the wave are called *transverse waves*. Light is a transverse wave. Waves that have displacements along the wave direction are termed *longitudinal*. Examples of this latter type are sound waves and compressional waves in coiled springs.

All the waves we consider have the important property of *linearity*. Linearity means that when two or more waves pass the same point, the resulting wave is the sum of the individual waves, and after passing, the waves continue along their paths as if no encounter had occurred. If one observes the circular waves traveling outward from separate points where two pebbles are dropped in still water, the peaks add up when they meet. Afterward, they travel along undisturbed. The description of such linear behavior is called the *principle of superposition*.

21.1 | THE REPRESENTATION OF WAVES

Figure 21.1 shows single wave *pulses* on a string and in a spring produced by a disturbance at their left ends and traveling toward the right. Figure 21.2 shows *periodic waves* produced by oscillatory motion of the left ends. The waves on the string are transverse, and the waves in the spring are longitudinal. Nevertheless, both types of waves can be represented symbolically by similar graphs.

Many kinds of wave phenomena occur in nature. Light is a transverse electromagnetic wave, with changing electric and magnetic fields at right angles to each other and to the direction of the light wave. Sound is an alternate compression and rarefaction of the medium along the direction of motion, so it is a longitudinal wave. Water waves are a mixture of longitudinal and transverse waves, with water molecules moving in roughly circular paths. All these wave phenomena can be represented by graphs similar to those for strings and springs and share many common properties (Fig. 21.3).

Periodic waves of any type are characterized by several quantities. The *frequency f* is the number of waves passing a point per second and is determined by the source of the waves. For example, in the string and spring of Fig. 21.2, the frequency is the rate at which the oscillations occur at the left end. The *period T* is the time between successive wave crests, or the inverse of the frequency; $T = 1/f$. The *velocity c* of a wave is the speed at which a wave peak travels. The *wavelength* λ of a periodic wave is the distance be-

455

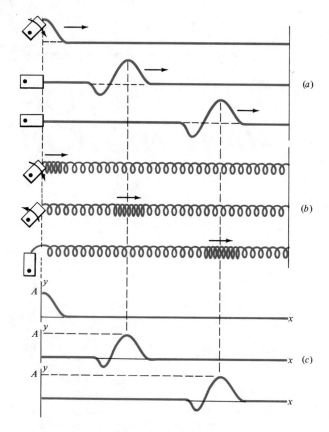

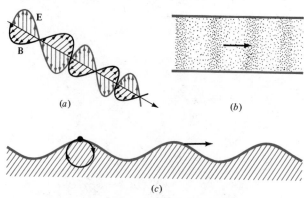

Figure 21.3. (a) An electromagnetic wave. (b) A sound wave. (c) A water wave. All these waves can be represented by the same kind of graphs as used for springs and strings.

tween successive wave peaks. Finally, the *amplitude A* is the maximum magnitude of the displacement; the displacement of a periodic wave varies back and forth between A and $-A$. Most of the periodic waves we consider are *sinusoidal* waves, and their graphs look like sine or cosine graphs.

The relationship among the frequency, wavelength, and velocity of a periodic wave can be found from Fig. 21.4. In one period T, the time required for one complete wave oscillation, the wave travels one wavelength λ. The velocity of the wave c is the distance traveled divided by the time, so $c = \lambda/T$. Since $T = 1/f$, this implies

$$f\lambda = c \qquad (21.1)$$

The following examples illustrate these relationships.

Example 21.1

A wave pulse on a string moves a distance of 10 m in 0.05 s. (a) What is the velocity of the pulse? (b) What is the frequency of a periodic wave on the same string if its wavelength is 0.8 m?

(a) The velocity of the pulse is $c = \Delta x/\Delta t$, where $\Delta x = 10$ m and $\Delta t = 0.05$ s, so

$$c = \frac{10 \text{ m}}{0.05 \text{ s}} = 200 \text{ m s}^{-1}$$

(b) The periodic wave has the same velocity, 200 m s^{-1}. Using $f\lambda = c$, the frequency of a 0.8-m wave is

$$f = \frac{c}{\lambda} = \frac{200 \text{ m s}^{-1}}{(0.8 \text{ m})} = 250 \text{ s}^{-1} = 250 \text{ Hz}$$

Example 21.2

A typical sound wave associated with human speech has a frequency of 500 Hz, while the frequency of yellow

Figure 21.1. (a) A single wave pulse is produced at the left end of a string. The appearance of the string is shown at three successive times. Note that the motion of the particles is transverse to the motion of the wave. (b) A similar wave pulse on a coiled spring. The spring is alternately compressed and extended along the direction of motion, so the wave is longitudinal. (c) Either wave pulse can be represented symbolically by the same graph. For the string, y is the displacement of the string from its undisturbed position; displacements above the equilibrium position are positive. For the spring, y is a measure of the compression or extension of the spring; a compression is regarded as a positive displacement.

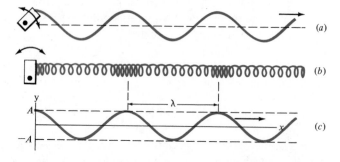

Figure 21.2. (a) A periodic disturbance produced in a string by an oscillating lever travels toward the right. The dashed line indicates the undisturbed position of the string. (b) A spring is alternately compressed and extended. (c) The same graph can represent either wave.

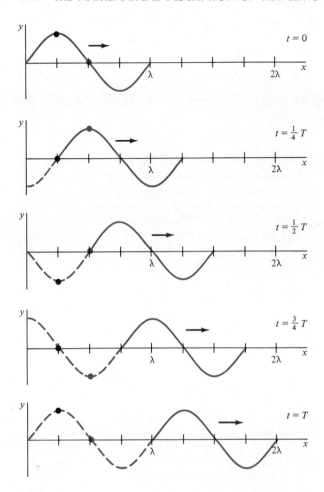

Figure 21.4. Graphs of the displacement at time intervals of one-quarter period. In one period T, the wave moves a distance equal to the wavelength λ, and the displacement at any point makes one full oscillation. For example, at $x = \frac{1}{4}\lambda$, the displacement is a maximum at $t = 0$ and again at $t = T$ (black dots). The displacement at $x = \frac{1}{2}\lambda$ has the same values but at times later by $\frac{1}{4}T$ (colored dots).

their wavelength. Such effects are often apparent with sound or water waves but are seldom noticed for light because the wavelength of visible light is so short. Light waves will bend around objects that are small enough.

21.2 | THE MATHEMATICAL DESCRIPTION OF TRAVELING WAVES

Waves can be described mathematically as well as graphically. A wave traveling in the $+x$ direction at a speed c has a displacement y that can be written as

$$y = f(x - ct) \qquad (21.2)$$

Here f is any function of the variable $(x - ct)$. To see that this is correct, suppose that the time t increases by an amount Δt, and x increases by $\Delta x = c\Delta t$. Then the difference $(x - ct)$ remains the same, and y is unchanged. That is, the point in the wave pattern with displacement y has moved a distance $\Delta x = c\Delta t$ in the positive x direction (Fig. 21.5). This is the definition of a *traveling wave*. Similarly, a wave moving in the $-x$ direction is described by

$$y = f(x + ct) \qquad (21.3)$$

Sinusoidal waves are an important type of traveling wave. When we speak of a 500-Hz sound wave or a beam of yellow light, we are referring to sinusoidal waves with a specific frequency or wavelength. Sinusoidal waves are special cases of Eqs. 21.2 and 21.3 where the function f is a sine or cosine. If the ampli-

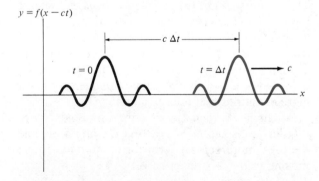

Figure 21.5. A wave pulse $y(x, t) = f(x - ct)$ is a traveling wave moving toward $+x$ at a velocity c. When t is increased by Δt and x by $c\Delta t$, the variable $(x - ct)$ stays the same. At $t = 0$ (black curve) and at $t = \Delta t$ (colored curve) the two waveforms are identical, but the wave at Δt has moved a distance $\Delta x = c\Delta t$ in the positive x direction. The variable $(x - ct)$ has the same value at the corresponding points on the two curves.

light is about 5×10^{14} Hz. In air, sound travels at 344 m s^{-1} and light at 3×10^8 m s^{-1}. Find the wavelengths of the two waves.

For both waves, $f\lambda = c$, so for the sound wave

$$\lambda = \frac{c}{f} = \frac{344 \text{ m s}^{-1}}{500 \text{ Hz}} = 0.688 \text{ m}$$

For the light wave,

$$\lambda = \frac{c}{f} = \frac{3 \times 10^8 \text{ m s}^{-1}}{5 \times 10^{14} \text{ Hz}} = 6 \times 10^{-7} \text{ m}$$

The fact that these characteristic wavelengths are so different is responsible for some of the apparent differences between the two types of waves. Waves tend to bend around obstacles comparable in size to

tude of the wave is A, then an example of a sinusoidal wave moving in the $+x$ direction is

$$y = A \sin k(x - ct) \quad (21.4)$$

Figure 21.4 illustrates a portion of such a wave. The sine function varies between -1 and $+1$, so y varies from $-A$ to $+A$.

The quantity k is called the *wave number*. We can relate it to the wavelength λ by noting that the sine repeats itself when the argument or *phase* increases by 2π radians or $360°$. (A graph illustrating this property is given in Fig. B.5 in Appendix B.) If x increases by λ, then the displacement y is also the same, and

$$y = A \sin k(x - ct) = A \sin k(x + \lambda - ct)$$

Since the sines are equal, their arguments differ by 2π. Thus $k\lambda = 2\pi$, or

$$k = 2\pi/\lambda \quad (21.5)$$

This is the desired relationship between the wave number and the wavelength. Since wavelengths are measured in length units or metres, wave numbers are measured in inverse metres, m^{-1}.

With the definition

$$\omega = kc \quad (21.6)$$

Eq. 21.4 can also be written in the form

$$y = A \sin (kx - \omega t) \quad (21.7)$$

When t increases by one period or by $T = 1/f$, the displacement y is the same. Thus we must have $\omega T = 2\pi$ or $\omega/f = 2\pi$, and

$$\omega = 2\pi f \quad (21.8)$$

The quantity ω is called the *angular frequency,* and it is given in radians per second. It has the same symbol and units as the angular velocity of a rotating object (Chapter Seven), but its meaning is different. The angular frequency is handy to use in discussing sinusoidal waves, although the ordinary frequency f has a more direct physical interpretation.

Equations 21.4 and 21.7 describe a sinusoidal wave that has a displacement $y = 0$ at the origin at time $t = 0$. A more general form for a sinusoidal wave moving in the $+x$ direction is

$$y = A \sin (kx - \omega t + \phi) \quad \text{(wave toward } +x) \quad (21.9)$$

Substituting $x = 0$ and $t = 0$, we have $y = A \sin \phi$. Thus, depending on the choice of the *phase constant* ϕ, the displacement at $x = 0$ at time $t = 0$ can have

any value between $-A$ and $+A$. For example, if $\phi = \pi/2$, $y = A \sin \pi/2 = A$. Also, any sinusoidal wave moving toward $-x$ can be represented by

$$y = A \sin (kx + \omega t + \phi) \quad \text{(wave toward } -x) \quad (21.10)$$

We use the expressions for sinusoidal waves in the following example.

Example 21.3

A sinusoidal wave on a string traveling in the $+x$ direction at 10 m s^{-1} has a wavelength of 2 m. (a) Find its wave number, frequency, and angular frequency. (b) If the amplitude is 0.1 m, and the point $x = 0$ on the string is at its equilibrium position ($y = 0$) at time $t = 0$, find the equation for the wave. (c) If instead the point at $x = 0$ is at its maximum displacement at $t = 0$, find the equation for the wave.

(a) The wave number is

$$k = 2\pi/\lambda = 2\pi/(2 \text{ m}) = 3.14 \text{ m}^{-1}$$

The frequency and angular frequency are

$$f = c/\lambda = (10 \text{ m s}^{-1})/(2 \text{ m}) = 5 \text{ Hz}$$
$$\omega = 2\pi f = 2\pi(5 \text{ Hz}) = 31.4 \text{ rad s}^{-1}$$

(b) Equation 21.7 is appropriate here, since it describes a wave in the $+x$ direction with a displacement $y = 0$ at the origin at $t = 0$. Thus, with $A = 0.1$ m and the results of part (a), we have

$$y = A \sin (kx - \omega t)$$
$$= (0.1 \text{ m}) \sin[(3.14 \text{ m}^{-1})x - (31.4 \text{ s}^{-1})t]$$

This equation tells us the displacement y at any point x on the string for all times.

(c) Here we must use the form of the wave in Eq. 21.9, $y = A \sin (kx - \omega t + \phi)$. At $x = 0$, $t = 0$, the wave has a maximum, and $y = A$. Thus we must have

$$A = A \sin (0 - 0 + \phi)$$

or $\sin \phi = 1$. The sine function is 1 at $90°$ or $\pi/2$ radians. Thus $\phi = \pi/2$ and

$$y = (0.1 \text{ m}) \sin[(3.14 \text{ m}^{-1})x - (31.4 \text{ s}^{-1})t + \pi/2]$$

Example 21.4

A sinusoidal wave on a string is represented by $y = A \sin (kx - \omega t)$. Each point on the string moves along the y or transverse direction from $y = -A$ to $y = +A$ and back to $y = -A$ once in each cycle. This motion is at right angles to the direction of the wave, which is along the string. (a) At a given point x on the string, find the velocity v associated with this transverse motion as a function of time. (b) Find the corresponding transverse acceleration.

(a) The position of the string relative to its equilibrium position is given by a function $y(t)$ for a fixed value of x. To find the corresponding velocity, we must evaluate the derivative of the position,

$$v = \frac{dy}{dt} = \frac{d}{dt}[A \sin(kx - \omega t)] \qquad \text{(fixed } x)$$

(Technically, the derivative with respect to t for a fixed value of x is a *partial derivative*.) Now with $u = (kx - \omega t)$ and x held constant, using the chain rule (Appendix B.8), we find

$$\frac{d}{dt}\sin u = \left[\frac{d}{du}\sin u\right]\frac{du}{dt} = (\cos u)(-\omega)$$

Thus

$$v = -\omega A \cos(kx - \omega t)$$

We see that the velocity also varies sinusoidally with time. However, since it varies as $-(\cos u)$ instead of as $\sin u$, its maxima and minima occur a quarter cycle before the maxima and minima in the displacement.

(b) The acceleration associated with the transverse motion is the derivative of the velocity,

$$a = \frac{dv}{dt} = \frac{d}{dt}[-\omega A \cos(kx - \omega t)] \qquad \text{(fixed } x)$$

With

$$\frac{d}{dt}\cos u = \left[\frac{d}{du}\cos u\right]\frac{du}{dt} = -(\sin u)(-\omega)$$

we have

$$a = -\omega^2 A \sin(kx - \omega t)$$

Note that the transverse acceleration a and the displacement y are related by $a = -\omega^2 y = -(2\pi f)^2 y$. As discussed in Chapter Nine, this is the defining equation for simple harmonic motion. This is the type of oscillatory motion frequently encountered when an object undergoes small oscillations about equilibrium. *Thus, each point on the string is undergoing simple harmonic motion transverse to the wave direction.*

21.3 | THE VELOCITY OF WAVES

While waves are common phenomena, each type of wave disturbance has its own specific physical origin. Accordingly, each type of wave has a characteristic velocity.

Electromagnetic waves are unique in that they require no medium in which to propagate. As we discussed in Chapter Twenty, these waves are due to mutually induced time varying electric and magnetic fields. In a vacuum, they have a velocity of 3×10^8 m s^{-1}; in matter, their velocity is always smaller.

Sound is a coordinated mechanical disturbance involving large numbers of molecules. These molecules move and collide as a wave disturbance passes, but on the average they suffer no net change in position. As we show in the next chapter, the velocity of sound in a substance depends on how the pressure changes when the density changes. The velocity of sound in air at $30°$ C is 344 m s^{-1}, but it is much higher in solids; in aluminum, for example, $c = 5000$ m s^{-1}. In solids there may also be transverse as well as longitudinal sound waves. The two types of sound waves have different velocities.

The velocity of a wave can be predicted from the physical laws that describe the specific wave phenomenon. For example, we saw in Chapter Twenty that Maxwell was able to derive an expression for the speed of an electromagnetic wave using the fundamental properties of electric and magnetic fields. Similarly, the velocities of the various kinds of mechanical waves, such as sound waves and waves in strings, springs, and water, can be predicted using Newton's laws of motion. Such calculations are mathematically complex and will not be carried out here. However, it is important to remember that the velocity of a wave depends on the type of wave, on the properties of the system in which the wave travels, and sometimes on the frequency. As an example, we describe the result for a wave on a string.

When a string is displaced, the restoring force is proportional to the tension in the string. Also, a thick string will respond more slowly to this restoring force than will a thin string. Hence, we expect that the wave velocity on a string will depend on the tension $\mathfrak{J}$ and the mass per unit length μ of the string. This is the case and the detailed analysis gives for the wave velocity c on a string,

$$c = \sqrt{\frac{\mathfrak{J}}{\mu}} \qquad (21.11)$$

This result has the features described above. The following example leads to a typical value for the wave velocity on a string.

Example 21.5
The tension on the longest string of a grand piano is 1098 N, and the mass per unit length is 0.065 kg m^{-1}. What is the velocity of a wave on this string?

Using the above result,

$$c = \sqrt{\frac{\Im}{\mu}} = \sqrt{\frac{1098 \text{ N}}{0.065 \text{ kg m}^{-1}}} = 130 \text{ m s}^{-1}$$

21.4 | WAVE INTERFERENCE AND STANDING WAVES

When two or more waves travel in a medium, the resulting wave is the sum of the displacements associated with the individual waves. This property is referred to either as *linearity* or as the *principle of superposition* and is applicable to all the waves we consider here. As we have noted, the resulting wave may have a very complex shape when the waves overlap, but each individual wave is unchanged and displays its original form when the waves separate. Since waves add algebraically, the resultant of overlapping waves can be larger or smaller than the individual waves, depending on their relative sign. This characteristic interaction of waves is termed *interference* and leads to many interesting and curious effects. In this section, we see that waves traveling in opposite directions can combine to form a wave disturbance that does not appear to travel at all and is therefore called a *standing wave*. Many other examples of interference are considered later.

To illustrate these ideas, suppose first that a string is held at each end by individuals who simultaneously produce wave pulses. The only difference between the pulses is that they are inverted mirror images of each other (Fig. 21.6). As the waves pass by the same points on the string, the resulting wave is found by adding the displacements of the individual waves.

An important feature of this particular experiment is that despite the fact that the wave shape becomes quite complex, there is one point on the string that is never displaced. This point is called a *node*. The string could actually be held fixed at this point without affecting the results. The pulses are said to be interfering with each other *destructively* at this point.

The same experiment can be performed with two periodic waves traveling in opposite directions but having the same amplitude and wavelength (Fig. 21.7). Now there are a number of nodes spaced one half wavelength apart. Midway between two successive nodes are points called *antinodes* where the string can experience a maximum displacement. At an antinode the waves add *constructively*.

Figure 21.8 shows a multiple exposure series of pictures of a string vibrating in this way. Notice that the

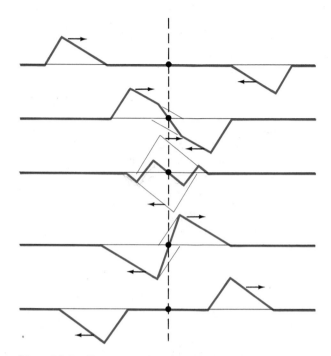

Figure 21.6. Two wave pulses that are inverted mirror images of each other are produced at the ends of a string. As the waves pass each other, the string has a complex shape but the one point on the dashed line is always at rest. The string shape is found by adding the displacements of the two pulses at every point.

traveling waves are no longer observed. What is seen is a pattern described as a *standing wave:* a wave with nodes and antinodes at fixed points.

The interference effects produced by waves depend on their *phases*. If two waves reaching a point have their maxima at the same time, they are *in phase* and add constructively. If a maximum of one wave coincides with a minimum of the other, they are a half wavelength out of phase and interfere destructively. Under these conditions the waves are said to be exactly out of phase. In general, there may be an arbitrary phase difference between two waves (Fig. 21.9).

Mathematical Description of Standing Waves |
The mathematical description of sinusoidal traveling waves leads directly to formulas for standing waves. Suppose that, as in Fig. 21.7, we have waves with the same frequency and equal amplitudes traveling in both the $+x$ and $-x$ directions.

$$y_1 = A \sin(kx - \omega t), \quad y_2 = A \sin(kx + \omega t) \quad (21.12)$$

Then the superposition principle states that the net

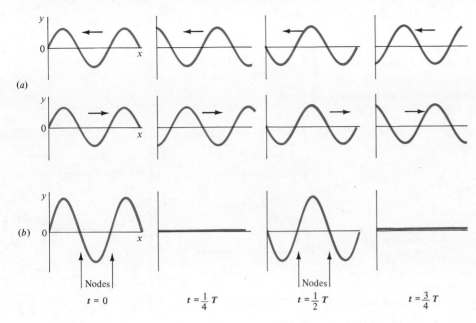

(a)

(b)

Nodes

Nodes

$t = 0$ $t = \frac{1}{4} T$ $t = \frac{1}{2} T$ $t = \frac{3}{4} T$

Figure 21.7. (a) Waves on a string at time intervals of $\frac{1}{4}T$. Top, a wave of the left; below, a wave to the right. (b) The appearance of the string if both waves are on the same string. The shape of the string changes with time, but its displacement is always zero at the nodes. A photograph taken with an exposure of several periods shows a blur corresponding to the maximum displacement at each point (Fig. 21.8).

displacement is their sum

$$y = A \sin (kx - \omega t) + A \sin (kx + \omega t)$$

Now we can expand the sines using Eq. B.34 from Appendix B.11,

$$\sin (a + b) = \sin a \cos b + \cos a \sin b$$
$$\sin (a - b) = \sin a \cos b - \cos a \sin b$$

With $kx = a$ and $\omega t = b$, we find the formula for a standing wave,

$$y = 2A \sin kx \cos \omega t \qquad (21.13)$$

The shape of a string described by this equation is a sinusoidal wave. Its displacement depends on the

time through the factor $\cos \omega t$. However, the locations of the nodes and antinodes are determined by the $\sin kx$ term and are fixed in time. The nodes occur in Fig. 21.7 at the places where $\sin kx = 0$. This happens where $kx = 0, \pi, 2\pi, \ldots$. Since $k = 2\pi/\lambda$, this

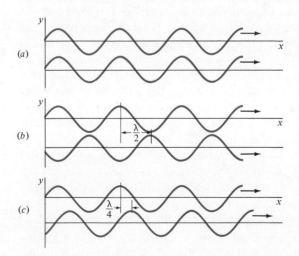

(a)

(b)

$\frac{\lambda}{2}$

(c)

$\frac{\lambda}{4}$

Figure 21.9. (a) Two waves exactly in phase. (b) Two waves exactly out of phase. Addition of these two waves results in complete destructive interference. (c) These two waves are out of phase by $\lambda/4$.

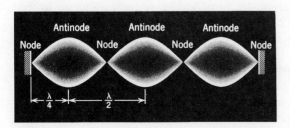

Antinode   Antinode   Antinode
Node   Node   Node   Node

$\frac{\lambda}{4}$ $\frac{\lambda}{2}$

Figure 21.8. The pattern formed by the string of Fig. 21.7 if viewed at many different times.

corresponds to $x = 0$, $\lambda/2$, λ, ... The antinodes occur where $|\sin kx| = 1$. This is at $kx = \pi/2$, $3\pi/2$, . . . , or $x = \lambda/4$, $3\lambda/4$, ...

Thus the superposition of two traveling waves produces a stationary pattern or standing wave, much like the string in Fig. 21.8. Standing waves with nodes at other locations and with a slightly different algebraic form result if y_2 has a different phase. This happens, for example, if the sign of y_2 is changed, or it is replaced by $A \sin(kx + \omega t + \phi)$.

21.5 | THE EFFECTS OF BOUNDARIES

Usually when a wave encounters a *boundary*, a point where the medium changes, part of the wave is reflected and part is absorbed or transmitted. The details depend on the nature of the boundary. In this section, we consider two special types of boundaries. The end of a string will either be held firmly fixed or will be free to move transversely. In both cases the wave is nearly completely reflected because almost no energy is lost from the system.

Figure 21.10 shows what happens to a wave pulse that reaches a fixed or a free end of a string. If the string end is fixed, the reflected wave is the *inverted* mirror image of the incident pulse; the wavelength and shape are unchanged. If the string end is free, the reflected wave is the mirror image of the incident wave but is *not inverted*.

The results can be understood from the principles of mechanics. When a pulse arrives at a fixed end of a string, it exerts an upward force on the support. The support will exert an *equal and opposite* force on the string, producing a reflected pulse opposite to the incident pulse or reversed in phase. If a pulse reaches the free end of a string, the particles of the string there acquire momentum in the upward direction. When the end of the string reaches the maximum height of the wave pulse, this momentum is not zero, and the string overshoots or continues upward. Now the free end of the string exerts a force on the remainder of the string and produces a reflected wave of exactly the same shape as the incident wave. No inversion takes place, and the phase is unchanged.

Once we know how a pulse behaves at a boundary, we can consider the effects of a boundary on a periodic wave. When a sinusoidal wave $y_1 = A \sin(kx - \omega t)$ reaches the fixed end of a string at $x = 0$, a reflected wave $y_2 = A \sin(kx + \omega t)$ is

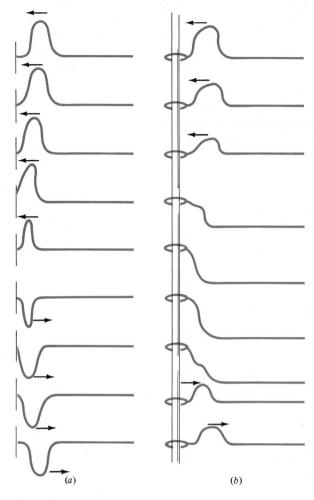

Figure 21.10. The reflection of a wave pulse on a string at (*a*) a fixed end boundary and (*b*) a free end boundary.

produced. We may verify that the reflected wave is inverted or reversed in phase by noting that at $x = 0$, $y_1 = A \sin(-\omega t)$, and $y_2 = A \sin(\omega t)$. The sine function is odd, so that $\sin(-\omega t) = -\sin(\omega t)$, and $y_2 = -y_1$ as required.

This means that two waves are present, traveling in opposite directions (Fig. 21.11). This is precisely the situation described in the preceding section by Eq. 21.12. Thus we know immediately that the incident and reflected waves interfere to produce a standing wave with a node at $x = 0$, the fixed end. There are additional nodes at $x = \lambda/2$, λ, ...

The reflection of sinusoidal waves at a free end is shown in Fig. 21.12. Here the reflected wave is not inverted. Adding the incident and reflected waves algebraically leads to the result for this situation that the string has an antinode at the free end.

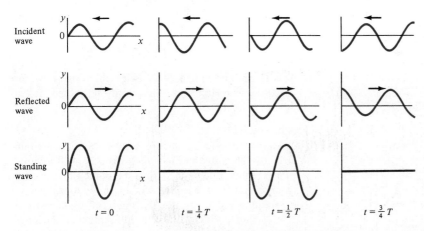

Figure 21.11. The top pictures show, at time intervals of $\frac{1}{4}T$, a wave approaching the fixed end of a string. The middle pictures show the reflected wave, which is inverted. In the actual string, both waves are present, and the result is a standing wave (bottom).

In summary, the reflection of sinusoidal waves at a boundary results in standing waves. On a string, a node appears at a fixed end and an antinode at a free end. Analogous standing waves are formed for all kinds of waves when reflections occur.

21.6 | RESONANT STANDING WAVES

In Chapter Nine on harmonic motion, we found that many structures have specific resonant frequencies at which large-amplitude vibrations are readily produced. Musical instruments and the voice involve systems that combine vibrating strings or air columns with structures that have certain resonant frequencies.

The fact that strings have specific resonant frequencies can be seen by tying one end of a rope to a post and shaking the other end. After a few tries, one finds standing waves are produced only at specific frequencies. At other frequencies the rope vibrates erratically and with a small amplitude.

To understand why only certain frequencies result in standing waves, consider a string of length l. With both ends rigidly fixed, only standing waves with nodes at each end can be produced. Figure 21.13 shows the five longest standing waves for a fixed end string, that is, the longest wave with nodes at both ends. The waves that fit on the string are called the *harmonics* of the string. The longest is called the *fundamental* or first harmonic; the first five harmonics of the fixed end string are shown in Fig. 21.13.

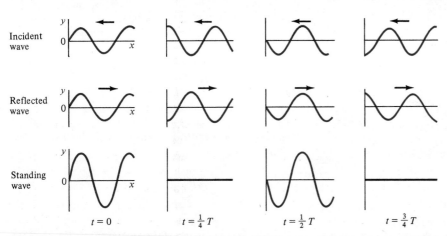

Figure 21.12. A wave approaching a free end of a string, its reflected wave (not inverted), and the resulting standing wave.

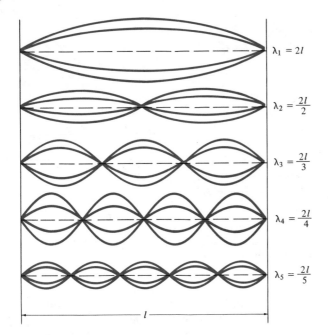

Figure 21.13. The five longest wavelength standing waves on a string of length l with fixed ends.

In Fig. 21.13 the wavelengths can be labeled with a subscript n, where n is a positive integer. For the nth harmonic, we see that

$$\lambda_n = \frac{2l}{n},$$

$$n = 1, 2, 3, \ldots \text{ (fixed end string)} \tag{21.14}$$

The corresponding frequencies are found from $f_n\lambda_n = c$, where c is the wave velocity on the string. Thus on a string of length l

$$f_n = \frac{n}{2l}c,$$

$$n = 1, 2, 3, \ldots \text{ (fixed end string)} \tag{21.15}$$

Since $c = \sqrt{\Im/\mu}$ for a string, we also have

$$f_n = \frac{n}{2l}\sqrt{\frac{\Im}{\mu}}$$

$$n = 1, 2, 3, \ldots \text{ (fixed end string)} \tag{21.16}$$

The following examples illustrate the use of these results.

Example 21.6

What are the frequencies of the first three harmonics of the longest string in a grand piano? The length is 1.98 m, and the velocity of the wave on the string is $c = 130 \text{ m s}^{-1}$.

Using the result for the fixed end string with $n = 1$,

$$f_1 = \frac{c}{2l} = \frac{130 \text{ m s}^{-1}}{2(1.98 \text{ m})} = 32.8 \text{ Hz}$$

The second harmonic is at twice this frequency, or $2(32.8) = 65.6$ Hz, and the third harmonic is at $3(32.8) = 98.4$ Hz.

Example 21.7

The wave velocity on the highest frequency violin string is 435 m s^{-1}, and its length l is 0.33 m. If a violin player lightly touches the string at a point a distance $l/3$ from an end, a node is produced there. What is the lowest frequency that can now be produced by the string?

Referring to Fig. 21.13, the third harmonic is the longest wavelength and lowest frequency standing wave with a node one-third of the distance from one end. Hence with $\lambda_3 = 2l/3$,

$$f_3 = \frac{c}{\lambda_3} = \frac{c}{\left(\dfrac{2l}{3}\right)} = \frac{3c}{2l} = \frac{3(435 \text{ m s}^{-1})}{2(0.33 \text{ m})} = 1977 \text{ Hz}$$

A violinist can, by lightly touching a string at some point along its length, produce a node on the string at that point so that only harmonics with nodes at that point are excited. However, if the string is pushed firmly against the finger board, its effective length is reduced, and all the harmonic frequencies are shifted upward.

The relationship between the frequency and mass per unit length of a string is discussed in the following example.

Example 21.8

The highest and lowest frequency strings of a piano are tuned to fundamentals of $f_H = 4186$ Hz and $f_L = 32.8$ Hz. Their lengths are 0.051 m and 1.98 m, respectively. If the tension in these two strings is the same, what is the ratio of the masses per unit length of the two strings?

Using Eq. 21.16 and solving for μ, we find for the fundamental, $n = 1$,

$$\mu = \frac{\Im}{(2lf_1)^2}$$

The ratio of μ_L for the low-frequency string to μ_H for the high-frequency string is

$$\frac{\mu_L}{\mu_H} = \frac{\dfrac{\Im}{(2l_Lf_L)^2}}{\dfrac{\Im}{(2l_Hf_H)^2}} = \frac{(l_Hf_H)^2}{(l_Lf_L)^2}$$

$$= \left[\frac{(0.051 \text{ m})(4186 \text{ Hz})}{(1.98 \text{ m})(32.8 \text{ Hz})}\right]^2 = 10.8$$

This is a large difference. It was found empirically by piano makers that a thick solid string does not produce a pleasing tone. This is apparently due to the great stiffness of such strings. The solution developed was to wind a wire in a tight coil about a straight wire (here of length 1.98 m). This increases the mass per unit length without changing the stiffness of the wire very much.

21.7 | COMPLEX WAVES AND BEATS

Often when a string is disturbed or standing waves are set up in air columns, not only is the fundamental frequency present but so also are some of the higher harmonics. The higher harmonics that are actually present are called *overtones*. If all the harmonics are present, the second harmonic is the first overtone, the third harmonic is the second overtone, and so on. In the next chapter, we discuss situations where every other harmonic is present. In this case, the third and fifth harmonics are the first and second overtones, respectively. The higher overtones are numbered similarly.

The higher harmonics are not all excited with the same amplitude; instead the higher harmonics usually have progressively smaller amplitudes. If several harmonics are simultaneously present, the actual shape of the string at any instant may be quite complex (Fig. 21.14). Any complex wave can be regarded as a sum of many sinusoidal waves.

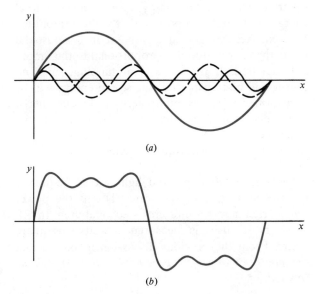

(a)

(b)

Figure 21.14. (a) The first, third, and fifth harmonics for a string fixed at both ends. (b) The addition of these harmonics produces the complex shape shown.

In Fig. 21.15, we show two traveling waves with slightly different wavelengths and frequencies. At points B and D the two waves add constructively. At the intermediate point C the waves interfere destructively. The resultant wave is a rapid oscillation that changes amplitude with time. At points A, C, and E the amplitude of the total wave is zero, and at points B and D the amplitude is a maximum. This phenomena is called *beating*. The frequency with which the nodes pass a given point on the x axis is called the *beat frequency*.

We can add the equations for the two waves to find the beat frequency. We are interested in the net wave at a single point as a function of the time, so at that point we write

$$y_1 = A \cos 2\pi f_1 t \qquad y_2 = A \cos 2\pi f_2 t \quad (21.17)$$

Here we write $2\pi f$ instead of ω and represent the sinusoidal waves by cosines instead of sines to obtain results that have a neater form. (This is permissible because the cosine and sine functions are the same except for a shift of 90°.) According to Eq. B.37 in Appendix B.11,

$$\cos a + \cos b = 2 \cos \frac{(a + b)}{2} \cos \frac{(a - b)}{2}$$

Thus the sum $y = y_1 + y_2$ can be written as

$$y = \left[2A \cos \frac{2\pi(f_1 - f_2)t}{2} \right] \cos \frac{2\pi(f_1 + f_2)t}{2} \quad (21.18)$$

The second cosine represents a wave with the average of the two frequencies, $\overline{f} = (f_1 + f_2)/2$. The first cosine varies with a frequency $(f_1 - f_2)/2$, which is very small if the two frequencies are nearly equal. Thus Eq. 21.18 describes a wave with the average frequency $\overline{f}$ and an amplitude given by the factor in

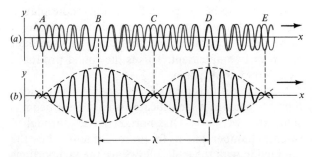

(a)

(b)

Figure 21.15. (a) Two waves of slightly different frequencies, f_1 (shown colored) and f_2, add to form the resultant wave in (b). The dashed lines show how the amplitude changes.

brackets that slowly varies in magnitude between 0 and 2A. The amplitude has a maximum whenever $\cos 2\pi(f_1 - f_2)t/2$ is $+1$ or -1. Each of these values occurs once in each cycle, so the beat frequency f_b is twice $(f_1 - f_2)/2$, or

$$f_b = f_1 - f_2 \qquad (21.19)$$

The beat frequency is equal to the difference in the frequencies of the two waves.

The human ear is sensitive to beats from nearly 0 to about 20 Hz, and the waves need not have exactly equal amplitudes. The following example illustrates how beats may be put to practical use.

Example 21.9

A tuning fork with a known frequency of 520 Hz is sounded at the same time that a violin string is plucked. Beats are heard with a frequency of 5 Hz. What is the frequency of the violin string?

From our discussion, we note that the difference between the frequencies of the tuning fork and violin string is 5 Hz. However, we cannot tell immediately which frequency is higher and which is lower. Since the fork has a frequency of 520 Hz, the violin string vibrates at either 525 Hz or 515 Hz.

We could determine which is the case by adjusting the tension in the violin string. From Eq. 21.16, the string frequency is proportional to the square root of the tension. If reducing the tension reduces the beat frequency, we know the violin was at 525 Hz. If the string were at 515 Hz, the tension would have to be increased to reduce the beat frequency.

Use is made of beats in tuning a piano. Most piano keys strike three strings that are supposed to produce identical notes. A modern piano tuner stops two of the three strings from vibrating and adjusts the tension of the third until it is synchronized with an electronic oscillator set at the desired frequency. Then the other two strings are released one at a time and adjusted until no beats are heard. A single note played on a poorly tuned piano has beats that are too rapid to be heard separately, but the sound is discordant and may be unpleasant. This is the sound produced by a honky-tonk piano.

Modern methods for studying large molecules with light use beats in a sophisticated way. A laser beam is split into two parts. One portion passes through a sample chamber, and its frequency is modified by the interaction with the molecules. Since the two portions of the beam now have different frequencies, they produce beats when recombined. Information about the molecules is obtained from the frequency changes.

21.8 | ENERGY AND MOMENTUM IN WAVES

Although it is not always apparent, waves of every kind carry energy and momentum obtained originally from their sources. This may be illustrated by some familiar examples. Light from the sun provides the energy that makes life possible on our planet. Ocean waves gradually transform the coastlines, and they can exert large forces on someone standing in shallow water. Intense sound waves can crack windows and cause other kinds of damage to mechanical structures including the human ear.

The specific expressions for the energy and momentum of a wave depend on the nature of the particular wave. However, it is always true that the energy and momentum stored in a sinusoidal wave are proportional to the square of its amplitude.

This fact can be understood if we make an analogy with the simple harmonic oscillator (Chapter Nine). If we observe a sinusoidal wave passing a given point, the displacement at that point varies with time as a sine or cosine. A simple harmonic oscillator, such as a mass on a spring, also has a displacement that varies sinusoidally with time. Hence, each point in a sinusoidal wave is undergoing simple harmonic motion. This was also shown in detail in Example 21.4. In an oscillator, the potential energy $\frac{1}{2}kx^2$ is proportional to the displacement squared and hence to the amplitude squared. The kinetic energy, $\frac{1}{2}mv^2$, and the total energy are also proportional to the amplitude squared. Accordingly, at any point in a sinusoidal wave, the energy stored is proportional to the amplitude squared. If the wave is a standing wave, no energy will be transported; the stored energy will be transformed from potential energy to kinetic energy and back again repeatedly, but it will remain in the same location. However, if the wave is moving in some direction, it will transport energy in that direction.

Discussions of sound and light waves frequently use the *intensity I* of the wave. This is the power transported across a unit cross-sectional area. In S.I. units, the intensity is measured in watts per square metre. From the previous discussion it follows that *the intensity of a wave is proportional to its amplitude squared,* or

$$I \propto A^2 \qquad (21.20)$$

The momentum carried by the wave also varies as the amplitude squared.

21.9 | THE POLARIZATION OF TRANSVERSE WAVES

Transverse waves have a property called *polarization* that is not shared by longitudinal waves. For example, in Fig. 21.16, a wave on a string is traveling along the x direction. The transverse wave disturbance can be along any line perpendicular to this direction of motion. If the disturbance is always along the same line, the wave is said to be *polarized* along that line. A wave whose direction of oscillation changes randomly from time to time or which is composed of many waves of random polarization is said to be *unpolarized*.

An unpolarized wave can be polarized by a *polarizer*. For the string shown in Fig. 21.16a, the polarizer is a slot. Beyond the slot the string can only vibrate in the y direction. In Fig. 21.16b, there is also a second slot. When the two slots are perpendicular the wave is totally suppressed.

The amplitude of the wave after passing through a polarizing structure is reduced, since only the *component* of the original wave parallel to the polarizer passes through. In Fig. 21.16a the amplitude of the wave on the left side of the slot is A_L and the intensity is $I_L \propto A_L^2$. The angle between the planes of polarization on the left side and right side of the slot is θ. The amplitude of the wave on the right side of the slot is $A_R = A_L \cos \theta$. The intensity is then

$$I_R = I_L \cos^2 \theta$$

In Fig. 21.16b the angle between the two slots is 90°, and since cos 90° = 0, the amplitude becomes zero.

Electromagnetic waves are transverse waves. We refer to the direction of the electric vector as the direction of polarization. Polarized sunglasses use the transverse property of light to reduce glare. Light that is reflected from smooth horizontal surfaces tends to be polarized in the horizontal direction. Polaroid sunglass lenses almost totally suppress the component of light along one direction and are oriented so that only the vertically polarized component is transmitted. If one rotates a Polaroid lens relative to a second, the transmitted light is seen to decrease nearly to zero when the polarizing axes are perpendicular. A more extensive discussion of light polarization is given in Chapter Twenty-three.

SUMMARY

Waves of all kinds share many common features. Periodic waves are characterized by their frequency f, wavelength λ, and velocity c, which are related by $f\lambda = c$. The velocity depends on the properties of the medium and in some cases on the frequency. The amplitude of a wave is the maximum magnitude of its displacement.

Waves can interfere with one another. When two waves are present at a point, the resulting wave is found by algebraically adding the displacements of the individual waves. Two waves in phase add constructively, while two waves a half wavelength out of

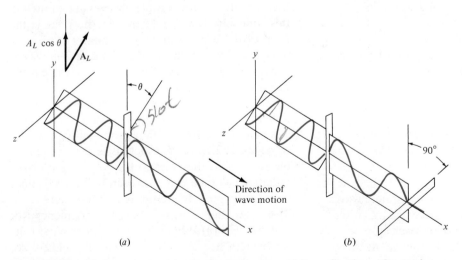

(a)   (b)

Figure 21.16. (a) A wave on a string is vibrating in some arbitrary direction. After passing through the slot, the wave is polarized in the y direction and has a smaller amplitude. (b) With the second slot at right angles to the first, the wave is completely suppressed. If the two slots were not perpendicular, the final wave would be polarized in the direction of the second slot.

phase interfere destructively. The property of combining waves by the addition of displacements is called the principle of superposition or linearity.

Waves incident on a boundary can be transmitted, absorbed, or reflected. Waves that reach a fixed end of a string are reflected as inverted mirror images. At a free end the reflected wave is not inverted.

Standing waves are formed when two waves of equal amplitude but opposite direction interfere. Standing waves on a string can persist only at certain special frequencies, called harmonics, which are determined by the length of the string and the velocity of the wave.

When two waves arrive at a point at slightly different frequencies, the intensity varies at the beat frequency, which is equal to the difference in frequencies of the two waves. Beats can be heard in sound waves at frequencies up to about 20 Hz.

The energy and momentum in a sinusoidal wave are proportional to the square of its amplitude. The intensity of a wave is the power transported per unit area and also varies as the amplitude squared.

If the disturbance is always along some particular axis perpendicular to the direction of propagation of a transverse wave, it is said to be polarized along that direction. If the direction of oscillation changes randomly over time, the wave is unpolarized.

Checklist

Define or explain:

wave pulse	node
periodic wave	standing waves
longitudinal waves	phase
wavelength	boundary
amplitude	fundamental frequency
sinusoidal waves	harmonics
wave velocity	overtones
wave number	beat frequency
angular frequency	intensity
superposition	polarized wave
destructive interference	

REVIEW QUESTIONS

Q21-1   For a given type of wave, periodic waves with a higher frequency will have _____ wavelengths than those with lower frequencies.

Q21-2   True or false: The velocities of all waves on all strings are the same.

Q21-3   True or false: A standing wave on a string occurs when all points on the string stand still.

Q21-4   Standing waves can be regarded as a _____ of two or more traveling waves.

Q21-5   True or false: If one end of a string is fixed, the possible resonant standing waves that can occur are the same whether the other end is fixed or free.

Q21-6   True or false: The wavelengths of the possible resonant standing waves on a string depend only on the length of the string and on the boundary conditions.

Q21-7   Beats occur because waves _____ with one another.

Q21-8   Two waves of the same type but with different frequencies f_1 and f_2 will produce beats at a frequency _____.

Q21-9   The energy carried by a wave is proportional to the _____ of the wave.

Q21-10   Only _____ waves may be polarized.

EXERCISES

Section 21.1 | The Representation of Waves

21-1   What is the wavelength of a sound wave with a frequency of 1000 Hz and a velocity 344 m s^{-1}?

21-2   What is the frequency of a wave of velocity 200 m s^{-1} and wavelength 0.5 m?

21-3   A radar antenna emits electromagnetic radiation ($c = 3 \times 10^8$ m s^{-1}) with a wavelength of 0.03 m for 0.5 s. (a) What is the frequency of radiation? (b) How many complete waves are emitted in this time interval? (c) After 0.5 s, how far is the front of the wave from the antenna?

21-4   A TV station broadcasts using 2-m waves. What is the frequency of the broadcast wave if the speed of the wave is 3×10^8 m s^{-1}?

21-5   A radio telescope is built to observe microwaves from interstellar hydrogen atoms at frequencies near 1.4×10^9 Hz. Since the ability of a telescope to distinguish fine details is determined by the ratio of its diameter to the wavelength of the radiation, its diameter is chosen to be 1000 times the wavelength. How large is the diameter?

21-6   A light wave has a frequency of 6×10^{14} Hz. (a) What is its period? (b) What is its wavelength in vacuum? (c) When the light wave enters water, its velocity decreases to 0.75 times its velocity in vacuum. What happens to the frequency and wavelength?

Section 21.2 | The Mathematical Description of Traveling Waves

21-7 Green light has a wavelength of 5×10^{-7} m. Find the (a) frequency; (b) angular frequency; (c) wave number.

21-8 Two waves are represented by $y_1 = A \sin (kx - \omega t)$ and by $y_2 = A \sin (kx - \omega t + \pi/4)$. At time $t = 0$, draw rough graphs of each wave for $x = 0$ to 3λ, where $\lambda = 2\pi/k$. Explain the relationship between the graphs for the two waves.

21-9 A wave is represented by $y = A \sin (kx + \omega t)$. Draw two cycles of the wave from $x = 0$ to $x = 2\lambda$ at (a) $t = 0$; (b) $t = T/4$, where $T = 1/f = 2\pi/\omega$.

21-10 Sinusoidal waves can be represented equally well by sines or by cosines. For example, $y = A \sin (kx - \omega t)$ can be rewritten as $y = A \cos (kx - \omega t + \phi)$. What is the phase angle ϕ?

21-11 A sinusoidal wave with a wavelength of 2 m and a velocity of 5 m s^{-1}. At $t = 0$ s, the left end is at its equilibrium position, $y = 0$, and the displacement is decreasing. The maximum displacement is 0.1 m. Taking $x = 0$ at the left end and $+x$ toward the right, find an equation of the form of Eq. 21.7 for the wave.

c21-12 A wave on a string is represented by $y = (0.05 \text{ m}) \sin [(10 \text{ m}^{-1})x - (50 \text{ s}^{-1})t]$. (a) Find the period T of the wave. (b) Find the transverse velocity at the point $x = 0$ at times $t = 0$ and at $T/4$. (c) Find the transverse acceleration at the same times.

Section 21.3 | The Velocity of Waters

21-13 A string on a steel guitar has a mass per unit length of 3×10^{-3} kg m^{-1}. If the tension in the string is 90 N, what is the velocity of a wave on the string?

21-14 The tension in a string is four times that in a second identical string. What is the ratio of the wave velocities of the strings?

21-15 The speed of a wave on a string is 160 m s^{-1} when the tension in the string is 100 N. To increase the speed to 200 m s^{-1}, to what value must the tension be increased?

21-16 When light enters heavy flint glass, its velocity is reduced from that in the vacuum by a factor of 1.647. What is its velocity in this kind of glass?

Section 21.4 | Wave Interference and Standing Waves

21-17 Two wave pulses on a string (Fig. 21.17) travel toward each other with speeds of 1 m s^{-1}. Sketch the shape of the string at $t = 1$, 1.25, and 1.5 s after the instant shown.

21-18 Two wave pulses travel toward each other on a string (Fig. 21.18). If the wave velocity is 2 m s^{-1}, sketch the string 1, 1.25, and 1.5 s after the instant shown.

21-19 Sketch the string of Fig. 21.19, 1, 1.25, and 1.5 seconds after the instant shown. The wave velocity is 1 m s^{-1}.

21-20 (a) Show that if $y_1 = A \sin (kx - \omega t)$ and $y_2 = -A \sin (kx + \omega t)$, the resulting wave $y = y_1 + y_2$ is a standing wave. (b) Where are its nodes and antinodes relative to those of the standing wave described by Eq. 21.13?

21-21 If $y_1 = A \sin (kx - \omega t)$ and $y_2 = 3A \sin (kx + \omega t)$, then the superposition $y = y_1 + y_2$ is a pure standing wave plus a traveling wave in the $-x$ direction. Find the amplitude of (a) the traveling wave; (b) the standing wave.

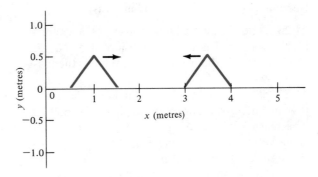

Figure 21.17. Exercise 21-17.

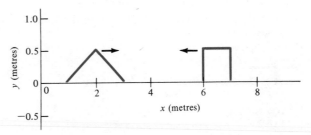

Figure 21.18. Exercise 21-18.

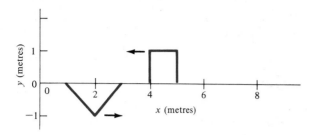

Figure 21.19. Exercise 21-19.

21-22 Sketch the standing wave represented by Eq. 21.13 from $x = 0$ to $x = 2\lambda$ at (a) $t = 0$; (b) $t = T/8$; (c) $t = T/4$.

21-23 For the standing wave represented by Eq. 21.13, at what values of x are (a) the nodes; (b) the antinodes?

Section 21.5 | The Effects of Boundaries

21-24 Sketch the shape of a square-wave pulse on a string (Fig. 21.20) if it is totally reflected from a fixed boundary.

21-25 Sketch the shape of the square-wave pulse on a string of Fig. 21.20 if it is totally reflected from a free end boundary.

Section 21.6 | Resonant Standing Waves

21-26 The lowest frequency string of a violin is 0.33 m long and is under a tension of 55 N. The fundamental frequency is 196 Hz. What is the mass per unit length of the string?

21-27 The E string of a violin has a length of 0.33 m and a fundamental frequency of 659 Hz. The tension of the string is 55 N. (a) What is the wave velocity of the string? (b) What is the mass per unit length of the string?

21-28 If the heaviest and lightest strings of a violin have masses per unit length of 3×10^{-3} kg m^{-1} and 2.9×10^{-4} kg m^{-1}, what is the ratio of the radii of these strings?

21-29 The fundamental frequency of a string with fixed ends is 100 Hz and the wave velociy is 350 m s^{-1}. (a) What is the wavelength of the funda-

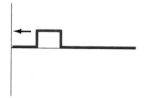

Figure 21.20. Exercises 21-24 and 21-25.

mental? (b) What is the length of the string?

21-30 If the wave velocity of a 0.5-m-long guitar string is 170 m s^{-1}, what is its fundamental frequency?

21-31 The highest C string of a piano has a frequency of 4186 Hz and a length of 0.051 m. The lowest C string has a frequency of 32.8 Hz. If these strings had the same tension and mass per unit length, what would be the length of the lowest C string?

21-32 The A string on a violin is 0.33 m long and tuned to a fundamental frequency of 440 Hz. How far from the end of the string should one press it against the finger board to obtain the same fundamental frequency, 659 Hz, as the E string?

21-33 A harp string of length 0.5 m is tuned to a fundamental frequency of 650 Hz. (a) What is the wavelength of the fourth harmonic of the string? (b) What is the wavelength of the sound produced in the air if the fourth harmonic is excited? (Use $c = 344$ m s^{-1} in air.)

21-34 If a string of an instrument is touched lightly at a point one-third of its length from one end, a node is formed there. Which harmonics of the string cn be excited?

Section 21.7 | Complex Waves and Beats

21-35 The first and second harmonics of a string are shown at a certain instant in Fig. 21.21. Sketch the actual shape of the string at this instant.

21-36 The fundamental frequency of the heaviest string on a cello is 65.4 Hz. What is the beat frequency of the third harmonic of this string with the 196-Hz fundamental of the heaviest string on a violin?

21-37 What beat frequency is heard if two tuning forks vibrate with frequencies $f_1 = 200$ Hz and $f_2 = 205$ Hz?

21-38 What are the possible frequencies of a tuning fork that produces a beat frequency of $f = 4$ Hz with a standardized tuning fork that has a frequency of 300 Hz?

Section 21.8 | Energy and Momentum in Waves, and Section 21.9 | The Polarization of Transverse Waves

21-39 A traveling wave on a rope has a vertical amplitude of 0.1 m. It passes through a slot that is tilted at 30° with the vertical. What is the amplitude of the wave after passing through the slot?

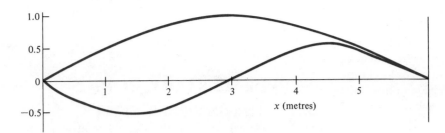

Figure 21.21. Exercise 21-35.

21-40 A traveling wave on a rope has a vertical amplitude of 0.2 m. It passes through two successive slots; the first is tilted at 45° with the vertical and the second is vertical. (a) What is the amplitude of the wave after it has passed through the first slot? (b) What is the amplitude of the wave after it has passed through both slots?

21-41 What is the ratio of the intensity of the wave of Exercise 21.40 after it has passed through both slots to its original intensity?

21-42 In Fig. 21.7*b*, the second and fourth pictures of the superposition of two waves, there is no displacement. What has happened to the energy of the wave?

21-43 A symmetrical wave pulse travels to the right on a string, and an identical pulse of opposite sign travels to the left. When they meet the waves cancel exactly, so the displacement is zero. What has happened to the energy, and why do the waves reappear shortly?

PROBLEMS

21-44 Using dimensional arguments, show that the velocity of a wave on a string with tension $\Im$ and mass per unit length μ is proportional to $\sqrt{\Im/\mu}$.

*21-45 The restoring force on long wavelength surface water waves is due to gravity as long as the water is deep and the bottom does not affect the wave. Using dimensional arguments and explaining your reasoning, show that the velocity of long water waves of wavelength λ is of the form

$$c = A \sqrt{g\lambda}$$

where g is the gravitational acceleration, and A is a numerical constant. (Exact calculation shows that $A = \sqrt{1/8\pi}$.

*21-46 Using dimensional reasoning, show that for very short wavelength surface waves on water the wave velocity depends on the surface tension and has the form

$$c = B \sqrt{\frac{\gamma}{\rho\lambda}}$$

Here γ is the surface tension, ρ is the density, and B is a numerical constant. (Exact calculations yield $B = \sqrt{9\pi/2}$.)

21-47 Using the results of Problem 21-45, what is the velocity of a water wave of wavelength 30 m?

21-48 The general expression for the velocity of surface water waves of wavelength λ is

$$c = \frac{1}{2} \sqrt{\frac{g\lambda}{2\pi} + \frac{18\pi\gamma}{\rho\lambda}}$$

where ρ is the density, g is the acceleration of gravity, and γ is the surface tension. At what wavelength λ does the transition from long to short waves take place? (This result is only valid if the water is very deep compared to the wavelength and crest height.)

21-49 How long will it take a 1-m bow wave from a boat to reach the short of a river 25 m away? (See Problem 21-45).

21-50 The 0.33-m-long A string of a violin is tuned to a fundamental frequency of 440 Hz. The 0.69-m-long A string of a cello is tuned to a fundamental frequency of 220 Hz. If both strings have the same tension, what is the ratio of the masses per unit length of these strings?

21-51 The E string of a violin is 0.33 m long, and the wave velocity of the string is 435 m s^{-1}. (a) What is the time required for the wave produced by plucking the string to make one complete circuit along the string and return to its original position? (b) Does this result have any bearing on the proper frequency of plucking or bowing the string?

21-52 If one of the three strings corresponding to middle C on a piano is adjusted to the proper frequency of 261.6 Hz and is struck simultaneously with an untuned string so that a beat frequency of 10 Hz is heard, (a) what two frequencies might the untuned string have? (b) If the tension in the tuned string is 1160 N, what are the possible tensions in the untuned string?

21-53 The amplitude of the horizontally polarized light on a beach is twice that of the vertical component. If a woman standing erect puts on Polaroid sunglasses, only a negligible portion of the horizontally polarized light reaches her eyes. (a) What is the percentage reduction of the light energy that reaches her eyes when she puts on the sunglasses? (b) If the woman is lying on her side, what is the percentage reduction in energy that reaches her eyes when she puts on the sunglasses? (*Hint:* The total amplitude of a wave is the vector sum of its horizontal and vertical components.)

21-54 A rope of length l has a wave velocity c. Find the characteristic standing wave frequencies if the rope is (a) tied at one end and free at the other; (b) free at both ends.

^c**21-55** A sinusoidal wave of amplitude A and angular frequency ω is traveling on a string. (a) What is the maximum transverse velocity of a point on the string? (b) What is the maximum kinetic energy of a small segment of the string of mass Δm? (c) Given the fact shown in Example 21.4 that the segment is undergoing simple harmonic motion, what is its maximum potential energy?

^c**21-56** A standing wave on a string is represented by Eq. 21.13. Show that each point on the string is undergoing simple harmonic motion transverse to the string.

21-57 The superposition of two sinusoidal waves traveling in the same direction with different phases, $y = A \sin(kx - \omega t) + A \sin(kx - \omega t + \phi)$, is a sinusoidal wave with an amplitude different from A. Find that amplitude. (*Hint:* Use Eq. B.36 from Appendix B.11.)

21-58 Show that beat frequency (Eq. 21.19) can also be derived by adding two waves at a point represented by $y_1 = A \sin 2\pi f_1 t$ and $y_2 = A \sin 2\pi f_2 t$. (*Hint:* Use Eq. B.36 from Appendix B.11.)

ANSWERS TO REVIEW QUESTIONS

Q21-1, shorter; **Q21-2,** false; **Q21-3,** false; **Q21-4,** superposition; **Q21-5,** false; **Q21-6,** true; **Q21-7,** inter-

fere; **Q21-8,** $f_1 - f_2$; **Q21-9,** square of the amplitude; **Q21-10,** transverse.

SUPPLEMENTARY TOPICS

21.10 | THE DOPPLER EFFECT

When a train passes by, the frequency at which its whistle is heard drops suddenly. This *Doppler effect* arises because as the source approaches a listener, more waves arrive each second than are emitted; conversely, as the source recedes, fewer waves arrive each second than are emitted. A similar effect is heard when the listener moves and the source is fixed.

All types of waves exhibit a Doppler effect, although the results for light waves are somewhat different than those for other types of waves. In this section, we describe the Doppler effect for sound and other mechanical waves in detail and merely state the results for light waves.

Consider first a sound source stationary relative to the air with frequency f and wavelength $\lambda = c/f$, and a listener moving away from the source at a velocity v_l (Fig. 21.22). The speed of the sound waves will appear to the listener to be $c' = c - v_l$. The distance between two successive wave maxima in the listener's moving frame of reference is the same as for the stationary frame of the source, λ. The apparent frequency f' heard by the listener then satisfies $f'\lambda = c'$, or $f' = c'/\lambda$. With $\lambda = c/f$ and $c' = c - v_l$, this becomes for a *listener moving away from a source at a velocity v_l*

$$f_r = f_0 \left(\frac{c - v_c}{c \times v_c} \right)$$

When the listener is approaching the source, v_l must be regarded as negative. Notice that the frequency heard is lower than the source frequency when the listener is moving away from the source. The perceived frequency is greater than the source frequency when the listener approaches the source.

The following example illustrates this type of Doppler effect.

Example 21.10
A stationary civil defense siren has a frequency of 1000 Hz. What frequency will be heard by drivers of cars moving at 15 m s^{-1} (a) away from the siren; (b) toward the siren? The velocity of sound in air is 344 m s^{-1}.

(a) With $v_l = 15$ m s^{-1}, drivers moving away from the siren hear a frequency

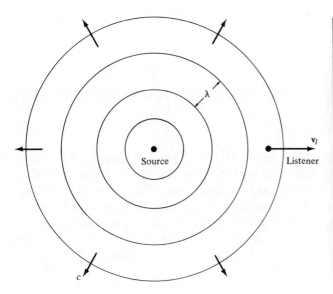

Figure 21.22. Waves leave the stationary source with a speed c but appear to the moving listener to have a speed of $c - v_l$.

$$f' = f\frac{(c - v_l)}{c} = (1000\ \text{Hz})\left(\frac{344\ \text{m s}^{-1} - 15\ \text{m s}^{-1}}{344\ \text{m s}^{-1}}\right)$$

$$= 956\ \text{Hz}$$

(b) Using $v_l = -15\ \text{m s}^{-1}$ for drivers moving toward the source,

$$f' = (1000\ \text{Hz})\left(\frac{344\ \text{m s}^{-1} + 15\ \text{m s}^{-1}}{344\ \text{m s}^{-1}}\right)$$

$$= 1044\ \text{Hz}$$

A different expression is obtained for a source moving at a velocity v_s away from a listener who is at rest relative to the air (Fig. 21.23). The sound waves are bunched together in the direction of motion and spread out in the opposite direction. If f_0 is the frequency at which the waves are emitted, then in the direction opposite to the motion, the distance between successive crests is increased from $\lambda = c/f$ to

$$\lambda' = \frac{c + v_s}{f}$$

The frequency heard by the stationary listener is determined by $f'\lambda' = c$, *so when the source is moving away from the listener at a velocity v_s,*

$$f' = \frac{c}{\lambda'} = f\left(\frac{c}{c + v_s}\right) \qquad (21.22)$$

If the source is approaching the listener, v_s is negative.

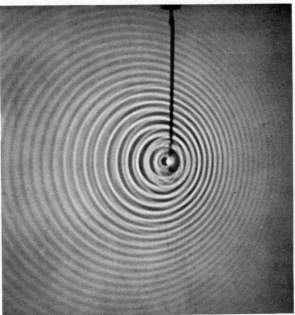

(a)

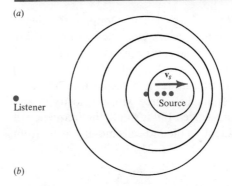

(b)

Figure 21.23. (a) A moving source of waves in a water-filled ripple tank produces waves that are bunched together in the direction of motion and spread out in the direction opposite the motion. (b) With a sound source moving with velocity v_s, a listener hears sound waves that have a longer wavelength than those from a source at rest. ((a) From Education Development Center, Inc., Newton, Massachusetts.)

The frequency heard is lower than the source frequency when the source moves away, and higher when it approaches the stationary listener. Both of these effects are seen in the next example.

Example 21.11

A police car with a 1000-Hz siren is moving at 15 m s^{-1}. What frequency is heard by a stationary listener when the police car is (a) receding from and (b) approaching the listener?

(a) With $c = 344\ \text{m s}^{-1}$ and $v_s = 15\ \text{m s}^{-1}$ for the receding source,

$$f' = f\left(\frac{c}{c + v_s}\right)$$

$$= (1000 \text{ Hz})\frac{(344 \text{ m s}^{-1})}{(344 \text{ m s}^{-1} + 15 \text{ m s}^{-1})}$$

$$= 958 \text{ Hz}$$

Note this is a slightly higher frequency than our earlier result of 956 Hz for a listener receding from a stationary siren with the same frequency and speed.

(b) When the source is approaching the listener, $v_s = -15 \text{ m s}^{-1}$, and

$$f' = f\left(\frac{c}{c + v_s}\right)$$

$$= (1000 \text{ Hz})\frac{(344 \text{ m s}^{-1})}{(344 \text{ m s}^{-1} - 15 \text{ m s}^{-1})}$$

$$= 1046 \text{ Hz}$$

Again this frequency is slightly greater than the frequency of 1044 Hz heard by a listener approaching a stationary siren.

Doppler Flow Meter

The Doppler frequency shift can be used to measure blood flow rates by using a high-frequency sound source on one side of a vessel and a detector on the other side (Fig. 21.24). The sound from the transmitter is reflected from red cells moving away from the source with a velocity v_c and detected at the receiver. If θ is small, the frequency of the sound incident on the red cells moving away from the source is given by Eq. 21.21,

$$f_1 = f\left(\frac{c - v_c}{c}\right)$$

c is the speed of sound in the bloodstream. The cells then act as a moving source with a frequency f_1. From

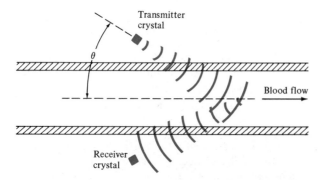

Transmitter
crystal

θ

Blood flow

Receiver
crystal

Figure 21.24. A schematic view of the use of the Doppler flow meter.

Eq. 21.22 the frequency f_r at the stationary receiver is then

$$f_r = f_1\left(\frac{c}{c + v_c}\right) = f\left(\frac{c - v_c}{c}\right)\left(\frac{c}{c + v_c}\right)$$

$$f_r = f\left(\frac{c - v_c}{c + v_c}\right) \qquad (21.23)$$

Doppler Shift for Light

A Doppler shift is also observed with light waves, but it differs somewhat from that for sound waves. This arises from the fact that no medium is necessary for light wave propagation. Therefore, there is no way to distinguish between motion of an observer relative to a stationary source or motion of a source relative to a stationary observer. One can only say that they are moving with respect to each other. This is called the *principle of relativity* and is discussed further in Chapter Twenty-five.

When the analysis is carried out it is found that if a light source of frequency f and an observer move away from each other with a relative velocity u, then the observed frequency is

$$f' = f\frac{c - u}{\sqrt{c^2 - u^2}} \qquad (21.24)$$

The velocity u should be interpreted as negative when the observer and source approach each other. This result is equivalent to our earlier result, Eq. 21.21, when u^2/c^2 is very small.

EXERCISES ON SUPPLEMENTARY TOPICS

Section 21.10 | The Doppler Effect

21-59 An automobile moves at 20 m s^{-1} toward a factory whistle that has a frequency of 1000 Hz. What is the frequency heard by a passenger?

21-60 A locomotive is moving at 40 m s^{-1}. Its whistle has a frequency of 2000 Hz. What is the frequency heard by a stationary observer when the locomotive is (a) approaching, and (b) moving away from the observer?

21-61 A source moves away from a stationary listener. If the frequency heard is 8 percent lower than the source frquency, what is the speed of the source?

21-62 The average velocity of blood flow in the aorta during systole is 1.5×10^{-2} m s^{-1}. What is the frequency shift in a Doppler flow meter with a source frequency of 10^5 Hz? (The speed of sound in blood is 1570 m s^{-1}.)

21-63 One of two identical 500-Hz tuning forks is at rest and the other is moving. These produce a beat frequency of 5 Hz. (a) What are the possible velocities of the moving fork? (b) Do you have enough information to determine whether the moving fork moves toward or away from the stationary fork? What additional information, if any, might be useful in this determination?

21-64 A galaxy is moving away from us with a speed of 3×10^7 m s^{-1}. What is the frequency of the light we would observe if it is emitted at 6×10^{14} Hz?

21-65 Light from distant galaxies that are moving away from us is said to be red-shifted; the wavelength observed is at longer wavelengths than that emitted by the galaxy. What is the observed wavelength shift for the galaxy of the preceding exercise?

PROBLEMS ON SUPPLEMENTARY TOPICS

21-66 A typical red blood cell has a radius of 5×10^{-6} m. Doppler shift flow meters depend on reflection from red cells and utilize ultrasonic frequencies. (a) If the source frequency is 10^7 Hz, how many red cells will "fit" into one wavelength of the sound? (b) Why are high frequencies necessary?

21-67 A tuning fork of frequency 500 Hz moves away from a stationary listener and toward a stationary wall at a speed of 2 m s^{-1}. (a) What is the frequency of the direct sound heard by the listener? (b) What is the detected frequency of the reflected wave? (c) How many beats per second are heard by the listener?

*21-68 The frequency of sound reflected from an object moving away from a detector was found in the text to be (Eq. 21.22)

$$f_r = f_0 \left(\frac{c - v_c}{c + v_c} \right)$$

(a) Show that if $v_c \ll c$, then

$$\Delta f = f_r - f_0 \simeq -\frac{2f_0 v_c}{c}$$

(b) If the object moves toward the detector, show that

$$\Delta f = f_r - f_0 \simeq \frac{2f_0 v_c}{c}$$

21-69 A bat emits squeaks of short duration at a frequency of 80,000 Hz. If the bat flies toward an obstacle at a speed of 20 m s^{-1}, what is the frequency of the reflected wave detected by the bat?

21-70 A bat travels toward a stationary obstacle. If emits sounds at a frequency of 50,000 Hz and detects a reflected sound of 51,000 Hz. How fast is the bat flying?

21-71 Show that the beat frequency detected when the reflected wave from an object moving toward or away from the source is combined with the unshifted wave of frequency f_0 is

$$f_B = \frac{2f_0 v}{c}$$

(Assume v, the speed of the reflecting object, is small compared to c.)

21-72 The average velocity of blood flow in a dog's artery with an inner radius of 4.0×10^{-3} m is 2.3×10^{-2} m s^{-1}. (a) What is the average frequency of the sound detected in a Doppler shift flow meter if the source frequency is 10^5 Hz? (b) What is the blood flow rate? (The speed of sound in blood is 1570 m s^{-1}.)

21-73 An average frequency shift of 100 Hz is detected in a Doppler shift flow meter with a source frequency of 5×10^6 Hz. What is the average blood flow velocity in the vessel being studied? (The speed of sound in blood is 1570 m s^{-1}.)

21-74 If one approaches a red light of wavelength 6.5×10^{-7} m, at a speed of 10^8 m s^{-1}, what is the observed wavelength?

21-75 Show that if $u^2/c^2 \ll 1$, the formula for the Doppler effect for light, Eq. 21.24, is the same as Eq. 21.21.

21-76 Two tuning forks produce beats at 5 Hz when at rest next to each other. The higher frequency fork has a frequency of 1000 Hz. (a) If the higher frequency fork is set into motion, in what direction and at what speed is it moving if no beats are heard by a stationary listener? (b) In what direction and at what speed should the lower frequency fork be moved, with the other at rest, so that beats are not heard by a stationary listener?

21-77 Two police cars with identical 1500-Hz sirens move south on a road, one at 80 km h^{-1} and the other at 70 km h^{-1}. At what frequency will a stationary observer further south hear beats?

21-78 A source of sound waves away from a listener at a speed v_s relative to the air, while the listener moves away from the source at a speed v_l relative to the air. If the source frequency is f_0, show that the listener hears a frequency

$$f = f_0 \left(\frac{c - v_l}{c + v_s} \right)$$

21-79 Using the results of the preceding problem, find the frequency of the sound heard by a bat when its squeaks are reflected by an insect approaching it at a speed relative to the air of 3 m s^{-1}. Assume the bat is flying toward the insect at 8 m s^{-1}, and that the squeaks are emitted at 50,000 Hz.

*21-80 A certain species of bat uses a pure sound wave of 83 kHz for detecting moving targets. When a Doppler shifted wave is detected, the bat lowers its own emitted frequency until the returning wave is at 83 kHz. What is the lowered emission frequency for a bat moving at 10 m s^{-1} toward a target that is moving toward the bat at 2 m s^{-1}?

Additional Reading

Sources marked with asterisks (*) are appropriate references for this and the next chapter.

R.A. Waldron, *Waves and Oscillations,* Momentum Series, D. Van Nostrand and Co., Princeton, N.J., 1964.

Willard Boscom, *Waves and Beaches: The Dynamics of the Ocean Surface,* Science Study Series, Doubleday and Co., Garden City, N.Y., 1964.

* Arthur H. Benade, *Horns, Strings and Harmony,* Anchor Books, Doubleday and Company, Inc., Garden City, N.Y., 1960.

* Alexander Wood, *The Physics of Music,* University Paperbacks, Chapman and Hall, London, 1975.

* Charles A. Culver, *Musical Acoustics,* McGraw-Hill Book Co., New York, 1957.

* C.A. Taylor, *The Physics of Musical Sounds,* American Elsevier Publishing Co., New York, 1965.

* John Backus, *The Acoustical Foundations of Music,* W.W. Norton and Co., Inc., New York, 1969.

Kenneth M. Baird, Frequency Measurements of Optical Radiation, *Physics Today,* January 1983, p. 52.

Scientific American articles:

J. Bernstein, Tsunamis, August 1954, p. 60.

Frank Press, Resonant Vibrations of the Earth, November 1965, p. 28.

John R. Percy, Pulsating Stars, June 1975, p. 66.

Eugene Helm, The Vibrating String of the Pythagoreans, December 1967, p. 92.

* Carleen Maley Hutchings, The Physics of Violins, November 1962, p. 79.

* Arthur H. Benade, The Physics of Brasses, July 1973, p. 24.

J.T. Gosling and A.J. Hundhausen, Waves in the Solar Wind, March 1977, p. 36.

* Carleen Maley Hutchins, The Acoustics of Violin Plates, October 1981, p. 170.

David K. Lynch, Tidal Bores, October 1982, p. 146. The effects of water waves.

Jearl Walker, Walking on the Shore, Watching the Waves and Thinking on How They Shape the Beach, *The Amateur Scientist,* August 1982, p. 144.

CHAPTER 22

SOUND

When a gas, liquid, or solid is mechanically disturbed, sound waves are often produced. In these waves the molecules of the substance vibrate and collide with one another but maintain the same average position. However, since their motions are coordinated, a wave results and energy is transmitted, even though no net particle displacement occurs.

The speed of sound depends on the physical properties of the substance in which it travels. When sound encounters a boundary between substances in which the sound speeds differ, some energy is transmitted and some is reflected. Thus a study of the characteristics of the production, propagation, detection, and uses of sound is necessarily a study of the transfer of mechanical energy.

Animals use sound for information exchange and for the detection and location of objects. Some bats and porpoises use sound for navigation and to locate food where inadequate light is present for vision. Humans also use sound as a substitute for light and even for X rays. *Sonar* is used for underwater navigation and observation, and *ultrasonic* or high-frequency sound is now commonly used for medical diagnosis and therapy. Very low-frequency sound is also used in geophysical studies.

The first part of this chapter deals with some basic properties of sound, including the speed of sound and the energy carried by sound waves. This is followed by a description of sound sources and detectors, including the human voice and ear. These are examples of *transducers,* devices that convert energy from one form to another. The remainder of the chapter is devoted to some special examples of man's use of sound.

22.1 | THE NATURE AND SPEED OF SOUND

In our discussion of the physics of fluids, we found that it is impractical to apply Newton's laws of motion directly to small segments of fluid or individual molecules. Instead, we introduced the density and pressure and explored their relationships. These same variables are useful in characterizing a sound wave, which has millions of molecules in a single wavelength.

Figure 22.1 illustrates the production of a sound wave in a medium by a piston that oscillates back and forth at a frequency f. When it moves forward, it compresses the medium, and a compressional wave moves outward. When the piston moves backward, there is a region of reduced pressure, a *rarefaction*. This disturbance also travels outward. Note that the individual molecules do not travel any appreciable distance but only oscillate about their average positions.

The speed c with which a sound wave travels in a medium is determined by the strength of the forces among the molecules. In a fluid, these forces are characterized at the macroscopic level by the *bulk modulus, K.* This quantity is a measure of how hard it is to compress a substance. When the pressure on an object is increased, its volume decreases; its density, which is the mass-to-volume ratio, increases. The

477

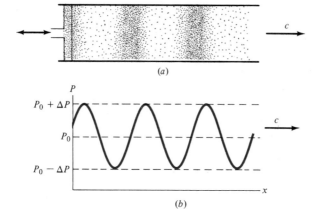

Figure 22.1. (a) The moving piston produces density and pressure changes that move to the right with a velocity c. (b) The graph of pressure versus position. P_0 is the equilibrium pressure.

bulk modulus relates the fractional density change $d\rho/\rho$ to the pressure change dP,

$$dP = K\frac{d\rho}{\rho} \qquad (22.1)$$

A fluid such as air that is easy to compress has a small bulk modulus. Water, which is hard to compress, has a large bulk modulus. In solids, the elastic moduli discussed in Chapter Eight characterize the molecular forces and determine the velocity of sound.

The velocity of sound in a fluid depends only on its bulk modulus K and density ρ. In Section 22.8 in the

Supplementary Topics, we show that applying Newton's laws of motion to a fluid subjected to a sudden compression leads to the result

$$c = \sqrt{\frac{K}{\rho}} \qquad \text{(fluid)} \qquad (22.2)$$

Note that the sound velocity is large for materials with large bulk moduli, which are stiff or hard to compress. This is reasonable if we consider the behavior of a simple harmonic oscillator (Chapter Nine). The period $T = (m/k)^{1/2}/2\pi$ decreases as the spring becomes stronger and the spring constant k increases. Thus the spring responds faster to a disturbance.

In Eq. 22.2 K is the *adiabatic bulk modulus*, not the isothermal bulk modulus sometimes tabulated. This is because, while temperature changes do accompany the compressions and rarefactions in a sound wave, they are so rapid that little heat flow occurs, and the motion is effectively adiabatic. Representative values of the density ρ and the longitudinal sound velocity c are given in Table 22.1.

The following example uses the difference in sound velocities in air and in iron.

Example 22.1

Two children are at opposite ends of an iron pipe. One strikes an end of the pipe with a stone. What is the ratio of times it takes the sound waves in air and in iron to reach the second child?

The time necessary for a sound wave of velocity c to

TABLE 22.1

Representative densities and sound velocities. For solids, the velocity given is the velocity for longitudinal waves in thin rods. The temperature is 20° C unless otherwise noted

Material	Density (kg m^{-3})	Sound Velocity (m s^{-1})
Air	1.20	344
Carbon dioxide (0° C)	1.98	259
Hydrogen (H_2)(0° C)	0.0899	1284
Alcohol (ethyl)	790	1207
Benzine	870	1295
Water (pure)	998	1498
Aluminum	2700	5000
Copper	8930	3750
Glass (Pyrex)	2320	5170
Iron	7900	5120
Blood (37° C)	1056	1570
Body tissue (37° C)	1047	1570

travel the length of the pipe d is found from $d = ct$ or $t = d/c$. Using Table 22.1, the ratio of the times for air and iron is

$$\frac{t_{air}}{t_{iron}} = \frac{\dfrac{d}{c_{air}}}{\dfrac{d}{c_{iron}}} = \frac{c_{iron}}{c_{air}} = \frac{5120 \text{ m s}^{-1}}{344 \text{ m s}^{-1}} = 14.9$$

Note that the sound travels faster in the denser material. Sound velocities are generally larger in solids and liquids than in gases.

As in all other waves, the velocity of sound, the frequency, and the wavelength are related by $f\lambda = c$. Typical wavelengths for sound waves in air are calculated in the next two examples.

Example 22.2

A typical young adult has a hearing range from 20 Hz to 20,000 Hz. What are the wavelengths of sound waves in air corresponding to these two frequencies?

For air, $c = 344 \text{ m s}^{-1}$, so with $f\lambda = c$ and $f = 20$ Hz,

$$\lambda = \frac{c}{f} = \frac{344 \text{ m s}^{-1}}{20 \text{ s}^{-1}} = 17.2 \text{ m}$$

At 20,000 Hz,

$$\lambda = \frac{344 \text{ m s}^{-1}}{20,000 \text{ s}^{-1}} = 0.0172 \text{ m} = 1.72 \text{ cm}$$

Example 22.3

A bat can hear sound at frequencies up to 120,000 Hz. What is the wavelength of sound in air at this frequency?

Again we use

$$\lambda = \frac{c}{f} = \frac{344 \text{ m s}^{-1}}{120,000 \text{ s}^{-1}} = 2.87 \times 10^{-3} \text{ m}$$
$$= 0.287 \text{ cm}$$

One may ask why a bat utilizes sound waves of such high frequencies and short wavelengths. This is because a wave will be disturbed only by objects comparable to or larger than a wavelength; it will pass by smaller objects with little effect. A bat is nearly blind and avoids obstacles and finds food using sound waves. The bat emits a series of high-frequency squeaks and senses the time it takes for the waves to return after being reflected by an object. The wavelength must be short enough so that reflections can occur from small objects.

Porpoises are sensitive to frequencies up to 2×10^5 Hz and use a similar system for underwater navigation and location.

Sound Waves in Solids | In solids, transverse as well as longitudinal sound waves occur. These transverse waves appear in solids because the forces among the ordered molecules not only act along the direction of the wave but also transverse to the wave; solids can oppose shear deformations. Thus a given molecule or atom can move as it does in a gas, back and forth along the wave direction. It can also move at right angles to the wave direction, as does a molecule on a string when a wave passes by. Because the restoring forces are weaker for this transverse motion, the velocity of the transverse wave is generally lower than for the longitudinal wave. This velocity difference makes it possible to determine the distance from the center of an earthquake to a seismograph by noting the time of arrival of the different kinds of seismic waves. In this chapter, we will limit our attention to longitudinal sound waves.

22.2 | STANDING SOUND WAVES

In the preceding chapter, we discussed standing waves on strings. However, standing waves can occur for any type of wave. In particular, standing waves are an important feature of sound-producing instruments.

We can discuss sound waves either in terms of the longitudinal displacements of the molecules or of the pressure variations. A detailed analysis shows that the displacement and pressure variations are a quarter wavelength out of phase, so that when one is at its maximum, the other is passing through a zero! Similarly, a standing wave node for one is located at an antinode for the other. Here it is convenient to think in terms of the displacement wave, but we use the amplitude of the pressure wave in the next section to discuss intensity.

We consider sound waves in long narrow tubes or air columns. The standing waves for sound are set up exactly as are those on a string; periodic waves are reflected at a boundary, and the incident and reflected waves combine to form a standing wave.

Sound waves can be reflected at either open or closed ends of air columns. At the closed end of an air column, the molecules cannot oscillate normally; those adjacent to the end do not move at all, and a displacement node appears at a closed end. At an open end, part of the wave is transmitted and part reflected. Since molecules at the open end can move

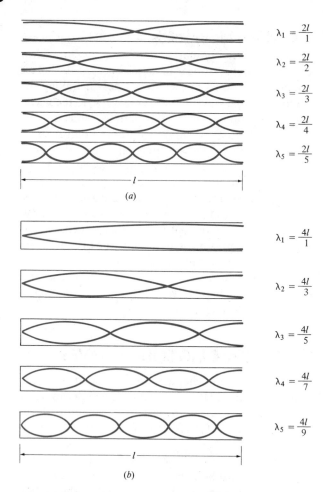

$$\lambda_1 = \frac{2l}{1}$$

$$\lambda_2 = \frac{2l}{2}$$

$$\lambda_3 = \frac{2l}{3}$$

$$\lambda_4 = \frac{2l}{4}$$

$$\lambda_5 = \frac{2l}{5}$$

(a)

$$\lambda_1 = \frac{4l}{1}$$

$$\lambda_2 = \frac{4l}{3}$$

$$\lambda_3 = \frac{4l}{5}$$

$$\lambda_4 = \frac{4l}{7}$$

$$\lambda_5 = \frac{4l}{9}$$

(b)

Figure 22.2. The first five resonant frequencies of an air column (a) with both ends open: (b) one end open.

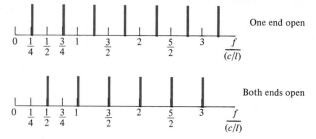

Figure 22.3. The lowest possible frequencies for air columns with one or both ends open.

open $\lambda_n = 4l/(2n - 1)$, where n is an integer. (Note that $2n - 1$ is always an odd integer.) The frequencies for this column are

$$f_n = \frac{c}{\lambda_n} = \frac{(2n - 1)c}{4l} \qquad (22.4)$$

$$n = 1, 2, 3, \ldots \qquad \text{(one open end)}$$

Only odd harmonics, odd multiples of the fundamental $f_1 = c/4l$, are possible for this type of column. The lowest frequencies for the two types of air column are illustrated in Fig. 22.3. These are the fundamentals and overtones for these columns.

The clarinet is an example of a standing wave instrument.

Example 22.4

Instruments such as the clarinet employ air columns with one open end. What is the effective length of a clarinet that has a fundamental frequency of 147 Hz?

The fundamental frequency is found from Eq. 22.4 with $n = 1$.

$$l = \frac{c}{4f_1} = \frac{344 \text{ m s}^{-1}}{4(147 \text{ Hz})} = 0.585 \text{ m}$$

Because the end of a clarinet is flared, its actual length is closer to 0.67 m. Similar differences occur in many instruments.

22.3 | THE INTENSITY OF SOUND WAVES

For most purposes the power per unit area or intensity of a sound wave is more important than the total energy carried over a period of time. For example, the noise of a jet engine will have very different effects on people and objects nearby than will the sound of a violin played for a much longer interval, even though the total sound energy produced might be the same in the two cases. However, a jet engine far away may be no louder than a nearby violin, since a sound wave spreads out as it travels (Fig. 22.4).

freely, the open end is a displacement antinode of the standing wave. The longest wavelength standing waves for two types of air columns are shown in Fig. 22.2.

For a column with two open ends, there is a displacement antinode at each end. The wavelengths λ_n are related to the length of the column by $\lambda_n = 2l/n$, where n is an integer (Fig. 22.2a). From $f_n\lambda_n = c$, we find that the standing waves have frequencies

$$f_n = \frac{c}{\lambda_n} = \frac{nc}{2l}, \qquad (22.3)$$

$$n = 1, 2, 3, \ldots \qquad \text{(two open ends)}$$

These are the same as the standing wave frequencies on a string fixed at both ends. For the string, however, the ends are nodes rather than antinodes.

For a column with one end closed and one end

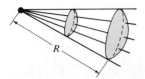

Figure 22.4. Sound waves spread out as they leave a source. Since the surface area of a sphere of radius R is $4\pi R^2$, the area the wave spreads over varies as R^2, and the intensity or power per unit area varies as $1/R^2$.

Also, some sound energy is dissipated by the internal friction or viscosity of the medium in which it is moving. Since this energy loss is usually small, we neglect it.

As noted in the preceding chapter, the intensity of any wave is proportional to its amplitude squared. The intensity of a sound wave with pressure amplitude ΔP is proportional to $(\Delta P)^2$ and is given exactly by

$$I = \frac{(\Delta P)^2}{2\rho c} \qquad (22.5)$$

The following example involves typical values of the pressure and intensity of audible sound waves.

Example 22.5

The maximum amplitude ΔP of a sound wave that is tolerable to a human ear is about 28 Pa. (a) What fraction is ΔP of normal atmospheric pressure? (b) What intensity of sound does ΔP correspond to in air at room temperature?

(a) Since normal atmospheric pressure is 1.013×10^5 Pa,

$$\frac{\Delta P}{P} = \frac{28 \text{ Pa}}{1.013 \times 10^5 \text{ Pa}} = 2.77 \times 10^{-4}$$

Thus even very loud sounds correspond to pressure fluctuations that are only a small fraction of a percent of atmospheric pressure.

(b) From Table 22.1, $\rho = 1.20$ kg m^{-3} and $c = 344$ m s^{-1}, so

$$I = \frac{(\Delta P)^2}{2\rho c} = \frac{(28 \text{ Pa})^2}{2(1.20 \text{ kg m}^{-3})(344 \text{ m s}^{-1})}$$

$$= 0.950 \text{ W m}^{-2}$$

Because the sound intensity decreases with the distance squared, the intensity of a sound source decreases rapidly as the distance from it increases. This is illustrated in the next example.

Example 22.6

The low-frequency speaker of a powerful stereo set has a surface area of 0.05 m^2 and produces 1 W of

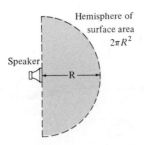

Figure 22.5. Example 22.6.

acoustical power. (a) What is the intensity at the speaker? (b) If the speaker projects sound uniformly into the forward hemisphere, at what distance from the speaker is the intensity 0.1 W m^{-2}?

(a) At the speaker, the intensity is

$$I = \frac{\mathcal{P}}{A} = \frac{1 \text{ W}}{0.05 \text{ m}^{-2}} = 20 \text{ W m}^{-2}$$

(b) At a distance R from the speaker, the wave has spread out over a hemisphere whose area is half the surface area of a sphere, or $\frac{1}{2}(4\pi R^2)$ (Fig. 22.5). Thus $I = \mathcal{P}/2\pi R^2$ or

$$R = \sqrt{\frac{\mathcal{P}}{2\pi I}} = \sqrt{\frac{1 \text{ W}}{2\pi(0.1 \text{ W m}^{-2})}} = 1.26 \text{ m}$$

22.4 | SOUND SOURCES

Many phenomena produce sound in an incidental but unavoidable fashion. For example, the combustion of fuel in an engine always produces some sound as a by-product. This sound is both annoying and wasteful of energy. However, there are many manmade and natural sources for which sound is the desired output. These usually have two primary components: a mechanism for producing a vibration and a *resonant structure*.

Musical instruments present a variety of arrangements for the production of sound. In a violin the strings vibrate, and their vibrations are efficiently transmitted to the air by the resonant hollow body of the instrument. In woodwinds and brasses the vibrations are produced by causing the air in the mouthpiece to puff, swirl, and eddy. This causes the reeds in woodwinds to vibrate. In brasses the lips themselves vibrate as air is blown into the mouthpiece. In both cases the oscillatory flow of air results in standing waves in the extended hollow body of the instrument, and the energy is then efficiently transmitted to the air outside. Similarly, the oral and nasal cavities in

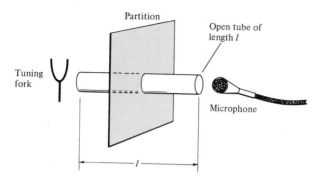

Figure 22.6. A tuning fork on the left of the partition produces sound at a single frequency. Sound can only reach the microphone by passing through the tube.

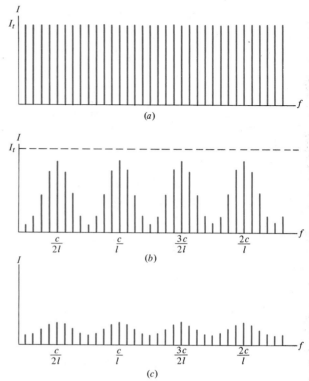

Figure 22.7. (*a*) A series of tuning forks used in the experiment of Fig. 22.6 produces sound of intensity I_t at the tube entrance at closely spaced frequencies. (*b*) The intensity reaching the microphone is greatest at the resonant frequencies of the tube. (*c*) The intensity received by the microphone through a narrower tube than that used in (*b*) is less sharply peaked about the resonant frequencies.

humans serve as resonant structures for vibrations produced by the vocal cords.

Some insight into these somewhat complicated systems can be obtained by considering the experiment illustrated in Fig. 22.6. We know that a cylindrical tube of length l that is open at each end will sustain resonant standing waves of frequency $f_n = nc/2l$, where n is an integer. The experiment is performed by using a number of tuning forks with different but closely spaced frequencies and monitoring the sound with a microphone on the opposite side of the tube. When tuning forks with frequencies equal to the resonant frequencies of the tube produce sound, standing waves are produced in the tube, and sound is transmitted efficiently to the microphone. Forks with other frequencies produce sound that does not result in standing waves. Much less sound reaches the microphone; the actual amount depends on how close one is to a resonant frequency (Fig. 22.7).

When sound enters the tube, the air inside vibrates, and the air near the tube walls loses some energy because of viscous forces. As we have seen with the harmonic oscillator and the *LRC* circuit, whenever the energy dissipation in a resonant system is increased, the intensity versus frequency curve or *spectrum* is lowered and proportionately broadened. If a narrower tube of the same length is used, the broadening is more pronounced because the relative amount of air near the walls is larger, and the viscous effects are proportionately greater (Fig. 22.7c). The sound energy not transmitted is partially reflected at the tube entrance and partially dissipated in the walls of the tube.

Thus, we conclude that the tube serves as a resonant structure that selectively transmits sound at and near specific resonant frequencies and suppresses all other frequencies. Similar selective transmission occurs in instruments such as the violin and in the human speech mechanism.

The Violin |
The body of a violin is a more complex resonant structure than the tube. When the strings of a violin are plucked or bowed, their vibrations are transferred to the body through the bridge (Fig. 22.8). Although the strings may vibrate with many different frequency components, the body resonates at and amplifies only certain frequencies.

The violin body vibrates so that its volume varies, and air is forced in and out through the f-holes. This is called the air resonance. The front and back plates of the body can also vibrate at characteristic frequencies called body resonances. Just as a string can vibrate at more than one frequency, so also can the violin plates, and several body resonances exist (Fig. 22.9). The frequencies of the air and body resonances

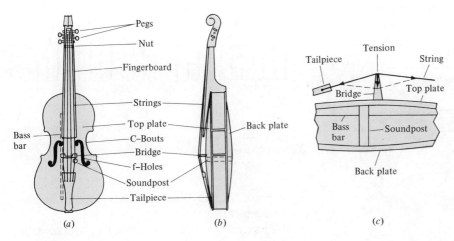

Figure 22.8. (a) Front and (b) side views of a violin. (c) An enlarged view of the bridge and part of the violin body.

should be at or near the fundamental string frequencies. If this is not achieved some notes will be muted or distorted.

The Human Voice

In human speech, the vocal cords initiate vibrations of the air; the throat and the nasal and oral cavities serve as resonant structures (Fig. 22.10). The remarkable ability of humans to produce so many different sounds stems

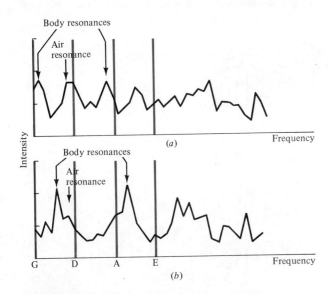

Figure 22.9. (a) The intensity-versus-frequency graph for a good violin. The vertical colored lines represent the characteristic frequencies of the four violin strings. Note that the resonances occur at nearly the same frequencies as those characteristic of the strings. (b) The spectrum of a poor violin.

from two facts. First, the vocal cord tension can be varied; hence the frequencies produced and the proportions of the harmonics present can be changed. Second, the resonant structures, particularly the oral cavity, can be changed in shape and dimensions to modify the frequency content of the amplified sounds. (These considerations apply to voiced sounds as opposed to nonvoiced sounds such as "sh" in "shut" and "f" in "feet," which do not require the use of the vocal cords.) Normally the vocal cords are relaxed and present no obstruction to air passing through the larynx. During preparation for speech the tension in the vocal cords increases and the larynx is closed (Fig. 22.11). The air pressure below the vocal cords increases until the cords are forced open. The air then rushes through the opening, causing the vocal cords to vibrate.

The vibrations of the vocal cords can be understood from Bernoulli's equation (Chapter Thirteen). As air rushes through the opening, its velocity is large, and hence the pressure is lowered, allowing the vocal cords to start to close. Once the opening becomes sufficiently small, the pressure below the vocal cords increases and forces the cords to spread apart again. This process is continuous as long as exhalation persists and the fundamental as well as many harmonics of the vocal cords are excited (Figs. 22.12 and 22.13).

As illustrated in Fig. 22.13, the spectrum of sound produced by the vocal cords is quite uniform up to about 3000 Hz, the range of frequencies used in speech. Without modification this would result sim-

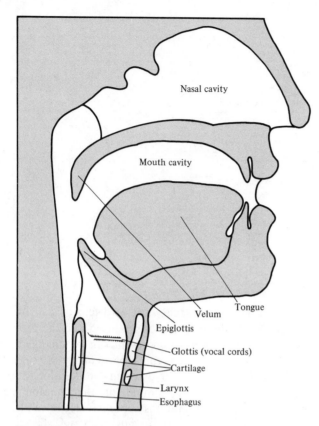

Figure 22.10. Essential components of the human speech mechanism.

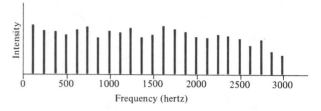

Figure 22.13. An example of the fundamental and harmonics produced by the vocal cords during speech. This spectrum will change with the tension in the vocal cords.

ply in noise. The modification takes place in the resonant structures.

The human oral cavity is a complex structure, but its properties are surprisingly close to those of a simple model. This model is a pipe 0.17 m long with one end open at the mouth and nose and the other end nearly closed at the vocal cords. The resonant frequencies of a pipe open at one end were found to be

$$f_n = \frac{(2n-1)c}{4l}, \qquad n = 1,2,3 \ldots \quad (22.4)$$

In speech the frequencies of importance are between 300 and 3000 Hz. Using $c = 344 \text{ m s}^{-1}$ and $l = 0.17$ m, we find that the fundamental and first two overtones of the pipe, which are near 500, 1500, and 2500 Hz, respectively, are within this range. If one analyzes a typical voiced sound, one finds that

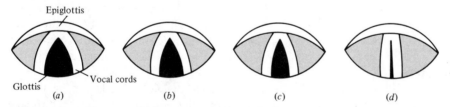

Figure 22.11. (a) The vocal cords are relaxed during normal breathing (a). When speech is begun, the vocal cord tension increases (b), (c), and the larynx is closed (d).

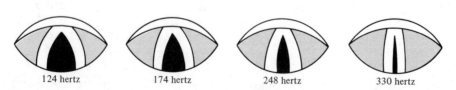

Figure 22.12. The vocal cords are shown for several fundamental frequencies during singing. Higher frequencies occur when the tension increases. Notice the stretching of the cords just as in a string.

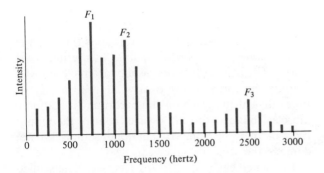

Figure 22.14. The intensity versus frequency spectrum of the sound "a" in "father." The formants are wide resonances labeled F_1, F_2, and F_3. The vertical lines show how the vocal cord frequencies shown in Figure 22.13 are selectively transmitted by the resonance structure.

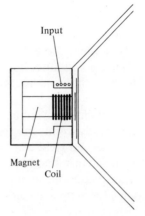

Figure 22.16. A speaker has a permanent magnet that is free to move along its length. One end of the magnet is attached to the speaker diaphragm and the magnet is surrounded by a current-carrying coil.

there usually are three resonant peaks that are close to the resonant frequencies of the model. These resonant peaks are referred to as *formants*. For example, the sound "ae" in "at" has formants at about 600, 1700, and 2300 Hz.

From Fig. 22.14, we see that there are many possible ways in which the formant composition may be varied to produce the different sounds characteristic of a language. The formants may be broad or narrow and may be shifted in frequency. In Fig. 22.15 the peak formant intensities of some English vowel sounds are shown.

The Loudspeaker | The reproduction and amplification of sound using electronic equipment

has revolutionized human communication. Sound that is modified by electronic systems is reconverted to sound by a speaker, the most common of which utilizes a permanent magnet surrounded by a current-carrying coil (Fig. 22.16). The coil carries a time varying current from the amplifier that produces a variable magnetic field inside the coil. This results in varying forces on the permanent magnet. As the magnet moves, so does the diaphragm, and its motion produces sound waves in the air.

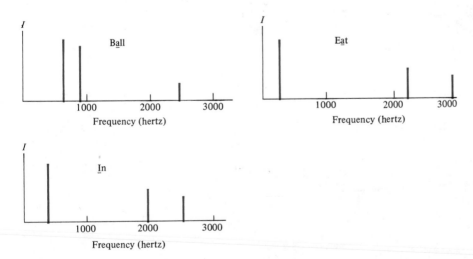

Figure 22.15. The position and relative intensity of the formants for some English vowel sounds as voiced by males.

22.5 | SOUND DETECTORS

Sound detection requires the conversion of the mechanical vibrations of sound waves into a form that permits the analysis of their frequency and intensity. Here, we discuss how this is done by the condenser microphone and the human ear.

The condenser microphone contains a flexible membrane that includes an electrically conducting layer. This membrane is close to a second rigid conductor, so together they form a capacitance that varies as the membrane vibrates in response to a sound wave. In a circuit, this variation of the capacitance causes a time variation in the current. Once this varying component of the current has been amplified electronically, it can drive a speaker or record information on magnetic tape.

Any useful sound detector must respond accurately to variations in the sound frequency and intensity. The human ear is remarkable, since it can do this well and is nevertheless virtually unaffected by motion and vibration of the body or by the sounds produced by the blood flow and in the internal organs. It also permits the listener to locate sound sources and to concentrate on specific sounds in a confused sound environment.

Figure 22.17a shows the anatomy of the human ear. The outer ear collects sound waves and transmits them to the eardrum. Bones called *ossicles* in the middle ear transmit the eardrum vibrations through the *oval window* to the *perilymph* fluid of the inner ear canals. The ossicles function as a hydraulic press and amplify the force on the oval window to nearly 15 times that of the force on the eardrum. In addition, muscles connected to the ossicles control the amplitude of their motion, so that loud sounds do not damage the sensitive inner ear.

The inner ear (Fig. 22.17b) has two canals filled with fluid. The *cochlear duct,* containing nerve endings in the *organ of Corti,* divides the two chambers except at the end farthest from the windows. The *oval window* is driven by one of the ossicles. A flexible *round window* in the other chamber flexes as the perilymph moves, so the volume of the inner ear stays constant. Because the cochlear duct is thicker near the narrow end of the cochlea, different frequencies of vibration of the perilymph cause the partition to flex at different points along its length. The flexing is sensed by the nerve hairs in the region of excitation, and nerve impulses travel toward the brain.

The mechanical operation of the ear is well understood; however, the perception of sound involves complex neurological processing of information that is not yet fully explained. An example is the use of

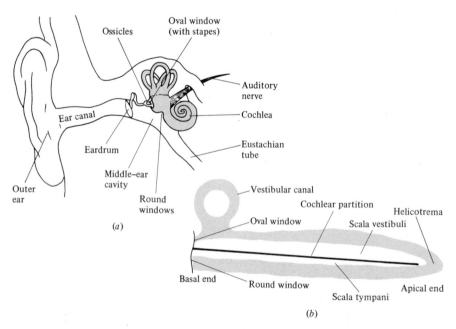

Figure 22.17. (a) Schematic diagram of the human ear. (b) The uncoiled inner ear.

earphones. Any sound with a wavelength much larger than the diameter of a speaker cannot be efficiently produced by that speaker. Earphone speakers are too small to reproduce the low-frequency fundamentals of many musical sounds, although they may accurately reproduce the overtones. Nevertheless, music heard on earphones sounds as though the fundamental were present because the brain, recognizing the harmonics, effectively supplies the fundamental.

22.6 | AUDITORY RESPONSE

The ears of animals are remarkable mechanical structures. For example the human ear can comfortably respond to intensities from as low as 10^{-12} W m^{-2} up to 1 W m^{-2}, 12 orders of magnitude. Without this range, many sounds in our natural environment would either be inaudible or unbearable.

Measurements of auditory response are somewhat subjective, but two objective characteristics have been fairly well established. One is the *threshold of hearing;* the minimum intensity that is just audible at a given frequency. This is represented by the lower curve of Fig. 22.18. The second is the *threshold of feeling.* At this high intensity, a tickling sensation is experienced when the ossicles vibrate so strongly that

they strike the middle ear wall. The normal hearing range lies between these two curves.

The ear is sensitive to such an enormous intensity range in part because muscles around the eardrum and the ossicles respond to neural feedback and modify the tension in these parts. Thus the eardrum is somewhat like a flexible drum head with an adjustable tension.

Because of the enormous intensity range of the ear, it is common to measure intensities in logarithmic units or *decibels* (dB). The *intensity level* is related to the intensity by

$$\beta = 10 \log \frac{I}{I_0} \qquad (22.6)$$

β is measured in decibels, I is the sound intensity, $I_0 = 10^{-12}$ W m^{-2} is an arbitrary reference level roughly equal to the lowest intensity normally audible, and log denotes the base 10 or common logarithm. (See Appendix B.10 for a review of logarithms.) The decibel range of hearing at 1000 Hz is from near 0 dB up to about 120 dB (Fig. 22.18). Table 22.2 lists the intensity levels of common sound sources.

The following is an example of the relationship between intensity level and perception.

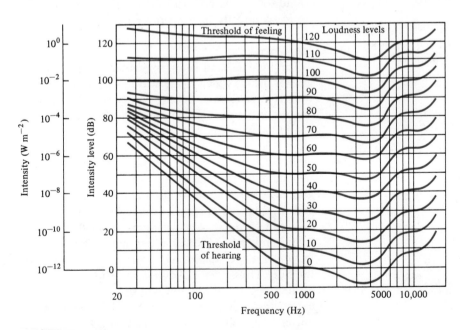

Figure 22.18. A logarithmic graph showing the range of audible intensities versus frequency. The human ear is most sensitive to frequencies near 3000 Hz; the audible range extends from about 20 to 20,000 Hz. The contours show the intensities required to produce the same subjective loudness levels for pure tones at various frequencies.

TABLE 22.2
The intensity level in decibels of common sound sources

Source	Intensity Level (in decibels)
Large orchestra (maximum)	98
Riveter	95
Bass drum (maximum)	94
Trumpet (maximum)	75
Busy street traffic	70
Clarinet (maximum)	67
Ordinary conversation	65
Quiet radio	40
Whisper	20

Example 22.7

If a 30-dB sound varies in frequency from 20 to 30,000 Hz, which frequencies will the normal ear hear?

From Fig. 22.18, we note that the horizontal line at 30 dB intersects the threshold of hearing curve at about 150 and 10,000 Hz. Thus, only frequencies within this range will be audible.

When more than one sound source is present, the intensity level does not usually increase markedly. This is shown in the next example.

Example 22.8

If the average intensity level of each of two radios is 45 dB, what is the average intensity level when both radios are on and tuned to different stations?

The intensity of one radio is I_r, where the corresponding intensity level is

$$\beta_1 = 45 \text{ dB} = 10 \log \frac{I_r}{I_0}$$

With both radios on, the intensity is $I_r + I_r = 2I_r$, so the intensity level is

$$\beta_2 = 10 \log \frac{2I_r}{I_0} = 10 \log 2 + 10 \log \frac{I_r}{I_0}$$
$$= 10 \log 2 + \beta_1$$

Since $10 \log 2 = 3$,

$$\beta_2 = 3 \text{ dB} + 45 \text{ dB} = 48 \text{ dB}$$

Notice that although the intensity I_1 doubles β increases by only 3 dB.

Subjective Factors | Properties such as *loudness, pitch,* and *quality* are often used to describe sound. These concepts are difficult to deal with because they are subjective. For example, a 20-dB sound at 1000 Hz will sound louder than a 20-dB sound at 400 Hz (Fig. 22.18). Thus loudness depends both on intensity and frequency. The pitch of a sound is closely related to the frequency. However, above about 3000 Hz, the pitch increases with intensity even if the frequency is constant. Below 2000 Hz the pitch decreases with increasing intensity.

The quality of a sound is even more ambiguous. We know that some types of sounds that have certain harmonics present with specific intensities sound pleasing. However, again we cannot characterize quality with a precise statement about measurable physical parameters.

The combination of a mechanical ear structure, neurological processing, and the psychological aspects of perception has so far allowed us to classify but not to understand the subjective qualities of sound. These aspects are under intensive study by those concerned with hearing defects and the science of musical sounds.

Auditory Localization | We can sometimes locate a sound source by turning our heads, since the ear nearest the source will often hear the loudest sound. However, people can often accurately locate sound sources without turning their heads, unless the source is directly in front of or behind them. There are different explanations for this at low and high frequencies.

The human head is approximately a sphere of diameter 0.2 m, which is the wavelength of a 1700 Hz sound wave in air. Sound waves of longer wavelengths or lower frequencies will pass around the head largely unaffected. If at some time a peak in the wave is at one ear, the wave has a different pressure at the other ear. The nerve impulses from the two ears then contain information about the relative pressures at the two ears that the brain uses to locate the source. This mechanism is most effective below 1000 Hz.

Sound waves with wavelengths much shorter than 0.2 m tend to be reflected when they strike the head, leaving a sound shadow with little or no sound on the far side. Thus, above 5000 Hz the source is usually located because the sound is distinctly louder in the ear nearest it. Between 1000 and 5000 Hz, both mechanisms are used, but the localization is less accurate.

An interesting correlation between auditory localization and the upper and lower frequency limits to hearing has been found for mammals of many different sizes. As we noted, the crossover frequency dividing two mechanisms of auditory localization is determined by the distance between the ears. More exactly, to take into account marine mammals, this frequency should be computed as the inverse of the distance between the ears divided by the speed of sound in the surrounding medium, be it air or water.

When this is done, it is found that mammals with a large ear separation (and hence a low crossover frequency) also have lower frequency limits to their hearing range. For example, the graph analogous to Fig. 22.18 for an elephant has very much the same shape but extends to lower frequencies on the left and only up to 10,000 Hz on the right. The reverse is found for small mammals; that is, they do not hear low frequencies as well but can hear up to higher frequencies. The average upper limits for several mammals reflect this; humans, 19,000 Hz; dogs, 44,000 Hz; rats, 72,000 Hz.

Thus there is a very strong correlation between how well a mammal can localize sounds at various frequencies, and what frequencies the mammals can even hear. It appears that a mammal with a given ear separation only hears a wide enough range of frequencies to efficiently localize the sound source. This empirically uniform trend even includes mammals making special use of sound, such as bats and porpoises.

SUMMARY

Sound waves are the result of mechanical disturbances in materials. The speed of sound depends on how much of a pressure change is necessary to produce a given density change. This relationship depends on the molecular properties of the material.

Standing sound waves are often produced in certain geometries and play an important role in the production and detection of sound. In sound sources, a resonant structure determines which standing waves are produced, amplifying certain frequencies and suppressing others. A sound receiver has similar characteristics.

The intensity of sound, the transported power per unit area, is proportional to the square of the amplitude of the pressure change in the wave. As a sound wave spreads out from a source, its intensity at any point on the wave decreases with the square of the distance from the source.

The intensity level of sound in decibels is often used when one is dealing with auditory response. This decibel or logarithmic scale is used because of the enormous range of ear sensitivities of humans and other animals.

Checklist

Define or explain:

pressure variations	loudspeaker
bulk modulus	middle ear
standing waves	inner ear
sound intensity	threshold of hearing
source of vibration	threshold of feeling
resonant structure	loudness
sound spectrum	pitch
formant	quality

REVIEW QUESTIONS

Q22-1 Why doesn't sound travel in a vacuum?

Q22-2 If a material is hard to compress, its velocity of sound is _____ .

Q22-3 High-frequency sound waves have _____ wavelengths.

Q22-4 Standing sound waves occur because of the presence of _____ in the region in which the wave travels.

Q22-5 Displacement nodes in standing sound waves occur at _____ ends of a tube and antinodes at _____ ends.

Q22-6 The intensity of a sound wave depends on the _____ of the _____ amplitude.

Q22-7 Instruments designed to produce sound usually contain a mechanism to produce vibrations and _____ .

Q22-8 The output of a sound source is often represented by a spectrum, a graph of _____ versus _____ .

Q22-9 Typically the spectrum of a voiced sound is composed of three _____ .

Q22-10 What plays the role of the resonant structure in the human ear?

Q22-11 In discussions of hearing, intensities are usually described in terms of _____ because the intensity range is so large.

EXERCISES

Section 22.1 | The Nature and Speed of Sound

22-1 A hi-fi speaker has a diameter of 0.3 m. What is the frequency of the sound produced if the wavelength is equal to the circumference of the speaker? (This is comparable to the resonant frequency of the speaker, and special enclosures must be designed so that most of the power is not emitted by the speaker at this frequency.)

22-2   A lightning bolt is seen by an observer. The accompanying thunder is heard 5 seconds later. What is the approximate distance of the observer from the storm cloud?

22-3   A stone is dropped into a well. The sound of the splash is heard 3 seconds after the stone is dropped. What is the depth of the well?

22-4   When one end of a copper pipe is struck, the time difference between the sounds heard in the copper and in air at the other end is 1 second. How long is the pipe?

22-5   The depth of a body of water can be found by emitting sound pulses at the surface and detecting the pulses reflected from the bottom. If the time interval from emission to detection is 2 seconds, what is the depth of the water?

22-6   A marching band takes a step each 0.8 second. At what distance will a marching band appear to be one half step out of time with the music being played?

22-7   When approaching an object, a bat decreases the duration of its chirps and also the time interval between chirps. When the chirps last 3×10^{-4} s, what is the minimum distance at which the first part of the echo overlaps with the end of the chirp?

22-8   The telephone is designed to transmit efficiently frequencies from about 50 to 3000 Hz. What are the wavelengths of sound in air corresponding to these two frequencies?

22-9   Bats utilize sound up to frequencies of about 1.2×10^5 Hz, while porpoises are sensitive to frequencies up to 2×10^5 Hz. Why might porpoises be expected to have a greater frequency range?

Section 22.2 | Standing Sound Waves

22-10   The fundamental frequency of the longest pipe on an organ is 16.35 Hz. If the pipe is open at both ends, how long is the pipe?

22-11   A bugle has the same characteristic frequencies as a cylindrical tube 1.3 m long with both ends open. What are the frequencies of the first four harmonics?

22-12   A clarinet has a fundamental frequency of 147 Hz and, when played, has one end closed.

(a) How many harmonics appear below 1350 Hz? (b) If an open end tube has the same fundamental frequency, how many harmonics appear below 1350 Hz?

22-13   A pipe organ has open end pipes and spans a frequency range from 65 to 2090 Hz. What are the lengths of the longest and shortest pipes of this organ?

22-14   The outer ear can be thought of as a pipe 2.7×10^{-2} m long with one closed end. (a) Using this model, predict what frequency sound would be most effectively detected by the ear. (b) How does this compare to the frequency of the minimum of the threshold of hearing curve (Fig. 22.18)?

Section 22.3 | The Intensity of Sound Waves

22-15   What is the pressure amplitude of thunder with an intensity of 0.1 W m^{-2}?

22-16   What is the total power output of a loudspeaker for which the intensity over the surface of a hemisphere 10 m away is 10^{-4} W m^{-2}?

22-17   Two sound waves of the same intensity travel in air and water. What is the ratio of the pressure amplitudes of the sounds in the two materials?

22-18   If the pressure amplitudes of sound in air and water are the same, what is the ratio of the two intensities?

22-19   If one sound wave has twice the pressure amplitude of another in the same medium, what is the ratio of the intensities of the two waves?

22-20   The ratio of the maximum intensities of a piano and a flute is 8. What is the ratio of the pressure amplitude produced by a piano to that of the flute?

22-21   The intensity of a large orchestra is equal to that of 216 trumpets. What is the ratio of the pressure amplitude of the orchestra to that of 1 trumpet?

22-22   What is the intensity of a sound wave in air with a pressure amplitude of 1 Pa?

22-23   What is the pressure amplitude of a sound wave in water of intensity $I = 10^{-12}$ W m^{-2}?

22-24   What is the pressure amplitude in air of a sound at the threshold of feeling, $I = 1$ W m^{-2}?

Section 22.4 | Sound Sources, and
Section 22.5 | Sound Detectors

22-25   A bat needs two ears for direction finding, just as humans do. If the distance between the ears is 0.01 m, what is the minimum frequency for which the ears are separated by at least one half wavelength? (Assume that the sound approaches directly from one side of the head.)

22-26 In hearing tests, a tone of a given frequency is gradually reduced in intensity until it becomes inaudible. Why are tones of different frequencies used during the test?

22-27 Sound can be heard by humans when the vibrations are transmitted to the inner ear via bone. The outer and middle ear play no role in such hearing. (a) Suggest a method of testing hearing such that conclusions may be drawn about whether damage is present in the inner ear only or if middle ear damage is also present. (b) For what types of hearing losses would hearing aids that transmit sounds to the skull bones be useful?

22-28 Bone is a better conductor of low-frequency vibrations than is air. Why does a person feel that his or her voice is richer and lower pitched than do listeners?

22-29 Why do sinus infections in which the eustachian tubes are infected often impair hearing?

Section 22.6 | Auditory Response

22-30 Two sound waves have intensities of 10^{-9} W m^{-2} and 5×10^{-8} W m^{-2}. What is the difference in the intensity levels of the two sounds?

22-31 Find the ratio of intensities of two sounds, one of which is 10 dB louder than the other.

22-32 What is the intensity level of ultrasound of intensity 25×10^4 W m^{-2}?

22-33 If the intensity level of one person speaking is 50 dB, what is the intensity level when 10 such people are speaking?

22-34 What is the intensity of a sound that is 5 dB louder than that of a sound of intensity 10^{-9} W m^{-2}?

22-35 The surface area of the eardrum is about 8×10^{-5} m^2. What is the power transmitted to the eardrum by a sound wave of 40 dB if no sound is reflected?

PROBLEMS

22-36 Bats, which use echo location for navigation, are observed to emit chirps lasting 2×10^{-3} s with 7×10^{-2} s of silence between. How close to an object can a bat be so that the reflected sound from the first part of the chirp is not masked by the emission of the first part of the next chirp?

22-37 Equation 22.2 for the sound velocity, $c = \sqrt{K/\rho}$, hides an understanding of the physics involved in sound propagation. In fact, c is larger for dense substances than for less dense materials (see Table 22.1). If dominoes are placed on end beside one another and the end one is toppled, the entire row falls. By noticing that the dominoes fall fastest when they are close together, discuss the dependence of the sound velocity on density.

22-38 During cruising flight the interval between chirps of a bat is 7×10^{-2} s. What is the maximum distance a bat can be from an object so that the complete reflected wave returns before the next chirp begins?

22-39 For an ideal gas the adiabatic bulk modulus is $K = \gamma P$, where P is the pressure, and γ is the ratio of the specific heat capacities at constant pressure and volume, $\gamma = c_P/c_V$. (a) Show that the ideal gas law can be written as $P/\rho = RT/M$, where ρ is the mass density, and M is the molecular weight of the gas. (b) Show that the sound velocity in an ideal gas is $c = \sqrt{\gamma RT/M}$.

22-40 In Problem 22-39 the velocity of sound in an ideal gas is found to be $c = \sqrt{\gamma RT/M}$. Using this expression, can one account for the difference in sound velocities in air and hydrogen, H$_2$?

22-41 A bat using echo location emits chirps at intervals of 5×10^{-3} s, each having a duration of 0.3×10^{-3} s. (a) How far from an object will the bat be so that the first part of the chirp is being reflected at the time the chirp ends? (b) How much time is there between the return of the complete reflected wave of part (a) and the beginning of the succeeding chirp?

22-42 A sonar source used for underwater detection emits pulses of duration 0.1 s once every T seconds. What is the minimum value for T if the echo and the next pulse are not to overlap from objects (a) 50 m away; (b) 1 km away?

22-43 A vertical tube of length 2 m can be filled to any level with water. Sound entering the open end of the tube is reflected at the water surface and this is the position of a node. (a) If the tube is filled with water to a depth of 1 m, what is the lowest frequency at which resonance will occur? (b) What is the depth of the water if the lowest frequency resonance occurs at 500 Hz?

22-44 The maximum tolerable pressure amplitude for the human ear in air is $\Delta P = 28$ Pa. What is the ratio of the corresponding density change to the mean density for sound waves in (a) air, and (b) water.

22-45 The acoustic intensity of an antiaircraft shell exploding at an altitude of 1000 m is 10^{-3} W m^{-2} at ground level. (a) What is the total acoustic power released? (b) What total acoustic energy is released during the explosion if it lasts 0.1 second?

22-46 If the air and body resonances of a violin had a very narrow frequency width or range, the instrument might not be effective. Explain why.

22-47 A jet plane flying at an altitude of 3000 m produces a sound of 40 dB at ground level. What would the intensity level be if the altitude were 1000 m? (Assume that the total sound energy produced does not change.)

22-48 An outdoor public address system is adjusted to a level of 70 dB for listeners 10 m distant. What intensity level is heard at 50 m?

22-49 The average intensity level of a radio is adjusted to 40 dB at a distance of 10 m. (a) What is the intensity in watts per square metre at this distance? (b) What is the intensity level in decibels 3 m from the radio? (c) What is the output power of the radio in watts if the sound is spread uniformly over the forward hemisphere?

22-50 The normal human ear can distinguish a difference in intensities of about 0.6 dB at a given frequency. What percentage of power increase is required to raise the intensity level 0.6 dB?

22-51 The tropical oilbird is often found flying in totally darkened caves. It uses sounds for obstacle avoidance, but the highest frequency it can emit and recognize is 8000 Hz. (a) Estimate the minimum-size object that the bird can detect. (b) It is observed that the actual minimum diameter of round objects that are easily avoided is 0.2 metre. What does this suggest about the actual frequency used by the bird?

22-52 The expected upper hearing limit of mammals may be compared if one assumes that it is determined by the separation of the ears. (a) Making this assumption and using the scaling model of Chapter Eight, $r \propto m^{3/8}$, what upper limit of hearing would one expect to find in elephants and rats whose body masses are typically 3×10^3 kg and 2×10^{-1} kg, respectively? Use the fact that a 70-kilogram human has an upper hearing limit of 19,000 Hz. (b) In fact, the upper limits of hearing for elephants and rats are 11,000 Hz and

72,000 Hz, respectively. Using the data for elephants and rats, find the experimental scaling exponent for the frequency-versus-body mass.

22-53 In Fig. 22.2a, assume that the two traveling waves are represented by $y_1 = A \cos (kx - \omega t)$ and $y_2 = A \cos (kx + \omega t)$, where $x = 0$ is at the left end of the tube, and $x = \ell$ is at the right end. (a) Using an identity from Appendix B.11, show that $y = y_1 + y_2$ is a standing wave that satisfies the appropriate boundary condition at $x = 0$. (b) Derive Eq. 22.3.

22-54 In Fig. 22.2b, assume that the two traveling waves are represented by $y_1 = A \sin (kx - \omega t)$ and $y_2 = A \sin (kx + \omega t)$, where $x = 0$ is at the left end of the tube, and $x = \ell$ is at the right end. (a) Using an identity from Appendix B.11, show that $y = y_1 + y_2$ is a standing wave that satisfies the appropriate boundary condition at $x = 0$. (b) Derive Eq. 22.4.

ANSWERS TO REVIEW QUESTIONS

Q22-1, sound is a mechanical disturbance with an actual movement of molecules and atoms; **Q22-2,** large; **Q22-3,** short; **Q22-4,** boundaries; **Q22-5,** closed, open; **Q22-6,** square, pressure; **Q22-7,** a resonant structure; **Q22-8,** intensity, frequency; **Q22-9,** formants; **Q22-10,** the outer ear canal; **Q22-11,** decibels.

SUPPLEMENTARY TOPICS
22.7 | ULTRASOUND

In this section, we describe some of the properties and medical applications of ultrasound, which is sound with frequencies above 20,000 Hz. Ultrasound can currently be produced at frequencies as high as somewhat more than 10^9 Hz. It is widely used as a diagnostic, therapeutic, and surgical tool in medicine, as well as in a widening variety of industrial applications.

Physical Principles and Limitations | In most applications, ultrasound is sent out in pulses that are partially reflected. Between pulses the transmitter acts as a receiver, detecting the reflected wave or echo. Accurately locating small objects at substantial distances requires narrow, short wavelength beams with sufficient range. These criteria tend to conflict, since short wavelengths (high frequencies) are strongly absorbed. For example, at a frequency of 1 megahertz (1 megahertz = 1 MHz = 10^6 Hz), the intensity drops 50 percent in 7 cm of soft animal tissue. Although in some situations the intensity is limited by the capacity of the apparatus, in medical applications the intensity limit is determined mainly by the destructive effects of intense ultrasound on animal tissues.

Ultrasound is useful because it is reflected from the boundaries between materials of nearly the same density and can be used without apparent harmful effects where X rays cannot. For example, ultrasonic scanning of the uterus during pregnancy is considered safe, while X rays must be avoided.

The fraction of the sound intensity reflected at a boundary can be found from a formula derived using advanced mathematical methods. Consider a sound wave passing from one medium of density ρ_1 and sound velocity c_1 to a second medium of density ρ_2 and sound velocity c_2. The ratio of the reflected to incident intensities when the wave travels perpendicular to the interface is

$$\frac{I_r}{I_i} = \left(\frac{\rho_1 c_1 - \rho_2 c_2}{\rho_1 c_1 + \rho_2 c_2}\right)^2 \qquad (22.7)$$

Since the sound velocity does not vary much, the amplitude of the reflected wave depends primarily on the density difference. It is this depenence that causes detectable reflections from boundaries even if the density change is small. This is illustrated by the next example.

Example 22.9

The densities of two types of muscle tissue are 1026 and 1068 kg m^{-3}. What is the ratio of the intensities of the reflected and incident waves if the wave passes from the more dense to the less dense medium? (Assume that the sound velocities are the same.)

Using the densities given and Eq. 22.7, the sound velocities factor out, and we have

$$\frac{I_r}{I_i} = \left(\frac{1068 \text{ kg m}^{-3} - 1026 \text{ kg m}^{-3}}{1068 \text{ kg m}^{-3} + 1026 \text{ kg m}^{-3}}\right)^2 = 0.00040$$

While this ratio is small the reflected sound is still measurable. The unreflected part of the wave continues on and further reflections at other boundaries can occur.

Equation 22.7 becomes simpler if we assume that the sound velocities c_1 and c_2 are equal, so we have

$$\frac{I_r}{I_i} = \frac{(\rho_1 - \rho_2)^2}{(\rho_1 + \rho_2)^2}$$

We see that the reflection is minimal when ρ_1 and ρ_2 are nearly equal. Conversely, if one of the densities is much larger than the other the reflection is comparatively large.

This result may be partially understood with the aid of a mechanical analog discussed in detail in Section 7.4. There we studied an elastic head-on collision of a moving object of mass m_1 with an object of mass m_2 that is initially at rest. We applied momentum and energy conservation to the case where they move away parallel to the original direction. In the present context, we can regard the energy acquired by the struck object as transmitted energy. We found that if the objects have equal mass—the billiards or pool situation—m_1 stops and m_2 has all the energy that m_1 originally had. Thus all the energy is transmitted, and none is reflected. We also saw that when the masses are not equal, m_2 still acquires some energy but the percentage becomes steadily smaller as the mismatch between the masses increases. Whether m_1 is much larger than m_2 or the other way around, most of the energy is retained by m_1 or reflected, and little is transmitted.

A situation comparable to the case of very unequal masses occurs for ultrasonic waves at an air-tissue boundary, where virtually all the sound is reflected because the density change is large. In most medical work this necessitates using a liquid to provide good transmission between the ultrasound source and the tissue. Nearly total reflection occurs at air-tissue boundaries in the body.

Destructive Effects | Intense ultrasound produces large density and pressure changes within each small wavelength. This results in large stresses, and molecules are forced to move rapidly. It also produces heat in most materials and *cavitation* in liquids and perhaps in tissue. Cavitation is the formation of bubbles of vapor caused by the mechanical fracture of the liquid in a region where the pressure is decreasing. These bubbles may then collapse violently. This catastrophic collapse can be used for cleaning, since the implosion scatters the liquid along with impurities. Cavitation is also used to produce minute droplets of liquid. Medication used in inhalation therapy is broken with ultrasound into droplets fine enough to enter the alveoli of the lungs.

Another violent result of ultrasound treatment is the mechanical rupture of cell membranes and break-

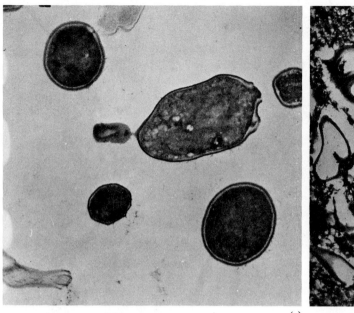
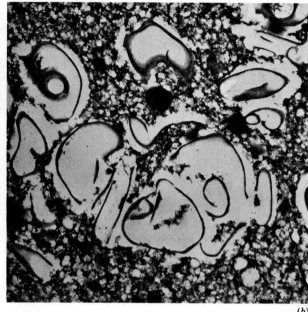

(a)

(b)

Figure 22.19. Electron photomicrographs of Baker's yeast cells (a) before and (b) after exposure to ultrasound. (Courtesy Heat Systems-Ultrasonics, Inc.)

age of cell constituents, such as chromosomes (Fig. 22.19). A major fraction of the energy lost in tissue is absorbed by proteins.

All these effects limit the therapeutic use of ultrasound to intensities below 3×10^4 W m^{-2}. This may be compared with surgical intensities of 25×10^4 W m^{-2}.

Instrumentation and Display

Ultrasound sources are usually crystals that are either *piezoelectric* or *magnetostrictive*. A piezoelectric crystal is one in which an applied electric field alters the molecular positions, producing stresses in the crystal. If the applied field is periodic, the crystal vibrates, producing sound. Magnetostrictive materials exhibit similar behavior in an applied magnetic field. Both types of transducers also act as receivers, since the mechanical vibrations produce electric and magnetic fields that can be detected and used to monitor incoming sound waves.

In diagnostic medical work, two techniques of scanning and display are common. In an A-scan the vertical deflection plates of an oscilloscope display voltages associated with the emission or reception of sound pulses. The horizontal motion of the beam is adjusted so that the sweep time is long enough to display the original pulse and its echo (Figs. 22.20 and 22.21).

A second technique, the B-scan, uses a movable transducer. In this case, the voltage due to the pulse or echo is added to the accelerating voltage in the electron gun of the cathode-ray tube. The oscilloscope is adjusted so that only when a pulse or echo voltage is present are the electrons fast enough to produce a bright spot on the screen (Fig. 22.22).

The echos then appear on the screen as bright spots whose brightness depends on the strength of the echo. Thus if the transducer were not moved, the peaks would appear only as bright spots along a line. When the transducer is moved parallel to the body surface the vertical deflection plates are adjusted so that the traces are displaced along horizontal lines one above another (Figs. 22.23 and 22.24). Newer instruments do not require movement of the transducer. The beam leaving the transducer can be swept across the region of interest. However, the principle of the display is the same.

Ultrasound is very useful for detecting motion. For example, the motion of the heart wall or even of the mitral valve can be detected (Fig. 22.25). Special source probes can even be used to do ultrasonic scanning from inside the heart. The Doppler effect with ultrasound can be used to detect fetal heartbeats and the pulsation of arterial walls.

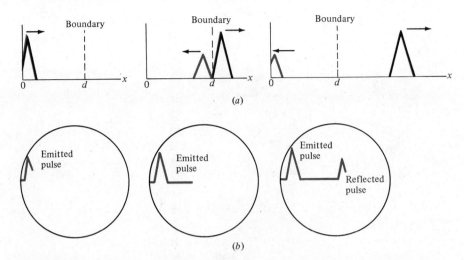

Figure 22.20. (*a*) A pulse is emitted and travels toward a tissue boundary a distance *d* away. Part of the pulse (shown colored) is reflected at the interface and returns to the transducer. (*b*) The oscilloscope trace monitors the emitted pulse and the reflected pulse a time $t = 2d/c$ later. Because these events are so rapid, the eye only sees the completed trace.

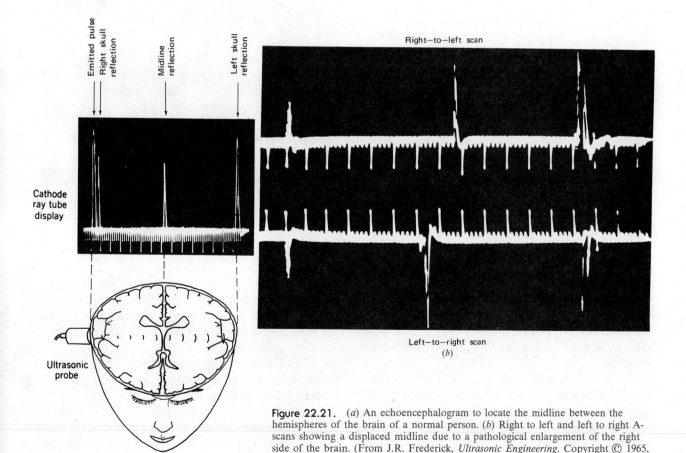

Figure 22.21. (*a*) An echoencephalogram to locate the midline between the hemispheres of the brain of a normal person. (*b*) Right to left and left to right A-scans showing a displaced midline due to a pathological enlargement of the right side of the brain. (From J.R. Frederick, *Ultrasonic Engineering.* Copyright © 1965, John Wiley & Sons, New York. Used with permission.)

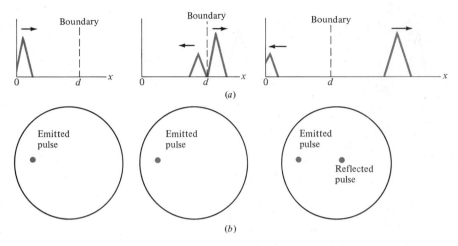

Figure 22.22.   The sequence of events in (*a*) appears on a single scan of the oscilloscope trace (*b*) as bright spots.

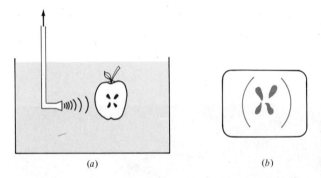

Figure 22.23.   A B-scan performed on an apple (shown in cross section). (*a*) The apple and transmitter-receiver are immersed in water to improve transmission. The ultrasonic beam is directed to the right, and partial reflection occurs at the apple surface and from the seeds. The receiver is moved upward during the scan producing the oscilloscope trace shown in (*b*).

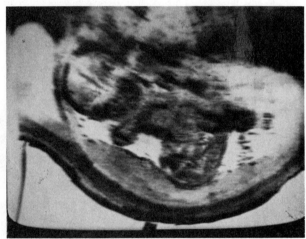

Figure 22.24.   B-scan of a fetus during the last trimester of gestation. The transducer is directed downward and is moved back and forth mechanically to obtain this result. (Photo courtesy of Picker Corporation.)

22.8 | THE SPEED OF SOUND

We now derive the speed of sound in a medium by applying Newton's second law of motion as stated in Section 7.1: The impulse imparted to an object equals its momentum change. We consider a sound wave in fluid in a cylindrical tube as in Fig. 22.26, although the result is more general. Here the object in the second law is a segment of the fluid, and the impulse is imparted by the piston.

In Fig. 22.26, the fluid has an initial density ρ and a pressure P. The piston is suddenly pushed to the right, compressing the adjacent fluid and increasing the pressure by ΔP. The piston moves at a velocity v and the collisions with the molecules in the fluid

nearby give them an average velocity v also. The disturbance in the fluid propagates forward at the sound velocity c. Note that the piston and sound velocities are different, just as the velocity of an individual falling domino differs from the rate at which a disturbance travels along a line of dominos. Normally the sound velocity c is much larger than v, the average velocity of the particles in the medium.

In a time Δt the piston moves a distance $v\Delta t$, while the front of the disturbed section of the fluid moves $c\Delta t$. The cross-sectional area of the piston and the

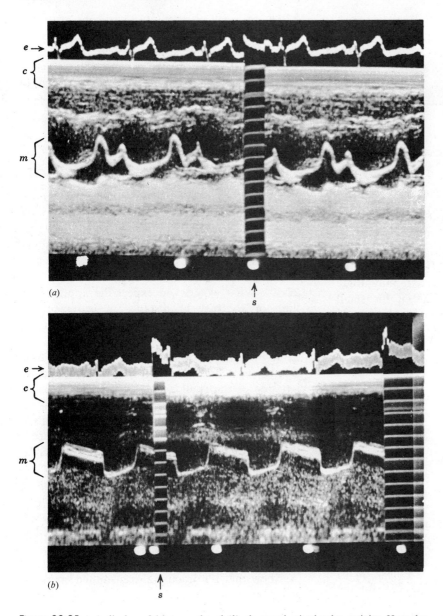

Figure 22.25. A display of (*a*) normal and (*b*) abnormal mitral valve activity. Here the transducer is held fixed and the horizontal axis is a time axis. A scale in centimeters appears at *s*. The reflections in region *c* are from the skin, those in region *m* are from the mitral valve, and the corresponding EKG trace is shown at *e*. (From J.R. Frederick, *Ultrasonic Engineering*. Copyright © 1965, John Wiley & Sons, New York. Used with permission.)

fluid segment is A. Thus this segment has a volume $A(c\Delta t)$, and its mass is the density ρ times the volume or $\rho(Ac\Delta t)$. The fluid within that segment is moving at a velocity v. Since the fluid is initially at rest, the magnitude of its momentum change $\Delta \mathbf{p} = \Delta(m\mathbf{v})$ is the product of its mass and its velocity, $(\rho Ac\Delta t)v$. The impulse is the product of the net force $A\Delta P$ on the fluid and the time it acts, $(A\Delta P)\Delta t$. Thus we have

$$A\Delta P\Delta t = \rho Ac\Delta t v$$

and the pressure change must satisfy

$$\Delta P = \rho c v \qquad (22.8)$$

Now from the definition of the bulk modulus (Eq. 22.1) the pressure change satisfies $\Delta P = K(\Delta \rho / \rho)$. The density of the fluid is its mass to volume ratio,

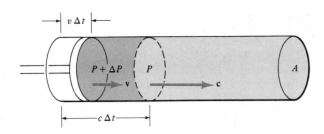

Figure 22.26. The fluid is initially at a uniform pressure P and density ρ. The piston is moved toward the right at a velocity v, and the disturbance it produces travels at the speed of sound c.

$\rho = m/V$. If the density increases slightly, there is a corresponding decrease in the volume of a given mass. (More formally, we note that if m is constant, $d\rho = d(m/V) = -mdV/V^2 = -\rho dV/V$.) Hence the fractional density change $\Delta\rho/\rho$ is equal in magnitude to the fractional volume change $\Delta V/V$ for small changes. The segment has an overall initial length $c\Delta t$ and is shortened by $v\Delta t$, so equating the magnitudes

$$\frac{\Delta\rho}{\rho} = \frac{\Delta V}{V} = \frac{Av\Delta t}{Ac\Delta t} = \frac{v}{c}$$

Thus $\Delta P = K\Delta\rho/\rho = Kv/c$. With Eq. 22.8, we have then

$$Kv/c = \rho cv$$

or

$$c = [K/\rho]^{1/2}$$

This is Eq. 22.2

When a thin rod is struck, a similar compressional pulse travels along the rod. However, the rod also expands transversely when this occurs. A detailed analysis shows that Young's modulus (Chapter Eight) then replaces the bulk modulus in Eq. 22.2 for the velocity of the longitudinal sound wave. In an extended solid medium, the sound velocity also depends on the shear modulus.

EXERCISES ON SUPPLEMENTARY TOPICS

Section 22.7 | Ultrasound

22-55 (a) If sound travels from air to water perpendicular to the interface, what percentage of the intensity is reflected? (b) If sound travels from water to air, what percentage of the intensity is transmitted?

22-56 A new ultrasound microscope uses frequencies of 3×10^9 Hz. What wavelength does this sound have in water at 20° C?

22-57 The energy loss of an ultrasonic wave is proportional to the square of the frequency. Find the ratio of the energy losses for waves at 1 MHz to those for waves at 3000 MHz.

PROBLEMS ON SUPPLEMENTARY TOPICS

22-58 The echo from the midline of the brain is detected on an oscilloscope 10^{-4} s after the source pulse. How far from the source is the midline? (Assume that the velocity of sound in brain tissue is 1540 m s^{-1}). Neglect corrections due to the skull bone.)

22-59 What would you expect the approximate limit of accuracy of distance measurements in human tissue to be when using sound of frequency 10^6 Hz?

22-60 In a brain scan the echos from the right side of the skull, the midline of the brain, and the left side of the skull are observed after times of 0.10×10^{-4}, 1.26×10^{-4}, and 2.40×10^{-4} s, respectively. If the sound velocity is 1540 m s^{-1}, (a) how far is the midline of the brain displaced? (b) Which hemisphere of the brain is enlarged?

22-61 An object of mass m_1 and velocity v collides elastically and head-on with one of mass m_2 that is initially at rest. We define the fractional transmitted energy as the ratio of the final kinetic energy of m_2 to that of m_1 before the collision. The fractional reflected energy is the ratio of the energies of mass m_1 after and before the collision. (a) Show that the fractional transmitted energy is given by

$$\frac{4m_1m_2}{(m_1 + m_2)^2}$$

(b) Show that the fractional reflected energy is given by

$$\left(\frac{m_1 - m_2}{m_1 + m_2}\right)^2$$

(c) What fraction of the energy is transmitted and reflected if $m_1 = m_2$, $m_1 = 10\,m_2$, and $m_2 = 100\,m_1$?

Additional Reading

See also the references marked with an asterisk (*) at the end of Chapter Twenty-one.

Winston E. Kock, *Sound Waves and Light Waves,* Science Study Series, Doubleday and Co., Garden City, N.Y., 1965.

William A. Van Bergeigk, John R. Pierce, and Edward B. David, Jr., *Waves and the Ear,* Science Study Series, Doubleday and Co., Garden City, N.Y., 1960.

Jess J. Josephs, *The Physics of Musical Sound,* D. Van Nostrand and Co., Princeton, N.J., 1967.

D.R. Griffin, How Bats Guide Their Flight by Supersonic Echoes, *American Journal of Physics,* vol. 12, 1944, p. 343.

F. S. Crawford, Singing Corrugated Pipes, *American Journal of Physics,* vol. 42, 1974, p. 278.

T.D. Rossing, Musical Acoustics (Resource Letter MA-1), *American Journal of Physics,* vol. 43, 1975, p. 944.

Arthur H. Benade, *Fundamentals of Musical Acoustics,* Oxford University Press, New York, 1976.

J.W. Coltman, Acoustics of the Flute, *Physics Today,* vol. 21, November 1968, p. 25.

H. Fletcher, The Pitch, Loudness, and Quality of Musical Tones, *American Journal of Physics,* vol. 14, 1946, p. 215.

J.L. Flanagan, *Speech Analysis: Synthesis and Perception,* 2nd ed., Springer-Verlag, Berlin, 1972.

Peter B. Denes and Elliot N. Pruson, *The Speech Chain,* Bell Telephone Laboratories, Murray Hill, N.J., 1963.

Dale Esnminger, *Ultrasonics; The Low and High Energy Applications,* Marcel Dekker, Inc., New York, 1973. Chapter 14 is a summary of the medical applications of ultrasound.

Julian R. Frederick, *Ultrasonic Engineering,* John Wiley and Sons, Inc., New York, 1965, chapters 7, 8, and 9.

P.N.T. Wells, *Physical Principles of Ultrasonic Diagnosis,* Academic Press, London and New York, 1969.

Thomas H. Maugh II, Acoustic Microscopy: A New Window to the World of the Small, *Science,* vol. 201, 1978, p. 1110.

Thomas H. Maugh II, Eavesdropping on Bones, *Science,* vol. 214, 1981, p. 172. Sounds from bones under stress diagnose bone fractures and monitor their healing.

James C. Smith, James T. Marsh, Steven Greenberg, and Warren S. Brown, Human Auditory Frequency-Following Responses to a Missing Fundamental, *Science,* vol. 201, 1978, p. 639.

Richard L. Popp and Albert Macovski, Ultrasonic Diagnostic Instruments, *Science,* vol. 210, 1980, p. 268.

Irwin Hersey, The Dangerous Decibels, *Engineering Opportunities,* August 1969, p. 9.

Beverly Karplus Hartline, Snow Physics and Avalanche Prediction, *Science,* vol. 203, 1979, p. 346.

Edgar A.G. Shaw, Noise Pollution—What Can be Done?, *Physics Today,* January 1975, p. 46.

Rickye Heffner and Henry Heffner, Hearing in the Elephant, *Science,* vol. 208, 1980, p. 518.

Gerhard Neuweiler, How Bats Detect Flying Insects, *Physics Today,* August 1980, p. 34.

James A. Simmons, M. Brock Fenton, and Michael J. O'Farrell, Echolocation and Pursuit of Prey by Bats, *Science,* vol. 203, 1979, p. 16.

Masakazu Konishi, How the Owl Tracks its Prey, *American Scientist,* vol. 61, 1973, p. 414.

Masakazu Konishi and Eric I. Knudsen, The Oilbird: Hearing and Echolocation, *Science,* vol. 204, 1979, p. 425.

Thomas D. Rossing, Physics and Psychophysics of High-Fidelity Sound, Part V, *The Physics Teacher,* vol. 22, 1984, p. 84. Earlier articles in the series are referenced.

Christine Shadle, Experiments on the Acoustics of Whistling, *The Physics Teacher*, vol. 21, 1983, p. 148.

Thomas D. Rossing, *The Science of Sound,* Addison-Wesley, Reading, Mass., 1982. An introductory text. Principles of acoustics, human voice, music, environmental noise.

Scientific American articles:

Maurice Ewing and Leonard Engel, Seismic Shooting at Sea, May 1962, p. 116.

Herbert A. Wilson, Jr., Sonic Boom, January 1962, p. 36.

Mark R. Rosenzweig, Auditory Localization, October 1961, p. 132.

Richard M. Warren and Roslyn P. Warren, Auditory Illusions and Confusions, December 1970, p. 30.

Robert C. Chanard, Aerodynamic Whistles, January 1970, p. 40.

Edward E. David, Jr., The Reproduction of Sound, August 1961, p. 72.

Victor E. Ragosine, Magnetic Recording, November 1969, p. 70.

Leo L. Beranck, Noise, December 1966, p. 66.

Arthur A. Few, Thunder, July 1975, p. 80.

Veru O. Knudsen, Architectural Acoustics, November 1963, p. 78.

Gordon S. Kiuo and John Shaw, Acoustic Surface Waves, October 1972, p. 50.

Georg von Békésy, The Ear, August 1957, p. 66.

James L. Flanagan, The Synthesis of Speech, February 1972, p. 48.

Adrian M. Wenner, Sound Communication in Honeybees, April 1964, p. 116.

Crawford Greenewalt, How Birds Sing, November 1969, p. 126.

Peter F. Ostwald, Acoustic Methods in Psychiatry, March 1965, p. 82.

Gerald Oster, Auditory Beats in the Brain, October 1973, p. 94.

Peter F. Ostwald and Phillip Peltzman, The Cry of the Human Infant, March 1974, p. 84.

F.A. Saunders, Physics and Music, July 1948, p. 33.

Arthur H. Benade, The Physics of Woodwinds, October 1960, p. 144.

E. Donnell Blackham, The Physics of the Piano, December 1965, p. 88.

John C. Schelleng, The Physics of the Bowed String, January 1974, p. 87.

B. Patterson, Musical Dynamics, November 1974, p. 78.

Diana Deutsch, Musical Illusions, October 1975, p. 92.

Johan Sundberg, The Acoustics of the Singing Voice, March 1977, p. 82.

G.E. Henry, Ultrasonics, May 1954, p. 54.

Klaus Dransfeld, Kilomegacycle Ultrasonics, June 1963, p. 60.

Kenneth D. Roeder, Moths and Ultrasound, April 1965, p. 94.

Calvin F. Quate, The Acoustic Microscope, October 1979, p. 62.

Gilbert B. Devey and Peter N.T. Wells, Ultrasound in Medical Diagnosis, May 1978, p. 98.

Eric I. Knudsen, The Hearing of the Barn Owl, December 1981, p. 112.

Jearl Walker, Some Whispering Galleries Are Simply Sound Reflectors, But Others Are More Mysterious, The Amateur Scientist, October 1978, p. 179.

A.J. Hudspeth, The Hair Cells of the Inner Ear, January 1983, p. 54. Mechanical forces are converted into electrical signals to the brain.

Neville H. Fletcher and Suzanne Thwaites, The Physics of Organ Pipes, January 1983, p. 94.

Jearl Walker, What Makes You Sound So Good When You Sing in the Shower?, The Amateur Scientist, May 1982, p. 170.

Thomas D. Rossing, The Physics of Kettledrums, The Amateur Scientist, November, 1982, p. 172.

CHAPTER 23

WAVE PROPERTIES OF LIGHT

Historically, the wave theory of light was not readily accepted because many observations seemed easier to explain with a model that treated a light beam as a stream of particles. Both particle and wave theories could adequately describe the *reflection* of light, as well as *refraction*, the bending of light as it crosses a boundary between two media. Waves tend to bend around obstacles, as is readily observed with water or sound waves. Since we cannot see around obstacles, this was taken as evidence against wave theories until experiments in the early part of the nineteenth century demonstrated interference effects. These experiments also showed that the bending of light around ordinary objects is difficult to observe because the wavelength of visible light is very short. Later in the century, Maxwell's work on electromagnetic radiation gave the wave model of light a firm theoretical foundation.

Despite the great success of Maxwell's electromagnetic wave theory in predicting many phenomena involving light, it failed to correctly describe some processes in which light is absorbed or emitted by matter. The modern *quantum theory*, developed early in the twentieth century, states that light is composed of *quanta* or *photons*. These are little packets or bundles of light waves, each with an energy proportional to the frequency. Photons have a particlelike aspect associated with the discreteness of their energy in addition to their wave attributes. The quantum theory of light will be discussed in Chapter Twenty-six, but for the situations considered in this unit, the classical electromagnetic wave theory is adequate.

In the first part of this chapter, we discuss the speed of light, reflection, and refraction. The next few sections cover some of the interference properties of light waves. The chapter concludes with a brief discussion of the polarization of light. Applications to optical instruments will be treated in the next chapter.

Although we concentrate mainly on visible light in this chapter, it should be noted that much of what we discuss applies also to other electromagnetic waves and to other wave phenomena, such as sound and water waves.

23.1 | THE INDEX OF REFRACTION

Electromagnetic waves of any frequency travel in a vacuum with the same velocity, $c = 3.00 \times 10^8$ m s^{-1}. In a material medium, the velocity v depends on the frequency of the wave, but it is never greater than the velocity c in a vacuum. The ratio of these velocities is the *index of refraction* of the medium,

$$n = \frac{c}{v} \qquad (23.1)$$

Since v is never greater than c, n is never less than 1. In comparing two media, the one with the larger refractive index is said to be *optically denser*. Representative indices of refraction are given in Table 23.1 for yellow light. The refractive index varies somewhat with the frequency. In a typical material the variation is 1 or 2 percent within the visible spectrum.

The frequency of a light wave is determined by its source and is unaffected by the medium. Since

TABLE 23.1

Indices of refraction of representative materials for yellow sodium light ($\lambda = 589$ nanometres) We use the approximate values $n = 1$ for air and $n = 4/3$ for water

Material	Index
Air	1.00029
Carbon dioxide	1.00045
Water	1.333
Ethyl alcohol	1.362
Benzene	1.501
Carbon disulfide	1.628
Glass, light crown	1.517
Glass, heavy flint	1.647
Fluorite	1.434
Diamond	2.417

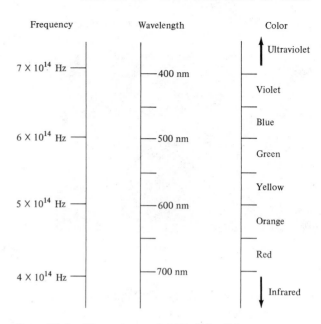

Figure 23.1. The spectrum of visible light. Color ranges are approximate.

$f\lambda = v = c/n$, the wavelength changes when the index of refraction changes. If a beam of light goes from a medium with index n_1 to a medium with index n_2, then $f\lambda_1 = c/n_1$, and $f\lambda_2 = c/n_2$. Dividing these two equations,

$$\frac{\lambda_2}{\lambda_1} = \frac{1/n_2}{1/n_1} = \frac{n_1}{n_2} \qquad (23.2)$$

This means that the wavelength is smaller in the optically denser medium, as is illustrated by the following example.

Example 23.1

Green light with a wavelength of 5×10^{-7} m in a vacuum enters a glass plate with refractive index 1.5. (a) What is the velocity of light in the glass? (b) What is the wavelength of the light in the glass?

(a) From the definition $n = c/v$, the velocity in the glass is

$$v = \frac{c}{n} = \frac{3.00 \times 10^8 \text{ m s}^{-1}}{1.5}$$
$$= 2.00 \times 10^8 \text{ m s}^{-1}$$

(b) In the vacuum, $v = c$ and $n = 1$. Thus

$$\lambda_2 = \lambda_1 \frac{n_1}{n_2} = (5 \times 10^{-7} \text{ m}) \left(\frac{1}{1.5}\right)$$
$$= 3.33 \times 10^{-7} \text{ m}$$

Several units of length are commonly employed in discussions of electromagnetic waves. Metres, centimetres, and millimetres are convenient for radio and microwaves. Discussions of optical instruments and

biological applications of visible light often use the *nanometre* (nm), where

$$1 \text{ nanometre} = 1 \text{ nm} = 10^{-9} \text{ m}$$

The visible spectrum for humans extends approximately from 400 nm (violet) to 700 nm (red) (Fig. 23.1). *X rays* are very-high-frequency electromagnetic waves. They have many diagnostic and therapeutic uses in medicine and also serve as a powerful probe of the structure of complex molecules. A typical X ray wavelength is about 0.1 nm $= 10^{-10}$ m.

23.2 | HUYGENS' PRINCIPLE

In 1678, almost two centuries before Maxwell's work on electromagnetic waves, Christian Huygens (1629–1695) proposed a wave theory of light. It is still very useful for understanding many properties of light and other waves, since it makes no reference to the physical nature of the wave phenomenon.

To discuss Huygens' idea, now called *Huygens' principle,* it is useful to introduce the concept of a *wave front.* We recall from Chapter Twenty-one that two waves that have their maximum amplitude at the same time are said to be in phase. Also, if one wave has a maximum amplitude when the other has a minimum, the two waves are exactly out of phase. *Wave*

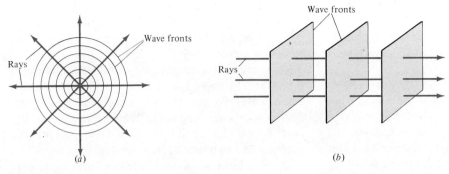

Figure 23.2. (a) The wave fronts corresponding to successive maxima in a spherical wave. (b) Wave fronts for a plane wave.

fronts are surfaces on which the waves at every point are in phase. They travel outward from the source with the speed of the wave. A useful picture of the wave is obtained by drawing several wave fronts (Fig. 23.2). A wave with spherical wave fronts is called a *spherical wave.* A wave that travels in a single direction has wave fronts that are planes, so it is called a *plane wave. A line perpendicular to the wave fronts is called a* ray *and indicates the direction of motion of the wave.* Often it is easier to draw only the rays and not the wave fronts.

Huygens' principle enables us to find the future shape and location of a wave front from its present shape and location. It states that *each point on a wave front can be considered as a source of small secondary spherical wavelets* (Fig. 23.3). The wave front at a later time is the surface tangent to the secondary wavelets, their *envelope.*

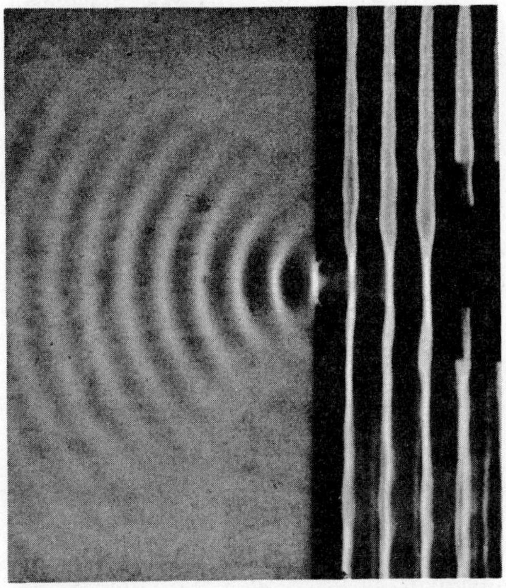

We can illustrate the use of Huygens' principle by considering a plane wave (Fig. 23.4). At several points on a wave front, we draw spheres of radius $r = ct$, representing the distance traveled by the secondary wavelets in time t. The tangent to these spheres is a plane displaced a distance ct from the original wave front. Thus the wave front has moved the expected distance in the time interval.

The original formulation of Huygens' principle stated that the wavelets radiate with equal intensity in all directions. This would imply that the waves travel backward as well as forward, contrary to what we see. In the nineteenth century, Fresnel and Kirch-

Figure 23.3. Water waves in a shallow ripple tank encounter a narrow slit in an obstacle. Circular wave fronts are produced to the left of the slit. (From Educational Development Center Inc., Newton, Massachusetts.)

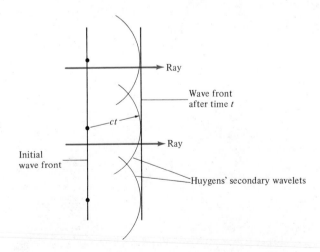

Figure 23.4. In a time t, the wave front moves a distance ct.

hoff put Huygens' ideas into a rigorous mathematical form, and showed that the intensity of the wavelets is a maximum in the forward direction and gradually decreases to zero in the backward direction. Thus there is no wave going backward.

In the next two sections we apply Huygens' principle to situations where light reaches a boundary between two media that extends over distances large compared to a wavelength. In such situations, the behavior of the light rays is simple to describe. Later in the chapter we consider light beams encountering objects comparable in size to a wavelength. Interference effects then can be quite complex.

23.3 | REFLECTION OF LIGHT

When a beam of light reaches the boundary between two media, some light is transmitted, some is absorbed, and the remainder is reflected. The smooth surface of a piece of glass or polished metal reflects light in a particular direction. This is called *specular* reflection. Sometimes the reflection is *diffuse,* and the reflected light travels in all directions. This happens when light strikes a surface such as a sheet of paper or a painted wall with random irregularities that are large compared to a wavelength. Each smooth section of such a surface produces specular reflection, but because of the varying orientations of the sections, the total reflected beam has no unique direction.

In specular reflection (Fig. 23.5), the directions of the incident rays relative to the normal to the surface are related very simply:

The reflected light rays are in the same plane as the incident rays and the normal, and make the same angle with the normal.

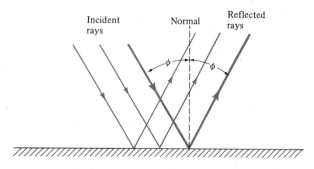

Figure 23.5. The reflected light rays are in the same plane as the incident rays and the normal and make the same angle with the normal.

This equality of the angles of incidence and reflection is a general property of waves. It follows from Huygens' principle and applies equally well to sound waves or water waves (Fig. 23.6).

Two additional properties of reflected light are important. First, we saw in Chapter Twenty-one that when waves on a string reach a fixed end, the reflected waves are *inverted* or *reversed in sign.* In this case, one says that the *phase is reversed.* This also happens to a light wave if it is incident on a boundary from an optically less dense medium moving toward a more dense medium. For example, a phase reversal occurs at an air-glass boundary when light is incident from the air. No phase reversal occurs when light is incident in the glass, which is the denser medium.

Second, the intensity I_r of the reflected wave is determined by the refractive indices n_1 and n_2 in the two media. If the incident intensity is I_0, then for normal incidence ($\phi = 0°$) in either direction

$$\frac{I_r}{I_0} = \left(\frac{n_2 - n_1}{n_2 + n_1}\right)^2 \quad \text{(normal incidence)} \quad (23.3)$$

This formula is also approximately correct if the angle of incidence is small but not zero. As the following example shows, significant amounts of light may be reflected at the air-glass surfaces of optical instruments.

Example 23.2
A lens is made of glass with $n = 1.5$. What fraction of the light is reflected at normal incidence?
With $n_1 = 1$ and $n_2 = 1.5$.

$$\frac{I_r}{I_0} = \left(\frac{n_2 - n_1}{n_2 + n_1}\right)^2 = \left(\frac{1.5 - 1}{1.5 + 1}\right)^2 = 0.040$$

Thus, 4 percent of the light is reflected. In a microscope or camera with several lenses, about 4 percent of the intensity is lost at each lens surface. Methods used to diminish these losses are discussed later in the chapter.

23.4 | REFRACTION OF LIGHT

When light rays go from one transparent medium to another with a different index of refraction, they are bent or *refracted* (Fig. 23.7). As in the case of reflection, it is possible to find a relationship between the two directions with the aid of Huygens' principle. This relationship is called *Snell's law,* after Willebrord Snell (1591–1626) who discovered it experimentally. If the indices of refraction of the media are n_1 and n_2, and the *angle of incidence,* ϕ_1, and the *angle*

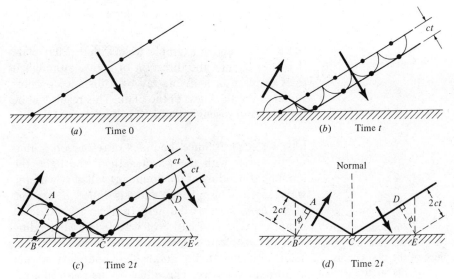

Figure 23.6. Huygens' principle applied to reflection. (*a*) A wave front reaches a reflecting surface. (*b*) After a short time *t* the secondary wavelets from five equally spaced points on the original wave front have spread a distance *ct*. The new wave front is found by drawing lines tangent to the wavelets. (*c*) The wavelets from points on a wave front in part (*b*) have spread a distance *ct*. (*d*) Two right triangles redrawn from (*c*) for clarity. The reflected wave has traveled from *B* to *A* in time 2*t*, and the incident wave will travel from *D* to *E* in an equal time. Hence *AB = DE*. Also, since the points shown on the wave front are equally spaced, *AC = CD*. This means that the triangles *ABC* and *CDE* have two sides and an included angle equal, so they are congruent and all their corresponding sides and angles are equal. Consequently the two wave fronts are at the same angle with the surface. This means that the two rays also make the same angle with the normal.

of refraction, ϕ_2, are measured relative to the normal direction, then Snell's law states

$$n_1 \sin \phi_1 = n_2 \sin \phi_2 \qquad (23.4)$$

As in the case of reflection, both rays and the normal are in the same plane.

Snell's law implies that if *n* increases, then $\sin \phi$ and consequently ϕ decrease. Thus, a ray bends toward the normal when it enters an optically denser medium ($n_2 > n_1$) and away from the normal when it enters a rarer medium ($n_2 < n_1$). Illustrations of Snell's law are given in the following two examples.

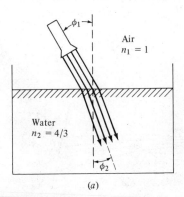

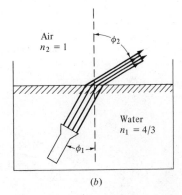

Figure 23.7. (*a*) Light bends toward the normal going from an optically rare to an optically dense medium. (*b*) Light bends away from the normal going from the denser to the rarer medium.

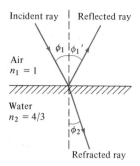

Figure 23.8.   Example 23.3.

Example 23.3

A beam of light is incident from air on water at an angle of 30°. Part of the light is reflected and part is refracted (Fig. 23.8). Find the angles of the two beams.

The incident and reflected rays form equal angles with the normal, so the angle of reflection is

$$\phi_1' = \phi_1 = 30°$$

By Snell's law, with $n_1 = 1$, $n_2 = 4/3$, and $\sin \phi_1 = \sin 30° = 0.5$, the angle of refraction is found from

$$\sin \phi_2 = \frac{n_1}{n_2} \sin \phi_1 = \frac{1}{4/3}(0.5) = 0.375$$

or $\phi_2 = 22°$.

Example 23.4

Light is incident in air at an angle ϕ_1 on a flat glass plate with index n_2 (Fig. 23.9). At what angle does the transmitted beam emerge from the other side?

At the first surface, with $n_1 = 1$, Snell's law gives

$$n_2 \sin \phi_2 = n_1 \sin \phi_1 = \sin \phi_1$$

Applying Snell's law at the second surface, with $n_3 = 1$,

$$n_2 \sin \phi_2 = n_3 \sin \phi_3 = \sin \phi_3$$

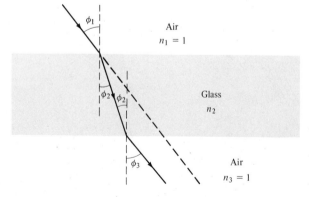

Figure 23.9.   Example 23.4.

Comparing these equations, we have $\sin \phi_3 = \sin \phi_1$, or $\phi_3 = \phi_1$.

From the above example we see that a flat plate does not change the direction of a ray, although it does displace or shift the ray by a distance proportional to the thickness of the plate. This result will be useful in our discussion of lenses in the next chapter.

Dispersion | Since the index of refraction of materials varies with the wavelength of the light, the amount of bending at the boundary will vary with the wavelength. This phenomenon is called *dispersion*. *White light,* such as the light from an incandescent lamp, contains a mixture of wavelengths extending over the visible range, so its various wavelengths or colors are separated at an air-glass boundary, except at normal incidence. This effect occurs twice in a triangular glass *prism,* producing a fan-shaped beam of light and a colorful pattern or *spectrum,* on a screen behind the prism (Fig. 23.10).

23.5 | TOTAL INTERNAL REFLECTION

Figure 23.11 shows what happens to rays from a light source inside a tank of water. When the rays reach the surface, some of the light is reflected and some is refracted. As the angle of incidence increases, the intensity of the reflected beam increases. The transmitted beam gradually becomes weaker, and its *intensity diminishes to zero as the angle of refraction reaches 90°.* The corresponding angle of incidence is called the *critical angle,* ϕ_c. If the angle of incidence exceeds ϕ_c, no refracted beam is observed, and all the light is reflected. This is called *total internal reflection.*

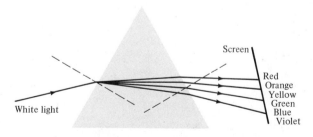

Figure 23.10.   The variation of the index of refraction with wavelength causes different colors in a beam of white light to be separated by a glass prism.

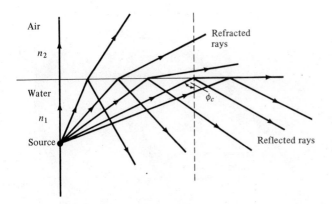

Figure 23.11. Total internal reflection occurs if the angle of incidence exceeds the critical angle ϕ_c.

The critical angle can be found from Snell's law by setting $\phi_2 = 90°$ or $\sin \phi_2 = 1$, corresponding to the maximum possible angle of refraction. This gives $n_1 \sin \phi_c = n_2$, or

$$\sin \phi_c = \frac{n_2}{n_1} \qquad (23.5)$$

If a value of ϕ_1 larger than ϕ_c is substituted in Snell's law, $\sin \phi_2$ turns out to be greater than 1. Since no angle has a sine greater than 1, this implies that there is no refracted beam. This is in agreement with the observations that at any angle equal to or greater than ϕ_c, the beam is totally reflected.

Example 23.5

What is the critical angle for light going from glass with refractive index 1.5 into air?

Using $n_1 = 1.5$ and $n_2 = 1$,

$$\sin \phi_c = \frac{n_2}{n_1} = \frac{1}{1.5} = 0.667$$

and $\phi_c = 42°$. Thus, light incident from the glass on a glass-air boundary will be completely reflected if the angle of incidence exceeds 42°. By contrast, light going from air into glass can enter at any angle of incidence.

In ordinary specular reflection the reflected beam is always weaker than the incident beam, even if the surface is highly polished. By contrast, *no loss of intensity occurs in total internal reflection.* For this reason, totally reflecting prisms are used rather than mirrors in binoculars, periscopes, and reflex cameras (Fig. 23.12). No loss of intensity is associated with the internal reflections in these prisms, although some light is reflected at the surfaces where the light enters or leaves.

The fact that no intensity loss occurs in total internal reflection is also the basis for *fiber optics,* a new and rapidly growing branch of optics. The basic principle of fiber optics is illustrated by a long transparent rod or *light pipe* (Fig. 23.13a). A ray of light entering the pipe is totally reflected if the angle of incidence is large enough when the ray reaches the surface. Even if the pipe curves gradually, light will travel its entire length and arrive unattenuated after many reflections. Consequently, a single light pipe can transmit light energy quite efficiently. However, the rays from different parts of an object are completely scrambled by the multiple reflections, so a single pipe cannot transmit an image. Images can be transmitted using bundles of fine glass or plastic fibers, since each fiber transmits rays from a very small region of the object. The quality of the image is largely determined by the diameter of the fibers, which can be as small as 10^{-6} m (Fig. 23.13b).

Medical applications of fiber optics include instruments used in urinary bladder examinations and in studying the bronchi. Light pipes are also used to transmit intense light beams for surgery. Fiber optics bundles have recently been developed that can transmit light a distance of several kilometres. These have made possible the construction of optical communications systems. Some systems similar to the one in Fig. 23.14 have been installed to link central offices within large cities in the United States, and a 700-mile Boston-to-Washington line is under construc-

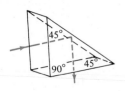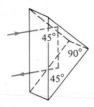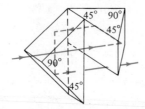

Figure 23.12. Totally reflecting prisms.

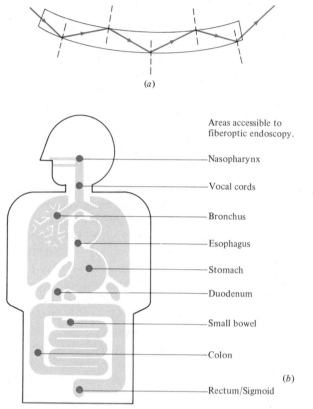

(a)

Areas accessible to
fiberoptic endoscopy.

— Nasopharynx

— Vocal cords

— Bronchus

— Esophagus

— Stomach

— Duodenum

— Small bowel

— Colon

(b)

— Rectum/Sigmoid

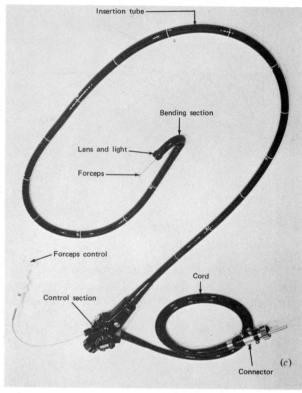

(c)

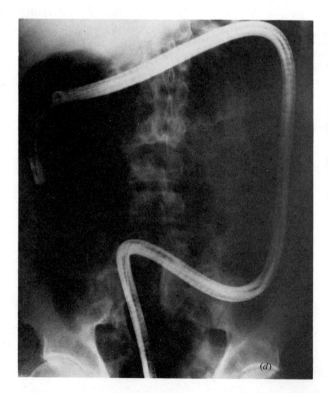

(d)

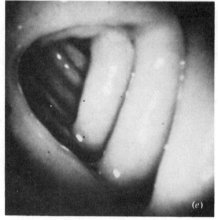

(e)

Figure 23.13. (a) In a light pipe, light suffers repeated total
internal reflection, so it is transmitted without any loss of intensity
(b) Bundles of very fine glass fibers can transmit a sharp image
even if the bundle is gradually bent. This permits a doctor to
visually inspect a variety of internal organs without resorting to
surgery. (c) A colonofiberscope. This instrument is fitted with a
light source and can be used to remove abnormal growths. (d) An
X-ray picture showing the colonofiberscope in use. (e) A
photograph of a normal colon as seen with the colonofiberscope
located as in (d). (Courtesy Olympus Corporation. Photos (d) and
(e) by Dr. Hiromi Shinya. Beth Israel Hospital, New York.)

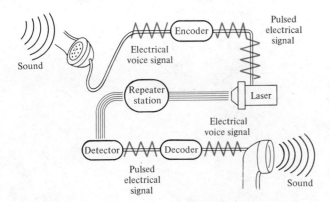

Figure 23.14. An experimental optical telephone system. Each thin glass fiber can carry up to 672 conversations simultaneously.

tion. In Canada, work has begun on a 19,000 mile network. Since visible light has a very high frequency compared to radio or microwaves, and since the information-handling capacity of a wave is proportional to its frequency, optical communication systems have a much greater capacity than those now in general use.

23.6 | YOUNG'S DOUBLE-SLIT INTERFERENCE EXPERIMENT

We have seen that when two waves are present in a string or in an acoustical medium, their interference results in a wave that is a superposition of the two waves. This can lead to standing waves and to beats. The existence of interference effects for light was first demonstrated in 1803 by Thomas Young (1773–1829), an English physician, physicist, and Egyptologist. In Young's experiment, single frequency or *monochromatic light* passes through two narrow slits spaced less than 1 mm apart. It then falls on a distant screen, forming a series of light and dark bands or *fringes*. These arise from the interference between the Huygens' wavelets originating at the two slits (Fig. 23.15).

Similar double-slit interference patterns can be observed with many kinds of waves, including water waves in a ripple tank (Fig. 23.16). The nodes correspond to points where the distances to the wavelet sources differ by a half wavelength, one and a half wavelengths, and so on. Exact cancellation or *destructive interference* occurs between the two out-of-phase waves. The wave maxima or antinodes occur when these two distances differ by an exact integer multiple of a wavelength. Then the waves are in phase, the amplitudes add, and maximum *constructive interference* occurs.

Huygens' principle can be used to locate the maxima and minima in Young's experiment. At an arbitrary point P on the screen in Fig. 23.17a the distances to the two slits are r_1 and r_2. The slit separation d is very small compared to the distance D to the screen. Consequently, if we draw a circle of radius r_1 with P as the origin, then the arc AB is approximately

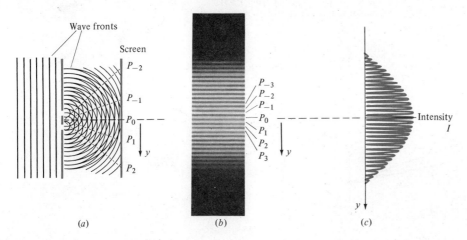

Figure 23.15. (*a*) Young's double-slit experiment. If the distances from a point on the screen to the two slits differ by an integer number of wavelengths, maximum constructive interference occurs there. (*b*) A photograph of the interference fringes produced on the screen. (*c*) A graph of the intensity I versus position y. ((*b*) From Sears, Zemansky and Young, fifth edition, 1976. Addison-Wesley, Reading, Mass.)

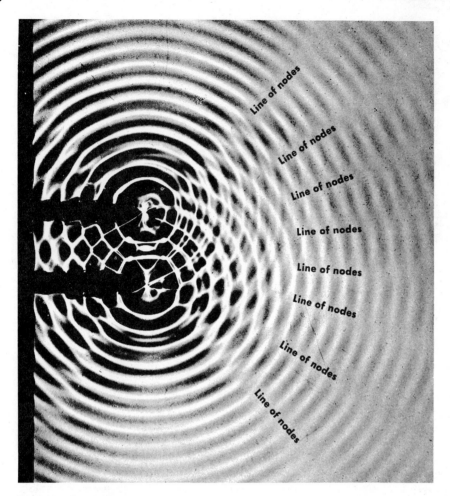

Figure 23.16. Interference of water waves in a ripple tank. The two waves are produced by identical synchronized vibrations. (Reproduced by permission of the publisher from *PSSC Physics,* fourth edition. © 1976, by D. C. Heath and Company, Lexington, Mass.)

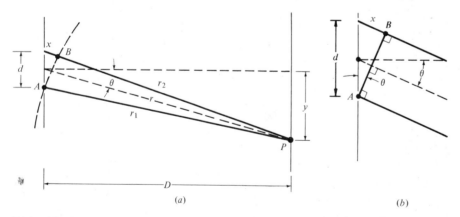

Figure 23.17. (*a*) The geometry of Young's double-slit apparatus. The scale is distorted for clarity; the screen distance D is actually very large compared to the slit spacing d. (*b*) Enlarged view of the region near the slits $x = d \sin \theta$.

THOMAS YOUNG
(1773–1829)

(Radio Times Hulton Picture Library.)

Thomas Young was an unusually talented English physician and physicist. At the age of 2, he could read fluently; at 4, he had read the Bible twice, and at 14, he knew eight languages. Throughout his adult life he was active in both medicine and physics, and he also found the time to be the first person to make major progress on the problem of deciphering Egyptian hieroglyphics.

Young worked on various topics in the physics of fluids and made significant contributions relating to surface tension, capillarity, and the tides. He was the first to use the concept of energy as the ability of a system to do work, and he argued against the caloric theory of heat. He also studied the elastic properties of materials; as we have seen in Chapter Eight, his name is still associated with the parameter characterizing elastic deformations. However, Young's most important research related to the properties of light.

While he was still a medical student, Young discovered how the lens of the eye changes shape in order to focus on objects at different distances. He received his medical degree from Göttingen in Germany in 1796 and went into practice in London in 1799. In 1801, he discovered that irregularities in the cornea are responsible for astigmatism. Some years later, he proposed a theory of color vision based on the idea that the perception of three basic colors would, in combination, generate all the many shades we see. This theory was later refined by Helmholtz and is referred to as the Young-Helmholtz theory.

Turning his attention from the eye to the nature of light itself, Young performed some experiments that were crucial in demonstrating its wave properties. In 1803, he showed that when light went through a narrow slit, bright bands appeared in a region that would be totally dark if light did not bend. Even more striking was his double-slit experiment, which showed that two light beams can combine to produce alternating light and dark bands. The dark bands indicated a cancellation of the two beams, which was easily understood with waves but rather hard to explain with a particle model.

Young's demonstration of the wave nature of light was not well received in England, since Newton and his spiritual descendents had argued for a particle model. As a result the work needed to extend his ideas was carried out by two French physicists, Fresnel and Arago. On the other hand, French physicists found it hard to abandon Lavoisier's "French" caloric theory. Several decades passed before this theory of heat was permanently put to rest.

a straight line perpendicular to both AP and BP. From Fig. 23.17b we see that the difference in path lengths $x = r_2 - r_1$ is then equal to $d \sin \theta$. Maximum constructive interference occurs when this difference is an exact integer number of wavelengths, or when

$$d \sin \theta = m\lambda \qquad m = 0, \pm 1, \pm 2, \dots \quad (23.6)$$

The maxima, or fringes, are located symmetrically about the central $m = 0$ maximum. The negative integers correspond to maxima above the axis, as indicated by the points $P_{-1}, P_{-2}, \dots$ in Fig. 23.15.

The positions of the fringes on the screen can easily be found. In Fig. 23.17a, the distance r from the center of the screen to the fringe is approximately equal to the distance D to the screen, since θ is small. Thus $\sin \theta = y/r \simeq y/D$. Using this result in Eq. 23.6, we have $d(y/D) = m\lambda$, and

$$y = m\lambda \frac{D}{d} \qquad m = 0, \pm 1, \pm 2, \dots \quad (23.7)$$

The intensity at the maxima is not simply the sum of the intensities of the waves from the two slits. The intensity of a wave is proportional to the square of its amplitude. If the amplitude of each wave is A, when the two waves are in phase the total amplitude of the combined wave is $2A$. The intensity at the maxima is then proportional to $(2A)^2 = 4A^2$, or four times the intensity from a single slit. When the doublt-slit interference pattern is averaged over both the maxima and the minima, it turns out that the resulting average intensity is twice that of a single slit. This is the result required by energy conservation, since the net energy reaching the screen is the sum of the energies from the two slits.

The position of a maximum is found in the next example.

Example 23.6

Find the angle of the $m = 3$ fringe if two slits 0.4 mm $= 4 \times 10^{-4}$ m apart are illuminated by yellow light of wavelength 600 nm.

Substituting $m = 3$ in $d \sin \theta = m\lambda$,

$$\sin \theta = \frac{m\lambda}{d} = \frac{3(600 \times 10^{-9} \text{ m})}{4 \times 10^{-4} \text{ m}}$$

$$= 4.50 \times 10^{-3}$$

When the sine is very small, we can use the approximation $\sin \theta = \theta$, where θ is in radians. Thus

$$\theta = 4.50 \times 10^{-3} \text{ rad} \left(\frac{180°}{\pi \text{ rad}} \right)$$

$$= 0.258°$$

This is a small angle, but the fringes can be seen readily if the screen is a metre or more from the slits.

When the slit separation is known, the interference pattern can be used to determine the wavelength of the light used.

Example 23.7

Two slits 4×10^{-4} m apart are 1 m from a screen. If the distance y between the central fringe and the $m = 1$ fringe is 1 mm $= 10^{-3}$ m, what is the wavelength of the light?

Solving $y = m\lambda D/d$ for λ, we have with $m = 1$

$$\lambda = \frac{yd}{mD} = \frac{(10^{-3} \text{ m})(4 \times 10^{-4} \text{ m})}{(1)(1 \text{ m})}$$

$$= 4 \times 10^{-7} \text{ m} = 400 \text{ nm}$$

In practice, it is difficult to measure the positions of the maxima in the double-slit pattern with great accuracy. The diffraction grating discussed in Section 23.8 is a much better way to use interference methods to determine the wavelength of a light source.

The interference fringes in the double-slit pattern (Fig. 23.15) diminish appreciably in intensity beyond the first few fringes on either side of the central maximum. This is due to the interference among wavelets produced in different parts of a *single* slit. This phenomenon, called *diffraction*, will be discussed later in this chapter.

23.7 | COHERENCE

In order to form fringes in a double-slit experiment, the light reaching the slits must originate from a *single* source. By contrast, if the slits are illuminated by light from separate lamps, only a relatively uniform level of brightness will appear on the screen, because the light waves from an ordinary source are not emitted continuously, but rather as short pulses at random intervals. Consequently the phases of the waves from two lamps frequently change, as do the positions of the maxima and minima of the interference pattern; no single interference pattern persists long enough to be detectable. The two lamps are said to be *incoherent* light sources. Only *coherent* waves, those that have a stable phase relationship, can produce interference effects.

Coherent and incoherent sources can be illustrated by two vibrators in a ripple tank. If the vibrators oscillate in unison, they produce coherent waves, and the pattern in Fig. 23.16 results. But suppose that once every few seconds one of the vibrators pauses for a brief but random period and then begins to vibrate again. The two sources will no longer be in phase. There will still be nodes and antinodes at the locations of maximum destructive and constructive wave interference. However, every time a vibrator pauses and restarts, these positions will shift. If these shifts occur frequently enough, no interference pattern will be discernable.

The electromagnetic waves produced by ordinary light sources such as heated wires or flames are emitted at random by single atoms or molecules. These act independently, much like the two vibrators in the ripple tank illustration of incoherent sources. Typically the atoms or molecules produce wave pulses of about 10^{-8}-second duration, or a few metres in length. Such a light source can produce interference effects if a colored filter and a small hole are placed directly in front of it. The filtered light is nearly monochromatic and has originated in a small portion of the source. This light is at least approximately coherent, and it will produce fringes in a double-slit apparatus.

In recent years, intense, coherent light sources called *lasers* have become available. In a laser the atoms emit light in phase, instead of randomly or incoherently. The light produced from a laser has an extremely small range of frequencies, so it is almost exactly monochromatic, and it is highly coherent. As a result, light emerging from the laser at two different points or at two slightly different times can be used to produce interference patterns.

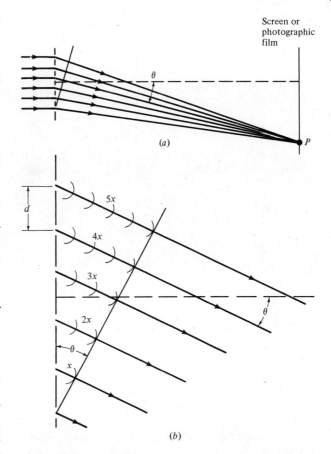

Figure 23.18. (a) Light rays reaching a point P on a screen from six closely spaced slits. (Light rays reaching other points are not shown.) (b) Enlarged view of the region near the slits.

23.8 | THE DIFFRACTION GRATING

The wavelength of a light beam can be found by measuring the spacing of the interference fringes in a double-slit arrangement. However, this is hard to do with accuracy, since the fringes are relatively wide and weak in intensity. A *diffraction grating*, which consists of many closely spaced slits, permits much more precise wavelength determinations. We will see that, as in the double slit, the *distance* between the slits in the grating determines the *locations* of the maxima; their *sharpness* depends on the *total number* of slits.

The principle of the diffraction grating is illustrated by considering the case of six slits (Fig. 23.18). If the difference $x = d \sin \theta$ of the distances from a point on the screen to any two adjacent slits is exactly an integral multiple of a wavelength, all the waves reaching

the screen are in phase. Thus a maximum in the intensity occurs when

$$d \sin \theta = m\lambda \qquad m = 0, \pm 1, \pm 2, \ldots \quad (23.8)$$

m is called the *order* of the maximum or spectrum produced by the grating.

This is the same formula as was found for the angular positions of the maxima in the two-slit apparatus. However, here the fringes are much more intense, since the amplitudes from each of the six slits add (Fig. 23.19). Equally important is the fact that the fringes are now much narrower. This happens because the waves from the six slits cancel almost completely as soon as the angle differs slightly from a value at which a maximum intensity occurs. Both the intensity increase and the line narrowing are accentuated as the number of slits is increased further.

Gratings are manufactured by cutting or ruling equally spaced parallel grooves or lines with an auto-

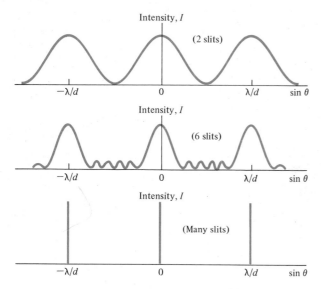

Figure 23.19. Central and $m = \pm 1$ maxima for 2, 6, and a very large number of slits. The intensity units in the three graphs are not to scale; the heights of the maxima are proportional to the square of the number of slits. Note that the fringes become more intense and narrower as the number of slits is increased. In the six-slit case, there is significant but not total cancellation among the waves from the various slits except at the peaks given by $d \sin \theta = m\lambda$. When there are a great many slits, this cancellation is almost exact.

matic ruling engine. Those used to measure wavelengths in the visible region usually have about 4000 to 12,000 lines per centimetre. Spacings not too much greater than a wavelength are employed so that the angular spacing of the maxima will be reasonably large. This makes it possible to measure the angles accurately. The large angles obtainable are illustrated by the next example.

Example 23.8

A grating has 4000 lines per centimetre. At what angles are maxima formed if it is illuminated with yellow light at 600 nm?

The slit spacing is

$$d = \frac{1}{4000} \text{ cm} = 2.5 \times 10^{-4} \text{ cm}$$

$$= 2.5 \times 10^3 \text{ nm}$$

Using $\lambda = 600$ nm, $d \sin \theta = m\lambda$ gives

$$\sin \theta = \frac{m\lambda}{d} = \frac{m(600 \text{ nm})}{(2.5 \times 10^3 \text{ nm})}$$

$$= m(0.240)$$

Substituting $m = 1$ gives $\sin \theta = 0.24$, or $\theta = 14°$; maxima are formed at $14°$ to either side of the central $m = 0$

maximum. Similarly, $m = 2$, 3, and 4 give maxima at $29°$, $46°$, and $74°$, respectively. There are no other maxima, since $m = 5$ requires $\sin \theta = 1.2$, and no angle has a sine greater than 1.

When a diffraction grating is used to measure wavelengths, a narrow slit is placed in front of the light source. If the source is monochromatic, the grating forms several bright lines on the screen, as in the previous example. If the light contains several specific frequencies, lines are formed for each frequency, and the pattern is called a *line spectrum* (Fig. 23.20). Such a spectrum results, for example, when atoms or molecules in a gas are disturbed by an electrical current. Measurements of the wavelengths of line spectra provide important atomic and molecular structure information. Other sources, such as incandescent lamps, produce white light, which is a continuous distribution of all frequencies in the visible region, and give rise with a grating to a *continuous spectrum*. The following example shows how rainbowlike patterns are formed when a grating is illuminated by white light.

Example 23.9

A grating with a slit spacing of 2.5×10^3 nm is illuminated with white light containing wavelengths from 400 to 700 nm. Describe the spectrum formed by the grating.

Using $m = 1$ and $\lambda = 400$ nm,

$$\sin \theta = \frac{m\lambda}{d} = \frac{(1)(400 \text{ nm})}{2.5 \times 10^3 \text{ nm}} = 0.16$$

and $\theta = 9°$. For $m = 1$ and $\lambda = 700$ nm, $\theta = 16°$. Thus the first-order ($m = 1$) spectrum extends from $9°$ to $16°$ on each side of a central white maximum, with the shortest wavelength (violet) at the smallest angle. In the same way, one finds that the second-order ($m = 2$) spectrum extends from $19°$ to $34°$, and the third-order spectrum, from $29°$ to $57°$. Note that the second- and third-order spectra overlap. In general, the first-order visible spectrum of a grating is isolated, but all the higher orders overlap.

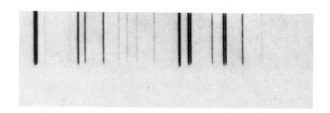

Figure 23.20. The line spectrum produced by iron atoms. This is a photographic negative; the dark lines seen here correspond to bright lines formed by the diffraction grating. (Courtesy of Lick Observatories.)

Angular Width of the Maxima; Resolution

We saw above that as a number of lines is increased, the maxima formed by a grating become progressively narrower. The widths of the maxima determine how accurately a wavelength can be measured with the grating and set a limit on the ability of the grating to separate or *resolve* two closely spaced wavelengths present in a beam of light.

The amplitude at a maximum is proportional to N, the number of slits, and the intensity varies therefore as N^2. Since the total power reaching the screen varies as the number of slits N, the widths of the maxima must vary as $1/N$.

Suppose now that we wish to use a grating to separate two light waves of nearly equal wavelength, λ and $\lambda' = \lambda + \Delta\lambda$. Clearly, this is possible only if the two maxima do not overlap too much. A conventional criterion for setting a limit on this overlap was introduced by Lord Rayleigh (1842–1919). *Rayleigh's criterion* is that one can just resolve two adjacent maxima if their intensities are each half their respective peak values at the point where the overlap is greatest (Fig. 23.21). Since the width of each peak is proportional to $1/N$, it is found that the peaks of the mth-order spectrum can just be distinguished when

$$\frac{\Delta\lambda}{\lambda} = \frac{1}{Nm} \tag{23.9}$$

The resolution obtainable with typical gratings is quite impressive, as can be seen in the following example.

Example 23.10

Trace amounts of sodium in a flame result in a characteristic bright yellow color due to a *doublet*, light at two nearby wavelengths, 589.59 and 589.00 nm. Can a grating with 10,000 lines resolve the two wavelengths in the first-order spectrum?

For the two wavelengths in the yellow sodium doublet,

$$\frac{\Delta\lambda}{\lambda} = \frac{(589.59 - 589.00)\ \text{nm}}{589\ \text{nm}} = 10^{-3}$$

The smallest wavelength difference that can be resolved by this grating in first order ($m = 1$) satisfies

$$\frac{\Delta\lambda}{\lambda} = \frac{1}{Nm} = \frac{1}{(10^4)(1)} = 10^{-4}$$

Since this is only one-tenth of the ratio $\Delta\lambda/\lambda$ for the sodium doublet, the grating readily resolves the two wavelengths.

Rayleigh's criterion for the minimum resolvable separation of two maxima is a good estimate of what can be achieved if a spectrum is examined visually. However, if an intensity pattern such as the one in Fig. 23.21b is carefully measured and analyzed, the minimum distinguishable wavelength separation can be greatly reduced.

23.9 | DIFFRACTION

Thus far, we have discussed examples of interference between waves from two discrete sources, as in a double-slit experiment, or from many discrete sources, as in the diffraction grating. However, parts of a wave

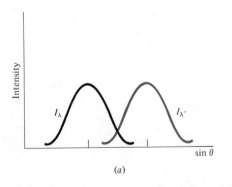

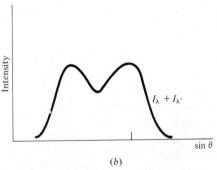

(a)   (b)

Figure 23.21. Intensity distributions in a very small angular region of a diffraction grating spectrum. (a) Light of wavelength λ produces the maximum shown by the black curve, while light with wavelength λ' produces the maximum shown by the colored curve. (b) When both waves are present, the intensity observed is the sum of the two curves in (a). In the case illustrated the two peaks are just barely discernable. According to Rayleigh's criterion, when the wavelengths λ and λ' are significantly closer, the two peaks cannot be resolved.

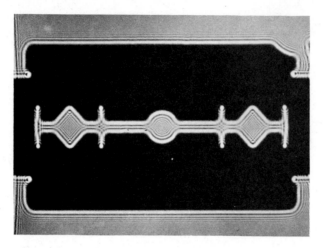

Figure 23.22. The shadow produced when a razor blade is illuminated by a monochromatic point source. (From Sears, Zemansky and Young, fifth edition, 1976, Addison-Wesley, Reading, Mass.)

from a single source will also interfere with each other. This causes effects such as the bending of waves around an obstacle. Single-source interference is referred to as *diffraction* and occurs for all kinds of waves.

Figure 23.22 shows the shadow of a razor blade placed between a point source of monochromatic light and a photographic plate. Some light is bent inside the *geometrical shadow,* the region that would be totally dark in the absence of any bending. Near the edge of the shadow, a diffraction pattern of alternate light and dark bands appears. Thus, even in this fairly simple situation, the diffraction pattern is quite complex.

Under ordinary conditions, we seldom notice the diffraction of light. Light sources such as incandescent lamps or the sun are not monochromatic point sources, and the diffraction patterns due to different parts of the source and to different wavelengths usually overlap and obscure each other. Nevertheless, diffraction patterns are visible when we look at a distant light source, such as a street lamp, through a crack between two fingers or through a cloth umbrella. By contrast, the diffraction of sound waves is hard to avoid. Sound readily bends around obstacles of ordinary size, such as furniture, and fills a room rather uniformly. This difference between the diffraction of visible light and of sound is due to their wavelengths, which are typically 5×10^{-7} m and 1 metre,

respectively. *Diffraction effects are large only when we deal with obstacles or apertures comparable in size to the wavelength.*

Diffraction by a Narrow Slit | Calculations

of diffraction patterns are usually complex. However, for the case of light diffracted by a narrow slit, some of the main features can be obtained without elaborate mathematics. Qualitatively similar results are found for other situations.

In Fig. 23.23, monochromatic light passes through a slit whose width a is comparable to the wavelength λ and falls on a screen or photographic plate at a large distance. The Huygens' wavelets from different parts of the slit interfere and produce the diffraction pattern shown in Fig. 23.24.

To locate the minima in this diffraction pattern, we start by dividing the slit into two halves (AB and BC in Fig. 23.25). The distance to the screen for the wavelet originating at A is greater than that for the wavelet from B by $x = \frac{1}{2}a \sin \theta$. Wavelets originating at any point on AB and at the corresponding point on BC will have the same path difference. Consequently, if x is exactly $\frac{1}{2}\lambda$, all the pairs of wavelets will be out of phase and cancel exactly, producing a minimum. Hence a minimum occurs when $x = \frac{1}{2}a \sin \theta = \frac{1}{2}\lambda$, *or*

$$a \sin \theta = \lambda$$

At this point one might suppose that a maximum occurs when $x = \lambda$, since the wavelets from A and B will then be exactly in phase. However, this is not the case, as can be seen by dividing the slit into four equal parts (Fig. 23.25). Wavelets from A and D will have a path difference which is $\frac{1}{2}x = \frac{1}{2}\lambda$, so they will be exactly out of phase and cancel completely. The same is also true for wavelets from any two points $a/4$ apart, so $x = \lambda$ also implies a minimum. Similarly,

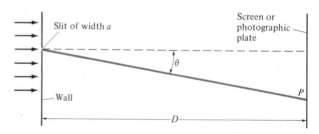

Figure 23.23. Geometry of the single-slit different experiment. The pattern seen on the screen is shown in Fig. 23.24.

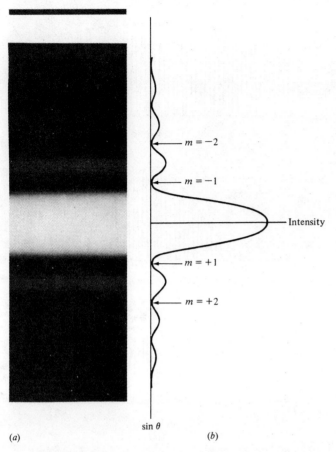

$m = -2$

$m = -1$

Intensity

$m = +1$

$m = +2$

$\sin \theta$

(a)   (b)

Figure 23.24. (a) Photograph of the diffraction pattern produced by a single slit illuminated by monochromatic light as in Fig. 23.23. (b) Intensity graph for this diffraction pattern. (From Sears, Zemansky and Young, fifth edition, 1976, Addison-Wesley, Reading, Mass.)

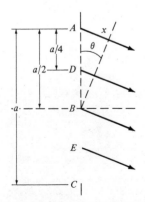

Figure 23.25. A diffraction minimum occurs when wavelets from different parts of the slit interfere destructively.

we can divide the slit into 6, 8, . . . parts and find that total destructive interference occurs when $x = \frac{1}{2}a \sin \theta = \frac{1}{2}\lambda, \lambda, \frac{3}{2}\lambda, 2\lambda, \dots$. Thus, in general, diffraction minima occur at angles satisfying

$$a \sin \theta = m\lambda \qquad m = \pm 1, \pm 2, \dots$$
$$\text{(diffraction minima)} \quad (23.10)$$

The negative integers correspond to minima above the axis in Fig. 23.24.

The maxima are located approximately (but not exactly) midway between the minima, or at $m = 0$, $\pm\frac{3}{2}\lambda, \pm\frac{5}{2}\lambda, \dots$. The central $m = 0$ maximum is very bright, since all the wavelets have nearly the same path length and are in phase. At the other maxima the intensity is much smaller, and it diminishes rapidly as m increases (Fig. 23.24). This happens because, except at $\theta = 0$, the wavelets from some parts of the slit cancel even at the maxima. This partial cancellation becomes more and more complete as the angle increases.

The angular width of the diffraction pattern depends on the size of the slit relative to the wavelength of light, as is seen in the next example.

Example 23.11

A slit is illuminated by light of wavelength λ. Find the angular position of the first diffraction minimum as the slit width expands from λ to 5λ and finally to 10λ.

Substituting $m = 1$ and $a = \lambda$ in $a \sin \theta = m\lambda$, we find

$$\sin \theta = m\frac{\lambda}{a} = (1)\frac{\lambda}{\lambda} = 1$$

so $\theta = 90°$. Similarly, $a = 5\lambda$ gives $\sin \theta = 0.2$ and $\theta = 12°$; $a = 10\lambda$ gives $\sin \theta = 0.1$ and $\theta = 6°$. Thus, as the slit becomes wider, the central diffraction peak becomes narrower. A similar narrowing of the diffraction pattern would be observed if the slit size were held fixed and the wavelength decreased.

Diffraction explains some of the features of the double-slit interference experiment that we neglected in Section 23.6. Earlier we supposed that all the light coming from a single slit had the same phase. This is equivalent to assuming that the slit is very narrow compared to a wavelength and that its diffraction pattern is very wide. However, if the slit widths are comparable to the wavelength, the waves reaching the screen from each slit will have an intensity that varies with position in accordance with the single-slit diffraction formulas. Thus the complete pattern for a double slit is actually formed from two superimposed diffraction patterns, and the double-slit interference maxima do not all have the same intensity. The single-slit diffraction pattern forms an *envelope* for the interference pattern so that the intensity of the peaks becomes smaller further away from the center of the pattern (Fig. 23.26).

Diffraction by a Circular Aperture | When a light wave enters an optical instrument with a circular opening, a diffraction pattern is produced by the interference of the wavelets originating at different points in the aperture. The diffraction pattern due to a distant point source of light has a bright central circular region, surrounded by concentric dark and light rings (Fig. 23.27). A detailed analysis shows that

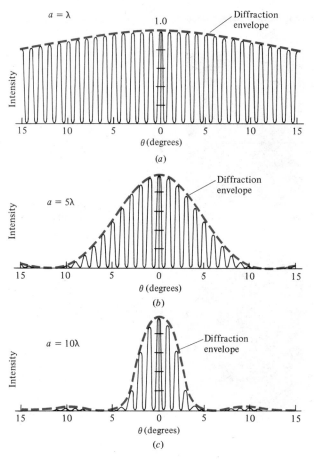

Figure 23.26. Interference patterns for two slits separated by $d = 50\lambda$. The fringes seen are centered at the peaks in the solid curve. Note how the pattern changes as the slit width a is increased and the single-slit diffraction envelope narrows. (From Halliday and Resnick, *Physics*, Part II, third edition, Copyright © 1978, John Wiley and Sons, New York.)

Figure 23.27. The image of a star formed on a photographic plate by a telescope. A central diffraction maximum and a circular secondary maximum are visible. Additional maxima also occur but their intensities are too small to be seen here. (From Halliday and Resnick, *Physics*, Part II, third edition, Copyright © 1978, John Wiley and Sons, New York.)

the first minimum for a circle with diameter d occurs when

$$\sin \theta = 1.22 \frac{\lambda}{d} \qquad (23.11)$$

This is similar to the formula for the first minimum of a slit, $\sin \theta = \lambda/a$, but the circular geometry results in the factor of 1.22.

We see in the next chapter how diffraction limits the sharpness of images formed by optical instruments and the human eye.

23.10 | POLARIZATION OF LIGHT

In Chapter Twenty-one, we noted that light, like any other transverse wave, can be polarized. We now examine this idea in more detail.

Electromagnetic waves have electric and magnetic fields oscillating at right angles to the direction of motion of the waves. If the electric field vector is always along a certain direction, the wave is said to be *linearly polarized* along this direction. The radiation from a single atom or molecule is polarized, but since the many individual atoms or molecules usually act randomly, the resultant light beam from most sources is unpolarized. This means that at a particular instant, the electric field is equally likely to point in any direction perpendicular to the direction of motion of the light wave. If we resolve this field into components along two convenient perpendicular axes, on the average the component along each axis will be equally large.

Polarized light can be produced from an unpolarized beam in several ways, including *absorption, reflection,* and *scattering.* The most familiar way, mentioned in Chapter Twenty-one, is by *absorption* in Polaroid filters. These filters contain long molecules that are aligned. When light has its electric field vector along the molecules, electric currents are set up, and the light is absorbed. However, little happens to a light beam polarized at right angles to the molecules. A similar affect can be demonstrated nicely with microwaves, which are electromagnetic waves with wavelengths of the order of centimetres (Fig. 23.28). Microwaves with vertical electric fields pass readily through horizontal wires. However, when the wires are vertical, the electric field sets up currents in the wires, and the waves are absorbed.

Light can also be polarized by *reflection* off the surface of a nonconducting material. Except at normal

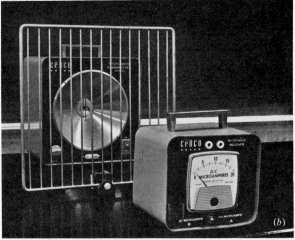

Figure 23.28. Microwaves polarized so that the electric field is vertical are beamed toward a detector. (*a*) Horizontal metal wires permit the waves to pass unimpeded. (*b*) Vertical wires absorb the waves, as shown by the reduced detector reading. (Courtesy of Worth Publishers, from P.A. Tipler, *Physics,* 1976. Photos by Larry Langrill.)

incidence, the fraction of the light reflected by a surface depends on its polarization. It was discovered in 1812 by Sir David Brewster (1781–1868) that the reflected beam is completely polarized at an angle of incidence, ϕ_p, which is determined by the indices of refraction n_1 and n_2 of the two media (Fig. 23.29). This angle is called *Brewster's angle* and satisfies

$$\tan \phi_p = \frac{n_2}{n_1} \qquad (23.12)$$

If an unpolarized beam is incident at this angle, the relatively weak reflected beam will be fully polarized

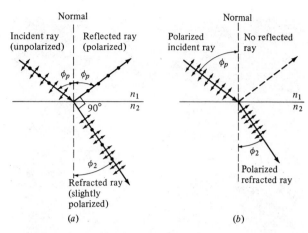

Figure 23.29.   Polarization by reflection. (*a*) On the average the unpolarized incident beam has equally large electric fields perpendicular to the plane of incidence (dots) and in the plane (arrows). When the wave is incident at Brewster's angle, the reflected beam is fully polarized. The transmitted beam is only slightly polarized, since only a small fraction of the light is reflected. (*b*) A wave polarized with its electric field in the plane of incidence is fully transmitted at Brewster's angle.

perpendicular to the *plane of incidence* defined by the incident ray and the normal. The stronger transmitted beam will be slightly polarized. The use of Brewster's angle is illustrated by the following common situation.

Example 23.12

Sunlight reflected off a still lake is fully polarized. At what angle is the light incident on the lake?

Since $n_1 = 1$ for air and $n_2 = 4/3$ for water, the reflected beam is polarized horizontally when

$$\tan \phi_p = \frac{n_2}{n_1} = \frac{4/3}{1} = 1.333$$

or $\phi_p = 53°$. Light reflected off the lake at angles near 53° will be mostly horizontally polarized.

Polarized light is also produced by *scattering*, which is the absorption and reradiation of light. An interesting demonstration of the polarization of scattered light can be done using a tank of water containing some powdered skim milk or soap. Using a Polaroid filter, one finds that the light scattered at right angles to the incident beam is polarized perpendicular to the plane of the incident and scattered rays (Fig. 23.30). If a Polaroid filter is used to polarize the incident beam vertically, no change is observed in the light scattered at right angles in the horizontal plane. However, the vertical beam now disappears.

These observations are explained by the fact that the incident light sets charges in the atoms into oscillatory motion along the direction of the electric field. These oscillating charges then emit radiation with an electric field that has a component along the direction of oscillation. Since the waves are transverse, this means that no radiation is emitted along the direction of oscillation. Accordingly the horizontal beam originates from vertical oscillators and is polarized vertically.

The daytime sky appears blue because short wavelength light is most readily scattered. The sunlight

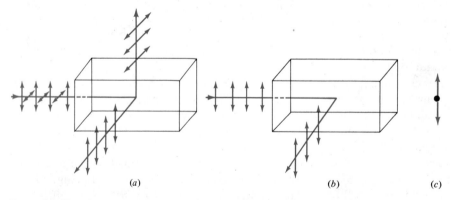

Figure 23.30.   (*a*) Light incident on water containing powdered skim milk or soap is scattered in all directions. Light scattered at right angles is fully polarized. (*b*) If the incident light is vertically polarized, the vertically scattered beam disappears. (*c*) A vertically polarized incident beam causes charges to oscillate vertically. These oscillating charges produce radiation with an electric field component along the direction of oscillation.

scattered at right angles is fully polarized, and all the scattered sunlight is at least partially polarized. Bees are able to sense this polarization and can use the information to determine the direction of the sun from a view of only a small part of the sky.

Polarizers and Analyzers

Several simple but fascinating experiments can be done with a few Polaroid filters. In Fig. 23.31, an unpolarized beam passes through two filters referred to as a *polarizer* and *analyzer,* respectively. The fields along the transmission axis of the polarizer (x axis) and along the direction perpendicular to it (y axis) have the same average amplitude, E_0. After the polarizer, the beam is polarized along the transmission axis with an amplitude E_0. Since the intensity is proportional to $E^2 = E_x{}^2 + E_y{}^2$, the intensity I_0 before the polarizer is proportional to $E_0{}^2 + E_0{}^2 = 2E_0{}^2$, while the intensity after it is proportional to $E_0{}^2$. *Hence the intensity after the polarizer is half the original intensity of the unpolarized beam:*

$$I_1 = \tfrac{1}{2}I_0 \qquad \text{(polarizer)} \qquad (23.13)$$

The transmission axis of the analyzer is at an angle θ to that of the filter. The electric field amplitude reaching the analyzer has a component $E_0 \cos \theta$ along its transmission axis. This component is permitted to pass through the analyzer, while the component at right angles is absorbed. Hence the intensity I_2 after the analyzer is proportional to $E_0{}^2 \cos^2 \theta$, and we have

$$I_2 = I_1 \cos^2 \theta \qquad \text{(analyzer)} \qquad (23.14)$$

This relationship, which was obtained in Chapter 21 for transverse waves on a string, was discovered experimentally by Étienne Louis Malus (1775–1812) in 1809 and is called *Malus's law.*

The following example shows how surprising results can be obtained.

Example 23.13

Unpolarized light of intensity I_0 is passed through three successive Polaroid filters. The second has its axis rotated 45° relative to the first, and the third has its axis rotated an additional 45°, so that it is at 90° to the first. (a) What is the final intensity after the last filter? (b) If the second filter is removed without disturbing the others, what is the final intensity?

(a) The first filter acts as a polarizer, so it reduces the intensity to $I_1 = \tfrac{1}{2}I_0$. The second acts as an analyzer, reducing the intensity by a factor of $\cos^2 \theta = \cos^2 45° = \tfrac{1}{2}$, so $I_2 = \tfrac{1}{2}(\tfrac{1}{2}I_0) = \tfrac{1}{4}I_0$. Similarly, the third filter reduces the intensity by an additional factor of $\cos^2 45° = \tfrac{1}{2}$, bringing the final intensity to $I_3 = \tfrac{1}{8}I_0$.

(b) With the second filter gone, the two remaining filters are at 90°. The first again reduces the intensity to $\tfrac{1}{2}I_0$, but now the next filter reduces the intensity by a factor of $\cos^2 90° = 0$. Eliminating the middle filter reduces the intensity to zero!

23.11 | X-RAY DIFFRACTION AND THE STRUCTURE OF BIOLOGICAL MOLECULES

Because waves are unaffected by objects that are small compared to a wavelength, wavelengths comparable to the interatomic spacing are needed to "see" the arrangement of atoms in a molecule. This spacing is typically a few tenths of a nanometre, where 1 nanometre = 10^{-9} metre. Visible light has a wavelength of 400 to 700 nm, so it cannot be used to study molecular structure. However, X rays, which

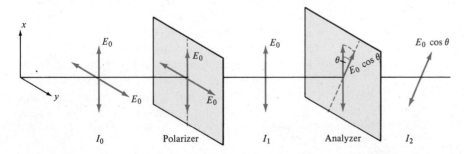

Figure 23.31. Unpolarized light passing through the polarizer emerges polarized along its transmission axis with an amplitude E_0. After the analyzer, the light is polarized along its axis and has an amplitude $E_0 \cos \theta$. The polarizer reduces the intensity by half, and the analyzer by $\cos^2 \theta$.

are electromagnetic waves with typically a 0.1-nm wavelength, are well suited to this task. A crystalline material, which consists of a regular array of atoms, serves for X rays as a three-dimensional analog to the diffraction grating. Information obtained from such X-ray diffraction studies has contributed greatly to unraveling the structure of complex biological molecules such as proteins and DNA, the carrier of the genetic code.

X rays were discovered accidentally by Wilhelm Konrad Roentgen in 1895 while he was studying the properties of cathode rays (electrons) in a gaseous-discharge tube. He observed that even though the tube was inside a box, a barium platinocyanide screen emitted light whenever the tube was on. He named the invisible radiation X rays because of their unknown nature. Within a very short time, their ability to penetrate matter was put to use in medicine as an invaluable diagnostic tool. However, the hazards

associated with X rays were discovered more gradually, as we see in Chapter Thirty-one.

When electrons from a heated filament are accelerated through a large potential difference and allowed to strike a metal target, X rays are produced with a continuous distribution of wavelengths (Fig. 23.32). If the X rays then strike a crystal, the reflected X rays form intense *Laue spots* on a screen or film (Fig. 23.33). These spots are due to the constructive interference of wavelets produced by many atoms.

To understand how one Laue spot is formed, consider a crystal made up of identical atoms in a cubic arrangement that is exposed to a beam of X rays containing many different wavelengths. We can imagine the atoms as forming a series of partially reflecting planes, so that some X rays are reflected by each plane (Fig. 23.34). If the distance between two successive planes is d, then the path lengths for waves reflected by two successive planes will differ by $2x = 2(d \sin \alpha)$, where α is the angle between the incident X-ray direction and the crystal plane. (Conventionally α is used in X-ray diffraction discussions rather than the angle between the beam and the normal direction.) Constructive interference occurs among the reflected beams from all the parallel planes if the path length difference is an integer number of wavelengths, or if

$$2d \sin \alpha = m\lambda, \qquad m = 1, 2, 3, \ldots \quad (23.15)$$

This equation was first obtained by Sir William Henry Bragg (1862–1942), a pioneer in X-ray research, and is called the *Bragg condition*.

Although the incident beam contains many wavelengths, for a given angle of incidence α, spots will be produced only for those wavelengths satisfying Eq. 23.15. The remainder of the beam, containing all other wavelengths, is either absorbed or transmitted. (Because there are many parallel reflecting planes, nearly total cancellation occurs among reflected beams if the Bragg condition is not satisfied.) Notice that spots, and not lines, are formed, because the incident beam is very narrow and would itself form a single spot if it were not deflected. The selective nature of the crystal diffraction is illustrated by the next example.

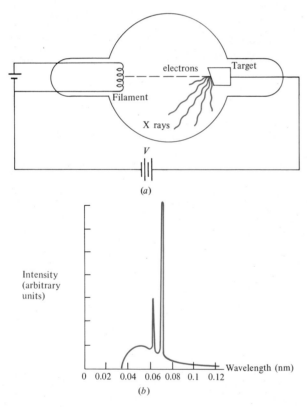

Figure 23.32. (*a*) An X-ray tube. Electrons from the heated filament are accelerated through the potential difference V, and produce X rays when they strike the metal target at the anode. (*b*) The spectrum of X rays produced by a typical X-ray tube.

Example 23.14

The interatomic spacing d in Fig. 23.34 is 0.2 nm, and the angle α is 10°. If the shortest wavelength in the X-ray

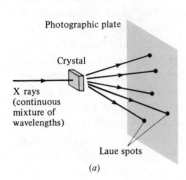

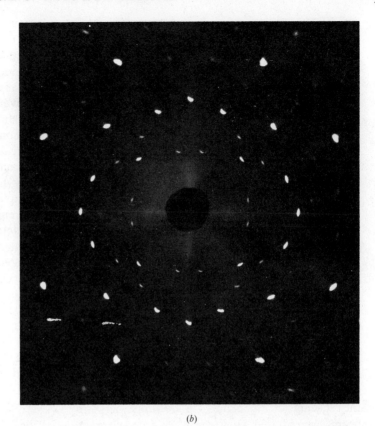

(b)

Figure 23.33. (a) A narrow beam of nonmonochromatic X rays strikes a crystal and forms a pattern of Laue spots on a photographic plate. (b) The Laue spots from a sodium chloride crystal. ((b) From Halliday and Resnick, *Physics*, Part II, third edition, Copyright © 1978, John Wiley and Sons, New York).

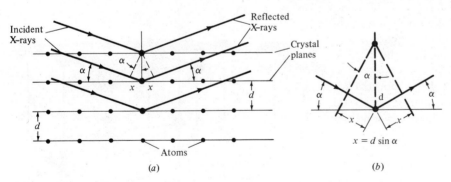

Figure 23.34. (a) X rays are reflected by parallel planes of atoms in a crystal. (b) An enlarged view of the geometric details.

beam is 0.04 nm, which wavelengths will be reflected strongly from the planes shown?

The wavelengths strongly reflected satisfy $m\lambda = 2d \sin \alpha$. Using $m = 1$ and $\sin 10° = 0.174$,

$$\lambda = \frac{2d \sin \alpha}{m} = 2 \frac{(0.2 \text{ nm})}{(1)} (0.174) = 0.0696 \text{ nm}$$

Using $m = 2$, we get half this wavelength, or 0.0348 nm, which is less than the 0.04-nm minimum wavelength of the beam. Hence only the 0.0696-nm X rays will interfere constructively, and the reflected beam will be monochromatic. We see then that reflection from crystals provides a way of making monochromatic X-ray sources.

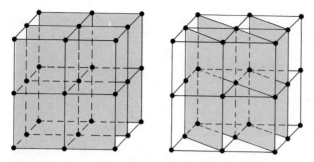

Figure 23.35. Two of the many sets of reflecting planes in a NaCl crystal.

The Laue pattern in Fig. 23.33 contains a large number of spots. This is because there are many sets of parallel reflecting planes in a crystal, with various orientations (Fig. 23.35). Each set of planes produces one or more spots in accordance with the Bragg condition.

The pattern of spots formed by a mixture of X-ray wavelengths cannot be used to determine the spacing of crystal planes, because there is no way to find the wavelength corresponding to a particular spot. Thus to obtain information about the crystal structure and the spacing of its planes, monochromatic X-ray beams are required. However, when monochromatic X rays are used, for most crystal orientations the Bragg condition is not satisfied for *any* set of planes and *no* spots are formed. To overcome this, the crystal is gradually rotated (Fig. 23.36). Spots are formed on the plate as sets of planes are brought to the required angles.

The locations and intensities of these spots contain remarkably detailed information about the crystal. Since a spot is formed whenever the Bragg condition

is satisfied for some set of planes, the spot *positions* depend only on the geometric structure of the crystal and the spacing of its planes. For example, sodium chloride (Fig. 23.37) and all other cubic crystals have the same characteristic X-ray diffraction pattern, with the overall scale of the pattern set by the atomic spacing. However, the *relative intensities* of the spots depend on the chemical composition of the crystal. This is because the X rays are actually reflected by the electronic clouds of the atoms, and heavy atoms, which have more electrons, are better reflectors. For example, in Fig. 23.37 different sets of planes contain different relative numbers of Na^+ and Cl^- ions, and this determines the intensity of the associated spots.

Although X-ray diffraction was originally used with relatively simple inorganic crystals, it has also been applied with spectacular success to biological molecules, such as proteins and nucleic acids, that can be put into crystalline form. For example, in the 1950s, Perutz compared the positions and intensities of thousands of spots for hemoglobin with those expected from various models of the molecule. He found that this oxygen-carrying protein of the blood consists of about 10,000 atoms assembled into four chains, each with a helical form and several bends. Combining the X-ray data with chemical determinations of the sequence of the amino acid building

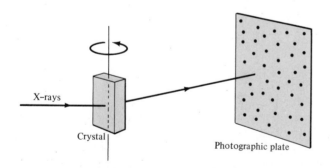

Figure 23.36. Monochromatic X rays are allowed to strike a crystal that is gradually rotated, forming a series of spots characteristic of the crystal structure.

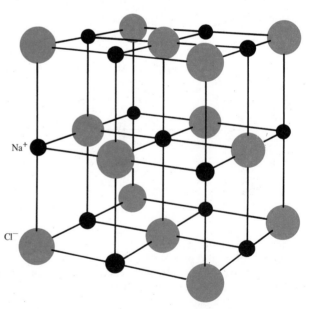

Figure 23.37. The crystal structure of sodium chloride. (From Halliday and Resnick, *Physics*, Part II, third edition, Copyright © 1978, John Wiley and Sons, New York.)

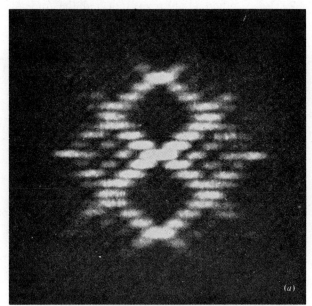

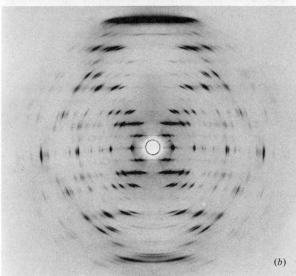

Figure 23.38. (*a*) An optical simulation of X ray diffraction from an array of helices. (*b*) The X ray diffraction pattern for a lithium salt of DNA. Helical structures always produce characteristic clear areas similar to those seen in both patterns. ((*a*) After A.R. Stokes, "The Theory of X-ray Fibre Diagrams," Prog. Biophysics vol. 5, 1955, p. 140. (*b*) Courtesy of Professor M.H.F. Wilkins, Biophysics Department of King's College, London.)

X-ray studies was the discovery by Crick and Watson in 1953 that the structure of the nucleic acid DNA is a double helix. The helical structure of DNA is apparent when its diffraction pattern is compared with that expected for an array of helices (Fig. 23.38).

SUMMARY

Electromagnetic waves of any frequency travel in a vacuum at a speed $c = 3.00 \times 10^8$ m s^{-1}. In a medium, the speed is $v = c/n$, where the index of refraction n is never less than 1.

The wave nature of light is demonstrated by many kinds of interference effects. For any wave, Huygens' principle states that each point on a wavefront can be considered as the source of small secondary wavelets. These wavelets can be used to predict many properties of light waves.

When light is reflected or refracted by objects large compared to its wavelength, the wave character of light plays a minor role and can usually be neglected. A light ray reflected from a smooth surface has its angle of reflection equal to the angle of incidence. When light enters a medium with a different index of refraction, it is bent or refracted; the angles of incidence and refraction are related by Snell's law,

$$n_1 \sin \phi_1 = n_2 \sin \phi_2$$

A ray bends toward the normal on entering an optically denser medium and away from the normal on entering a rarer medium. Total internal reflection will occur in the latter case if Snell's law gives $\sin \phi > 1$ for the refracted ray.

For a pair of narrow slits illuminated by a coherent beam of monochromatic light, Huygens' principle predicts a pattern of alternating dark and bright fringes on a distant screen. These appear at the angles of maximum destructive and constructive interference, respectively. *Maxima* appear at angles satisfying

$$d \sin \theta = m\lambda \qquad m = 0, \pm 1, \pm 2, \ldots$$

In a diffraction grating, which has many very closely spaced slits, nearly total destructive interference occurs, except in very narrow angular regions. The maxima are given by the same formula, but they are much sharper and further apart.

Secondary Huygens' wavelets from different parts of a single wavefront also interfere with each other. This diffraction property gives rise to a small but detectable bending of light around objects of ordinary

blocks in the hemoglobin molecule, he was able to construct a detailed picture of the location and spacing of the atoms. Perutz also showed how the shape of hemoglobin changes when it gains or loses oxygen.

Another major advance in biology based in part on

size and is a limiting factor in the resolution obtained with optical instruments. When light passes through a narrow slit, diffraction *minima* appear at

$$a \sin \theta = m\lambda \qquad m = \pm 1, \pm 2, \ldots$$

The maxima are approximately halfway between the minima.

When the electric field vector of an electromagnetic wave is always directed along some line, the wave is said to be linearly polarized in that direction. Light can be polarized by various means, including absorption, reflection, and scattering. A polarizer reduces the intensity of an unpolarized beam by a half. If it is followed by an analyzer with its axis rotated by an angle θ, then the intensity is reduced further by a factor of $\cos^2 \theta$ (Malus's law).

A crystal serves as a diffraction grating for X rays. If α is the angle between the beam and the crystal planes, maxima occur when the Bragg condition is satisfied,

$$2d \sin \alpha = m\lambda \qquad m = 1, 2, 3, \ldots$$

Checklist

Define or explain:

index of refraction	light pipe
nanometre	critical angle
Huygens' principle	coherence
wave front	diffraction grating
envelope	Rayleigh criterion
ray	resolution
secondary wavelets	line, continuous spectra
diffuse, specular	diffraction
reflection	Brewster's angle
Snell's law	scattering
angles of incidence,	polarizer
reflection, refraction	analyzer
white light	Malus's law
dispersion	Bragg condition
total internal reflection	Laue spots

REVIEW QUESTIONS

Q23-1 If the index of refraction is 2, the speed of light is _____ times the speed in the vacuum.

Q23-2 The wavelength of a beam of light _____ when it goes from air into water, and the frequency _____.

Q23-3 On a wave front, the waves at every point have _____.

Q23-4 When a light ray is reflected, the angle of reflection equals _____.

Q23-5 When light goes from an optically denser medium to an optically rarer medium, the rays are bent _____.

Q23-6 Total internal reflection can occur when light goes from a _____ dense to a _____ dense region. It occurs when the sine of the angle of refraction would be _____.

Q23-7 In a double-slit interference pattern, the bright fringes occur where the waves from the two slits are _____.

Q23-8 In order for interference effects to be observed, the light sources must be _____.

Q23-9 As more slits are added to a grating, the lines formed become _____.

Q23-10 The diffraction pattern formed by light passing through a small opening arises from interference among _____.

Q23-11 Light can be polarized by _____, _____, and _____.

Q23-12 The spacing of atoms in a crystal is comparable to the wavelength of _____.

EXERCISES

Section 23.1 | The Index of Refraction

23-1 Find the speed of light in water.

23-2 What is the speed of light in diamond?

23-3 A lamp emits yellow light with wavelength 600 nm in air. (a) What is the wavelength of the light in water? (b) What color will the lamp appear to be to a diver who is not wearing a mask? Explain.

23-4 A beam of light has a wavelength of 640 nm in glass with a refractive index of 1.5. What is the frequency of the light?

23-5 Light with a wavelength of 500 nm in air enters water. What is its wavelength in the water?

23-6 A lamp emits light at a frequency of 5×10^{14} Hz. (a) Find the wavelength in air. (b) If the light enters glass with an index of refraction equal to 1.5, find the frequency and wavelength.

Section 23.3 | Reflection of Light

23-7 What fraction of the light intensity is reflected when light is normally incident in air on water?

23-8 In which direction must light initially travel

to have a phase reversal at (a) an air-water suface; (b) a water-glass surface?

23-9 Light is incident in water along the normal to a glass plate with refractive index 1.5. What fraction of the light intensity is transmitted into the glass?

23-10 Light is normally incident in air on a glass lens with an index of refraction of 1.6. (a) What fraction of the light is reflected? (b) What fraction is transmitted? (c) If the light emerges from the second surface along the normal direction, what fraction of the original intensity leaves the lens on this side?

23-11 When light is normally incident in air on a surface, 6 percent of the intensity is reflected. What is the index of refraction of the material?

Section 23.4 | Refraction of Light

23-12 A ray of light is incident on a glass-water surface from the glass at an angle of 45°. Find the angle of refraction if the index of refraction of the glass is 1.5.

23-13 Light is incident in air on water at an angle of 15° to the normal. (a) At what angle is the reflected ray? (b) At what angle is the refracted ray?

23-14 Light is incident in water on an air-water boundary at an angle of 30° to the normal. What are the angles of reflection and refraction?

23-15 When light is incident in air at 30° to the normal of a surface of an unknown material, the angle of refraction is 25°. What is the refractive index of the material?

23-16 A student notes in his lab notebook that a light beam has an angle of incidence in air of 40° and an angle of refraction in a plastic slab of 50°. Is this reasonable? Explain.

23-17 The index of refraction of air depends on its density. How does this fact explain the shimmering or watery appearance of distant parts of black roads on hot, sunny days?

Section 23.5 | Total Internal Reflection

23-18 What is the critical angle for total internal reflection in diamond? (Assume the diamond is in air.)

23-19 A glass light pipe in air will totally internally reflect a light ray if its angle of incidence is at least 39°. What is the minimum angle for total internal reflection if the pipe is in water?

23-20 The critical angle for light going from glass to air is found to be 36°. What is the refractive index of the glass?

Section 23.6 | Young's Double-Slit Interference Experiment

23-21 Two narrow slits are illuminated with light of wavelength 500 nm. Adjacent maxima near the center of the interference pattern are separated by 1.5°. How far apart are the slits?

23-22 Two narrow slits 0.2 mm = 2×10^{-4} m apart are illuminated with red light of wavelength 700 nm. At what angles are the five maxima closest to the center of the pattern formed?

23-23 A double-slit apparatus is illuminated with yellow sodium light ($\lambda = 589$ nm). The maxima on a screen 1 m away are 1 cm apart. Find the slit separation.

23-24 Two slits separated by 10^{-4} m are illuminated with monochromatic light and form a pattern on a screen 2 m away. The fifth maximum, not counting the one at the center of the pattern, is 6 cm from the center of the screen. What is the wavelength of the light?

Section 23.7 | Coherence

23-25 A double-slit apparatus is illuminated with a narrow beam of white light from a small portion of a lamp. A filter that transmits only red light is placed just before one slit and a similar green filter before the other. Will a double-slit pattern be observed on the screen? Explain.

Section 23.8 | The Diffraction Grating

23-26 A diffraction grating with 5000 lines per centimetre is illuminated with yellow sodium light of wavelength 589 nm. (a) What is the angular position of the $m = 1$ line? (b) How many lines can be seen?

23-27 A grating has 4000 lines per centimetre. What is the longest wavelength for which the fourth-order line can be observed?

23-28 A source emits a red doublet near 656 nm with a separation of 0.2 nm. (a) At what angles will lines be seen if the diffraction grating has 8000 lines per centimetre? (b) If the two first-order lines are just resolved, how many lines does the grating have?

23-29 What is the minimum number of lines needed in a diffraction grating to resolve in second order the sodium doublet at 589.59 and 589.00 nm?

23-30 A diffraction grating with 8000 lines per centimetre is illuminated with light from a hydrogen lamp. In the first-order spectrum, what is the angular separation between the 656 and 410 nm lines emitted by atomic hydrogen in the lamp?

23-31 Explain why diffraction gratings have (a) closely spaced slits; (b) a large number of slits.

23-32 A grating has 8000 lines per centimetre, and it is 0.5 m from a screen. When it is illuminated with light of wavelength 550 nm, how far from the center of the screen will the first order line be located?

23-33 A grating has 6000 lines per centimetre. It is located 0.7 m from a screen, and the first order line appears 0.32 m from the center of the screen. What is the wavelength of the light?

Section 23.9 | Diffraction

23-34 A narrow slit is illuminated with white light. What will be seen on a screen beyond the slit?

23-35 A narrow slit is illuminated with yellow light of wavelength 589 nm. If the central diffraction maximum extends from 0° to 40°, how wide is the slit?

23-36 A slit of width 1000 nm is illuminated by light of wavelength 600 nm. At what angle is the first diffraction minimum?

23-37 A slit of width 1600 nm is 0.5 m from a screen. It is illuminated by light with a wavelength of 400 nm. What is the distance between the first minima on either side of the bright central maximum?

23-38 A circular aperture of radius 10^{-5} m is illuminated with light of wavelength 500 nm. At what angle is the first diffraction minimum?

23-39 A circular aperture of radius 800 nm is illuminated by light of wavelength 600 nm. The first diffraction minimum appears as a circle on a screen 0.3 m away. What is the radius of that circle?

Section 23.10 | Polarization of Light

23-40 Light is incident from air on glass with an index of refraction 1.5. At what angle will the reflected light be fully polarized?

23-41 When light goes from air into a plastic, the reflected beam is completely polarized if the angle of incidence is 60°. If the light goes from the plastic into the air, at what angle of incidence will the reflected beam be completely polarized?

23-42 A light beam is initially unpolarized and is passed through a Polaroid filter. If the beam emerging from the filter has an intensity of 10 W m^{-2}, what is the intensity of the incident beam?

23-43 A light beam is partially polarized so that it has an average amplitude E_0 along one axis and an average amplitude $2E_0$ along the axis at right angles. If a Polaroid filter is rotated in front of the beam, find (a) the minimum fraction of the intensity transmitted; (b) the maximum fraction transmitted.

23-44 Unpolarized light passes through three filters. The first has its axis vertical, the second at 30° to the vertical, and the third at 60° to the vertical, or at 30° to the second. (a) What fraction of the intensity is transmitted through the three filters? (b) Suppose the second and third filters are interchanged, without altering the orientation of their axes. What fraction of the intensity is transmitted now?

23-45 When unpolarized light passes through two successive Polaroid filters, its intensity is reduced by 90 percent. What is the angle between the transmission axes of the filters?

Section 23.11 | X-ray Diffraction and the Structure of Biological Molecules

23-46 It is found that X rays are reflected strongly from crystal planes 0.2 nm apart when they are incident at 20° from the planes. What is their longest possible wavelength?

23-47 A set of planes in a crystal has a spacing of 0.3 nm. If a beam of monochromatic X rays of wavelength 0.1 nm is incident on the crystal, what is the smallest angle relative to the planes at which constructive interference occurs?

23-48 X rays are reflected from crystal planes separated by 0.3 nm. They are incident at an angle of 12° to the planes. If the beam contains wavelengths as short as 0.2 nm, what wavelengths will be strongly reflected?

PROBLEMS

23-49 A monochromatic light ray enters the prism shown in Fig. 23.39 at an angle of 30° to the normal. The refractive index is 1.5. At what angle θ does the ray leave the prism?

23-50 A beam of light enters the top of a glass cube at an angle greater than zero to the normal direction, and then hits a side of the cube. If the

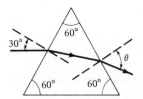

Figure 23.39. Problem 23-49.

refractive index of the cube is 1.5, can the beam emerge from the side into the air? Explain your answer with the aid of a diagram.

23-51 A small lamp is 10 m below the surface of a lake. It emits light in all directions. A boat goes from directly over the lamp to the point where it can no longer be seen. How far does it go?

23-52 A glass plate with refractive index 1.5 is inserted into a beam so that the angle of incidence is 30°. (a) What is the direction of the refracted ray inside the plate? (b) What is the direction of the ray emerging from the far side of the plate? (c) How far is the ray shifted at right angles to its original path by the plate if it is 0.02 m thick?

*23-53 An object 1 m under water is observed from the air. (a) What is the angle of refraction for a light ray from the object reaching the surface at an angle of 10° to the normal? (b) Where does the object appear to be? (*Hint:* Calculate where the refracted ray appears to intersect a ray normal to the surface as in Fig. 23.40. Q is the actual position and Q′ is the apparent position.)

*23-54 Snell's law of refraction can be derived from Huygens' principle by using a procedure similar to that used in Fig. 23.6 to derive the law of reflection. (a) Draw a series of diagrams showing how a wave front changes as it travels from a medium with light velocity v_1 into a medium with velocity v_2. (b) Using similar triangles derive Snell's law.

23-55 Obtain a formula for the angular positions of the minima in a double-slit apparatus.

23-56 One does not normally detect interference effects between the sound coming from two loudspeakers of a stereo system. Explain why.

23-57 Show how one could devise a simple experiment to demonstrate the interference between sound waves from two sources.

23-58 What will be observed in a double-slit experiment if white light is used? Explain.

*23-59 A coherent light beam is produced by placing a filter and a slit of adjustable width in front of an incandescent lamp. The light then falls on a double-slit apparatus and forms an interference pattern on a screen. Describe what happens as the source slit is gradually increased in width. (Ignore the effects of source-slit diffraction.)

23-60 A grating with 6000 lines per centimetre is illuminated with white light. (a) What is the highest order spectrum that includes the entire visible spectrum? (b) What is the highest order spectrum that includes any portion of the visible spectrum?

23-61 A grating with 4000 lines per centimetre forms a line at 45°. (a) What are the possible wavelengths of the light illuminating the grating? (b) How could one determine the actual wavelength?

23-62 When a small opaque circular disk is illuminated by a coherent plane wave, a bright spot is seen at the center of the shadow formed on a distant screen. Explain why.

*23-63 Suppose that light is incident on a diffraction grating at an angle ϕ to the normal direction (Fig. 23.41). (In the discussion in this chapter, ϕ was always chosen to be zero.) Show that the maxima occur at angles θ such that $d(\sin \phi + \sin \theta) = m\lambda$, where $m = 0, 1, 2, \ldots$, and d is the slit separation.

23-64 A slit of width 10^{-5} m is illuminated with

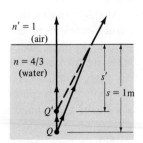

Figure 23.40. Problem 23-53.

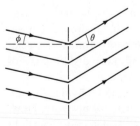

Figure 23.41. Problem 23-63.

light of wavelength 500 nm. How wide is the central diffraction peak on a screen 0.5 m away?

23-65 A lens when properly focused on a distant-point light source of wavelength 550 nm forms a circular spot 0.1 mm = 10^{-4} m in diameter on a screen 0.2 m from the lens. If the spot size is entirely due to diffraction by the lens, what is the diameter of the lens?

23-66 Why are high-frequency loudspeakers (tweeters) more directional than low-frequency speakers (woofers)? (*Hint:* Regard the speaker as a round hole through which the sound waves pass.)

23-67 Using Snell's law, show that when light is incident at Brewster's angle, the reflected and refracted beams are at right angles.

23-68 Suppose a sodium chloride crystal is ground into a powder, so that it becomes a large number of randomly oriented small crystals. If monochromatic X rays are incident on the powder, what kind of interference pattern will be formed? (*Hint:* Consider what will happen to the spots in Fig. 23.33 if the axis of rotation is changed.)

ANSWERS TO REVIEW QUESTIONS

Q23-1, 0.5; **Q23-2,** decreases, stays the same; **Q23-3,** the same phase; **Q23-4,** the angle of incidence; **Q23-5,** away from the normal; **Q23-6,** more, less, greater than 1; **Q23-7,** in phase; **Q23-8,** coherent; **Q23-9,** sharper; **Q23-10,** different parts of the wave; **Q23-11,** absorption, reflection, scattering; **Q23-12,** X rays.

SUPPLEMENTARY TOPICS

23.12 | HOLOGRAPHY

One of the most intriguing and versatile advances of recent years is *holography.* A *hologram* is a photograph of the interference pattern produced when monochromatic light reflected or transmitted by an object interferes with a coherent *reference* light beam. When a hologram is illuminated by a coherent light source, an image of the original object is formed. Unlike an ordinary two-dimensional photograph, this *reconstructed* image is three dimensional. Consequently, a moving viewer observes *parallax:* the relative displacement of nearer and more distant parts of the object (Fig. 23.42).

Holography was invented in 1947 by D. Gabor, who received the 1971 Nobel prize for his work. However, the lack of sufficiently strong coherent light sources limited the usefulness of holography until the advent of the laser in 1960. Since then, increasingly varied and ingenious applications of holography have been developed.

In order to understand how holography works, it is useful to first consider a circular arrangement of concentric transparent and opaque rings called a *zone plate* (Fig. 23.43) illuminated by coherent monochromatic light of wavelength λ. Going outward from the center of the plate, the average distance from a ring to a particular point P on the axis is one wavelength greater than from the preceding ring. Accordingly the Huygens' wavelets from all the transparent rings arrive approximately in phase and interfere constructively. The opaque rings block out the wavelets that would interfere destructively with those from the transparent rings.

Because of the opaque rings, about half of the incident light is blocked by the zone plate. Much of the remaining light travels straight ahead. However, the Huygens' wavelets from the transparent rings arrive at P approximately in phase, forming a bright spot due to the constructive interference (Fig. 23.44*a*). Wavelets spreading outward as though they had originated at point P', a distance d to the left of the zone plate, are also in phase. If they are eventually brought together by a lens, as in the eye or a camera, they interfere constructively. Hence, if one looks through a zone plate toward a light source, one sees a bright spot at P', even though no light actually originates there (Fig. 23.44*b*). This is called a *virtual image,* while the bright spot actually formed at P is called a *real image.*

The hologram produced by a single-point object is very similar to a zone plate. In the idealized arrangement of Fig. 23.45*a*, a laser beam is incident on a pointlike object, producing spherical Huygens' wavelets. At some places on the photographic plate, the spherical wave from the object and the undeflected reference wave are exactly in phase, and maximum constructive interference occurs. At other places the waves are out of phase, and destructive interference occurs. This produces a pattern that looks like a zone plate, although the abrupt changes from opaque to transparent rings are replaced by a more gradual variation in intensity. If the developed photographic plate is later illuminated by a laser beam, virtual and real images are formed of the point object just as they were for the zone plate (Fig. 23.45*b*).

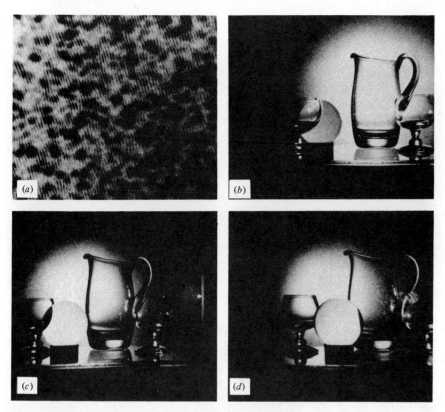

Figure 23.42. (*a*) A hologram. (*b*), (*c*), and (*d*) are photographs taken of the reconstructed images taken at slightly different angles. The parallax shows that holography reconstructs the images in three dimensions. (From Howard K. Smith, *Principles of Holography,* second edition, Copyright © 1975, John Wiley and Sons, New York.)

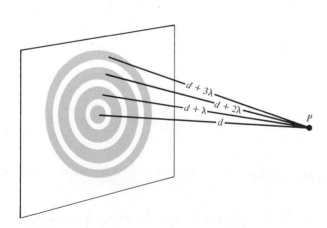

Figure 23.43. A zone plate has alternate opaque and transparent rings. All the wavelets reaching point P are in phase and interfere constructively.

When a hologram is made of a complex object, each point on the object produces spherical waves that interfere with the reference wave and form a series of zone platelike rings on the photographic plate. These rings are superimposed, so the resulting hologram is an incredibly complex pattern of lines and swirls. Nevertheless, it represents a coded diffraction pattern containing the information about all the points on the object necessary to reconstruct the wave fronts when the hologram is illuminated by a coherent source (Fig. 23.46).

Applications of Holography Holography makes it possible to overcome a basic limitation of the microscope. If the magnification is large, the range of depths over which an object is in focus at any microscope setting is very small. Biological specimens often are suspended in a fluid and tend to drift about.

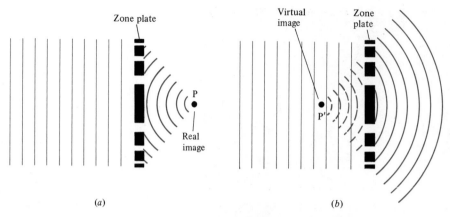

Figure 23.44. A coherent monochromatic wave incident on a zone plate has alternate zones blocked by the plate. Some transmitted light goes straight through, some forms a real image at *P* as shown in (*a*), and some forms a virtual image at *P'* as shown in (*b*).

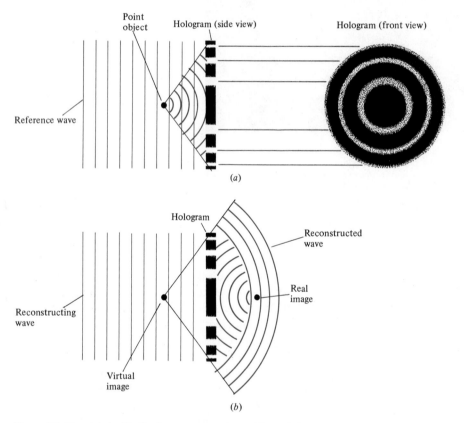

Figure 23.45. (*a*) An idealized arrangement for making a hologram of a point object. (*b*) The corresponding reconstruction of the virtual and real images.

Hence they move in and out of focus, unless the motion is restricted by putting the specimen into a solid form or by making it very thin. Either procedure may alter the specimen. However, if one makes a holographic snapshot with the microscope, this "freezes" the motion while preserving all the three-dimensional information. The reconstructed three-dimensional image can then be examined at leisure with a microscope, and the instrument can readily be adjusted to focus on successive layers. Similarly, if a sample

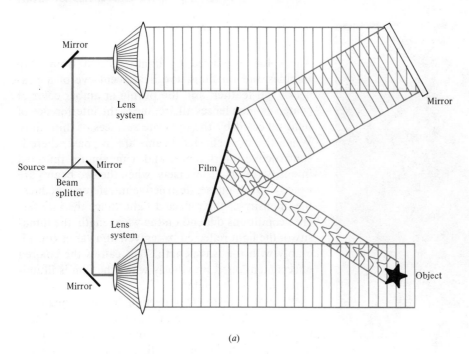

(a)

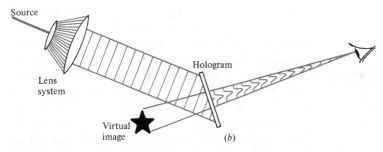

(b)

Figure 23.46. (a) Production of a hologram for a complex object. A single highly coherent laser beam is split into two by a partially silvered mirror. Lenses are used to broaden the beams without altering their coherence properties. One beam illuminates the object, while the other acts as a reference beam. When the reference beam and the reflected beam interfere at the photographic plate, a hologram is formed. (b) The reconstruction stage. Only the virtual image is shown, although a real image is also formed.

changes gradually, a series of holograms can be made that, when reconstructed, allows one to study the sample both at various depths and times.

Holography can be used in various ways to detect minute changes that have occurred in an object. Suppose an object and its hologram made at an earlier time are returned to precisely the positions they had when the hologram was originally made. If the same laser again is used to illuminate them, then the wave from the object and the reconstructed wave from the hologram will interfere. Any changes that have occurred in the appearance of the object will produce

light and dark interference fringes, which directly pinpoint changes not readily found by visual inspection. Differences between two carefully machined parts can be determined in a similar way.

A closely related procedure is to make two successive holograms on a single photographic plate by a double exposure. When the image is reconstructed interference effects will occur if the object has changed in any way. A remarkable photograph made from a double exposed hologram is shown in Fig. 23.47.

So far we have discussed cases in which the original

Figure 23.47. Two holograms were made on a single photographic plate by illuminating a subject with two short laser pulses 1.5×10^{-4} s apart. This photograph shows the interference fringes produced when the double hologram was illuminated and the wave fronts were reconstructed. Destructive interference occurs where the subject's chest has moved far enough so that the two reconstructed waves are out of phase. The fringes are most closely spaced in regions where the motion is most rapid. This man has had his left lung removed, and is inhaling. The distortion in his muscular action is readily apparent. (Courtesy of The Athletic Institute, from an article by S.M. Zivi in *Biomechanics,* 1971.)

23.13 | INTERFERENCE EFFECTS IN THIN FILMS

The colors produced by light reflected from soap bubbles and oil films, the irridescent eye of a peacock's tail feather, and the purple or amber color of coated camera lenses all are due to the interference of light reflected by the opposite surfaces of thin films. When the two reflected beams are in phase, there is constructive interference, and a maximum reflected intensity results. Conversely, when the two beams are exactly out of phase, destructive interference reduces the intensity of the reflected light. Since the interference conditions depend on the wavelength, the intensity of the light reflected by a given film varies considerably with the wavelength. This causes the colored effects mentioned previously when the film is illuminated with white light.

We now investigate reflection by films quantitatively. Fig. 23.48 shows monochromatic light incident in air on a thin film of refractive index n and thickness d. At normal incidence, light reflected by the second surface travels a distance $2d$ further than light reflected by the first surface. If the wavelength in vacuum or air is λ, then the wavelength in the film is $\lambda' = \lambda/n$. The light reflected at the first surface is reversed in phase. However, the light reflected at the second surface is not, since the phase reversal occurs only when the light is incident from the less dense medium. Thus, if $2d$ is exactly one wavelength, the waves are exactly out of phase and interfere destructively. In general, destructive interference occurs if $2d$

light source and the reconstructing beam have the same wavelength. However, this is not necessary. If a shorter wavelength is used in the reconstruction, this will have the effect of increasing the apparent depth of the object. Also, it is not necessary that the original and reconstructing waves have the same physical nature. Using special photographic techniques, holograms have been made from the interference between reference and reflected or transmitted beams of ultrasonic waves. Holograms are also made with X rays. In both cases visible light is used for reconstructing the image. The ability of ultrasound and X rays to penetrate where light may not is thereby combined with the great flexibility of image manipulation afforded by lenses for visible light.

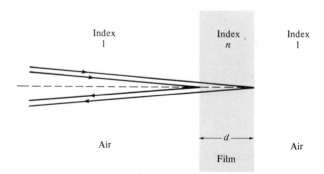

Figure 23.48. When light is incident on a thin film, some of the light is reflected at each surface, although most of it is transmitted (not shown). At normal incidence the light reflected by the second surface travels a distance $2d$ further than the light reflected by the first surface.

is any integer multiple of λ', or if $2d = m\lambda' = m\lambda/n$. Thus the reflected intensity is a minimum if

$$2nd = m\lambda, \qquad m = 0, 1, 2, \ldots \quad (23.16)$$

Similarly, the constructive interference and the reflected intensity are a maximum when

$$2nd = (m + \tfrac{1}{2})\lambda, \qquad m = 0, 1, 2, \ldots \quad (23.17)$$

These two equations also hold true if the film is less dense than the media on either side. However, the conditions for the maxima and minima are reversed when the film is intermediate in density between the media on either side, since then either both reflected beams or neither is reversed in sign. The following example illustrates the reflecting properties of a typical film.

Example 23.15

A film of soap solution ($n = 1.33$) is just thick enough to cause the maximum reflection of red light of wavelength 700 nm at normal incidence. (a) How thick is the film? (b) The film is illuminated with white light at close to normal incidence. If an observer sees the reflected light, what color does the film appear to be?

(a) Using $m = 0$ in Eq. 23.17 for the maximum reflected intensity,

$$d = \frac{(m + \tfrac{1}{2})\lambda}{2n} = \frac{(\tfrac{1}{2})(700 \text{ nm})}{2(1.33)} = 132 \text{ nm}$$

(b) The distance across the film and back is a half wavelength at 700 nm, so it is a full wavelength at 350 nm. This is the longest wavelength at which the reflected intensity is a minimum, and it is just beyond the limit of visible light, 400 nm. This means that partial destructive interference also occurs at wavelengths in the visible near 400 nm, so relatively little light is reflected at the shorter wavelengths. Conversely, the reflection at wavelengths close to the long wavelength or red end of the spectrum will be enhanced. Thus if the observer sees light reflected at or near normal incidence, the film will appear red. However, if the film is part of a curved surface or a bubble, the angle of incidence and path length will vary over the film, and various colors will be seen in different parts of the film.

Nonreflective Coatings | We saw in Section 23.3

that about 4 percent of the light is reflected at each of the several air-glass surfaces in an optical instrument such as a camera or a microscope. Hence the reduction of these intensity losses made possible by nonreflective coatings is of considerable practical value.

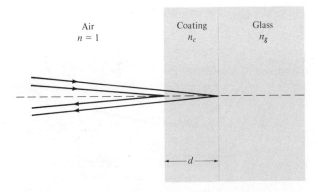

Figure 23.49. A nonreflecting coating on glass. The refractive index of the coating is less than that of the glass.

A nonreflective coating is illustrated in Fig. 23.49. A thin film of a material with index n_c is evaporated onto glass with index n_g. Since n_c is chosen to be less than n_g, the waves reflected from *both* sides of the film are reversed in sign. Hence, maximum destructive interference and minimal reflection occurs when the path lengths differ by a half wavelength. For normal incidence this happens if the film thickness d is a quarter wavelength. For light with wavelength λ in air, the wavelength in the coating is λ/n_c, and $d = \lambda/4n_c$. This result is illustrated by the following example.

Example 23.16

A camera lens has a coating of magnesium fluoride, which has a refractive index $n_c = 1.38$. If the reflected intensity at normal incidence of blue light of wavelength 500 nm is a minimum, how thick is the coating?

From the previous discussion, the thickness of the coating is

$$d = \frac{\lambda}{4n_c} = \frac{500 \text{ nm}}{4(1.38)} = 90.6 \text{ nm}$$

Total cancellation or zero reflected light intensity can occur only if the two reflected beams have the same intensity. It follows from Eq. 23.3 that the intensities are equal if $n_c = \sqrt{n_g}$. For glass with index 1.5, this means that the coating should have an index of $\sqrt{1.5} = 1.22$. However, suitable materials with the necessary hardness are not available with this index, and magnesium fluoride, which has an index equal to 1.38, is used as a compromise.

A single coating on a lens reduces the reflected light intensity to varying degrees throughout the visible range. Camera lenses designed for color film are

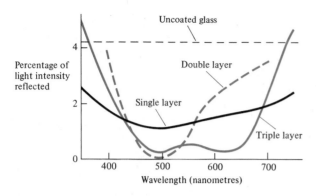

Figure 23.50. Percentage of light intensity reflected by glass with one, two, or three coating layers.

least reflective in the blue, and they appear amber because the reflection is greater at longer wavelengths. Lenses designed primarily for black and white film are least reflective for green light. They reflect more red and violet light, which gives them a purple appearance. In recent years, techniques have been developed to place two or even three layers on lenses. Although difficult and expensive, this process further reduces the reflective intensity (Fig. 23.50).

EXERCISES ON SUPPLEMENTARY TOPICS

Section 23.12 | Holography

23-69 An object is located so that the light diffracted from any one point on the object interferes with the reference beam at every point on a photographic plate. (a) If the resulting hologram is cut in half and one half is illuminated in the usual way with a laser beam, what portion of the original object will be seen? (b) Compare what happens here to the case where a diffraction grating is cut in half, parallel to the lines.

Section 23.13 | Interference Effects in Thin Films

23-70 A lens with refractive index 1.6 is coated with magnesium fluoride with index 1.38. (a) If it is designed to be nonreflecting for green light of wavelength 550 nm, how thick is the film? (b) If this lens is used in water, will it be nonreflecting for green light? Explain.

23-71 A soap film with refractive index 1.33 is just thick enough to produce destructive interference of the shortest wavelength visible light

(400 nm) at normal incidence. How thick is the film?

23-72 Room acoustics can be modified by placing slots in the walls. (a) How deep should the slots be to minimize sound reflections at 500 Hz? (Use 344 m s^{-1} as the speed of sound.) (b) What effect will such slots have at 1000 Hz?

PROBLEMS ON SUPPLEMENTARY TOPICS

23-73 The zone plate in Fig. 23.43 is constructed so that the center of the mth dark ring is at a distance $d_m = d + m\lambda$ from point P. (a) Show that the distance r_m from the center of the plate to the center of the mth ring satisfies $r_m{}^2 = d_m{}^2 - d^2$. (b) Show that when d is large compared to λ, $r_m{}^2 = 2md\lambda$ is a good approximation.

*23-74 If a hologram is made with ultrasonic waves and reconstructed with visible light, the depth of the image is highly distorted. (a) Use the result of the preceding problem to explain this observation. (b) Make a numerical estimate of the effect.

*23-75 A zone plate is constructed to produce a bright spot 0.1 m from the plate when illuminated with red light of wavelength 700 nm. Using the result of Problem 23-73, determine what will happen when the plate is illuminated with yellow light of wavelength 600 nm.

*23-76 Using the result of Problem 23-73, describe what will happen when a zone plate is illuminated with white light.

23-77 Explain why a soap film appears black when illuminated with white light if it is much thinner than the average wavelength of visible light.

*23-78 Show that if a lens with refractive index n_g is coated with a film of index n_c, the intensity ratio I_r/I_0 is the same for the light reflected from the two sides of the film when $n_c{}^2 = n_g$. (Assume that the lens is used in a vacuum and that the light is incident normally.)

23-79 Explain why the reflections from thin oil layers are brightly colored but those from thick ones are not.

*23-80 Two glass plates are placed in contact, and a sheet of paper is inserted at one end so that an angle of 0.0005 rad is made by the plates. If the plates are illuminated normally from above with light of wavelength 600 nm, what is the spacing of the observed bright and dark fringes?

Additional Reading

J.B. Cohen, The First Explanation of Interference, *American Journal of Physics,* vol. 8, 1940, p. 99.

R.W. Pohl, Discovery of Interference by Thomas Young, *American Journal of Physics,* vol. 28, 1960, p. 530.

R.B. Setlow and E.C. Pollard, *Molecular Biophysics,* Addison-Wesley Publishing Co., Reading, Mass., 1962, chapter 5. X-ray diffraction.

Eugene Ackerman, *Biophysical Science,* Prentice-Hall, Inc., Englewood Cliffs, N.J., 1962, chapter 15. X-ray diffraction.

J.F. Mulligan and D.F. McDonald, Recent Determinations of the Speed of Light, *American Journal of Physics,* vol. 20, 1952, p. 165; vol. 25, 1957, p. 180.

Winston E. Kock, *Radar, Sonar, and Holography,* Academic Press, New York, 1973.

B.G. Ponseggi and B.J. Thompson (eds.), Holography, *Proceedings of the Society of Photo-Optical Instrumentation Engineers,* vol. 15, 1968.

Winston E. Kock, Sound Visualization and Holography, *The Physics Teacher,* January 1975, p. 14.

A.G. Porter and S. George, An Elementary Introduction to Holography, *American Journal of Physics,* vol. 43, 1975, p. 954.

R.A.R. Tricker, *Introduction to Meteorological Optics,* American Elsevier Publishing Co., New York, 1970.

Lightwave Communications, a special issue of *Physics Today,* vol. 29, number 5, May 1976.

The Light Fantastic, *Science 84,* May, p. 26. Photo essay on interference effects in thin films in nature.

James H. Underwood and David T. Attwood, The Renaissance of X-Ray Optics, *Physics Today,* April 1984, p. 44.

Phillip H. Abelson, Glass Fiber Communication, *Science,* vol. 220, p. 463.

Scientific American articles:

J.H. Rush, The Speed of Light, August 1955, p. 62.

The entire issue of September 1968 is devoted to the subject of light.

J.A. Giordmaine, The Interaction of Light with Light, April 1964, p. 38.

A.G. Ingalls, Ruling Engines, June 1952, p. 45.

Karl H. Drexhage, Monomolecular Layers and Light, March 1970, p. 108.

Phillip Baumeister and Gerald Pincus, Optical Interference Coatings, December 1970, p. 59.

Eric Deuton, Reflectors in Fish, January 1971, p. 64.

H. Moyses Nussensvieg, The Theory of the Rainbow, April 1977, p. 116.

Richard J. Wurtman, The Effects of Light on the Human Body, July 1975, p. 69.

N.S. Kapany, Fiber Optics, November 1960, p. 72.

J.S. Cook, Communication by Optical Fibers, November 1978, p. 28.

Henri Birsigonies, Communication Channels, September 1972, p. 99.

W.S. Boyle, Light-Wave Communications, August 1977, p. 40.

Arthur Ashkin, The Pressure of Laser Light, February 1972, p. 62.

Victor Vali, Measuring Earth Stresses by Laser, December 1969, p. 88.

Stewart E. Miller, Communication of Laser, January 1966, p. 19.

Glenn L. Berge and George A. Seilestad, The Magnetic Field of the Galaxy, June 1965, p. 46.

Rüdiger Wehner, Polarized Light Navigation by Insects, July 1976, p. 106.

Don R. Sullenger and C.H. Keunard, Boron Crystals, July 1966, p. 96.

M.F. Perutz, The Hemoglobin Molecule, November 1964, p. 64.

J.C. Kendrew, The Three-Dimensional Structure of a Protein Molecule, December 1961, p. 96.

D.C. Phillips, The Three-Dimensional Structure of an Enzyme Molecule, November 1966, p. 78.

Sir Laurence Bragg, X-ray Crystallography, July 1968, p. 58.

Edward A. Stern, The Analysis of Materials by X-Ray Absorption, April 1976, p. 96.

Emmett N. Leith and Juris Upatnieks, Photography by Laser, June 1965, p. 24.

S. Henman, How to Make Holograms, The Amateur Scientist, February 1967, p. 122.

Keith S. Pennington, Advances in Holography, February 1968, p. 40.

Alexander F. Metherell, Acoustical Holography, October 1969, p. 36.

Emmett N. Leith, White Light Holograms, October 1976, p. 80.

David K. Lynch, Atmospheric Halos, April 1978, p. 144.

A.D. Moore, Henry Rowland, February 1982, p. 159. The scientist who developed techniques for ruling high-precision diffraction gratings.

Anthony C. S. Readhead, Radio Astronomy By Very-Long-Baseline Interferometry, June 1982, p. 52.

Jearl Walker, The Bright Colors in a Soap Film Are a Lesson in Wave Interference, The Amateur Scientist, August 1978, p. 232.

Jearl Walker, Studying Polarized Light with Quarter-Wave and Half-Wave Plates of One's Own Making, The Amateur Scientist, December 1977, p. 172; More about Polarizers and How to Use Them, Particularly for Studying Polarized Sky Light, January 1978, p. 132; The Physics of the Patterns of Frost on a Window, Plus an Easy-to-Read Sundial, December 1980, p. 230.

Jearl Walker, Mysteries of Rainbows, Notably Their Rare Supernumerary Arcs, The Amateur Scientist, June 1980, p. 174.

Jearl Walker, Dazzling Laser Displays That Shed Light on Light, The Amateur Scientist, August 1980, p. 158; More About Edifying Visual Spectacles Produced by Laser, January 1981, p. 164.

Jearl Walker, The "Speckle" on a Surface Hit by Laser Light Can Be Seen With Other Kinds of Illumination, The Amateur Scientist, February 1982, p. 82.

Richard E. Dickerson, The DNA Helix and How It Is Read, December 1983, p. 94. X-ray analysis.

Jearl Walker, Simple Optical Experiments in Which Spatial Filtering Removes the "Noise" From Pictures, *The Amateur Scientist*, November 1982, p. 194.

Jearl Walker, What Causes the Color in Plastic Objects Stressed Between Two Polarizing Filters?, *The Amateur Scientist*, June 1983, p. 146.

Jearl Walker, In Which a Lifesaver Lights Up in the Mouth and Light Takes Funny Bounces Through a Lens, *The Amateur Scientist*, July 1982, p. 146.

Jearl Walker, What Is a Fish's View of a Fisherman and the Fly He Has Cast on the Water?, *The Amateur Scientist,* March 1984, p. 138.

Dina F. Mandoli and Winslow R. Briggs, Fiber Optics in Plants, August 1984, p. 90.

CHAPTER 24

MIRRORS, LENSES, AND OPTICAL INSTRUMENTS

Cameras, microscopes, telescopes, and the human eye are examples of optical instruments that employ lenses and, in some cases, mirrors. Ordinarily lenses and mirrors are large compared to the wavelengths of visible light, so their principal effects on beams of light can be discussed without reference to interference or diffraction phenomena. However, as expected from the preceding chapter, these wave phenomena do play a role in limiting the resolution and sharpness of the images formed by optical instruments.

The properties of mirrors and lenses are discussed in the first part of this chapter. The later sections cover applications to specific optical instruments.

24.1 | MIRRORS

When we stand a metre away from a plane mirror and look into it, we see someone apparently standing a metre behind the mirror who looks much like us but whose hair is parted on the wrong side. The light reaching our eyes seems to originate at a place behind the mirror called the *image*. The image is *virtual* rather than *real,* since the light never actually passes through that location. It is called an *erect* image, since we do not seem to be standing upside down, or *inverted.*

It is useful to infer these observed properties of the image formed by a mirror from the equality of the angles of incidence and reflection for a light ray, since similar reasoning can be applied to the more complex problem of image formation by a lens. Figure 24.1a shows two of the many light rays produced by a point source of light or a point *object* at O, a distance d from the mirror. The ray incident along the normal to the mirror is reflected directly back along the normal, so it appears to have originated behind the mirror on

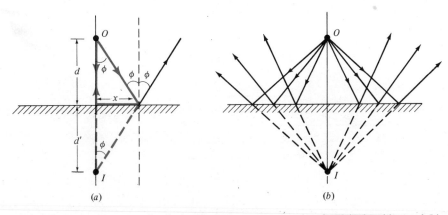

Figure 24.1. (a) The two colored triangles are congruent, so $d = d'$. (O is the object and I is its image.) (b) All the light reflected by the mirror appears to originate at the image I.

Figure 24.2. In a plane mirror, the image of each point on an object is directly in front of it. (K. Bendo.)

sometimes said to be reversed side to side, but not the normal. The light ray that strikes the mirror at an angle ϕ to the normal is reflected at an equal angle and appears to come from somewhere behind the mirror along the dashed projection of the outgoing ray. The two projected paths cross behind the mirror at point I, the image position. In Fig. 24.1a, the two colored right triangles have a common side x and an equal angle ϕ. Hence the triangles are congruent, and all their angles and sides are equal. *This means that the object distance d and the image distance d' are the same.*

When we have a complex illuminated object instead of a point, the image of each point on the object is directly in front of it (Fig. 24.2). This image is vertically; that is, the image is erect rather than inverted. This remark is not quite correct, as can be seen if we lie on our sides in front of a mirror. Then we perceive our images as reversed vertically but not horizontally!

Spherical mirrors are discussed in Secion 24.8 in the Supplementary Topics.

24.2 | LENSES

A lens is a piece of transparent material that can focus a transmitted beam of light so that an image is formed. The lenses in man-made optical instruments are usually manufactured from glass or plastic, while the lens in the human eye is formed by a transparent membrane filled with a clear fluid. For our purposes, it is sufficient to consider *thin, spherical lenses.* These have two spherical surfaces or a spherical and a plane surface and a thickness that is small compared to the radii of the surfaces.

We can categorize all lenses as either *converging* or *diverging.* A converging lens is thicker at its center than at the edge, while the opposite is true for a diverging lens (Fig. 24.3). A converging lens bends light rays toward its *axis,* the line through its centers of curvature, so that a beam of parallel rays converges at a point (Fig. 24.4). For example, in bright sunlight, a converging lens may produce a spot of light intense enough to ignite paper. A diverging lens bends rays outward from its axis.

(a)   (b)

Figure 24.3. (*a*) Converging lenses. (*b*) Diverging lenses.

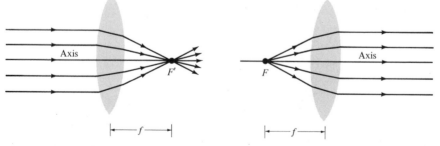

Figure 24.4. Focal points of a converging lens. The focal length f is positive for a converging lens.

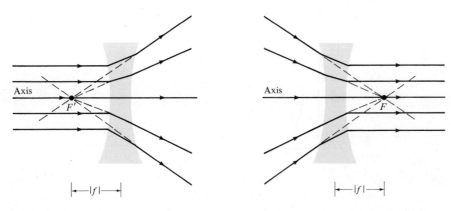

Figure 24.5. Focal points of a diverging lens. The focal length *f* is negative for diverging lenses.

Suppose a very distant object is on the axis of a lens, so that the light rays are nearly parallel to the axis when they reach the lens. A converging lens will refract these parallel rays so that they meet or form an *image* at a *focal point F'* beyond the lens, while a diverging lens will bend the rays outward so that they *appear to have come from* a focal point *F'* before the lens (Figs. 24.4 and 24.5). The distance from the center of the lens to the focal point is called the *focal length f.* Conventionally *f* is taken to be positive for converging lenses and negative for diverging lenses.

Lenses also have a second focal point. If an object is placed at the point *F*, which is a distance *f* in front of a converging lens, then light rays from the object will emerge parallel to the axis (Fig. 24.4). Similarly, light rays aimed at the point *F* a distance |*f*| beyond a diverging lens emerge parallel to the axis (Fig. 24.5).

The effect of a lens is determined by its focal length; a lens with a short focal length is stronger and bends light rays more than one with a long focal length. The focal length depends on the index of refraction of the lens and on its shape, that is, on the *radii of curvature* of its surfaces. Lens surfaces may be *convex, concave,* or *plane.* (A convex surface bulges outward, like the outside of a spoon; a concave surface bulges inward, like the inside of a spoon.) We use the following conventions in characterizing lens surfaces:

1 A convex surface has a positive radius of curvature.

2 A concave surface has a negative radius of curvature.

3 A plane surface has an infinite radius of curvature.

These conventions are illustrated by Fig. 24.6.

The focal length of a lens is related to its index of refraction *n* and the radii of curvature R_1 and R_2 of its surfaces by a formula that can be derived using Snell's law and the approximation that the angles of incidence are small. Since the derivation is lengthy, we only present the result here. The focal length of a lens with index *n* in a medium of index 1 is

$$\frac{1}{f} = (n - 1)\left(\frac{1}{R_1} + \frac{1}{R_2}\right) \qquad (24.1)$$

This is called the *lensmaker's equation.* Its use is illustrated by the following example.

Example 24.1

Lenses similar to those in Fig. 24.6a and 24.6c are made from glass with a refractive index of 1.5. Find the focal length if there are (a) two convex surfaces with radii of curvature 0.1 m and 0.2 m; (b) one plane surface and one concave surface of radius 4 m.

(a) According to our conventions, convex surfaces have positive radii of curvature, so $R_1 = 0.1$ m and $R_2 = 0.2$ m. Thus, the lensmaker's equation gives

$$\frac{1}{f} = (n - 1)\left(\frac{1}{R_1} + \frac{1}{R_2}\right)$$
$$= (1.5 - 1)\left(\frac{1}{0.1 \text{ m}} + \frac{1}{0.2 \text{ m}}\right)$$
$$= (0.5)(10 + 5)\text{m}^{-1} = 7.5 \text{ m}^{-1}$$

or *f* = 0.133 m.

(b) A plane surface has an infinite radius of curvature,

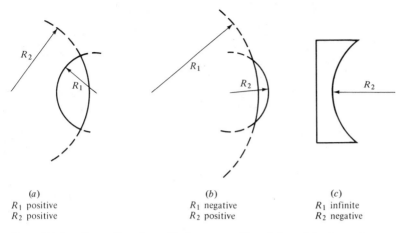

(a)
R_1 positive
R_2 positive

(b)
R_1 negative
R_2 positive

(c)
R_1 infinite
R_2 negative

Figure 24.6. Conventions for radii of curvature. From left to right, the surfaces are (a) convex, convex; (b) concave, convex; (c) plane, concave.

so $1/R_1 = 1/\infty = 0$. A concave surface has a negative radius of curvature. Thus, with $R_2 = -4$ m,

$$\frac{1}{f} = (1.5 - 1)\left[0 + \frac{1}{(-4 \text{ m})}\right]$$
$$= -0.125 \text{ m}^{-1}$$

and $f = -8$ m. The minus sign indicates that this is a diverging lens.

The lensmaker's equation in the form given in Eq. 24.1 assumes that the lens is in a medium with index of refraction one. If the lens is not used in air or vacuum, then the symbol n in the equation must be interpreted as the *relative* index of refraction, $n(\text{lens})/n(\text{medium})$. Lenses have longer focal lengths in water than in air, as seen in this example.

Example 24.2

What is the focal length of the lens in Example 24.1a if it is placed in water?

Since the index of refraction of water is 1.333, we should use $n = 1.5/1.333 = 1.125$ in the lensmaker's equation. Then with $R_1 = 0.1$ m and $R_2 = 0.2$ m, we find

$$\frac{1}{f} = (n - 1)\left(\frac{1}{R_1} + \frac{1}{R_2}\right)$$
$$= (1.125 - 1)\left[\frac{1}{0.1 \text{ m}} + \frac{1}{0.2 \text{ m}}\right]$$

and $f = 0.533$ m. This is four times the focal length of the lens in air as determined in the preceding example.

The increase in the focal length of a lens that occurs when it is placed in water explains why we see so poorly under water. Our eyes contain fluids whose indices of refraction are close to that of water. Light bends appreciably when it enters the eye through the curved transparent *cornea,* but it bends very little when it enters from water. Consequently the eye forms a badly focused image when it is in contact with water. When a person wears goggles or a face mask, light passes through the glass into an air layer at or near normal incidence without appreciable deflection. The light is then refracted in the usual fashion as it enters the eye from the air, so that vision is improved.

24.3 | IMAGE FORMATION

We saw earlier that light rays from a very distant point on a lens axis arrive parallel to the axis and meet to form an image at the focal point. Rays from other points form images whose locations can be found graphically or algebraically if the focal length of the lens is known.

Images may be *real* or *virtual.* A real image is one that is actually formed; if a screen is placed at that location, the image appears on the screen (Fig. 24.7). When there is a virtual image, the outgoing light seems to be coming from a place before the lens. However, if a screen is put there, no image is seen (Fig. 24.8a). Most objects are real; the light rays actually diverge from points at the object location. Virtual objects sometimes occur in multilens systems when the converging rays from one lens pass through a second lens (Fig. 24.8b).

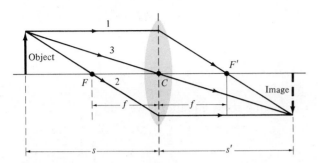

Figure 24.7. Three rays are used to locate an image graphically.

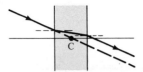

Figure 24.9. A ray passing through a flat plate is displaced but not deflected. Accordingly, a ray passing through or near the center of a thin lens is displaced very slightly.

We use the following conventions in the graphical or *ray-tracing* approach to locating an image:

1 Light always goes from left to right.
2 Real objects are to the left of the lens, and real images to the right.
3 Virtual images are to the left of the lens, and virtual objects to the right.

Three of the many rays emanating from a point on an object not on the axis have readily predicted paths. Their intersection determines the location of the image. Two rays are actually sufficient to find the image location; the third serves as a check. This ray-tracing procedure is illustrated in Fig. 24.7, where an illuminated arrow serves as a real object located at an *object distance s* from the lens. The lens forms a real image at an *image distance s′*. The three numbered rays in the diagram are drawn from the arrowhead as follows:

1 The ray leaving the arrowhead parallel to the axis is deflected by the lens so that is passes through the focal point F′, in accordance with the definition of the focal point.

2 The ray going through the focal point F emerges from the lens parallel to the axis.

3 The ray directed at the center of the lens is undeflected. This happens because at that point the two sides of the lens are almost parallel, so a ray is effectively going through a flat plate (Fig. 24.9). Since the lens is thin, the ray is displaced from its original path by an amount that is negligibly small.

Points below the top of the arrow at the same object distance s will have images at the same image distance s′. Thus, once we have located the image of the top, we can sketch in the entire image of the arrow.

Ray tracing is illustrated again in Fig. 24.8 for situations involving virtual images and virtual objects. In each case, the rays through F′, F, and the center are followed and their intersection located. We use ray tracing in several applications later in this chapter.

Although ray tracing provides good qualitative insight into the formation of an image by a lens or system of lenses, numerical work is best done with algebraic formulas. In order to develop and apply these formulas, we adopt the following sign conventions for the quantities shown in Fig. 24.10:

1 s is positive for a real object, negative for a virtual object.

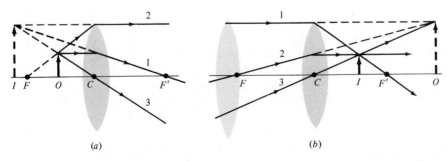

(a)   (b)

Figure 24.8. (a) Real object, virtual image. (b) Virtual object, real image. The virtual object is formed by the lens shown in color.

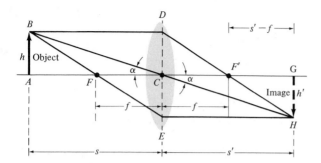

Figure 24.10. The object distance s, the image distance s', and the focal length f are related by the thin lens formula.

2 s' is positive for a real image, negative for a virtual image.

3 The *object height h* is positive if it points above the axis and negative if it points below the axis.

4 The *image height h'* is positive if it points above the axis and negative if it points below the axis.

The *linear magnification m* is the ratio of the image and object heights, h'/h. The linear magnification is negative when the image is inverted as in Fig. 24.7; it is positive when the image is erect, as in Fig. 24.8a.

We can derive formulas relating s, s', f, and m using similar triangles. Two triangles are similar, and their corresponding sides are therefore proportional, if two of their angles are equal. In Fig. 24.10, the triangles ACB and GCH each have a right angle and the same angle α, and so they are similar. Thus

$$\left| \frac{h'}{h} \right| = \left| \frac{s'}{s} \right|$$

With our sign conventions, h' is negative in Fig. 24.10, so the *linear magnification m* can be written as

$$m = \frac{h'}{h} = -\frac{s'}{s} \qquad (24.2)$$

Also, triangles CDF' and GHF' are similar, so in the same way

$$m = \frac{h'}{h} = -\left(\frac{s' - f}{f} \right) \qquad (24.3)$$

Comparing these equations, we find

$$\frac{s'}{s} = \frac{s' - f}{f}$$

Dividing by s' and rearranging, we find the *thin lens formula*,

$$\frac{1}{s} + \frac{1}{s'} = \frac{1}{f} \qquad (24.4)$$

We can check to see that this equation agrees with our earlier discussion. For a very distant object, s is infinite and $1/s$ is zero, so the thin lens formula gives $1/s' = 1/f$, or $s' = f$, as expected from our definition of the focal point. Similarly, for a very distant image, s' is infinite and $s = f$. Note that for a given focal length, the image distance s' depends only on the object distance s and not on the height h of the object. *This means that all object points at a distance s from the lens have their images on the same plane.* The following examples further illustrate the thin lens formula.

Example 24.3

A lens has a focal length of $+0.1$ m. Find the image distance when the object distance is (a) 0.5 m; (b) 0.08 m.

(a) Using the thin lens formula,

$$\frac{1}{s'} = \frac{1}{f} - \frac{1}{s} = \frac{1}{0.1 \text{ m}} - \frac{1}{0.5 \text{ m}} = 8 \text{ m}^{-1}$$

and $s' = 0.125$ m. Note that s' is positive, corresponding to a real image (Fig. 24.11a).

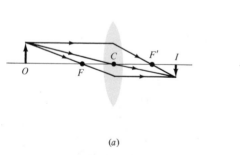

(a)

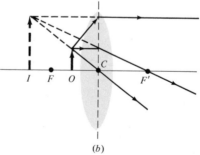

(b)

Figure 24.11. (a) When the distance from a real object to a converging lens is greater than the focal length, a real image is formed. (b) When the object is between the focal point and the converging lens, the image is virtual.

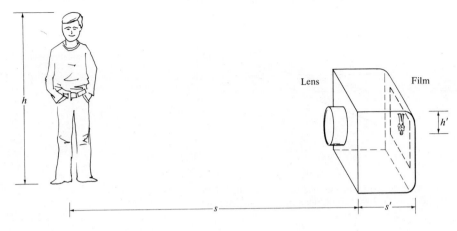

Figure 24.12.   Principle of the camera.

(b) Again using the thin lens formula,

$$\frac{1}{s'} = \frac{1}{f} - \frac{1}{s} = \frac{1}{0.1 \text{ m}} - \frac{1}{0.08 \text{ m}} = -2.5 \text{ m}^{-1}$$

and $s' = -0.4$ m. Here s' is negative, so the image is virtual (Fig. 24.11b).

Example 24.4

A camera lens has a focal length of $+0.1$ m. (a) If the camera is focused on a child 2 m from the lens, what is the distance from the lens to the film (Fig. 24.12)? (b) If the child has a height of 1 m, how tall is the image on the film?

(a) If a sharp image is to be formed, the distance from the lens to the film must equal the image distance s'. Using the thin lens formula

$$\frac{1}{s'} = \frac{1}{f} - \frac{1}{s} = \frac{1}{0.1 \text{ m}} - \frac{1}{2 \text{ m}} = 9.5 \text{ m}^{-1}$$

and $s' = 0.105$ m. Thus the film should be at a distance from the lens slightly greater than the focal length, 0.1 m. Except when extreme closeups are taken, the object distance is always large compared to f, so the image and the correct film location are just beyond the focal point.

(b) The image height can be found from the object height, $h = 1$ m, and the linear magnification, $m = h'/h = -s'/s$. Approximating s' by f in accordance with our discussion above,

$$m = -\frac{s'}{s} \simeq -\frac{f}{s} = \frac{-0.1 \text{ m}}{2 \text{ m}} = -0.05$$

Hence

$$h' = mh = (-0.05)(1 \text{ m}) = -0.05 \text{ m}$$

The height of the image is 0.05 m. The minus sign indicates that the image is inverted.

Since the magnification is approximately proportional

to the focal length, cameras are often equipped with interchangeable lenses of varying focal lengths. The film size remains constant, so the field of view decreases as the focal length and magnification increase.

Example 24.5

A diverging lens has a focal length of -0.4 m. (a) Find the image location for an object placed 2 m from the lens. (b) If there is a real image 1 m from the lens, where is the object?

(a) Since the object is real, $s = +2$ m. Thus

$$\frac{1}{s'} = \frac{1}{f} - \frac{1}{s} = \frac{1}{(-0.4 \text{ m})} - \frac{1}{2 \text{ m}} = -3 \text{ m}^{-1}$$

and $s' = -0.333$ m. The image is virtual, and it is located between the focal point F' and the lens (Fig. 24.13a).

(b) With $s' = +1$ m,

$$\frac{1}{s} = \frac{1}{f} - \frac{1}{s'} = \frac{1}{(-0.4 \text{ m})} - \frac{1}{(1 \text{ m})} = -3.5 \text{ m}^{-1}$$

so $s = -0.286$ m. The object is virtual, and therefore must be due to another lens. The object is located between the lens and the focal point F (Fig. 24.13b).

24.4 | THE POWER OF A LENS; ABERRATIONS

In discussing lenses, it is often more convenient to deal with the reciprocal of the focal length, which is called the *power* of the lens:

$$P = \frac{1}{f} \tag{24.5}$$

It is clear from this definition that the meaning of the word power in optics is unrelated to its meaning in mechanics, work per unit time.

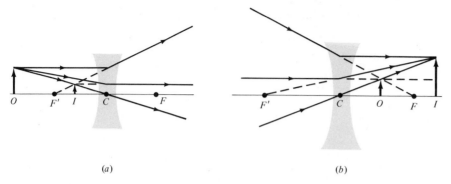

Figure 24.13. (a) A diverging lens always forms a virtual image of a real object. (b) When there is a virtual object between the lens and F formed by another lens (not shown), a diverging lens forms a real image.

If the focal length f is measured in metres, then P is measured in *diopters;* 1 diopter = 1 m^{-1}. For example, a lens with a focal length of -0.4 m has a power $P = 1/(-0.4 \text{ m}) = -2.5$ diopters. A short-focal-length lens, which bends light through large angles, has a large power.

It is left as a problem (Problem 24.43) to show that two thin lenses with focal lengths f_1 and f_2 placed next to each other are equivalent to a single lens with a focal length f satisfying

$$\frac{1}{f} = \frac{1}{f_1} + \frac{1}{f_2} \qquad (24.6)$$

Alternatively, with $P_1 = 1/f_1$ and $P_2 = 1/f_2$, the power of the pair of lenses is

$$P = P_1 + P_2 \qquad (24.7)$$

The powers of lenses in contact are simply added to find the net power. Thus, using powers instead of focal lengths avoids a good deal of arithmetic involving fractions. For example, an ophthalmologist placing 3-diopter and 0.25-diopter lenses in front of a patient's eye immediately knows that the combination is equivalent to a single 3.25-diopter lens.

We see in the following subsection how the additivity of the powers can be used to treat a lens configuration designed to minimize aberrations.

Aberrations | No matter how perfectly spherical its surfaces, any lens suffers from various kinds of *aberrations,* which limit the sharpness of its images independently of diffraction effects. Since the index of refraction of glass varies with the wavelength of the light, the focal length of a lens also varies with the

wavelength. When an object is illuminated with white light, if its image on a screen is in focus for one color component, it will be slightly out of focus for the others. This is *chromatic aberration.* Also, the lensmaker's equation is derived using small angle approximations. Corrections to this formula show that rays parallel to the axis have image locations that vary slightly with their distance from the axis. For this reason, a parallel beam of light actually forms an image of finite size rather than a true point image (Fig. 24.14). Aberrations of this and similar types occur even for light of a single wavelength and are called *monochromatic aberrations.* Complex lens systems with several elements are designed so that their separate aberrations tend to cancel (Fig. 24.15).

We can illustrate these cancellations by considering a doublet, two lenses in contact (Fig. 24.16). Lens 1 has two convex sides, and is made from crown glass. Lens 2 has one flat side and one concave side, and is made from flint glass. All the curved surfaces have

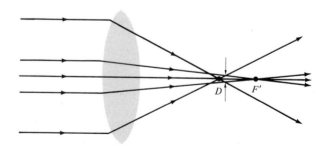

Figure 24.14. Monochromatic light rays near the axis are focused at F', while those near the edge of the lens meet at D. The arrows show where the bundle of rays has the smallest diameter.

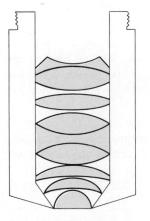

Figure 24.15. A modern multielement microscope lens.

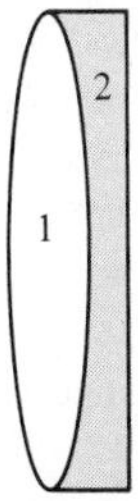

	656 nm (red)	589 nm (yellow)	486 nm (blue)
Refractive indices:			
Crown glass	1.517	1.520	1.527
Flint glass	1.644	1.650	1.664
Powers:			
P_1 (crown)	10.34	10.40	10.54
P_2 (flint)	−6.44	−6.50	−6.64
$P = P_1 + P_2$	3.90	3.90	3.90

Figure 24.16. A doublet made up of two different kinds of glass and designed to minimize chromatic aberration. The curved surfaces all have 10-cm radii of curvature. Note that the combined power is independent of the wavelength.

radii of curvature of 10 cm. In both types of glass, the refractive index varies about 1 percent over the visible spectrum. The powers P_1 and P_2 can be calculated using the lensmaker's equation,

$$P = \frac{1}{f} = (n - 1)\left(\frac{1}{R_1} + \frac{1}{R_2}\right)$$

As is seen in Fig. 24.16, P_1 varies by 2 percent over the spectrum, and P_2 by 3 percent. However, when the lenses are in contact, the effective power $P = P_1 + P_2$ is constant! Thus the doublet is free of chromatic aberration.

24.5 | THE SIMPLE MAGNIFIER

The normal human eye can just barely distinguish two well-illuminated point objects with an angular separation $\theta_0 \simeq 5 \times 10^{-4}$ rad $\simeq 0.03°$. This minimum angluar separation is called the *visual acuity* and can, in effect, be reduced with a *simple magnifier* or *magnifying glass*.

To see fine details, a person holds an object as close to the eye as possible, or at the *near point:* the closest point at which one can focus comfortably. For a normal young adult the distance x_n to the near point is about 0.25 m (Fig. 24.17a). At the near point, two points a small distance y apart have an angular separation small enough so that $\theta \simeq \tan \theta = y/x_n$ is a good approximation. If θ is $\theta_0 = 5 \times 10^{-4}$ rad, then

$$y = x_n\theta = (0.25 \text{ m})(5 \times 10^{-4})$$
$$= 1.25 \times 10^{-4} \text{ m}$$
$$= 0.125 \text{ mm}$$

Thus the finest details discernable to the naked eye have a size of about 0.1 mm.

The simple magnifier is a converging lens that allows the object to be brought closer to the eye so that it subtends a larger angle and permits one to see finer details. Usually the object under study is placed just inside the focal point of the lens, which is held close to the eye (Fig. 24.17b). The resulting virtual image is far from the eye and thus can be viewed comfortably. The angle subtended by the image is

$$\theta' \simeq \tan \theta' = \frac{y}{f}$$

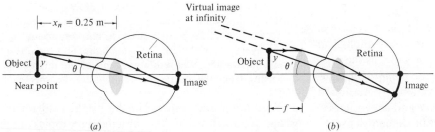

Figure 24.17. The simple magnifier. The lens permits the object to be brought closer to the eye, thereby subtending a larger angle.

The *angular magnification M* is the ratio (Fig. 24.17*b*)

$$M = \frac{\theta'}{\theta} = \frac{y/f}{y/x_n} = \frac{x_n}{f}$$

or with $x_n = 0.25$ m,

$$M = \frac{0.25 \text{ m}}{f} \qquad (24.8)$$

The following example illustrates the use of a typical magnifier.

Example 24.6

A collector uses a lens with a focal length of 0.1 m to examine a stamp. What angular magnification is provided by the lens?

The angular magnification is

$$M = \frac{0.25 \text{ m}}{f} = \frac{0.25 \text{ m}}{0.1 \text{ m}} = 2.5$$

Details will appear 2.5 times larger with this magnifier.

Lenses with focal lengths much shorter than the 0.1 m of the previous example would give much larger magnifications, but they would require that the object under study be held very close to the lens. Also, they will have large aberrations. Thus simple magnifiers usually have an angular magnification of only two or three. Magnifiers containing two or more lenses can have considerably greater useful magnification, since the aberrations can be partially corrected.

24.6 | THE BRIGHT-FIELD LIGHT MICROSCOPE

Although the microscope is one of the oldest and most widely used physical instruments in biology and medicine, new types of microscopes have been developed in recent decades. These make possible more detailed study of cellular structures and sometimes avoid the need for destructive methods in observing living cells.

Figure 24.18 shows a bright-field light microscope, the ordinary microscope found in every biological laboratory. The lenses in the *condenser* focus the incident light on the specimen, and the *diaphragm* regulates the intensity. The magnification is determined by the focal lengths of the *objective* and *ocular* lenses. In practice, both of these are multielement lenses.

The principle of the microscope is shown in Fig. 24.19. The object under study is placed just beyond

the focal point of the objective, so $s_1 \simeq f_1$. Its image is real and inverted, and it is much larger than the object. The linear magnification is $m_1 = -s_1'/s_1 \simeq -s_1'/f_1$, and it is typically around 50. This image then serves as the object for the ocular, which acts as a simple magnifier and provides an enlarged virtual image at a comfortable distance for viewing.

Since the angular magnification of the ocular is $M_2 = (0.25 \text{ m})/f_2$, the overall magnification of the microscope is the product of the two magnifications,

$$M = m_1 M_2 = \frac{-s_1' \times 0.25 \text{ m}}{f_1 f_2} \qquad (24.9)$$

This result is illustrated by the following example.

Example 24.7

The focal length of a microscope objective is $0.4 \text{ cm} = 4 \times 10^{-3}$ m, and the focal length of the ocular is $3.2 \text{ cm} = 10^{-2}$ m. The image formed by the objective is 0.2 m from the objective. (a) Where is the object under study? (b) What is the angular magnification? (c) Neglecting any diffraction effects, what is the smallest separation between two points that can be resolved by the eye with this instrument?

(a) From the thin lens formula, the object distance s_1 satisfies

$$\frac{1}{s_1} = \frac{1}{f_1} - \frac{1}{s_1'} = \frac{1}{4 \times 10^{-3} \text{ m}} - \frac{1}{0.2 \text{ m}} = 245 \text{ m}^{-1}$$

so $s_1 = 4.08 \times 10^{-3}$ m. The object is just outside of the focal point of the objective, since $f_1 = 4 \times 10^{-3}$ m.

(b) The angular magnification is

$$M = \frac{-s_1' \times 0.25 \text{ m}}{f_1 f_2}$$

$$= \frac{-(0.2 \text{ m})(0.25 \text{ m})}{(4 \times 10^{-3} \text{ m})(3.2 \times 10^{-2} \text{ m})}$$

$$= -391$$

The minus sign means that the image is inverted.

(c) The smallest resolvable separation of two points for an unaided eye is about $0.1 \text{ mm} = 10^{-4}$ m. If the angular size is increased by a factor of $391 \simeq 400$, then the minimum resolvable separation becomes $10^{-4} \text{ m}/400 = 2.5 \times 10^{-7} \text{ m} = 250 \text{ nm}$, or about half the average wavelength of visible light. This separation is comparable to the resolution limit of the microscope arising from the diffraction of light by the specimen.

Resolution and contrast in the bright field microscope are discussed in the supplementary topics at the end of this chapter, along with polarizing, interference, and phase contrast microscopes.

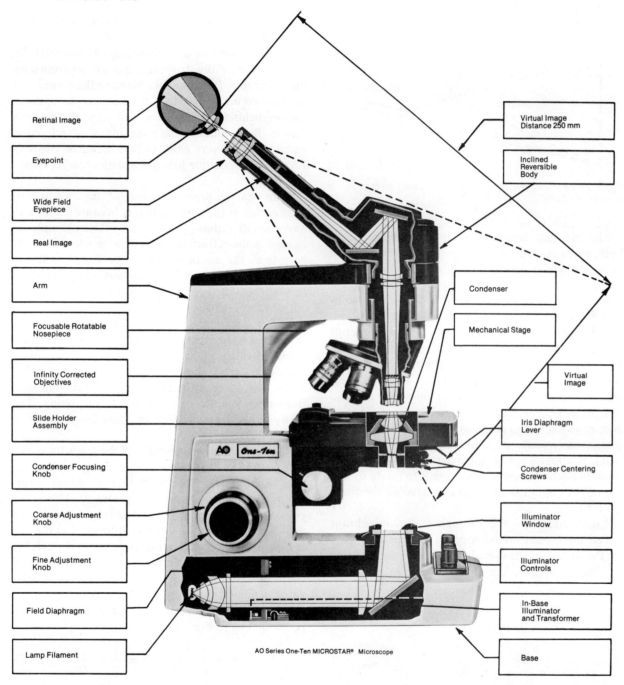

Figure 24.18. The ordinary bright-field light microscope. (Courtesy of American Optical Corporation.)

24.7 | THE HUMAN EYE

The human eye is a remarkable evolutionary achievement. It has an intensity range of 10^9, covers a field of view of over 180°, can rapidly shift its focus from very short distances to infinity, and has a resolution close to the limit imposed by diffraction. Also, as we see in Chapter Twenty-six, its threshold sensitivity is comparable to the theoretical limit imposed by the quantum properties of light.

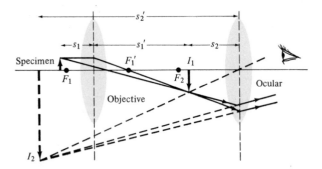

Figure 24.19. The microscope. The ocular and objective in practice are multielement lenses.

The eye and camera have many similarities. In both, a lens system forms an inverted real image on a light-sensitive surface. The eyeball is approximately spherical in shape with a diameter of about 2.3 cm (Fig. 24.20). Its outer covering is a nearly opaque fibrous layer called the *sclera*. Inside this is a dark membrane, the *choroid* which, like the black interior of a camera, absorbs stray light. The inner surface of the eyeball is the *retina*, a membrane containing numerous nerves and blood vessels. The nerve fibers terminate at *rods* or *cones* in the retina which respond to light by generating electrical nerve pulses. The eye is most sensitive at a small retinal depression, the *yellow spot* or *macula;* its central portion, the *fovea centralis,* is about $\frac{1}{4}$ mm in diameter and contains only densely packed cones. The eye tends to rotate so that the object under examination is imaged on the fovea centralis.

Light enters the eye through a thin membrane called the *cornea,* which covers a transparent bulge on the surface of the eyeball. The *iris* is a colored ring behind the cornea; like a camera diaphragm, it ad-

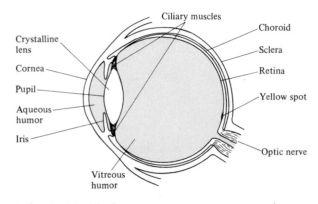

Figure 24.20. The human eye.

justs in size and aids in regulating the amount of light entering the eye through the *pupil*. The *crystalline lens* is composed of a fibrous, jellylike material. Its shape and therefore its focal length are controlled by the *ciliary muscles*. The space between the cornea and the lens contains a watery fluid called the *aqueous humor;* behind the lens is a thin jelly, the *vitreous humor*. Both humors have an index of refraction equal to 1.336, very close to that of water, which is 1.333. The crystalline lens has a slightly larger index, 1.437.

Somewhat surprisingly, most of the bending of light occurs at the cornea. This is because the cornea has a small radius of curvature (0.8 cm), and the change in the refractive index is large when the light goes from the air ($n = 1$) into the aqueous humor ($n = 1.336$). The function of the lens is to provide the fine adjustments needed to focus on objects at different distances.

If the ciliary muscles are relaxed, the front surface of the lens is kept relatively flat and light from distant objects is focused on the retina. When the ciliary muscles contract, the lens assumes a more rounded shape, and its focal length decreases, bringing light from nearby objects into focus on the retina. The ablity of the lens to adjust its focal length is called *accommodation.*

The *power of accommodation* of the eye is the maximum variation of its power for focusing on near and distant objects. Suppose, with the eye relaxed, a person sees objects at a distance x_f clearly. This is the person's far point, and it is at infinity for someone with normal vision. The image distance s' to the retina is somewhat less than the diameter of the eye. We will use an image distance $D = 2$ cm $= 0.02$ m for simplicity in the arithmetic, although the correct value is slightly smaller; the value used is not important for our considerations. At the far point, the power P_f of the eye is

$$P_f = \frac{1}{f} = \frac{1}{s} + \frac{1}{s'} = \frac{1}{x_f} + \frac{1}{D} \qquad (24.10)$$

For a person with normal vision, $x_f = \infty$, and this gives $P_f = 1/0.02$ m $= 50$ diopters. When the eye adjusts its focal length so that it focuses on an object at the near point, the object distance is $s = x_n$. Since again $s' = D$, the power of the eye is now

$$P_n = \frac{1}{f} = \frac{1}{s} + \frac{1}{s'} = \frac{1}{x_n} + \frac{1}{D} \qquad (24.11)$$

For a young adult with normal vision, $x_n = 0.25$ m, and $P_n = (1/0.25 \text{ m}) + (1/0.02 \text{ m}) = 54$ diopters. The power of accommodation is the difference,

$$A = P_n - P_f \qquad (24.12)$$

For a young adult with normal vision, $A = (54 - 50)$ diopters $= 4$ diopters. Young children have a much greater power of accommodation, and often can read books held quite close to their eyes. The accommodation decreases with aging, and most people find their near point gradually recedes until they cannot read comfortably without corrective glasses. Corrective glasses are discussed in the supplementary topics.

Acuity

We noted in our discussion of the magnifier that the visual acuity of a typical person is about 5×10^{-4} rad; objects with a smaller angular separation cannot be distinguished. It is reasonable to inquire whether this limit is due to diffraction effects. According to Chapter Twenty-three, when light from a distant source passes through a small circular aperture of diameter d, the first diffraction minimum is at $\sin \theta = 1.22 \lambda/d$. To estimate what this implies for the eye, let us take the iris diameter to be 5 mm $=$ 5×10^{-3} m, and the wavelength to be 500 nm $=$ 5×10^{-7} m. Since θ is small, to one significant figure

$$\theta \simeq \sin \theta = 1.22 \frac{\lambda}{d}$$

$$= 1.22 \frac{(5 \times 10^{-7} \text{ m})}{5 \times 10^{-3} \text{ m}}$$

$$= 10^{-4} \text{ rad}$$

According to the Rayleigh criterion, two objects will just barely be resolvable if they are separated by this angle (Fig. 24.21). Experiments show that while a few people under optimum conditions have an acuity of twice the diffraction limit, or 2×10^{-4} rad, nobody can reach the limit of 10^{-4} rad.

An explanation for this failure of the eye to quite match the diffraction limit is provided by the structure of the retina. From Fig. 24.21a, the radius of the image circle due to diffraction is

$$r = D \tan \theta \simeq D\theta$$
$$= (2.3 \times 10^{-2} \text{ m})(10^{-4})$$
$$= 2.3 \times 10^{-6} \text{ m}$$

This is about equal to the separation of the cones in the fovea, which is the most sensitive region of the retina and contains only cones and not rods. Now the

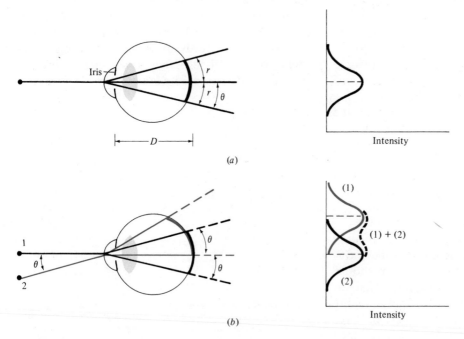

(a)

(b)

Figure 24.21. (a) The minimum image size for a point object due to diffraction at the iris. (b) If two point objects are separated by θ, their images will be barely distinguishable.

Figure 24.22.   When our neurological decoding of a visual image contradicts our experience, we find it disturbing. (M.C. Escher, "Waterfall," Escher Foundation—Haags Gemeentemuseum—The Hague.

best resolution observed is about 2×10^{-4} rad, corresponding to diffraction centers separated by about 4.6×10^{-6} m, or by two cones. *Thus, it appears that to distinguish two small objects, at least one unexcited cone must intervene between the excited cones.*

The interpretation of the images detected on the retina as meaningful patterns involves complex neurological processing and is a fascinating problem inviting computer simulations and theories of pattern recognition. Given a complex scene, the brain suddenly locks into an interpretation that looks "correct" (Fig. 24.22).

Sensitivity | The *minimum* or *threshold intensity*
needed to see a flash of light depends on the wavelength. The cornea is opaque to wavelengths shorter than 300 nm, and the crystalline lens to wavelengths below 380 nm, so ultraviolet light does not normally contribute to vision. However, some persons have had their lenses surgically removed because they have become opaque or developed *cataracts*. These

people have poor acuity and no accommodation, but to them objects illuminated with ultraviolet light alone are visible and appear violet.

One long wavelength limit on the sensitivity of the eye is set by the strong absorption of light by water in the cornea and the aqueous humor at wavelengths above 1200 nm. However, the sensitivity of the eye goes to zero rapidly above 700 nm. It is thought that the photosensitive molecules in the rods and cones do not respond to the longer wavelengths.

Figure 24.23 shows the variation with wavelength of the sensitivity, which is the inverse of the threshold intensity, for *dark-adapted* and *light-adapted* eyes. When we go from daylight into a dimly lit room, our eyes gradually adapt and reveal initially invisible details. This adaptation takes over half an hour to be complete. Since the dark-adapted eye is much more sensitive, the two curves have been adjusted to have the same maximum value.

The cones are active only in light-adapted vision, while the rods are always active. Accordingly, the fovea, which lacks rods but has the greatest concentration of cones, is the most sensitive part of the retina of the light-adapted eye, while the edges of the retina, which have more rods, are more sensitive in the dark-adapted eye. The visual acuity of the dark-adapted eye is much poorer than that of the light-adapted eye.

From Fig. 24.23 it can be seen that the sensitivity is greatest near 500 and 550 nm for dark- and light-adapted eyes, respectively. Both wavelengths correspond to green light. Often, green glass is used in sunglasses and in tinted windows. Since green glass

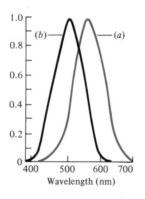

Figure 24.23.   Relative sensitivity curves for the (*a*) light-adapted and (*b*) dark-adapted eye. Although the dark-adapted eye is more sensitive, both curves are drawn with the same maximum values. (Adapted from Eugene Ackerman, *Biophysical Science,* Prentice-Hall, Englewood Cliffs, N.J., 1962.)

absorbs green light less than other wavelengths, it provides the most useful illumination while reducing the total transmitted intensity.

The dark-adapted eye responds to a large range of wavelengths, but the sensation of color is perceived only in the light-adapted eye. Color perception will be discussed in the supplementary topics.

SUMMARY

A plane mirror forms a virtual image whose distance from the mirror is equal to the object distance. The image of each point on a complex object appears directly in front of that point.

A converging lens will bend rays parallel to its axis so that they meet at the focal point, while a diverging lens will bend such rays outward so that they appear to have come from a focal point before the lens. The focal length is determined by the index of refraction of the lens and the radii of curvature of its surfaces with the aid of the lensmaker's equation,

$$\frac{1}{f} = (n - 1)\left(\frac{1}{R_1} + \frac{1}{R_2}\right)$$

The radius of curvature is positive for a convex surface, and negative for a concave surface.

The intersection of three special rays determines the location of the image of a point on an object. The ray parallel to the axis is deflected so it passes through the focal point F', the ray going through the focal point F emerges parallel to the axis, and the ray through the center is undeflected. Alternatively, the image and object distances are related by the thin lens formula,

$$\frac{1}{s} + \frac{1}{s'} = \frac{1}{f}$$

The linear magnification of a thin lens is

$$m = \frac{h'}{h} = \frac{-s'}{s}$$

A simple magnifier produces an angular magnification that is approximately

$$M = \frac{0.25 \text{ m}}{f}$$

Multiple lens systems such as microscopes are analyzed by considering the image of one lens as the object of the next. The common bright-field light microscope consists of an objective and an ocular. The objective forms a real image just inside the focal point of the ocular. The ocular then forms an enlarged virtual image, which is seen by the viewer. For a microscope, the overall angular magnification is

$$M = \frac{-s_1' \times 0.25 \text{ m}}{f_1 f_2}$$

In the human eye most of the bending of light occurs at the cornea. The lens of the eye provides the additional adjustments needed to focus on objects at varying distances. The limit of resolution of the eye corresponds to the requirement that the diffraction circles corresponding to the images of two object points must fall on cones in the retina separated by at least one unexcited cone.

Checklist

Define or explain:

object, image	radius of curvature
virtual, real	linear magnification
erect, inverted	thin lens formula
converging, diverging	aberration
lenses	angular magnification
focal length	visual acuity
lens axis	accommodation
lensmaker's equation	power

REVIEW QUESTIONS

Q24-1 The image distance for a plane mirror equals the _____.

Q24-2 Converging lenses have _____ focal lengths; diverging lenses have _____ focal lengths.

Q24-3 The focal point F' is the place where rays _____ meet after passing through a converging lens.

Q24-4 A plane lens surface has a radius of curvature equal to _____.

Q24-5 When a lens is moved from air to water, the light rays bend _____, and the focal length is _____.

Q24-6 A ray through the center of the lens emerges _____.

Q24-7 A ray passing through the focal point F emerges from the lens _____.

Q24-8 If the image distance is twice the object distance and of the same sign, the image is _____ and has a linear magnification of _____.

Q24-9 The variation of the refractive index of glass with wavelength is responsible for _____ aberration.

Q24-10 When you use a lens as a simple magnifier, the object under study is placed _____.

Q24-11 If a lens has a focal length of −2 m, its power is _____.

Q24-12 For the eye to distinguish two small objects, at least _____ must intervene between two excited cones.

Q24-13 The eye is most sensitive to wavelengths corresponding to _____ light.

EXERCISES

Section 24.1 | Mirrors

24-1 A man stands 2 m from a vertical plane mirror. What is the distance from the man to his image?

24-2 A cat sees its image in a plane mirror. The cat is 1 m from the mirror. (a) Where is the image? (b) The cat lunges at the mirror at 2 m per second. How fast does the cat approach its image?

24-3 A nearsighted man cannot see objects clearly beyond 40 cm from his eyes. How close must he stand to a mirror in order to see what he is doing when he shaves?

Section 24.2 | Lenses

24-4 An eyeglass lens has one concave surface with a radius of 0.5 m and a convex surface with a radius of 0.7 m. If the index of refraction of the lens is 1.6, find the focal length.

24-5 A lens is made from a plastic with a refractive index of 1.4. One side is flat and the other is concave with a radius of 0.2 m. Find the focal length of the lens.

24-6 A lens is made of glass with $n = 1.5$. One side is convex and has a radius of curvature equal to 0.1 m. Find the radius of curvature of the other surface and draw a sketch of the lens if (a) $f = 0.15$ m; (b) $f = 0.1$ m; (c) $f = −0.15$ m.

24-7 A lens made of glass with refractive index 1.6 has a focal length of 0.5 m in air. What is the focal length in water?

24-8 Crown glass has a refractive index of 1.523 for blue light and an index of 1.517 for red light. If a crown glass lens has a focal length of 1 m for red light, what is its focal length for blue light?

24-9 Flint glass has a refractive index of 1.645 for blue light and an index of 1.629 for red light. A lens made of flint glass has two convex surfaces of radius of curvature 0.1 m. Find its focal lengths for blue and red light.

Section 24.3 | Image Formation

24-10 A lens has a focal length of 0.2 m. A real object is placed 0.08 m from the lens. (a) Locate the image approximately using graphical methods. (b) Locate the image algebraically. (c) What is the magnification?

24-11 An object is placed 1 m from a lens with a focal length of −0.5 m. Locate the image position (a) graphically, and (b) algebraically.

24-12 A lens 0.1 m from a lamp forms a real image of this lamp that is 10 times larger. What is the focal length of the lens?

24-13 A lens of focal length 0.1 m is held 0.08 m from an insect. (a) Where is the image of the insect? (b) What is the magnification of the image? Is it erect or inverted?

24-14 An eyeglass lens has a focal length of −2 m. If it is 4 m from a book, where is the image of the book?

24-15 A camera is focused on a group of people 3 m from the lens, which has a focal length of 50 mm = 0.05 m. (a) What is the lens-to-film distance? (b) What is the linear magnification? (c) If the film height is 24 mm = 0.024-m, what is the maximum height of a person whose image can completely fit on the film?

24-16 Accessories are available for some cameras which permit the lens-to-film distance to be increased when extreme close-ups are taken of very small objects. Why are these accessories needed?

24-17 A photographer replaces a lens with a focal length of 50 mm by a lens with a focal length of 200 mm. What happens to the image size for distant objects?

Section 24.4 | The Power of a Lens; Aberrations

24-18 A doctor finds that a nearsighted person needs one lens with a power of −8 diopters and one with a power of −6 diopters. Find the focal lengths of the lenses.

24-19 Find the focal length of a lens with a power of 4 diopters.

24-20 A person wears reading glasses with a focal length of 2 m. What is their power?

24-21 An optometrist fitting a person for eyeglasses places a 0.25-diopter lens directly in front of a 4.5-diopter lens. (a) What is the power of the combination? (b) What is the effective focal length of the combination?

24-22 A lens has a focal length of 2 m. When a

second lens is placed in contact the pair has a focal length of 1.5 m. What is the focal length of the second lens?

24-23 A 50-year-old woman who is nearsighted wears eyeglasses with a power of −5.5 diopters for distance viewing. Her doctor prescribes a correction of +2 diopters in the close-vision section of her bifocals. This is measured relative to the main part of the lens. (a) What is the focal length of her distance-viewing part of the lens? (b) What is the focal length of the close-vision section of the lens?

Section 24.5 | The Simple Magnifier

24-24 A lens of focal length 0.1 m is used as a simple magnifier. What is its magnification?

24-25 A farsighted woman has a near point 1 m from her eyes. If her visual acuity is 10^{-3} rad, what is the smallest separation between two objects that she can distinguish?

24-26 A lens used as a simple magnifier gives an angular magnification of 6. What is its focal length?

Section 24.6 | The Bright-Field Light Microscope

24-27 An insect of length 2 mm = 2×10^{-3} m is observed through a microscope that has an angular magnification of 100. How large an "insect" would one have to find to see its details as well with the unaided eye?

24-28 A microscope objective of focal length 5 mm forms an image 150 mm from the objective. If the ocular has a focal length of 28 mm, find the magnification of the microscope.

24-29 A microscope has an objective with a 4-mm focal length and an ocular with a 30-mm focal length. The two lenses are separated by 0.16 m, and the final image is formed 0.25 m from the ocular. (a) Where is the image formed by the objective? (b) Where is the specimen relative to the objective? (c) What is the magnification of the microscope?

24-30 A microscope provides an angular magnification of 150. What is the smallest separation that can be distinguished with this instrument?

Section 24.7 | The Human Eye

24-31 Explain why faint stars are best seen by looking "out of the corner of the eye" rather than directly at the stars.

24-32 Why is it believed that color perception is due only to cones and not to the rods?

24-33 (a) At approximately what wavelengths is the sensitivity of the light-adapted eye half its max-

imum value? (b) What are the corresponding wavelengths for the dark-adapted eye?

24-34 A bright star can have a sufficiently intense diffraction pattern on the retina so that the second ring will be above the threshold intensity. What effect will this have on its apparent size?

24-35 (a) Under optimum conditions, the smallest black dot that can be seen subtends an angle of 2.3×10^{-6} rad. If a dot is viewed at a distance of 0.25 m, the near point of a normal adult, what is the smallest diameter it can have and still be seen? (b) The maximum resolution is obtained when the image falls on the fovea centralis. At 10° away from this region, the acuity is 10 times poorer. What is the minimum size spot that can be seen at that angle under these conditions?

24-36 It is possible for a trained person to align two straight lines on a slide rule or a vernier scale to within 9×10^{-6} rad, much less than the minimum separation needed to resolve two points. How large is the alignment error if the instrument is held at the near point, 0.25 m from the eye?

24-37 Images on the retina are inverted. Explain why this must be true and suggest how we see things properly.

PROBLEMS

24-38 A girl 1.5 m tall stands in front of a vertical mirror that is just large enough so that she can see her entire body. How tall is the mirror?

24-39 A man holds a plane mirror of height 0.1 m vertically at a distance of 0.25 m from his eyes and observes that the image of a building just fills the height of the mirror. If the building is 200 m from the mirror, what is its height?

24-40 Prove that a ray of light travels the shortest possible distance in going from a light source at point P to point Q after reflection at the mirror (Fig. 24.24). (*Hint:* Draw a straight line from Q to the image of P.)

24-41 For a lens with a positive focal length f, (a) calculate the image position for $s = 0$, $f/2$, f,

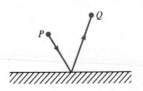

Figure 24.24. Problem 24-40.

$3f/2$, and $3f$. (b) Sketch a graph of the image position as s varies from 0 to $+\infty$.

24-42 A lens has a focal length of -1 m. (a) Calculate the image position when the object is placed at $s = 0$, 0.5, 1, 1.5, and 3 m from the lens. (b) Sketch a graph of the image position as s varies from 0 to $+\infty$.

24-43 Show that if two thin lenses with focal lengths f_1 and f_2 are in contact, they are equivalent to a single lens with focal length f satisfying

$$\frac{1}{f} = \frac{1}{f_1} + \frac{1}{f_2}$$

(*Hint:* The image formed by the first lens is the object for the second, so $s_2 = -s'_1$.)

24-44 An object is 1 m from a screen. At what points may a 0.05-m focal length lens be placed so as to produce a sharp image on the screen? What are the corresponding magnifications?

24-45 A lens of index 1.5 has one plane side and one concave side of radius 0.2 m. The lens is placed horizontally with the concave side up and is filled with water. What is the focal length of the water-glass system? (*Hint:* Treat the system as a pair of thin lenses in contact.)

24-46 A slide projector lens is 3 m from the screen and has a focal length of 0.08 m. (a) Where is the slide located when the projector is in focus? (b) A child has a height on the screen of 10 cm. What is her height on the slide? (c) If we want to double the size of images on the screen, where should we put the projector? (d) If we wish to double the image size without moving the projector, what focal length lens must we use?

24-47 A converging lens has a focal length f. At what object location is the magnification equal to -1?

24-48 The radius of curvature of the cornea is typically 7.7 mm. If we construct a lens with this radius of curvature on one side, a plane surface on the other, and a refractive index of 1.37 (the index of the cornea), what power lens do we obtain? (This calculation shows that most of the focusing power of the eye comes from the cornea.)

24-49 The focal length of a converging lens can easily be found by seeing where a distant object is imaged. Explain how this can be done for a diverging lens with the aid of an additional converging lens of known power.

24-50 An ocular consists of two identical thin lenses of focal length 5 cm separated by a distance of 2.5 cm. Where is light from a distant point source focused?

ANSWERS TO REVIEW QUESTIONS

Q24-1, object distance; **Q24-2,** positive, negative; **Q24-3,** initially parallel to the axis; **Q24-4,** infinity; **Q24-5,** less, longer; **Q24-6,** undeflected; **Q24-7,** parallel to the axis; **Q24-8,** inverted, 2; **Q24-9,** chromatic; **Q24-10,** just inside the focal point; **Q24-11,** -0.5 diopters; **Q24-12,** one unexcited cone; **Q24-13,** green.

SUPPLEMENTARY TOPICS

24.8 | SPHERICAL MIRRORS

Spherical mirrors are encountered in applications as varied as medical instruments and truck rearview mirrors. Although their effects are quite different, spherical mirrors can be analyzed using concepts and equations very similar to those we developed for spherical lenses.

If light rays are incident on a concave mirror parallel to its axis, they are reflected and meet at a common point F (Fig. 24.25*a*). Thus a concave mirror is converging, and it can form a real image. Since light rays from a very distant source are parallel when they reach the mirror, F is the focal point. Similarly, parallel rays incident on a convex mirror are reflected so that they diverge as though they had come from a focal point behind the mirror (Fig. 24.25*b*)

As for lenses, we draw our diagrams with the light incident from the left, and real objects to the left of the mirror. Real objects are at positive object distances s, and real images are at positive image distances s'. However, since mirrors reverse the ray directions, real images are also to the left of a mirror, and virtual objects and images are to the right. The distances s and s' are measured from the *vertex* of the mirror, point V in Fig. 24.25.

The focal length f is related to the radius of curvature R of the mirror by

$$f = R/2 \qquad (24.13)$$

This equation assumes these sign conventions for R:

1 R is positive for a concave (converging) mirror
2 R is negative for a convex (diverging) mirror

Like the lensmaker's formula, $f = R/2$ holds exactly only in the small angle limit, that is, when all the

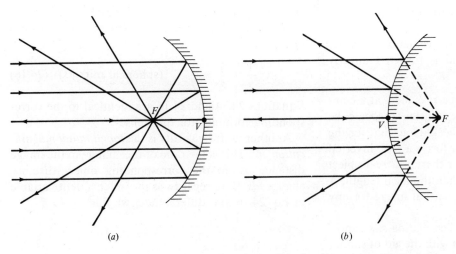

Figure 24.25. Focal points of (a) a converging mirror; (b) a diverging mirror. V is the vertex.

rays are almost normally incident on the mirror. However, it is easier to derive. In Fig. 24.26, C is the center of curvature. The radius CA is perpendicular to the mirror, so by the law of reflection, the incident and reflected rays each make the same angle ϕ with CA. Since the incident ray is parallel to the axis, θ is also equal to ϕ. Thus the triangle CFA is isosceles, and the sides CF and FA are equal. In the small angle limit, CF and FA are each half of CA, so $(R - f) = R/2$, or $R = f/2$.

As in the case of a lens, three special rays from a point on the object (Fig. 24.27) can be used to find the position of an image:

1 The ray parallel to the axis is reflected through the focal point F of a concave mirror, and appears to come from the focal point of a convex mirror.

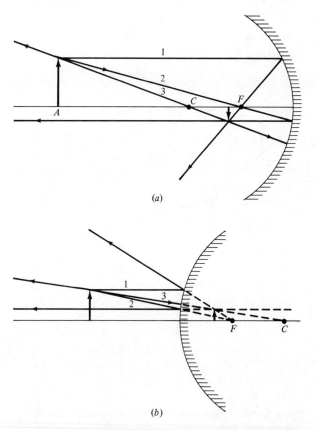

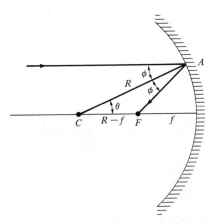

Figure 24.26. Since the angle of incidence equals the angle of reflection, both rays make the same angle ϕ with the radius CA. θ is also equal to ϕ, so the triangle CFA is isosceles. If ϕ is small, the sides CF and FA are each approximately half of CA.

Figure 24.27. Three rays used to locate an image formed by a spherical mirror. (a) A concave mirror. (b) A convex mirror.

2 The ray passing through the focal point is reflected parallel to the axis.

3 The ray passing through the center of curvature is normally incident and is reflected directly backward.

Neglecting aberrations, all three rays meet at a common image point. Two rays are sufficient to locate the image; the third serves as a check. Note that for the case shown the concave mirror forms a real, inverted image. This happens whenever there is a real object at an object distance greater than the focal length. A convex mirror forms an erect, virtual image for any real object.

We can again derive formulas relating s, s', and f and the linear magnification m with the aid of similar triangles. In Fig. 24.28, the ray striking the vertex V reflects at the angle of incidence ϕ and passes through the image. Thus the right triangles ABV and HGV are similar. Since the image is inverted, its height h' is negative, and the linear magnification is

$$m = \frac{h'}{h} = \frac{-s'}{s} \qquad \text{(linear magnification)} \quad (24.14)$$

A second expression for m follows from the similar triangles ABC and HGC:

$$m = \frac{h'}{h} = \frac{-(R - s')}{(s - R)} \qquad (24.15)$$

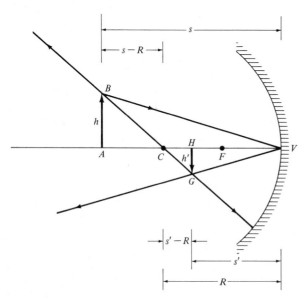

Figure 24.28. Triangles ABV and HGV are similar, as are triangles ABC and HGC.

We can equate the right sides of these expressions for m. With the relation $f = R/2$, we find after a few lines of algebra that

$$\frac{1}{s} + \frac{1}{s'} = \frac{1}{f} \qquad \text{(spherical mirrors)} \quad (24.16)$$

Equations 24.14 and 24.16 are identical to the corresponding formulas for thin lenses.

Another useful quantity is the *longitudinal magnification, m'*. This is the ratio of the change in the image distance ds' to the corresponding object distance change, ds. If we regard s as the independent variable in Eq. 24.16 and differentiate, we find

$$\frac{-1}{s^2} - \frac{1}{s'^2}\frac{ds'}{ds} = 0$$

Solving for $m' = ds'/ds$, we obtain

$$m' = -\frac{s'^2}{s^2} = -m^2 \qquad (24.17)$$

This result states that the longitudinal magnification is minus the square of the linear magnification. Since m^2 is always positive, m' is always negative. Suppose, for example, that the object approaches the mirror, so that ds is negative. Then ds' is positive, and the image moves toward the object.

In many applications, such as the convex rearview mirrors used on vehicles to give a large field of view, the linear magnification m is small compared to 1. The longitudinal magnification $m' = -m^2$ is then very small. This makes it difficult to judge the distances to other vehicles.

The following examples illustrate the use of these equations.

Equation 24.8

A convex truck rearview mirror has a radius of curvature of 0.4 m. A car is 30 m from the mirror, and its height is 1.2 m. (a) Where is the image of the car? (b) How large is this image? (c) If the car decreases its distance by 5 m, how far does the image move?

(a) The mirror is convex, so its radius of curvature is negative. The focal length of the mirror is $f = R/2 = -(0.4\text{ m})/2 = -0.2$ m. With $s = 30$ m, $1/s + 1/s' = 1/f$ becomes

$$\frac{1}{30\text{ m}} + \frac{1}{s'} = \frac{1}{-0.2\text{ m}}$$

Solving for s', we find $s' = -0.1987$ m. The image is virtual, and it is very close to the focal point, since the object is very far away compared to the focal length.

(b) The image size is found from the linear magnification,

$$m = h'/h = -s'/s = -(-0.1987 \text{ m})/30 \text{ m} = 0.00662$$

$$h' = 0.00662 \, h = (0.00662)(1.2 \text{ m})$$

$$= 0.00785 \text{ m} = 0.785 \text{ cm}$$

The image is less than 1-cm tall, so it is not easy to see.

(c) The longitudinal magnification is

$$m' = -m^2 = -(0.00662)^2 = -4.38 \times 10^{-5}$$

Thus if the object distance changes by $\Delta s = -5$ m, the image distance changes by

$$\Delta s' = \frac{ds'}{ds}\Delta s = m'\Delta s = (-4.38 \times 10^{-5})(-5 \text{ m})$$

$$= 0.000219 \text{ m} = 0.219 \text{ mm}$$

Since $\Delta s'$ is positive, s' has become less negative. That is, the image has moved toward the mirror. However, the position of the image has changed by a mere two-tenths of a millimetre. Thus it is extremely hard to judge the location of the car from the position of its image. A better clue is provided by the size of the image, since this changes to a good approximation in proportion to the object distance.

Example 24.9

A concave makeup mirror has a radius of 0.5 m. (a) If it is held 0.2 m from a woman's face, where is her image and how large is it? (b) Where is the image of a light bulb 3 m from the mirror? What will it look like?

(a) Using $1/s + 1/s' = 1/f$, with $f = R/2 = (0.5 \text{ m})/2 = 0.25$ m,

$$\frac{1}{0.2 \text{ m}} + \frac{1}{s'} = \frac{1}{(0.25)}$$

$$s' = -1 \text{ m}$$

The image of her face is 1 metre behind the mirror. The linear magnification is $m = -s'/s = -(-1 \text{ m})/(0.2 \text{ m}) = 5$, so her face is enlarged by a factor of 5.

(b) Again applying $1/s + 1/s' = 1/f$,

$$\frac{1}{3 \text{ m}} + \frac{1}{s'} = \frac{1}{(0.25 \text{ m})}$$

$$s' = 0.273 \text{ m}$$

s' is positive, so the image of the light bulb is real; it is located just beyond the focal point. The linear magnification is

$$m = -s'/s = -(0.273 \text{ m})/(3 \text{ m}) = -0.091$$

Thus a bright spot with a radius less than one-tenth that of the bulb will be seen if a hand or sheet of paper is placed at the image location.

24.9 | THE CAMERA

The essential elements of any camera are a light-tight box, a converging lens, a shutter that can open briefly, and a film that records the image. In order to admit enough light to permit short exposures, a lens with a large opening or aperture must be used. This, in turn, requires that in a high-quality camera, complex multielement lens designs be used in order to reduce aberrations.

Camera lenses are specified by two quantities. One of these is the focal length, f. As we saw in Example 24.4, the image size is approximately proportional to the focal length. The other quantity is the diameter d of the lens. Usually this is expressed in terms of the f-number. For example, an $f/8$ lens has a diameter $d = f/8$, which is one-eighth its focal length. The light-gathering power of a lens is proportional to its area or to its diameter squared. All but the simplest camera lenses have a diaphragm that can be adjusted to vary the aperture or effective diameter of the lens. For example, changing from $f/8$ to $f/16$ means halving the diameter and reducing the area by a factor of $2^2 = 4$. The shutter time must be increased by a factor of 4 to admit the same total light energy.

Because of lens aberrations, a point on the object being photographed will always have an image of finite size no matter how well the lens is positioned or focused. This image size will be reduced if the effective lens diameter or aperture is reduced with a diaphragm. Reducing the aperture also increases the *depth of field,* the range of distances over which object points are imaged with satisfactorily small circles (Fig. 24.29). Thus, a photographer must compromise between the need for a short exposure times to stop the motion of a scene and the need for small apertures to reduce aberrations and increase the depth of field.

When the lens aperture is small, diffraction effects may be more important than lens aberrations in limiting the sharpness of photographs. We saw in Chapter Twenty-three that when light passes through a small circular aperture, a diffraction pattern is formed consisting of a bright central circle and weak concentric rings. If the diameter of the aperture is d, the bright central region extends to the first diffraction minimum at $\sin \theta = r/R \simeq r/f$ (Fig. 24.30). With $\sin \theta = 1.22\lambda/d$, we have

$$r = 1.22\lambda \frac{f}{d} \qquad (24.18)$$

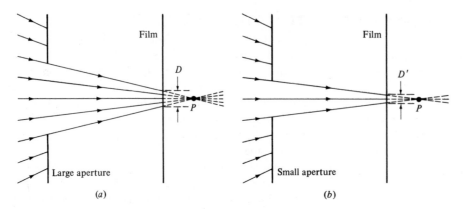

Figure 24.29. The effect of aperture size on the depth of focus. The lens (not shown) of a camera is positioned so that distant objects are imaged sharply on the film. (*a*) Light rays from a point object closer to the lens converge toward a point *P* behind the film. An image circle of diameter *D* is formed on the film. (Diffraction and aberration effect would make the actual diameter larger.) (*b*) When the aperture is made smaller, the beam of light rays is narrower and the diameter *D′* of the image circle of the film is smaller. Consequently the blurring is reduced, and objects located over a larger range of distances appear adequately focused. However, the light intensity reaching the film is reduced, and a longer exposure time is required.

This result indicates that when a camera is properly focused on a point object, the image formed on the film will be a circle of radius *r* even in the absence of aberrations. (The weak concentric rings are not normally noticeable.) Since this radius is proportional to f/d, the effects of diffraction grow as the diaphragm is closed (Fig. 24.31). This sets an upper limit on the useful aperture, as can be seen from the following example.

Example 24.10

If the lens of a 35-mm camera is stopped down to $f/22$, and a photograph is taken, what are the effects due to diffraction? (The film size in a 35-mm camera is 35 × 24 mm.)

In accordance with our discussion, each point in the scene being photographed will have an image that, at best, is a circle whose radius *r* is determined by diffraction. If the scene is illuminated with white light, the wavelengths range from 400 to 700 nm, averaging about 550 nm. Using this wavelength and $f/d = 22$,

$$r = 1.22\lambda \frac{f}{d} = 1.22(550 \times 10^{-9} \text{ m})(22)$$
$$= 1.5 \times 10^{-5} \text{ m}$$
$$= 0.015 \text{ mm}$$

This result means that all details on the film smaller than about 0.015 mm are hopelessly blurred by the overlapping diffraction patterns. Enlarging the photo 10 times produces a picture 35 by 24 cm. It also increases the diffraction circles to 0.15 mm in radius, making them large

enough to be discernable to someone looking carefully at the photograph. For this reason, apertures smaller than $f/22$ are not normally provided on 35-mm cameras. Instead, filters are used if the light intensity must be reduced further.

24.10 | RESOLUTION AND CONTRAST IN MICROSCOPES

When the separation between two points in a microscopic specimen is comparable to the wavelength λ of the light, we would expect diffraction effects to be important. A detailed analysis shows that the mini-

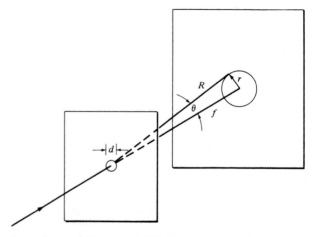

Figure 24.30. The central diffraction circle for light passing through a smaller circular lens aperture of diameter *d*.

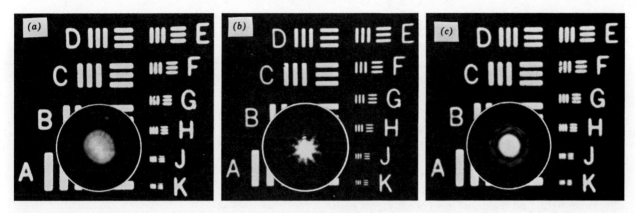

Figure 24.31. In these enlargements, aberrations limit image quality at large apertures (a) and diffraction sets the limit at small apertures (b, c). (Courtesy of Eastman Kodak.)

mum separation d that can be resolved by a microscope is

$$d = \frac{\lambda}{2n \sin \theta} \qquad (24.19)$$

Here, λ is the wavelength in air, n is the refractive index of the medium between the objective lens and the object under study, and θ is the angle subtended by the objective lens (Fig. 24.32). The product $n \sin \theta$ is called the *numerical aperture* and is sometimes marked on the instrument. If two points in a specimen are separated by less than d, their diffraction patterns overlap so much that their images cannot be distinguished.

A typical value of this minimum separation is found in the next example.

Example 24.11

What is the minimum resolvable separation for objects in air illuminated by green light ($\lambda = 500$ nm) if the angle subtended by the objective is 90°?

With $n = 1$ and $\sin 90° = 1$,

$$d = \frac{\lambda}{2n \sin \theta} = \frac{500 \text{ nm}}{2(1)(1)} = 250 \text{ nm}$$

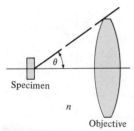

Figure 24.32. The numerical aperture of this lens is $n \sin \theta$, where n is the refractive index of the medium between the lens and the specimen.

We saw in Example 24.7 that a magnification of 400 leads to a minimum resolvable separation of about 250 nm if diffraction is ignored. However, we see now that diffraction cannot be ignored, since it independently limits the resolution of any light microscope to about 250 nm. Consequently, the useful magnification of the ordinary light microscope is limited to roughly 400. Larger magnifications may make viewing more comfortable, but they reveal no additional details. Since a typical bacterial cell has a diameter of about 1000 nm, it is not possible to make very detailed studies of bacterial structures with light microscopes.

Equation 24.19 suggests two ways to improve the resolution of a microscope: use shorter wavelength light and media with larger indices of refraction. With *oil immersion objectives,* some improvement is obtained by immersing the object in a medium such as oil of cedar, which has $n = 1.4$. *Ultraviolet microscopes* use light with wavelengths somewhat shorter than visible light. In addition to having a smaller minimum resolvable separation, ultraviolet microscopes are useful because substances such as nucleic acids and proteins strongly absorb ultraviolet light. This leads to very good contrast, which is also necessary to obtain good resolution.

Much greater resolution can be achieved with *electron microscopes,* in which electrons are accelerated by a potential difference and focused by magnetic fields. Although electrons behave in many ways as particles, they also have wave attributes that are discussed in Chapter Twenty-seven. The wavelength associated with electrons accelerated through 50,000 V is 5×10^{-3} nm or about $1/10^5$ times that of visible light. In practice the resolution is limited to about

0.2 nm, which is about 1000 times better than can be achieved with light microscopes.

Contrast | To distinguish an object from its surroundings, there must be sufficient contrast or variation in light intensity. Without good contrast, the actual resolution achieved will be much less than that implied by the design of the microscope. *Staining* with dyes that are absorbed differently in various parts of an object is sometimes used to improve contrast. Other dyes are used to make a sample *fluoresce*. In this case the sample is illuminated with intense ultraviolet light, and the fluorescent constituents emit light at a longer wavelength. This light is detected after the ultraviolet light is removed from the beam by a filter.

24.11 | POLARIZING, INTERFERENCE, AND PHASE CONTRAST MICROSCOPES

We now briefly describe *polarizing, interference,* and *phase contrast* microscopes. These instruments ingeniously exploit the wave properties of light in improving the contrast of transparent structures.

A *polarizing microscope* uses polarized light to illuminate an object. When the object has a random or uniform character, the polarization of the light passing through it is unaffected. If this light then passes through an analyzer set at 90° to the original polarization direction, no light is transmitted. However, if the sample contains *birefringent* materials, whose indices of refraction depend on the direction of the light beam and on the electric field direction, then the polarization vector of the light is rotated and some light passes through the analyzer (Fig. 24.33). Proteins and nucleic acids are birefringent and can be seen in a polarizing microscope.

In an *interference microscope,* the illuminating light is split into two beams (Fig. 24.34). One beam passes through a sample whose refractive index varies with the position. The phases of the different parts of this beam have corresponding variations after it leaves the sample. The second beam travels through an identical optical path, except that it does not go through the sample. When the two beams are recombined, interference between them produces intensity variations. Thus, many kinds of structures can be made visible, even though the sample is completely transparent.

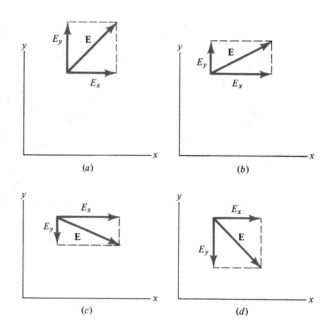

Figure 24.33. The polarization of an electromagnetic wave is parallel to the electric field vector **E**. In a birefringent medium, waves with electric fields along different crystal axes travel with different velocities. (a) Traveling into the page, the wave enters the birefringent medium. The initial electric field has x and y components that now travel at different velocities. (b) At a point inside the medium, E_x has a maximum at some instant. The wave with an electric field in the y direction travels slower, so E_y is not yet a maximum. Hence the **E** vector is rotated somewhat from its original direction. (c) Further into the medium, E_y is still negative when E_x is a maximum. (d) After an additional distance, E_y has a maximum negative value when E_x is a maximum. The polarization has now been rotated 90° from its original direction by the birefringent material.

The *phase contrast microscope,* invented in 1932 by Fritz Zernike, is an ingenious instrument that employs interference to enhance contrast but involves only a single beam of light. Consequently, it is less

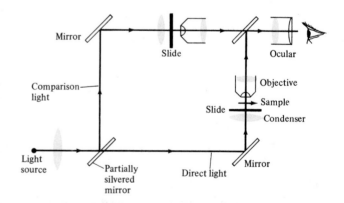

Figure 24.34. Interference microscope. The elements are described in the text.

costly and complicated than the interference microscope.

Since the operation of this microscope depends on diffraction effects, we first consider some aspects of diffraction. We have seen in Chapter Twenty-three that when a parallel beam of monochromatic light passes through a narrow slit, an undeviated bright central line is formed on a screen, with progressively weaker lines on either side. (Similarly, when light goes through a circular aperture, a bright central circular spot and weak concentric rings are seen.) The central line and the lines immediately adjacent on either side can be shown to be exactly a quarter wavelength out of phase. Now suppose the screen is removed, and the bright central beam passes through a partially absorbing coated glass plate so that its amplitude becomes equal to that of an adjacent diffracted beam (Fig. 24.35a). The latter passes through a thicker transparent glass plate. The beam travels slower in the glass than in air, so it emerges an *additional* quarter wavelength out of phase with the central beam. These two beams are now a half wavelength out of phase, or exactly opposite in sign, and equal in amplitude. Hence, they interfere destructively when they are recombined with a lens.

Essentially, this idea is applied to a rather different geometry in the phase contrast microscope (Fig. 24.35b). A transparent ring is placed in front of the light source, producing a hollow cone of light with its point at the specimen. If there is no specimen, this light forms a bright circular image just beyond the focal point of the objective. The same thing happens if there is a completely uniform transparent specimen. However, suppose the specimen contains a small transparent structure. This structure will have a diffraction pattern consisting of an undeviated bright circle and weak diffracted concentric circles (Fig. 24.35c). The adjacent diffracted circles are again a quarter wavelength out of phase with the undeviated beam. A *phase plate* placed at the location of this

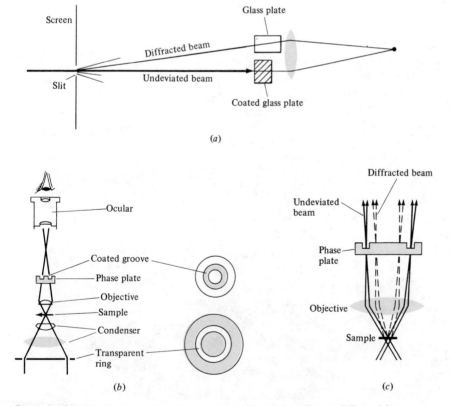

(a)

(b)

(c)

Figure 24.35. (a) The undeviated beam from a slit and the adjacent diffracted beam are a quarter wavelength out of phase. If the phase of the diffracted beam is shifted an additional quarter wavelength by a glass plate, it is exactly out of phase with the undeviated beam. (b) A phase contrast microscope. (c) An expanded view near the objective.

image has a circular groove corresponding to the bright circle. The groove is coated so that it absorbs much of the light passing through this part of the plate, reducing its amplitude to that of the adjacent diffracted light. The latter passes through a thicker portion of the phase plate, emerging exactly out of phase with the light passing through the groove, so that destructive interference occurs when the two beams are later recombined. In this way, a dark spot is formed due to a structure with a slightly different refractive index than the remainder of a transparent object (Fig. 24.36).

24.12 | OPTICAL DEFECTS OF THE EYE

Four common optical defects of the eye can be corrected by the use of eyeglasses. In three of these defects, the glasses are used to shift the apparent position of an object, so that the defective eye is able to focus properly. In the last, *astigmatism,* the glasses correct for a distortion produced by the eye.

In *myopia* or nearsightedness, parallel light from a distant object is focused by the relaxed eye at a point in front of the retina (Fig. 24.37). Consequently a nearsighted person cannot focus clearly on objects farther away than the far point located at a distance x_f. This problem arises because the power of the eye is too great; either the cornea has an excessive curvature or the eye is longer than normal. Diverging lenses with negative powers will compensate for this defect.

To calculate the required correction, we find the power of the eye at its far point, and then select a lens of the correct power to move the far point to infinity. We use the thin lens formula, $1/f = 1/s + 1/s'$, rewritten in terms of the power $P = 1/f$:

$$P = \frac{1}{s} + \frac{1}{D} \qquad (24.20)$$

Here D is the image distance in the eye, approximately 0.02 m. The procedure is illustrated by the following example.

Example 24.12

A nearsighted man has a far point at a distance of 0.2 m. His power of accommodation is 4 diopters. (a) What power lenses does he need to see distant objects? (b) What is his near point without the glasses? (c) What is his near point with the glasses?

At the far point, $s = 0.2$ m. Thus the power of his eye when fully related is, using Eq. 24.20,

$$P_f = \frac{1}{0.2 \text{ m}} + \frac{1}{0.02 \text{ m}} = 55 \text{ diopters}$$

To have his far point at infinity, he needs a power of

$$P_f' = \frac{1}{\infty} + \frac{1}{0.02 \text{ m}} = 50 \text{ diopters}$$

When he wears glasses, the sum of the powers of the lens and of his eye determines the effective power. Hence if he wears a lens of power $(50 - 55) = -5$ diopters, he will have a net power of 50 diopters when his eye is relaxed, and will see distant objects clearly.

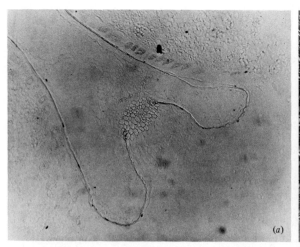 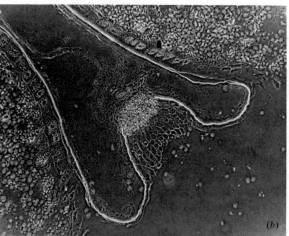

(a)   (b)

Figure 24.36. (*a*) An unstained cross section of a grain of wheat as seen with a conventional bright-field microscope. (*b*) The same scene as viewed with a phase contrast microscope. (Courtesy of Bausch & Lomb.)

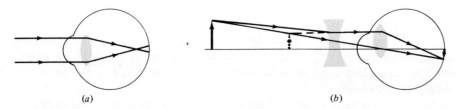

(a)   (b)

Figure 24.37. (a) Parallel light rays from a distant object are focused at a point before the retina of a nearsighted or myopic eye. (b) A divergent lens bends the rays so that they appear to come from a closer location and enables the eye to image the rays on the retina.

(b) Since his power of accommodation is $A = P_n - P_f = 4$ diopters, $P_n = P_f + A = (55 + 4)$ diopters = 59 diopters. He will focus at a point $s = x_n$ satisfying

$$P_n = \frac{1}{x_n} + \frac{1}{D}$$

$$59 \text{ diopters} = \frac{1}{x_n} + \frac{1}{0.02 \text{ m}} = \frac{1}{x_n} + 50 \text{ diopters}$$

$$x_n = 0.11 \text{ m}$$

(c) With the glasses, $P'_n = P'_f + A = (50 + 4)$ diopters = 54 diopters. Then

$$P_n = \frac{1}{x'_n} + \frac{1}{D}$$

$$54 \text{ diopters} = \frac{1}{x'_n} + \frac{1}{0.02 \text{ m}} = \frac{1}{x'_n} + 50 \text{ diopters}$$

$$x'_n = 0.25 \text{ m}$$

This is the near point distance for a person with normal vision and average accommodation. The lenses have fully corrected his vision.

Hypermetropia, or farsightedness, is the opposite of myopia. Light from an object close to the eye is focused toward a point behind the retina, even when the lens is adjusted by the ciliary muscles to have its

maximum power (Fig. 24.38). Eyeglasses with converging lenses supply the additional focusing power needed, as in the next example.

Example 24.13

A woman has her near point 1 m from her eyes. What power glasses does she require to bring her near point to 0.25 m from her eyes?

Using Eq. 24.20, at her near point the power of her eye is

$$P_n = \frac{1}{1 \text{ m}} + \frac{1}{0.02 \text{ m}} = 51 \text{ diopters}$$

To focus at 0.25 m, she would need a power

$$P'_n = \frac{1}{0.25 \text{ m}} + \frac{1}{0.02 \text{ m}} = 54 \text{ diopters}$$

A lens with a power of +3 diopters will give her the required power to bring her near point to the normal location.

Presbyopia, the reduction in accommodation that occurs with age, is the result of a gradual weakening of the ciliary muscles and diminishing flexibility of the lens. The far point of a person who has normal vision as a young adult eventually recedes far enough so that converging lenses are needed for close work or reading, much like the person with hypermetropia.

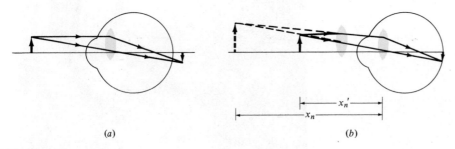

(a)   (b)

Figure 24.38. (a) Rays from a nearby object are bent so they would focus at a point beyond the retina of a farsighted or hypermetropic eye. (b) A converging lens bends the rays so they appear to come from a more distant location. The distance x'_n to the near point with the glasses is less than the distance x_n without the glasses.

Many people eventually require bifocal lenses. In the most common type of bifocals, the upper portion of the lens is used for distant vision, and the lower part is used for close work. For example, an elderly myopic person will wear bifocals with a diverging lens in the upper part and a weaker diverging lens below.

A person with *astigmatism* cannot simultaneously focus on both horizontal and vertical lines. Usually this is due to a cornea that is not perfectly spherical, so that it has different curvatures in different directions. Occasionally, astigmatism is caused by irregularities elsewhere in the eye. Astigmatism can be corrected by cylindrical lenses oriented to compensate for the distortion (Fig. 24.39). If a correction for myopia or hypermetropia is also needed, then lenses shaped like the outer surface of a doughnut are used. The maximum and minimum radii of curvature are chosen to correct both defects.

Contact lenses are an alternative to ordinary eyeglasses and have many advantages (Fig. 24.40) Hard contact lenses are made of rigid plastic and are about 1 mm thick and 1 cm in diameter. Placed directly on the eye, they are separated from the cornea by a layer of tears, *effectively replacing the cornea as the front of the eye.* Since the light now enters the eye through a spherical lens surface, astigmatism due to corneal irregularities is automatically corrected. Also, the lenses move with the eyes, avoiding the aberrations that distort the view near the edges of strong eyeglass lenses. Soft contact lenses are somewhat larger than hard contact lenses. They conform to the shape of the cornea and are often more comfortable. Since most soft contacts do not maintain a spherical surface when placed on an irregular cornea, they do not correct for astigmatism. Toric soft contacts, which are relatively expensive and sometimes less comfortable, are shaped to compensate for corneal irregularity, so they do correct for astigmatism.

Surgeons in Colombia, the Soviet Union, and more recently the United States, have corrected the vision of nearsighted and farsighted patients by reshaping the cornea. In one procedure in use since the 1960's, the front of the cornea is removed and reshaped on the inside by a computer-controlled lathe. For a farsighted patient, tissue is removed from the edges, increasing the curvature and power. Alternatively, tissue from a donor's cornea is properly shaped and placed under the center of the cornea before it is replaced to increase the curvature. Similarly, for a nearsighted person, tissue is removed from the center to flatten the cornea and decrease the power. In a newer operation for nearsightedness, 16 or so shallow radial cuts are made from the outer edge of the cornea toward the center, rather like the spokes of a wheel. This stretches and flattens the nicked regions. Both procedures are still considered experimental and there is limited data on long-term benefits and problems.

24.13 | COLOR PERCEPTION AND MEASUREMENT

Scientists, artists, and paint manufacturers have long been intrigued by our perception of color, and by the

Figure 24.39. A cylindrical lens is used in eyeglasses to correct for astigmatism.

Figure 24.40. A soft contact lens being inserted. (Courtesy Bausch & Lomb, Soflens Division.)

complex relationship between the physical properties of light and its perceived attributes of *hue, brightness,* and *saturation. Hue* is what we colloquially call color; it includes the basic spectral colors corresponding to single wavelengths—red, yellow, blue, and so forth—as well as some *extraspectral* hues, such as crimson and purple, formed by mixing violet and red lights from the opposite ends of the visible spectrum. *Brightness* is the subjective impression of intensity, but differs from it because of the wavelength variation of the intensity discussed in Sec. 24.7. *Saturation* is the purity of depth of a color; when mixed with a neutral color (white, gray, black), a color becomes less saturated. For example, white light added to saturated red results in pink. The hue, brightness, and saturation can all be changed somewhat by variations in any one of the physical parameters, the wavelength, intensity, and spectral composition, as well as by changes in the surroundings. Clearly, color perception is a very complex phenomenon.

Color perception is made possible by the existence of three kinds of cones in the retina, each with its own photosensitive pigment. Each pigment can absorb light over a large range of wavelengths, but the peak sensitivities occur at different wavelengths: 445 nm, 535 nm, and 575 nm (Fig. 24.41). Light at a particular wavelength will excite all three kinds of cones to some

extent, depending on how close it is to the corresponding peaks. Although the existence of three different kinds of detectors was confirmed in recent years by direct experiments, it had been suggestyed in the last century by Young, Helmholz, and Maxwell because of the observed properties of mixtures of light beams with different wavelengths.

Every color has a *complementary color,* which when mixed with it in the right proportion produces the same sensation of white as a beam of sunlight containing the whole visible spectrum. For example, red (R) plus cyan (C) (a blue green) produces the sensation of white (W). Symbolically, we can write $W = R + C$, or $C = W - R$. That is, the complementary color is obtained if we subtract the original color from white by absorbing it with a suitable pigment.

Any color that is not too saturated can be matched in appearance by a suitable mixture of *primaries,* a set of three saturated colors well spread across the spectrum but otherwise arbitrary, such as red, green, and blue. This is why color television screens use three kinds of colored dots, and color photography employs three emulsions. Mixing colors always results in some desaturation, so a saturated color that is not selected as a primary cannot be exactly matched by such a mixture. However, if enough of one of the primaries

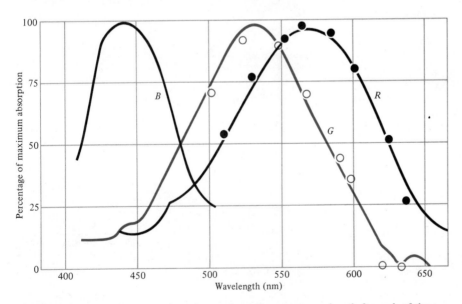

Figure 24.41. The percentage of maximum absorption versus wavelength for each of three cone pigments. The curves are the results of measurements on single cones taken from human and monkey retinas. The open and filled circles are data obtained by studies of the absorption of light in the living human eye. (Adapted from W.A.H. Rushton, Visual Pigments and Color Blindness, *Scientific American,* vol, 232, March 1975.)

is added to this color as a *desaturant,* the resulting mixture can be matched by a mixture of the two remaining primaries. Thus, experiments show that *any perceived color can be represented mathematically by a mixture of three primaries,* provided that a desaturant is treated as a negative component.

A crude circular model that displays all these general features of color mixing is shown in Fig. 24.42. Three primaries *A, B,* and *C* define a triangle that represents in an approximate way the colors that can be obtained by mixing the primaries. It also allows the identification of complementary colors. Saturated colors, such as *C* and *E,* which are at the ends of a line through *W,* are complementary.

The representation of colors in terms of three primaries can be cast into a compact, symbolic form. The chosen primaries are designated as *X, Y,* and *Z.* The fractional amount of each primary is given by the *chromaticity coordinates x, y,* and *z*; since they are fractional parameters, their sum is one:

$$x + y + z = 1 \qquad (24.21)$$

If *x* and *y* are known, *z* is equal to $(1 - x - y)$ and is also known. Thus a color represented by the chromaticity coordinates (x, y, z) can be located on an *xy* plane or *chromaticity diagram.*

A standard chromaticity diagram has been adopted for the specification of colors (Fig. 24.43). It uses imaginary primaries to avoid negative values of the chromaticity coordinates corresponding to desaturants. The heavy colored line shows the saturated colors. Moving inward from this line, the saturation decreases, reaching pure white at the point indicated; the intensity units for the primaries have been adjusted so that $x = 1/3$, $y = 1/3$, $z = 1/3$ produces the sensation of white. Effects of color matching are readily found with this diagram. For example, mixing the saturated colors *A* and *B* in varying proportions yields matches to unsaturated colors along the line *AB*. At point *P*, $x = 0.2$, $y = 0.4$, and $z = (1 - 0.2 - 0.4) = 0.4$. The same perceived color is produced equally well by this mixture of primaries or by a mixture of *A* and *B* with relative proportions $5:3$, which is the inverse of the ratio of the distances found with a ruler from *P* to *A* and from *P* to *B*. The

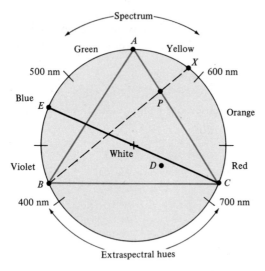

Figure 24.42. A simplified model of color mixing that displays many of the observed properties. Points on the perimeter are fully saturated; the center is pure white, and saturation decreases as one approaches the center. Mixing the three primaries *A, B,* and *C* in varying proportions leads to colors matching any color such as *D* within the triangle they define. The saturated color *X* is outside this triangle and cannot be matched. However, if enough *B* is admixed to *X,* they produce a mixture which falls at point *P,* and this color can also be produced by mixing *A* and *C.* The complement to *C* is *E*; when these colors are combined, they produce the sensation of white.

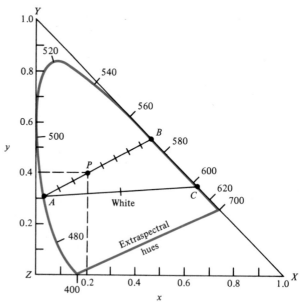

Figure 24.43. The standard chromaticity diagram adopted in 1931 by the International Commission on Illumination. A point on the diagram corresponds to a color with chromaticity coordinates *x, y,* and $z = (1 - x - y)$. These coordinates specify the relative amounts of the imaginary primaries *X, Y,* and *Z.* The points on the colored curve are labeled with wavelengths in nanometres and correspond to the saturated spectral hues. *C* is the complement of *A.*

matches predicted by this diagram hold over a large range of brightness levels, even though in some cases the perceived colors will vary as the brightness changes.

By itself, the existence of three kinds of detectors does not explain some color-vision phenomena, such as the specific colors seen by color-blind people lacking one of the pigments, and the occurrence of negative afterimages, the sensation of the complementary color that occurs upon the removal of a color after a long exposure. Intricate, only partially understood neurological processing is involved, so many questions remain to be answered in this fascinating field.

EXERCISES ON SUPPLEMENTARY TOPICS

Section 24.8 | Spherical Mirrors

24-51 A concave spherical mirror with a 3-m radius of curvature is used as a lunar telescope. Find the radius of the image it forms of a full moon, using the information given in the table of solar and terrestrial data on the inside back cover.

24-52 A peacock standing 2 m from the highly polished curved metal of a car sees "another bird" (its own image) 0.5 m behind the metal. (a) What is the radius of curvature of this part of the car? (b) If the peacock moves 0.01 m closer to the car, where does the other bird move?

24-53 A concave shaving mirror with a radius of 0.3 m is held so that the image of a man's face is three times as large as his face. How far is the mirror from his face?

24-54 A silvered sphere used as a Christmas tree ornament has a diameter of 0.08 m. A girl holds the ornament 0.2 m from her eye. (a) Where does she see the image of her eye? (b) What is the ratio of the height of this image to the height of her eye? (c) Is the image erect or inverted?

24-55 Use graphical methods to find the location and magnification of the image formed by a concave (converging) mirror if the object distance is (a) $2R$; (b) R; (c) $R/4$.

24-56 Use graphical methods to find the location and magnification of the image formed by a convex (diverging) mirror if the object distance is (a) $2|R|$; (b) $|R|$; (c) $|R|4$.

Section 24.9 | The Camera

24-57 An $f/1.4$ camera lens has a focal length of 50 mm = 0.05 m. What is the lens diameter?

24-58 With a certain set of lighting conditions, a photographer finds that an exposure of 1/25 second is needed at $f/16$. What exposure is needed at $f/8$?

24-59 The diaphragm of a lens is stopped down from $f/2$ to $f/16$. By what factor is the light intensity reduced?

24-60 Why do telephoto (long focal length) lenses usually have larger f-numbers than normal camera lenses of similar cost and quality?

24-61 A camera lens is focused on a distant scene illuminated with white light. Find the radii of the diffraction circles on the film at f-numbers of (a) $f/2$, and (b) $f/32$.

Section 24.10 | Resolution and Contrast in Microscopes

24-62 An oil immersion objective is immersed in oil of cedar with a refractive index 1.4. If the objective subtends an angle of 80°, find the (a) numerical aperture, and (b) minimum resolvable separation at a wavelength of 400 m.

Section 24.12 | Optical Defects of the Eye

24-63 An elderly man has his near point 2 m from his eyes. What power lenses does he need to read comfortably at 0.25 m?

24-64 A woman has a far point 0.5 m from her eyes. (a) If she is to see distant objects clearly, what focal length lenses does she require? (b) If her power of accommodation is 4 diopters, where is her near point without the glasses? (c) Where is her near point with the glasses?

24-65 A farsighted person with an accommodation of 3 diopters has a near point 2 m from the eyes. (a) What power eyeglasses are needed to move the near point to 0.25 m from the eyes? (b) Where is the far point with these eyeglasses?

24-66 A very nearsighted man has his near point at 0.1 m. His power of accommodation is 4 diopters. (a) Where is his far point? (b) What power lens does he require? (c) What is his near point when wearing the glasses?

24-67 A nearsighted man wears glasses with a correction of −6 diopters to see distant objects clearly. Where is his far point without the glasses?

24-68 A farsighted woman wears glasses with a power of +3 diopters to read books at a distance of 0.25 m. Where is her near point without glasses?

Section 24.13 | Color Perception and Measurement

24-69 (a) In Fig. 24.41, at what wavelength will the pigments labeled B and G have equal percentages of their maximum absorption? (b) What is the corresponding percentage for the pigment labeled R at that wavelength?

24-70 In Fig. 24.41, what are the percentages of maximum absorption for the pigments labeled G and R if light with a wavelength of 600 nm is incident on the retina?

24-71 What are the chromaticity components and wavelength of the complementary color of (a) spectral 480 (blue); (b) spectral 520 (green)? (Use Fig. 24.43).

24-72 (a) What are the chromaticity coordinates of a mixture of equal parts of spectral 500 (green) and spectral 580 (yellowish orange)? (b) What spectral color should be mixed with white and in what proportions to give the same result? (Use Fig. 24.43).

PROBLEMS ON SUPPLEMENTARY TOPICS

24-73 Show in detail that equating the expressions for m in Eqs. 24.14 and 24.15 leads to $1/s + 1/s' = 1/f$.

24-74 A boy walks toward a convex spherical mirror at 2 m s^{-1} and sees his image walk towards him at 0.5 m s^{-1}. If the radius of curvature of the mirror is 2 m, how far is he from the mirror?

24-75 A nearsighted middle-aged man has a near point of 0.1 m and a power of accommodation of 2 diopters. Find his far point (a) without glasses, and (b) with the correct glasses to move his near point to 0.25 m.

*24-76 A refracting telescope (Fig. 24.44) focused on an astronomical object forms an image at infinity. (a) If the focal lengths of the objective and ocular are f_1 and f_2, respectively, show that the separation between the lenses is $f_1 + f_2$. (b) Using small angle approximations, show that the angular magnification $M = \theta_2/\theta_1$ satisfies $M = -f_1/f_2$.

24-77 A telescope is constructed using lenses of focal lengths 2 m and 0.1 m. Using the result of the preceding problem, find its angular magnification.

24-78 The ocular of a refracting telescope has a focal length of 0.1 m. If the length of the telescope is 3 m, what is its angular magnification? (Use the results of Problem 24-76.)

24-79 The discussion of color mixing in the text is based on mixing lights of different wavelengths and perceived colors. The colors of surfaces follow quite different rules, because surfaces absorb particular parts of the visible spectrum, and transmit or reflect the remainder. Thus a pigment or dye that absorbs one color strongly will make the surface appear to have the complementary color. Suppose we take our primaries to be red (R), green (G), and blue (B). The complementary colors are cyan (C) (a blue green), magenta (M), and yellow (Y), respectively. (a) What primaries are reflected by magenta paints? (b) What primaries are reflected by yellow paints? (c) What primary will be reflected if the paints are mixed? (*Hint:* White (W) = $R + G + B$.)

Additional Reading

The following books have chapters on microscopes and the eye:

R.B. Setlow and E.C. Pollard, *Molecular Biophysics,* Addison-Wesley Publishing Co., Reading, Mass., 1962, chapters 12 and 14.

Simon G.G. MacDonald and Desmond M. Burns, *Physics for the Life and Health Sciences,* Addison-Wesley Publishing Co., Reading, Mass., 1975, chapters 20, 21, and 23.

D. Ackerman, *Biophysical Science,* Prentice-Hall, Inc., Englewood Cliffs, N.J., 1962, chapters 2, 7, 19, and 29.

John R. Cameron and James G. Skofronick, *Medical Physics,* John Wiley and Sons, Inc., New York, 1978, chapters 14 and 15.

William Hughes, *Aspects of Biophysics,* John Wiley and Sons, Inc., New York, 1979, chapter 15.

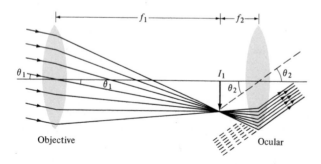

Figure 24.44. A refracting telescope. Problem 24-76.

Other references on the material of this chapter:

Michael J. Ruiz, Camera Optics, *The Physics Teacher,* vol. 20, 1982, p. 372.

S.F. Jacobs and A.B. Stewart, Chromatic Aberration in the Eye, *American Journal of Physics,* vol. 20, 1952, p. 247.

M.H. Pirenne, *Vision and the Eye,* Chapman and Hall, London, 1967.

S.M. Luria, Color Vision, *Physics Today,* March 1968, p. 35.

Joseph J. Sheppard, Jr., *Human Color Perception,* American Elsevier Publishing Co., New York, 1968.

Kenneth Neil Ogle, *Optics: An Introduction for Ophthalmologists,* Charles C Thomas, Springfield, Ill., 1968.

Gerald S. Wasserman, *Color Vision: An Historical Introduction,* John Wiley and Sons, Inc., New York, 1978.

Timothy H. Goldsmith, Hummingbirds See Near Ultraviolet Light, *Science,* vol. 207, 1980, p. 786.

Michael J. Ruiz, The Physics of Visual Acuity, *The Physics Teacher,* vol. 18, 1980, p. 457.

Thomas H. Maugh III, A New Microscopic Tool For Biology, *Science,* vol. 206, 1979, p. 918. Near-infrared microscopy.

Michael L. Brines and James L. Gould, Bees Have Rules, *Science,* vol. 206, 1979, p. 573. Honey bee dances correlate with patterns of polarized skylight.

Karl Von Frisch, *The Dance Language and Orientation of Bees,* Belknap Press of Harvard University Press, Cambridge, Mass., 1967.

Scientific American Articles:

F. Dow Smith, How Images are Formed, September 1968, p. 97.

R. Clark Jones, How Images are Detected, September 1968, p. 111.

P.K. Tien, Integrated Optics, April 1974, p. 28.

Ivo Kohler, Experiments with Goggles, May 1962, p. 62.

P. Connes, How Light is Analyzed, September 1968, p. 72.

W.R.A. Muntz, Vision in Frogs, March 1964, p. 110.

Talbot H. Waterman, Polarized Light and Animal Navigation, July 1955, p. 88.

John F. Harris and Igor Gamow, The Infrared Receptors of Snakes, May 1973, p. 94.

Eric A. Newman and Peter H. Hartline, The Infrared "Vision" of Snakes, March 1982, p. 106.

George Wald, Eye and Camera, August 1950, p. 32.

Charles R. Michael, Retinal Processing of Visual Images, May 1969, p. 105.

Derek H. Fender, Control Mechanisms of the Eye, July 1964, p. 24.

Ulric Neisser, The Processes of Vision, September 1968, p. 204.

G. Adrian Horridge, The Compound Eye of Insects, July 1977, p. 108.

Edwin H. Land, Experiments in Color Vision, May 1959, p. 84.

Edwin H. Land, The Retinex Theory of Color Vision, December 1977, p. 108.

George Wald, Life and Light, October 1959, p. 92.

W.A.H. Rushton, Visual Pigments and Color Blindness, March 1975, p. 64.

W.A. Rushton, Visual Pigments in Man, November 1962, p. 120.

Jacob Bech, The Perception of Surface Color, August 1975, p. 62.

Bela Julesz, Experiments in the Visual Perception of Textures, April 1975, p. 34.

Robert Sekular and Eugene Levinson, The Perception of Moving Targets, January 1977, p. 60.

Gunnar Johannson, Visual Motion Perception, June 1975, p. 76.

Robert H. Wurtz, Michael E. Goldberg, and David Lee Robinson, Brain Mechanisms of Visual Attention, June 1982, p. 124.

Olga Eizner Favreau and Michael E. Corballis, Negative Aftereffects in Visual Perception, December 1976, p. 42.

Edward F. MacNichol, Jr., Three-Pigment Color Vision, December 1964, p. 48.

Joseph S. Levine and Edward F. MacNichol, Jr., Color Vision in Fishes, February 1982, p. 140.

John I. Yellott, Jr., Binocular Depth Inversion, July 1981, p. 148.

Michael Menaker, Nonvisual Light Reception, March 1972, p. 22.

Alistair B. Fraser and William H. Mack, Mirages, January 1976, p. 102.

Leonard A. Herzenberg, Richard G. Sweet, and Leonore A. Herzenberg, Fluorescence-Activated Cell Sorting, March 1976, p. 108.

William H. Price, The Photographic Lens, August 1976, p. 72.

David Emil Thomas, Mirror Images, December 1980, p. 206.

Eberhard Spiller and Ralph Feder, The Optics of Long-Wavelength X Rays, November 1978, p. 70.

Jearl Walker, Experiments with Edwin Land's Method of Getting Color Out of Black and White, The Amateur Scientist, June 1979, p. 189.

Jearl Walker, Anamorphic Pictures: Distorted Views from

Which Distortion Can Be Removed, The Amateur Scientist, July 1981, p. 176.

Jearl Walker, Interference Patterns Made by Motes on Dusty Mirrors, The Amateur Scientist, August 1981, p. 146.

Jeremy M. Wolfe, Hidden Visual Processes, February 1983, p. 94.

Donald D. Hoffman, The Interpretation of Visual Illusions, December 1983, p. 154.

Robert H. Wurtz, Michael E. Goldberg and David Lee Robinson, Brain Mechanisms of Visual Attention, June 1982, p. 124.

Tomaso Poggio, Vision by Man and Machine, April 1984, p. 106.

John N. Bahcall and Lyman Spitzer, Jr., The Space Telescope, July 1982, p. 40.

Eitan Abraham, Colin T. Seaton and S. Desmond Smith, The Optical Computer, February 1983, p. 85.

UNIT SEVEN

MODERN PHYSICS

Albert Einstein (1879–1955) (The Bettmann Archive).

I N the latter part of the nineteenth century, physics was considered by many to be a completed science. The fundamental laws of motion and electromagnetism, including light waves, were well understood, and it seemed that only calculational and experimental difficulties would hinder further progress. However, after only 20 years of the twentieth century, this description of the physical world was severely shaken. The very few unsolved problems that existed in 1900 proved to be explainable only by drastic assumptions that had no historical precedents. The illusion of a complete science proved to be a result of a lack of experience with atomic-size particles and with objects that move at nearly the speed of light.

By 1912 the work of Max Planck, Niels Bohr, and Albert Einstein had given us a new picture of the world. These three set forth the tentative but crucial ingredients of what is now known as quantum mechanics. During the same period, Einstein also developed the special theory of relativity. Quantum mechanics and relativity are our present basis for understanding nature. Since their predictions agree with many kinds of experiments, they must be considered not as speculative but as a largely correct description of nature.

The three chapters of this unit set forth our fundamental concepts of special relativity, the particle nature of light, and the wave nature of matter. These concepts are the foundations of the remaining chapters in this book, which cover atoms, molecules, and atomic nuclei.

It should be recognized at the outset that while modern physics has forced us to change our philosophy of natural processes, the procedures of the old or classical physics are still essentially correct under certain conditions. Those conditions are that the objects under consideration are large compared to atomic sizes and that their velocities are much less than the speed of light. In nearly all the situations that we have considered up to now, these conditions are well satisfied.

CHAPTER 25

SPECIAL RELATIVITY

At the end of the nineteenth century, several fundamental problems existed in physics. One set of problems eventually led to the development of the revolutionary concepts of quantum mechanics, although it took several decades and many people to develop them. A second set of problems was solved in a single stroke by Einstein when his theory of special relativity was published in 1905.

The problem solved by Einstein was almost obscure enough to be largely ignored, but his solution has had a profound effect on our conception of the physical world. The *special theory of relativity* relates time and length measurements made by inertial observers moving at constant velocities with respect to one another. The *general theory of relativity,* which Einstein began to develop in 1911, deals with accelerated systems and gravitational forces. However, in this book, we only discuss the special theory of relativity.

The problem that Einstein originally addressed was an apparent violation of a fundamental rule of physics. This rule is called the principle of relativity and was developed first by Galileo. It states that the laws of physics should be the same for all observers moving at constant velocities with respect to each other. We describe this idea further in the next section. Here, we need only remark that Newton's laws of mechanics seemed in agreement with this rule, but the laws of electricity and magnetism did not. It appeared that the effects of moving charges on each other depended on which ones were moving. If this were the case, there would be different results measured by different moving observers.

Einstein found that electromagnetism would be a consistent theory if one insisted that the speed of light in the vacuum is the same for all observers and is independent of the motion of the source and the detector. In this way, light waves are different from sound or water waves, whose velocities are determined by their media. Einstein's view, which is now well supported by experiment, is that there is no medium required for the propagation of electromagnetic waves.

If one accepts Einstein's principles that the laws of physics and the speed of light are the same for all inertial observers, one is led directly to some remarkable conclusions. Moving clocks run slow, and moving objects are shortened. Thus, according to an earth observer, a fast-moving astronaut will live longer, and his rocket ship is shorter than before blastoff. Also, events that appear simultaneous according to the astronaut are not simultaneous as viewed by the earth observer.

Because our ideas of space and time are altered for rapidly moving objects, the laws of mechanics are also affected. The definitions of energy and momentum must be changed if these quantities are to obey the usual conservation laws. If an object has a mass m and a velocity v, its relativistic energy turns out to be

$$E = \frac{mc^2}{\sqrt{1 - v^2/c^2}}$$

When the velocity is zero, the energy becomes the *rest energy*

$$E_0 = mc^2$$

This is the most famous equation of twentieth-century physics. It asserts that mass (or matter) and energy are equivalent. Mass is converted into energy in the nuclear reactions that provide the energy of the sun and in nuclear explosives and reactors. If just 1 kg of matter is converted into energy, the energy released is $mc^2 = (1 \text{ kg})(3 \times 10^8 \text{ m s}^{-1})^2 = 9 \times 10^{16}$ J. This is comparable to the total energy used per day to generate electricity in the United States.

25.1 | THE FUNDAMENTAL PRINCIPLES OF SPECIAL RELATIVITY

We saw in Chapter Three that Newton's laws of motion apply only to measurements made with respect to inertial reference frames. These are frames in which Newton's first law holds true: an object with no net force on it remains at rest or in motion with a constant velocity. For example, observers in accelerating vehicles or on rotating merry-go-rounds cannot apply Newton's laws directly to their observations. Special relativity also deals only with measurements made by observers in inertial reference frames.

The theory of special relativity is based on two fundamental principles.

1 *All the laws of physics (and nature) have exactly the same form in all inertial reference frames. This is the principle of relativity.*
2 *The speed of light in free space is the same for all observers in inertial reference frames.*

The full implications of these ideas will become gradually clearer in this chapter, but we can immediately see some of the consequences. For example, the first principle implies that two people moving at a constant velocity with respect to each other can never decide which is moving and which is at rest. This is because all physical laws are the same in both systems, and they do not depend on any absolute velocity. Thus the observers can only determine their relative motion. For example, a woman on a train moving with constant velocity relative to the ground might say that she is stationary and the earth is moving. No experiment will decide whether this is true or false. The assumption that the earth is at rest and the train is moving is no more and no less valid.*

*Actually the earth is not exactly an inertial frame; as it rotates about the sun and spins on its axis, objects on the earth experience a centripetal acceleration. In most situations the effects of this acceleration are small, so we assume in this chapter that the earth is an inertial frame.

Turning to the second principle, it is apparent that if the speed of light is the same for all inertial observers, light must be different from all other waves we have studied. For example, a listener moving toward a sound source will observe a higher speed of sound than someone at rest relative to the air. Nothing like this happens for light.

It was long thought that there was some medium, the *ether,* in which light moved at a speed c. This ether was the counterpart of the medium in which sound moves. Thus if one were moving relative to the ether, the speed of light would appear different from c. However, no experiment ever detected the ether or measured changes in the speed of light due to motion relative to the ether, and as far as we know now, there is no ether. In fact the strongest evidence we have that there is no ether is the success of the theory of relativity. The theory of relativity could not be correct if the ether existed.

Einstein's motivation for the theory of relativity was to remove the contradictions then present in the laws of mechanics and electromagnetism. To the objection that the results were disturbing to our commonsense view of things, he answered, "Common sense is that layer of prejudices laid down in the mind prior to the age of eighteen."

25.2 | MOVING CLOCKS AND TIME DILATION

Einstein's theory predicts that if a clock is moving with respect to an observer in an inertial reference frame, he will observe it to be running slower than a clock at rest relative to the observer. The latter clock is said to be in the observer's *rest frame.* In order to show how this follows directly from the principles of special relativity, we first consider a particular type of clock called a *light clock.*

This clock is a stick of length l with a mirror R and a photodetector P at opposite ends (Fig. 25.1). A flash of light emitted at one end will be reflected by the mirror at the other end and return to the photodetector next to the light source. Each time a light flash is detected, the clock "ticks" and emits another flash.

We can easily relate the time t between ticks of the clock to the length l when the clock is at rest in a laboratory. From Fig. 25.1a the total distance traveled by a light pulse is $x = 2l$, and the speed of the light is c. Thus $x = vt$ becomes $2l = ct$, or $l = ct/2$, where t is the time between ticks.

When the clock is moving with a speed u relative to

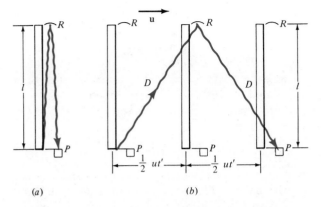

Figure 25.1. (*a*) Each time a flash of light is reflected from the mirror *R* and returns to the photodetector *P* beside the source, the light clock "ticks" and emits a new pulse. (*b*) The same clock moving to the right at a constant velocity **u**. The light pulse must travel the distance 2*D* to return to the photodetector *P*.

the laboratory, the speed of light is again *c* (principle 2), but the distance it must travel between ticks is 2*D* (Fig. 25.1*b*). With the Pythagorean theorem, we have

$$D^2 = l^2 + \left(\frac{1}{2}ut'\right)^2$$

where *t'* is the time for the light to return to the photodetector of the moving clock. But we also know that $2D = ct'$ or $D = ct'/2$ and that $l = ct/2$. Substituting these for *D* and *l*, we find

$$\left(\frac{ct'}{2}\right)^2 = \left(\frac{ct}{2}\right)^2 + \left(\frac{ut'}{2}\right)^2$$

Solving for *t'*,

$$t' = \frac{t}{\sqrt{1 - u^2/c^2}} \qquad (25.1)$$

The time *t'* is greater than *t* because the factor $\sqrt{1 - u^2/c^2}$ is less than 1. (We see later that *u* must be less than *c*.) This result is referred to as *time dilation*. It states that as measured by the observer at rest in the laboratory, time does not pass as rapidly in the moving frame. Note that we have *not* concluded that an observer moving with the clock would see any difference in the interval between ticks of his clock. He would, however, observe that a clock at rest in the laboratory has slowed down by exactly the same time dilation factor. *Each observer observes that the other's clocks are slowed down.* To illustrate the remarkable implications of time dilation, consider an imaginary world in which the effects of special relativity are very pronounced. We do this by pretending that the speed

of light is not 3×10^8 ms^{-1}, but it is instead comparable to everyday speeds.

Example 25.1

Assume for this example that the speed of light is 100 km h^{-1}. A woman relates the story of the delivery of her baby. She says that she drove herself at 80 km h^{-1} to the hospital, which was 100 km away according to the highway signs. She also says that according to her watch the baby was born 1 hour after she left. Was the baby born in the hospital?

If we use our usual reasoning, we would say no, since it would take $t = x/u = 100$ km/80 km h$^{-1} = 1.25$ hours to get to the hospital. However, because of time dilation, once she is traveling at 80 km h^{-1}, we see that her watch slows down. Thus when her watch indicates 1 hour, our watches indicate that the elapsed time is.

$$t' = \frac{t}{\sqrt{1 - u^2/c^2}} = \frac{1\,\text{h}}{\sqrt{1 - \left(\frac{80\,\text{km h}^{-1}}{100\,\text{km h}^{-1}}\right)^2}}$$

$$= \frac{1\,\text{h}}{0.6} = 1.67\,\text{h}$$

Since all clocks, *including biological ones,* must behave in the same way, in 1 hour of her time she can drive (80 km h^{-1})(1.67 h) = 133.6 km. The hospital is only 100 km away, so she must have arrived with time to spare.

In this example one might object to the idea that the woman's wristwatch and her biological processes obey the same law that we derived for a light clock. However, *all* clocks and processes must obey this law; otherwise, the principle of relativity would be violated. For example, we can synchronize a light clock and several other types of clocks when they are all at rest. Now when they are set in motion with a common velocity, all the clocks must behave the same way. If some but not all apparently slow down, those that do not could be said to operate correctly in only one reference frame.

There is one more point to clarify about this example. The woman can consider herself to be at rest and the other people in motion. *Accordingly, their clocks will appear to be ticking slowly to her.* Nevertheless, when she arrives, her watch reads an earlier time than the hospital clocks. Since her watch agreed with the clocks where she started, she concludes that the hospital clocks have not been properly synchronized with those at her starting point. We can interpret this to mean that *clocks synchronized by observers at rest in one inertial frame are not synchronized in another.* Note, however, the difference between her experience and that of someone who did not make the trip. Since she accelerated and decelerated, she was not always in the same inertial frame. We explore these questions more fully in Section 25.5 in the Supplementary Topics.

The violent disagreement between the conclusions of this example and our expectations derives from the fact that in our world speeds are normally very small compared to that of light. Our intuition leads us astray when speeds approach c. An experiment that represents one of many direct confirmations of special relativity is described in the next example.

Example 25.2

Short-lived subatomic particles called mu mesons are created by cosmic rays in the upper atmosphere, approximately 10,000 m above us. They travel at $0.999\,c$ ($c = 3 \times 10^8$ m s^{-1}). In laboratory experiments with mu mesons at rest, they have an average lifetime of 2.2×10^{-6} s. (a) How long will the moving particles appear to live to an observer on the earth? (b) Will the average meson reach the earth?

(a) The average lifetime of the moving meson will appear increased to the earth observer. The average lifetime will appear to be

$$t' = \frac{t}{\sqrt{1 - u^2/c^2}} = \frac{2.2 \times 10^{-6}\,\text{s}}{\sqrt{1 - (0.999c/c)^2}}$$
$$= 4.92 \times 10^{-5}\,\text{s}$$

This lifetime is more than 20 times the lifetime of a stationary mu meson.

(b) Using this as the average lifetime, the earth observer sees the average meson travel a distance of

$$D = vt'$$
$$= (0.999)(3 \times 10^8 \text{ m s}^{-1})(4.92 \times 10^{-5} \text{ s})$$
$$= 14,700 \text{ m}$$

This is more than the 10,000-m height at which they were produced, so the average meson does reach the earth. If the lifetime were 2.2×10^{-6} s, a meson would travel only 660 m, and these particles would not be observed at the surface of the earth. Since mu mesons are observed at the earth's surface, this represents a direct confirmation of time dilation.

25.3 | LENGTH CONTRACTION

An object in motion is shorter than it is at rest. This is called *length contraction* and is an immediate consequence of the time dilation effect. To take a concrete situation, consider the woman in Example 25.1. She and an observer stationary with respect to the road

must agree on their relative velocity, in accordance with the principle of relativity. As far as she can tell, her clocks are working normally. Therefore, she will reason that she was able to reach the hospital safely because the distance she had to travel was less than the posted distance of 100 km. In other words, as measured by this moving observer, the length of the road was reduced.

According to a stationary observer, the woman's velocity was $u = l/t'$, where t' is the time as measured by *a clock at rest relative to the road*, and l is the length of the road. The woman measures the same velocity. As we saw in the preceding section, according to *her clock* the time elapsed, t, is less than t' by the time dilation factor, so

$$t = t'\sqrt{1 - \frac{u^2}{c^2}}.$$

Hence the length l' of the road as observed by the woman is $l' = ut = (l/t')(t'\sqrt{1 - u^2/c^2})$, or

$$l' = l\sqrt{1 - \frac{u^2}{c^2}} \tag{25.2}$$

This means that the road is measured to be shorter by the factor $\sqrt{(1 - u^2/c^2)}$ by an observer moving relative to the road. *An object is shortened or contracted according to an observer moving along its length.* Let us again examine the mu meson example from this point of view.

Example 25.3

A mu meson approaches the earth from a height of 10,000 m at a speed of $0.999c$. According to an observer moving with the velocity of the meson, what is the height of the atmosphere?

Using the length contraction formula, the height as measured by this observer is

$$l' = l\sqrt{1 - \frac{u^2}{c^2}}$$
$$= (10^4 \text{ m})\sqrt{1 - (0.999)^2}$$
$$= 447 \text{ m}$$

Since the atmosphere rushes by at $0.999c$, the meson reaches the ground in $(447 \text{ m})/(0.999 \times 3 \times 10^8 \text{ m s}^{-1}) = 1.49 \times 10^{-6}$ s, which is less than its average lifetime of 2.2×10^{-6} s. Thus the average meson reaches the earth's surface, in agreement with our earlier conclusion.

Contemporary spaceships travel at speeds small compared to the speed of light. However, exploration beyond our solar system involves distances so large

ALBERT EINSTEIN (1879–1955)

(California Institute of Technology Archives.)

By the end of 1905, at the age of 26, Albert Einstein had developed the special theory of relativity and explained the photoelectric effect. He was also well on his way to formulating the general theory of relativity. This precocious activity was not preceded by exhibitions of genius. On the contrary, Einstein had dropped out of secondary school in his native Germany and later returned to school only because he could not pass the university entrance examinations without further preparation. He was graduated from the University of Zurich in 1900 and settled for a position in the Swiss Patent Office in Bern, since his mediocre performance at Zurich precluded his obtaining an academic position.

The Patent Office was quiet and afforded him the time to do research in theoretical physics. Because of his isolation from others working on contemporary problems, some of his work was a duplication of that already done. However, by 1905, Einstein was working on unsolved problems and was having immense success in developing new ideas and concepts. The work on the photoelectric effect was later cited in his Nobel prize award. The special theory of relativity not only explained a number of fundamental problems in physics but also changed the way in which we regard space and time. In this same year, Einstein finished his doctoral dissertation, and in 1907, he joined the faculty at the University of Zurich. By 1913, his work had brought great professional praise, an important position at the University of Berlin, and increasing fame among nonphysicists. Popular descriptions of the theory of relativity captured the imagination of his time.

The general theory of relativity was published in 1916. It predicted the deflection of light in a gravitational field, which was confirmed in England in 1919. The emotions of World War I affected these scientific achievements of a German-born physicist. In Einstein's words, "Today in Germany I am called a German man of science and in England I am represented as a Swiss Jew. If [my theory is overthrown] the description will be reversed, and I shall become a Swiss Jew for the Germans and a German for the English."

While the theory never was overthrown, the situation did change. As the Nazis gained power, his Jewish heritage and his advocacy of world government and disarmament led to vitriolic personal attacks on him. Consequently, he emigrated to the United States in 1933 and never returned to his homeland.

Although we will see that he also helped to initiate the development of quantum mechanics, it evolved in a way that he found unsatisfying. Modern quantum mechanics holds that only the relative likelihood of various outcomes from an experiment may be predicted and not the result of a specific measurement. Einstein did not like this notion of "God playing dice with the Universe." While he played an important role as a consultant and sympathetic protagonist in the development of quantum mechanics, in his later years Einstein's research concentrated on attempts at a unified theory of gravitation and electromagnetism. His philosophy and work took him out of the mainstream of physics, although his early contributions continued to play an important role in the theory of modern physics.

Einstein's reputation was greater and more lasting than that of any other scientist of this century. His name was and is still used as a synonym for the revolutionary developments of modern natural science.

---

that it would be practical only at speeds comparable to c. Relativistic effects such as length contraction would then be important. Note that distances in astronomy are usually not measured in metres but in *light-years*. A light-year is the distance a light pulse travels in one year:

$$D = vt = c \, (1 \text{ year}) = 1 \text{ light-year}$$

The following example illustrates these remarks.

Example 25.4

Suppose that on January 1, 2000 a spaceship leaves the earth at $0.6\,c$ and heads towards planet X orbiting a star 12 light-years distant according to earth observers. (a) How far away from the earth is this planet according to the astronauts? (b) When does the spaceship reach planet X according to earth observers and to the astronauts?

(a) Using the length contraction formula, in the rest frame of the spaceship the distance to be travelled is

$$l' = l \sqrt{1 - u^2/c^2}$$
$$= (12 \text{ light-years}) \sqrt{1 - (0.6c)^2/c^2}$$
$$= (12 \text{ light-years})(0.8) = 9.6 \text{ light-years}$$

(b) According to earth observers, the time required is

$$t_E = l/v = (12 \text{ light-years})/(0.6c) = 20 \text{ years}$$

The ship arrives on January 1, 2020 according to earth observers.

The astronauts observe the time required is

$$t_s = l'/v = (9.6 \text{ light-years})/(0.6c) = 16 \text{ years}$$

and arrive on January 1, 2016 according to their clocks. This is 4 years before the date indicated by the earth clocks.

As in the case of the woman traveling to the hospital, earth observers find that the moving spaceship clocks tick slowly, while the astronauts moving relative to the earth and planet X see a shortened separation between them. Both sets of observers agree that the spaceship clocks indicate an elapsed time of 16 years, 4 years less than on earth clocks, although they differ as to why this has happened.

No change in length is seen by an observer moving at right angles to the length of an object. It is left as a problem (Prob. 25-24) to show this is a direct consequence of the principle of relativity.

25.4 | MOMENTUM AND ENERGY

In Chapter Seven, we saw that the momentum $p = mv$ is a useful quantity in ordinary nonrelativistic mechanics because the momentum of a system remains constant in a collision. When collisions occur between objects moving at speeds approaching that of light, it is found that momentum is not conserved if the nonrelativistic definition is used. However, the momentum can be redefined so that momentum conservation does hold true in collisions. Specifically, it is found by considering collisions that the correct definition of the relativistic momentum of an object of mass m and velocity v is

$$p = \frac{mv}{\sqrt{1 - v^2/c^2}} \qquad (25.3)$$

Note that the square root factor becomes 1 at low velocities, so we recover the nonrelativistic expression for the momentum.

In this equation, as well as everywhere else in this chapter, m is the ordinary mass of the object as measured by an observer in its rest frame. (Some books refer to this quantity as the rest mass and also define a velocity-dependent mass. We do not do this.)

In nonrelativistic mechanics, Newton's second law is $\mathbf{F} = m\mathbf{a} = md\mathbf{v}/dt$, or with $\mathbf{p} = m\mathbf{v}$

$$\mathbf{F} = \frac{d\mathbf{p}}{dt} \qquad (25.4)$$

The force is the rate of change of the momentum. In this form, Newton's second law applies in relativistic mechanics as well.

If an object is initially at rest, the work done on it as it accelerates is equal to its final kinetic energy. The work done in an infinitesimal displacement dx parallel to the force is the product, $dW = Fdx = (dp/dt)dx$. By definition, the velocity of the object is $v = dx/dt$, so $dx = vdt$. Thus $dW = (dp/dt)vdt = vdp$, and the final kinetic energy is found by evaluating

$$K = \int_0^v v\,dp$$

If we differentiate Eq. 25.3, we find after some algebra that

$$dp = \frac{dp}{dv}dv = \frac{mdv}{(1 - v^2/c^2)^{3/2}}$$

Then we obtain an expression which integrates readily,

$$K = \int_0^v \frac{mvdv}{(1 - v^2/c^2)^{3/2}} = mc^2\left[\frac{1}{(1 - v^2/c^2)^{1/2}}\right]_0^v$$

Thus, we find that the kinetic energy of an object with mass m and velocity v is

$$K = mc^2\left(\frac{1}{\sqrt{1 - v^2/c^2}} - 1\right) \qquad (25.5)$$

If the denominator is expanded in powers of v^2/c^2, this reduces at low velocities to the nonrelativistic kinetic energy, $\frac{1}{2}mv^2$. However, the kinetic energy of a rapidly moving object is much larger than $\frac{1}{2}mv^2$.

The following example shows that it takes much more work to accelerate an object to high velocities than predicted by nonrelativistic mechanics.

Example 25.5

An object of mass m has a speed $v = 0.9c$. What is the ratio of the nonrelativistic kinetic energy of this object to its relativistic kinetic energy?

The nonrelativistic kinetic energy is

$$K_n = \frac{1}{2}mv^2 = \frac{1}{2}m(0.9\ c^2) = 0.405mc^2$$

The relativistic kinetic energy is

$$K_r = mc^2\left(\frac{1}{\sqrt{1 - v^2/c^2}} - 1\right)$$

$$= mc^2\left(\frac{1}{\sqrt{1 - (0.9)^2}} - 1\right)$$

$$= 1.29mc^2$$

Thus the relativistic energy is more than three times the nonrelativistic value. This means that the work necessary to accelerate an object from rest to a velocity $v = 0.9c$ is more than three times greater than we would have predicted nonrelativistically. As the velocity approaches the velocity of light, the kinetic energy of the object increases very rapidly. Hence, it takes proportionately more work to achieve these speeds. In fact, it would require an infinite amount of work to have an object achieve a speed of c. This is one way of seeing that *the velocity of light cannot be attained by objects with a nonzero mass.*

The *rest energy* of an object of mass m is defined to be

$$E_0 = mc^2 \qquad (25.6)$$

This says that the mass m of an object is equivalent to an amount of energy mc^2. The *total energy* of the object is then the sum of the rest energy and the kinetic energy, $E = E_0 + K$, or

$$E = \frac{mc^2}{\sqrt{1 - v^2/c^2}} \qquad (25.7)$$

This is a statement that mass and energy are two forms of the same thing, and that one can be converted into the other. This was a revolutionary idea when Einstein introduced it, but it is now very well supported by experiment.

Before we examine some of the experimental evidence for the equivalence of mass and energy, let us consider an imaginary experiment in which two iden-

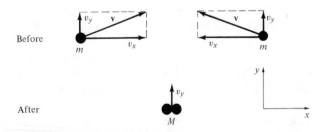

Figure 25.2. Two objects each with mass m collide and stick together.

tical masses m collide and stick together (Fig. 25.2). They form a single object of mass M moving with a velocity v_y, which does not change, since by symmetry neither object can exert a vertical force on the other. We can determine the mass of this object using momentum conservation.

The x component of the total momentum of the system is zero after and before the collision. Equating the y components after and before, we have

$$\frac{Mv_y}{\sqrt{1 - v_y^2/c^2}} = \frac{2mv_y}{\sqrt{1 - v^2/c^2}}$$

Dividing out the factor of v_y and multiplying by c^2,

$$\frac{Mc^2}{\sqrt{1 - v_y^2/c^2}} = \frac{2mc^2}{\sqrt{1 - v^2/c^2}}$$

This equation must hold for *any* value of v_y. In particular, if v_y is very small, then this becomes

$$Mc^2 = \frac{2mc^2}{\sqrt{1 - v^2/c^2}}$$

This is a remarkable result. It states that the rest energy of the combined object is equal to the total energy—rest plus kinetic—of the original two objects; *kinetic energy has been converted into mass.* Note that this result has followed directly from the conservation of the relativistic momentum. Thus the equivalence of mass and energy is an automatic consequence of momentum conservation.

There are many interesting situations in which the opposite of this experiment occurs, and mass is converted into energy. Every *particle* in nature has its *antiparticle,* which is identical in mass but opposite in charge. Our world is made up of particles, but a few particle-antiparticle pairs may be created by very energetic cosmic rays entering the atmosphere or by an accelerator beam striking a target. If, for example, an *antiproton* with a charge $-e$ encounters a proton with charge $+e$, their entire mass may be converted into gamma rays, very-high-frequency electromagnetic waves. Both the proton and antiproton cease to exist.

Other examples of the conversion of matter into energy are provided by *nuclear fission* and *fusion.* In nuclear fission, a uranium nucleus splits into two smaller nuclei whose total mass is less than that of the uranium nucleus by about 0.25 u. This is approximately 0.1 percent of the original mass and is equivalent to an amount of energy

$$E_0 = mc^2 = (0.25 \text{ u})(1.66 \times 10^{-27} \text{ kg u}^{-1})$$
$$\times (3 \times 10^8 \text{ m s}^{-1})^2$$
$$= 3.7 \times 10^{-11} \text{ J} = 2.3 \times 10^8 \text{ eV}$$

This is about 10^8 times the energy released in a typical chemical reaction, which is a few electron volts. We can see that nuclear fission reactors and explosives employ huge energy sources.

Nuclear fusion, in which two light nuclei fuse to form a larger nucleus, releases even more energy in proportion to the mass of the constituents. This is the source of energy in the stars and may eventually provide us with almost unlimited amounts of energy. Both fission and fusion are discussed further in Chapter Thirty.

Mass is also converted into energy in chemical reactions. However, because the energies released are very small compared to the rest energies of the atoms, the mass changes are too small to be directly measured.

Relation Between Energy and Momentum

If one squares the defining equations for the relativistic energy E and momentum p, one finds that

$$E = \sqrt{p^2c^2 + m^2c^4} \qquad (25.8)$$

When v/c is small, pc is small compared to mc^2, and this equation can be expanded to give $E = mc^2 + p^2/2m$. The first term is the rest energy. The second term can be rewritten as $(mv)^2/2m = mv^2/2$, which is the usual nonrelativistic kinetic energy.

If the mass m is zero, which is the case for light, Eq. 25.8 becomes

$$E = pc \qquad (m = 0) \qquad (25.9)$$

The energy equals the momentum times the speed of light. Exactly this result had been obtained decades earlier by Maxwell, when he calculated the energy and momentum carried by an electromagnetic wave. This is another indication that Einstein's theory of relativity is in full agreement with Maxwell's theory of electromagnetism.

SUMMARY

The special theory of relativity is based on two fundamental facts about our natural world. First, there is no way for inertial observers moving with a constant relative velocity to determine which is at rest and which is in motion, since all the laws of nature are the same for both. Second, the speed of a light beam in a vacuum is $c = 3 \times 10^8$ m s^{-1}, no matter which observer makes the measurement. These two facts are our clues to the nature of space and time and lead to a number of interesting results.

If two observers are moving with a constant relative velocity, each will observe that objects used by the other are foreshortened in the direction of motion. An observer in motion relative to an object sees it foreshortened according to

$$l' = l\sqrt{1 - \frac{u^2}{c^2}}$$

Each of the two observers will also say that the time t between ticks of the other's clocks is longer and is

$$t' = \frac{t}{\sqrt{1 - \frac{u^2}{c^2}}}$$

The total energy of an object is the rest energy

$$E_0 = mc^2$$

plus the kinetic energy, $E = E_0 + K$. For an object with velocity v,

$$E = \frac{mc^2}{\sqrt{1 - \frac{v^2}{c^2}}}$$

The modifications of our ideas of space and time require corresponding changes in our identification of momentum and energy. The work necessary to increase the energy of an object by a given amount increases with the speed of the object. This increase is such that an infinite amount of energy is necessary to cause an object with nonzero mass to travel with the speed of light.

Checklist

Define or explain:

inertial reference frames	relativistic momentum
principle of relativity	relativistic kinetic
constancy of the speed	energy
of light	rest energy
time dilation	energy-momentum
length contraction	relationship

REVIEW QUESTIONS

Q25-1 An inertial reference frame is one that moves with _____ velocity.

Q25-2 One postulate of special relativity states that the speed of light is the same for all observers in _____ .

Q25-3 An observer A, moving at a constant, nonzero velocity with respect to another observer B, observes that B's clocks are running slow compared to A's. B, in turn, observes that A's clocks are running _____ with respect to B's.

Q25-4 True or false? All clocks, no matter how constructed, should keep the same time if in the same inertial reference frame.

Q25-5 True or false? An observer at rest observes that the dimensions of a moving object are lengthened along the direction of motion.

Q25-6 True or false? The work required to increase the kinetic energy of an object of mass m from zero to the speed of light is $mc^2/2$.

Q25-7 The idea that mass and energy are different forms of the same entity is embodied in the equation $E_0 = $ _____ .

Q25-8 Why can't an object of mass m travel faster than the speed of light?

Q25-9 Why is $E_0 = mc^2$ called the rest energy of an object?

EXERCISES

Use $c = 3 \times 10^8$ m s^{-1} unless the problem specifies otherwise.

Section 25.1 | The Fundamental Principles of Special Relativity

25-1 An observer at rest on the ground sees a jet plane fly overhead at a constant speed of 2000 km h^{-1}. At what speed does the pilot see the ground moving?

25-2 A spaceship is traveling away from the earth at a constant speed of $c/2$. A light pulse is emitted by a lamp on the earth and travels toward the rocket. Find the speed of the light pulse according to observers on (a) the earth, and (b) the rocket.

25-3 A spaceship traveling away from the earth at a speed of $c/4$ emits a light pulse when it is a distance of 3×10^{10} m from the earth as measured by earth observers. How long will it take for the light pulse to reach the earth according to earth clocks?

Section 25.2 | Moving Clocks and Time Dilation

25-4 A person on a spaceship travels for 1 year (earth time) at a speed of $0.95c$. (a) How much time will have elapsed for the traveler? (b) How far will the traveler observe he has traveled?

25-5   A beam of pi mesons has a velocity of $0.6c$. Their average lifetime when at rest is 3×10^{-8} s. (a) On the average, how long will the moving pi mesons appear to live to an observer at rest? (b) How far will they travel in this time?

25-6   Two identical twins A and B are each 20 years old when B starts on a round trip that lasts 20 years according to A. If B travels at a speed of $u = 0.99c$ except during brief acceleration periods, what are the ages of the twins when B returns?

Section 25.3 | Length Contraction

25-7   A pi meson is moving at $0.9c$ relative to a magnet. If the magnet has a length of 2 m in its rest frame, how long is it in the frame of the meson?

25-8   (Assume that the speed of light is $c = 100$ m s^{-1} for this exercise.) Suppose you are driving a racing car at Indianapolis. The homestretch is 1300 m long according to the spectators and you are driving at 90 m s^{-1}. (a) How long does the homestretch appear to you? (b) How long do you think it takes you to drive this far? (c) How long do the spectators think it took you to drive that far?

25-9   At what speed, would a moving metrestick appear to be 0.5 m long to an observer at rest?

25-10   Suppose we see a pole passing us, moving parallel to its length at $0.6c$. What length would we measure it to be if its length in its rest frame is 20 m?

25-11   A star is 10 light-years from the earth according to earth observers. If a spaceship travels toward it from the earth at $0.6c$, how far from the earth is the star according to its occupants?

25-12   How many metres are there in a light-year?

Section 25.4 | Momentum and Energy

25-13   The kinetic energy and rest energy of a particle are equal. What is the speed of the particle?

25-14   The energy obtained by burning one gram of coal is about 3×10^4 J. What is the ratio of this energy to the rest energy of the 1 g of coal?

25-15   The energy released in chemical reactions is on the order of 5 eV per molecule. (a) Estimate the mass change in atomic mass units (1 u $= 1.66 \times 10^{-27}$ kg). (b) If the mass of the constituent atoms is 30 u, what is the fractional mass change?

25-16   A nuclear power plant produces 10^9 W of electrical power and simultaneously releases 2×10^9 W to the environment as waste heat. (a) At what rate is mass converted to energy? (b) How much mass is used up in 1 year? (1 year $= 3.16 \times 10^7$ seconds.) (c) What fraction is this of the 10^5 kg of uranium oxide in the reactor core?

25-17   In the decay of a neutron at rest into a proton, an electron e^-, and an antineutrino $\bar{\nu}$, one observes a total kinetic energy release of 1.25×10^{-13} J. What is the difference between the mass of a proton plus electron and the mass of a neutron? (The antineutrino has zero mass.)

25-18   Although our world is made up of matter, it has been conjectured that some galaxies may be composed of antimatter. If a 10-kg meteor of antimatter struck the earth, (a) how much mass would be converted into energy? (b) How much energy would be released?

25-19   A proton has a rest energy of 938 MeV. (1 MeV $= 10^6$ eV.) When its velocity is $0.9c$, what is its (a) total energy, and (b) kinetic energy?

25-20   An electron has a rest energy of 0.51×10^6 eV. (a) What is its total energy if it is accelerated from rest through a potential difference of 12×10^6 V? (b) What is its velocity?

PROBLEMS

25-21   A person on a rocket notes that she is traveling toward a star 1 light-year away from the earth with a speed u. A person on earth sees the rocket traveling with speed u but says the star is 10 light-years away. (a) How long will the traveler say the trip takes? (b) How long does the earth observer say the trip takes? (c) Discuss who is correct.

25-22   If the speed of light were $c = 100$ m s^{-1}, what would a sprinter say about his time needed to run the 100-m dash? (The officials say the time is 10 seconds.)

25-23   Consider two islands 10 km apart on a straight river flowing at 3 km h^{-1}. Using a boat that moves at 5 km h^{-1} with respect to the water, (a) how long does the upstream trip from island to island take? (b) How long does the downstream trip take? (c) How long does the total round trip take? (d) Compare the duration of the trip with the time it would take in still water. (e) Describe the analogy between the results of this problem and the derivation leading to Eq. 25.1.

25-24 Two identical sticks both have pins at their ends and will leave scratch marks on each other if they come into contact. Suppose they are set into relative motion perpendicular to their length as in Fig. 25.3. Show that it follows from the principle of relativity that they cannot change in length.

25-25 The energy flux from the sun on the earth averages about 1 kW m^{-2} (24 hours per day). How much of the sun's mass reaches each square metre of the earth each year in the form of energy? (1 year = 3.15×10^7 seconds.)

25-26 (a) Using the relativistic expressions for the energy and momentum show that the velocity of a particle satisfies $v = pc^2/E$. (b) Show how this goes over into a correct nonrelativistic formula when v/c is small.

25-27 An electron is accelerated from rest through a potential difference of 60,000 V. (a) What is the ratio of the total energy of the moving electron to its rest energy? (b) What is the final speed of the electron? (c) What is the final speed of the electron according to nonrelativistic mechanics?

25-28 Verify that $E = \sqrt{p^2c^2 + m^2c^4}$ follows from the definitions of E and p.

25-29 From the binomial theorem,

$$(1 - x)^{-1/2} = 1 + \frac{1}{2}x + \frac{3}{8}x^2 + \cdots$$

Using this to show that the relativistic expression for the kinetic energy reduces to the nonrelativistic formula when v/c is small.

ᶜ25-30 Using Eq. 25.3, show that $dp/dv = m/(1 - v^2/c^2)^{3/2}$.

ᶜ25-31 Verify the integration in the equation preceding Eq. 25.5.

25-32 Tom flies on an airplane at 1000 km h^{-1}. How many hours must he fly to be 1 second younger than his twin Harry who stays on the ground? [*Hint:* Because u/c is very small, you cannot simply plug the velocities into a calculator; instead, use the binomial expansion $(1 - u^2/c^2)^{1/2} = 1 - (1/2)u^2/c^2 + \cdots$.]

25-33 Suppose one uses the nonrelativistic approximation for the kinetic energy. The fractional error resulting from the use of this approximation is $|K_{\text{exact}} - K_{\text{approx}}|/K_{\text{exact}}$. Find the fractional error when v/c is (a) 0.1; (b) 0.5.

ANSWERS TO REVIEW QUESTIONS

Q25-1, constant; **Q25-2,** inertial reference frames; **Q25-3,** slow; **Q25-4,** true; **Q25-5,** false; **Q25-6,** false; **Q25-7,** mc^2; **Q25-8,** it would require an infinite amount of energy to reach the speed of light; even this is impossible; **Q25-9,** the total energy is $E = mc^2/\sqrt{1 - v^2/c^2}$, which is mc^2 when $v = 0$, and the object is at rest.

SUPPLEMENTARY TOPICS

25.5 | THE PROBLEM OF SIMULTANEOUS EVENTS; TWIN PARADOX

We have seen that Einstein's postulates led to the conclusion that the time elapsed between events is not an absolute quantity; rather, it depends on the motion of an observer. The mu meson observations described in Example 25.2 represent just one of many experimental verifications of this prediction.

A further consequence of Einstein's postulates is that observers may disagree on the question of whether two spatially separated events occur simultaneously. To see this qualitatively, suppose that an observer S standing at the center of a room sends two light pulses simultaneously toward the front and back of the room. Since the distances to the two walls are the same, the pulses strike the walls simultaneously according to this observer. If an observer M passes S at a velocity u just as the pulses are emitted, he too sees the two pulses traveling at a speed c (principle 2) toward the walls. However, he sees one wall approaching at a speed u and the other wall receding at the same rate. Consequently, M sees the pulse arrive at the approaching wall first, and he disagrees with S's observation that the two pulses arrive simultaneously at the walls.

To consider this question of simultaneity quantitatively, we consider what happens when an observer S

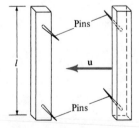

Figure 25.3. Two identical sticks with length l in their rest frames are in relative motion perpendicular to their length (Problem 25-24).

synchronizes two clocks at different locations. Suppose S is stationary relative to a stick of length l with a clock at each end. To synchronize the clocks so that they both read $t = 0$ at the same time, S has a light pulse emitted at one clock at $t = 0$. This pulse will travel the distance l to the second clock in a time $t = l/c$. If the second clock reads $t = l/c$ at the instant the pulse arrives, S will conclude that both his clocks read $t = 0$ when the pulse was emitted and that they are synchronized (Fig. 25.4).

We now ask how an observer M moving at a speed u relative to the stick observes these events (Fig. 25.5). He agrees with S that a light pulse is emitted when S's clock at A reads $t = 0$ and that the pulse reaches clock B when that clock reads $t = l/c$. According to S, the time it takes for the light pulse to travel from clock A to clock B is $t = l/c$. M disagrees, since he observes the stick to have a foreshortened length $l' = l\sqrt{1 - u^2/c^2}$ and also sees clock B moving toward the pulse. According to M, the pulse must travel a distance $ct' = l' - ut'$. Thus

$$t' = \frac{l'}{c + u} = \frac{l\sqrt{1 - u^2/c^2}}{c(1 + u/c)}$$

This is the transit time t' for the pulse as observed by M *using his own clocks*. It is less than the time interval l/c measured by S. Now suppose M looks at the *clocks belonging to S*. According to M, these clocks run slower by the time dilation factor. Hence, when a time interval t' has elapsed on M's clocks, M will observe that S's clocks indicate an interval t given by $t = t'\sqrt{1 - u^2/c^2}$. Using our result for t', this becomes

$$t = \frac{l}{c}\frac{1 - u^2/c^2}{1 + u/c} = \frac{l}{c}\left(1 - \frac{u}{c}\right)$$

This is the transit time for the pulse as observed by M using S's clocks. Thus, M concludes that even according to S's clocks, the light pulse does not take as long in transit as the time l/c measured by S. In fact, M reasons that since the pulse arrives at clock B when it reads $t = l/c$, then when the pulse was emitted, clock B read $t = lu/c^2$, not $t = 0$ as S concluded! While S says the clocks read $t = 0$ simultaneously, M says that when clock A read $t = 0$, clock B read $t = lu/c^2$. This is a real difference; neither M nor S is wrong.

This result can be restated in more general terms. Suppose two clocks a distance l apart in their rest

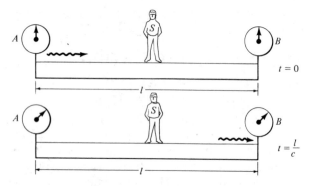

Figure 25.4. An observer S is stationary with respect to the stick and clocks shown. He will say that the clocks are synchronized if clock A reads $t = 0$ when a light pulse is emitted and clock B reads $t = l/c$ when the pulse arrives.

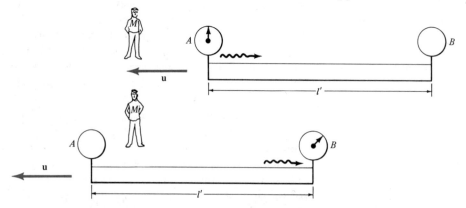

Figure 25.5. An observer M in motion relative to the stick sees the stick and clocks of Figure 25.4 moving past him with a speed u. M sees the pulse emitted when clock A reads $t = 0$ and sees the pulse arrive when clock B reads $t = l/c$. When M measures the length of the moving stick, he finds it has a contracted length $l' = l\sqrt{1 - u^2/c^2}$.

frame are synchronized according to an observer in that frame. An observer who sees them in motion with a speed u sees the *trailing clock set ahead* by

$$\Delta t = \frac{lu}{c^2} \quad \text{(trailing clock leads)} \quad (25.10)$$

This is the time difference she sees on the *moving clocks*. Because of time dilation, her own clocks indicate a larger difference $\Delta t'$ in when the two clocks read $t = 0$:

$$\Delta t' = \frac{\Delta t}{(1 - u^2/c^2)^{1/2}} \quad \begin{array}{l}\text{(trailing clock} \\ \text{leads)}\end{array} \quad (25.11)$$

We now apply this result to the spaceship clock problem of Example 25.4.

Example 25.6

Astronauts travel at $0.6c$ from the earth to planet X located 12 light-years away, as measured by earth observers. X clocks have been synchronized with earth clocks by light pulses. The trip starts on January 1, 2000; it lasts 20 years according to earth clocks and 16 according to spaceship clocks. (a) What date do the astronauts expect to see on X clocks upon arrival? (b) Explain why they do not see what they expected.

(a) The astronauts find X clocks read January 1, 2020 upon their arrival. But the astronauts see the earth and X clocks in motion, and therefore ticking slowly. Thus they had expected to arrive after (16 years) $(1 - 0.6^2)^{1/2} = 12.8$ years or in October, 2012 according to these clocks. The astronauts therefore infer that the clocks on planet X were not correctly synchronized with earth clocks. Instead, they conclude that these clocks were set ahead by $(20 - 12.8) = 7.2$ years in their own inertial frame.

(b) The astronauts see the earth and planet X moving past their ship at $0.6c$. If their clocks have been synchronized in their rest frame, the astronauts will see the X clocks (trailing clocks) set ahead of the earth clocks by

$$\Delta t = \frac{lu}{c^2} = \frac{(12 \text{ light-years})(0.6c)}{c^2} = 7.2 \text{ years}$$

This is precisely the difference in clock settings according to *earth clocks*, as seen by the astronauts.

Another rocket problem illustrates the relationship between time differences as seen in two inertial frames.

Example 25.7

People in a rocket of length $l = 30$ m traveling toward the earth at a speed of $u = 0.6c$ turn on lights in the front and back of the rocket simultaneously (Fig. 25.6). What will observers on earth see?

Observers on earth see the trailing light B turned on Δt before light F, where

$$\Delta t = \frac{lu}{c^2} = \frac{(30 \text{ m})(0.6c)}{c^2}$$

$$= \frac{(30 \text{ m})(0.6)}{3 \times 10^8 \text{ m s}^{-1}} = 6 \times 10^{-8} \text{ s}$$

This is the time measured according to *rocket clocks*. According to *earth clocks*, the light B will have been turned on earlier than light F by

$$\Delta t' = \frac{\Delta t}{\sqrt{1 - u^2/c^2}} = \frac{6 \times 10^{-8} \text{ s}}{\sqrt{1 - (0.6c/c)^2}}$$

$$= \frac{6 \times 10^{-8} \text{ s}}{0.8}$$

$$= 7.5 \times 10^{-8} \text{ s}$$

The results we have obtained in this section can sometimes lead to very intriguing apparent paradoxes. The following example which takes place in an imaginary world where c is comparable to everyday speeds illustrates one of these.

Example 25.8

(For the purposes of this example, assume that the speed of light is $c = 5$ m s^{-1}.) Three people own some boards 5 m long and a barn 4 m wide. They know that a moving object is contracted in length. Hence, they decide to move the boards at a speed $u = 3$ m s^{-1}, so that the length of the boards will be contracted to

$$l' = l\sqrt{1 - \frac{u^2}{c^2}} = 5 \text{ m} \sqrt{1 - \left(\frac{3}{5}\right)^2}$$

$$= (5 \text{ m})(0.8) = 4 \text{ m}$$

Thus the boards will fit into the barn. They station one person (B) at the back door of the barn and another (F) at the front door. The third (M) will run with the board (Fig. 25.7). F and B are to close the doors simultaneously when the board is exactly inside the barn.

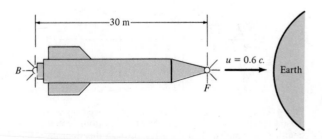

Figure 25.6. Observers on a rocket turn on lights at F and B simultaneously.

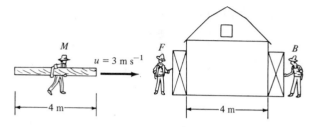

Figure 25.7. The barn and board *as measured by the observers F and B who are at rest relative to the barn.* F and B plan to close the doors of the barn simultaneously when the board is inside. When the board is at rest it has a length of 5 m.

The apparent paradox is that M, moving with the board, sees the board as 5 m long and the barn as having a contracted width of only $(4 \text{ m})(0.8) = 3.2$ m. How can the two sets of observations be compatible?

We first consider the situation as measured by F and B who are at rest with respect to the barn. F and B observe:

1 The barn and board are the same length.
2 When the board is just inside the barn they both close the doors.
3 They both observe the board crashing into the closed back door just after both doors are closed.

However, M's measurements indicate that (Fig. 25.8):

1 The barn is 3.2 m long and the board is 5 m long.
2 B does shut his door just as the front of the board reaches him, but the back of the board is still $5 \text{ m} - 3.2 = 1.8$ m outside the barn.
3 F does *not* shut his door when B does. According to F and B's clocks, M sees F shut his door

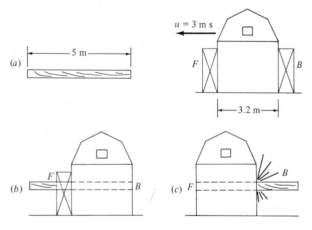

Figure 25.8. (*a*) The board and barn as viewed by M, the person moving with the board. (*b*) The back door is shut as the board reaches it. The board crashes through this door an instant later. (*c*) The front door is shut later, when the trailing end of the board is just inside the barn.

$$\Delta t = \frac{lu}{c^2} = \frac{(4 \text{ m})(3 \text{ m s}^{-1})}{(5 \text{ m s}^{-1})^2} = 0.48 \text{ s}$$

later than B. In M's frame, this is

$$\Delta t' = \frac{0.48 \text{ s}}{\sqrt{1 - (3/5)^2}} = 0.6 \text{ s}$$

Now, in 0.6 s, M sees the barn move toward him a distance of $u \Delta t' = (3 \text{ m s}^{-1})(0.6 \text{ s}) = 1.8$ m. This is just the distance that places the back of the board inside the barn when F shuts his door.

We see that all observers agree that the back door was shut just before the board reached it and the front door was shut just after the end of the board was inside the door. What the observers do not agree on was when the board crashed through the back door. F and B say it happened just after they both shut their doors, while M says it happened just after B shut his door but before F shut his. This disagreement is real and is typical of those that can arise when two sets of observers view events that occur at different positions.

The Twin Paradox | We saw in Examples 25.4 and 25.6 that a spaceship traveling at $0.6c$ reaches a planet 12 light-years away in 20 years, according to earth observers. However, because of time dilation, the elapsed time according to the astronauts is only 16 years. The return trip takes equally long. Thus, on her return, astronaut Angela has aged 32 years, while her stay-at-home identical twin Susan has aged 40 years. However, from Angela's point of view, she has been at rest. Susan and the earth were in motion, and *their* clocks therefore ticked slowly, with only $32(1 - 0.6^2)^{1/2} = 25.6$ years passing; hence Susan must be younger! They can't both be right, so which twin has actually aged less? This is the *twin* or *clock paradox*. It is resolved by looking critically at the statements we have just made.

The twin who has aged less is astronaut Angela: She has lived 32 years, while Susan has lived 40. The basic point is that Angela cannot consider herself to be at rest throughout her trip, because she was not in an inertial frame during the periods of acceleration and deceleration at the trip's beginning, midpoint, and ends. The lack of uniformity in her motion removes the symmetry between her experience and Susan's. Furthermore, we can account for the difference in their ages using the clock synchronization results above.

To see this in detail, suppose Angela starts her voyage with a very brief (and uncomfortable) acceleration from rest to $0.6c$. She then travels at a constant velocity toward planet X, so she is in an inertial frame. Upon arrival she sees the trailing clock on X set ahead by 7.2 years (Example 25.6). After a rapid deceleration and acceleration, she is heading back toward the earth in a *different* inertial frame. According to observers in this frame, the trailing clock is set ahead by 7.2 years, but now this is the clock on the *earth*. These two adjustments are the price Angela must pay for switching inertial frames! If Angela adds 2(7.2 years) = 14.4 years to the 25.6 years that she sees elapse on the clocks in the earth's rest frame, her sum is 40 years. This agrees with Susan's observation of the duration of the trip on earth clocks.

Let's consider another and probably more convincing way for Angela and for us to see that Susan has lived 40 years while Angela was away. Suppose that on each birthday Susan sets off a fireworks display. During the outward trip, the earth is moving away at $0.6c$, and explosions occur every $1/(1 - 0.6^2)^{1/2} = 1.25$ years according to Angela's clocks. Furthermore, to reach the spaceship, successive flashes must travel an additional distance (0.6 light-year)/$(1 - 0.6^2)^{1/2} = 0.75$ light-year. (The distance is *longer* in the rest frame of the spaceship than in the moving frame of the earth.) Adding the time between flashes to the extra travel time gives $(1.25 + 0.75) = 2$ years. Angela sees flashes once every 2 years, or 8 flashes during the 16 outbound years. On the return trip, successive flashes travel distances shorter by 0.75 light-years, so they arrive every $(1.25 - 0.75) = 0.5$ years. During the 16 years of the return trip, Angela sees 32 flashes. All together she sees $(8 + 32) = 40$ flashes, and she knows that Susan has celebrated 40 birthdays!

25.6 | THE ADDITION OF VELOCITIES

According to special relativity, two observers in relative motion disagree about the properties of their metresticks and clocks. Accordingly, when they measure the velocity of an object moving relative to both of them, their observations are related in a way that reflects these disagreements.

The derivation of this relationship is similar to those of previous sections but lengthy, so we simply state the results. Consider two observers with a relative velocity u. An object moves along the x direction with a velocity v' according to O' (Fig. 25.9). According to O, the velocity of the object is

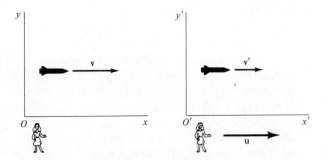

Figure 25.9. Observer O' moves relative to O at a velocity $\mathbf{u}$. An object has a velocity $\mathbf{v'}$ according to O' and $\mathbf{v}$ according to O.

$$v = \frac{v' + u}{1 + \dfrac{uv'}{c^2}} \qquad (25.12)$$

This is called the *velocity addition formula*.

When the velocities are small, we can neglect u/c and v'/c. This gives the usual nonrelativistic result, $v = v' + u$. However, when v' is comparable to or equal to c, the results are quite different. In particular, it turns out that if v' and u are less than c, but $v' + u$ is greater than c, then v is still less than c. This means that nothing can be observed to move faster than the speed of light. The case $v' = c$ is considered in the next example.

Example 25.9

An observer moving away from the earth at $0.5c$ sends a light pulse along his direction of motion and measures its speed to be c. How fast does the pulse travel according to the earth observer?

Here $v' = c$ and $u = 0.5c$. Using the velocity addition formula, the velocity according to an earth observer is

$$v = \frac{v' + u}{1 + \dfrac{v'u}{c^2}} = \frac{c + 0.5c}{1 + \dfrac{(c)(0.5c)}{c^2}} = c$$

Thus both observers measure the same speed of light, in agreement with Einstein's second principle.

When the motion of the object is not along the x axis, the addition of velocities involves more complicated expressions. However, these expressions again ensure that no object can be observed to travel faster than the speed of light.

EXERCISES ON SUPPLEMENTARY TOPICS

Section 25.5 | The Problem of Simultaneous Events

25-34 An observer on an earth satellite traveling at 8×10^3 m s^{-1} relative to the earth sees rocket launchings occurring in New York and in Los Angeles, which he judges to be simultaneous. (a) The distance between the two cities is 4.5×10^6 m. What will the time interval between these launchings be according to observers on earth? (The satellite is moving directly from Los Angeles toward New York.) (b) In which city does the rocket launching occur first according to earth observers?

25-35 An observer on a train sees two lights flash on the station platform. Light A is seen to flash 0.8 second after light B (Fig. 25.10), and this observer measures the lights as being 100 m apart. If the train is moving at 16 m s^{-1} and the speed of light is $c = 20$ m s^{-1}, (a) what is the distance between the lights as measured by an observer at rest on the platform? (b) Which light flashes first according to an observer at rest on the platform?

25-36 According to Susan, the stay-at-home twin in The Twin Paradox discussion, how many flashes will reach astronaut Angela on (a) the outward part of the trip; (b) the inward leg?

25-37 Assume for this exercise that $c = 100$ km h^{-1}. A woman drives to a hospital at 80 km h^{-1}. According to the highway signs, the distance is 100 km. When she arrives, she finds the hospital clocks are set ahead of the clocks she saw on the road at her starting point. By how much does she say they are ahead according to (a) clocks in the hospital; (b) her watch?

Section 25.6 | The Addition of Velocities

25-38 In a large-particle accelerator facility, two beams of protons are directed toward each other, each with a velocity of $0.8c$ as measured in the laboratory. What is the speed of one beam as measured by an observer at rest with respect to the other beam?

25-39 A space ship approaching the earth at $0.6c$ fires a rocket at a speed of $0.6c$ relative to the space ship. According to earth observers, what is the speed of the rocket if it is fired (a) toward the earth; (b) away from the earth?

25-40 A pi meson decays into a mu meson and a neutrino. In the rest frame of the pi meson, the velocity of the mu meson is always found to be $0.507c$. Suppose now the pi meson is moving relative to the laboratory at $0.6c$ when it decays, emitting the mu meson along its direction of motion. What will be the speed of the mu meson relative to the laboratory?

25-41 Particle A decays into particles B and C. When the A particle is at rest, B is always found to have a speed $0.6c$. If the A particle is moving at $0.4c$ relative to the laboratory when it decays, find the velocity of the B particle as measured in the laboratory frame if the B particle moves (a) parallel to the direction of the A particle; (b) opposite to the direction of the A particle.

PROBLEMS ON SUPPLEMENTARY TOPICS

25-42 An observer on earth sees a rocket P traveling away from the earth at 2×10^8 m s^{-1}. P is overtaking a second rocket Q, which is traveling away from the earth at 1.5×10^8 m s^{-1}. Find the velocity of (a) P measured by observers in Q, and (b) Q measured by observers in P.

25-43 A rocket ship moves at $0.6c$ past two observers, A and B, at rest on a platform. They arrange to fire guns simultaneously, striking the rocket at points A' and B' (Fig. 25.11). (a) How far apart are A and B as measured by those on the rocket? (b) Observers on the platform and on the rocket agree that B fires at $t = 0$. According to observers on the rocket looking at their own clocks, when does A fire? (c) Using the results found in (a) and (b), determine the distance between points A' and B' as measured by people on the rocket. (d) Verify that the distances between A' and B' as measured by observers on the rocket and on the platform are consistent with the length contraction formula.

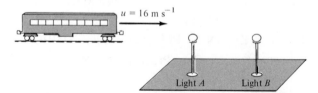

Figure 25.10. Exercise 25-35.

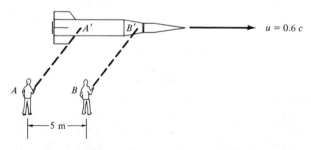

Figure 25.11. Problem 25-43.

25-44 Two observers O and O' are in relative motion along their x axes at a speed u. If O' observes an object in motion along the y' axis with a speed v', then O will observe it to have both x and y velocity components, where

$$v_x = u, \qquad v_y = v' \sqrt{1 - \frac{u^2}{c^2}}$$

If $v' = c$, find $v = \sqrt{v_x{}^2 + v_y{}^2}$.

25-45 Observer O' observes a rocket ship traveling along the y direction at a speed $0.9c$. O' is moving at $0.8c$ along the x direction relative to observer O. Using the formulas in the preceding problem, find the components of the rocket ship velocity as observed by O and show that v is less than c.

Additional Reading

N. David Mermin, *Space, Time, and Relativity,* McGraw-Hill Book Co., New York, 1965.

Hermann Bondi, *Relativity and Common Sense,* Science Study Series, Doubleday and Co., Garden City, N.Y., 1964.

Scientific American articles:

R.S. Shankland, The Michelson-Morley Experiment, November 1964, p. 107.

J. Bronowski, The Clock Paradox, February 1963, p. 134.

Milton A. Rothman, Things That Go Faster Than Light, July 1960, p. 142.

V.L. Ginzburg, Artificial Satellites and Relativity, May 1959, p. 149.

Gerard K. O'Neill, Particle Storage Rings, November 1966, p. 107.

Albert B. Stewart, The Discovery of Stellar Aberration, March 1964, p. 100.

CHAPTER 26

PARTICLE PROPERTIES OF LIGHT: THE PHOTON

As physics moved into the twentieth century, just a very few physical phenomena appeared unexplained. However, these few observations proved to be the tip of the iceberg of results that led to quantum mechanics. At that time, the wave theory of light and electromagnetic radiation was on a particularly firm basis due to the work of Maxwell and others. It was only when certain experiments involving the interaction of light with matter were performed that the theory appeared incomplete. In these experiments light interacted with single atoms, molecules, or electrons. Such experiments became possible only toward the end of the last century, and hence were initially few in number.

While the experiments were few, they proved to be explainable only by the development of the idea that light sometimes behaves like a wave and sometimes like a collection of particlelike objects called *photons*. These photons are small bundles of light with discrete energies. Although this idea was radical, it was soon accepted because it explained the new observations so completely. Furthermore, as a complete quantum theory developed in the first decades of the new century, photons were a natural part of that theory.

In retrospect, it is easy to understand why electromagnetic radiation or light was considered to consist of continuous waves. Any particlelike behavior was masked by the enormous number of light quanta usually present. The situation is analogous to that encountered in electrostatics where the amount of charge present is usually very large. The fact that charge comes in discrete amounts is irrelevant under these conditions.

In this chapter we describe the particle like properties of light and see how these ideas explain several crucial experiments.

26.1 | THE PHOTOELECTRIC EFFECT

An early experiment that most clearly illustrates the particle properties of light is also the one that prompted Einstein to develop the concept of the photon. A description of the experiment follows.

The photoelectric effect apparatus is shown in Fig. 26.1. Monochromatic light incident on the metal plate yields sufficient energy to allow electrons to escape from the metal. Some of these emitted electrons reach a collection plate, and the ammeter measures the resulting *photoelectric current*. The intensity and

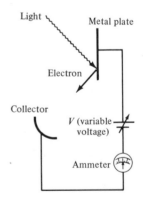

Figure 26.1. The photoelectric effect apparatus. Light incident on the metal plate causes electrons to be emitted. These can travel to the collector producing a current.

frequency of the incident light can be varied, as can the potential difference V between the metal plate and the collector.

The conduction electrons in a metal move in the attractive electric field of the stationary positive ions of the lattice. While electrons move relatively freely inside, it requires a minimum energy to actually pull them from inside to outside the metal. This minimum energy is called the *work function W* and depends on the properties of the metal and of the surface. If an electron is given an energy E larger than W, it can escape the metal and will have a maximum kinetic energy

$$\tfrac{1}{2}mv_{max}^2 = E - W$$

The results of the photoelectric experiment can be summarized as follows:

1 When $V = 0$, photoelectrons are detected whenever the metal is illuminated by light at a frequency f, which is greater than a critical or threshold frequency f_0. However, no matter what the intensity of the light, no current is observed if the frequency is below f_0.

2 At each light frequency above the threshold, the potential V can be increased until, at some value V_0, the current becomes zero. If the emitter and collector are made of the same material, this occurs when the potential energy difference eV_0 of an electron with charge $-e$ is just equal to the *maximum* kinetic energy of the emitted electrons

$$eV_0 = \tfrac{1}{2}mv_{max}^2$$

A graph of V_0, the stopping potential, versus the frequency of the incident light appears in Fig. 26.2a.

3 Above the threshold frequency, an increase in intensity results in an increase in the number of photoelectrons, but the maximum kinetic energy of the electrons does not change.

These observations are in direct conflict with predictions based on the wave picture of light. If light is a classical wave, the electrons should absorb energy continuously, and at any intensity it should be merely a matter of time until an electron has sufficient energy to escape. Hence there would be no threshold frequency, although there would be a delay in production of photoelectrons at low intensities until enough light energy had been absorbed by the mate-

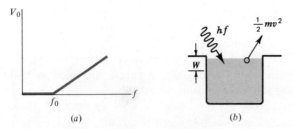

Figure 26.2. (a) The stopping voltage V_0 versus incident light frequency. If f is less than f_0 no current is observed, even with zero stopping potential. Above f_0 the stopping potential increases linearly with f. This graph is valid for *all* nonzero light intensities. (b) Electrons in a metal have energies which vary over the colored region. The minimum energy required to remove an electron is the work function, W. Thus if an electron at the highest energy (top of the colored region) absorbs a photon of energy hf, it acquires a kinetic energy $\tfrac{1}{2}mv_{max}^2 = eV_0 = hf - W$. Other electrons acquire smaller kinetic energies when they absorb photons.

rial. Furthermore, at high intensities the electrons should receive more energy, so the stopping potential should be greater. No such effects are observed.

Einstein's Explanation | In 1905, Einstein discovered that the photoelectric effect experiments could be explained in a straightforward way if one assumed that the energy carried by the incoming light came in discrete amounts, rather than continuously. Furthermore, he suggested that the amount of energy in each *light quantum* or photon depends only on the frequency of the light, and not on its intensity. The intensity of a beam of light is determined by the number of photons present; the energy of each photon is determined by the frequency.

The light quanta behave like particles that travel at the speed of light. If the light has a frequency f and wavelength $\lambda = c/f$, the photons each have an energy

$$E = hf \quad \text{(photon energy)} \quad (26.1)$$

The intensity of monochromatic light is proportional to the number of photons present. A beam of white light contains photons of many different energies. The quantity h is a proportionality constant and is fit to experiment. In fact, h was first introduced in 1900 by Max Planck (1858–1947) in a less fully developed theory of discrete radiation; for this reason, h is known as Planck's constant. Its value is

$$h = 6.625 \times 10^{-34}\,\text{J s} = 4.135 \times 10^{-15}\,\text{eV s}$$

The small size of this constant is our first hint that classical physics may fail when extremely small energies are important.

The photon theory of light offers a complete explanation of the photoelectric effect. An electron will leave the metal only if it absorbs a photon of energy equal to or greater than the work function W (Fig. 26.2b). The threshold corresponds to the frequency at which the photon energy hf_0 equals W or

$$f_0 = \frac{W}{h} \qquad (26.2)$$

If the photon frequency is above threshold, the excess energy appears as kinetic energy of the photoelectrons. Since W is the minimum energy necessary to remove an electron, the maximum kinetic energy is

$$\tfrac{1}{2}mv_{max}^2 = hf - hf_0 = hf - W \qquad (26.3)$$

Increasing the intensity with f constant will result in more electrons being emitted, but with the same maximum kinetic energy.

Since the stopping voltage V_0 is adjusted so that $eV_0 = \tfrac{1}{2}mv_{max}^2$, we see that $eV_0 = hf - W$, or

$$V_0 = \frac{h}{e}f - \frac{W}{e} \qquad (26.4)$$

This is an important result, because it predicts that the slope of the V_0 versus f curve is h/e. Planck's constant h and the electronic charge e were known, and the slope in Fig. 26.2 turned out just as predicted. This convincing demonstration was one of the first in a series of discoveries that led to the acceptance of a particle or photon description of electromagnetic radiation.

The following example illustrates the ideas we have discussed.

Example 26.1

Light is incident on the surface of a metal for which the work function is 2 eV. (a) What is the minimum frequency the light can have and cause the emission of electrons? (b) If the frequency of the incident light is 6×10^{14} Hz, what is the maximum kinetic energy of the electrons?

(a) At the threshold frequency, the energy of the photon equals the work function, so $hf = W$ or

$$f = \frac{W}{h} = \frac{(2 \text{ eV})}{(4.135 \times 10^{-15} \text{ eV s})} = 4.84 \times 10^{14} \text{ Hz}$$

(b) The energy of a photon of frequency 6×10^{14} Hz is

$$E = hf = (4.135 \times 10^{-15} \text{ eV s})(6 \times 10^{14} \text{ Hz}) = 2.48 \text{ eV}.$$

Hence the maximum kinetic energy is the difference

$$\tfrac{1}{2}mv_{max}^2 = (2.48 \text{ eV}) - (2 \text{ ev}) = 0.48 \text{ eV}$$

26.2 | THE PHOTON

As we saw in the preceding section, a light beam is composed of quanta. For monochromatic light of frequency f, each quantum of light or photon has an energy hf. The number of photons present determines the intensity. The following example shows that under usual conditions the number of photons present is so large that their quantum nature may be overlooked.

Example 26.2

Monochromatic green light of frequency 6×10^{14} Hz is produced by a laser. The power emitted is 2×10^{-3} W. (a) What is the energy of a photon in the beam? (b) How many photons per second pass a point in the beam?

(a) Each photon has an energy

$$E = hf = (6.63 \times 10^{-34} \text{ J s})(6 \times 10^{14} \text{ s}^{-1})$$
$$= 3.98 \times 10^{-19} \text{ J}$$

(b) Let N be the number of photons passing a point in the beam each second. The power $\mathcal{P}$ transmitted in the beam must be equal to N times the energy per photon, $\mathcal{P} = NE$. Thus

$$N = \frac{\mathcal{P}}{E} = \frac{2 \times 10^{-3} \text{ W}}{3.98 \times 10^{-19} \text{ J}}$$
$$= 5.03 \times 10^{15} \text{ photons per second}$$

Blackbody Radiation | As we mentioned earlier, Max Planck was the first to suggest that light has discrete energies proportional to the frequency. Planck was trying to explain the observed relation between the power radiated by a hot object versus the wavelength (and frequency) at which the radiation appears.

In Chapter Twelve we found that the power radiated from a heated object depends on the wavelength of the radiation. The details of how this power varies with wavelength and temperature usually depend on the characteristics of the object. However, the radiation emitted through a small opening from a cavity does not depend either on the material of its walls or on the shape of the cavity and the hole. The cavity radiation obeys a universal law of nature and acts as a perfect blackbody: a perfect absorber and emitter with emissivity equal to 1 (Fig. 26.3).

Many attempts were made to derive the blackbody

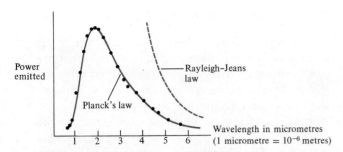

Figure 26.3. Blackbody radiation at 1600 K. The curve predicted by Planck's law passes through the data points. The Rayleigh-Jeans prediction becomes very large at short wavelengths.

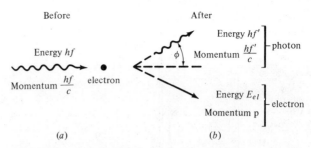

Figure 26.4. (a) A photon incident on a stationary electron is scattered (b) through an angle ϕ. The energy and momentum lost by the photon are taken up by the electron.

radiation curve (Fig. 26.3) using classical electromagnetic theory, but all of them failed. The most famous of these, called the Rayleigh-Jeans law, was correct at long wavelengths, but at short wavelengths the theoretical power emitted became infinite. This failure was referred to as the ultraviolet (short wavelength) catastrophe.

Planck found that two radical assumptions could be combined with classical theory to give an excellent fit to experiment. These assumptions are equivalent to the photons that Einstein used to describe the photoelectric effect.

Planck assumed that the radiation in the cavity was emitted and absorbed by atomic oscillators in the walls of the cavity. The oscillators had two unique characteristics:

1 The oscillators could only have energies given by

$$E = nhf, \qquad n = 0, 1, 2, 3, 4, \ldots \quad (26.5)$$

where f is the oscillator frequency and the proportionality factor h is called *Planck's constant.*

2 The oscillators could emit energy only in discrete or quantized amounts corresponding to a change in n of one unit:

$$E_{n+1} - E_n = (n + 1)hf - nhf = hf$$

We see that an oscillator is releasing a photon when it loses energy.

Planck was able to derive a formula for the power radiated with these assumptions. This formula fit the experimental results (Fig. 26.3) very well and led to the first determination of h.

The Compton Effect | Several further experiments indicated convincingly that the photon picture of light is correct. One, performed in 1923 by A. H.

Compton (1892–1962), involved the observation of X rays scattered from free electrons. Compton found that the scattered X rays have a reduced frequency. He showed that this implies that a photon-electron collision is much like an elastic collision of two billiard balls (Fig. 26.4). When a photon of energy hf collides with an electron at rest, the electron recoils with an energy E_{el}. The scattered photon has an energy hf', where f' must be less than f since energy conservation requires

$$hf' = hf - E_{\text{el}}$$

The analogy of a billiard ball collision implies that the total momentum of the photons and the electron is conserved. Using classical electromagnetic theory, Maxwell showed that the energy E carried by a light wave is related to the momentum of the wave by $E = pc$, where c is the velocity of light. (This is also the result we obtained from relativity for a particle with zero mass.) Thus the photon momentum is given by

$$p = \frac{E}{c} = \frac{hf}{c} \qquad (26.6)$$

When momentum and energy conservation are applied to the collision of Fig. 26.4, the predicted dependence of the frequency change $f - f'$ on the angle θ is found to be in very good agreement with experiment. Thus, Compton's experiment provides further evidence of the particle nature of light.

X-Ray Spectrum | X rays provided another significant verification of the photon theory. We saw in Chapter Twenty-three that when an electron beam is accelerated in a vacuum and allowed to strike a target, X rays are emitted with various frequencies up to some maximum frequency, f_{max} (Fig. 26.5). It was

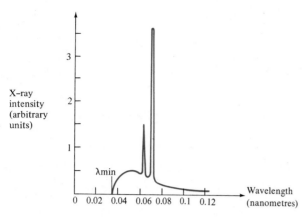

Figure 26.5.   X rays produced by the electron bombardment of a target result in X ray photons of a maximum frequency $f_{max} = eV/h$ and a minimum wavelength $\lambda_{min} = c/f_{max}$. The sharp peaks appear at different wavelengths for different target materials. They are due to the rearrangement of electrons disturbed from their normal atomic orbits.

found that f_{max} depends only on the accelerating potential difference V and is independent of the number of electrons in the beam. Classically, these observations are difficult to comprehend, but the photon theory gives a ready explanation. If the electrons are accelerated from rest through a potential difference V, their kinetic energy is eV. The electrons lose part or all of their kinetic energy when they interact with an atom in the target. The highest frequency photon will be produced when all the kinetic energy of one electron is converted into a single photon, or when

$$hf_{max} = eV \qquad (26.7)$$

This equation is in excellent agreement with the observed frequency maxima.

Example 26.3

What are (a) maximum frequency and (b) minimum wavelength of X rays produced by 40,000 V electrons?

(a) Using Eq. 26.7 for the photon frequency,

$$f_{max} = \frac{eV}{h} = \frac{(1.60 \times 10^{-19}\ \text{C})(4 \times 10^4\ \text{V})}{6.63 \times 10^{-34}\ \text{J s}}$$
$$= 9.65 \times 10^{18}\ \text{Hz}$$

(b) The corresponding minimum wavelength is

$$\lambda_{min} = \frac{c}{f_{max}} = \frac{3 \times 10^8\ \text{m s}^{-1}}{9.65 \times 10^{18}\ \text{Hz}}$$
$$= 3.11 \times 10^{-11}\ \text{m} = 0.0311\ \text{nm}$$

26.3 | WAVE-PARTICLE DUALITY

Having discussed a number of situations where the particle or quantum nature of light is important, we recall the wide variety of phenomena that were described by a wave theory of light. The two pictures seem very contradictory.

Some reconciliation of these two ideas can be achieved if we view the photon as a small *wave packet* (Fig. 26.6). This illustration is a representation of the mathematical formulation of quantum mechanics. It shows a bundle of waves traveling at the speed of light. The total energy of the bundle or packet is hf. If the frequency f is increased, so is the energy of the packet; the energy contained in a packet is determined by the frequency. The packet is a wave disturbance, but it is localized in a small region of space, much like a particle. Large numbers of photons at a single frequency behave in a way similar to a continuous wave when they pass through slits, reflect off objects, and so on.

The dual wave-particle nature of light is now a basic part of our theory of light and matter. The theory as formulated mathematically contains both aspects of light, and the outcome of any experiment can in principle be predicted. On the other hand, a verbal description of the behavior of light is not so unambiguous. As a tentative rule, we may say that light behaves as a wave as long as no absorption or emission of light occurs. Whenever light is absorbed, its quantum or particle nature may be expected to be evident.

26.4 | PHOTONS AND VISION

Light energy is received by the visual receptors of the eye, that is, the rods and cones, in discrete amounts as photons are absorbed. Hence, there is necessarily a minimum number of photons needed for vision. This

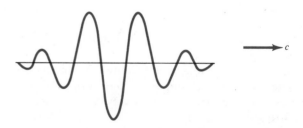

Figure 26.6.   The photon pictured as a wave packet traveling at the speed of light. It is a bundle of electromagnetic energy with energy hf.

minimum number is called the *absolute threshold of vision.*

A distinction must be made between the number of photons incident on the eye and the number of photons causing excitation of the receptors. About 90 percent of the photons incident on the eye are absorbed or scattered in the lens and fluid of the eye. Furthermore, less than 40 percent of the photons reaching the retina are actually absorbed by the receptors. Thus the number of photons incident on the eye is far larger than the number of photons used by the receptors for visual perception. Our discussion will involve only photons absorbed by the receptors.

We might guess that a single photon is sufficient for vision. However, theoretical and experimental evidence seems to rule this out. Theoretically we know that molecules in the eye have an average thermal energy that is quite small. However, there is a small probability that an occasional atom or molecule will have an energy large enough to emit a photon that could cause excitation of the receptor. Multiplying this small probability by the large number of molecules present and the large number of receptors available indicates that about one photon per second might be detected. These photons have nothing to do with the external visual scene, and hence their perception would be a constant source of annoyance and confusion. This *background* would be eliminated by the requirement that *coincident* or nearly simultaneous absorption of more than one photon must occur to stimulate the visual process.

Experimentally, coincident absorption of photons apparently must take place in a region where the neurons are interconnected. This region includes about 100 rods in the dark-adapted eye. In this region, if the coincidence does not occur during the eye's "storage time" of 0.2 second, the coincidence is not counted. Thus the visual memory of the absorption of one photon lasts about 0.2 second.

The actual experiment is performed with pulses that illuminate about 100 rods and last 0.2 second. Subjects are asked to say when they see a pulse. The fraction of positive responses is then plotted versus the average pulse energy $\bar{E}$ (Fig. 26.7). The average number of photons $\bar{n}$ absorbed by the rods is proportional to $\bar{E}$,

$$\bar{n} = A\bar{E} \qquad (26.8)$$

Here A is a constant that must be determined from the experiment.

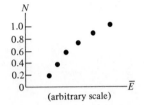

Figure 26.7. The experimental data indicating the fraction of positive responses to light pulses of varying average energy $\bar{E}$.

A light source emits photons randomly in time, so not every pulse in a group with average energy $\bar{E}$ will result in exactly $\bar{n}$ absorbed photons. For each pulse of the group with an average of $\bar{n}$ absorbed photons, there is some probability, denoted by $P(k, \bar{n})$, that at least k photons are actually absorbed.

This is analogous to the following situation. When rain falls on a sidewalk divided into squares, the average number of drops per square $\bar{n}$ equals the total number of drops divided by the number of squares. However, examination will show that some squares have fewer and some have more than the average number of drops. The number of squares receiving k or more drops divided by the total number of squares is $P(k, \bar{n})$.

The visual threshold experiment effectively measures $P(k, \bar{n})$, where k is the minimum number of photons necessary for vision. That is, there should be a correspondence between the graph of Fig. 26.7 and a graph of the theoretical value of $P(k, \bar{n})$ versus $\bar{n}$ (Fig. 26.8).

The problem is that neither k nor the quantity A in Eq. 26.8 is known. By plotting the number of positive

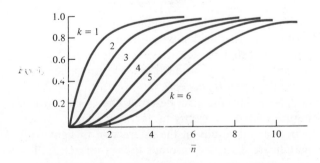

Figure 26.8. Graphs of the predicted value of $P(k, \bar{n})$ versus $\bar{n}$ for $k = 1, \ldots, 6$. One of the curves of this type should correspond to the results shown in Figure 26.7. (Adapted from R.K. Clayton, *Light and Living Matter, Volume II: The Biological Part,* McGraw-Hill, New York, 1971. Used with permission.)

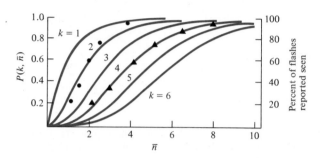

Figure 26.9. The curves showing the fractional probability of absorption of at least k photons versus the average number of photons $\bar{n}$. The symbols ● and ▲ represent two attempts to fit a single set of data to the curves for $k = 2$ and $k = 4$. The fits are achieved by adjusting the constant A in Eq. 26.8. (Adapted from Clayton.)

responses versus $\bar{n}$, using A as an adjustable parameter, N of Fig. 26.7 can be compared with $P(k, \bar{n})$. Typical results are shown in Fig. 26.9. It is found that an approximate fit to the data can be obtained for k between 2 and 6. Hence at least 2 but fewer than 7 photons are apparently necessary for vision.

SUMMARY

The wave nature of light was well established by the end of the nineteenth century. However, several experiments, including blackbody radiation and the photoelectric effect, could not be explained in terms of such a picture. Max Planck made the first tentative assertion that vibrating atoms emitted light energy in specific amounts, and Einstein extended this idea to say that light itself is composed of quanta called photons. These carry a specific amount of energy equal to Planck's constant times the light frequency,

$$E = hf$$

Compton's studies of photon-electron collisions provided further evidence that light has a particlelike nature and that photons have a definite energy and momentum. X-ray production was also understood using the photon concept.

Present-day theories describe the wave-particle nature of light correctly as far as we know. When experimental results of the interaction of light with matter are discussed in other than a formal mathematical way, the nature of the experiment will determine whether a wave or particle description is best suited for understanding the results. The eye is an excellent example of this. The gathering and focusing of light is well described in terms of waves, but the ultimate absorption of light in rods and cones involves the absorption of photons.

Checklist

Define or explain:

photon
Planck's constant
photoelectric effect
work function
blackbody radiation

Planck's law
Compton effect
X-ray spectrum
absolute threshold of
vision

REVIEW QUESTIONS

Q26-1 True or false? If light behaved only as a wave, electrons would be emitted from a metal for any frequency.

Q26-2 The maximum energy of an electron emitted in the photoelectric effect is determined by the _____ of the absorbed photon.

Q26-3 The intensity of a beam of light is determined by the _____ of photons in the beam.

Q26-4 The frequency of the light with photons of energy E is _____.

Q26-5 When a blackbody emits radiation, the oscillators in the material lower their energy by emitting _____.

Q26-6 The collision of an electron and a photon can be analyzed by requiring that the total _____ and _____ are conserved.

Q26-7 When photons are produced by electrons colliding with atoms, a _____ frequency photon is produced when all of the electron kinetic energy is converted into a single photon.

EXERCISES

Section 26.1 | The Photoelectric Effect

26-1 When light of frequency 7×10^{14} Hz shines on a metal surface, electrons with a maximum speed of 6×10^{5} m s^{-1} are emitted. What is the photoelectric threshold frequency of the sample?

26-2 The work function of metallic sodium is 2.3 eV. What is the maximum wavelength of the incident light for which photoemission of electrons will occur?

26-3 The work function of a metal is 6.4×10^{-19} J. What is the minimum (threshold) frequency for photoemission of electrons?

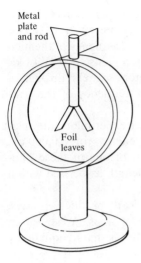

Figure 26.10. When light strikes the plate, the leaves of an electroscope spread apart at the bottom.

26-4 The maximum energy for photoelectrons emitted from a metal with work function 3 eV is 20 eV. What are the (a) maximum energy, and (b) maximum frequency of the incoming photons?

26-5 An electroscope has two metal foil leaves that normally hang side by side. They are both connected to a metal plate (Fig. 26.10). If light of a sufficiently high frequency is shone on the plate, the leaves are observed to spread apart. Describe what causes this.

Section 26.2 | The Photon

26-6 A radio antenna radiates 10^4 W of power at 9.2×10^5 Hz. How many photons per second are emitted?

26-7 A sodium vapor lamp emits 10 W of light at $\lambda = 590$ nm uniformly in all directions. How far from the lamp will the number of photons per square metre per second be equal to 10^{15}?

26-8 When the sun is directly overhead, the power incident on the earth at ground level is about 10^3 W m^{-2}. Assuming the average wavelength of the light is 550 nm, how many photons per square metre are incident on the earth per second?

26-9 What are the energies in electron volts of photons at the ends of the visible spectrum, (a) $\lambda = 400$ nm; (b) $\lambda = 700$ nm?

26-10 In a Compton scattering experiment, the initial and final photon frequencies are

2.4×10^{20} Hz and 1.6×10^{20} Hz, respectively. What is the recoil energy of the scattering electron in electron volts?

26-11 In order to disintegrate a deuteron (heavy hydrogen nucleus) into a proton and neutron, a γ ray must have an energy of at least 2.2 MeV. (γ rays are very energetic photons; 1 MeV $= 10^6$ eV $= 1.60 \times 10^{-13}$ J.) Find the minimum γ-ray frequency for photodisintegration of the deuteron.

26-12 In a monochromatic beam of light of wavelength 500 nm, what is the photon (a) energy, and (b) momentum?

26-13 The maximum frequency of X-ray photons from a certain apparatus is 9.5×10^{18} Hz. What is the accelerating voltage of the X-ray tube?

26-14 The electrons in an X-ray tube are accelerated from rest through a potential difference of 50,000 V. (a) What is the maximum frequency X-ray photon that is produced? (b) What is the wavelength of this photon?

26-15 A 1-kg mass is suspended from a spring with constant $k = 16$ N m^{-1}. The amplitude of vibration is 0.01 m. (a) If the energy is quantized as Planck suggested, what is the quantum number associated with the energy of the spring? (b) If the quantum number changes by one unit, what is the ratio of the energy change to the total energy of the spring?

Section 26.4 | Photons and Vision

26-16 The center-to-center distance between rods is 5×10^{-6} m in one portion of the eye. Estimate the area on the retina taken up by 100 rods.

26-17 A minimum of 100 photons incident on 2.5×10^{-9} m^2 of the pupil of the eye is necessary for vision. These must be incident during the eye's storage time of 0.2 second. (a) What is the minimum intensity of the light if $\lambda = 500$ nm? (b) If the source is 2 m from the eye and radiates uniformly in all directions, what is the power of the source?

26-18 A candle emits energy at a rate of 1 W at an average wavelength of 550 nm. (a) How many photons are emitted per second? (b) How far from the eye can the candle be seen if the threshold intensity is 10^{-7} W m^{-2}? (c) How many photons per square metre per second is this?

26-19 How far from a lamp emitting 5 W of power at a wavelength of 590 nm can one be and still see it with the naked eye? (The threshold intensity is 10^{-7} W m^{-2}.)

PROBLEMS

26-20 The vibrational motions of the atoms of a diatomic molecule are like Planck's oscillators. If we assume that the two atoms are connected by a spring with constant k, the vibrational energy of the molecule is $E = nhf$, where $f = \sqrt{k/\mu}/2\pi$. μ is called the *reduced mass* and for two atoms of masses m_1 and m_2,

$$\mu = \frac{m_1 m_2}{m_1 + m_2}$$

For the diatomic molecule H_2, $m_1 = m_2 = 1.67 \times 10^{-27}$ kg, and the measured energy-level spacing is 0.55 eV $= 8.74 \times 10^{-20}$ J. (a) What is the reduced mass of the molecule? (b) What is the spring constant k for the molecule?

26-21 The minimum frequency of photons emitted in vibrational transitions of the HCl molecule is 8.97×10^{13} Hz. The masses of the two atoms are 1.67×10^{-27} kg and 5.85×10^{-26} kg for hydrogen and chlorine, respectively. (a) What is the reduced mass of the molecule? (b) What is the spring constant of the molecule? (See Problem 26-20.)

26-22 The masses of O and H in the OH group are 2.67×10^{-26} kg and 1.67×10^{-27} kg, respectively. The effective spring constant is 50.5 N m^{-1}. (a) What is the reduced mass of the group? (b) What is the characteristic vibrational frequency of the group? (See Problem 26-20.)

26-23 The work function of tungsten is 4.49 eV. (a) Find the threshold wavelength for photoemission. (b) Ultraviolet light of wavelength 250 nm falls on the surface. What is the maximum kinetic energy of the emitted electrons? (c) What is the stopping potential?

*26-24 Light of intensity 10^{-2} W m^{-2} falls on a metallic surface. The energy required to emit a photoelectron is 3 eV (1 eV $= 1.60 \times 10^{-19}$ J). If a single electron absorbes light from a surrounding area 10 atoms in radius (10^{-9} m), how long will it take the electron to absorb enough energy to be emitted if light is assumed to be a classical wave?

26-25 In a Compton effect experiment, the incident X-ray photons have an energy of 10^5 eV $= 1.6 \times 10^{-14}$ J. (a) What is the frequency of the incident photons? (b) An electron gains 4000 eV of kinetic energy when a photon scatters through a certain angle. What is the frequency of the scattered photons?

26-26 In a Compton effect experiment, the incident photons have an energy of 20,000 eV. (a) What is the momentum of one of these photons? (b) What is the wavelength associated with a photon?

26-27 Nuclear reactions can produce γ rays, which are very energetic photons. When a π° meson decays at rest, 135 MeV is shared equally by two γ rays. Find the wavelength of the γ rays. (1 MeV $= 10^6$ eV $= 1.60 \times 10^{-13}$ J.)

ANSWERS TO REVIEW QUESTIONS

Q26-1, true; **Q26-2,** frequency; **Q26-3,** number; **Q26-4,** E/h; **Q26-5,** photons; **Q26-6,** energy, momentum; **Q26-7,** maximum.

Additional Reading

J. Andrade, E. Silva, and G. Lochak, *Quanta,* World University Library, McGraw-Hill Book Co., New York, 1969. Paperback.

A.H. Compton. The Scattering of X rays as Particles, *American Journal of Physics,* December 1961, p. 817.

Roderick K. Clayton, *Light and Living Matter, Volume I: The Physical Part; Volume II: The Biological Part,* McGraw-Hill Book Co., New York, 1971.

Albert Rose, Quantum Effects in Human Vision, in *Advances in Biological and Medical Physics,* vol. V., Academic Press, New York, 1957, p. 211.

Scientific American articles:

Karl K. Darrow, The Quantum Theory, March 1952, p. 47.

Richard Gordon, Image Reconstruction From Projections, October 1975, p. 56.

CHAPTER 27

WAVE PROPERTIES OF MATTER

The photon theory of radiation did not resolve all the difficulties facing physicists in the early part of the twentieth century. In many experiments, it was noted that matter absorbed and emitted radiation only at certain frequencies. For example, it had long been known that if hydrogen gas is placed in an electrical discharge tube, only a specific set of wavelengths is observed when the light emitted is analyzed with a diffraction grating. Furthermore, every pure gas had its own characteristic frequencies of absorption and emission. By contrast, all classical models of atoms predicted a continuous range of light frequencies.

By 1911, experiments had also shown conclusively that atoms have a positively charged but very small nucleus. The nucleus accounts for most of the mass of the atom, and the negatively charged electrons occupy a region outside the nucleus. This immediately suggests a "planetary" model of atoms with electrons orbiting about the nucleus and the charges held together by the electric forces among them. The electrons must move or they will be pulled into the nucleus, just as the earth would fall into the sun if it stopped moving. However, according to classical physics, accelerating charges produce electromagnetic radiation and hence lose energy. For example, if electrons are in circular orbits in atoms, they have a centripetal acceleration. Consequently, they must radiate energy and very soon fall into the nucleus. Thus the experimental structure of the atom and classical physics led to the unacceptable result that all atoms are unstable.

In discussing the resolution of these and other difficulties, we focus our attention on the work of two physicists. In 1912, Niels Bohr proposed a model of the one-electron atom that was successful in accounting for many observations. In a more refined form, many of the ideas of Bohr's model are contained in our present-day theories. Also, in 1924, Louis de Broglie hypothesized that electrons and all other "particles" possess wave attributes. This was strikingly confirmed when it was found that electrons reflected from a crystal form a pattern similar to that produced when X rays are diffracted by a crystal.

By about 1930, most of the fundamental puzzles involving the wave-particle behavior of light and matter had been resolved. Although the sequence of events that occurred in the unraveling of these problems is extremely interesting, we do not present them in chronological order here. Rather, we take a retrospective view of how the ideas and experiments were made compatible and how a solid foundation for our modern picture of nature was established. Specifically, we are concerned in this chapter with the wave properties of matter which, along with the photon theory of light, give us a consistent view of nature.

27.1 | FAILURES OF CLASSICAL PHYSICS

In the 1880s, experiments showed that when electrodes of opposite polarity are immersed in a gas, a current may be produced in the gas. In 1897, J.J. Thomson (1856–1940) showed that this current is composed of negatively charged particles now called electrons. These particles can be deflected by a magnetic field and the charge-to-mass ratio, e/m, meas-

603

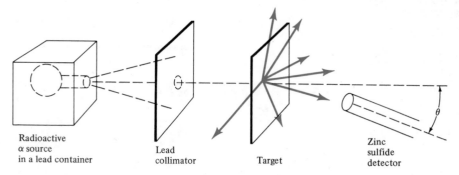

Figure 27.1. In Rutherford's experiments, α particles are emitted by a radioactive source. The lead collimator absorbs all the α particles except those that pass through a tiny hole and form a beam. The α particles scatter in all directions after striking the target and are detected at different angles by a moveable zinc sulfide detector.

ured. The same particles appear no matter what gas or electrode material is used. Hence, Thomson concluded that these electrons are a fundamental constituent of matter.

In 1906, Thomson proposed that atoms consist of electrons imbedded in a positively charged background. The total charge added up to zero because atoms are normally electrically neutral. This model could be tested by bombarding materials with charged particles and observing the way in which they are deflected.

Radioactivity was receiving a great deal of attention at this time, and one of those most prominent in that work was Ernest Rutherford (1871–1937). One of his most important early discoveries was the spontaneous emission by some heavy radioactive elements of *alpha* (α) *particles*. These particles have a positive charge that is twice the magnitude of the electronic charge; they are about 7000 times as massive as an electron. Starting in 1907, Rutherford began an intensive study of scattering α particles by various targets. The scattered α particles were detected by observing the flashes of light they produced upon reaching a zinc sulfide screen (Fig. 27.1).

According to Thomson's atomic model, the α particles should not be disturbed much by the diffuse positive charge, and the light electrons should also deflect them only slightly. Thus, most of the α particles should pass almost straight through the very thin gold foil target (Fig. 27.2a).

When the experiment was performed, it was found that most of the α particles were scattered through small angles, but a significant fraction were scattered through much larger angles. And occasionally an α

particle was stopped and sent back out toward the source! This remarkable fact was described by Rutherford: "It was almost as incredible as if you had fired a 15-inch shell at a piece of tissue paper and it came back and hit you."

The only explanation consistent with Rutherford's data is that of a small positively charged nucleus (of radius on the order of 10^{-14} m) with the electrons occupying the remainder of the atom. (Atoms were known to be on the order of 10^{-10} m in radius, a factor of 10,000 times larger than the nucleus.) The α particles experience a repulsive force from the heavy, positively charged nuclei. Occasionally, in a head-on collision, an α particle is stopped close to a nucleus and scattered directly back toward the source (Fig. 27.2b).

Rutherford's nuclear atom was an important step forward, but it presented serious problems. It now

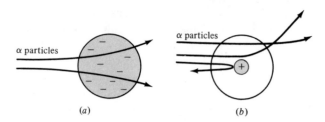

Figure 27.2. (a) In Thomson's model of the atom all the α particles would travel through the foil, deflected slightly by the diffuse positive charge (shown colored) and by the light electrons. (b) Rutherford explained the relatively frequent large angle scatterings as due to the repulsive electric force of a small positively charged nucleus. The light electrons occupying the remainder of the atom would not scatter the α particles significantly.

appeared that the electrons must be in some sort of orbit in the attractive electric field of the nucleus. The energy of an electron was the sum of its kinetic energy in its orbit plus the potential energy due to the positive nucleus. If the energy were to decrease, the electron would move closer to the nucleus. The problem was that accelerated charges were known to radiate energy as electromagnetic waves. For example, an electron in a circular orbit has a centripetal acceleration and should radiate energy. Its energy should decrease, and eventually the electron should fall into the nucleus. All classical theories of this nature led to unstable pictures of atoms. It remained for Bohr and those who followed to resolve these problems through the development of quantum mechanics.

Discrete Spectra | A second important body of

data existed during the same period that showed that atoms emitted and absorbed light only at specific frequencies. For illustrative purposes, we describe the data for hydrogen.

In examining a flame or an electric discharge tube containing hydrogen through a diffraction grating, only certain characteristic frequencies and wavelengths of emitted light were observed (Fig. 27.3). Jo-

hann Balmer (1825–1898) discovered in 1884 that the wavelengths of the visible and near ultraviolet lines of hydrogen obey a remarkably simple formula almost exactly,

$$\frac{1}{\lambda} = R_{H}\left(\frac{1}{2^2} - \frac{1}{n^2}\right) \qquad n = 3, 4, 5, \ldots \quad (27.1)$$

Here $R_H = 1.097 \times 10^7 \text{ m}^{-1}$ is called the *Rydberg constant*.

The longest visible wavelength in the hydrogen spectrum is found in the next example.

Example 27.1

What is the longest wavelength light predicted by the Balmer formula?

The longest wavelength is found using Balmer's formula for $n = 3$,

$$\frac{1}{\lambda} = R_{H}\left(\frac{1}{2^2} - \frac{1}{3^2}\right) = \frac{5}{36}R_{H}$$

Then

$$\lambda = \frac{36}{5R_{H}} = \frac{36}{5(1.097 \times 10^7 \text{ m}^{-1})}$$
$$= 6.56 \times 10^{-7} \text{ m} = 656 \text{ nm}$$

which is the longest visible wavelength observed (Fig. 27.3).

Other spectral lines were also observed including some not in the visible spectrum. The wavelengths of *all* the observed lines could be predicted using

$$\frac{1}{\lambda} = R_{H}\left(\frac{1}{n_f^2} - \frac{1}{n_i^2}\right) \qquad (27.2)$$
$$n_i = n_f + 1, n_f + 2, \ldots$$

The visible lines correspond to $n_f = 2$, the ultraviolet spectrum has $n_f = 1$, and the infrared spectrum corresponds to $n_f \geq 3$.

Using energy conservation, when a photon of energy hf is emitted by an atom the internal energy of the atom must decrease. Because the observed photons only appear at certain frequencies, the atoms must only change their energies by fixed amounts. This too contradicts a classical picture of electrons orbiting a nucleus at an arbitrary radius and a corresponding arbitrary energy.

From the point of view of classical physics, it seemed impossible for Rutherford's nuclear atom to be able to change its energy in discrete jumps. In the remainder of this chapter we describe the new ideas needed to explain the observations we have discussed.

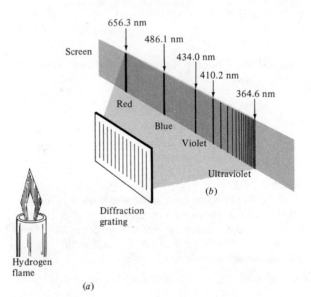

Figure 27.3. (a) When the light from a flame containing hydrogen passes through a diffraction grating and is projected on a screen, discrete lines are seen corresponding to the emission of monochromatic light at several different wavelengths. (b) The visible and near ultraviolet lines of hydrogen.

27.2 | THE de BROGLIE WAVE HYPOTHESIS

In 1924 Louis de Broglie suggested that matter as well as light might have wave properties. This would give both light and matter a dual wave-particle nature and put them both on the same footing. De Broglie also suggested a formula relating the wavelength of any object to its momentum. This formula was rapidly verified by experimental measurements.

De Broglie noted that photons are zero-mass particles and their energy and momentum are related by $E = hf = pc$, where h is Planck's constant and c is the speed of light (Chapter Twenty-six). Using $f\lambda = c$, the relation between the photon momentum and wavelength is $hc/\lambda = pc$ or

$$\lambda = \frac{h}{p} \qquad (27.3)$$

Using this as a guide to the possible dual nature of particles, de Broglie suggested that particles of momentum p should also have a wavelength λ associated with them, given by exactly the same relation.

This idea was tested beautifully in 1926 in an experiment performed by C. Davisson (1881–1958) and L.H. Germer (1896–1971). They directed a beam of electrons at a crystal and observed that the electrons scattered in various directions for a given crystal orientation (Fig. 27.4). In this experiment the pattern formed by the electrons reflected from the crystal lattice of aluminum is almost identical to that produced by X rays (Fig. 27.5). This strongly suggests that the electrons have a wavelength λ associated with them and that the Bragg condition for X-ray diffraction

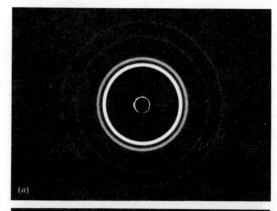

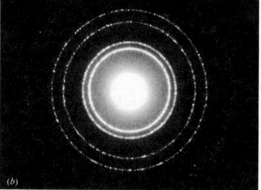

(a)

(b)

Figure 27.5. The diffraction pattern in aluminum obtained with (a) X rays, and (b) electrons. (From Educational Development Center, Inc., Newton, Mass.)

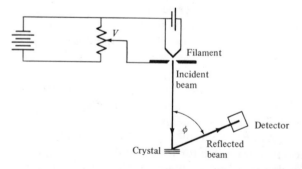

Figure 27.4. The apparatus of Davisson and Germer. Electrons emitted from a hot filament are accelerated through a potential difference V. They strike a crystal, and the scattered electrons are detected at an angle ϕ with the incident beam.

discussed in Chapter Twenty-three holds true for electrons also,

$$m\lambda = 2d \sin \theta \qquad m = 1, 2, 3, \ldots$$

The following examples illustrate how the wave character of electrons is predicted by de Broglie's hypothesis.

Example 27.2

What is the wavelength associated with the electrons that scatter as shown in Fig. 27.6?

From the figure $\theta = 90° - \frac{1}{2}(52°) = 64°$, and $\sin 64° = 0.899$. With $d = 0.09$ nm $= 9 \times 10^{-11}$ m, the Bragg condition with $m = 1$ gives

$$\lambda = 2d \sin \theta = 2(9 \times 10^{-11} \text{ m})(0.899)$$
$$= 1.62 \times 10^{-10} \text{ m}$$
$$= 0.162 \text{ nm}$$

Example 27.3

What is the momentum and corresponding de Broglie wavelength of the electrons scattered in Fig. 27.6?

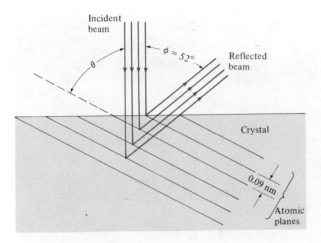

Figure 27.6. When the electron accelerating voltage is 56 V, electrons scattered from atomic planes 0.09 nm apart in the crystal appear mostly at $\phi = 52°$.

The momentum of electrons accelerated through a 56-V potential difference can be found by noting that their kinetic energy is

$$\tfrac{1}{2}mv^2 = eV$$

where m is the electron mass and e is the magnitude of the electronic charge. Using $p = mv$, $\tfrac{1}{2}mv^2 = p^2/2m = eV$, so

$$p = \sqrt{2emV}$$
$$= \sqrt{2(1.6 \times 10^{-19}\,\text{C})(9.1 \times 10^{-31}\,\text{kg})(56\,\text{V})}$$
$$= 4.04 \times 10^{-24}\,\text{kg m s}^{-1}$$

Using de Broglie's hypothesis,

$$\lambda = \frac{h}{p} = \frac{6.63 \times 10^{-34}\,\text{J s}}{4.04 \times 10^{-24}\,\text{kg m s}^{-1}}$$
$$= 1.64 \times 10^{-10}\,\text{m} = 0.164\,\text{nm}$$

This value agrees with that found from experiment (Example 27.2).

As a result of this and other experiments, we now accept de Broglie's idea as one of the cornerstones of modern physics. The diffraction effects of electrons and neutrons are now important tools to the scientist.

Neutron Diffraction | The wave properties of

matter provide us with ways to study microscopic objects on a much smaller scale than is possible with the ordinary microscope. Both neutrons and electrons have been used extensively for this purpose.

No analog to the lens is available for neutrons, so it is not possible to construct a neutron microscope.

However, a crystal serves as a three-dimensional diffraction grating for neutrons, just as it does for electrons and X rays. In many situations, neutron diffraction provides molecular structure information that cannot be obtained in other ways.

The reactions occurring in a nuclear reactor produce many neutrons with a wide range of energies. If neutrons collide with several atoms before or after leaving the reactor, their average kinetic energy is comparable to the average thermal energy of the atoms. When the temperature is T, this energy is $\tfrac{3}{2}k_BT$. Thus the average energy and de Broglie wavelength of such *thermal neutrons* can be controlled by varying the temperature of the materials.

The wavelength of a thermal neutron is approximately the size of an atom, as is shown in the next example.

Example 27.4

What is the wavelength of a neutron with kinetic energy $\tfrac{3}{2}k_BT$, when $T = 300$ K?

The neutron kinetic energy is $\tfrac{1}{2}mv^2 = \tfrac{3}{2}k_BT$ so the momentum $p = mv$ is found from $p^2 = (mv)^2 = 3k_BTm$ or

$$p = \sqrt{3mk_BT}$$

Using $m = 1.67 \times 10^{-27}$ kg and $T = 300$ K,

$$\lambda = \frac{h}{p} = \frac{h}{\sqrt{3mk_BT}}$$
$$= \frac{6.63 \times 10^{-34}\,\text{J s}}{\sqrt{3(1.67 \times 10^{-27}\,\text{kg})(1.38 \times 10^{-23}\,\text{J K}^{-1})(300\,\text{K})}}$$
$$= 1.46 \times 10^{-10}\,\text{m} = 0.146\,\text{nm}$$

This is a typical atomic or molecular distance.

Electron Microscope | The electron micro-

scope is an astonishing improvement over the optical microscope. The shortest useful wavelength and resolution for optical work is about 200 nm, but electrons produced by using accelerating voltages of 50 kV have a de Broglie wavelength of 0.0055 nm. Theoretically, this would be the resolution of an electron microscope. In practice, electrons must be focused using electric and magnetic fields as lenses, and these limit the resolution to about 0.2 nm. However, even this is 1000 times better than that obtained with optical instruments and allows the study of minute cellular constituents and molecules.

Two types of electron microscopes are in common use. The *transmission microscope* requires the use of

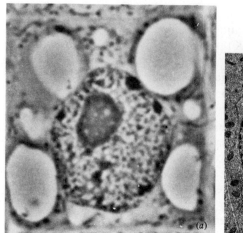
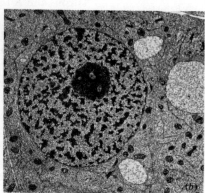
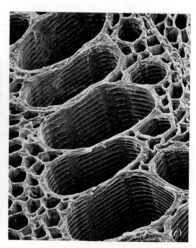

Figure 27.7.   (a) Light microscope picture of a plant cell. (b) Transmission electron microscope picture of the same type of cell. (c) A scanning electron microscope picture of a cross section of the leaf stalk of a fern. The large hollow cells are xylem tracheids and function as water conducting tissues. The three-dimensional structure is clearly evident. ((a) and (b) courtesy William A. Jensen, from *Cell Ultrastructure* by W.A. Jensen and R.B. Park, Wadsworth, 1967; (c) courtesy John R. Troughton, from *Plants* by J.R. Troughton and F.B. Sampson, John Wiley & Sons, 1973.)

samples less than 100 nm thick. If the sample is too thick, the energy lost by off-axis electrons is much greater than that lost by electrons going directly through the sample. The wavelengths are then sufficiently varied so that the electrons cannot all be focused simultaneously. Although the resolution of these microscopes is excellent, sample preparation limits their usage somewhat, and the depth of field is very small. Thus, three-dimensional structures will not be properly focused. These disadvantages are overcome with a *scanning microscope,* but at the price of a reduction of the resolution to about 10 nm. Live samples can be used, and the results show three-dimensional structures clearly (Fig. 27.7).

The transmission microscope is schematically much like an optical microscope. Produced at a hot cathode, the electrons pass through the sample and are focused into a real image (Fig. 27.8a). In the scanning microscope the electrons are focused to a very small spot, which is swept across the sample. The image is formed by monitoring secondary electrons knocked out of the sample by the main beam. These secondary electrons are collected and accelerated in a cathode-ray tube. The cathode-ray beam is swept across the screen at the same rate as the main beam is swept across the sample. The number of secondary electrons varies with the composition and orientation

of the surfaces of the sample, and the brightness of the trace on the cathode-ray screen varies accordingly (Fig. 27.8b, 27.9).

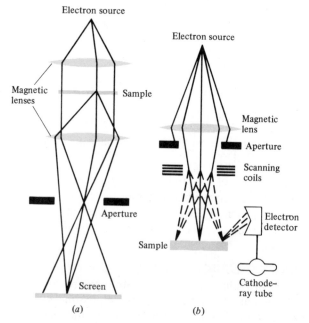

Figure 27.8.   (a) Transmission and (b) scanning electron microscopes.

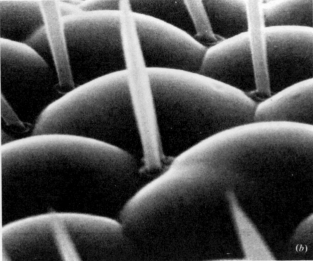

Figure 27.9. Scanning electron microscope pictures of (*a*) a fruit fly and (*b*) a portion of the eye alone. ((*a*) Photo from Thomas Eisner, Cornell University; (*b*) Photo from H. Hartman and T.L. Hayes, *Journal of Heredity*, Vol. 62, p. 41, 1971. Copyright 1971 by the American Genetic Association.)

27.3 | THE BOHR ATOM

In the remaining sections of this chapter, we describe the resolution of the atomic spectra and nuclear atom problems discussed earlier. The wave-particle nature of matter provides the basis for these discussions.

In 1913, well before de Broglie postulated the wave nature of matter, Niels Bohr (1885–1962) proposed a model of one-electron atoms. The immediate purpose of the model was to explain the observed emission spectrum of hydrogen. Although the model was successful in this respect, it is now known to be quite incomplete and oversimplified. However, it does provide a very simple picture that we sometimes use to visualize complex atomic processes.

Bohr adopted Rutherford's nuclear atom model with electrons moving in the electric field of a small massive nucleus. He also suggested that certain orbits are stable; the accelerating electrons do not radiate energy when in these orbits. We can find these orbits by attributing a wavelength $\lambda = h/p$ to the electron, although Bohr used a different approach.

We consider an atom composed of a heavy nucleus containing Z protons and a single electron in a circular orbit of radius r around the nucleus. Since the nuclear mass is much larger than that of the electron,

we assume that the nucleus is at rest. The electric force between the positive nucleus (with charge Ze) and the electron (with charge $-e$) holds the electron in orbit (Fig. 27.10). As we show below, the energy of the electron depends on the radius of the circular orbit. Classical physics predicts that all radii are possible.

To restrict the possible orbits, Bohr postulated that the orbital angular momentum of the electron $L = rmv$ is *quantized,* that is, it can have only certain values,

$$L = rmv = n \frac{h}{2\pi} = n\hbar \qquad n = 1, 2, 3, \ldots \quad (27.4)$$

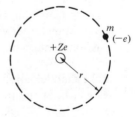

Figure 27.10. An electron of mass m and charge $-e$ in a circular orbit around a nucleus with charge Ze. The massive nucleus is assumed fixed in place.

NIELS HENRIK DAVID BOHR
(1885–1962)

(American Institute of Physics.)

The years from 1910 to 1930 were among the most exciting in physics. A unique generation of physicists participated in the conception and development of the revolution called quantum physics. Most of this excitement centered on three places, Rutherford's laboratory in Manchester, England, the University of Göttingen in Germany, and the Institute for Theoretical Physics at the University of Copenhagen. The center of attraction in Copenhagen was Niels Bohr. In addition to his abilities as a physicist, Bohr was able to cultivate an atmosphere in which incredible progress and understanding was possible by physicists from all over the world. The flavor of this period is illustrated by the following story told by George Gamow*:

> The evening work in the Institute's library was often interrupted by Bohr, who would say that he was very tired and would like to go to the movies. The only movies he liked were wild Westerns (Hollywood style), and he always needed a couple of students to go with him and explain the complicated plots. . . . But his theoretical mind showed even in these movie expeditions. He developed a theory to explain why although the villain always draws first, the hero is faster and manages to kill him. This Bohr theory was based on psychology. Since the hero never shoots first, the villain has to decide when to draw, which impedes his action. The hero, on the other hand, acts according to a conditional reflex and grabs the gun automatically as soon as he sees the villain's hand move. We disagreed with this theory, and the next day I went to a toy store and bought two guns in Western holsters. We shot it out with Bohr, he playing the hero, and he "killed" all his students.

Bohr was the son of a physiology professor. He received his doctorate in physics in Copenhagen in 1911 and then spent three years in England, two of them in Manchester with Rutherford. It was during this period that Bohr took the revolutionary steps described now as the Bohr model for the hydrogen atom.

Bohr's model combined elements of classical physics with ideas and postulates which, at the time, were not verifiable nor even readily believable. For example, Max Planck, whose work Bohr used as a point of departure, was very reluctant to accept Bohr's ideas. It was the task of physicists during the next two decades to fill in the missing pieces and put the entire picture together in a

*George Gamow, *Thirty Years That Shook Physics,* Anchor Books, New York, 1966.

coherent way. In this process, Bohr represented the generation that had to cast off the old and press forward with the new. The generation he helped educate was not so tied to classical physics and felt less need to look back.

It is important to realize that during this period it was not simply new physics problems that were being solved; rather, an entirely new view of natural law was at issue. During this time, one needed not only new equations but also new philosophies and new pictures of a world that was beyond the direct reach of the senses but not beyond experimental reach. Bohr's role in forging the physics and philosophy of this age was immense.

By 1930 the quantum picture of atoms was fairly firm, and Bohr joined others in trying to understand the atomic nucleus. This led to Bohr's interest in fission. In 1939, in a paper with G. V. Wheeler, Bohr showed how many one may understand the role of neutrons in causing the ^{235}U nucleus to break apart. This work was fundamental in the process that ultimately led to nuclear chain reactions.

In 1943, Bohr's ever-present concern for others resulted in his having to flee Denmark to avoid arrest by the Nazis. He was flown to England from Sweden in a British fighter plane. Given an oxygen mask that was too small, he was saved when the pilot noticed that Bohr had lost consciousness, and he reduced the altitude.

From England, Bohr went to the United States and participated in the atomic bomb development program. After the war, he became an early and persistent voice in calling for international reason in the face of the tremendous power he had helped to release. He also played an important role in developing international and Danish facilities for atomic and nuclear research. Having given birth to a model that was awesome in both its simplicity and its novel character, Bohr was an active participant of the scientific and political events that followed for another half-century.

Since the combination $h/2\pi$ appears frequently, it is denoted by

$$\hbar = \frac{h}{2\pi} = 1.055 \times 10^{-34} \text{ J s}$$

and is read as "h bar."

This quantization postulate does not contain any obvious reference to the wave property of the electron. However, if the electron has a wavelength $\lambda = h/p$ as later suggested by de Broglie, then it should "fit" into the orbit. If it does not "fit," it will interfere with itself (Fig. 27.11). Notice that this idea is similar to the one that we used for waves on strings. Resonant standing waves occur when the wavelength "fits" on a string. Here the wavelength must fit into the orbit.

With this assumption, we must fit an integer number of waves of wavelength $\lambda = h/p = h/mv$ into the orbital circumference $2\pi r$. Thus

$$2\pi r = n\lambda = n\frac{h}{mv}$$

or

$$rmv = n\frac{h}{2\pi} = n\hbar \qquad n = 1, 2, 3, \ldots \quad (27.5)$$

This is exactly Bohr's criterion for the angular momentum! Thus our assumption is equivalent to Bohr's postulate.

We now use Bohr's postulate plus Newton's laws to obtain equations for the radii and energies of the electron orbits. If the electron is in a circular orbit of radius r, the acceleration is $a = v^2/r$. Using Newton's second law, $\mathbf{F} = m\mathbf{a}$, where $\mathbf{F}$ is the electric force with magnitude kZe^2/r^2,

$$k\frac{Ze^2}{r^2} = \frac{mv^2}{r} \qquad (27.6)$$

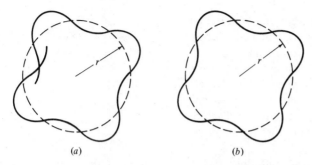

Figure 27.11. (a) The wavelength associated with the electron does not fit into the orbit. In (b) the wave "fits" correctly. The orbital circumference is $2\pi r$.

From this we see that the kinetic energy of the electron is

$$K = \frac{1}{2} mv^2 = k\frac{Ze^2}{2r} \qquad (27.7)$$

From Eq. 27.5, we have $v = n\hbar/mr$. If we substitute this expression for v in Eq. 27.7, we can solve for the radius r associated with n wavelengths. Labeling this radius with a subscript n, we find

$$r_n = \frac{n^2\hbar^2}{kZme^2} = \frac{n^2}{Z}a_0, \qquad n = 1, 2, 3, \ldots \,(27.8)$$

Here $a_0 = \hbar^2/kme^2 = 5.29 \times 10^{-11}$ m is called the *Bohr radius*. There are an infinite number of allowed values of r_n corresponding to all the integer values of n. The electrons can only be in these orbits. This is quite different from the classical idea that any value of r is possible.

The potential energy of the electron is $\mathcal{U} = -kZe^2/r$, so the total energy for an orbit of radius r is, with Eq. 27.7,

$$E = K + \mathcal{U} = -\frac{kZe^2}{2r}$$

Using our result for the possible radii r_n, this becomes

$$E_n = -\frac{kZ^2e^2}{2a_0n^2} = -\frac{Z^2}{n^2}E_0 \qquad n = 1, 2, 3, \ldots \,(27.9)$$

Here

$$E_0 = \frac{ke^2}{2a_0} = 13.6 \text{ eV} = 2.18 \times 10^{-18} \text{ J}$$

E_0 is the lowest possible energy of the electron. The corresponding orbit is called the *ground state*.

This is the fundamental result that Bohr obtained.

The electrons can only occupy orbits with certain energies. The allowed values of the electron energy E_n are termed *energy levels,* and n is called a *quantum number.*

Bohr's result is in close agreement with experiment for *any* atom with one electron, such as hydrogen ($Z = 1$), singly ionized helium, He$^+$($Z = 2$), and doubly ionized lithium, Li^{++} ($Z = 3$). The lowest levels for He$^+$ are found in the next example.

Example 27.5

(a) What are the lowest three energy levels of singly ionized helium? (b) What are the radii associated with these levels?

(a) Helium has two protons in the nucleus ($Z = 2$), and when it is singly ionized, it has one electron. Using $Z = 2$ and $n = 1$,

$$E_1 = -\frac{Z^2}{n^2}E_0 = -\frac{(2)^2}{(1)^2}(13.6 \text{ eV})$$

$$= -54.4 \text{ eV}$$

If instead we use $n = 2$, the result obtained is smaller by $1/2^2$, and is -13.6 eV; for $n = 3$, the energy is -6.04 eV.

(b) Using Eq. 27.8, the $n = 1$ radius is

$$r_1 = \frac{n^2}{Z}a_0 = \frac{(1)^2}{2}(5.29 \times 10^{-11} \text{ m})$$

$$= 2.65 \times 10^{-11} \text{ m}$$

Similarly, $r_2 = 10.6 \times 10^{-11}$ m, and $r_3 = 23.8 \times 10^{-11}$ m (Fig. 27.12).

Bohr also postulated that electrons in allowed orbits do not radiate energy, even though they are ac-

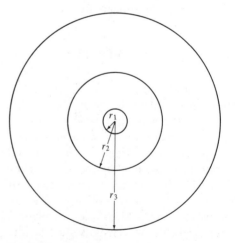

Figure 27.12. The radii predicted by the Bohr model for the three lowest energy orbits in a one-electron atom. Here $r_3 = 9r_1$ and $r_2 = 4r_1$.

celerated. Electrons can only lower their energy by dropping into lower energy orbits and emitting a photon in the process. If the initial and final energies of the electron are $E_{n,i}$ and $E_{n,f}$, from energy conservation the photon energy hf must be

$$hf = E_{n,i} - E_{n,f}$$

Using our result for the energy levels, this gives for hydrogen

$$hf = \frac{hc}{\lambda} = hcR_H \left(\frac{1}{n_f^2} - \frac{1}{n_i^2} \right) \quad (27.10)$$

$$n_i = n_f + 1, n_f + 2, \ldots$$

where the theoretical value of R_H is

$$R_H = \frac{mk^2e^4}{2\hbar^2hc} = 1.097 \times 10^7 \text{ m}^{-1}$$

This is exactly the empirical result found by Balmer and others and described in the preceding section.

Ordinarily the electrons in hydrogen atoms are in the energy level with the lowest or most negative energy, the $n = 1$ level or ground state. In a flame or an electrical discharge tube, many of the atoms are in *excited states,* with the electrons in less tightly bound orbits that have larger values of n. An electron may return to the lowest level in a single step, emitting one photon, or in a series of steps from higher to lower energy levels, emitting photons at each step. Figure 27.13 is an energy-level diagram that shows the allowed energy levels for the electron and the transitions that may occur.

An atom will also absorb photons with these same frequencies and wavelengths. In absorption, the energy provided by the photon must be just equal to the energy required to raise an electron from one stable level to a higher one. The following example illustrates photon absorption.

Example 27.6

What energy photon is necessary to raise an electron in hydrogen from the $n = 1$ to the $n = 3$ energy level?

From Fig. 27.13, we see that this is the reverse of the process indicated by the second line in the Lyman series.

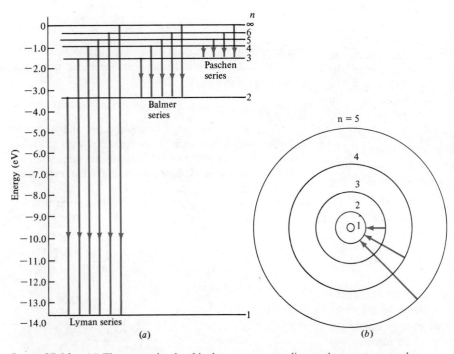

Figure 27.13. (a) The energy levels of hydrogen corresponding to the quantum numbers $n = 1$ through 6 and the highest level ($n = \infty$) which has zero energy. The vertical arrows show possible electron transitions that are accompanied by photon emission. For example, the longest wavelength line in the Balmer series corresponds to an electron dropping from the $n = 3$ to the $n = 2$ energy level in its hydrogen atom. The three series shown involve ultraviolet (Lyman), visible (Balmer), and infrared (Paschen) photons. (b) The transitions corresponding to the first three lines of the Balmer series.

The electron increases its energy from E_1 to E_3, and the photon energy must be

$$hf = E_3 - E_1 = -\frac{1}{3^2} E_0 - \left(-\frac{1}{1^2} E_0\right)$$

$$= \left(1 - \frac{1}{9}\right) E_0$$

With $E_0 = 13.6$ eV, the photon energy is

$$hf = \frac{8}{9}(13.6 \text{ eV}) = 12.1 \text{ eV}$$

Bohr's results are correct for any one-electron atom, despite the fact that the model on which they are based is now considered to be a great oversimplification. However, we do see how the two problems described earlier are explained. The fact that the electron is not a particle as we usually think of one classically leads to quantized energy levels. The discrete spectra observed are the result of electrons moving from one orbit to another. The wavelike behavior of the electron also means that it is not an object that radiates energy when accelerated. The orbital radii we have calculated represent the average distances of electrons from the nucleus but are not actually the radii of electronic trajectories.

Despite the success of Bohr's model for one-electron atoms, most attempts to extend these ideas to multielectron atoms and molecules were unsuccessful. The requirement that the angular momentum is quantized is insufficient to determine the orbits in these complex systems.

In 1925, two detailed theories of *quantum mechanics* were proposed, one by Erwin Schroedinger (1887–1961) and the other by Werner Heisenberg (1901–1975). Although the two theories were very different in appearance, they were soon recognized to be equivalent in content. Unlike Bohr's theory, these are complete theories which, in principle, can be applied to all physical systems, including atoms, molecules, and macroscopic objects. We discuss Schroedinger's form of quantum mechanics in the next chapter.

We have seen that Bohr's model predicts that there is a lowest energy level. It is interesting to ask, why can't the electron go into a deeper energy level, one closer to or even inside the nucleus? In nature, systems tend to stay in the lowest energy state whenever possible, and the electron could reduce its energy enormously by moving much closer to the nucleus. This question can be partially answered using the *uncertainty principle*.

27.4 | THE UNCERTAINTY PRINCIPLE

Both Schroedinger's and Heisenberg's formulations of quantum mechanics implicitly contain the de Broglie hypothesis of the wave character of matter. They also contain the uncertainty principle, which was first discussed by Heisenberg in 1927.

The uncertainty principle describes limits imposed by nature on the precision of *simultaneous* measurements of the position and momentum of an object. Mathematically stated, if an object is said to be at a position x within an uncertainty of Δx, then any simultaneous measurement of the x component of momentum must have an uncertainty Δp_x consistent with

$$\Delta x \, \Delta p_x \geq \hbar \qquad (27.11a)$$

This means that an increase in the accuracy of a position measurement must be accompanied by a decrease in the accuracy of the momentum measurement. The best that can be done even with idealized experiments is to have the equality hold, $\Delta x \, \Delta p_x = \hbar$. The x direction here is chosen arbitrarily; the uncertainty relation could equally well be stated as

$$\Delta y \, \Delta p_y \geq \hbar \qquad (27.11b)$$

Before pursuing the ramifications of these statements, we may ask just how important are these naturally occurring limitations.

Example 27.7

Suppose the velocities of an electron and of a rifle bullet of mass 0.03 kg are each measured with an uncertainty of $\Delta v = 10^{-3}$ m s^{-1}. What are the minimum uncertainties in their positions according to the uncertainty principle?

Using $\Delta p_x = m \, \Delta v_x$ for each, the minimum position uncertainty satisfies $\Delta x \, m \, \Delta v_x = \hbar$. For the electron $m = 9.11 \times 10^{-31}$ kg, so

$$\Delta x = \frac{\hbar}{m \, \Delta v_x} = \frac{1.055 \times 10^{-34} \text{ J s}}{(9.11 \times 10^{-31} \text{ kg})(10^{-3} \text{ m s}^{-1})}$$

$$= 0.116 \text{ m}$$

For the bullet,

$$\Delta x = \frac{\hbar}{m \, \Delta v_x} = \frac{(1.055 \times 10^{-34} \text{ J s})}{(0.03 \text{ kg})(10^{-3} \text{ m s}^{-1})}$$

$$= 3.5 \times 10^{-30} \text{ m}$$

We can see from the preceding example that for normal macroscopic objects such as the bullet the uncertainty principle does not impose any effective limit on experimental measurements because errors in position measurements are always very much larger than 10^{-30} m. However, the opposite is true of objects as small as electrons. For example, since atoms in a solid are about 10^{-9} m apart, a position measurement with an uncertainty of about 0.1 m means that the electron could be anywhere among billions of atoms!

The uncertainty principle leads to the conclusion that, in any experiment, one cannot simultaneously observe the wave and particle properties of light or matter. This can be illustrated with a double-slit interference experiment performed with electrons. Consider a beam of monoenergetic electrons incident on two slits of width $d/4$ a distance d apart (Fig. 27.14). Since the electrons have a wavelength $\lambda = h/p$ associated with them, an interference pattern will be detected at the screen. From Chapter Twenty-three we know that there will be an intensity maximum at $\theta = 0$, and the first minimum occurs when $\sin \theta = \lambda/2d$. As long as $\lambda/2d$ is small, we can use $\sin \theta \simeq \theta$, so the first minimum occurs at

$$\theta_m = \frac{\lambda}{2d}$$

Thus a measurement of θ_m can be used to determine the wavelength λ. This value can be compared with de Broglie's prediction to test his hypothesis.

The interference pattern is a manifestation of the wave properties of electrons. However, electrons also have particle properties. If they are regarded as particles, we should be able to say which slit each electron passes through. This might be done by having a beam of photons passing to the right of the slits, so that

each time an electron goes through a slit, it will collide with a photon. The electron proceeds toward the screen, but the scattered photon can be detected and the area of the collision located. For example, if the collision takes place above the centerline of Fig. 27.14, we would say the electron went through the upper slit, and its position uncertainty at the slits would be, at most, $\Delta y = d/4$, the slit width.

We must find out what our measurements mean in the light of the uncertainty principle. The uncertainty principle requires that $\Delta y \, \Delta p_y \geq \hbar$. Since $\hbar = h/2\pi$,

$$\Delta p_y \geq \frac{h}{2\pi \, \Delta y} = \frac{h}{2\pi d/4} = \frac{2h}{\pi d}$$

The angular deflection of the electron is then at least

$$\theta = \frac{\Delta p_y}{p} = \frac{2h/\pi d}{h/\lambda} = \frac{2\lambda}{\pi d}$$

But this deflection is larger than $\theta_m = \lambda/2d$, the angular position of the interference minimum. This means that the interference pattern will be hopelessly blurred if we determine which slit the electron passes through. The act of measurement that pinpoints the electron as a particle going through one slit also destroys the evidence of the wave property of the electron!

Viewed in this way, the uncertainty principle is nature's way of ensuring that the wave and particle properties of an object cannot be simultaneously observed. We also face the extraordinary fact that *the measurement process itself affects the results*. In the case illustrated, the measurement of the electron position deflects the electron enough to destroy the interference pattern.

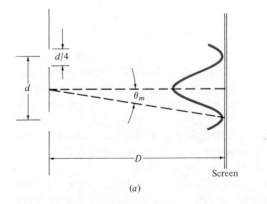
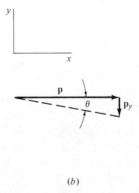

(a)   (b)

Figure 27.14. (a) Electrons of wavelength $\lambda = h/p$ are incident from the left. After passing through the slits, they form an interference pattern at the screen. (b) The slit area is shown much enlarged; in practice $D \gg d$ and θ is very small.

We conclude this section by considering the question raised in the previous section: Why doesn't an electron radiate energy and drop into the nucleus where its potential energy would be much lower than it is in the first Bohr orbit? Suppose an electron is within a distance r of the nucleus. Then we could say that we know its position within an uncertainty $\Delta x = r$. From the uncertainty principle, the electron momentum cannot be zero, and it must be at least $p = \Delta p_x = \hbar/r$. The total energy of the electron is then at least (for $Z = 1$)

$$E = \frac{1}{2}mv^2 - \frac{ke^2}{r} = \frac{p^2}{2m} - \frac{ke^2}{r} = \frac{\hbar^2}{2mr^2} - \frac{ke^2}{r}$$

If the electron approaches the nucleus, r becomes small; the potential energy $-ke^2/r$ becomes more negative, but the kinetic energy $\hbar^2/2mr^2$ increases (Fig. 27.15). We would expect the electron to stay at an average radius r such that its total energy is a minimum. This occurs when r equals the Bohr radius, $a_0 = \hbar^2/kme^2$. Thus the lowest energy state of an atom is not one with the electron in the nucleus. The uncertainty principle requires that the kinetic energy becomes large as the position uncertainty decreases, more than offsetting the reduction in the potential energy.

We see from this discussion that the stability of atoms and of matter depends on the uncertainty principle.

SUMMARY

Matter, as well as light, has both wave and particle characteristics. The wavelength is associated with the momentum of an object according to de Broglie's relationship,

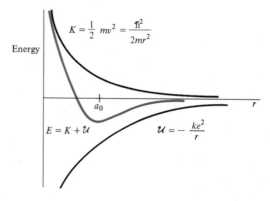

Figure 27.15. The kinetic and potential energy contributions to give the total energy $E = K + \mathcal{U}$. The total energy is a minimum at $r = a_0$.

$$\lambda = \frac{h}{p}$$

where h is Planck's constant. This concept is one of the fundamental ideas of quantum theory.

By combining the wave properties of electrons with the picture of an atom as a massive, positively charged nucleus with orbiting electrons, Bohr was able to find the energy levels for an atom with one electron. The allowed energy levels have energies

$$E_n = -\frac{Z^2}{n^2}E_0, \qquad n = 1, 2, 3 \ldots$$

where $E_0 = 13.6$ eV. Photons with specific energies are emitted or absorbed when an electron goes from one energy level to another.

The uncertainty principle describes a natural limit to the precision of simultaneous measurements of the position and momentum of particles,

$$\Delta x\, \Delta p_x \geq \hbar$$

It also implies that either wave or particle properties, but not both, may be observed in a given experiment.

Checklist

Define or explain:

nuclear atom	Bohr radius
discrete spectra	ground state
de Broglie wavelength	excited state
electron diffraction	energy levels
neutron diffraction	quantum number
Bohr atom	uncertainty principle
quantized angular	wave-particle duality
momentum	

REVIEW QUESTIONS

Q27-1   Rutherford's experiments led to the idea that an atom was composed of a massive _____ nucleus with _____ orbiting outside it.

Q27-2   True or false? When the light from a discharge in hydrogen gas is passed through a diffraction grating, a continuous spectrum of colors is seen.

Q27-3   De Broglie postulated that there is a relationship between the _____ of an object and the wavelength associated with it.

Q27-4   Electron diffraction is an example of the appearance of the _____ properties of what were originally considered "particles."

Q27-5   The de Broglie wavelength of large objects is irrelevant because it is so _____.

Q27-6   In Bohr's model, it is the _____ force that holds the atom together.

Q27-7   The incorporation by Bohr of angular momentum quantization in an otherwise classical theory is equivalent to the idea that the electron _____ "fits" into the allowed orbits.

Q27-8   The spectral lines discovered for hydrogen are explained by the _____ or _____ of photons as electrons move from one allowed energy level to another.

Q27-9   The uncertainty principle explains why an electron does not get arbitrarily _____ to the nucleus to lower its energy.

Q27-10   The uncertainty principle plays no significant role in classical phenomena because the predicted uncertainties are _____.

EXERCISES

Section 27.1 | Failures of Classical Physics

27-1   A 4.8-MeV (1 MeV = 10^6 eV) α particle makes a head-on collision with a gold nucleus ($Z = 79$ for gold). What is the distance of closest approach of the α particle and the gold nucleus at which the α particle has zero kinetic energy? (Neglect the kinetic energy transferred to the gold nucleus.)

27-2   In hydrogen, the proton radius is 0.8×10^{-15} m and the average distance of the electron from the proton center is 5.3×10^{-11} m. If the proton is represented by an orange 0.06 m in radius, how far away will the electron be?

27-3   What is the wavelength of the visible light of atomic hydrogen corresponding to $n = 4$?

27-4   What is the wavelength of the ultraviolet radiation from hydrogen corresponding to $n_i = 2$ and $n_f = 1$?

Section 27.2 | The de Broglie Wave Hypothesis

27-5   An electron and a photon each have a wavelength of 0.1 nm. What are their momenta?

27-6   Compare the de Broglie wavelengths of an electron and a neutron each having energy $\frac{3}{2}k_B T$, where $T = 300$ K.

27-7   What is the de Broglie wavelength associated with the earth in its motion about the sun? (The earth's mass is 6×10^{24} kg, and its orbital radius is 1.5×10^{11} m.)

27-8   An oxygen atom has a mass of 16 u, where 1 u $= 1.66 \times 10^{-27}$ kg. If the atom has a kinetic energy of 1 eV, what is its de Broglie wavelength?

27-9   A 1-kg mass has a velocity of 1 m s^{-1}.

(a) What is its de Broglie wavelength? (b) Should one expect to see interference effects from these waves? Explain.

27-10   The protons and neutrons in the atomic nucleus are typically 10^{-15} m apart, so any probe of nuclear structure must have a de Broglie wavelength of this order of magnitude. (a) What momentum must a probe have if its de Broglie wavelength is 10^{-15} m? (b) If the probe is an alpha particle of mass 6.64×10^{-27} kg, what is its kinetic energy in electron volts?

27-11   Electrons incident on a set of crystal planes at an angle of 20° to the planes interfere constructively and form a spot on a photographic plate. If the crystal planes are 0.12 nm apart, find (a) the wavelength of the electrons; (b) the energy of the electrons.

27-12   Neutrons with a de Broglie wavelength of 0.3 nm are needed to perform a molecular structure experiment. (a) What is the velocity of these neutrons? (b) A monoenergetic neutron pulse can be made by opening a nuclear reactor port and using only those neutrons that reach the target within a certain time interval. How long will it take the neutrons in part (a) to travel 10 m?

Section 27.3 | The Bohr Atom

27-13   Can a hydrogen atom absorb a photon of energy greater than 13.6 eV? Explain.

27-14   In an electrical discharge tube an electron in a hydrogen atom is excited from the $n = 1$ to the $n = 4$ level. (a) How much energy has the atom absorbed? (b) Sketch an energy-level diagram showing all the transitions that the electron may make in returning to the $n = 1$ level. (c) What is the wavelength of the most energetic photon that may be emitted?

27-15   The binding energy of a one-electron atom is the energy necessary to completely remove the electron from the atom. What are the binding energies in electron volts of (a) hydrogen; (b) singly ionized helium; (c) doubly ionized lithium?

27-16   How many wavelengths of an electron "fit" into the orbit if it is in the $n = 3$ energy level in hydrogen?

27-17   If a photon of energy 2.55 eV is emitted when an electron drops out of the $n = 4$ level of hydrogen, into which level will it drop?

27-18   What frequency photon is absorbed when an electron in hydrogen jumps from the $n = 2$ to the $n = 4$ energy level?

Section 27.4 | The Uncertainty Principle

27-19 What is the position uncertainty of a 0.1 kg stone if its velocity is uncertain by 0.003 m s^{-1}?

27-20 What is the minimum kinetic energy of a particle of mass 10^{-29} kg if its position uncertainty is 10^{-10} m?

27-21 Suppose that an electron and a stone of mass 0.2 kg each have the same position uncertainty, 10^{-10} m. Find for each (a) the minimum momentum uncertainties; (b) the minimum velocity uncertainties.

PROBLEMS

27-22 If the angular momentum of the earth in its motion about the sun satisfies Bohr's angular momentum hypothesis, $L = rmv = n\hbar$, what is the quantum number associated with the earth's orbit? (The earth's mass is 5.98×10^{24} kg, and the orbital radius is 1.50×10^{11} m.)

*27-23 Bohr's theory can be used for positronium, a negative electron and a positron (positive electron) orbiting around one another. (*Note:* Since both particles have the same mass, they circle about the center of mass, halfway between them. The atom must be analyzed as if one particle with half the mass of an electron orbits around the stationary second particle.) (a) What is the binding energy of positronium in its ground state? (b) What is the radius of the lowest energy orbit?

*27-24 The dimensionless number $\alpha = ke^2/\hbar c$ is called the *fine structure constant* and is almost exactly 1/137. (*c* is the velocity of light.) (a) What is the velocity of an electron in the lowest energy level of a one-electron atom with atomic number Z, expressed as a fraction of c, the velocity of light? (b) The model of a one-electron atom given in this chapter does not take relativistic effects into account. Thus, we would expect that for atoms with v/c fairly large the numerical predictions of the model will fail. For what value of Z would the model fail completely? Explain your reasoning.

27-25 A negative *mu meson* (μ^-) has a mass 207 times that of an electron and charge $-e$. It can orbit a nucleus much like an electron. If a μ^- orbits a sulfur nucleus ($Z = 16$), find (a) the lowest energy level, and (b) the corresponding average radius. (c) Find the ratio of this orbital radius to that of the sulfur nucleus, 4×10^{-15} m.

*27-26 The conduction electrons in a metal may be considered as free particles in a box the size of the sample. As a start to such a model, consider an electron moving in one dimension with momentum p. The de Broglie wavelength must "fit" between walls a distance D apart with a node at each wall. (a) Draw the longest and next to longest waves that will "fit" between the walls. (b) What is the momentum associated with the longest wave that will "fit" between the walls? (c) What are all of the allowed values of the kinetic energy of an electron? (d) If $D = 0.01$ m, what is the spacing between the lowest two energy levels?

27-27 Hydrogen atoms initially in the ground state are exposed to photons of energies up to 13 eV. What energy photons will be produced as the atoms are excited and return to their ground states? (Neglect the recoil of the atoms during the absorption and emission processes.)

27-28 What are the average velocities of electrons in the lowest energy states of (a) hydrogen, and (b) singly ionized helium? (c) Would you expect relativistic effects to be important in either case?

*27-29 Bohr's quantum condition states that the angular momentum L must be some integer times $\hbar$. This condition can also be applied to the rotation of a diatomic molecule. If we assume that a diatomic molecule is like a dumbbell (Fig. 27.16) with a moment of inertia I, its angular momentum is $L = I\omega$, where ω is the rotational angular velocity of the molecule. The kinetic energy is $K = \frac{1}{2}I\omega^2 = L^2/2I$. (a) Apply Bohr's quantum condition to the rotating molecule and find a general expression for the energy levels. (b) Sketch an energy-level diagram for the four lowest levels. (The complete quantum theory gives the result that $L^2 = n(n+1)\hbar^2$. The energy levels of the rotating diatomic molecule are then correctly given by $n(n+1)\hbar^2/2I$.)

27-30 The moment of inertia of the O_2 molecule is 1.92×10^{-46} kg m^2. (a) What is the energy of the

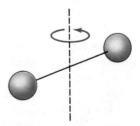

Figure 27.16. A dumbbell model of a diatomic molecule (Problem 27-29).

photon emitted during a rotational transition of the molecule from the $n = 2$ to the $n = 0$ level? (See Problem 27-29.) (b) What is the wavelength of this photon?

27-31 Hydrogen atoms in their ground states bombarded by an electron beam absorb up to 12.5 eV and emit photons. What photon energies are observed?

27-32 A double-slit interference experiment is performed with electrons. No attempt is made to determine which slit each electron passes through, so the position uncertainty of the electrons equals the slit separation d. Show that the uncertainty in the angle through which the electrons are deflected is less than $\theta_m = \lambda/2d$.

*27-33 Assume that the smallest average distance of an electron from the proton in hydrogen is $a_0 + \varepsilon$, where a_0 is the Bohr radius, and ε is much smaller than a_0. Show, using the uncertainty principle argument of Section 27.4, that the energy of the electron is smallest for $\varepsilon = 0$.

27-34 Before Chadwick discovered the neutron in 1932, it was suggested that atomic nuclei were composed of protons and electrons. For example, a carbon nucleus has a mass of about 12 proton masses, but only 6 units of charge, so it was considered to have 12 positive protons and 6 negative electrons. (a) Show that, according to the uncertainty principle, the minimum kinetic energy for an electron confined to a box of size 10^{-14} m (a typical nuclear size) is much larger than the typical 1-MeV $= 10^6$-eV spacing of nuclear energy levels. (b) Is this true also for neutrons? (This problem should actually be solved using the theory of special relativity, but the qualitative conclusion is the same as found using $E = p^2/2m$.)

ANSWERS TO REVIEW QUESTIONS

Q27-1, positively charged, electrons; **Q27-2,** false; **Q27-3,** momentum; **Q27-4,** wave; **Q27-5,** small; **Q27-6,** electric; **Q27-7,** wave; **Q27-8,** absorption, emission; **Q27-9,** close; **Q27-10,** very small.

Additional Reading

C.E. Behrens, Atomic Theory from 1904–1913, *American Journal of Physics*, vol. 11, 1943, p. 60; The Early Development of the Bohr Atom, p. 135; Further Developments of Bohr's Early Atomic Theory, p. 272.

T.H. Osgood and H.S. Hirst, Rutherford and His Alpha Particles, *American Journal of Physics*, vol. 32, 1964, p. 681.

G.P. Thomson, J.J. Thomson and the Discovery of the Electron, *Physics Today*, August 1956, p. 19.

M.A. Medicus, Fifty Years of Matter Waves, *Physics Today*, vol. 27, February 1974, p. 38.

Bruce R. Wheaton, Louis de Broglie and the Origins of Wave Mechanics, *The Physics Teacher*, vol. 22, 1984, p. 297.

C. Jönsson, Electron Diffraction at Multiple Slits, *American Journal of Physics*, vol. 42, 1974, p. 4.

George Gamow, *Mr. Tomkins in Wonderland: Stories of c, G, and h*. The Macmillan Co., New York, 1940. A world in which h is so large that quantum mechanics is part of daily life.

George Gamow, *Thirty Years That Shook Physics*, Anchor Books, New York, 1966.

John L. Heilbron, J.J. Thomson and the Bohr Atom, *Physics Today*, vol. 30, April 1977, p. 23.

F.P. Ottensmeyer, Scattered Electrons in Microscopy and Microanalysis, *Science*, vol. 215, January 29, 1982, p. 461.

Ferdinand G. Brickwedde, Harold Urey and the Discovery of Deuterium, *Physics Today*, September 1982, p. 34.

Scientific American articles:

Karl K. Darrow, The Quantum Theory, March 1952, p. 47.

R. Furth, The Limits of Measurement, July 1950, p. 48.

George Gamow, The Principle of Uncertainty, January 1958, p. 51.

Karl K. Darrow, Davisson and Germer, May 1948, p. 50.

Erwin Schroedinger, What Is Matter?, September 1953, p. 52.

E.N. DaCosta Andrade, The Birth of the Nuclear Atom, November 1956, p. 93.

Albert V. Crewe, A High Resolution Scanning Electron Microscope, April 1971, p. 26.

Thomas E. Eberhart and Thomas L. Hayes, The Scanning Electron Microscope, January 1972, p. 54.

Urve Essmann and Hermann Träuble, The Magnetic Structure of Superconductors, March 1971, p. 74.

F. Reif, Quantized Vortex Rings in Superfluid Helium, December 1964, p. 116.

Lester H. Germer, The Structure of Crystal Surfaces, March 1965, p. 32.

Donald M. Engelman and Peter B. Moore, Neutron Scattering Studies of the Ribosome, October 1976, p. 44.

S.W. Hawking, The Quantum Mechanics of Black Holes, January 1977, p. 34.

Phillip Ekstrom and David Wineland, The Isolated Electron, August 1980, p. 104.

UNIT EIGHT

ATOMS AND MOLECULES

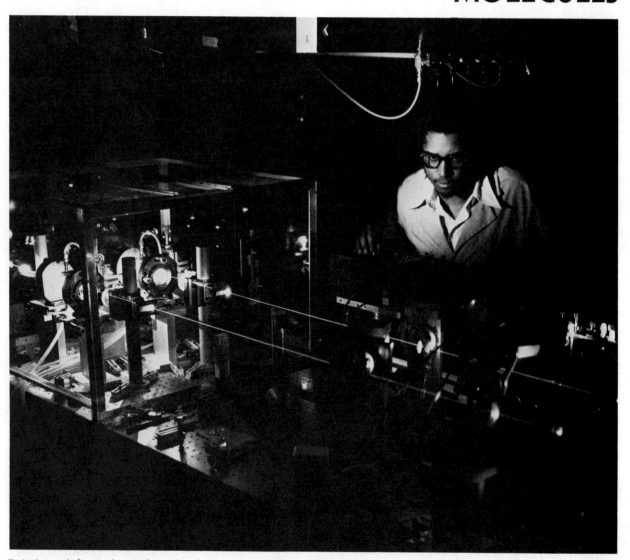

Twin Lasers being used to study semiconductor surfaces. (Courtesy Bell Labs).

E have seen that the failure of classical physics to explain several crucial experiments led to fundamental changes in scientists' view of the physical world. Specifically, physicists now believe that wave-particle duality, and in some cases special relativity, must be built into any theory of microscopic phenomena that can hope to be successful. The modern quantum mechanics developed by Schroedinger and others has extended the basic ideas of Planck, Einstein, de Broglie, and Bohr into a more general and complete theory of matter and energy. In this unit, we study the application of quantum mechanics to the structure of atoms and molecules, including such topics as lasers and methods of proving the structure of complex biological molecules.

CHAPTER 28

QUANTUM MECHANICS AND ATOMIC STRUCTURE

In 1925, Erwin Schroedinger proposed an equation whose solutions would represent the matter waves associated with electrons or other "particles." This *wave equation* and some related concepts introduced soon afterward comprise what has come to be called quantum mechanics. Quantum mechanics has been highly successful in predicting and correlating a vast amount of information about atoms. It is also able to deal successfully with aggregates of atoms and with atomic nuclei, as we will see in later chapters.

Schroedinger noticed that starting from Newton's laws, it is possible to obtain an equation for the disturbance associated with a wave traveling along a string or in an acoustical medium. Similarly, from the laws of electricity and magnetism, one can find an equation for the electric and magnetic fields in an electromagnetic wave. Thus, for any classical wave there is a wave equation whose solutions are formulas for the possible wave disturbances or *wave functions*.

Schroedinger reasoned that de Broglie's matter wave hypothesis might explain why atoms have discrete energy levels. While waves generally can have any wavelength, boundary conditions can restrict the possible values. For example, sound waves in a pipe or waves on a string with fixed ends form resonant standing waves only when they "fit." If the matter waves in an atom could have only specific wavelengths, the corresponding momenta and energies would also be discrete.

Accordingly, Schroedinger developed a procedure for constructing the wave equation for a particle or system of particles if the force or forces acting among

them are known. For example, in a hydrogen atom consisting of a proton and an electron, this force is the electrical attraction given by Coulomb's law. The equation relates derivatives of the wave function and solving it in most situations involves mathematical complexities beyond the level of this book for most applications. However, we do summarize the results for the hydrogen atom in this chapter. We also develop and apply Schroedinger's equation for a few idealized situations in the Supplementary Topics.

Schroedinger's equation determines the quantum mechanical wave function $\Psi(x)$ (psi), the analog for a matter wave in an atom or molecule of the displacement $y(x)$ for a wave on a string. Its absolute value squared $|\Psi(x)|^2$ is proportional to the probability of finding the particle at the point x.

In classical physics, the *state* of a system at any instant in time is defined by giving the positions and velocities of all the objects making up the system. The uncertainty principle tells us that such a detailed description is impossible at the microscopic level. Classically, we can predict with arbitrary accuracy the results of a measurement on a system whose state we know. However, in quantum mechanics the specification of the state of a physical system provided by the wave function only enables us to make probabilistic or statistical statements about the outcomes of many kinds of experiments.

The wave function itself cannot be directly measured or observed. However, if we know the wave function, we can compute measurable quantities such as the energy of an atom or a molecule. Also, from the appropriate wave functions, we can predict the outcome of experiments, such as the double-slit electron interference experiment discussed in the previous chapter. The results of all such analyses are consistent with the uncertainty principle and with de Broglie's wave hypothesis. Thus, these two concepts are embedded firmly in quantum mechanics.

A remarkable example of the parallel development of major scientific ideas is the fact that Werner Heisenberg developed an entirely different form of quantum mechanics at almost the same time that Schroedinger did his initial work. Heisenberg represented observable quantities, such as the position or momentum of an electron, by mathematical quantities called matrices and proposed certain equations relating these matrices. Although no wave functions appeared in his theory, it was soon realized that its predictions are always identical to those of Schroedinger's wave theory. Hence the matrix and wave forms of quantum mechanics are completely equivalent. We restrict our discussions to Schroedinger's approach, although Heisenberg's methods are sometimes more convenient for detailed calculations.

28.1 | THE OUTLINE OF QUANTUM MECHANICS

In this chapter, we discuss those aspects of Schroedinger's theory that are important in understanding atomic structure. This will also set the foundation for our picture of how atoms combine to form molecules.

Some important ideas that we use can be described in terms of the results of the preceding chapter. In Bohr's model of one-electron atoms, we found that the electron has certain allowed orbital radii labeled by a *quantum number*. Each value of the quantum number designates a unique configuration or quantum state of the system. There are as many quantum states as there are values of the quantum number, and each state has a characteristic energy. The electron can be in any one of these states and can change configurations only by moving from one allowed state to another.

In Schroedinger's theory the full specification of an atomic state requires additional quantum numbers. Some of these are associated with the *orbital angular momentum* corresponding to the electronic motion about the nucleus. Others are associated with the *spin angular momentum* of the electrons or, more concisely, the electron *spin*. This spin can be pictured as due to a rotation somewhat like that of the earth about its axis. However, this classical analogy to the

quantum mechanical concept of spin is of only limited applicability, since electrons have no discernable structure or size. Furthermore, all electrons have a spin angular momentum of exactly the same magnitude. Spin appears to be an intrinsic property of all the basic constituents of matter, the so-called *elementary particles*. These include not only the electron, proton, neutron, and photon but also some more recently discovered particles, such as neutrinos, mesons, and hyperons. The fact that all particles of a given type have the same spin cannot be explained classically.

In addition to lacking a classical analog, the spin angular momentum does not appear in the original form of Schroedinger's theory. Although spin can be inserted into this theory, it appears naturally in the theory only when special relativity is included at the outset. The existence of the electron spin was first found experimentally in studies of the properties of atoms and plays a crucial role in our understanding of the structure of matter.

In the next two sections, we describe the quantum numbers and wave functions found from Schroedinger's theory for the hydrogen atom. We will then be able to make the step to atoms with more than one electron.

28.2 | HYDROGEN ATOM QUANTUM NUMBERS

Schroedinger's theory requires the solution of an equation for a wave function ψ. It is found that each allowed configuration of any system has a unique wave function labeled by certain values of the quantum numbers. We describe the quantum numbers for atomic hydrogen in this section and the wave functions in the following section. The quantum numbers for hydrogen are n, l, m_l, and m_s. These are associated with the energy, the total orbital angular momentum, a component of the orbital angular momentum, and a component of the spin angular momentum, respectively.

The Principal Quantum Number, n | In Schroedinger's original theory of hydrogen, the electronic energy levels are completely determined by a *principal quantum number, n,* and satisfy the Bohr formula

$$E_n = -\frac{Z^2}{n^2} E_0, \qquad E_0 = \frac{ke^2}{2a_0} \qquad (28.1)$$

Using $Z = 1$ and $E_0 = 13.6$ eV,

$$E_n = -\frac{13.6}{n^2} \text{ eV}, \qquad n = 1, 2, 3, \ldots$$

This formula is only approximately correct. The more complete relativistic theory of hydrogen developed shortly after Schroedinger's original work includes the small magnetic forces associated with the motion of the charges in the atom. This complete theory leads to energy levels that have a slight dependence on other quantum numbers. The energy levels in multi-electron atoms depend even more strongly on other quantum numbers because of the additional forces among the electrons.

The Angular Momentum Quantum Number, l

In the Bohr model, one quantum number labeled both the angular momentum and the energy of each state. However, in solving Schroedinger's equation, it is found that the magnitude L of the total orbital angular momentum $\mathbf{L}$ is related to a separate quantum number l by,

$$L = \sqrt{l(l + 1)}\,\hbar, \qquad l = 0, 1, 2, \ldots, n - 1 \quad (28.2)$$

Two important results should be noted. First, the angular momentum may be zero for any value of n. A classical picture of an electron with zero angular momentum would be one in which the electron moves back and forth directly through the nucleus! The second point is that for each value of n, l may be any integer up to and including $n - 1$.

A somewhat archaic nomenclature is still used to designate states with certain values of n and l. Table 28.1 shows the letters assigned to states with $l = 0$ through 6. With this notation, the state with $n = 1$, $l = 0$ is said to be the $1s$ state, while $n = 2$, $l = 0$ is the $2s$ state, and $n = 2$, $l = 1$ is the $2p$ state. Other states are denoted in a similar way.

The z Component of the Angular Momentum

The direction of the angular momentum of the one electron in hydrogen becomes important when a magnetic field is applied to the

atom. An electron with an orbital angular momentum $\mathbf{L}$ is effectively a current loop or magnetic dipole. The energy of the loop depends on its orientation relative to the magnetic field $\mathbf{B}$ (Fig. 28.1).

In Section 19.5 on magnetic dipoles, we found that an electron of charge $-e$ and mass m moving in an orbit with angular momentum $\mathbf{L}$ has a magnetic dipole moment that is proportional to $\mathbf{L}$. In vector form, this relationship is

$$\boldsymbol{\mu} = -\frac{e}{2m}\mathbf{L} \qquad (28.3)$$

The minus sign is needed because the electronic charge is negative. The energy of the dipole in a magnetic field is $E = -\mu B \cos \theta$. If $\mathbf{B}$ is in the z direction, this can also be written in terms of the z component of $\boldsymbol{\mu}$ (Fig. 28.1),

$$E = -\mu_z B \qquad (28.4)$$

Thus, we expect that the energy of an electron due to a magnetic field can have any value between $-\mu B$ and $+\mu B$.

The electron can change its energy by emitting or absorbing a photon. Since its energy can vary between $-\mu B$ and $+\mu B$, the photons observed might be expected to have a spread of energies; but in experiments only certain photon energies are seen. The analysis of the experiments and Schroedinger's theory both show that L_z, the z component of $\mathbf{L}$, can have only certain values:

$$L_z = m_l \hbar$$
$$m_l = -l, -l + 1, \ldots, 0, \ldots l - 1, l \qquad (28.5)$$

Thus m_l, the quantum number denoting the z compo-

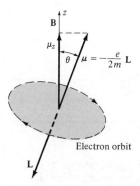

Figure 28.1. An electron with orbital angular momentum $\mathbf{L}$ has a magnetic dipole moment $\boldsymbol{\mu}$ opposite to $\mathbf{L}$ because the electron charge is negative. The energy of the dipole in the field $\mathbf{B}$ is $E = -\mu_z B$.

TABLE 28.1

Spectroscopic notation for the first seven angular momentum states in atoms. The sequence is alphabetic for $l \geq 3$.

$l =$	0	1	2	3	4	5	6
Notation	s	p	d	f	g	h	i

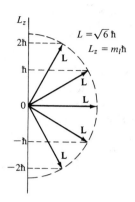

Figure 28.2. The allowed orientations of **L** for $l = 2$ correspond to five values of $L_z = m_l h$. Here m_l can have any of the values 0. ± 1. ± 2.

nent of **L**, can have any of $2l + 1$ integer values for a given value of l (Fig. 28.2).

Since L_z can only have certain values, μ_z also has specific allowed values. The resulting energy levels are given (Fig. 28.3) by

$$E_{ml} = -\mu_z B = -\left(-\frac{e}{2m}\right) m_l \hbar B = m_l \frac{e\hbar}{2m} B \quad (28.6)$$

The following example illustrates the enumeration of states possible for $n = 2$.

Example 28.1

What are the allowed values of l and m_l for states of the hydrogen atom with $n = 2$?

For $n = 2$, the orbital angular momentum number l can be either 0 or 1. For $l = 0$, we can only have $m_l = 0$; for $l = 1$, m_l can be -1, 0, or $+1$. In tabular form,

State (n, l, m_l)	n	l	m_l
2s state (2, 0, 0)	2	0	0
2p states (2, 1, −1)	2	1	−1
(2, 1, 0)	2	1	0
(2, 1, 1)	2	1	1

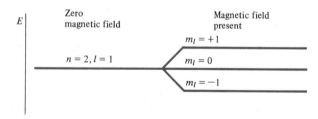

Figure 28.3. In the presence of a magnetic field the 2p state is broken up into three energy levels corresponding to the three allowed values of m_l. In the $m_l = +1$ state the magnetic dipole is parallel to the magnetic field: for $m_l = -1$, it is antiparallel, and for $m_l = 0$, it is perpendicular to the field.

We see that for $n = 2$, there are four possible combinations of the quantum numbers l and m_l. However, we see in the following that an additional quantum number is needed to fully specify all of the $n = 2$ states.

Spin Angular Momentum | When experiments were performed on atoms in a magnetic field the energy levels were found to be split, but more levels were observed than anticipated. Furthermore, even when no external magnetic field was present, it was found that many of the observed spectral lines were, in fact, closely spaced pairs of lines or *doublets*. These energy-level splittings led to the conclusion that the electron has not only an orbital angular momentum but also an intrinsic spin angular momentum with an associated magnetic dipole moment. The observed effects were explained in terms of both the orbital and spin magnetic moments.

The magnitude of the spin angular momentum **S** of a particle is found to be (Fig. 28.4)

$$S = \sqrt{s(s + 1)}\,\hbar \quad (28.7)$$

where s is the *spin quantum number*. For electrons, $s = \frac{1}{2}$, as it is for protons and neutrons; for photons, $s = 1$. These particles are said to have spin one-half and spin one, respectively.

In atomic structure, it is the z component of the spin angular momentum S_z that is important. For a spin one-half particle, S_z is

$$S_z = m_s \hbar, \qquad m_s = \frac{1}{2}, -\frac{1}{2} \quad (28.8)$$

Despite the fact that **S** never points directly along the z axis, the two states $m_s = \pm\frac{1}{2}$ are often referred to as

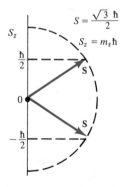

Figure 28.4. For electrons, with spin $\frac{1}{2}$, the z component of the spin angular momentum is either $+\hbar/2$ or $-\hbar/2$. **S** is always at an angle with the z axis.

spin up ($m_s = \frac{1}{2}$) and spin down ($m_s = -\frac{1}{2}$). Thus, for every set of values for n, l, and m_l, there are two allowed values of m_s. This means that the results of Example 28.1 are incomplete. There are two $n = 1$ states and eight, rather than four, $n = 2$ states.

The electron spin accounts for the extra observed spectral lines. The electron has an orbital and a spin magnetic moment, and these two dipoles may have two possible relative orientations corresponding to the two values of S_z. Hence, all the energy levels are split into two, giving rise to the doublet spectral lines observed. This is called the *spin-orbit splitting* or *fine structure*.

Later Developments | In the two decades following Schroedinger's work, theory and experiment were found to be in excellent agreement for hydrogen and for more complex atoms. However, after World War II, totally new tools were applied to the study of atomic structure. Most notable was the use of short-wavelength electromagnetic radiation (micro-waves) employing technology developed in wartime radar research. It was found that the existing theory failed to predict small but significant details of the measurements.

Feynman, Schwinger, Tomonaga, and others showed that agreement with experiment could be restored if several new processes were taken into account. These processes included the emission and reabsorption of photons by electrons in atoms and the spontaneous production of short-lived pairs of electrons and *positrons*. The positron is a particle with the same mass as the electron but with a positive charge $+e$. This new theory is called *quantum electrodynamics* and is sufficient to explain all features of atomic structure that do not depend on the internal structure of the atomic nucleus. For most applications of quantum mechanics to problems in chemistry and biochemistry, quantum electrodynamic effects are negligibly small. Consequently, we do not consider them further.

In summary, the complete solution to the Schroedinger equation for hydrogen shows that neglecting spin, the energy levels are given correctly by the Bohr model if no external fields are present. The multiplicity of the states for a given energy level can be found from the splitting in a magnetic field. In the next section, we examine the wave functions corresponding to different sets of quantum numbers.

28.3 | HYDROGEN ATOM WAVE FUNCTIONS

As we mentioned earlier, the wave functions used to describe the states available to electrons are not measurable quantities. However, these wave functions are used to calculate measurable quantities such as the energy and angular momentum of a state. The wave function, denoted by ψ, is the analog of the displacement in the standing waves on strings and in air columns. In quantum mechanics the wave function is most directly related to the probability of finding an electron at a certain position when it is in a given state.

The square of the absolute value of the wave function, $P = |\Psi|^2$, is a measure of the probability of finding an electron at a given point in space. (Sometimes Ψ contains $\sqrt{-1}$ and is therefore a complex quantity. However, $|\Psi|^2$ is a positive, real number as we would expect for a probability. If Ψ is real, then $|\Psi|^2 = \Psi^2$.) Graphs of the wave functions and of $|\Psi|^2$ for the $n = 1$ and 2 states of hydrogen are shown in Fig. 28.5. Three-dimensional representations of $|\Psi|^2$ (Fig. 28.6) can be thought of as superpositions of many photographs of the electron position. An electron is most likely to be found in the darker areas.

What is the probability of locating an electron at a given radius r? The probability of finding an electron in a given small volume ΔV is proportional to ΔV; the larger the detector, the greater the chance of finding the electron. Thus the probability of finding the electron in ΔV is determined by the product $|\Psi|^2 \Delta V$. A sphere of radius r has a surface area $A = 4\pi r^2$, and a thin spherical shell of thickness Δr has a volume $\Delta V = A\Delta r = 4\pi r^2 \Delta r$, which varies as r^2. (For example, an orange peel of a given thickness has a volume that is proportional to the square of its radius.) Hence the probability of finding an electron within a spherical shell of radius r is proportional to the *radial probability*, $r^2 P = r^2 |\Psi|^2$. These radial probabilities for the $n = 1$ and 2 states are shown in Fig. 28.7.

From the $n = 1$, $l = 0$ (1s) graph in Fig. 28.5, we see that Ψ and $|\Psi|^2$ are largest at $r = 0$, suggesting a relatively large probability of finding a 1s electron near the nucleus. However, because of the r^2 factor, the radial probability is zero at $r = 0$; its greatest value occurs at $r = a_0$ (Fig. 28.7). The nuclear radius is about 10,000 times smaller than the Bohr radius. On the graphs, the nuclear radius falls within the

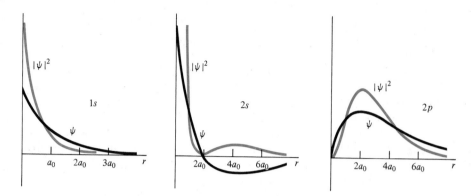

Figure 28.5. Graphs of the radial variation of Ψ and $|\Psi|^2$ for the $n = 1$ and $n = 2$ states of hydrogen. The p state wave functions vary with the direction as well. a_0 is the radius of the first Bohr orbit. The graphs are drawn with an arbitrary vertical scale for clarity.

width of the line representing the vertical axis. Thus the probability is very small of finding a $1s$ electron actually within the tiny nuclear volume. The $2s$ and all the higher s states also have wave functions that are largest at $r = 0$. With the r^2 factors included, their radial probabilities for being found at small r values diminish rapidly as n increases. The s states are all spherically symmetric (Fig. 28.6).

Now consider the $2p$ wave functions. It is clear from Fig. 28.6 that these are dependent on the direction, unlike the s states. The wave functions are zero at $r = 0$ (Fig. 28.5), and the radial probabilities are small even at $r = a_0$ (Fig. 28.7); the peaks occur at $r = 4a_0$. Thus $2p$ electrons are less likely to be found near the nucleus than are $1s$ or even $2s$ electrons. This has important implications for multielectron atoms which we explore in Section 28.5.

The wave functions we have described are those that are most useful for atomic physics, where the magnetic interactions between the spin and orbital magnetic moments play a role. However, when atoms bind to one another, it is more convenient to use a different set of wave functions. In Fig. 28.8, we show representations of the three $2p$ states for hydrogen which are used in discussions of molecular binding. The wave functions labeled p_x, p_y, and p_z are superpositions of the three $2p$ states of Fig. 28.6. Both sets of wave functions are solutions of Schroedinger's equation and the choice of which set to use depends on whether one is interested in magnetic effects or molecular structure. We use the states of Fig. 28.8 in discussing molecular binding in the next chapter.

Explicit formulas for the hydrogen atom wave functions and examples of their use are given in the Supplementary Topics in Section 28.12.

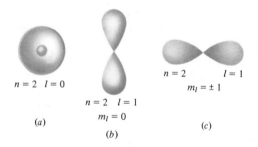

Figure 28.6. The representations of $|\psi|^2$ for the $n = 2$ states of hydrogen. (a) The $l = 0$ cloud is darkest at $r = 0$. (b) For $l = 1$ and $m_l = 0$ the representation of $|\psi|^2$ appears as two droplets above and below the origin. For this state $|\psi|^2 = 0$ at $r = 0$. (c) For $l = 1$, $m_l = \pm 1$. $|\psi|^2$ is doughnut shaped. (Here it is viewed in cross section from the side). Again $|\psi|^2 = 0$ at $r = 0$ for this state. While the pictures for $m_l = +1$ and $m_l = -1$ appear the same, the angular momentum is opposite. Thus one may think of the electron in the doughnut as circulating one way for $m_l = 1$ and oppositely for $m_l = -1$.

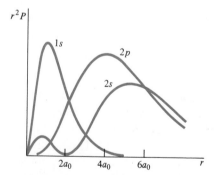

Figure 28.7. The radial probability distributions $r^2P = r^2|\Psi|^2$ for the $n = 1$ and 2 states. The maximum in r^2P is at $r = a_0$ for the $1s$ state, at $4a_0$ for the $2p$, and at a slightly larger radius for the $2s$.

28.4 | THE PAULI EXCLUSION PRINCIPLE

We noted that all elementary particles have an intrinsic spin angular momentum. The spin plays a crucial role in atomic and molecular structure and in the behavior of collections of particles. For example, the two isotopes of helium, ^{3}He, with two protons and a neutron in the nucleus, and ^{4}He, with two protons and two neutrons, have very different properties at low temperatures because the net spins of the two nuclei are different. In ^{4}He the two protons have equal but oppositely directed spins, as do the two neutrons. Thus the net spin angular momentum is zero. This substance, which at atmospheric pressure is a liquid even at the lowest temperatures so far achieved, becomes a *superfluid* below about 2 K. Among other extraordinary characteristics, this superfluid will flow through openings that no other liquid or gas will. ^{3}He displays equally remarkable but very different behavior below about 5×10^{-3} K. Here the net nuclear spin is one-half due to the single unpaired neutron.

Another example of the effects of spin is observed in metals. At low temperatures the free electrons, each having a spin of one-half, match up to form pairs with a net spin of zero. When this happens the metal becomes a superconductor; its electrical resistance becomes zero!

All these characteristics depend on two observations:

1   *The Pauli exclusion principle.* Two or more indistinguishable or identical particles with spin one-half cannot be in the same quantum state. This means for example that in atoms with many electrons, no more than one electron can be in a level denoted by a given set of quantum numbers, n, l, m_l, and m_s.

2   *Indistinguishable or identical particles with a spin of zero or one* can be put in any quantum state whether it is occupied or not.

The word "indistinguishable" in these rules is important because in classical physics, individual particles are usually considered as being identifiable and distinguishable. That is, if we know the position and momentum of a particle at a certain moment, then using Newton's laws we can, in principle, follow the motion of that particle as long as we wish. In quantum systems, the uncertainty principle precludes an accurate determination of the position and momentum, so at different times we cannot tell if we are seeing the original particle or another that looks just like it. In the latter case the particles must be considered as indistinguishable. Atoms provide many illustrations of the effects of the Pauli exclusion principle.

Example 28.2

How many electrons can be in the $n = 2$ hydrogenlike states of an atom?

In Example 28.1, we found that there were four possible values of l and m_l for $n = 2$. However, for each of these states, m_s can be $+\frac{1}{2}$ and $-\frac{1}{2}$. There are then a total of eight $n = 2$ states. The Pauli exclusion principle states that only one electron can be in each state, so the $n = 2$ states can accommodate eight electrons.

28.5 | ATOMIC STRUCTURE AND THE PERIODIC TABLE

We now have sufficient information to gain some insight into atoms with more than one electron. We will see that many of the physical and chemical properties of atoms can be explained by the *atomic shell model,* which is based on the idea that the electrons in atoms are in hydrogenlike states.

In the ground state of an atom, the electrons occupy the lowest states allowed by the Pauli exclusion principle. When all the states corresponding to the same or nearly the same energy have been filled, an *electron shell* is said to be completed or *closed.* Atoms with all their electrons in closed shells are similar to each other in their physical and chemical properties and are very stable. For example, the minimum energy needed to remove one electron from a neutral ground state atom, the *ionization energy,* is very large in these closed shell atoms (Fig. 28.9). The regularities associated with successive shell closures are responsible for similarities among the elements in the same column of the periodic table (Appendix A).

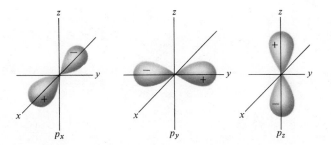

Figure 28.8.   The three $2p$ wave functions of hydrogen most often used in chemistry. These wave functions are positive on one side of the origin and negative on the other side.

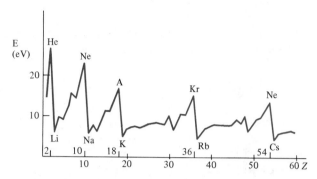

Figure 28.9. The ionization energy, the minimum energy required to remove one electron from a neutral atom in its ground state, versus Z for atoms up to $Z = 60$. Closed shell inert gases have $Z = 2, 10, 18, 36,$ and 54.

We now consider individual atoms in more detail, starting with $Z = 2$.

Helium ($Z = 2$)

Helium has two electrons and a nucleus with a charge $2e$. It is not currently possible to find exact solutions to the Schroedinger equation when there are three or more objects present, although very accurate approximate solutions to the helium atom problem can be obtained. These solutions show, as one might expect, that in the helium ground state the two electrons are both in $n = 1$, $l = 0$ states. In accordance with the Pauli principle, one electron has its spin directed up ($m_s = \frac{1}{2}$) and the other down ($m_s = -\frac{1}{2}$).

The fact that two electrons are present affects the energies of their states. Each electron has a probability distribution $|\psi|^2$ similar to that of Fig. 28.5a, so the electrons partially shield each other from the nucleus. Roughly half the time, each electron "sees" an attractive nuclear charge $Z = 2$, and half the time, it is shielded and sees a net charge $Z = 1$. Thus, we might suppose that in the Bohr formula we should use an effective nuclear charge $Z_{eff} = \frac{1}{2}(2 + 1) = 1.5$. However, the mutual repulsion of the electrons pushes them apart and reduces the shielding effect, so that Z_{eff} increases to approximately 1.7. Hence, when they are in $n = 1$ orbitals, using Eq. 28.1 with $Z = Z_{eff}$, the two electrons have a total energy

$$E_2 = -2\frac{Z_{eff}^2}{n^2}E_0 = -2\frac{(1.7)^2}{1^2}(13.6 \text{ eV})$$
$$= -79 \text{ eV}$$

If one electron is removed, then the remaining electron experiences the full nuclear charge $Z = 2$. The energy of this electron is then

$$E_1 = -\frac{2^2}{1^2}(13.6 \text{ eV}) = -54 \text{ eV}$$

The difference $E_1 - E_2 = -54 - (-79) = 25$ eV is the energy needed to remove one electron from the helium atom. This ionization energy is greater than the 13.6 eV needed to ionize hydrogen. We see then that the two electrons in helium are in the two $1s$ states, and despite their mutual repulsion, they are each more tightly bound than is the one electron in hydrogen. Helium is the lightest closed shell atom, since it has the $n = 1$ levels filled; chemically, it is the lightest inert or *noble gas*. In lithium, with $Z = 3$, the third electron will have to go into the higher energy $n = 2$ orbital.

Lithium—Neon ($Z = 3$-10)

From our discussion of the hydrogen energy states, the $n = 2$ level contains a total of eight states with different values of l, m_l, and m_s. The $n = 2$ levels are significantly higher in energy than the $n = 1$ levels but are still well below the $n = 3$ levels (Fig. 28.10). In lithium ($Z = 3$) two electrons occupy the $n = 1$ states and one electron goes into a $n = 2$ level. Because this third electron is further from the nucleus on the average, it is shielded from the nuclear charge by the two inner electrons. This shielding is less complete in the $n = 2$, $l = 0$ state than in the $n = 2$, $l = 1$ state, since in the $l = 0$

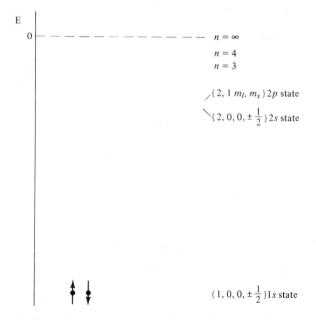

Figure 28.10. A schematic energy-level diagram for the neutral helium atom. The two electrons have opposite spins and are both in $1s$ states. The states are labeled by $\{n, l, m_l, m_s\}$.

state the electron has a higher probability of being close to the nucleus. Accordingly the energy is somewhat lower in the $l = 0$ state, and that level will be filled in the ground state of the atom. We find the effective nuclear charge for the $n = 2$, $l = 0$ electron in the next example.

Example 28.3

From the Bohr model, an $n = 2$ electron in the presence of an effective nuclear charge $Z_{eff} = 1$ will have an energy of $E_2 = -13.6/(2)^2$ eV $= -3.4$ eV. In lithium the energy of this electron is -5.39 eV. What is the effective nuclear charge for this state?

From the Bohr model with $Z = Z_{eff}$,

$$E_2 = -5.39 \text{ eV} = -13.6 \frac{(Z_{eff})^2}{(2)^2}$$

Solving, we find $Z_{eff} = 1.26$. Thus, because the electron can be close to the nucleus, it sees an effective nuclear charge that is 26 percent greater than that of a completely shielded electron.

For $Z = 4$, beryllium, the fourth electron can also be in an $n = 2$, $l = 0$ state with its spin opposite to that of the third electron. These two electrons shield each other somewhat. However, the nuclear charge is now 4 and the inner electrons only shield two of the protons. Thus the two outer electrons see an effective nuclear charge of between 1 and 2. The ionization energy is found to be 9.32 eV for beryllium.

With $Z = 5$, boron, the $n = 2$, $l = 1$ levels begin to fill. There are six $n = 2$, $l = 1$ states and electrons fill these states in going from boron to carbon ($Z = 6$), nitrogen ($Z = 7$), oxygen ($Z = 8$), fluorine ($Z = 9$), and finally neon ($Z = 10$). With neon, the $n = 2$ levels are filled, and the shell is completed; it is an inert gas. We note that for neon, the six outer electrons are competing with one another to share the portion of the nuclear charge that is not shielded by the four inner electrons. Because each of these six electrons can sample a nuclear charge as large as six part of the time, their effective nuclear charge is relatively large. For neon the ionization energy has the large value of 21.56 eV. This corresponds to a Z_{eff} of about 2.5.

The competition between the attractive effects of an effective nuclear charge and the repulsion of the electrons from each other is very sensitive to the number of electrons present. For example, fluorine, with five $n = 2$, $l = 1$ electrons, is one electron short of having a filled shell. The effect of adding an additional electron to fluorine to obtain the negative ion F^- is to cause six outer electrons to compete for five unshielded nuclear charges. The energy gained in this

sharing is sufficient to offset the repulsive effects of the added electron, and F^- is quite stable. Its ionization energy is 4.2 eV. Fluorine is said to have a high *electron affinity*.

In a crude model of the molecule NaF, the fluorine atom takes one of the electrons from sodium. The molecule is then held together by the attractive electrical force between Na^+ and F^-. This is called *ionic binding*. Fluorine is the lightest of the *halogens*, which include chlorine, bromine, iodine, and astatine, all of which are one electron short of having a closed shell. All are highly reactive as they try to gain an electron.

We can also see how carbon ($Z = 6$), a constituent of so many organic compounds, combines with other elements. The $n = 2$ level of carbon has two electrons in the $l = 0$ states and two in the $l = 1$ states. Four more are needed to completely fill the $n = 2$ shell. Since atoms are most stable when they have closed shells, carbon will usually try to "share" four electrons from other elements. This is called *covalent binding*.

$Z = 11\text{-}18$ | As Z increases from 11, the 3s and

then the 3p levels fill just as the 2s and 2p states were filled. The chemical characteristics of these elements are very similar to the corresponding elements with Z between 3 and 10. At first glance, we would expect that the next closed shell inert gas would appear at $Z = 28$, when all the $n = 3$ levels are filled (Table 28.2). However, the electrons in the $l = 2$ states spend much less time near the nucleus than those in $l = 1$ or 0 states and are therefore substantially higher in energy. In fact, the 4s levels ($n = 4$, $l = 0$) are also below the 3d levels ($n = 3$, $l = 2$). Accordingly, the 3s and 3p states form a complete shell by themselves.

TABLE 28.2

Atomic states for $n = 3$

State	l	m_l	m_s	
3s	0	0	$\pm\frac{1}{2}$	} 2 states
3p	1	-1	$\pm\frac{1}{2}$	
		0	$\pm\frac{1}{2}$	} 6 states
		$+1$	$\pm\frac{1}{2}$	
3d	2	-2	$\pm\frac{1}{2}$	
		-1	$\pm\frac{1}{2}$	
		0	$\pm\frac{1}{2}$	} 10 states
		$+1$	$\pm\frac{1}{2}$	
		$+2$	$\pm\frac{1}{2}$	

Argon ($Z = 18$) has filled $3s$ and $3p$ states and is therefore an inert gas. Similarly, chlorine ($Z = 17$) is a halogen like fluorine, and potassium ($Z = 19$) is an *alkali metal* similar to lithium and sodium, which also have one electron outside a closed shell.

$Z \geq 19$ |

Above $Z = 18$ the order in which levels are filled becomes complicated. The zero angular momentum states in a shell are always filled first because electrons in these states penetrate closest to the nucleus. This effect is so strong that for $Z = 19$ the $n = 4$, $l = 0$ states begin to fill even though the $n = 3$, $l = 2$ states are empty. Thus in Fig. 28.11 we see that for intermediate values of Z, closed shells are not always associated with a single value of n. The one-to-one correspondence between shells and the quantum number n reappears for large values of Z.

Two general features of atomic structure are important. First, there is no obvious reason that elements with very large values of Z could not exist. However naturally occurring or artificially produced elements have only been found for values of Z up to slightly over 100. Very heavy atoms do not exist because their nuclei tend to break up.

A second notable feature of atoms is the remarkable similarity in their radii. The alkali atoms, with a single, weakly bound s electron, have the largest radii. But the largest alkali atom, cesium, has a volume that is only about twice that of helium. This happens because, in heavy atoms, the innermost electrons see a very large Z_{eff} and consequently are much closer to the nucleus than in light atoms. This permits the outer electrons to be closer to the nucleus as well, so that the average atomic radius does not increase much with Z.

28.6 | ATOMIC EMISSION AND ABSORPTION SPECTRA

We described earlier how the electrons in an atom can change their energies by moving from one state to another. This process is accompanied by the emission or absorption of photons.

The photon has a spin of 1. When an atom absorbs

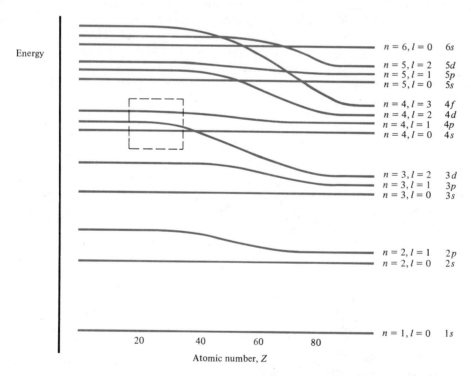

Figure 28.11. The positions of the energy levels of atoms versus the atomic number Z. The dotted square indicates the levels filled for $Z = 19$ through 36. The shells contain different levels in different parts of the periodic table. For example for $Z = 30$ the $3s$ and $3p$ levels comprise a shell, but for $Z = 70$ the $3s$, $3p$, and $3d$ levels form a shell.

or emits a photon, the *total* angular momentum of the atom and photon must be constant. Thus, because the photon carries angular momentum, the electron that emits or absorbs that photon must change its angular momentum when making a transition.

Usually the electrons *change their orbital angular momentum quantum number l by* 1 when making transitions. For example, suppose a hydrogen atom in its ground state, which is a $l = 0$ state, absorbs a photon. It can only do so if its final orbital angular momentum quantum number is $l = 1$. During a downward transition the electron must also change its angular momentum by 1. Figure 28.12 shows these *allowed* transitions among the lowest five energy levels of hydrogen.

When atoms are excited, the electrons usually return to their ground states via allowed transitions. These excited states are very short-lived, lasting no more than about 10^{-8} second. However, when the electrons cannot return to their ground states via allowed transitions, they may do so by the emission of two photons. This is a relatively slow process, and such *metastable* states may last for times of the order of a second.

SUMMARY

The structure of atoms with more than one electron can be largely understood in terms of electronic energy levels similar to those of the hydrogen atom. The positions of the electronic energy levels are determined by the attractive electric force due to the nucleus and the repulsive electric forces among the electrons.

The Pauli exclusion principle requires that only one electron can be in a given state. These states are labeled by quantum numbers that indicate the energy, orbital angular momentum, and spin angular momentum of an electron in that state. In the ground states of atoms, electrons occupy the lowest energy states available consistent with the exclusion principle.

The square of the wave function is an indication of the position probability of an electron in a given state. This probability distribution can be used to understand the screening of the nucleus by other electrons and the relatively low energy of the zero angular momentum states.

The observed atomic spectra of complex atoms depend not only on the positions of the energy levels but also on angular momentum conservation.

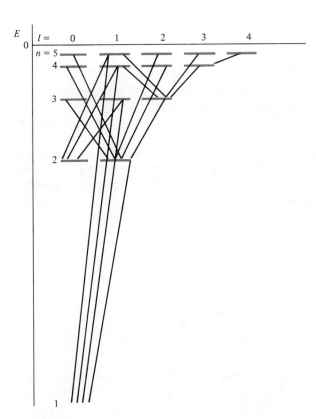

Figure 28.12. Some of the allowed transitions for the electron in hydrogen. These are accompanied by the emission or absorption of one photon. The $n = 2$, $l = 0$ state is metastable because the electron cannot lower its energy by direct one-photon emission.

Checklist

Define or explain:

wave function	ionization energy
principal quantum number	effective nuclear charge
	electron affinity
quantum number l	ionic binding
quantum number m_l	halogen
spin angular momentum	covalent binding
$|\psi|^2$	alkali metal
s states, p states	allowed transition
Pauli exclusion principle	metastable states
closed electron shell	

REVIEW QUESTIONS

Q28-1 In the hydrogen atom, the energy is specified by the _____ quantum number.

Q28-2 The orbital angular momentum quantum number may be any _____ up to _____.

Q28-3 The quantum number determining the z component of the orbital angular momentum can have any integer value between _____ and _____.

Q28-4 In units of $\hbar$, the z component of the spin angular momentum of an electron has a magnitude of _____.

Q28-5 The probability of finding an electron at a certain position is determined by the square of _____.

Q28-6 The probability of finding an electron near the nucleus is greatest for an $l =$ _____ state.

Q28-7 Two identical spin one-half particles may not be in the same _____.

Q28-8 The minimum energy needed to remove an electron from an atom in its ground state is the _____.

Q28-9 An atom with all the states corresponding to approximately the same energy filled has a closed _____ and is very _____.

Q28-10 When an excited atom emits a photon, in addition to energy the process must conserve _____.

EXERCISES

Section 28.1 | The Outline of Quantum Mechanics, and Section 28.2 | Hydrogen Atom Quantum Numbers

28-1 An electron in hydrogen is in an $n = 4$ state. (a) What are the possible values of the orbital angular momentum quantum number l? (b) What is the spectroscopic notation for each state? (c) What is the maximum magnitude of the z component of the orbital angular momentum for each of these values of l?

28-2 Sketch a figure similar to Fig. 28.2 showing the possible orientations of the z component of the angular momentum for an $l = 3$ state. Label the values of m_l for each of the allowed orientations of **L**.

28-3 (a) What are the possible energy states of an electron in an $l = 2$ state of hydrogen when a magnetic field B is applied? (Neglect all effects of spin in this exercise.) (b) If the magnetic field has a magnitude of 10 T, what is the energy difference between the $m_l = 1$ and $m_l = 0$ states? (c) If an electron makes a transition from the $m_l = 0$ to the $m_l = 1$ state, is a photon emitted or absorbed?

28-4 A spinning charge in a magnetic field is effectively a current loop. Quantum mechanics shows that an electron in a magnetic field B with a z component of spin S_z has an energy $\mathcal{U} = 2\mu_B B m_s$, where $\mu_B = e\hbar/2m$ is called the Bohr magneton; $\mu_B = 0.93 \times 10^{-23}$ A m^2. What is the energy difference between the spin up and spin down states of an electron in a magnetic field of magnitude 1 T?

28-5 How many $n = 4$ states are there in hydrogen? (Include spin effects.)

28-6 How many states in hydrogen have $n = 3$ and (a) $l = 0$; (b) $l = 1$; (c) $l = 2$? (Include spin effects.)

Section 28.5 | Atomic Structure and the Periodic Table

28-7 Why is potassium chemically similar to sodium and lithium? ($Z = 19$ for potassium.)

28-8 In $Z = 56$, barium, the $n = 6$, $l = 0$ state is filled and in succeeding atoms the electrons start filling the $n = 4$, $l = 3$ subshell. How many states are available in this subshell? (These are the rare-earth elements.)

28-9 Using the ideas of screening described for the helium atom, estimate the energy required to remove the second electron from lithium, leaving Li^{++}. Compare your estimate of Z_{eff} with $Z = 3$, the nuclear charge of lithium.

28-10 One of the two electrons in a helium atom is excited to an $n = 2$ state while the other electron remains in the $n = 1$ state. Will the $n = 2$ electron have the same energy in the $l = 1$ and $l = 0$ states? If not, which state has the lower energy and why?

28-11 The alkali metals are strongly reactive; they readily interact with other atoms, particularly those that attract electrons. Why are these metals so chemically active?

28-12 In the water molecule, H$_2$O, the oxygen atom shares the electrons of the hydrogen atoms. Why is this such a stable compound?

28-13 Hydrogen combines with the halogens to form such acids as HF, HCl, HBr, and HI. Why do these compounds form so readily?

28-14 The Nobel prize work of Tomonaga, Feynman, and Schwinger on quantum electrodynamics was prompted, in part, by some unexplained features of atomic spectra. Techniques developed in the late 1940s enabled experimenters to observe that the splitting of energy levels in a magnetic field B due to the electron spin is not $\Delta\mathcal{U} = 2\mu_B B$, but is, instead, $\Delta\mathcal{U} = 2\mu_B(1 + \alpha/2\pi)B$. ($\alpha = ke^2/\hbar c = 1/137$ is the fine structure constant.

$\mu_B = e\hbar/2m = 0.93 \times 10^{-23}$ A m^2. See Exercise 28-4.) (a) In a magnetic field of 10 T, what is the magnitude of the discrepancy in the spectrum? (b) What is the percentage of the shift from the value $\Delta \mathcal{U} = 2\mu_B B$?

28-15 What are the quantum numbers for the outer electron in the ground state of sodium ($Z = 11$)?

28-16 Estimate the energy of a 1s electron in a lead atom ($Z = 82$).

Section 28.6 | Atomic Emission and Absorption Spectra

28-17 The effective nuclear charge for a 3s ($n = 3$, $l = 0$) electron is $Z_{eff} = 3.1$, and for the same electron in an excited state $n = 3$, $l = 1$, it is $Z_{eff} = 1.5$. What is the energy of the photon emitted when the electron returns to the 3s state?

28-18 When one of the innermost electrons of an atom is captured by the nucleus, an X ray is emitted. Further X rays are emitted as electrons from higher levels drop down to fill the vacant level or hole left by the captured electron. In copper, $Z = 29$. (a) Estimate Z_{eff} for the 2p and 1s states in copper. Explain your reasoning. (b) Estimate the wavelength of the X ray emitted when an electron drops from the 2p state to the 1s state.

PROBLEMS

28-19 The energy necessary to remove a 1s electron in nitrogen ($Z = 7$) is 540 eV. (a) Using $Z_{eff} = 6.5$, estimate the energy needed to remove a 1s electron. (b) Why does this calculation underestimate the energy required? (c) Might the presence of the 2s electrons affect the energy of the 1s state? Why?

28-20 Beryllium ($Z = 4$) has two 1s and two 2s electrons in its ground state. The ionization energy for the first 2s electron is 9.32 eV, and for the second it is 18.12 eV. (a) What is Z_{eff} for the single 2s electron of Be$^+$? (b) What is Z_{eff} for the two 2s electrons in the neutral ground state of Be?

28-21 In singly ionized neon Ne$^+$ ($Z = 10$), the five outer 2p electrons see an effective nuclear charge as high as 6 and as low as 2, depending on the shielding due to the other four 2p electrons. By a simple averaging process we find $Z_{eff} = (6 + 5 + 4 + 3 + 2)/5 = 4$. (a) What is the total energy of the five outer electrons of Ne$^+$ using this value of Z_{eff}? (b) The measured ionization energy of neon is

21.56 eV. Assuming that the Z_{eff} of part (a) is correct, compute Z_{eff} for the six 2p electrons for neutral neon in its ground state.

28-22 Potassium ($Z = 19$) has one outer 4s electron and an ionization energy of 4.3 eV. (a) What is Z_{eff} for this electron in potassium? (b) Calcium ($Z = 20$) has two outer 4s electrons. A rough estimate for the effective nuclear charge for the single 4s electron of Ca$^+$ would be $Z_{eff} + 1$, where Z_{eff} is the result from (a). The measured ionization energy of calcium is 6.09 eV. Estimate Z_{eff} for the two 4s electrons of neutral calcium.

ANSWERS TO REVIEW QUESTIONS

Q28-1, principal; **Q28-2**, integer, $(n - 1)$; **Q28-3**, $-l$, $+l$; **Q28-4**, $\frac{1}{2}$; **Q28-5**, absolute value of the wave function; **Q28-6**, zero; **Q28-7**, quantum state; **Q28-8**, ionization energy; **Q28-9**, electron shell, stable; **Q28-10**, angular momentum.

SUPPLEMENTARY TOPICS

28.7 | MASERS AND LASERS

The ability to produce intense, coherent electromagnetic waves at nearly a single frequency has made possible much of modern communication. Masers and lasers now produce very intense coherent monochromatic microwave and visible radiation in a narrow beam. The word maser stands for *Microwave Amplification by Stimulated Emission of Radiation*. The laser, which relies on the same principles, produces visible light rather than microwaves. In this section, we consider some general features common to all masers and lasers and then describe the helium-neon gas laser.

Two important conditions must occur in any maser or laser. One is called *population inversion* and the other is called *stimulated emission*. We describe these ideas in terms of atoms that have two energy levels E_1 and E_2 (Fig. 28.13).

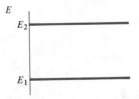

Figure 28.13. Two energy levels of an atom. Under normal conditions, a system of atoms will have more electrons in the lower energy state, E_1. Heating will at best result in the nearly equal population of the two states.

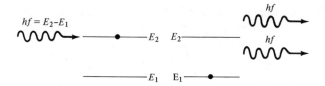

Figure 28.14. A photon with energy $hf = E_2 - E_1$ is incident on an atom in the state E_2. The incident photon stimulates the atom to drop into the lower energy state E_1, emitting a second photon with the same energy and coherent with the first.

Under normal conditions most of the atoms will be in the lower energy state, with a few in the upper level. Atoms that are in the excited state spontaneously emit photons and drop into the lower state. This process is random, and the emitted photons are not coherent, that is, they are not in phase with one another. In a laser, another type of emission process, called stimulated emission, plays an important role. If one atom emits a photon of energy $E_2 - E_1$, that photon may collide with a second atom that is in the state E_2. When this happens the second atom is stimulated to also emit a photon of the same energy. *These two photons are coherent with each other* (Fig. 28.14).

Stimulated emission may occur under normal conditions but is a rather minor effect because so few atoms are in the higher energy state. However, in lasers the upper energy level is caused to be overpopulated, with more atoms in the upper than the lower state, and *population inversion* is said to occur. Under these conditions the probability that stimulated emission occurs is large, and many coherent photons can be obtained.

We can now see how intense coherent radiation may be produced. If a system of atoms or molecules can be forced into having the higher of two energy states overpopulated, the emission of a single photon during a transition to the lower state by one atom will trigger the emission of photons by many more atoms. In practice, this is achieved by an arrangement like that in Fig. 28.15.

The laser diagrammed is said to be a two-state or pulsed laser. After the process of stimulated emission is completed and most of the atoms in the material have been deexcited, energy must again be fed into the system to invert the population. This energy is usually supplied by electromagnetic radiation.

Lasers that produce continuous radiation rather

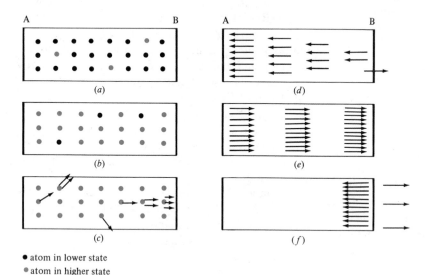

● atom in lower state

● atom in higher state

Figure 28.15. A schematic diagram of a laser. A material is placed between two parallel walls A and B. Wall A is a perfect reflector and B is 99% reflecting. (*a*) The material is in its normal state with most atoms, black circles, in a lower energy state and some, colored circles, in a higher energy state. (*b*) Population inversion is achieved. This is done by feeding energy into the system and is often called *pumping*. (*c*) Photons are emitted by random atoms. Some photons are lost through the sides, but one is shown initiating stimulated emission perpendicular to the ends. (*d*) After partial reflection the buildup of coherent photons continues. (*e*) The buildup continues after total reflection from wall A and in (*f*) 1% of the very large number of photons present escape through wall B. The photons inside continue to sweep back and forth until the population becomes normal and all the photons escape.

than a pulsed beam require the presence of at least three, rather than two, energy states. In a continuous laser, atoms are continuously pumped from a low energy state to a higher one. These excited atoms then do not return to the initial state, but instead fall into and overpopulate a third state of intermediate energy. The laser action then takes place between this third level and some lower state.

One continuous laser, producing light at 632.8 nm, contains a mixture of helium and neon gases. These two atoms have filled $n = 1$ and $n = 2$ electronic shells, respectively. When these atoms are excited, the electrons must return to a 1s state in helium and to a 2p state in neon to reestablish the ground state (Fig. 28.16).

The 2s ($n = 2$, $l = 0$) level in helium is 20.61 eV above the ground state and is lower in energy than the 2p ($n = 2$, $l = 1$) state. When an electric discharge occurs in the gas, electrons that end up in the

2s state in helium cannot return to the 1s ground state because the orbital angular momentum of the two states is the same; the 2s state is metastable. However, the 5s state of neon happens to be 20.66 eV above its ground state. Thus one way that the electrons in helium can return to their ground states from the 2s state is by colliding with a neon atom and exciting a neon electron from the 2p state into the 5s state. The slight energy difference is made up by the thermal kinetic energy of the helium atom. In effect, the continuous fund of 2s electrons in helium is pumping electrons into the 5s state of neon. The laser action then occurs when the electrons of the 5s state of neon falls into the 3p state, emitting photons with wavelength 632.8 nm. Other coherent transitions can also occur, but none are in the visible spectrum.

A variety of lasers and masers are now available that produce radiation in large portions of the visible, infrared, and microwave regions. Although not all wavelengths in these regions are available, some devices are tunable; that is, their output wavelengths may be changed. Masers and lasers have already found a wide range of applications and more are being developed.

In surveying and construction work the intense narrow laser beam can be used as a reference for accurate alignment. Holography is almost totally reliant on the use of lasers. In medicine, the laser is used for repair of retinal damage and as a surgical tool. Laser light scattering is an excellent tool for studying molecular structure and motion.

The same characteristics that make lasers a useful tool also pose hazards. While the total energy output of a laser is sometimes small, the rate at which it is released and the narrowness of the beam often result in extremely intense radiation. Available pulsed lasers can have power outputs of more than 10^{13} W with a beam of cross-section less than 1 mm^2 = 10^{-6} m^2. However, the pulse may be as brief as about 10^{-12} second. Laser beams can vaporize metals and may cause severe damage to biological systems and tissue. For example, the retina is a strong absorber of light. Thus, direct visual observation of even a relatively weak laser beam can cause severe retinal damage.

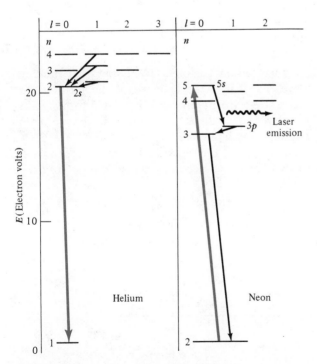

Figure 28.16. If electrons in helium are excited they may return to the ground state emitting photons in the process. However, many electrons end up in the metastable 2s state of helium. When such a helium atom collides with a neon atom, the electron in helium drops into the ground state (colored line) and an electron in neon is raised into its 5s state (also shown colored). Laser action occurs as the electron in neon drops into the 3p state.

28.8 | SCHROEDINGER'S EQUATION

Schroedinger's equation, like Newton's laws of motion or the basic laws of electricity and magnetism, cannot be derived from more fundamental concepts. Its validity rests on the successful testing of its predictions over a wide range of phenomena. We show how to construct the equation in this section, and examine its solutions for some simple situations in the following sections.

Applying Newton's laws to sound waves or waves on a string leads to a *wave equation* involving second derivatives of the displacement. Maxwell's equations lead to a similar equation for electromagnetic waves. Thus, Schroedinger guessed that matter waves also satisfy a wave equation containing second derivatives of Ψ.

Schroedinger developed a procedure or "recipe" for constructing the wave equation for any system. It starts from the total energy of the system written in terms of the momenta and positions of its particles. For example, suppose there is just one particle of mass m, and it can only move along the x axis. The particle has a kinetic energy $mv^2/2 = p^2/2m$, and a potential energy $\mathcal{U}(x)$ that depends on the nature of the forces acting on it. The total energy is

$$\frac{p^2}{2m} + \mathcal{U}(x) = E$$

The equation for Ψ is constructed by making the replacement

$$p^2 \rightarrow -\hbar^2 \frac{d^2}{dx^2} \qquad (28.9)$$

Then Schroedinger's equation for this case is

$$\left[\frac{p^2}{2m} + \mathcal{U}(x) \right] \Psi(x) = E\Psi(x) \qquad (28.10)$$

or, with Eq. 28.9,

$$\frac{-\hbar^2}{2m} \frac{d^2}{dx^2} \Psi(x) + \mathcal{U}(x)\Psi(x) = E\Psi(x)$$

$$\text{(Schroedinger's equation)} \qquad (28.11)$$

This equation does not contain the time, and is sometimes called the *time-independent* Schroedinger equation. A related *time-dependent* Schroedinger equation, which we will not consider, determines how the wave function evolves in time. Also, if the particle can move in two or three dimensions, or if there are more particles, the energy must include additional terms, and Schroedinger's equation is then more complicated.

28.9 | PARTICLE IN A ONE-DIMENSIONAL BOX

To see how you can solve Schroedinger's equation and interpret the resulting wave function, we consider a particle trapped in a one-dimensional "box." This idealized situation contains many of the features of more realistic but complex problems.

In Fig. 28.17, the particle experiences no force as long as it is the region between $x = 0$ and $x = a$. Thus the potential energy $\mathcal{U}$ is constant in that region, and can be set to the convenient value $\mathcal{U} = 0$. At $x = 0$ and $x = a$, infinitely large forces keep the particle in the box. If the forces were finite but large, the potential energy would be rising steeply at these points. Here, since it would take an infinite amount of work to go beyond these limits, the potential energy becomes infinite at the ends of the box.

Inside the box, with $\mathcal{U} = 0$, Schroedinger's equation is

$$\frac{-\hbar^2}{2m} \frac{d^2}{dx^2} \Psi(x) = E\Psi(x) \qquad (28.12)$$

We try a solution of the form

$$\Psi(x) = A \sin kx \qquad (28.13)$$

Now we saw in Chapter Twenty-one that standing waves have an x dependence of the form $\sin kx$. There the wave number k was related to the wavelength λ by $k = 2\pi/\lambda$. Equation 28.13 therefore represents a standing wave with a wavelength $\lambda = 2\pi/k$.

According to Eqs. B.26 and B.27 in Appendix B, $(d/dx) \sin kx = k \cos kx$, and $(d/dx) \cos kx =$

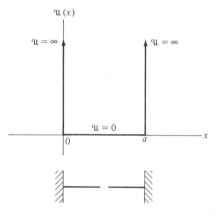

Figure 28.17. The potential energy function $\mathcal{U}(x)$ for a particle in a one-dimensional "box." There is no force on the particle inside the box, so its potential energy is constant. At the ends of the box, there are infinitely large forces, and the potential energy jumps to infinity.

$-k \sin kx$. Hence substituting $\Psi(x) = A \sin kx$ in Eq. 28.12 leads to

$$\frac{\hbar^2 k^2}{2m} (A \sin kx) = E \, (A \sin kx)$$

Since $(A \sin kx)$ is a common factor on both sides, it can be divided out. Thus we have a solution to Eq. 28.12 provided that the wave number k satisfies

$$\hbar^2 k^2/2m = E \qquad (28.14)$$

or

$$\hbar k = (2mE)^{1/2} \qquad (28.15)$$

Now $E = p^2/2m$, or $2mE = p^2$, so this equation reduces to $\hbar k = p$. Furthermore, with $\hbar = h/2\pi$ and $k = 2\pi/\lambda$, we have

$$p = \hbar k = \frac{h}{2\pi} \frac{2\pi}{\lambda} = \frac{h}{\lambda}$$

This is de Broglie's hypothesis! Schroedinger's recipe produces waves with de Broglie's wavelength. Note that if we replaced the sine by a cosine in Eq. 28.13, we would again find a solution.

Now let's consider the effect of the "walls." If a classical particle is moving within a box, and its energy is less than the potential energy at the top of the walls, it cannot get out. The probability of finding the particle outside the box is zero. In quantum mechanics, for this case of an infinite potential energy or "infinitely high walls," the probability of the particle being found outside the box is also zero. (This isn't true, however, for walls of finite height, as we will see in Section 28.10.) Since $|\Psi(x)|^2$ is proportional to that probability, $\Psi(x)$ must be zero outside the box. If we require that the wave functions inside and outside match at the walls, we have the condition that $\Psi(x) = 0$ at $x = 0$ and at $x = a$. *This is the boundary condition for a particle in a box:* $\Psi(x)$ *vanishes at the walls.*

At the point $x = 0$, $\sin kx = \sin 0 = 0$ as required. (Note that $\cos kx = \cos 0 = 1$ at $x = 0$, which rules out a cosine solution.) At $x = a$, $\sin kx = \sin ka$. This is zero only if ka equals one of the arguments for which the sine is zero. These are $0, \pi, 2\pi, \ldots$, corresponding to $0°, 180°, 360°, \ldots$. Thus $ka = n\pi$, where n is an integer. The case $ka = 0$ is uninteresting, since it implies $k = 0$ and $\Psi(x) = A \sin 0 = 0$ for all x. Hence k is restricted to the values

$$k_n = \frac{n\pi}{a} \qquad n = 1, 2, 3, \ldots \qquad (28.16)$$

Using $E = \hbar^2 k^2/2m$, the corresponding energies are

$$E_n = \frac{n^2 \hbar^2 \pi^2}{2ma^2} \qquad n = 1, 2, 3, \ldots \quad (28.17)$$

The energies E_n *are discrete.* There is a lowest or ground state energy $E_1 = \hbar^2 \pi^2/2ma^2$, a first excited state with $E_2 = 2^2 E_1$, and so on (Fig. 28.18). The *quantum number* n labels the energy levels E_n and the associated wave functions $\Psi_n(x)$.

Since the wavelength λ is related to k by $\lambda = 2\pi/k$, and k can have the values $n\pi/a$, the possible wavelengths are $\lambda = 2\pi(a/n\pi) = 2a/n$. Thus $n\lambda/2 = a$, or *an integer number of half wavelengths must "fit" into the box.* This is just the condition for resonant waves on a fixed end string or in a closed pipe. Clearly we could have obtained these results without a detailed mathematical discussion by simply requiring the de Broglie waves to fit within the box.

However, our primary goal is not to solve the particle in a box problem, but rather to illustrate how we solve quantum mechanical problems in general. We construct the wave equation using Schroedinger's recipe and find solutions consistent with the boundary conditions.

An atom is not a box with infinitely high walls, but an electron without sufficient energy to leave the atom is kept within the general neighborhood of the nucleus. *Thus the boundary condition for an atomic wave function is that* Ψ *must go to zero when an electron is far from the nucleus, or when* $r \to \infty$. This requirement implies that the constants analogous to k in the wave function must have special values. The corresponding energies are then discrete.

The following example illustrates that the solutions to Schroedinger's equation are consistent with the uncertainty principle, as expected.

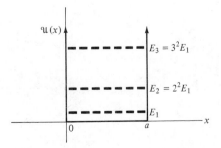

Figure 28.18. The first three energy levels for the particle in a one-dimensional box.

Example 28.4

Show that the ground state energy is consistent with the minimum energy predicted by the uncertainty principle.

If we guess that the particle is at the midpoint of the box, we cannot be off by more than $a/2$. Hence we take $\Delta x \simeq a/2$. Then the uncertainty principle requires $\Delta x \Delta p \geq \hbar$, so $\Delta p \geq \hbar/\Delta x = 2\hbar/a$. The momentum is at least as large as its uncertainty, so its minimum value is $p = \Delta p = 2\hbar/a$. The corresponding kinetic energy is

$$E = \frac{p^2}{2m} = \frac{4\hbar^2}{2ma^2}$$

The ground state energy is actually $\pi^2\hbar^2/2ma^2$, which is larger than this minimum by a factor of $\pi^2/4 \simeq 2.5$.

Interpretation of the Wave Function | We have noted that for matter waves, $|\Psi(x)|^2$ is a measure of the probability of finding a particle at x. More precisely, $|\Psi|^2$ is a *probability density*. Thus, in one dimension $P(x) = |\Psi(x)|^2$ is a *probability per unit length*. The probability of finding the particle between x and $x + dx$ is

$$P(x)dx = |\Psi(x)|^2 dx \qquad (28.18)$$

In regions where Ψ is large, the odds are high of finding the particle. It is less likely to be found where Ψ is small.

For example, Fig. 28.19a shows Ψ and $|\Psi|^2$ for the $n = 1$ or ground state of a particle in a one-dimensional box. At the center, $x = a/2$, $|\Psi|^2$ has its greatest value. We can't say where the particle is, except in a probabilistic way. However, if we measure its position, the chances are best of finding it near $a/2$. If we have many identical particles in their ground states in identical boxes, and we measure their positions, a graph of the relative frequencies of the observed positions will look like the $|\Psi(x)|^2$ plot. Figure 28.19b shows Ψ and $|\Psi|^2$ for the first excited state, $n = 2$. For this state the probability density is zero at the center, and it peaks at $a/4$ and $3a/4$. The probability density $P(x)$ is very different for the $n = 1$ and $n = 2$ states.

The total probability of finding the particle somewhere in the box must be 1, since it is somewhere!

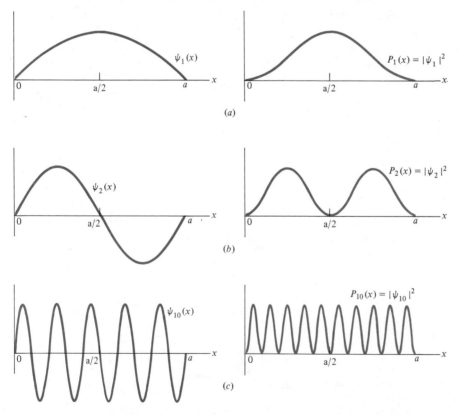

Figure 28.19.   Graphs of $\Psi(x)$ and $|\Psi(x)|^2$ for a particle in a one-dimensional box. (a) Ground state ($n = 1$). (b) First excited state ($n = 2$). (c) $n = 10$ state.

This means that if we integrate $P(x)$ over all regions where the wave function is not zero, we must find $\int P\,dx = 1$, or

$$\int |\Psi(x)|^2 dx = 1 \quad \text{(normalization condition)} \quad (28.19)$$

This is called the *normalization condition*. If the constant factor A in $\Psi = A \sin kx$ is chosen so that Eq. 28.19 is satisfied, Ψ is said to be *normalized*, and A is called the *normalization constant*. Thus we must have

$$\int_0^a |\Psi(x)|^2 dx = \int_0^a A^2 \sin^2 kx\, dx = 1 \quad (28.20)$$

It is left as an exercise to verify that $A^2 = 2/a$, so that the normalized wave function is

$$\Psi_n(x) = \left[\frac{2}{a}\right]^{1/2} \sin \frac{n\pi x}{a} \quad \text{(particle in a box)}$$

$$(28.21)$$

Since classical physics correctly describes the world at the macroscopic level, a quantum description must approach the classical predictions in suitable limiting cases. This is referred to as the *correspondence principle*. The next example shows that this occurs in the limit of large energies or quantum numbers for a particle in a box.

Example 28.5

Show that the probability distribution for a particle in a box approaches the classical distribution for large values of the quantum number n.

If a classical ball is moving back and forth within a one-dimensional box, it is equally likely to be found at any point in the box. Stated mathematically, the classical probability distribution $P_{\text{C.M.}}(x)$ is the same everywhere in the box. In the quantum mechanical case, in the ground state the probability density is greatest at the center (Fig. 28.19a), and

$$P(x) = |\Psi(x)|^2 = (2/a) \sin^2 k_1 x = (2/a) \sin^2 \pi x/a$$

Thus the classical and quantum probability distributions are very different for a particle with the ground state energy.

However, when n is large, $|\Psi_n|^2 = (2/a) \sin^2 k_n x$ oscillates very rapidly (Fig. 28.19c). Now consider the probability of finding the particle somewhere within a narrow "bin" of x values comparable to the width of the $\sin^2 k_n x$ oscillations but still small compared to a. The odds of finding the particle are the same for the bins everywhere in the box. As the energy E_n and the quantum number n become larger, the bins become smaller. Thus for all practical purposes the probability distribution becomes constant, as in the classical case. This is what is expected from the correspondence principle.

28.10 | THE SQUARE WELL; TUNNELING

A ball or a car rolling up a hill can travel uphill only until its kinetic energy decreases to zero. However, we see now with the aid of an idealized problem called a *square well*, that quantum mechanics predicts that particles can in fact sometimes be found in or pass through places forbidden by energy considerations.

Figure 28.20 shows a square well. The potential energy $\mathcal{U}$ equals zero inside, and $\mathcal{U}_0$ outside. Suppose we have a particle with a total energy $E = K + \mathcal{U}$. As it leaves the well, its potential energy increases by $\mathcal{U}_0$. Since E is constant, its kinetic energy decreases by an equal amount.

According to classical physics, the particle cannot leave the well if $E < \mathcal{U}_0$, since it would then have a negative kinetic energy, which is impossible. ($K = p^2/2m$ can never be negative.) What does quantum mechanics say about this situation? Inside the well, just as in the box problem of Sec. 28.9, Schroedinger's equation predicts sinusoidal waves of the form $\sin kx$ or $\cos kx$, where again $\hbar k = (2mE)^{1/2}$. Outside, we have

$$\frac{-\hbar^2}{2m} \frac{d^2}{dx^2} \Psi(x) + \mathcal{U}_0 \Psi(x) = E\Psi(x)$$

or

$$\frac{-\hbar^2}{2m} \frac{d^2}{dx^2} \Psi(x) = (E - \mathcal{U}_0)\,\Psi(x) \quad (28.22)$$

This is similar to Eq. 28.12 for the particle in a box, but with $(E - \mathcal{U}_0)\Psi$ replacing $E\Psi$. However,

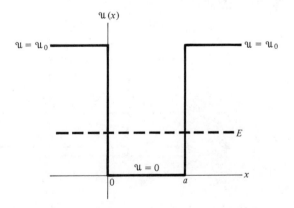

Figure 28.20. A one-dimensional square well. If the total energy E is less than $\mathcal{U}_0$ as shown here, the kinetic energy $K = E - \mathcal{U}_0$ is negative for $x < 0$ or $x > a$. According to classical physics the particle cannot be found in these regions.

$(E - \mathfrak{U}_0)$ is negative, since $E < \mathfrak{U}_0$. Because of this, a sine or a cosine no longer is a solution. Instead, we need exponentials. Specifically, we can try the solutions

$$\Psi(x) = Be^{+Kx} \quad \text{or} \quad \Psi(x) = Ce^{-Kx}$$

Using Eq. B.25 in Appendix B, $(d/dx)e^{ax} = ae^{ax}$, we find with either function $(d^2/dx^2)\Psi = k^2\Psi$. Thus either is a solution if $(\hbar K)^2/2m = (E - \mathfrak{U}_0)$ or

$$\hbar K = [2m(E - \mathfrak{U}_0)]^{1/2} \qquad (28.23)$$

Just as in the box problem, boundary conditions limit the acceptable solutions of the Schroedinger equation. First, we must reject as unphysical any wave functions that become infinite as we go far from the well. The wave function becomes infinite, for example, if Ψ varies as e^{+Kx} for $x \rightarrow \infty$, or as e^{-Kx} for $x \rightarrow -\infty$ (Fig. 28.21a). Thus the solutions outside the well must be of the form (Fig. 28.21b)

$$\Psi(x) = Be^{+Kx} \qquad x < 0$$
$$\Psi(x) = Ce^{-Kx} \qquad x > a$$

In addition to behaving properly at infinity, the solutions in the three regions must join smoothly. More precisely, the wave function Ψ and its derivative $d\Psi/dx$ must be the same or "match" as we approach an edge of the well from either side. (Otherwise we could not compute meaningful values of $(d^2/dx^2)\Psi$ to substitute into the Schroedinger equation.) These boundary conditions serve to limit the possible energies to discrete values, just as the requirement $\Psi = 0$ at the ends of the box limits the energy values in that situation.

Specifically, if the energy E is picked arbitrarily, we can always adjust the normalization constants B and C to make Ψ match at both edges, but $d\Psi/dx$ will not be the same on both sides except for very special values of E. Thus we must look systematically for those special energy values that permit both Ψ and $d\Psi/dx$ to be matched.

Carrying out this calculation in detail is messy, and the results can only be stated in numerical or graphical form. Some of the allowed energy values and the corresponding wave functions are sketched in Fig.

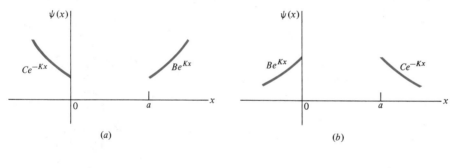

(a)

(b)

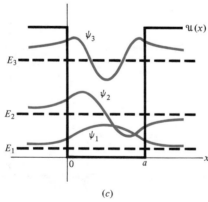

(c)

Figure 28.21. (a) Wave functions that grow rapidly as we move away from the well are not physically meaningful. (b) The acceptable wave functions decrease exponentially outside the well. (c) The first three energy levels. The corresponding wave functions are shown, using the energy levels as axes for convenience.

28.21c. The wave function is sinusoidal inside the well, but it has exponential "tails" extending outside. Thus the wave function and the probability density are not zero in the region where the kinetic energy is negative! There is a nonzero probability of finding the particle in places where it could never be according to classical physics. Note that as $\mathcal{U}_0$ becomes larger relative to the energy E, the tails become smaller; the lowest energy level has the smallest tails. In the limit of infinitely high walls, $\mathcal{U}_0 \rightarrow \infty$, the tails are zero. This is why we used the boundary condition $\Psi = 0$ at the walls of a box.

The energy values for the square well correspond closely to those for the particle in a box, but they are systematically a bit lower. This happens because the wave function for the particle is now spread over a larger region than the box size a, so its wavelength is longer. Thus its momentum $p = h/\lambda$ is smaller, as is the kinetic energy $p^2/2m$.

If the energy of a particle is greater than the depth of the well, the particle is not "bound" or attached to the well, and it may travel anywhere. It has a sinusoidal wave function with different wavelengths inside and outside the well.

The following example illustrates some of these ideas.

Example 28.6

Suppose the square well in Fig. 28.20 has only two bound states. Using the approximation that the square well energy levels are the same as for a box, find the range of possible values for $\mathcal{U}_0 a^2$.

For a particle in a box, Eq. 28.17 gives the possible energies as $E_n = n^2 \hbar^2 \pi^2 / 2ma^2$. Thus the first three energy levels are

$$E_1 = \hbar^2 \pi^2 / 2ma^2$$
$$E_2 = 2^2 E_1 = 4\hbar^2 \pi^2 / 2ma^2$$
$$E_3 = 3^2 E_1 = 9\hbar^2 \pi^2 / 2ma^2$$

If there are only two bound states, this means that the height $\mathcal{U}_0$ of the well is more than E_2 and less than E_3. Thus

$$4\hbar^2 \pi^2 / 2ma^2 < \mathcal{U}_0 < 9\hbar^2 \pi^2 / 2ma^2$$

or

$$4\hbar^2 \pi^2 / 2m < \mathcal{U}_0 a^2 < 9\hbar^2 \pi^2 / 2m$$

In the approximation that the square well levels are the same as those for the box, this is the range of $\mathcal{U}_0 a^2$ values for which the well has two bound states. If a^2 is small, $\mathcal{U}_0$ is large; the well is then very deep, but it

extends over a small region of space. Conversely, a large value of a^2 implies a shallow well. Allowing for the fact that the actual energies for a square well are lower than for a box would slightly lower the limits.

Tunneling | The penetration of the wave function into classically inaccessible regions has remarkable effects. Suppose, for example, that a proton is directed at an atomic nucleus. Since both particles are positively charged, their electrical repulsion limits how close the proton can come to the nucleus. The nuclear force plays no role at large distances; the proton must be nearly at the nucleus before its effect is noticeable. Thus, if the kinetic energy of the proton is small, it cannot get close enough to the nucleus for a nuclear interaction to occur. Nevertheless, because of this penetration of the wave function, the wave function of the proton can overlap with the nucleus at quite low energies. Hence nuclear fusion reactions can occur in stars at much lower energies and temperatures than would otherwise be required.

Tunneling through a barrier is an even more striking quantum effect. We can illustrate what it means by thinking about a ball on a hill. Objects tend to move toward the place where their potential energy is lowest, and if it is on the slope, the ball tends to roll downhill. This won't happen, however, if the ball is stuck inside a hole on the hillside without enough energy to get out.

Now consider an electron or an alpha particle instead of a macroscopic object. Then classical concepts do not apply, and we must turn to quantum mechanics. What we find is that there is a chance that the electron or alpha particle will be found outside the hole, "rolling" downhill! It is as though it had tunneled through the region between the bottom of the hole and the adjacent slope.

To see this in a little more detail, consider Fig. 28.22. It is a quantum mechanical analog of the ball in a hole on a hill. To the left of the barrier the wave function is sinusoidal and within the barrier Ψ is decreasing exponentially. To the right Ψ again is oscillatory, although it is not just a simple sine or cosine in this region where $\mathcal{U}$ is varying. The requirement of matching smoothly at each edge of the barrier leads to the wave function shown, with a small but nonzero wave function on the right. If a particle is initially in the region to the left, this means it has a finite probability of later being found on the right of the barrier where its potential energy is lower. Effectively, it is

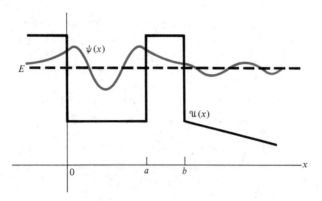

Figure 28.22. Inside the well, $\psi(x)$ is sinusoidal. In the classically inaccessible regions, it is exponentially decreasing. Beyond the barrier, $x > b$, $\psi(x)$ is again oscillatory rather than decreasing. Each time the particle hits the barrier, it has a small but nonzero chance of tunneling through.

repeatedly striking the barrier as it moves back and forth, and eventually it will escape and go lower down the hill. This tunneling phenomenon is responsible for nuclear alpha decay (Chapter Thirty) and for many processes that occur in solid state physics.

28.11 | APPLICATION TO ATOMS, MOLECULES, AND NUCLEI

The basic ideas developed in our discussion of idealized one-dimensional problems apply to atoms, molecules, and nuclei. A Schroedinger equation is constructed and solved either exactly or approximately for the wave functions. The acceptable bound state wave functions are limited by the boundary conditions, so that only special energy values are allowed. The allowed energies and the associated wave functions are characterized by several quantum numbers, rather than by one quantum number as in the one-dimensional case. Again, the wave functions typically oscillate in classically allowed regions, and decrease rapidly in classically inaccessible regions.

Carrying out the actual calculations for even the simplest atom, hydrogen, requires more space and more advanced mathematics than is appropriate here. Textbooks on modern physics, quantum mechanics, and quantum chemistry go into these problems in detail. Some of the most basic results of these calculations for hydrogen and larger atoms were discussed earlier in this chapter. In the next section, we give explicit formulas for the hydrogen wave functions and show how to use them. Molecules and atomic nuclei are considered in later chapters.

28.12 | EXPLICIT FORM OF THE HYDROGEN WAVE FUNCTIONS

Although the Schroedinger equation for the hydrogen atom is too complicated for us to solve, we can write down and discuss the wave functions it predicts. A number of interesting and important results can be obtained in this way.

Since the electron in a hydrogen atom moves in three dimensions, its wave function depends on three coordinates. These may be chosen to be the Cartesian coordinates x, y, and z. However, the spherical symmetry of the Coulomb force between the electron and the nucleus is reflected in the wave function. Hence the spherical coordinates r, θ, and ϕ are often more convenient.

The two sets of coordinates are related by (Fig. 28.23a)

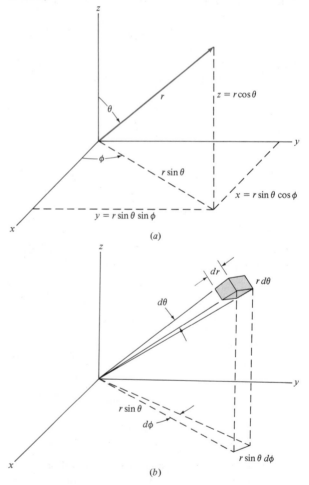

Figure 28.23. (a) Relation between spherical and Cartesian coordinates. (b) The volume element formed when the coordinates change by dr, $d\theta$, and $d\phi$.

$$x = r \sin\theta \cos\phi$$
$$y = r \sin\theta \sin\phi \qquad (28.24)$$
$$z = r \cos\theta$$

Another useful relationship is

$$r^2 = x^2 + y^2 + z^2 \qquad (28.25)$$

An infinitesimal element of volume in spherical coordinates is (Fig. 28.23b)

$$dV = r^2 \sin\theta \; d\theta \; d\phi \; dr \qquad \text{(volume element)} \quad (28.26)$$

Table 28.3 lists the $n = 1$ and $n = 2$ hydrogen atom wave functions. The ground state (1s) wave function is the simplest, and all the s states are spherically symmetric, or independent of the angles θ and ϕ. As n or l increases, the formulas contain more and more terms; these give rise to additional oscillations and nodes in the wave functions (Figs. 28.5 and 28.6).

In three dimensions, the probability density $|\Psi(x,y,z)|^2$ is the *probability per unit volume*. This means that if the electron wave function is Ψ, the probability of finding it in the infinitesimal box of volume $dV = dx \; dy \; dz$ centered at the point x,y,z is

$$P(x,y,z) \; dV = |\Psi(x,y,z)|^2 \; dx \; dy \; dz \quad (28.27)$$

If the wave function is expressed in spherical coordinates, then the probability of finding the electron in a small box centered at r,θ,ϕ is

$$P(r,\theta,\phi) \; dV = |\Psi(r,\theta,\phi)|^2 \; r^2 \sin\theta \; d\theta \; d\phi \; dr \quad (28.28)$$

As in one dimension, the probability of finding the electron *somewhere* must be one, so the wave function must be normalized:

$$\int |\Psi|^2 dV = 1 \qquad \text{(normalization)} \quad (28.29)$$

The integral is taken over all space. Thus in Cartesian coordinates, x, y, and z each go from $-\infty$ to $+\infty$; in spherical coordinates, r varies from 0 to ∞, θ from 0 to π, and ϕ from 0 to 2π. The wave functions in Table 28.3 have normalization constants chosen so that this condition holds. For example, the 1s state normalization constant is

$$N_{1s} = (1/\pi a_0^3)^{1/2}$$

As we noted in Sec. 28.3, the s state wave functions are largest at $r = 0$, so there is a relatively large probability of finding an s electron near the nucleus. The wave functions for all $l > 0$ states are zero at the origin, so electrons in these states are less likely to be found at small values of r. The chance of finding an s state electron at a very small value of r is reduced, however, by the factor of r^2 in Eq. 28.28; spherical shells with small radii have very small volumes. We apply some of these ideas in the following example.

Example 28.7

For the 1s state, find the maximum value of the radial probability density r^2P.

From Table 28.3, the wave function can be written as

$$\Psi_{1s} = N_{1s} e^{-r/a_0}$$

Thus the radial probability density is

$$r^2P = r^2 |\Psi_{1s}|^2 = N_{1s}^2 r^2 e^{-2r/a_0} \qquad (i)$$

To find the maximum value of r^2P, we set its derivative $d(r^2P)/dr$ equal to zero:

$$N_{1s}^2 e^{-2r/a_0}(2r - 2r^2/a_0) = 0$$

This has solutions at $r = 0$ and $r = \infty$, which are not relevant since they correspond to minima, and at $r = a_0$;

TABLE 28.3

The $n = 1$ and $n = 2$ wave functions for atomic hydrogen. Note that some of the Ψ's involve $i = (-1)^{1/2}$, and are complex numbers. For one electron atoms with $Z > 1$, the wave functions are obtained from those below if a_0 is replaced by a_0/Z.

1s	$\dfrac{1}{\pi^{1/2} a_0^{3/2}} e^{-r/a_0}$
2s	$\dfrac{1}{2\pi^{1/2}(2a_0)^{3/2}} \dfrac{(2-r)}{a_0} e^{-r/2a_0}$
2p $(m = 0)$	$\dfrac{1}{2\pi^{1/2}(2a_0)^{3/2}} \dfrac{r}{a_0} e^{-r/2a_0} \cos\theta$
2p $(m = \pm1)$	$\dfrac{1}{\pi^{1/2}(4a_0)^{3/2}} \dfrac{r}{a_0} e^{-r/2a_0} \sin\theta \, (\cos\phi \pm i \sin\phi)$

the last solution corresponds to the desired maximum. Thus, as stated earlier, the electron is most likely to be found at a radius equal to the Bohr radius. Substituting $r = a_0$ in Eq. i, the probability of finding the electron in that region is proportional to

$$r^2 P = r^2 |\Psi_{1s}|^2 = N_{1s}^2 a_0^2 e^{-2}$$

If this calculation is repeated for the $2p$ states, you find that the radial probability density peaks at $4a_0 = 2^2 a_0$, the radius predicted by the Bohr model for $n = 2$. The $2s$ state has its greatest radial probability density at a somewhat larger radius.

A particle orbiting around a force center has a total energy $E = K + \mathcal{U}(r)$ which is constant. The kinetic energy, $K = E - \mathcal{U}(r)$, varies with position, and it is zero at the radius r_t where

$$E - \mathcal{U}(r_t) = 0 \qquad \text{(classical turning point)} \quad (28.30)$$

In classical physics, this is the maximum radius or *classical turning point* for the orbiting particle. The particle can never go beyond this point, since it cannot have a negative kinetic energy. As we have seen, in quantum mechanics particles can sometimes be found in the classically forbidden region. This is true for electrons in atoms, as we see in the next example.

Example 28.8

(a) For the $1s$ state, find the classical turning point r_t.
(b) Find the value of $r^2 P$ at r_t and at $2r_t$.

(a) The energy of the $1s$ state is, from Eq. 28.1,

$$E_1 = \frac{-ke^2}{2a_0}$$

The potential energy is

$$\mathcal{U} = \frac{-ke^2}{r}$$

Then $K = E - \mathcal{U} = 0$ when $r = r_t$, where r_t is found by solving

$$\frac{-ke^2}{2a_0} + \frac{ke^2}{r_t} = 0$$

This gives $r_t = 2a_0$. Thus the classical turning point is at twice the Bohr radius. Note that, in Fig. 28.7, $r^2 P$ is nonzero for the $1s$ state well beyond $r_t = 2a_0$.

(b) The radial probability at $r = r_t = 2a_0$ is

$$r^2 P = r^2 |\Psi_{1s}|^2 = N_{1s}^2 r^2 e^{-2r/a_0} |_{r=2a_0}$$

$$= 4N_{1s}^2 a_0^2 e^{-4}$$

We saw in Example 28.7 that for the $1s$ state, the maximum radial probability density occurs at a_0 and is

$N_{1s}^2 a_0^2 e^{-2}$. Thus, at the turning point, $r^2 P$ is smaller by a factor of $4e^{-2} = 0.54$.

In the same way, at $r = 2r_t = 4a_0$, we find

$$r^2 P = 16 N_{1s}^2 a_0^2 e^{-8}$$

At $r = 2r_t$, $r^2 P$ is smaller than at r_t by a factor of $4e^{-4} = 0.073$, and smaller than its maximum value at a_0 by a factor of $16e^{-6} = 0.040$. At twice the classical turning point, the chance of finding the electron is 4 percent as great as at its most probable radius.

The total probability of finding the electron at radii larger than the classical turning radius is obtained by evaluating $\int P\,dV$ over all points in space with $r \geq r_t$. This probability turns out to be 0.24. Thus, the electron spends almost a quarter of its time in the region where classical physics says it can never be! As we have noted, the fact that in quantum mechanics particles can penetrate into such regions is responsible for many remarkable phenomena in atoms, molecules, and nuclei.

We noted in Sec. 28.3 that in molecular structure discussions it is convenient to use $2p$ wave functions that are superpositions of those shown in Table 28.3. In the $2p$ wave functions, the angular factors can be rewritten with the aid of Eqs. 28.24 as follows:

$$m = 0: \qquad \cos\theta \rightarrow z/r$$
$$m = \pm 1: \qquad \sin\theta\,(\cos\phi \pm i\sin\phi) \rightarrow (x + iy)/r$$

Thus adding and subtracting the $m = +1$ and $m = -1$ wave functions yeilds wave functions proportional to x and to y, respectively:

$$\Psi_x = (\Psi_{2p,m=+1} + \Psi_{2p,m=-1})/2^{1/2} \propto x/r$$
$$\Psi_y = (\Psi_{2p,m=+1} - \Psi_{2p,m=-1})/2^{1/2}i \propto y/r \qquad (28.30)$$

These are the p_x and p_y wave functions, respectively. The factors in the denominators are chosen to give real, normalized wave functions. Also, we define the p_z wave function by

$$\Psi_z = \Psi_{p,m=0} \qquad (28.31)$$

These wave functions are illustrated in Fig. 28.9. Because of the factor of x in Ψ_x, at any value of r the probability density $|\Psi_x|^2$ is greatest along the x axis. Similarly, $|\Psi_y|^2$ is greatest along the y axis, and $|\Psi_z|^2$ along the z axis.

EXERCISES ON SUPPLEMENTARY TOPICS

Section 28.7 | Masers and Lasers

28-23 The angular spread of a particular laser beam is 10^{-5} rad. What is the diameter of the spot formed on the moon's surface if the laser is directed toward the moon from the earth? (The earth-to-moon distance is 3.8×10^5 km.)

28-24 (a) If a laser emits 10 J of energy in a pulse lasting 5×10^{-11} s, what power is emitted? (b) What is the intensity of the beam if it is 2×10^{-6} m² in area?

28-25 What is the length of a laser pulse in a vacuum if it is emitted in 10^{-11} s?

28-26 A ruby laser emits light at 693.4 nm. If the energy released in each 10^{-11} second pulse is 0.1 J, how many photons are in a pulse?

Section 28.9 | Particle in a One-Dimensional Box

ᶜ**28-27** Verify that the wave function in Eq. 28.21 is normalized, so that Eq. 28.20 holds.

28-28 The discreteness of energy levels is unobservably small for macroscopic objects. Illustrate this point by finding the separation in joules of the $n = 1$ and $n = 2$ levels for a 1-g object in a 1-cm box.

28-29 In atomic nuclei, the separations between energy levels decrease as the nuclei increase in size. Consider a neutron trapped in a one-dimensional box. Find the difference in energy in MeV of the $n = 1$ and $n = 2$ states if the length of the box is (a) 3 fm; (b) 8 fm. (1 MeV $= 10^6$ eV; 1 fm $= 10^{-15}$ m; the neutron mass is listed on the inside rear cover. The lengths chosen correspond roughly to the sizes of carbon and lead nuclei.)

28-30 A model for an atom is an electron in a one-dimensional box of length 10^{-10} m. (a) What is the energy difference in electron volts between the ground and first excited states of this system? (b) Find the ratio of the calculated energy difference to the actual energy difference in atomic hydrogen.

Section 28.10 | The Square Well; Tunneling

28-31 The heavy hydrogen or deuterium nucleus consists of one proton and one neutron. Their mutual attraction is just barely strong enough so that there is one bound state. A simple model for this system is a neutron moving in a square well of length 1 fm $= 10^{-15}$ m. Assuming that the square well energy levels are the same as in a box, what is the minimum well depth in MeV that will permit a bound state? (1 MeV $= 10^6$ eV; the neutron mass is listed on the inside rear cover.)

Section 28.12 | Explicit Form of the Hydrogen Wave Functions

ᶜ**28-32** Verify that the 1s state hydrogen atom wave function is normalized.

ᶜ**28-33** Verify that the 2p, $m = 0$ hydrogen atom wave function is normalized.

ᶜ**28-34** Find the radius at which the radial probability density $r^2 P$ is a maximum for the 2s state of hydrogen.

ᶜ**28-35** (a) Find the radius at which the radial probability density $r^2 P$ is a maximum for the 2p states of hydrogen. (b) At what radius is P largest for these states?

28-36 What is the radial probability density for a hydrogen atom 2s electron at its classical turning point?

ᶜ**28-37** (a) Show that since the atomic nucleus is very small compared to the Bohr radius, the wave function of a hydrogen atom 1s state electron is essentially constant at its $r = 0$ value within the nucleus. (b) If the nuclear radius is 1 fm $= 10^{-15}$ m, what is the probability of finding a 1s hydrogen atom electron within the nucleus?

PROBLEMS ON SUPPLEMENTARY TOPICS

ᶜ**28-38** A particle of mass m moves in one dimension. It is attached to a spring with spring constant k, so its potential energy is $\frac{1}{2}kx^2$. (a) What is the Schroedinger equation for this system? (b) What boundary conditions must Ψ satisfy for $x \to \pm\infty$? (c) Show that

$$\Psi(x) = Ae^{-ax^2}$$

is a satisfactory solution of the Schroedinger equation if a is suitably chosen. (This is actually the ground state wave function.) (a) What is the energy corresponding to this wave function?

ᶜ**28-39** A particle of mass m is in a one-dimensional box with its walls at $x = -a$ and $x = +a$. (a) What are the boundary conditions the wave functions must satisfy? (b) Show that both sine and cosine solutions are acceptable wave functions if k has suitable values. (c) Find the normalized wave functions and the corresponding energies.

ᶜ**28-40** A particle in a one-dimensional box extending from 0 to a is in its ground state. Calculate the probability of finding it in the region $0 < x < a/4$. [*Hint:* $\int \sin^2 bx \, dx = x/2 - (\sin 2bx)/4b$]

c28-41 Find the total probability that the electron in a hydrogen atom 1s state is at a radius greater than the maximum radius permitted by classical physics.

c28-42 Find the probability that a 2s electron in a hydrogen atom is at a radius smaller than the Bohr radius. [*Hint:* integrate the probability density over the appropriate volume.]

c28-43 The average value of r^n is denoted by $<r^n>$. If an electron has a wave function Ψ, then $<r^n> = \int r^n |\Psi|^2 dV$. For the 1s state in hydrogen, find the average values of (a) r; (b) $1/r$. (c) Discuss the reasons why $1/<r> \neq <1/r>$.

c28-44 The average value of z^2 is $<z^2> = \int z^2 |\Psi|^2 dV$, and the average value of r^2 is $<r^2> = \int r^2 |\Psi|^2 dV$. Evaluate both of these quantities for the 2s state in hydrogen.

c28-45 Verify that the $2p_x$ hydrogen atom wave function is normalized. (Use $|a + ib|^2 = a^2 + b^2$.)

c28-46 Repeat the previous problem for the 2p, $m = 1$ state.

Additional Reading

R. Riuker, The Safe Use of Lasers, *Physics Teacher,* vol. 11, 1973, p. 455.

H. Weichel, W.A. Danne, and L.S. Pedroth, Laser Safety in the Laboratory, *American Journal of Physics,* vol. 42, 1974, p. 1006.

J.M. Coakley, Probability of Laser Injury to the Eye—Further Considerations, *Physics Teacher,* vol. 13, 1975, p. 388.

Arthur L. Schawlow, Lasers and Physics: A Pretty Good Hint, *Physics Today,* December 1982, p. 46.

Phillip Sprangle and Timothy Coffey, New Sources of High-Power Coherent Radiation, *Physics Today,* March 1984, p. 44. Free-electron lasers and cyclotron-resonance masers.

Mitja Kregar and Victor Weisskopf, Ionization Energies and Electron Affinities of Atoms up to Neon, *American Journal of Physics,* vol. 50, March 1982, p. 213.

James J. Wayne, Current Trends in Atomic Spectroscopy, *Physics Today,* November 1982, p. 52.

Bernd Crasemann and Francois Weilleumeir, Atomic Physics with Synchrotron Radiation, *Physics Today,* June 1984, p. 34.

Scientific American articles:

Leslie Holliday, Early View on Forces Between Atoms, May 1970, p. 116.

H.R. Crane, The g Factor of the Electron, January 1968, p. 77.

Vernon W. Hughes, The Muonium Atom, April 1966, p. 93.

O.R. Frisch, Molecular Beams, May 1965, p. 58.

William C. Livingston, Magnetic Fields on the Quiet Sun, November 1966, p. 54.

J.P. Gordon, The Maser, December 1958, p. 42.

Arnold L. Bloom, Optical Pumping, October 1960, p. 72.

Arthur L. Schawlow, Optical Masers, June 1961, p. 52.

Arthur L. Schawlow, Advances in Optical Masers, July 1963, p. 34.

S.E. Miller, Communication by Laser, January 1966, p. 19.

George C. Pimentel, Chemical Lasers, April 1966, p. 32.

Alexander Lempick and Harold Samelson, Liquid Lasers, June 1967, p. 80.

Donald F. Nelson, The Modulation of Laser Light, June 1968, p. 17.

C.K.N. Patel, High Power Carbon Dioxide Lasers, August 1968, p. 22.

Arthur L. Schawlow, Laser Light, September 1968, p. 120.

Donald R. Herriott, Applications of Laser Light, September 1968, p. 141.

Peter Sorokin, Organic Lasers, February 1969, p. 30.

M.W. Berus and D.E. Rounds, Cell Surgery by Laser, February 1970, p. 98.

M.S. Field and V.S. Letokhov, Laser Spectroscopy, December 1970, p. 69.

M.B. Parish and I. Hayashi, A New Class of Diode Lasers, July 1971, p. 32.

Richard N. Zare, Laser Separation of Isotopes, February 1977, p. 86.

Daniel Kleppner, Michael G. Littman, and Myron L. Zimmerman, Highly Excited Atoms, May 1981, p. 28.

R.I.G. Hughes, Quantum Logic, October 1981, p. 202.

Aldo V. La Rocca, Laser Applications in Manufacturing, March 1982, p. 94.

Jearl Walker, The Spectra of Streetlights Illuminate Basic Principles of Quantum Mechanics, *The Amateur Scientist,* January 1984, p. 138.

Isaac F. Silvera and Jook Walraven, The Stabilization of Atomic Hydrogen, January 1982, p. 66.

Marie-Anne Bouchiat and Lionel Pottier, An Atomic Preference Between Left and Right, June 1984, p. 100.

CHAPTER 29

THE STRUCTURE OF MATTER

Molecules and bulk matter are aggregates of atoms held together by electrical forces. If the electrons and nuclei of a molecule have a lower energy than that of the separate neutral atoms, the molecule will be stable. Similarly in metals and semiconductors, the energy of the aggregate is less than that of its isolated constituents. The problem of understanding the structure of matter is one of finding the arrangement that has the lowest energy or greatest binding energy.

Fortunately, it is not necessary to solve Schroedinger's equation for all the electrons and nuclei of a system, because this can only be done approximately and is generally difficult. Instead, we use simplified models based on our knowledge of atoms that are very successful in accounting for the structure and properties of molecules and bulk materials. There are three models of primary interest to us here, which describe *ionic, covalent,* and *metallic* binding, respectively. Weaker bonds such as *van der Waal's* and *hydrogen* bonds are also important where the stronger bonding mechanisms are not effective.

29.1 | IONIC BINDING

When some molecules and crystals are formed, one or more electrons from one atom are completely transferred to another atom. The atom losing the electrons acquires a net positive charge, and the atom accepting these electrons becomes negatively charged. These two ions are then held together by the electric forces between them. This *ionic binding* usually involves one atom with one or more loosely bound electrons and a second with a nearly full outer shell. For

example, the alkali metals, such as sodium and potassium with one valence electron, readily combine with the halides, such as chlorine and fluorine, that have one vacancy in their outer shells. This type of binding is responsible both for the formation of single molecules and of bulk crystals.

Ionic binding is illustrated by the ionic molecule potassium chloride, KCl. In order to understand why it is energetically favorable for potassium and chlorine atoms to form KCl molecules, we break up the formation of the molecule into three stages. These are the removal of an electron from the potassium atom, its binding to the chlorine atom, and the attraction of the resulting K^+ and Cl^- ions. These steps are not necessarily followed in the actual formation of the molecule.

1 The outer electron of potassium is removed leaving the K^+ ion and an electron. This requires an energy equal to the ionization energy of potassium which is 4.34 eV.

2 The extra electron binds to the neutral chlorine atom yielding Cl^-. This releases energy called the energy of formation or *electron affinity*. For chlorine this is 3.82 eV. The entire process of transferring one electron from potassium to chlorine then requires that $(4.34 - 3.82) = 0.52$ eV of energy be supplied.

3 The oppositely charged ions approach each other. Because of their mutual attraction, their potential energy initially decreases as they approach. However, when they are so close that their electron clouds begin to overlap, there is a

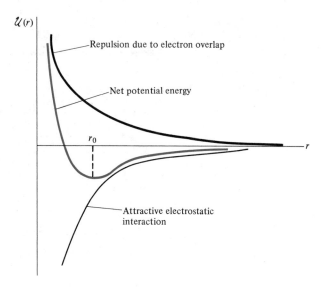

Figure 29.1. The potential energy of two ions versus the distance between their centers. The repulsive part due to the overlap of core electrons, which is positive, plus the attractive electrostatic part, which is negative, equals the net potential energy (shown colored). This energy is a minimum at a spacing r_0.

repulsive force and the potential energy begins to increase (Fig. 29.1). The observed equilibrium separation of the ions in KCl is $r_0 = 2.79 \times 10^{-10}$ m. The electrical potential energy at this separation is

$$E = -\frac{ke^2}{r_0}$$

$$= -(9 \times 10^9 \text{ N m}^2 \text{ C}^{-2}) \frac{(1.6 \times 10^{-19} \text{ C})^2}{2.79 \times 10^{-10} \text{ m}}$$

$$= -8.26 \times 10^{-19} \text{ J} = -5.16 \text{ eV}$$

Ignoring other contributions to the potential energy momentarily, this means that the molecule has an energy of $(0.52 - 5.16)$ eV $= -4.64$ eV compared to the separate neutral atoms. Conversely, 4.64 eV of energy would be required to break up the molecule into neutral potassium and chlorine. The measured value is 4.40 eV. The difference in the two results can be attributed to the repulsion of the inner electron clouds.

In ionic crystals the ions form a lattice. A number of crystal structures are possible, but the common feature of all of them is that ions of one charge have closest neighbors that are ions of the opposite charge. The electric attraction between neighbors is responsible for the binding of the crystals.

29.2 | COVALENT BINDING

The vast majority of molecules are formed by sharing the outermost or *valence* electrons of the constituent atoms. In such *covalent* binding, the distribution of electrons in the molecule may be quite different from that in the separated atoms. These electronic distributions and other molecular properties are studied by interpreting detailed experimental results with the aid of theoretical models. These models rely on two important concepts described in the preceding chapter.

1 The square of a wave function $|\psi|^2$ indicates the relative probability of an electron being at a particular location. Equivalently, $|\psi|^2$ determines how often an electron is at a given place. When two atoms share an electron, it effectively belongs to each atom part of the time. Thus we expect that in molecules the wave functions of the electronic states should overlap so that the shared electron is favorably positioned with repect to both atoms.
2 When the atoms combine, the Pauli principle requires that each spatial state contains at most two electrons, one with spin up and one with spin down.

We can illustrate these ideas with the diatomic hydrogen molecule, H_2. Neutral hydrogen atoms have one electron in the $1s$ shell, although this state could hold two electrons if their spins were opposite (Fig. 29.2a). When two such atoms move together, one of two things will occur. If the electron spins are parallel, the Pauli principle prevents the electrons from being very close together since they cannot be in the same spatial state. The electrons repel each other and their wave functions are distorted (Fig. 29.2b). Because the electrons are far from each other most of the time, the nuclear charges also repel each other, and the whole arrangement is energetically unfavorable.

On the other hand, if the spins are opposite each electron "fits" into the $1s$ state of the other atom. The electrons can now be close to each other, and the molecular wave function looks like the sum of the two $1s$

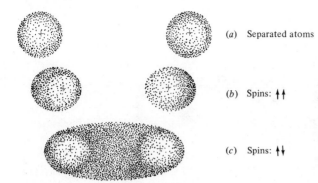

(a) Separated atoms

(b) Spins: ↑↑

(c) Spins: ↑↓

Figure 29.2. Representation of $|\psi|^2$ for the $1s$ states of two hydrogen atoms when they are far apart. (a) Contours representing the electron probability. (b) When the electrons have parallel spins, they repel each other because of the exclusion principle and the molecule does not form. (c) When the spins are opposite the two $1s$ electronic wave functions add. In a sense, each atom has a filled $1s$ shell. The energy of the H_2 molecule is lower than that of two separate neutral atoms and the molecule is stable.

wave functions of each atom (Fig. 29.2c). The electrons have a high probability of being between the two nuclei where they attracted by both protons. This reduces the total energy of the molecule below that of the separated atoms and results in a stable molecule. The characteristic elongated wave function of Fig. 29.2c is called a σ (sigma) *orbital*. This bond is said to be a σ bond.

Hybridization | An approximate approach to the structure of complex covalent molecules that has proved very useful is called *hybridization*. This method employs wave functions that are combinations or *hybrids* of atomic states, such as the $3s$ and $3p$ states of the atoms, that have slightly different energies. These energy differences can usually be ignored in molecules because they are small compared to the molecular binding energy. To illustrate the method of hybridization we describe the scheme for magnesium fluoride, MgF_2.

Magnesium ($Z = 12$) has closed $n = 1$ and $n = 2$ shells, and two $3s$ valence electrons; its six $3p$ levels are empty. Fluorine ($Z = 9$) is one electron short of having a filled $n = 2$ shell and hence attracts electrons. Since the outer $3s$ wave functions of magnesium are spherically symmetric, their overlap with the $2p$ wave functions of fluorine is small (Fig. 29.3a). However, using combinations of magnesium wave

functions, the overlap can be greatly increased, resulting in stronger binding (Fig. 29.3b, c).

We saw in the last chapter that the three p wave functions have pairs of lobes oriented along coordinate axes. If a p wave function and an s wave function are added and subtracted, the result is a pair of sp *hybridized* wave functions, or *orbitals*, each with a large lobe in one direction and a small lobe opposite (Fig. 29.3b). The two magnesium valence electrons can be thought of as being in these states, which overlap well with the single vacant p state of a fluorine atom. Consequently MgF_2 is a linear molecule with the magnesium atom between two fluorine atoms. Note that the sp orbital is a mixture of two atomic orbitals with slightly different energies, so a free magnesium atom would not ordinarily be found in such a state. However, the energy needed to occupy this sp orbital is much smaller that that gained in the formation of the molecule, so in the molecule it is energetically favorable for the electrons to be in this configuration.

In sp^3 *hybridization*, we consider combinations of one s state and all three p states. We illustrate this procedure for the water molecule H_2O. Oxygen ($Z = 8$) has two $2s$ electrons and four $2p$ electrons in its $n = 2$ shell. Thus it is two electrons short of having a closed $n = 2$ shell. Figure 29.4a shows the four sp^3 orbitals that can be formed from suitable combinations of the s and p states of oxygen. These states have a maximum angular separation, 109.5°. When occupied by electrons, this will minimize their repulsion.

Each sp^3 orbital can accommodate two electrons with opposite spins. In the water molecule, four of the $n = 2$ oxygen electrons fill two of the sp^3 orbitals. Each of the other two orbitals has one electron from the oxygen and one from a hydrogen atom (Fig. 29.4b and 29.4c).

A reminder that the hybridization model for water is not exact is provided by the experimental bond angle, 104.5°, which is slightly less than the predicted angle of 109.5°. The difference is due to the effects of electrical forces among the bond electrons and the hydrogen nuclei in the asymmetric water molecule.

The structure of the water molecule makes it a *polar* molecule, one with a permanent electric dipole moment. The oxygen nucleus attracts the bond electrons somewhat more strongly than do the hydrogen nuclei. The result is an effective excess of positive charge near the hydrogen nuclei and an excess of

(Wide World.)

LINUS CARL PAULING
Born: 1901

Pauling belonged to the generation that developed the methods of quantum physics and applied them to atoms and molecules. After receiving a B.S. in chemical engineering from Oregon State College, Pauling obtained his Ph.D. in Chemistry at the California Institute of Technology in 1925. He has been a professor there and at Stanford University since 1931.

By the mid-1920s the quantum theory of atoms was relatively well developed, and enough was known so that one could try to understand molecules and molecular binding. The models then existing depicted molecules as composed of atoms containing fixed charges, somewhat like knobs on the surface of a spherical nucleus. Atoms combined when these knobs fit into place, held together by the electric attraction between the various electrons and nuclei. This model was unsatisfactory for several reasons and was completely incompatible with the wave-particle description of electrons in atoms.

From 1928 to 1932, Pauling published a series of papers in which he developed the quantum mechanical theory of molecular bonding. He was responsible for many of the ideas that are regarded today as fundamental. These included the hybridization scheme described in this chapter, the idea of resonance bonds such as those in benzene (see Section 29.10), and the correlation of interatomic distances with the electronic structure of molecules. The culmination of this work was the publication of *The Nature of the Chemical Bond* in 1939, a book that is still required reading for anyone doing research on the subject. Pauling received the Nobel prize in chemistry for this work.

In the mid-1930s Pauling turned his interest to biochemistry. He was the first to suggest and analyze the helical structure of some proteins. His work at the molecular level led to the idea that sickle-cell anemia is caused by a genetic defect in hemoglobin and also to a molecular theory of anesthesia.

Since World War II, Pauling has been an outspoken critic of nuclear testing and of the large nuclear arsenals of the major powers. This ongoing activity resulted in his receiving the 1963 Nobel peace prize.

Pauling, who once picketed in front of the White House during the afternoon and then dined inside with the president and other Nobel Laureates in the evening, has had a profound effect on physics, physical chemistry, and biochemistry. In his efforts for nuclear disarmament, Pauling has become a thoughtful and important advocate.

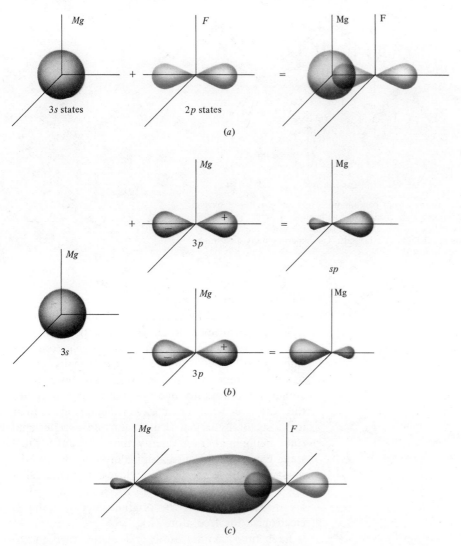

Figure 29.3. (*a*) The 3*s* wave functions in magnesium and the 2*p* in fluorine have a small overlap area. (*b*) Adding and subtracting 3*s* and 2*p* magnesium wave functions produces *sp* wave functions with large lobes. (*c*) The *sp* wave function has a large overlap with the 2*p* fluorine wave function. In the MgF_2 molecule, a second fluorine atom binds on the left with the other *sp* wave function, forming a linear, symmetric molecule.

negative charge (electrons) near the oxygen nucleus. This permanent electric dipole moment plays an important role in the properties of water. For example, water as a solvent breaks ionic salts up into ions, as in $NaCl \rightarrow Na^+ + Cl^-$ (Fig. 29.5).

Carbon | Carbon ($Z = 6$), because it has only four out of a possible eight $n = 2$ electrons, can par-

ticipate in covalent bonds in a variety of ways. When the *s* and *p* states are combined to form four *sp*^{3} orbitals, carbon can, for example, bond with four hydrogen atoms forming methane CH_4. Each orbital has an electron from carbon and a second from a hydrogen atom. The carbon wave functions can also be hybridized to form two *sp* orbitals or three *sp*^{2} orbitals. The *sp*^{2} orbitals are used for molecules such as ethylene $CH_2{=}CH_2$ (Fig. 29.6).

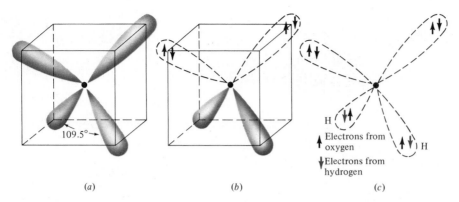

Figure 29.4. (a) The four sp^3 orbitals formed from combinations of s and p states of oxygen. (b) Two of the orbitals are full, containing a pair of electrons with opposite spins. The remaining two orbitals each have one of the oxygen electrons. (c) In the water molecule the hydrogen atoms provide the electrons to fill these two orbitals. The experimentally determined angle between the two hydrogen bonds is 104.5°.

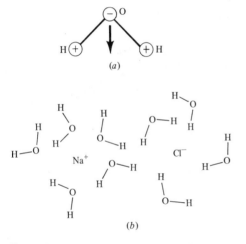

Figure 29.5. (a) The water molecule is electrically neutral, but the electrons are closer to the oxygen nucleus. The result is a permanent electric dipole in the molecule. Water acts as a solvent because its dipole field can weaken and break the ionic bond in molecules such as NaCl. (b) The water molecules tend to surround the dissociated ions so that they are nearly electrically neutral and do not recombine.

29.3 | THE METALLIC BOND

Covalent bonds occur in molecules containing two to several thousand atoms. By contrast, metallic binding holds billions of atoms together. Usually these substances are solid, with ionized atoms forming a rigid lattice and electrons contributed by the atoms that are free to move over the entire crystal. These delocalized electrons are the charge and energy carriers in metals and are responsible for their large electrical and thermal conductivities.

To see how the electrons become delocalized, consider lithium atoms, which have one $2s$ electron outside a filled $1s$ shell. This electron can have its spin up or down. When two lithium atoms are brought together, there are only two possible energy levels for the outer electrons. Their wave functions are similar to those for hydrogen (Fig. 29.2), the parallel spin state being higher in energy than the antiparallel state. Now, when a third atom is brought close to the first two, three separate energy levels become available for the outer electrons (Fig. 29.7). The energy levels are clustered near the original energy of the $2s$ electron in the free atom (Fig. 29.8).

Each time another atom is added, a new energy level results. However, the net effect of each new atom becomes smaller as the number of atoms already close together increases. Finally, with N atoms present, we have a *band* of N closely spaced energy levels. This band can accommodate $2N$ electrons. For example, for $N = 3$, there are three energy levels and six states (Fig. 29.8). The electronic wave functions are spread over large distances and the electrons in this *conduction band* are free to move over the entire crystal.

Metals are good electrical and thermal conductors because of the presence of the conduction electrons. There are many unoccupied states in the conduction band into which electrons can move with only a small change in energy. Thus, if a metal is heated at one end, electrons in this region move into higher energy

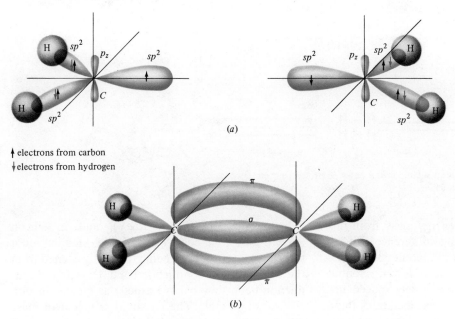

electrons from carbon

electrons from hydrogen

Figure 29.6. (*a*) The ethylene molecule $CH_2 = CH_2$ is formed by attaching two hydrogen atoms along two of the three sp^2 orbitals of each carbon atom and then (*b*) bringing the two complexes together. The σ bonds are formed along the C—C and C—H directions, and a bond is also formed by the overlap of the p_z wave functions of the two carbon atoms. The overlapping p_z states of the two complexes are called a π bond.

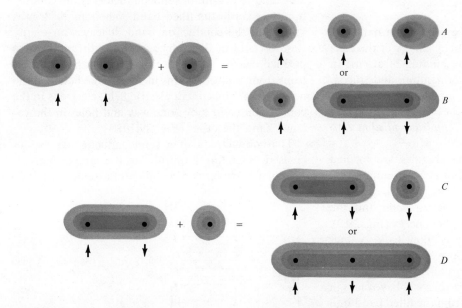

Figure 29.7. Two lithium atoms have their outer electron spins either parallel (above left) or antiparallel (below left). When a third atom is brought near, configurations like those at the right result. We expect that the configurations represented schematically as *B* and *C* would have the same energy. Reversing all the spins in *A* or *D* produces two more configurations *A′* and *D′* with the same energies as *A* and *D*. Thus the picture suggests three energy levels, each having two states. In fact this picture is oversimplified. Combining the wave functions of several particles is complicated and this diagram is only suggestive. Extending this type of diagram to four or more particles can lead to erroneous conclusions.

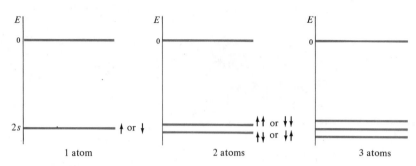

Figure 29.8. The lowest energy levels available to the outer electrons from one, two, and three lithium atoms placed close together.

states and transport the heat energy away from this region. Similarly, a small voltage difference along a metal will lead to an increase in the energy of electrons, which then move to regions of lower potential energy. It is the presence of the closely spaced unfilled energy levels of the band that facilitates these conduction processes. In a filled band in a semiconductor or insulator, electrons cannot easily change their states and conduction is much reduced.

29.4 | INSULATORS AND SEMICONDUCTORS

In metals the formation of the conduction band, resulting in a decrease in the kinetic energy of the electrons, plays a vital role in the binding of atoms. In materials such as diamond, germanium, and silicon, bands are also formed when a large number of atoms are assembled. However, in these cases the valence electrons form a *band that is completely filled* at zero temperature.

Additional bands at higher energies are formed from higher atomic orbitals, but at zero temperature these are completely empty. There is an *energy gap* between the fully occupied *valence band* and the empty conduction band at a higher energy (Fig. 29.9).

In diamond the energy gap is about 6 eV. A temperature of 300 K, which is approximately room temperature, implies a mean thermal energy of $\frac{3}{2}k_B T = 0.04$ eV. Thus, virtually no electrons will have enough thermal energy to be excited from the filled band into the conduction band at any reasonable temperature. The motion of electrons due to electrical forces or to temperature gradients requires that they increase their kinetic energy. Since the electrons in the filled band cannot change energy states, conduction is impossible and diamond is an excellent insulator.

In germanium $(E_g = 0.72$ eV), and in silicon $(E_g = 1.1$ eV), the energy gap is much smaller and a few electrons succeed in being thermally excited from the valence to the conduction band. Once in the conduction band they behave exactly as the conduction electrons in metals. The resistivity is higher in these *semiconductors* than in metals, because the number of electrons free to move is much smaller than in metals. For example, the room temperature resistivities of metals, semiconductors, and insulators are on the order of 10^{-8}, 10^2, and 10^{13} ohm m, respectively.

A unique feature of semiconductors is the conduction by *holes* in the filled band. When an electron is excited to the conduction band, it leaves an empty state or hole in the valence band with an effective positive charge. If an electric field is applied, electrons from adjacent atoms jump into the hole and the hole moves. Thus, in an electric field, electrons in the conduction band move one way and holes in the valence band the other (Fig. 29.10).

The conductivity of a semiconductor can be increased by adding impurities to the sample. For example, if a few parts per million of arsenic are added

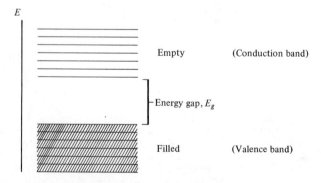

Figure 29.9. Energy bands of insulators and semiconductors.

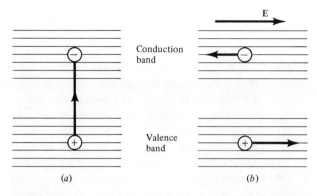

Figure 29.10. (a) An electron excited from the filled valence to the conduction band leaves a positively charged hole behind. (b) In an electric field both move, producing a net current.

to germanium, the conductivity increases a thousand-fold. Arsenic has one more electron in its outer shell than does germanium. When added as an impurity, it replaces a germanium atom, and its extra electron is freed for conduction. Similarly, gallium has one less outer electron than germanium, and when present as an impurity, it takes one electron from germanium, leaving a hole.

Semiconductors have many practical uses. A semiconductor can serve as a photocell for use in photography or in automatic door openers. The quanta of visible light have energies of about 2 or 3 eV, more than enough to excite electrons from the valence to the conduction band. In an electric field the semiconductor current increases greatly when the material is exposed to light and the number of charge carriers increases. Different types of semiconductors in contact are used in a wide variety of transistors and other devices. These devices have all but replaced the older vacuum tubes in electronics. They are desirable because they have low power requirements, no heating problems, are rugged, and last almost indefinitely.

29.5 | WEAKER BONDS

The bonds we have discussed typically have energies of a few electron volts. However, two bonds with energies of at most one-tenth of an electron volt play an important role in nature: the *van der Waals bond* and the *hydrogen bond*. These bonds are especially important when the stronger types of bonds are not present. For example, the inert gases such as helium, neon, argon, and krypton condense to liquids and solids only because of attractive van der Waals forces

among their atoms. Hydrogen bonding is often responsible for important structural features of molecules and solids such as the helical and pleated sheet structures of some organic molecules. In some instances, both types of bonding are present simultaneously.

van der Waals Attraction | The van der Waals attraction of molecules or atoms for each other is electric in nature just as are other types of bonds. However, this attraction is due to the presence of electric dipoles, either permanent or induced.

Molecules such as water have a permanent electric dipole moment corresponding to an effective separation of charge within the molecule (Fig. 29.5). As a result these polar molecules tend to cluster together (Fig. 29.11). This clustering is opposed by the normal thermal motion of the molecules. The motion diminishes as the temperature decreases, so the van der Waals attraction becomes more effective and may cause condensation or solidification.

A polar molecule can induce an electric dipole moment in a nearby nonpolar molecule. The positive end of the polar molecule repels the nucleus and attracts the electrons of the nonpolar molecule. Once this induced dipole moment is formed, the two molecules attract each other just as do two polar molecules.

A van der Waals attraction even occurs between nonpolar atoms and molecules. Although the centers of positive and negative charge are coincident on the average in a nonpolar atom, the electrons are constantly moving. At any instant the atom may have a temporary electric dipole moment in some random direction. This moment will induce a dipole moment in a neighboring atom, and the two dipoles are always aligned, so there is a net attraction. On the aver-

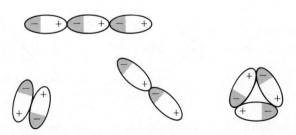

Figure 29.11. Polar molecules tend to cluster together due to the electric attraction of the positive and negative ends of neighboring molecules.

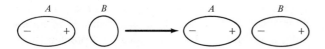

Figure 29.12.   The instantaneous presence of a dipole moment in molecule *A* induces a dipole moment in molecule *B*, and the two nonpolar molecules attract one another.

age then there is a net van der Waals attraction (Fig. 29.12).

Van der Waals forces are responsible for the attraction between inert gas atoms and for the behavior of liquids in capillary tubes. For example, water rises in a glass tube because the van der Waals attraction between the water and glass molecules is stronger than that among the water molecules themselves.

The Hydrogen Bond | The hydrogen atom
plays a unique role in compounds because of its small size and its single electron. A classic example of a hydrogen bond (H-bond) is found in ice. In liquid form, the primary attraction among water molecules is due to van der Waals forces among the polar molecules, with H-bonding playing a secondary role. However, in the solid phase the H-bond is responsible for the specific structure of the ice crystal.

In water, two of the hybrid orbitals of oxygen hold paired spin electrons and do not participate in the binding of the molecule (Fig. 29.13). We refer to them

as nonbonding orbitals. Since the bonding electrons are mainly between the oxygen and hydrogen nuclei, the region near the hydrogen nucleus is electrically positive. Each nonbonding orbital has a pair of electrons that is attracted to the positive hydrogen nucleus on an adjacent molecule. The hydrogen nucleus then is positioned with two electrons on either side (Fig. 29.13*b*). In ice, we find the hydrogen nuclei situated between the bonding and nonbonding orbitals of adjacent molecules. This is the hydrogen bond.

Hydrogen bonding plays an important role in the spatial arrangement of atoms in large organic molecules. For example, in DNA there are two helical strands of alternate sugars and phosphate groups. Each sugar residue has a side chain, and the side chain strands are hydrogen bonded to each other. Thus the strands of the double helix have hydrogen bonds as their weakest link (Fig. 29.14).

SUMMARY
Molecules and bulk matter represent energetically favorable configurations of electrons and nuclei held together by electric forces. Ionic bonds form between atoms that have completely transferred an electron. The electric potential energy gained by having the ions close together more than compensates for the energy required to transfer the electron.

In covalent bonds, electrons are shared. In general, this sharing can be understood by examining the

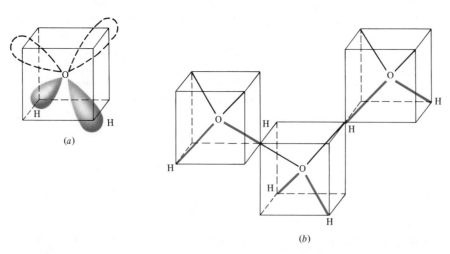

Figure 29.13.   (*a*) The water molecule. Each orbital has a pair of electrons. The colored orbitals form the covalent bond with the hydrogen atoms. (*b*) In the crystal, water molecules join so that a hydrogen atom from one molecule connects to a nonbonding orbital (represented by a single black line) of an adjacent molecule. Bonding orbitals are shown as colored lines.

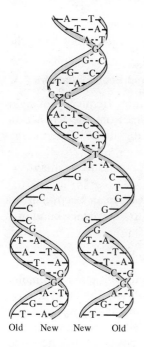

Old New   New   Old

Figure 29.14. The DNA structure during replication A, T, G, and C label side chain components attached to opposing side chains by hydrogen bonds as shown by the dashes. The double helix separates and rejoins at the hydrogen bonds.

electronic structure of the constituent atoms. The wave functions of the atoms should overlap well, so that the electrons are favorably positioned with respect to the nuclei. Usually, such bonds contain two opposite spin electrons satisfying the Pauli principle. The pairs of bonding electrons tend to be as far apart as possible so the electric repulsion between electrons in different bonds is minimized. A model of the molecular wave function treats the constituent atoms as having hybrid electronic states in covalent bonding.

In metals the electrons become the common property of the metal as a whole. They are completely delocalized. In semiconductors in their ground state, the electrons remain bound to the nuclei, although they may be shared by neighboring atoms. An electron must increase its energy by at least a minimum amount to move into a conduction band. That is, excitation of an electron is equivalent to taking it away from an atom and allowing it to be free to move throughout the sample, just as in a metal. This excitation can be produced by heating or by the absorption of light.

Weak hydrogen bonds and van der Waals bonds often act between or within molecules. The van der

Waals bond, which is caused by the mutual attraction of permanent or induced electric dipole moments, is responsible for the existence and properties of many liquids and solids. Hydrogen bonds are formed when a covalently bonded hydrogen atom attracts the electrons in a nearby orbital. This weak bond is responsible for much of the detailed structure in some solids and many organic molecules.

Checklist
Define or explain:

ionic bond	conduction band
electron affinity	valence band
covalent binding	insulator
Pauli principle	semiconductor
σ orbital	energy gap
hybridization, *sp*, *sp²*,	holes
sp³	van der Waals forces
directional *p* state	hydrogen bond
electric dipole moment	

REVIEW QUESTIONS

Q29-1   In _____ binding, one or more electrons is transferred from one atom to another.

Q29-2   In _____ binding, electrons are shared by the atoms.

Q29-3   If electron spins are parallel, the _____ prevents them from being very close to each other.

Q29-4   A combination of 3*s* and 3*p* wave functions is called a _____ wave function.

Q29-5   A polar molecule has a _____.

Q29-6   Electrons in a conduction band in a metal can move _____.

Q29-7   In an insulator, the valence electrons form a band that is _____.

Q29-8   Semiconductors have much smaller _____ than insulators.

Q29-9   A hole is a _____.

Q29-10   When a polar molecule is near a nonpolar molecule, it can induce an _____.

EXERCISES

Section 29.1 | Ionic Binding

29-1   The ionization energy of neutral sodium (Na) is 5.12 eV, and the electron affinity of chlorine

(Cl) is 3.82 electron volts. (Neglect the repulsions of the core electrons.) (a) How much energy is required to transfer one electron from Na to Cl? (b) If the ionic separation in the molecule NaCl is 2.36×10^{-10} m, what is the electric potential energy of the molecule? (c) What is the predicted energy required to separate NaCl into its separate neutral constituents, Na and Cl?

29-2 The ionization energy of lithium atoms is 5.39 eV, and the interatomic distance in the lithium fluoride molecule, LiF, is 1.51×10^{-10} m. The electron affinity of fluorine is 3.51 eV. (Neglect the repulsions of the core electrons.) (a) What is the potential energy of the ions Li^+ and F^- in the molecule? (b) What is the binding energy of LiF?

29-3 The binding energy of sodium bromide (NaBr) is 3.77 eV, the interatomic spacing is 2.50×10^{-10} m, and the ionization energy of Na is 5.12 eV. (Neglect the repulsions of the core electrons.) (a) What is the potential energy of the ions Na^+ and Br^- in the molecule? (b) What is the electron affinity of bromine?

Section 29.2 | Covalent Binding

29-4 Describe how magnesium and chlorine form the molecule $MgCl_2$.

29-5 Does the molecule H_2 have a permanent electric dipole moment? Explain.

29-6 The distance between the protons in a singly ionized H_2^+ molecule is 1.06×10^{-10} m. If the remaining electron is midway between the protons, what is its electric potential energy?

29-7 Describe the formation and structure of the methane molecule CH_4 using the sp^3 hybridization scheme for carbon.

29-8 What is the theoretical angle between the hydrogen-carbon bonds at one end of the ethylene molecule?

Section 29.3 | The Metallic Bond

29-9 When N atoms are brought together to form a certain material there are $2N$ electronic energy-levels formed in a band. If each atom contributes one valence electron to the band the band will be half filled and the material will behave like a metal. (a) Describe the occupation of the band for another material in which the atoms contribute two valence electrons. (b) How might the electrical and thermal conduction processes of this second material compare with those of the metal? Explain.

29-10 If the conduction band electrons are free to move about in the metal, why don't they leave the metal entirely?

Section 29.4 | Insulators and Semiconductors

29-11 What is the minimum frequency of light necessary to excite electrons from the valence to the conduction band in pure silicon? (The energy gap is 1.1 eV.)

29-12 Explain how a semiconductor might be used as a thermometer.

29-13 Is energy conserved when an electron drops from the conduction band to the valence band in a semiconductor? Explain.

Section 29.5 | Weaker Bonds

29-14 In Fig. 29.12, molecule A is polar and molecule B is nonpolar. Molecule A induces an electric dipole moment in B. Explain qualitatively why the net force between the molecules is attractive. Include the effects of the positive and negative charges of both molecules in your discussion.

29-15 Draw a schematic diagram of the boundary between a horizontal water surface and a vertical piece of glass. Describe qualitatively in terms of the van der Waals forces the shape of the water surface near the boundary.

29-16 Figure 29.15 shows a methane molecule. Would you expect that hydrogen bonding is as important in solid methane as it is in ice? Explain.

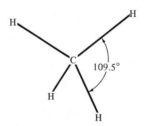

Figure 29.15. The structure of the methane molecule (Exercise 29-16).

PROBLEMS

29-17 The electric dipole moment of a molecule is the charge separation distance times the charge itself. What is the electric dipole moment of KCl? (The separation of the K^+ and Cl^- ions is 2.79×10^{-10} m.)

29-18 The neutral molecule H_3 is not stable although H_3^+ is. Using the exclusion principle, explain these results.

29-19 The ammonia molecule (NH_3) is usually described by an sp^3 hybridization of the nitrogen wavefunctions. Starting from the fact that the neutral nitrogen atom has two $2s$ and three $2p$ electrons in its $n = 2$ shell, describe the structure of NH_3. Do you expect this molecule to have a permanent electric dipole moment? Explain.

29-20 The hybridization model is a highly simplified one that sometimes leads to results that are not in agreement with experiment. For example, hydrogen sulfide (H_2S) might be expected to have the same shape as H_2O, since oxygen and sulfur are both two electrons short of having filled outer shells. However, the bond angle in H_2S is 92°, which is quite different from the bond angle in H_2O of 104.5°. Show that H_2S may be expected to have 90° bond angles if the hydrogen atoms overlap with the pure p states of sulfur. Sulfur has two $3s$ and four $3p$ electrons.

*29-21 Two common forms of pure solid carbon are graphite and diamond (Fig. 29.16). (a) Describe the hybridization scheme appropriate to the two structures. (b) Graphite can be used as a lubricant, while diamond is extremely hard. Can this difference in properties be explained by their difference in crystal structure?

*29-22 In methane (CH_4) the carbon atom shares an electron from each of four hydrogen atoms. These four electrons would be sufficient to fill the $n = 2$ shell of carbon. Chlorine atoms need a p state electron to complete a shell. Carbon tetrachloride (CCl_4) is a covalently bound molecule.

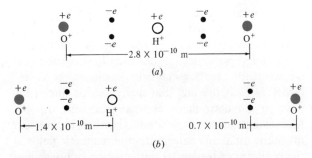

Figure 29.17. (*a*) A crude model of a hydrogen bond in ice. The oxygen nuclei (shown colored) have effective charges $+e = 1.6 \times 10^{-19}$ C. The hydrogen nucleus (open circle) is midway between with a charge $+e$. The electrons with charges $-e$ are midway between the oxygen and hydrogen nuclei. (*b*) The same charges grouped as shown but separated by a large distance (Problem 29-23).

(a) What is the structure of CCl_4? (b) Does this molecule have a permanent electric dipole moment? (c) How might the positions of the bonding electrons differ from those in methane?

*29-23 Figure 29.17*a* shows a crude model of a hydrogen bond. Compute the difference between the total electric potential energy of the charges in Fig. 29.17*a* and Fig. 29.17*b*. (Do not include the repulsion between electrons that are in the same side of the hydrogen nucleus).

ANSWERS TO REVIEW QUESTIONS

Q29-1, ionic; **Q29-2,** covalent; **Q29-3,** Pauli principle; **Q29-4,** hybridized; **Q29-5,** permanent electric dipole moment; **Q29-6,** over the entire crystal; **Q29-7,** completely filled; **Q29-8,** energy gaps; **Q29-9,** missing electron; **Q29-10,** electric dipole moment.

SUPPLEMENTARY TOPICS

29.6 | NUCLEAR MAGNETIC RESONANCE

A number of experimental techniques originally used by physicists are now in wide use by biochemists and others to study complex molecules. These studies yield information about the structure of molecules and their role in biological processes.

The experimental information necessary for any description of molecules includes the identification of the types of atoms present, their positions in the molecule, and the properties of the molecule as a whole. Usually the experimental evidence is acquired in bits

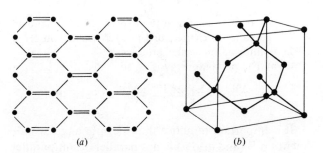

Figure 29.16. The positions of carbon atoms in (*a*) graphite and (*b*) diamond. The structure of graphite is nearly two dimensional, with each carbon bound to three neighbors in the same plane. In diamond, each carbon atom has four equidistant nearest neighbors (Problem 29-21).

and pieces and is very difficult to interpret, except in terms of models such as those of this chapter.

We now describe one of the most commonly applied and versatile techniques used to study molecular structure: *nuclear magnetic resonance or NMR*. NMR exploits the fact that many atomic nuclei behave in magnetic fields as small magnetic dipoles. Measuring the energy needed to reorient the nuclear magnetic moments determines the magnetic fields at the positions of the nuclei in the molecule For example, if an external magnetic field $\mathbf{B}_e$ is applied to a molecule, the actual magnetic field at a dipole is the external field plus the fields due to the electrons and nuclei in the neighborhood of the dipole. We call this the total field $\mathbf{B}_T$. The experiment is performed to measure the difference $\mathbf{B}_T - \mathbf{B}_e$. As we will see, this yields detailed information about the structure of the molecule. NMR is also used in a new type of imaging system (Sec. 31.5) that supplements ultrasonic scans and X rays.

NMR can be used on any molecule that has nuclei with nonzero magnetic dipole moments. The single proton in a hydrogen nucleus has such a moment, and the presence of hydrogen in so many organic molecules makes them excellent subjects for study. Our discussion in the following sections focuses on the origin of this proton magnetic moment and the methods of NMR.

29.7 | THE BEHAVIOR OF A MAGNETIC DIPOLE IN A MAGNETIC FIELD

In Section 19.5, we found that a current loop can be considered as a magnetic dipole. The magnetic dipole moment was found to be the current times the area of the loop. In the case of a particle moving in a circular orbit this magnetic dipole moment turns out to be proportional to the particle's orbital angular momentum.

A spinning charge also has a magnetic dipole moment $\boldsymbol{\mu}$ proportional to its spin angular momentum, $\mathbf{S}$. Experimentally the relationship between these quantities for a proton is found to be

$$\boldsymbol{\mu} = 2.79 \frac{e}{m_p} \mathbf{S} \qquad (29.1)$$

where m_p is the proton mass. In the remainder of this section, we describe how the energy of a proton depends on the magnetic field it experiences. We also

find that the proton moment experiences a torque due to the magnetic field. This torque causes the proton moment and spin to precess. This motion is quite analogous to that of a precessing top (Section 7.8).

We saw in Section 19.5 that the energy of a dipole at an angle θ to a magnetic field $\mathbf{B}_T$ is $\mathfrak{U} = -\mu B_T \cos\theta$. When the total magnetic field $\mathbf{B}_T$ is in the z direction, the potential energy of the dipole is $\mathfrak{U} = -\mu_z B_T$. If S_z is the z component of the spin angular momentum, this energy is

$$\mathfrak{U} = -2.79 \frac{e}{m_p} S_z B_T \qquad (29.2)$$

For the proton $S_z = +\hbar/2$ or $-\hbar/2$. It is conventional to define the *nuclear magneton* by

$$\mu_N = \frac{e\hbar}{2m_p} = 5.05 \times 10^{-27} \text{ A m}^2$$

so the potential energy becomes

$$\mathfrak{U} = \pm 2.79 \, \mu_N B_T \qquad (29.3)$$

The minus sign holds when S_z and μ_z are parallel to the field (spin up) and the plus sign when they are opposite (spin down). The following example illustrates these energy relationships.

Example 29.1

A photon is absorbed when a proton moves from the spin up to the spin down state in a magnetic field of 1.4 T. (a) What is the energy of the absorbed photon in electron volts? (b) What is the photon frequency?

(a) Using $\mathfrak{U} = -2.79 \, \mu_N B_T$ for the spin up state and $\mathfrak{U} = +2.79 \, \mu_N B_T$ for the spin down state, the change in energy is

$$\begin{aligned}
\Delta\mathfrak{U} &= 2(2.79)\mu_N B_T \\
&= 2(2.79)(5.05 \times 10^{-27} \text{ A m}^2)(1.4 \text{ T}) \\
&= 3.94 \times 10^{-26} \text{ J} = 2.47 \times 10^{-7} \text{ eV}
\end{aligned}$$

We see that the energy of the photon absorbed is small compared to the changes of 1 to 10 eV usually observed in atomic and molecular processes.

(b) The photon energy is hf, so

$$f = \frac{\Delta\mathfrak{U}}{h} = \frac{2.47 \times 10^{-7} \text{ eV}}{4.14 \times 10^{-15} \text{ eV s}} = 5.96 \times 10^7 \text{ Hz}$$

It is important to notice that the spin angular momentum $\mathbf{S}$ (and also $\boldsymbol{\mu}$) is not parallel or antiparallel to $\mathbf{B}_T$. This follows from Chapter Twenty-eight where we found that the magnitude of the spin angular momentum $\mathbf{S}$ for spin $\frac{1}{2}$ particles is given by $S = \sqrt{s(s+1)}\,\hbar = \sqrt{\frac{1}{2}(\frac{1}{2}+1)}\,\hbar = \sqrt{3}\,\hbar/2$. Since $S_z =$

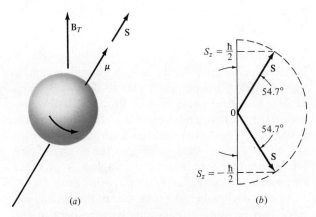

Figure 29.18. (a) A classical model of a spinning charge in a magnetic field $\mathbf{B}_T$. The spin angular momentum $\mathbf{S}$ and the magnetic moment $\boldsymbol{\mu}$ are parallel if the charge is positive. (b) The spin of a proton has a magnitude of $S = \sqrt{3}\,\hbar/2$, and the z component of $\mathbf{S}$ is $S_z = \pm\hbar/2$.

$\pm\hbar/2$, $\mathbf{S}$ must always be at an angle with the z axis (Fig. 29.18b).

In a magnetic field, the proton moment experiences a torque $\boldsymbol{\tau}$ (Section 19.5) given by

$$\boldsymbol{\tau} = \boldsymbol{\mu} \times \mathbf{B}_T \tag{29.4}$$

This torque is perpendicular to $\boldsymbol{\mu}$, $\mathbf{B}_T$, and also to $\mathbf{S}$ (Fig. 29.19a). The torque produces a change in the angular momentum $\mathbf{S}$ according to $\boldsymbol{\tau} = \Delta\mathbf{S}/\Delta t$ (Chapter Seven) and the result is a precession of $\mathbf{S}$

and $\boldsymbol{\mu}$ around the direction of the field $\mathbf{B}_T$ (Fig. 29.19b). We can see this more clearly if we write $\Delta\boldsymbol{\mu} = 2.79\,(e/m_p)\,\Delta\mathbf{S}$, so

$$\boldsymbol{\tau}\,\Delta t = \Delta\mathbf{S} = \frac{m_p}{2.79\,e}\,\Delta\boldsymbol{\mu} \tag{29.5}$$

Since $\boldsymbol{\tau}$ is always perpendicular to $\boldsymbol{\mu}$, the magnitude of $\boldsymbol{\mu}$ is constant, but its direction changes. The vector $\boldsymbol{\mu}$ is "pulled" around by the torque and $\boldsymbol{\mu}$ precesses around $\mathbf{B}_T$ (Fig. 29.19b). This precession is analogous to the precession of a spinning top.

The rate of precession is an important parameter in a resonance experiment. From Fig. 29.20, we see that the z component of $\boldsymbol{\mu}$ is constant, but the component of $\boldsymbol{\mu}$ perpendicular to $\mathbf{B}_T$, $\mu\sin\theta$, turns through a small angle $\Delta\Omega$ in a time Δt. Since the arc AB is approximately equal to the chord $\Delta\mu$, $\Delta\mu = \mu\sin\theta\,\Delta\Omega$.

The torque is in the same direction as $\Delta\boldsymbol{\mu}$, and from Fig. 29.19, we see that its magnitude is $\tau = \mu B_T\sin\theta$. Using these results for $\Delta\mu$ and τ in Eq. 29.5, we find

$$\mu B_T\sin\theta\,\Delta t = \frac{m_p}{2.79\,e}\,\mu\sin\theta\,\Delta\Omega$$

The precession frequency f_p is defined as $\Delta\Omega/2\pi\,\Delta t$, so we find

$$f_p = \frac{1}{2\pi}\frac{\Delta\Omega}{\Delta t} = 0.444\,\frac{e}{m_p}\,B_T \tag{29.6}$$

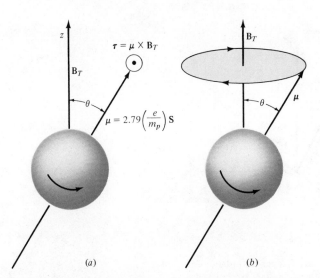

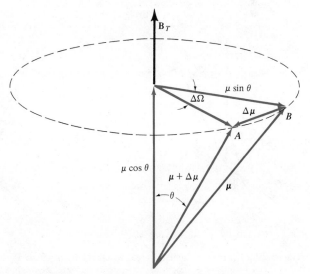

Figure 29.19. (a) The magnetic dipole moment $\boldsymbol{\mu}$ experiences a torque $\boldsymbol{\tau}$ due to the field $\mathbf{B}_T$. Here $\boldsymbol{\tau}$ is directed out of the page. (b) The magnetic moment $\boldsymbol{\mu}$ precesses around $\mathbf{B}_T$; the tip of the arrow representing $\boldsymbol{\mu}$ traces out a circle.

Figure 29.20. The vector $\boldsymbol{\mu}$ precesses around $\mathbf{B}_T$. The component of $\boldsymbol{\mu}$ perpendicular to $\mathbf{B}_T$ changes by an amount $\Delta\mu = \mu\sin\theta\,\Delta\Omega$ in a time Δt.

Note that the precession frequency is proportional to the magnetic field. This is illustrated in the next example.

Example 29.2

Calculate the precession frequency of a proton in a total magnetic field of 1.4 T.

Using Eq. 29.6, we have

$$f_p = 0.444 \frac{e}{m_p} B_T$$

$$= 0.444 \frac{(1.6 \times 10^{-19} \text{ C})}{(1.67 \times 10^{-27} \text{ kg})} (1.4 \text{ T})$$

$$= 5.96 \times 10^7 \text{ Hz}$$

This is a typical radio wave frequency.

In an NMR experiment the proton moments of the sample precess at a frequency $f_p = 0.444 \, eB_T/m_p$, where B_T is the total field at the position of the proton. A free proton precesses at a frequency $f_e = 0.444 \, eB_e/m_p$, where B_e is the magnitude of the external magnetic field. The difference between the two precession frequencies is

$$\Delta f = f_p - f_e = 0.444 \frac{e}{m_p} (B_T - B_e) \qquad (29.7)$$

The difference of $\mathbf{B}_T$ from the external field $\mathbf{B}_e$ indicates the effects of the surroundings. This information is obtained by measuring Δf.

29.8 | MEASURING THE PRECESSION FREQUENCY

The measurement of $\Delta f = f_p - f_e$ is the object of NMR experiments. To do this, a second magnetic field $\mathbf{B}_\perp$ is applied perpendicular to $\mathbf{B}_e$ and $\mathbf{B}_T$ (Fig. 29.21). This field changes with time, oscillating between the positive and negative x directions with a frequency $f_\perp$, and also produces a torque on the proton. When $f_\perp = f_p$, the resulting torque $\boldsymbol{\tau}_\perp$ on the proton moment tends to increase the angle θ. At all other frequencies the effect of $\boldsymbol{\tau}_\perp$ averages to zero.

As we noted earlier, the magnetic moment can only have two possible orientations with respect to the magnetic field $\mathbf{B}_T$. From Eq. 29.3, the energy difference between these states is $\Delta\mathcal{U} = 2(2.79)\mu_N B_T = 5.58 \, \mu_N B_T$. When the magnetic field $\mathbf{B}_\perp$ is applied at the precession frequency, the torque $\boldsymbol{\tau}_\perp$ will cause the dipole to "flip" (Fig. 29.22). Energy $\Delta\mathcal{U}$ is absorbed by the dipole, and this absorption can be detected. The frequency at which energy absorption occurs determines the precession frequency. (*Note:* The magnitude of $\mathbf{B}_\perp$ is always much smaller than that of $\mathbf{B}_T$, and is thus neglected in the calculation of $\Delta\mathcal{U}$.)

To summarize, when an external magnetic field is applied to a sample, the proton moments precess at the frequency f_p. The perpendicular time varying magnetic field acts to flip the dipoles over when $f_\perp = f_p$ (Fig. 29.22b).

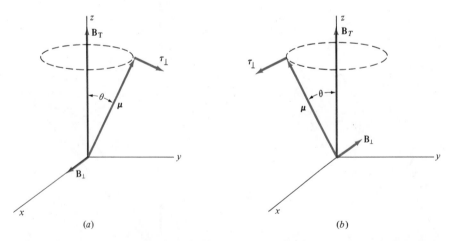

Figure 29.21. (*a*) An oscillating magnetic field $\mathbf{B}_\perp$ oriented as shown at a certain instant, produces a torque $\boldsymbol{\tau}_\perp = \boldsymbol{\mu} \times \mathbf{B}_\perp$ that tends to increase the angle θ. (*b*) If the field $\mathbf{B}_\perp$ varies at a frequency $f_\perp$, which is just equal to the precession frequency f_p, the torque $\boldsymbol{\tau}_\perp$ again tends to increase θ. If $f_\perp \neq f_p$, the effect of $\mathbf{B}_\perp$ averages to zero.

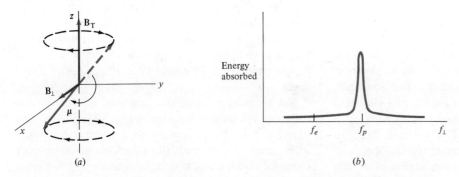

Figure 29.22. (a) A dipole precessing at a frequency f_p in a magnetic field $\mathbf{B}_T$ can absorb energy and "flip" over if a second field $\mathbf{B}_\perp$ is properly applied. (b) Absorption will occur when the frequency $f_\perp$ of $B_\perp$ is equal to f_p.

29.9 | THE NMR APPARATUS

The NMR apparatus has four fundamental components:

1 A permanent magnet or electromagnet used to produce the large external field $\mathbf{B}_e$. Standard instruments produce a field of 1.4 T with the field uniform over the region of the sample to one part in 10^8.

2 A method for varying $\mathbf{B}_e$ over a small range. In Fig. 29.23, the sweep coils fill this function.

3 A radio frequency (R.F.) oscillator that produces electromagnetic radiation at a fixed frequency, normally $f_\perp = 6 \times 10^7$ Hz. The magnetic field of this radiation plays the role of $\mathbf{B}_\perp$.

4 A radio frequency receiver used as a detector for the absorption of energy from the oscillator. Sometimes a single coil serves as both the oscillator and detector. When energy is absorbed an induced EMF is produced in the coil. This induced EMF is slightly out of phase with the oscillator voltage and can be electronically detected.

Since the R.F. oscillator has a fixed frequency, variations in f_p are produced by varying B_e. This is done by altering the current in the sweep coils in Fig. 29.23. Resonance occurs when the external field is adjusted so that $f_p = f_\perp = 6 \times 10^7$ Hz. Each time a resonance appears energy is absorbed. As we see in the following sections, there are often many resonant frequencies for a given molecule.

The sample region is quite small, usually a tube of radius 0.2 cm filled about 3 cm deep. This is done to minimize the expense of the high-precision magnet. Liquid solutions are usually used with the solvents chosen so that their resonant absorption frequencies do not overlap with those of the sample. An internal reference is commonly employed, normally tetramethylsilane (TMS), a molecule that does not interact strongly with other molecules. It has a single resonant frequency that shows up on the absorption graph and is used to calibrate the system.

Having established the physical basis for NMR measurements, we now consider what information can be obtained.

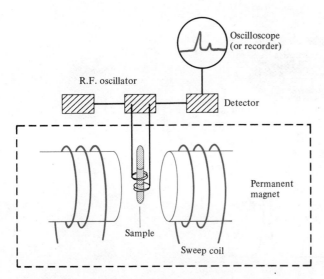

Figure 29.23. A schematic diagram of an NMR apparatus. The electronic instruments outside the box produce and monitor the energy absorbed. The sweep coils are used to make small variations in the field $\mathbf{B}_e$ produced by the permanent magnet. Other features are described in the text.

29.10 | THE CHEMICAL SHIFT

The *chemical shift* is the observed shift of the resonant frequency of protons in a given molecule from that of free protons in a field $\mathbf{B}_e$. For example, in benzene a single resonant frequency and chemical shift are observed. The measured frequency f_p is greater than f_e, so $\Delta f = f_p - f_e$ is positive. We can understand this result by examining the benzene molecule. Benzene has six carbon atoms in a ring with hydrogen atoms attached to them (Fig. 29.24a). Above and below the carbon ring, electrons are free to move in doughnut-shaped regions (Fig. 29.24b). As an external magnetic field $\mathbf{B}_e$ is applied, the electrons circulate in accordance with Lenz's law (Chapter Twenty). They form a current whose magnetic field opposes the initial increase in $\mathbf{B}_e$, reducing the flux inside the ring (Fig. 29.24). However, *outside the ring*, this induced current produces an increase in the magnetic field, and the total magnetic field there is greater than B_e by ΔB. Since the hydrogen atoms in benzene are outside the ring, they have a precession frequency that is greater than f_e by Δf.

Notice that at least two qualitative pieces of information about benzene have been confirmed. First, there exists a closed path around which the electrons can circulate and second, the hydrogen nuclei are outside this path. If the hydrogen nuclei were inside the ring, the resonant frequency would be less than f_e; the chemical shift would be in the opposite direction.

Subtle differences in the positions of electrons in different bonds can be seen using NMR. When an external field $\mathbf{B}_e$ is applied to a neutral hydrogen atom in its ground state, the s electron has its energy changed slightly, and it circulates according to Lenz's law (Fig. 29.25). The local field at the proton is less than B_T and the precession frequency is smaller than f_e. This is called the *diamagnetic* effect.

When hydrogen occurs in a covalent bond, the electron wave function is no longer symmetric about the nucleus. The more the electron is pulled away from the hydrogen nucleus, the smaller the diamagnetic effect, since its ability to produce an effective induced field at the nucleus is reduced. For example, in an OH group the oxygen atom strongly attracts the bond electron from the hydrogen atom and the precession frequency of the proton in hydrogen is only slightly smaller than f_e. The bond electrons in CH_2 are not as far from the hydrogen as in OH. Hence the diamagnetic effect is larger and the precession frequency of these two protons is less than in OH. Fi-

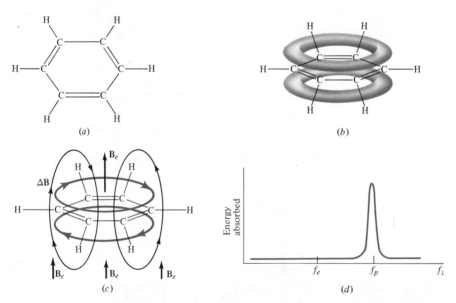

Figure 29.24. (a) The chemical diagram of benzene. (b) Electrons in benzene move freely in two rings above and below the plane of the molecule. (c) When $\mathbf{B}_e$ is applied, an induced electron current produces a magnetic field that opposes $\mathbf{B}_e$ inside the rings and adds to it outside the rings. (d) The field at the protons is larger than $\mathbf{B}_e$, so the resonant frequency is larger than f_e.

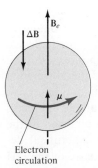

Figure 29.25. When B_e is applied to a hydrogen atom, the electron circulates as shown, producing the induced field ΔB. The field at the proton is then $B_T = B_e - \Delta B$.

29.11 | SPIN-SPIN SPLITTING

The same proton magnetic moments that precess and absorb energy in molecules give rise to magnetic fields within the molecule. These dipole magnetic fields produce modifications in the electronic configuration of the molecule which, in turn, affect other dipoles. The net result is that the orientation of each dipole may influence the energy of the other dipoles.

When the effects of interactions among spins within a group such as CH_3 are taken into account, the energy required to flip any one spin does not depend on the orientation of the others. On the other hand, the effect of proton spins on adjacent groups is significant and measureable. Consider first the effect of the CH_3 spins on the protons of an adjacent group.

The three proton spins in CH_3 can have four possible types of spin configurations:

1. ↑ ↑ ↑
2. ↑ ↑ ↓   or ↑ ↓ ↑   or ↓ ↑ ↑
3. ↓ ↓ ↑   or ↓ ↑ ↓   or ↑ ↓ ↓
4. ↓ ↓ ↓

The configurations with one spin opposite to the other two are three times as probable as those where all the spins are parallel.

Each of the configurations shown produces a different effect on dipoles of a neighboring group. Thus the protons of the group adjacent to CH_3 will experience four slightly different magnetic fields and in a sample four slightly different precession frequencies will be observed. The fourfold splitting of the CH_2 line in Fig. 29.27 shows this. It is also seen that the

nally, electrons in CH_3 are only weakly attracted by the carbon so the diamagnetic effect and chemical shifts are even larger. The precession frequency of these three protons is well below f_e. Thus in ethanol, CH_3—CH_2—OH, we see three chemical shifts (Fig. 29.26).

The identification of the three resonance peaks in ethanol is also facilitated by the fact that for every proton in an OH group that absorbs energy, there are two protons in CH_2 and three in CH_3. Energy is absorbed in the ratio of 1 to 2 to 3 in the three groups, as seen in Fig. 29.26b.

When an NMR experiment of higher resolution is performed on ethanol, the broad spectrum of Fig. 29.26 is resolved into a number of sharp peaks (Fig. 29.27). The explanation of these peaks is the subject of the next section.

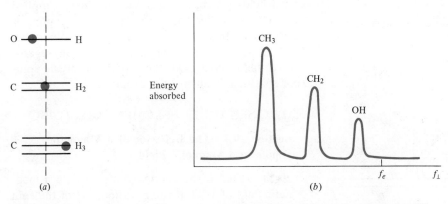

Figure 29.26. (a) Schematic diagram of the center of charge of bond electrons in the OH, CH_2, and CH_3 groups of ethanol. (b) The changes in the electron density and current at the hydrogen nucleus are evidenced by the different chemical shifts or precession frequencies of each group.

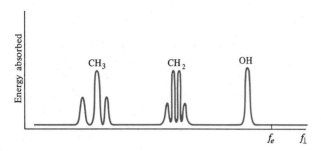

Figure 29.27. The NMR absorption spectrum of ethanol (CH_3—CH_2—OH) under high resolution at room temperature. (The effect of temperature will be discussed later.)

two central peaks are about three times as high as the outermost peaks, corresponding to the relative frequency of occurrence of the spin configurations in CH_3.

The two proton spins in CH_2 can have three different configurations:

1. ↑ ↑
2. ↑ ↓ or ↓ ↑
3. ↓ ↓

These three configurations produce three different fields in adjacent groups, and the CH_3 resonance is split into three peaks. The central peak is very close to twice as intense as the two side peaks (Fig. 29.27). Using the same reasoning we expect to see a threefold splitting of the OH resonance due to the adjacent CH_2 group. This splitting is not seen in the room temperature spectrum of Fig. 29.27. However, if the temperature of the sample is lowered, the threefold splitting is finally observed (Fig. 29.28). Furthermore, the CH_2 resonance undergoes an additional splitting so that eight peaks are seen.

The disappearance of this additional structure at

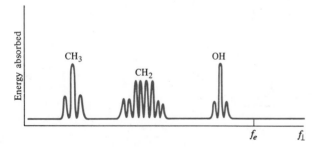

Figure 29.28. At low temperatures, further structure is observed in the spectrum of ethanol. There are eight peaks near the CH_2 resonance, and the OH line has three peaks.

room temperature is explained by *chemical exchange.* The hydroxyl group OH is rather weakly bound in ethanol and can be pictured as moving from molecule to molecule, binding briefly at each site and then moving again. This process is so rapid that during the time it takes for the resonant absorption of energy to occur, about the time it takes for the dipole to precess once, several OH groups have been attached to the same molecule. Thus the resonance experiment detects an average effect from the presence of the OH groups and the detailed structure is not seen. As the temperature is lowered, the thermal energy that facilitates this random exchange motion is decreased, and the OH group remains attached to each molecule for a longer period of time. When the exchange frequency is lower than the precession frequency, the effects of the OH group are seen. The OH resonance splits into three peaks due to the spin-spin coupling with the adjacent CH_2 group, and there are two possible spin alignments of OH:

1. ↑
2. ↓

These lead to a twofold splitting of the CH_2 resonance. Since the CH_2 resonance already had four peaks, it now has eight.

When experiments are performed on complex molecules, it may be difficult to distinguish the resonance of a given group from the spin-spin splitting of another group. In such cases the experiment can be performed at several different values of the external field. The size of the chemical shift varies with the applied field, but the spin-spin coupling does not. The chemical shift depends on the magnitudes of induced fields which, in turn, depend on the applied field. On the other hand, the spin-spin coupling is produced by the effects of spins on each other. Because these spins have the same magnitude, regardless of the applied field, the separation of the spin-spin splittings is always the same.

EXERCISES ON SUPPLEMENTARY TOPICS

Section 29.7 | The Behavior of a Magnetic Dipole in a Magnetic Field

29-24 What is the energy of a proton in a magnetic field of 1.2 T if the z component of the spin angular momentum is (a) along the field; (b) opposite to the field?

29-25 A photon of frequency 5×10^7 Hz is emit-

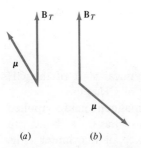

(a)   (b)

Figure 29.29. At the instant shown, the proton magnetic moment μ is in the plane of the page as is $\mathbf{B}_T$. Exercise 29-26.

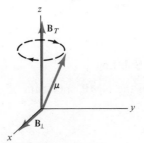

Figure 29.30. At $t = 0$, μ and $\mathbf{B}_T$ are in the yz plane, and $\mathbf{B}_\perp$ is along the x axis. Exercise 29-33.

ted when a proton spin flips. (a) What is the orientation of the proton spin angular momentum in the final state after the spin flips? (b) How large is the magnetic field?

29-26 Indicate the direction of the torque and the path of precession of the proton magnetic moment μ in the field $\mathbf{B}_T$ of Fig. 29.29a and 29.29b.

29-27 What is the precession frequency f_p of a proton in a magnetic field $B_T = 1.2$ T?

29-28 When a magnetic field is applied to protons in a molecule, the difference between the precession frequencies $\Delta f = f_p - f_e$ is found to be 300 Hz. What is the difference between the actual and applied magnetic fields at the position of the dipole?

29-29 Show that the energy absorbed when a proton spin flips in a magnetic field $\mathbf{B}_T$ can be written as $\Delta \mathcal{U} = hf_p$, where f_p is the precession frequency.

29-30 (a) Using the results of Exercise 29-29, compute the frequency of the photons absorbed by protons in a magnetic field of 5 T. (b) What is the energy change when absorption occurs?

Section 29.8 | Measuring the Precession Frequency

29-31 In a typical NMR experiment the difference between the external field and the field at the position of the proton is $B_e - B_T = 1.5 \times 10^{-6}$ T. What is the measured frequency shift Δf?

29-32 A free proton precesses with a frequency f_e in an external field B_e. At resonance in a molecule the total field is $B_T > B_e$. Is the frequency at which the perpendicular magnetic field is applied greater or less than f_e?

29-33 In Fig. 29.30 the proton magnetic moment precesses with a frequency f_p. The period, the time necessary for the dipole to precess once, is $T = 1/f_p$. At resonance the perpendicular magnetic field $\mathbf{B}_\perp$ has a maximum magnitude and is oriented as

shown at $t = 0$. Sketch the field $\mathbf{B}_\perp$ and show the direction of the torque on μ due to $\mathbf{B}_\perp$ at $t = 0$, $T/4$, $T/2$, $3T/4$, and T.

Section 29.9 | The NMR Apparatus

29-34 Standard instruments produce a field $\mathbf{B}_\perp$ at a frequency of $f_\perp = 6 \times 10^7$ Hz. At what value of $\mathbf{B}_e$ will resonance occur for a free proton?

29-35 Describe briefly what might be observed if the applied field $\mathbf{B}_e$ in an NMR apparatus did not have exactly the same value everywhere in the sample.

Section 29.10 | The Chemical Shift

29-36 Sketch a graph of energy absorbed versus $f_\perp$ for a molecule shaped like benzene but with the hydrogen atom protons located inside the carbon ring.

29-37 In the compound CH_3—CH_2—CH_2—I, the two CH_2 groups experience slightly different local fields because their environments in the molecule are not identical. (a) Assuming the chemical shift in the CH_2 group beside the iodine (I) atom is smaller than in the adjacent CH_2 group, sketch a graph of energy absorbed versus $f_\perp$ for this molecule analogous to that in Fig. 29.26. Do not include the effects of spin-spin coupling. (b) What is the predicted ratio of the peak heights of the resonances that appear for this molecule?

29-38 When molecules such as ethanol, CH_3—CH_2—OH, are in certain solutions, the hydrogen in the OH group may form the center for a hydrogen bond with a solvent molecule. In such a bond, the hydrogen proton has four electrons near it rather than two as when the hydrogen bond is not present. The OH resonance when hydrogen bonds are present is observed to shift so that it appears between the CH_3 and CH_2 resonances in Fig. 29.26. Describe why this shift occurs in terms of the diamagnetic effect.

Section 29.11 | Spin-Spin Splitting

29-39 Sketch and explain the NMR spectrum of 1,1,2-trichloroethane,

$$Cl—CH_2—CH \begin{array}{c} Cl \\ \\ Cl \end{array}$$

(The chemical shift of CH is smaller than that of CH_2.)

29-40 Figure 29.31 shows the NMR absorption spectrum of a molecule containing one carbon atom, one oxygen atom, and four hydrogen atoms. What is the formula for the molecule?

PROBLEMS ON SUPPLEMENTARY TOPICS

29-41 Figure 29.32 shows the circulating electron current of the C—O bond in aldehyde. What will the effect of this induced current be on the precession frequency of the proton dipole moment in the hydrogen atom?

29-42 Sketch the NMR absorption spectrum of acetaldehyde:

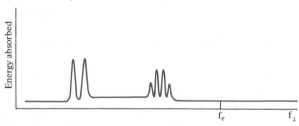

Figure 29.31. Exercise 29-40.

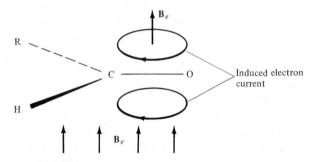

Figure 29.32. An electron current is induced in the C—O group of the aldehyde molecule when an external field is applied. R stands for the remainder of the molecule. Problem 29-41.

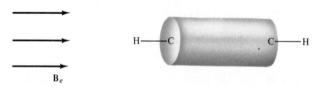

(Use the fact that the chemical shift of the CH group is smaller than that of CH_3.)

29-43 When an external magnetic field is applied to acetylene, the electrons of the C—C bond can circulate around the surface of a cylinder (Fig. 29.33). (a) What is the direction of the induced electron current if viewed looking along the $\mathbf{B}_e$ direction? (b) What is the effect of this current on the precession frequency of the protons in the hydrogen atoms?

29-44 The NMR absorption spectrum of $CH_3—CH_2—CH_2—I$ is shown in Fig. 29.34. The different chemical shifts of the two CH_2 groups are due to the different environments of the two groups in the molecule. (a) Explain the threefold splitting of the CH_3 peak. (b) Why is the central CH_2 line split into 12 peaks?

Additional Reading

Linus Pauling, *The Nature of the Chemical Bond,* Cornell University Press, Ithaca, New York, 1960.

Alan Holden, *Bonds Between Atoms,* Bell Telephone Laboratories, Murray Hill, N.J., 1966.

Gordon M. Barrow, *Physical Chemistry,* 3rd ed., McGraw-Hill Book Co., New York, 1973. Schematic representations of atomic and molecular orbitals.

Figure 29.33. Problem 29-43.

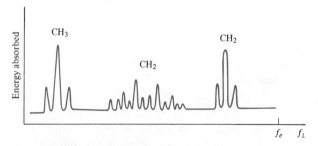

Figure 29.34. Problem 29-44.

R.B. Setlow and E.C. Pollard, *Molecular Biophysics,* Addison-Wesley Publishing Co., Reading, Mass., 1962, chapter 7.

John R. Dyer, *Applications of Absorption Spectroscopy of Organic Compounds,* Prentice-Hall, Inc., Englewood Cliffs, N.J., 1965. Paperback.

Dudley H. Williams and Ian Fleming, *Spectroscopic Methods in Organic Chemistry,* McGraw-Hill Publishing Co., London, 1966. Paperback.

Joe Demuth and Phaedon Avouris, Surface Spectroscopy, *Physics Today,* November 1982, p. 62. Information from emitted electrons.

Edward Edelson, Scanning the Body Magnetic, *Science 83,* July/August, p. 60. NMR scans.

D. Chapman and P.D. Magnus, *Introduction to Practical High Resolution Nuclear Magnetic Resonance Spectroscopy,* Academic Press, London and New York, 1966. Paperback.

Stanley J. Opella, Biological Nuclear Magnetic Resonance, *Science,* vol. 198, 1977, p. 158.

Jean L. Marx, NMR Researchers Embark on New Enterprise, *Science,* vol. 213, 1981, p. 425. Clinical applications of phosphorus NMR methods.

G.C. Levy and D.J. Craik, Recent Developments in Nuclear Magnetic Resonance, *Science,* vol. 214, 1981, p. 291.

G.L. Brownell *et al.*, Positron Tomography and Nuclear Magnetic Resonance Imaging, *Science,* vol. 215, 1982, p. 619.

J. Jonas, Nuclear Magnetic Resonance at High Pressure, *Science,* vol. 216, 1982, p. 1179. Application to the study of liquids.

Scientific American articles:

Boris V. Derjaguim, The Force Between Molecules, July 1960, p. 47.

Arnold C. Wahl, Chemistry by Computer, April 1970, p. 54.

Harry Selig *et al.*, The Chemistry of the Noble Gases, May 1964, p. 66.

Ronald Breslow, The Nature of Aromatic Molecules, August 1972, p. 32.

The entire September 1967 issue is devoted to *Materials,* with a number of articles related to this chapter.

Marvin L. Cohen, Volker Heine, and James C. Phillips, The Quantum Mechanics of Materials, June 1982, p. 82.

Gerald L. Pollack, Solid Nobel Gases, October 1966, p. 64.

L.K. Rumels, Ice, December 1966, p. 118.

J.D. Bernal, The Structure of Liquids, August 1960, p. 125.

A.R. Mackintosh, The Fermi Surface of Metals, July 1963, p. 110.

N.W. Ashcroft, Liquid Metals, July 1968, p. 72.

N.Ya. Azbel', M.I. Kaganov, and I.M. Lifshitz, Conduction Electrons in Metals, January 1973, p. 88.

Robert Gomer, Surface Diffusion, August 1982, p. 98. Metallic surfaces.

Gordon A. Thomas, An Electron-Hole Liquid, June 1976, p. 28.

Bernard Bertram and Robert A. Guyer, Solid Helium, August 1967, p. 84.

James L. Fergason, Liquid Crystals, August 1964, p. 76.

Raymond Bowers, Plasmas in Solids, November 1963, p. 46.

R.D. Parks, Quantum Effects in Superconductors, October 1965, p. 57.

F. Reif, Superfluidity and "Quasi-Particles," November 1960, p. 138.

N. David Mermin and David M. Lee, Superfluid Helium 3, December 1976, p. 56.

Richard E. Dickerson, The DNA Helix and How It Is Read, December 1983, p. 94. X-ray analysis.

Klaus Bechgaard and Denis Jerome, Organic Superconductors, July 1982, p. 52.

Donald H. Levy, The Spectroscopy of Supercooled Gases, February 1984, p. 96. Molecular energy levels.

Ednor M. Rowe and John H. Weaver, The Uses of Synchrotron Radiation, June 1977, p. 32.

David Adler, Amorphous-Semiconductor Devices, May 1977, p. 36.

James L. Dye, Anions of the Alkali Metals, July 1977, p. 92.

Kurt Nassau, The Causes of Color, October 1980, p. 124. Absorption of light by molecules.

Jearl Walker, The Physics and Chemistry of a Failed Sauce Béarnaise, The Amateur Scientist, December 1979, p. 178. van der Waals forces.

George B. Benedek, Magnetic Resonance at High Pressure, January 1965, p. 102.

Ian L. Pykett, NMR Imaging in Medicine, May, 1982, p. 78.

R.G. Shulman, NMR Spectroscopy of Living Cells. January 1983, p. 86.

UNIT NINE

THE ATOMIC NUCLEUS

A Tokamak, an experimental fusion machine, at the Princeton Plasma Laboratory. (Courtesy Princeton Plasma Physics Laboratory).

 E have seen that the development of quantum physics in the early decades of this century made possible an understanding of atomic and molecular phenomena. By the early 1930s the frontier of physics and the search for the ultimate structure of matter had moved on to the much smaller scale of the atomic nucleus.

Until 1939, nuclear physics was a scholarly activity of little practical importance. In that year O. Hahn (1879–1968) and F. Strassman, working in Germany, discovered that a uranium nucleus can fission into two lighter nuclei. This process releases millions of times more energy than any chemical reaction among atoms or molecules and represents an energy source of unprecedented magnitude.

The great contemporary impact of nuclear physics is demonstrated by a few examples. Nuclear weapons brought a gruesome but rapid end to World War II, and the threat of their possible use has become a central factor in international politics. The world's reserves of fossil fuels are rapidly diminishing, and of all the possible alternative energy sources, only nuclear fission reactors have as yet become available for widespread use. Many kinds of radioactive materials have been made available; their use in medicine has saved many lives, and important applications have been developed in research, agriculture, and industry. Major advances in geology and archeology have resulted from radioactive dating and other nuclear physics tools.

This unit is divided into two chapters. The first covers the fundamental physics of the nucleus, and the second deals with the effects and uses of ionizing radiation.

CHAPTER 30

NUCLEAR PHYSICS

The atomic nucleus is a very small, dense object made up of two kinds of nucleons: protons and neutrons. A proton has a positive electrical charge equal in magnitude to the electronic charge, and a mass about 1840 times that of the electron. Neutrons are about 0.1 percent more massive than protons. As their name suggests, they bear no electrical charge.

A nucleus is specified by its *atomic number Z* and its *mass number A*. *Z* is the number of protons, and *A* is the total number of nucleons, so the *neutron number N* equals $A - Z$. The standard notation for nuclei is illustrated by $^{238}_{92}U$. This nucleus has 238 nucleons, of which 92 are protons and $238 - 92 = 146$ are neutrons. U is the chemical symbol for the 92nd element, uranium. Sometimes the atomic number is omitted, since it is implicitly given by the name of the element. The notation U-238 is occasionally used when one needs to specify the isotope as well as the element.

Nuclear species, or *nuclides,* which have the same atomic number but different neutron numbers are called *isotopes.* Because the electronic structure of atoms depends mainly on the total positive charge of the nucleus, different isotopes of an element are nearly identical chemically. Slightly over 100 naturally occurring or artificially produced elements and about 300 stable nuclides are known.

Three distinct types of forces play important roles in nuclei. Nuclei are held together by very strong, short-ranged *nuclear forces* among the nucleons. *Electrical forces* are smaller in magnitude but they become progressively more important as the number of protons in the nucleus increases. The *weak interactions* are much weaker than either the strong nuclear forces or the electromagnetic interactions, but they are responsible for the *beta decay* processes in which, for example, neutrons in nuclei are converted into protons as they emit electrons and neutrinos. Gravitational forces are weaker still, and are unimportant in nuclear physics.

The detailed motions of the individual nucleons inside a nucleus are very complex. However, their average orbits and the resulting nuclear structure can be fairly well described by a nuclear shell model, which closely resembles the atomic shell model.

In this chapter, we first explore some basic properties of nuclei and then discuss nuclear forces and the nuclear shell model. We conclude with an examination of nuclear decay processes. Nuclear energy is discussed in the supplementary topics.

30.1 | RADIOACTIVITY

Antoine Henri Becquerel (1852–1908) accidentally made the first observation of a purely nuclear phenomenon in 1896, 15 years before Rutherford inferred the existence of the nucleus. Becquerel noted that uranium compounds produce invisible rays or *radiation* that can penetrate an opaque container and expose a photographic emulsion. Soon thereafter, Pierre and Marie Curie showed that uranium ores also contain traces of polonium ($Z = 84$) and radium ($Z = 88$), both much more intensely *radioactive* than uranium. Many other radioactive nuclear species or *radionuclides* were subsequently found.

Before long, some important properties of the radiation were discovered. A lead plate an inch or so thick

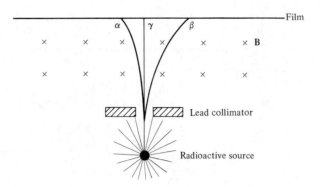

Figure 30.1. Radiation from a radioactive source splits into three components in a magnetic field. The α particles are $_2^4$He nuclei, the β particles are electrons, and the uncharged γ rays are photons.

stops most of the radiation from a uranium source, so a plate with a small hole can be used to form a narrow collimated beam of radiation. In the presence of a magnetic field, this beam splits into three components, labeled *alpha* (α), *beta* (β), and *gamma* (γ) Fig. (30.1). *Alpha particles* are positively charged and have a very short range in matter; they are now known to be helium nuclei ($_2^4$He). The negatively charged *beta particles* have a longer range in matter and are electrons. The neutral *gamma rays* penetrate furthest in matter. They are photons whose energies are usually greater than those of X rays.

The energies of α, β, and γ radiation are as much as several million electron volts (MeV) per particle. Since atomic and molecular processes typically involve energies of a few electron volts, radioactivity represented a totally new kind of phenomenon and suggested the existence of forces much stronger than electrical forces.

In 1903, Rutherford and Soddy showed that when a uranium nucleus emits an alpha particle, it is changed into a thorium nucleus. Symbolically,

$$^{238}_{92}\text{U} \rightarrow {}^{234}_{90}\text{Th} + {}^4_2\text{He}$$

Thus a transmutation of the elements occurs in α *decay processes*. This is a change that cannot be produced by chemical means, with or without medieval incantations. Similar transmutations occur in β *decays,* but not in γ *decays.*

Along with the early discoveries of the basic physical properties of radiation, there was progress in understanding the biological effects of radiation. By the start of this century, it was realized that both X rays and nuclear radiation can cause skin burns. However,

it was not then known that they also can induce cancer. Research scientists, physicians, and industrial workers often received massive radiation doses, and many developed malignancies, sometimes decades after the exposure had ceased. Not until the 1920s were the first governmental limitations on radiation exposures formulated.

30.2 | HALF-LIFE

Consider a large group of students at a dull lecture. If they all toss coins, saying "heads I stay, tails I leave," about half will remain after one toss. Roughly half of these will remain after the next toss, and so on. We cannot predict in any way when a specific student will depart, and the process is said to be *random*. Because about half the students leave after each toss, the time between tosses is called the *half-life* of the class.

A nuclear decay is a similar random process and is characterized by a half-life T, the time required for half the nuclei present to decay. If at time $t = 0$, there are N_0 nuclei, then one half-life later at $t = T$, an average of $N_0/2$ will remain. At $t = 2T$, when two half-lives have elapsed, half of these, or $N_0/4$, nuclei will be left; at $t = 3T$, $N_0/8$ will be left, and so on. Depending on the nuclide, the half-life may vary from a small fraction of a second to billions of years.

To find the number of nuclei remaining when the elapsed time is not an exact integer multiple of the half-life, we must use a formula that we now derive. The change ΔN in the number of nuclei present during a short time Δt is proportional to N and to Δt, so

$$\Delta N = -\lambda N \Delta t$$

The minus sign is needed because N is decreasing and ΔN is negative; the proportionality constant λ is called the *decay constant*. Dividing by Δt and taking the limit as Δt approaches zero, we find

$$\frac{dN}{dt} = -\lambda N \qquad (30.1)$$

We must solve Eq. 30.1 for the case where the initial number of nuclei at time $t = 0$ is a given value, N_0. The correct solution is $N = N_0 e^{-\lambda t}$ or

$$N/N_0 = e^{-\lambda t} \qquad (30.2)$$

At $t = 0$, this exponential decay formula reduces to $N = N_0$ as required; it is also a solution of Eq. 30.1, as can be verified by differentiation.

The decay constant λ is related to the half-life, T. Suppose one half-life has elapsed, so $N/N_0 = \frac{1}{2}$. Substituting $t = T$ in Eq. 30.2, we have $e^{-\lambda T} = \frac{1}{2}$, or

$$e^{\lambda T} = 2$$

Taking the base e logarithm (the natural logarithm) of this equation, we find $\lambda T = \ln 2 = 0.693$, or

$$\lambda = \frac{\ln 2}{T} = \frac{0.693}{T} \tag{30.3}$$

We see that a short half-life implies a large decay constant.

The exponential decay formula is plotted versus time in Fig. 30.2. When $t = T$, $N/N_0 = \frac{1}{2}$, in accordance with our discussion of the half-life; when $t = 2T$, $N/N_0 = \frac{1}{4}$, and so forth. Values of N/N_0 for any time can be read from the graph or calculated using an electronic calculator or tabulated values of e^{-x}, as in the next example.

Example 30.1

Iodine 131 is used in the treatment of thyroid disorders. Its half-life is 8.1 days. If a patient ingests a smll quantity of ^{131}I and none is excreted from the body, what fraction N/N_0 remains after 8.1 days, 16.2 days, 60 days?

Since 8.1 days is the half-life, the fraction remaining at this time is $\frac{1}{2}$. Similarly, 16.2 days is $2\,T$, so $(\frac{1}{2})(\frac{1}{2}) = \frac{1}{4}$ remains. Sixty days is not an exact integer multiple of the half-life, so to find the fraction remaining we use the exponential decay formula:

$$\frac{N}{N_0} = e^{-\lambda t} = e^{-0.693\,t/T}$$
$$= e^{-0.693(60\,\text{d})/8.1\,\text{d}}$$
$$= e^{-5.13} = 0.0059$$

Thus only 0.59 percent of the radioactive iodine remains after 60 days.

The assumption made in the above example that no ^{131}I is lost from the body by biological processes is not quite correct; ^{131}I is excreted steadily but slowly with a biological half-life of 180 days. Thus if nonradioactive iodine were ingested, only $\frac{1}{2}$ would remain in the body after 180 days, $\frac{1}{4}$ after 360 days, and so on. The effective half-life T_{eff} is obtained by combining the biological half-life T_b and the radioactive or physical half-life T_p according to the formula

$$\frac{1}{T_{\text{eff}}} = \frac{1}{T_b} + \frac{1}{T_p} \tag{30.4}$$

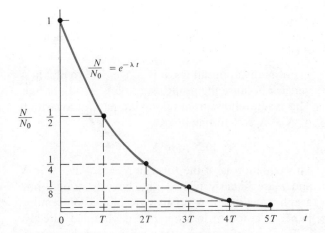

$$\frac{N}{N_0} = e^{-\lambda t}$$

Figure 30.2. The fraction of radioactive nuclei N/N_0 remaining after time t. N_0 nuclei are present at $t = 0$, and T is the half-life.

Some representative half-lives are given in Table 30.1.

Example 30.2

^{59}Fe is administered to a patient to diagnose blood anomalies. Find its effective half-life.

Using Table 30.1, $T_b = 65$ days and $T_p = 46.3$ days. Thus

$$\frac{1}{T_{\text{eff}}} = \frac{1}{T_b} + \frac{1}{T_p} = \frac{1}{65\ \text{d}} + \frac{1}{46.3\ \text{d}}$$
$$= 0.037\ \text{d}^{-1}$$
$$T_{\text{eff}} = 27\ \text{days}$$

TABLE 30.1

Half-lives of some radionuclides used in medicine and biology

Nuclide	Organ Where Concentrated	Half-Life (days)	
		Physical	Biological
$^{3}_{1}$H	Total body	4.6×10^3	19
$^{14}_{6}$C	Fat	2.09×10^6	35
	Bone	2.09×10^6	180
$^{24}_{11}$Na	Total body	0.62	29
$^{32}_{15}$P	Bone	14.3	1200
$^{35}_{16}$S	Skin	87.1	22
$^{36}_{17}$Cl	Total body	1.6×10^8	29
$^{42}_{19}$K	Muscle	0.52	43
$^{45}_{20}$Ca	Bone	152	18,000
$^{59}_{26}$Fe	Blood	46.3	65
$^{64}_{29}$Cu	Liver	0.54	39
$^{131}_{51}$I	Thyroid	8.1	180

Note that the effective half-life is shorter than either the biological or physical half-life. This happens because both processes are depleting the supply of the radionuclide.

For many applications, it is convenient to plot ln N versus t because the resulting graph is a straight line. This result follows from taking the natural logarithm of $N = N_0 e^{-\lambda t}$ to obtain

$$\ln N = \ln N_0 - \lambda t$$

This equation is of the form $\ln N = a + bt$, so $\ln N$ and t are linearly related and have a straight-line graph. If values of N versus t are plotted on semilog paper, a graph of ln N versus t is obtained directly.

This kind of graph is useful when we wish to determine the half-life of a radioactive sample. We can measure its activity with one of several detectors described in the next chapter. From Eq. 30.1, *the number of decays observed per second, the count rate dN/dt, is proportional to the number N of radioactive nuclei present and will diminish at the same rate.* Consequently, we can deduce the half-life from observations of the count rate by noting when the count rate drops to about half its initial value. However, a better value is obtained by plotting the data on semilog paper and drawing a straight line through the points. This procedure is more accurate since it uses all the observations, not just two data points. It is illustrated by the following example.

Example 30.3

The measured count rate for a radioactive sample is given in Table 30.2. The initial count rate is 400 per second, the rate after 2 minutes is 336 per second, and so on. What is the half-life of the sample?

Inspecting the table, we can immediately estimate the half-life by noting that the count rate dropped to 200 per second or half the initial rate of 400 per second sometime between $t = 6$ minutes and $t = 8$ minutes. Hence, T is between 6 and 8 minutes. To find a more exact value, we plot the data in the table on semilog paper as in Fig. 30.3. The straight line drawn through the data

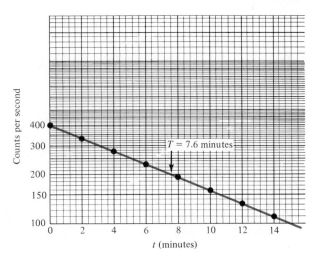

Figure 30.3.   Data of Table 30.2 plotted on semilogarithmic paper.

points crosses 200 counts per second at 7.6 minutes, so this is the half-life.

It is worth noting that the half-life is determined without using any information about the absolute number of nuclei present. Also, one does not need to know what fraction of the radiation is observed by the detector as long as it is a constant fraction.

30.3 | DATING IN ARCHAEOLOGY AND GEOLOGY

The ratio of the original and present-day quantities of a radionuclide in an object indicates the time elapsed since that object was created or formed. While the present-day concentration of a radionuclide may be measured directly, the original amount of the radionuclide must be determined indirectly.

Carbon-14 Dating | Radiocarbon dating of seeds, wooden artifacts, human and animal remains, and other objects containing plant or animal materials provides a reliable way of dating events occurring within the past 60,000 to 70,000 years. Developed by Willard Libby in the 1940s, this technique has had a tremendous impact on archaeology and related fields.

Carbon-14, which has a half-life of 5730 years, is always present in the environment because of the effects of cosmic-ray particles arriving in the atmosphere from outer space. These energetic particles interact with the atomic nuclei of the upper atmosphere to produce neutrons (n) that subsequently collide

TABLE 30.2

t (min)	Count Rate (s^{-1})	t (min)	Count Rate (s^{-1})
0	400	8	194
2	336	10	162
4	280	12	131
6	230	14	110

with nitrogen nuclei to produce ^{14}C and protons (p) in the reaction

$$n + \,^{14}_{7}N \rightarrow \,^{14}_{6}C + p$$

The radiocarbon mixes thoroughly with the ordinary carbon in the environment and is ingested by all living organisms. Once an organism dies, the input of radiocarbon stops, and the ratio of radiocarbon to ordinary carbon decreases steadily as the ^{14}C decays. Thus the quantity of ^{14}C remaining indicates the date of death.

Although this basic concept is simple, the actual application of radiocarbon dating is complex. One reason for this is that the ^{14}C is a minute fraction of the total carbon, only about one part in 10^{12}. Consequently the radioactivity produced by the ^{14}C as it beta decays to ^{14}N is very small compared to the total background radiation from sources normally present in the environment. To reduce this background, it is necessary to perform a careful chemical separation of the carbon and to use elaborate shielding arrangements during the counting. A second complication arises from uncertainties about the levels of radiocarbon in antiquity. The plausible assumption that radiocarbon levels have remained constant leads to predicted radioactivities that are close to the measured values for objects of known age (Fig. 30.4). However, studies made on tree rings have shown that minor variations have occurred in the ^{14}C level. These studies make use of the fact that only the outer portion of a tree is living material, so the ring formed in any one year records the radiocarbon level at that time. Using older trees that have long since fallen, but whose lifetimes overlapped appropriately, tree ring chronologies have been constructed extending back some 8000 years. These data permit small corrections to be made in doing accurate work.

The fraction of ^{14}C in the atmosphere decreased by about 3 percent over the past century because of the additional carbon dioxide produced by the large-scale burning of fossil fuels that no longer contain ^{14}C. Hydrogen bomb testing, which began in 1954, reversed the trend, doubling the radiocarbon level by 1963. For example, whiskies made after 1954 can be dated accurately by their radiocarbon levels.

Direct Detection With Accelerators

An accelerator-based technique involving the direct detection of radionuclide atoms has recently opened the doors to a wide range of new dating applications. It

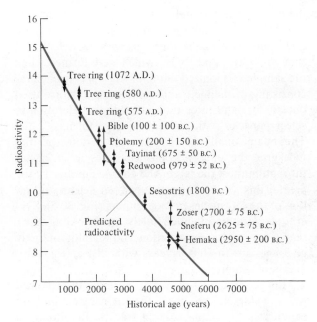

Figure 30.4. Predicted and observed radioactivities of samples of known age, assuming radiocarbon levels were constant in past centuries. "Bible" refers to Dead Sea scrolls, "Tayinat" to wood from the floor of a Syrian palace, and the other names refer to wooden objects from Egyptian tombs. (Adapted from W. F. Libby. *Radiocarbon Dating,* 2nd ed., University of Chicago Press, Chicago, 1955.)

was first applied to radiocarbon dating, but the method is being extended to other cosmic-ray-produced radionuclides.

For each decay per minute occurring in a sample containing ^{14}C, there are actually 4×10^9 such atoms present. If these could be detected directly, without waiting for them to decay, the sensitivity would be much greater. Attempts to do this with mass spectrometers, which measure the charge to mass ratio with magnetic fields (Chapter Nineteen), failed because the tiny quantities of ^{14}C were masked by ^{14}N, which has nearly the same mass and is always present in large amounts because it is the main constituent of the atmosphere.

Although some direct detection work is done with cyclotrons (Chapter Nineteen), most of the experiments use tandem Van de Graaff accelerators. In these machines, electrons are attached to ^{14}C atoms in a sample to form negative $^{14}C^-$ ions, which are then accelerated electrostatically down a long tube. Since negative nitrogen ions are unstable and fall apart, the ^{14}N contamination is mostly eliminated at the start. Partway through the acceleration process, the beam passes through a thin foil. This foil removes electrons, converting the particles into positive ions, and breaking up molecular ions with masses of 14 u such as $^{12}CH_2^-$ and $^{13}CH^-$ which were formed when the sample was ionized. Simple "mass spectrometers" consisting of magnets and collimating slits are used once before and three times after the acceleration to select particles with the right charge to mass ratio. The beam finally passes through a detector that measures the energy a particle loses as it passes through matter; the larger the atomic number Z, the greater this loss is. The result is a very clean separation of the ^{14}C nuclei from all others (Fig. 30.5). Several accelerators intended solely for this kind of research are under construction and should routinely give ages up to 60,000 years with just a few milligrams of material.

Minute amounts of other cosmic-ray-produced radionuclides are present in the environment and are candidates for similar dating studies. Beryllium 10 has a half-life of 1.5×10^6 years and accumulates in ocean sediments; it can be used to date sedimentary rocks. Since one decay per minute corresponds to 10^{12} atoms of ^{10}Be, direct detection offers a vast improvement over radioactive decay detection. Studies of the circulation of deep ocean layers and of underground water reservoirs have been done with ^{14}C and ^{36}Cl (half-life 30,000 years); other promising radionuclides for these applications with shorter half-lives are ^{32}Si (650 years) and ^{39}Ar (269 years).

Geochronology | Most of the radionuclides used for dating rocks have half-lives comparable to geological times (Table 30.3). The oldest rocks on the surface of the earth are 3.3 billion years old, while the earth itself is estimated to be about 4.5 billion years old.

The techniques used in geochronology depend on the particular kind of rock or mineral under study. For example, ordinary lead of nonradioactive origin

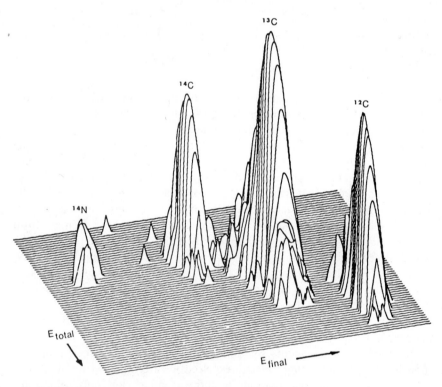

Figure 30.5. A display of the particles passing through the Van de Graaff accelerator system for direct detection of ^{14}C nuclei. E_{total} is the energy of the particles as they emerge from the last magnet. The spread in E_{final} reflects the increasing energy loss with Z as particles pass through matter. The small ^{14}N peak arises from $^{14}NH^-$ molecules. (From C. L. Bennett *et al.*, *American Scientist,* vol. LXVII, 1979, p. 456.)

TABLE 30.3

Radionuclides used in geochronology

Natural Radionuclide	Stable Nuclide Produced	Half-Life (Billion Years)
^{238}U	^{206}Pb	4.49
^{235}U	^{207}Pb	0.71
^{232}Th	^{208}Pb	14.1
^{87}Rb	^{87}Sr	50
^{40}K	^{40}A	1.3

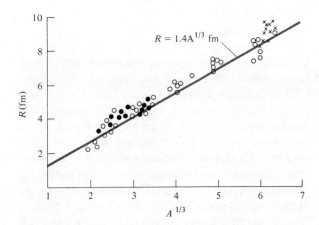

Figure 30.6. Experimental values of the radius of the nucleus versus $A^{1/3}$.

is a mixture of ^{204}Pb, ^{206}Pb, ^{207}Pb, and ^{208}Pb. From Table 30.3, we see that the radioactive decays of the uranium isotopes and thorium produce all of these isotopes except ^{204}Pb. If the lead in a sample does not contain any ^{204}Pb, this indicates that the lead present was produced by radioactive decay, and the sample can be used for dating. According to the table, ^{238}U decays into ^{206}Pb with a half-life of 4.49 billion years. Suppose a sample with no ^{204}Pb contains equal numbers of ^{238}U and ^{206}Pb nuclei. Then exactly one half-life or 4.49 billion years must have elapsed since the specimen was formed. The ^{232}Th to ^{208}Pb ratio can be employed in the same way. Alternatively the ^{206}Pb/^{207}Pb ratio can be used, since ^{235}U and ^{238}U decay at different rates. The ^{87}Rb/^{87}Sr and ^{40}K/^{40}Ar ratios are also sometimes employed.

30.4 | NUCLEAR SIZES

Beginning in 1907, Rutherford conducted a series of experiments in which he bombarded various atoms with alpha particles. As we saw in Chapter Twenty-seven, he found that an atom contains a small positive nucleus with a radius of less than 10^{-14} m, which is about 10^{-4} times the radius of an atom. From further experiments with alpha particles, nucleons, and other projectiles, considerable information has emerged about the spatial distribution of matter in the nucleus.

Roughly speaking, a nucleus containing A nucleons is a uniformly dense sphere of radius (Fig. 30.6.)

$$R = 1.4\,A^{1/3} \times 10^{-15}\text{ m}$$
$$= 1.4\,A^{1/3}\text{ femtometres} \qquad (30.5)$$

(1 femtometre = 1 fm = 10^{-15} m). The radius is proportional to the cube root of the mass number, and therefore increases very slowly as is seen in this next example.

Example 30.4

Find the nuclear radii of ^{27}Al and ^{64}Zn.
The radius of ^{27}Al is

$$R = 1.4\,A^{1/3}\text{ fm} = 1.4(27)^{1/3}\text{ fm}$$
$$= (1.4)(3)\text{ fm}$$
$$= 4.2\text{ fm}$$

Similarly, the radius of ^{64}Zn is

$$R = 1.4(64)^{1/3}\text{ fm} = (1.4)(4)\text{ fm}$$
$$= 5.6\text{ fm}$$

The radius has increased only by a third although A has more than doubled.

^{238}U is the largest naturally occurring nuclide, with a radius of about 9×10^{-15} m. By comparison, atomic radii are about 10^{-10} m, more than 10,000 times larger. The fraction of the atomic volume occupied by the nucleus is much smaller than the corresponding fraction occupied by the sun in the solar system.

Because the nuclear radius varies as $A^{1/3}$, it follows that the nuclear volume $\frac{4}{3}\pi R^3$ varies as $(A^{1/3})^3$ or A. Thus the volume is proportional to the number of nucleons. This is similar to the situation in ordinary bulk matter; if we double the number of water molecules, we double the volume of water present.

30.5 | PROTONS AND NEUTRONS

In 1921, Rutherford produced hydrogen nuclei or protons ($^1_1$H or p) by bombarding nitrogen with alpha particles ($^4_2$He or α). This reaction,

$$^4_2\text{He} + ^{14}_7\text{N} \rightarrow ^{17}_8\text{O} + ^1_1\text{H}$$

was the first artificially induced transmutation of elements, and suggested that protons are a constituent of nuclei. Since it was known that electrons are emitted by nuclei in β decay, it seems plausible that a nucleus has A protons and $A - Z$ electrons.

Two problems arose with this idea. First, it was found that $^{14}_{7}N$ has spin 1. However, 14 protons and 7 electrons, each with spin $\frac{1}{2}$, can only combine to give a *half integer* spin $(\frac{1}{2}, \frac{3}{2}, \ldots)$. Second, if an electron is confined to a region of nuclear dimensions, the uncertainty princple can be used to show that its energy levels must differ by roughly 100 MeV. However, the energy needed to excite a nucleus is usually 5 MeV or less. Thus the observed spacing of energy levels is much smaller than would be predicted by having electrons in the nucleus.

James Chadwick's (1891–1974) discovery of the neutron in 1932 began the modern era of nuclear physics (Fig. 30.7). Bombarding beryllium with alphas, he produced neutrons in the reaction

$$^{4}_{2}He + ^{9}_{4}Be \rightarrow ^{12}_{6}C + n$$

Neutrons have spin $\frac{1}{2}$, no electric charge, and a mass about 0.1 percent greater than the proton mass. This discovery suggested that nuclei are made up of protons and neutrons, which has now been well verified.

We have mentioned that nuclei that differ only in their neutron number are called isotopes and are chemically very similar. For example, hydrogen exists in three forms: ordinary hydrogen, $^{1}_{1}H$; deuterium or heavy hydrogen, $^{2}_{1}H$; and tritium, $^{3}_{1}H$. The first two nuclei are stable, but tritium β decays into

$^{3}_{2}He$ with a half-life of 12.3 years. For a given atomic number, it is found that from zero to nine stable isotopes may exist, and several naturally occurring or artificially produced radioactive isotopes may also exist. Each chemical element as it is found on the earth normally contains a mixture of its stable isotopes with nearly constant relative abundances.

30.6 | NUCLEAR MASSES AND BINDING ENERGIES

The masses of many nuclei have been accurately measured (Table 30.4) using mass spectrometers (Chapter Nineteen). A little arithmetic shows that the mass of a nucleus is less than the sum of the masses of its constituents. For example, $6m_p + 6m_n + 6m_e = 12.0989$ u, while a $^{12}_{6}C$ atom has a mass of only 12.000 u. This *mass defect* tends to increase with the mass number A.

The significance of the mass defect is made clear by Einstein's principle of the equivalence of mass and energy (Chapter Twenty-five). For an object at rest,

$$E = mc^2 \qquad (30.6)$$

In words, a mass m of matter can be converted into an amount of energy, E; the quantity c is the speed of light in a vacuum. The mass of a ^{12}C nucleus is less than that of its constituent nucleons because it is a bound system. One must supply energy equal to the *nuclear binding energy* to pull apart the protons and neutrons. According to Einstein's principle, this energy is equal to the mass defect times c^2.

To relate the mass defect and binding energy, we first calculate the energy associated with a mass of 1 u = 1.66×10^{-27} kg:

$$\begin{aligned} E &= (1\ u)(c^2) \\ &= (1.66 \times 10^{-27}\ kg)(3 \times 10^8\ m\ s^{-1})^2 \\ &= 1.49 \times 10^{-10}\ J \end{aligned}$$

Since 1 eV = 1.602×10^{-19} J, this energy can be expressed in electron volts as 931×10^6 eV = 931 MeV. Effectively,

$$1\ u = 931\ MeV \qquad (30.7)$$

The ^{12}C mass defect of 0.0989 u corresponds to a total binding energy of 0.0989×931 MeV = 92.1 MeV. Dividing by the mass number $A = 12$, the *binding energy per nucleon* in ^{12}C is 7.7 MeV.

The binding energy per nucleon for the stable nuclei is plotted versus A in Fig. 30.8. It is about 8 MeV

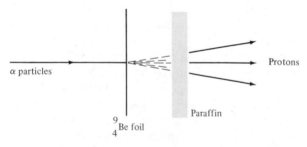

α particles

Protons

$^{9}_{4}$Be foil

Paraffin

Figure 30.7. Chadwick's discovery of the neutron. A beam of α particles struck a thin beryllium foil, producing uncharged particles (shown dashed). When a block of paraffin (a hydrocarbon containing many hydrogen nuclei or protons) was placed in front of the foil, protons were ejected. From measurements of the energy and momentum of the protons, he found the velocity and mass of the neutral particles striking them. The mass of these neutral particles turned out to be slightly greater than that of the proton.

TABLE 30.4

Atomic masses in u. Except for e, p, and n, the masses given are for the *neutral atom and include the electronic masses*. The atomic mass unit is defined so that the mass of a ^{12}C atom is exactly 12 u

Nuclide	Mass, m (u)	$Zm_p + Nm_n + Zm_e$ (u)	Difference (u)	Binding Energy per Nucleon (MeV)
e	5.48×10^{-4}			
p	1.00728			
n	1.00866			
^1_1H	1.00783			
^2_1H	2.0141	2.0165	0.0024	1.1
^4_2He	4.0026	4.0330	0.0304	7.1
$^{12}_6C$	12.0000	12.0989	0.0989	7.7
$^{13}_6C$	13.0034	13.1078	0.1044	7.5
$^{56}_{26}Fe$	55.9349	56.4633	0.5284	8.8
$^{238}_{92}U$	238.0508	239.9845	1.9337	7.6

per nucleon except for the lightest nuclei. There is a broad maximum in the region of medium-size nuclei, with a peak of 8.8 MeV per nucleon at ^{56}Fe. Above $A \simeq 100$, the curve gradually declines, reaching 7.6 MeV per nucleon for uranium.

The initial increase and later decrease in the binding energy per nucleon can readily be explained. The strong *nuclear forces* among the nucleons that hold the nucleus together have a very short range; these

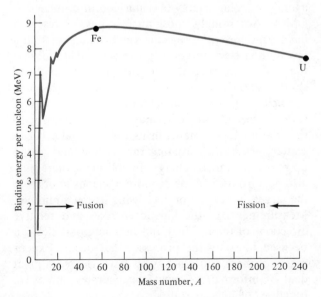

Figure 30.8. Binding energy per nucleon versus A for stable nuclei.

forces are zero at distances greater than a few femtometres. Accordingly a nucleon is attracted only to its closest neighboring nucleons. A nucleon near the surface has fewer neighbors than one in the interior of the nucleus and is less tightly bound. This *surface energy* effect implies that the binding energy per nucleon will rise as the nucleus increases in size and proportionately fewer nucleons are near the surface. This explains the initial rise in the binding energy per nucleon for the light elements.

To explain the decrease in the binding energy per nucleon for large A, we must take into account the fact that the electrical repulsions among the protons are proportional to the number of proton pairs, or to Z^2. The potential energy of two protons at a separation r is ke^2/r, so the total electrical potential energy due to the proton charges varies as Z^2e^2/R, where R is the nuclear radius. This energy grows rapidly as the number of protons increases. In the region near ^{56}Fe, the changes in the surface and electrical energies are roughly equal in magnitude but opposite in sign. Above $A \simeq 100$, the electrical repulsion gradually outstrips the surface effects, leading to the observed gradual decline in the binding energy per nucleon.

The fact that intermediate-size nuclei have the greatest binding energy per nucleon has some important consequences. If a heavy nucleus splits or *fissions* into two intermediate-size nuclei, the binding energy increases by close to 1 MeV per nucleon. The extra energy is released as kinetic energy of the fission products or as γ rays. Similarly, if two very light nuclei such as ²H or ³H combine, this *fusion* releases several MeV. Fission and fusion will be discussed in the supplementary topics at the end of this chapter, but we can already see that both processes release large amounts of energy.

30.7 | NUCLEAR FORCES

We now discuss the strong nuclear forces that hold the nucleus together. We have seen that the nuclear volume is proportional to A; as more nucleons are added, the nucleus grows in size but its density remains constant. We also saw that the binding energy of each additional nucleon is roughly constant at 8 MeV. These properties both occur because the *nuclear forces are short ranged* and are zero at distances greater than a few femtometres. Note that if the forces were long ranged as is true for electrical and gravitational forces, then every pair of nucleons would interact, leading to an increase in the density and mean binding energy with A. These do not happen because each nucleon interacts with only a few neighbors.

Complex nuclei are tightly bound despite the large electrical repulsions among the pairs of closely spaced protons. This indicates that nuclear forces are much stronger than electric forces. They must also be attractive, at least at the average internucleon separation in nuclei. However, it is found by studying collisions between nucleons that the force actually becomes repulsive at very short internucleon distances (Fig. 30.9). It has also been established that, except

for the electrical repulsion between two protons, the proton-proton, proton-neutron, and neutron-neutron forces are the same; *nuclear forces do not depend on the electric charge.*

Since the nuclear radius is $1.4\,A^{1/3}$ fm, the average internucleon separation turns out to be over a femtometre. At this distance the nucleon-nucleon potential energy is of moderate size and slowly varying. Consequently, the nucleus should not be viewed as a rigid arrangement of nucleons. To a surprisingly good approximation, each nucleon in the interior of the nucleus can be viewed as moving fairly freely with a constant potential energy arising from the effects of its neighbors. This idea is the basis for the nuclear shell model, which is discussed in the next section.

30.8 | NUCLEAR ENERGY LEVELS AND NUCLEAR STABILITY

A nucleus, like an atom, has a ground state and excited states. These states can be studied by bombarding nuclei with energetic protons, alpha particles, and other projectiles from particle accelerators. Many of the observed energy levels and other properties of nuclei can be predicted using the *nuclear shell model*. Initially developed in 1949 by Maria Mayer (1906–1972) and J.H. Jensen, this model resembles the atomic shell model.

Successive nuclear energy levels are usually separated by a few million electron volts or less, and the average binding energy of a nucleon in a nucleus is 8 MeV. Accordingly, most nuclear physics research has been done with projectile kinetic energies from a few million electron volts up to about 1000 MeV. At the higher energies the de Broglie wavelength is shorter, so the nucleus can be probed on a finer scale.

Nuclear processes studied with accelerators are of several types. A projectile may scatter *elastically,* so that the nucleus remains in its original state. It can scatter *inelastically,* exciting the nucleus and giving up part of its kinetic energy. In collisions, there may also be a *transfer* of one or more nucleons to or from the target, increasing or decreasing its mass number. Measurements of the kinetic energies and relative directions of the incident and outgoing particles can be used to study the processes that occur. Experiments of these and other types have yielded a great deal of information about the energies and wave functions of nuclear states.

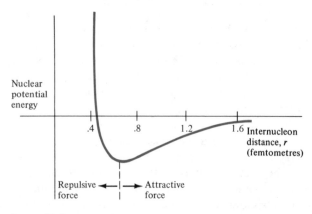

Figure 30.9. A rough graph of the nucleon-nucleon potential energy. The force becomes large and repulsive for $r \lesssim 0.4$ fm.

Nuclear Shell Model

In the preceding section, we noted that when a nucleon is deep inside the nucleus, its potential energy is approximately constant. However, a nucleon near the edge of a nucleus experiences a net inward attraction, since there are fewer neighboring nucleons on the side away from the nuclear center. Thus near the edge of a nucleus the potential energy increases as r increases (Fig. 30.10). Such a potential energy curve is referred to as a *potential well*.

The possible energy levels for a nucleon are obtained by solving the basic equation of quantum mechanics, the Schroedinger equation. Its solutions for stable states correspond to fitting waves into the potential well. The longest waves that fit into the potential well have nodes at $r = 0$ and at the edge. Now using the de Broglie relation $p = h/\lambda$, the kinetic energy of a nucleon $\frac{1}{2}mv^2 = p^2/2m$ can be rewritten as $h^2/2m\lambda^2$. Thus the longest wavelength wave functions represent the lowest energy states. The longest waves fitting into the well are shown in Fig. 30.10. When angular momentum is taken into account, the level scheme is more complex, but it is found that, as in atoms, the levels fall into closely spaced groups or shells.

Nucleons are spin $\frac{1}{2}$ particles, so the Pauli principle applies, and two identical nucleons cannot occupy a single quantum state. Any energy level may contain at most two protons, one with spin up and one with spin down, the two possible orientations for a spin $\frac{1}{2}$ particle. Since there can also be spin up and spin

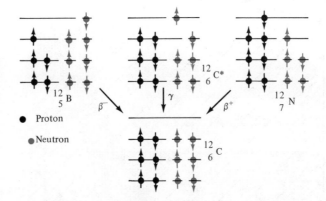

● Proton
● Neutron

Figure 30.11. Lowest states of three $A = 12$ nuclei, and the first excited state of $^{12}_{6}$C. The $^{12}_{5}$B, $^{12}_{6}$C*, and $^{12}_{7}$N nuclei have nearly the same energy. All three decay to the ground state of $^{12}_{6}$C as described in the text. (*Indicates excited state.)

down neutrons in an energy level, it can have at most four nucleons.

The lowest energy configuration for a given number of protons and neutrons is obtained by filling the lowest level with two protons and two neutrons, then the next level, and so on, until all the nucleons are used. A schematic diagram for three $A = 12$ nuclei is given in Fig. 30.11. Because of the Pauli principle, the lowest energy ground state occurs for the $Z = N$ case, $^{12}_{6}$C. Neglecting the small differences due to the electrical repulsions among the protons and to the neutron-proton mass difference, the *excited state* of $^{12}_{6}$C and the *ground states* of $^{12}_{5}$B and $^{12}_{7}$N have the same energy. The latter two nuclei will tend to β decay into the ground state of $^{12}_{6}$C, since that is a state of lower total energy. (In β^+ decay, a proton is converted into a neutron and in β^- decay, a neutron is converted into a proton. See Section 30.9.) Also, the excited $^{12}_{6}$C state will tend to emit a γ ray and undergo a transition to the ground state.

In general, the ground state of a nucleus is found by filling the lowest states in accord with the Pauli principle. Excited states correspond to one or more nucleons in higher states.

From our $A = 12$ example, we expect the stable nuclei to have $N \simeq Z$, which is in fact observed for nuclei with values of Z up to about 20 (Fig. 30.12). This equality maximizes the number of nucleons in the lowest nuclear energy states. If N differs substantially from Z, the nucleus tends to β decay. As Z increases beyond 20, the known stable and radioactive nuclei tend to have increasing neutron excesses. A

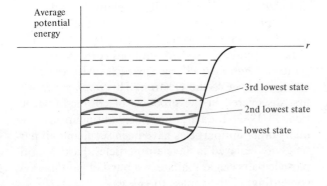

Average potential energy

r

3rd lowest state
2nd lowest state
lowest state

Figure 30.10. Average potential energy of a nucleon in the nucleus versus the distance r from the nuclear center. The possible nucleon energy levels are shown along with the wave functions for the lowest three states.

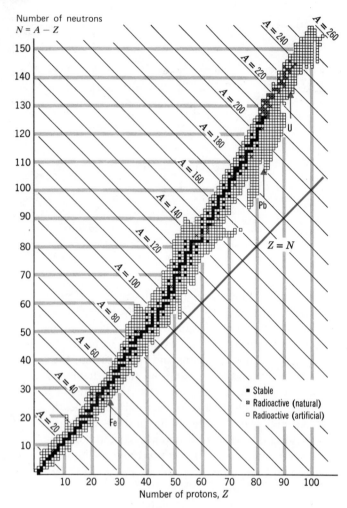

Figure 30.12. The known stable and radioactive nuclei. Nuclei beyond those shown in most cases rapidly break up into fragments.

30.9 | RADIOACTIVE DECAYS

At the beginning of this chapter, we noted that many nuclei undergo alpha, beta, or gamma decay. In these decays, as in all nuclear processes, the following quantities must be conserved:

1 energy (including mass energy)
2 momentum (both linear and angular)
3 electric charge
4 number of nucleons

Notice in particular that the total charge and number of nucleons does not change. We now discuss these decays individually.

γ Decay |

Gamma rays are electromagnetic quanta or photons emitted when a nucleus undergoes a transition from a higher to a lower energy level. They are completely equivalent to the light quanta and X rays emitted by excited atoms, but their energies are usually much greater. The half-lives for γ decay are usually very short, typically 10^{-13} s; however, in a few special cases the half-lives may be as long as several years.

Closely related to γ decay is *internal conversion*. Here an excited nucleus gives up its excess energy to an electron in one of the inner atomic shells, ejecting it from the atom. No gamma ray is emitted. Some radionuclides used in nuclear medicine decay in this fashion.

β Decay |

In β decay, an electron (e^-) or in a few cases a positron (e^+) is emitted by a nucleus. In 1928, P.A.M. Dirac proposed a theory of the electron that included the effects of special relativity and accurately predicted some fine details of the hydrogen atom spectrum not given by Schroedinger's equation. Dirac's equations also had solutions corresponding to positrons, *positively* charged particles with the same mass and spin as the electron. This *antiparticle* was first discovered in the products of cosmic-ray reactions in 1932. Today, antiprotons, antineutrons, and many other antiparticles have been seen, and all particles are believed to have antiparticles. Particle-antiparticle pairs can annihilate in a burst of gamma rays, converting all their mass to energy.

Some typical β decays are listed in Table 30.5. The half-lives are very long compared to γ decay half-lives, varying from seconds to many years. This indicates that the force responsible for the β decay is

neutron excess occurs because the energy that would be released by converting a neutron to a proton ($β^-$ decay) and placing it in a lower nuclear energy level is less than the increase that would occur in the electrical potential energy as a result of the increase in Z. Since the electrical energy is proportional to Z^2e^2/R, it becomes increasingly important as Z becomes large. Consequently, in the heaviest nuclei there are about 50 percent more neutrons than protons.

We saw in Chapter Twenty-eight that atoms with a completely filled outermost shell are inert gases and are very stable. Similarly, nuclei with closed outer shells are more tightly bound than neighboring nuclei, and are called *magic nuclei*. Nuclei with both proton and neutron shells closed are especially stable.

TABLE 30.5

Typical beta decays

Decay	Half-Life	Maximum Kinetic Energy (MeV)
$^3_1H \rightarrow \, ^3_2He + e^-$	12.3 years	0.186
$^{14}_6C \rightarrow \, ^{14}_7N + e^-$	5730 years	0.156
$n \rightarrow p + e^-$	11.0 minutes	0.782
$^{34}_{15}P \rightarrow \, ^{34}_{16}S + e^-$	12.4 seconds	5.1
$^{22}_{11}Na \rightarrow \, ^{22}_{10}Ne + e^+$	2.60 years	0.546
$^{13}_7N \rightarrow \, ^{13}_6C + e^+$	9.99 minutes	1.19

weak compared to the electromagnetic forces that are responsible for γ decay.

A striking feature of β decay is that the emitted betas are variable in kinetic energy (Fig. 30.13). For example, in $^3_1H \rightarrow \, ^3_2He + e^-$, the mass energy of the ^3_1H nucleus exceeds that of a ^3_2He nucleus plus an electron by 0.0186 MeV, the maximum electron kinetic energy. However, sometimes the electron energy is less than 0.0186 MeV. What has happened to the missing energy?

Enrico Fermi (1901–1954) supplied the answer in 1933. He proposed that when a nucleus β decays, it creates not only an electron but also a *neutrino* (ν): a massless, uncharged, spin $\frac{1}{2}$ particle. Then the ^3_1H decay is more completely written as

$$^3_1H \rightarrow \, ^3_2He + e^- + \nu$$

Since the decaying nucleus emits two particles they can share the decay energy in various combinations.*

The $\beta^\pm$ decay processes *in nuclei* result in the following conversions

$$n \rightarrow p + e^- + \nu \qquad (\beta^- \text{ decay}) \qquad (30.8)$$
$$p \rightarrow n + e^+ + \nu \qquad (\beta^+ \text{ decay}) \qquad (30.9)$$

Both processes tend to occur whenever they produce a nucleus with more binding energy. The β^+ decay can never happen for a free proton, since the proton mass is less than that of the neutron. However, the free neutron does β^- decay with a half-life of 11 minutes.

Another closely related way in which protons are changed into neutrons in some nuclei is *electron capture*. Here an atomic electron and a proton interact via

$$p + e^- \rightarrow n + \nu \qquad (\text{electron capture}) \qquad (30.10)$$

*According to the modern form of Fermi's theory, an *antineutrino* ($\overline{\nu}$) is emitted in a β^- decay, and a *neutrino* (ν) in a β^+ decay. We will ignore the differences between the neutrino and antineutrino.

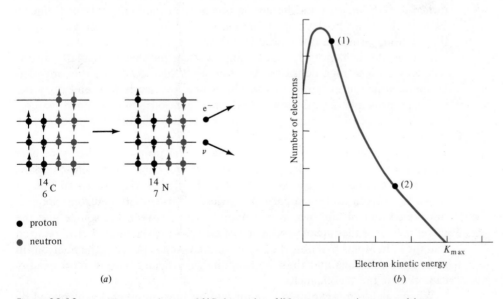

● proton

● neutron

Figure 30.13. (*a*) The ground state of $^{14}_6C$ decays into $^{14}_7N$ as a neutron is converted into a proton. An electron and a neutrino are emitted, and they share the energy given up as the nucleus goes into a lower energy state. (*b*) The number of electrons emitted versus kinetic energy for a large number of decays. Point (1) represents a decay in which the electron has less energy than that represented by point (2). The total energy available is constant so the neutrino in decay (1) has correspondingly more energy than in decay (2).

(Los Alamos Laboratory.)

ENRICO FERMI
(1901 – 1954)

Physics has become sufficiently specialized so that most physicists concentrate their efforts either on experimental work or theoretical calculations. Enrico Fermi was a notable exception to this rule. He made many important contributions of both types to physics.

Born in Rome, Fermi received his doctorate in 1922 from the University of Pisa for his research on X rays. As a student he taught himself the new quantum theories that were being developed elsewhere but were not yet known in Italy. His efforts to explain these concepts to his fellow students and his professors helped to establish modern physics in Italy and also developed his abilities as a teacher.

After a brief period of further study in Germany and Holland, Fermi returned to Italy in 1924 and became a professor at the University of Rome in 1926. In that year, he developed the theory of an ideal gas whose atoms obey the exclusion principle introduced in 1925 by Pauli. He found striking departures from the behavior predicted using classical physics, particularly at low temperatures and high densities. An example of such a *Fermi gas* is provided by the conduction electrons in a metal, which are nearly free and obey the exclusion principle. Fermi showed that many previously unexplained features of the electrical and thermal properties of metals were correctly predicted by this theory.

Turning his attention to nuclear physics, Fermi proposed a theory of beta decay in 1933, which in a slightly modified form remains our present-day basis for understanding this process. In 1934, he began a series of experiments in which he systematically bombarded a variety of targets with neutrons. Soon he discovered that placing water or a hydrocarbon between the source and the target increased the rate of production of artificial radioactivity. Fermi realized that the light atoms had absorbed some of the kinetic energy of the neutrons in a series of collisions and that the resulting slow neutrons were more readily captured by the target nuclei.

When a neutron is captured by a nucleus, its mass number A is increased by one; if a subsequent β^- decay occurs, the atomic number Z also is increased by one. Fermi and his collaborators therefore tried, in 1934, to go beyond the last known element by bombarding uranium ($Z = 92$) with neutrons. They thought mistakenly that they succeeded in producing the first transuranic element ($Z = 93$) when in fact they had caused uranium nuclei to fission. This was not realized until further work was done by Hahn and Strassman in Germany in 1939 that conclusively identified the fission products. Shortly before this discovery, Fermi and his family had fled the Fascist regime in Italy by traveling to Stockholm, where he accepted the Nobel prize, and then proceeding to New York. It was there that he learned of the work done in Germany.

Fermi immediately realized the importance of the discovery of fission and the possibility of a nuclear chain reaction. Working initially at Columbia University and later at the University of Chicago, he supervised the construction of the first nuclear reactor. When it was first operated on December 2, 1942, a now famous but then secret telegram announced that "the Italian navigator has entered the new world."

Fermi worked on the development of the atomic bomb during World War II, and then returned to academic life at the University of Chicago. In 1949, he joined several other leading scientists in opposing the development of the hydrogen bomb on ethical grounds. His postwar research centered on neutron studies, the properties of newly discovered particles called pi mesons, and the origin of cosmic rays. Shortly after his death in 1954, the artificially produced element with atomic number 100 was named fermium in his honor.

Two additional aspects of beta decay are worth noting. First, the forces responsible for the beta decay processes, including electron capture, are very weak compared both to the strong nuclear forces and to the electromagnetic forces among the nucleons. Accordingly, these β decay forces are referred to as *weak interactions;* they are the weakest of the fundamental forces in nature, except the gravitational forces. Second, as we mentioned earlier, beta decay is responsible for the fact that in stable nuclei the proton and neutron levels are filled to about the same maximum energy. When electrical forces are taken into account, this implies $N \simeq Z$ for light nuclei and $N \simeq 1.5\,Z$ for the heaviest nuclei.

α Decay | Unlike β decay, α decay always produces particles with the same kinetic energy. The emitted particle is a $_2^4$He nucleus, with two protons and two neutrons. Alpha decay is usually observed in the heavier unstable nuclei. All known nuclei above $Z = 83$ are unstable; those that do not β decay have been observed to α decay with half-lives ranging from about 10^{-3} s to 10^{10} years.

In β decay, the weak interaction determines the decay time. We might expect α decay to be much faster, since it is due to the strong nuclear interaction. However, typical decay processes are much slower than β decays. Why then does α decay take so long if it is energetically favorable for the strong nuclear forces to eject four nucleons in the form of an α particle?

The explanation of why this process is so slow involves a quantum mechanical phenomenon called *tunneling.* In classical physics, a ball bouncing elastically inside an open container can bounce over the walls and escape from the container only if its energy is great enough. Specifically, its energy must exceed the gravitational potential energy at the walls (Fig. 30.14). Quantum mechanically, if a particle reaches a region of high potential energy without sufficient energy to pass over this *barrier,* it has a small but nonzero probability of passing through the barrier (Fig. 30.15). This means that each time a particle strikes the barrier, it has a small probability of getting past the barrier. After enough impacts, the odds are that it will have passed through the barrier and escaped. It is then said to have *tunneled* through the region where, according to classical physics, it lacks sufficient energy to ever penetrate.

To see how this argument applies to α decay, we must consider the origin of the energy barrier for α particles. We do this by examining the potential energy curve for an alpha particle and a nucleus (Fig. 30.16). If we bring an α particle up to a nucleus, we

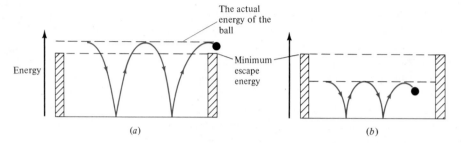

Figure 30.14.   (a) The ball has enough energy and bounces high enough so that it can escape from the container. (b) The ball is trapped according to classical mechanics.

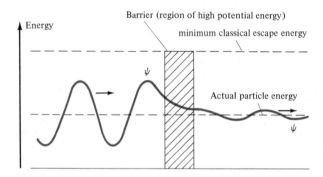

Figure 30.15.   The quantum mechanical wave associated with a particle approaching a barrier, a region where the potential energy exceeds the energy of the particle. The wave functon ψ is small but nonzero to the right of the barrier, so the probability $|\psi|^2$ that the particle may be found beyond the barrier is greater than zero. This probability diminishes rapidly as the barrier becomes taller or wider.

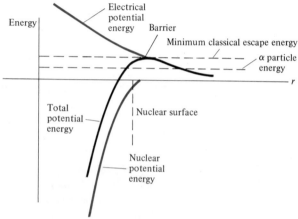

Figure 30.16.   The potential energy of an α particle and a nucleus.

must do work to overcome the electrical repulsion. Thus the potential energy increases as the particles approach and r decreases. At the nuclear surface, the α particle begins to feel the attractive nuclear force, which corresponds to a negative potential energy. The sum of the two potential energies has a positive maximum value somewhere near the nuclear surface. This peak corresponds to the barrier in the preceding discussion.

When the Schroedinger equation for the α particle inside this potential well is solved, it is found that the lowest α particle state has an energy greater than zero but below the barrier peak. Thus the alpha can reach a lower energy state by leaving the nucleus, and if it strikes the barrier often enough, it will eventually escape. This will take a long time if the barrier is very high or wide. Detailed calculations based on this theory are in excellent agreement with the measured α decay half-lives and energies.

SUMMARY

A nucleus is specified by its mass number A, atomic or proton number Z, and neutron number $N = A - Z$. After one half-life has elapsed, about half the unstable or radioactive nuclei present will remain. The decay constant is related to the half-life by $\lambda = 0.693/T$. After a time t, the fraction of radioactive nuclei remaining is

$$N/N_0 = e^{-\lambda t}$$

The effective half-life of a radionuclide in an organism is determined by the physical and biological half-lives. If the original amount of a radioactive substance in a sample is known, the amount remaining at the present time indicates its age.

Nuclei are approximately spheres of radius $1.4\,A^{1/3}$ fm. The mass of a nucleus is less than the mass of its constituent protons and neutrons by an amount equal to its mass defect. According to the Einstein mass-energy relationship, this represents the nuclear binding energy.

Using beams of particles produced by accelerators, the energy levels and structure of nuclei can be studied in detail. The nuclear shell model, which is similar to the atomic shell model, accounts for much of these data. Light nuclei either have $Z \simeq N$ or are unstable against β decay; because of electrical forces, very heavy stable nuclei have $N \simeq 1.5\ Z$. Excited states of nuclei tend to γ decay to their ground states. All nuclei above $Z = 83$ are unstable; those that do not β decay are all α particle emitters, and some will also spontaneously fission.

Checklist
Define or explain:

nuclide	binding energy
mass number	fission
atomic number	fusion
neutron number	nuclear force
isotope	nuclear shell model
alpha, beta, gamma	potential well
radiation	magic nuclei
half-life	internal conversion
nuclear radius	positron
mass defect	electron capture
equivalence of mass	tunneling
and energy	

REVIEW QUESTIONS

Q30-1 If a nucleus has 10 protons and 11 neutrons, what is its mass number?

Q30-2 Nuclear species with the same atomic number but different neutron numbers are called _____.

Q30-3 Alpha particles are _____, beta particles are _____, and gamma rays are _____.

Q30-4 The energy associated with nuclear processes is typically _____.

Q30-5 If there are 1000 nuclei initially, about _____ remain after one half-life, and about _____ after two half-lives.

Q30-6 Radiocarbon in living things originates in the atmosphere as a result of reactions caused by _____.

Q30-7 The volume of a nucleus is proportional to the number of _____.

Q30-8 The difference between the mass of a nucleus and the total mass of its constituents is its _____.

Q30-9 Compared to electrical forces, nuclear forces have a _____ magnitude and a _____ range.

Q30-10 Nucleons, like electrons, obey the Pauli principle, which prevent two protons or two neutrons from occupying the _____.

Q30-11 A positron has the same _____ as the electron, but the opposite _____.

Q30-12 An alpha particle can escape from a nucleus by _____, even though it lacks the minimum energy required according to classical physics.

EXERCISES

Section 30.2 | Half-Life

30-1 After 24 hours the radioactivity of a nuclide is 1/8 times its original level. What is its half-life?

30-2 How many half-lives are required for the activity of a radionuclide to decrease by a factor of 64?

30-3 Using the data in Table 30.1, find the effective half-life for ^{35}S.

30-4 A radionuclide with a physical half-life of 10 days is observed to have an effective half-life of 6 days when administered to a patient. What is the biological half-life of the nuclide?

30-5 Estimate the half-life of the radioactive substance whose count rate is given in Table 30.6 by inspection of the table.

30-6 Find the half-life of the radioactive sub-

TABLE 30.6

Time (days)	Count Rate (counts per minute)
0	455
1	402
2	356
3	315
4	278
5	246
6	218
7	193
8	171
9	151
10	133

stance whose count rate is given in Table 30.6 by plotting the data on semilog paper.

30-7 A radionuclide has a half-life of 10 hours. What percentage of a sample remains after 24 hours?

30-8 A patient is administered ^{35}S in a diagnostic study. (a) Using the data in Table 30.1, find the effective half-life. (b) What percentage of this radionuclide remains in the body after 22 days?

Section 30.3 | Dating in Archaeology and Geology

30-9 A wooden bowl has one-fourth the ^{14}C activity observed in contemporary wooden objects. Estimate its age. (Assume ^{14}C levels in the atmosphere have remained the same.)

30-10 A rock contains three ^{207}Pb nuclei for each ^{235}U nucleus. How old is the rock, assuming all the ^{207}Pb is from the uranium decay?

30-11 Why is there no radiocarbon in fossil fuels?

30-12 Carbon from living organisms contains ^{14}C at about the level of 1 part in 10^{12}. What is the corresponding number for a sample 40,000 years old?

30-13 Water taken from a deep well is found to have one-fourth the amount of ^{32}Si found in surface water. How long does it take for the source of the water to be replenished? (The half-life of ^{32}Si is 650 years.)

Section 30.4 | Nuclear Sizes

30-14 (a) Find the nuclear radii for ^4He, ^{27}Al, ^{64}Cu, ^{125}I and ^{216}Po. (b) Draw a rough graph of the nuclear radius versus A.

30-15 What fraction of the volume of a helium atom is occupied by its nucleus, assuming an atomic radius of 10^{-10} m?

30-16 (a) Calculate the density of an oxygen nucleus in kilograms per cubic metre. (b) Find the ratio of this density to the density of water, 10^3 kg m^{-3}.

30-17 *Neutron stars* are believed to have a mass similar to that of the sun, but a density comparable to that of atomic nuclei. Estimate the radius of such a star.

30-18 Find the volume of 1 mole of carbon nuclei. (1 mole = 6.02×10^{23} particles)

Section 30.5 | Protons and Neutrons

30-19 How many neutrons are there in $^{14}_{6}C$, $^{36}_{17}Cl$, $^{64}_{29}Cu$, and $^{208}_{82}Pb$?

30-20 (a) Of the nuclides ^1_1H, ^2_1H, ^3_1H, ^3_2He, and ^4_2He, which have the same neutron numbers? (b) Which nuclides have similar chemical properties?

Section 30.6 | Nuclear Masses and Binding Energies

30-21 The atomic mass of $^{208}_{82}Pb$ is 207.9766 u. What is its average binding energy per nucleon?

30-22 The atomic mass of $^{207}_{82}Pb$ is 206.9759 u, and the atomic mass of $^{208}_{82}Pb$ is 207.9766 u. (a) Find the difference of their mass defects. (b) What is the minimum energy needed to remove a neutron from $^{208}_{82}Pb$?

30-23 A nucleus has a mass defect of 1.5 u. (a) What is its binding energy in MeV? (b) If the mass number is 200, find the binding energy per nucleon.

Section 30.7 | Nuclear Forces

30-24 (a) Find the magnitude of the electric force between two protons separated by a typical nuclear distance, 10^{-15} m. (b) A typical atomic dimension is about 10^{-10} m. How large is the electrical force between a proton and an electron with this separation? (c) Calculate the ratio of the two forces in parts (a) and (b).

30-25 When a proton moves toward a nucleus, its initial kinetic energy must be great enough to overcome the electrical repulsion if it is to experience nuclear forces. For the largest naturally occurring nucleus, $^{238}_{92}U$, this minimum energy is about 15 MeV. Find the corresponding energies for incident (a) alpha particles (^4_2He nuclei); (b) neutrons.

30-26 All nuclei above $Z = 83$ are radioactive; if they do not β decay, they eventually α decay or fission. What is the origin of this instability?

30-27 (a) What is the electrical potential energy in MeV of two protons separated by 1 fm? (b) How much does this contribute to the mass defect if the two protons are in a nucleus?

Section 30.8 | Nuclear Energy Levels and Nuclear Stability

30-28 What is X in the following reactions? (d = ^2_1H) (a) p + $^{12}_{6}C \rightarrow$ d + X. (b) ^3_2He + $^3_2He \rightarrow$ ^4_2He + p + X. (c) p + $^{14}_{7}N \rightarrow$ $^{12}_{7}N$ + X.

30-29 What is X in the following reactions? (d = ^2_1H, $\alpha = ^4_2He$) (a) d + d $\rightarrow X$ + p. (b) ^6_3Li + $X \rightarrow \alpha + \alpha$. (c) X + $^{16}_{8}O \rightarrow$ $^{19}_{9}F$ + p.

30-30 In Fig. 30.11, the ground states of $^{12}_5$B and $^{12}_7$N and an excited state of $^{12}_6$C have the same states filled. (a) Which of these nuclides has the greatest binding energy if electrical forces are taken into account? (b) What effect does the neutron-proton mass difference have on the relative energies? Explain.

Section 30.9 | Radioactive Decays

30-31 An excited nucleus decays, emitting a 2-MeV gamma ray. Find (a) the frequency of the gamma-ray photon emitted; (b) the wavelength of the photon.

30-32 Complete the following decay processes by adding the missing decay particles (α, γ, or $\beta^\pm$ + ν): (a) $^{11}_6$C $\rightarrow$ $^{11}_5$B + ? (b) $^{32}_{15}$P $\rightarrow$ $^{32}_{16}$S + ? (c) $^{12}_6$C* $\rightarrow$ $^{12}_6$C + ? (d) $^{240}_{94}$Pu $\rightarrow$ $^{236}_{92}$U + ?

PROBLEMS

30-33 How many half-lives have elapsed if the activity of a radionuclide has diminished to 1 percent of its initial value?

30-34 An excavated wooden beam has 20 percent of the ^{14}C found in atmospheric carbon. How old is it? (Assume the atmospheric ^{14}C levels have remained the same.)

30-35 ^{90}Sr has a half-life of 28 years. How many years must this material be stored before its activity drops to $1/e$ times its original value?

30-36 A radioactive material contains two radionuclides, one with a half-life of 1 day and the other with a half-life of 8 days. Initially the radioactivity of the short-lived nuclide is $2^7 = 128$ times greater than that of the long-lived nuclide. When will their activities be equal?

*30-37 In the preceding problem, suppose one cannot distinguish the radiation from the two radionuclides and plots the *total* count rate on semilog paper. (a) What is the shape of the graph for the first few days? (b) Why does the shape change when the two activities are nearly equal? (c) What is the shape of the graph when the short-lived nuclide has largely disappeared? Is there any difference from the initial curve?

30-38 The μ meson (muon) is a short-lived particle which, like the electron, has electrical interactions with nuclei but does not experience the strong nuclear force. The muon mass is 207 times that of an electron. Both $\mu^\pm$ have been observed and the magnitude of their charge is equal to that of the electron. (a) What is the radius of the first Bohr orbit of a negative muon in orbit around a $^{238}_{92}$U nucleus? (b) Find the ratio of that radius to the nuclear radius. (Actually the radii and energy levels of muonic atoms are changed somewhat from the Bohr values because a significant fraction of the wave function is inside the nucleus for low-lying atomic states. These changes provide information about the distribution of protons in the nucleus.)

30-39 The energy of the sun is derived from reactions that convert four protons into an alpha particle ($^4_2$He) with the emission of two positrons and two massless neutrinos. (a) How much energy in MeV is released in this process? (See Table 30.4 for the required atomic masses.) (b) When H_2 and O_2 gases combine to produce H_2O, 6.2 eV is released for each molecule formed. What mass of hydrogen would have to be burned to equal the energy released by the nuclear fusion of 1 kg of hydrogen into alphas?

30-40 How can reactions in which a projectile transfers nucleons to a target nucleus be used to find nuclear energy levels?

30-41 $^9_4$Be is stable but $^9_5$B is unstable as is $^9_3$Li. (a) Draw nuclear energy-level diagrams similar to those of Fig. 30.11 for these three $A = 9$ nuclides. (b) Suggest why two of these three nuclides are unstable. (c) What decay process would be expected for $^9_3$Li?

30-42 After a heavy nucleus α decays, a β^- decay often follows, but never a β^+ decay. What is the reason for this?

30-43 We can estimate the electrostatic energy converted into kinetic energy in the fission of ^{235}U with a simple model. (a) What is the radius of a ^{235}U nucleus? (b) Suppose half the 92 protons in the nucleus are, on the average, separated from the other half by a nuclear diameter. What is the potential energy in MeV corresponding to the electrical repulsion of the two groups of protons?

30-44 Just as the mass of a nucleus is less than that of its constituents because of the binding energy, the mass of an atom is less than the mass of its nucleus and its electrons. Proportionately the effect is much smaller. (a) What is the ratio of the binding energy of a typical nucleus with a binding energy of 8 MeV per nucleon to its total mass energy? (b) A hydrogen atom consists of a proton and an electron bound by electrical forces with an energy of 13.6 eV. Find the corresponding ratio.

30-45 When a particle and its antiparticle meet,

they can annihilate into a burst of gamma rays. However, other final products may result if they are consistent with the conservation laws. For example, pi mesons are much less massive than nucleons, and can be produced when a proton and antiproton collide. If an antiproton and proton with negligible kinetic energy annihilate forming two positive pi mesons and two negative pi mesons, how much kinetic energy is shared by the four pi mesons? (The rest energy of a positive or negative pi meson is 139.6 MeV.)

*30-46 ^3_1H is unstable; it beta decays into ^3_2He, even though the electrical potential energy of the two protons is present in ^3_2He and not in ^3_1H. The extra mass of the neutron that is converted into a proton is sufficiently large to make ^3_2He more stable. Use this information to find a minimum average separation of the protons in ^3_2He.

ANSWERS TO REVIEW QUESTIONS

Q30-1, 21; **Q30-2,** isotopes; **Q30-3,** helium nuclei, electrons, energetic photons; **Q30-4,** several MeV; **Q30-5,** 500, 250; **Q30-6,** cosmic rays; **Q30-7,** nucleons; **Q30-8,** mass defect; **Q30-9,** greater, shorter; **Q30-10,** same quantum state; **Q30-11,** mass, charge; **Q30-12,** tunneling through a barrier.

SUPPLEMENTARY TOPICS

30.10 | NUCLEAR FISSION

We saw earlier that the binding energy per nucleon decreases gradually as A increases beyond 100. Consequently, about 1 MeV per nucleon is released if a large nucleus such as uranium splits into two smaller fragments. Such fission processes are the energy sources in nuclear power reactors and in the fission weapons first developed during World War II.

A simple model of the $^{235}_{92}U$ nucleus shows how fission occurs. The model pictures a $^{235}_{92}U$ nucleus as composed of two parts, each containing a large number of protons and neutrons. The potential energy curve for the two parts is much like that of an α particle and a nucleus. If the two positively charged parts are initially far apart and gradually approach, their electrical potential energy rises. When they become close enough so the strong attractive nuclear forces play a role, the total potential energy begins to diminish (Fig. 30.17).

The ground state of the ^{235}U nucleus turns out to

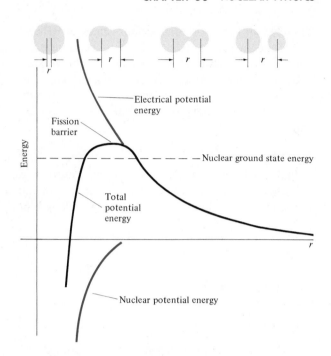

Figure 30.17. The potential energy of the uranium nucleus versus the separation r of the two parts of the nucleus. The corresponding shape of the nucleus is shown above the graph.

have an energy considerably greater than zero, but the system is trapped in its bound state below the peak in the potential energy curve. The region that is inaccessible according to classical physics is called the *fission barrier*. As in α decay, tunneling through the barrier makes it possible for this nucleus to *spontaneously fission*. As the two parts of the nucleus separate, a large amount of electrical potential energy is converted into kinetic energy. Spontaneous fission occurs in both ^{235}U and ^{238}U, but much less frequently than α decay.

Some nuclides can also undergo *induced fission* when they are bombarded by *thermal neutrons,* very slow neutrons with kinetic energies comparable to kT (less than 1 eV). Thermal neutrons have a large de Broglie wavelength and therefore a large effective size for an interaction. Many nuclei have high probabilities (large *cross sections*) for capturing thermal neutrons that pass close by and incorporating them into their structure. When this happens, a nucleus with A nucleons is transformed into one with $A + 1$ nucleons. The average binding energy of the last nucleon in a heavy nucleus is roughly 7 MeV. Hence when the neutron is captured the $A + 1$ particle nucleus will have an energy excess of about 7 MeV, and it will be in an excited state. If this excitation energy is sufficient to put the nucleus above the fission barrier, it will immediately fission, releasing a large amount of energy.

Precisely this happens when ^{235}U is bombarded by thermal neutrons. An excited state of ^{236}U is formed, with an energy 6.8 MeV above the ground state. Since the ^{236}U fission barrier height is also 6.8 MeV, the nucleus immediately fissions. On the other hand, ^{238}U does not fission when it captures a slow neutron, because the resulting excitation energy in the ^{239}U nucleus is 5.3 MeV, while the fission barrier is 7.1 MeV. Consequently, induced fission can occur with ^{238}U only if the incoming neutron has at least 1.8 MeV of kinetic energy. Such fast neutrons have shorter de Broglie wavelengths and smaller effective sizes and are hard to capture. The fast neutron fission cross section for ^{238}U is 1/2000 times the thermal neutron fission cross section for ^{235}U. For this reason, ^{235}U is much more useful in a fission energy source.

Natural uranium is 99.3 percent ^{238}U and only 0.7 percent ^{235}U. However, it is necessary to use nuclear fuels with a larger percentage of ^{235}U in most types of power reactors and in nuclear weapons. Since separation of these isotopes is costly and difficult, this both retards nuclear power development and tends to stabilize some aspects of international politics.

The energy released in nuclear fission can be estimated by noting that the binding energy per nucleon is 7.6 MeV for ^{235}U and about 8.5 MeV for the two $A \simeq 100$ fission fragments. (See Fig. 30.8.) Consequently the energy released is 235(8.5 MeV − 7.6 MeV) $\simeq$ 200 MeV per nucleus. To put that in perspective, we may note that chemical reactions typically liberate about 1 eV per atom, so nuclear fission is 200 million times more powerful! If all the nuclei in one kilogram of ^{235}U fission, the energy released is equivalent to that of 20,000 tons of TNT. The World War II fission bombs released approximately this much energy.

Usually the fission fragments have proportionately too many neutrons for medium A nuclei, so they eject one or more *prompt neutrons* almost immediately in 10^{-13} s or less (Fig. 30.18). They then undergo a series of three or four β^- decays, each of which converts one neutron into a proton and brings the nucleus closer to stability. Accordingly, the first β^- decay proceeds in a few seconds, while the next has a half-life of minutes or hours, and the last may take days or years. It is these relatively long-lived radioactive nuclides that comprise the *radioactive wastes* produced in reactors and that must be safely stored until the radioactivity is negligible. Sometimes a *delayed neutron* is ejected after the first β^- decay. On the average, there are 2.6 prompt neutrons for each fission; about 1 percent of the fission events lead to the emission of a delayed neutron after a delay averaging some 10 seconds.

The neutrons released in an induced ^{235}U fission can be captured by other ^{235}U nuclei, causing them to fission in turn in a *chain reaction*. Suppose, on the average, the 2.6 neutrons from one fission induce more than one additional fission. The *fission ratio* is then greater than 1, and the reaction will grow exponentially. Before long an explosion will occur if the growth proceeds unchecked. On the other hand, if enough neutrons escape from the ^{235}U or are absorbed by other materials, the fission ratio is less than 1. The process will then not be self-sustaining and will die out after being initiated by a stray neutron. When each fission produces exactly one more fission, the fission ratio is one and the reaction continues at a constant rate. The ^{235}U is then said to have a *critical mass*. In a nuclear reactor the fission ratio is controlled so that fission reactions occur at a desired rate.

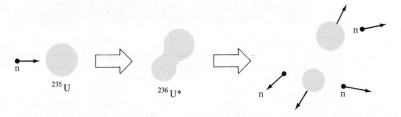

Figure 30.18. Induced fission of ^{235}U. A thermal neutron approaching a ^{235}U nucleus is captured, producing an excited ^{236}U nucleus that fissions. In this example the two fission fragments that are produced emit three prompt neutrons.

30.11 | FISSION REACTORS AND EXPLOSIVES

To achieve a critical mass, explosives and most fission reactors require enriched uranium containing more than the 0.7 percent of ^{235}U found in natural uranium. In the most common reactors, enrichment to 3 percent is required, while explosives use about 90 percent enriched material. The two uranium isotopes cannot be separated chemically, since their chemical properties are almost identical. However, the average molecular kinetic energy is determined by the temperature, so molecules containing a heavier isotope move more slowly on the average. Hence physical processes, such as diffusion, that depend on the average velocity can be used to separate isotopes. The masses of the uranium isotopes are so similar that although the separation may be accomplished by diffusion, it is slow and costly. Electromagnetic separators, which are similar to mass spectrometers, are sometimes used to further concentrate partially enriched uranium.

Reactors | All the commercial electric power reactors in the United States employ thermal neutrons to induce fission and have several common features. They have many *fuel rods,* tubes containing enriched uranium, along with a *moderator* and *control rods.*

The moderator serves to slow down the neutrons. This is necessary because the neutrons are produced with an initial energy of several MeV, and the ^{235}U fission cross section is largest below 1 eV. The control rods are made of a substance such as boron, which has a large neutron capture probability, and are used to control the fission rate. As these rods are withdrawn from the reactor assembly, the chain reaction grows until the desired rate is reached. The approximately 1 percent of the neutrons that are delayed play a key role in this process. They permit the operators to gradually adjust the control rods and to make readjustments if the reaction grows too rapidly.

The majority of the commercial reactors in operation or under construction in the United States are *boiling water reactors* (Figs. 30.19 and 30.20). The water in the reactor serves as the moderator; neutrons slow down when they collide with the hydrogen nuclei (protons) of the water molecules. The energy released in the fission process is converted into heat when the fast moving fission fragments collide with other atoms in the fuel rods. The heated water carries off this heat energy and then in the form of steam is used to drive turbines and produce electricity.

An important feature of boiling water reactors is

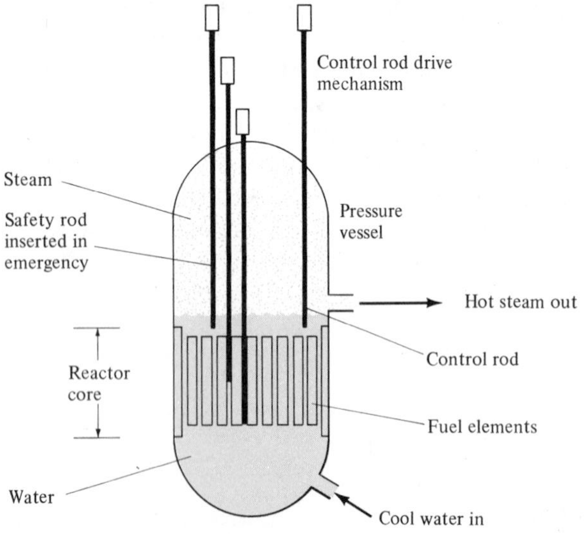

Figure 30.19. The boiling water reactor. The water serves as the moderator and also carries off the heat energy produced. Steam heated in the reactor drives a turbine and is then condensed before returning to the reactor.

Figure 30.20. The reactor at Vernon, Vermont, is a typical boiling water reactor. Located on the Connecticut River, its cooling system normally requires 23 m^3 of water per second, which exceeds the flow of the river during dry periods. Hence, it is equipped with cooling towers that require evaporation of much smaller quantities of water. The reactor is designed to produce 515 MW of electrical power, which is about 0.1% of the U.S. electrical generating capacity. It has 62,000 kilograms of uranium enriched to 2.7% ^{235}U contained in 23,000 fuel rods. (ERDA photo.)

their intrinsic stability against runaway chain reaction accidents. Any unintended rise in temperature increases the rate of boiling and bubble formation, thereby reducing the amount of water available to moderate the neutrons and slowing the chain reaction. Thus, such a reactor cannot explode like a nuclear bomb, and only an energy equivalent to a minor chemical explosion could be released in an accident. The controversy surrounding the use of nuclear power reactors centers not so much on the risk of a large-scale explosion as on the possible accidental dispersal of radioactive wastes into the environment.

Breeder Reactors

If the number of nuclear power reactors expands rapidly, uranium supplies are likely to be depleted in a few decades. A possible solution to this problem is the *breeder reactor,* which produces more fissionable material than it consumes. One type of breeder reactor produces ^{239}Pu, which can be fissioned like ^{235}U by thermal neutrons. In this reactor a fast neutron causes the reaction

$$^{238}_{92}U + n \rightarrow ^{239}_{92}U + \gamma$$

This is folowed by two β^- decays, which produce first neptunium and then plutonium:

$$^{239}_{92}U \rightarrow ^{239}_{93}Np + \beta^-, \qquad ^{239}_{93}Np \rightarrow ^{239}_{94}Pu + \beta^-$$

The plutonium α decays with a half-life of 24,000 years, so it is relatively stable. It is a suitable material for thermal neutron reactors as well as for bomb construction, but has the unfortunate property of being highly toxic.

Since fast neutrons are needed for breeding, a breeder reactor contains no moderator. The design currently receiving the greatest study in the United States uses liquid sodium metal, which is extremely reactive chemically, as a coolant. The reactor fuel is highly enriched, and the reactor core is very compact and hot. This places very great demands on the components and materials.

Fission Explosives

In a fission or atomic bomb, a subcritical sphere of ^{235}U or ^{239}Pu is surrounded by a chemical explosive. When this is detonated, shock waves implode the sphere, substantially reducing its volume and increasing its density. This makes the fission ratio greater than 1, and the chain reaction grows rapidly. Because of the inertia of the imploding material, the fission ratio stays above 1 long enough for a large total energy release to occur before the bomb material disperses. Such a bomb is limited in size by the requirements that it must not be critical before it is detonated, and that it must be brought to critical size quickly and held together long enough for a substantial amount of fission to occur. The largest fission explosives are equivalent to about 250 kilotons of TNT, and are about 12 times the size of the World War II atomic bombs.

30.12 | NUCLEAR FUSION

Nuclear fusion is potentially a much greater source of energy than fission because the supplies of suitable materials are nearly inexhaustible. Fusion is also attractive because its final products are stable and do not present radioactive waste disposal problems. Nevertheless, the achievement of controlled fusion and the extraction of useful power is a very difficult challenge.

The source of the difficulty is seen by considering two fusion reactions among the hydrogen isotopes, deuterium (d $= ^2_1$H) and tritium (t $= ^3_1$H):

$$d + d \rightarrow t + p + (4 \text{ MeV kinetic energy})$$
$$t + d \rightarrow ^4_2He + n + (17.6 \text{ MeV kinetic energy})$$

In both reactions, there is a strong electrical repulsion keeping the two positively charged nuclei apart. Unless their total initial kinetic energy is at least 0.1 MeV, they cannot come close enough together for the nuclear force to cause them to fuse. It is easy to supply 0.1 MeV or more to a nucleus in an accelerator, but such a machine consumes much more energy than is released, since fusion occurs in only a relatively few nuclei. Thus the only practical way to attain large-scale fusion is in a *thermonuclear reaction;* that is, by heating the materials until at least a small fraction of the nuclei have enough kinetic energy to fuse. This requires temperatures of about one million degrees Celsius.

The necessary temperatures are attained in a fission-fusion bomb by using a fission bomb to heat the hydrogen isotopes. Tritium is desirable in a thermonuclear device, since the d-t fusion process starts at a lower temperature than does the d-d reaction. However, tritium is radioactive, with a half-life of 12 years, so it is costly to produce and difficult to work with. These problems are avoided in hydrogen bombs by effectively storing the needed hydrogen isotopes in a solid, nonradioactive form in the chemical compound ^{6}Li d, lithium deuteride. When struck by a fast neutron from the fission explosion, the ^{6}Li nuclide produces tritium in the reaction

$$^6_3\text{Li} + \text{n} \rightarrow {}^4_2\text{He} + \text{t} + (4.8 \text{ MeV kinetic energy})$$

The tritium produced in this reaction then undergoes fusion with the deuterium. The required ^{6}Li is separated from natural lithium, which also contains the more abundant ^{7}Li, at a much lower cost than that of producing tritium.

The basic problem in achieving controlled fusion is containing the materials, since no solid can exist at the required temperatures. Since the hot deuterium or tritium gases are completely ionized, they form an electrically conducting mixture of positive and negative charges, which is called *plasma* and can be contained by magnetic fields. After the plasma is heated by bombardment from an electron gun, the magnetic field is suddenly increased. This compresses the plasma, raises its temperature still further, and causes some nuclei to fuse.

Although thermonuclear reactions have been produced in this way, it is not yet possible to produce them for a long enough time and in a large enough volume to be useful. It may turn out that an approach quite different from the magnetic containment schemes will work best. For example, considerable progress has been made recently using intense laser beams or electron beams focused on tiny spheres containing hydrogen isotopes. In general, progress has been slower than was hoped when the fusion program began in the 1950s, and practical power seems to be decades away at best. However, deuterium, unlike fossil fuels and uranium, is available in almost unlimited amounts. This is because one hydrogen nucleus out of 20,000 is a deuteron, making the oceans giant reservoirs of this fuel. Thus the potential benefits from the successful generation of power from deuterium fusion are very great.

30.13 | QUARKS

In the nineteenth century, scientists discovered that the many different substances that occur in nature or can be made in a laboratory are all composed of ninety-odd kinds of atoms. Early in this century, it was learned that atoms consist of electrons bound to nuclei containing protons and neutrons. Thus atoms are not the basic constituents of matter; rather they are built up from more elementary objects—electrons and nucleons. Now it appears that nucleons are also composites, and contain objects called *quarks*.

The first evidence that nucleons might be formed

from still smaller objects came with the development of high-energy accelerators and the discovery in the 1950s and 1960s of an embarrassingly large number of *hadrons*. Hadrons are particles that interact with each other via the strong nuclear forces, also known as the strong interactions. By now, physicists have observed *hundreds* of kinds of short-lived hadrons. Some of these particles are *mesons*, particles with integer spins (0, 1, . . .). The lowest mass meson is the *pi meson* or *pion,* which has spin 0 and comes in three charge states, π^+, π°, π^-; pion masses are about $\frac{1}{7}$ times that of a nucleon. The other new hadrons are *baryons,* with half integer spins ($\frac{1}{2}, \frac{3}{2}, . . .$), and masses greater than the nucleon mass; these particles can decay into nucleons, which are the lowest mass baryons, by emitting one or more mesons or photons. Regularities in the masses, lifetimes, magnetic moments, and other properties of the hadrons are reminiscent of the relations among the ground and excited states of molecules, atoms, and nuclei, all of which are composite structures.

Violent collisions between electrons and protons or between two protons provided additional evidence that nucleons are composite objects. In 1911, Rutherford discovered that alpha particles passing through thin foils are sometimes deflected through large angles. This was hard to understand in terms of a distributed charge model of an atom, but rather natural if most of the atomic mass is concentrated in a small nucleus. Similarly, these new experiments produced far more frequent large-angle events than would be expected with a diffuse distribution of matter in the nucleon, suggesting that the nucleon contains discrete objects.

In the early 1960s, many hadronic properties were correlated using a branch of mathematics called *group theory* and employing the concept of *broken symmetries.* To understand what this term means, think of an atom. In the absence of an externally applied magnetic field, the orientation of the atom is irrelevant, and its energy is the same for any value of the z component of the angular momentum. However, if a magnetic field is turned on, defining a special direction in space, this *removes or breaks the rotational symmetry* of the system. The energy levels now depend on the component of the angular momentum along the field, and the energy levels are split apart. In particle physics, a fairly small portion of the strong interactions is believed to lead to analogous splittings among the mass energies of related hadrons.

Murray Gell-Mann and George Zweig noted in

1964 that the mathematical relations among the hadronic properties followed very naturally if there are three *quarks* in a baryon, and a quark-antiquark pair in a meson. Quarks are spin $\frac{1}{2}$ objects with electric charges equal to $2e/3$ or $-e/3$, where e is the charge on a proton. (The word quark is from "Three Quarks for Muster Mark," in *Finnegan's Wake* by James Joyce.) In the full quantum mechanical formulation of atomic physics, quantum electrodynamics, an atom is held together because the electrons and protons exchange photons, which are massless, electrically neutral, spin 1 particles. Coulomb's law represents a first approximation to this description. Similarly, hadrons are held together because the quarks exchange massless, electrically neutral, spin 1 particles called *gluons*.

Originally, many physicists regarded quarks as a handy mnemonic device for an abstract mathematical description of hadronic properties. This attitude was due in part to the failure to observe fractionally charged particles. However, physicists now generally feel that quarks are, in fact, real physical entities. The failure to find free quarks is an indication that they are bound by forces that do not weaken as the separation grows, so that they cannot be pulled apart with a finite amount of energy.

Flavors and Colors | All the hadrons discovered before 1974 could be built up from three kinds or *flavors* of quarks, labeled u, d, and s, for *up*, *down*, and *strange*. These carry electric charges $2e/3$, $-e/3$, and $-e/3$, respectively; the antiquarks, $\bar{u}$, $\bar{d}$, and $\bar{s}$, carry charges of equal magnitude and opposite sign. Thus a proton has charge e and is (uud); a neutron has no charge and is (udd); π^+ is $(u\bar{d})$; π° is $(u\bar{u})$ or $(d\bar{d})$; and π^- is $(d\bar{u})$. The strange quark was needed to construct *strange particles*, hadrons that can decay into nucleons or pions only via the weak interactions, and hence have comparatively long lifetimes. The discovery in 1974 of a very massive, relatively long-lived meson called the J/Ψ confirmed the suspected existence of a fourth *charmed* quark, c. A fifth quark (*bottom, b*) was discovered more recently, and it is expected that a sixth (*top, t*) will be found.

Quarks are spin $\frac{1}{2}$ particles, so only one quark can occupy a given quantum state in accordance with the Pauli principle. A given spatial state can contain at most one spin up and one spin down quark of a given type. However, as many as three spin up quarks of a given flavor are found in a single spatial state. This implied that quarks have an additional quantum number that specifies their state. This quantum number has been named *color*, although, of course, it has nothing to do with color in the conventional sense. Each quark flavor comes in three colors, usually taken to be red, yellow, and green. Only "colorless" or "white" combinations of quarks correspond to physically observable hadronic states.

In analogy with quantum electrodynamics (QED), the theory of the interactions of quarks and gluons is called *quantum chromodynamics* (QCD). Many exact and approximate results have been obtained from this theory, but the mathematics of QCD are quite difficult, and many questions remain. Both theoretical and experimental progress in this field are quite rapid, and many new ideas are likely to appear before long. However, the basic elements of the quark picture are likely to remain with us, just as Rutherford's ideas about the atom still have relevance today.

EXERCISES ON SUPPLEMENTARY TOPICS

Section 30.10 | Nuclear Fission, and
Section 30.11 | Fission Reactors and Explosives

30-47 Why are β^-'s and γ's emitted in the decay of fission fragments but not β^+'s?

30-48 If control rods are inserted so that the chain reaction in a reactor stops, the reactor still produces a substantial amount of heat for a long time. Why?

30-48 Geologists spend a great deal of time inspecting potential nuclear reactor sites. What do you think they are looking for?

30-50 It is easier to separate fissionable ^{239}Pu bred in a reactor from the uranium and other materials present than to separate ^{235}U from the more abundant ^{238}U. Why? (This plutonium isotope is suitable for bombmaking. One reason for international inspection of reactors sold to the nonnuclear powers is that any thermal reactor breeds a small amount of ^{239}Pu. Over a period of years a reactor could make enough plutonium to assemble a bomb.)

30-51 Suppose an object of mass m_1 strikes a stationary object of mass m_2 squarely so that object 2 moves straight ahead. If the collision is elastic the final and initial kinetic energies of object 1 satisfy

$$K_f = \left(\frac{m_1 - m_2}{m_1 + m_2}\right)^2 K_i$$

Use this result to explain why the hydrogen nuclei

are more effective than the oxygen nuclei in water in slowing down neutrons.

Section 30.12 | Nuclear Fusion

30-52 The $_1^3$H-$_1^2$H fusion process releases 17.4 MeV, while $_{92}^{235}$U fission releases about 200 MeV. Find the energy released per nucleon for each process.

30-53 The reaction ^{6}Li + d → α + α + 22 MeV occurs in a hydrogen bomb. Why does this reaction need higher temperatures to occur than does the d + d or d + t fusion reactions? (d = $_1^2$H; t = $_1^3$H.)

30-54 2.2-MeV γ rays are produced when a proton captures a slow neutron in the reaction p + n → d + γ. (d = $_1^2$H.) Why can this fusion process occur at very low temperatures? (This process is not usable as a fusion power source since suitable neutron sources do not exist.)

PROBLEMS ON SUPPLEMENTARY TOPICS

30-55 (a) Calculate the energy released in joules when 1 kg of ^{235}U fissions, assuming 200 MeV per nucleus is released. (b) The electric power capacity of the United States is about 5×10^{11} W. Assuming 30 percent efficiency in producing electricity from nuclear energy, at what rate must ^{235}U be consumed to provide this power?

30-56 In a pressurized water reactor, as in a boiling water reactor, water cools the reactor core and serves as a moderator. However, the water is kept at high pressure so that it is well below its boiling point. Why is this reactor less stable than a boiling water reactor? (*Hint:* What is the effect, if any, of a local increase in the water temperature?)

30-57 (a) At what temperature is the average thermal kinetic energy of a nucleus, $\frac{3}{2}kT$, equal to 0.1 MeV? (b) Why is it not necessary to achieve this temperature to initiate a fusion reaction that requires a kinetic energy of 0.1 MeV?

Additional Reading

W.A. Blanpied, *Physics: Its Structure and Evolution,* Ginn Blaisdell, Waltham, Mass., 1969, chapters 20–23.

W.F. Libby, *Radiocarbon Dating,* 2nd ed., University of Chicago Press, Chicago, 1955.

I. Perlman, F. Asara, and H.V. Michel, Nuclear Applications in Art and Archaeology, *Annual Reviews of Nuclear Science,* vol. 22, 1972, p. 383.

D.E. Nelson, R.G. Korteling, and W.R. Stott, Carbon 14: Direct Detection at Natural Concentrations, *Science,* vol. 198, 1977, p. 507.

L. Paul Knauth and Madhurendu B. Kumar, Uranium Series Dating of Human Skeletal Remains from the Del Mar and Sunnyvale Sites, California, *Science,* vol. 213, 1981, p. 1003.

C.L. Bennett *et al.,* Radiocarbon Dating Using Electrostatic Accelerators: Negative Ions Provide the Key, *Science,* vol. 198, 1977, p. 508.

Richard A. Muller, Radioisotope Dating With Accelerators, *Physics Today,* vol. 32, February 1979, p. 23.

G.M. Raisbeck, F. Yiou, M. Fruneau, and J.M. Loiseaux, Beryllium-10 Mass Spectroscopy With a Cyclotron, *Science,* vol. 202, 1978, p. 215.

D.R. Inglis, *Nuclear Energy: Its Physics and Its Social Challenge,* Addison-Wesley Publishing Co., Reading, Mass., 1973.

B.N. DaCosta Andrade, *Rutherford and the Nature of the Atom,* Science Study Series, Doubleday and Company, Garden City, N.Y., 1964.

Enrico Fermi, The Nucleus, *Physics Today,* vol. 5, March 1952, p. 6.

C.F. Tsang, Superheavy Elements, *Physics Teacher,* vol. 13, 1975, p. 279.

G.T. Seaborg, W. Loveland, and D.J. Morrissey, Superheavy Elements: A Crossroads, *Science,* vol. 203, 1979, p. 711.

Frederick Reines, The Early Days of Neutrino Physics, *Science,* vol. 203, 1979, p. 11.

R.F. Post and F.L. Ribe, Fusion Reactors as Future Energy Sources, *Science,* vol. 186, 1974, p. 397.

Robert G. Sachs, Maria Goeppert-Mayer, *Physics Today,* vol. 35, 1982, p. 46.

Tracey Kidder, Taming a Star, *Science 82,* vol. 3, 1982, p. 54.

Otto R. Frisch and John A. Wheeler, the Discovery of Fission, *Physics Today,* vol. 20, Nov. 1967, p. 43.

F.L. Culler, Jr. and W.O. Harms, Energy from Breeder Reactors, *Physics Today,* vol. 25, May 1972, p. 28.

Anthony V. Nero, Jr., *A Guidebook to Nuclear Reactors,* University of California Press, Berkeley, 1979. A detailed but highly readable description of specific reactor designs and systems.

Charles M. Koplik, Maureen F. Kaplan, and Benjamin Ross, The Safety of Repositories for Highly Radioactive Wastes, *Reviews of Modern Physics,* vol. 54, 1982, p. 269.

K.G. McNeill, Photonuclear Reactions in Medicine, *Physics Today,* vol. 27, April 1974, p. 75. See also the printing corrections in vol. 27, May 1974, p. 71.

Laurie M. Brown and Lillian Hoddeson, The Birth of Elementary-Particle Physics, *Physics Today,* vol. 35, April 1982, p. 36.

Fred A. Donath and Robert O. Pohl, A Debate on Radioactive Waste Disposal, *Physics Today*, December 1982, p. 36.

John H. Nuckols, The Feasibility of Inertial-Confinement Fusion, *Physics Today*, September 1982, p. 24.

D. Allan Bromley, Neutrons in Science and Technology, *Physics Today*, December 1983, p. 30.

Fay Ajzenberg-Selove and Ernest K. Warburton, Nuclear Spectroscopy, *Physics Today*, November 1982, p. 26.

Sallie A. Watkins, Lise Meitner: The Making of a Physicist, *The Physics Teacher*, vol. 22, 1984, p. 12. Short biography of a pioneering nuclear scientist.

O. Lewin Keller Jr., Darlene Hoffman, Robert A. Penneman, and Gregory R. Choppin, Accomplishments and Promise of Transplutonium Research, *Physics Today*, March 1984, p. 34.

H.E. Gove, A New Accelerator—Based Mass Spectrometry, *The Physics Teacher*, vol. 21, 1983, p. 237.

William A. Fowler, Experimental and Theoretical Nuclear Astrophysics: the Quest for the Origin of the Elements, *Reviews of Modern Physics,* vol. 56., 1984, p. 149. Nobel lecture.

Nathan Isgur and Gabriel Karl, Hadron Spectroscopy and Quarks, *Physics Today*, November 1982, p. 36.

Edward Whitten, New Ideas About Neutrino Masses, *The Physics Teacher*, vol. 21, 1983, p. 78.

Richard C. Henry, Particle Physics Meets Cosmology— The Search for Decaying Neutrinos, *The Physics Teacher*, vol. 20, 1982, p. 531.

Paul Forman, The Fall of Parity, *The Physics Teacher,* vol. 20, 1982, p. 281.

John S. Laughlin, History of Medical Physics, *Physics Today*, July 1983, p. 26.

Paul R. Moran, R. Jerome Nickles, and James A. Zagzebski, The Physics of Medical Imaging, *Physics Today*, July 1983, p. 36.

Scientific American articles:

O. Hahn, The Discovery of Fission, February 1958, p. 76.

Lawrence Bodash, How the Newer Alchemy Was Received, August 1966, p. 88.

Victor F. Weisskopf, The Three Spectroscopies, May 1968, p. 15.

G.T. Seaborg and A.R. Fritsch, The Synthetic Elements: III, April 1963, p. 68.

Glenn T. Seaborg and Justin L. Bloom, The Synthetic Elements: IV, April 1969, p. 56.

Clyde E. Wiegand, Exotic Atoms, November 1972, p. 102.

Vernon D. Barger and David B. Cline, High-Energy Scattering, December 1967, p. 76.

G. Feinberg and M. Goldhaber, Conservation Laws, October 1963, p. 36.

M.D. Kameu, Tracers, January 1949, p. 31.

C. Emiliani, Ancient Temperatures, February 1958, p. 54.

C. Renfrew, Carbon 14 and the Prehistory of Europe, October 1971, p. 63.

R.K. O'Nions, P.J. Hamilton, and Norman M. Evensen, The Chemical Evolution of the Earth's Mantle, May 1980, p. 120. Geochronology.

Stephen Moorbath, The Oldest Rocks and the Growth of Continents, March 1977, p. 92.

J.D. Macdougall, Fission-Track Dating, December 1976, p. 114.

Maria G. Mayer, The Structure of the Nucleus, March 1951, p. 228.

Hans Bethe, What Holds the Nucleus Together, September 1953, p. 201.

Victor F. Weisskopf and E.P. Rosenbaum, A Model of The Nucleus, December 1955, p. 261.

R.H. Hofstadter, The Atomic Nucleus, July 1956, p. 55.

R.B. Peierls, Models of the Nucleus, January 1959, p. 235.

Robert B. Marshak, The Nuclear Force, March 1960, p. 98.

Lawrence Cranberg, Fast-Neutron Spectroscopy, March 1964, p. 79.

Michael Baranger and Raymond A. Sorenson, The Size and Shape of Atomic Nuclei, August 1969, p. 58.

H.W. Kendall and W. Panofsky, The Structure of the Proton and the Neutron, June 1971, p. 66.

Chris D. Zafiratus, The Texture of the Nuclear Surface, October 1972, p. 100.

Joseph Cerny and Arthur M. Poskanzer, Exotic Light Nuclei, vol. 238, p. 60.

Alan D. Krisch, The Spin of the Proton, May 1979, p. 68.

D. Allan Bromley, Nuclear Molecules, December 1978, p. 58.

S.B. Treiman, The Weak Interactions, March 1959, p. 72.

Sheldon Penman, The Muon, July 1961, p. 46.

Leon M. Lederman, The Two-Neutrino Experiment, March 1963, p. 60.

W.B. Fowler and N.P. Samios, The Omega-Minus Experiment, October 1964, p. 36.

John N. Bahcall, Neutrinos From the Sun, July 1969, p. 29.

Steven Weinberg, Unified Theories of Elementary Particle Interactions, July 1974, p. 50.

Sidney D. Drell, Electron-Positron Annihilation and the New Particles, June 1975, p. 50.

Sheldon Lee Glashow, Quarks with Charm and Color, October 1975, p. 38.

David B. Cline, Alfred K. Mann, and Carlo Rubbia, The

Search for New Families of Elementary Particles, January 1976, p. 44.

David B. Cline, Carlo Rubbia and Simon van der Meer, The Search for Intermediate Vector Bosons, March 1982, p. 48.

John G. Learned and David Eichler, A Deep-Sea Neutrino Telescope, February 1981, p. 138.

Howard Georgi, A Unified Theory of Elementary Particles and Forces, April 1981, p. 48.

Lewis P. Fulcher, Johann Rafelski, and Abraham Klein, The Decay of the Vacuum, December 1979, p. 150.

Martin L. Perl and William T. Kirk, Heavy Leptons, March 1978, p. 50.

Gerard 't Hooft, Gauge Theories of the Forces Between Elementary Particles, June 1980, p. 104.

Frank Wilczek, The Cosmic Asymmetry Between Matter and Antimatter, December 1980, p. 82.

Kenneth A. Johnson, The Bag Model of Quark Confinement, July 1979, p. 112.

Elliot D. Bloom and Gary J. Feldman, Quarkonium, May 1982, p. 66.

D.J. Hughes, The Reactor as a Research Instrument, August 1953, p. 23.

A.M. Weinberg, Power Reactors, December 1954, p. 33.

Alvin M. Weinberg, Breeder Reactors, January 1960, p. 82.

R.B. Leachman, Nuclear Fission, August 1965, p. 49.

Glenn T. Seaborg and Justin L. Bloom, Fast Breeder Reactors, November 1970, p. 13.

Hugh C. McIntyre, Natural-Uranium Heavy-Water Reactors, October 1975, p. 17.

George A. Cowan, A Natural Fission Reactor, July 1976, p. 36.

H.A. Bethe, The Necessity of Fission Power, January 1976, p. 21. See also April 1976, pp. 8–21 for letters to the editor.

Bernard L. Cohen, The Disposal of Radioactive Wastes from Fission Reactors, June 1977, p. 21. See also October 1977, p. 7, for letters to the editor about this article.

David J. Rose and Richard K. Lester, Nuclear Power, Nuclear Weapons, and Nuclear Stability, April 1978, p. 45.

Harold W. Lewis, The Safety of Fission Reactors, March 1980, p. 53. The lessons of the Three Mile Island accident.

Harold M. Agnew, Gas-Cooled Nuclear Power Reactors, June 1981, p. 55.

Kenneth S. Deffeyes and Ian D. MacGregor, World Uranium Resources, January 1980, p. 66.

Steven A. Fetter and Kosta Tsipis, Catastrophic Releases of Radioactivity, April 1981, p. 48.

William P. Bebbington, The Reprocessing of Nuclear Fuels, December 1976, p. 30.

George A. Vendryes, Superphénix: A Full Scale Breeder Reactor, March 1977, p. 26.

R.F. Post, Fusion Power, December 1957, p. 73.

T.K. Fowler and R.F. Post, Progress Toward Fusion Power, December 1966, p. 21.

Francis F. Chen, The Leakage Problem in Fusion Reactors, July 1967, p. 76.

William C. Gough and Bernard J. Eastlund, The Prospects of Fusion Power, February 1971, p. 50.

Moshe J. Lubin and Arthur P. Fraas, Fusion by Laser, June 1971, p. 21.

Bruno Coppi and Jan Rems, The Tokamak Approach in Fusion Research, July 1972, p. 65.

John L. Emmett, John Nuckolls, and Lowell Wood, Fusion Power by Laser Implosion, July 1975, p. 24.

Gerold Yonas, Fusion Power With Particle Beams, November 1978, p. 50.

Harold P. Furth, Progress Toward a Tokamak Fusion Reactor, August 1979, p. 50.

Haim Harari, The Structure of Quarks and Leptons, April 1983, p. 56.

Kenzo Ishikawa, Glueballs, November 1982, p. 142.

Claudio Rebbi, The Lattice Theory of Quark Confinement, February 1983, p. 54.

Duane A. Dicus, John R. Letaw, Doris C. Teplitz, and Vigdor L. Teplitz, The Future of the Universe, March 1983, p. 90.

Nariman B. Mistry, Ronald A. Poling, and Edward H. Thorndike, Particles With Naked Beauty, July 1983, p. 106.

George F. Bertsch, Vibrations of the Atomic Nucleus, May 1983, p. 62.

Wm. C. McHarris and John O. Rasmussen, High Frequency Collision Between Atomic Nuclei, January 1984, p. 58.

Robert W. Conn, The Engineering of Magnetic Fusion Reactors, October 1983, p. 60.

CHAPTER 31

IONIZING RADIATION

The radioactive decay of nuclei produces several kinds of *ionizing radiation* with energies that are typically several million electron volts per particle or quantum. When this radiation passes through matter, it leaves a trail of ionized atoms along its path. Even a small amount of ionization can seriously disrupt a sensitive system such as a living cell or a transistor.

The term "ionizing radiation" includes both nuclear radiation and atomic X rays. The less energetic quanta of lower frequency electromagnetic waves such as visible light and microwaves do not ordinarily cause appreciable ionization. In general, radiation refers only to ionizing radiation in this chapter.

Radiation is an excellent example of an area of science that has been studied intensively by physicists because of its intrinsic interest and that has also become invaluable in applications to many other fields including biology and medicine. Radiation also illustrates with unusual clarity how a scientific advance may, despite its great benefits, have a very large potential for harm. For example, X-ray pictures are often essential in diagnosing a serious illness, but even one X-ray exposure slightly increases the chance of developing cancer. Consequently, all those working with radiation, especially those in the health sciences, have an obligation to understand the physics and biology of radiation and to use it wisely and carefully.

In this chapter, we first consider the interaction of radiation with matter. We then discuss common radiation sources and the biological effects of radiation. We conclude with a discussion of some applications where radiation is used.

31.1 | THE INTERACTION OF RADIATION WITH MATTER

There are four major categories of radiation of interest to us. In order of increasing range in matter, these are:

1 positive ions, such as alpha particles
2 electrons and positrons
3 photons (gamma rays and X rays)
4 neutrons

Positive Ions | Alpha particles, protons, and other positive ions have very short ranges in matter. Roughly speaking, the average range or stopping distance varies inversely with the density of the medium, so a 5-MeV alpha that can travel about 4 cm in air cannot penetrate a sheet of paper or a layer of skin (Fig. 31.1).

A fast alpha particle undergoes frequent collisions with atomic electrons as it passes through matter, leaving a wake of excited and ionized atoms. About 100 eV is transferred in a single collision, so many collisions occur as the alpha slows. When its kinetic energy decreases to about 1 MeV, the alpha acquires two electrons and becomes a neutral helium atom. The neutral atom comes to rest in a short distance after a few more collisions.

Because an alpha is so much more massive than an electron, it is scarcely deflected in the collisions, and its path is nearly a straight line. Although individual alphas do not all travel exactly the same distance before stopping, the spread in stopping distances or

703

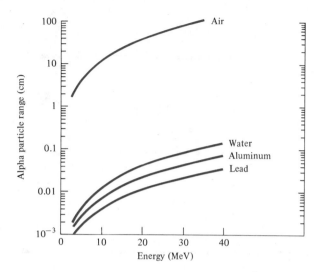

Figure 31.1. Range of alpha particles versus energy for air, water, aluminum, and lead. Water is a good approximation to soft animal tissue. (Adapted from G.S. Hurst and J.E. Turner, *Elementary Radiation Physics,* Copyright © 1970, John Wiley and Sons, New York.)

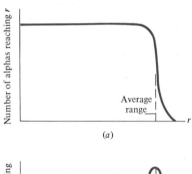

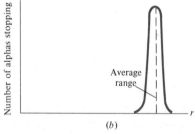

Figure 31.2. Alpha particles from a monoenergetic point source enter a medium. (*a*) The number of alphas passing a point a distance *r* into the medium. (*b*) The number of alphas stopping versus distance *r* traveled. Note most alphas travel a distance within a few percent of the average range.

straggling is only a few percent of the average range (Fig. 31.2).

Energy Loss Rate | The fact that alpha particles have a short range in matter compared to electrons of the same energy can be understood from

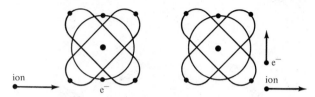

Figure 31.3. (*a*) An ion approaches an atomic electron that is moving slowly compared to the ion. (*Ab*) The ion collides with the electron, transferring momentum and energy to it. In this case the electron is removed, so the atom it came from is now ionized. Collisions can also excite atoms without ionizing them.

basic physical principles. An ion traveling through a medium collides repeatedly with atomic electrons and gradually loses energy (Fig. 31.3). We show in Section 31.8 that in a given medium, the energy ΔK lost per unit distance by an ion with charge q and velocity v is proportional to q^2/v^2, or

$$\Delta K \propto -\frac{q^2}{v^2} \qquad (31.1)$$

The minus sign indicates that the ion is losing kinetic energy. ΔK is also approximately proportional to the density of atomic electrons in the medium. It is convenient to rewrite this result for the rate of energy loss with distance in terms of the kinetic energy instead of v^2. If the ion is moving slowly compared to the speed of light, its kinetic energy is $K = \frac{1}{2}mv^2$, where m is the mass of the ion. Thus $v^2 = 2K/m$ and Eq. 31.1 can be rewritten as

$$\Delta K \propto \frac{-mq^2}{2K} \qquad (v \ll c) \qquad (31.2)$$

This shows that for a given kinetic energy, the energy loss rate is proportional to ion mass, so that massive particles such as alphas lose energy rapidly and come to rest in a short distance.

These equations are in good agreement with observed energy loss rates except at very low ion velocities. They show that the rate of energy loss increases as the ion slows and is a maximum near the end of the path. To the extent that they are correct, these equations allow one to compare the energy loss rates of various ions in a given medium. This is illustrated in the following example.

Example 31.1

Compare the rates of energy loss in a given material for protons and alphas having the same initial kinetic energies, assuming $v \ll c$.

From Eq. 31.2, the energy loss rate for a given particle is proportional to mq^2. A proton ($_1^1H$) has a charge e, while an alpha ($^4_2$He) has a charge $2e$. Thus q^2 for the two particles is related by $q_p^2 = q_\alpha^2/4$. The mass of a proton is close to 1 u and that of an alpha is close to 4 u, so $m_p = m_\alpha/4$. The energy loss rate for the proton is then proportional to

$$m_p q_p^2 = \frac{m_\alpha}{4} \frac{q_\alpha^2}{4} = \frac{m_\alpha q_\alpha^2}{16}$$

Thus the rate of energy loss with distance for a proton is 1/16 times that of an alpha. This suggests that a proton has a range 16 times that of an alpha with the same initial energy. Experimentally, this is only roughly correct; the proton range is typically about 10 times the alpha range. This happens because, as noted, Eq. 31.2 does not hold for very low velocities.

Electrons and Positrons |

These products of nuclear beta decays have ranges that are typically a hundred times greater than those of alpha particles. For example, a 1-MeV electron has a range in water or soft tissues of 0.4 cm (Fig. 31.4). Like positive ions, electrons lose energy mainly by ionizing or exciting atoms. However, for a given kinetic energy the velocity will be much larger than for a proton or an alpha, because the electron mass is so small. Thus the rate of energy loss given by Eq. 31.1 will be much smaller. This accounts for the large range of the electrons. Also, because of the small electron mass, a large deflection occurs in each collision with an atomic electron. Hence the electrons do not travel in a straight

line, but instead wander randomly, and the straggling is large.

The range of positrons is approximately the same as that of electrons. Eventually a positron slows and comes close enough to an electron so that they annihilate, producing gamma rays.

Photons |

Gamma rays and X rays are both electromagnetic quanta or photons, but since the gamma rays originate in nuclear rather than atomic processes, they typically have more energy. Photons do not produce appreciable ionization directly; instead, they lose energy to electrons which, in turn, cause ionization. Consequently, photons have a long range in matter. For example, a 1-MeV photon in water has a mean range of roughly 10 cm.

Photons transfer energy to electrons by three processes (Fig. 31.5). At energies below 0.1 MeV, the *photoelectric effect* is most important. Here a photon is absorbed by an atom, and an atomic electron is ejected. This process is most likely for inner shell electrons and for atoms with large atomic numbers. At about 1 MeV, *Compton scattering* dominates; this is photon-electron scattering, with the photon transferring some but not all of its energy to an atomic electron. At higher energies, it becomes possible for a photon to produce an *electron-positron pair*. This can occur near a nucleus when the photon energy is greater than the total mass energy of the two particles, $2m_e c^2 = 1.02$ MeV. Above a few million electron volts this is the most likely fate of a photon.

Photon absorption probabilities usually diminish as the energy rises. Consequently, as the photon energy increases, the absorption becomes more gradual, and the radiation becomes more penetrating or *harder*.

Neutrons |

Neutrons are uncharged and produce ionization only indirectly. Since they interact primarily with the small atomic nuclei rather than the atomic electrons, they have a very long range in matter. Neutrons with energies of a few million electron volts may travel a metre or so in water or in animal tissues.

Neutrons are slowed by elastic scattering from nuclei and by nuclear reactions. Some of these reactions lead to proton or gamma-ray emission; the protons in particular are significant in causing biological effects. Once a neutron has slowed to thermal energies—less than 1 eV—it has a high probability of being cap-

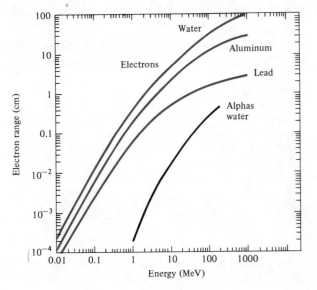

Figure 31.4. Average range of electrons in water, aluminum, and lead. The range of alpha particles in water is shown again for comparison. (Adapted from Hurst and Turner.)

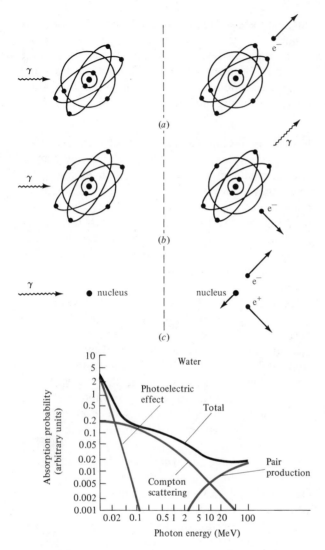

Figure 31.5. Energy loss mechanisms for photons.
(a) Photoelectric effect. The photon is absorbed by an atom, and an inner shell electron is ejected. (b) Compton scattering. Some energy and momentum are transferred to an atomic electron. (c) Pair production. The photon disappears and an electron-positron pair is created. Some momentum is transferred to a nucleus. (d) Relative importance in water of the three mechanisms versus energy. [(d) is adapted from Gloyna and Ledbetter.]

tured by a nucleus. Such a capture is often followed by gamma-ray emission.

31.2 | RADIATION UNITS

Four types of radiation measurements are used in various applications: *source activity, exposure, absorbed dose,* and *biologically equivalent dose.* We dis-cuss each of these briefly and define their most com-mon units.

The *source activity* $\mathfrak{A}$ is the disintegration rate of a radioactive material, or the rate of decrease in the number of radioactive nuclei present. It is measured in *curies,* where the curie is defined by the *exact* rela-tionship

$$1 \text{ curie} = 1 \text{ Ci}$$
$$= 3.7 \times 10^{10} \text{ disintegrations}$$
$$\text{per second} \qquad (31.3)$$

One gram of radium has approximately 3.7×10^{10} disintegrations per second, so that 1 Ci of radium has a mass of about 1 g.

The S.I. source activity unit is the *becquerel* (Bq), which is one disintegration per second. It is expected to gradually replace the curie.

The activity of a sample is related to its half-life, T. From Chapter Thirty, $\Delta N = -\lambda N \Delta t$ and $\lambda = 0.693/T$, so

$$\mathfrak{A} = \frac{-\Delta N}{\Delta t} = \lambda N = \frac{0.693}{T} N$$

The minus sign is needed because ΔN is the change in the number of nuclei present and is negative, while the disintegration rate or activity is positive. If there are n moles in the sample, then the number of atoms is $N = nN_A$, where Avogadro's number $N_A = 6.02 \times 10^{23}$ is the number of particles in a mole. The activity of n moles of a sample is then

$$\mathfrak{A} = \frac{0.693}{T} nN_A \qquad (31.4)$$

Note that the activity varies inversely with the half-life. This result is used in the following example.

Example 31.2
^{60}Co beta decays with a half-life of 5.27 years = 1.66×10^8 seconds into ^{60}Ni, which then promptly emits two gamma rays. These gamma rays are widely used in treating cancer. What is the mass of a 1000-Ci cobalt source?

Solving Eq. 31.4 for the number of moles and using $\mathfrak{A} = 1000$ Ci,

$$n = \frac{\mathfrak{A}T}{0.693 \, N_A} = \frac{1000(3.7 \times 10^{10} \text{ s}^{-1})(1.66 \times 10^8 \text{ s})}{0.693(6.02 \times 10^{23} \text{ mole}^{-1})}$$
$$= 0.0147 \text{ mole}$$

Since a mole of ^{60}Co has a mass of 60 g, the mass of the sample is

$$m = (0.0147 \text{ mole})(60 \text{ g mole}^{-1}) = 0.882 \text{ g}$$

Exposure and Absorbed Dose | *Exposure*
indicates the amount of radiation reaching a material, while *absorbed dose* indicates the energy absorbed in the material from the beam. Thus the absorbed dose depends on the properties of the material and the beam, while the exposure depends on the characteristics of the beam alone.

Exposure is defined only for X rays and gamma rays with energies up to 3 MeV and not for other forms of radiation. It is defined as the amount of ionization produced in a unit mass of dry air at standard temperature and pressure (STP), 1 atmosphere, and 0° C. The conventional unit is

$$1 \text{ roentgen} = 1 \text{ R}$$
$$= 2.58 \times 10^{-4} \text{ coulomb per kilogram} \quad (31.5)$$

Thus 1 roentgen of X rays will produce 2.58×10^{-4} coulomb of positive ions in a kilogram of air at STP, and an equal amount of negative ions.

The *absorbed dose* is the energy imparted by ionizing radiation to a unit mass of absorbing tissue. It is measured in *rads,* where

$$1 \text{ rad} = 0.01 \text{ joule per kilogram} \quad (31.6)$$

A 1 roentgen exposure to X rays or gamma rays produces a soft tissue absorbed dose of approximately 1 rad. The *gray* (Gy), which is 1 joule per kilogram or 100 rads, is the official S.I. unit and is gradually coming into use.

Unlike exposure, the absorbed dose is used with all kinds of ionizing radiation, not just X rays and gamma rays with energies below 3 MeV. Its use is illustrated in the following example, which also shows the great effectiveness of energy in the form of ionizing radiation in altering biological structures.

Example 31.3
Living tissues exposed to 10,000 rads are completely destroyed. By how much will this absorbed dose raise the temperature of the tissues if none of the heat is lost? (Assume that the specific heat of the tissue is the same as that of water, $c = 4180 \text{ J kg}^{-1} \text{ K}^{-1}$.)

According to Chapter Twelve, the heat ΔQ needed to produce a temperature change ΔT in a mass m is $\Delta Q = mc \, \Delta T$. Now 10,000 rads corresponds to an absorbed energy per unit mass $\Delta Q/m = 10,000(0.01 \text{ J kg}^{-1}) = 100 \text{ J kg}^{-1}$. Thus

$$\Delta T = \frac{\Delta Q}{m} \frac{1}{c} = (100 \text{ J kg}^{-1}) \left(\frac{1}{4180 \text{ J kg}^{-1} \text{ K}^{-1}} \right)$$
$$= 0.0239 \text{ K}$$

Such a small temperature rise would have a negligible effect if it were achieved by simply heating the tissue. Radiation is so lethal to living tissue because it does not deposit the energy uniformly at all points. Instead it imparts energy in relatively large amounts to single atoms at random locations, and this may disrupt critical biological molecules.

Biological Quantities | The absorbed dose refers to a physical effect: the transfer of energy to a material. However, the effects of radiation on biological systems also depend on the type of radiation and its energy. The quality factor (QF) of a particular radiation is defined by comparing its effects to those of a standard kind of radiation, which is usually taken to be 200-keV X rays. For example, fast neutrons (with energies above 0.1 MeV) have a QF of about 10 for causing cataracts. Hence the absorbed dose (in rads) of 200-keV X rays needed to produce cataracts is 10 times the dose required for neutrons. The QF varies with the radiation type and energy, with the animal species, and the biological effect under consideration (Table 31.1). Positive ions, which deposit more energy per unit length than beta or gamma rays, generally do more biological damage than the same absorbed dose of betas or gammas. However, their effects are often limited to the surface tissue because they have short ranges.

The *rem* and the millirem $= 10^{-3}$ rem are the units used in discussions of biological effects. For example, in the case of cataract formation, 1 rad of 200-keV X rays and 0.1 rad of fast neutrons each produce 1 rem of damage. In any situation the *biologically equivalent dose* (in rems) is the physical absorbed dose (in rads) times the QF. *One rem of any kind of radiation* produces the same biological effect, namely the effect of *one rad of 200-keV X rays.* In S.I. units,

TABLE 31.1
Typical QF values. By definition, the QF is exactly 1 for 200-keV X rays

Radiation	Typical QF
^{60}Co γ rays (1.17 and 1.33 MeV)	0.7
4-MeV γ rays	0.6
β particles	1.0
Protons (1 to 10 MeV)	2
Neutrons	2–10
α particles	10–20

TABLE 31.2

Radiation units

	Unit	Definition
Source activity	curie (Ci)	3.70×10^{10} disintegrations per second
	becquerel (Bq)	1 disintegration per second
Exposure (X and γ rays)	roentgen (R)	2.58×10^{-4} C kg^{-1} in dry air at STP
Absorbed dose	rad	0.01 J kg^{-1}
	gray (Gy)	1 J kg^{-1}
Biologically equivalent dose	rem	QF $\times$ (dose in rads)
	sievert (Sv)	QF $\times$ (dose in grays)

the biologically equivalent dose in *sieverts* (Sv) equals the dose in grays times the QF.

A summary of various radiation units is given in Table 31.2. An illustration of the use of these units is given in the next example.

Example 31.4

A cancer is irradiated with 1000 rads of ^{60}Co gamma rays, which have a QF of 0.7. Find the exposure in roentgens and the biologically equivalent dose in rems.

For gamma rays, the soft tissue dose in rads is approximately equal to the exposure in roentgens, so the exposure is approximately 1000 R. The biologically equivalent dose is the product of the QF and the dose in rads, or (0.7)(1000) = 700 rems.

31.3 | HARMFUL EFFECTS OF RADIATION

When radiation passes through living cells, it can alter or damage the structure of important molecules. This may lead to the malfunctioning or death of the cells and ultimately to the death of the organism.

Usually immature cells and the cells that are growing or dividing most rapidly are the most radiosensitive. Often cancer cells are rapidly growing; in such cases, they are highly vulnerable to radiation. Similarly, fetuses and infants are much more readily harmed by radiation than are adults. For example, in one study it was found that children whose mothers received pelvic X rays while pregnant had a 30 to 40 percent increase in the incidence of cancer.

The limited knowledge of the immediate effects on humans of large radiation doses comes from studies of victims of atomic bomb explosions and of occasional accidents. Whole body doses under 25 rems have no observable effect. As the dose increases above 100 rems, damage to the blood-forming tissues becomes evident, and above 800 rems severe gastro-

intestinal disorders occur. Death usually follows in a period of days or weeks if the dose is much more than about 500 rems.

Sublethal, short-term doses and doses acquired gradually over a long period may lead to cancer after a latent period of many years during which no ill effects are discernible. The chance of dying of cancer is doubled by a dose somewhere between 100 and 500 rems. Over a wide range of exposures, both in animal experiments and in human data, the increase in the cancer rate is directly proportional to the total accumulated dose.

Studies of low-level doses are difficult and inconclusive, because of the much larger incidence of cancer from other causes. Consequently, it is possible that the damage resulting from doses below some threshold is repaired, so that no increase in the cancer rate occurs. For many years, most experts favored the more conservative *linear hypothesis,* which assumes that the effects of radiation in causing cancer are proportional to the dose at all levels (Fig. 31.6). How-

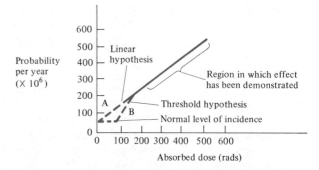

Figure 31.6. Leukemia rates in humans. Both the linear and threshold hypotheses are consistent with the high-dose data. Measurements of the very small rates at low doses such as 1 to 10 rads would require huge populations to be statistically meaningful. (Adapted from Hurst and Turner.)

ever, in 1980 the Advisory Committee on the Biological Effects of Ionizing Radiations (the BEIR committee) of the National Academy of Sciences found that the linear hypothesis probably overestimates the effects of low-level exposures. If this is correct, the public health implications of occupational exposures or of nuclear accidents may be less serious than previously believed. Nevertheless, since it is simple to apply and definite in its predictions, we will use the linear hypothesis in this chapter.

Genetic Effects | Although genetic mutations have helped humans to evolve to their present state, most mutations are harmful. Any increase in the mutation rate means more prenatal deaths and more people born with serious defects. In addition to the personal tragedy this implies, there is also the problem that the increasing ability of medical science to keep such people alive until they reproduce tends to keep defective genes in the human genetic pool.

Mutations can be increased above the normal rate by elevated temperatures, some chemicals, and by ionizing radiation. The mutations caused by radiation are similar to those occurring naturally. It is generally believed that their rate is proportional to the absorbed dose, no matter how small, and that there is no threshold or repair mechanism. The dose that will double the mutation rate is probably between 25 and 150 rems.

31.4 | CHRONIC RADIATION EXPOSURE

Living things have always been exposed to chronic low-level radiation arising from cosmic rays and from radionuclides naturally present in the environment. To this natural background, we have added a nearly equal amount of artificially produced radiation, mostly from medical diagnostic X rays (Table 31.3). Some individuals receive much more than the average chronic doses from natural sources. For example, some villages in Brazil and India happened to be built on soil with a high thorium content; doses in these villages are up to 100 times the norm. Also, the water from some wells and mineral springs in Italy, Austria, and elsewhere contains up to a million times the usual radioactivity. In these localities the cancer rates are abnormally high.

The National Council on Radiation Protection establishes a *maximum permissible dose* (MPD) for radiation workers and the general public from all artificial radiation sources (Fig. 31.7) *except* medical diagnosis and treatment. (The council does not regulate medical exposures.) The MPD has been lowered repeatedly as the hazards of radiation have become better understood. For example, in the late 1930s the MPD was 0.1 rem per day. The current MPD for radiation workers is a seventh of this, or 5 rems per year (Table 31.4). The average permitted for the entire population averaged over all individuals is a thirtieth of the occupational MPD, or 170 millirems per year.

Since any radiation exposure implies some hazard, these standards represent a compromise involving judgments of risks versus benefits. No single radiation-producing device or activity, such as watches, TV sets, or nuclear reactors, is permitted to expose the public to more than a small fraction of the MPD. In general, the Council requires that exposures must be "as low as reasonably achievable." Total present-day general population exposures are much less than the limit.

Nevertheless, the 1972 National Academy of Sciences report quoted in Table 31.4 urged that the allowed doses be lowered substantially. It is estimated that *if* everyone received the present MPD from nonmedical sources, between 3000 and 15,000 *additional* U.S. cancer deaths would occur annually. To see how this estimate comes about, let us assume that 250 rems will double the cancer rate. At 170 millirems per year, the average American who is about 30 years old will have accumulated 30(170 mrem) $\simeq$ 5000 mrem = 5 rems. This dose is 5/250 or 2 percent of the amount needed to double the cancer rate. The total death rate is roughly the U.S. population of 220 million divided by 70 years, or about 3 million, of which one-tenth or 300,000 deaths are due to cancer. A 2 percent increase would mean 6000 more deaths per year. (The range of additional deaths quoted in the Academy report corresponds to the uncertainty in the doubling dose, which is somewhere between 100 and 500 rems.) At present, nonmedical artificial radiation sources other than bomb testing contribute less than 1 percent of the MPD. Thus the actual number of deaths due to these sources is small.

Medical diagnostic X rays are currently the major source of artificial radiation exposure, and they are increasing annually. These average 40 percent of the nonmedical MPD and are estimated to cause 1500 to 3000 cancer deaths annually, plus a possibly greater number of genetic deaths in future generations via

TABLE 31.3

Chronic sources of radiation in the United States. All values are approximate or average (1 millirem = 10^{-3} rem)

Source	Dose (millirems per year except as noted)	
Cosmic rays:		
Sea level	41	
Denver (5000 ft)	70	
Leadville, Colorado, area (10,500 ft)	160	
20,000 ft	400	
Commercial jet (35,000 ft)	0.7 millirems per hour	
U.S. average from cosmic rays		44
γ Rays from rocks, soil (Ra, U, Th, K, etc.):		
Atlantic Coastal Plains	22.8	
Colorado Front Range	89.7	
U.S. average from external radionuclides	40	
Calculation of gonadal dose:		
Correction factor due to housing shielding	0.8	
Correction factor due to biological shielding	0.8	
U.S. average gonadal dose from external radionuclides	(40)(0.8)(0.8)	26
Internal radionuclides:		
^{40}K	16	
^{14}C, Ra, and decay products	2	
Total		18
Total U.S. environmental average gonadal dose: (Oakley, 1972)		88 ± 11
Fallout (1970)	4	
Nuclear power	0.003	
Medical diagnostic	72	
Medical radiopharmaceuticals	1	
Occupational	0.8	
Miscellaneous	2	

(Sources: Natural Radiation Exposure in the U.S., by D. T. Oakley, U.S. Environmental Protection Agency, June 1972; National Academy of Sciences Committee on Biological Effects of Ionizing Radiation Report, November 1972.)

lethal mutations. It is estimated that *medical exposures could be greatly reduced* without any loss of the benefits of diagnostic X rays. Clearly the medical and dental professions have a great responsibility to minimize exposures. This means avoiding unneeded diagnostic X rays and assuring the use of modern, well-shielded machines and high-sensitivity films. It also means educating the public, so that people do not equate diagnostic X rays with good medical practice in evaluating the care they receive.

The importance of using the proper X-ray devices and techniques is illustrated by the range of doses received from chest X rays. The best procedures give only 6 millirems, although the average is 200 millirems. By contrast, the X-ray machines used in mobile units to screen large populations for tuberculosis, which are relatively inexpensive to operate, gave about 1000 millirems to the chest area. Public health authorities no longer recommend their general use, since they believe the risks of radiation exposure out-

Figure 31.7. An example of the misuse of X rays was the shoe-fitting fluoroscopy machine, a popular selling feature of many children's shoe stores from 1946 until the late 1950s. The first machines were largely unshielded. Doses to the feet sometimes were as high as 40 rems per minute and averaged 8 rems per minute. The machines were eventually eliminated by legislation that restricted X-ray use to licensed medical practitioners. Here, an expert is measuring X-ray levels while a little girl watches the bones in her toes wiggle. (Courtesy of Ira A. Paul, Bureau of Radiation Control, New York City Department of Health.)

31.5 | RADIATION IN MEDICINE

Despite its hazards, the use of ionizing radiation in medical research, diagnosis, and therapy has been invaluable and has saved many lives. Within a few weeks of their discovery by Roentgen in 1895, the value of X rays for medical diagnosis was realized. By early in this century, X rays and naturally occurring radionuclides (radium, radon) were being used in cancer therapy. Today, many different radionuclides are produced in reactors and accelerators and are available for use in nuclear medicine.

Medical Research | Amino acids, sugars, DNA, and penicillin are a few of the hundreds of biological compounds commercially available today containing ^{14}C, ^{3}H, ^{35}S, ^{32}P, or other radionuclides. The radioactivity of these *tracers* makes it possible to follow their pathways and metabolism very accurately and conveniently. The active transport of sodium in nerve fibers, the metabolism of starches and sugars, the building up of proteins from amino acids, and the action of hormones and drugs are examples of fundamental biological processes that have been studied with tracers.

Diagnosis | Most diagnostic studies done with radionuclides share some common features. A radioactively labeled compound that is absorbed in the organ of interest is administered, and a detector outside the body measures the radioactivity at various times or locations. The radiation doses are normally comparable to those from diagnostic X rays. To minimize the dose, short-lived nuclides are selected that

weigh those of lung diseases for the average person.

To summarize, any radiation exposure carries with it a small but very real risk. The benefits of such exposures must be weighed against the hazards, both for the individual and for society overall.

TABLE 31.4

Maximum permissible doses (MPD) for whole body exposures as revised in 1972 by the National Council on Radiation Protection. Larger doses are permitted for some parts of the body

	MPD
General Population	
MPD for any individual[a]	500 millirems per year
Average for whole U.S. population[b]	170 millirems per year
Radiation Workers	
Annual	5000 millirems per year
3-month period	1250 millirems per 3 months
Pregnant workers	500 millirems per 9 months

[a] Facilities using radiation may occasionally expose some individuals to this dose.
[b] This is the average for everyone, including radiation workers.

emit gamma rays with just enough energy to be detected. For many purposes, a long-lived excited state of technicium ($Z = 45$) referred to as ^{99m}Tc is almost ideal. Formed when ^{99}Mo beta decays, this nuclide decays to its stable ground state ^{99}Tc with a half-life of 6.0 hours by emitting a 140-keV gamma ray. Its chemical properties permit it to be incorporated into many kinds of molecules, so it can be directed at a large number of organs.

A *dynamic function study* measures the rate at which a thyroid, kidney, or other organ absorbs or eliminates a material. For example, the amount of a ^{99m}Tc labeled compound taken up by the thyroid as measured by a counter held against the throat is an indicator of its condition. Similarly, when diuretics containing radioactive compounds are concentrated in the kidneys, alterations from the usual patterns of uptake and excretion signal possible abnormalities.

In a *gamma-ray-imaging study* or *scan,* a picture is produced by detecting the radiation from an absorbed radionuclide. The radiation is detected with the aid of a *scintillating crystal,* which emits flashes of light when beta or gamma rays pass through it. (Scintillation detectors are discussed in more detail in the supplementary topics at the end of this chapter.) The scintillation counter is equipped with a lead collimator containing several tapered holes that transmit the gamma rays from a small region of the organ that has absorbed the radionuclide. The counter is slowly moved or *scanned* over the region of interest, and the pulses generated are amplified and recorded on paper, photographic film, or video tape (Fig. 31.8).

For example, a thyroid scan is done by administering a suitable compound containing ^{99m}Tc. Since nonfunctioning tissue in the thyroid does not absorb these materials, it appears as a less radioactive region in the thyroid. Areas with above normal metabolic activity also show up with high levels of radioactivity. Other common scans include those of the liver, kidneys, and brain.

The *Anger* or *gamma camera* is an ingenious device developed by H.O. Anger in 1957. It measures the radioactivity everywhere in a large area simultaneously and displays the results on an oscilloscope screen as well as recording them on film or tape. This makes it possible to perform dynamic studies in which one observes a radionuclide enter or leave an organ. It also permits static examinations to be completed in a minute or so. For this reason, the Anger camera has largely replaced the older scanners, which required up to an hour.

The gamma camera has a single, large, thin scintillating crystal shielded by a lead collimator containing hundreds of holes (Fig. 31.9a). This collimator is placed over the organ under study. Nineteen light-sensitive *photomultiplier tubes* behind the crystal observe flashes of light produced when a gamma ray passes through the crystal and send signals to a small computer. The amount of light entering a particular tube depends on its distance from the point where the gamma ray passes through the crystal. The computer locates this point from the relative light intensities and sends appropriate signals to the oscilloscope plates. The oscilloscope then displays a two-dimensional picture whose intensity at each point is proportional to the level of radioactivity at the corresponding point in the organ. Gamma cameras are used to diagnose diseases of the thyroid, liver, brain, kidneys, lungs, spleen, heart, and circulatory system.

In the 1970s, the availability of high speed digital computers facilitated the development of several powerful new medical imaging systems, including the now widely used *computer-assisted-tomography* (CAT). "Tomography" comes from the Greek words *tomos* (slice) and *graph* (picture); a tomograph looks like a slice through the body. CAT-scans provide important information that cannot be obtained from conventional X-ray pictures.

Conventional X-ray pictures are the oldest and most common use of ionizing radiation. A beam from an X-ray tube passes through the body and strikes a fluorescent screen or a sensitive film, producing im-

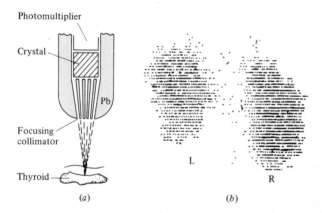

Figure 31.8.   Diagnostic scans. (*a*) A scintillation counter shielded with a lead collimator. (*b*) A kidney scan. Note the difference in appearance between the healthy right kidney and the diseased left kidney.

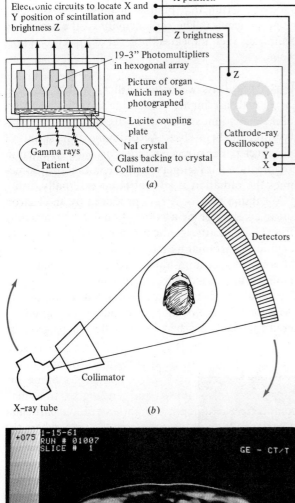

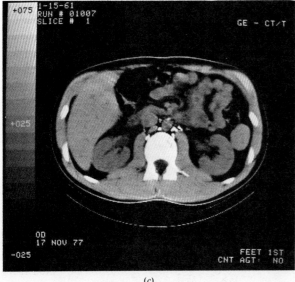

Figure 31.9. (*a*) The Anger or gamma camera. (*b*) A CAT-scanner. Each of the hundreds of detectors is connected to a computer. (*c*) The computer processes the information to produce a cross-sectional view in this case of the abdomen. (Courtesy of General Electric.)

ages with excellent spatial resolution. This system works well for examining bone fractures, for example, because bones differ greatly from the body fluids and soft tissues in their density and X-ray-absorption rate. However, the contrast is very poor when one is studying soft tissues with similar characteristics. For example, a 1-cm brain tumor with a difference in X-ray-absorption rates from the surrounding tissues of only 1% produces an intensity difference of only 0.2% after the beam has passed through the whole head. Since intensities must vary by at least 2 or 3% to be distinguishable, such a tumor cannot be detected.

In a CAT-scan, an X-ray beam is collimated into a thin fan, 1.5 to 10 mm thick (Fig. 31.9*b*). Hundreds of ionization detectors feed a computer their measurements of the X-ray transmission along each narrow pathway. The source and the detector array rotate about the body axis, and measurements are made at hundreds of angles over a total period of a few seconds. This procedure provides depth information to the computer, which unfolds a two-dimensional picture of a cross section of the body with great contrast and sensitivity (Fig 31.9*c*). CAT-scans can distinguish tissues with density differences of a fraction of 1%, and can readily display tumors which cannot be detected with the older techniques.

Another new imaging system is *positron-emission-tomography* (PET), which is useful in metabolic studies. Compounds labeled with positron-emitting radionuclides are administered to the patient. The positrons travel only a short distance before they stop and annihilate with electrons, producing two 511-keV gamma-ray photons traveling in nearly opposite directions. These photons are observed by a ring of detectors placed around the patient. Again a computer unfolds the accumulated data to construct an image of the distribution of the absorbed radionuclide in the organ of interest.

The newest imaging technique employs nuclear magnetic resonance (NMR) instead of ionizing radiation. NMR probes the magnetic environments of the nuclei of specific atomic nuclei, such as hydrogen or phosphorus (Sec. 29.6). NMR imaging systems offer the possibility of distinguishing normal and abnormal tissues, of measuring the flow of blood in capillary beds, and of doing chemical analyses in the body. As in ultrasonic scans (Sec. 22.7), no exposure to ionizing radiation is involved, which is a major advantage. However, so far the spatial resolution achieved has been relatively poor.

Radioimmunoassay (RIA) | A technique with increasingly widespread applications in research and in medical diagnosis, RIA can detect minute quantities of *antibodies,* hormones, and other complex molecules. An antibody is a protein molecule manufactured by the body to combat a specific *antigen.* For example, someone who inhales ragweed pollen produces a particular antibody that binds to the pollen.

In one version of RIA, a known quantity of radioactively labeled antigen is added to a sample of blood serum. (Serum is a clear yellowish fluid obtained by removing the cells in the blood with a centrifuge.) Some of the labeled antigen binds to antibodies in the serum, forming antigen-antibody complexes. These complexes are separated out using electrophoresis (Section 17.11) or other methods. The radioactivity of the separated complexes indicates the fraction of the labeled antigen that has been bound in competition with the unlabeled antigen originally in the serum. Comparison with known calibration samples then determines the amount of antigen originally present.

In another RIA procedure, labeled antibodies are added to serum that bind with human growth hormone molecules. The radioactivity of the separated complexes then indicates the level of this crucial hormone.

Therapy | As we noted earlier, rapidly dividing cancer cells are highly vulnerable to radiation. Also, unlike many kinds of normal cells, they lack the ability to repair themselves. Accordingly, at least half of all cancer patients receive *radiation therapy,* often in combination with surgery and chemotherapy. Sometimes the radiation is administered externally, using ^{60}Co gamma rays, or X rays produced by an electron linear accelerator or a conventional X-ray machine (Fig. 31.10). Large medical centers frequently have facilities for irradiating patients with electron beams. Alternatively, radium-filled needles, small seeds containing radon gas, or wires or ribbons containing artificial radionuclides may be implanted in the tumor. Occasionally, radioactive compounds are administered that are absorbed by the affected part of the

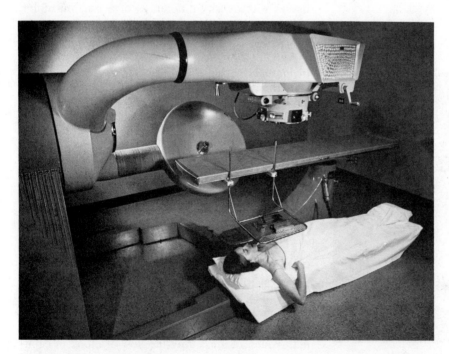

Figure 31.10. A facility for ^{60}Co radiation therapy. The cobalt is in a heavily shielded container that can be moved near various parts of the patient's body. Since the cobalt produces gamma rays that have a greater energy than is available with standard X-ray machines, they penetrate further into the body and are more effective in treating many forms of cancer. (Courtesy of the University of Texas, Houston, M.D. Anderson Hospital & Tumor Institute.)

body. Still in the experimental stage is the use of beams of neutrons and other particles with fairly high energies. Because these particles deposit much of their energy in a small region near the end of their range, it is hoped that they will do less damage to the tissues surrounding the cancer being irradiated and therefore permit larger doses.

31.6 | OTHER USES OF RADIATION

Radiation has hundreds of uses in agriculture and industry. Sometimes radiation offers the cheapest or most convenient way to perform some function, while in other situations, no other technique is available. For example, there is often no substitute for tracer studies of metabolic processes or complex reactions.

In agriculture, tracers have been used to study how fertilizers, hormones, weed killers, and pesticides perform their functions. With labeled growth hormones or pesticides, it is even possible to measure residues in the final food products that are too minute to be detected chemically.

Radiation has been used to induce plant mutations, which have led to improved varieties of many crops, such as wheat, peas, and beans, with higher yields and more resistance to diseases. It has also been used to eliminate a major agricultural pest, the screwworm fly, without employing pesticides. This was done by using radiation to sterilize hordes of male flies and releasing them to compete with normal males for mates. The females mate only once, and unfertilized eggs do not hatch. Hence, repeatedly saturating an area with millions of sterile males soon wiped out the population completely.

The range of uses that radiation has found in industry can be seen with some typical examples (Fig. 31.11). Gamma rays or X rays are used much like medical diagnostic X rays to reveal potentially dangerous but invisible defects in metal castings and welds. The thickness of rolled sheets of plastic, paper, or metal, or of thin coatings, can be controlled by measuring the transmission of radiation. Densities and fluid levels are easily measured. Tracers are used to locate leaks in underground pipelines, to measure the rate of wear of automobile tires and engine parts, and to determine the effectiveness of detergents.

SUMMARY

Charged particles passing through matter ionize atoms along their paths by colliding with atomic elec-

trons. For a given energy, the rate of energy loss with distance is greatest for the most massive particles. Hence, at energies of a few MeV, alpha particles penetrate about 0.01 cm in water, while electrons travel about 1 cm. Photons lose energy to electrons via the photoelectric, Compton, and pair production processes; at 1 MeV their range is about 10 cm. Neutrons lose energy only in nuclear processes, and their ranges are of the order of a metre.

The source activity refers to the number of nuclear disintegrations occurring per second in a source and is measured in curies. The exposure indicates the number of photons in an X-ray or gamma-ray beam and is measured in roentgens. The absorbed dose for any kind of radiation is the energy absorbed per unit mass of material; its unit is the rad. One roentgen of X rays or gamma rays produces a dose of about one rad. The biologically equivalent dose in rems is the absorbed dose in rads times the quality factor, which is defined to be one for 200 keV X rays.

Radiation doses from cosmic rays and naturally occurring radionuclides average about 88 millirems per year in the United States. Medical diagnostic X rays contribute almost as large a dose on the average. The effects of low-level radiation in causing cancer and inducing mutations are apparently proportional to the total dose accumulated. Immature rapidly growing cells are highly radiosensitive, so many kinds of cancer cells can be killed with radiation.

Checklist

Define or explain:

ionizing radiation	rad
range	quality factor
energy loss rate	rem, sievert
photoelectric effect	doubling dose
Compton scattering	maximum permissible
pair production	dose
source activity	linear hypothesis
curie, becquerel	latent period
exposure	tracers
roentgen	Anger camera
absorbed dose	CAT-scan

REVIEW QUESTIONS

Q31-1 The rate at which a charged particle loses energy in matter is _____ to the mass for a given kinetic energy.

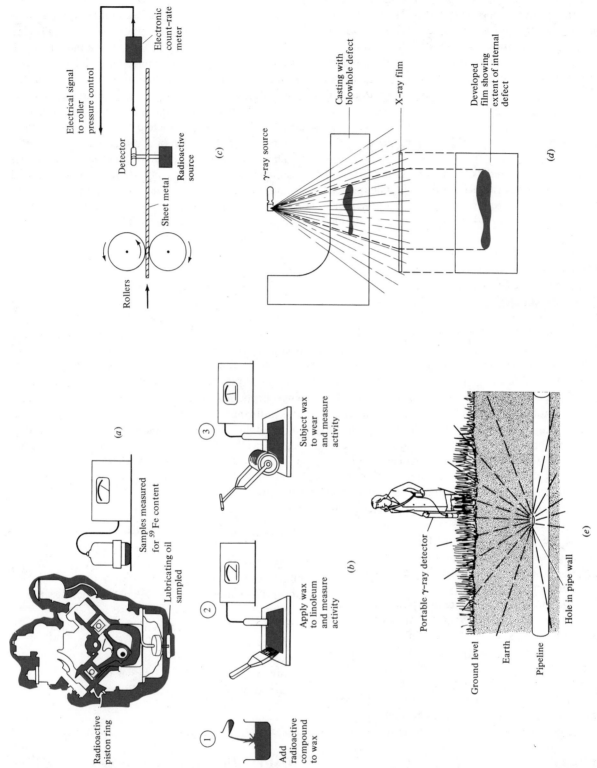

Figure 31.11. Typical industrial uses of radiation. Measurement of wear rates of (a) engine parts and (b) floor wax. (c) A thickness gauge. (d) Inspection of castings. (e) Locating pipeline leaks with tracers.

Q31-2 Photons lose energy by _____, _____, and _____.

Q31-3 Which of the following has the longest range: protons, alpha particles, positrons, neutrons? Which has the shortest range?

Q31-4 The source activity varies _____ with the half-life.

Q31-5 Absorbed dose is the energy absorbed per _____.

Q31-6 A 1-roentgen exposure to X rays gives an absorbed dose of about _____.

Q31-7 Radiation is so lethal because it can disrupt critical _____.

Q31-8 The major source of artificial radiation exposure is _____.

Q31-9 Prolonged exposure to low-level radiation can cause _____ and _____.

Q31-10 Rapidly growing cells are most _____.

EXERCISES

Section 31.1 | The Interaction of Radiation with Matter

31-1 If a microwave beam is sufficiently intense, it can heat a material sufficiently to ionize some of its atoms. How does this differ qualitatively from the way gamma-ray photons cause ionization?

31-2 The rate of energy loss for a charged particle in a medium with atomic number Z is almost exactly proportional to nZ, where n is the number of atoms per unit volume. Give an argument for this.

31-3 In a laboratory experiment to study the properties of radioactive sources, a student finds that the count rate drops sharply to zero when a Geiger-Mueller counter in the air is moved from 9 to 10 cm from the source. (a) What is the most likely nature of the radiation? Explain. (b) Estimate the range of this radiation in aluminum, assuming the range varies inversely with the density. (The density of aluminum is 2700 kg m^{-3}, and that of air is 1.29 kg m^{-3}.)

31-4 Using Fig. 31.4, find the ranges of 10-MeV alpha particles and 10-MeV electrons in water.

31-5 In the photoelectric process, a gamma ray is absorbed and an electron is ejected from an atom. Usually X rays are then produced. Explain why.

31-6 When a gamma ray is absorbed in a photo-electric process, an electron is ejected from an atom, and the gamma ray disappears. Is the kinetic energy of the electron equal to the gamma-ray energy? Explain your answer.

31-7 In electron-positron pair production by gamma rays, some momentum is transferred to a nearby nucleus. Why is the energy transferred to the nucleus very small?

31-8 According to Fig. 31.5, when gamma rays are absorbed in water, at what energy are pair production and Compton scattering equally important?

31-9 Protons and deuterons ($^1_1$H and $^2_1$H) lose the same amount of energy in a thin sheet of material. How are their energies related?

31-10 If protons and deuterons ($^1_1$H and $^2_1$H) have the same kinetic energy when they enter a thin sheet of material, how are the energies they lose related?

31-11 How could thin foils and a proton detector be used to determine the energy of a beam of protons produced by an accelerator?

31-12 Suppose ^{14}C and ^{14}N nuclei are both accelerated to an energy of 40 MeV and are then allowed to pass through a thin foil. If the ^{14}C nuclei lose 2 MeV, how much energy will the ^{14}N nuclei lose? (Some of the dating studies based on the direct detection of radioactive nuclei with accelerators use this technique to sort out the various nuclei with nearly equal masses. See Section 30.3.)

31-13 Using Fig. 31.4, estimate the range of a 10-MeV electron in (a) water; (b) lead.

Section 31.2 | Radiation Units

31-14 What is the mass of a 1-microcurie (10^{-6}-Ci) ^{131}I source? (The half-life of ^{131}I is 8.1 days.)

31-15 How many positive ions are produced in 1 kg of air at standard conditions by a 1-R X-ray exposure? (Assume each ion has one electronic charge, $+e$.)

31-16 A 10^{-6}-Ci source has a half-life of 8 hours. How many moles of the radioactive material are present in the source?

31-17 When ^{99m}Tc decays, it emits a 140-keV gamma ray. (a) How many photons per second are emitted by a 10^{-6}-Ci ^{99m}Tc source? (b) How much energy, in keV, is carried by these photons? (c) Find the power in joules per second or watts associated with the photons.

31-18 A well-insulated tank of water exposed to

radiation has its temperature increased by 30 K. How many rads were absorbed?

31-19   A beam of gamma rays is found to produce 10^{16} positive ions per kilogram in dry air at STP. What is the corresponding exposure if each ion has a charge $+e$?

31-20   Treating fish or meat with 200,000 rads kills many of the bacteria present and increases the refrigerated shelf life five to seven times. (a) How much energy is absorbed per kilogram of food? (b) Neglecting heat losses, estimate the temperature rise. (Assume that the specific heat is equal to that of water.)

31-21   A ^{60}Co source produces an absorbed dose of 4000 rads per hour in tissue. The QF is 0.7 for cobalt gamma rays. (a) How much time is required for an absorbed dose of 300 rads? (b) How much time is required for a biologically equivalent dose of 300 rems?

31-22   A mouse receives an absorbed dose of 200 rads of 10-MeV protons, which have a QF of 2. What is the biologically equivalent dose?

Section 31.3 | Harmful Effects of Radiation, and Section 31.4 | Chronic Radiation Exposure

31-23   A nuclear radiation technologist receives a whole body dose of 0.1 rem every time she loads a radium source. How many times is she permitted to load the source in (a) a quarter; (b) a year?

31-24   The lethal dose of radiation for mammals and birds is under 1000 rads, while it takes about 100,000 rads to kill amoebas and over a million rads to kill viruses. What is the likely reason for this?

31-25   Suppose 100,000 radiation workers each receive 5 rems. Is this more or less important genetically than the entire world population of about 3 billion people receiving 5 millirems? Why?

31-26   A stewardess flies at an average height of 11 km for 20 hours per week. If she receives 0.7 millirems per hour, what is her annual biologically equivalent dose? Compare this with the MPD for the general population and for radiation workers.

31-27   Estimate the death rate due to diagnostic X rays, assuming the U.S. cancer death rate is 300,000 per year, that the doubling dose for cancer is 250 rems, and that the average person of age 30 has received 72 millirems per year.

31-28   A radiation worker accidentally receives 50 rems. Using the assumptions in the preceding exercise, estimate his chances of developing cancer because of this accident.

31-29   The earliest safety rules for radiological personnel limited their dose to the equivalent of 100 rems per year, a level low enough to avoid skin burns. Estimate the cancer risk and genetic effects of such doses.

Section 31.5 | Radiation in Medicine

31-30   In *rotation therapy* the patient or the radiation source is rotated about an axis through the tumor being irradiated. What is the advantage of this procedure?

31-31   How do you explain the apparent paradox that X rays induce cancer and that X rays are used to treat cancer?

Section 31.6 | Other Uses of Radiation

31-32   How can you use radioactively labeled dirt to test the effectiveness of a detergent?

PROBLEMS

31-33   X-ray machines produce a continuous distribution of X-ray photon energies up to a maximum determined by the voltage of the machine. A thin sheet of aluminum will absorb the lowest energy X rays without appreciably absorbing the higher energy photons. (a) What is the reason for this? (b) What is the advantage of such an aluminum filter in medical uses of X rays?

31-34   A 1000-Ci alpha emitter is in a lead container. (a) How many disintegrations occur per second? (b) The alpha particles have an energy of 2.5 MeV, and all the alphas are stopped in the lead. At what rate is energy absorbed in the lead?

31-35   Cobalt 60 produces two gamma rays per disintegration. (a) How many gammas are emitted per second by a 10-Ci source? (b) What is the flux of gammas per square metre at a distance of 1 m from the source? (Assume the radiation is uniform in all directions.)

31-36   A 10-Ci β^- emitter is 2 m from your hand. How many electrons strike each square centimetre of your hand per second? (Neglect absorption in air.)

31-37   A radiation worker is accidentally exposed to 100 R of 200-keV X rays. Estimate the absorbed and biologically equivalent doses.

31-38   One milligram of radium produces a dose

of 8.2 rads per hour at a distance of 0.01 m. The QF for radium gamma rays is 0.965 for the tissue irradiated. (a) Explain why the exposure varies inversely with the square of the distance from the source. (b) How long is required for a 60-mg radium source to give a dose of 100 rems at a distance of 0.01 m? (c) How long would this take at 0.05 m?

31-39 A small quantity of radioactive serum albumin is administered to a patient, and an equal quantity is placed in 2000 cm^3 of water to serve as a standard. After 10 minutes, a blood sample is taken from the patient and the red cells removed by centrifugation, leaving the serum. The activity of 10 cm^3 of the serum is measured with a scintillation counter and found to be 2600 counts per minute. A 10-cm^3 sample from the standard gives 1892 counts per minute. In both cases, 155 counts per minute are due to background from cosmic rays and other environmental sources. What is the total volume of the patient's blood serum?

31-40 A laboratory experiment in a physics class uses a 10-microcurie ^{137}Ce source. Each decay emits a 0.66-MeV gamma ray. (a) How many decays occur per hour? (b) A student standing close to the source absorbs a small fraction of the gamma rays. Assume that fraction is 10 percent. How much energy does the student absorb in an hour? (c) If the student has a mass of 60 kg, what is the absorbed dose in rads? (d) Find the biologically equivalent dose in millirems, if the quality factor is 0.8. (e) Calculate the ratio of the biologically equivalent dose to the average annual dose from natural causes, 88 mrem.

31-41 Suppose a young child accidentally swallowed the radioactive source described in the preceding problem. (a) Assuming that half the gamma rays are absorbed by the child, how much energy does the child absorb in a year? (b) If the mass of the child is 20 kg, what is the absorbed dose in rads in a year? (c) Using a quality factor of 0.8, find the ratio of the biologically equivalent dose to the average annual dose from natural causes, 88 mrem. (This problem and the preceding one illustrate that radioactive sources used in introductory physics experiments are safe when used properly, but must be handled correctly to avoid potentially serious problems.)

31-42 After it has been in operation for a year or so, a large nuclear reactor has accumulated about 10^{10} Ci of radioactive wastes from the fission reac-

tions. If the reactor is shut down by inserting the control rods, this radioactive material continues to emit decay products for a long time. (a) How many kilograms of radium would be needed to produce as many decays per second? (b) Assuming a typical decay releases 1 MeV, at what rate in watts does the reactor continue to produce heat? (c) Find the ratio of this power to the rate heat is generated when the reactor is in operation, 3 × 10^9 watts.

31-43 A recent article on the safety of nuclear reactors stated that if many people are each exposed to a small dose of radiation, the number of latent fatal cancers induced is one per 10,000 man-rems of exposure. (The number of man-rems of exposure is the average exposure in rems times the number of people exposed.) (a) According to the estimates made in this chapter, how many fatal cancers would be induced by 10,000 man-rems? (b) In March 1979, an accident occurred at the reactor at Three Mile Island near Middletown, Pennsylvania. According to the article, ^{133}Xe, which has a half-life of about 5 days, was released in sufficient quantity to cause a total exposure to the public over a large region of approximately 4000 man-rems. Using the 10,000 man-rems per cancer ratio, how many extra cancer deaths are likely to occur because of the accident? (c) Could these predictions be verified with public health data? Explain.

31-44 Estimate the percentage of all deaths that are due to cancers caused by radiation from natural sources.

31-45 If a large population consumes milk containing radioactive ^{131}I, latent thyroid cancers will be induced at approximately 63 cancers per 10^6 man-rads of thyroid dose. (The number of man-rads equals the average dose in rads times the number of people exposed.) In 1976, fallout from nuclear explosions in the atmosphere detonated by the People's Republic of China resulted in an average thyroid dose in the United States of 3.1 × 10^{-4} rad. The total U.S. population is about 2.2 × 10^8. (a) How many extra thyroid cancers were induced in the United States by these explosions? (b) These cancers will appear over a period of 45 years. The present rate of thyroid cancers in the United States is 8400 per year. Find the fractional increase in this rate. (c) Infants received thyroid doses as large as 0.02 rad. What chance does such an infant have of developing thyroid cancer because of the fallout? (In order to carry out the calculation, assume that the risk for a given dose to an infant is the same as for the overall population. The risk is, in fact, larger.)

ANSWERS TO REVIEW QUESTIONS

Q31-1, proportional; **Q31-2,** photoelectric effect, Compton scattering, pair production; **Q31-3,** neutrons, alpha particles; **Q31-4,** inversely; **Q31-5,** unit mass; **Q31-6,** 1 rad; **Q31-7,** biological molecules; **Q31-8,** diagnostic X rays; **Q31-9,** cancer, genetic mutations; **Q31-10,** radiosensitive.

SUPPLEMENTARY TOPICS

31.7 | RADIATION DETECTION AND MEASUREMENT

Radiation is detected by observing the ionization it produces in matter. In this section, we describe the radiation detectors most commonly used in biomedical applications of radiation.

Gas Ionization Counters | These detectors, which are easy to use, are very sensitive to beta particles. They can also be used for gamma rays, but are not very effective in stopping this more penetrating radiation. A typical arrangement has argon gas in a cylinder with a fine wire along its axis (Fig. 31.12). When ionization is produced by radiation, a brief current pulse results that is metered or recorded.

Depending on the applied voltage V_0, the counter operates in basically different ways (Fig. 31.13). A small voltage drives all the ions and electrons produced to the cathode or anode, causing a small current pulse. Increasing the voltage further, we reach

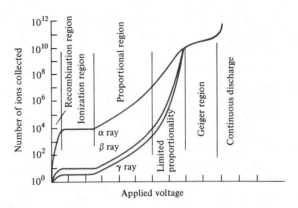

Figure 31.13. Number of ions collected versus applied voltage in a gas ionization counter.

the *proportional region,* where each electron released in the original ionization process acquires sufficient kinetic energy to ionize one or more additional atoms before it reaches the anode. The total charge in the current pulse now is proportional to the initial amount of ionization, and it indicates the energy of the beta particle or gamma ray that passed through the detector. Often these current pulses are fed to a *multichannel analyzer,* which records the number of pulses produced versus the total charge in the pulse. This information can be used to identify the nuclide producing the radiation (Fig. 31.14).

If the applied voltage is increased still more, the counter operates in the *Geiger region.* Here each electron produces several secondary electrons, which in turn produce others, and so on. A current pulse develops that is large enough so that it can be observed or recorded with very simple circuits. Such a detector is called a *Geiger-Mueller counter,* and it can be quite compact and portable. However, the total amount of charge in a pulse is the same for any incident ionizing particle, since the tube becomes fully ionized. Consequently, a Geiger-Mueller counter cannot be used to identify the nature or energy of the radiation being detected.

Scintillation Counters | These counters have relatively high efficiencies for detecting gamma rays and can be used to measure their energies. They are the detectors most widely used in biomedical applications.

A scintillation counter consists of a crystal such as sodium iodide and a *photomultiplier.* When it is irradiated, the crystal *scintillates* or emits flashes of visi-

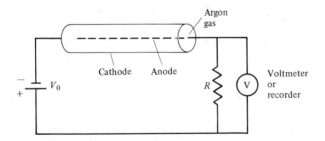

Figure 31.12. Gas ionization counter. When radiation ionizes one of the argon atoms, the electrons are attracted to the positively charged wire (anode) and the positive ions to the negatively charged outer cylinder (cathode). This produces a current through R and a voltage drop across it that is measured by the voltmeter.

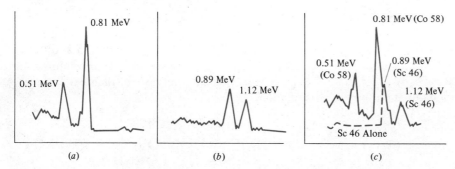

Figure 31.14. Gamma-ray spectra obtained with a multichannel analyzer. The gamma-ray energy is the horizontal coordinate, and the vertical coordinate gives the intensity at that energy. (a) ^{58}Co. (b) ^{46}Sc. (c) A mixture of ^{58}Co and ^{46}Sc. Mixtures of radionuclides can be analyzed if the energy resolution of the apparatus is good enough to distinguish the characteristic peaks. Sometimes samples are made radioactive artificially by exposing them to beams of neutrons, gamma rays, or charged particles that induce various nuclear reactions. Such *activation* techniques require only small samples and can be used to detect and measure extremely small traces of many elements.

ble light as atoms excited by the radiation emit photons. The photons enter the photomultiplier and strike a light-sensitive cathode, causing the emission of electrons (Fig. 31.15). These electrons are accelerated by a potential difference and strike an electrode called a *dynode,* which emits approximately four secondary electrons for each incident electron. A series of 10 dynodes gives an amplification factor of 4^{10} or approximately a million. As in the proportional gas counter, the total charge in the current pulse is proportional to the amount of ionization, and a multichannel analyzer can be used to identify the source.

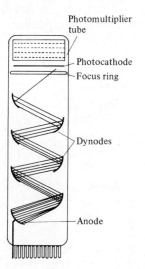

Figure 31.15. Scintillation counter.

Semiconducting Detectors

When radiation deposits energy in a semiconducting material, electrons can be excited from filled valence bands to conduction bands. This produces pairs of conduction electrons and electron vacancies or *positive holes*. If a potential difference is maintained across the semiconductor, these charge carriers will move, and a current pulse will be observed. The energy needed to produce an electron-hole pair is typically about 1 eV, considerably less than the energy needed to ionize gas atoms or to excite atoms in a scintillating crystal. Accordingly, a relatively large number of charge carriers is produced for a given energy loss. The statistical fluctuations in the number of atoms excited or ionized limit the energy resolution of any detector. The large number of charge carriers produced in the semiconductor reduces these statistical effects and enables such a detector to determine energies very accurately when used with a multichannel analyzer.

At the present time, these detectors are expensive and difficult to make unless they are quite small. Small semiconducting detectors have the disadvantage of low gamma-ray efficiencies.

Photographic Emulsions

Film badges are used to monitor the beta- and gamma-ray exposure of radiation personnel. Thin metal strips shield parts of the film from beta rays so that they indicate the gamma flux; the unshielded area gives the total dose from beta and gamma rays. The radiation exposure is determined by examining the processed film. Neutron doses can be measured if the film is impregnated or covered with a material having a high probability for neutron reactions.

Thermoluminescent dosimeters or TLDs

These are widely used in calibrating radiation sources in hospitals and are rapidly replacing film badges in personnel monitoring. A TLD is a crystal such as lithium fluoride (LiF) or calcium fluoride (CaF_2) containing trace impurities. After a TLD is exposed to radiation, it will emit visible light upon being heated. The total amount of light emitted is directly proportional to the dose received by the crystal, even if the crystal is stored for many months after exposure. TLDs are more accurate than film badges and are usable over a much larger range of radiation levels.

The principle of the TLD is illustrated in Fig. 31.16, which shows the energy levels of a LiF crystal. A perfect crystal will have a completely filled valence band and an empty conduction band. Electrons cannot be in energy levels between these bands. Electrons can be excited from the valence band to the conduction band, leaving empty valence levels or holes if they absorb enough energy from the radiation passing through the crystal. The impurity atoms present in the crystal add isolated energy levels to this band structure, which lie just below the conduction band. If an electron is excited to the conduction band and falls into one of these isolated levels, it is "trapped." It cannot migrate and thus cannot return to an empty valence level. When the crystal is heated at a later time, the trapped electron can receive enough thermal energy to return to the conduction

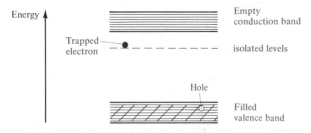

Figure 31.16. Energy levels of a LiF crystal. A perfect crystal has a filled band and an empty conduction band. Isolated levels below the conduction band are caused by impurities. If radiation excites an electron to the conduction band, leaving a valence band hole, the electron can fall into an isolated level and become trapped. Heating will provide the energy needed for the electron to return to the conduction band. It will emit visible light when it drops to the vacant valence band level.

band and migrate until it reaches a hole. When the electron loses energy by filling the hole, it emits a photon of visible light. The total light emitted as the crystal is heated is a measure of the number of trapped electrons and therefore of the total absorbed radiation.

31.8 | DERIVATION OF THE ENERGY LOSS RATE FORMULA

We now derive Eq. 31.1 for the rate of energy loss with distance for an ion in matter.

The velocity $\mathbf{v}_e$ of an atomic electron in the medium is negligible compared to its final velocity $\mathbf{v}'_e$ after it is struck by the ion. The momentum transferred by the ion to the electron is therefore $\mathbf{F}\,\Delta t = m_e(\mathbf{v}'_e - \mathbf{v}_e) \simeq m_e\mathbf{v}'_e$, where $\mathbf{F}$ is the force exerted by the ion and Δt is the time duration of the collision. The energy transferred to the electron is approximately $\frac{1}{2}m_e v'^2_e$, which is proportional to $(\mathbf{F}\,\Delta t)^2$. Now the electrical force between the ion and the electron is proportional to the ion charge q, and the collision time varies inversely with the ion velocity v. Thus $\mathbf{F}\,\Delta t$ must vary as q/v, and the energy transfer as q^2/v^2, or as we stated earlier,

$$\Delta K \propto -\frac{q^2}{v^2} \qquad (31.1)$$

EXERCISES ON SUPPLEMENTARY TOPICS

Section 31.7 | Radiation Detection and Measurement

31-46 In gamma-ray scans, the scintillation counter pulses are sometimes simply counted, and the count rate is then used to construct a picture. Alternatively, the pulses may be processed by a multichannel analyzer before constructing a picture. How can this latter procedure reduce the effects of background radiation from other sources?

31-47 Given a Geiger-Mueller counter and several thin slabs of a material containing a large amount of hydrogen, such as a hydrocarbon, how can you tell whether a given radioactive source is emitting neutrons?

Additional Reading

Steward C. Barshong, Radiation Exposure in Our Daily Lives, *The Physics Teacher,* vol. 15, 1977, p. 135. An up-to-date survey of environmental, occupational, and medical exposures.

G.S. Hurst and J.E. Turner, *Elementary Radiation Physics,* John Wiley and Sons, Inc., New York, 1970. An excellent brief introduction at the level of this book.

E.F. Gloyna and J.E. Ledbetter, *Principles of Radiological Health,* Marcel Dekker, Inc., New York, 1969. Emphasis on simplified calculational methods and engineering estimates.

Louis E. Etter (ed.), *The Science of Ionizing Radiation,* Charles C Thomas, Springfield, Ill., 1965. Interesting chapters on history and on medical and other applications.

J.N. Gregory, *The World of Radioisotopes,* Angus and Robertson, Sydney, 1966. Directed at nontechnical audiences; many interesting applications.

H.E. Johns and J.R. Cunningham, *The Physics of Radiology,* 3rd ed., Charles C Thomas, Springfield, Ill., 1969.

Victor Arena, *Ionizing Radiation and Life,* The C.V. Mosby Co., St. Louis, 1971.

Bacq and Alexander, *Fundamentals of Radiobiology,* Pergamon Press, New York, 1961.

P.N. Goodwin, E.H. Quimby, and R.H. Morgan, *Physical Foundations of Radiology,* 4th ed., Harper & Row, New York, 1970.

G.E.M. Jauncey, The Early Days of Radioactivity, *American Journal of Physics,* vol. 14, 1946, p. 226.

Gordon L. Brownell and Robert J. Shalek, Nuclear Physics of Medicine, *Physics Today,* August 1970, p. 33.

Advisory Committee on the Biological Effects of Ionizing Radiations, *The Effects on Populations of Exposure to Low Levels of Ionizing Radiation,* National Academy of Sciences, National Research Council, Washington, D.C., 1972.

Advisory Committee on the Biological Effects of Ionizing Radiation, *The Effects on Populations of Exposure to Low Levels of Ionizing Radiation,* National Research Council, National Academy of Sciences, National Academy Press, Washington, D.C., 1980.

Charles M. Koplile, Maureen F. Kaplan, and Benjamin Rass, The Safety of Repositories for Highly Radioactive Wastes, *Reviews of Modern Physics,* vol. 54, 1982, p. 269.

William J. Schull, Masanori Otake, and James V. Neel, Genetic Effects of the Atomic Bombs: A Reappraisal, *Science,* vol. 213, 1981, p. 1220. Descendents of atomic bomb survivors show genetic effects at a rate indicating a doubling dose of 156 rems, some four times larger than the doubling dose suggested by experiments on mice.

J. Michael Smith, Jon A. Broadway, and Ann B. Strong, United States Population Dose Estimates for Iodine-131 in the Thyroid After the Chinese Atmospheric Nuclear Weapons Tests, *Science,* vol. 200, 1978, p. 44.

High Background Radiation Research Group, Health Survey in High Background Radiation Areas in China, *Science,* vol. 209, 1980, p. 877. Studies of people living in areas with above average natural radioactivity showed no discernible health differences from control groups.

Charles E. Land, Estimating Cancer Risks from Low Doses of Ionizing Radiation, *Science,* vol. 209, 1980, p. 1197.

Rosalyn S. Yallow, Radioimmunoassay: A Probe for the Fine Structure of Biologic Systems, *Science,* vol. 200, 1978, p. 1236.

Arthur L. Robinson, Position-Sensitive Detectors: An "Electronic Film" for X rays, *Science,* vol. 199, 1978, p. 39.

Maurice J. Cotter and Kathleen Taylor, Neutron Activation Analysis of Paintings, *The Physics Teacher,* vol. 16, May 1978, p. 263.

C.L. Melcher and D.W. Zimmerman, Thermoluminescent Determination of Prehistoric Heat Treatment of Chert Artifacts, *Science,* vol. 197, 1977, p. 1359.

The October 1978 issue of *Physics Today* is devoted to articles on new types of particle detectors.

Arthur Robinson, Image Reconstruction (I) Computerized X-ray Scanners, *Science,* vol. 190, 1975, p. 542; (II) Computerized Scanner Explosion, *Science,* vol. 190, 1975, p. 647.

Rowland W. Redington and Walter H. Berninger, Medical Imaging Systems, *Physics Today,* vol. 34, August 1981, p. 5.

G.L. Brownell *et al.,* Positron Tomography and Nuclear Magnetic Resonance Imaging, *Science,* vol. 215, 1982, p. 619.

Scientific American articles:

S.A. Korff, Counters, July 1950, p. 40.

G.B. Collins, Scintillation Counters, November 1953, p. 36.

D.A. Glaser, The Bubble Chamber, February 1955, p. 46.

H. Yagoda, The Tracks of Nuclear Particles, May 1956, p. 40.

Gerard K. O'Neill, The Spark Chamber, August 1962, p. 36.

Olexa-Myron Bilamick, Semiconductor Particle Detectors, October 1962, p. 78.

Jearl Walker, A Radiation Detector Made Out of Aluminum Foil and a Tin Can, The Amateur Scientist, September 1979, p. 234.

R.L. Fleischer, P.B. Price, and R.M. Walker, Nuclear Tracks in Solids, June 1969, p. 30.

W.H. Wahl and H.H. Kramer, Neutron Activation Analysis, April 1967, p. 68.

The September 1959 issue was devoted to ionizing radiation.

H.J. Muller, Radiation and Human Mutations, November 1955, p. 58.

E.F. Knipling, The Eradication of the Screw-Worm Fly, October 1960, p. 54.

Theodore T. Puck, Radiation and the Human Cell, April 1960, p. 142.

Renato Baserga and W.B. Kisielski, Autobiographies of Cells, August 1963, p. 103.

Richard Gordon, Gabor T. Herman, and Steven A. Johnson, Image Reconstruction from Projection, October 1975, p. 56.

Michel M. Ter-Pogossian, Marcus E. Raichle, and Burton E. Sobel, Positron-Emission Tomography, October 1980, p. 170.

Ian L. Pykett, NMR Imaging in Medicine, May 1982, p. 78.

Harold W. Lewis, The Safety of Fission Reactors, March 1980, p. 3.

Steven A. Fetter and Kosta Tsipis, Catastrophic Releases of Radioactivity, April 1981, p. 48.

Arthur C. Upton, The Biological Effects of Low-Level Ionizing Radiation, February 1982, p. 41.

Edward R. Landa, The First Nuclear Industry, November 1982, p. 180. Radium production and uses.

PHYSICS AND THE FUTURE

Many textbooks simply stop at a suitable point, without an epilogue or a postscript or sometimes even a concluding paragraph. However, because physics is still a rapidly developing science with many exciting discoveries being made, we would like to conclude with some brief remarks about the directions in which physics seems to be going.

We have at least touched on some of the currently active subfields of physics, such as atomic, nuclear, and solid-state physics. Physicists working in these areas are continuing to explore new frontiers. For example, nuclear physicists are trying to produce *superheavy nuclei,* with mass and atomic numbers much greater than those currently known. Solid-state physicists are studying the curious behavior of many materials at temperatures as low as a millidegree (0.001 K). It has been known since 1956 that the weak interactions responsible for processes such as nuclear beta decay violate the principle called *parity conservation.* This means that a physical process would not be the same in a mirror-image world as in our world if weak interactions are involved. Atomic and nuclear physicists are performing a variety of experiments designed to probe this and other symmetry breakdowns that have been observed.

A subfield that currently involves a large number of research physicists, high-energy or particle physics, deals with phenomena at the scale of the nucleon diameter (about 1 fm = 10^{-15} m) or smaller. When two particles collide at high speeds, some of their kinetic energy may be converted into mass energy with the creation of other particles. Using particle accelerators of ever-increasing energy, a vast number of short-lived strongly interacting particles (hadrons) have

been created, and their masses, lifetimes, decays, and interactions have been studied. As we saw in Chapter Thirty, this research has lead to the idea that hadrons are made up of fractionally charged quarks that are held together by the exchange of gluons. Progress has also been made recently in unifying the four fundamental forces. (In order of increasing strength, these are the gravitational, weak, electromagnetic, and strong interactions.) It has been found that the weak and electromagnetic interactions are most likely two manifestations of the same underlying forces. Furthermore, in some more speculative Grand Unified Theories (GUTS, colloquially), these two are also connected to the strong interactions. According to GUTS, the proton is not truly stable, but instead decays into less massive particles with a very long lifetime of about 10^{31} years. For comparison, the age of the universe is about 10^{10} years. Experiments are in progress to test some of these ideas.

Because much of particle physics is tentative and remote from applications, we have not discussed particle physics in detail in this book. For similar reasons, we have excluded any general discussion of astrophysics, even though many striking discoveries have occurred in recent years. Many of these have been made using *radio telescopes,* which detect the electromagnetic waves emitted at radio frequencies by stars just as ordinary telescopes detect the light they emit. Among the remarkable objects found with radio telescopes are *pulsars,* which emit rapid pulses in the radio and optical regions. Pulsars are believed to be *neutron stars,* objects first predicted theoretically, with about the same mass as our sun but a radius of only 10 kilometres! Also, there are several

725

double star systems in which one member may be a *black hole:* an object so large and dense that the gravitational force bends all light rays back into it, and no light ever escapes! Since new kinds of astronomical tools—millimetre radio wave, infrared, and x-ray telescopes—are just now being developed, we can expect more exciting advances in astronomy and astrophysics in the decade ahead.

Returning to the earth, physicists are currently active in many applied and interdisciplinary fields. One of these is plasma physics, the physics of highly ionized matter. Thermonuclear reactions occur in a plasma, so advances in this field are crucial for achieving controlled fusion. Increasing numbers of physicists are working in biophysics and medical physics, using physical tools to probe basic biological processes and to develop new kinds of instrumentation. For example, light scattering using laser sources is now being used to study aspects of nerve conduction. Light scattering is also employed in a system being developed that will automatically sort biological cells.

To summarize, physicists are continuing the process of learning the basic laws of the universe, as they have for centuries. Although we cannot predict the future course with any certainty, we can be sure that many surprises lie ahead. We can also be confident that physics will continue to make important contributions to many fields, including the life sciences.

PERIODIC TABLE OF ELEMENTS

The symbol for each element is preceded by the atomic number; below is the atomic mass of the element as it occurs naturally on the earth. The atomic mass unit (u) is defined so the mass of a ^{12}C atom is exactly 12 u. The mass of carbon is listed as 12.01115 u because naturally occurring carbon is 98.89% ^{12}C and 1.11% ^{13}C. For artificially produced elements, the approximate atomic mass of the most stable isotope is given in brackets.

Period	Series	I	II	III	IV	V	VI	VII	VIII			O
1	1	1 H 1.00797										2 He 4.0026
2	2	3 Li 6.939	4 Be 9.0122	5 B 10.811	6 C 12.01115	7 N 14.0067	8 O 15.9994	9 F 18.9984				10 Ne 20.183
3	3	11 Na 22.9898	12 Mg 24.312	13 Al 26.9815	14 Si 28.086	15 P 30.9738	16 S 32.064	17 Cl 35.453				18 A 39.948
4	4	19 K 39.102	20 Ca 40.08	21 Sc 44.956	22 Ti 47.90	23 V 50.942	24 Cr 51.996	25 Mn 54.9380	26 Fe 55.847	27 Co 58.9332	28 Ni 58.71	
4	5	29 Cu 63.54	30 Zn 65.37	31 Ga 69.72	32 Ge 72.59	33 As 74.9216	34 Se 78.96	35 Br 79.909				36 Kr 83.80
5	6	37 Rb 85.47	38 Sr 87.62	39 Y 88.905	40 Zr 91.22	41 Nb 92.906	42 Mo 95.94	43 Tc [99]	44 Ru 101.07	45 Rh 102.905	46 Pd 106.4	
5	7	47 Ag 107.870	48 Cd 112.40	49 In 114.82	50 Sn 118.69	51 Sb 121.75	52 Te 127.60	53 I 126.9044				54 Xe 131.30
6	8	55 Cs 132.905	56 Ba 137.34	57–71 Lanthanide series*	72 Hf 178.49	73 Ta 180.948	74 W 183.85	75 Re 186.2	76 Os 190.2	77 Ir 192.2	78 Pt 195.09	
6	9	79 Au 196.967	80 Hg 200.59	81 Tl 204.37	82 Pb 207.19	83 Bi 208.980	84 Po [210]	85 At [210]				86 Rn [222]
7	10	87 Fr [223]	88 Ra [226.05]	89–Actinide series**								
		104 [261]	105 [262]	106 [263]								

*Lanthanide series:

| 57 La 138.91 | 58 Ce 140.12 | 59 Pr 140.907 | 60 Nd 144.24 | 61 Pm [145] | 62 Sm 150.35 | 63 Eu 151.96 | 64 Gd 157.25 | 65 Tb 158.924 | 66 Dy 162.50 | 67 Ho 164.930 | 68 Er 167.26 | 69 Tm 168.934 | 70 Yb 173.04 | 71 Lu 174.97 |

**Actinide series:

| 89 Ac [227] | 90 Th 232.038 | 91 Pa [231] | 92 U 238.03 | 93 Np [237] | 94 Pu [242] | 95 Am [243] | 96 Cm [247] | 97 Bk [247] | 98 Cf [249] | 99 Es [254] | 100 Fm [257] | 101 Md [256] | 102 No [253] | 103 Lw [260] |

MATHEMATICAL REVIEW

This appendix reviews topics covered in high school or introductory college mathematics courses that are needed in various parts of this book. Students who are somewhat rusty in basic algebra, geometry, and trigonometry may find it helpful to study Sections B.1 to B.6 in detail; the remaining sections are provided mainly for reference at specific points in the book. A list of paperback mathematics review books is given at the end of the appendix for students who need additional preparation in this area. Answers to all the review problems are also given at the end of this appendix.

B.1 | POWERS AND ROOTS

A quantity x multiplied by itself n times is written as x^n. For example, $(2)(2)(2) = 2^3$; in words, this is 2 raised to the *exponent* or *power* 3. The basic rule in manipulating powers of a given number is that exponents add. For example, $(2^2)(2^3) = (2)(2) \cdot (2)(2)(2) = 2^5$. In symbols, the rule is

$$(x^n)(x^m) = x^{n+m} \qquad \text{(B.1)}$$

From this rule, we see that $x^n x^0 = x^n$, so $x^0 = 1$ for any value of x. Also, $x^n x^{-n} = x^0 = 1$, so x^{-n} is the inverse of x^n:

$$x^{-n} = \frac{1}{x^n} \qquad \text{(B.2)}$$

For example, $10^{-2} = 1/10^2 = 1/100 = 0.01$. Another useful rule is

$$(x^n)^m = x^{nm} \qquad \text{(B.3)}$$

For example, $(10^2)^3 = (10)(10) \cdot (10)(10) \cdot (10)(10) = 10^6$; similarly, $(10^2)^{-3} = 10^{-6}$.

When two numbers are raised to the same power, their products and quotients obey simple rules:

$$(x^n)(y^n) = (xy)^n \qquad \text{(B.4)}$$

$$\frac{x^n}{y^n} = \left(\frac{x}{y}\right)^n \qquad \text{(B.5)}$$

For example, $(4)^3(2)^3 = (4 \cdot 2)^3 = (8)^3$, and $(4)^3(2)^{-3} = (4/2)^3 = (2)^3$.

A fractional exponent means that a root of a number is involved. For example, $x^{1/2}x^{1/2} = x^1 = x$, so $x^{1/2}$ is the square root of x. The quantity $x^{1/n}$ is the nth root of x:

$$x^{1/n} = \sqrt[n]{x} \qquad \text{(B.6)}$$

For example, $(64)^{1/3} = \sqrt[3]{64} = 4$. More complicated fractional powers can be evaluated with the aid of Eq. B.3. For example, $(27)^{2/3} = (27^{1/3})^2 = (3)^2 = 9$. All the other rules listed above also apply to fractional powers.

REVIEW PROBLEMS

Evaluate or simplify the following quantities:

1. 2^4
2. 3^2
3. $(2^2)(2^3)$
4. $(x^5)(x^3)(x)$
5. 5^{-2}
6. $(5^{-3})(5^4)$
7. $(x^4)(x)(x^{-3})$
8. x^4/x^2
9. x^4/y^4
10. $(a^2x^4)^{1/2}$
11. $(a^3x^6)^{1/2}$
12. $(x^2y^6)^{1/2}$
13. $(x^4y^4)^{-1/2}$
14. $(1000)^{1/3}$

15. $(10,000)^{-1/4}$

16. $x^2(x^6)^{-1/3}$

17. $(125)^{-1/3}$

18. $\left(\dfrac{x^2}{64}\right)^{1/2}$

19. $(x^4y^{-8})^{1/2}$

20. $(10^4)^{3/4}$

B.2 | SCIENTIFIC NOTATION

A number is said to be in scientific notation when it is written as a number between 1 and 10 times a power of 10. For example, 376 can be written as $3.76 \times 100 = 3.76 \times 10^2$, since $10^2 = 10 \times 10 = 100$. One advantage of this notation is compactness; 376,000,000 can be written as 3.76×10^8. Note that the power of 10 is the number of places the decimal point has been shifted to the left. Similarly, $0.0000376 = 3.76 \times 0.00001 = 3.76 \times 10^{-5}$. Here the number in this negative exponent indicates how many places the decimal point has been shifted to the right.

Scientific notation facilitates many kinds of numerical calculations. It is especially useful in manipulations involving very large or small numbers. As an illustration, consider 2×10^{20} times 3×10^{-15} divided by 8×10^8:

$$\frac{(2 \times 10^{20})(3 \times 10^{-15})}{8 \times 10^8} = \frac{(2)(3)}{8} \times 10^{20-15-8}$$
$$= 0.75 \times 10^{-3} = 7.5 \times 10^{-4}$$

The use of scientific notation also aids in the evaluation of roots, as in the following illustrations:

$$(2.32 \times 10^8)^{1/2} = (2.32)^{1/2}(10^8)^{1/2} = \sqrt{2.32} \times 10^4$$
$$= 1.52 \times 10^4$$
$$(2.32 \times 10^8)^{1/3} = (232 \times 10^6)^{1/3}$$
$$= (232)^{1/3}(10^6)^{1/3}$$
$$= \sqrt[3]{232} \times 10^2 = 6.14 \times 10^2$$
$$(9.37 \times 10^{-4})^{1/3} = (937 \times 10^{-6})^{1/3}$$
$$= (937)^{1/3}(10^{-6})^{1/3}$$
$$= \sqrt[3]{937} \times 10^{-2}$$
$$= 9.79 \times 10^{-2}$$

In the last two examples, we have rewritten the power of 10 so that it leads to an integer power of 10 when the root is calculated. Notice also from the first two examples that the cube root of a number greater than 1 is less than the square root; the fourth root is smaller still. The converse statement holds for numbers less than 1.

REVIEW PROBLEMS

Write the following numbers in scientific notation:

21. 27,631

22. 2,763,100

23. 15,000

24. 0.000000034

25. 1,600

26. 4,329.76

27. 0.003902

28. 0.08002

Express the following numbers in ordinary notation:

29. 2.34×10^{-3}

30. 1.76×10^6

31. 5.799×10^{-5}

32. 4.5×10^7

33. 0.067×10^4

34. 27.2×10^5

35. 0.0272×10^8

Evaluate the following expressions:

36. $(3 \times 10^6)(5 \times 10^4)$

37. $\dfrac{4 \times 10^8}{8 \times 10^6}$

38. $(5 \times 10^{10})(3 \times 10^{-8})(4 \times 10^6)$

39. $\dfrac{(4.4 \times 10^6)(3 \times 10^3)^2}{6 \times 10^{-4}}$

40. $\dfrac{(8.25 \times 10^4)(3.14)(5.2 \times 10^3)^2}{(6.25 \times 10^{-3})}$

41. $(4 \times 10^4)^{1/2}$

42. $(90,000)^{1/2}$

43. $(2.7 \times 10^7)^{1/3}$

44. $(8,000)^{1/3}$

45. $(4 \times 10^{-6})^{1/2}$

46. $(160,000)^{1/4}$

47. $(10^{10})^{1/2}$

48. $(10^{10})^{-1/2}$

49. $(10^{10})^{1/3}$

50. $(3.2 \times 10^8)^{1/3}$

B.3 | SIGNIFICANT FIGURES

The accuracy of any measurement is limited by errors of various types (see Section 1.1). It is important to keep track of these errors at least approximately in using or manipulating experimentally determined numbers. This is accomplished most readily with the rules for significant figures.

The principle involved is illustrated by the problem of determining the area A of a rectangular sheet of paper using a ruler whose smallest spacing is 0.1 cm. If we place one end of the ruler at the edge of the paper, the other edge might lie between the markings indicating 8.4 and 8.5 cm. We can, at best, then judge its position to one-tenth of a spacing, so we might report our reading as 8.43 cm. However, a more elaborate measuring arrangement might well give a length closer to 8.44 or 8.42 cm; the last digit we report is somewhat uncertain. The number 8.43 is said to have three *significant figures*. In the same way we might find 6.77 cm for the other dimension of the rectangle. The area is then the product

$$A = (8.43 \text{ cm})(6.77 \text{ cm}) = 57.0711 \text{ cm}^2 = 57.1 \text{ cm}^2$$

Each of the factors in the product is uncertain in the third place, so only three places on the right have any meaning. Hence, A is given to three significant figures. To clarify the reason for this, suppose the first factor is found to be closer to 8.42 cm when more careful measurements are made. Then the area becomes $A = (8.42 \text{ cm})(6.77 \text{ cm}) = 57.0034 \text{ cm}^2$, and the digits beyond 57.0 are changed. Clearly, these digits in the product are meaningless, and the area A is somewhat uncertain in the third digit. Note that our answer for A has been rounded up from 57.07 $\cdots$ to 57.1; a number below 57.05 would be rounded down to 57.0.

In all computations involving multiplication and division, the factor with the fewest significant figures determines the number of significant figures in the answer. For example, in

$$\frac{(8.2239)(2.7)(98.35)\pi^2}{2764} = 7.797899 \cdots$$

the first three factors in the numerator have five, two, and four significant figures, respectively; $\pi^2 = (3.1415926 \cdots)^2$ is known to an arbitrarily great accuracy; and the denominator is known to four significant figures. Accordingly, the answer obtained for this expression should be rounded to two figures, that is, to 7.8. However, it is a good idea to retain one or more extra places in *intermediate steps* of the calculation in order to avoid introducing additional errors in the process of rounding off the numbers. This is important in complex multistep calculations and is easy to do with an electronic calculator.

The significant figures procedure used in addition and subtraction differs from that for multiplication and division. It is illustrated by the sum

$$\begin{array}{r} 45.76 \\ + \; 0.123 \\ \hline 45.883 \end{array}$$

Here, the 6 in the first number is somewhat uncertain, and the next place is completely unknown. Accordingly, the 3 in the sum is meaningless, and the answer is rounded to 45.88. *The answer contains as many*

places relative to the decimal point as the "least accurate number" in the sum. Notice that in this example the least accurate number that limits the accuracy is 45.76, which has four significant figures; 0.123 has only three significant figures but is more accurate in the sense meant here.

Since the same ideas apply to subtraction, the difference of two nearly equal numbers may have very few significant figures. For example, consider

$$\begin{array}{r} 35.179 \\ - 35.17813 \\ \hline 0.001 \end{array}$$

This result has essentially no accuracy, since it is uncertain by approximately 1 in the last place. If a new set of measurements changed the numbers slightly, their difference could well be 0.002 or -0.001.

Adding or subtracting numbers expressed in scientific notation requires that they be written with the same power of 10. For example,

$$\begin{aligned} 2.25 \times 10^6 + 6.4 \times 10^7 &= 2.25 \times 10^6 + 64 \times 10^6 \\ &= 66.25 \times 10^6 \\ &= 6.6 \times 10^7 \end{aligned}$$

Note that we have rounded 66.25 to 66 in accordance with the rules.

Significant Zeros | The number 1200 may have two, three, or four significant figures, depending upon whether the zeros represent measurements or are merely used to locate the decimal point. Scientific notation avoids this ambiguity; 1.2×10^3, 1.20×10^3, and 1.200×10^3 have two, three, and four significant figures, respectively.

REVIEW PROBLEMS

Round off the following quantities to three significant figures and write them in scientific notation:

51. 27632.0   **54.** 3.33333
52. 0.3729   **55.** 2.45558×10^4
53. 4.6667   **56.** 0.000034567

How many significant figures are there in the final result for each of the following expressions?

57. (3.2)(8.67)/(3.008)
58. (0.0002)(45.6)
59. $(2.0 \times 10^5)(3.777 \times 10^{-4})$
60. 17.2 + 2.35 + 4.3333
61. 88.45 + 9.24 − 6.05043
62. 186.45 − 186.12

Evaluate the following expressions in accordance with the rules for significant figures:
63. $3.28 \times 10^5 + 4.25 \times 10^7$
64. $3.7 \times 10^6 + 2.91 \times 10^7$
65. $1.91 \times 10^{-3} - 1.7 \times 10^{-5}$

B.4 | SOLUTION OF ALGEBRAIC EQUATIONS

The application of physical laws often leads to one or more algebraic equations that must be solved for the desired quantities. To do this, we must have as many equations as there are unknowns; for example, if we want to find two forces acting on an object, we must have two different equations relating them.

Equations in One Unknown | The basic rule in manipulating any algebraic equaiton is that both sides of the equation must be treated in the same way. If we add a number to one side, or multiply it by some factor, or square it, we must do the same thing to the other side.

To illustrate how a simple equation is solved, we first consider a *linear* equation, one in which the unknown quantity x appears only to the first power:

$$5x - 10 = 30$$

To solve for x, we first add 10 to both sides, giving

$$5x = 40$$

Dividing by 5 leads to the desired solution.

$$x = 8$$

A *quadratic* equation is one in which the highest power of the unknown is 2. Quadratic equations may or may not have a term linear in the unknown. An example of the latter situation is

$$16t^2 = 64$$

Dividing by 16 gives $t^2 = 4$. Taking the square root then gives two possible answers, $t = +2$ and $t = -2$, since either of these numbers squared is equal to 4. (Quadratic equations generally have two solutions.) Since t represents some number such as a time that is measured, it can only have one correct value. The circumstances of the specific problem will indicate which is the appropriate solution. For example, in Chapter One, in determining when a ball thrown straight up is at a specific height, we encounter equations in the time variable t. The two solutions of these

equations correspond to the time when the ball is at that height on the way up and on the way down.

The quadratic equation

$$t^2 - 6t + 8 = 0$$

is somewhat more complicated than the preceding one, since it contains t to the first power as well as t^2. It can be solved by *factoring*, rewriting the equation as a product of two factors that is equal to zero:

$$(t - 2)(t - 4) = 0$$

This factoring can be verified by noting that when we multiply out the two factors, we obtain four terms, $t^2 - 4t - 2t + 8 = t^2 - 6t + 8$. Clearly the product can be zero only if either the first *or* the second factor is 0. Thus

$$t - 2 = 0, \quad t = 2$$

or

$$t - 4 = 0, \quad t = 4$$

The two solutions of this equation are $t = 2$ and $t = 4$.

The disadvantage of factoring is that one must guess the factors somehow. The *quadratic formula* is a general solution that can always be used without any guesswork. The equation

$$at^2 + bt + c = 0$$

has two solutions,

$$t = \frac{-b + \sqrt{b^2 - 4ac}}{2a}$$

and

$$t = \frac{-b - \sqrt{b^2 - 4ac}}{2a}$$

(B.7)

This can be applied to the example $t^2 - 6t + 8 = 0$ by setting $a = 1$, $b = -6$, and $c = 8$. We find then

$$t = \frac{-(-6) + \sqrt{(-6)^2 - 4(1)(8)}}{2(1)}$$

$$= \frac{6 + \sqrt{4}}{2} = 4$$

and

$$t = \frac{-(-6) - \sqrt{(-6)^2 - 4(1)(8)}}{2(1)}$$

$$= \frac{6 - \sqrt{4}}{2} = 2$$

These are the same solutions as found before.

Simultaneous Equations | Two different equations containing the same two unknowns are called *simultaneous equations*. The unknowns are found by combining the equations in such a way that a single equation is obtained that contains just one unknown. For example, consider these two equations for a force F and an acceleration a, $F - 6a = 20$ and $-F + 8a = 0$. If we add these equations, the F and $-F$ cancel:

$$F - 6a = 20$$
$$\underline{-F + 8a = 0}$$
$$F - F - 6a + 8a = 20 + 0$$

This reduces to

$$2a = 20$$

or

$$a = 10$$

F is then found by substituting the value found for a into the first equation:

$$F - 6(10) = 20, \qquad F = 20 + 60 = 80$$

Sometimes one equation must be multiplied by a factor before the equations are added or subtracted in order to eliminate one unknown. For example, consider $x + 3y = 6$ and $2x - y = 5$. Neither the x's nor y's will completely cancel if these equations are added. However, if we multiply the first equation by 2 and the second by -1, we find upon adding them

$$2x + 6y = 12$$
$$\underline{-2x + y = -5}$$
$$7y = 7$$

and $y = 1$. Substituting this value of y into either equation then gives $x = 3$.

The procedure just used is readily extended to three simultaneous equations containing three unknowns, x, y, and z. Two of the equations are combined to yield an equation containing only two unknowns, say x and y, and another combination of two of the original equations is formed that again contains only x and y. The problem has then been reduced to two equations in two unknowns, and one then proceeds as in the examples just given to find x and y.

REVIEW PROBLEMS

Solve the following equations for the unknown quantities.

66. $x - 7 = 3$
67. $3x + 7 = 4 + 6x$
68. $1 + 0.2x = 7$
69. $x^2 + 4 = 13$
70. $x^{1/2} + 4 = 13$
71. $-4x + 7 = 2x + 15$
72. $(x/3)^{1/2} = 2$
73. $0 = 64 - 16t^2$
74. $x^3 - 1 = 63$
75. $(x + 2)(x + 4) = 0$
76. $x^2 + 3x + 2 = 0$
77. $3x^2 + 2x - 5 = 0$
78. $x^2 + 4 = -4x$
79. $2x^2 = -3x$
80. $-3x + 2x^2 - 5 = 0$
81. $x + y = 5, \ x - y = 1$
82. $2 - T = 3a, \ T = 4a$
83. $x + 3y = 9, \ x - 2y = 10$
84. $2x - y = 10, \ x + y = 6$
85. $3x - 7y = 2, \ 3x - 2y = 4$

B.5 | GRAPHS

Just as pictures are often more informative than words, graphs are often more useful than algebraic formulas in understanding what is happening in a physical system. To illustrate how a graph is constructed, consider the equation for the position coordinate x of an object moving at a constant velocity at a time t:

$$x = 5 + 2t$$

We start by making a table of the values of x obtained from this equation for several values of t between -4 and $+4$:

time, t	-4	-2	0	2	4
position, x	-3	1	5	9	13

Using ordinary (Cartesian) graph paper, we draw a horizontal axis for the independent variable, t, and a vertical axis for the dependent variable, x. The tabu-

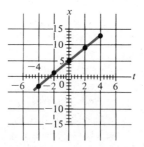

Figure B.1. The graph of $x = 5 + 2t$.

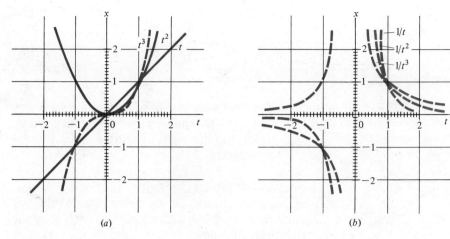

Figure B.2. (a) Graphs of $x = t$, $x = t^2$, and $x = t^3$. (b) Graphs of $x = 1/t$, $x = 1/t^2$, and $x = 1/t^3$. Notice that $1/t^3$ becomes very small most rapidly as t increases and also grows most rapidly as t approaches zero.

lated values are then used to locate the points marked on the graph (Fig. B.1).

In this example, all the points fall on a single straight line because t appears only to the first power. If other powers are present, the graph of the equation will be a curve. Some examples of such curves are shown in Fig. B.2.

The choice of the dependent and independent variables is not fixed, and it depends on what one is trying to do. For example, the equation

$$x = 16t^2$$

gives the position of an object dropped from $x = 0$ at time $t = 0$. If we want to know when it will be at a position x, we must solve for t in terms of x:

$$t = \tfrac{1}{4}\sqrt{x}$$

Now x has become the independent variable. The graphs of the two forms of the equation look quite different (Fig. B.3), although both are curves rather than straight lines. We can also get a straight-line graph if we consider t^2 rather than t as a variable. This is particularly useful in analyzing experimental measurements, since it is easy to see if the data points all fall on a straight line as expected.

REVIEW PROBLEMS

Draw the graphs of the following equations:
86. $y = 3x - 7$
87. $x = 2t^2$
88. $y = 2x^4 - 3$

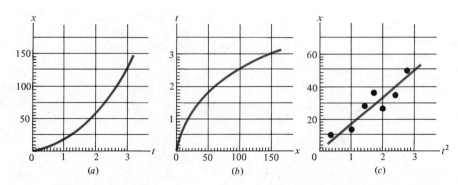

Figure B.3. (a) A graph of $x = 16t^2$. (b) A graph of $t = \tfrac{1}{4}\sqrt{x}$. (c) Measured pairs of x and t^2 values are plotted. They fall close to a straight line drawn through them with a ruler. On a plot of x versus t, they would fall near a curved line that would be harder to draw accurately and to analyze numerically.

B.6 | PLANE GEOMETRY AND TRIGONOMETRIC FUNCTIONS

Plane Geometry | The following results from plane geometry are often useful.

1 The sum of the internal angles of any triangle is 180°. In a right triangle, where one angle is 90°, the other two angles must add up to 90°.

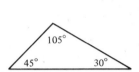
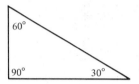

2 Two triangles are *similar* if two of their angles are equal. The corresponding sides of similar triangles are proportional. For example, $a/A = b/B$ in this figure:

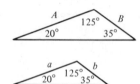

3 Two angles are equal if their sides are parallel.

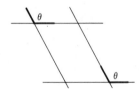

4 Two angles are equal if their sides are mutually perpendicular. For example, the angle between the weight vector **w** and the line perpendicular to the inclined plane equals the angle between the plane and the horizontal direction.

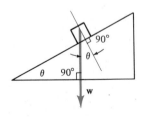

5 Two angles are equal if they are *vertical* angles.

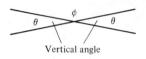

Vertical angle

6 Two angles are said to be *complementary* if they add up to 90° and *supplementary* if they add up to 180°. In the previous figure, θ and ϕ are supplementary.

Trigonometric Functions | The sine, cosine, and tangent of an angle are abbreviated as sin, cos, and tan, respectively, and are defined in terms of the right triangle below.

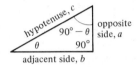

$$\sin \theta = \frac{\text{opposite side}}{\text{hypotenuse}} = \frac{a}{c} \qquad (B.8)$$

$$\cos \theta = \frac{\text{adjacent side}}{\text{hypotenuse}} = \frac{b}{c} \qquad (B.9)$$

$$\tan \theta = \frac{\text{opposite side}}{\text{adjacent side}} = \frac{a}{b} \qquad (B.10)$$

The Pythagorean theorem states that

$$a^2 + b^2 = c^2 \qquad (B.11)$$

Dividing by c^2,

$$\frac{a^2}{c^2} + \frac{b^2}{c^2} = 1$$

or

$$\sin^2 \theta + \cos^2 \theta = 1 \qquad (B.12)$$

The trigonometric functions of angles larger than 90° may be either positive or negative, depending on the angle. In the notation of the diagram below, they are defined by

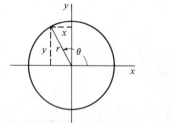

$$\sin \theta = \frac{y}{r}$$

$$\cos \theta = \frac{x}{r}$$

$$\tan \theta = \frac{y}{x}$$

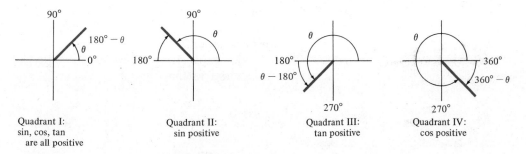

Quadrant I:
sin, cos, tan
are all positive

Quadrant II:
sin positive

Quadrant III:
tan positive

Quadrant IV:
cos positive

Figure B.4. A summary of the rules for finding the trigonometric functions. For example, if θ is in the third quadrant, a plus sign is used for $\tan \theta$, and a minus sign for $\sin \theta$ and $\cos \theta$.

By convention, r is always positive. However, for the angle shown, which is said to be in the second *quadrant,* x is negative and y is positive. Accordingly the sine is positive, and the tangent and cosine are negative. In the third quadrant only the tangent is positive, and in the fourth only the cosine is positive.

Many books contain a table of the trigonometric functions for angles up to 90°. For angles between 90° and 180°, one looks up the supplement, $180° - \theta$. Between 180° and 270°, one looks up $\theta - 180°$; and from 270° to 360°, one looks up $360° - \theta$. Minus signs are inserted wherever required in accordance with the discussion above. For example, $\sin 150° = \sin 30°$, $\cos 150° = -\cos 30°$, and $\tan 150° = -\tan 30°$. These rules are summarized in Fig. B.4. Many pocket calculators have these rules built in.

Graphs of the sine, cosine, and tangent (Fig. B.5) are useful in understanding the general behavior of these quantities. They all repeat after one full circle or cycle; their *period* is 360° or 2π radians. (Measuring angles in radians is discussed in Chapter Five.)

The average of the sine or cosine over a full cycle is zero, since for every positive value, there is a corresponding negative value. Notice that the $\sin \theta$ and $\cos \theta$ curves are identical if the $\cos \theta$ curve is shifted 90° to the right. The same shift also makes the $\sin^2 \theta$ and $\cos^2 \theta$ curves (Fig. B.6.) identical, so the average values of these quantities over a full cycle must be equal. Denoting the average by a bar, this means

$$\overline{\sin^2 \theta} = \overline{\cos^2 \theta}$$

Since we know from Eq. B.12 that $\sin^2 \theta + \cos^2 \theta = 1$, it follows that

$$\overline{\sin^2 \theta} + \overline{\cos^2 \theta} = 1$$

Therefore

$$\overline{\sin^2 \theta} = \overline{\cos^2 \theta} = \tfrac{1}{2} \tag{B.13}$$

Also, we can see from Fig. B.6 that the product, $\sin \theta \cos \theta$, has a corresponding negative value for each positive value, so averaging the product over a full cycle must give zero:

$$\overline{\sin \theta \cos \theta} = 0 \tag{B.14}$$

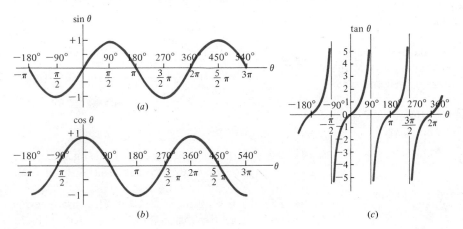

Figure B.5. Graphs of (a) $\sin \theta$; (b) $\cos \theta$; (c) $\tan \theta$.

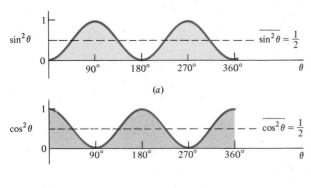

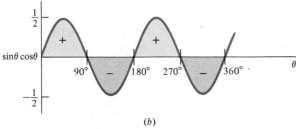

Figure B.6.   (a) $\sin^2 \theta$ and $\cos^2 \theta$ have the same average value, $\frac{1}{2}$. (b) The average value of $\sin \theta \cos \theta$ is zero.

REVIEW PROBLEMS

89. If two angles of a triangle are 29° and 111°, what is the third angle?

90. Find sin 120°; cos 120°; tan 120°.

91. Find sin 270°; cos 270°; tan 270°.

92. At what angle or angles between 0° and 360° is $\sin \theta$ equal to 0, +1, −1? Give the corresponding angles for the cosine and tangent.

B.7 | SERIES EXPANSIONS

We are often interested in the value of a trigonometric function or algebraic expression when the variable is much less than 1. In these cases, it is often convenient to approximate the exact formula by a series expansion involving successively higher powers of the variable, since these terms quickly become very small.

As an example, we may consider the series expansion for $(1 - x)^{-1}$ derived in algebra texts:

$$\frac{1}{1 - x} = 1 + x + x^2 + x^3 + \cdots$$

When $x = 0.1$, $(1 - x)^{-1} = 1/0.9 = 1.11$ to two decimal places. Since $x = 0.1$, $x^2 = 0.01$, and $x^3 = 0.001$. Thus

$$1 + x + x^2 = 1.11$$

is sufficient to approximate the formula to two decimal places. If $x = 0.01$, then $1 + x$ is sufficient to give two place accuracy.

Many series expansions involve the quantity $n!$ (read "n factorial") defined by

$$n! = n(n - 1)(n - 2) \cdots (2)(1)$$

For example, $4! = 4 \times 3 \times 2 \times 1 = 24$. By definition,

$$1! = 1, \qquad 0! = 1$$

The following series expansions are often useful:

$$(1 \pm x)^{-1} = 1 \mp x + x^2 \mp x^3 + \cdots \qquad (-1 < x < 1) \quad \text{(B.15)}$$

$$(1 \pm x)^n = 1 \pm nx + \frac{n(n - 1)x^2}{2!} \pm \frac{n(n - 1)(n - 2)x^3}{3!} + \cdots \qquad (-1 < x < 1) \quad \text{(B.16)}$$

$$(1 \pm x)^{-n} = 1 \mp nx + \frac{n(n + 1)x^2}{2!} \mp \frac{n(n + 1)(n + 2)x^3}{3!} + \cdots \qquad (-1 < x < 1) \quad \text{(B.17)}$$

$$e^x = 1 + x + \frac{x^2}{2!} + \frac{x^3}{3!} + \cdots \qquad \text{(B.18)}$$

In the following series for the trigonometric functions, the angles must be measured in radians, where 1 rad = $180°/\pi = 57.3°$.

$$\sin x = x - \frac{x^3}{3!} + \frac{x^5}{5!} - \cdots \qquad \text{(B.19)}$$

$$\cos x = 1 - \frac{x^2}{2!} + \frac{x^4}{4!} - \cdots \qquad \text{(B.20)}$$

$$\tan x = x + \frac{x^3}{3} + \frac{2x^5}{15} + \frac{17x^7}{314} + \cdots \quad \text{(B.21)}$$

REVIEW PROBLEMS

93. What is the percentage error in using $\sin x = x$, when $x = 10° = 0.1745$ rad? What is the error at 30°?

94. For $x = 0.1$, $e^x = e^{0.1} = 1.105$. How many terms in the series expansion for e^x are needed to obtain this accuracy?

95. (a) Write out the first three terms in the series for

$(1 + x)^{1/2} = \sqrt{1 + x}$. (b) What does this approximation give for $x = 0.1$? Compare this with the exact answer to five decimal places, 1.04881.

B.8 | DERIVATIVES

Derivatives are used in this text in various places. The most frequently used formulas are listed below; others may be found in calculus texts. Note that the derivative of a constant is zero. Also, when an expression is multiplied by a constant, its derivative is multiplied by the same constant. For example,

$$\frac{d}{dt}(3t^2) = 3\frac{d}{dt}(t^2)$$

In the following expressions a and n are constants.

$$\frac{d}{dt}(t^n) = nt^{n-1} \qquad \text{(B.22)}$$

$$\frac{d}{dt}\left(\frac{1}{t}\right) = -\frac{1}{t^2} \qquad \text{(B.23)}$$

$$\frac{d}{dt}\left(\frac{1}{t^n}\right) = \frac{-n}{t^{n+1}} \qquad \text{(B.24)}$$

$$\frac{d}{dt}(e^{at}) = ae^{at} \qquad \text{(B.25)}$$

$$\frac{d}{dt}\sin at = a\cos at \qquad \text{(B.26)}$$

$$\frac{d}{dt}\cos at = -a\sin at \qquad \text{(B.27)}$$

If y depends on a variable u, and u depends in turn on t, then the *chain rule* states that

$$\frac{dy}{dt} = \frac{dy}{du}\frac{du}{dt} \qquad \text{(chain rule)}$$

REVIEW PROBLEMS

Find the derivatives of the following expressions.
96. $3t + 7$
97. $4t^3$
98. $1 - 1/t$
99. $4e^{-3t}$
100. $10\sin 2\pi t$

B.9 | AREAS AND VOLUMES

At various points in this book, we have to make use of the formulas for the areas and volumes of simple shapes. They are listed here for reference.

Circle
>   radius = r
>   diameter = $2r$
>   circumference = $2\pi r$
>   area = πr^2

Square
>   side = a
>   area = a^2

Triangle
>   area = $\frac{1}{2}$(base)(height)

Cube
>   side = a
>   surface area = $6a^2$
>   volume = a^3

Sphere
>   radius = r
>   surface area = $4\pi r^2$
>   volume = $4\pi r^3/3$

Cylinder
>   radius = r, length = ℓ
>   area of curved surface = $2\pi r\ell$
>   area of each end = πr^2
>   volume = $\pi r^2\ell$

B.10 | THE EXPONENTIAL FUNCTION; LOGARITHMS

Many times the rate at which a quantity changes is proportional to the quantity present. For example, the rate at which a population of bacteria increases is directly proportional to the size of the population itself, as is the rate of growth in the funds in a savings account. Similarly, the rate of change of the charge on a capacitor is sometimes proportional to the charge present. Students of calculus will recognize these as situations where the quantity, call it y, depends on time as

$$y = Cb^{Dt}$$

Here b and D are numbers and C is a constant which is determined by the conditions of the situation when $t = 0$. This is called an exponential dependence of y on t.

The choice of b is arbitrary, although it will affect the value of D when it is determined. However, there is a particular choice of b that can be made that greatly simplifies the manipulations. This choice is $b = e = 2.718....$ For this seemingly unlikely choice, the rate of change of $y = Ce^t$ with t is exactly equal to y itself. For any other choice of b, the rate of change of $y = Cb^t$ with t is proportional to but not equal to y. Thus the function $y = e^t$ is very common in many scientific studies. The decreasing function $e^{-t} = 1/e^t$ is of similar importance when the rate of *decrease* of a quantity is proportional to the present amount of that quantity. This happens, for example, in the decay of radioactive nuclei.

The exponential function e^t is built into many pocket calculators, sometimes as the inverse of the *natural logarithm*. If $x = 10^t$, then the more familiar *common logarithm* (log, or logarithm to the base 10) is the power of 10, or t in this case. Symbolically,

$$\log x = \log(10^t) = t \qquad (B.28)$$

Conversely, the inverse common logarithm of t is 10^t.

Similarly, the natural logarithm (ln, or logarithm to the base e) is defined as the power of e needed to obtain a given quantity. Thus if $y = e^t$,

$$\ln y = \ln(e^t) = t \qquad (B.29)$$

The inverse natural logarithm of t (INV ln on many calculators) is then e^t.

Some simple rules can be obtained from the definitions (B.28) and (B.29) for the logarithms of products, quotients, and powers. The following rules are given for natural logarithms but apply equally well to common logarithms:

$$\ln(xy) = (\ln x) + (\ln y) \qquad (B.30)$$
$$\ln(x^n) = n(\ln x) \qquad (B.31)$$
$$\ln(1/x^n) = -n(\ln x) \qquad (B.32)$$
$$\ln(x/y) = (\ln x) - (\ln y) \qquad (B.33)$$

Note that there is no simple rule for the logarithms of sums or differences of quantities.

Using either tabulated values of e^t or values found with a calculator, we can construct a plot of e^t (Fig. B.7). Notice how it grows very rapidly as t increases, much more rapidly than any power of t. Also, the plot of e^{-t} diminishes very rapidly for large t values, although it never reaches zero.

Several special values of e^{-t} are used in this book. To three significant figures, $e^{-1} = 0.368$ and $e^{-2} = 0.135$. Also, $e^{-t} = 0.500$ when $t = 0.693$.

A series expansion for e^x is given in Eq. B.18 above, and the derivative of e^x is given in Eq. B.25.

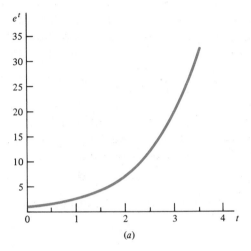

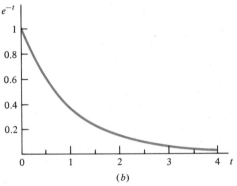

Figure B.7.   (*a*) The graph of e^t. (*b*) The graph of e^{-t}. Note the difference in vertical scales in the two graphs.

REVIEW PROBLEMS

Evaluate the following quantities with the aid, where needed, of tables or a pocket calculator.

101. Find e^2, e^3, e^4.

102. Find $e^{0.5}$, $e^{1.5}$.

103. Find $\ln e^3$, $\ln 10$, $\ln 4$, $\ln 0.75$, $\ln 2$.

104. What is x if its natural logarithm is 0.5, 0.1, 1, -1, 2, 2.5?

105. What are the common logarithms of 10, 100, 1000, 5, 0.5, e?

106. What is x if its common logarithm is 0.5, 0, 1, -1, 2, 2.5?

Simplify the following expressions:

107. $\log(x^4)$

108. $\ln(x/y^2)$

109. $\ln(\sqrt{x})$

110. $\log(1/x^3)$

111. $\ln[x(a + b)]$

B.11 | TRIGONOMETRIC IDENTITIES

There are a variety of identities relating trigonometric quantities. The following formulas are used in this text.

$$\sin (a + b) = \sin a \cos b \pm \cos a \sin b \qquad (B.34)$$

$$\cos (a + b) = \cos a \cos b \mp \sin a \sin b \qquad (B.35)$$

$$\sin a + \sin b = 2 \sin \tfrac{1}{2}(a + b) \cos \tfrac{1}{2}(a - b) \qquad (B.36)$$

$$\cos a + \cos b = 2 \cos \tfrac{1}{2}(a + b) \sin \tfrac{1}{2}(a - b) \qquad (B.37)$$

B.12 | INTEGRALS

Integrals are used in several sections. The following brief list contains most of the *indefinite integrals* used in the text or needed for the problems. Definite integrals are discussed below.

$$\int t^n \, dt = t^{n+1}/(n + 1) \quad (n \neq -1) \qquad (B.38)$$

$$\int dt/t = \ln t \qquad (B.39)$$

$$\int e^{at} dt = \frac{1}{a} e^{at} \qquad (B.40)$$

$$\int \sin at \, dt = -(1/a) \cos at \qquad (B.41)$$

$$\int \cos at \, dt = (1/a) \sin at \qquad (B.42)$$

$$\int \frac{t \, dt}{(a^2 + t^2)^{1/2}} = (a^2 + t^2)^{1/2} \qquad (B.43)$$

$$\int \frac{dt}{(a^2 + t^2)} = \frac{1}{a} \tan^{-1} \frac{t}{a} \qquad (B.44)$$

(Here $\tan^{-1}$ is the *arctangent;* if $y = \tan^{-1} x$, then $x = \tan y$)

$$\int \frac{t \, dt}{(a^2 + t^2)^{3/2}} = \frac{-1}{(a^2 + t^2)^{1/2}} \qquad (B.45)$$

$$\int \frac{a^2 \, dt}{(a^2 + t^2)^{3/2}} = \frac{t}{(a^2 + t^2)^{1/2}} \qquad (B.46)$$

$$\int u dv = uv - \int v du \qquad \text{(integration by parts)} \qquad (B.47)$$

$$\int t^n e^{at} dt = (t^n/a) - (n/a) \int t^{n-1} e^{at} dt \qquad (B.48)$$

$$\int t^n e^{-at} dt = -(t^n/a) + (n/a) \int t^{n-1} e^{-at} dt \qquad (B.49)$$

A *definite integral* is evaluated by substituting the values of the variable at the ends of the integration range into the indefinite integral. If $g(t)$ is the indefinite integral of $f(t)$, then

$$\int_a^b f(t) dt = g(t) \Big|_a^b = g(b) - g(a) \qquad (B.50)$$

For example, according to Eq. B.38, $\int t dt = t^2/2$. Then

$$\int_a^b t dt = \frac{t^2}{2} \Big|_a^b = \frac{b^2}{2} - \frac{a^2}{2}$$

Also, repeated application of Eq. B.49 shows that

$$\int_0^\infty t^n e^{-at} dt = n!/a^{n+1} \qquad (B.51)$$

(Here $n! = n(n - 1)(n - 2) \cdots (2)(1)$ is *n-factorial.*)

Additional Reading

Students needing additional mathematical preparation may find the following paperback books helpful.

Jerry B. Marion and Ronald C. Davidson, *Mathematical Preparation for General Physics,* W. B. Saunders Co., Philadelphia, 1972.

Clifford E. Swartz, *Used Math,* Prentice-Hall, Inc., Englewood Cliffs, N.J., 1973.

Michael Ram, *Essential Mathematics for College Physics,* John Wiley & Sons, New York, 1982.

———, *Essential Mathematics for College Physics With Calculus,* John Wiley & Sons, New York, 1984.

Answers to Review Problems in Appendix B

1. 16
2. 9
3. $2^5 = 32$
4. x^9
5. $\frac{1}{5^2} = 0.04$
6. 5
7. x^2
8. x^2
9. $\left(\dfrac{x}{y}\right)^4$
10. ax^2
11. $a^{3/2}x^3$
12. xy^3
13. $\dfrac{1}{x^2y^2}$
14. 10
15. 0.1
16. 1
17. $\frac{1}{5} = 0.2$
18. $\dfrac{x}{8}$
19. $x^2 y^{-4} = \dfrac{x^2}{y^4}$
20. 10^3
21. 2.7631×10^4
22. 2.7631×10^6
23. 1.5×10^4
24. 3.4×10^{-8}
25. 1.6×10^3
26. 4.32976×10^3
27. 3.902×10^{-3}

28. 8.002×10^{-2}
29. 0.00234
30. $1{,}760{,}000$
31. 0.00005799
32. $45{,}000{,}000$
33. 670
34. $2{,}720{,}000$
35. $2{,}720{,}000$
36. 1.5×10^{11}
37. 50
38. 6×10^{9}
39. 6.6×10^{16}
40. 1.12×10^{15}
41. 200
42. 300
43. 300
44. 20
45. 2×10^{-3}
46. 20
47. 10^{5}
48. 10^{-5}
49. 2.15×10^{3}
50. 684
51. 2.76×10^{4}
52. 3.73×10^{-1}
53. 4.67
54. 3.33
55. 2.46×10^{4}
56. 3.46×10^{-5}
57. two
58. one
59. two
60. three
87.

61. four
62. two
63. 4.28×10^{7}
64. 3.28×10^{7}
65. 1.89×10^{-3}
66. $x = 10$
67. $x = 1$
68. $x = 30$
69. $x = \pm 3$
70. $x = 81$
71. $x = -\frac{4}{3}$
72. $x = 12$
73. $t = \pm 2$
74. $x = 4$
75. $x = -2,\ x = -4$
76. $x = -1,\ x = -2$
77. $x = 1,\ x = -\frac{5}{3}$
78. $x = -2$
79. $x = -\frac{3}{2},\ x = 0$
80. $x = -1,\ x = \frac{5}{2}$
81. $x = 3,\ y = 2$
82. $a = \frac{2}{7},\ T = \frac{8}{7}$
83. $y = -\frac{1}{5},\ x = 9\frac{3}{5}$
84. $x = 5\frac{1}{3},\ y = \frac{2}{3}$
85. $y = \frac{2}{5},\ x = 1\frac{3}{5}$
86.

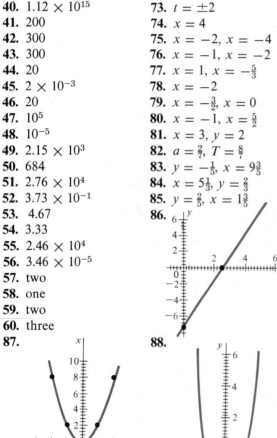

88.

89. $40°$
90. $\sin 120° = 0.866$, $\cos 120° = -0.5$,
   $\tan 120° = 1.732$
91. $\sin 270° = -1$, $\cos 270° = 0$, $\tan 270° = \infty$
92. $\sin \theta = 0$ at $0°$, $180°$, $360°$
   $\sin \theta = 1$ at $90°$
   $\sin \theta = -1$ at $270°$
   $\cos \theta = 0$ at $90°$, $270°$
   $\cos \theta = 1$ at $0°$, $360°$
   $\cos \theta = -1$ at $180°$
   $\tan \theta = 0$ at $0°$, $180°$, $360°$
   $\tan \theta = 1$ at $45°$, $225°$
   $\tan \theta = -1$ at $135°$, $315°$
93. 0.52 percent at $10°$, 4.7 percent at $30°$
94. Three terms
95. (a) $1 + \frac{1}{2}x - \frac{1}{8}x^2$
   (b) 1.04875, which is 0.006 percent less than
      the exact answer.
96. 3
97. $12t^2$
98. $\dfrac{1}{t^2}$
99. $-12e^{-3t}$
100. $20\pi \cos 2\pi t$
101. 7.39, 20.1, 54.6
102. 1.65, 4.48
103. 3, 1, 1.39, -0.288, 0.693
104. 1.65, 1, 2.72, 0.368, 7.39, 12.2
105. 1, 2, 3, 0.699, -0.301, 0.434
106. 3.16, 1, 10, 0.1, 100, 316
107. $4 \log x$
108. $\ln x - 2 \ln y$
109. $\dfrac{1}{2} \ln x$
110. $-3 \log x$
111. $\ln x + \ln (a + b)$

SYSTEMS OF UNITS

In this text we have used Systeme Internationale (S.I.) units almost exclusively. This is in keeping with a worldwide effort to develop a single set of units for scientific, commercial, and social use. Because this effort is far from complete and because historical material will always be important, a variety of units will be with us for a long time. In this appendix we describe the *c.g.s.* and *British* systems of units to assist the reader in relating units in these systems to S.I. units. Some units that do not fall clearly into any of these systems are also discussed.

We assume the reader is familiar with the general scheme of conversion of units described in Chapter One of this textbook. A table of useful conversion factors can be found inside the front cover. Multiples and submultiples of S.I. units are constructed using prefixes given on the right front endpaper of this text.

C.1 | TIME, LENGTH, AND MASS

The single quantity with the same units in all systems is time. The basic unit of time is the *second*, with the *minute* and *hour* in common use. In scientific work, submultiples of the second are often convenient and the prefixes of the S.I. system are useful for these cases.

The basic units of length are the *metre, centimetre,* and *foot* in the S.I., c.g.s., and British systems, respectively. There are 100 centimetres and 3.281 feet in 1 metre. Areas and volumes have the obvious units of m², cm², and ft² and of m³, cm³, and ft³, respectively, in the three systems and require the use of conversion factors either two or three times. In the British sys-

tem, the *inch* (= 1/12 foot) is often used, as is the *mile* (= 5280 feet).

The volume measures of liquids and gases has led to common use of the litre (= 10^{-3} m³ = 1000 cm³) and a plethora of units in the British system including the *gallon, quart, pint, cup,* and various sizes of spoons. The conversions between these are most readily found in cookbooks and dictionaries. The *gallon* is equivalent to 3.786 litres.

The units of mass in the three systems are the *kilogram, gram,* and *slug.* The relationship between the kilogram and gram is obvious from the prefix, and the slug is equal to 14.59 kilograms. The atomic mass unit, denoted u, is accepted for use with the S.I. system; 1 u = 1.66×10^{-27} kg.

The acceleration of gravity, g, is important in all three systems:

$$g = 9.8 \text{ m s}^{-2} = 980 \text{ cm s}^{-2} = 32 \text{ ft s}^{-2}$$

Until recently, two length units were very commonly used in studies of light and in atomic physics. The *angstrom* and the *micron* are equal to 10^{-10} metres and 10^{-6} metres, respectively, and are simply names given to convenient submultiples of the metre.

In the same spirit, the *fermi*, equal to 10^{-15} metres, was used for years as the convenient length unit in nuclear studies. Similarly the *barn*, equal to 10^{-28} m², is a convenient measure of cross-sectional areas in subatomic particle collisions.

C.2 | UNITS OF FORCE, WEIGHT

The *newton*, the S.I. unit of force, is equivalent to 1 kg m s⁻², as can be seen from Newton's second law,

741

$F = ma$. In the c.g.s. system, the force unit is the *dyne;* by the same reasoning, it is equal to 1 gm cm s^{-2}. Similarly, the *pound,* the British unit of force, is 1 slug ft s^{-2}. The numerical conversion factors are

$$1 \text{ N} = 10^5 \text{ dynes} = 0.2248 \text{ pounds}$$

These can be simply obtained from $F = ma$ where $a = g$ and the force is interpreted as the weight.

Notice that these units are names given to the force in the fundamental law $F = ma$. Thus the force units are derived from those for mass, length, and time given in Section C.1. In fact, with the exception of the units of electric charge, all the units of this section and those following are derived from those of mass, length, and time.

C.3 | OTHER UNITS OF MECHANICS

The derived S.I. unit of energy is the *joule.* This can be related to the units of mass, length, and time using $W = Fs$, that is work equals force times distance, or from the units of kinetic energy, $mv^2/2$. Thus the joule is equal to 1 N m = 1 kg m^2 s^{-2}. Similarly, the c.g.s. energy unit, the *erg*, is 1 dyn cm = 1 gm cm^2 s^{-2}. The British energy unit is the *foot-pound* = 1 ft lb = 1 slug ft^2 s^{-2}.

Unfortunately, the complete energy unit story is more complex. Historically, thermal energy was not recognized as the same entity as mechanical energy. Hence we have inherited a separate set of energy units from theoretical studies that are still in common use. A *calorie* is the heat necessary to raise the temperature of 1 gram of water by 1 K and is equal to 4.184 joules or 4.184 × 10^7 ergs. The British Thermal Unit, the BTU, had an origin similar to the calorie, but involved raising the temperature of a pound of water by 1° F. It is equal to 777.2 ft lb = 1.054 × 10^3 joules.

In atomic physics, a commonly used energy unit is the *electron volt*. This is a small energy unit equal to 1.602 × 10^{-19} joule. It is accepted for use with the S.I. system.

There is no defined c.g.s. unit of power; it is just the erg s^{-1}. In the British system, the logical power unit is the foot-pound per second. However, the more common British unit is the *horsepower,* equal to 550 ft lb s^{-1}.

There is a wide variety of pressure units in common use. The standard units in the three systems are the *pascal,* dyn cm^{-2}, and lb in^{-2}. Atmospheric pressures are often measured in terms of the *standard atmosphere* = 1.013 × 10^5 pascals = 1.013 × 10^6 dyn cm^{-2} = 14.7 lb in^{-2}. Pressures are also often given in terms of the height of a column of liquid that is supported, usually water or mercury. In fact, the pressure of 1 mm Hg (one millimetre of mercury) is given the name of *torr*; 1 torr = 133.3 pascals. One inch of water (1 in. H$_2$O) is equal to a pressure of 249.1 pascals.

In addition to the variety of systems of units, many combinations of units are used because they are numerically convenient. As an example, consider the conversion between two units of flow resistance, 1 Pa s m^{-3} = 7.50 × 10^{-9} torr s cm^{-3}. The smaller units are convenient in biological studies.

C.4 | ELECTRICAL UNITS

In the study of electricity and magnetism, a large number of units is introduced. However, if the unit of charge, the coulomb, is regarded as a fundamental unit, as are mass, length, and time, then all other units are derived from these four units.*

For example, the *volt* is a joule per coulomb and, as described in the preceding section, one joule is equivalent to one kg m^2 s^{-2}. Fortunately, S.I. units are widely used in electric and magnetic studies and conversions from other units systems are seldom required. One exception to this is the c.g.s. unit of magnetic field strength, the *gauss*, which equals 10^{-4} *tesla*. For example, the gauss is widely used in studies of the earth's magnetic field where the numbers are of a convenient size.

C.5 | SUMMARY

This appendix does not by any means exhaust all the systems of units or all of the combinations of units that are in use. However, if the reader will keep in mind that all units can be reduced to a combination of the units of mass, length, time, and charge, the task of relating and converting units will be greatly eased.

---

* If, in Coulomb's law, $\mathbf{F} = kQ_1Q_2/r^2$, the constant k were chosen to be dimensionless, the coulomb could be expressed in terms of units of mass, length, and time. This is a perfectly valid procedure, and it is employed in some c.g.s. systems of electrical units. However, in this text (and in most other modern books), we assign units to k and regard the coulomb as a fundamental unit, and not one expressed in units of mass, length, and time.

ANSWERS TO ODD NUMBERED EXERCISES AND PROBLEMS

Chapter 1

1-1 4.05×10^3 m^2

1-3   3.79 litres

1-5   (a) 7×10^{-9} m   (b) $7 \times 10^{-3} \mu$m

1-7   40.5 hectares

1-11   40 km h^{-1}

1-15   80 km h^{-1}

1-17   (a) 10.2 m s^{-1}

1-19   (a) -9.8 m s^{-1} (b) -29.4 m

1-21   (a) 10 m s^{-1} (b) 5 m s^{-1} (c) 0

1-23   6.94 km h^{-1}

1-27   positive at 0, T; negative at $T/2$

1-29 $\bar{v} = 10.1$ m s^{-1}; $\bar{v} < v(\max)$

1-31   (a) -98 m s$^-$ (b) -98 m s^{-1}

1-35   (a) $T_0 < t < 3T_0$ (b) $0 < t < T_0$, $3T_0 < t < 5T_0$ (c) $t > 5T_0$

1-37   (a) $-v_0/T$ (b) $-v_0/T$ (c) equal for constant a

1-39   (a) 50 m (b) 20 m s^{-1}

1-41   (a) -0.6 m s^{-2} (b) 750 m

1-45   (a) -113 m s^{-2} (b) -11300 m s^{-2}

1-47   400 m s^{-2}

1-49   45.9 m

1-51   (a) 400 m   (b) 2440 m   (c) 46.7 s

1-53   (a) 11.5 m   (b) 1.53 s   (c) 0.473 s, 2.59 s

1-55   (a) 39.4 m s^{-1} (b) 74.1 m

1-57   2.5 m

1-59   (a) 0.204 s   (b) 38.1 m   (c) 38.5 m

1-63   7 m s^{-1}

1-65   (a) 5 km h^{-1} (b) 1 km h^{-1}

1-71   6 s

1-73   (a) 4.52 m s^{-2} (b) 2.55 s   (c) 11.5 m s^{-1} (d) 200, yes; 1000-m, no

1-75   Closest approach is 10 m

1-77   (a) 9.7 m s^{-2} (b) 30 m s^{-2} (c) 60 m should have been 20 m

1-79   1720 m

1-81   (b) A (c) $(2\pi A/T) \cos 2\pi t/\tau$ (d) $-(2\pi/T)^2 A \sin 2\pi t/T$ (e) $(2\pi/T)^2$

1-83   (a) 1.84 m   (b) 1.22 s

1-85 $v = 2.42$ m s^{-1}, $a = 98.0$ m s^{-2}

1-87   62.5 m

1-89 $bT^3/3$

1-91 $v_f t + (v_f/b)(e^{-bt} - 1)$

1-93 $v_0 e^{-ct}$

Chapter 2

2-1   (a) **D** (b) **G** (c) **C** (d) 0 (e) **A** (f) **G** (g) **C** (h) **D**

2-3   (b) 10.4, 17° above $-x$ axis

2-5   (a) 7.21, 56° above **A** (b) 7.21, 56° above $-$**A**   (c) 7.21, 56° below **A**

2-7   (a) 24.9, 36° above $-x$ axis

2-9   (a) 224 km, 27° N of W

2-11   (a) 8.54, 69° above **C** (b) 17.0, 28° below **A**

2-13   (a) 39.3 s   (b) 12.7 m s^{-1}

2-15   (a) $v_x = 28.2$ m s^{-1}, $v_y = 10.3$ m s^{-1}

2-17 p, 2qt

2-19   (a) 78.5 s   (b) 0.510 m s^{-2} toward S

2-21   (a) 41,700 m s^{-2} (b) $a_x = 36,100$ m s^{-2}, $a_y = 20,800$ m s^{-2}

2-23   (a) 7.50 m s^{-1} (b) g

2-25   (a) 1.79×10^5 m s^{-2} (b) $a_x = 1.55 \times 10^5$ m s^{-2} $a_y = 8.93 \times 10^4$ m s^{-2}

2-27   (a) 1.33 s   (b) 11.5 m from building

2-29   (a) 141 m   (b) 4.08 s

2-31   (a) 0.0304 m s^{-2}

2-33   (a) The second   (b) Speeds are equal

2-35 (*a*) 0.25 h (*b*) 1 km downstream

2-37 0.56 m; does not clear net

2-39 (*a*) 1.03 s (*b*) 30.8 m

2-41 $V_{ox} = 6.41$ m s^{-1}, $V_{oy} = 15.3$ m s^{-1}

2-45 (*a*) (50 m) sin (0.16 s^{-1})t, (50 m) cos (0.16 s^{-1})t
(*b*) (8 m s^{-1}) cos (0.16 s^{-1})t, $-$(8 m s^{-1}) sin (0.16 s^{-1})t
(*c*) (1.28 m s^{-2}) sin (0.16 s^{-1})t,
$-$(1.28 m s^{-2}) cos (0.16 s^{-1})t

2-47 (*a*) $pt\hat{\mathbf{x}} + (qt^2/2)\hat{\mathbf{y}}$ (*b*) $\mathbf{r} = (pt^2/2)\hat{\mathbf{x}} + (qt^3/6)\hat{\mathbf{y}}$

2-49 (*a*) 2.04 s (*b*) 2.89 s (*c*) 3.53 s

2-51 40.8 m

2-53 8.85 m s^{-1}

2-55 20.2 m s^{-1}

2-57 26.1 m s^{-1}

2-59 71.4°

2-63 (*a*) 2.97 m s^{-1} (*b*) 0.45 m

2-65 30 m

Chapter 3

3-1 37° below the 20 N force, 25 N

3-3 15° from either force, 19.3 N

3-5 490 N

3-7 (*a*) 4.90 N (*b*) 1.10 lb

3-9 54.5 kg

3-11 1.0595 kg

3-13 (*a*) 1420 kg m^{-3} (*b*) Density near edge of sun is low compared to center

3-15 (*a*) 1.27×10^{17} kg m^{-3} (*b*) 1.27×10^{14} kg m^{-3}

3-17 1.002

3-19 127 kg

3-21 0.4515 kg

3-23 (*a*) 11.3 (*b*) 11.3 kg

3-25 No, no

3-27 (*a*) 0 (*b*) 19,600 N

3-29 *w*

3-31 Stable, cables restore position

3-33 No, car is accelerated, net force is not zero.

3-39 (*a*) 1490 N (*b*) 3.04

3-41 3000 N

3-43 11,500 N

3-45 2.45 m s^{-2}

3-47 (*a*) 3320 m s^{-2} (*b*) 339

3-49 0.86 m s^{-2}

3-51 652.5 N

3-53 0.444

3-55 296 N

3-57 $m_E/8$

3-59 5.98×10^{24} kg

3-61 2*w*, upward

3-63 (*a*) 3 m s^{-2} (*b*) 1.045 mg, 17.0° to vertical

3-65 0.867

3-67 30 N

3-69 Increase maximum frictional force

3-71 (*a*) 300 N (*b*) 150 N

3-75 1125 N

3-77 (*a*) 81.6 kg (*b*) 10 m s^{-2} (*c*) 816 N

3-79 (*a*) 5122 N (*b*) 1914 N

3-81 (*a*) 1.23 m s^{-2} (*b*) 32,000 N (c) 16,000 N

3-83 (*a*) 0.647 m s^{-2} (*b*) 30.6 kg

3-85 8.85 m s^{-2}

3-87 (*a*) 4.105×10^{12} N (*b*) 2.93×10^{16} N

3-89 5.12 m s^{-2}

3-91 (*a*) 1.55 *mg*, 1.225 *mg* (*b*) 0.225 *g*

3-93 (*a*) 23.5 N (*b*) 1.96 m s^{-2} (*c*) 0.98 m s^{-1}, 0.245 m

3-95 190 kg

3-97 (*a*) 6.24 m s^{-2} (*b*) 22.3 m s^{-1}

3-99 3.20 m s^{-2}

Chapter 4

4-1 $\mathbf{w}_1$, 0; $\mathbf{w}_2$, -6 N m; $\mathbf{w}_3$, -40 N m; $\mathbf{w}_4$, -75 N m

4-3 (*a*) $\mathbf{A} \times \mathbf{A}, \mathbf{A} \times \mathbf{C}$ (*b*) $\mathbf{A} \times \mathbf{D}, \mathbf{A} \times \mathbf{E}$
(*c*) $\mathbf{A} \times \mathbf{B}$ (*d*) $\mathbf{A} \times \mathbf{D}, \mathbf{A} \times \mathbf{E}$

4-5 4 m, 1.732 m

4-7 (*a*) 30 N m, into page; 24 N m, into page; 21.2 N m, out of page; (*b*) position (*a*)

4-9 $T_1 = 25.7$ N, $T_2 = 40.3$ N

4-11 $F_1 = 0.75$ N, $F_2 = 0.25$ N

4-13 (*a*) $w + w_1 + w_2$ (*b*) w_1/w_2

4-15 3.06 m

4-17 0.175 m

4-19 1.31 m

4-21 27°

4-25 500 N

4-27 0.667

4-31 (*a*) I (*b*) III (*c*) Splenius

4-33 193 N

4-35 (*a*) 4000 N (*b*) 3606 N, 16° below horizontal

4-37 24.1 N

4-39 $X = 0$, $Y = 2mL/(M + 2m)$

4-41 36°

4-43 3.33 kg

4-45 (*a*) $T = 264$ N, $R_x = 251$ N, $R_y = 46.7$ N
(*b*) 0.133

4-49 (*a*) $T = 2020$ N, $R_x = 1980$ N, $R_y = 70$ N
(*b*) $T = 3220$ N, $R_x = 3150$ N, $R_y = -5$ N

4-51 $4L/9$

4-53 $3h/4$

4-55 (*a*) 0.075 m (*b*) 3 N

4-57 3 m from scale 1

4-59 $X = 0.229h, \ Y = 0.443h$

4-61 75°

4-63 42.4 N

Chapter 5

5-1 0.64 m s^{-2}

5-3 367 m

5-5 (a) 19,700 N (b) 2010 kg

5-7 5100 m

5-9 0.546

5-11 31.3 m s^{-1}

5-15 17°

5-17 Car may slide downward

5-19 Up the embankment

5-21 42,300 rev min^{-1}

5-23 (a) 1.58 m s^{-2} (b) 370 N (c) 512 N

5-25 (a) 60° (b) 135° (c) 405°

5-27 (a) Up (b) Down

5-29 (a) 12.5 rad s^{-1}; along axle away from us
(b) 2.5 rad s^{-2}

5-31 7670 rad s^{-1}, 73,200 rev min^{-1}

5-33 (a) No (b) Possibly (c) Yes

5-35 (a) 66.7 rad s^{-1} (b) 4.44 rad s^{-1} (c) 500 rad

5-37 0.245 kg m^2

5-39 0.289 l

5-41 (a) 0.00640 kg m^2 (b) 0.0960 N m

5-43 $g/4$

5-45 $7ml^2/48$

5-47 $2mR^2$

5-49 6.25×10^{12} must be removed

5-51 (a) 3.33×10^{-6} C (b) 2.08×10^{13}

5-53 (a) -9.63×10^7 C (b) $+9.63 \times 10^7$ C
(c) 8.35×10^{25} N

5-55 9.22×10^{-10} N

5-57 (a) 18.9 m s^{-2} (b) 1.06

5-59 (a) 22.8 m s^{-1} (b) 57.8 m s^{-1}, 0

5-61 (a) 0.0338 m s^{-2} (b) 698 N (c) Further reduces
weight

5-63 16.0 s

5-65 (a) $4w$ (b) $w\sqrt{10}$

5-67 $(31/32)\pi\rho aR^4$

5-69 5.39 s

5-71 7.54 N m

5-73 (a) 0 (b) $\sqrt{3}kqQ/4a^2$, away from line joining
$+q$ charges

5-75 (a) $mb^2/6$ (b) $2mb^2/3$

5-79 $3.5d$

5-81 $27y$

5-83 $7.81h$

5-85 0.539 m

5-87 $T = Cr^2$

Chapter 6

6-1 56.4 J

6-3 76°

6-5 80 J

6-7 2.21×10^5 J

6-9 12.5 J

6-11 (a) 20,000 J (b) 20,000 N

6-13 (a) 2500 J (b) 2500 J

6-15 13.1 J

6-17 4230 N

6-19 Less energy per second required

6-21 24.2 m s^{-1}

6-23 0

6-25 31.3 m s^{-1}

6-27 35.9 m s^{-1}

6-29 8.82×10^4 J

6-31 0.408

6-33 0.592

6-35 (a) 49,000 J (b) 6.45×10^{-3} kg

6-37 2.35×10^{12} J

6-39 4.32×10^{14} J

6-41 (a) 29.4 J (b) 4.85 m s^{-1}

6-43 39,220 J

6-45 (a) $GM_s m/2R$ (b) $GM_E m/2R_E$ (c) 28.3

6-47 -6.60×10^{-8} J

6-49 3.17×10^4 m s^{-1}

6-51 2.37×10^3 m s^{-1}

6-53 1.227×10^7 m s^{-1}

6-55 8464

6-57 (a) 2.915×10^{-15} J (b) 1.822×10^4 eV

6-59 40 W

6-61 209 W

6-63 (a) 1.80 W (b) 5.14 N (c) 1080 J

6-65 24 cents

6-67 (a) 200 W (b) 800 W

6-69 6.53×10^4 W

6-71 (a) 200 m^2 (b) Comparable to roof of large
one-story house

6-73 0.790 J

6-75 (a) 48.0 J (b) 96.0 J

6-77 (a) 1350 J (b) 2.149 N m (c) Converted into
thermal energy

6-79 $(4gd/5)^{1/2}$

6-81 (a) 36.3 J kg^{-1}, 132 W kg^{-1} (b) 3.23 J kg^{-1},
131 W kg^{-1}

6-85 $mgl \sin \theta$

6-87 (a) 1.28 m (b) 0.573 m s^{-1} (c) 0.724 m
(d) 3.77 m s^{-1}

6-89 (a) 6×10^{15} kg (b) 311 GW

6-91 (a) 1.529×10^{15} J (b) 3.54×10^{10} W (c) 1.77

6-93 (a) 1.25×10^{18} J (b) 1.45×10^{13} W (c) 1.45

6-95 (a) 498 W (b) 8.53 km

6-97 1.46°

6-101 Earth, 11,200 m s^{-1}, Sun, 42,100 m s^{-1}

6-103 (a) $GM_s^2/4R_s$ (b) 9.50×10^{40} J
(c) 7.93×10^6 y

6-105 5.27×10^{-3} s

6-111 $ma^2t^2/2$

6-113 (a) 718 N (b) 5740 W

6-115 6.09 m s^{-1}

6-117 (a) 198 m (b) 58.0 s

Chapter 7

7-1 69.1 m s^{-1}

7-3 (a) 36,000 kg m s^{-1} (b) 36.0 m s^{-1}

7-5 $2mv/\Delta t$ to left

7-11 $mv_0/(m + M)$

7-13 (a) 8.26 m s^{-1} (b) 1650 N

7-15 0.005 m s^{-1}

7-17 $\phi = \theta$

7-19 (a) 3.34×10^{-11} m s^{-1} (b) 1.12×10^{-15}

7-21 $3\mathbf{v}/2$

7-23 (a) Earth-moon center of mass (b) 4.66×10^6 m

7-25 (a) None (b) 1.33×10^5 J

7-27 1/2

7-29 (a) 16.7 rad s^{-1} (b) 4.33 kg m^2 s^{-1}

7-31 0.251 kg m^2 s^{-1}

7-33 To increase I, hence stability

7-37 500 s

7-39 2.05 rad s^{-1}

7-41 $v/2$

7-43 (a) Car, 33.3 m s^{-1}; truck 6.67 m s^{-1}
(b) 6.67 m s^{-1}

7-45 4.62 m s^{-1}

7-49 (a) 2.48 m s^{-1} (b) 0.314 m

7-51 (b) and (c)

7-53 m_2/m_1

7-55 $m/(m + M)$

7-57 (a) 18.1 m s^{-1}, 83.7° south of west (b) 3.62×10^5 J

7-59 0.2 rev s^{-1}

7-61 It will lengthen

7-63 $n^2h^2/8\pi^2I$

7-65 (a) 1.71 rad s^{-1} (b) 1.71 rad s^{-1}

7-71 1470 N

7-75 (a) 114 J (b) 55.3 J

7-81 Increase rate

Chapter 8

8-1 1000 N m^{-2}

8-3 0.0125

8-5 1.56×10^{-4} m

8-7 4.84×10^{-3} m

8-9 7.07×10^4 N

8-11 (a) 9800 N m^{-2}, 4.90×10^{-8} (b) 9.80×10^{-8} m
(c) 5.10×10^6 kg

8-13 77,400 N

8-15 (a) 3.18×10^6 N m^{-2}, 1.59×10^{-5}
(b) 4.77×10^{-5} m

8-17 2.25×10^7 N m^{-1}

8-19 78.5 N m

8-21 (a) 2.13×10^{-7} m^4 (b) 8.53×10^{-7} m^4, 5.33×10^{-8} m^4 (c) Board B, $\perp$ to 2-cm dimension
(d) Board A

8-23 (a) Yes (b) Weight, normal force (c) No;
torques add to zero

8-25 They produce a large torque with respect to the base

8-27 8.73 m

8-29 37.6 m, compared to 8.73 m

8-31 (a) 10^6 N m^{-2}, 1.19×10^{-5} (b) 1.19×10^{-7} m

8-33 2.83×10^4 N

8-35 (a) 1.25×10^7 N m^{-2} (b) 0.125

8-37 (a) 615 m^3 (b) 0.005 m

8-39 (a) 3.89×10^{-3} m^2 (b) 2×10^{-4} m

8-41 16

8-43 (a) 8.80×10^{-5} m^4 (b) 7.41×10^{-7} m^4

8-45 R(solid)/R(hollow) = 0.0914

8-47 1.73 cm

8-49 Areas would scale with weight

8-51 $m^{5/8}$

8-53 (a) $m^{-1/3}$ (b) $m^{-1/4}$

8-55 Independent of mass, as in Chapter 6

8-57 25 m

Chapter 9

9-1 (a) 1/48 s, 1/32 s, 1/24 s, 1/16 s
(b) -8.70 m s^{-1}, -10.05 m s^{-1}, -8.70 m s^{-1}, 0

9-3 0.0477 m

9-5 -148 m s^{-2}

9-7 4

9-9 7.40 N m^{-1}

9-11 1.36 Hz

9-13 0.248 m

9-15 12.2 m

9-17 1.49 m

9-19 4

9-21 1.050 Hz

9-23 (a) 4.90 N m^{-1} (b) 1.58 Hz (c) 0.635 s
(d) 6.13 × 10^{-3} J

9-25 0.0707 m

9-27 (a) ±0.447 m s^{-1} (b) ±0.387 m s^{-1}

9-29 (a) 0.769 s (b) 4.61 s

9-31 (a) x_0, 0, $-\omega^2 x_0$ (b) $-x_0$, 0, $\omega^2 x_0$

9-33 $x = R\sin(2\pi ft)$, $v = (2\pi f)R\cos(2\pi ft)$,
$a = -(2\pi f)^2 R\sin(2\pi ft)$

9-35 (a) 8.88 N m^{-1} (b) 7.5 Hz

9-37 (a) 19.7 N m^{-1} (b) 0.995 m

9-39 (a) ≃0.7 Hz (b) ≃4 km

9-41 5.95 m s^{-1}

9-43 (a) 147 J (b) 1.176 × 10^5 N m^{-1} (c) 7.72 Hz

9-45 3.40 m

9-47 (a) $\theta_0 \sqrt{g/l}$

9-51 (b) determines phase (point in cycle at $t = 0$)

9-55 ≃10^5 N m^{-1}

9-57 (a) 98,000 N m^{-1} (b) 7.05 Hz (c) No

9-59 0.0621 m

9-61 (a) 22.2 Hz (b) Well

Chapter 10

10-1 37.8°C

10-3 −40°

10-5 36.461 u

10-7 4.032 grams

10-9 1.806 × 10^{24}

10-11 1.58 moles

10-13 127 grams

10-15 5.07 × 10^8 N

10-17 2030 N; no

10-19 (a) ≃10^5 Pa (b) ≃1000 Pa

10-21 5

10-23 1.37

10-25 10,970 K

10-27 1.05 atm

10-29 10,100 m

10-31 Pressure is doubled

10-33 0.9957

10-35 (a) 4.25 × 10^{-21} J (b) 205 K

10-37 746 s

10-39 3.74 × 10^6 Pa = 37.0 atm

10-41 196 moles m^{-3}

10-43 303 u

10-45 (a) 0.3 m^3 (b) 37.5 min

10-47 5.12%

10-49 9750 K

10-51 (a) 0.45 s (b) 4.5 times as large

10-53 7.88 atm

10-55 12.5 J

Chapter 11

11-1 (a) $P_1(V_3 - V_1)$ (b) $-P_1(V_3 - V_1)$

11-3 1.4 J

11-5 3.24 × 10^5 J

11-7 25 W

11-9 Faster increase with piston fixed. All heat goes into internal energy change

11-13 (a) No (b) Increased

11-15 (a) All tails (heads) 1 way; 1 head (tail) and 5 tails (heads)—6 ways; 2 heads (tails) and 4 tails (heads)—15 ways; 3 heads and 3 tails—20 ways. (b) 3 heads and 3 tails most probable.

11-17 (a) 58.1% (b) 3.3 × 10^6 J (c) 41.8%

11-19 500 K, 429 K

11-21 (a) 9.31 (b) 25 W

11-23 (a) 70,000 J (b) −80,000 J (c) 20,000 J

11-25 50%

11-27 (a) 2 or 12—1 way; 3 or 11—2 ways; 4 or 10—3 ways; 5 or 9—4 ways; 6 or 8—5 ways; 7—6 ways (b) 7

11-29 (a) 148 s

11-31 (a) 43.9% (b) 1.52

11-33 January, 1.07%; July, 6.08%

11-37 (a) 7.43 × 10^{-3} litres s^{-1} (b) 214 litres

11-39 (a) 8.82 litres (b) 34.5 litres

11-41 (a) 25% (b) 231 W

11-43 1.24 × 10^6 J

11-45 2.68 grams

11-47 No

11-49 0.375 W kg^{-1}

11-51 18.5d

Chapter 12

12-1 5.08 × 10^{-3} m

12-3 4.6 × 10^{-4} m

12-5 Lid expands more than glass

12-7 0.02991 m

12-9 46.6 kJ

12-11 2.48 kJ kg^{-1} K^{-1}

12-13 10,000 kJ

12-15 3.54 × 10^3 kJ

12-17 (a) No (b) 0°C

12-19 93.5 min

12-21 315 W

12-23 53.3 W

12-25 (a) 0.263 m^2 K W^{-1} (b) 3.95 m^2 K W^{-1}

12-27 79.999°C

12-29   21.2°C
12-31   1.5
12-33 2.28×10^{-6} m
12-35   23.4%
12-37   (a) 895 W   (b) 240 W
12-39   55.5 W
12-41   (a) 3.1×10^{-4} m   (b) Slow   (c) 1.27×10^{-4}
         (d) 11.0 s
12-43 3.41×10^{-3} K^{-1}
12-45   775 W
12-47   1.2 K h^{-1}
12-49   (a) 62.4 kJ   (b) 104 kJ
12-51   714 J
12-53   (a) 102°C   (b) 3200 W
12-55   100.8°C
12-57   22%
12-59   (a) 418 W m^{-2}
12-61   (a) 317 W   (b) No
12-63 8.74×10^{-3} kg
12-65   (a) 2.65×10^{6} J   (b) 0.39 cents h^{-1}
12-67   (a) 79.997°C   (b) 71.6 W
12-69   0.314 V_0
12-75   0.260 kg h^{-1}

Chapter 13

13-1   0.075 m^3
13-3   265 N
13-5   0.01 m^3
13-7   1 m s^{-1}
13-9   No, flow is turbulent
13-11   No, work done against gravity
13-13   20.8 k Pa
13-15   20.4 m
13-17   1170 kg m^{-3}
13-19   9.34 k Pa = 70.1 torr
13-21   25.8 k Pa
13-23   21.6 m s^{-2}
13-25   No
13-27   (a) 75 Pa   (b) 0.4 m s^{-1}
13-31   750 kg m^{-3}
13-33   Fallen
13-37   99.986 atm
13-39   0.193 m
13-41   (a) 3.83 m s^{-1} (b) Yes, until water level falls
         0.25 m   (c) 3.43 m s^{-1}
13-43   (a) Yes, increases   (b) Decreases
13-47   High takeoff velocity
13-49   (a) 48 m s^{-1} (b) Doubtful validity
13-51 $l^{7/2}$
13-53 $l^{3/4}$

Chapter 14

14-1   Halfway
14-3   1.8 W
14-5   (a) 1.13×10^{-7} m^3 s^{-1} (b) 0.0720 m s^{-1}
14-7   (a) 0.0314 N   (b) 4.71×10^{-4} W
14-9   (a) 0.354 m s^{-1} (b) 632 Pa
14-11   (a) 7.96 m s^{-1} (b) Turbulent   (c) No
14-13   (a) 0.983 m s^{-1} (b) 1.24×10^{-5} m^3 s^{-1}
14-15 $2R^2/r^2$
14-17   (a) 1.49×10^{10} kPa s m^{-3} (b) $8.72 \times$
         10^{-11} m^3 s^{-1}
14-19   (a) 8.38×10^{-3} N   (b) 8.33×10^{-2} N   (c) Oil
         inhibits rust
14-21   (a) 8×10^{-3} m^3 s^{-1} (b) 6.37 m s^{-1} (c) $N_R =$
         2.5×10^5, so flow is turbulent
14-23   900 W
14-25   409.6P_0
14-27   (a) 11.6 kPa, 0.873 kPa   (b) Right ventrical does
         less work than left
14-29   (a) $R_f = \Delta P/Q$ (b) Increase substantially
14-31   (a) 100 kPa s m^{-3} (b) 10 W   (c) No
14-33   (a) 3.32×10^{14} kPa s m^{-3} (b) 3.08×10^{10}
14-35   (a) $7\pi\Delta PR^4/128\eta l$ (b) 7/16   (c) 1/4
14-37   (a) 4.33×10^{-10} m s^{-1} (b) 50 m s^{-1}
14-39 4.23×10^{-8} m
14-41 $\simeq$200 N
14-43   (a) 4.23   (b) 3.90 m s^{-2}, 45.8 m s^{-2}
14-45   (a) 3.25×10^{-16} N   (b) 2.41×10^{-16} N
14-47   805 s
14-49 3.39×10^7 u
14-51 10^{-5} m s^{-1}
14-57   (a) 2.29×10^{-24} m^3 (b) 8.18×10^{-9} m
         (c) 1.54×10^{-7} m   (d) No, probably oblong
14-59   Carries material to and from walls
14-61 B does 21% more work

Chapter 15

15-1   0.02 N m^{-1}
15-3   Minimize film area outside
15-5 -0.0535 m
15-7   0.0703 m
15-9   0.025 N m^{-1}
15-11 3.35×10^{-3} N m^{-1}; 7%
15-13   4 Pa
15-17   21%
15-19   Yes; negative pressure situation is destroyed
15-21 -0.934 atm
15-23 $w + 4\pi r\gamma$
15-29 3.86×10^{-2} N m^{-1}

15-31 (*a*) 41.5 J (*b*) 5.03×10^{-4} J
15-33 (*a*) 0.514 m (*b*) No
15-35 0.0495 m

Chapter 16

16-1 $(2kQ^2/9b^2)\hat{\mathbf{y}}$
16-3 $-(kQ^2/2b^2)\hat{\mathbf{y}}$
16-5 4.36×10^{-9} N
16-7 (*a*) 1.32×10^{13} N C^{-1} outward (*b*) 2.11×10^{-6} N, toward nucleus
16-9 5.69×10^{-4} N C^{-1}, opposite to **a**
16-11 $-(kQ/2b^2)\hat{\mathbf{y}}$
16-13 -1.77×10^{-8} C
16-15 (*a*) 10^{-6} C m^{-1} (*b*) 1.8×10^5 N C^{-1}
16-17 (*a*) 0.0050 (*b*) 0.41
16-19 (*a*) 86.4 V (*b*) -1.38×10^{-17} J
16-21 $-2a$
16-23 (*a*) 0 (*b*) $4\sqrt{2}kQ/a$
16-25 (*a*) Toward positive plate (*b*) Toward negative plate (*c*) Acquire some energy (*d*) 42.9
16-27 $E = 0$
16-29 (*a*) 1.127×10^{-12} J (*b*) 3.67×10^7 m s^{-1}
16-31 3.13×10^{-11} m
16-33 $2qE$
16-35 (*a*) 0 (*b*) 1.6×10^{-23} N m (*c*) 0
16-37 10^{-7} F
16-39 3.23 m^2
16-41 4×10^8 V
16-43 (*a*) 0.325 m^2 (*b*) 1.44 cm
16-45 (*a*) 60 μF (*b*) 0.060 C
16-47 (*a*) 8.85×10^{-8} C (*b*) 4.43×10^{-5} J (*c*) 100 V, 4.43×10^{-6} J
16-49 (*a*) 4.79×10^{-19} C (*b*) 1890 V
16-51 720 N C^{-1}, toward plate
16-53 (*a*) 1.76×10^{14} m s^{-2} (*b*) 10^{-8} s (*c*) 8.8×10^{-3} m (*d*) 5°
16-55 $mv^2/2qE$
16-59 (*a*) $-2kqa\hat{\mathbf{y}}/(x^2 + a^2)^{3/2}$ (*b*) 1/2
16-65 (*a*) $2kp_1p_2/R^3$ (*b*) kp_1p_2/R^3
16-67 (*b*) 0.316 e
16-69 5 capacitors in parallel
16-71 1.33 μF
16-75 (*a*) 0 (*b*) $kQ\hat{\mathbf{r}}/r^2$ (*c*) 0
16-77 (*a*) at the surface (*b*) 9×10^7 N C^{-1} (*c*) 0.141 m, 0.05 m
16-79 (*a*) $4\pi kQ$ (*b*) No, not enough symmetry to do integral
16-83 No. Charges are in motion, not in static equilibrium.

16-85 $3kQ/2R$
16-87 (*a*) $2k\lambda/r$

Chapter 17

17-1 6.25×10^{12}
17-3 3.43×10^{-4} m s^{-1}
17-5 4.75×10^{-4} kg
17-7 2.5 ohms
17-9 34.2 ohms
17-11 1.05×10^{-3} m
17-13 11.5 ohms
17-15 20 ohms
17-17 (*a*) 6 A (*b*) 60 C (*c*) -720 J (*d*) 720 J (*e*) 0 (*f*) 720 J (*g*) Chemical energy in the battery
17-19 (*a*) 0.15 A (*b*) -1.5 V, 1.5 V, 1.5 V, -0.6 V, -0.9 V
17-21 (*a*) 3000 A, 3.6×10^4 W (*b*) 0.146 ohm (*c*) 934 W (*d*) 25.6 W
17-23 (*a*) 144 ohms (*b*) 0.833 A
17-25 (*a*) 0.4 A (*b*) 4.8 W, -3.2 W (*c*) 0.64 W, 0.96 W
17-27 64.8 cents
17-29 (*a*) 4.17 A (*b*) 1.15 ohms
17-31 460
17-33 (*a*) 240 ohms (*b*) 40 ohms
17-35 1 ohm
17-37 (*a*) 1.5 A (*b*) 3 V (*c*) 1 A
17-39 90-ohm series resistor
17-41 (*a*) 0.1 ohm (*b*) 10^{-7} s
17-43 0.02 s
17-45 (*a*) 0.012 A (*b*) 0.005 s
17-47 (*a*) 10 V (*b*) 100 V
17-51 (*a*) 7200 C (*b*) 4.50×10^{22} (*c*) 6.19×10^{23}
17-53 (*a*) 2.09×10^{-3} ohm (*b*) 0.00162 m (*c*) $w(\mathrm{Al})/w(\mathrm{Cu}) = 0.462$
17-55 (*a*) 4600 W (*b*) 37.7 min
17-57 (*a*) 1.51×10^4 J (*b*) 5.16 K
17-59 (*a*) Positive (*b*) 241 W (*c*) 12.06 V
17-61 (*a*) $\mathcal{E}/r$ (*b*) $\mathcal{E}/2r$
17-63 (*a*) 0.1 C (*b*) 10 ohms (*c*) 37 A
17-67 $\mathcal{E} = \mathcal{E}_1 R/(R + R_1)$
17-69 $(\ln^2)/2 = 0.347$
17-71 1.295 ohm m
17-73 (*a*) 13.6 ohm m (*b*) 7.37×10^{-2} ohm^{-1} m^{-1}
17-75 (*a*) 1 A (*b*) 10 ohms (*c*) 14 V
17-77 3.427 W

Chapter 18

18-1 10^6
18-3 1.59×10^9 ohms

18-5 (a) 1.26×10^{-9} F (b) 1.59×10^{6} ohms

18-7 (a) 1.2×10^{7} N C^{-1}, into axon (b) 8.48

18-9 (a) 4.5×10^{-6} C m^{-2} (b) 9.0×10^{-4} C m^{-2}

18-11 $10 \, \mu$m

18-13 $0.8 \, \mu$m

18-15 8.17×10^{12}

18-17 74.1 mV

18-19 (a) 0.0289 A (b) 0.00260 W

18-21 1.5 mm

18-23 (a) -86.3 mV (b) -88.6 mV

18-25 (a) 0.2 s (b) 0.01 s

18-27 (a) 2.67×10^{7} ohm m (b) 1.46×10^{10} m^{2}
(c) $8.28 \, \mu$m

18-29 $\rho_a = 0.812$ ohm m, $\rho_{if} = 0.599$ ohm m

18-31 (a) 3.9×10^{-10} C (b) 2.44×10^{9}

18-37 (a) 75 (b) 75,000

Chapter 19

19-1 (a) P_4 (b) P_1

19-3 No; can have $\mathbf{v} \| \mathbf{B}$

19-5 (a) 0 (b) 0 (c) qvB/m, into page (d) qvB/m, out of page (e) qvB/m, $+y$ direction (f) qvB/m, $-y$ direction

19-7 (a) 5×10^{-4} N, into page (b) 0.05 m s^{-2}, into page

19-9 0.5 T

19-11 (a) 30 N

19-13 (a) 2 N, 2 N, 0 (b) 0

19-15 (a) 0.1 A m^{2} (b) 0.01 N m (c) $\mathbf{B}$ in plane of loop

19-17 Yes; $-x$ direction

19-19 3.77×10^{-7} T

19-21 1.257×10^{-3} T

19-23 6.28×10^{-3} T

19-25 (a) 8×10^{-6} T (b) 8×10^{-7} T

19-27 7.80×10^{-7} A

19-29 (a) 2×10^{-4} m (b) Yes

19-31 0.25 A, parallel

19-33 (a) ev_dB, upward (b) v_dB, upward (c) av_dB
(d) av_dB (e) 2×10^{-5} V (f) Out of page
(g) Opposite in sign for a given current direction

19-35 (a) 0.24 A m^{2} (b) 0.12 N m (c) $\boldsymbol{\mu} \| \pm \mathbf{B}$

19-37 2×10^{-6} N, toward left

19-39 (a) 10^{-5} T, into page (b) 1.67×10^{-5} T, into page (c) 2.33×10^{-5} T, out of page (d) $x = 0.0333$ m

19-41 15.9 A

19-43 (a) 1.09×10^{-3} A (b) 13.4 T

19-45 $\pi k'I/a$

19-47 (a) 9.6×10^{-20} kg m s^{-1} (b) 5.75×10^{7} m s^{-1}

19-49 5 MeV

19-51 (a) D (b) D (c) D

19-53 (a) 1.67×10^{5} m s^{-1} (b) 8.66×10^{-3} m

19-55 (a) $evB/2m$ (b) $0.866\,v$, 0 (c) $v/2$, $mv/2eB$
(d) $1.732\pi mv/eB$ (e) Helix

19-59 (a) $2k'Ir/a^2$ (b) $2k'I/r$
(c) $2k'I(r^2 - b^2)/[r(c^2 - b^2)]$ (d) 0

Chapter 20

20-1 (a) Yes (b) No (c) Yes

20-3 (a) Same (b) Large in copper, almost zero in rubber

20-5 3.14 A

20-7 (a) 48 V (b) 0

20-9 Clockwise

20-11 $2\omega BA/\pi$

20-13 150 V

20-15 0.075

20-17 1

20-19 Induced EMF can lead to sparks

20-21 (a) Clockwise (b) Counterclockwise

20-27 0.32 H

20-31 (a) qvB (b) $qvB\ell$ (c) $-B\ell v$; same

20-33 Viewed from above: (a) Counterclockwise (b) Zero (c) Clockwise

20-35 (a) Proportional to ω (b) Reduces it (c) Increases I (d) I^2R becomes large

20-39 (a) $4\pi K_m k'IN/\ell$ (b) $2\pi K'_m N^2 I^2/\ell^2$
(c) $1/(8\pi k')$

20-43 (a) 1.333 H (b) 6 H

20-45 (a) $(Li_0/T)e^{-t/T}$ (b) $\frac{1}{2}Li_0^2 e^{-2t/T}$

20-47 (a) 1.2 A (b) 0.02 s (c) 0.759 A (d) 1.2 A

20-49 2.392 J

20-53 28.3 A

20-55 (a) 200 W (b) 1.67 A (c) 2.36 A

20-57 Ground plus either "hot" line

20-59 (a) 159 Hz (b) 70.7 V

20-61 (a) $39.8 \, \mu$F (b) 66.7 ohms (c) 240 V

20-63 100

20-65 (a) 0.796 A (b) 1.125 A

20-67 (a) 75.4 ohms (b) 125 ohms (c) 1.92 A

20-69 (a) 101 ohms (b) 1.19 A

20-71 (a) 1.592×10^{8} Hz (b) 1.885 m

20-73 250 V

20-75 (a) 13 ohms (b) 426 W

20-77 (a) 15.8 ohms (b) 2 W

20-79 R_v should be much larger to avoid altering currents and voltages significantly

20-81 9.9×10^{-3} ohms

20-83 5.77 s

20-85 (a) 1.29×10^{-9} F (b) 1.75 ohms (c) 0.452 ohms

20-87 (a) $\mathcal{P}_s/\mathcal{P}_l = 1$ (b) $\mathcal{P}_s/\mathcal{P}_l \simeq 0$

Chapter 21

21-1 0.344 m

21-3 (a) 10^{10} Hz (b) 5×10^9 waves (c) 1.5×10^8 m

21-5 214 m

21-7 (a) 6×10^{14} Hz (b) 3.77×10^{15} rad s^{-1} (c) 1.257×10^7 m^{-1}

21-11 (0.1 m) $\sin[(3.14 \text{ m}^{-1})x - (15.7 \text{ s}^{-1})t]$

21-13 173 m s^{-1}

21-15 156 N

21-21 (a) 2A (b) $2A \cos \omega t$

21-23 (a) $0, \pi/k, 2\pi/k, \ldots$ (b) $\pi/2k, 3\pi/2k, \ldots$

21-27 (a) 435 m s^{-1} (b) 2.91×10^{-4} kg m^{-1}

21-29 (a) 3.5 m (b) 1.75 m

21-31 6.51 m

21-33 (a) 0.25 m (b) 0.132 m

21-37 (a) 5 Hz

21-39 0.0866 m

21-41 0.25

21-43 All kinetic energy; string is still moving

21-47 3.42 m s^{-1}

21-49 40.0 s

21-51 (a) 1.52×10^{-3} s

21-53 (a) 80% (b) 20%

21-55 (a) ωA (b) $\omega^2 A^2 \Delta m/2$ (c) $\omega^2 A^2 \Delta m/2$

21-57 $2A \cos \phi/2$

21-59 1058 Hz

21-61 29.9 m s^{-1}

21-63 8.99×10^4 Hz

21-65 (a) $+3.47$ m s^{-1}, -3.41 m s^{-1} (b) No

21-67 5.28×10^{-8} m

21-69 (a) 497.1 Hz (b) 502.9 Hz (c) 5.8 Hz

21-73 0.0157 m s^{-1}

21-77 13.7 Hz

21-79 53,300 Hz

Chapter 22

22-1 365 Hz

22-3 40.7 m

22-5 1498 m

22-7 0.0516 m

22-9 In water, c is larger, λ larger

22-11 132 Hz, 265 Hz, 397 Hz, 529 Hz

22-13 2.65 m, 0.0823 m

22-15 9.09 Pa

22-17 0.0166

22-19 4

22-21 14.7

22-23 1.73×10^{-3} Pa

22-25 17,200 Hz

22-31 10

22-33 60 dB

22-35 8×10^{-13} W

22-41 (a) 0.103 m (b) 4.4×10^{-3} s

22-43 (a) 86.0 Hz (b) 1.828 m

22-45 (a) 1.26×10^4 W (b) 1.26×10^3 J

22-47 49.5 dB

22-49 (a) 10^{-8} W m^{-2} (b) 50.5 dB (c) 6.28×10^{-6} W

22-51 (a) 0.0430 m (b) 1720 Hz

22-55 (a) 99.9% (b) 0.11%

22-57 9×10^6

22-59 $\simeq 2$ mm

Chapter 23

23-1 2.25×10^8 m s^{-1}

23-3 (a) 450 nm (b) Yellow

23-5 375 nm

23-7 0.0204

23-9 0.997

23-11 1.649

23-13 (a) 15° (b) 11°

23-15 1.183

23-17 Air near road is less dense, has smaller n; sunlight bends upward

23-19 57°

23-21 1.91×10^{-5} m

23-23 5.98×10^{-5} m

23-25 No, two beams are not coherent

23-27 625 nm

23-29 500

23-31 (a) To separate lines (b) To sharpen lines

23-33 693 nm

23-35 916 nm

23-37 0.258 m

23-39 0.154 m

23-41 30°

23-43 (a) 0.2 (b) 0.8

23-45 63.4°

23-47 10°

23-49 77°

23-51 11.34 m

23-53 (a) 13.4° (b) $s' = 0.74$ m

23-55 $d \sin \theta = (m + \frac{1}{2})\lambda$

23-61 (a) $(1768/m)$nm (b) Look for lines at smaller angles

23-65 2.68×10^{-3} m

23-69 (a) All of it (b) Resolution is reduced in both cases

23-71   150 nm

23-75   Spot is 0.117 m from plate

23-77   All wavelengths interfere destructively

Chapter 24

24-1   4 m

24-3   20 cm

24-5 -0.5 m

24-7   1.50 m

24-9   Blue, 0.0775 m; red, 0.0795 m

24-11 -0.333 m

24-13   (a) -0.4 m   (b) 5; erect

24-15   (a) 0.0508 m   (b) -0.0169 (c) 1.42 m

24-17   Four times larger

24-19   0.250 m

24-21   (a) 4.75 diopters   (b) 0.211 m

24-23   (a) -0.1818 m   (b) -0.286 m

24-25 10^{-3} m

24-27   0.2 m

24-29   (a) 0.133 m from objective   (b) 0.00412 m
(c) -277

24-31   Edges of retina have most rods

24-33   (a) 510 nm, 620 nm   (b) 450 nm, 560 nm

24-35   (a) 5.75×10^{-7} m   (b) 5.75×10^{-6} m

24-37   Real images are inverted; processing in nervous system

24-39   80.1 m

24-41   (a) 0, $-f$, ∞, $3f$, $3f/2$

24-45 -1.20 m

24-47 $2f$

24-51 6.80×10^{-3} m

24-53   0.1 m away

24-57   0.0357 m

24-59   64

24-61   (a) 1.34×10^{-6} m   (b) 2.15×10^{-5} m

24-63   3.5 diopters

24-65   (a) 3.5 diopters   (b) 1 m

24-67   (a) 0.1667 m

24-69   (a) 480 nm   (b) 28%

24-71   (a) 0.52, 0.48, 0; 580nm   (b) 0.44, 0.13, 0.43; extraspectral

24-75   (a) 0.125 m (b) 0.5 m

24-77 -20

24-79   (a) Red, blue (b) Red, green (c) Red

Chapter 25

25-1   2000 km h^{-1}

25-3   100 s

25-5   (a) 3.75×10^{-8} s   (b) 6.75 m

25-7   0.872 m

25-9 $0.866c$

25-11   8 light-years

25-13 $0.866c$

25-15   (a) 5.35×10^{-9} u   (b) 1.78×10^{-10}

25-17 8.37×10^{-4} u

25-19   (a) 2150 MeV   (b) 1210 MeV

25-21   (a) 1.005 y   (b) 10.05 y   (c) Both

25-23   (a) 5 h   (b) 1.25 h   (c) 6.25 h   (d) 4 h

25-25 3.50×10^{-7} kg

25-27   (a) 1.12   (b) 1.35×10^{8} m s^{-1}
(c) 1.45×10^{8} m s^{-1}

25-33   (a) 0.0075   (b) 0.192

25-35   (a) 167 m   (b) A flashes first

25-37   (a) 0.8 h   (b) 1.333 h

25-39   (a) $0.822c$ (b) 0

25-41   (a) $0.806c$ (b) $-0.263c$

25-43   (a) 4 m   (b) 1.25×10^{-8} s later   (c) 6.25 m

25-45 $v_x = 0.8c$, $v_y = 0.54c$, $v = 0.965c$

Chapter 26

26-1 4.53×10^{14} Hz

26-3 9.65×10^{14} Hz

26-5   Removal of electrons leaves positive charge

26-7   48.6 m

26-9   (a) 3.10 eV   (b) 1.77 eV

26-11 5.32×10^{20} Hz

26-13   39,300 V

26-15   (a) 1.895×10^{30} (b) 5.276×10^{-31}

26-17   (a) 7.95×10^{-8} W m^{-2} (b) 4.00×10^{-6} W

26-19   1995 m

26-21   (a) 1.62×10^{-27} kg   (b) 515 N m^{-1}

26-23   (a) 277 nm   (b) 0.48 eV   (c) 0.48 eV

26-25   (a) 2.42×10^{19} Hz   (b) 2.32×10^{19} Hz

26-27 1.84×10^{-14} m

Chapter 27

27-1 4.73×10^{-14} m

27-3   486 nm

27-5 6.63×10^{-24} kg m s^{-1} (for both)

27-7 3.70×10^{-63} m

27-9   (a) 6.63×10^{-34} m   (b) No, λ is too small

27-11   (a) 0.821 nm   (b) 224 eV

27-13   Yes

27-15   (a) 13.6 eV   (b) 54.4 eV   (c) 122 eV

27-17 $n = 2$

27-19 $\gtrsim 3.52 \times 10^{-31}$ m

27-21   (a) 1.06×10^{-24} kg m s^{-1} (b) 1.16×10^{6} m s^{-1}
5.28×10^{-24} m s^{-1}

27-23   (a) 6.8 eV   (b) $2a_0 = 1.06 \times 10^{-10}$ m

27-25   (a) -7.21×10^{5} eV   (b) 1.60×10^{-14} m   (c) 4

27-27 12.75 eV, 12.1 eV, 10.2 eV, 2.55 eV, 1.89 eV, 0.661 eV

27-29 (a) $n^2\hbar^2/2I$, $n = 0, 1, 2, \ldots$

27-31 12.1 eV, 10.2 eV, 1.89 eV

Chapter 28

28-1 (a) 0,1,2,3 (b) $4s, 4p, 4d, 4f$ (c) $0, \hbar, 2\hbar, 3\hbar$

28-3 (a) $0, \pm e\hbar B/2m, \pm 2e\hbar B/2m$ (b) 5.79×10^{-4} eV (c) Absorbed

28-5 32

28-7 All have one s electron outside closed shells

28-9 48 eV

28-11 Small ionization energies

28-13 H electron closes shell in halogen atoms

28-15 $n = 3, l = 0, m_l = 0, s = 1/2, m_s = \pm 1/2$

28-17 11.1 eV

28-19 (a) 483 eV (b) Mutual repulsion of 1s electrons (c) Yes; additional shielding

28-21 (a) -272 eV (b) 3.79

28-23 3.8 km

28-25 0.003 m

28-29 (a) 68.3 MeV (b) 9.6 MeV

28-31 205 MeV

28-35 (a) $4a_0$ (b) $2a_0$

28-37 (b) 9.01×10^{-15}

28-39 (a) $\Psi(a) = \Psi(-a) = 0$ (c) $a^{-1/2} \cos n\pi x/2a$, $n = 1, 3, 5, \ldots$; $a^{-1/2} \sin n\pi x/2a$, $n = 2, 4, 6, \ldots$; $E_n = n^2\hbar^2\pi^2/8ma^2$

28-41 $13e^{-4} = 0.238$

28-43 (a) $1.5\,a_0$ (b) $1/a_0$ (c) r and $1/r$ are large in different regions of space

Chapter 29

29-1 (a) 1.30 eV (b) -6.10 eV (c) 4.80 eV

29-3 (a) -5.76 eV (b) 3.13 eV

29-5 No; it is symmetrical

29-9 (b) 4 (c) 2

29-11 2.66×10^{14} Hz

29-13 Yes; photons are emitted or lattice excited

29-17 4.46×10^{-29} C m

29-19 Yes; nonbonding orbital is negative, H nuclei positive

29-21 (a) Graphite, sp^2; diamond, sp^3 (b) Yes; no bonding between planes in graphite

29-23 12.0 eV

29-25 1.18 T

29-27 5.11×10^7 Hz

29-31 63.8 Hz

29-35 Broader peaks

29-37 3 to 2 to 2

29-41 $f_p > f_e$

29-43 (a) Counterclockwise (b) Reduces f_p

Chapter 30

30-1 8 h

30-3 17.6 d

30-5 5.6 to 5.7 d

30-7 18.95%

30-9 11,500 y

30-11 Too old

30-13 1300 y

30-15 1.1×10^{-14}

30-17 1.48×10^4 m

30-19 8,19,35,126

30-21 7.86 MeV

30-23 (a) 1397 MeV (b) 6.98 MeV

30-25 (a) 30 MeV (b) 0

30-27 (a) 1.44 MeV (b) -1.547×10^{-3} u

30-29 (a) ^3_1H (b) ^2_1H (c) α

30-31 (a) 4.84×10^{20} Hz (b) 6.20×10^{-13} m

30-33 6.64

30-35 40.4 y

30-37 (a) Straight line (b) Decay rate is not a simple exponential (c) Straight line, less rapid decrease

30-39 (a) 24.7 MeV (b) 1.99×10^6 kg

30-41 (b) ^9_5B and ^9_5Li have unfilled lower levels (c) β^-

30-43 (a) 8.64 fm (b) 176.3 MeV

30-45 1319 MeV

30-47 Fragments have too many neutrons

30-49 Geological stability (against earthquakes)

30-53 Electrical repulsion is three times larger

30-55 (a) 8.21×10^{13} J kg^{-1} (b) 0.0203 kg s^{-1}

30-57 (a) 7.72×10^8 K (b) Some atoms have above average energies

Chapter 31

31-1 A single γ can ionize atoms; one microwave quantum cannot

31-3 (a) α particles (b) 0.0045 cm

31-7 Nuclear mass is much larger than electron mass

31-9 $K(\text{d}) = 2K(\text{p})$

31-11 Determine range

31-13 (a) 5 cm (b) 0.5 cm

31-15 1.61×10^{15}

31-17 (a) 3.7×10^4 (b) 8.29×10^{-10} J (c) 8.29×10^{-10} W

31-19 6.20 R

31-21 (a) 4.50 min (b) 6.43 min

31-23 (a) 12 (b) 50

31-25 Less important because of smaller total dose to genetic pool

31-27 $2600 \, \text{y}^{-1}$

31-29 1 to 5 years doubles cancer rate, 0.25 to 1.5 years doubles mutation rate

31-33 (a) Low energy X rays have shorter ranges (b) Reduce skin dose

31-35 (a) $7.40 \times 10^{11} \, \text{s}^{-1}$ (b) $5.89 \times 10^{10} \, \text{m}^{-2} \, \text{s}^{-1}$

31-37 100 rad, 100 rem

31-39 $1420 \, \text{cm}^3$

31-41 (a) 0.616 J (b) 3.08 rad (c) 28

31-43 (a) 2 (b) 0.4 (c) No; masked by naturally occurring cancers

31-45 (a) 4.30 (b) 1.14×10^{-5} (c) 1.26×10^{-6}

31-47 Count rate increases as slabs are added

INDEX

FUNDAMENTAL CONSTANTS

The numerical values of most constants have been rounded off to three significant figures for convenience.

Quantity	Symbol	Numerical Value
Speed of light (in vacuum)	c	3.00×10^8 m s^{-1}
Gravitational constant	G	6.67×10^{-11} N m^2 kg^{-2}
Avogadro's number	N_A	6.02×10^{23} molecules mole^{-1}
Universal gas constant	R	8.31 J K^{-1} mole^{-1}
Boltzmann constant	k_B	1.38×10^{-23} J K^{-1}
		8.62×10^{-5} eV K^{-1}
Stefan's constant	σ	5.67×10^{-8} W m^{-2} K^{-4}
Atomic mass unit	u	1.66×10^{-27} kilograms
Coulomb constant	k	9.00×10^9 N m^2 C^{-2}
	$\epsilon_0 = 1/4\pi k$	8.85×10^{-12} C^2 N^{-1} m^2
Biot-Savart constant	k'	10^{-7} T m A^{-1}
Electron charge	$-e$	-1.60×10^{-19} coulombs
Electron mass	m_e	9.11×10^{-31} kilograms
Proton charge	e	1.60×10^{-19} coulombs
Proton mass	m_p	1.673×10^{-27} kilograms
Neutron mass	m_n	1.675×10^{-27} kilograms
Planck's constant	h	6.63×10^{-34} J s
		4.14×10^{-15} eV s
	$\hbar = h/2\pi$	1.055×10^{-34} J s
		6.58×10^{-16} eV s
Rydberg constant	R_H	1.10×10^7 metres^{-1}
Bohr radius	a_0	5.29×10^{-11} metres
Bohr magneton	μ_B	9.27×10^{-24} J T^{-1}

SOLAR AND TERRESTRIAL DATA

The numerical values given have been rounded off to three significant figures for convenience.

Standard atmospheric pressure	1 atm
	1.013×10^5 Pa
	1.013 bars
	760 mm Hg
	760 torr
Acceleration of gravity, g	9.81 m s^{-2}
Magnetic field (Washington, D.C.)	5.7×10^{-5} teslas
Speed of sound (dry air, 20°C)	344 m s^{-1}
Mass of earth	5.98×10^{24} kilograms
Volume of earth	1.09×10^{21} m^3
Mean radius of earth	6.38×10^6 metres
Mean density of earth	5.52×10^3 kg m^{-3}
Mean angular rotational speed of earth	7.29×10^{-5} rad s^{-1}
Earth to sun, mean distance	1.50×10^{11} metres
Earth to moon, mean distance	3.84×10^8 metres
Mean orbital speed of earth about sun	2.98×10^4 m s^{-1}
Sun, mean radius	6.95×10^8 metres
mass	1.99×10^{30} kilograms
Moon, mean radius	1.74×10^6 metres
volume	2.20×10^{19} m^3
mass	7.35×10^{22} kilograms
mean density	3.34×10^3 kg m^{-3}
acceleration of gravity	1.62 m s^{-2}